Nano-Bio- Electronic, Photonic and MEMS Packaging

C. P. (Ching-Ping) Wong • Kyoung-sik (Jack) Moon

Editors

Nano-Bio- Electronic, Photonic and MEMS Packaging

Second Edition

Editors
C. P. (Ching-Ping) Wong
School of Materials Science & Engineering
Georgia Institute of Technology
Atlanta, Georgia, USA

Kyoung-sik (Jack) Moon
School of Materials Science & Engineering
Georgia Institute of Technology
Atlanta, Georgia, USA

Yi Li
Intel Corp.
Chandler, Arizona, USA

ISBN 978-3-030-49993-8 ISBN 978-3-030-49991-4 (eBook)
https://doi.org/10.1007/978-3-030-49991-4

This Springer imprint is published by the registered company Springer Nature Switzerland AG
The registered company address is: Gewerbestrasse 11, 6330 Cham, Switzerland

Preface

In the past decade, we have witnessed tremendous advances in nano- and bio-related science, with enormous technical information and literature published worldwide. However, limited numbers of commercialized and industrialized items have appeared. Reproducibility, low yield, and difficulties in packaging, which are critically needed to manufacture usable devices and systems, are issues that researchers strive to resolve.

This book provides comprehensive reviews and overviews on the latest developments and cutting edge on nano- and biopackaging technologies and their science, including nano- and biomaterials, devices and thermal issues for nanobiopackaging, and the molecular or atomistic scale modeling to predict those small world phenomena.

This book comprises 20 chapters written by world-renowned experts in this field. Chapters 1, 2, 3, 4, 8, and 10 review various nanomaterials for nanopackaging technologies and their most recent research including nanomaterials for electrical and thermal interconnections, and nanosurface manipulation for nanopackaging. Chapter 5 addresses a novel combustion method for nanoparticle synthesis and nanosurface coating. Novel nanomaterials for renewable energy and energy conversion devices are reviewed in Chaps. 6 and 7. Chapter 9 addresses passive devices by using nanomaterials, and Chap. 11 reviews structural analysis of nanoelectronics and optical devices. Latest reviews on nanothermal science are presented in Chaps. 12 and 13. Some of the hot issues in the healthcare arena are the biomedical device and NEMS packaging, and the biosensor device and materials. Their recent development and research are reviewed in Chaps. 14, 15, 16 and 17. Computational molecular and atomistic scale simulations could provide us with a profound insight into the understanding of nano- and bio-related systems, which are difficult to realize without expensive equipment. These modeling efforts are reviewed in Chaps. 18, 19 and 20.

We are indebted to the contributions and efforts from all of the authors who shared their expertise in these vital areas in delivering this book to our readers.

Atlanta, Georgia, USA

C. P. (Ching-Ping) Wong
Kyoung-sik (Jack) Moon

Contents

Part X Computational

Contributors

Gang Bao Department of Biomedical Engineering, Georgia Institute of Technology, Atlanta, GA, USA

Avram Bar-Cohen Department of Mechanical Engineering, University of Maryland, College Park, MD, USA

Kuo-Ning Chiang Department of Power Mechanical Engineering, National Tsing Hua University, Hsinchu, Taiwan

Jaephil Cho School of Energy Engineering, Ulsan National Institute of Science & Technology, Ulsan, Korea

Ruo Li Dai Department of Mechanical and Automation Engineering, The Chinese University of Hong Kong, Shatin, NT, Hong Kong

B. Dang IBM Corporation, Armonk, NY, USA

J. J. Fan College of Mechanical and Electrical Engineering, Hohai University, Changzhou, China

X. J. Fan Department of Mechanical Engineering, Lamar University, Beaumont, TX, USA

Jan Felba Institute of Microsystem Technology, Wrocław University of Technology, Wrocław, Poland

Claire Gu Department of Electrical Engineering, University of California, Santa Cruz, CA, USA

Cuiping Han College of Materials and Engineering, Shenzhen University, and Shenzhen Key Laboratory of Special Functional Materials, Shenzhen, China

Peter J. Hesketh School of Mechanical Engineering, Georgia Institute of Technology, Atlanta, GA, USA

Jeffery A. Hinkley Langley Research Center, NASA, Hampton, VA, USA

J. L. Huang EEMCS Faculty, Delft University of Technology, Delft, The Netherlands

Andrew Hunt nGimat Co., Atlanta, GA, USA

Karl I. Jacob Department of Polymer, Textile and Fiber Engineering, Georgia Institute of Technology, Atlanta, GA, USA

Dasharatham G. Janagama Package Research Center (PRC), Georgia Institute of Technology, Atlanta, GA, USA

Hongjin Jiang Intel Corporation, Chandler, AZ, USA

Yongdong Jiang nGimat Co., Atlanta, GA, USA

Ranjith John College of Engineering, University of Arkansas, Fayetteville, AR, USA

Y. Joshi School of Mechanical Engineering, Georgia Institute of Technology, Atlanta, GA, USA

Chunho Kim Medtronic plc, Phoenix, AZ, USA

K. Kota School of Mechanical Engineering, Georgia Institute of Technology, Atlanta, GA, USA

Jan Krajniak School of Chemical and Biomolecular Engineering, Georgia Institute of Technology, Atlanta, GA, USA

Shi-Wei Ricky Lee Department of Mechanical Engineering, Hong Kong Institute of Science and Technology, Kowloon, Hong Kong

Wei-Hsin Liao Department of Mechanical and Automation Engineering, The Chinese University of Hong Kong, Shatin, NT, Hong Kong

Chun-Te Lin Department of Power Mechanical Engineering, National Tsing Hua University, Hsinchu, Taiwan

Wei Lin School of Materials Science and Engineering, Georgia Institute of Technology, Atlanta, GA, USA

Xiaohui Lin School of Mechanical Engineering, Georgia Institute of Technology, Atlanta, GA, USA

Yao Li Department of Polymer, Textile and Fiber Engineering, Georgia Institute of Technology, Atlanta, GA, USA

Yi Li Intel Corporation, Chandler, AZ, USA

Hang Lu School of Chemical and Biomolecular Engineering, Georgia Institute of Technology, Atlanta, GA, USA

Ajay P. Malshe College of Engineering, University of Arkansas, Fayetteville, AR, USA

Kyoung-sik (Jack) Moon School of Materials Science and Engineering, Georgia Institute of Technology, Atlanta, GA, USA

John H. L. Pang School of Mechanical & Aerospace Engineering, Nanyang Technological University, Singapore, Singapore

Edward S. Park School of Chemical and Biomolecular Engineering, Georgia Institute of Technology, Atlanta, GA, USA

C. Qian School of Reliability and Systems Engineering, Beihang University, Beijing, China

Won Jong Rhee Department of Biomedical Engineering, Georgia Institute of Technology, Atlanta, GA, USA

Herbert Robert George W. Woodruff School of Mechanical Engineering,Georgia Institute of Technology, Atlanta, GA, USA

Ephraim Suhir Department of Electrical Engineering, University of California, Santa Cruz, CA, USA

Bell Laboratories, Physical Sciences and Engineering, Research Division, Murray Hill, NJ, USA

Department of Mechanical Engineering, University of Maryland, College Park, MD, USA

ERS Co., Los Altos, CA, USA

Y. F. Sun School of Mechanical & Aerospace Engineering, Nanyang Technological University, Singapore, Singapore

Andrew Tsourkas Department of Bioengineering, University of Pennsylvania, Philadelphia, PA, USA

Rao R. Tummala Package Research Center (PRC), Georgia Institute of Technology, Atlanta, GA, USA

Ganesh Venugopal nGimat Co., Atlanta, GA, USA

Peng Wang Department of Mechanical Engineering, University of Maryland, College Park, MD, USA

Xudong Wang School of Materials Science and Engineering, Georgia Institute of Technology, Atlanta, GA, USA

Zhong Lin Wang School of Materials Science and Engineering, Georgia Institute of Technology, Atlanta, GA, USA

X. Wei IBM Corporation, Armonk, NY, USA

C. P. (Ching-Ping) Wong School of Materials Science and Engineering, Georgia Institute of Technology, Atlanta, GA, USA

Yonghao Xiu School of Materials Science and Engineering, Georgia Institute of Technology, Atlanta, GA, USA

Woon-Hong Yeo George W. Woodruff School of Mechanical Engineering, Wallace H. Coulter Department of Biomedical Engineering, Georgia Institute of Technology, Atlanta, GA, USA

G. Q. Zhang EEMCS Faculty, Delft University of Technology, Delft, The Netherlands

Yi Zhang Department of Electrical Engineering, University of California, Santa Cruz, CA, USA

Zhiyong Zhao nGimat Co., Atlanta, GA, USA

Part I

Electronics (Electrical Interconnections and Thermal Management)

Some Nanomaterials for Microelectronics and Photonics Packaging

1

C. P. (Ching-Ping) Wong and Kyoung-sik (Jack) Moon

1.1 Introduction

Microelectronics, photonics, and bio-packaging technology trends have been always relying on increasing the packaging density, resulting from meeting consumer demands for lightweight, compact, highly reliable and multifunctional electronic, biomedical, or communication devices. To address such demanding ultra-high-density packaging technology, nanotechnology, nanoscience, and bioengineering play an important role in this arena. For over a decade, nano-/biotechnologies such as carbon nanotubes (CNTs), nanoparticles, molecular self-assembly, nanoimprints, and sensors have been explored and studied for mechanical, physical, chemical, photonic, electronic, and biological properties. The packaging scientists prospect beyond nanomaterials or devices, aiming at system-level components (chip-to-chip or chip-to-package interconnection, thin-film components, dielectrics, passives, encapsulants, coatings, power sources, etc.) leading to nano-/biomodules. This emphasis is expected to lead to a number of commercial applications during the next decades. However, the implementation of nanotechnology in the system-level applications requires a thorough understanding of nano-/bioscience and engineering.

In this chapter, several fundamental researches related to nanomaterials for microelectronics, photonics, and bio-packaging are discussed, including carbon nanotubes (CNTs) for electrical/thermal interconnects, nanosolder materials with melting point depression, molecular wires for high-performance electrical interconnects, high-*k* nanocomposites, high thermally conductive nanocomposites, and bio-mimic self-cleaning nanosurfaces.

C. P. Wong (✉) · K. Moon
School of Materials Science and Engineering, Georgia Institute of Technology, Atlanta, GA, USA
e-mail: cp.wong@mse.gatech.edu; jack.moon@gatech.edu

1.1.1 Interfaces of Carbon Nanotubes (CNTs)/Substrates for Electrical and Thermal Interconnects

Carbon nanotubes (CNTs) have attracted much interest due to their extraordinary electrical, thermal, and mechanical properties and their wide range of potential applications [1–4]. The quality control of the CNTs grown, e.g., purity (catalyst residual and amorphous carbon), diameter, length, and defects, directly affect the final properties of CNTs. Unfortunately, the results reported so far vary much, and reproducible growth of high-quality CNTs is still an issue. Aligned CNT (ACNT) structures make the best use of the 1-D transport properties of individual CNTs and have been proposed as the promising candidate for thermal interface materials (TIM) and electrical interconnect materials, gas sensors and electrodes, etc. Chapter 3 will address the carbon materials in more detail.

However, there are challenges in the chemical vapor deposition (CVD) for ACNT fabrication, including its high growth temperature and few suitable substrate materials for the CVD. To address these key issues, three low-temperature ACNT transfer techniques to position ACNTs onto various surfaces are reported by Georgia Tech team, as shown in Fig. 1.1, which use in situ open-ended and surface functionalized, aligned multiwall-CNTs via (1) solder, (2) polymer adhesives, and (3) self-assembled monolayer as transfer media [4, 5]. Introducing a trace amount of oxidants during the CVD process can eliminate amorphous carbons on CNT surfaces, open the CNT tips, and controllably functionalize the tips and walls. Consequently, the CNTs are of high purity and superior electrical conductivity. A number of research papers on CNTs (recently on graphene – a monolayer of graphite, another emerging carbon material) have been published in the past decade resulting in the tremendous amount of findings on their material properties, structures, fabrications, manipulation, and applications. Nevertheless, due to a lack of knowledge on the defect control

C. P. Wong et al. (eds.), *Nano-Bio- Electronic, Photonic and MEMS Packaging*, https://doi.org/10.1007/978-3-030-49991-4_1

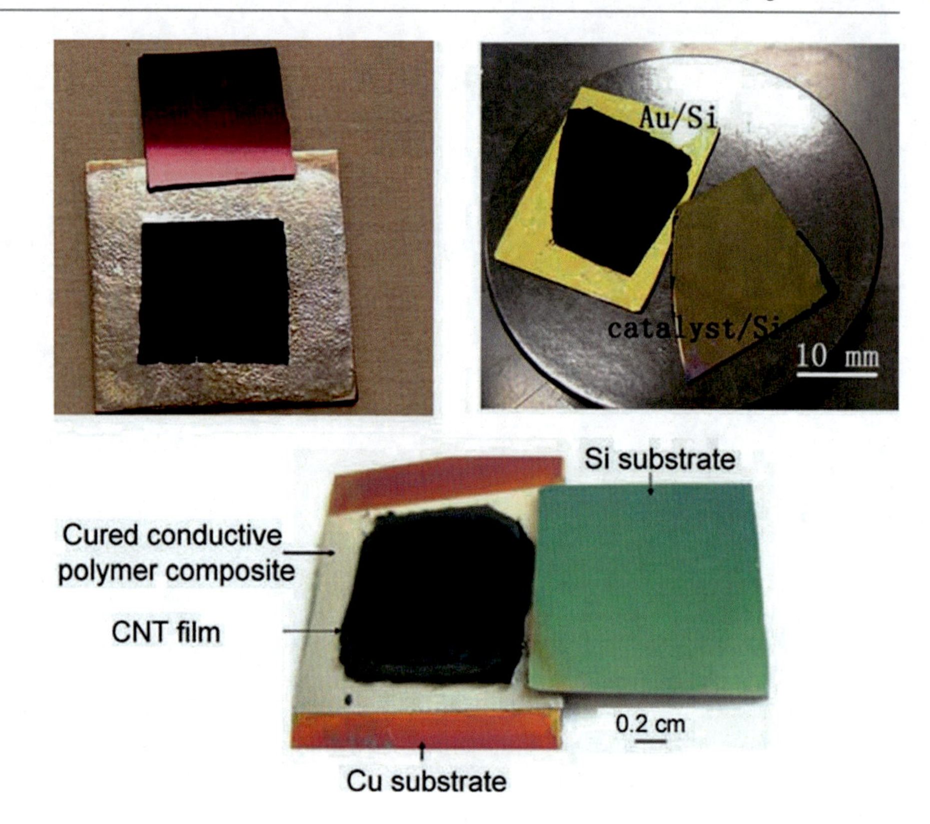

Fig. 1.1 Low-temperature ACNT transfer technology: solder transfer, adhesive transfer, and chemical transfer (from left to right in sequence)

and their interface manipulation, there are still barriers in using CNTs and graphenes in device-level applications. Therefore, more understanding on defects and interfaces should be made in the near future in order to fully utilize their exotic material properties.

1.1.2 Tin/Silver Alloy Nanoparticle Pastes for Low-Temperature Lead-Free Interconnect Applications

Tin/silver (96.5Sn3.5Ag) is one of the promising lead-free alternatives for a Sn/Pb solder that has been a de facto standard bonding material in microelectronics packaging [6]. However, the melting point (*T*m) of 96.5Sn3.5Ag alloy (221 C) is around 40°C higher than that of eutectic Sn/Pb solders. The higher *T*m requires the higher reflow temperature in the electronic manufacturing process, which has adverse effects not only on energy consumption but also on the package reliability. Therefore, studies on lowering the melting point of lead-free metals have been drawing much attention of electronic packaging engineers.

The Georgia Tech team has reported that different sizes of SnAg alloy nanoparticles were successfully synthesized by the chemical reduction method. An organic surfactant was used to not only prevent aggregation but also protect the SnAg alloy nanoparticles from oxidation. Both the particle size-dependent melting point depression and latent heat of fusion have been observed (Fig. 1.2). It has already been found that surface melting of small particles occurs in a continuous manner over a broad temperature range, whereas the homogeneous melting of the solid core occurs abruptly at the critical temperature *T*m [7]. For smaller-sized metal nanoparticles, the surface melting is strongly enhanced by curvature effects. Therefore with the decreasing of particle size, both the melting point and latent heat of fusion will decrease. Among all the synthesized particles, the 5 nm (radius) SnAg alloy nanoparticles have a melting point at ~194.3°C, which will be a good candidate for the low-melting temperature lead-free solders and can solve the issues from the high-temperature reflow for the micron-sized lead-free particles. The wetting properties of the (32 nm in radius) SnAg alloy nanoparticle pasted on the cleaned copper surface were studied in a 230°C oven with an air atmosphere. The cross section of the sample after reflow is shown in Fig. 1.3. It was observed that the SnAg alloy nanoparticles completely melted and wetted on the cleaned copper foil surface. The energy-dispersive spectroscopy (EDS) results revealed the formation of the intermetallic compound (Cu_6Sn_5), which showed scallop-like morphologies in Fig. 1.2.

This study has demonstrated the feasibility of the SnAg alloy nanoparticle pastes as a candidate for the low processing temperature lead-free interconnect applications.

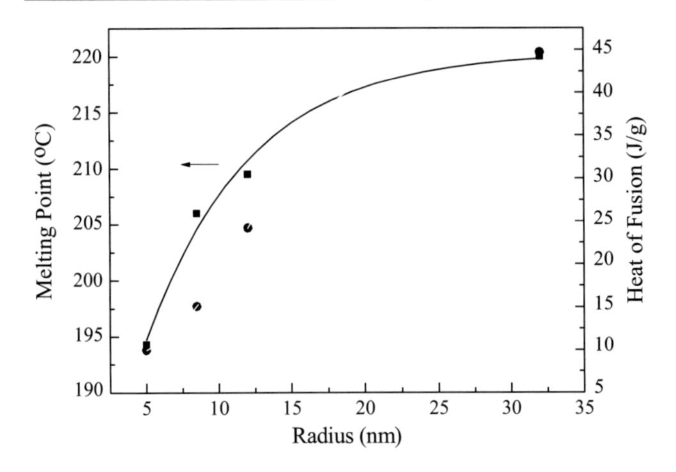

Fig. 1.2 Relationship between the radius of the as-synthesized SnAg alloy nanoparticles and their corresponding melting points and heat of fusion, respectively

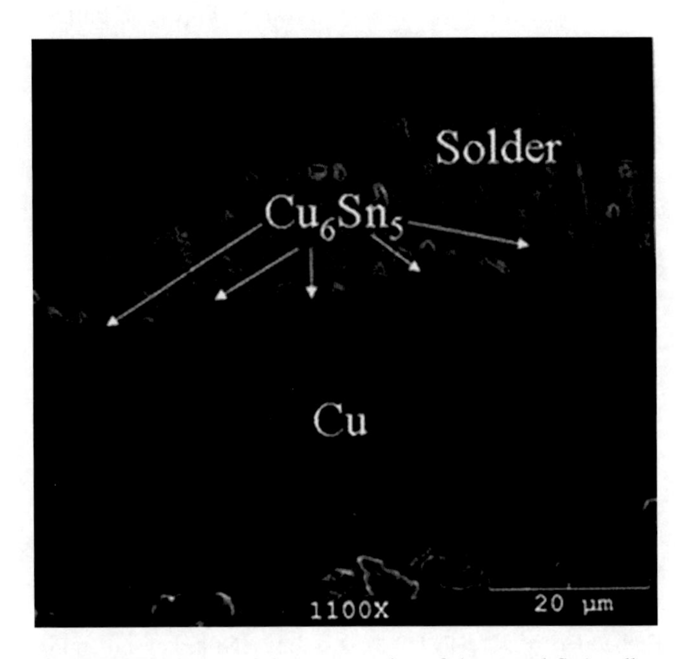

Fig. 1.3 SEM image of the cross section of the wetted SnAg alloy nanoparticles (32 nm in radius) on the cleaned copper foil

The melting point can be dramatically decreased when the size of substances is reduced to nanometer size [8]. To date, the size-dependent melting behavior has been proved both theoretically and experimentally [9–11]. The high ratio of the surface area to volume of nanoparticles has been known as one of the driving forces for the size-dependent melting point depression, which is employed currently in various areas, in particular, where low-temperature bonding, high-thermal or electrical conductivity is required. Chapter 5 addresses this as well.

1.1.3 Enhanced Electrical Properties of Anisotropically Conductive Adhesive with π -Conjugated Molecular Wire Junctions for Enhanced Electrical Properties

As microelectronic requirements are driven toward smaller, higher-density, and more cost-effective solutions, anisotropically conductive adhesives (ACAs) are widely used in electronics industry [12, 13]. The ACA joint is established by trapping conductive particles dispersed in the polymer matrix in between integrated circuit and conductive pads when heat and pressure are applied. ACAs can provide electrical conductivity in *Z*-direction as well a mechanical interconnect between integrated circuit and conductive pads. The most significant advantage of ACAs is fine pitch capability. However, the ACA joints have lower electrical conductivity and poor current-carrying capability due to the restricted contact area and poor interface between conductive particles and IC/conductive pads, compared to the metallurgical joint of the metal solders. In order to enhance the electrical performance of ACA joints, the Georgia Tech team has introduced π-conjugated self-assembled molecular wire junctions to improve the electrical properties of anisotropically conductive adhesives (ACAs) [14] and nonconductive films (NCFs) [15] for the electronic interconnects in previous work. Crucial for successful applications of self-assembly monolayers (SAMs) in ACAs and NCFs are interfacial properties (chemisorption or physisorption), the characteristics of the molecules (conjugated or unconjugated, thermal stability, and so on), surface binding geometry, and packing density. Therefore, current work investigates the effects of those factors on the performance of ACAs in detail by using a series of molecules with different functional groups. The electrical properties of an anisotropic conductive adhesive (ACA) joint were investigated using a submicron-sized (~500 nm in diameter) silver (Ag) particle as conductive filler with the effect of π-conjugated self-assembled molecular wires (a). The ACAs with submicron-sized Ag particles have higher current-carrying capability (~3400 mA) than those with micro-sized Au-coated polymer particles (~2000 mA) and Ag nanoparticles (~2500 mA). More importantly, by construction of π-conjugated self-assembled molecular wire junctions between conductive particles and integrated circuit (IC)/substrate, the electrical conductivity has increased by one order of magnitude, and the current-carrying capability of ACAs has improved by 600 mA (Fig. 1.4).

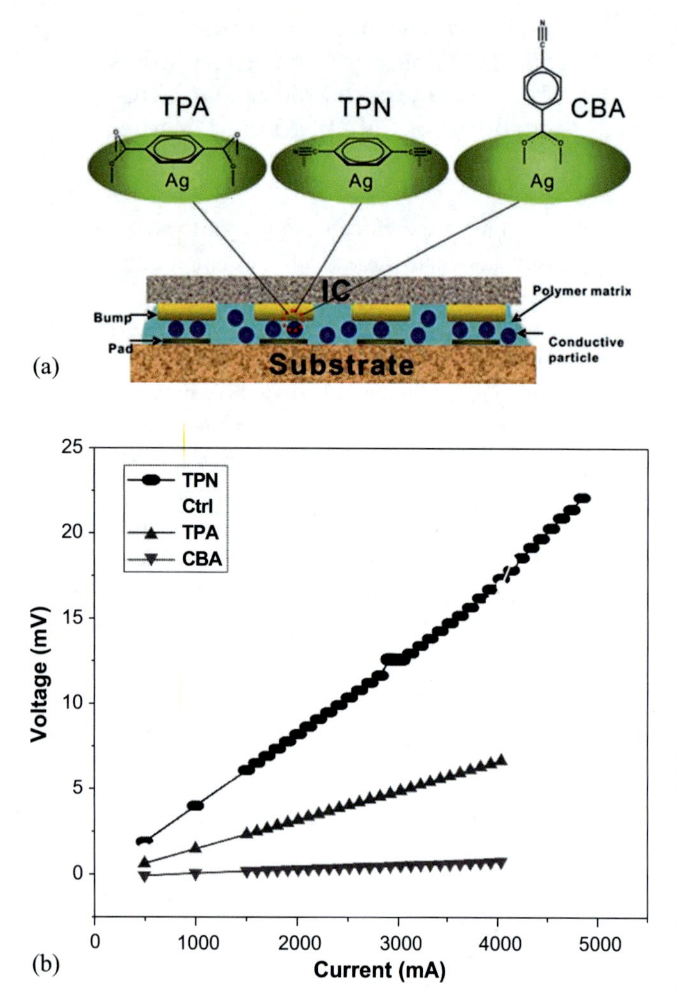

Fig. 1.4 (**a**) Cross section of ACA joints with π-conjugated molecular wire junctions; (**b**) *I–V* curve of ACA filled with Ag particles (Ctrl), TPN-treated Ag particles, TPA-treated Ag particles, and CBA-treated particles

1.1.4 Low-Stress and High-Thermal Conductive Underfill for Cu/Low-k Application

The underfill materials have been used to improve the flip-chip device reliability by effectively distributing thermal stress of the package evenly to all solder joints/balls versus concentrating on the farthest to the distance neutral point (DNP), resulting in dramatically enhancing solder joint fatigue life [16–18]. In addition to using low modulus and CTE underfills for reducing the stress, researches also show the thermally conductive underfills can enhance the reliability as well, because of its thermal dissipative capability [18]. Therefore, the high-thermal conductive underfill material is expected to play an important role to improve the Cu/low-*k* device reliability, as the thermal management is becoming the bottleneck of electronic packaging technology due to the reduced device profile, increased package density (e.g., 3D packaging), and application of porous materials in Cu/low-*k* device. However, there is no underfill material available, which can absorb such high-thermal stress and dissipate heat to protect the next-generation Cu/low or ultra-low-*k* and high-power devices. Consequently, a novel high-performance underfill/polymer composite material is required for the highly reliable next-generation fine pitch Cu/low-*k* devices.

Optimizing the thermal conductivity and mechanical properties of these high-performance underfill materials by surface modification of high-thermal conductive inorganic filler has been focused. SiC particles are high thermally conductive but have relatively low amounts of the hydroxyl group on the particle surfaces, hence the difficulty to be involved in the surface reaction and poor compatibility with the epoxy resin. The SiC particle surfaces were thermally coated with a nanolayer of SiO2 by oxidizing Si atoms on the surface at high temperature, as shown in Fig. 1.5a. Further surface modification of SiC by γ-glycidoxypropyl-trimethoxysilane (GOP) (Fig. 1.5b) was applied to graft epoxide groups onto SiC surface. The surface modification improves the chemical bonding density between filler and polymer matrix and reduces the surface phonon scattering, resulting in the thermal conductivity improvement in epoxy/SiC composites. Moreover, results showed that the added silane improved the adhesion strength between the underfill and the silicon wafer. Also, mechanical property measurements indicated the relatively low moduli and low coefficient of thermal expansion, which will help to reduce the global thermal stress of the underfilled flip-chip structure.

1.1.5 High-Dielectric Constant (k) Polymer Nanocomposites for Embedded Capacitor Applications

Embedded passive technology is one of the key emerging techniques for realizing the system integration such as system-in-package (SiP) and system-on-package (SOP), which offers various advantages over traditional discrete components, including higher component density, increased functionality, improved electrical performance, increased design flexibility, improved reliability, and reduced unit cost. Novel materials for embedded capacitor applications are in great demand, for which high dielectric constant (*k*), low dielectric loss, and process compatibility with printed circuit boards (PCBs) are the most important material prerequisites. Chapter 6 covers the embedded component packaging in more detail.

Dielectric nanocomposites based on nanoparticles with controlled parameters were designed and developed to fulfill the balance between sufficiently high-*k* and low dielectric loss and satisfy the requirements to be a feasible option for

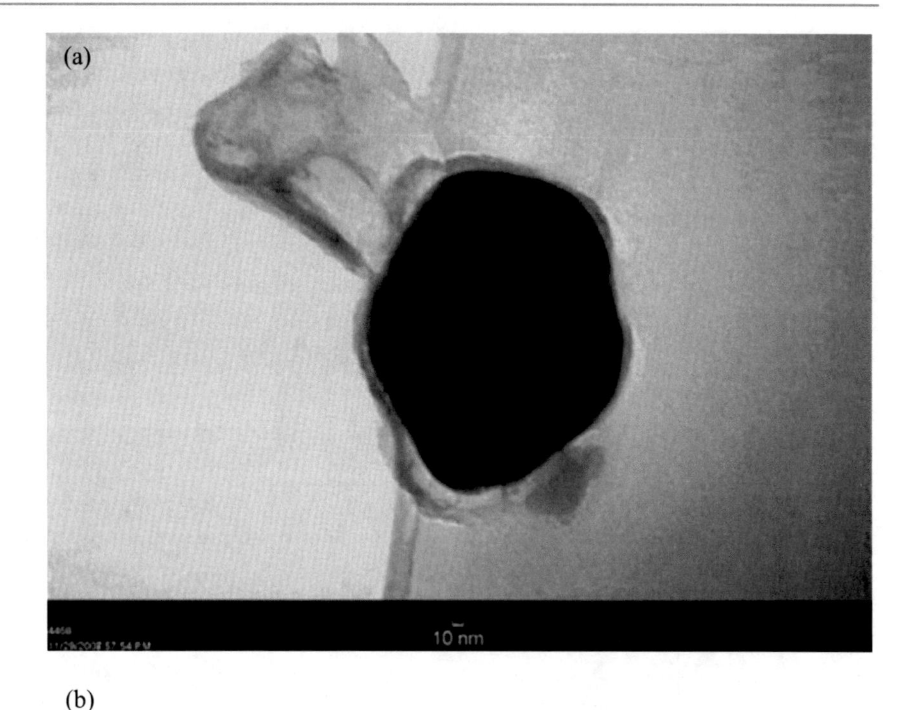

Fig. 1.5 (a) TEM image of a SiC particle thermally coated by a nanolayer SiO2; (b) surface modification of thermally coated SiC particle by silane

embedded capacitor applications [19]. To this end, people have explored the use of material and process innovations to reduce the dielectric loss sufficiently while maintaining the high-*k* of the nanocomposites [20, 21]. They take advantage of the high-*k* polymer matrix and the interparticle barrier layer formed intrinsically by an insulating shell around the conducting metal core to achieve a simultaneously high-*k* and low dielectric loss conductive filler/polymer composite. First, a silver (Ag)-epoxy mixture was prepared via an in situ photochemical method, in which Ag nanoparticles were generated by chemical reduction of a metallic precursor upon UV irradiation. Figure 1.6 displays TEM micrograph and histogram of Ag nanoparticles synthesized via an in situ photochemical reduction in epoxy resin. The particle size analysis results show that well-dispersed Ag nanoparticles with the size mostly smaller than 15 nm in polymer matrix were obtained. The average size was calculated as 5.2 nm [22].

The as-prepared Ag-epoxy mixture was then utilized as a high-*k* polymer matrix to host self-passivated aluminum (Al) fillers for preparation of Al/Ag-epoxy composites as embedded capacitor candidate materials. Figure 1.7 shows the dielectric properties of Al/epoxy composites and Al/Ag-epoxy composites as a function of Al filler loading at a frequency of 10 kHz. The Al/Ag-epoxy composites exhibit more than 50% increase in *k* values compared with Al/epoxy composites with the same filler loading of Al, which demonstrated that the incorporation of well-dispersed Ag nanoparticles in the polymer matrix delivers a huge impact on the dielectric constant enhancement. It is also notable that the dielectric loss tangent values (see the inset of Fig. 1.7) for Al/Ag-epoxy composites with different Al filler loadings are all below 0.1, which is tolerable for some applications such as decoupling capacitors. The low dielectric loss can be attributed to the good dispersion of Ag nanoparticles in the polymer matrix together with the thin self-passivated aluminum oxide layer forming an insulating boundary outside of the Al particles. The results suggest that the metal-polymer nanocomposites via an in situ photochemical approach can be employed as a high-*k* polymer matrix to enhance the *k* value of the composites while maintaining the relatively low dielectric loss tangent.

The trends of the dielectric properties with the variation of filler loadings behaved differently for Al/epoxy composites and Al/Ag-epoxy composites. This phenomenon can be explained in terms of the origins and mechanisms of the

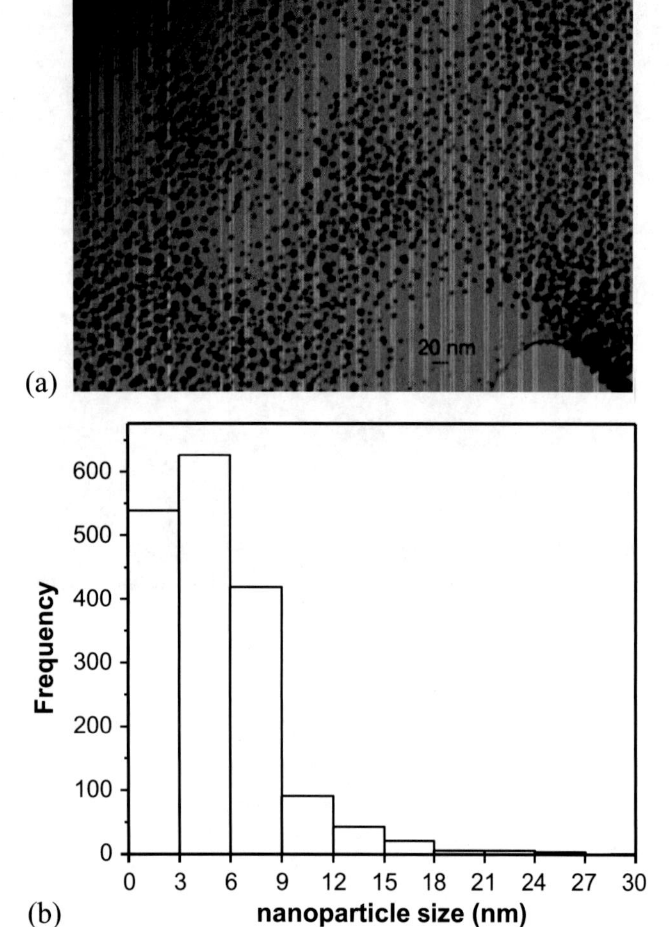

Fig. 1.6 (**a**) TEM micrograph and (**b**) histogram of Ag nanoparticles synthesized by an in situ photochemical reduction in epoxy resin

dielectric loss of the heterogeneous conductive filler/polymer composite material. The concentration of silver nanoparticles in the polymer matrix plays a significant role in determining the dielectric properties of the high-*k* nanocomposites. At low Ag concentration, the dielectric behaviors of Al/Ag-epoxy composites are mainly determined by interfacial polarization, while conduction and electron transport of Ag dominate the Al/Ag-epoxy composites at higher Ag concentration [22].

1.1.6 Bio-Mimetic Lotus Surface

The effect of surface structures on superhydrophobicity (SH) is of much interest due to the dependence of a structure on the attainment of a high water contact angle (>150°) and reduced contact angle hysteresis (<10°). Superhydrophobicity was first observed on lotus leaves where high water contact angle and low hysteresis cause water droplet that falls on the surface to bead up and roll off the surface, thereby leading to water repellency and self-cleaning characteristics [23–25]. Such surface properties are also critical for antistiction in microelectromechanical system (MEMS) [26], friction reduction [27], and anticorrosion [28] applications. In addition, superhydrophobic surfaces offer much promise for the formation of high-performance micro-/nanostructured surfaces with multifunctionality that can be used in optoelectronic [29, 30], photoelectric [31], microelectronic, catalytic, and biomedical applications. In particular, its antireflection application in photovoltaic (PV) solar cell devices for enhancing the efficiency has been considered very important since a PV cell is a critical

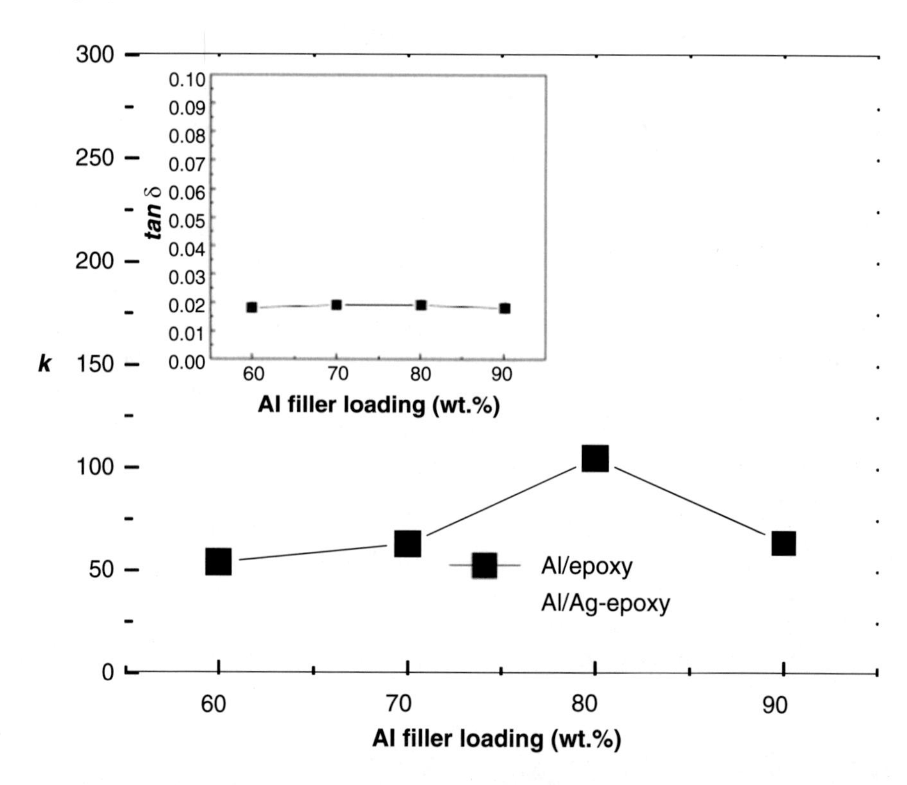

Fig. 1.7 Dielectric constant (*k*) values of Al/epoxy and Al/Ag-epoxy composites as a function of Al filler loading (at 10 kHz). Dielectric loss tangent (tan δ) values of Al/epoxy and Al/Ag-epoxy composites are displayed in the inset

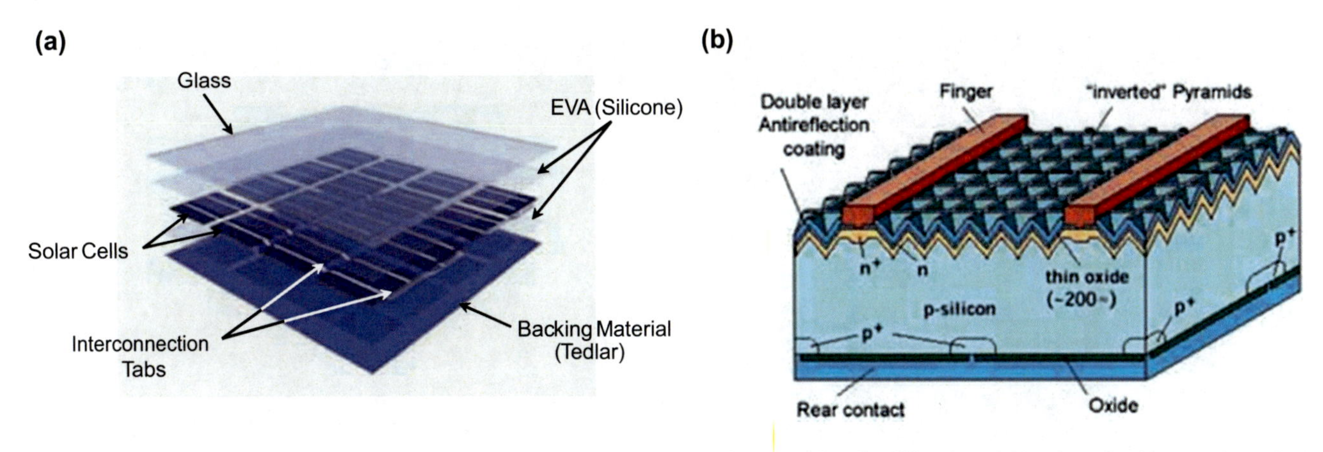

Fig. 1.8 (a) General structure of a PV module; (b) structure of passivated emitter, real locally diffused (PERL) solar cell with near theoretical efficiency

Fig. 1.9 Schematics of two-tier surface roughness fabrication: (a) only micron-sized roughness and (b) additional nano-sized surface roughness

component for renewable energy developments. Figure 1.8 shows the typical PV module package structures where incident sunlight transmits through a cover glass and a polymer layer and arrives at the top surface of a silicon cell. Light reflection at the silicon top surface can significantly reduce the cell efficiency.

The Georgia Tech team used a metal-assisted etching technique to form nanoscale roughness and thereby form hierarchical structures by metal-assisted etching of micron-size pyramid-textured surfaces (Fig. 1.9). These structures yield superhydrophobic surfaces, which may have important application in light harvesting, antireflection, and light-trapping properties of solar cells.

A SiN coating on the pyramid-textured silicon surface has been employed to reduce the reflection. Recently, a nanosurface with two-tier roughness was found to provide not only superhydrophobicity with self-cleaning capability but also very low light reflectivity on a silicon cell even without SiN coating [32]. Because of the presence of surface nanostructures, the surface absorbs most of the incident light, thus reducing reflection, especially in the 300–1000 nm wavelength regime. The nanotextured surface may also increase the path length of light as it travels through the cell, which allows thinner solar cells with reduced cost; furthermore, the surface may trap the weakly absorbed light reflected from the back surface by total internal reflection at the front surface/air interface. On the micro-/nanostructured two-scale surfaces (Fig. 1.10), the weighted light reflectance is further reduced to 3.8%, as shown in Table 1.1. More studies on nanostructure surfaces are introduced in Chap. 15.

1.1.7 Molecular Dynamic (MD) Simulations in Nanomaterial Study

Although a variety of nanomaterials have been studied and considered in many applications, a number of fundamental questions on their individual behavior have been asked, such

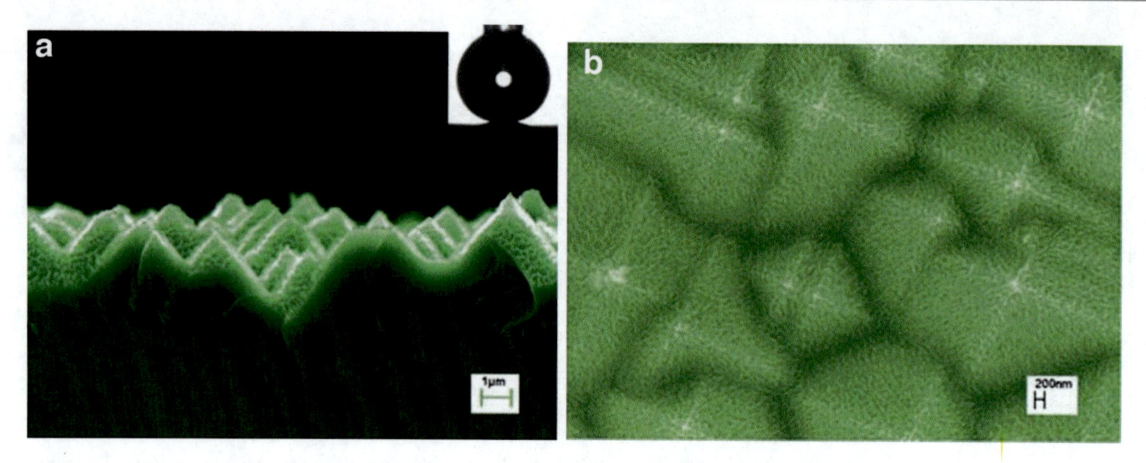

Fig. 1.10 (**a**) Cross section and (**b**) top view of Si pyramid surfaces etched made by Georgia Tech team

Table 1.1 Weighted reflectance on different textured surfaces

Sample	Weighted reflectance %
Flat Si surface	37.3
Pyramid-textured surface	12.3
Nanotextured surface	6.4
Two-scale textured surface	3.8

as thermal behavior during cooling or heating, how they react with different organic molecules in specific solvents, and thermal transport at complicated interfaces, because the deep understanding of their individual behavior enables scientists and engineers to manipulate the nanomaterials appropriately and utilize their properties fully. To answer those questions, many researchers have been reporting their experimental results, which have been obtained and analyzed by sophisticated equipments. However, those experimental approaches still have technical limitations such as characterization resolutions, noise, and reproducibility.

In particular, for thermal behavior, various equipments such as transmission electron microscopy and electron diffraction [33, 34], x-ray diffraction (XRD) [35], and differential scanning calorimetry (DSC) [36, 37] have been applied to monitor the melting behavior of the nano-sized materials. At the melting point, there are changes in the diffraction pattern associated with the disordering of the structure; thus XRD and electron diffraction have been used to investigate the melting of the nanomaterials. Since melting is associated with the heat of a system, DSC is also used to directly measure the thermal properties of the nanomaterials. However, experimental techniques are normally incapable of resolving the melting temperature of a single nanoparticle, and the distribution in size of nanoparticles often obscures the physics behind melting. Also, experimental measurements have another limitation, i.e., the melting of particles is influenced by the type of the substrate on which the nanoparticles are placed. Moreover, experiments cannot give detailed information on the atomic-level behavior at elevated temperatures since direct observations are extremely difficult. As such, experimental methods have some technical limitations when studying the melting behavior of nanoparticles.

Theoretical models based on the kinetics, thermodynamics, and mechanical properties of solids that are typically bulk crystals have been employed. Those approaches also have limitations since they do not have the information about the individual atoms comprising the nanoparticle [38]. For the nano-scale substances, a more atomistic approach is necessary. In particular, molecular dynamic (MD) simulation is an effective way of investigating the coalescence or melting behavior at an atomic level. It is generally more difficult and more expensive to study the properties of a microstructure experimentally than to perform the corresponding simulation. As such, the MD simulation technique has been applied by several researchers to explore the melting behavior of nanoparticles [39–41], including the melting point depression phenomenon and the role of extrinsic defects (grain boundaries and free surfaces) in the melting process. It was found that disorder and the melting transition were initiated at the interface in the simulations. More details on MD simulation will be shown in Chap. 24.

The thermal behavior and interaction of single elements or alloy nanoparticles on certain substrates [42–48] are very interesting topics because various applications have already been employing nanomaterials in their systems to improve electrical and thermal properties. For instance, how the individual nanoparticles behave or interact with adjacent particles during heating and cooling as a function of time can be beautifully visualized as shown in Fig. 1.11.

The modified embedded atom method (MEAM) is popularly used in conjunction with MD simulations to investigate the effect of different particle sizes and temperature ramp-up rates on the melting behavior of nanoparticles. The MEAM is an empirical extension of the embedded atom method

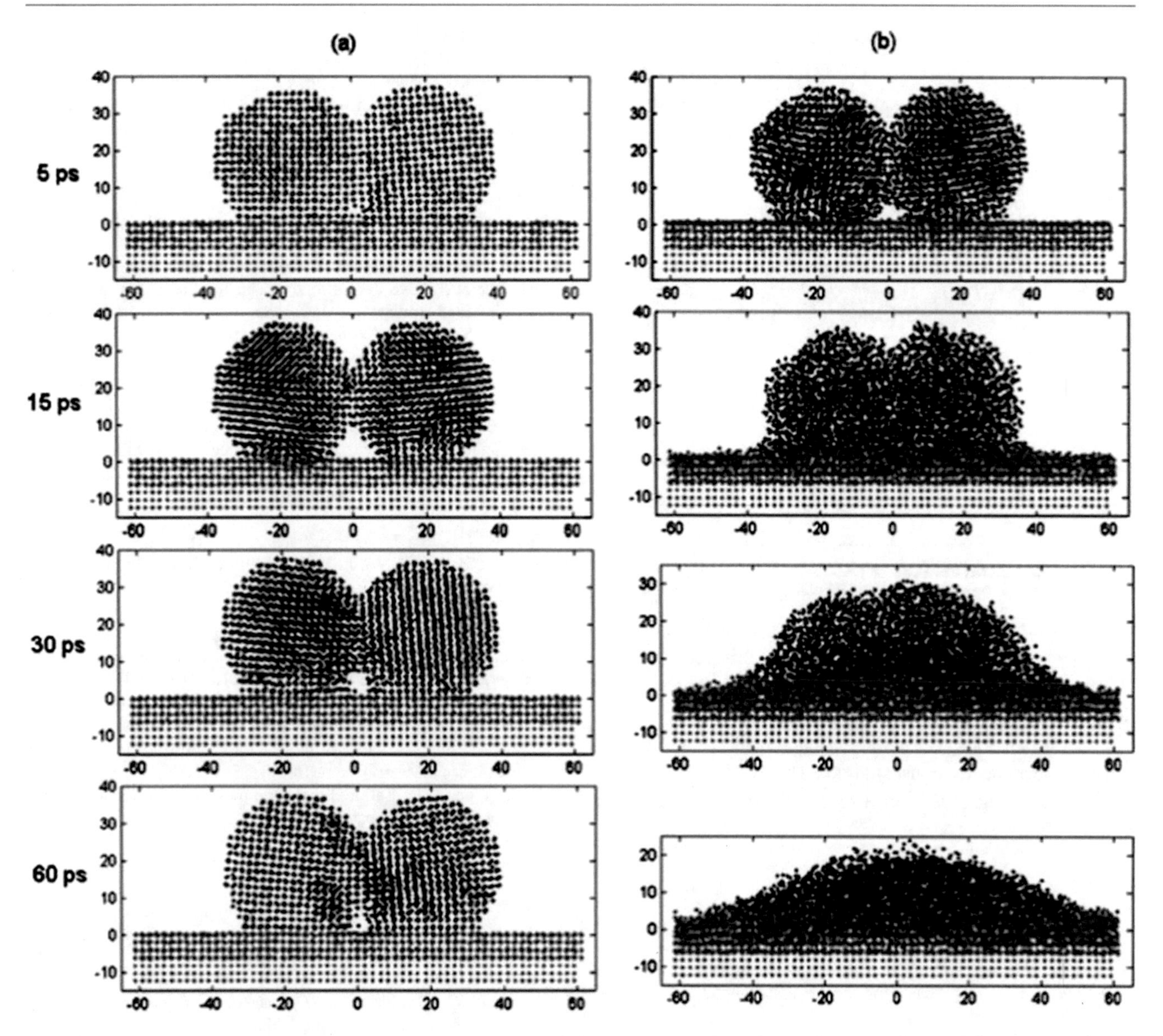

Fig. 1.11 The *x-z* plane projection of the system with simulation temperatures of (**a**) 400 K and (**b**) 1000 K and simulation time of 5, 15, 30, and 60 ps, respectively [44]

(EAM), which can be applied to both metallic and covalent bonding by including angular forces [49, 50]. In those modeling studies, it is not simple to define what temperature will be the melting point of particles, and a heating rate is another of this complication. Thus, the evolution of potential energy with respect to temperature is used to obtain the melting temperature of nanoparticles, and the two-phase equilibrium method is used to identify the melting points of nanoparticles with different sizes [46]. Also the pair correlation function (PCF) is used to monitor the crystal regularity of particles and study the structure differences of shells and cores of a particle [51]. Comparisons of these researches with thermodynamic models [52] showed that the thermodynamic models were not able to correctly predict the melting points of nanoparticles with very small sizes.

Figure 1.12 demonstrated that atoms located in the outer shell of the particle lost their structural order before the atoms in the core region did. This result confirmed the surface melting behavior of nanoparticles of this size. The MD simulation has been proved as a very powerful tool for exploring nanomaterial systems at atomistic scale, by itself as well as combined with other simulation methods and has been demonstrating its strong potential in this arena. We can expect that a numerous current and future questions on such a small-scale world could be answered through the powerful computational studies together with sophisticated experimental methods.

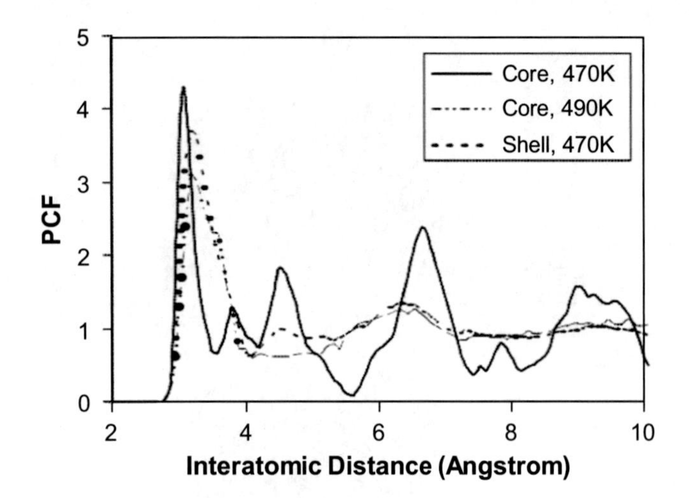

Fig. 1.12 PCF of a Sn particle; heating rate 0.1 K/ps, diameter 8 nm, shell width is 1 nm, core radius is 3 nm [46]

1.2 Conclusive Remarks

While a variety of researches on nanomaterials for nano-bio devices and systems have been carried out, this field is still at its infancy. However, there are many potential applications in electronic, photonic, and bio-medical applications, which will benefit our society. Therefore, much research and development work need to be studied. This is an area where scientists and engineers need to work together to make this goal a reality.

References

1. Zhu, L.B., Wong, C.P.: Well-aligned open-ended carbon nanotube architectures: an approach for device assembly. Nano Lett. **6**, 243 (2006)
2. Zhu, L.B., Wong, C.P.: Aligned carbon nanotube stacks by water-assisted selective etching. Nano Lett. **5**, 2641 (2005)
3. Jiang, H.J., Moon, K., Wong, C.P.: The preparation of stable metal nanoparticles on carbon nanotubes whose surfaces were modified during production. Carbon. **45**, 655 (2007)
4. Jiang, H.J., Moon, K., Wong, C.P.: Low temperature carbon nanotube film transfer via conductive polymer composites. Nanotechnology. **18**, 125203 (2007)
5. Lin, W., Xiu, Y., Jiang, H., Zhang, R., Hildreth, O., Moon, K., Wong, C.P.: Self-assembled monolayer-assisted chemical transfer of in situ functionalized carbon nanotubes. J. Am. Chem. Soc. **30**, 9636–9637 (2008)
6. Artaki, I., Jackson, A.M., Vianco, P.T.: Evaluation of lead-free solder joints in electronic assemblies. J. Electron. Mater. **23**(8), 757 (1994)
7. Hu, W.Y., Xiao, S.G., Yang, J.Y., Zhang, Z.: Melting evolution and diffusion behavior of vanadium nanoparticles. Eur. Phys. J. B. **45**(4), 547 (2005)
8. Pawlow, P.Z.: The dependency of melting point on the surface energy of a solid body. (Supplement). Phys. Chem. **65**(5), 545 (1909)
9. Lai, S.L., Guo, J.Y., Petrova, V., Ramanath, G., Allen, L.H.: Size-dependent melting properties of small tin particles: nanocalorimetric measurements. Phys. Rev. Lett. **77**(1), 99 (1996)
10. Bachels, T., Guntherodt, H.J., Schafer, R.: Melting of isolated tin nanoparticles. Phys. Rev. Lett. **85**(6), 1250 (2000)
11. Zhao, S.J., Wang, S.Q., Cheng, D.Y., Ye, H.Q.: Three distinctive melting mechanisms in isolated nanoparticles. J. Phys. Chem. B. **105**(51), 12857 (2001)
12. Li, Y., Moon, K., Wong, C.P.: Electronics without lead. Science. **308**(141), 9 (2005)
13. Li, Y., Wong, C.P.: Recent advances of conductive adhesives as a lead-free alternative in electronic packaging: materials, processing, reliability and applications. Mater. Sci. Eng. R. Rep. **R51**, 1 (2006)
14. Li, Y., Moon, K., Wong, C.P.: Adherence of self-assembled monolayers on gold and their effects for high-performance anisotropic conductive adhesives. J. Electron. Mater. **34**, 266 (2005)
15. Li, Y., Yim, M., Wong, C.P.: High performance nonconductive film with π-conjugated self-assembled molecular wires for fine pitch interconnect applications. J. Electron. Mater. **36**, 549 (2007)
16. Tsukada Y.: Surface laminar circuit packaging. Proceedings of the 42nd Electronic Components and Technology Conference, p. 22 (1992)
17. Han, B., Guo, Y.: J. Electron. Packag. **117**, 185 (1995)
18. Lee, W., Han, I.Y., Yu, J., Kim, S.J., Byun, K.Y.: Thermal characterization of thermally conductive underfill for a flip-chip package using novel temperature sensing technique. Thermochim. Acta. **455**, 148–155 (2007)
19. Ulrich, R.K., Schaper, L.W.: Integrated Passive Component Technology. IEEE Press, Wiley-Interscience, Hoboken (2003)
20. Lu, J., Moon, K., Xu, J., Wong, C.P.: Synthesis and dielectric properties of novel high-K polymer composites containing in-situ formed silver nanoparticles for embedded capacitor applications. J. Mater. Chem. **16**(16), 1543 (2006)
21. Lu J., Wong C.P.: Tailored dielectric properties of high-k polymer composites via nanoparticle surface modification for embedded passives applications. IEEE Proceedings of the 57th Electronic Components and Technology Conference, Reno, NV, pp. 1033–1039 (2007)
22. Lu, J., Moon, K., Wong, C.P.: Silver/polymer nanocomposite as a high-k polymer matrix for dielectric composites with improved dielectric performance. J. Mater. Chem. https://doi.org/10.1039/B807566B (2008)
23. Neinhuis, C., Barthlott, W.: Characterization and distribution of water-repellent, self- cleaning plant surfaces. Ann. Bot. **79**, 667 (1997)
24. Barthlott, W., Neinhuis, C.: The purity of sacred lotus or escape from contamination in biological surfaces. Planta. **202**, 1 (1997)
25. Neinhuis, C., Koch, K., Barthlott, W.: Movement and regeneration of waxes through plant cuticles. Planta. **213**, 427 (2001)
26. Bhushan, B., Jung, Y.C.: Micro- and nanoscale characterization of hydrophobic and hydrophilic leaf surfaces. Nanotechnology. **17**, 2758 (2006)
27. Cottin-Bizonne, C., Barrat, J.-L., Bocquet, L., Charlaix, E.: Low-friction flows of liquid at nanopatterned interfaces. Nat. Mater. **2**, 237 (2003)
28. Wang, S.T., Feng, L., Jiang, L.: One-step solution-immersion process for the fabrication of stable bionic superhydrophobic surfaces. Adv. Mater. **18**, 767 (2006)
29. Chattopadhyay, S., Li, X., Bohn, P.W.: In-plane control of morphology and tunable photoluminescence in porous silicon produced by metal-assisted electroless chemical etching. J. Appl. Phys. **91**, 6134 (2002)
30. Campbell, P., Green, M.A.: Light trapping properties of pyramidally textured surfaces. J. Appl. Phys. **62**, 243 (1987)

31. Koynov, S., Brandt, M.S., Stutzmann, M.: Black nonreflecting silicon surfaces for solar cells. Appl. Phys. Lett. **88**, 203107.1–203107.3 (2006)
32. Xiu, Y., Zhang, S., Yelundur, V., Rohatgi, A., Hess, D., Wong, C.P.: Superhydrophobic and low light reflectivity silicon surfaces fabricated by hierarchical etching. Langmuir. **24**, 10421–10426 (2008)
33. Yasuda, H., Mitsuishi, K., Mori, H.: Particle-size dependence of phase stability and amorphouslike phase formation in nanometer-sized Au-Sn alloy particles. Phys. Rev. B. **64**, 094101 (2001)
34. Lee, J., Mori, H.: Solid-liquid two-phase structures in isolated nanometer-sized alloy particles. Phys. Rev. B. **70**, 144105 (2004)
35. Shi, F.G.: Size dependent thermal vibrations and melting in nanocrystals. J. Mater. Res. **9**, 1307 (1994)
36. Koch, C.C., Jang, J.S.C., Gross, S.S.: The melting point depression of tin in mechanically milled tin and germanium powder mixtures. J. Mater. Res. **4**, 557 (1989)
37. Zhang, M., Yu, M., Schiettekatte, F., Olson, E., Kwan, A., Lai, S., Wisleder, T., Greene, J., Allen, L.: Size-dependent melting point depression of nanostructures: nanocalorimetric measurements. Phys. Rev. B. **62**, 10548 (2001)
38. Zhao, M., Zhou, X.H., Jiang, Q.: Comparison of different models for melting point change of metallic nanocrystals. J. Mater. Res. **16**, 3304 (2001)
39. Phillpot, S.R., Lutsko, J.F., Wolf, D., Yip, S.: Molecular-dynamics study of lattice-defectnucleated melting in silicon. Phys. Rev. B. **40**, 2831 (1989)
40. Nguyen, T., Ho, P.S., Kwok, T., Nitta, C., Yip, S.: Grain-boundary melting transition in an atomistic simulation model. Phys. Rev. Lett. **57**, 1919 (1986)
41. Zhao, S.J., Wang, S.Q., Zhang, T.G., Ye, H.Q.: A cluster-induced structural disorder and melting transition in grain-boundary of B2 NiAl: a molecular-dynamics simulation on parallel computer. J. Phys. Condens. Matter. **12**, L549 (2000)
42. Dong, H., Zhang, Z., Wong, C.P.: Molecular dynamics study of a nano-particle joint for potential lead-free anisotropic conductive adhesives applications. J. Adhes. Sci. Technol. **19**(2), 87–94 (2006)
43. Dong, H., Moon, K., Wong, C.P.: Molecular dynamics study on coalescence of Cu nanoparticles and their deposition on Cu substrate. J. Electron. Mater. **33**, 1326–1330 (2004)
44. Dong, H., Moon, K.S., Wong, C.P.: Molecular dynamics study of nanosilver particles for low-temperature lead-free interconnect applications. J. Electron. Mater. **34**(1), 40 (2005)
45. Dong, H., Moon, K., Jiang, H., Wong, C.P., Hua, F.: Simulation study of nanoparticle melting behavior for lead free nano solder application. 11th International Symposium on Advanced Packaging Materials: Processes, Properties and Interface, 15–17 March 2006, p. 107
46. Dong, H., Moon, K.-S., Jiang, H., Baskes, M.I., Hua, F., Wong, C. P.: Computer simulation of nano tin melting behavior: Effect of particle size temperature ramping up rate. American Chemical Society National Meeting & Exposition, March 29, 2006 (Invited)
47. Alavi, S.: Molecular dynamics simulations of the melting of aluminum nanoparticles. J. Phys. Chem. A. **110**(4), 1518–1523 (2006)
48. Yang, Z., Yang, X., Xu, Z.: Molecular dynamics simulation of the melting behavior of Pt−Au nanoparticles with core−shell structure. J. Phys. Chem. C. **112**(13), 4937–4947 (2008)
49. Baskes, M.I.: Modified embedded-atom potentials for cubic materials and impurities. Phys. Rev. B. **46**, 2727 (1992)
50. Ravelo, R., Aguilar, J., Baskes, M.I.: Molecular-dynamics studies of thin films of Sn on Cu. Mater. Res. Soc. Symp. Proc. **492**, 43 (1998)
51. Baskes, M.I.: Determination of modified embedded atom method parameters for nickel. Mater. Chem. Phys. **50**, 152 (1997)
52. Hanszen, K.: Theoretische Untersuchungen über den Schmelzpunkt kleiner Kügelchen. Z. Phys. **157**, 523–553 (1960)

2 Nano-conductive Adhesives for Nano-electronics Interconnection

Yi Li, Kyoung-sik (Jack) Moon, and C. P. (Ching-Ping) Wong

2.1 Introduction

Miniaturization of electronic and photonic devices is a demanding trend in current and future technologies including consumer products, military, biomedical, and space applications. In order to achieve this miniaturization, new packaging technologies have emerged such as embedded active and passive components and 3D packaging, where new packaging designs and materials are required.

Fine-pitch interconnection is one of the most important technologies for the device miniaturizations. Metal solders including eutectic tin/lead solder and lead- free alloys are the most common interconnect materials in the electronic/ photonic packaging areas.

Conventional eutectic tin/lead solder or lead-free solders cannot meet the requirements for fine-pitch assembly due to their stencil printing resolution limit and bridging issues of solders in fine-pitch bonding, which is an intrinsic characteristic of metal solders.

Recently, polymer-based nanomaterials have been intensively studied to replace the metal-based interconnect materials, and many research efforts to enhance material properties of the electrically conductive adhesives (ECAs) are ongoing. Furthermore, incorporating nanotechnology into the ECA systems not only enables the ultra-fine-pitch capability but also enhances the electrical properties such as reducing the percolation threshold and improving the electrical conductivity.

In this chapter, recent research and study trends on ECAs and their related nanotechnologies are discussed, and the future of the nanomaterial-based polymer composites for fine-pitch interconnection is reviewed.

Electrically conductive adhesives (ECAs) are composites of polymeric matrices and electrically conductive fillers. The polymeric resin, such as an epoxy, a silicone, or a polyimide, provides physical and mechanical properties such as adhesion, mechanical strength, impact strength, and the metal filler (such as silver, gold, nickel, or copper) conducts electricity. Metal-filled thermoset polymers were first patented as electrically conductive adhesives in the 1950s [1–3]. Recently, ECA materials have been identified as one of the major alternatives for lead-containing solders for microelectronics packaging applications. ECAs offer numerous advantages over conventional solder technology, such as environmental friendliness, mild processing conditions (enabling the use of heat-sensitive and low-cost components and substrates), fewer processing steps (reducing processing cost), low stress on the substrates, and fine-pitch interconnect capability (enabling the miniaturization of electronic devices) [4–9]. Therefore, conductive adhesives have been used in flat panel displays such as LCD (liquid crystal display) and smart card applications as an interconnect material and in flip-chip assembly, CSP (chip-scale package), and BGA (ball grid array) applications in replacement of the solder. However, no currently commercialized ECAs can replace tin-lead metal solders in all applications due to some challenging issues such as lower electrical conductivity, conductivity fatigue (decreased conductivity at elevated temperature and humidity aging or normal use condition) in reliability testing, limited current-carrying capability, and poor impact strength. Table 2.1 gives a general comparison between tin-lead solder and a generic commercial ECA [6].

Depending on the conductive filler loading level, ECAs are divided into isotropically conductive adhesives (ICAs), anisotropically conductive adhesives (ACAs), and nonconductive adhesives (NCAs). For ICAs, the electrical conductivity in all *x*-, *y*-, and *z*-directions is provided due to high filler content exceeding the percolation threshold. For ACAs or NCAs, the electrical conductivity is provided only in *z*-direction between the electrodes of the assembly. Figure 2.1 shows the schematics of the interconnect structures and typical cross-sectional images of flip-chip joints by ICA, ACA,

Y. Li (✉) · K. Moon · C. P. Wong
Intel Corporation, Chandler, AZ, USA
e-mail: yi.li@intel.com

C. P. Wong et al. (eds.), *Nano-Bio- Electronic, Photonic and MEMS Packaging*, https://doi.org/10.1007/978-3-030-49991-4_2

Table 2.1 Conductive adhesives compared with solder

Characteristic	Sn/Pb solder	ECA (ICA)
Volume resistivity typical junction R	0.000015 Ω cm 10–15 mW	0.00035 Ω cm <25 mW
Thermal conductivity	30 W/m-deg.K	3.5 W/m-deg.K
Shear strength	2200 psi	2000 psi
Finest pitch Minimum processing temperature	300 μm 215 °C	<150–200 μm <150–170 °C
Environmental impact	Negative	Very minor
Thermal fatigue	Yes	Minimal

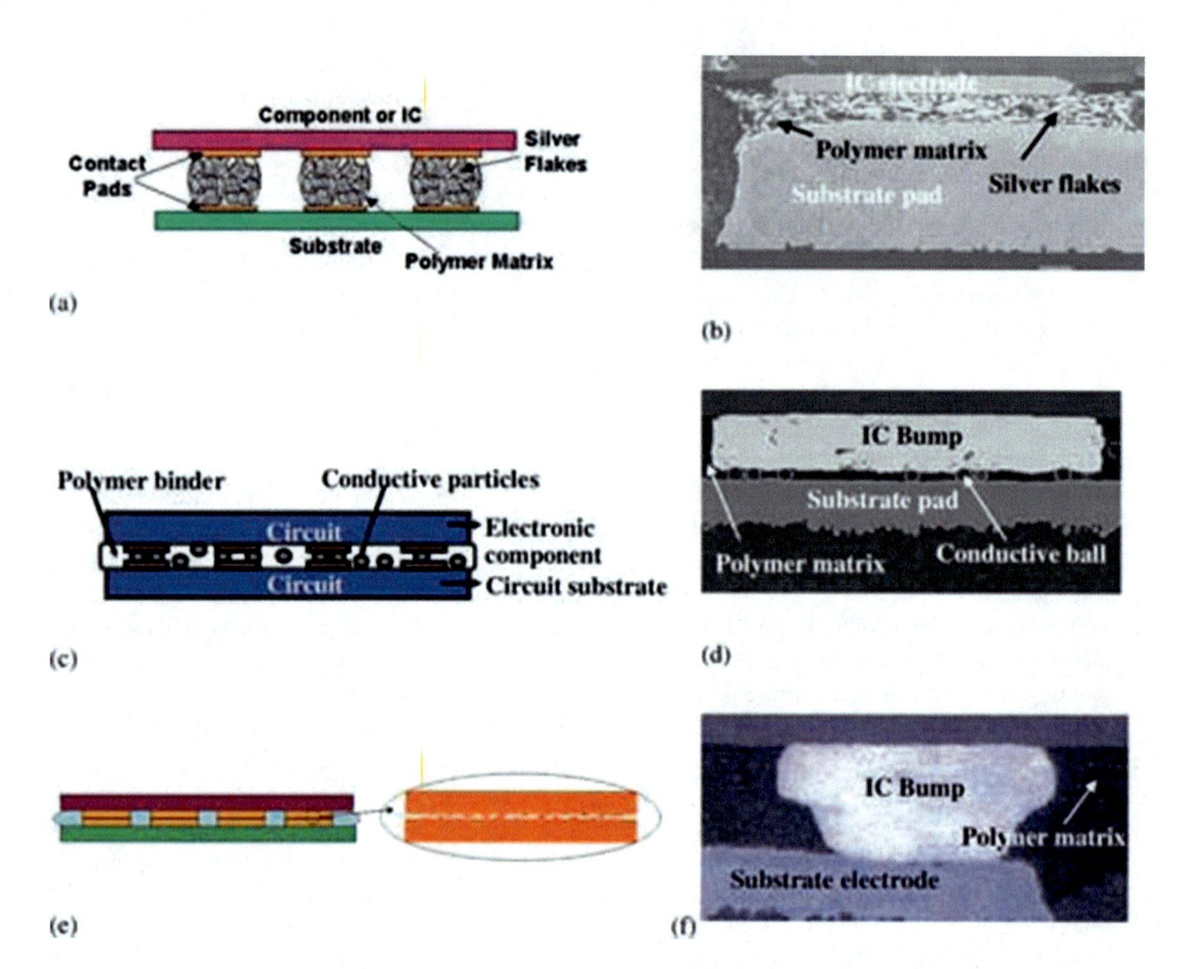

Fig. 2.1 Schematic illustrations and cross-sectional views of (**a**, **b**) ICA, (**c**, **d**) ACA, and (**e**, **f**) NCA flip-chip bondings

and NCA materials illustrating the bonding mechanism for all three adhesives.

Isotropic conductive adhesives, also called as "polymer solder," are compos- ites of polymer resin and conductive fillers. The adhesive matrix is used to form a mechanical bond for the interconnects. Both thermosetting and thermoplastic materials are used as the polymer matrix. Epoxy, cyanate ester, silicone, polyurethane, etc. are widely used thermosets, and phenolic epoxy, maleimide acrylic preimidized polyimide, etc. are the commonly used thermoplastics. An attractive advantage of thermoplastic ICAs is that they are reworkable, i.e., can easily be repaired. A major drawback of thermoplastic ICAs, however, is the degradation of adhesion at high temperature. A drawback of polyimide-based ICAs is that they generally contain solvents. During heating, voids are formed when the solvent evaporates. Most of commercial ICAs are based on thermosetting resins. Thermoset epoxies are by far the most common binders due to the superior balanced properties, such as excellent adhesive strength, good chemical and corrosion resistances, and low cost, while thermoplastics are usually added to allow toughening and rework under moderate heat. The conductive fillers provide the composite with electrical conductivity through physical contact between the conductive particles. The possible conductive fillers include silver (Ag), gold (Au), nickel (Ni), copper (Cu), and carbon in various forms (graphites, carbon nanotubes, etc.), sizes, and shapes. Among different metal particles, silver flakes are the most commonly used conductive fillers for current commercial ICAs because of the high conductivity and the maximum contact between flakes. In addition, silver is unique among all the cost-effective metals by nature of its conductive oxide. Oxides of most common metals are good electrical insulators. Copper powder, for example, becomes a poor conductor after oxidation/aging. Nickel and copper-based conductive adhesives generally do not have good resistance stability. As such, electrical reliability issues of ECA joints, in particular on non-noble metal finishes at elevated temperature and high relative humidity, are of critical concern for implementing ECAs into the high-performance device interconnects. Intensive studies on this ECA joint failure led to understanding of the failure mechanisms that corrosion at the ECA joints was the underlying mechanism rather than a simple oxidation process of the non-noble metal used [10, 11] (Fig. 2.2). Under the elevated temperature/humidity (e.g., 85 °C/85%RH), the non-noble metal acts as an anode, and is oxidized by losing electrons, and then turns into metal ion ($M–ne^- = M^{n+}$). The noble metal acts as a cathode, and its reaction generally is $2H_2O + O_2 + 4e^- = 4OH$. Then M^{n+} combines with OH^- to form a metal hydroxide, which further oxidizes to metal oxide. After corrosion, a layer of metal hydroxide or metal oxide is formed at the interface. Because this layer is electrically insulating, the contact resistance increases dramatically.

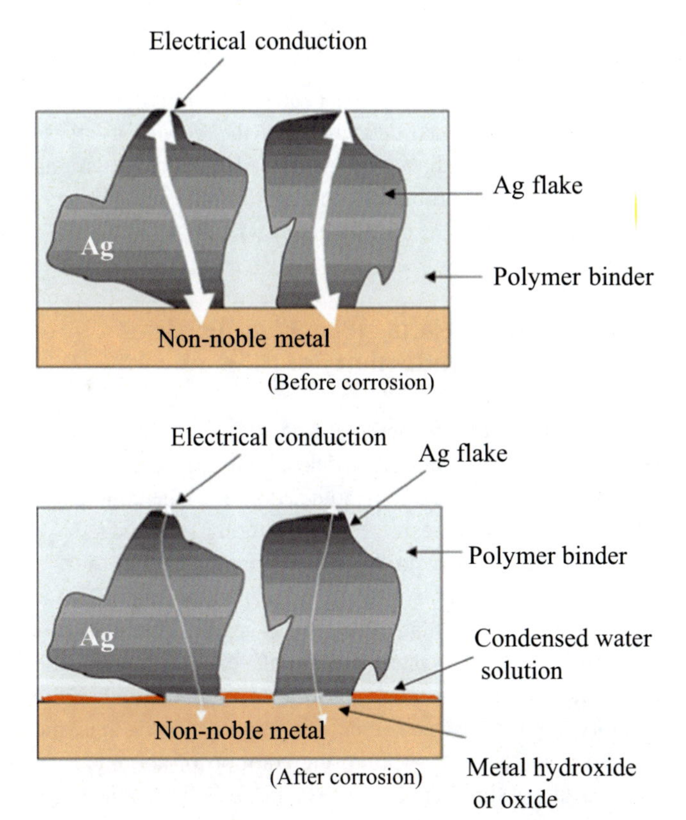

Fig. 2.2 Metal hydroxide or oxide formation after galvanic corrosion

Since this underlying mechanism was reported, many research activities for reliable ECAs have been intensively conducted via various approaches, such as using low moisture absorption resins, incorporating corrosion inhibitors or oxygen scavengers, and using sacrificial anode materials and oxide-penetrating sharp particles [10, 12–15]. In addition, many approaches to improve the electrical conductivity of ECAs have been reported, such as increasing the cure shrinkage of the matrix resin, replacing a lubricant of Ag flake with a shorter chain length carboxylic acid, adding an in situ reducing agent for preventing the Ag flake-surface oxidation, and adding low melting point alloys (LMA) for enhancing conductivity [16–21]. With the help of those intensive researches on enhancing the ECA performance, recently, ICAs have also been considered as an alternative to tin/lead solders in surface mount technology (SMT) [22, 23], flip chip, [24] and other applications.

Anisotropic conductive adhesives (ACAs) or anisotropic conductive films (ACFs) provide unidirectional electrical conductivity in the vertical or *Z*-axis. This directional conductivity is achieved by using a relatively low-volume loading of conductive fillers (5–20 vol.%). The low-volume loading is insufficient for inter- particle contact and prevents conductivity in the *X-Y* plane of the adhesives. The *Z*-axis adhesive, in film or paste form, is interposed between the surfaces to be connected. Heat and pressure are simultaneously applied to this stack-up until the particles bridge the two conductor surfaces.

The ACF bonding process is a thermo-compression bonding as shown in Fig. 2.3. In case of tape-carrier packages (TCP) bonding, the ACF material is attached on a glass substrate and pre-attached. Final bonding is established by the thermal cure of the ACF resin, typically at 150–220 °C, 3–20 s, and 100–200 MPa, and the conductive particle deformation between the electrodes of TCP and the glass substrate by applied bonding pressure.

Interconnection technologies using ACFs are major packaging methods for flat panel display modules to be of high resolution, light weight, thin profile, and low power consumption [25] and are already successfully implemented in the forms of outer lead bonding (OLB), flex to PCB bonding (FPCB), reliable direct chip attach such as chip-on-glass (COG), and chip-on-film (COF) for flat panel display modules [26–29], including liquid crystal display (LCD), plasma display panel (PDP), and organic light-emitting diode display (OLED). As for the small and fine-pitched bump of driver ICs to be packaged, fine-pitch capability of ACF interconnection is much more desired for COG, COF, and even OLB assemblies. There have been advances in

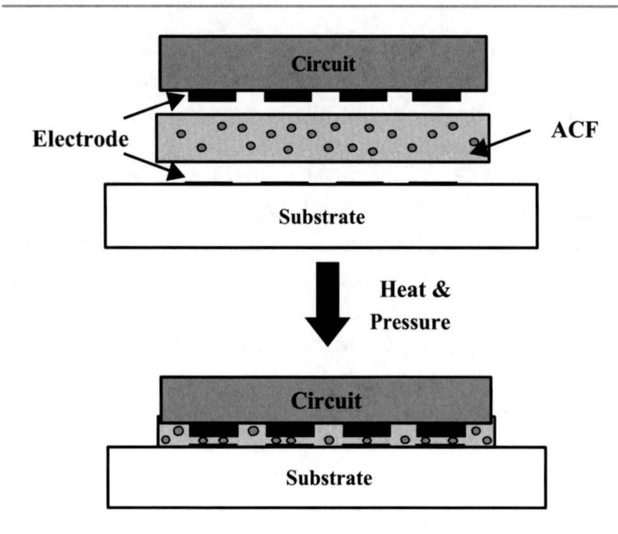

Fig. 2.3 Thermo-compression bonding using ACF

development works for improved material system and design rule for ACF materials to meet fine-pitch capability and better adhesion characteristics of ACF interconnection for flat panel displays.

In addition to the LCD industry, ACA/ACF is now finding various applications in flex circuits and surface mount technology (SMT) for chip-scale package (CSP), application-specific integrated circuit (ASIC), and flip-chip attachment for cell phones, radios, personal digital assistants (PDAs), sensor chip in digital cam- eras, and memory chip in laptop computers. In spite of the wide applications of ACA/ACF, there are some key issues that hinder their implementations as high-power devices. The ACA/ACF joints generally have lower electrical conductivity, poor current-carrying capability, and electrical failure during thermal cycling.

To meet the requirements for future fine-pitch and high-performance interconnects in advanced packaging, ECAs with nanomaterial or nanotechnology attract more and more interests due to the specific electrical, mechanical, optical, magnetic, and chemical properties. There has been extensive research on nano-conductive adhesives, which contain nano-filler such as nano-conductive particles, nanowires, or carbon nanotube (CNT). This chapter will provide a comprehensive review of the most recent research results on nano-conductive adhesives.

2.2 Recent Advances on Nanoisotropic Conductive Adhesive (Nano-ICA)

2.2.1 ICAs with Silver Nanowires

Wire-type nano1D materials (such as silver nanowires and CNT) have been addressed for their interconnect applications because of the flexibility and lowering the percolation threshold in composites. Those advantages bring stress absorbing as well as light-weight interconnections. Silver nanowires are one of the important 1D materials for interconnect due to their high electrical conductivity. An ICA filled with nano-silver wires has been reported and compared to two other ICAs filled with micrometer-sized (roughly 1 μm and 100 nm, respectively) silver particles for the electrical and mechanical properties [30]. It was found that, at a low filler loading (e.g., 56 wt%), the bulk resistivity of ICA filled with the Ag nanowires was significantly lower than the ICAs filled with 1 μm or 100 nm silver particles. The better electrical conductivity of the ICA-filled nanowires was contributed to the fewer number of contact (lower contact resistance) between nanowires, more stable conductive network, and more significant contribution from the tunneling effects among the nanowires.

It was also found that, at the same filler loading (e.g., 56 wt%), the ICAs filled with Ag nanowires showed similar shear strength to those of the ICAs filled with 1 μm and 100 nm, respectively. However, to achieve the same level of electrical conductivity, the filler loading must be increased to at least 75 wt% for the ICA filled with micro-sized Ag particles, and the shear strength of these ICAs is then decreased (lower than that of the ICA filled with 56 wt% nanowires) due to the higher filler loading.

Chen et al. reported synthetic routes of Ag nanowires for conductive adhesive applications and demonstrated that the electrical conductance of a multicomponent system (mixture of different geometry fillers) was better than that of a single-component system (e.g., sphere or flake themselves). A possible mechanism was described in the report as these nanowires act as bridges to establish perfect linkage among particles, and the probability of contact and contact area becomes more than the cases without wires [31].

2.2.2 Effect of Nano-sized Silver Particles to the Conductivity of ICAs

Nano-sized fillers are being introduced into the ECA to replace the micro-sized silver flakes in recent research. Lee et al. studied the effects of nano-sized filler to the conductivity of conductive adhesives by substituting micro-sized Ag particles with nano-sized Ag colloids either in part or as a whole to a polymeric system (polyvinyl acetate, PVAc) [32]. It was found that, when nano-sized silver particles were added into the system at 2.5 wt% each increment, the resistivity increased in almost all the cases, except when the quantity of micro-sized silver was slightly lower than the threshold value. At that point, the addition of about 2.5 wt % brought a significant decrease in resistivity. Near the percolation threshold, when the micro-sized silver particles were still not connected, the addition of a small amount of

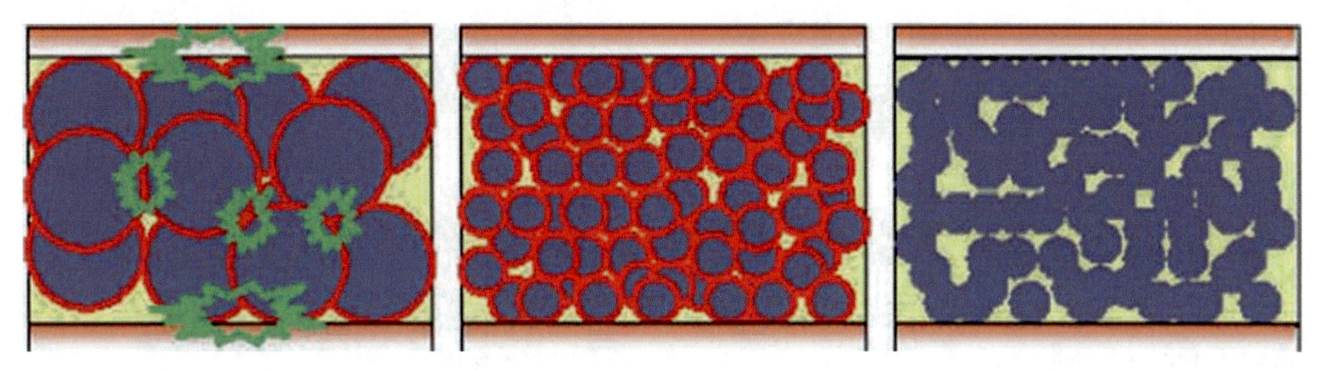

Fig. 2.4 Schematic illustration of particles between the metal pads [35]

nano-sized silver particles helped to build the conductive network and lowers the resistivity of the composite. However, when the filler loading was above the percolation threshold and all the micro-sized particles were connected, the addition of nanoparticles seemed only to signify the relative contribution of contact resistance between the particles. Due to its small size, for a fixed amount of addition, the nano-sized silver colloid contained a larger number of particles when compared with micro-sized particles. This large number of particles should be beneficial to the connection between particles. However, it also inevitably increased the contact resistance. As a result, the overall effect was an increase in resistivity upon the addition of nano-sized silver colloids. They also studied the effects of temperature on the conductivity of ICAs. Heating the composite to a higher temperature could reduce the resistivity significantly. This is likely due to the high surface activity of nano-sized particles. For micro-sized paste, this temperature effect was considered negligible. The interdiffusion of silver atoms among nano-sized particles helped to reduce the contact resistance quite significantly and the resistivity reached 5×10^{-5} Q cm after treatment at 190 °C for 30 min.

Ye et al. reported that the addition of nanoparticles showed a negative effect on electrical conductivity [33]. They proposed two types of contact resistance, i.e., restriction resistance due to small contact area and tunneling resistance when nanoparticles are included in the system. It was believed that the conductivity of micro-sized Ag particle-filled adhesives could be dominated by constriction resistance, while the nanoparticle containing conductive adhesives is controlled by tunneling and even thermionic emission. Fan et al. also observed a similar phenomenon (adding nano-sized particles reduced both electrical and thermal conductivities) [34].

The overall resistance (R_{totai}) of the isotropic conductive adhesive (ICA) formulation is the sum of the resistance of fillers ($R_{fillers}$), the resistance between fillers ($R_{btw\ fillers}$), and the resistance between filler and pads ($R_{filler\ to\ bond\ pad}$) (Eq. 2.1). In order to decrease the overall contact resistance, the reduction of the number of contact points between the particles may be obviously effective. Incorporation of nanofillers increases the contact resistance and reduces the electrical performance of the ICAs. The number of contacts between the small particles is larger than that between the large particles as shown in Fig. 2.4a, b. A recent study by Jiang et al. showed that nano-silver particles could exhibit sintering behavior at curing temperature of ICAs [35, 36]. If nanoparticles are sintered together, then the contact between fillers will be fewer. This will lead to smaller contact resistance (Fig. 2.4c). By using effective surfactants for the dispersion and effective capping those nano-sized silver fillers in ECAs, obvious sintering behavior of the nano-fillers can be achieved. The sintering of nano-silver fillers improved the interfacial properties of conductive fillers and polymer matrices and reduced the contact resistance between fillers. The dispersion and interdiffusion of silver atoms among nano-sized particles could be facilitated, and the resistivity of ICA could be reduced to 5×10^{-6} Ω cm.

$$R_{total} = R_{btw\ fillers} + R_{filler\ to\ bond\ pad} + R_{filler} \quad (2.1)$$

2.2.3 ICA Filled with Aggregates of Nano-sized Ag Particles

To improve the mechanical properties under thermal cycling conditions while still maintaining an acceptably high level of electric conductivity, Kotthaus et al. studied an ICA material system filled with aggregates of nano-sized Ag particles [37]. The idea was to develop a new filler material which did not sacrifice the mechanical property of the polymer matrix to such a great extent. A highly porous Ag powder was attempted to fulfill these requirements. The Ag powder was produced by the inert gas condensation method. The powders consist of sintered networks of ultrafine particles in the size range 50–150 nm. The mean diameter of these aggregates could be adjusted down to some microns. The as-sieved powders were characterized by a low level of impurity content, an internal porosity of about 60%, and a good ability for resin infiltration.

Using Ag IGC instead of Ag flakes is more likely to retain the properties of the resin matrix because of the infiltration of the resin into the pores. Measurements of the shear stress-strain behavior indicated that the thermomechanical properties of bonded joints may be improved by a factor up to two, independent of the chosen resin matrix.

Resistance measurements on filled adhesives were performed within a temperature range from 10 to 325 K. The specific resistance of a Ag IGC-filled adhesive is about 10^{-2} Ω cm and did not achieve the typical value of commercially available adhesives of about 10^{-4} Ω cm. The reason may be that Ag IGC particles are more or less spherical, whereas Ag flakes are flat. So, the decrease of the percolation threshold because of the porosity of Ag IGC is overcompensated by the disadvantageous shape and the intrinsically lower specific conductivity. For certain applications where mechanical stress plays an important role, this conductivity may be sufficient, and therefore the porous Ag could be suitable as a new filler material for conductive adhesives.

2.2.4 Nano-Ni Particle-Filled ICA

Sumitomo Electric Industries, Ltd. (SEI) has developed a new liquid-phase deposition process using plating technology. This new nanoparticle fabrication process achieves purity greater than 99.9% and allows easy control of particle diameter and shape. The particles' crystallite size calculated from the result of X-ray diffraction measurement is 1.7 nm, which leads to an assumption that the particle size of primary particles is extremely small. When the particle size of nickel and other magnetic metals becomes smaller than 100 nm, they are changed from the multidomain particles to the single-domain particles, and their magnetic properties change. That is, if the diameter of nickel particles is around 50 nm, each particle acts like a magnet that has only a pair of magnetism and magnetically connects with each other to form the chain-like clusters. When the chain-like clusters are applied to conductive paste, electrical conduction of the paste is expected to be better than the existing paste. The developed chain-like nickel particles were mixed with pre-defined amount of polyvinylidene difluoride (PVdF) that acts as an adhesive. Then, *N*-methyl-2-pyrrolidone was added to this mixture to make conductive paste. This paste was applied on a polyimide film and then dried to make a conductive sheet. Specific volume resistivity of the fabricated conductive sheet was measured by the quadruple method. The same measurement was also conducted on the conductive sheet that used the paste made of conventional spherical nickel particles. Measurement of the sheet resistance immediately after paste application defined that the developed chain-like nickel powders had low resistance of about one-eighth of that of the conventionally available spherical nickel particles. This result showed that, when the newly developed chain-like nickel particles were applied as conductive paste, high conductivity can be achieved without pressing the sheet.

2.2.5 Nano-conductive Adhesives for Via-Filling Applications in Organic Substrates

The demand of higher density organic substrates is increasing recently. Traditionally, greater wiring densities are achieved by reducing the dimensions of vias, lines, and spaces, increasing the number of wiring layers, and utilizing blind and buried vias. However, each of these approaches possess inherent limitations, for example, those related to drilling and plating of high aspect ratio vias, reduced conductance of narrow circuit lines, and increased cost of fabrication related to additional wiring layers. One method of extending wiring density beyond the limits imposed by these approaches is to form more metal-to-metal Z-axis interconnection of subcomposites during lamination to form a composite structure. Conductive joints can be formed during lamination using an electrically conductive adhesive. As a result, one is able to fabricate structures with vertically terminated vias of arbitrary depth. Replacement of conventional plated through holes with vertically terminated vias opens up additional wiring channels on layers above and below the terminated vias and eliminates via stubs, which cause reflective signal loss.

Conductive adhesives usually have high filler loading that weaken the overall mechanical strength. Therefore, reliability of the conductive joint formed between the conductive adhesive and the metal surface to which it is mated is of prime importance. Conductive adhesives can have broad particle size distributions. Larger particles can be a problem when filling smaller holes (e.g., diameter of 60 μm or less), resulting in voids.

Das et al. developed conductive adhesives using controlled-sized particles, ranging from nanometer scale to micrometer scale, and used them to fill small diameter holes to fabricate *Z*-axis interconnections in laminates for interconnect applications [38]. A variety of metals, including Cu and Ag, and low melting point (LMP) alloys with particle sizes ranging from 80 nm to 15 μm were used to make epoxy-based conductive adhesives. Among all, silver-based adhesives showed the lowest volume resistivity and highest mechanical strength. It was found that with increasing curing temperature, the volume resistivity of the silver-filled paste decreased due to sintering of metal particles. Sinterability of silver adhesive was further evaluated using high-temperature/pressure lamination and shows a continuous metallic network when laminated at 365 °C.

As a case study, they used a silver-filled conductive adhesive as a Z-axis interconnecting construction for a flip-chip plastic ball grid array package with a 150 μm die pad pitch. High aspect ratio, small diameter (~55 μm) holes were successfully filled. Silver-filled adhesives were electrically and mechanically better than Cu and LMP-filled adhesives. All adhesives maintained high tensile strength even after 1000 cycles thermal cycling (−55 to 120 °C). Conductive joints were stable after three times reflowing, 1000 cycles thermal cycling, pressure cooker test (PCT), and solder shock. The adhesive-filled joining cores were laminated with circuitized subcomposites to produce a composite structure. High-temperature/pressure lamination was used to cure the adhesive in the composite and provide stable, reliable Z-interconnections among the circuitized subcomposites.

2.2.6 Nano-ICAs Filled with CNT

2.2.6.1 Electrical and Mechanical Characterization of CNT-Filled ICAs

Carbon nanotubes are a new form of carbon, which was first identified in 1991 by Sumio Iijima of NEC, Japan [39]. Nanotubes are sheets of graphite rolled into seamless cylinders. Besides growing single-wall nanotubes (SWNTs), nanotubes can also have multiple walls (MWNTs) – cylinders inside the other cylinders, which act as multichannel transport of electrical and thermal for thermal interface materials (TIMs) applications. The carbon nanotube can be 1–50 nm in diameter and 10–100 μm or up to a few centimeters in length, with each end "capped" with half of a fullerene dome consisting of five- and six-member rings. Along the sidewalls and cap, additional molecules can be attached to functionalize the nanotube to adjust its properties. CNTs are chiral structures with a degree of twist in the way that the graphite rings join into cylinders. The SWCNT chirality determines whether a nanotube will conduct in a metallic or semiconducting manner. Carbon nanotubes possess many unique and remarkable properties. The measured electrical conductivity of metallic carbon nanotubes is in the order of 104 S/cm. The thermal conductivity of carbon nanotubes at room temperature can be as high as 6600 W/m K [40]. The Young's modulus of carbon nanotube is about 1 TPa. The maximum tensile strength of carbon nanotube is close to 30 GPa, some reported at TPa [41]. Since carbon nanotubes have very low density and long aspect ratios, they have the potential of reaching the percolation threshold at very low weight percent loading in the polymer matrix.

Epoxy-based conductive adhesives filled with MWNTs have been recently developed [42]. It was found that ultrasonic mixing process helped disperse CNTs in the epoxy more uniformly and made them contact better, and thus lower electrical resistance was achieved. The contact resistance and volume resistivity of the conductive adhesive decreased with increasing CNT loading. The percolation threshold for the MWNTs used was less than 3 wt%. With 3 wt% loading, the average contact resistance was comparable with solder joints. It was also found that the performance of CNTs-filled conductive adhesive was comparable with solder joints in high frequency. By replacing metal particle fillers with CNTs in the conductive adhesive, a higher percentage of mechanical strength was retained. For example, with 0.8 wt% of CNT content, the 80% shear strength of the polymer matrix was retained, while conventional metal-filled conductive adhesives only retain less than 28% of shear strength of the polymer matrix.

Experiments conducted by Qian et al. [43] show 36–42% and 25% increases in elastic modulus and tensile strength, respectively, in polystyrene (PS)/CNT composites. The TEM observations in their experiments showed that cracks propagated along weak CNT-polymer interfaces or relatively low CNT density regions and caused failure. If the outer layer of MWNTs can be functionalized to form strong chemical bonds with the polymer matrix, the CNT-polymer composites can be further reinforced in mechanical strength and have controllable thermal and electrical properties.

2.2.6.2 Effect of Adding CNT to the Electrical Properties of ICAs

Effects of adding CNT to the electrical conductivity of silver-filled conductive adhesive with various CNT loadings were reported [44]. It was found that CNT could greatly enhance the electrical conductivity of the conductive adhesives when the silver filler loading was still below percolation threshold. For example, 66.5 wt% silver-filled conductive adhesive without CNT had a resistivity of 10^4 Ω cm but showed a resistivity of 10^{-3} Ω cm after adding 0.27 wt% CNT. Therefore, it is possible to achieve the same level of electrical conductivity by adding a small amount of CNT to replace the silver fillers.

2.2.6.3 Composites Filled with Surface-Treated CNTs

Although CNTs have exceptional physical properties, incorporating CNTs into other materials has been inhibited by the surface chemistry of carbon. Problems such as phase separation, aggregation, poor dispersion within a matrix, and poor adhesion to the host must be overcome. Zyvex claimed that they have overcome these restrictions by developing a new surface treatment technology that optimizes the interaction between CNTs and the host matrix [45]. A multifunctional bridge was created between the CNT sidewalls and the host material or solvent. The power of this bridge was demonstrated by comparing the fracture behavior of the composites filled with untreated and surface-treated nanotubes. It was observed that the untreated nanotubes interacted poorly with the polymer matrix and thus left behind voids in the matrix after fracture. However, for composite filled with treated nanotubes, the nanotube remained in the matrix even after the fracture, indicating strong CNT interaction with the matrix. Due to their superior dispersion in the polymer matrix, the treated nanotubes achieved the same level of electrical conductivity at much lower loadings than the untreated nanotubes [45].

2.2.7 Inkjet Printable Nano-ICAs and Inks

Areas for printing very-fine-pitch matrix (e.g., very-fine-pitch paths, antennas) are very attractive. But there are special requirements for inkjet printing materials, namely, the most important ones are low viscosity and very homogenous structure (such as molecular fluid) to avoid sedimentation and separation during the process. Additionally, for electrical conductivity of printed structure, the liquid has to contain conductive particles, with nano-sized dimension to avoid the printing nozzle blocking and prevent sedimentation phenomenon. The nano-sized silver seems to be one of the best candidates for this purpose, especially when its particle size dimensions will be less than 10 nm.

Inkjetting is an accepted technology for dispensing small volumes of material (50–500 pl). Currently, traditional metal-filled conductive adhesives cannot be processed by inkjetting (due to their relatively high viscosity and the size of filler material particles). Smallest droplet size achievable by traditional dispensing techniques is in the range of 150 μm, yielding proportionally larger adhesive dots on the substrate. Electrically conductive inks are available on the market with metal particles (gold or silver) <20 nm suspended in a solvent at 30–50 wt%. After deposition, the solvent is evaporated, and electrical conductivity is enabled by a high metal ratio in the residue. Some applications include a sintering step [46, 47].

There are many requirements for an inkjettable, Ag particle-filled conductive adhesive. The silver particles must not exceed a maximum size determined by the diameter of the injection needle used. At room temperature, the adhesive should resist sedimentation for at least 8, preferably 24 h. A further requirement by the end user on the adhesive's properties was a two-stage curing mechanism. In the first curing step, the adhesive surface is dried. In this state, the product may be stored for several weeks. The second curing step involves glueing the components with the previously applied adhesive. By heating and applying pressure, the adhesive is cured. Thus, the processing operation is similar to that required for soldering. A conductivity in the range of 10^{-4} Ω cm in the bulk material is required. An adhesive less prone to sedimentation was formulated by using suitable additives. Furthermore, the formation of filler agglomerations during deflocculation and storage was reduced. This effect was achieved by making the additives adhere to the filler particle surfaces. This requires a very sensitive balance. If the insulation between individual silver particles becomes very strong, overall electrical conductivity is significantly reduced.

Jana Kolbe et al. demonstrated feasibility of an inkjettable, isotropically conductive adhesive in the form of a silver-loaded resin with a two-step curing mechanism [48, 49]. In the first step, the adhesive was dispensed (jetted) and pre-cured leaving a "dry" surface. The second step consisted of assembly (wetting of the second part) and final curing. The attainable droplet sizes were in the range of 130 μm but could be further reduced by using smaller (such as 50 μm) and more advanced nozzle shapes.

Nano-Ag particles for nano-ink applications are coated with capping agents for dispersion, which should be removed before the solidification process such as sintering. Otherwise, the particles cannot fuse or sinter each other. The capping agent removal usually requires heating the printed pattern to evaporate solvent and induce sintering. Wakuda et al. recently demonstrated room temperature sintering Ag nanoparticles that are protected by a dispersant in an air atmosphere, where the dodecylamine dispersant was removed in methanol and the Ag nanoparticles were densely sintered. The sintered Ag wires showed low resistivity of 7.3×10^{-5} Ω cm, which can be used for inkjet printing [50].

2.3 Recent Advances of Nano-ACA/ACF

2.3.1 Low-Temperature Sintering of Nano-Ag-Filled ACA/ACF

One of the concerns for ACA/ACF is the higher joint resistance since interconnection using ACA/ACF relies on mechanical contact, unlike the eutectic solder, which forms the metallurgical bonding during reflow. An approach to minimize the joint resistance of ACA/ACF is to make the conductive fillers fuse each other and form metallic joints such as metal solder joints. However, to fuse metal fillers in polymers does not appear feasible, since a typical organic printed circuit board (T_g~125 °C), on which the metal filled polymer is applied, cannot withstand such a high temperature; the melting temperature (T_m) of Ag, for example, is around 960 °C. Studies have showed that T_m and sintering temperatures of materials could be dramatically reduced by decreasing the size of the materials [51–53]. It has been reported that the surface pre-melting could be a primary mechanism of the T_m depression of the fine nanoparticles (<50 nm). For nano-sized particles, sintering behavior could occur at much lower temperatures; as such, the use of the fine metal particles in ACAs would be promising for high electrical performance of ACA joints by eliminating the interface between metal fillers. The application of nano-sized particles can also increase the number of conductive fillers on each bond pad and result in more contact area between fillers and bond pads as illustrated in Fig. 2.5. For the sintering reaction in a certain material system, temperature and duration are the most important parameters, in particular, the sintering temperature.

Current-resistance (*I–R*) relationship of the nano-Ag-filled ACA joint is shown in Fig. 2.6. As can be seen from the figure, with increasing curing temperatures, the resistance of the ACA joints decreased significantly, from 10^{-3} to 5×10^{-5} Ω. Also, ACAs cured at higher temperature exhibited higher current-carrying capability than ACAs cured at low temperature. This phenomenon suggested that more sintering of nano-Ag particles and subsequently superior interfacial properties between fillers and metal bond pads were achieved at higher temperatures [53], yet the *x-y* direction of the ACF maintains an excellent insulation property for electrical insulation.

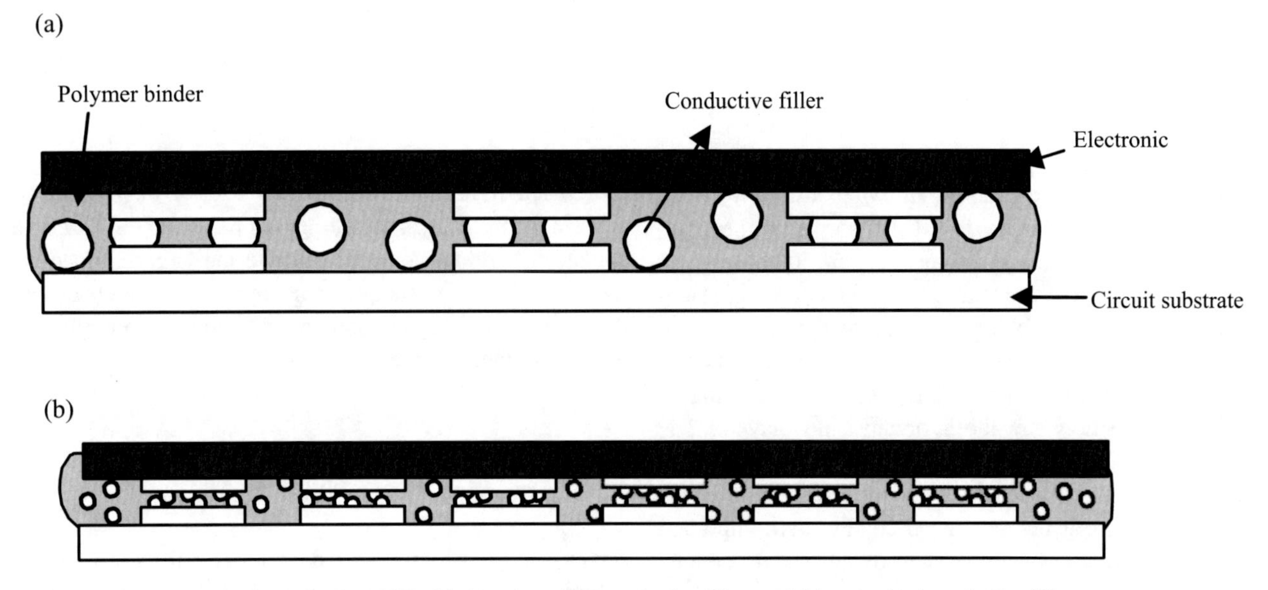

Fig. 2.5 Schematic illustrations of ACA/ACF with (**a**) micro-sized conductive fillers and (**b**) nano-sized conductive fillers

Fig. 2.6 Current-resistance (*I–R*) relationship of nano-Ag-filled ACAs with different curing temperatures [53]

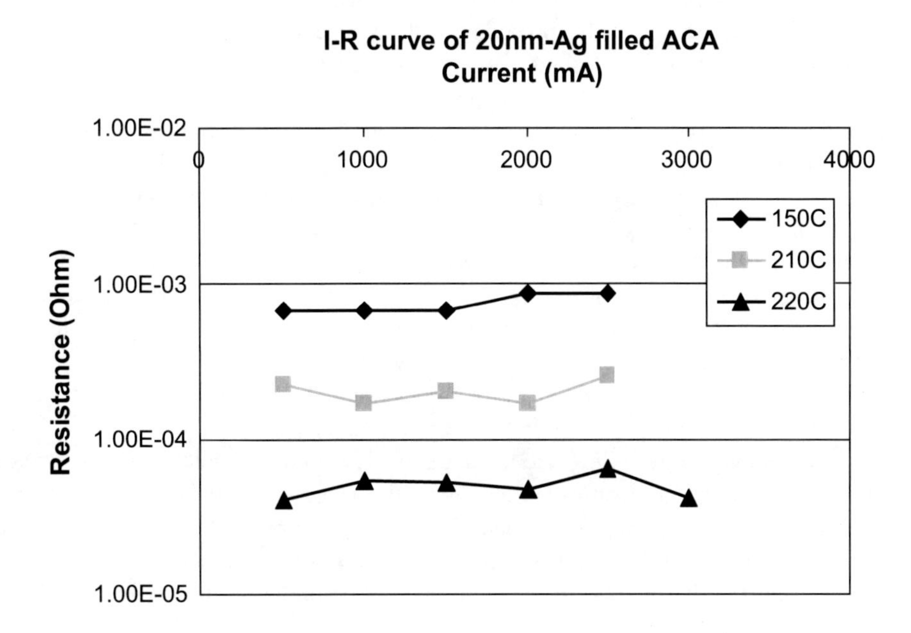

Table 2.2 Potential organic monolayer interfacial modifiers for different metal finishes

Formula	Compound	Metal finishes
R-S-H	Thiols	Au, Ag, Cu, Pt, Zn
R-COOH	Carboxylates	Fe/FexOy, Ti/TiO2, Ni, Al, Ag
R-C ≡ N	Cyanides	Au, Ag, Pt
R-N=C=O	Isocyanates	Pt, Pd, Rh, Ru
(imidazole ring: N, NH)	Imidazole and triazole derivatives	Cu
R-SiOH	Organosilane derivatives	SiO_2, Al_2O_3, qu, artz, glass, mica, GeO_2, etc.

R denotes alky and aromatic groups

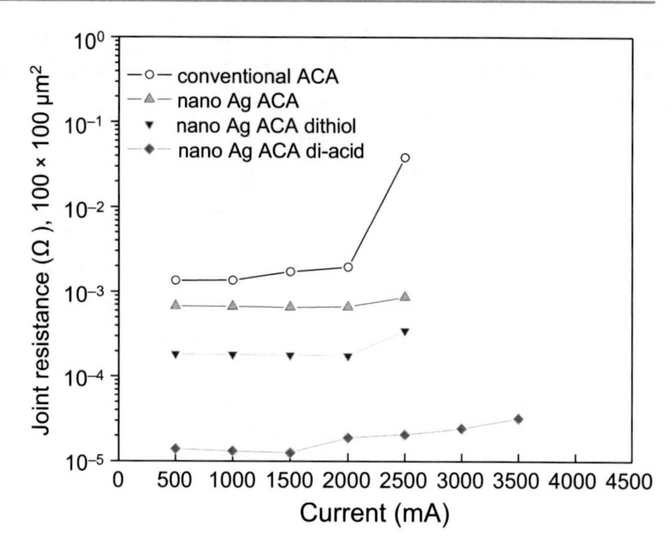

Fig. 2.7 Electrical properties of nano-Ag-filled ACA with dithiol or dicarboxylic acid [57]

2.3.2 Self-Assembled Molecular Wires for Nano-ACA/ACF

To enhance the electrical performance of ACA/ACF materials, self-assembled molecular wires (SAMW) have been introduced into the interface between metal fillers and metal-finished bond pad of ACAs [54, 55]. These organic molecules adhere to the metal surface and form physico-chemical bonds, which allow electrons to flow; as such, it reduces electrical resistance and enables a high current flow. The unique electrical properties are due to their tuning of metal work functions by those organic monolayers. The metal surfaces can be chemically modified by the organic monolayer, and the reduced work functions can be achieved by using suitable organic monolayer coatings. An important consideration when examining the advantages of organic monolayers pertains to the affinity of organic compounds to specific metal surfaces. Table 2.2 gives the examples of molecules preferred for maximum interactions with specific metal finishes; although only molecules with symmetrical functionalities for both head and tail groups are shown, molecules and derivatives with different head and tail functional groups are possible for interfaces concerning different metal surfaces.

Different organic molecular wires, dicarboxylic acid, and dithiol have been introduced into ACA/ACF joints. For SAM-incorporated ACA with micron-sized gold/polymer or gold/nickel fillers, lower joint resistance and higher maximum allowable current (highest current applied without inducing joint failure) was achieved for low-temperature curable ACA (<100 °C). For high curing temperature ACA (150 °C), however, the improvement was not as significant as low curing temperature ACAs, due to the partial desorption/degradation of organic monolayer coating at the relatively high temperature [56]. However, when dicarboxylic acid or dithiol was introduced into the interface of nano-silver-filled ACAs, significantly improved electrical properties could be achieved at high-temperature curable ACA/ACF, suggesting the coated molecular wires did not suffer degradation on silver nanoparticles at the curing temperature (Fig. 2.7). The enhanced bonding could attribute to the larger surface area and higher surface energy of nanoparticles, which enabled the monolayers to be more readily coated and relatively thermally stable on the metal surfaces [57].

2.3.3 Silver Migration Control in Nano-silver-Filled ACA

Silver is the most widely used conductive filler in ICAs and exhibits exciting potentials in nano-ACA/ACF due to its many unique advantages. Silver has the highest room temperature electrical and thermal conductivity among all the conductive metals. Silver is also unique among all the cost-effective metals by nature of its conductive oxide (Ag_2O). In addition, silver nanoparticles are relatively easy to be formed into different sizes (a few nanometers to 100 nm) and shapes (such as spheres, rods, wires, disks, and flakes) and well dispersed in a variety of polymeric matrix materials. Also the low-temperature sintering makes silver one of the promising candidates as conductive fillers in nano-ACA/ACF. However, silver migration has long been a reliability concern in electronic industry. Metal migration is an electrochemical process, whereby metal (e.g., silver), in contact with an insulating material, in a humid environment, and under an applied electric field, leaves its initial location in ionic form and deposits at another location [58]. It is considered that a threshold voltage exists above which the migration starts. Such migration may lead to a reduction in electrical spacing or cause a short circuit between interconnections. The migration process begins when a thin continuous film of water forms on an insulating material between oppositely charged electrodes. When a potential is applied across the electrodes, a chemical reaction takes place at the positively biased electrode where positive metal ions are formed. These ions, through ionic conduction, migrate toward the negatively charged cathode, and over time, they accumulate to form metallic dendrites. As the dendrite growth increases, a reduction of electrical spacing occurs. Eventually, the dendrite silver growth reaches the anode and creates a metal bridge between the electrodes, resulting in an electrical short circuit [59].

Although other metals may also migrate under specific environment, silver is more susceptible to migration, mainly due to the high solubility of silver ion, low activation energy for silver migration, high tendency to form dendrite shape, and low possibility to form a stable passivation oxide layer [60–62]. The rate of silver migration is increased by (1) an increase in the applied potential; (2) an increase in the time of the applied potentials; (3) an increase in the level of relative humidity; (4) an increase in the presence of ionic and hydroscopic contaminants on the surface of the substrate; and (5) a decrease in the distance between electrodes of the opposite polarity.

To reduce silver migration and improve the reliability, several methods have been reported. The methods include (1) alloying the silver with an anodically stable metal such as palladium [61] or platinum [63] or even tin [64]; (2) using hydrophobic coating over the PWB to shield its surface from humidity and ionic contamination [65], since water and contaminates can act as a transport medium and increase the rate of migration; (3) plating of silver with metals such as tin, nickel, or gold, to protect the silver filler and reduce migration; (4) coating the substrate with polymer [66]; (5) applying benzotriazole (BTA) and its derivatives [67]; (6) employing siloxane epoxy polymers as diffusion barriers due to the excellent adhesion of siloxane epoxy polymers to conductive metals [68]; and (7) chelating silver fillers in ECAs with molecular monolayers [69]. As an example shown in Fig. 2.8 [70], with carboxylic acids and forming chelating compounds with silver ions, the silver migration behavior (leakage current) could be significantly reduced and controlled.

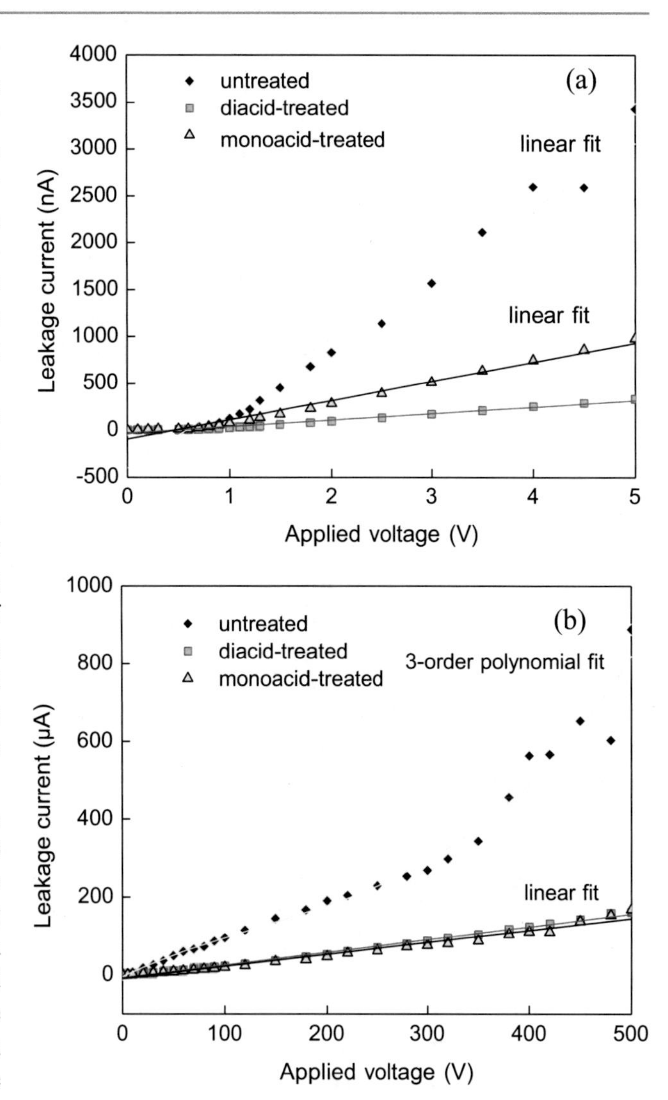

Fig. 2.8 Leakage current-voltage relationship of nano-Ag conductive adhesives at (**a**) low voltages and (**b**) high voltages [71]

2.3.4 ACF with Straight-Chain-Like Nickel Nanoparticles

Sumitomo Electric recently developed a new concept ACF using nickel nanoparticles with a straight-chain-like structure as conductive fillers [72]. They applied the formulated straight-chain-like nickel nanoparticles and solvent in a mixture of epoxy resin on a substrate film. Then the particles were made to orient toward the vertical direction of the film surface and fixed in a resin by evaporating the solvent, and this approach is similar to the elastomeric conductive polymer interconnects (ECPI), dispersed Ni-coated glass spheres in a silicone matrix, which have been developed by AT&T Bell Laboratory [73]. In the estimation using 30 μm-pitch IC chips and glass substrates (the area of Au bumps was 2000 μm^2; the distance of space between neighboring bumps was 10 μm), the new ACF showed excellent reliability of electrical connection after high-temperature, high-humidity (60 °C/90%RH) test and thermal cycle test (between −40 and 85 °C). The samples were also exposed to high temperature and high humidity (60 °C/90%RH) for insulation ability estimation. Although the distance between two electrodes was only 10 μm, ion migration did not occur, and insulation resistance has been maintained at over 1 GΩ for 500 h. This result showed that the new ACF has superior insulation reliability and has the potential to be applied in very fine interconnections.

2.3.5 Nanowire ACF for Ultrafine-Pitch Flip-Chip Interconnection

To satisfy the reduced I/O pitch and avoid electric shorting, a possible solution is to use high-aspect-ratio metal post. Nanowires exhibit high possibilities due to the small size and extremely high aspect ratio. In literature, nanowires could be applied in FET (field effective transistor) sensor for gas detection, magnetic hard disk, nanoelectrodes for electrochemical sensor, thermal-electric device for thermal dissipation and temperature control, etc. [71, 74, 75]. To prepare nanowires, it is important to define nanostructures on the photoresist. Many expensive methods such as e-beam, X-ray, or scanning probe lithography have been used, but the length of nanowires cannot be achieved to micro-meter size. A less expensive alternative is electrodeposition of metal into nano-porous template such as anodic aluminum oxide (AAO) [76] or block-copolymer self-assembly template. The disadvantages of block-copolymer template include thin thickness (that means short nanowires), non-uniform distribution, and poor parallelism of nano-pores. However, AAO has benefits of higher thickness (>10 μm), uniform pore size and density, larger size, and very parallel pores. Lin et al. [77] developed a new ACF with nanowires. They used AAO templates to obtain silver and cobalt nanowires array by electrodeposition. And then low-viscosity polyimide (PI) was spread over and filled into the gaps of nanowires array after surface treatment. The bi-metallic Ag/Co nanowires could remain parallel during fabrication by magnetic interaction between cobalt and applied magnetic field. The silver and cobalt nanowires/polyimide composite films could be obtained with a diameter of about 200 nm for nanowire and maximum film thickness up to 50 μm. The X-Y insulation resistance is about 4–6 GΩ, and Z-direction resistance including the trace resistance (3 mm length) is less than 0.2 Ω. They also demonstrated the evaluation of this nanowire composite film by stress simulation. They found that the most important factor for designing nanowire ACF was the volume ratio of nanowires. However, actually the ratio of nanowires cannot be too small to influence the electric conductance. They concluded that it is important to get a balance between electric conductance and thermal-mechanical performance by increasing film thickness or decreasing modulus of polymer matrix.

2.3.6 An In Situ Formation of Nano-conductive Fillers in ACA/ACF

One of the challenging issues in the formation of nano-filler ACA/ACF is the dispersion of nano-conductive fillers in ACA/ACF. A lot of research has been going on in recent years to address the dispersion issue of nano-composite because nano-fillers due to their high surface reactivity and high electrostatic force tend to agglomerate. For the fine-pitch electronic interconnects using nano-ACA/ACF, the dispersion issues need to be solved. The efforts usually include the physical approaches such as sonication and chemical approaches such as surfactants. Recently, a novel ACA/ACF incorporated with in situ formed nano-conductive particles is proposed for the next-generation high-performance fine-pitch electronic packaging applications [78, 79]. This novel interconnect adhesive combines the electrical conduction along the *z*-direction (ACA-like) and the ultrafine- pitch (<100 nm) capability. Instead of adding the nano-conductive fillers in the resin, the nanoparticles can be in situ formed during the curing/assembly process. By using in situ formation via chemical reduction of nanoparticles, during the polymer curing process, the filler concentration and dispersion could be controlled, and the drawback of surface oxidation of the nano-fillers could be easily overcome [80].

2.3.7 CNT-Based Conductive Nanocomposites for Transparent, Conductive, and Flexible Electronics

The electrically conductive and optically transparent coating is an upcoming potential for various applications such as transistors [81–84], diodes [85], sensors [86, 87], optical modulators [88], or conductive backbone for electrochemical polymer coating [89], smart windows [90], photovoltaics [91], solar cells [92, 93], and organic light-emitting diode (OLED) [94, 95].

With the popularity of flex circuits or substrates, in particular for large-scale flexible flat panel displays, there are demands for flexible interconnect materials, especially under highly mechanical bending conditions of printed electronics. ITO (indium tin oxide), fluorine tin oxide (FTO), and conductive polymers such as polyaniline (PANI) have been de facto options as transparent and conductive coating materials. However, those materials have no flexibility in nature, which is critical for flexible devices such as bendable displays and sensors. Thus, materials with mechanical flexibility as well as electrical conductivity and transparency can find various applications. The CNT incorporated polymer composite is one of the candidates because its tremendously high aspect ratio can provide a significantly low percolation threshold in electrical conduction. Its tiny size (<10 nm in diameter (SWCNT), smaller than 1/4λ of visible light) lets light pass through, and the polymer matrix renders flexibility. Therefore, intensive efforts to materialize transparent, conductive, and flexible CNT-based polymer composite have been dedicated.

Kaempgen et al. found that single-wall CNTs (SWCNTs) are more suitable for transparent conductive coatings than multi-wall CNTs (MWCNTs) [96]. This is likely due to a significantly higher diameter of each MWCNT, which increases the light absorption (Fig. 2.9). The longer CNTs increase the conductivity for the same transparency with the decreased number of contacts because the electrical transport is dominated by contact resistances between the CNTs [97–101]. The uniform dispersion is a key factor to manufacture reproducible coatings or composite films.

Some CNT-based polymer composites already met the requirements for various applications. Figure 2.10 compares the resistivity (1 kΩ/sq) of the CNT coating with 90% light transmittance to those of various target applications [96], where it can be seen that currently 90% transparent CNT coating can meet the requirements for many applications such as touch screen and electrostatic discharge, although only shielding of electromagnetic interference (EMI) is out of reach at least for transparent coatings. There are trade-off values between transparency and surface resistance of the CNT composites with regard to their applications. Recently, a CNT film on polymer or glass with 80% light transmittance exhibited a resistance of 20 Ω/sq. [102]. With extensive research and development of CNT nanocomposite, more reliable conductive and transparent adhesives will find versatile applications in microelectronics.

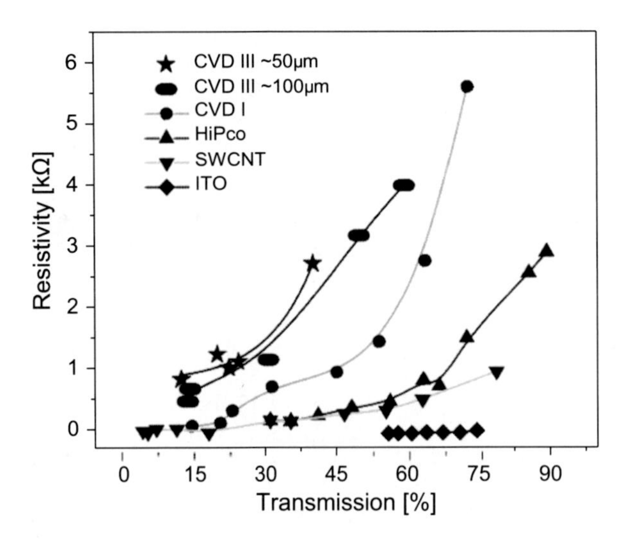

Fig. 2.9 Transmittance vs. resistivity plot for sprayed CNT layers and ITO [96]

2.4 Concluding Remarks

In this chapter, recent advances in nano-electrically conductive adhesives (ECAs) were reviewed, and in particular, innovative approaches by using cutting-edge nanotechnologies were presented. The nanotechnology-based ECAs will be playing a very important role in future nanoelectronics, nanophotonics, nano-biomedical devices, etc.

Although Ag filler is known as one of best conductive materials for ECA, its high cost and intrinsic propensity for silver migration are of concern for low-cost consumer products or high-performance device-packaging interconnects. As such, copper will be a silver alternative as ECA filler, as copper is the second highest electrically conductive element after silver and has less migration tendency. However, copper is very vulnerable to oxygen and corrosion, and unlike Ag, its oxide is a poor electrical conductor. Therefore, in order to achieve all the benefits from copper, much research effort on preventing copper from oxidation/

Fig. 2.10 Required surface resistivity in typical applications for transparent conductive coatings compared to the performance of thin conductive CNT composites [96]

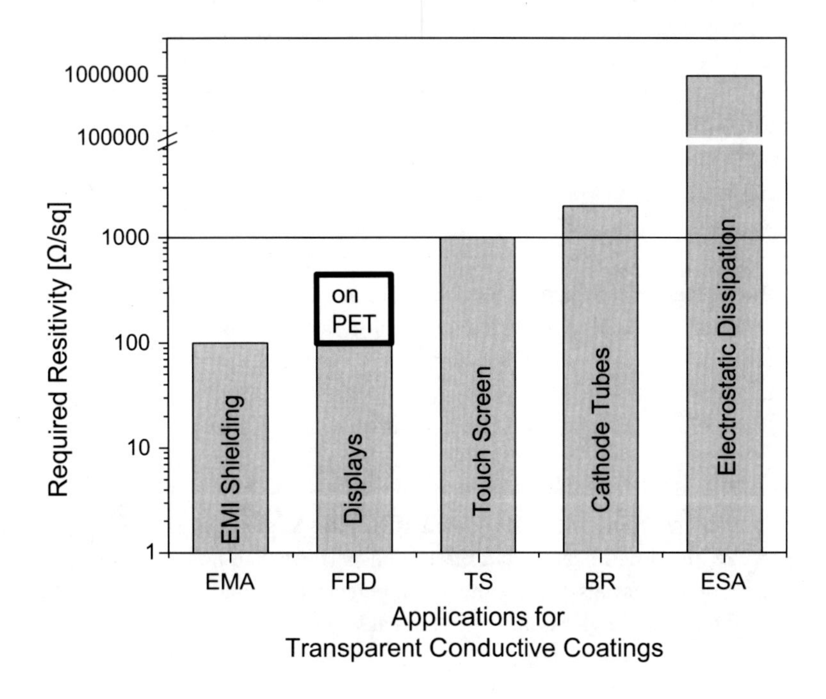

corrosion is needed, and the use of SAM/surfactant as mentioned in Table 2.2 may provide some solutions.

Tremendous evolution in display technologies has been demonstrated over the past decades, and future display technologies need very flexible and high- performance interconnect materials for nano- or biomedical device packaging. In general, bending or physical deformation of ECAs can degrade the electrical conductivity. Therefore, novel flexible electrically conductive materials to respond to repeated bending or high elongations will be required for future flexible display technologies. In addition, transparent conductive adhesive is needed for some applications such as organic LED (OLED) application. Ouyang et al. [103] have demonstrated a transparent conducting polymer glue and the feasibility of fabricating organic electronic devices through a lamination process by using the PEDOT/PSS electric glue. By combining this technology with the large-area continuous coating process of organic films, it is possible to fabricate large-area organic electronic devices through a continuous roll-to-roll coating process. This electric glue made from a conducting polymer provides new applications for conducting polymers; it will revolutionize the fabrication process of flexible electronic devices.

Although some researches on enhancing current-carrying capability of ECAs have been carried out, there is no commercial ECA material to replace metallic solders yet. Thus, increasing current-delivering capability of the ECA will be a continuing research effort. Employing molecular wires in the ECA joints or conductive elements and manipulating conductive joint work functions with high thermal stability are some of the promising approaches for the high-current ECA challenges.

With particle size of ECAs becoming smaller in particular for nano-ACAs, there is serious implication from the point of view of the manufacturing process. When ACAs are used to attach the metal pads on two mating surfaces, the warpage and co-planarity of these surfaces need to be carefully controlled to ensure all the metal pads on the mating surfaces are making contact to the conductive spheres in the ACA simultaneously. Large and compliant (deformable) conductive particles will ease the manufacturing process because these particles can accommodate the warpage and non-co-planarity. With conductive particles becoming even smaller in nano-ACAs, the manufacturing will become more challenging.

The ECAs will be continuing to evolve to meet future technological needs; as such, much attention is needed on the breakthrough technology developments by engineers and scientists including materials scientists, chemists, physicists, etc. We all should bear in mind that high-performance interconnection materials for advanced nanoelectronics, photonics, biomedical devices, and systems packaging are the key for future microelectronics.

References

1. Wolfson, H., Elliot, G.: Electrically conducting cements containing epoxy resins and silver. US Patent. **2**(774), 747 (1956)
2. Matz, K.R.: Electrically conductive cement and brush shunt containing the same. US Patent. **2**(849), 631 (1958)
3. Beck, D.P.: Printed electrical resistors. US Patent 2,866,057, 1958
4. Li, Y., Moon, K., Wong, C.P.: Science. **308**, 1419–1420 (2005)
5. Li, Y., Wong, C.P.: Mater. Sci. Eng. R. **51**, 1–35 (2006)
6. Liu, J. (ed.): Conductive Adhesives for Electronics Packaging. Electrochemical Publications, Isle of Man, Chapter 1 (1999)
7. Hwang, J.S. (ed.): Environment-Friendly Electronics: Lead-free Technology. Electrochemical Publications, Port Erin, Isle of Man, Chapter 1, pp. 4–10 (2001)
8. Murray, C.T., Rudman, R.L., Sabade, M.B., Pocius, A.V.: Mater. Res. Bull. **28**, 449–454 (2003)
9. Lau, J., Wong, C.P., Lee, N.C., Lee, S.W.R.: Electronics Manufacturing: with Lead-Free, Halogen-Free, and Conductive-Adhesive Materials. McGraw-Hill, New York (2002)
10. Lu, D., Tong, Q.K., Wong, C.P.: IEEE Trans. Compon. Packag. Manuf. Technol. Part C. **22**(3), 228–232 (1999)
11. Persson, K., Nylund, A., Liu, J., Olefjord, I.: Proceedings of the 7th European Conference on Applications of Surface and Interface Analysis, Gothenburg (June 1997)
12. Lu, D., Wong, C.P.: J. Appl. Polym. Sci. **74**, 399–406 (1999)
13. Li, Y., Moon, K., Wong, C.P.: J. Adhes. Sci. Technol. **19**(16), 1427–1444 (2005)
14. Matienzo, L.J., Egitto, F.D., Logan, P.E.: J. Mater. Sci. **38**, 4831–4842 (2003)
15. Li, H., Moon, K., Wong, C.P.: J. Electron. Mater. **33**, 106–113 (2004)
16. Lu, D., Tong, Q.K., Wong, C.P.: IEEE Trans. Compon. Packag. Manuf. Technol. Part C. **22**, 223–227 (1999)
17. Li, Y., Moon, K., Wong, C.P.: IEEE Trans. Compon. Packag. Manuf. Technol. **29**(1), 173–178 (2006)
18. Li, Y., Moon, K., Whitman, A., Wong, C.P.: IEEE Trans. Compon. Packag. Manuf. Technol. **29**(4), 758–763 (2006)
19. Gallagher, C., Matijasevic, G., Maguire, J.F.: Proceedings of 47th IEEE Electronic Components and Technology Conference, pp. 554–560 (1997)
20. Roman, J.W., Eagar, T.W.: Proceedings of the International Society for Hybrids and Microelectronics Society, San Francisco, CA, p. 52 (1992)
21. Li, Y., Moon, K., Li, H., Wong, C.P.: Proceedings of 54th IEEE Electronic Components and Technology conference, Las Vegas, Nevada, June 1–4, 2004 (1959–1964)
22. Cavasin, D., Brice-Heams, K., Arab, A.: Proceedings of 53rd Electronic Components and Technology Conference, pp. 1404–1407 (2003)
23. Kisiel, R.: J. Electron. Packag. **124**, 367 (2002)
24. de Vries, H., van Delft, J., Slob, K.: IEEE Trans. Compon. Packag. Manuf. Technol. **28**, 499 (2005)
25. Watanabe, I., Gotoh, Y., Kobayashi, K.: Proceedings of the Asia Display/IDW, Nagoya, Japan, pp. 553–556 (2001)
26. Nishida, H., Sakamoto, K., Ogawa, H.: IBM J. Res. Dev. **42**, 517 (1998)
27. Williams, D.J., Whalley, D.C., Boyle, O.A., Ogunjimi, A.O.: Solder. Surf. Mount Technol. **5**, 4 (1993)
28. Liu, J., Tolvgard, A., Malmodin, J., Lai, Z.: IEEE Trans. Compon. Packag. Manuf. Technol. **22**, 186 (1999)
29. Clot, P., Zeberli, J.F., Chenuz, J.M., Ferrando, F., Styblo, D.: Proceedings of the International Electronics Manufacturing Technology Symposium 24th IEEE/CPMT, Austin, TX, 36 (1999)

30. Wu, H., Wu, X., Liu, J., Zhang, G., Wang, Y., Zeng, Y., Jing, J.: Development of a novel isotropic conductive adhesive filled with silver nanowires. J. Compos. Mater. **40**(21), 1961–1968 (2006)
31. Chen, C., Wang, L., Li, R., Jiang, G., Yu, H., Chen, T.: J. Mater. Sci. **42**(9), 3172 (2007)
32. Lee, H.S., Chou, K.S., Shih, Z.W.: Effect of nano-sized silver particles on the resistivity of polymeric conductive adhesives. Int. J. Adhes. Adhes. **25**, 437–441 (2005)
33. Ye, L., Lai, Z., Liu, J., Tholen, A.: Effect of Ag particle size on electrical conductivity of isotropically conductive adhesives. IEEE Trans. Electron. Packag. Manuf. **22**(4), 299–302 (1999)
34. Fan, L., Su, B., Qu, J., Wong, C.P., Electrical and thermal conductivities of polymer composites containing nano-sized particles. In: Proceedings of Electronic Components and Technology Conference. IEEE, NJ, pp. 148–154 (2004)
35. Jiang, H.J., Moon, K., Lu, J., Wong, C.P.: J. Electron. Mater. **34**, 1432–1439 (2005)
36. Jiang, H.J., Moon, K., Li, Y., Wong, C.P.: Surface functionalized silver nanoparticles for ultrahigh conductive polymer composites. Chem. Mater. **18**(13), 2969–2973 (2006)
37. Kotthaus, S., Günther, B.H., Haug, R., Schafer, H.: Study of isotropically conductive bondings filled with aggregates of nano-sized Ag-particles. IEEE Trans. Compon. Packag. Manuf. Technol. Part A. **20**(1), 15–20 (1997)
38. Das, R.N., Lauffer, J.M., Egitto, F.D.: Proceedings of Electronic Components and Technology Conference. IEEE, pp. 112–118 (2006)
39. Iijima, S.: Nature. **354**, 56 (1991)
40. Berber, S., Kwon, Y.K., Tomànek, D.: Unusually high thermal conductivity of carbon nanotubes. Phys. Rev. Lett. **84**(20), 4613–4616 (2000)
41. Yu, M.F., Files, B.S., Arepalli, S., Ruoff, R.S.: Tensile loading of ropes of single wall carbon nanotubes and their mechanical properties. Phys. Rev. Lett. **84**(24), 5552–5555 (2000)
42. Li, J., Lumpp, J.K.: Electrical and mechanical characterization of carbon nanotube filled conductive adhesive. In: Proceedings of Aerospace Conference. IEEE, NJ, pp. 1–6 (2006)
43. Qian, D., Dickey, E.C., Andrews, R., Rantell, T.: Load transfer and deformation mechanisms in carbon nanotube polystyrene composites. Appl. Phys. Lett. **76**, 2868 (2000)
44. Lin, X.C., Lin, F.: Improvement on the properties of silver-containing conductive adhesives by the addition of carbon nanotube. In: Proceedings of High Density Microsystem Design and Packaging. IEEE, NJ, pp. 382–384 (2004)
45. Rutkofsky, M., Banash, M., Rajagopal, R., Chen, J.: Using a carbon nanotube additive to make electrically conductive commercial polymer composites. Zyvex Corporation Application Note 9709. http://www.zyvex.com/Documents/9709.PDF 28 (2006)
46. Kamyshny, A., Ben-Moshe, M., Aviezer, S., Magdassi, S.: Ink-jet printing of metallic nanoparticles and microemulsions. Macromol. Rapid Commun. **26**, 281–288 (2005)
47. Cibis, D., Currle U.: Inkjet printing of conductive silver paths. In: 2nd International Workshop on Inkjet Printing of Functional Polymers and Materials, Eindhoven (2005)
48. Kolbe, J., Arp, A., Calderone, F., Meyer, E.M., Meyer, W., Schaefer, H., Stuve, M.: Inkjettable conductive adhesive for use in microelectronics and microsystems technology. In: Proceedings of IEEE Polytronic Conference. IEEE, NJ, pp. 1–4 (2005)
49. Moscicki, A., Felba, J., Sobierajsk,i T., Kudzia, J., Arp, A., Meyer, W.: Electrically conductive formulations filled nano-size silver filler for ink-jet technology. In: Proceedings of IEEE Polytronic Conference. IEEE, NJ, pp. 40–44 (2005)
50. Wakuda, D., Hatamura, M., Suganuma, K.: Chem. Phys. Lett. **441**, 305–308 (2007)
51. Moon, K., Dong, H., Maric, R., Pothukuchi, S., Hunt, A., Li, Y., Wong, C.P.: J. Electron. Mater. **34**, 132–139 (2005)
52. Matsuba, Y.: Erekutoronikusu Jisso Gakkaishi. **6**(2), 130–135 (2003)
53. Li, Y., Moon, K., Wong, C.P.: J. Appl. Polym. Sci. **99**(4), 1665–1673 (2006)
54. Li, Y., Moon, K., Wong, C.P: Proceedings of 54th IEEE Electronic Components and Technology Conference. IEEE, NJ, pp. 1968–1974 (2004)
55. Li, Y., Wong, C.P: Proceedings of 55th IEEE Electronic Components and Technology Conference. IEEE, NJ, pp. 1147–1154 (2005)
56. Li, Y., Moon, K., Wong, C.P.: J. Electron. Mater. **34**(3), 266–271 (2005)
57. Li, Y., Moon, K., Wong, C.P.: J. Electron. Mater. **34**(12), 1573–1578 (2006)
58. Davies, G., Sandstrom, J.: Circuits Manuf. 56–62 (1976)
59. Harsanyi, G., Ripka, G.: Electrocomp. Sci. Technol. **11**, 281–290 (1985)
60. Giacomo, G.A.: In: McHardy, J., Ludwig, F. (eds.) Electrochemistry of Semiconductors and Electronics: Processes and Devices, pp. 255–295. Noyes Publications, Park Ridge (1992)
61. Manepalli, R., Stepniak, F., Bidstrup-Allen, S.A., Kohl, P.A.: IEEE Trans. Adv. Packag. **22**, 4–8 (1999)
62. Giacomo, D.: Reliability of Electronic Packages and Semiconductor Devices. McGraw- Hill, New York, Chapter 9 (1997)
63. Wassink, R.: Hybrid. Circ. **13**, 9–13 (1987)
64. Shirai, Y., Komagata, M., Suzuki, K: 1st International IEEE Conference on Polymers and Adhesives in Microelectronics and Photonics. IEEE, NJ, pp. 79–83 (2001)
65. Der Marderosian, A.: Ratheon Co. Equipment Division, Equipment Development Laboratories, pp. 134–141 (1978)
66. Schonhorn, H., Sharpe, L.H.: Prevention of surface mass migration by a polymeric surface coating. US Patent 4377619 1983
67. Brusic, V., Frankel, G.S., Roldan, J., Saraf, R.: J. Electrochem. Soc. **142**, 2591–2594 (1995)
68. Wang, P.I., Lu, T.M., Murarka, S.P., Ghoshal, R.: US Pending Patent 20050236711 2005
69. Li, Y., Wong, C.P.: US Patent Pending, 2006
70. Li, Y., Wong, C.P.: Monolayer-protection for eletrochemical migration control in silver nanocomposite. Appl. Phys. Lett. **81**, 112112 (2006)
71. Prinz, G.A.: Science. **282**, 1660 (1998)
72. Toshioka, H., Kobayashi, M., Koyama, K., Nakatsugi, K., Kuwabara, T., Yamamoto, M., Kashihara, H.: SEI Tech. Rev. **62**, 58–61 (2006)
73. Chang, D.C., Smith, E.L.: US patent #5206585 1993
74. Lieber, C.M.: Nanowire nanosensors for high sensitive and selective detection of biological and chemical species. Science. **293**, 1289–1292 (2001)
75. Martin, C.R., Menon, V.P.: Fabrication and evaluation of nanoelectrode ensembles. Anal. Chem. **67**, 1920–1928 (1995)
76. Xu, J.M.: Appl. Phys. Lett. **79**, 1039–1041 (2001)
77. Russell, T.P.: Ultra-high density nanowire array grown in self-assembled di-block copolymer template. Science. **290**, 2126–2129 (2000)
78. Li, Y., Moon, K., Wong, C.P.: Proceedings of 56th IEEE Electronic Components and Technology Conference. IEEE, NJ, pp.1239–1245 (2006)
79. Li, Y., Zhang, Z., Moon, K., Wong, C.P.: Ultra-fine pitch wafer level ACF (anisotropic conductive film) interconnect by in-situ formation of nano fillers with high current carrying capability. US Patent Pending 2006
80. Moon, K., Jiang, H., Zhang, R., Wong, C.: In-situ formed nanoparticles in polymer matrix for low pressure bonding. GTRC Invention Disclosure No. 4443 (2008)
81. Snow, E.S., Novak, J.P., Lay, M.D., Houser, E.H., Perkins, F.K., Campbell, P.M.: J. Vac.Sci. Technol. B. **22**(4), 1990 (2004)
82. Lay, M.D., Novak, J.P., Snow, E.S.: Nano Lett. **4**(4), 603 (2004)

83. Shiraishi, M., Takenobu, T., Iwai, T., Iwasa, Y., Kataura, H., Ata, M.: Chem. Phys. Lett. **394**, 110 (2004)
84. Meitl, M.A., Zhou, Y., Gaur, A., Jeon, S., Usrey, M.L., Strano, M. S., Rogers, J.A.: Nano Lett. **4**(9), 1643 (2004)
85. Zhou, Y., Gaur, A., Hur, S., Kocabas, C., Meitl, M.A., Shim, M., Rogers, J.A.: Nano Lett. **4**(10), 2031 (2004)
86. Li, Z., Dharap, P., Nagarajaiah, S., Barrera, E.V., Kim, J.D.: J. Adv. Mat. **16**, 640 (2004)
87. Abraham, J.K., Philip, B., Witchurch, A., Varadan, V.K., Reddy, C.C.: Smart Mater. Struct. **13**, 1045 (2004)
88. Wu, Z., Chen, Z., Du, X., Logan, J.M., Sippel, J., Nikolou, M., Kamaras, K., Reynolds, J.R., Tanner, D.B., Hebard, A.F., Rinzler, A.G.: Science. **305**, 1273 (2004)
89. Ferrer-Anglada, N., Kaempgen, M., Skákalová, V., Dettlaf-Weglikowska, U., Roth, S.: Diam. Relat. Mater. **13**(2), 256 (2004)
90. Hu, L., Zhoa, Y.-L., Ryu, K., Zhou, C., Stoddart, J.F., Grüner, G.: Adv. Mater. **20**, 5939–5946 (2008)
91. Gruner, G.: J. Mater. Chem. **16**, 3533–3539 (2006)
92. Ago, H., Petritsch, K., Shaffer, M.S.P., Windle, A.H., Friend, R.H.: Adv. Mater. **11**, 1281 (1999)
93. Du Pasquier, A., Unalan, H.E., Kanwal, A., Miller, S., Chhowalla, M.: Appl. Phys. Lett. **87**, 203511 (2005)
94. Shan, B., Cho, K.: Phys. Rev. Lett. **94**, 236602 (2005)
95. Mitschke, U., Bauerle, P.: J. Mater. Chem. **10**, 1471 (2000)
96. Kaempgen, M., Duesberg, G.S., Roth, S.: Appl. Surf. Sci. **252**, 425–429 (2005)
97. Kaiser, A.B., Dusberg, G., Roth, S.: Phys. Rev. B. **57**, 1418 (1998)
98. Liu, K., Roth, S., Duesberg, G.S., Kim G.-T., Schmid M.: Progress in molecular nanostructures.In: Kuzmany H., Fnk J., Mehring M., Roth S. (eds.) American Institute of New York, AIP Conference Proceedings, vol. 442, pp. 61–64 (1998)
99. Shirashi, M.: Synth. Met. **9198**, 1 (2002)
100. Duesberg, G.S., Graham, A.P., Kreupl, F., Liebau, M., Seidel, R., Under, E., Hönlein, W.: Diam. Rel. Mater. **13**, 354 (2004)
101. Lu, K.L., Lago, R.M., Chen, Y.K., Green, M.L.H., Harris, P.J.E., Tsang, S.C.: Carbon. **34**, 814 (1996)
102. Trancik, J.E., Barton, S.C., Hone, J.: Nano Lett. **8**(4), 982–987 (2008)
103. Ouyang, H.Y., Yang, Y.: Adv. Mater. **18**, 2141–2144 (2006)

Applications of Carbon Nanomaterials as Electrical Interconnects and Thermal Interface Materials

3

Wei Lin and C. P. (Ching-Ping) Wong

3.1 Introduction

As we have been experiencing, revolutionary increase in speed and reliability of microprocessors has been successfully achieved in the past 60 years. The faster and higher performance of microprocessors is based on increased transistor density. As originally proposed, Moore's law stated that the number of transistors in semiconductor devices or integrated circuits (ICs) would double approximately every 2 years [1]. This prediction has been realized, largely due to device scaling, characteristic of fine-pitch (pitch = distance between two adjacent electrodes or electrical traces) interconnects. Copper interconnects, introduced in 1998, are now routinely used with the minimum feature size down to 65 nm; 45-nm node can be found in some commercial devices; new technologies for smaller nodes are being intensively studied. However, the electrical resistivity of copper interconnects increases with a decrease in dimensions due to grain-boundary and electron surface scattering [2]. As current density further increases, the electromigration issue for metal interconnects becomes more severe. In these regards, carbon nanotubes (CNTs) have been proposed as a future interconnecting material due to their ultra-high current-carrying capacity (10^9 A/cm^2), thermal stability, and high resistance to electromigration [3–11]. Today, the main challenges of CNT interconnects (circuits) are (1) purification of metallic or semiconducting CNTs; (2) selective positioning of CNTs; and (3) effective and reliable contacts at CNT junctions [12, 13]. Although the purification issue has almost been addressed by recent development of various CNT separation methods, CNT positioning and contact reliability issues are still unaddressed [14–18]. Appropriately, graphene, an alternative to CNTs in the family of carbon nanomaterials, has been proposed to be capable of offering a solution to the problem of selective positioning for interconnects.

In spite of electrical performances, increasing microprocessor performance is associated with an increased cooling demand; in other words, more heat dissipation is required. It has been reported that a reduction in the device operation temperature corresponds to an exponential increase in reliability and life expectancy of a device [19]. To control device temperature within operation limits is critical. Thermal design power (TDP, the maximum sustained power dissipated by the microprocessor) has in the past increased steadily with increasing microprocessor performance [20]. Although multicore microprocessors should alleviate the growth in TDP with increased performance, thermal nonuniformity, usually referred to as the "hot spot" issue, where the local power density could be >300 W/cm^2, must be paid more attention to in circuit design and operation [20]. In fact, heat flux from a chip is not uniform; more than 90% of the total power is dissipated from the core [21]. Although microprocessors, ICs, and other sophisticated electronic components are designed to operate at temperatures above ambient temperature (generally in the range of 60–100 °C), the thermal problem for chip cooling is not about maintaining the average temperature of the chip below the design point but is about maintaining the temperature of the hottest spot below the design point. The hot spot issue makes the heat dissipation near the chip more difficult.

Effective heat dissipation has become a key issue for further development of high-performance semiconductor devices. Efforts on improving thermal management have led to new cooling technology developments, including turbulator-enhanced air cooling, hybrid air–water cooling, water cooling, direct liquid cooling, etc. [22]. As chip power keeps increasing, novel cooling techniques will be needed to maintain chip temperatures within functional temperature limits. The development of liquid cooling technologies such as liquid microchannel–minichannel cooling represents a major evolution in this field

W. Lin (✉) · C. P. Wong
School of Materials Science and Engineering, Georgia Institute of Technology, Atlanta, GA, USA
e-mail: wlin31@gatech.edu

C. P. Wong et al. (eds.), *Nano-Bio- Electronic, Photonic and MEMS Packaging*, https://doi.org/10.1007/978-3-030-49991-4_3

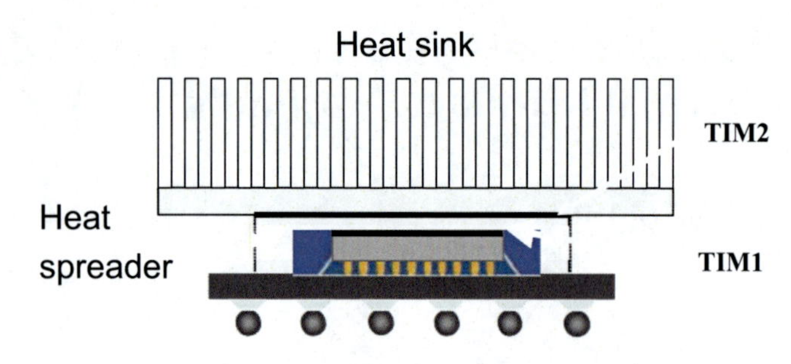

Fig. 3.1 Schematic illustration of TIM architecture in flip-chip bonding

[19, 23]. No matter what kind of cooling systems are used, usually heat sinks are necessarily implemented into the packaging, either directly on the backside of a chip or on the heat spreader which sits between the chip and the heat sink, disperses heat, and averages heat fluxes [21]. However, surface asperities greatly limit the actual contact between solid surfaces (e.g., die-heat spreader and heat spreader–heat sink), undermining the thermal conduction. Therefore, a thin layer of thermal interface materials (TIMs), e.g., TIM1 between the chip and the heat spreader and TIM2 between the heat spreader and the heat sink, has been introduced to fill the gap between the asperities to reduce the contact thermal resistance. Schematic illustration of a typical TIM assembly architecture is shown in Fig. 3.1.

Unavoidably, a portion of the allowable temperature drop occurs when heat flows from the chip to the heat spreader through TIM1 and from the heat spreader to the heat sink through TIM2. Such temperature drops might not a big problem in advanced cooling systems but are crucial for air cooling systems. We note that thermal management is more than a technical challenge; economic challenge is often the reason why new cooling technologies find barriers against introduction. The cost, simplicity, and fit considerations of convection methods underscore the desire to cool systems and components with air convection. In an air cooling system, heat dissipation from the chip is by conduction through the many solid layers and interfaces of the package into the heat sink, conduction through the base and into the fins of the heat sink, and finally convection from the fins of the heat sink to the cooling airstream. The temperature drop caused by thermal conduction between the chip and heat spreader is predicted to become comparable to the maximum allowed temperature drop (temperature difference at the chip and heat sink). In other words, the chip-heat spreader thermal resistance will be comparable to the total allowed resistance by 2010, eliminating the possibility for a temperature drop at the heat sink [24]. Heat spreaders are usually high thermal conductivity materials, such as Cu, diamond, AlN, and SiC. Thus, the spreaders themselves cause only a very small fraction of the total thermal resistance. Effective reduction of resistance between the chip and the spreader (the largest possible fraction of system resistance in the near future) is therefore limited to the development of high thermal conductivity TIMs and the suppression of the interfacial thermal resistance between TIMs and mating surfaces. Development of novel TIMs is thus crucial to meet thermal performance requirements for future generations of high-performance IC chips. Heat dissipation challenges create opportunities for fundamental research in materials and thermal management strategies. Specifically, it has been suggested that future cooling approaches may be based on micro- and nanotechnologies [25]. For thermal management applications, the distinctive thermal properties of carbon nanotubes (CNTs) and graphene attract much attention and give rise to new opportunities in thermal management of microelectronic devices and packaging systems.

In this chapter, a brief review is first given on understandings of CNT structure, electrical property, heat transport, and synthesis methods (Sects. 3.2 and 3.3). Next, CNT applications as interconnects (Sect. 3.4) and TIMs (Sect. 3.5) are presented, followed by CNT integration into circuits and packaging, with focus on CNT-transfer technology (Sect. 3.6). Based on the basic understanding of CNT nanoelectronics, a bright future of graphene nanoelectronics is foreseen in Sect. 3.7. In comparison with CNT-polymer composites, nanographite-polymer composites are more promising for TIM applications, which will be discussed in Sect. 3.8. Section 3.9 gives a brief summary of current thermal measurement techniques, commercialized and non-commercialized. Finally, in Sect. 3.10, future research needs for CNT and graphene applications as interconnects and TIMs in nanoelectronics are discussed.

3.2 Structure and Properties of Carbon Nanotubes

3.2.1 Carbon Nanotube Structure

Carbon nanotubes (CNTs) are carbon allotropes, with cylindrical hollow structures and extremely high length–diameter aspect ratio of up to 10^7 [26, 27]. Generically, CNTs can be divided into two distinct types – single-walled CNT (SWNT) and multi-walled CNT (MWNT), depending on the number of their walls. An SWNT has only one shell with a small diameter (usually <2 nm). An MWNT, basically similar to an SWNT in the individual wall structure, consists of two or more concentric cylindrical shells of rolled graphene sheets coaxially arranged around a central hollow with a constant separation between the layers [28–30]. Typically, MWNT diameters are in the range of 2–30 nm.

An SWNT is conveniently illustrated by rolling a graphene sheet of honeycomb structure along the chiral vector C_h [31, 32]. The circumference of any SWNT can be described in terms of the chiral vector:

$$C_h = m\alpha_1 + n\alpha_2 \tag{3.1}$$

where α_1 and α_2 are unit vectors and m and n are numbers of steps along the corresponding unit vectors. The chiral angle, θ, determined relative to the direction defined by α_1, uniquely defines an SWNT structure. Chiral nanotubes have general (n, m) values and a chiral angle between 0 and 30°. Zigzag nanotubes correspond to $(m, 0)$ or $(0, n)$ and have a chiral angle of 0°; armchair nanotubes have (n, n) and a chiral angle of 30°. Both zigzag and armchair SWNTs have a mirror plane and thus are considered as achiral. The diameter of an SWNT is given by

$$d = \frac{|C_h|}{\pi} = \frac{\sqrt{3}a_{\mathrm{c-c}}\sqrt{m^2 + mn + n^2}}{\pi} \tag{3.2}$$

where $a_{\mathrm{c-c}}$ is the C–C bond length (1.42 Å). Figure 3.2 shows the vector C_h constructed for $(m, n) = (3, 1)$.

The formation of CNTs is determined by free energy. Carbon has four electrons in total in its valence orbitals, which construct an sp^2-hybridized conjugated structure in the honeycomb-like graphene sheets. However, graphene sheets of a finite size have many edge atoms with dangling bonds possessing high-energy states. When graphene sheets wrap up to become CNTs (although this may not be the real mechanism of CNT synthesis), curvature-induced strain energy is compensated by the reduction in energy by eliminating edge dangling bonds [33]. Although perfect open-ended SWNTs are seamless and fully consist of six-membered rings, real CNTs are far from being exempt of local defects. Typically, synthesized CNTs are capped with high strain energy at the tips. Hexagonal graphitic unit can be locally replaced with heptagonal and pentagonal units, corresponding to positive and negative Gaussian curvatures in the two-dimensional (2D) graphene lattice [34]. In MWNTs, the most distinct defect is the lattice misfit between adjacent walls, also called interfacial dislocation [35, 36]. The consequence of this interplanar stacking disorder weakens the electronic coupling between the layers compared with graphite stacks.

3.2.2 Electronic Structure and Electrical Properties

Recall that a simple approximation based on a tight-binding calculation for the electronic structure of a 2D graphene sheet is expressed as [33]

$$E_{2\mathrm{D}}(k_x, k_y) = \pm\gamma_0\left\{1 + 4\cos\left(\frac{\sqrt{3}k_x a}{2}\right)\cos\left(\frac{k_y a}{2}\right) + 4\cos^2\left(\frac{k_y a}{2}\right)\right\}^{1/2} \tag{3.3}$$

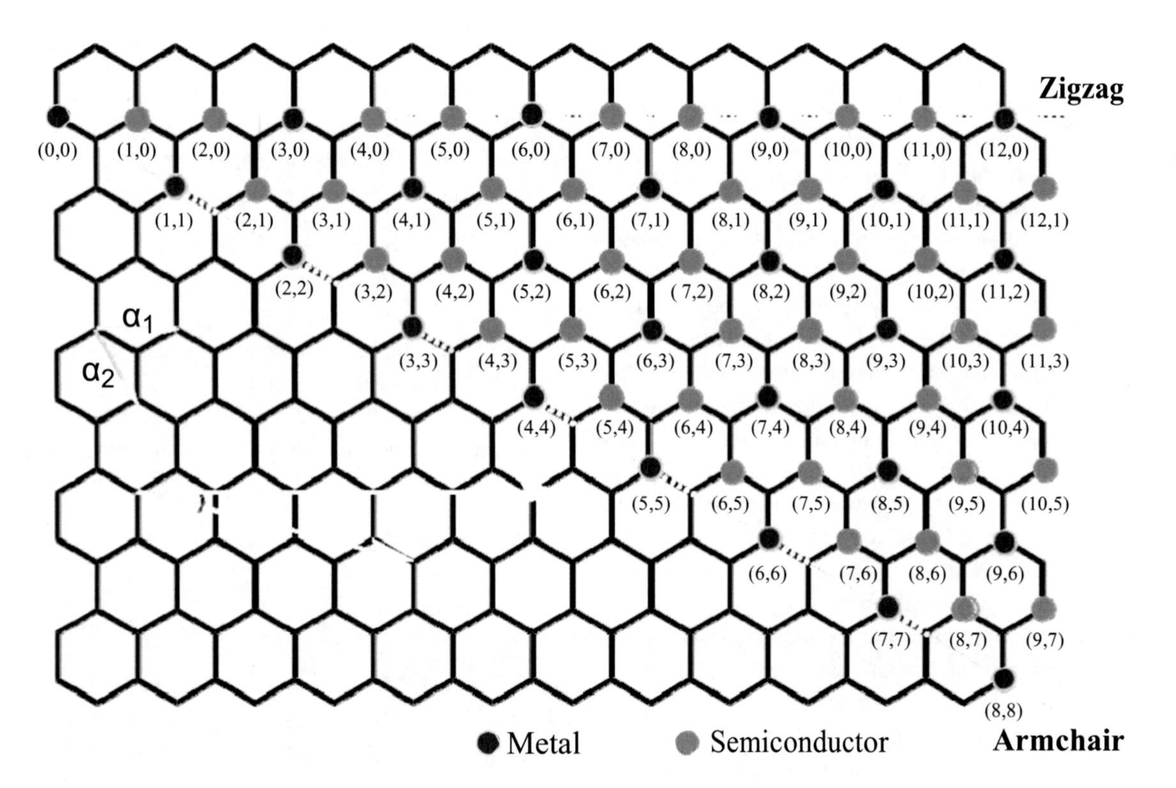

Fig. 3.2 Formation of a carbon nanotube from a 2D graphene sheet along the vector that specifies a chiral nanotube

where k_x, k_y, and γ 0 are wave vectors and the nearest-neighbor overlap integral, respectively; $a = 0.246$ nm is the in-plane lattice constant. Based on this calculation, the valence and conductance bands touch and are degenerated at the six K (k_F) points at the 2D Brillouin zone corner where the Fermi level in reciprocal space is located. Therefore, graphene is a zero-gap semiconductor. This has proved to be in good agreement with the first-principles results [37].

When graphene is rolled over to form a CNT, the periodic boundary condition of $C_h \cdot k = 2\pi\ q$ is imposed to eliminate the wave vectors $k = (k_x, k_y)$, where q is an integer. In this way, 1D energy bands can be obtained by slicing the 2D energy dispersion relations [31]. From tight-binding approximations, all CNTs are conductive but have different gap widths. SWNTs with

$$(n - m) = 3q \tag{3.4}$$

(where q is an integer) are metallic and have band gap as small as a few meV, while the remainders with

$$(n - m) = 3q \pm 1 \tag{3.5}$$

are semiconducting and have a higher band gap energy. This suggests that one third of the SWNTs grown are metallic, while two thirds are semiconducting. The band gap for a semiconducting nanotube is given by [38]

$$E_g = 2a_{\mathrm{c-c}}\gamma_0/D \tag{6}$$

where D is the nanotube diameter. Therefore, the band gap of a 1-nm semiconducting nanotube is between 0.7 and 0.9 eV, which is in good agreement with scanning tunneling microscopy (STM) measurements [38, 39].

MWNTs can also be metallic or semiconducting, analogous to SWNTs, and determined by the band gap width. Electron energy loss spectroscopy (EELS) has been experimentally used to probe the local electronic structure of MWNTs. The low electron energy loss spectroscopy (LEELS) contains information about the joint density of states (DOS) related to the individual and collective excitation (plasmons) processes of the valence electrons [29]. For MWNTs, smaller tubes show lower carrier density and greater average carrier localization due to the increasing curvature. Therefore, MWNTs, which usually have large diameters, are metallically conductive.

Fortunately, owing to the strong C–C bond and rolled graphitic structure, CNTs with even extremely small diameters as 1 nm do not show the Peierls instability – a common phenomenon for most of the 1D materials, which exhibit a metal-to-insulator transition when they come down to nanoscale [40]. When the length of a 1D conductor is smaller than its electron mean-free path, the electron transport is ballistic. This has become the most significant property of CNTs. For a perfect metallic SWNT with 100% transparent contacts, low-bias transport (e.g., $V_b < 0.1$ V) is ballistic over length scales that can reach micrometers, due to extremely weak acoustic phonon scattering in the 1D conductor. At high bias, strong excitation of optical phonons by electrons leads to energy dissipation in the tube and current saturation.

For ballistic transport, neglecting electron spin and sublattice degeneracy, the quantum resistance for a single channel is a constant:

$$R_Q = h/2e^2 \tag{3.7}$$

where h is the Planck's constant and e is the electron charge. Taking into account electron spin and sublattice degeneracy, the corresponding conductance is $G_0 = 2e^2/h$, which is frequently used for metallic SWNTs in literatures.

A single metallic SWNT can carry up to 25 μA of current, which corresponds to a current density of 10^9 A/cm^2 [41, 42]. MWNTs have similar conductance and current density when only the outer wall–shell contributes to the current transport [43–45]. However, multichannel quasiballistic conduction with extremely high conductance of 460 G_0 and large current-carrying capacity was found in an open-ended MWNT [46]. This implies that CNT conductance and current-carrying capacity will be dramatically improved by integrating open-ended MWNTs which have multichannels in metallic contact with the electrodes.

It was proposed that the conductance of an MWNT or an SWNT is determined by two factors: the conducting channels per shell and the number of shells. The number of shells for an individual MWNT is diameter-dependent [47]:

$$N_{\mathrm{shell}} = 1 + \frac{D_{\mathrm{outer}} - D_{\mathrm{inner}}}{2\delta} \tag{3.8}$$

where $\delta = 0.34$ nm is the intershell distance, corresponding to the interlayer distance in graphite, and D_{outer} and D_{inner} are the external and the internal tube diameter, respectively. Assuming that the metallic to semiconducting tube ratio is 1/2, the number of conducting channels per shell can be written as [48]

$$\mathrm{N}_{\mathrm{chan/shell}} = \begin{cases} (ad + b)/3, & d > 6\,\mathrm{nm} \\ 2r, & d < 6\,\mathrm{nm} \end{cases} \tag{3.9}$$

where $a = 0.1836$ nm^{-1}, $b = 1.275$, r is the metallic nanotube ratio, and d is the diameter of the nanotube shell. One conducting channel provides either quantized conductance

G_0 or ohmic conductance ($G_i = G_0\, l/\lambda$), depending upon the tube length l. For the low-bias situation that is suitable for interconnect applications, the diameter-dependent channel conductance for one shell is [47]

$$G_{\text{shell}}(d,l) = \begin{cases} G_0 N_{\text{chan/shell}}, & l \leq \lambda \\ G_i N_{\text{chan/shell}}, & l > \lambda \end{cases} \tag{3.10}$$

where λ is the electron mean-free path. For example, the electrical conductance of a metallic SWNT ($N_{\text{chan/shell}} = 2$) is

$$G_{\text{shell}}(d,l) = \begin{cases} 2G_0, & l \leq \lambda \\ 2G_i, & l > \lambda \end{cases} \tag{3.11}$$

An MWNT consists of several shells, each with its own d, λ, and $\widetilde{N}_{\text{chan/shell}}$. Therefore the total conductance comes from the contribution by each shell:

$$G_{\text{MWNT}} = \sum_{N_{\text{shell}}} G_{\text{shell}}(d,l) \tag{3.12}$$

An SWNT rope or MWNT can be viewed as a parallel assembly of individual SWNTs. Naeemi et al. derived physical models for the conductivity of MWNT interconnects [48]. The results indicate that for long interconnects (hundreds of micrometers), MWNTs may have conductivities several times larger than that of copper or even SWNT bundles. Furthermore, the more robust structure of MWNTs renders them better mechanical stability, which is an advantage for interconnect applications over SWNTs.

As mentioned before, a CNT structure is not perfect. When a few local defects are present, electron scattering is distinctly increased. In actual operation, conductance may be even lower due to electron–electron coupling, intertube coupling, and contact resistance at CNT-substrate interface. Therefore, it is considered promising to use nanotube bundles/arrays aligned in parallel for interconnects.

3.2.3 Separation of Metallic and Semiconducting SWNTs

In terms of CNT nano devices, distinct properties of metallic and semiconducting SWNTs lead to distinct applications. For example, metallic SWNTs can function as leads in a nanoscale circuit, while semiconducting SWNTs could perform as Schottky-type field-effect transistors [43, 49–53]. Since SWNT samples produced via various techniques are generally mixtures of metallic and semiconducting SWNTs, large-scale separation of metallic and semiconducting SWNTs in a nanotube sample is important and has attracted much recent attention [43, 54–70]. Current separation techniques can be divided roughly into the following three categories:

1. *Destructive separations*. In 2001, Collins et al. found that Joule heat could burn off metallic SWNTs in an SWNT fabrication, generating entire arrays of nanoscale field-effect transistors based solely on the fraction of semiconducting SWNTs [43]. Miyata et al. reported selective oxidation of SWNTs by hydrogen peroxide. An enrichment in metallic SWNTs of higher than 80% was achieved in the final product [68]. Recently, Huang et al. found that upon laser irradiation in air, metallic SWNTs in a carbon nanotube thin film can be destroyed in preference to their semiconducting counterparts when the wavelength and power intensity of the irradiation are appropriate and the carbon nanotubes are not heavily bundled [67].
2. *Nondestructive physical separations*. Krupke et al. took advantage of the difference in the relative dielectric constants of metallic and semiconducting SWNTs with respect to the solvent and used an alternating current dielectrophoresis process to separate metallic from semiconducting SWNTs [55].
3. *Nondestructive chemical separations*. Selective chemical functionalization of SWNT sidewalls, covalently or non-covalently, is of particular interest since it facilitates the solubilization/suspension of the functionalized SWNTs in certain organic solvents during the final separation step [54]. Sun's group reported an effective separation process based on the selective interaction of porphyrins toward semiconducting SWNTs [56]. Dyke et al. realized the separation by a controlled reaction process. 4-*tert*-butylbenzenediazonium salt and 4-(2^{t}-hydroxyethyl)-benzenediazonium salt were used to react with metallic and semiconducting SWNTs in a dispersion, respectively, and sequentially [64]. The as-functionalized SWNTs were finally separated based on a chromatographic separation process. Maeda et al. achieved a metallic SWNT-enriched (87%) tetrahydrofuran solution based on selective attachment of amine to metallic SWNT surfaces [57]. This process is characteristic of a large-scale separation; however, the enrichment of the metallic SWNTs needs further improvement. Toyoda et al. synthesized a long-alkyl-chain benzenediazonium compound that reacted to a metallic SWNT surface selectively and then precipitated the unreacted semiconducting SWNTs with an organic solvent [58]. This technique was able to separate semiconducting from metallic SWNTs, whereas it required a careful kinetic control over the chemical reaction. Mioskowski's group reported separation of semiconducting from metallic SWNTs by selective functionalization of semiconducting SWNTs with azomethine ylides [66]. Bao's group reported a

spin-assisted process to prepare self-sorted, aligned SWNT networks for thin-film transistors [65]. The selective deposition was based on selective interaction between the semiconducting SWNTs and the functional groups, i.e., amine groups, on the silane monolayer on a silicon wafer. Recently, Tanaka et al. reported an interesting and effective approach of separating semiconductor from metallic SWNTs. They froze, thawed, and squeezed a piece of gel containing SWNTs and sodium dodecyl sulfate. The solution squeezed out contained 70% pure metallic SWNTS, and the gel left contained 95% pure semiconducting SWNTs [70]. Field-effect transistors constructed from the separated semiconducting SWNTs were shown to function without any electrical breakdown.

3.2.4 Heat Transport

Great progress has been achieved on understanding the extraordinarily high thermal conductivity of individual SWNTs, both theoretically and experimentally [71–76]. Molecular dynamics simulations of an SWNT by Berber et al. indicated that the thermal conductivity of an SWNT can be as high as 6600 W/m-K at room temperature [71]. Pop et al. presented a method for extracting the thermal conductivity of an individual SWNT from high-bias electrical measurements in the temperature range from 300 to 800 K by reverse fitting the data to an existing electrothermal transport model [76]. Their results showed that the thermal conductivity for an SWNT of 2.6 μm long and 1.7 nm diameter was nearly 3500 W/m-K at room temperature.

The high thermal conductivity of SWNTs is based on the large phonon mean-free path, i.e., on the order of micron at low temperatures [72, 73, 77]. Although both phonon and electron contribute to thermal conductance, for SWNTs at low temperatures or even at room temperature, electron contribution is negligible compared to that from phonons.

For graphite and related systems, the heat capacity can be written as

$$C = C_{\mathrm{ph}} + C_e \tag{3.13}$$

where C_{ph} and C_e are the contributions from phonons and electrons, respectively. If $T << \hbar v/\kappa_B R$, C_{ph} can be written as [78]

$$C_{\mathrm{ph}} = \frac{3L\kappa_B^2 T}{\pi \hbar v} \times 3.292 \tag{3.14}$$

where κ_B is the Boltzmann's constant, v is the velocity of sound ($\sim 10^4$ m/s), and L is the tube length. The electronic-specific heat for a metallic SWNT is also linear with temperature for $T << \hbar v_F/\kappa_B R$:

$$C_{\mathrm{e}} = \frac{4\pi L\kappa_B^2 T}{3\hbar v_F} \tag{3.15}$$

where v_F is the Fermi velocity ($\sim 10^6$ m/s).

Therefore, the ratio between phonon and electron contributions to the specific heat is

$$\frac{C_{\mathrm{ph}}}{C_{\mathrm{e}}} \approx \frac{v_F}{v} = 100 \tag{3.16}$$

Experiments showed a linear increment of thermal conductance of SWNTs with increased temperature at low temperatures [74, 77]. However, at high temperatures, e.g., from 300 to 800 K, the thermal conductivity decays with temperature due to strong scattering. The length effect of the thermal conductivity of SWNTs has been measured by a 3ω method, showing that thermal conductivity of SWNTs increases with length, which is consistent with theoretical predictions [79].

For MWNTs, heat transport is complex due to the intershell interaction. The strong phonon coupling between the MWNT shells should cause a behavior similar to that of graphite. A thermal relaxation technique has been used to measure the specific heat of MWNTs from 0.6 to 210 K; results indicate that MWNTs have a similar specific heat to that of SWNT ropes and graphite [80]. Thermal conductivity of MWNTs has been experimentally measured [81–83]. Kim et al. developed a microfabricated suspended device hybridized with MWNTs (~1 μm) to allow the study of thermal transport where no substrate contact was involved [81]. The thermal conductivity and thermoelectric power of a single CNT were measured, and the observed thermal conductivity was >3000 W/m-K at room temperature. Choi et al. obtained the thermal conductivity of individual MWNTs (outer diameter of ~ 45 nm) by employing a 3ω method [82, 83]. The thermal conductivity was reported to be 650–830 W/m-K at room temperature. Given that MWNTs are more easily synthesized and assembled and possess higher rigidity, MWNTs have been considered more promising, compared with SWNTs, to be applied as TIMs in next-generation electronic packaging.

In an SWNT rope, phonons propagate along individual tube axes as well as between parallel tubes, leading to phonon dispersion in both longitudinal and transverse directions. The net effect of dispersion is a significant reduction of the specific heat at low temperatures compared to an isolated nanotube [84]. Yang et al. investigated the thermal conductivity of MWNT films prepared by microwave CVD using a pulsed photothermal reflectance technique [85]. The average thermal conductivity of carbon nanotube films, with film thickness from 10 to 50 μm, was ~15 W/m-K at room temperature and independent of tube length. However, by taking into account a small volumetric fraction of CNTs, the effective nanotube thermal conductivity can reach 200 W/m-K.

Finally, we should note that the influence of structural defects on the heat transport in the 1D structure of CNTs could be anomalously high, much larger than in bulk materials. Yi et al. measured the thermal conductivity of millimeter-long aligned MWNTs [86]. At room temperature, the thermal conductivity of these samples was only ~ 25 W/m-K, due to a large number of CNT defects. However, thermal conductivity recovered to a high value after the aligned MWNTs were annealed at 3000 °C. Further discussion about MWNT films for TIMs will be presented in Sect. 3.5.

3.3 CNT Growth

Reliable synthesis techniques capable of yielding high-purity CNTs in desirable quantities and at selective positions on growth substrates are critical to realize CNT applications in electronic packaging. Till now, various synthesis techniques have been developed to produce CNTs for specific research uses or applications, mainly including arc-discharge, laser ablation, and various chemical vapor deposition (CVD) processes, each of them having its own advantages and disadvantages. Since there are a large number of scientific literatures dealing with various CNT syntheses, the readers are referred to other monographs on CNTs and the related references in this section for detailed information. Only a brief summary and comparison is presented in this section.

3.3.1 Arc-Discharge and Laser Ablation Methods

In 1992, Ebbesen and Ajayan adapted the standard arc-discharge technique used for fullerene synthesis to a gram-level synthesis of MWNTs under a helium atmosphere [28]. In this process, carbon atoms are evaporated with inert gas plasma characterized by high electric currents passing between opposing carbon electrodes (cathode and anode). Usually the carbon anode contains a small percentage of metal catalyst such as cobalt, nickel, or iron. The results show that the purity and yield depend sensitively on the gas pressure in the reaction vessel. The length of the synthesized MWNTs is several micrometers with diameters ranging from 2 to 20 nm. However, such synthesized CNTs are inevitably accompanied by the formation of carbon particles that are attached to the nanotube walls. A subsequent purification process is necessary to achieve high-purity nanotubes. In 1993, Bethune and coworkers reported a large-scale synthesis of SWNTs by an arc-discharge method using a carbon electrode that contained ~ 2 at .% cobalt [87]. At high temperatures, carbon and metal catalysts are co-vaporized into the arc, leading to the formation of carbon nanotubes with very uniform diameter (~1.2 nm). However, fullerenes (by-products of the arc-discharge process) also form readily in this process. In order to obtain high-purity SWNTs, a purification process is therefore necessary. In 1996, the Smalley group reported the synthesis of high-quality SWNTs with yields >70% using a laser ablation method [88]. This method utilized double pulse lasers to evaporate graphite rods doped with 1.2 at.% of a 50:50 mixture of Co and Ni powder, which was placed in a tube furnace heated to 1200 °C in flowing argon at 500 Torr; this process was followed by heat treatment in vacuum at 1000 °C to sublime C60 and other small fullerenes. The resulting SWNTs were quite uniform, had a diameter of ~1.38 nm, and formed ropes consisting of tens of individual SWNTs closely packed into hexagonal crystal structures that were stabilized through van der Waals forces.

The success in producing large quantities of CNTs by arc-discharge offers wide availability of CNTs for fundamental studies and exploration of potential applications. However, there are several concerns associated with these two growth methods: first, energy consumption is large; second, only powdered samples with tangled CNT bundles are produced, thereby making purification, manipulation, and assembly difficult; third, by-products are usually unavoidable, including fullerenes, graphitic polyhedrons, and amorphous carbon in the form of particles or overcoats on nanotubes; and fourth, patterned growth of CNTs is a great challenge [89].

3.3.2 Chemical Vapor Deposition (Common Thermal CVD, Plasma-Enhanced CVD, and Liquid Injection CVD)

Both MWNT and SWNT syntheses have been well-developed using chemical vapor deposition (CVD) processes. Generally, a CVD process for CNT synthesis involves decomposition of carbon sources, e.g., gaseous precursors or volatile compounds of carbon, catalyzed by metal nanoparticles at high temperatures. The mechanism for CNT growth has been generally assumed to be a dissociation-diffusion-precipitation process during which elemental carbon is formed on the surface of a metal particle followed by diffusion and precipitation in the form of cylindrical graphite [90, 91]. However, the detailed nanotube growth mechanisms are still not well understood.

CVD processes have great advantages in controllable tuning of CNT length, diameter, morphology, alignment, doping, and CNT adhesion on substrates. This is why CVD is the most attractive for CNT applications in nanoscale electronics. Various CVD processes have been developed, including common thermal CVD [92], plasma-enhanced CVD (PECVD) [93, 94], floating catalyst CVD (e.g., HiPCO) [95], liquid injection CVD (LJ-CVD) [96], hot filament CVD (HFCVD) [97], etc. A brief comparison between thermal CVD, PECVD, and LJ-CVD is listed in Table 3.1.

The critical parameters in CVD growth of CNTs are the carbon precursor, catalyst composition, reactor chamber pressure, and growth temperature. The effects of these parameters on CNT growth have been investigated extensively. Growth temperature, typically ranging from 400 °C [98, 99] to 3600 °C [100], has a significant effect on CNT growth. Generally, PECVD can lower the growth temperature, compared to thermal CVD, because plasma dissociates hydrocarbon molecules more efficiently and promotes surface diffusion of carbon on the catalyst droplet, thereby allowing CNT growth to occur at lower temperatures that are possible with thermal CVD [101, 102]. Interestingly, a thermal CVD process for CNT growth (even SWNT growth) at as low as 450 °C was recently reported [103]. PECVD also has particular advantages in good alignment of CNT arrays, which is significant for CNT application as vertical interconnects [104]. One drawback of the current synthesis of aligned CNT (ACNT) arrays by thermal CVD or PECVD methods is the relatively low ACNT packing density on substrates. This density limitation mainly comes from the low density of catalyst nanoparticles deposited on a substrate. Usually, a thin layer of catalyst is deposited on a substrate surface, e.g., by an e-beam evaporation process, for CVD or PECVD synthesis. The thickness of the catalyst layer after deposition is typically 0.2–6 nm. A proper thickness of the catalyst layer is one prerequisite for formation and separation of catalyst particles at high temperatures with small distribution in diameter. For thermal CVD process, ACNT packing density on silicon substrate typically lies between 300 and 1000 per μm^2. This means that, for ACNTs with average diameter of 15 nm, ACNT coverage on silicon substrate is <10%. Although such a surface coverage is sufficient to, in theory, provide excellent electrical or thermal conductance, defects are unavoidable in real CNTs, which greatly degrade the intrinsic properties of ACNTs. Therefore, for electrical or thermal applications, higher packing density of ACNTs on substrates is desired.

The advantage of liquid injection CVD (LJ-CVD) over thermal CVD and PECVD is to grow ACNT arrays with a higher packing density [96]. LJ-CVD does not need a catalyst layer on the substrate; instead, the catalyst is generated in situ in the reaction chamber by decomposition of a metal catalyst precursor, i.e., ferrocene dissolved in xylene. Shown in Fig. 3.3a and b are the SEM (scanning electron microscope)

Table 3.1 A brief comparison between CNTs synthesized by thermal CVD, PECVD, and LJ-CVD

	Growth rate	Quality	Packaging density	Cost
Thermal CVD PECVD LJ-CVD	Low Medium High	High Medium Low	Low Medium High	High Medium Medium or low

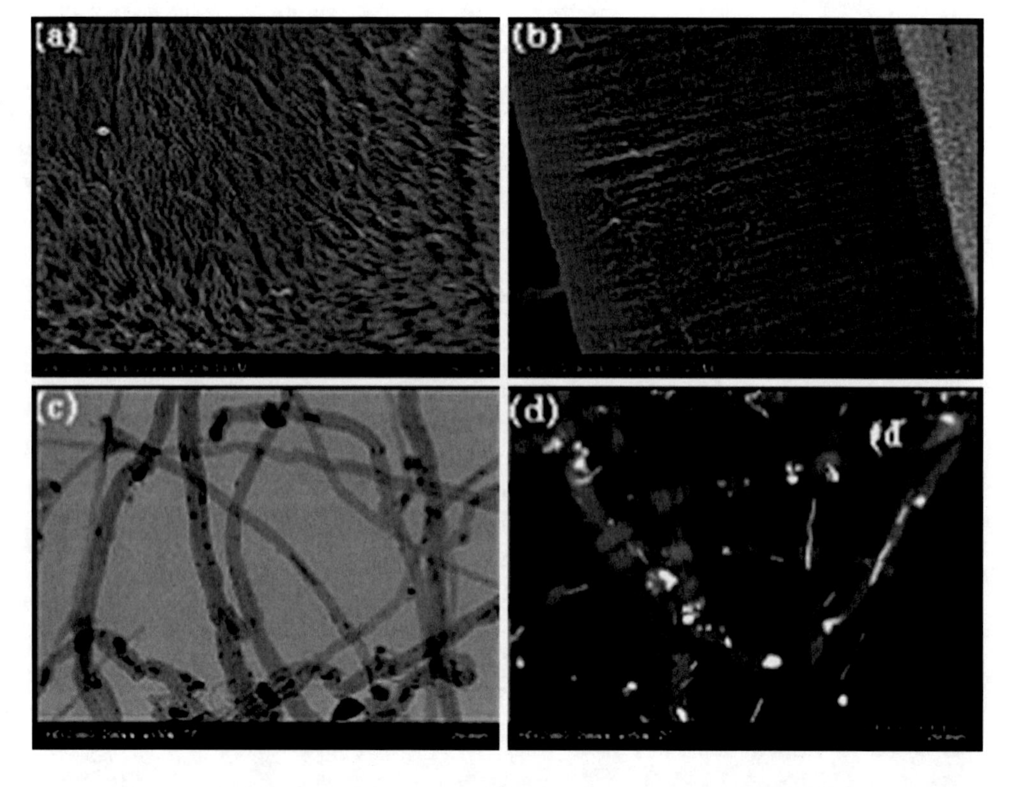

Fig. 3.3 SEM images of aligned CNT arrays synthesized by LJ-CVD: (**a**) side view; (**b**) top view; (**c**) TEM image of randomly dispersed individual CNTs; (**d**) Z-scan TEM image of randomly dispersed individual CNTs with bright spots indicating metal catalysts

images of ACNTs synthesized by LJ-CVD. A TEM image in Fig. 3.3c shows the diameters of CNTs ranging from 20 to 30 nm. Since the metal catalyst was formed continuously during the CNTs growth in LJ-CVD, it is randomly embedded in the cavity of CNTs, as evidenced by the Z-scan TEM image (Fig. 3.3d). From TGA results, the average metal content in CNTs arrays is about 10–15 wt%. Currently, helical CNT arrays have been synthesized by LJ-CVD, indicating more flexibility of LJ-CVD in controlling CNT morphology [105].

3.4 Carbon Nanotubes for Interconnect Applications

The International Technology Roadmap for Semiconductors (ITRS) predicts that in 2010, the current density will reach the order of magnitude of 10^6 A/cm^2, a value which can only be supported by CNTs. Wei et al. [42] measured the reliability of two individual MWNTs as electrical interconnects, by forcing a constant current through the CNTs at 250 °C. The CNTs showed no degradation after 350 h at current densities exceeding 10^{10} A/cm^2. Naeemi and Meindl [106] suggested that mono- or bilayer metallic SWNTs may be promising candidates for short local interconnects. We should note that, at small bias voltages, i.e., for interconnect applications, mean-free path in CNTs is large; however, at large bias voltages, i.e., transistor applications, electron-phonon scatterings in CNTs degrade the transconductance [7]. Therefore, transistor applications require assembly of ultra-short CNTs, which is still a big challenge for GSI. In this section, a brief review is given on current progresses in vertical ACNT interconnects.

Vertical interconnects pass through vias or contact holes in the dielectrics to connect horizontal interconnects to the source, drain, or gate metallization of transistors. Researchers from Fujitsu and Infineon have investigated vertical CNT interconnects extensively [3, 8, 107–110]. In ref. 3, holes were formed using conventional photolithography and subsequent wet etching with a buffered HF solution. Catalyst was selectively introduced into the vias by deposition followed by a lift-off process. However, CNTs grown inside the trenches are of low packing density and entangled. Therefore, Li et al. [4] introduced an alternative bottom-up approach to separate ACNTs, as shown in Fig. 3.4. They grew vertical ACNTs on patterned catalyst spots and then used SiO_2 to fill the gaps. After chemical mechanical polishing, the ACNT tips shot out and were ready for next step of metallization. In order to improve the CNT-substrate interfacial strength after CVD growth and simultaneously achieve low-resistance ohmic contact, Nihei et al. [5] modified the catalyst composition by using a layer of titanium below the nickel catalyst layer. It was postulated that titanium carbide formed during the growth at the CNT-Ti interface. Actually, conductive contacts between the graphitic network of CNTs and metallic electrodes are of great importance in electronic devices. To date, most results indicate that the properties of CNT-interconnected devices are dominated by the electronic behavior at the CNT-metal contacts [111]. Yang et al. [85] reported using intense electron beam irradiation to form covalent bonding between CNTs and metal particles; metallic conductivity of the junctions with low resistance was formed. Nihei et al. [5] has also predicted that uniformly small diameters of CNTs and a higher CNT packaging density would be beneficial. In this regard, a conventional CVD process needs improvement. Normally, catalyst particles for CVD growth of CNT are formed from a deposited thin catalyst film by a high-temperature treatment that causes segregation of the film into small particles or islands. There is no guarantee that the CNTs would be

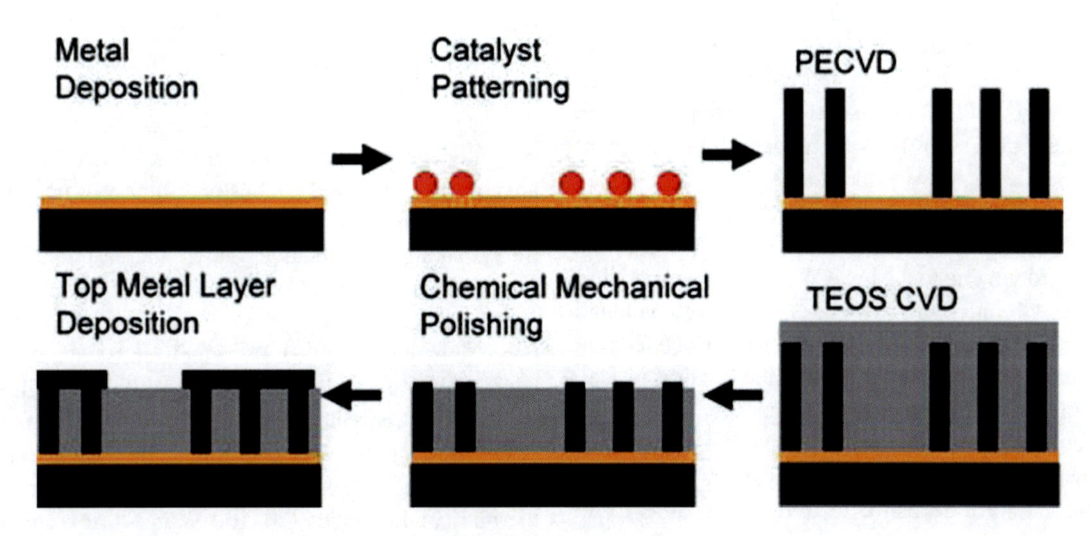

Fig. 3.4 Schematic process of the bottom-up approach for ACNT interconnects [4]

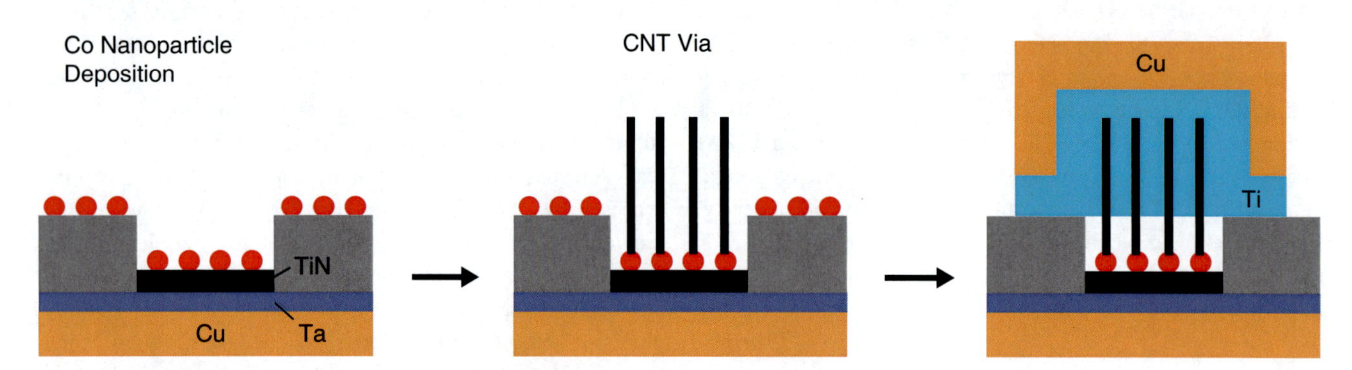

Fig. 3.5 Schematic process of ACNT via fabrication [109]

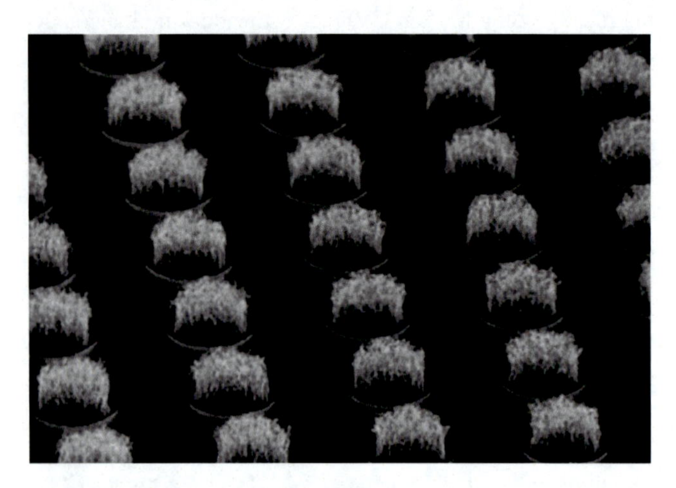

Fig. 3.6 A SEM image of the ACNT via array produced by the size-classified catalyst nanoparticles

uniformly small in diameter with high packing density on growth substrates. Awano et al. [8, 109, 112] have reported a size-classified catalyst nanoparticle CVD process for controlled growth of CNTs with uniform diameter and high density at low temperatures. Laser ablation is used to evaporate catalyst particles from a catalyst metal target. The resulting nanoparticles are size-separated by a low-pressure differential mobility analyzer in terms of their movement within an electric field. The size-classified nanoparticles are then deposited onto substrates in a deposition chamber. Finally, the substrates with deposited nanoparticles are placed in a hot-filament CVD chamber for nanotube growth. The nanoparticles present will not coalesce on the substrate during CNT growth due to the low CNT growth temperature.

ACNT growth on copper substrates was a big challenge. Awano et al. [109] put a 5-nm Ta barrier layer between the catalyst layer and the copper wiring film and used the process described in ref. 8 to grow ACNTs on the Cu film for interconnection. A schematic of a CNT via fabrication process is shown in Fig. 3.5. The top contact layers of 50-nm Ti and 300-nm Cu were deposited to form connections to the CNTs. These metal layers were then patterned, and the Co particles on the SiO_2 layer were simultaneously removed during the etching step. Figure 3.6 shows a SEM image of CNT vias with a diameter of 2 μm grown by a size-classified catalyst nanoparticle CVD process [8]. Via resistance for a 2-μm-diameter CNT structure was measured with a four-point probe using Kelvin patterns and a TiN contact layer. The measured resistance of CNT vias is ~0.59 Q at room temperature. This value is of the same order as the theoretical value of W plugs but is still one order of magnitude higher than the theoretical value of Cu vias. Further improvement is needed on CNT quality and packing density control.

3.5 Carbon Nanotubes as Thermal Interface Materials (TIMs)

3.5.1 A Brief Review of TIMs and Requirements for Next-Generation TIMs

TIMs play a key role in developing thermal solutions and have been extensively researched in the past decades [21, 113–131]. Developing new TIMs is the route toward higher packaging system reliability. Commonly used TIMs include polymers filled with fillers of high thermal conductivity, solders, and filled phase change materials (PCMs). Although polymer-based TIMs, e.g., boron nitride (BN)/polyethylene glycol (PEG) and BN/silicone, have their advantages in good adhesion, high compliance, easy processing, and low cost, their thermal conductivities at low filler loadings are relatively low. Increasing filler loading improves the thermal conductivity of TIM layer itself (~7 W/m-K) but brings down its compliance and increases bond line thickness (BLT), which consequently increases contact thermal resistance. Additionally, reliability is also an issue for polymer-based TIMs because compliance of polymers degrades with operation time. In comparison, solder TIMs such as In, AuSn, InPb, and InSnBi have high thermal conductivities (50–100 W/m-K), low melting point

Table 3.2 Summary of published data on effective thermal conductivity of CNT-polymer composites

Matrix	CNT	CNT loading	CNT length (nm)	Outer wall diameter (nm)	K_e/K_m
Epoxy	SWNT	1 wt%	100–2000	Broad distribution	2.25
Epoxy	SWNT	1 vol%	77–257	1.1	< 1.08
Epoxy	SWNT	0.5 vol%	77–257	1.1	1.30
Silicon	SWNT	3.8 wt%	–	–	1.65
Silicon	MWNT	0.4 vol%	300	12	2.16
Epoxy	MWNT	0.2 vol%	~ 5000	< 2	1.06
Epoxy	SWNT	0.2 vol%	~ 1000	15	1
Epoxy	MWNT	0.2 vol%	~ 5000	< 2	1.01

K_e and K_m refer to the thermal conductivities of the composite and the polymer matrix, respectively

(<170 °C), and high compliance and are now actively being pursued as TIMs. Currently, thermal resistance of a solder TIM layer (R_{TIM}, including the resistance at the two contact interfaces) is typically 5–20 mm^2-K/W. However, they have the drawbacks such as high CTE (coefficient of thermal expansion), low compliance, high cost, relatively low fatigue strength, voids issue, and complexity in application. PCMs have high fluidity during processing, making use of the latent heat of phase change during heating up, which dissipates much more heat generated by the die. However, PCMs, compared with solder TIMs, are not so popular due to a reliability issue because sometimes PCMs leak (flow out) from the interfaces during operation. In general, none of the above TIMs meet the requirements for next-generation TIMs; implementation of nanotechnology seems unavoidable.

3.5.2 Carbon Nanotube-Polymer Composites

Amorphous polymers such as epoxy resin have extremely low thermal conductivities (usually ~0.2 W/m-K). One challenging issue in polymer TIM composites is that thermal conductivity of the composite should be enhanced with limited filler loading levels. CNT, as the filler, theoretically is able to effectively enhance the thermal conductivity of polymer composites. Great efforts have been dedicated, however, with only moderate success [132–140]; representative data from the references are summarized in Table 3.2, showing that no significant enhancement has been achieved yet. Researchers have been trying to understand such a phenomenon [138, 141–148]. One important factor is structural defects in CNTs that act as scattering centers for phonons. The other widely accepted finding is that a large interfacial thermal resistance across the nanotube-polymer interface causes a significant degradation in the thermal conductivity enhancement.

Thermal resistance at an interface – Kapitza resistance, R_K – represents a barrier to heat flow associated with the differences in phonon spectra of two phases in contact. For high structural mismatch and weak contact at an interface, R_K is large.

In thermal transport, R_K is defined by

$$J_Q = \Lambda \Delta T \quad (3.17)$$

$$R_K = \frac{1}{\Lambda} = \frac{a_K}{K_m} \quad (3.18)$$

where J_Q, Λ, R_K, K_m, and a_K are heat flux, interfacial thermal conductance, interfacial thermal resistance, thermal conductivity of the matrix, and Kapitza radius, respectively.

In a CNT-polymer nanocomposite system, the R_K plays a much more important role than in common composites with micro-sized fillers, not only because the interface phase accounts for as high as 15 vol% but also because the a_K is around 5–20 nm, comparable to or even larger than the tube diameter [149, 150].

Experimental measurements showed the R_K in CNT suspension is around 8′ 10^{-8} m^2-K/W [150]; this value is of the same order of magnitude as those in other composite materials [149]. Such a large R_K arises from the weak coupling between the rigid tube structure and the soft polymer matrix [143]. For a nanotube, there are four acoustic modes: a longitudinal mode, corresponding to motion of atoms along the tube axis; a twist mode, corresponding to torsion around the tube axis; and two degenerate transverse modes – bending and squeezing modes, corresponding to atomic displacements perpendicular to the nanotube axis. The longitudinal and twisting modes involve only the movement of carbon atoms tangential to the tube surface and thus are expected to couple very weakly to the soft material around the tube, while the bending and squeezing modes involve out-of-plane motion and thus are more strongly coupled with the soft material around. Due to the weak coupling, sharp contrast in the phonon spectra at the interface between CNTs and an amorphous polymer matrix leads to strong backscattering of thermal energy waves at the interface.

Based on the theory above, an effective approach to reducing R_K is to provide a stronger coupling between CNTs and a polymer matrix by covalent bonding. Therefore, proper surface functionalization of CNTs should help improve the interaction between the CNT fillers and the polymer matrix and consequently result in a significant increase in the interfacial thermal conductance [144], as shown in Fig. 3.7a.

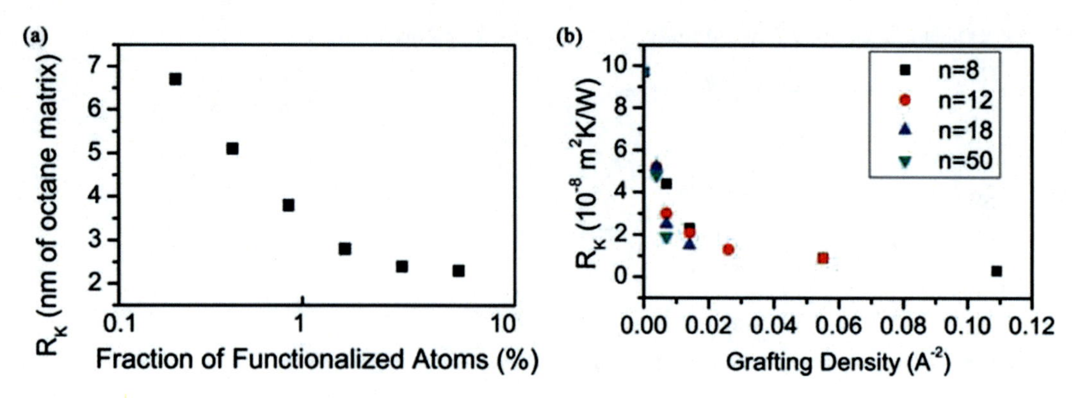

Fig. 3.7 Effects of surface functionalization (**a**) and grafting density of organic chains (**b**) on the interfacial thermal resistance of CNT-polymer matrix [146]

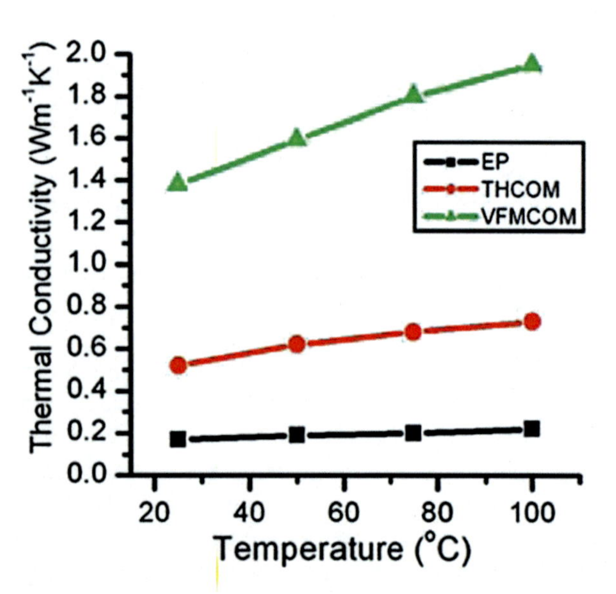

Fig. 3.8 Thermal conductivities of pure epoxy, ACNT-epoxy composite cured by thermal heating (THCOM), and ACNT-epoxy composite cured in a VFM chamber (VFMCOM)

Atomistic-based modeling was conducted to calculate the R_K and found that grafting organic chains or polymer chains to the surface of CNTs with covalent chemical bonds increase the conductance across the interface [145]. R_K was found to be a function of several parameters, e.g., grafting density and length of the linear chain covalently bonded to the CNT surface. Increasing chain length and grafting density lowers the interfacial thermal resistance, as shown in Fig. 3.7b.

Some modeling and experiment have been performed to examine the conclusions [138, 146]. However, conventionally, length and diameter of functionalized CNTs are not controllable. Evaluation of the effects of surface functionalization on the R_K, and consequently on the overall effective thermal conductivity of the composite, should exclude the influence brought about by the uncontrollable length and diameter of CNTs. This problem becomes a barrier for understanding the thermal transport in CNT-polymer systems, especially the R_K.

Progress has recently been made in understanding the influence of interface bonding between CNTs and a polymer matrix on the overall thermal conductivity of the composite. Lin et al. embedded ACNTs into an epoxy matrix and used variable frequency microwave (VFM) radiation to assist the curing process; the interface between the ACNTs and the epoxy matrix was enhanced [151]. In fact, microwave irradiation has attracted much interest in synthetic organic chemistry due to its special role in dramatically increasing reaction rates and the capability of inducing chemical reactions which cannot proceed by thermal heating alone [152–156]. It is postulated that during the curing process in the VFM chamber, the interfacial bonding between the CNTs and the epoxy matrix was dramatically improved. The enhanced interface effectively improved the overall thermal conductivity of the composite, as shown in Fig. 3.8.

3.5.3 Aligned Carbon Nanotube (ACNT)-Based Thermal Interface Materials (TIMs)

Vertically aligned CNTs utilize the superior longitudinal thermal conductivity of individual nanotubes and exhibit the overall thermal conductivities of ~ 80 W/m-K or higher [116, 157–162]. Fabrications and characterizations of ACNT TIMs have been a recent research focus. Hu et al. used a 3ω method to test the thermal contact resistance between a 13-µm-thick ACNT array and the surface of a free mating substrate. The results showed that the contact resistances were 17 and 15 mm^2-K/W, respectively, under the pressures of 0.040 and 0.100 MPa [163]. Wang et al. used a noncontact technique (photothermal) to measure the thermal conductivity of MWNT arrays but revealed only ~ 27 W/m-K for MWNTs [164]. Ngo et al. used electrodeposited copper as a

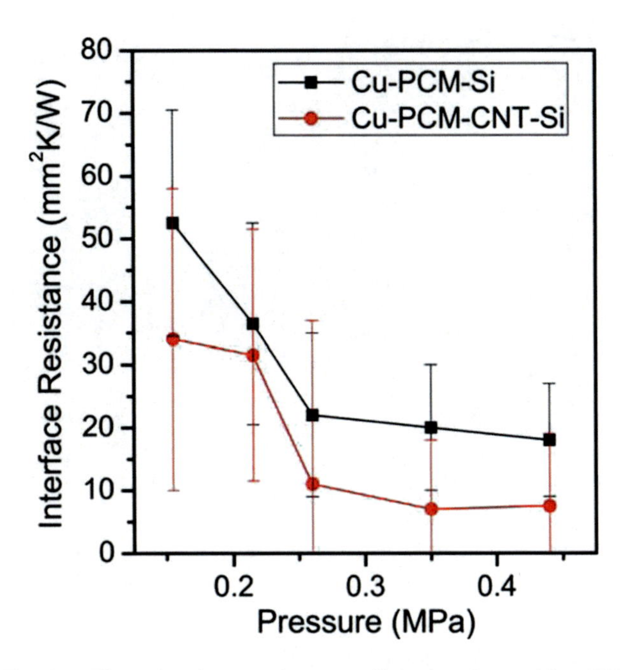

Fig. 3.9 Thermal resistance of copper–silicon interfaces with a CNT array (Cu-PCM-CNT-Si), compared with PCM (Cu-PCM-Si), as a function of pressure [116]

gap filler to enhance the stability and thermal transport of carbon nanofibers [165]. They reported the interfacial thermal resistance of 25 mm^2-K/W under a pressure of 0.414 MPa with a 1D reference bar method. Using a similar characterization method, Xu et al. reported a minimum thermal interface resistance of 19.8 mm^2-K/W for an ACNT array on a silicon substrate [116]. By using phase change material (PCM) with ACNTs, they produced a minimum resistance of 5.2 mm^2-K/W, as shown in Fig. 3.9. Xu et al. used a photothermal metrology to evaluate the thermal conductivity of aligned CNT arrays grown on silicon substrates by PECVD [166]. The effective thermal resistance was 12–16 mm^2-K/W, which is comparable to the resistance of commercially available thermal grease. Tong et al. used a phase-sensitive transient thermoreflectance technique to measure the thermal resistance of the two interfaces on each side of a vertically aligned CNT array as well as the ACNT itself [167]. They concluded that the interface between the free-end CNT tips and the opposing substrate in contact dominated the interfacial resistance. Cola et al. fabricated an interface material comprised of a metal foil with CNTs synthesized on both sides of the foil. This fabrication lowered the interfacial resistance to less than 10 mm^2-K/W [160]. Furthermore, they grew ACNTs on Si and copper substrates and fabricated them to make a two-sided ACNT TIM layer. By this assembly, they reported a minimum resistance of 4 mm^2-K/W.

A brief comparison between the thermal interfacial resistance for currently used TIMs and ACNT TIMs is listed in Table 3.3. Although solder TIMs have the lowest thermal

Table 3.3 A comparison of the best data reported on TIM resistance, reduced to the equivalent BLT of 1 mil

Polymer composites	Filled PCMs	Solders	Graphite paper	ACNT
7	8	2	10	4–20

resistivity among all the commonly used TIMs so far, bond line thickness (BLT) of solder TIMs cannot be as small as polymer-based TIMs due to the high CTE and processing complexity; therefore, the final total thermal interface resistance of solder TIMs is not so low as expected, which hinders application of solder TIMs as next-generation TIMs. The best data reported so far on ACNT TIMs, by Fisher's research group at Purdue University, lie in the range of 4–20 mm^2-K/W [116, 157, 160, 163, 168]. However, it should be further understood why different synthesis methods or catalyst compositions give the same CNT quality and why the same synthesis method gives totally different CNT diameter, diameter distribution, and substrate coverage. In fact, a high reproducibility in ACNT synthesis is a prerequisite for ACNT TIM applications, which requires further understanding on CNT growth mechanism and surface science involved in the growth substrate during a CVD process. This will be discussed in detail in Sect. 5.5.

What indeed determines the interfacial thermal resistance for an ACNT TIM? First, the intrinsic thermal resistance of the ACNTs is an important factor.

Theoretically, a longer CNT possesses a better thermal conductivity. However, in reality, this may not be true. Unavoidably, more defects present in a long CNT will lead to thermal conductance degradation. Therefore, the shorter the ACNT array, the better the thermal conductance. There is no doubt that improving CNT structural perfection by high-temperature annealing would be beneficial. However, common growth substrates such as silicon, copper, and quartz cannot sustain an effective annealing temperature as high as, for instance, 2400 °C.

Therefore, microwave annealing has been considered an alternative method of CNT post-growth annealing [169].

Second, look at the question again: "What is a TIM?" The answer seems so simple: a layer between two mating substrates (let us say "A" and "B") for more effective heat dissipation. Yes. Then "what is the picture of the real microstructure at the A–TIM–B interface?" Or let us ask the question in a different way: "Should A and B be separated by the TIM layer or not?" As a matter of fact, an ideal TIM should go only into the asperities created by the surface roughness of A and B while maintaining as much direct A–B solid contacts as possible. The direct A–B contacts are the main paths where heat flux flows. In this sense, no matter how low the thermal resistance of the TIM is, since an ACNT TIM unavoidably

Fig. 3.10 An illustration of the ACNT-EP TIM status at the mating surfaces

blocks part of the direct A–B contacts, it introduces extra resistance to the direct A–B contacts. This is evident when we refer to a series model for thermal resistance calculations. Of course, ACNTs filled in the asperities reduce the resistance at the noncontact sites. In these regards, a short ACNT TIM layer is preferred. Too small a thickness is, however, not desirable, not only because of the difficulty to grow uniform ACNTs with such a small length but also because of the surface waviness and the contact pressure. During the pressure-involved TIM assembling, ACNTs become cranked under the pressure at the contact interface with dramatically increased CNT coverage [94]. Figure 3.10 shows an ideal CNT status at the interface. For a properly rough (waved) surface, when a proper pressure is imposed, more actual contact area between the CNTs and the mating surface can be achieved and consequently better thermal conductance.

Third, there are still chances that a direct A-B contact exhibits a higher resistance than the case with a TIM in between that separates A and B; this means that a classical series model for resistance calculation fails. This is possible when we consider that fundamental of interfacial thermal resistance. The thermal resistance at the A-B contact sites originates from phonon spectrum mismatch. Different A-B combinations show different thermal resistance. In this sense, a short ACNT TIM layer between A and B may play a role of mitigating the A-B interfacial phonon scattering by avoiding a sharp A-B contact. In this case, the interfacial modification at the ACNT-substrate interface is important and will be discussed in detail in Sect. 3.6.2.

3.5.4 ACNT TIM Synthesis on Bulk Copper Substrates

Although ACNT synthesis on silicon as the growth substrate has been widely investigated, direct synthesis of a ACNT layer on the backside of a silicon device is not compatible with a current front-end semiconductor fabrication process due to the high temperature (typically >650 °C) required for ACNT syntheses by, for example, chemical vapor deposition (CVD). Great efforts have been made on low-temperature ACNT syntheses; however, overall speaking, the CNT qualities by low-temperature CVD processes are low [98, 101–103, 170–209]. Thus, a silicon substrate is not the growth substrate of choice for real-life ACNT TIM applications; instead, ACNT synthesis on a bulk copper substrate, i.e., a copper lid, is preferred. A few papers have reported ACNT syntheses on various metal substrates; however, the direct ACNT synthesis on a bulk copper substrate is still a great challenge [210–216]. Here, we emphasize "bulk" copper because ACNT synthesis on a thin copper coating on a silicon wafer has been demonstrated for interconnects as discussed before. However, a thin copper coating by, for example, e-beam evaporation, DC sputtering, and electroplating on a smooth substrate is different from a bulk copper substrate in terms of purity, roughness, and crystallinity. Wang et al. developed a process to grow ACNTs on a copper foil surface by water-vapor-assisted CVD [217]. However, the quality and the structure of the "ACNTs" on copper were not clearly described. Recently, Yin et al. reported growing multi-walled CNTs on an oxygen-free copper substrate, however, with a poor CNT alignment [218]. Integrations of ACNTs grown on copper into thermal management were reported by Fisher's research group [116, 157]. Nevertheless, the ACNTs in these two references exhibited different average diameters, diameter distributions, and CNT coverage on the growth substrates, even though the same synthetic conditions were used. Awano et al. synthesized ACNT arrays on TiN-coated copper plates for interconnect demonstration; however, from Fig. 3.6, we see that the as-synthesized ACNTs show a poor alignment and uniformity. Moreover, Raman spectra of the ACNTs were not given. Recently, Lin and Wong [219] reported a remarkable progress on rapid synthesis of high-quality ACNTs on bulk copper substrates through a common thermal CVD process. By introducing a well-controlled conformal Al2O3

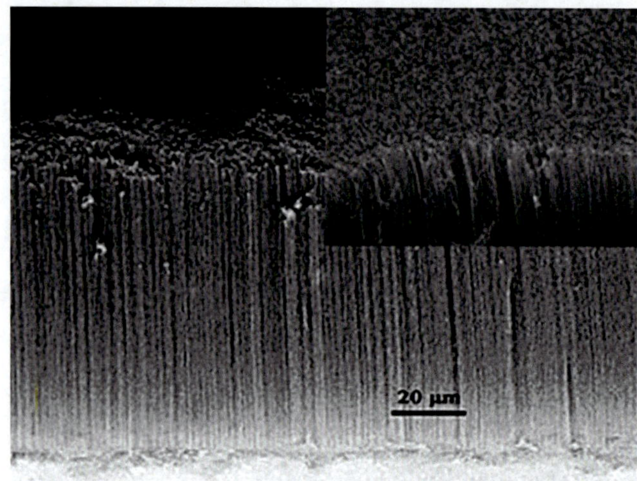

Fig. 3.11 A side-view SEM image of the ACNT TIM synthesized on a polished bulk copper substrate (*inset*: a further magnification)

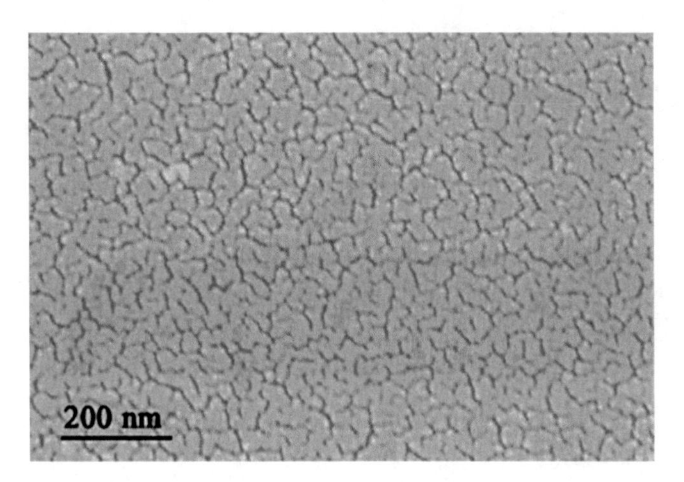

Fig. 3.12 Surface morphology of a 10-nm gold layer on Si (100) deposited by e-beam evaporation

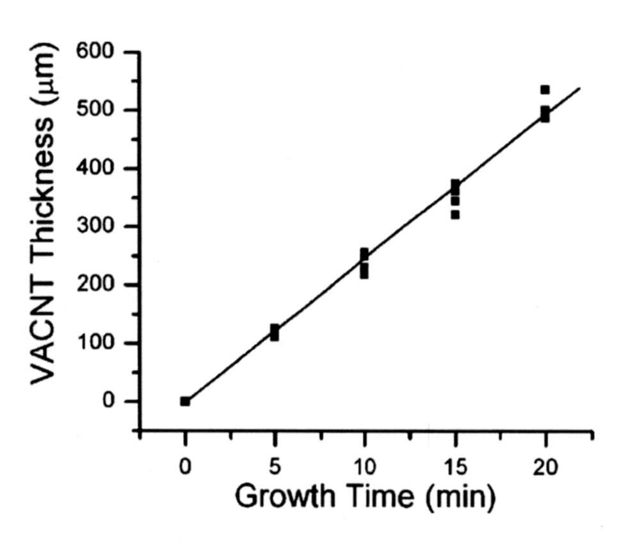

Fig. 3.13 A plot of ACNT thickness versus growth time, indicating kinetics-controlled ACNT growth on the bulk copper substrate

support layer on the bulk copper substrate by atomic layer deposition (ALD) prior to the deposition of the iron catalyst layer, they reproducibly synthesized ACNTs of good alignment and high quality on the copper substrate. A SEM image of the synthesized ACNT on copper is shown in Fig. 3.11.

Al_2O_3 support layers of 10- to 15-nm thick have been used to accelerate ACNT growth due to their capability of decomposing hydrocarbons and the interfacial interactions between Al_2O_3 and iron nanoparticles during the CVD syntheses of ACNTs [92, 220, 221]. Generally, literatures describe the conditions for the formation of an Al_2O_3 deposit layer on a target substrate by e-beam evaporation; however, the continuity of the deposit has rarely been mentioned. During an e-beam evaporation process, a continuous deposit layer forms on a substrate surface through nucleation and growth steps. At an early stage of deposition, heterogeneous nucleation may become rate controlling. This nucleation mechanism leads to isolated aggregates of condensate on the substrate when the deposit layer is thin [222]. This discontinuity is briefly discussed by de los et al. [222, 223]. In fact, the discontinuity of the deposit is the right reason why a thin layer of gold deposit on a bare silicon surface by e-beam evaporation was recently employed to selectively etch silicon wafers (Fig. 3.12) [224]. Therefore, when the Al_2O_3 deposit layer is too thin to become continuous, isolated nanoaggregates are formed [223, 225]. In this case, the bottom copper surface is partially exposed to the top iron catalyst and/or the carbon source at the inter-grain voids during the CVD process, and the ACNT synthesis fails. There are at least two reasons that account for the synthesis failure. First, copper deactivates the CNT growth. Pure copper does not catalyze decomposition of hydrocarbons as the interaction of hydrocarbons with copper surface does not lead to the rupture of the carbon–carbon bonds [226]. In addition, solubility of carbon in solid copper is extremely low [227, 228]. Thus, when the copper surface is exposed to the top iron particles during the CVD process, the local copper-rich catalyst composition is capable of decomposing hydrocarbons, however, without giving rise to ordered graphitic structures [226, 229]. Second, iron diffuses into copper when the iron particles are in contact with the copper surface at the inter-grain voids. Iron diffusion coefficient in single-crystalline copper is higher than 10^{-12} cm^2/s at 1000 K, at least three orders of magnitude higher than that in silicon dioxide ($< 10^{-15}$ cm^2/s) [230, 231]. Moreover, it is expected that a faster iron diffusion through enrichment at the grain boundaries in the polycrystalline copper has been used in Lin's study. In comparison with e-beam evaporation, ALD, also known as atomic layer epitaxy, is a special modification of CVD with the capability of producing films with excellent conformality and precisely controlled thicknesses [232–234]. Furthermore, unlike the thickness monitoring during e-beam evaporation, the thickness of the Al_2O_3 layer by ALD is accurate and insensitive to the surface roughness of copper substrates; as such, it guarantees a reproducible ACNT synthesis.

In comparison with the diffusion-controlled regime (>740 °C) of ACNT synthesis on SiO_2-Si surface [235], kinetics-controlled growth on the copper substrate was observed (Fig. 3.13).

3.6 Assembling Technologies of ACNTs

Generally, integration of CNTs into microsystems includes vertical and horizontal positioning. For horizontal interconnects, CNTs must be grown horizontally. However, there has been limited success in achieving controlled

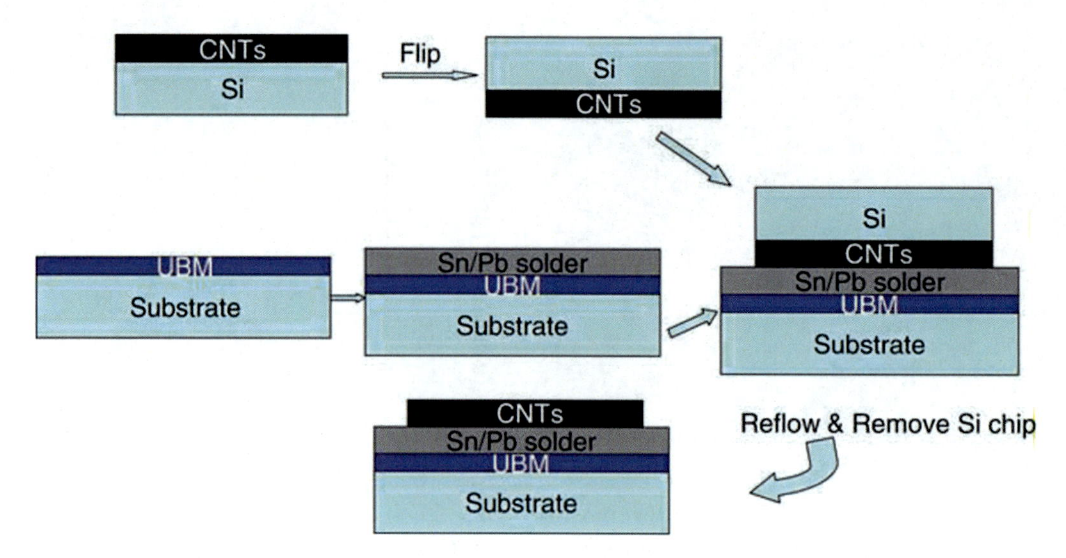

Fig. 3.14 An illustration of the solder transfer process

horizontal assembly or growth of CNTs [14, 15]. In this section, low-temperature assembling technologies for vertical positioning of ACNTs are discussed.

3.6.1 Physical Transfer/Anchoring

Transfer of ACNT films has been reported using aqueous treatments to debond the films from the growth substrates and place them onto another substrate surface [236]. Although this transfer separates the growth and assembly of ACNTs, effective contact of ACNTs on the substrate is limited, and the interfacial strength is weak. Zhu et al. proposed a solder transfer methodology for ACNTs, similar to flip-chip technology as illustrated schematically in Fig. 3.14 [220]. The substrates for this process can be copper or FR-4 boards coated with copper. A thin layer of solder paste, e.g., SnPb and SnAgCu, is stencil-printed on the substrate surface. After reflow at a proper temperature, the solidified solder layer is polished to ~ 30 μm. The ACNTs on the growth substrate are then flipped to the solder-coated substrate and reflowed to form electrical and mechanical connections. This process is straightforward to implement and overcomes the serious obstacles of integrating ACNTs into electronic packaging by offering low process temperatures and improved adhesion of ACNTs to substrates. This CNT transfer process can be expanded to assemble fine-pitch CNT bundles. Figures 3.15 and 3.16 show the transferred CNT film and bundles, respectively, onto copper substrates. Kordas et al. demonstrated a simple and scalable assembly process similar to the above solder transfer technique to fabricate ACNT microfins onto the chip for an air cooling system, as shown in Fig. 3.17 [161]. For a 1-mm^2 test chip to reach the same temperature, the applied power can be ~1 W larger when the ACNT sink is implemented than that of the situation with a bare chip. Testing indicated that the CNT fin structure allowed heat dissipation of ~30 and ~100 W/cm^2 more power at 100 °C from a hot chip for the cases of natural and forced convections, respectively. The cooling performance of the nanotube fin structures combined with their low weight, mechanical robustness, and ease of fabrication makes them possible candidates for on-chip thermal management applications. Kim et al. used a low-temperature ceramic cement to anchor ACNTs on silicon substrates with a mild curing temperature [237]. By filling the cement with gold particles, conductivity was enhanced, although it was not clearly determined whether the contact between the ACNTs

Fig. 3.15 A CNT film anchored on a copper plate by the solder transfer process

and the gold-filled cement was ohmic. They also demonstrated the transfer of ACNTs onto silicon surface coated with a thin gold layer. The high baking temperature, e.g., 800 °C, however, was not favored by electronic packaging. Sunden et al. reported a microwave-assisted transfer of ACNTs onto thermoplastic polymer substrates [156]. Jiang et al. proposed a low-temperature transfer process using conductive polymer composites [238]. An ohmic contact was formed between a CNT film and a highly conductive polymer composite, while a semiconductor joint was formed between a CNT film and a high-resistivity polymer composite. Recently, Gan et al. reported a large-scale bonding of ACNTs to metal substrates with the assistance of high-frequency induction heating. This technique provides a potential approach to reproducible large-scale fabrication of ACNTs for various applications [239].

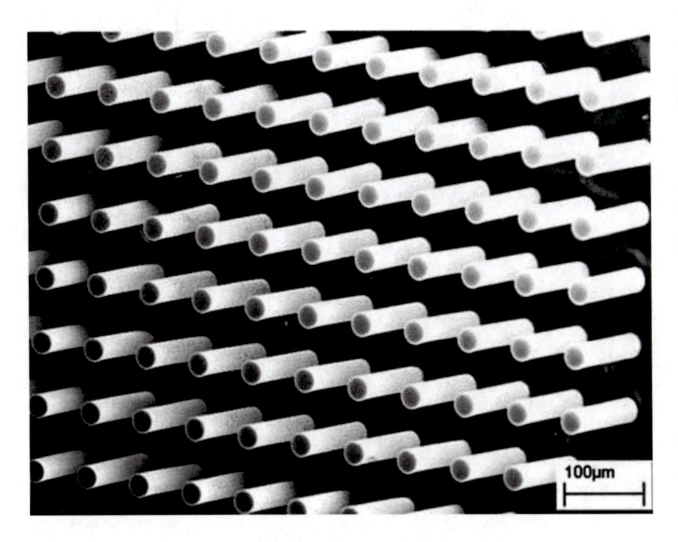

Fig. 3.16 CNT bundles anchored on a copper plate by the solder transfer process

3.6.2 Chemical Transfer

Recently, a chemical transfer technology was developed by Lin et al. (Fig. 3.18) [240]. They in-situ functionalized the ACNT surface and open tips with certain functional groups during the CVD process by introducing a trace amount of H_2O_2 as the oxidant into a reaction chamber. The synthesis

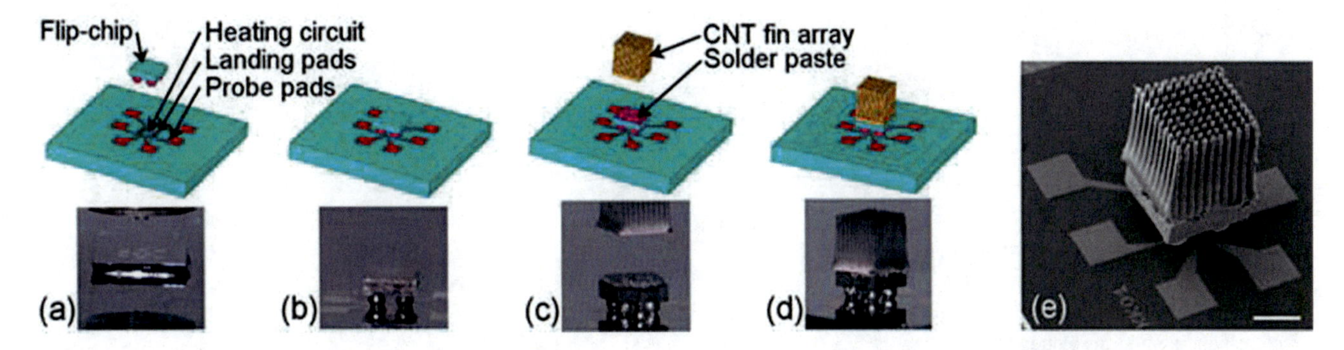

Fig. 3.17 Schematic illustration of the ACNT microfins fabrication process (**a–d**) and the image of the fabricated microfins as a heat sink on a flip-chip assembly [161]

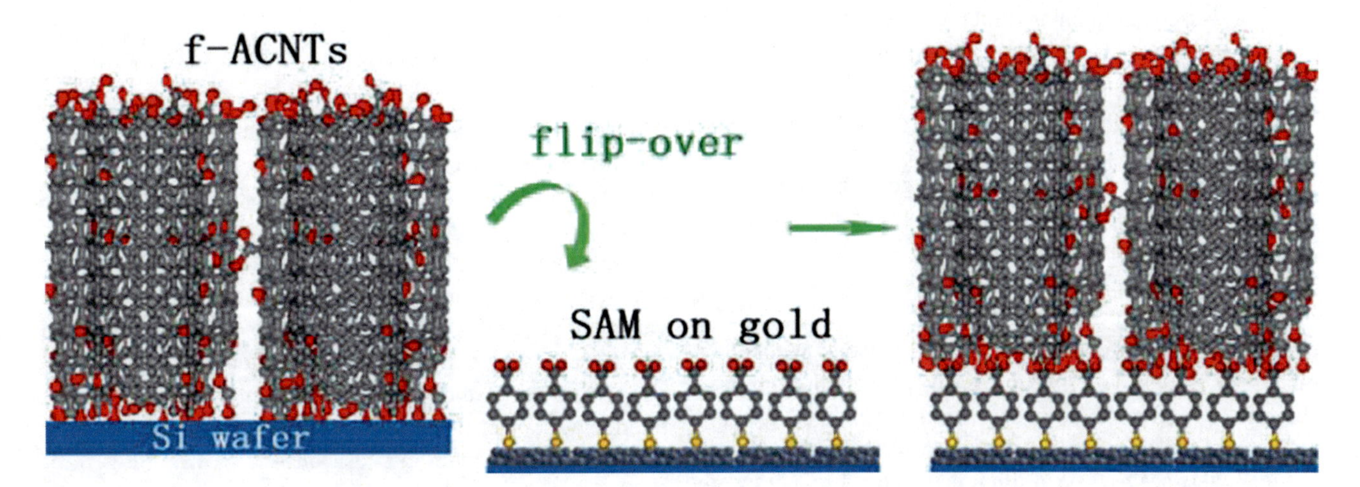

Fig. 3.18 Schematic illustration of the chemical transfer process

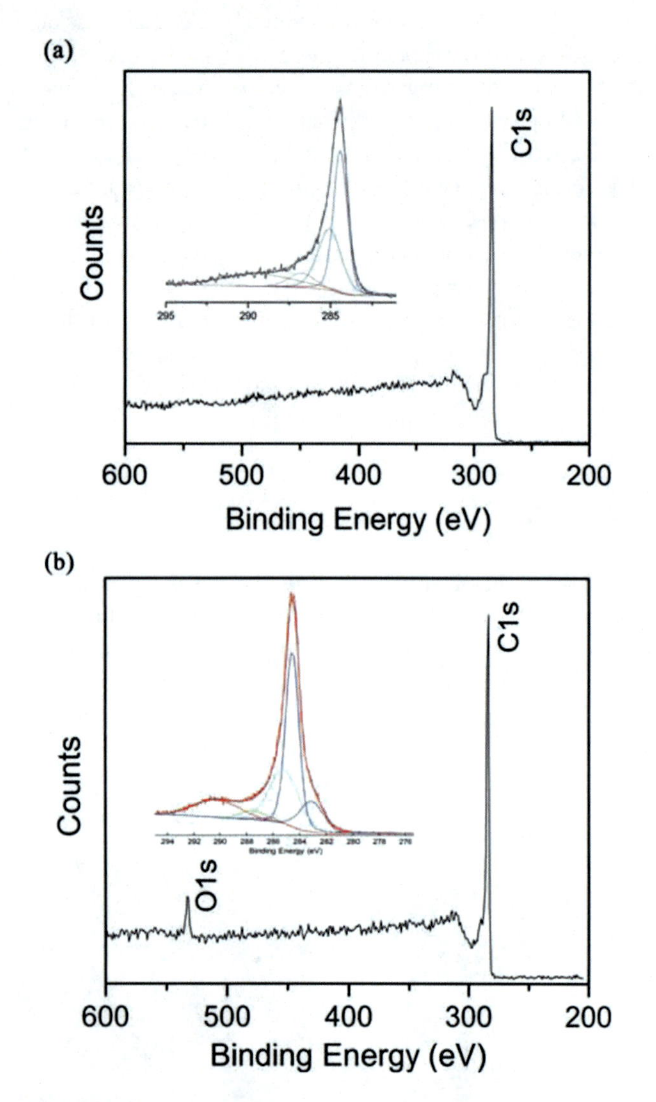

Fig. 3.19 XPS spectra of pristine ACNT (**a**) and f-ACNT (**b**). Inserted are corresponding high-resolution C 1 s spectra

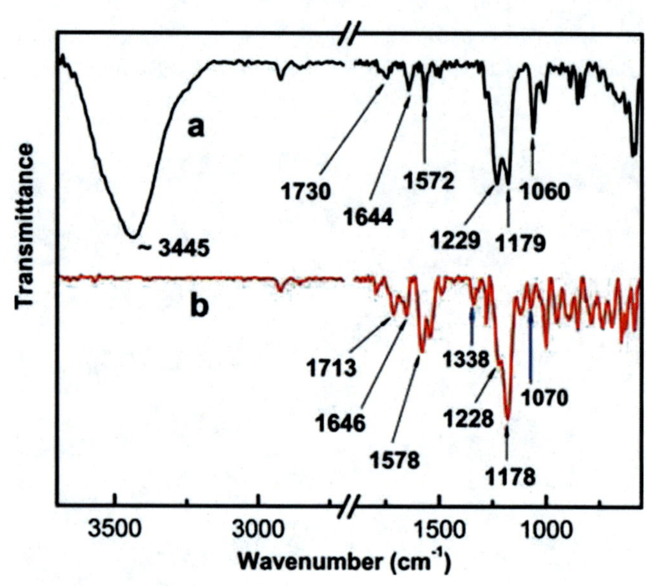

Fig. 3.20 FTIR spectra of the f-ACNTs (**a**) and the s-CNTs (**b**)

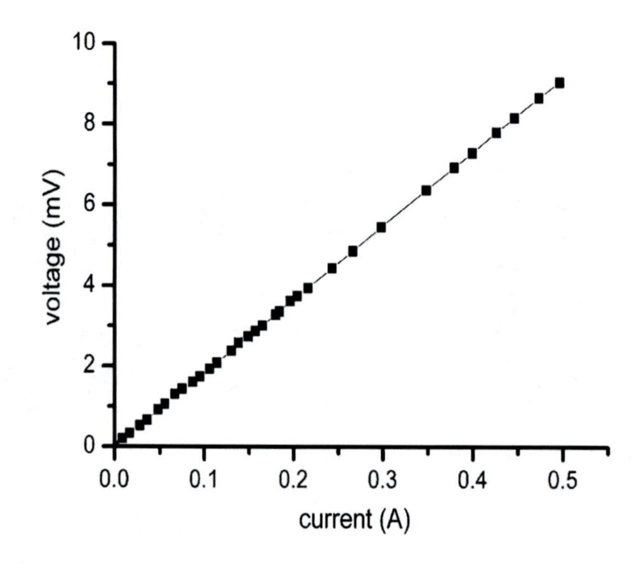

Fig. 3.21 The room-temperature *I–V* curve of the f-ACNT-gold interconnect via the chemical transfer

temperature could be lowered by 50 °C (700 °C). X-ray photoelectron spectroscopy (XPS) scans indicate the existence of oxygen-containing functional groups on ACNTs (Fig. 3.19). Figure 3.20 (trace a) shows the FTIR spectrum of the in situ functionalized ACNTs. The peak at ~ 1572 cm^{-1} is attributed to the C=C asymmetric stretching in graphite-like CNT structure.

Consistent with the XPS results, oxygen-containing functional groups were found. Peaks at 3445, 1229, 1179, and 1060 cm^{-1} are attributed to O–H, C–O, C–O–C, and C–OH stretchings, respectively. These oxygen-containing functional groups, particularly the hydroxyl groups and the carboxylic acid groups, make ACNTs reactive with other functional groups. To prove the reactivity, they used thionyl chloride to react with the in situ functionalized ACNTs, the FTIR spectrum of the resultant product s-CNT being shown in Fig. 3.20 (trace b). The in situ functionalized ACNTs were directly transferred to a gold-coated Si surface (or quartz surface) coated with a self-assembled monolayer of conjugated molecules as the bridging agent for mechanical anchoring and electron transport. This transfer technology has numerous benefits in that it covalently bonds the CNTs to the substrate, minimizes the damage to the CNTs during functionalization, can be tailored to attach CNTs to a wide variety of substrates, is carried out at low temperatures, and can be easily reworked. More importantly, the chemical bonds formed at the CNT-substrate interfaces reduce the interfacial contact resistance. Ohmic contact at the ACNT-substrate interface with low resistance is shown in Fig. 3.21. Preliminary data show that the transfer process dramatically reduces the thermal interfacial resistance as well.

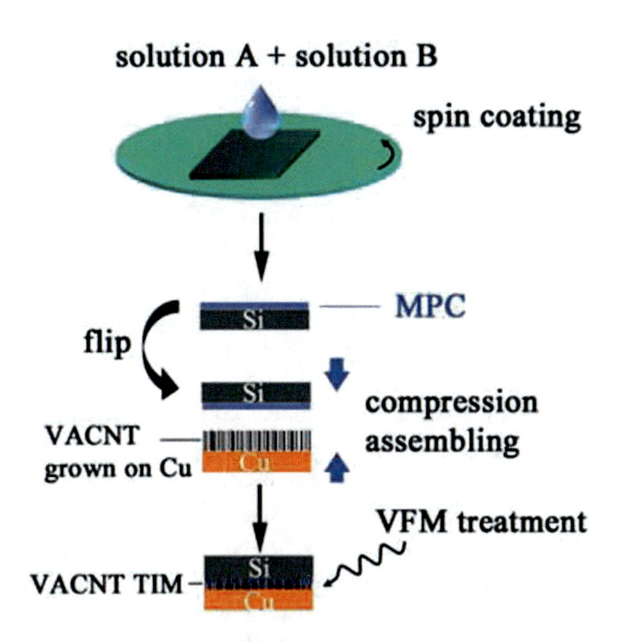

Fig. 3.22 Schematic illustration of the Si/ACNT TIM/Cu assembling process

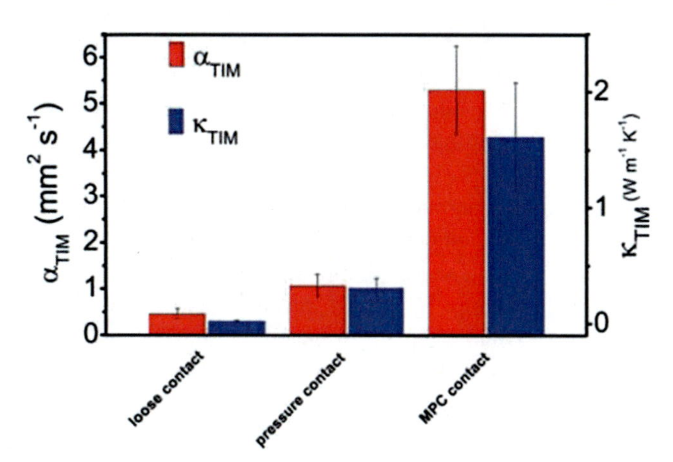

Fig. 3.23 A comparison of the thermal diffusivities (α_{TIM}) and the equivalent thermal conductivities (κ_{TIM}) among the loose contact, the pressure contact, and the MPC contact ACNT TIM assemblies

In our previous discussion, we summarized representative work on ACNT TIMs that have been reported so far. There are at least two main issues in common that inhibited real-life applications of the ACNT "TIMs": (1) the thermal contact resistance at the CNT-mating substrate interface was large and dictated the overall thermal resistance of the ACNT TIM [241], even when a relatively high pressure was imposed, and (2) the interfacial adhesion was extremely weak. One reason for the large contact resistance is mainly that, according to the very recent report by Panzer, the fraction of CNT-substrate contacts is low [159]. More importantly, in the molecular scale or the atomic scale, the CNTs are not in "contact" with the mating substrate surfaces because there is no chemical bonding or special association, which is also the reason for the weak interfacial adhesion. Thus, a challenge arises as how can people modify the CNT-mating substrate interface to effectively reduce the thermal contact resistance and simultaneously improve the interfacial adhesion? When heat transfers at the interface between a nanotube and a planar substrate, constriction resistance is developed due to the nanometer-scale contact area and enhanced phonon-boundary scattering at the nanocontacts [242]. Diao et al. [243] used MD simulations to study the ACNT-Si contact resistance. Their results showed that open-ended ACNTs at the interface reduced contact resistance; contact area and the number of bonds formed between the CNT and the Si substrate are the key to the interfacial thermal conductance. Therefore, Lin et al. extended their chemical transfer (or "chemical anchoring") technology to ACNT TIM assembling [244]. They grew ACNTs on a copper substrate and introduced a thermally stable cross-linked molecular phonon coupler (MPC) to chemically bond the ACNTs to the mating Si substrate. The ACNT TIM assembling process is schematically shown in Fig. 3.22. The MPC at the mating interface facilitates phonon transport across the ACNT-Si interface. The developed ACNT TIM assembling process not only improves the equivalent thermal conductivity of the ACNT TIM by two orders of magnitude (Fig. 3.23) but also dramatically enhances the interfacial adhesion.

3.7 Graphene Nanoelectronics

Semiconductor technology has pushed the size of transistors and electrical interconnects toward a few nanometers, at which scale quantum effects become important. Take the physical limits on CMOS scaling, for example. Gate tunneling current, subthreshold channel leakage current, device parameter variability/tolerances, interconnect latency, static and dynamic power dissipation, and so on are great challenges. Although some innovations have, to a certain extent, addressed some of these challenges such as strain in channel, three to five channel FETs, high-permittivity gate dielectrics (e.g., HfSiON), metal gates (e.g., TaSiN), and nanolithography techniques, a revolutionary resolution is considered necessary. Graphene, due to its intrinsic peculiar electronic properties, quantum effects, and extremely high in-plan thermal conductivity, has shown potential advantages to bring such a revolution to conventional Si semiconductors. Graphene, a single atomic layer of coplanarized hexagonally bonded carbon atoms, has recently generated enormous interests in academy and industry [245–259]. It has been proposed that graphene nanoelectronics may provide a solution to the critical challenges faced by current and future semiconductor technology due to its unusual properties that, so far as we know, include an (1) intrinsically zero band gap but nonzero (tunable) band gap due to influences of external factors such as the growth substrate and the shape effect [260–265]; (2) a large mean-free path of carrier transport comparable to that of SWNTs, allowing potential applications as ballistic transistors [266–275]; (3) an extraordinarily high carrier mobility on the order of 10,000 cm^2/V-s even at room temperature [274, 276–278]; (4) constant velocity of holes and electrons that does not depend on their kinetic energy [248]; (5) quantum Hall effect [278–280]; (6) extremely high in-plane thermal conductivity [281, 282]; and (7) 3D scalability compared with CNTs.

For potential application of graphene interconnects, research work reported is still quite limited. Recently, fabricating electronic circuits consisting of both transistors and interconnects in the same continuous graphene sheet has been proposed. Naeemi and Meindl used compact physical models to study the electrical conductance of graphene nanoribbons as functions of chirality, ribbon width, Fermi level, and the type of electron scatterings at the edges [283]. It was proposed that for widths below 8 nm, single-layer graphene nanoribbons can potentially outperform copper wires with unity aspect ratio. Shao et al. reported the high-temperature electrical resistance of graphene [284]. It was found that as temperature increases, the electrical resistance of graphene interconnects drops. They explained this phenomenon by the thermal generation of electron–hole pairs at elevated temperatures and carrier scattering by acoustic phonons.

3.8 Graphite Nanosheet-Polymer Composites: Very Promising TIM Composite Materials

As usual, like any advanced material discovered and prepared, one of its potential applications will be to incorporate it into polymer matrices to make high- performance composites [285–287]. Exfoliated graphite, a stack of tens to hundreds of graphene layers, has been used as an effective filler in polymer matrices for thermal conductivity improvement [288–291]. Exfoliated graphite with a few layers of graphene and a high aspect ratio, also named "graphite nanosheet (GNS)," is of special interest [117, 255]. Recently, Sun's group reported their successful preparation of GNS with a high aspect ratio and incorporation of the GNS into an epoxy (EP) matrix [292]. The as-prepared GNS-EP composite displayed an extremely high in-plane thermal conductivity ($\sim$ 80 W/m-K at 33 vol% GNS loading). Compared with carbon nanotube-epoxy nanocomposites, there is no doubt that the GNSs are considered more effective fillers for improving thermal conductivity. This great enhancement, hypothetically, resulted from the reduced GNS-polymer interfacial thermal resistance. However, this hypothesis is contrary to the results by Hung et al. where a strong influence of interfacial resistance was estimated [291]. Lin et al. used a conventional effective medium model to analyze Sun's results and showed good validity of the model in predicting thermal conductivities (both in-plane and through-thickness) of the GNS-polymer nanocomposites [293]. A high in-plane thermal conductivity, i.e., 1500 W/m-K, of a GNS was chosen in their modeling, on the basis of the following considerations. The in-plane thermal conductivity of a single graphene sheet was estimated to be as high as $\sim$5000 W/m-K [281, 282]. Given the fact that a multi-walled carbon nanotube (MWNT), which is treated as a concentrically wrapped GNS [29], possesses a similar longitudinal thermal conductivity (>3000 W/m-K) [81] to a single-walled carbon nanotube ($\sim$3500 W/m-K) [76], it was believed that a GNS possesses a high *K*f11 in the same order of magnitude with *K*f11 of a single graphene sheet. Previous experimental data on carbon nanotubes (CNTs) also showed that the longitudinal thermal conductivities of MWNT arrays were $\sim$80 W/m-K [116, 157, 161]. Since the packing densities of these MWNT arrays are $\sim$5% and tube–tube interactions degrade the effective thermal conductivities [294], they estimated the equivalent longitudinal thermal conductivity of an individual MWNT to be at least 1600 W/m-K. Therefore, the GNS, an unrolled form of an MWNT, should possess a *K*f11 of approximately 1600 W/m-K. Further proof came from the in-plane thermal conductivity of the commercial pyrolytic graphite sheets (PGS) provided by Panasonic, Inc. As the PGS density approaches the theoretical mass density of graphite, the in-plane thermal conductivity of the PGS becomes 1500 and 1700 W/m-K. To be conservative, 1500 W/m-K was taken in their modeling. The modeling results indicated the influence of GNS-polymer interfacial thermal resistance on the overall thermal transport, in analogy to CNT-polymer composites. Based on their modeling results, it was proposed that further exfoliation of GNS to achieve a large aspect ratio and GNS-polymer interfacial modification would lead to further enhancement of the thermal transport of GNS-polymer nanocomposites.

3.9 Thermal Property Measurement Techniques

Various thermal property measurement techniques have been applied to evaluate ACNT TIMs, including laser flash technique, steady-state TIM testers based on ASTM 5470, thermoreflectance technique, photoacoustic technique, photothermal technique, and 3ω technique. For conventional TIMs, that is, TIMs with relatively high thermal resistance, the ASTM-based steady-state measurement technique is the most popular and well-accepted technique. It has almost become a standard in industry. However, generally, the ASTM-based technique has very limited capability of measuring CNT TIMs with small thickness (e.g., a few microns) and low thermal resistance (e.g., < 10 mm^2-K/W). Moreover, CNT-substrate contact resistance is needed for scientific understanding of interfacial phonon transport, which cannot be measured directly by the ASTIM-based technique. Therefore, the other measurement techniques have been employed to study ACNT TIMs; we will go through them briefly in this section.

3.9.1 Laser Flash Technique (LFA 447, NETZSCH, Inc.)

Laser flash technique (LFA) is a sensitive, noncontact light flash method and has been widely used for measuring thermal diffusivity and specific heat of dense materials [295–299]. In an LFA setup, a high-performance xenon flash lamp, a sample, and a liquid-nitrogen-cooled Sb infrared detector are vertically arranged. During testing, the entire bottom surface of the sample is illuminated homogeneously by a pulsed laser produced by the flash lamp. Both the released energy of flash lamp and the length of the heating pulse can be adjusted via the 32-bit MS windows software named Nanoflash. An integrated furnace and a thermocouple positioned in the sample carrier are used for accurate control of sample temperatures. The temperature on the top surface of the sample is monitored by the infrared detector versus time after the light pulse. The thermal diffusivity of the sample is calculated by fitting the as-obtained data (temperature-time) with various models integrated in the analysis software of the LFA device. The thermal diffusivity measuring range is 0.01–1000 mm^2/s for LFA 447, with a reproducibility of approximately ±3%. In addition to the thermal diffusivity, by employing a comparative method, the specific heat of the sample can also be determined, with a reproducibility of approximately ±5%.

Typically, a thin layer of gold (100 nm) is sputtered onto the top and bottom surfaces. The gold layer on the bottom surface is to ensure a fast heat homogenization within the bottom surface so as to avoid measurement errors introduced by the hot spot issue. The gold layer on the top surface is to make sure that the infrared signal collected by the detector comes only (or mostly) from the very top surface. For thermal measurement of CNT-polymer composites, which are dense materials, a well-polished sample with a proper size, surface flatness, and roughness is required. For an ACNT TIM sample, since it is porous rather than dense, LFA could not be used to measure the intrinsic thermal properties of the ACNT array. Instead, a substrate/ACNT-TIM/substrate trilayer assembly is used. The measurement result is, therefore, the equivalent thermal diffusivity of the ACNT TIM layer that consists of the ACNT array and the two contact interfaces.

3.9.2 Steady-State Measurement

One-dimensional steady-state measurement is based on ASTM 5470, in which, in simple terms, the TIM is filled between two blocks with coplanar surfaces [116, 119, 300–303]. One block is heated and the other cooled. Therefore, a constant heat flux is supplied through the copper (or aluminum) blocks across the TIM sample. The overall interface thermal conductance (θ_{TIM}) is determined by extrapolating the temperature jump across the sample. ASTM 5470-based TIM testers are already commercialized, e.g., TIM tester 1300 from Analysis Tech Inc. Two advantages of steady-state measurements over other techniques are as follows:

(1) It is convenient to control the pressure (compression) imposed during the testing.
(2) It is suitable for TIM reliability evaluation.

However, in the steady-state measurement method, it is difficult to determine the thermal resistance between the ACNTs and the substrates. Similar to the laser flash technique, only a combined resistance is measured. The contact resistance may be obtained by extrapolating the resistance-TIM thickness curve toward zero thickness; however, the precision is low.

3.9.3 Thermoreflectance Technique

Phase-sensitive transient thermoreflectance technique was first developed by Ohsone et al. to determine the thermal conductance of the interface between a silicon substrate and the thermally grown silicon dioxide on the silicon substrate [304]. Thermoreflectance technique has been used in the nanosecond regime to characterize the thermal properties of thin films [305, 306], in the picosecond regime to measure both the thermal properties of thin films and their interfaces [307–309], and in the femtosecond regime to measure electron and phonon processes [310, 311]. Experimentally, a sample is heated by a high-power pulsed heating laser with intensity sinusoidally modulated at an angular frequency [125, 159]. A second low-power probe laser is focused on one surface of the sample. The heat flux oscillation caused by the heating laser propagates through the sample resulting in a temperature oscillation in the sample and at the interfaces. The reflected laser beam of the probe laser is captured by a photodetector, and the intensity signal is sent to an amplifier to extract the signal oscillation frequency. On the basis that the intensity and the phase of the reflected beam are dependent on the temperature of the reflecting surface, the phase shift and/or the intensity provides a dynamic measurement of the temperature response of the sample. A multiparameter search algorithm based on a least-square fit to the experimental data is used to obtain the thermal resistances/conductances of the interfaces and the thermal diffusivity of the TIM itself. The key advantage of this transient technique is its capability of distinguishing the interface thermal conductances from various interfaces and TIM itself.

3.9.4 Photoacoustic Technique

An alternative technique to measure the thermal resistances of thin films and the various interfaces discretely in a multi-layered structure is photoacoustic (PA) technique. Experimentally, a sample is mounted at the bottom of a PA cell, which is typically made of sapphire [157, 160, 168, 312, 313]. Sapphire has low reflectance and high transmittance for the infrared laser used; therefore, most of the reflected laser energy from the sample surface transmits out of the cell. A modulated heating laser is directed onto the sample surface. An oscillatory change of the temperature in the gas cell is induced due to the heat conduction through the contact with the sample surface and, as well, the mechanical work imposed on the gas medium due to the vibration of the sample surface. The oscillatory temperature change causes an oscillatory pressure change of the gas, which can be sensed as an acoustic wave. A condenser microphone mounted near the inside wall of the cell senses the acoustic signal and transfers it to a lock-in amplifier, where the amplitude and phase of the acoustic signal are measured. Because the transient temperature in the gas is related to the thermal properties of the sample, measuring the pressure allows determination of the thermal quantities such as the interfacial thermal resistance and the thermal diffusivity. Though it is thought that PA technique provides high accuracy and a convenient measurement procedure, PA devices are still not commercially available.

Other techniques such as photothermal technique and 3ω technique are not introduced here; readers are recommended to refer to related literature for details.

3.10 Future Needs

From the discussions above, we expect promising applications of carbon nanotubes as vertical electrical interconnects and TIMs for microelectronic packaging. However, a number of issues need to be addressed:

1. *Structural defect control* of ACNTs during CVD syntheses or post-growth treatments.
2. Effective increase in CNT *packing density*. Recent progress has been reported in increased ACNT packing density, i.e., from $\sim$1 to $\sim$20%, and consequently increased current-carrying capability [314]. However, a densification process with less complexity is in need for real-life applications. The influence of densification on thermal dissipation needs to be investigated.
3. For ACNT TIMs, ACNT-substrate *interface modification* is significant for improving phonon transport by reducing phonon scattering. The abovementioned MPC-assisted chemical modifications need to be optimized and are underway.
4. Although graphene nanoelectronics is approaching conceptually, a number of technological issues will emerge and need to be resolved effectively, such as 3D interconnection, controlled doping, band gap tuning, and gate fabrications.
5. Undoubtedly, a great deal of attention will be paid to GNS-polymer composites or graphene-polymer composites for applications as TIMs.

References

1. Moore, G.E.: Progress in digital integrated electronics. In: International Electron Devices Meetings, pp. 11–13 (1975)
2. Steinlesberger, G., Engelhardt, M., Schindler, G., Steinhogl, W., von Glasow, A., Mosig, K., Bertagnolli, E.: Electrical assessment of copper damascene interconnects down to sub-50 nm feature sizes. Microelectron. Eng. **64**, 409–416 (2002)
3. Kreupl, F., Graham, A.P., Duesberg, G.S., Steinhogl, W., Liebau, M., Unger, E., Honlein, W.: Carbon nanotubes in interconnect applications. Microelectron. Eng. **64**, 399–408 (2002)
4. Li, J., Ye, Q., Cassell, A., Ng, H.T., Stevens, R., Han, J., Meyyappan, M.: Bottom-up approach for carbon nanotube interconnects. Appl. Phys. Lett. **82**, 2491–2493 (2003)
5. Nihei, M., Horibe, M., Kawabata, A., Awano, Y.: Simultaneous formation of multiwall carbon nanotubes and their end-bonded ohmic contacts to Ti electrodes for future ULSI interconnects. Jpn. J. Appl. Phys. Part 1 Regul. Pap. Short Notes Rev. Pap. **43**, 1856–1859 (2004)
6. Naeemi, A., Sarvari, R., Meindl, J.D.: Performance comparison between carbon nanotube and copper interconnects for gigascale integration (GSI). IEEE Electron. Device Lett. **26**, 84–86 (2005)
7. Naeemi, A., Meindl, J.D.: Impact of electron-phonon scattering on the performance of carbon nanotube interconnects for GSI. IEEE Electron. Device Lett. **26**, 476–478 (2005)
8. Awano, Y.: Carbon nanotube technologies for LSI via interconnects. IEICE Trans. Electron. **E89C**, 1499–1503 (2006)
9. Park, M., Cola, B.A., Siegmund, T., Xu, J., Maschmann, M.R., Fisher, T.S., Kim, H.: Effects of a carbon nanotube layer on electrical contact resistance between copper substrates. Nanotechnology. **17**, 2294–2303 (2006)
10. Nieuwoudt, A., Massoud, Y.: Understanding the impact of inductance in carbon nanotube bundles for VLSI interconnect using scalable modeling techniques. IEEE Trans. Nanotechnol. **5**, 758–765 (2006)
11. Xu, T., Wang, Z., Miao, J., Chen, X., Tan, C.M.: Aligned carbon nanotubes for through-wafer interconnects. Appl. Phys. Lett. **91**, 042108 (2007)
12. Kaushik, B.K., Goel, S., Rauthan, G.: Future VLSI interconnects: optical fiber or carbon nanotube – a review. Microelectron. Int. **24**, 53–63 (2007)
13. Graham, A.P., Duesberg, G.S., Seidel, R., Liebau, M., Unger, E., Kreupl, F., Honlein, W.: Towards the integration of carbon

nanotubes in microelectronics. Diam. Relat. Mater. **13**, 1296–1300 (2004)
14. Huang, S.M., Cai, X.Y., Liu, J.: Growth of millimeter-long and horizontally aligned single-walled carbon nanotubes on flat substrates. J. Am. Chem. Soc. **125**, 5636–5637 (2003)
15. Zhang, Y.G., Chang, A.L., Cao, J., Wang, Q., Kim, W., Li, Y.M., Morris, N., Yenilmez, E., Kong, J., Dai, H.J.: Electric-field-directed growth of aligned single-walled carbon nanotubes. Appl. Phys. Lett. **79**, 3155–3157 (2001)
16. Chai, Y., Xiao, Z.Y., Chan, P.C.H.: Electron-shading effect on the horizontal aligned growth of carbon nanotubes. Appl. Phys. Lett. **94**, 043116 (2009)
17. Lan, C., Zakharov, D.N., Reifenberger, R.G.: Determining the optimal contact length for a metal/multiwalled carbon nanotube interconnect. Appl. Phys. Lett. **92**, 213112 (2008)
18. Nihei, M., Horibe, M., Kawabata, A., Awano, Y.: Simultaneous formation of multiwall carbon nanotubes and their end-bonded ohmic contacts to Ti electrodes for future ULSI interconnects. Jpn. J. Appl. Phys. **43**, 1856–1859 (2004)
19. Schmidt, R.: Challenges in electronic cooling – opportunities for enhanced thermal management techniques – microprocessor liquid cooled minichannel heat sink. Heat Transf. Eng. **25**, 3–12 (2004)
20. Mahajan, R., Chiu, C.P., Chrysler, G.: Cooling a microprocessor chip. Proc. IEEE. **94**, 1476–1486 (2006)
21. Prasher, R.: Thermal interface materials: historical perspective, status, and future directions. Proc IEEE. **94**, 1571–1586 (2006)
22. Chu, R.C.: The challenges of electronic cooling: past, current and future. J. Electron. Packag. **126**, 491–500 (2004)
23. Azar, K.: Power consumption and generation in the electronics industry – a perspective. In: 20th IEEE SEMI-Therm symposium. San Jose, CA, pp. 201–212 (2004)
24. Prasher, R.S.: Nano and micro technology based next generation package-level cooling solutions. Int. Technol. J. **9**, 285–292 (2005)
25. Schelling, P.K., Shi, L., Goodson, K.E.: Managing heat for electronics. Mater. Today. **8**, 30–35 (2005)
26. Iijima, S.: Helical microtubules of graphitic carbon. Nature. **354**, 56–58 (1991)
27. Ajayan, P.M., Iijima, S.: Smallest carbon nanotube. Nature. **358**, 23 (1992)
28. Ebbesen, T.W., Ajayan, P.M.: Large-scale synthesis of carbon nanotubes. Nature. **358**, 220–222 (1992)
29. Ajayan, P.M., Ebbesen, T.W.: Nanometre-size tubes of carbon. Rep. Prog. Phys. **60**, 1025–1062 (1997)
30. Ajayan, P.M.: Nanotubes from carbon. Chem. Rev. **99**, 1787–1799 (1999)
31. Hamada, N., Sawada, S., Oshiyama, A.: New one-dimensional conductors – graphitic microtubules. Phys. Rev. Lett. **68**, 1579–1581 (1992)
32. Dresselhaus, M.S., Dresselhaus, G., Saito, R.: Carbon-fibers based on C-60 and their symmetry. Phys. Rev. B. **45**, 6234–6242 (1992)
33. Dresselhaus, M.S., Dresselhaus, G., Avouris, P.: Carbon Nanotubes – Synthesis, Structure, Properties and Applications. Springer, New York (2001)
34. Mackay, A.L., Terrones, H.: Diamond from graphite. Nature. **352**, 762 (1991)
35. Zhang, X.F., Zhang, X.B., Vantendeloo, G., Amelinckx, S., Debeeck, M.O., Vanlanduyt, J.: Carbon nano-tubes – their formation process and observation by electron-microscopy. J. Cryst. Growth. **130**, 368–382 (1993)
36. Tsang, S.C., Chen, Y.K., Harris, P.J.F., Green, M.L.H.: A simple chemical method of opening and filling carbon nanotubes. Nature. **372**, 159–162 (1994)
37. Mintmire, J.W., Dunlap, B.I., White, C.T.: Are fullerene tubules metallic. Phys Rev Lett. **68**, 631–634 (1992)
38. Odom, T.W., Huang, J.L., Kim, P., Lieber, C.M.: Atomic structure and electronic properties of single-walled carbon nanotubes. Nature. **391**, 62–64 (1998)
39. Wildoer, J.W.G., Venema, L.C., Rinzler, A.G., Smalley, R.E., Dekker, C.: Electronic structure of atomically resolved carbon nanotubes. Nature. **391**, 59–62 (1998)
40. Peierl, R.F.: Quantum Theory of Solids. Clarendon, Oxford (1995)
41. Yao, Z., Kane, C.L., Dekker, C.: High-field electrical transport in single-wall carbon nanotubes. Phys. Rev. Lett. **84**, 2941–2944 (2000)
42. Wei, B.Q., Vajtai, R., Ajayan, P.M.: Reliability and current carrying capacity of carbon nanotubes. Appl. Phys. Lett. **79**, 1172–1174 (2001)
43. Collins, P.C., Arnold, M.S., Avouris, P.: Engineering carbon nanotubes and nanotube circuits using electrical breakdown. Science. **292**, 706–709 (2001)
44. Bachtold, A., Strunk, C., Salvetat, J.P., Bonard, J.M., Forro, L., Nussbaumer, T., Schonenberger, C.: Aharonov-Bohm oscillations in carbon nanotubes. Nature. **397**, 673–675 (1999)
45. McEuen, P.L., Bockrath, M., Cobden, D.H., Yoon, Y.G., Louie, S. G.: Disorder, pseudospins, and backscattering in carbon nanotubes. Phys. Rev. Lett. **83**, 5098–5101 (1999)
46. Li, H.J., Lu, W.G., Li, J.J., Bai, X.D., Gu, C.Z.: Multichannel ballistic transport in multiwall carbon nanotubes. Phys. Rev. Lett. **95**, (2005)
47. Haruehanroengra, S., Wang, W.: Analyzing conductance of mixed carbon-nanotube bundles for interconnect applications. IEEE Electron. Device Lett. **28**, 756–759 (2007)
48. Naeemi, A., Meindl, J.D.: Compact physical models for multiwall carbon-nanotube interconnects. IEEE Electron. Device Lett. **27**, 338–340 (2006)
49. Tans, S.J., Verschueren, A.R.M., Dekker, C.: Room-temperature transistor based on a single carbon nanotube. Nature. **393**, 49–52 (1998)
50. Bachtold, A., Hadley, P., Nakanishi, T., Dekker, C.: Logic circuits with carbon nanotube transistors. Science. **294**, 1317–1320 (2001)
51. Heinze, S., Tersoff, J., Martel, R., Derycke, V., Appenzeller, J., Avouris, P.: Carbon nanotubes as Schottky barrier transistors. Phys. Rev. Lett. **89**, 4 (2002)
52. Soh, H.T., Quate, C.F., Morpurgo, A.F., Marcus, C.M., Kong, J., Dai, H.J.: Integrated nanotube circuits: controlled growth and ohmic contacting of single-walled carbon nanotubes. Appl. Phys. Lett. **75**, 627–629 (1999)
53. Strano, M.S., Dyke, C.A., Usrey, M.L., Barone, P.W., Allen, M.J., Shan, H.W., Kittrell, C., Hauge, R.H., Tour, J.M., Smalley, R.E.: Electronic structure control of single-walled carbon nanotube functionalization. Science. **301**, 1519–1522 (2003)
54. Campidelli, S., Meneghetti, M., Prato, M.: Separation of metallic and semiconducting single- walled carbon nanotubes via covalent functionalization. Small. **3**, 1672–1676 (2007)
55. Krupke, R., Hennrich, F., von Lohneysen, H., Kappes, M.M.: Separation of metallic from semiconducting single-walled carbon nanotubes. Science. **301**, 344–347 (2003)
56. Li, H.P., Zhou, B., Lin, Y., Gu, L.R., Wang, W., Fernando, K.A.S., Kumar, S., Allard, L.F., Sun, Y.P.: Selective interactions of porphyrins with semiconducting single-walled carbon nanotubes. J. Am. Chem. Soc. **126**, 1014–1015 (2004)
57. Maeda, Y., Kimura, S., Kanda, M., Hirashima, Y., Hasegawa, T., Wakahara, T., Lian, Y.F., Nakahodo, T., Tsuchiya, T., Akasaka, T., Lu, J., Zhang, X.W., Gao, Z.X., Yu, Y.P., Nagase, S., Kazaoui, S., Minami, N., Shimizu, T., Tokumoto, H., Saito, R.: Large-scale separation of metallic and semiconducting single-walled carbon nanotubes. J. Am. Chem. Soc. **127**, 10287–10290 (2005)
58. Toyoda, S., Yamaguchi, Y., Hiwatashi, M., Tomonari, Y., Murakami, H., Nakashima, N.: Separation of semiconducting

single-walled carbon nanotubes by using a long-alkyl-chain benzenediazonium compound. Chem-Asian J. **2**, 145–149 (2007)
59. Wang, W., Fernando, K.A.S., Lin, Y., Meziani, M.J., Veca, L.M., Cao, L., Zhang, P., Kimani, M.M., Sun, Y.P.: Metallic single-walled carbon nanotubes for conductive nanocomposites. J. Am. Chem. Soc. **130**, 1415–1419 (2008)
60. Mattsson, M., Gromov, A., Dittmer, S., Eriksson, E., Nerushev, O. A., Campbell, E.E.B.: Dielectrophoresis-induced separation of metallic and semiconducting single-wall carbon nanotubes in a continuous flow microfluidic system. J. Nanosci. Nanotechnol. **7** (10), 3431–3435 (2007)
61. Tanaka, T., Jin, H.H., Miyata, Y., Kataura, H.: High-yield separation of metallic and semiconducting single-wall carbon nanotubes by agarose gel electrophoresis. Appl. Phys. Express. **1**, (2008)
62. Ghosh, S., Rao, C.N.R.: Separation of metallic and semiconducting single-walled carbon nanotubes through fluorous chemistry. Nano Res. **2**, 183–191 (2009)
63. Green, A.A., Duch, M.C., Hersam, M.C.: Isolation of single-walled carbon nanotube enantiomers by density differentiation. Nano Res. **2**, 69–77 (2009)
64. Dyke, C.A., Stewart, M.P., Tour, J.M.: Separation of single-walled carbon nanotubes on silica gel. Materials morphology and Raman excitation wavelength affect data interpretation. J. Am. Chem. Soc. **127**, 4497–4509 (2005)
65. LeMieux, M.C., Roberts, M., Barman, S., Jin, Y.W., Kim, J.M., Bao, Z.N.: Self-sorted, aligned nanotube networks for thin-film transistors. Science. **321**, 101–104 (2008)
66. Menard-Moyon, C., Izard, N., Doris, E., Mioskowski, C.: Separation of semiconducting from metallic carbon nanotubes by selective functionalization with azomethine ylides. J. Am. Chem. Soc. **128**, 6552–6553 (2006)
67. Huang, H.J., Maruyama, R., Noda, K., Kajiura, H., Kadono, K.: Preferential destruction of metallic single-walled carbon nanotubes by laser irradiation. J. Phys. Chem. B. **110**, 7316–7320 (2006)
68. Miyata, Y., Maniwa, Y., Kataura, H.: Selective oxidation of semiconducting single-wall carbon nanotubes by hydrogen peroxide. J. Phys. Chem. B. **110**, 25–29 (2006)
69. Arnold, M.S., Green, A.A., Hulvat, J.F., Stupp, S.I., Hersam, M.C.: Sorting carbon nanotubes by electronic structure using density differentiation. Nat. Nanotechnol. **1**, 60–65 (2006)
70. Tanaka, T., Jin, H.H., Miyata, Y., Fujii, S., Suga, H., Naitoh, Y., Minari, T., Miyadera, T., Tsukagoshi, K., Kataura, H.: Simple and scalable gel-based separation of metallic and semiconducting carbon nanotubes. Nano Lett. **9**(4), 1497–1500 (2009)
71. Berber, S., Kwon, Y.K., Tomanek, D.: Unusually high thermal conductivity of carbon nanotubes. Phys. Rev. Lett. **84**, 4613–4616 (2000)
72. Yu, C.H., Shi, L., Yao, Z., Li, D.Y., Majumdar, A.: Thermal conductance and thermopower of an individual single-wall carbon nanotube. Nano Lett. **5**, 1842–1846 (2005)
73. Mingo, N., Broido, D.A.: Carbon nanotube ballistic thermal conductance and its limits. Phys. Rev. Lett. **95**, (2005)
74. Yamamoto, T., Watanabe, K.: Nonequilibrium Green's function approach to phonon transport in defective carbon nanotubes. Phys. Rev. Lett. **96**, (2006)
75. Hone, J., Llaguno, M.C., Biercuk, M.J., Johnson, A.T., Batlogg, B., Benes, Z., Fischer, J.E.: Thermal properties of carbon nanotubes and nanotube-based materials. Appl. Phys. A Mater. Sci. Process. **74**, 339–343 (2002)
76. Pop, E., Mann, D., Wang, Q., Goodson, K., Dai, H.J.: Thermal conductance of an individual single- wall carbon nanotube above room temperature. Nano Lett. **6**, 96–100 (2006)
77. Hone, J., Whitney, M., Piskoti, C., Zettl, A.: Thermal conductivity of single-walled carbon nanotubes. Phys. Rev. B. **59**, R2514–R2516 (1999)
78. Benedict, L.X., Louie, S.G., Cohen, M.L.: Heat capacity of carbon nanotubes. Solid State Commun. **100**, 177–180 (1996)
79. Wang, Z.L., Tang, D.W., Li, X.B., Zheng, X.H., Zhang, W.G., Zheng, L.X., Zhu, Y.T.T., Jin, A.Z., Yang, H.F., Gu, C.Z.: Length-dependent thermal conductivity of an individual single-wall carbon nanotube. Appl. Phys. Lett. **91**, (2007)
80. Mizel, A., Benedict, L.X., Cohen, M.L., Louie, S.G., Zettl, A., Budraa, N.K., Beyermann, W.P.: Analysis of the low-temperature specific heat of multiwalled carbon nanotubes and carbon nanotube ropes. Phys. Rev. B. **60**, 3264–3270 (1999)
81. Kim, P., Shi, L., Majumdar, A., McEuen, P.L.: Thermal transport measurements of individual multiwalled nanotubes. Phys. Rev. Lett. 8721 (2001)
82. Choi, T.Y., Poulikakos, D., Tharian, J., Sennhauser, U.: Measurement of thermal conductivity of individual multiwalled carbon nanotubes by the 3-omega method. Appl. Phys. Lett. **87**, (2005)
83. Choi, T.Y., Poulikakos, D., Tharian, J., Sennhauser, U.: Measurement of the thermal conductivity of individual carbon nanotubes by the four-point three-omega method. Nano Lett. **6**, 1589–1593 (2006)
84. Hone, J., Batlogg, B., Benes, Z., Johnson, A.T., Fischer, J.E.: Quantized phonon spectrum of single-wall carbon nanotubes. Science. **289**, 1730–1733 (2000)
85. Yang, D.J., Zhang, Q., Chen, G., Yoon, S.F., Ahn, J., Wang, S.G., Zhou, Q., Wang, Q., Li, J.Q.: Thermal conductivity of multiwalled carbon nanotubes. Phys. Rev. B. **66**, (2002)
86. Yi, W., Lu, L., Zhang, D.L., Pan, Z.W., Xie, S.S.: Linear specific heat of carbon nanotubes. Phys. Rev. B. **59**, R9015–R9018 (1999)
87. Bethune, D.S., Kiang, C.H., Devries, M.S., Gorman, G., Savoy, R., Vazquez, J., Beyers, R.: Cobalt- catalyzed growth of carbon nanotubes with single-atomic-layerwalls. Nature. **363**, 605–607 (1993)
88. Thess, A., Lee, R., Nikolaev, P., Dai, H.J., Petit, P., Robert, J., Xu, C.H., Lee, Y.H., Kim, S.G., Rinzler, A.G., Colbert, D.T., Scuseria, G.E., Tomanek, D., Fischer, J.E., Smalley, R.E.: Crystalline ropes of metallic carbon nanotubes. Science. **273**, 483–487 (1996)
89. Liu, J., Rinzler, A.G., Dai, H.J., Hafner, J.H., Bradley, R.K., Boul, P.J., Lu, A., Iverson, T., Shelimov, K., Huffman, C.B., Rodriguez-Macias, F., Shon, Y.S., Lee, T.R., Colbert, D.T., Smalley, R.E.: Fullerene pipes. Science. **280**, 1253–1256 (1998)
90. Amelinckx, S., Zhang, X.B., Bernaerts, D., Zhang, X.F., Ivanov, V., Nagy, J.B.: A formation mechanism for catalytically grown helix-shaped graphite nanotubes. Science. **265**, 635–639 (1994)
91. Amelinckx, S., Bernaerts, D., Zhang, X.B., Vantendeloo, G., Vanlanduyt, J.: A structure model and growth-mechanism for multishell carbon nanotubes. Science. **267**, 1334–1338 (1995)
92. Hata, K., Futaba, D.N., Mizuno, K., Namai, T., Yumura, M., Iijima, S.: Water-assisted highly efficient synthesis of impurity-free single-waited carbon nanotubes. Science. **306**, 1362–1364 (2004)
93. Kong, J., Soh, H.T., Cassell, A.M., Quate, C.F., Dai, H.J.: Synthesis of individual single-walled carbon nanotubes on patterned silicon wafers. Nature. **395**, 878–881 (1998)
94. Qu, L., Dai, L.: Gecko-foot-mimetic aligned single-walled carbon nanotube dry adhesives with unique electrical and thermal properties. Adv. Mater. **19**, 3844–3849 (2007)
95. Nikolaev, P., Bronikowski, M.J., Bradley, R.K., Rohmund, F., Colbert, D.T., Smith, K.A., Smalley, R.E.: Gas-phase catalytic growth of single-walled carbon nanotubes from carbon monoxide. Chem. Phys. Lett. **313**, 91–97 (1999)
96. Andrews, R., Jacques, D., Rao, A.M., Derbyshire, F., Qian, D., Fan, X., Dickey, E.C., Chen, J.: Continuous production of aligned carbon nanotubes: a step closer to commercial realization. Chem. Phys. Lett. **303**, 467–474 (1999)
97. Xu, Y.Q., Flor, E., Kim, M.J., Hamadani, B., Schmidt, H., Smalley, R.E., Hauge, R.H.: Vertical array growth of small

diameter single-walled carbon nanotubes. J. Am. Chem. Soc. **128**, 6560–6561 (2006)
98. Shyu, Y.M., Hong, F.C.N.: Low-temperature growth and field emission of aligned carbon nanotubes by chemical vapor deposition. Mater. Chem. Phys. **72**, 223–227 (2001)
99. Shyu, Y.M., Hong, F.C.N.: The effects of pre-treatment and catalyst composition on growth of carbon nanofibers at low temperature. Diam. Relat. Mater. **10**, 1241–1245 (2001)
100. Laplaze, D., Alvarez, L., Guillard, T., Badie, J.M., Flamant, G.: Carbon nanotubes: dynamics of synthesis processes. Carbon. **40**, 1621–1634 (2002)
101. Choi, Y.C., Bae, D.J., Lee, Y.H., Lee, B.S., Park, G.S., Choi, W.B., Lee, N.S., Kim, J.M.: Growth of carbon nanotubes by microwave plasma-enhanced chemical vapor deposition at low temperature. J. Vac. Sci. Technol. A Vac. Surf. Films. **18**, 1864–1868 (2000)
102. Hofmann, S., Ducati, C., Robertson, J., Kleinsorge, B.: Low-temperature growth of carbon nanotubes by plasma-enhanced chemical vapor deposition. Appl. Phys. Lett. **83**, 135–137 (2003)
103. Devaux, X., Vergnat, M.: On the low-temperature synthesis of SWCNTs by thermal CVD. Phys. E. **40**(7), 2268–2271 (2008)
104. Bower, C., Zhu, W., Jin, S.H., Zhou, O.: Plasma-induced alignment of carbon nanotubes. Appl. Phys. Lett. **77**, 830–832 (2000)
105. Wang, W., Yang, K.Q., Gaillard, J., Bandaru, P.R., Rao, A.M.: Rational synthesis of helically coiled carbon nanowires and nanotubes through the use of tin and indium catalysts. Adv. Mater. **20**, 179–182 (2008)
106. Naeemi, A., Meindl, J.D.: Monolayer metallic nanotube interconnects: promising candidates for short local interconnects. IEEE Electron. Device Lett. **26**, 544–546 (2005)
107. Nihei, M., Kawabata, A., Kondo, D., Horibe, M., Sato, S., Awano, Y.: Electrical properties of carbon nanotube bundles for future via interconnects. Jpn. J. Appl. Phys. Part 1 Regul. Pap. Short Notes Rev. Pap. **44**, 1626–1628 (2005)
108. Hoenlein, W., Kreupl, F., Duesberg, G.S., Graham, A.P., Liebau, M., Seidel, R., Unger, E.: Carbon nanotubes for microelectronics: status and future prospects. Mater. Sci. Eng. C Biomimetic. Supramol. Syst. **23**, 663–669 (2003)
109. Awano, Y., Sato, S., Kondo, D., Ohfuti, M., Kawabata, A., Nihei, M., Yokoyama, N.: Carbon nanotube via interconnect technologies: size-classified catalyst nanoparticles and low-resistance ohmic contact formation. Phys. Status Solidi A. **203**, 3611–3616 (2006)
110. Graham, A.P., Duesberg, G.S., Hoenlein, W., Kreupl, F., Liebau, M., Martin, R., Rajasekharan, B., Pamler, W., Seidel, R., Steinhoegl, W., Unger, E.: How do carbon nanotubes fit into the semiconductor roadmap? Appl. Phys. A Mater. Sci. Process. **80**, 1141–1151 (2005)
111. Rodríguez-Manzo, J.A., Banhart, F., Terrones, M., Terrones, H., Grobert, N., Ajayan, P.M., Sumpter, B.G., Meunier, V., Wang, M., Bando, Y., Golberg, D.: Heterojunctions between metals and carbon nanotubes as ultimate nanocontacts. Proc. Natl. Acad. Sci. U. S. A. **106**, (2009)
112. Sato, S., Kawabata, A., Nihei, M., Awano, Y.: Growth of diameter-controlled carbon nanotubes using monodisperse nickel nanoparticles obtained with a differential mobility analyzer. Chem. Phys. Lett. **382**, 361–366 (2003)
113. Gwinn, J.P., Webb, R.L.: Performance and testing of thermal interface materials. Microelectron. J. **34**(3), 215–222 (2003)
114. Prasher, R.S.: Surface chemistry and characteristics based model for the thermal contact resistance of fluidic interstitial thermal interface materials. J. Heat Transf. Trans. ASME. **123**, 969–975 (2001)
115. Prasher, R.S., Shipley, J., Prstic, S., Koning, P., Wang, J.L.: Thermal resistance of particle laden polymeric thermal interface materials. J. Heat Transf. Trans. ASME. **125**, 1170–1177 (2003)
116. Xu, J., Fisher, T.S.: Enhancement of thermal interface materials with carbon nanotube arrays. Int. J. Heat Mass Transf. **49**, 1658–1666 (2006)
117. Yu, A.P., Ramesh, P., Itkis, M.E., Bekyarova, E., Haddon, R.C.: Graphite nanoplatelet-epoxy composite thermal interface materials. J. Phys. Chem. C. **111**, 7565–7569 (2007)
118. Aoyagi, Y., Leong, C.K., Chung, D.D.L.: Polyol-based phase-change thermal interface materials. J. Electron. Mater. **35**, 416–424 (2006)
119. Chung, D.D.L.: Thermal interface materials. J. Mater. Eng. Perform. **10**, 56–59 (2001)
120. Howe, T.A., Leong, C.K., Chung, D.D.L.: Comparative evaluation of thermal interface materials for improving the thermal contact between an operating computer microprocessor and its heat sink. J. Electron. Mater. **35**, 1628–1635 (2006)
121. Liu, Z.R., Chung, D.D.L.: Calorimetric evaluation of phase change materials for use as thermal interface materials. Thermochim. Acta. **366**, 135–147 (2001)
122. Maguire, L., Behnia, M., Morrison, G.: Systematic evaluation of thermal interface materials - a case study in high power amplifier design. Microelectron. Reliab. **45**, 711–725 (2005)
123. Prasher, R.S.: Rheology based modeling and design of particle laden polymeric thermal interface materials. IEEE Trans. Compon. Packag. Technol. **28**(2), 230–237 (2005)
124. Singhal, V., Siegmund, T., Garimella, S.V.: Optimization of thermal interface materials for electronics cooling applications. IEEE Trans. Compon. Packag. Technol. **27**, 244–252 (2004)
125. Tong, T., Zhao, Y., Delzeit, L., Kashani, A., Meyyappan, M., Majumdar, A.: Dense, vertically aligned multiwalled carbon nanotube arrays as thermal interface materials. IEEE Trans. Compon. Packag. Technol. **30**, 92–100 (2007)
126. Abadi, P., Leong, C.K., Chung, D.D.L.: Factors that govern the performance of thermal interface materials. J. Electron. Mater. **38** (1), 175–192 (2009)
127. Aoyagi, Y., Chung, D.D.L.: Antioxidant-based phase-change thermal interface materials with high thermal stability. J. Electron. Mater. **37**(4), 448–461 (2008)
128. De Mey, G., Pilarski, J., Wojcik, M., Lasota, M., Banaszczyk, J., Vermeersch, B., Napieralski, A.: Influence of interface materials on the thermal impedance of electronic packages. Int. Commun. Heat Mass Transf. **36**, 210–212 (2009)
129. Kanuparthi, S., Subbarayan, G., Siegmund, T., Sammakia, B.: An efficient network model for determining the effective thermal conductivity of particulate thermal interface materials. IEEE Trans. Compon. Packag. Technol. **31**, 611–621 (2008)
130. Lin, C., Chung, D.D.L.: Graphite nanoplatelet pastes vs. carbon black pastes as thermal interface materials. Carbon. **47**, 295–305 (2009)
131. Liu, X., Zhang, Y., Cassell, A.M., Cruden, B.A.: Implications of catalyst control for carbon nanotube based thermal interface materials. J. Appl. Phys. **104**, (2008)
132. Thostenson, E.T., Ren, Z.F., Chou, T.W.: Advances in the science and technology of carbon nanotubes and their composites: a review. Compos. Sci. Technol. **61**, 1899–1912 (2001)
133. Borca-Tasciuc, T., Mazumder, M., Son, Y., Pal, S.K., Schadler, L. S., Ajayan, P.M.: Anisotropic thermal diffusivity characterization of aligned carbon nanotube-polymer composites. J. Nanosci. Nanotechnol. **7**, 1581–1588 (2007)
134. Choi, E.S., Brooks, J.S., Eaton, D.L., Al-Haik, M.S., Hussaini, M. Y., Garmestani, H., Li, D., Dahmen, K.: Enhancement of thermal and electrical properties of carbon nanotube polymer composites by magnetic field processing. J. Appl. Phys. **94**, 6034–6039 (2003)
135. Velasco-Santos, C., Martinez-Hernandez, A.L., Fisher, F.T., Ruoff, R., Castano, V.M.: Improvement of thermal and mechanical properties of carbon nanotube composites through chemical functionalization. Chem. Mater. **15**, 4470–4475 (2003)

136. Liu, C.H., Huang, H., Wu, Y., Fan, S.S.: Thermal conductivity improvement of silicone elastomer with carbon nanotube loading. Appl. Phys. Lett. **84**, 4248–4250 (2004)
137. Huang, H., Liu, C.H., Wu, Y., Fan, S.S.: Aligned carbon nanotube composite films for thermal management. Adv. Mater. **17**, 1652–1656 (2005)
138. Bryning, M.B., Milkie, D.E., Islam, M.F., Kikkawa, J.M., Yodh, A.G.: Thermal conductivity and interfacial resistance in single-wall carbon nanotube epoxy composites. Appl. Phys. Lett. **87**, (2005)
139. Gojny, F.H., Wichmann, M.H.G., Fiedler, B., Kinloch, I.A., Bauhofer, W., Windle, A.H., Schulte, K.: Evaluation and identification of electrical and thermal conduction mechanisms in carbon nanotube/epoxy composites. Polymer. **47**, 2036–2045 (2006)
140. Bonnet, P., Sireude, D., Garnier, B., Chauvet, O.: Thermal properties and percolation in carbon nanotube-polymer composites. Appl. Phys. Lett. **91**, (2007)
141. Shenogina, N., Shenogin, S., Xue, L., Keblinski, P.: On the lack of thermal percolation in carbon nanotube composites. Appl. Phys. Lett. **87**, (2005)
142. Nan, C.W., Liu, G., Lin, Y.H., Li, M.: Interface effect on thermal conductivity of carbon nanotube composites. Appl. Phys. Lett. **85**, 3549–3551 (2004)
143. Shenogin, S., Xue, L.P., Ozisik, R., Keblinski, P., Cahill, D.G.: Role of thermal boundary resistance on the heat flow in carbon-nanotube composites. J. Appl. Phys. **95**, 8136–8144 (2004)
144. Shenogin, S., Bodapati, A., Xue, L., Ozisik, R., Keblinski, P.: Effect of chemical functionalization on thermal transport of carbon nanotube composites. Appl. Phys. Lett. **85**, 2229–2231 (2004)
145. Clancy, T.C., Gates, T.S.: Modeling of interfacial modification effects on thermal conductivity of carbon nanotube composites. Polymer. **47**, 5990–5996 (2006)
146. Liu, C.H., Fan, S.S.: Effects of chemical modifications on the thermal conductivity of carbon nanotube composites. Appl. Phys. Lett. **86**, (2005)
147. Nan, C.W., Shi, Z., Lin, Y.: A simple model for thermal conductivity of carbon nanotube-based composites. Chem. Phys. Lett. **375**, 666–669 (2003)
148. Ju, S., Li, Z.Y.: Theory of thermal conductance in carbon nanotube composites. Phys. Lett. A. **353**, 194–197 (2006)
149. Nan, C.W., Birringer, R., Clarke, D.R., Gleiter, H.: Effective thermal conductivity of particulate composites with interfacial thermal resistance. J. Appl. Phys. **81**, 6692–6699 (1997)
150. Huxtable, S.T., Cahill, D.G., Shenogin, S., Xue, L.P., Ozisik, R., Barone, P., Usrey, M., Strano, M.S., Siddons, G., Shim, M., Keblinski, P.: Interfacial heat flow in carbon nanotube suspensions. Nat. Mater. **2**, 731–734 (2003)
151. Lin, W., Moon, K.S., Wong, C.P.: A combined process of in-situ functionalization and microwave treatment to achieve ultra-small thermal expansion of aligned carbon nanotube/polymer nanocomposites for thermal management. Adv. Mater. **21**, 2421–2424 (2009)
152. Young, D.D., Nichols, J., Kelly, R.M., Deiters, A.: Microwave activation of enzymatic catalysis. J. Am. Chem. Soc. **130**, 10048–10049 (2008)
153. Noh, H.S., Moon, K.S., Cannon, A., Hesketh, P.J., Wong, C.P.: Wafer bonding using microwave heating of parylene intermediate layers. J. Micromech. Microeng. **14**, 625–631 (2004)
154. Jiang, H.J., Moon, K.S., Zhang, Z.Q., Pothukuchi, S., Wong, C.P.: Variable frequency microwave synthesis of silver nanoparticles. J. Nanopart. Res. **8**, 117–124 (2006)
155. Moon, K.S., Li, Y., Xu, J.W., Wong, C.P.: Lead-free interconnect technique by using variable frequency microwave. J. Electron. Mater. **34**, 1081–1088 (2005)
156. Sunden, E., Moon, J.K., Wong, C.P., King, W.P., Graham, S.: Microwave assisted patterning of vertically aligned carbon nanotubes onto polymer substrates. J. Vac. Sci. Technol. B. **24**, 1947–1950 (2006)
157. Cola, B.A., Xu, J., Cheng, C.R., Xu, X.F., Fisher, T.S., Hu, H.P.: Photoacoustic characterization of carbon nanotube array thermal interfaces. J. Appl. Phys. **101**, (2007)
158. Huang, H., Liu, C.H., Wu, Y., Fan, S.S.: Aligned carbon nanotube composite films for thermal management. Adv. Mater. **17**, 1652–1653 (2005)
159. Panzer, M.A., Zhang, G., Mann, D., Hu, X., Pop, E., Dai, H., Goodson, K.E.: Thermal properties of metal-coated vertically aligned single-wall nanotube arrays. J. Heat Transf. Trans. ASME. **130**, (2008)
160. Cola, B.A., Xu, X.F., Fisher, T.S.: Increased real contact in thermal interfaces: a carbon nanotube/foil material. Appl. Phys. Lett. **90**, (2007)
161. Kordas, K., Toth, G., Moilanen, P., Kumpumaki, M., Vahakangas, J., Uusimaki, A., Vajtai, R., Ajayan, P.M.: Chip cooling with integrated carbon nanotube microfin architectures. Appl. Phys. Lett. **90**, (2007)
162. Zhu, L.B., Hess, D.W., Wong, C.P.: Assembling carbon nanotube films as thermal interface materials. Electron.Compon. Technol. Conf. IEEE. 2006–2010 (2007)
163. Hu, X.J., Padilla, A.A., Xu, J., Fisher, T.S., Goodson, K.E.: 3-omega measurements of vertically oriented carbon nanotubes on silicon. J. Heat Transf. Trans. ASME. **128**, 1109–1113 (2006)
164. Wang, X.W., Zhong, Z.R., Xu, J.: Noncontact thermal characterization of multiwall carbon nanotubes. J. Appl. Phys. **97**, (2005)
165. Ngo, Q., Cruden, B.A., Cassell, A.M., Sims, G., Meyyappan, M., Li, J., Yang, C.Y.: Thermal inter- face properties of Cu-filled vertically aligned carbon nanofiber arrays. Nano Lett. **4**, 2403–2407 (2004)
166. Xu, Y., Zhang, Y., Suhir, E., Wang, X.W.: Thermal properties of carbon nanotube array used for integrated circuit cooling. J. Appl. Phys. **100**, (2006)
167. Tong, T., Zhao, Y., Delzeit, L., Majumdar, A., Kashani, A.: Third ASME Integrated Nanosystems Conference. New York (2004)
168. Amama, P.B., Cola, B.A., Sands, T.D., Xu, X.F., Fisher, T.S.: Dendrimer-assisted controlled growth of carbon nanotubes for enhanced thermal interface conductance. Nanotechnology. **18**, (2007)
169. Moon, K.S., Lin, W., Jiang, H.J., Ko, H., Zhu, L.B., Wong, C.P.: Surface treatment of MWCNT array and its polymer composites for TIM application. Electron.Compon. Technol. Conf. 234–237 (2008)
170. Amama, P.B., Ogebule, O., Maschmann, M.R., Sands, T.D., Fisher, T.S.: Dendrimer-assisted low-temperature growth of carbon nanotubes by plasma-enhanced chemical vapor deposition. Chem. Commun. 2899–2901 (2006)
171. Bae, E.J., Min, Y.S., Kang, D., Ko, J.H., Park, W.: Low-temperature growth of single-walled carbon nanotubes by plasma enhanced chemical vapor deposition. Chem. Mater. **17**, 5141–5145 (2005)
172. Chen, K.C., Chen, C.F., Chiang, J.S., Hwang, C.L., Chang, Y.Y., Lee, C.C.: Low temperature growth of carbon nanotubes on printing electrodes by MPCVD. Thin Solid Films. **498**(1–2), 198–201 (2006)
173. Chen, K.C., Chen, C.F., Lee, J.H., Wu, T.L., Hwang, C.L., Tai, N. H., Hsiao, M.C.: Low-temperature CVD growth of carbon nanotubes for field emission application. Diam. Relat. Mater. **16**, 566–569 (2007)
174. Chiang, W.H., Sankaran, R.M.: Synergistic effects in bimetallic nanoparticles for low temperature carbon nanotube growth. Adv. Mater. **20**, 4857–4861 (2008)
175. Dai, L.M.: Low-temperature, controlled synthesis of carbon nanotubes. Small. **1**, 274–276 (2005)

176. Dubosc, M., Minea, T., Besland, M.P., Cardinaud, C., Granier, A., Gohier, A., Point, S., Torres, J.: Low temperature plasma carbon nanotubes growth on patterned catalyst. Microelectron. Eng. **83**, 2427–2431 (2006)
177. Iwasaki, T., Zhong, G.F., Kawarada, H.: Low-temperature growth of vertically aligned single-walled carbon nanotubes by radical CVD. New Diam. Front Carbon Technol. **16**, 177–184 (2006)
178. Jayatissa, A.H., Guo, K.: Synthesis of carbon nanotubes at low temperature by filament assisted atmospheric CVD and their field emission characteristics. Vacuum. **83**, 853–856 (2009)
179. Kim, S.M., Zhang, Y., Wang, X., Teo, K.B.K., Gangloff, L., Milne, W.I., Wu, J., Eastman, M., Jiao, J.: Low-temperature growth of single-wall carbon nanotubes. Nanotechnology. **18**, (2007)
180. Kondo, D., Sato, S., Awano, Y.: Low-temperature synthesis of single-walled carbon nanotubes with a narrow diameter distribution using size-classified catalyst nanoparticles. Chem. Phys. Lett. **422**, 481–487 (2006)
181. Kyung, S.J., Lee, Y.H., Kim, C.W., Lee, J.H., Yeom, G.Y.: Field emission properties of carbon nanotubes synthesized by capillary type atmospheric pressure plasma enhanced chemical vapor deposition at low temperature. Carbon. **44**, 1530–1534 (2006)
182. Li, C.H., Liu, H.C., Tseng, S.C., Lin, Y.P., Chen, S.P., Li, J.Y., Wu, K.H., Juang, J.Y.: Enhancement of the field emission properties of low-temperature-growth multi-wall carbon nanotubes by KrF excimer laser irradiation post-treatment. Diam. Relat. Mater. **15**(11–12), 2010–2014 (2006)
183. Liao, K.H., Ting, J.M.: Characteristics of aligned carbon nanotubes synthesized using a high-rate low-temperature process. Diam. Relat. Mater. **15**, 1210–1216 (2006)
184. Min, Y.S., Bae, E.J., Oh, B.S., Kang, D., Park, W.: Low-temperature growth of single-walled car- bon nanotubes by water plasma chemical vapor deposition. J. Am. Chem. Soc. **127**, 12498–12499 (2005)
185. Rummeli, M.H., Gruneis, A., Loffler, M., Jost, O., Schonfelder, R., Kramberger, C., Grimm, D., Gemming, T., Barreiro, A., Borowiak-Palen, E., Kalbac, M., Ayala, P., Hubers, H.W., Buchner, B., Pichler, T.: Novel catalysts for low temperature synthesis of single wall carbon nanotubes. Phys. Status Solidi B. **243**(13), 3101–3105 (2006)
186. Tam, E., Ostrikov, K.: Plasma-controlled adatom delivery and (re)distribution: enabling uninterrupted, low-temperature growth of ultralong vertically aligned single walled carbon nanotubes. Appl. Phys. Lett. **93**, (2008)
187. Uchino, T., Bourdakos, K.N., de Groot, C.H., Ashburn, P., Kiziroglou, M.E., Dilliway, G.D., Smith, D.C.: Metal catalyst-free low-temperature carbon nanotube growth on SiGe islands. Appl. Phys. Lett. **86**, (2005)
188. Yu, G.J., Gong, J.L., Zhu, D.Z., He, S.X., Zhu, Z.Y.: Synthesis of carbon nanotubes over rare earth zeolites at low temperature. Carbon. **43**, 3015–3017 (2005)
189. Chen, M., Chen, C.M., Shi, S.C., Chen, C.F.: Low-temperature synthesis multiwalled carbon nanotubes by microwave plasma chemical vapor deposition using CH4-CO2 gas mixture. Jpn. J. Appl. Phys. **42**, 614–619 (2003)
190. Choi, Y.C., Bae, D.J., Lee, Y.H., Lee, B.S., Han, I.T., Choi, W.B., Lee, N.S., Kim, J.M.: Low temperature synthesis of carbon nanotubes by microwave plasma-enhanced chemical vapor deposition. Synth. Met. **108**, 159–163 (2000)
191. Honda, S., Katayama, M., Lee, K.Y., Ikuno, T., Ohkura, S., Oura, K., Furuta, H., Hirao, T.: Low temperature synthesis of aligned carbon nanotubes by inductively coupled plasma chemical vapor deposition using pure methane. Jpn. J. Appl. Phys. Part 2 Lett. **42**, L441–L443 (2003)
192. Jeong, H.J., Jeong, S.Y., Shin, Y.M., Han, J.H., Lim, S.C., Eum, S. J., Yang, C.W., Kim, N.G., Park, C.Y., Lee, Y.H.: Dual-catalyst growth of vertically aligned carbon nanotubes at low temperature in thermal chemical vapor deposition. Chem. Phys. Lett. **361**, 189–195 (2002)
193. Kang, H.S., Yoon, H.J., Kim, C.O., Hong, J.P., Han, I.T., Cha, S. N., Song, B.K., Jung, J.E., Lee, N.S., Kim, J.M.: Low temperature growth of multi-wall carbon nanotubes assisted by mesh potential using a modified plasma enhanced chemical vapor deposition system. Chem. Phys. Lett. **349**, 196–200 (2001)
194. Lee, C.J., Park, J., Kim, J.M., Huh, Y., Lee, J.Y., No, K.S.: Low-temperature growth of carbon nanotubes by thermal chemical vapor deposition using Pd, Cr, and Pt as co-catalyst. Chem. Phys. Lett. **327**, 277–283 (2000)
195. Lee, C.J., Son, K.H., Park, J., Yoo, J.E., Huh, Y., Lee, J.Y.: Low temperature growth of vertically aligned carbon nanotubes by thermal chemical vapor deposition. Chem. Phys. Lett. **338**, 113–117 (2001)
196. Li, M.W., Hu, Z., Wang, X.Z., Wu, Q., Chen, Y., Tian, Y.L.: Low-temperature synthesis of carbon nanotubes using corona discharge plasma at atmospheric pressure. Diam. Relat. Mater. **13**, 111–115 (2004)
197. Li, Y.L., Yu, Y.D., Liang, Y.: A novel method for synthesis of carbon nanotubes: low temperature solid pyrolysis. J. Mater. Res. **12**, 1678–1680 (1997)
198. Liao, H.W., Hafner, J.H.: Low-temperature single-wall carbon nanotube synthesis by thermal chemical vapor deposition. J. Phys. Chem. B. **108**, 6941–6943 (2004)
199. Lin, Y.H., Cui, X.L., Yen, C., Wai, C.M.: Platinum/carbon nanotube nanocomposite synthesized in supercritical fluid as electrocatalysts for low-temperature fuel cells. J. Phys. Chem. B. **109**, 14410–14415 (2005)
200. Maruyama, S., Kojima, R., Miyauchi, Y., Chiashi, S., Kohno, M.: Low-temperature synthesis of high-purity single-walled carbon nanotubes from alcohol. Chem. Phys. Lett. **360**, 229–234 (2002)
201. Shao, M.W., Wang, D.B., Yu, G.H., Hu, B., Yu, W.C., Qian, Y.T.: The synthesis of carbon nanotubes at low temperature via carbon suboxide disproportionation. Carbon. **42**, 183–185 (2004)
202. Shiratori, Y., Hiraoka, H., Yamamoto, M.: Vertically aligned carbon nanotubes produced by radio-frequency plasma-enhanced chemical vapor deposition at low temperature and their growth mechanism. Mater. Chem. Phys. **87**, 31–38 (2004)
203. Simon, F., Kuzmany, H., Rauf, H., Pichler, T., Bernardi, J., Peterlik, H., Korecz, L., Fulop, F., Janossy, A.: Low temperature fullerene encapsulation in single wall carbon nanotubes: synthesis of N@C-60@SWCNT. Chem. Phys. Lett. **383**, 362–367 (2004)
204. Ting, J.M., Liao, K.H.: Low-temperature, nonlinear rapid growth of aligned carbon nanotubes. Chem. Phys. Lett. **396**, 469–472 (2004)
205. Vohs, J.K., Brege, J.J., Raymond, J.E., Brown, A.E., Williams, G. L., Fahlman, B.D.: Low-temperature growth of carbon nanotubes from the catalytic decomposition of carbon tetrachloride. J. Am. Chem. Soc. **126**, 9936–9937 (2004)
206. Wang, W.L., Bai, X.D., Xu, Z., Liu, S., Wang, E.G.: Low temperature growth of single-walled carbon nanotubes: small diameters with narrow distribution. Chem. Phys. Lett. **419**, 81–85 (2006)
207. Wang, W.Z., Kunwar, S., Huang, J.Y., Wang, D.Z., Ren, Z.F.: Low temperature solvothermal synthesis of multiwall carbon nanotubes. Nanotechnology. **16**, 21–23 (2005)
208. Wang, X.H., Hu, Z., Wu, Q., Chen, Y.: Low-temperature catalytic growth of carbon nanotubes under microwave plasma assistance. Catal. Today. **72**(3–4), 205–211 (2002)
209. Zhong, D.Y., Liu, S., Zhang, G.Y., Wang, E.G.: Large-scale well aligned carbon nitride nanotube films: low temperature growth and electron field emission. J. Appl. Phys. **89**, 5939–5943 (2001)

210. Wang, B.A., Liu, X.Y., Liu, H.M., Wu, D.X., Wang, H.P., Jiang, J.M., Wang, X.B., Hu, P.A., Liu, Y.Q., Zhu, D.B.: Controllable preparation of patterns of aligned carbon nanotubes on metals and metal-coated silicon substrates. J. Mater. Chem. **13**, 1124–1126 (2003)
211. Xu, F.S., Liu, X.F., Tse, S.D.: Synthesis of carbon nanotubes on metal alloy substrates with voltage bias in methane inverse diffusion flames. Carbon. **44**, 570–577 (2006)
212. Hofmeister, W., Kang, W.P., Wong, Y.M., Davidson, J.L.: Carbon nanotube growth from cu-co alloys for field emission applications. J. Vac. Sci. Technol. B. **22**, 1286–1289 (2004)
213. Karwa, M., Iqbal, Z., Mitra, S.: Selective self-assembly of single walled carbon nanotubes in long steel tubing for chemical separations. J. Mater. Chem. **16**, 2890–2895 (2006)
214. Talapatra, S., Kar, S., Pal, S.K., Vajtai, R., Ci, L., Victor, P., Shaijumon, M.M., Kaur, S., Nalamasu, O., Ajayan, P.M.: Direct growth of aligned carbon nanotubes on bulk metals. Nat. Nanotechnol. **1**, 112–116 (2006)
215. Gao, L.J., Peng, A.P., Wang, Z.Y., Zhang, H., Shi, Z.J., Gu, Z.N., Cao, G.P., Ding, B.Z.: Growth of aligned carbon nanotube arrays on metallic substrate and its application to supercapacitors. Solid State Commun. **146**, 380–383 (2008)
216. Singh, M.K., Singh, P.P., Titus, E., Misra, D.S., LeNormand, F.: High density of multiwalled carbon nanotubes observed on nickel electroplated copper substrates by microwave plasma chemical vapor deposition. Chem. Phys. Lett. **354**, 331–336 (2002)
217. Wang, H., Feng, J.Y., Hu, X.J., Ng, K.M.: Synthesis of aligned carbon nanotubes on double-sided metallic substrate by chemical vapor deposition. J. Phys. Chem. C. **111**, 12617–12624 (2007)
218. Yin, X.W., Wang, Q.L., Lou, C.G., Zhang, X.B., Lei, W.: Growth of multi-walled CNTs emitters on an oxygen-free copper substrate by chemical-vapor deposition. Appl. Surf. Sci. **254**, 6633–6636 (2008)
219. Lin, W., Wong, C.P.: Synthesis of Vertically Aligned Multi-Walled Carbon Nanotubes on Copper Substrates for Applications as Thermal Interface Materials, MRS Spring Meeting (2009)
220. Zhu, L.B., Sun, Y.Y., Hess, D.W., Wong, C.P.: Well-aligned open-ended carbon nanotube architectures: an approach for device assembly. Nano Lett. **6**, 243–247 (2006)
221. Zhu, L.B., Xiu, Y.H., Hess, D.W., Wong, C.P.: Aligned carbon nanotube stacks by water-assisted selective etching. Nano Lett. **5**, 2641–2645 (2005)
222. Powell, C.F., Oxley, J.H., Johan, M., Blocher, J.: Vapor deposition. Wiley, New York (1966)
223. de los Arcos, T., Vonau, F., Garnier, M.G., Thommen, V., Boyen, H.G., Oelhafen, P., Duggelin, M., Mathis, D., Guggenheim, R.: Influence of iron-silicon interaction on the growth of carbon nanotubes produced by chemical vapor deposition. Appl. Phys. Lett. **80**, 2383–2385 (2002)
224. Xiu, Y.H., Zhang, S., Yelundur, V., Rohatgi, A., Hess, D.W., Wong, C.P.: Superhydrophobic and low light reflectivity silicon surfaces fabricated by hierarchical etching. Langmuir. **24**, 10421–10426 (2008)
225. Pisana, S., Cantoro, M., Parvez, A., Hofmann, S., Ferrari, A.C., Robertson, J.: The role of precursor gases on the surface restructuring of catalyst films during carbon nanotube growth. Phys. E. **37**, 1–5 (2007)
226. Rodriguez, N.M.: A review of catalytically grown carbon nanofibers. J. Mater. Res. **8**, 3233–3250 (1993)
227. Deck, C.P., Vecchio, K.: Prediction of carbon nanotube growth success by the analysis of carbon-catalyst binary phase diagrams. Carbon. **44**, 267–275 (2006)
228. Deng, W.Q., Xu, X., Goddard, W.A.: A two-stage mechanism of bimetallic catalyzed growth of single-walled carbon nanotubes. Nano Lett. **4**, 2331–2335 (2004)
229. Krishnankutty, N., Park, C., Rodriguez, N.M., Baker, R.T.K.: The effect of copper on the structural characteristics of carbon filaments produced from iron catalyzed decomposition of ethylene. Catal. Today. **37**, 295–307 (1997)
230. Almazouzi, A., Macht, M.P., Naundorf, V., Neumann, G.: Diffusion of iron and nickel in single-crystalline copper. Phys. Rev. B. **54**, 857–863 (1996)
231. Kononchuk, O., Korablev, K.G., Yarykin, N., Rozgonyi, G.A.: Diffusion of iron in the silicon dioxide layer of silicon-on-insulator structures. Appl. Phys. Lett. **73**, 1206–1208 (1998)
232. Puurunen, R.L.: Surface chemistry of atomic layer deposition: a case study for the trimethylaluminum/water process. J. Appl. Phys. **97**, (2005)
233. Rahtu, A., Alaranta, T., Ritala, M.: In situ quartz crystal microbalance and quadrupole mass spectrometry studies of atomic layer deposition of aluminum oxide from trimethylaluminum and water. Langmuir. **17**, 6506–6509 (2001)
234. Groner, M.D., Elam, J.W., Fabreguette, F.H., George, S.M.: Electrical characterization of thin Al2O3 films grown by atomic layer deposition on silicon and various metal substrates. Thin Solid Films. **413**, 186–197 (2002)
235. Zhu, L.B., Hess, D.W., Wong, C.P.: Monitoring carbon nanotube growth by formation of nanotube stacks and investigation of the diffusion-controlled kinetics. J. Phys. Chem. B. **110**, 5445–5449 (2006)
236. Murakami, Y., Maruyama, S.: Detachment of vertically aligned single-walled carbon nanotube films from substrates and their re-attachment to arbitrary surfaces. Chem. Phys. Lett. **422**, 575–580 (2006)
237. Kim, M.J., Nicholas, N., Kittrell, C., Haroz, E., Shan, H.W., Wainerdi, T.J., Lee, S., Schmidt, H.K., Smalley, R.E., Hauge, R. H.: Efficient transfer of a VA-SWNT film by a flip-over technique. J. Am. Chem. Soc. **128**, 9312–9313 (2006)
238. Jiang, H.J., Zhu, L.B., Moon, K.S., Wong, C.P.: Low temperature carbon nanotube film transfer via conductive polymer composites. Nanotechnology. **18**, 4 (2007)
239. Gan, Z., Liu, S., Song, X., Chen, M., Lv, Q., Yan, H., Cao, H.: Large-scale Bonding of Aligned Carbon Nanotube Arrays onto Metal Electrodes (2009)
240. Lin, W., Xiu, Y.G., Jiang, H.J., Zhang, R.W., Hildreth, O., Moon, K.S., Wong, C.P.: Self-assembled monolayer-assisted chemical transfer of in situ functionalized carbon nanotubes. J. Am. Chem. Soc. **130**, 9636–9637 (2008)
241. Son, Y., Pal, S.K., Borca-Tasciuc, T., Ajayan, P.M., Siegel, R.W.: Thermal resistance of the native interface between vertically aligned multiwalled carbon nanotube arrays and their SiO2/Si substrate. J. Appl. Phys. **103**, (2008)
242. Yu, C.H., Saha, S., Zhou, J.H., Shi, L., Cassell, A.M., Cruden, B. A., Ngo, Q., Li, J.: Thermal contact resistance and thermal conductivity of a carbon nanofiber. J. Heat Transf. Trans. ASME. **128**, 234–239 (2006)
243. Diao, J., Srivastava, D., Menon, M.: Molecular dynamics simulations of carbon nanotube/silicon interfacial thermal conductance. J. Chem. Phys. **128**, (2008)
244. Lin, W., Zhang, R.W., Moon, K.S., Wong, C.P.: Molecular phonon couplers at carbon nanotube/substrate interface to enhance interfacial thermal transport. Carbon. (2009). https://doi.org/10.1016/j.carbon.2009.08.033
245. Berger, C., Song, Z.M., Li, T.B., Li, X.B., Ogbazghi, A.Y., Feng, R., Dai, Z.T., Marchenkov, A.N., Conrad, E.H., First, P.N., de Heer, W.A.: Ultrathin epitaxial graphite: 2D electron gas properties and a route toward graphene-based nanoelectronics. J. Phys. Chem. B. **108**, 19912–19916 (2004)
246. Freitag, M.: Graphene – nanoelectronics goes flat out. Nat. Nanotechnol. **3**, 455–457 (2008)

247. Lee, B.K., Park, S.Y., Kim, H.C., Cho, K., Vogel, E.M., Kim, M. J., Wallace, R.M., Kim, J.Y.: Conformal Al2O3 dielectric layer deposited by atomic layer deposition for graphene-based nanoelectronics. Appl. Phys. Lett. **92**, (2008)
248. Westervelt, R.M.: Applied physics – graphene nanoelectronics. Science. **320**, 324–325 (2008)
249. Xuan, Y., Wu, Y.Q., Shen, T., Qi, M., Capano, M.A., Cooper, J.A., Ye, P.D.: Atomic-layer-deposited nanostructures for graphene-based nanoelectronics. Appl. Phys. Lett. **92**, (2008)
250. Chen, Z.H., Lin, Y.M., Rooks, M.J., Avouris, P.: Graphene nanoribbon electronics. Phys. E. **40**(2), 228–232 (2007)
251. Eisberg, N.: Electronics – graphene set to replace, silicon. Chem. Ind. 11–11 (2008)
252. Hass, J., Feng, R., Li, T., Li, X., Zong, Z., de Heer, W.A., First, P. N., Conrad, E.H., Jeffrey, C.A., Berger, C.: Highly ordered graphene for two dimensional electronics. Appl. Phys. Lett. **89**, (2006)
253. Wu, X.S., Sprinkle, M., Li, X.B., Ming, F., Berger, C., de Heer, W. A.: Epitaxial-graphene/graphene-oxide junction: an essential step towards epitaxial graphene electronics. Phys. Rev. Lett. **101**, (2008)
254. Berger, C., Song, Z.M., Li, X.B., Wu, X.S., Brown, N., Naud, C., Mayou, D., Li, T.B., Hass, J., Marchenkov, A.N., Conrad, E.H., First, P.N., de Heer, W.A.: Electronic confinement and coherence in patterned epitaxial graphene. Science. **312**, 1191–1196 (2006)
255. Geim, A.K., Novoselov, K.S.: The rise of graphene. Nat. Mater. **6**, 183–191 (2007)
256. Katsnelson, M.I., Novoselov, K.S., Geim, A.K.: Chiral tunnelling and the Klein paradox in graphene. Nat. Phys. **2**, 620–625 (2006)
257. Novoselov, K.S., Geim, A.K., Morozov, S.V., Jiang, D., Katsnelson, M.I., Grigorieva, I.V., Dubonos, S.V., Firsov, A.A.: Two-dimensional gas of massless dirac fermions in graphene. Nature. **438**, 197–200 (2005)
258. Ohta, T., Bostwick, A., Seyller, T., Horn, K., Rotenberg, E.: Controlling the electronic structure of bilayer graphene. Science. **313**, 951–954 (2006)
259. Son, Y.W., Cohen, M.L., Louie, S.G.: Half-metallic graphene nanoribbons. Nature. **444**, 347–349 (2006)
260. Han, M.Y., Ozyilmaz, B., Zhang, Y.B., Kim, P.: Energy band-gap engineering of graphene nanoribbons. Phys. Rev. Lett. **98**, (2007)
261. Son, Y.W., Cohen, M.L., Louie, S.G.: Energy gaps in graphene nanoribbons. Phys. Rev. Lett. **97**, (2006)
262. Castro, E.V., Novoselov, K.S., Morozov, S.V., Peres, N.M.R., Dos Santos, J., Nilsson, J., Guinea, F., Geim, A.K., Neto, A.H.C.: Biased bilayer graphene: semiconductor with a gap tunable by the electric field effect. Phys. Rev. Lett. **99**, (2007)
263. Giovannetti, G., Khomyakov, P.A., Brocks, G., Kelly, P.J., van den Brink, J.: Substrate-induced band gap in graphene on hexagonal boron nitride: ab initio density functional calculations. Phys. Rev. B. **76**, (2007)
264. McCann, E.: Asymmetry gap in the electronic band structure of bilayer graphene. Phys. Rev. B. **74**, (2006)
265. Zanella, I., Guerini, S., Fagan, S.B., Mendes, J., Souza, A.G.: Chemical doping-induced gap opening and spin polarization in graphene. Phys. Rev. B. **77**, (2008)
266. Gunlycke, D., Lawler, H.M., White, C.T.: Room-temperature ballistic transport in narrow graphene strips. Phys. Rev. B. **75**, (2007)
267. Huard, B., Sulpizio, J.A., Stander, N., Todd, K., Yang, B., Goldhaber-Gordon, D.: Transport measurements across a tunable potential barrier in graphene. Phys. Rev. Lett. **98**, (2007)
268. Katsnelson, M.I., Geim, A.K.: Electron scattering on microscopic corrugations in grapheme. Philos. Trans. R Soc. A. **366**, 195–204 (2008)
269. Wang, X.R., Ouyang, Y.J., Li, X.L., Wang, H.L., Guo, J., Dai, H. J.: Room-temperature all- semiconducting sub-10-nm graphene nanoribbon field-effect transistors. Phys. Rev. Lett. **100**, (2008)
270. Fiori, G., Iannaccone, G.: Simulation of graphene nanoribbon field-effect transistors. IEEE Electron. Device Lett. **28**, 760–762 (2007)
271. Liang, G.C., Neophytou, N., Lundstrom, M.S., Nikonov, D.E.: Ballistic graphene nanoribbon metal-oxide-semiconductor field-effect transistors: a full real-space quantum transport simulation. J. Appl. Phys. **102**, (2007)
272. Liang, G.C., Neophytou, N., Nikonov, D.E., Lundstrom, M.S.: Performance projections for ballistic graphene nanoribbon field-effect transistors. IEEE Trans. Electron. Devices. **54**, 677–682 (2007)
273. Liang, X., Fu, Z., Chou, S.Y.: Graphene transistors fabricated via transfer-printing in device active-areas on large wafer. Nano Lett. **7**, 3840–3844 (2007)
274. Obradovic, B., Kotlyar, R., Heinz, F., Matagne, P., Rakshit, T., Giles, M.D., Stettler, M.A., Nikonov, D.E.: Analysis of graphene nanoribbons as a channel material for field-effect transistors. Appl. Phys. Lett. **88**, (2006)
275. Yan, Q.M., Huang, B., Yu, J., Zheng, F.W., Zang, J., Wu, J., Gu, B.L., Liu, F., Duan, W.H.: Intrinsic current-voltage characteristics of graphene nanoribbon transistors and effect of edge doping. Nano Lett. **7**, 1469–1473 (2007)
276. Hwang, E.H., Adam, S., Das Sarma, S.: Carrier transport in two-dimensional graphene layers. Phys. Rev. Lett. **98**, (2007)
277. Lemme, M.C., Echtermeyer, T.J., Baus, M., Kurz, H.: A graphene field-effect device. IEEE Electron. Device Lett. **28**, 282–284 (2007)
278. Zhang, Y.B., Tan, Y.W., Stormer, H.L., Kim, P.: Experimental observation of the quantum hall effect and Berry's phase in graphene. Nature. **438**, 201–204 (2005)
279. Gusynin, V.P., Sharapov, S.G.: Unconventional integer quantum hall effect in graphene. Phys. Rev. Lett. **95**, (2005)
280. Novoselov, K.S., McCann, E., Morozov, S.V., Fal'ko, V.I., Katsnelson, M.I., Zeitler, U., Jiang, D., Schedin, F., Geim, A.K.: Unconventional quantum hall effect and Berry's phase of 2 pi in bilayer graphene. Nat. Phys. **2**, 177–180 (2006)
281. Ghosh, S., Calizo, I., Teweldebrhan, D., Pokatilov, E.P., Nika, D. L., Balandin, A.A., Bao, W., Miao, F., Lau, C.N.: Extremely high thermal conductivity of graphene: prospects for thermal management applications in nanoelectronic circuits. Appl. Phys. Lett. **92**, (2008)
282. Balandin, A.A., Ghosh, S., Bao, W.Z., Calizo, I., Teweldebrhan, D., Miao, F., Lau, C.N.: Superior thermal conductivity of single-layer graphene. Nano Lett. **8**, 902–907 (2008)
283. Naeemi, A., Meindl, J.D.: Conductance modeling for graphene nanoribbon (GNR) interconnects. IEEE Electron. Device Lett. **28**, 428–431 (2007)
284. Shao, Q., Liu, G., Teweldebrhan, D., Balandin, A.A.: High-temperature quenching of electrical resistance in graphene interconnects. Appl. Phys. Lett. **92**, (2008)
285. Lan, T., Pinnavaia, T.J.: Clay-reinforced epoxy nanocomposites. Chem. Mater. **6**, 2216–2219 (1994)
286. Moniruzzaman, M., Winey, K.I.: Polymer nanocomposites containing carbon nanotubes. Macromolecules. **39**, 5194–5205 (2006)
287. Stankovich, S., Dikin, D.A., Dommett, G.H.B., Kohlhaas, K.M., Zimney, E.J., Stach, E.A., Piner, R.D., Nguyen, S.T., Ruoff, R.S.: Graphene-based composite materials. Nature. **442**, 282–286 (2006)
288. Debelak, B., Lafdi, K.: Use of exfoliated graphite filler to enhance polymer physical properties. Carbon. **45**, 1727–1734 (2007)
289. Fukushima, H., Drzal, L.T., Rook, B.P., Rich, M.J.: Thermal conductivity of exfoliated graphite nanocomposites. J. Therm. Anal. Calorim. **85**, 235–238 (2006)

290. Ganguli, S., Roy, A.K., Anderson, D.P.: Improved thermal conductivity for chemically functionalized exfoliated graphite/epoxy composites. Carbon. **46**, 806–817 (2008)
291. Hung, M.T., Choi, O., Ju, Y.S., Hahn, H.T.: Heat conduction in graphite-nanoplatelet-reinforced polymer nanocomposites. Appl. Phys. Lett. **89**, (2006)
292. Veca, L.M., Meziani, M.J., Wang, W., Wang, X., Lu, F., Zhang, P., Lin, Y., Fee, R., Connell, J.W., Sun, Y.P.: Carbon nanosheets for polymeric nanocomposites with high thermal conductivity. Adv. Mater. **21**, 1–5 (2009)
293. Lin, W., Zhang, R.W., Wong, C.P.: High effective thermal conductivity of graphite nanosheet composites. J. Electron. Mater. (2009) in press
294. Prasher, R.: Thermal boundary resistance and thermal conductivity of multiwalled carbon nanotubes. Phys. Rev. B. **77**, (2008)
295. Sihn, S., Ganguli, S., Roy, A.K., Qu, L.T., Dai, L.M.: Enhancement of through-thickness thermal conductivity in adhesively bonded joints using aligned carbon nanotubes. Compos. Sci. Technol. **68**, 658–665 (2008)
296. Lin, W., Moon, K.S., Wong, C.P.: A combined process of in-situ functionalization and microwave treatment to achieve ultra-small thermal expansion of aligned carbon nanotube/polymer nanocomposites: toward applications as thermal interface materials. Adv. Mater. **21**, (2009)
297. Schriemp, J.T.: Laser flash technique for determining thermal diffusivity of liquid-metals at elevated-temperatures. Rev. Sci. Instrum. **43**, 781–786 (1972)
298. Tada, Y., Harada, M., Tanigaki, M., Eguchi, W.: Laser flash method for measuring thermal- conductivity of liquids – application to low thermal- conductivity liquids. Rev. Sci. Instrum. **49**, 1305–1314 (1978)
299. Shaikh, S., Li, L., Lafdi, K., Huie, J.: Thermal conductivity of an aligned carbon nanotube array. Carbon. **45**, 2608–2613 (2007)
300. Zhang, K., Chai, Y., Yuen, M.M.F., Xiao, D.G.W., Chan, P.C.H.: Carbon nanotube thermal interface material for high-brightness light-emitting-diode cooling. Nanotechnology. **19**, (2008)
301. Wang, S.R., Liang, Z.Y., Gonnet, P., Liao, Y.H., Wang, B., Zhang, C.: Effect of nanotube functionalization on the coefficient of thermal expansion of nanocomposites. Adv. Funct. Mater. **17**, 87–92 (2007)
302. Chen, C.I., Ni, C.Y., Chang, C.M., Liu, D.S., Pan, H.Y., Yuan, T. D.: Thermal characterization of thermal interface materials. Exp. Tech. **32**, 48–52 (2008)
303. Chen, C.I., Ni, C.Y., Pan, H.Y., Chang, C.M., Liu, D.S.: Practical evaluation for long-term stability of thermal interface material. Exp. Tech. **33**, 28–32 (2009)
304. Ohsone, Y., Wu, G., Dryden, J., Zok, F., Majumdar, A.: Optical measurement of thermal contact conductance between wafer-like thin solid samples. J. Heat Transf. Trans. ASME. **121**, 954–963 (1999)
305. Chu, D.C., Touzelbaev, M., Goodson, K.E., Babin, S., Pease, R.F.: Thermal conductivity measurements of thin-film resist. J. Vac. Sci. Technol. B Microelectron. Nanometer. Struct. **19**(6), 2874–2877 (2001)
306. Kading, O.W., Skurk, H., Goodson, K.E.: Thermal conduction in metallized silicon-dioxide layers on silicon. Appl. Phys. Lett. **65**, 1629–1631 (1994)
307. Smith, A.N., Hostetler, J.L., Norris, P.M.: Thermal boundary resistance measurements using a transient thermoreflectance technique. Microscale Thermophys. Eng. **4**, 51–60 (2000)
308. Miklos, A., Lorincz, A.: Transient thermoreflectance of thin metal-films in the picosecond regime. J. Appl. Phys. **63**, 2391–2395 (1988)
309. Paddock, C.A., Eesley, G.L.: Transient thermoreflectance from thin metal-films. J. Appl. Phys. **60**, 285–290 (1986)
310. Schoenlein, R.W., Lin, W.Z., Fujimoto, J.G., Eesley, G.L.: Femtosecond studies of nonequilibrium electronic processes in metals. Phys. Rev. Lett. **58**, 1680–1683 (1987)
311. Elsayedali, H.E., Norris, T.B., Pessot, M.A., Mourou, G.A.: Time-resolved observation of electron- phonon relaxation in copper. Phys. Rev. Lett. **58**, 1212–1215 (1987)
312. Hu, H.P., Wang, X.W., Xu, X.F.: Generalized theory of the photoacoustic effect in a multilayer material. J. Appl. Phys. **86**, 3953–3958 (1999)
313. Wang, X.W., Hu, H.P., Xu, X.F.: Photo-acoustic measurement of thermal conductivity of thin films and bulk materials. J. Heat Transf. Trans. ASME. **123**, 138–144 (2001)
314. Wardle, B.L., Saito, D.S., Garcia, E.J., Hart, A.J., de Villoria, R.G., Verploegen, E.A.: Fabrication and characterization of ultrahigh-volume-fraction aligned carbon nanotube-polymer composites. Adv. Mater. **20**, 2707–2714 (2008)

Nanomaterials via NanoSpray Combustion Chemical Vapor Condensation and Their Electronic Applications

4

Andrew Hunt, Yongdong Jiang, Zhiyong Zhao, and Ganesh Venugopal

4.1 Introduction to Nanomaterials and Their Synthesis

The term nanomaterials is commonly used to refer to materials that have at least one dimension of less than 100 nm [1]. It is important to consider in the definition that there be some changes in properties due to this small dimension. In this manuscript, the term will be used interchangeably with recently coined terms like nanoparticles and nanopowders. Nanomaterials are not new; some have existed in nature forever; and man-made versions have been around for more than a 100 years. Volcanoes emit more nanoparticulates than nanomaterials' manufacturers might ever dream of making [2]. This has occurred before life even began and has continued to happen ever since. Carbon black, fumed silica, alumina and titania, and nano-silver (photographs) are a few examples of man-made nanomaterials that have been in use for several generations now. Tires are one of the oldest and most widely used nanomaterial-enhanced products, with the wear and strength of the "rubber" coming predominantly via incorporation of nanosilica and carbon. Renewed interest in these materials over the last couple of decades stems from the correlation of the reduction in dimension to unique characteristics that were observed in the nanomaterial when compared to larger-sized particles of the same composition [3]. Nanomaterials have also spawned advances in other multicomponent, multiphase materials like dispersions or composites, by enabling optimization of structure, property, and processing characteristics.

Nanomaterial fabrication methods can be broadly classified into three categories: mechanical, physical, and chemical means. For most of the physical and mechanical methods, the composition of the nanomaterial has already been determined via an upstream chemical synthesis process (by man or nature). The final fabrication is usually a conversion process, whereby the larger particles are sized down to the nanometer scale. The mechanical approach is frequently referred to as the "top-down" approach [4]. For example, a mechanical process would involve mechanical milling either in a planetary or rotating ball mill.

Physical nano-methods typically involve a change in the material from one phase to another without a chemical change or just a reduction in size from grinding or other physical process. For example, in physical precipitation, a material may be dissolved in a solvent and converted to a nanosized form via controlled physical precipitation of the material-induced via temperature/pressure change or via addition of another component to reduce the solubility limit. Alternatively, drying of the solvent in a spray-drying process could also lead to the formation of the nanomaterial. While it is difficult to achieve nanosized material by grinding or pulverization, such top-down (size reduction) processes are used but can result in hard agglomerated material and yield a wide size distribution. Physical methods also include vapor condensation methods such as thermal or plasma processing and laser ablation techniques [5, 6]. These techniques have been widely utilized for making nanomaterials of metals.

Chemical nano-methods are defined here as a processing method where the nanomaterial is synthesized from precursor materials directly in nanoscale form. Chemical approaches are frequently referred to as being "bottom-up" approaches [4]. Some post-processing could take place to further reduce or even increase the size of the particles to a desired value. Perhaps the most widely used chemical process for making nanomaterials is flame pyrolysis. Since the turn of the twentieth century, this method has been used to make carbon black by the pyrolysis of liquid or gaseous hydrocarbons [7]. Carbon black has since seen widespread application as a pigment and a reinforcement agent for plastics and elastomers. In addition to carbon black, flame pyrolysis can also be used to make fumed silica, fumed titania, and fumed

A. Hunt (✉) · Y. Jiang · Z. Zhao · G. Venugopal
nGimat Co., Atlanta, GA, USA
e-mail: ahunt@ngimat.com; yjiang@ngimat.com; zzhao@ngimat.com; gvenugopal@ngimat.com

C. P. Wong et al. (eds.), *Nano-Bio- Electronic, Photonic and MEMS Packaging*, https://doi.org/10.1007/978-3-030-49991-4_4

alumina [8–10]. In spite of its popularity, flame pyrolysis is limited in terms of the compositional control and compound flexibility in the nanomaterials that it can produce. This is largely because limited number of high-volatility precursors are available that are suitable for flame pyrolysis. Care needs to be taken, or flame pyrolysis may promote formation of >100 nm secondary particles through aggregation [11]. Other chemical methods include variations of the flame pyrolysis method, such as spray pyrolysis and wet-chemical methods such as sol–gel synthesis (chemical precipitation).

The line between physical and chemical processing can get blurred. There is physical vaporization combined with gas phase reaction to form compounds such as oxides, carbides, nitrides, and borides. This is the same physical vapor process used to make metal nanomaterials but with a reaction gas metered in during processing to facilitate a chemical change. Surface adsorbed monolayers are another example of nanomaterials. These are mostly homogenous materials, except for where the monolayer is chemically or physico-chemically bonded to the surface of a substrate. Although the primary material being deposited stays the same, the reaction at the substrate leads to modification of the properties of the surface and the adsorbed species.

A large number of nanomaterial fabrication techniques currently exist, and new ones continue to develop. While it is not possible to review each of these techniques in this chapter, it is important to make the distinction between the fabrication of nanoparticles or nanopowders and other nanostructured materials that may exhibit nanoscale (<100 nm) features, but are not necessarily distinct particles. For example, lithographic patterning and vacuum deposition techniques like physical vapor deposition (PVD) and chemical vapor deposition (CVD) can produce nanostructured films, layers, or three-dimensional (3D) structures [12]. However, they are not very efficient for producing nanoparticles or nanopowders. Some of these processes have been referred to as "form-in-place" techniques for making nanostructured materials.

In the following section, we will focus on the combustion chemical vapor condensation method for making nanomaterials as a comparison with other processes discussed in this section. A brief mention of a related NanoSpray Combustion technique, called combustion chemical vapor deposition, will also be made. This method can be used to make nanostructured thin films and coatings.

4.2 NanoSpray Combustion Processing

NanoSpray Combustion processes to form nanomaterials have subsets depending on whether the material forms independently (homogeneous nucleation) or onto an existing solid (heterogeneous nucleation). Combustion chemical vapor condensation (nCCVC) is a homogeneous process for synthesizing powdered material by chemical change via combustion of precursor gases or solutions in droplet spray form, followed by the condensation of the material into ultrafine or nanopowder. nGimat Co. uses a proprietary atomizer technology, called the Nanomiser^{QR}, to produce submicron liquid aerosol droplets (NanoSpray), which allows the vapor formation of the nanoparticles or nanopowder after combustion. Large volume commercial CCVC exists via gas precursor's combustion in forming carbon black, fumed silica, and titania. The process of heterogeneously producing nanomaterial films via the combustion of a spray of submicron droplet has been patented by Georgia Tech [13] and licensed exclusively to nGimat, which formed the basis of the NanoSpray Combustion^{QR} technology umbrella. Thin-film coatings can be deposited onto substrates in the open atmosphere by way of combustion chemical capor deposition (CCVD) process [14]. Figure 4.1 shows a schematic of the nCCVC system utilizing a Nanomiser device to produce nanopowders. In this chapter, the acronyms CCVC and nCCVC may be used interchangeably. Similarly, CCVD may be used instead of nCCVD and vice versa.

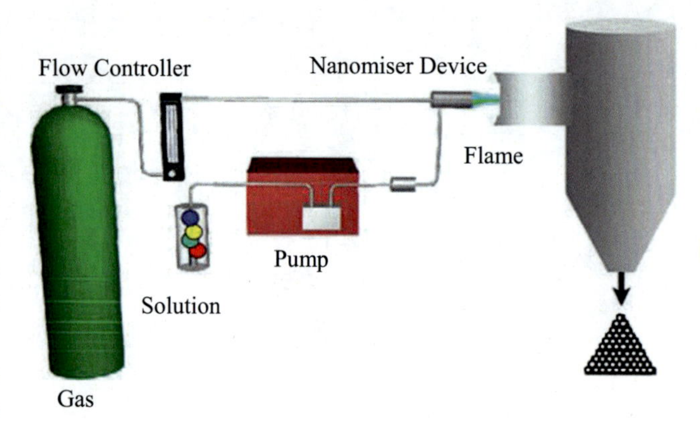

Fig. 4.1 Schematic drawing of an nCCVC system comprising a Nanomiser device for NanoSpray Combustion processing

The key advantage of the NanoSpray Combustion process is its ability to condense nanomaterials as nanoparticles or thin films in atmospheric pressure using low-cost functional precursor chemicals in solution. This eliminates the need for costly high vapor pressure precursors, furnaces, vacuum equipment, and reaction chambers. Typical precursors, such as metal nitrates or metal carboxylates, are dissolved in a solvent, which typically also acts as the combustible fuel. Water-soluble precursors may be dissolved in water and then mixed with a suitable liquid or gas fuel. The solution is atomized to form submicron droplets, and these nanodroplets are then conveyed by an oxygen-containing stream to the flame where they combust in a manner similar to premixed gas fuels. In nCCVC of nanopowders, Fig. 4.1, the flame flash vaporizes the solvent and precursors, which decompose to yield metal and oxide vapors that consist of gaseous atoms, ions, and molecular-oxide species (e.g., M-O "monomers," <5 Å in size). These vapor species then condense to form atomic clusters (e.g., (M-O) x, where $x \approx 10$ and the cluster is <5 nm in size) that then may coalesce to form nanoparticles with a desired size distribution in any range from 2 to 100 nm. The synthesized nanoparticles are then captured in the dry form in bag house filters or via wet collections methods.

The number and the size of nuclei formed in the flame are a function of the flow rates, solution density, temperatures, time at temperature, and concentration. Depending on flame temperature and position in the flame where powder is collected, non-agglomerated particles can be readily produced. Atmospheric pressure flame temperature is in the range of 1200–2400 °C, depending on oxygen ratio, flame size, solvent, precursor loading, and Nanomiser device atomization setting. In contrast to conventional spray pyrolysis from larger-sized droplets, NanoSpray Combustion processing technology is able to produce nCCVC nanopowders with grain sizes less than 10 nm. By adjusting precursor solution concentrations and constituents, a wide range of nanopowder stoichiometries and compositions can be achieved.

In summary, the NanoSpray Combustion processing technologies, nCCVC, and nCCVD, offer the following capabilities for producing nanopowders or thin films:

- *Inexpensive precursor requirements:* Soluble precursors of any element are used instead of limiting, expensive, high vapor pressure metal organics employed in traditional flame nanopowder technologies.
- *High degree of composition control:* Solution properties are adjustable to allow formation of a wide variety of complex, multicomponent compounds with precisely targeted stoichiometries.
- *Suitable for continuous production:* The equipment enables a robust production system shown to operate without failure around the clock for several days.
- *No need for specialized equipment:* CCVC nanopowders are formed at ambient conditions without the need for expensive, specialized equipment, thereby reducing downtime and associated costs.
- *Accelerated development cycle for new applications:* Development of CCVC nanopowders for specific applications is achieved more rapidly than with nanopowders produced by traditional technologies; a large series of samples are quickly prepared for iterative testing and optimization.
- *Environmentally friendly:* Relatively safe chemical precursors (nontoxic and halogen-free) are used, resulting in benign by-products, thus reducing workplace hazards. By using biosolvents such as alcohols, the process is nearly carbon neutral.

4.3 Overview of Nanomaterials Capabilities

The NanoSpray Combustion process is ideally suited for producing oxides of metals. More than 150 compositions have been made, and these have been further varied over a range of stoichiometries. Most of the elements in the periodic table can be combined in different ratios, thereby enabling nanomaterials with novel properties. Phosphates and sulfates can also be produced, with the addition of stable precursors for phosphorus and sulfur. For sulfates, the process has to be controlled so that sulfur does not exit the nCCVC system as sulfur dioxide gas. By maintaining a reducing atmosphere in the combustion chamber downstream from the flame, metal nanopowders of moderate to low oxidation potential elements, like nickel, copper, tin, silver, and palladium, can also be produced. A reducing atmosphere could also promote the formation a carbon coating on the surface of the nanoparticle. This surface treatment can be carried out both on reduced (metal) and oxidized (oxides, phosphates, etc.) nanoparticles. Unlike other competing processes, the surface modification takes place simultaneously with nanomaterial synthesis. This allows for low-cost production of complex nanomaterial compositions.

4.3.1 Single Metal Oxides and Phase Control

The CCVC process has been used to synthesize a wide variety of single metal oxides. Typical examples include MgO, Al2O3, CaO, TiO2, Fe2O3, Fe3O4, ZnO, ZrO2, SnO2, and CeO2. Single metal oxides are the simplest nanomaterials that can be made by nCCVC since metal stoichiometry is not an issue. Stable oxidation states such as the +2 state for MgO and CaO and the +4 state for ZrO2 and SnO2 are most easily achieved. Figure 4.2 shows the X-ray diffraction (XRD) pattern and transmission electron microscopy (TEM) micrograph for an SnO2 nanopowder sample made using by NanoSpray Combustion of a tin carboxylate precursor solution in an nCCVC system. Average particle sizes are less than 100 nm, and the morphology of the particle is faceted.

The precursor material composition and the process parameters are controlled carefully to ensure that carbonates and hydroxides are not formed. Some elements, like Ca and Ba, are more susceptible to the formation of these additional by-products than others. Tight control of materials and process parameters are also required during synthesis of less stable oxidation states of specific metals. For example, iron oxides can be formed as FeO (iron (II) oxide), Fe2O3 (iron (III) oxide), or the mixed valence Fe3O4 (iron (II, III) oxide). While Fe2O3 is relatively simple to make by nCCVC, FeO and Fe3O4 require careful control of the

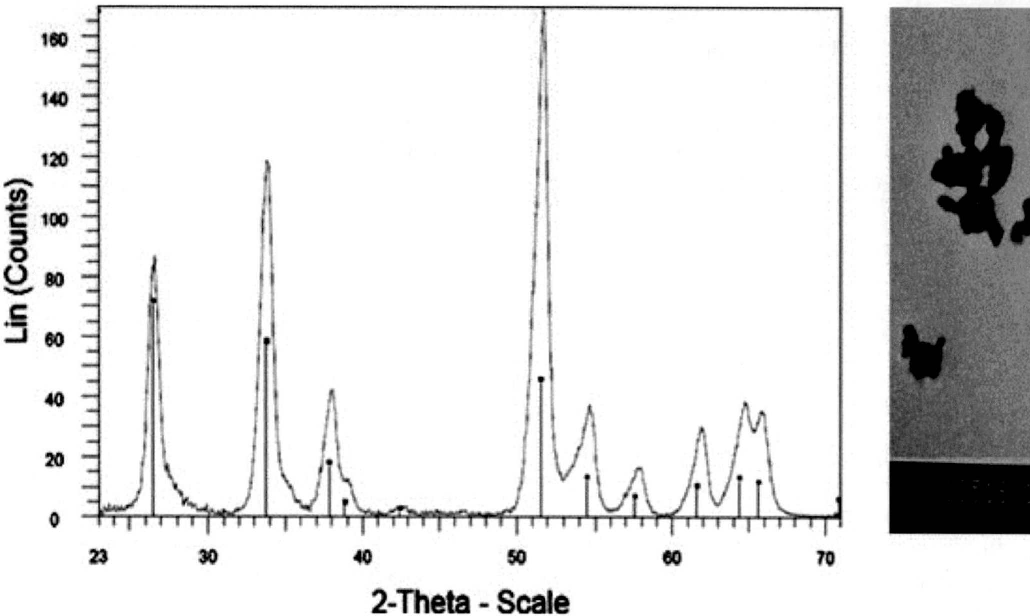

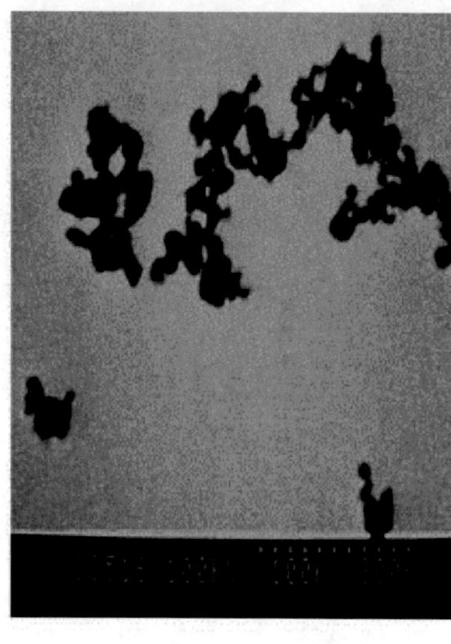

Fig. 4.2 X-ray diffraction pattern (*left*) and transmission electron micrograph (*right*) for tin oxide nanopowder made by NanoSpray Combustion of a tin precursor solution in a CCVC system

combustion process, so that a slightly reducing atmosphere is maintained to promote formation of the lower oxidation states of the metal.

For some metals, the oxide may be capable of forming different phases, even though the oxidation state of the metal is unchanged. Titanium is a typical example, where the oxide (TiO2) can form the rutile, the anatase, or the brookite phase. Anatase and brookite are known to change into the rutile phase upon heating. Certain applications however require the anatase phase preferentially. This is the case with dye-sensitized solar cells, where high efficiency is achieved when the TiO2 is in the anatase phase [15]. Both the rutile and the anatase phase of TiO2 can be selectively produced using nCCVC. For anatase, the precursor material solution and the process parameters are chosen so that the oxide forms under relatively cool conditions. Figure 4.3 shows the XRD curves for rutile, mostly anatase, and rutile with anatase forms of TiO2 made by nCCVC operating under different materials and processing conditions.

4.3.2 Multi-Metal Oxides

Synthesis of multi-metal oxides can be achieved by the combustion of precursor solutions containing multiple metal compounds in the desired stoichiometry. Precise formulation of stable precursor solutions is critical to achieving high-quality nanopowders. More than 100 multi-metal oxide nanopowders have been synthesized by nCCVC. Figure 4.4 shows the XRD and TEM (center) for as-synthesized Ni–YSZ nanopowder. In most cases, the material will form a single phase of similar stoichiometry to that of the feed solution, except when there are components that are not phase compatible as is the case with Ni and YSZ. The XRD pattern shows well-formed NiO and YSZ phases, even in as-synthesized form. The TEM shows mostly spherical particles with average particle size <50 nm. The TEM on the right-hand side of the same figure is for Cu/Sm/Ni/Gd-doped CeO2 nanopowders, a complex metal oxide containing five metals synthesized by nCCVC.

In most cases, nCCVC of a precursor solution containing multiple metals leads to the formation of either an amorphous mixture or a compound phase of the individual oxides of the component metal. Such is the case with some yttrium aluminum oxide materials, where an almost amorphous mixture is formed in the as-synthesized powder. Subsequent heating of the composite material at around 1700 °C leads to the formation of the crystalline garnet YAG phase. The XRDs for the nanopowder before and after heat treatment are shown in Fig. 4.5. Many elements can be combined in each grain, and this has been demonstrated in a multi-metal oxide, Cu/Sm/Ni/Gd-doped CeO2, containing five metals. Both the CeO2- and the YSZ-based materials discussed above play an important role in the electrolytes and electrodes that make up solid-oxide fuel cells [16] and as reforming catalysts for hydrogen generation [17]. By using nanopowders to fabricate the ceramic bodies in these applications, researchers have realized lower sintering temperatures, higher conductivities, and higher efficiencies in their respective end applications [16].

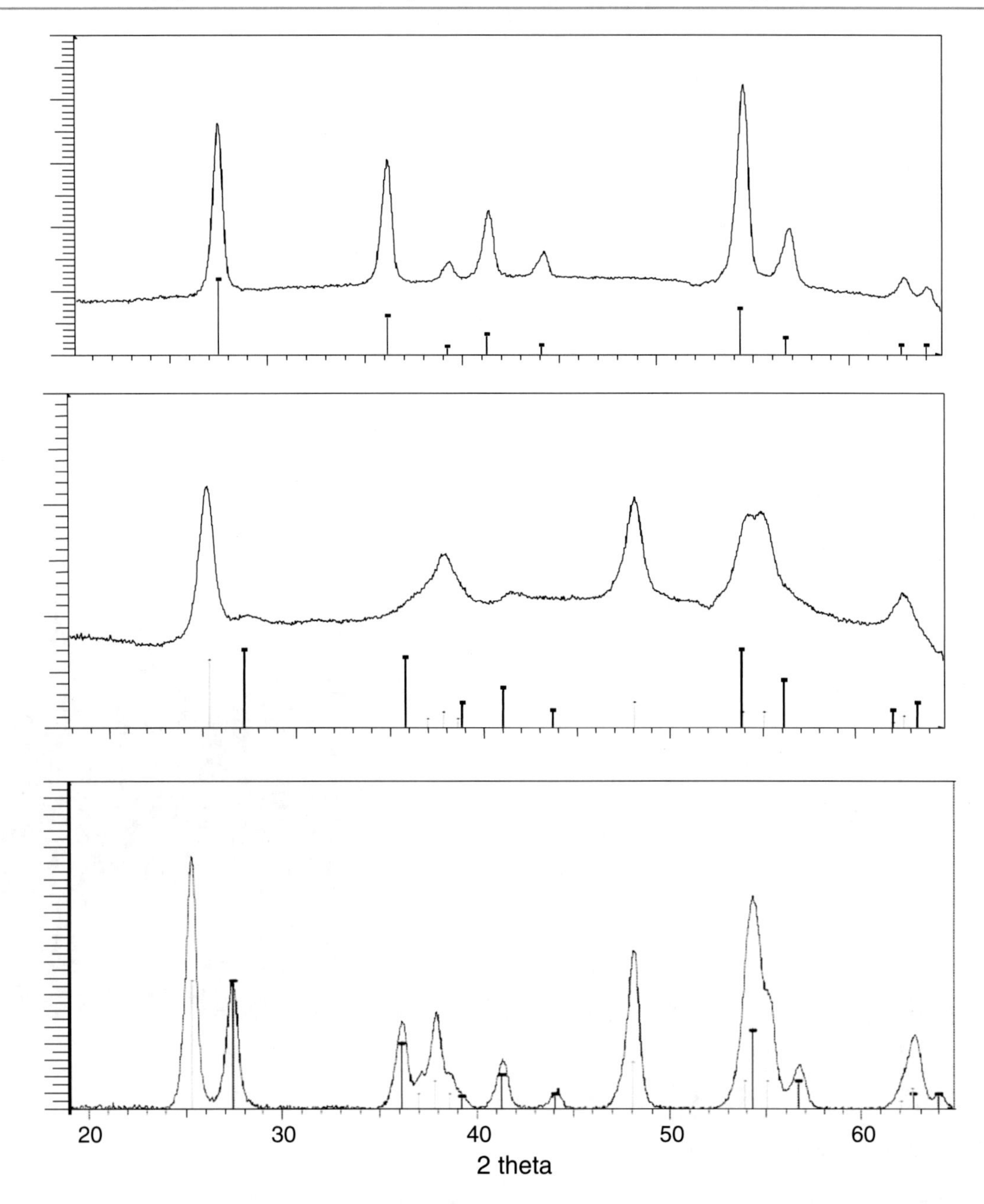

Fig. 4.3 X-ray diffraction patterns for rutile (*top*), mostly anatase (*center*), and rutile + anatase (*bottom*) forms of TiO2 made using nCCVC

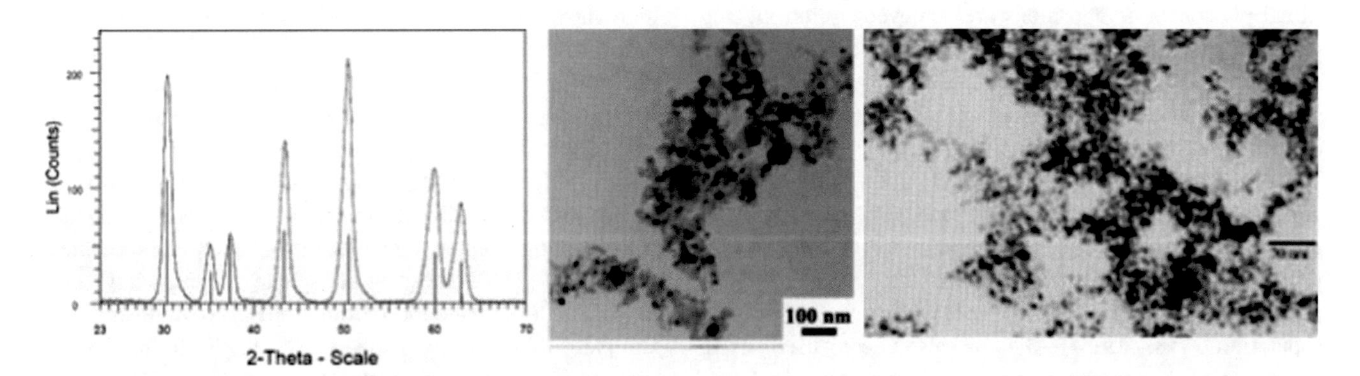

Fig. 4.4 XRD and TEM (*center*) for Ni–YSZ nanopowder made using CCVC; the TEM micrograph on the right is for Cu/Sm/Ni/Gd-doped CeO2 (scale bar = 50 nm)

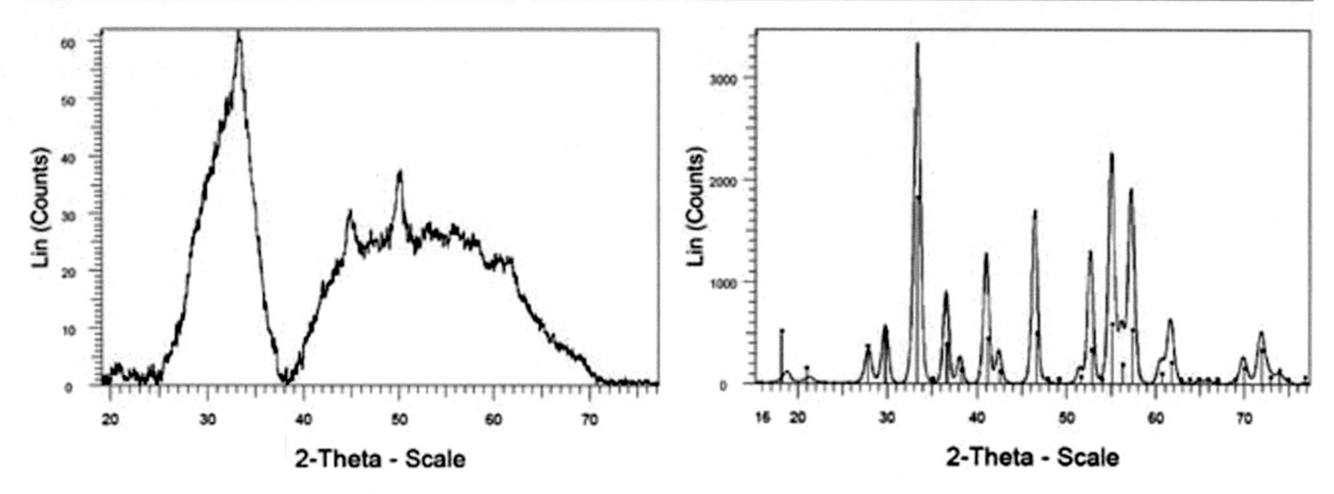

Fig. 4.5 XRD for yttrium aluminum oxide nanopowder before and after heat treatment at around 1700 °C, leading to the formation of the yttrium aluminum garnet (YAG) phase

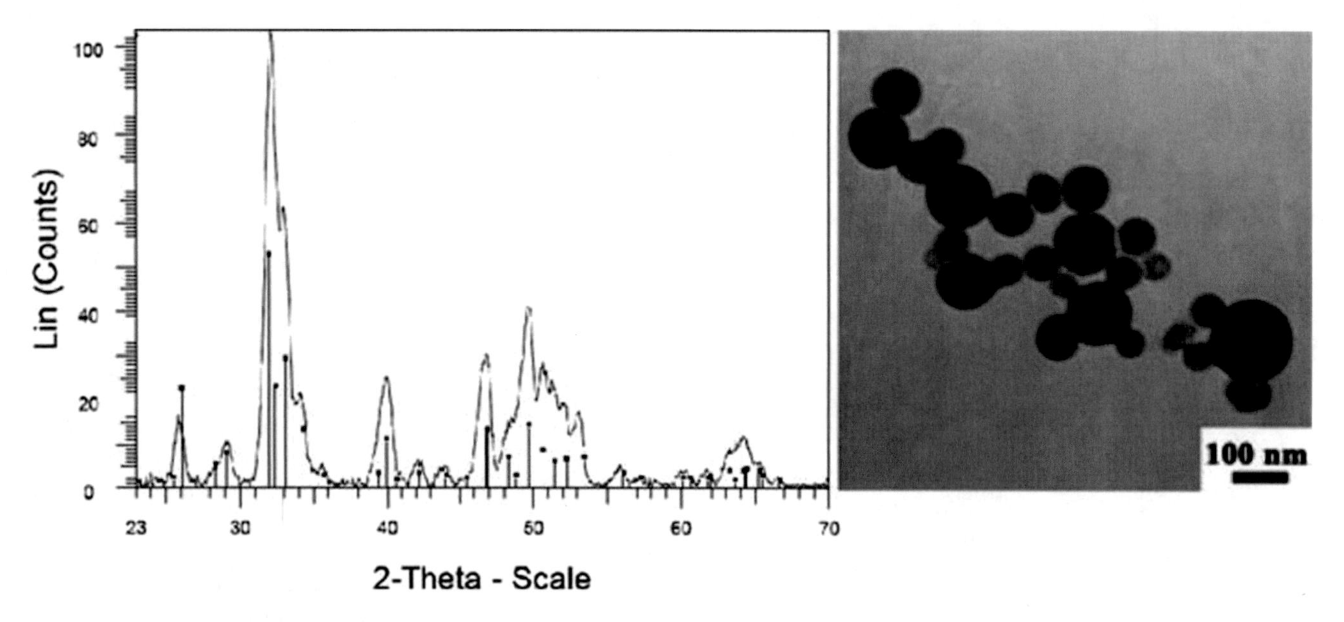

Fig. 4.6 XRD and TEM of synthetic hydroxylapatite nanopowder

4.3.3 Metal Phosphates

Metal phosphate compositions are achieved by introducing phosphorus compounds into the metal precursor solution mix. nCCVC has been used to produce a variety of metal phosphates when a stable phase exists, including those of calcium, lanthanum, and cesium. Precursor selection and solution formulation are very important for stoichiometric control. Figure 4.6 shows the XRD and TEM for a synthetic nanopowder version of the bone mineral hydroxylapatite (HAP). Synthetic HAP is used for making bone cements for repairing fractured bone and as a sorbent for chromatographic separations [18, 19]. HAP has also been used as a cytocompatible growth promoter of bone tissue in tissue engineering applications [20]. For applications such as these, the higher surface area offered by nanosized particles could allow for faster growth and tissue regeneration.

Through careful control of precursor materials and process conditions, it has also been possible to produce other HAP-related calcium phosphate materials by CCVC. Examples of these are tricalcium phosphate, amorphous calcium phosphate, and silicon-doped HAP. Synthetic versions of HAP with dopants have garnered a lot of attention recently for improving the biocompatibility of HAP in biomedical applications [21]. Nanosized phosphate materials have also attracted a lot of attention from the battery industry. For example, lithium iron phosphate particles in the nanometer size range have enabled the fabrication of high-power lithium-ion batteries that could be used for electric vehicle applications [22]. Other nanomaterials that serve the lithium-ion application are discussed in detail in a following section.

4.3.4 Metals

Since nCCVC is a combustion technology, particular attention has to be paid to the process conditions while using this method to make metal nanopowders. The list of metals that can be produced is limited to those that have moderate to low propensity to oxidize. More specifically, the metal has to have less oxidation potential than water or carbon dioxide. In combustion flames, water and carbon dioxide are the primary reactants. An element must have lower oxidation potential, be less competitive, to hydrogen and carbon to form as a metal in the high-energy flame environment. This list includes noble metals like palladium, silver, and platinum. Other base metals such as iron, copper, and nickel have also been produced in nanopowder form by nCCVC, as has tin. Alloys of some of these metals such as silver–copper, iron–nickel, and tin–copper have also been produced. Figure 4.7 shows the XRD pattern and the TEM for silver metal nanopowder. Later in this chapter, we discuss in greater detail the application of silver nanopowder in conductive adhesives for interconnect applications.

4.3.5 Nanocomposites

Nanocomposites are engineered multiphase materials that contain two or more distinct constituent materials, with at least one of the constituents having a dimension less than 100 nm. The resulting material has some physical, mechanical, or chemical property that is better than either of the constituents alone. The simplest nanocomposite consists of matrix material, the major phase, which uniformly dispersed another material, and the minor phase, which may have at least one dimension that is <100 nm.

Nanocomposites relevant to this discussion can be made in one of the several ways. The minor-phase material could be an nCCVC synthesized nanopowder that is dispersed in the matrix phase in a separate fabrication process, for example, extrusion, casting, or spraying. Alternatively, the minor phase could be formed in situ by nCCVC during the nCCVD of the matrix inorganic material. Examples of each of these nanocomposites and their applications are discussed in the applications section.

4.3.6 Nanodispersions (Not Solutions)

In this context, the term nanodispersion refers to a suspension of solid nanomaterial in a liquid medium. Typical solids loadings for the dispersions are in the 5–30% range, although compositions outside this range are also feasible. The stability (length of settling time) of a nanodispersion depends on the surface characteristics of the nanoparticles, size, density, and also on the chemical composition and viscosity of the liquid medium. From this standpoint, nCCVC is capable of producing nanomaterials with specific surface characteristics that could compatibilize the nanoparticles with the liquid medium of choice. If surface modification does not address the desired compatibility between solid and liquid, a compatibilizer such as a surfactant or stabilizer might be added to achieve the desired result. Some people use the term "solution" incorrectly for dispersion. A true solution is molecular in scale which is smaller than solid nanoparticles. nCCVC-derived nanomaterials have been used to produce many dispersions including cerium oxide, zirconium oxide, silicon dioxide, hydroxylapatite, and cobalt aluminate.

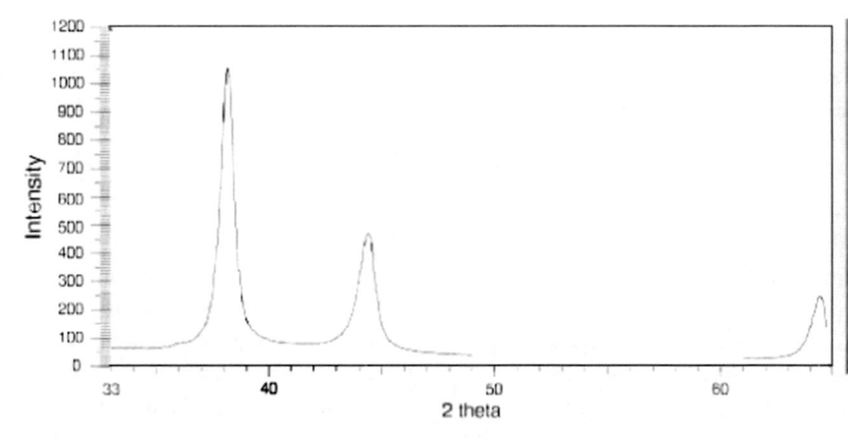

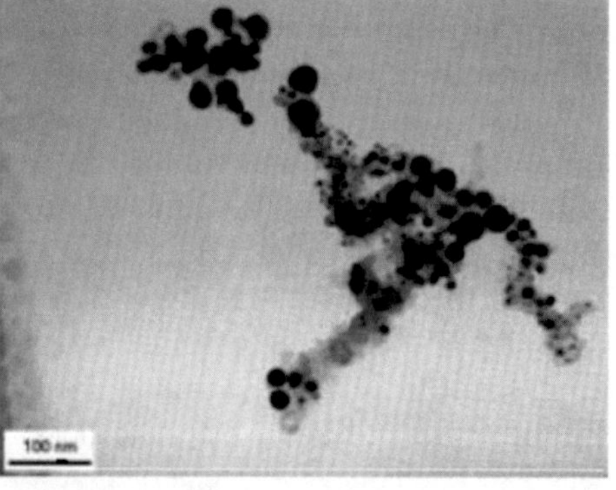

Fig. 4.7 XRD and TEM pattern for silver nanopowder

4.4 Applications of Nanomaterials Made by CCVC

In addition to the dimensional reference, the term nanomaterial implies that the material in question also has a unique performance characteristic that is not exhibited by larger-sized particles or bulk materials of the same composition [3]. The emergence of nanotechnology as an interdisciplinary science in recent years has mostly come about because of this realization. The performance characteristics that are being explored by research groups around the world are as varied as the materials themselves. Nanomaterials have been widely investigated for energy storage and energy conversion applications, where they are known to significantly improve power density in batteries or conversion efficiency in solar cells. For catalysis, nanomaterials can achieve similar efficiencies as micron-sized materials at lower activation temperatures or higher efficiencies at similar temperatures. This behavior has a huge impact on myriad applications, from petroleum and petrochemical production to fuel cell electrodes. Nanomaterials also offer unique fabrication advantages, particularly for those related to ceramics processing. Small particle sizes enable the fabrication of highly dense ceramic bodies at low processing temperatures that can have novel crystal structures. This is of particular importance to fuel cell electrolytes and laser host systems.

Whether in particle form, as made with nCCVC, or in thin-film form, as made by nCCVD, nanomaterials could be instrumental in the fabrication of miniaturized micro- or nanodevices. NanoSpray Combustion processing can enable a wide variety of the materials compositions for the components that make up these devices, in a form that can be easily integrated into the fabrication process. This opens the door for future application of these devices in electronics, electromechanical, and biomedical areas. For example, in the area of biomedical device packaging, nanomaterials can enable novel biosensors and micro-fluidic devices that would have superior capability compared to what would otherwise be achieved with larger-scale materials [23].

Nanocomposites have been used in structural applications to provide improved physical and mechanical properties. In addition to serving to improve the performance of conventional composites with larger-sized reinforcing agents, nanocomposites are also being used in biomedical applications. A typical example is bone cement, made from synthetic hydroxylapatite nanopowders, which is used to repair broken or defective bone. Applications for nanodispersions have also emerged at a rapid pace. Typical examples include printing inks for printed electronics and pigments for coloring high-temperature-stable ceramic tiles.

4.4.1 Conductive Adhesives as Electronic Interconnects

Continuous improvement in electronic interconnect technology has played a very important role in enabling sophisticated electronic devices that are reliable and environmentally friendly. Nanomaterials will play an important role in advancing future improvements in interconnect technology at several levels. Some examples have already started to emerge. Solder paste containing nanosized solder particles can be reflowed at lower temperatures than paste composed of conventional micron-sized solder particles [24]. This is of particular importance for lead-free solders, like tin–silver (Sn–Ag) and tin–silver–copper (Sn–Ag–Cu), where high melting points of the alloys require high reflow temperatures that potentially stress heat-sensitive components on the board. Similarly, nanomaterials with the right formulations can enhance performance in other areas of interconnect technology such as conductive adhesives [25], printing inks, underfill, thermal interface materials [26], and board technology.

Conductive adhesives can be used for connecting electronic components to rigid or flexible printed wiring (PWB) substrates instead of solder. They are particularly useful when one requires a low-temperature solder alternative for connecting temperature-sensitive components to conventional PWB substrates, conventional components to low-temperature stable boards, or a combination of both. When properly designed, conductive adhesives could also enable flexible electronic devices, since the adhesive composition can be tailored to impart the requisite mechanical properties in the interconnect material.

Silver-based nanopowders have been utilized to prepare conductive adhesive formulations [27]. In these formulations, the conductive component was composed of a mixture of micron-sized and nanosized silver particles. Silver nanopowders were made by nCCVC under oxidizing or reducing conditions. Figure 4.8 shows the thermogravimetric analysis (TGA) for two samples, representative of each of those production conditions, respectively. TGA experiments were done in air. Samples made under oxidizing conditions showed a weight loss of around 1%, leading up to approximately 100 °C. This is believed to be due to the loss of adsorbed surface species. These samples also showed another pronounced weight loss of around 1%, leading up to approximately 210 °C. This is attributed to decomposition of surficial silver oxide in the sample [28]. Since XRD shows only Ag metal peaks, the amount of oxide is nominal. Above 210 °C, these samples are very stable and do not exhibit any further weight loss. Samples made under reducing conditions showed very little, if any, weight loss below 100 °C and a relatively small loss (0.5%) at 210 °C. These

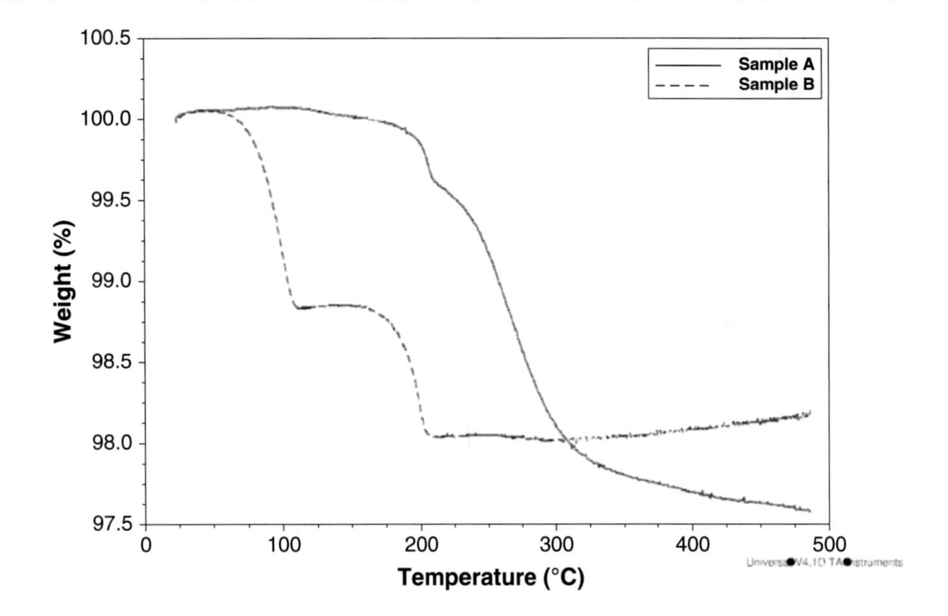

Fig. 4.8 Thermogravimetric analysis (TGA) curves for silver nanopowders made under relatively oxidizing (sample B – dashed curve) and relatively reducing (sample A – solid curve) conditions in the CCVC system

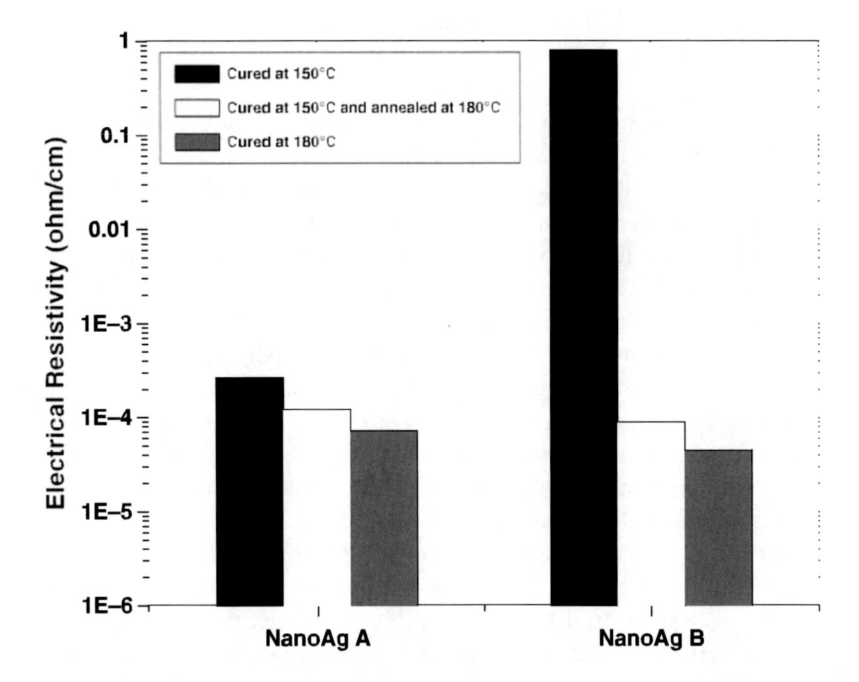

Fig. 4.9 Electrical resistivity measurements of conductive adhesives containing silver nanopowder made by CCVC under relatively oxidizing and relatively reducing conditions

samples loose approximately 2% of their weight leading up to 300 °C. This is indicative of the formation of carbonaceous surface coatings, as is to be expected of sample made under reducing conditions. The small loss leading up to 200 °C is indicative of a relatively small amount of surface silver oxide being formed as a result of the reducing conditions used for this run. Furthermore, the negligible loss at temperatures below a 100 °C suggests that the carbonaceous coating may be discouraging the adsorption of undesired chemical species on the surface of the nanomaterials.

Electrical resistivity measurements have been made on conductive adhesive formulations containing an epoxy matrix and a combination of micron-sized and nanosized silver particles. These measurements are shown graphically in Fig. 4.9. Samples with carbonaceous surface coatings yielded low resistivities ($\sim 2 \times 10^{-4}$ Ωcm) even at 150 °C. Similar formulations consisting of silver nanopowders made under oxidizing conditions showed resistivity values $>2 \times 10^{-2}$ Ωcm, indicating that the surficial silver oxide may be disrupting electron flow at the particle surface. Upon curing at 180 °C, resistivities of all of the conductive adhesives dropped to lower than 1×10^{-4} Ωcm, with the formulations containing the silver nanopowders made under oxidative conditions showing the lowest values. This indicates that at higher temperatures, the silver oxide layer is decomposed, thereby leading to lower resistivities. In

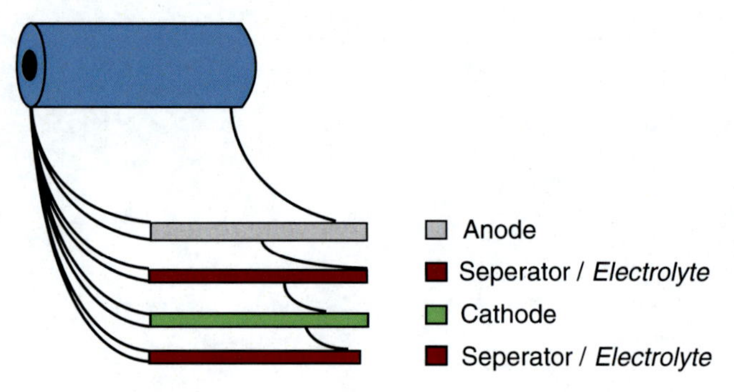

Fig. 4.10 Schematic of a cylindrical lithium-ion cell in unwound form showing anode, separator, and cathode layers

summary, nanomaterial fabrication and the resulting particle characteristics play an important role in determining the performance of the material in an end-use application.

4.4.2 Lithium-Ion Battery Electrodes for Energy Storage

Lithium-ion (Li-ion) batteries have been the preferred choice for powering portable electronic devices such as laptop computers and cell phones for over a decade and a half. Figure 4.10 shows a schematic of a cylindrical lithium-ion cell in unwound form, consisting of anode, cathode, and separator layers. Around a decade after its introduction into cellular phones, lithium-ion batteries completely replaced nickel-based chemistries. The selection of active (anode and cathode) materials for these energy-hungry applications has hence leaned toward high-capacity energy storage electrodes with layered structures such as LiCoO2 and (for cathodes) and graphite (for anodes). These traditional materials are not the best in class when it comes to other characteristics such as power density, low temperature performance, thermal stability, calendar life, and cost. For this reason, Li-ion has found limited applications to date in areas such as hybrid electric vehicles (HEV), power tools, and backup/storage, where traditional chemistries such as nickel metal hydride (NiMH), nickel cadmium (NiCd), and lead acid batteries are still preferred.

By sacrificing some electrode capacity, emerging Li-ion chemistries are capable of making significant improvements in the areas of power density (fast charge and discharge), cycle life, calendar life, thermal stability, and cost. This expands the utility of Li-ion batteries from portable electronic to aforementioned applications like HEV, power tools, and battery energy storage systems. Of these emerging materials classes, one that is considered a promising candidate is lithium titanate (Li4Ti5O12, LTO). The 3D spinel structure of the LTO provides a more stable framework for the lithium ions to move back and forth during the charge discharge cycle, thus enhancing the cycle life and thermal stability [29]. The higher voltage of the LTO with respect to Li will eliminate the need to form a secondary interface (SEI) layer that is a prerequisite for stable graphite anodes [30]. Furthermore, due to the higher voltage of the LTO half-cell with respect to the Li/Li+ electrode couple, the possibility of lithium plating is eliminated [30]. Both of these features render LTO-based batteries more thermally stable and safer compared to conventional Li-ion cells. The rate capability of conventional micron-sized LTO materials can be improved dramatically by synthesizing these electrode materials in the nanometer size range and/or coating with carbon. Nanosized particles allow shorter diffusion lengths for the lithium-ions in going from the bulk to the surface, and they allow high contact area between electrode and electrolyte, both requirements for high-power capability [31]. The benefits of coating LTO with carbon to improve the conductivity of the LTO and thereby the power capability of the resulting batteries have also been demonstrated [32]. While both of the aforementioned efforts, producing nanosized LTO and coating it with carbon, have been successful in synthesizing improved LTO materials, they have relied on tedious multi-step processing techniques. The versatility of the nGimat CCVC process simplifies what has otherwise been a laborious process for achieving carbon coating of nanometer-sized LTO, thereby enabling high-power capable LTO at low cost. Figure 4.11 shows the XRD pattern for LTO nanopowder made using CCVC of a solution of lithium carboxylate and titanium alkoxide precursors. The pattern is in good agreement with the spinel phase of LTO (Li1.33Ti1.66O4) published in the literature [33]. Lower-cost lithium nitrate precursors can also be used to introduce lithium into the compound and form the spinel structure.

Average particle size of the LTO was typically maintained to less than 100 nm. Particle morphology is determined by process conditions. For example, nanopowders formed at higher temperatures tended to form faceted morphologies, whereas particles that were formed at relatively low temperatures demonstrated spherical morphologies. This trend is shown in the TEM micrographs of the LTO samples in Fig. 4.12.

By adjusting the combustion environment and conditions, we have also been able to demonstrate the synthesis of carbon-coated LTO using a single-step process. This is an important process advantage offered by NanoSpray Combustion, since most commercial processes utilize at least two separate steps to manufacture carbon-coated electrode materials. The first step(s) in a commercial process typically involves the synthesis of the electrode powder (e.g., metal oxide), while the final step is dedicated to a carbon coating. Figure 4.13 shows a picture of three nanopowders that were made under oxidizing, slightly reducing and highly reducing conditions.

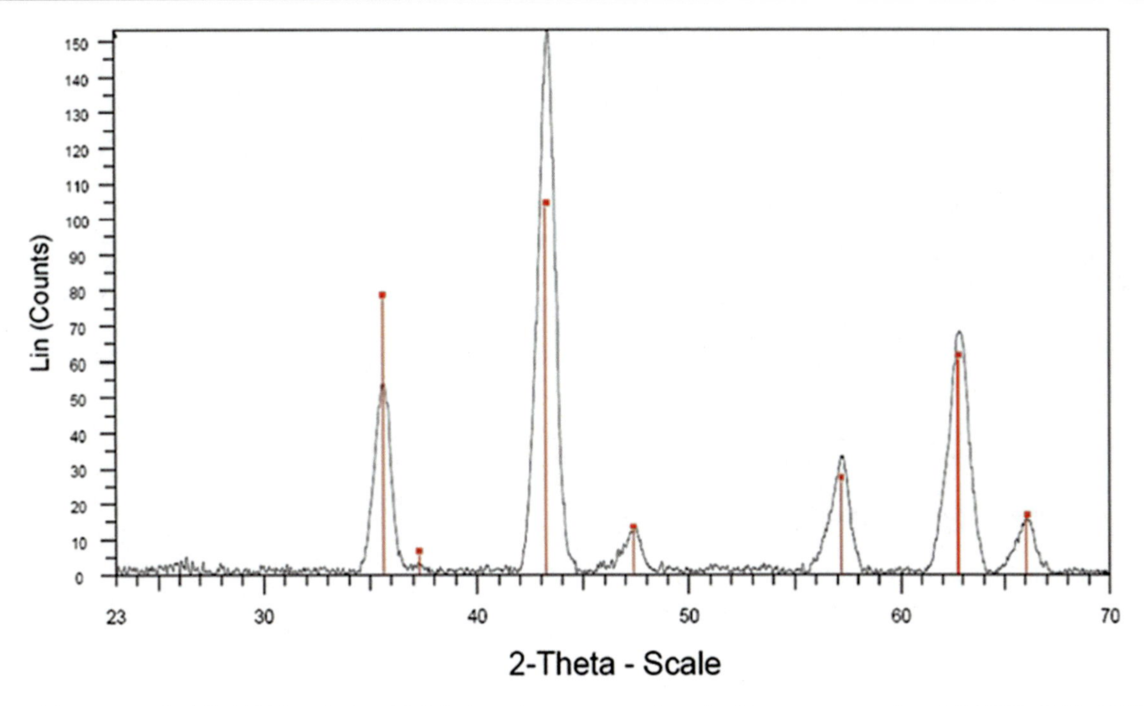

Fig. 4.11 XRD pattern for Li4Ti5O12 made by nCCVC, the reference pattern from the library, corresponds to Li1.33Ti1.66O4. (PDF number 26–1198)

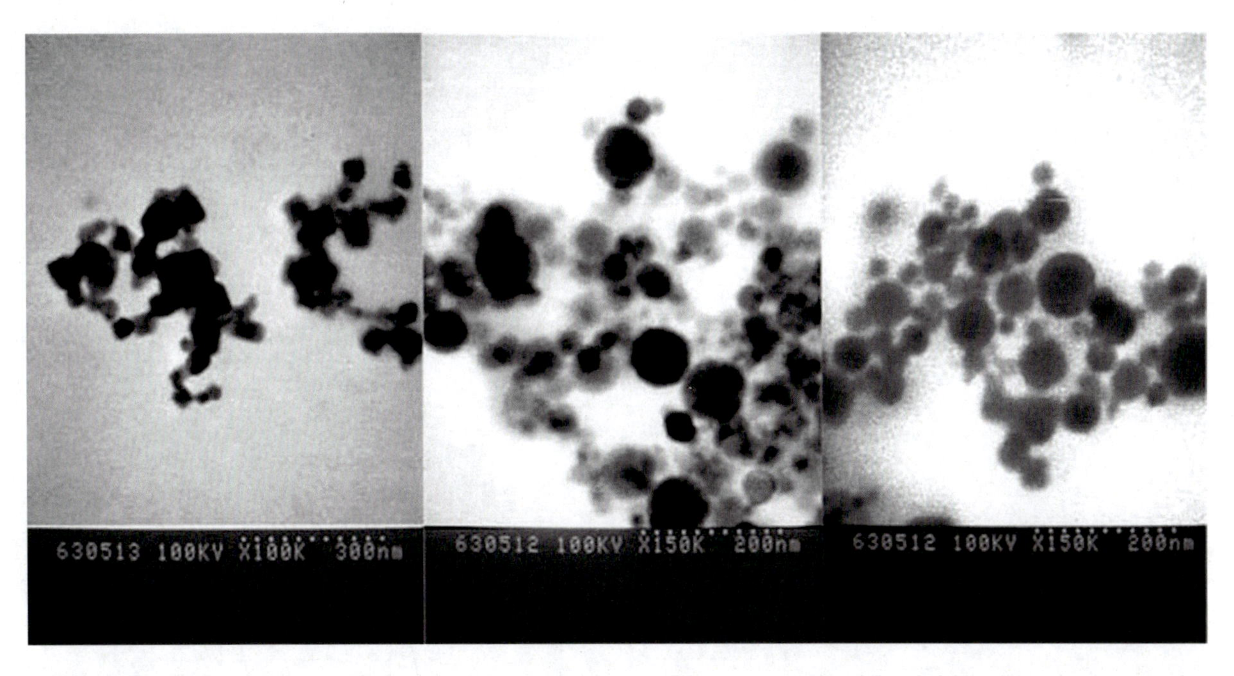

Fig. 4.12 TEM for LTO made using different processing conditions and different precursors: *left*, relatively high temperature, carboxylate precursor; *center*, relatively low temperature, carboxylate precursor; *right*, relatively low temperature, nitrate precursor

Thermogravimetric analysis (TGA) results for the carbon-coated materials indicated that the amount of carbon could be varied from 0.5 to 20%. Figure 4.14 compares the TGA scan for one typical carbon-coated LTO nanopowder sample with an LTO sample without any coating. Both scans were those of raw powder as removed from the nCCVC system prior to further treatment. The initial weight loss, up to 350 °C is expected to be due to surface adsorbed species that will need to be removed by heat treatment. This peak is visible even for uncoated LTO. The weight loss between 400 and 500 °C is attributed to the decomposition of the carbon layer that is coated on the LTO. Around 6 wt% of carbon was coated in this case.

The best capacity for prototype cells containing nCCVC LTO nanopowder was approximately 143 mAh/g. Literature values have ranged anywhere between 120 and 155 mAh/g [32, 34]. The lower capacity, compared to the best in class, is attributed to surface-induced capacity losses due to high-surface area of the as-synthesized nanopowders that were used for the battery tests. Creating secondary particles that have larger particle size but at the same time retain some of nanostructure necessary to provide high rate capability could reduce these surface-induced losses. Nanostructured electrode powders of this kind have been shown by the DOE to provide optimum power and energy for LTO-based Li-ion batteries [35].

Fig. 4.13 Picture showing three nanopowders that were made under oxidizing (*left*), slightly reducing (*center*), and highly reducing (*right*) conditions

Figure 4.15 shows the 1C–1C charge–discharge curves for the first five cycles of an LTO I Li half-cell. Even though the prototype cell is a half-cell that has not been optimized, it demonstrates relatively stable cycling behavior at 1C.

Rate capability studies were carried out at rates of 1C, 5C, and 10C. Figure 4.16 shows the discharge rate capability plot for a typical LTO I Li T cell. Normalized to a 1C capacity of 100%, the 5C capacity was at around 80%, and the 10C capacity was at around 60%. This performance is quite

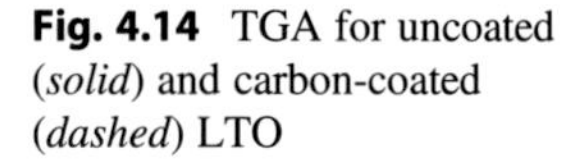

Fig. 4.14 TGA for uncoated (*solid*) and carbon-coated (*dashed*) LTO

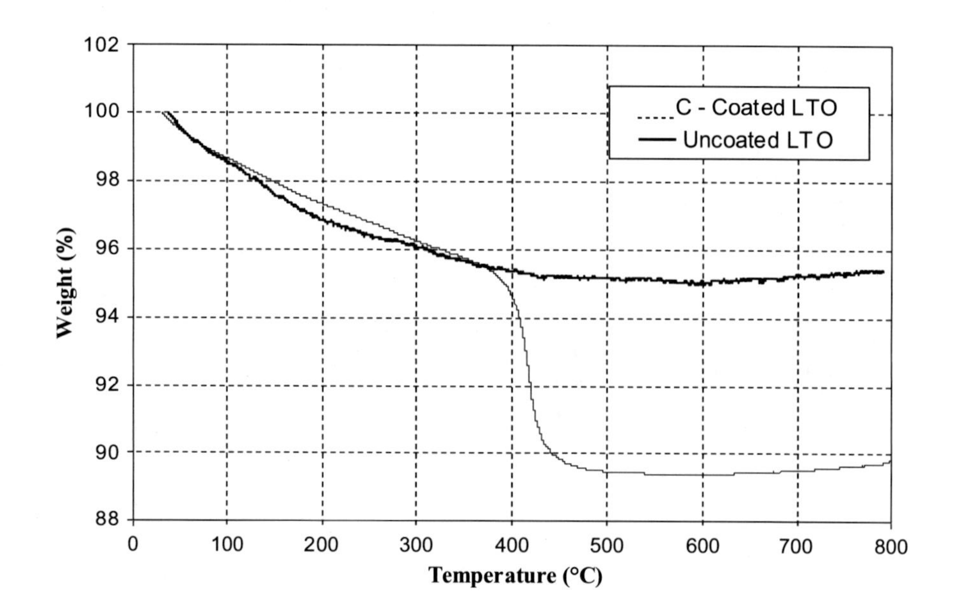

Fig. 4.15 1C–1C discharge–charge curves for an LTO I Li-metal T cells over five cycles

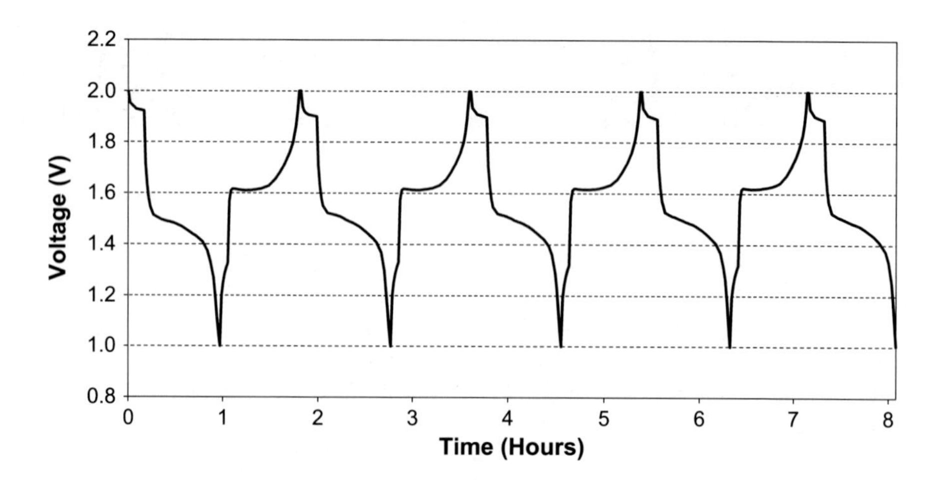

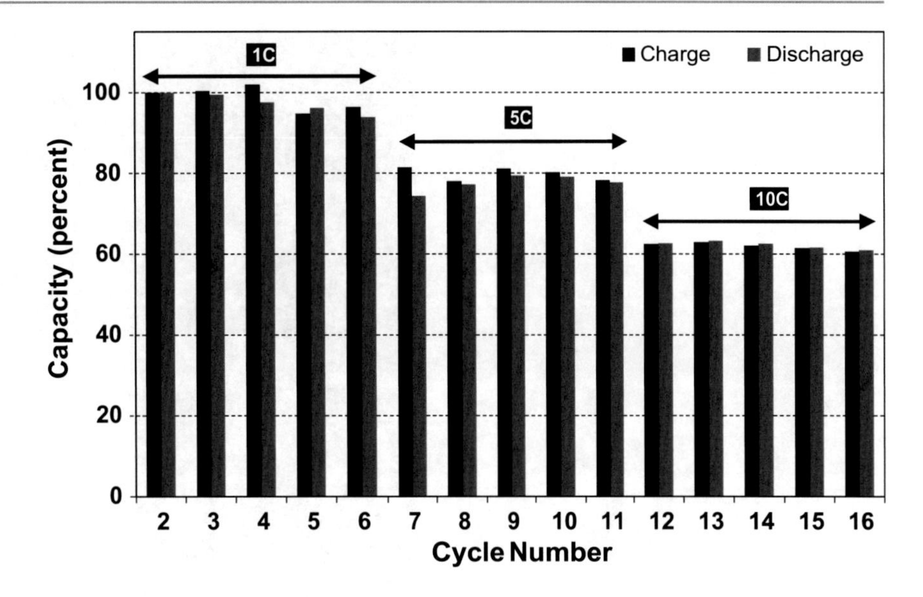

Fig. 4.16 Rate capability tests for nCCVC LTO | Li-metal T cells at 1C, 5C, and 10C discharge rates. Charge rate was 1C for all. The left bar in each cycle corresponds to the charge capacity, while the right bar in each cycle corresponds to the discharge capacity

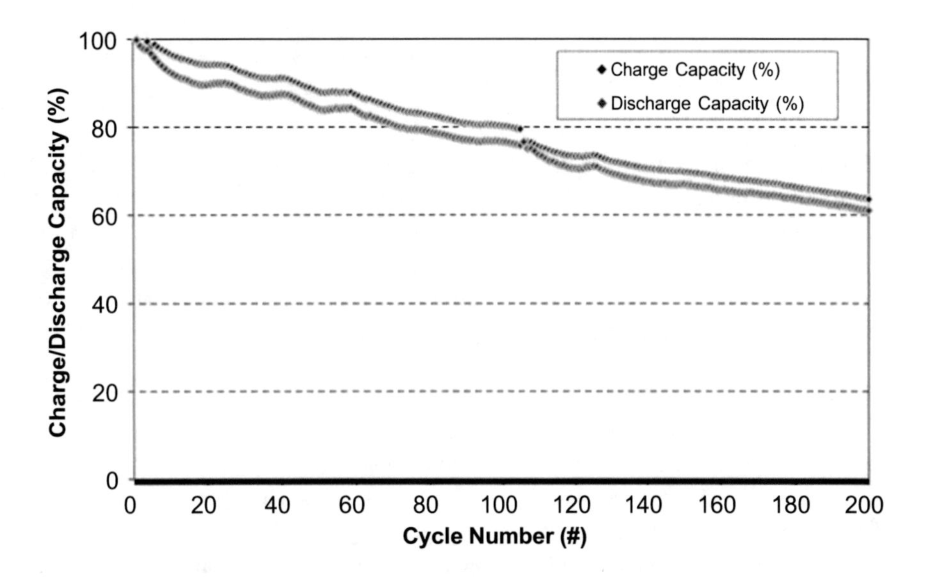

Fig. 4.17 Cycle life plot for a prototype Li-ion T cell consisting of an LTO cathode and a lithium-metal anode. The upper trace represents the charge capacity, and the lower trace represents the discharge capacity

impressive, considering that the materials were tested in an unoptimized T-cell format. Most portable electronic applications can operate in the 1C–5C power range. However, higher rate capabilities are required for emerging lithium-ion batteries applications like power tools and electric vehicles.

Some of the cells were cycled for an extended period to investigate preliminary cycle life capability. Figure 4.17 shows the chart for one such cell that was cycled to over 200 cycles at 1C charge and discharge rate. Prior to the test, this cell was cycled for five cycles in order to stabilize capacity. The sixth cycle capacity was set to 100%, and the capacities of all subsequent cycles were normalized with respect to the sixth cycle. Around 75% of the capacity was retained after 100 cycles and around 60% after 200 cycles. Manufactured cells normally perform better than laboratory cells due to better elimination of oxygen and moisture.

4.4.3 Polymer Nanocomposites for Capacitors

Development of electronic systems with reduced size, weight, cost, and improved functionality and performance will enable advances in applications such as telecommunications, network systems, automotive, and computer electronic devices. Electronic systems are composed of active components such as integrated circuits (ICs) and passive components including resistors, capacitors, and inductors. Currently more than 98% of passives used in the electronic industry are discrete passives. In a typical electronic system, discrete passives outnumber the active ICs by several times and occupy more than 70% real estate of the substrate. Discrete passives, especially capacitors, have already become the major barrier of the electronic systems miniaturization. Therefore, the development of integral passives is of much importance. Moreover, integral passives have the advantage

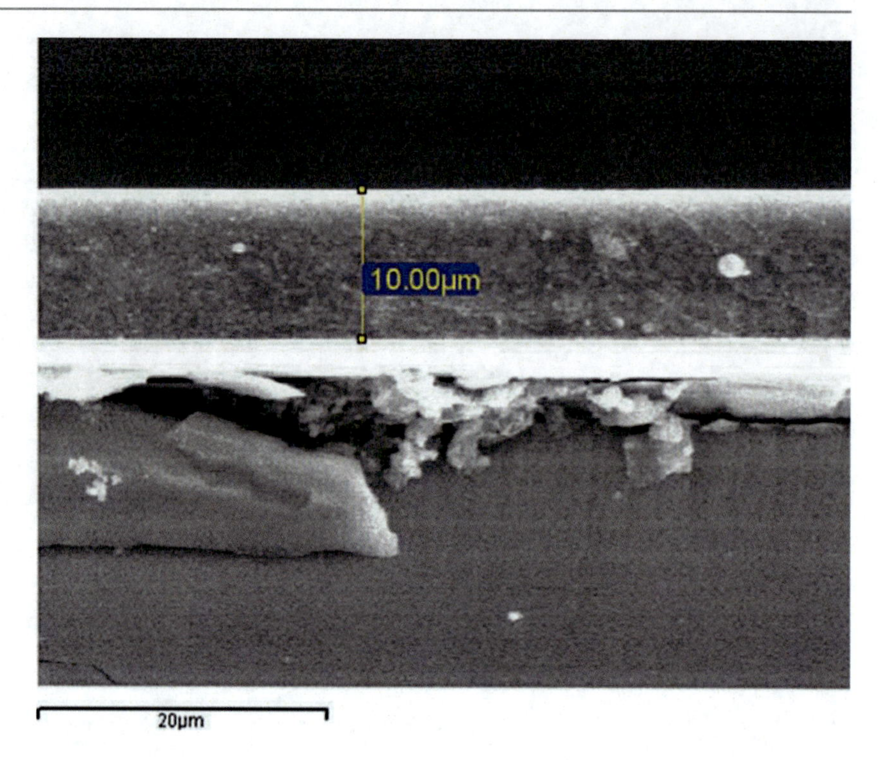

Fig. 4.18 SEM cross section of cyclotene/(Ag+BTO) nanocomposites with 5 vol.% Ag and 20 vol.% BTO

of no separate interconnects to the substrate, reduced parasitic inductance, improved electrical performance, lower cost, and easier processing. Among passive components, special interest is focused on embedded capacitors because they are used for various functions, such as decoupling, bypassing, filtering, and timing capacitors. Important requirements for embedded capacitor materials include high dielectric constant, low capacitance tolerance, good processability, and low cost [36–41]. Polymer/ceramic nanocomposites have great potential to meet these practical requirements and are under extensive investigations [42–44]. However, to obtain a dielectric constant of 100 or higher, ceramic filler is generally required to be higher than 50 vol.%, which results in limited flexibility and processability. Meanwhile, polymer-metal nanocomposites have also attracted great attention for embedded capacitor applications [45–48]. Although polymer-metal composites have shown promise both experimentally and theoretically, it is worthwhile to point out that the high dielectric constant of polymer-metal composites is usually achieved near the percolation threshold (the particle volume fraction at which particles have enough connectivity to create continuous paths through the structure) of metal particles. At these particle loadings, the composites have high leakage current and dissipation factors and low breakdown voltage and hence low performance and reliability.

In order to utilize the advantages of each materials system, nGimat has developed three-phase nanocomposites, in which polymer matrix dispersed with both ceramic and metal nanoparticles, which show promise to achieve high dielectric constant while retaining their flexibility and processability. In this study, the polymer matrix is cyclotene with a dielectric constant of about 2.6. All ceramic and metal nanoparticles were synthesized by the nCCVC technique. The nanocomposites were deposited by the NanoSpray technique onto platinized Si wafers. Capacitance and quality factor were measured by an HP 4285a precision LCR meter at 1 MHz.

As an example, Fig. 4.18 shows SEM cross-section view of a typical cyclotene/(Ag+BTO) nanocomposite film with 5 vol.% Ag and 20 vol.% of barium strontium titanate (BTO) nanoparticles. The film is dense and uniform with nanometer-sized particulate features that are likely BTO particles. It is approximately 10 μm thick.

Capacitance and quality factor were measured at 1 MHz. Dielectric constant was calculated based on capacitance, film thickness, and electrode area. Figure 4.19 shows dielectric constant and quality factor of the nanocomposite films as a function of Ag loading and BTO loading at different Ag loading levels. For cyclotene/Ag nanocomposites (no BTO), dielectric constant increases, and quality factor (Q) decreases with the increase of Ag loading, which suggests that Ag nanoparticles are partially percolated at relatively low loadings (2 vol.% or more). This is likely due to silver's propensity to diffuse at relatively low temperatures. Though the dielectric losses were large ($Q < 5$) for these partially percolated nanocomposites, impressive dielectric constants and capacitance densities of greater than 3000 and 15 μF/in^2, respectively, were measured. Alternatively, for cyclotene/BTO nanocomposites (no Ag), the dielectric constant increases slightly as BTO loading is

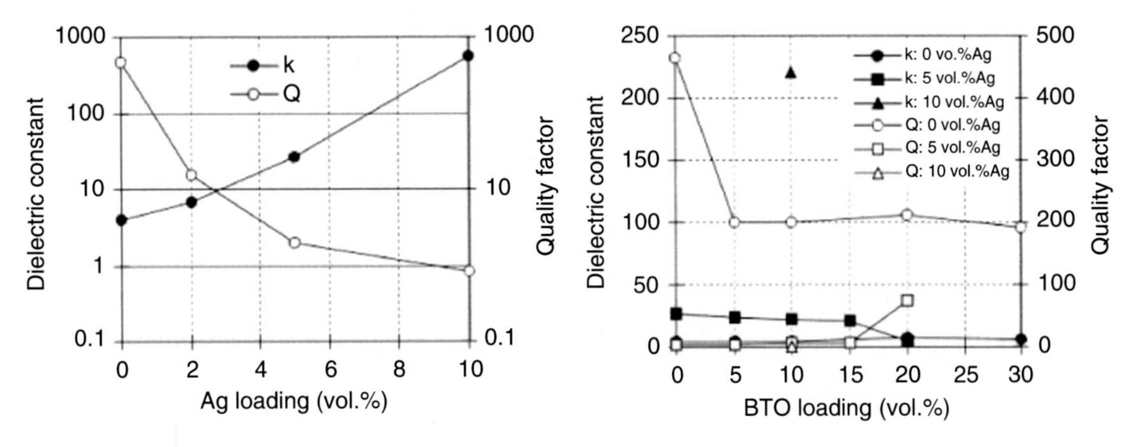

Fig. 4.19 Dielectric constant and quality factor as a function of (**a**) Ag loading and (**b**) BTO loading at different Ag loading levels

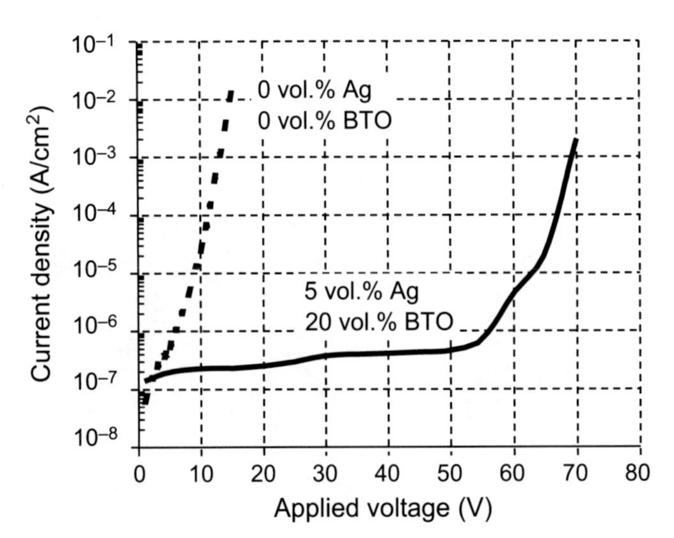

Fig. 4.20 Leakage current density as a function of applied voltage for pure cyclotene and cyclotene/(Ag+BTO)

increased. Dielectric constant increases from 4 to 7 at 20 vol.% BTO. Quality factor decreases from 465 to 200 with increasing BTO loading from 0 to 5 vol.% and then remains approximately 200 at BTO loadings higher than 5 vol.%. For the cyclotene/(Ag+BTO) nanocomposites with 5 vol.% Ag, dielectric constant decreases, and quality factor increases slightly with the increase of BTO loading. For example, the dielectric constant and quality factor are 27 and 2 at 0 vol.% BTO and 21 and 5 at 15 vol.% BTO, respectively. A sharp decrease in dielectric constant and increase in quality factor were observed at 20 vol.% BTO, showing that Ag nanoparticles are not percolated at this point.

Figure 4.20 shows leakage current density as a function of applied voltage for pure cyclotene and cyclotene/(Ag+BTO) with 5 vol.% Ag and 20 vol.% BTO. The leakage current density of the pure cyclotene film increases rapidly with the increase of applied voltage. The leakage current density of the cyclotene/(Ag+BTO) film increases slightly with increasing applied voltage below 55 V. Above 55 V, it increases rapidly. At 5 V, leakage current density of the pure cyclotene and cyclotene/(Ag+BTO) with 5 vol.% Ag and 20 vol.% BTO is 5.3×10^{-7} and 2.0×10^{-7} A/cm^2, respectively. The breakdown voltage of the nanocomposite film is about 55 V, while that of pure cyclotene film is about 5 V, corresponding to breakdown strengths of 5.5 and 6.5 MV/m. Please note that the cyclotene/(Ag+BTO) nanocomposite is approximately ten times thicker than the pure cyclotene film.

In addition, nanocomposites of cyclotene embedded with Cu and BTO, cyclotene embedded with Cu and calcium copper titanate (CCT), and cyclotene embedded with Pt and BTO have also been developed. The nanocomposites were also scaled up onto large size copper foils for practical applications.

4.4.4 Inorganic Nanocomposites for Nonlinear Optical Materials

Nonlinear optical (NLO) effects play a major role in the development of high-speed data processing and communication systems. Materials with third-order optical nonlinearity ($\chi^{(3)}$) and fast response time are essential for light-controlled phase and refractive index modulation, which can replace electronic devices for next-generation applications in switching, routing, and signal processing. Future opportunities in photonic switching and information processing will depend critically on the development of improved photonic materials with enhanced third-order nonlinearity and reduced absorption coefficient (α).

It has been demonstrated that semiconductor or metal-dispersed dielectric thin films exhibited high third-order nonlinear properties which could be of great importance in future optical data processing. To obtain a high $\chi^{(3)}$ NLO material, a larger volume fraction of metallic particles with controllable shape and size in the dielectric matrix is desired [49]. Currently, techniques used for producing the nanocluster (NC) embedded composite materials include sol–gel, sputtering, ion implantation, and pulsed laser deposition (PLD). Most of these techniques are multistep processes, in which post-heat treatments are often needed to attain the desired properties. However, these treatments can also alter the average size, size dispersion, and spatial arrangement of the NCs, which could have unfavorable effects on the nonlinear properties. Furthermore, the vacuum-based methodologies are expensive, capital intensive, batch-process limited, and suffer from low throughput. nGimat has successfully synthesized Pt nanoclusters embedded BaTiO3 thin films, which exhibited high $\chi^{(3)}$ ($\sim$4.8 $\times$ 10^{-10} esu) with a low α ($\sim 10^4$ cm^{-1}). This is the first time, to the best of our knowledge, that Pt was used as a metal dopant to improve NLO properties.

Since there is some mobility of Pt at high deposition temperatures, we deposited a Pt seed layer and then coated with BT using the CCVD at 600–1000 °C. It was predicted that the mobility of the Pt would enable it to diffuse into the dielectric matrix to form a nanocomposite, which it did.

Figure 4.21 shows the area detector XRD spectra of Pt seed layer and BT-coated Pt. The Pt seed layer is polycrystalline (slightly preferred). After BT deposition, the Pt diffraction peaks become sharper, which is contributed to possible Pt particle growth. This was found to be the case for Au seed layer as well. We theorize that the migrated Pt crystallites remain on the nanoscale after BT deposition (Fig. 4.21c).

Figure 4.22 shows the optical transmission spectra of Pt and BT/Pt. It is apparent that after BT deposition, the transmission increases, suggesting some Pt has migrated into small nanoparticles within the BT matrix. These Pt nanosized particles do not perturb light with wavelengths greater than 300 nm and therefore do not contribute to light scattering and absorption, resulting in higher transmission.

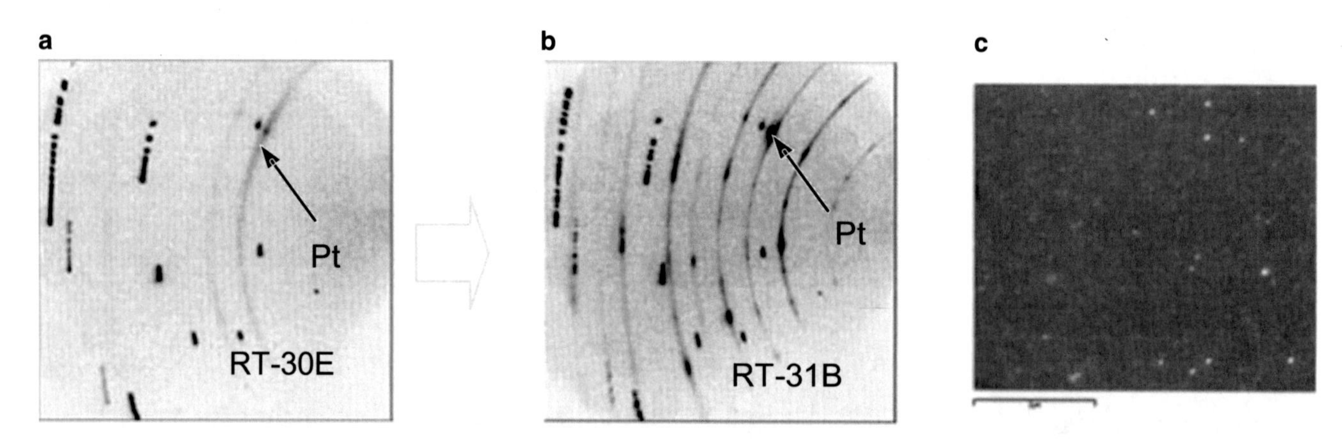

Fig. 4.21 XRD spectra of (**a**) Pt, (**b**) BT-coated Pt, and (**c**) plain-view SEM micrograph of Pt seed layer

Fig. 4.22 Optical transmission spectra of Pt (RT-30C and RT-25C) and BT/Pt (RT-31A and RT-31D) layers. *Arrows* indicate the spectra of the same sample after BT deposition. In order of increasing transmission at 800 nm, the curves are RT-30C, RT-25C, RT-31A, and RT31-D

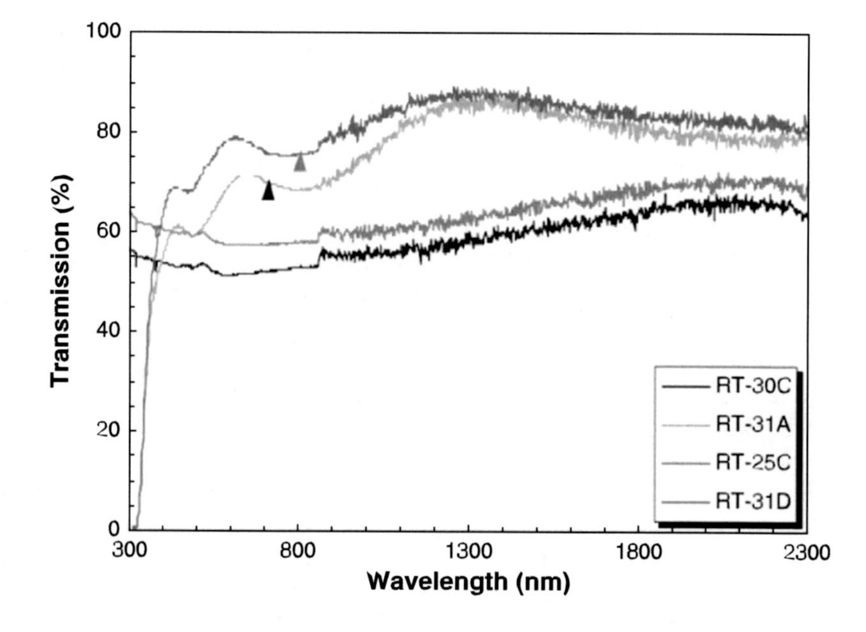

The z-scan technique [50] has been used to determine the sign and magnitude of nonlinear refractive index ($n2$) of the nanocomposite films. The optical source uses a ~200-fs Ti-sapphire laser which permits peak power densities of ~1 GW/cm^2 requiring 50 mW average powers. The following parameters were given quantitatively or qualitatively through the z-scan measurement: the relative strength of the nonlinear index ($n2$), the apparent two-photon absorption (TPA), which is an increased absorption with increasing intensity.

A pure $n2$ response is indicated by a bipolar response. For example, a positive $n2$ is indicated by a decrease in signal followed by an increase in signal strength. The signature is thus antisymmetric around the z position corresponding to having the focus at the sample. Two-photon absorption is indicated by a decrease in signal strength and is symmetric about the focus location. Bleaching is indicated by an increase in signal strength and is also symmetric about the focus location.

As a baseline, z-scan of a plain c-sapphire is shown in Fig. 4.23. Displayed characteristic nonlinear refraction coefficient of sapphire was on the order of ~3 × 10^{-16} cm^{2}w^{-1}. Numerical estimation of z-scan response based upon measured experimental parameters corresponds very well to measured response. The z-scan peak-to-peak amplitude was $\Delta I/I$~0.1%. z-Scan shows negligible nonlinear absorption.

Typical results of the z-scan measurement of BT and BT–Pt nanocomposite films are shown in Fig. 4.24 a (BT) and b

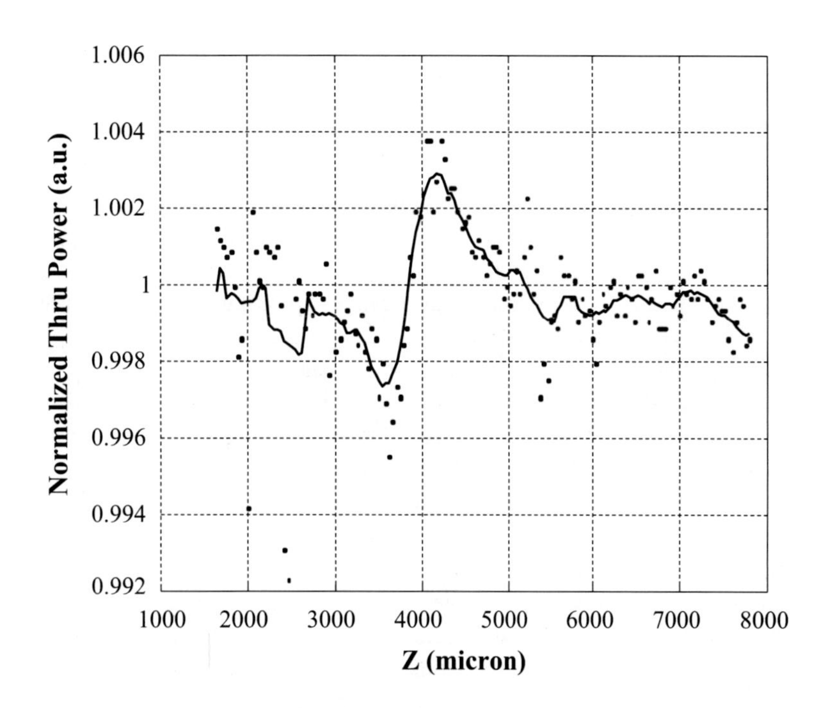

Fig. 4.23 z-Scan result of c-sapphire substrate

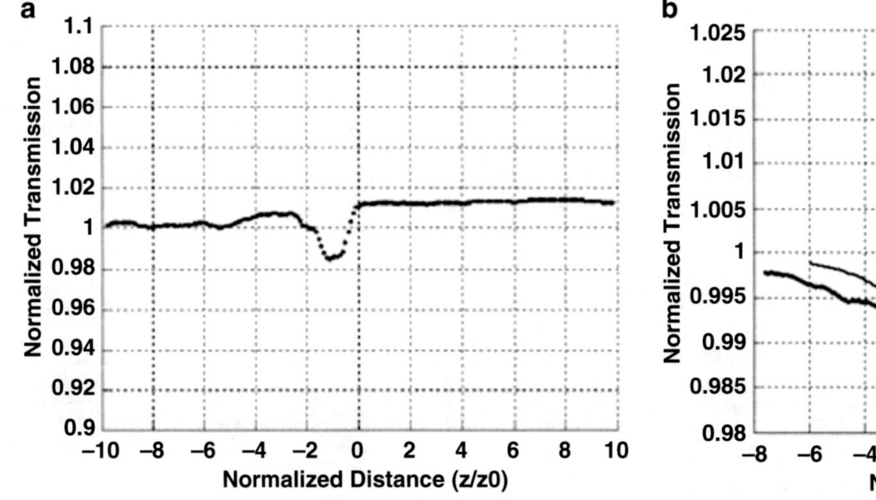

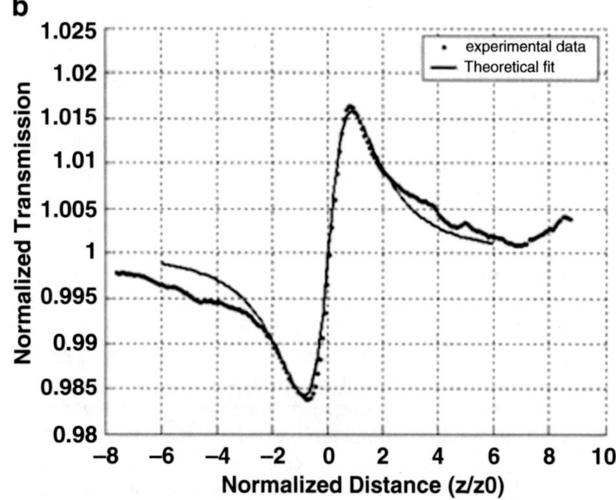

Fig. 4.24 z-Scan results of (**a**) BT film and (**b**) BT on Pt/sapphire

(BT on Pt/sapphire with a thickness of 250 nm). The dots correspond to the far-field transmission data as a function of its distance (z) to the lens focus; the solid curve of Fig. 4.24b corresponds to the theoretical fit. The curve of Fig. 4.24b exhibits a valley to peak of ~3.5% and indicates a positive value of the nonlinear refractive index, $n2$. The sapphire substrate (0.47 mm thick) has a very small nonlinear optical response at 810 nm as shown above; thus the high nonlinear optical properties observed here result from the BaTiO3/Pt films. Using the z-scan analysis, the nonlinear refractive index of BaTiO3 on Pt layer is determined to be $\sim 4.8 \times 10^{-10}$ esu.

4.5 Conclusions

The versatility of NanoSpray Combustion technology is made evident by the variety of applications that are served by the nanomaterials made using both nCCVC and nCCVD. Nanopowders, nanocomposites, nanodispersions, and thin films of metals, metal oxides, mixed metal oxides, and metal phosphates have applications in a number of areas, including energy, electronics, biomedical, and printing. The ability to tailor the composition, particle size, particle morphology, and surface characteristics of these nanomaterials in a cost-effective process is critical for achieving superior applications' performance and therefore successful commercialization. NanoSpray CCVC and CCVD have demonstrated this ability to provide tailored materials and are poised to be the process of choice for the mass production of next-generation nanomaterials. As new applications for nanomaterials continue to emerge, there is an urgent need for a scalable and low-cost process to manufacture these materials in large volumes so that these technologies can be commercialized. The CCVC and CCVD processes offer the scaling efficiencies demanded by these applications and do so for a broad range of chemistries and compositions. Compared to other nanomaterials manufacturing schemes, like those involving plasma, laser, and vacuum processes, it can also claim to be significantly lower in power consumption characteristics. CCVC and CCVD can also be easily adapted to use biosolvents and chemicals derived from biomass sources, further reducing the carbon footprint of the process.

Acknowledgments We thank all of the nGimat team members, past and present, who have contributed to the work discussed in this manuscript. In particular, we are grateful for the help provided by Dr. Elena Krumenaker and Mr. Michael Sapp. We would also like to thank Professor Steve Ralph at the Georgia Institute of Technology for performing z-scan measurements and Professor C.P. Wong for his many discussions and collaborations. Last but not least, we are grateful for the funding provided by the US Department of Energy, Department of Defense, and the National Science Foundation that helped defray the cost of carrying out the research and development efforts described in this chapter.

References

1. ASTM Standards Document E 2456-06 (2006)
2. Buzea, C., Pacheco, I.I., Robbie, K.: Nanomaterials and nanoparticles, sources and toxicity. Biointerphases. **2**(4), M17–M71 (2007)
3. Rempel, A.A.: Nanotechnologies. Properties and applications of nanostructured materials. Russ. Chem. Rev. **76**(5), 435–461 (2007)
4. Brechignac, C., Houdy, P., Lahmani, M. (eds.): Nanomaterials and Nanochemistry. Springer, Berlin (2008)
5. Vollath, D.: Plasma synthesis of nanopowders. J. Nanopart. Res. **10**, 39–57 (2008)
6. Swihart, M.T.: Vapor-phase synthesis of nanoparticles. Curr. Opin. Colloid Interf. Sci. **8**, 127–133 (2003)
7. Carbon Black Users Guide.: International Carbon Black Association. http://www.carbon-black.org/carbonblackuserguide.pdf (2004). Accessed 4 May 2009
8. Aerosil: Fumed Silica for Batteries.: Technical Bulletin 2125. www.aerosil.com (2007). Accessed 4 May 2009
9. Aeroxide and Aeroperl: Titanium Dioxide as a Photocatalyst, Technical Bulletin 1243. www.aerosil.com (2005). Accessed 4 May 2009
10. Lin, C., Chung, D.D.L.: Nanostructured fumed metal oxides for thermal interface pastes. J. Mater. Sci. **42**, 9245–9255 (2007)
11. Pitkethly, M.J.: Nanomaterials – the driving force. NanoToday. **7** (December issue), 20–28 (2004)
12. Rolland, J.P., Hagberg, E.C., Denison, G.M., Carter, K.R., De Simone, J.M.: High-resolution soft lithography: enabling materials for nanotechnologies. Angew. Chem. (43), 5796–5799 (2004)
13. Hunt, A.T., et al.: US Patents: 5652021, 5858465, 5863604 and 6013318
14. Hunt, A.T., et al.: Combustion chemical vapor deposition: A novel thin‐film deposition technique. Appl. Phys. Lett. **63**(2), 266 (1993)
15. Gratzel, M.: Photoelectrochemical cells. Nature. **414**, 338–344 (2001)
16. Hui, S., Roller, J., Sin, Y., Zhang, X., Deces-Petit, C., Xie, Y., Maric, R., Ghosh, D.: J. A brief review of the ionic conductivity enhancement for selected oxide electrolytes. Power Sour. **172**, 493–502 (2007)
17. Cheekatamarla, P.K., Finnerty, C.M.: J. Reforming catalysts for hydrogen generation in fuel cell applications. Power Sour. **160**, 490–599 (2006)
18. Fu, Q., Zhou, N., Huang, W., Wang, D.: Preparation and characterization of a novel bioactive bone cement: glass based nanoscale hydroxyapatite bone cement. J. Mater. Sci. Mater. Med. **15**, 133–1338 (2004)
19. Atkinson, A., et al.: Large-scale preparation of chromatographic grade hydroxylapatite and its application in protein separating procedures. J. Appl. Chem. Biotechnol. **23**, 517–523 (1973)
20. Schnieder, O.D.: Cotton wool-like nanocomposite biomaterials prepared by electrospinning: In vitro bioactivity and osteogenic differentiation of human mesenchymal stem cells. J. Biomed. Mater. Res. B Appl. Biomater. **84B**, 350–362 (2008)
21. Iliescu, M., et al.: Morphological and structural characterisation of osseointegrable Mn^{2+} and $CO3^{2-}$ doped hydroxylapatite thin films. Mater. Sci. Eng. **27**(1), 105–109 (2007)
22. Chung, S.-Y., Bloking, J.T., Chiang, Y.M.: Electronically conductive phospho-olivines as lithium storage electrodes. Nat. Mater. **1**, 123–128 (2002)
23. Ferrari, M.: BioMEMS and Biomedical Nanotechnology. Springer, New York (2006)
24. Wang, X.D., et al.: Nanomaterials and nanopackaging. In: Lu, D., Wong, C.P. (eds.) Materials for Advanced Packaging Reference, pp. 503–545. Springer, New York (2009)

25. Lu, D., Wong, C.P.: Electrically conductive adhesives. In: Lu, D., Wong, C.P. (eds.) Materials for Advanced Packaging Reference, pp. 365–405. Springer, New York (2009)
26. Prasher, R., Chiu, C.-P.: Thermal interface materials. In: Lu, D., Wong, C.P. (eds.) Materials for Advanced Packaging Reference, pp. 437–458. Springer, New York (2009)
27. Zhang, R., et al.: Georgia Institute of Technology, Manuscript in preparation
28. Lide, D.R. (ed.): Handbook of Chemistry and Physics, 87th edn. CRC Press, Boca Raton (2007)
29. Nazri, G.A., Pistoia, G. (eds.): Lithium Batteries: Science and Technology. Kluwer Academic Publishers, Boston/ Dordrecht/New York/London (2004)
30. Belharouak, I., et al.: On the safety of the Li4Ti5O12/ LiMn2O4 lithium-ion battery system. J. Electrochem. Soc. **154**, A1083–A1087 (2007)
31. Manthiram, A., et al.: Nanostructured electrode materials for electrochemical energy storage and conversion. Energy Environ. Sci. (2008). https://doi.org/10.1039/b811802g32
32. Amine, K., et al.: US Patent 2007/014845A1 (2007)
33. X-ray Diffraction Particle Data File # 26–1198
34. Thackeray, M., et al.: US Patent 7,452,630 B2 (2008)
35. Energy Storage R&D Annual Report, DOE Vehicle Technologies Program, (2007–2008)
36. Rao, Y., Wong, C.P.: Material characterization of a high�dielectric� constant polymer–ceramic composite for embedded capacitor for RF applications. J. Appl. Polym. Sci. **92**, 2228 (2004)
37. Rao, Y., Ogitani, S., Kohl, P., Wong, C.P.: Novel polymer–ceramic nanocomposite based on high dielectric constant epoxy formula for embedded capacitor application. J. Appl. Polym. Sci. **83**, 1084 (2002)
38. Kuo, D.H., Chang, C.C., Su, T.Y., Wang, W.K., Lin, B.Y.: Dielectric properties of three ceramic/epoxy composites. Mater. Chem. Phys. **85**, 201 (2004)
39. Bhattacharya, S.K., Tummala, R.R.: Integral passives for next generation of electronic packaging: application of epoxy/ceramic nanocomposites as integral capacitors. Microelectron. J. **32**, 11 (2001)
40. Windlass, H., Markondeya Raj, P., Balaraman, D., Bhattacharya, S. K., Tummala, R.R.: Colloidal processing of polymer ceramic nanocomposite integral capacitors. IEEE Trans. Adv. Packag. **26**, 10 (2003)
41. Pothukuchi, S., Li, Y., Wong, C.P.: Development of a novel polymer–metal nanocomposite obtained through the route of in situ reduction for integral capacitor application. J. Appl. Polym. Sci. **93**, 1531 (2004)
42. Dang, Z.M., Wang, L., Wang, H.Y., Nan, C.W., Xie, D., Yin, Y., Tjong, S.C.: Giant dielectric permittivities in functionalized carbon-nanotube/electroactive�polymer nanocomposites. Appl. Phys. Lett. **86**, 172905 (2005)
43. Kakimoto, M., Takashi, A., Tsurumi, T., Hao, J.J., Li, L., Kikuchi, R., Miwa, T., Oono, T., Yamada, S.: Polymer-ceramic nanocomposites based on new concepts for embedded capacitor. Mater. Sci. Eng. B. **132**, 74 (2006)
44. Wu, C.C., Chen, Y.C., Su, C.C., Yang, C.F.: The chemical and dielectric properties of epoxy/(Ba0.8Sr0.2)(Ti0.9Zr0.1) O3 composites for embedded capacitor application. Eur. Polym. J. **45**, 1442 (2009)
45. Li, Y., Pothukuchi, S., Wong, C.P.: Formation and dielectric properties of a novel polymer-metal nanocomposite [for embedded capacitor application] Proc. 9th Symp. Adv. Pack. Mat. p175, (2004)
46. Sakzm, D., Warjm, N., Baalmann, A., Simon, U., Jaeger, N.: Metal clusters in plasma polymer matrices. Phys. Chem. Chem. Phys. **4**, 2438 (2002)
47. Kelly, J.M., Stenoien, J.O., Isbell, D.E.: Wave-guide measurements in the microwave region on metal powders suspended in paraffin wax. J. Appl. Phys. **24**, 258 (1953)
48. J. Xu, K. Moon, C. Tison, C. P. Wong.: A Novel Aluminum-filled Composite Dielectric for Embedded Passive Applications. IEEE Trans. Adv. Packag. **29**, 2, 295–306 (2006)
49. Fukumi, K., Chayahara, A., Kadono, K., Sakaguchi, T., Horino, Y., Miya, M., Fujii, K., Hayakawa, J., Satou, M.: Gold nanoparticles ion-implanted in glass with enhanced nonlinear-optical properties. J. Appl. Phys. **75**, 3075 (1994)
50. Sheik-Bahae, M., Said, A.A., Wei, T.H., Hagan, D.J., Van Stryland, E.W.: Sensitive measurement of optical nonlinearities using a single beam. IEEE J. Quantum Electron. **26**, 760 (1990)

5 Nanolead-Free Solder Pastes for Low Processing Temperature Interconnect Applications in Microelectronic Packaging

Hongjin Jiang, Kyoung-sik (Jack) Moon, and C. P. (Ching-Ping) Wong

Solder has long played an important role in the assembly and interconnection of integrated circuit (IC) components on substrates, i.e., ceramic or organic printed circuit boards. This type of alloy can provide electrical, thermal, and mechanical connections in electronic assemblies. Although the electronics industry has made considerable advances over the past few decades, the essential requirements of communications among all types of components in all electronic systems remain unchanged. Components need to be electrically connected for power and ground as signal transmissions. Tin/lead (SnPb) solder alloy has been the de facto interconnect material in most areas of electronic packaging, including interconnection technologies such as pin-through hole (PTH), surface-mount technology (SMT), ball grid array (BGA) (Fig. 5.1), chip scale packaging (CSP), and flip chip.

There are increasing concerns nowadays about the use of tin/lead alloy solders. Lead, a major component in solder, has long been recognized as a health threat to human beings. The major concern is that lead from discarded consumer electronics products in landfills could leach into underground water and eventually into drinking water system [1]. According to the latest report (US Geological Survey, Jan 2009), the total lead consumption by the US industries was 1,620,000 tons in 2008, a significant amount of which was used to produce alloy solders. Worst of all, most of the electronic products have a very short service life (e.g., cell phones, pagers, electronic toys, PDA), which often end up in landfills in just a few months or years. Recycling of lead-containing consumer electronic products has been proven to be very difficult, compared to that of lead-containing batteries or cathode-ray tube (CRT) displays.

The attempt to ban lead from electronic solder was initiated in the US Congress in 1990; however, the lead-free movement has advanced much more rapidly in Japan and Europe. The leading Japanese original equipment manufacturers (OEMs) have already introduced products that contain no lead in the interconnection. Although the WEEE (Waste Electrical and Electronic Equipment) Directive has put the European Union's lead-free legislation back to 2006, many companies such as Ericsson, Nokia, and Philips have already implemented "green" marketing strategies into their new products. These differences in lead-free progress have triggered great concerns about maintaining business opportunity among users of lead-containing solders, as such; it further expedites the advancement of lead-free soldering programs. Hence, lead-free electronics are not only perceived as a health issue, but their development is also driven by governmental and commercial interests. Many IC chip and board manufacturers including Intel, the largest microprocessor manufacturer, claim that their lead-containing products would be phased out around 2006.

A variety of lead-free solder alloys have been investigated as potential replacements for tin/lead solders. The main requirement for a lead-free solder alloy are

1. Low melting point: The melting point should be low enough to avoid thermal damage to the packages and high enough for the solder joint to bear the operating temperature. The solder should retain adequate mechanical properties at these temperatures:
2. Wettability: The solder materials should wet the base metal properly to provide the electrical, thermal, and mechanical connections between chips and substrates.
3. Availability and cost: There should be adequate supplies and at the same time the cost should be reasonable. The microelectronics industry is extremely cost conscious. The electronic manufactures are unlikely to change to an alternative solder material with an increased cost unless it has

H. Jiang (✉)
Intel Corporation, Chandler, AZ, USA

K. Moon · C. P. Wong
School of Materials Science and Engineering, Georgia Institute of Technology, Atlanta, GA, USA
e-mail: jack.moon@gatech.edu; cp.wong@mse.gatech.edu

C. P. Wong et al. (eds.), *Nano-Bio- Electronic, Photonic and MEMS Packaging*, https://doi.org/10.1007/978-3-030-49991-4_5

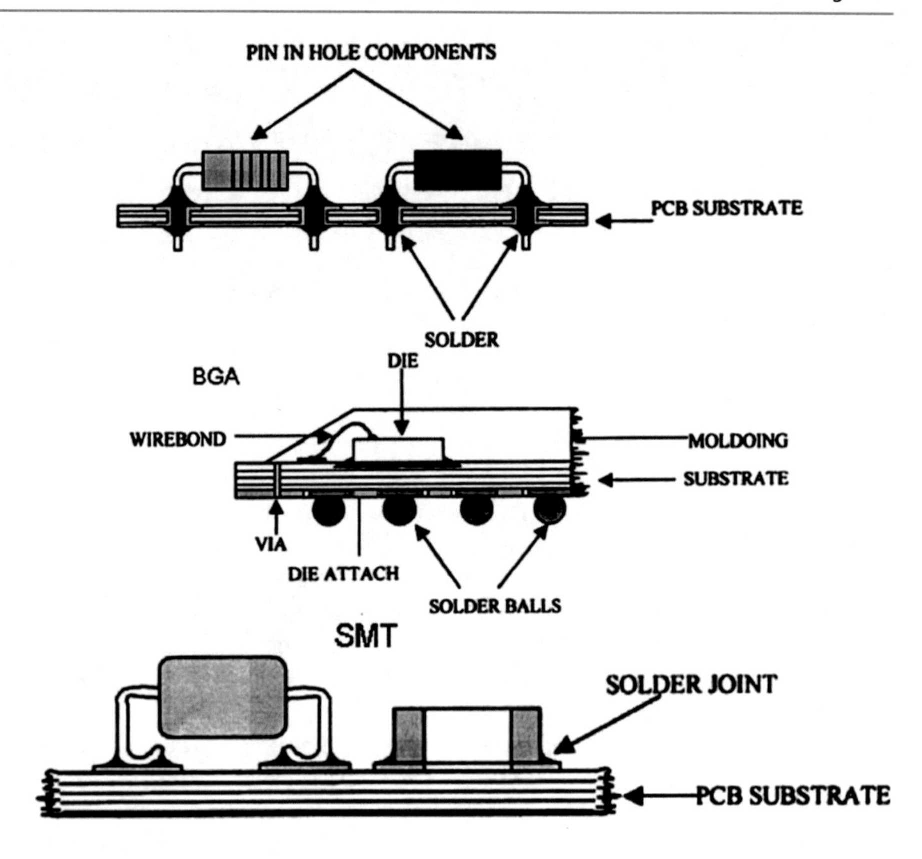

Fig. 5.1 Schematic structures of pin-through hole (PTH), ball grid array (BGA), and surface-mount technology (SMT) packages using solder interconnects

demonstrated better properties or there is legislative pressure to do so.

Although several commercial and experimental Sn-based lead-free solder alloys exit, none can meet all the above requirements, especially if the melting point of the candidate alloys is required to be very close to that of the SnPb. Lead-free candidates and their respective melting points are listed in Table 5.1. Two alloy families, tin/silver/copper (SnAgCu) and tin/copper (SnCu), seem to be generating the most interest. SnAgCu alloy composition (with or without the addition of a fourth element) appears to be the most popular replacement and has been chosen to be the benchmark, with SnPb being the baseline, that all other potential alloys for the industry to be tested against. Concerns with this alloy family include higher processing temperatures, poorer wettability due to their higher surface tension, and their compatibility with lead-bearing finishes. The SnCu alloy composition is a low-cost alternative for wave soldering and is compatible with most lead-bearing finishes. Process considerations must be addressed with this alloy due to its higher melting temperature than most SnAgCu alloys.

As shown in Table 5.1, the melting points of the SnAgCu and SnCu alloys are more than 30–40 °C higher than that of the eutectic SnPb alloy requiring the reflow.

The temperature for soldering components on assembly boards should be raised by 30–40 °C. The higher reflow temperature leads to a number of undesirable consequences such as higher residual stress of components and substrates, which adversely affects their reliability. It is also considered likely to have an increased tendency of the "pop-corning" found in the plastic-molded/plastic-encapsulated ICs during the reflow process. The high required temperature can potentially create serious warpage in organic board substrates. Furthermore, heat-sensitive components such as electrolytic capacitors might not survive the high process temperatures of lead-free assembly. Solders tend to be re-oxidized at high reflow temperatures, and the wettability becomes worsened

Table 5.1 Lead-free alloys

Alloy	Melting point
Sn96.5Ag3.5	221 °C
Sn96Ag3.5Cu0.5	217 °C
Sn20Au80	280 °C (mainly used in interconnects for optoelectronic packaging)
Sn99.3/Cu0.7	227 °C
SnAgCuX(Sb, in)	Ranging according to compositions, usually above 210 °C
SnAgBi	Ranging according to compositions, usually above 200 °C
Sn95Sb5	232–240 °C
Sn91Zn9	199 °C
SnZnAgAlGa	189 °C
Sn42Bi58	138 °C

unless expensive nitrogen reflow is used. Therefore, industry's attentions have been paid on lowering the processing temperature of the lead-free metals.

The melting point of many materials can be dramatically reduced by decreasing the size of the materials. The melting and freezing behaviors of finite systems have been of considerable theoretical and experimental interests for many years. As early as 1888, J.J. Thomson suggested that the freezing temperature of a finite particle depends on the physical and chemical properties of the surface. It was not until 1909, however, that an explicit expression for a size-dependent solid–liquid coexistence temperature first appeared. By considering a system consisting of small solid and liquid spheres of equal mass in equilibrium with their common vapor, it was shown that the temperature of the triple point was inversely proportion to the particle size. A similar conclusion was later reached based on the conditions for equilibrium between a solid spherical core and a thin surrounding liquid shell. Systematic experimental studies of the melting and freezing behavior of small particles began to appear in the late 1940s and the early 1950s: first in a series of experiments on the freezing behavior of isolated micrometer-sized metallic droplets and second in an electron diffraction study of the melting and freezing temperatures of vapor-deposited discontinuous films consisting of nano-sized islands of Pb, Sn, and Bi. These studies demonstrated that small molten particles could often be dramatically undercooled and that solid particles melted significantly below their bulk melting temperature. To date, the melting point of substances can be dramatically decreased when their size is reduced to nanometer size [2–12]. This is due to the high surface area-to-volume ratio for nanoparticles, which as a consequence of the higher surface energy substantially affects the interior "bulk" properties of the material, resulting in a decrease of the melting point. The outer surface of the nanoparticles plays a relevant role in the melting point depression. And the surface premelting process has been suggested as one of the sources of the melting point depression of the nanoparticles [13].

Transmission electron microscopy (TEM) and nanocalorimeter have been used to study the melting behavior of a single Sn nanoparticle or cluster. At the melting point, the diffraction pattern of the crystal structure in TEM exhibited the order–disorder transition [14]. However, unlike calorimeters, TEM cannot measure the latent heat of fusion (*f..H*m), which is very useful to understand the thermodynamics of a finite material system. Lai et al. investigated the melting process of supported Sn clusters by nanocalorimeter [3] and found that the melting point depended nonlinearly on the inverse of the cluster radius R, which was in contrast to the traditional description of the melting behavior of small

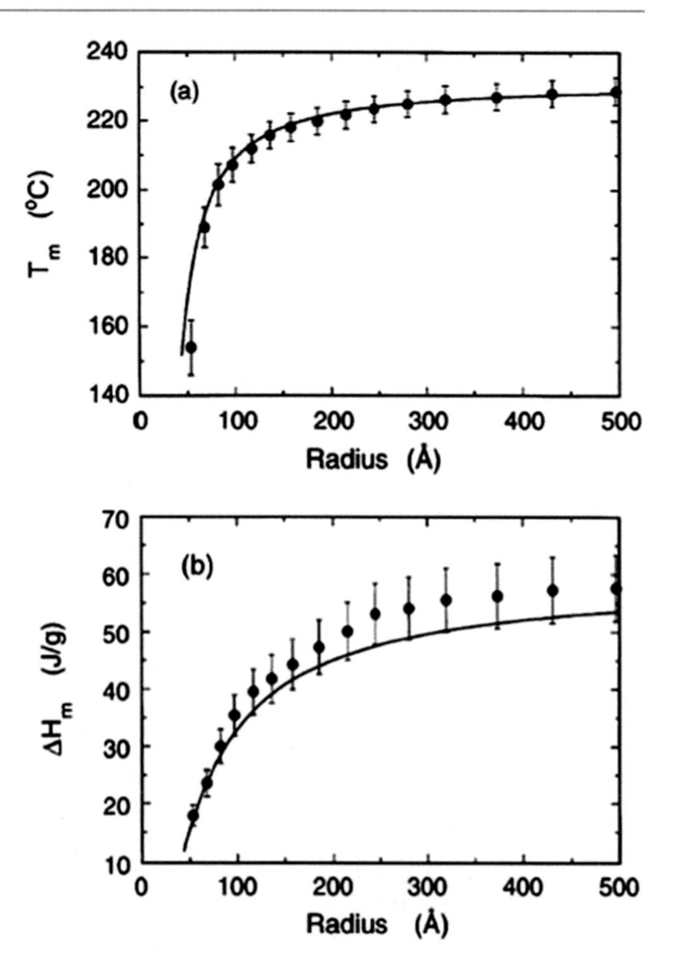

Fig. 5.2 (**a**) Size dependence of the melting points of Sn nanoparticles. The *solid line* is calculated in terms of Eq. (5.1). (**b**) Size dependence of the normalized heat of fusion. The *solid line* is calculated in terms of Eq. (5.2)

particles (Fig. 5.2a). They first reported a particle size-dependent reduction of *f..H*m for Sn nanoparticles (Fig. 5.2b). Bachels et al. studied the melting behavior of isolated Sn nanoparticles or clusters by nanocalorimeter as well [4]. The melting point of the investigated Sn clusters was found to be lowered by 125 K, and the latent heat of fusion per atom was reduced by 35% compared to bulk Sn. However, for both the TEM and nanocalorimetric measurements, the single element of Sn nanoparticle or the cluster was synthesized by the deposition method inside the measurement equipments in order to prevent the particles from oxidation.

Although this synthesis method was sufficiently sophisticated for the pure observation, the real-world synthesis and the characterization possess more experimental parameters to consider such as oxidation during moving samples and differences in melting/wetting behavior on different substrate materials.

$$T_m = 232 - 782\left[\frac{\sigma_s}{15.8(r - t_0)} - \frac{1}{r}\right] \quad (5.1)$$

$$\Delta H_m = \Delta H_0\left(1 - \frac{t_0}{r}\right)^3 \quad (5.2)$$

Various approaches to synthesize the single-element nanoparticles have been reported [41–43], which can be largely categorized into "bottom-up" (chemical reduction) and "top-down" methods (physical method). The chemical reduction methods include the inert gas condensation, sol–gel, aerosol, micelle/reverse micelle, and irradiation by UV, γ-ray, microwave, etc. [15–23]. For bimetallic or multicomponent nanoparticles, chemical "bottom-up" and physical "top-down" methods have also been used as well. The chemical methods use the coreduction of dissimilar metal precursors or successive reduction of two-metal salts, which is usually carried out to prepare a core–shell structure of bimetallic nanoparticles. The binary alloys reported in the form of nanoalloys or core–shell structures such as Ag–Au are an alloy system that forms solid solutions and their structures which can be controlled by the reduction order. On the other hand, in the case of alloys that do not favor the formation of solid solutions such as eutectic alloys (Sn-based alloys), little research has been reported because more sophisticated synthetic methods are required due to their oxidative nature. The physical method can be used to synthesize monometallic [24] and bimetallic nanoparticles [25]. Using this technique, nanoparticles can be synthesized in gram quantities directly from the bulk materials without complex reaction procedures. This is suitable for the low melting point metal precursors and their alloys.

Duh et al. have reported Sn–3.5Ag–xCu ($x = 0.2, 0.5, 1.0$) nanoparticle synthesis for the lead-free solder application [26], where the differential scanning calorimetry (DSC) profile showed the melting (endothermic) peak of their SnAgCu alloy nanoparticles at ~216 °C, which is the melting point of the micron meter-sized SnAgCu alloy powders. No obvious melting point depression from their particles might be due to the surface oxidation or heavy agglomeration of the nanoparticles.

Tin and its alloys are easily oxidized due to their low chemical potential. For nano-sized tin and its alloys, oxidation more easily happens due to the higher surface area-to-volume ratio of nanoparticles. However, the oxides of Sn and its alloys cannot form an interconnection between chips and substrates. Therefore, trapping each nanoparticle is critical for the prevention of oxidation, and capping agents/surfactants can cover the particle surfaces to serve as an effective barrier against the penetration of oxygen [44, 45]. Our research has found that some surfactant, such as 1,10-phenanthroline, is an effective capping agent on preventing tin and its alloy nanoparticles from oxidation. When Sn, SnAg, or SnAgCu alloy nanoparticles are formed, they are instantly coordinated to 1,10-phenanthroline through the two lone pair of chelating nitrogen donor sites adjoining to the two heterocyclic aromatic rings.

5.1 Size-Dependent Melting Point of Tin Nanoparticles

Figure 5.3a shows the TEM and inserted HRTEM images of Sn nanoparticles synthesized by using 2.1×10^{-4} mol tin (II) acetate as a precursor in the presence of 0.045 mol surfactants. The average particle size calculated from the TEM picture was around 61 nm, and the obvious lattice fringes in HRTEM image imply the crystalline structure. The interplanar spacing was about 0.29 nm which corresponds to the orientation of (200) atomic planes of the

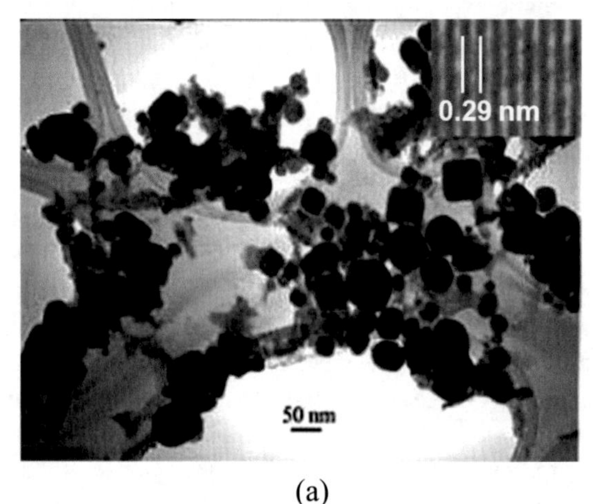

(a)

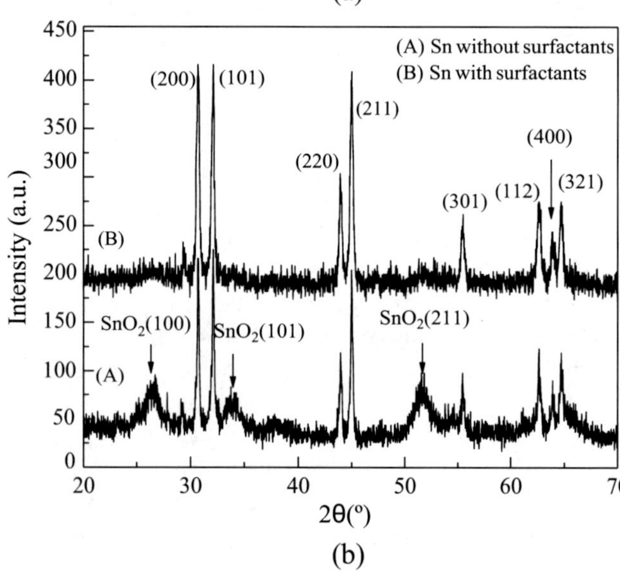

(b)

Fig. 5.3 (**a**) TEM and inserted HRTEM images; (**b**) XRD of Sn nanoparticles which were synthesized by using 2.1×10^{-4} mol tin (II) acetate as a precursor in the presence of 0.045 mol surfactants

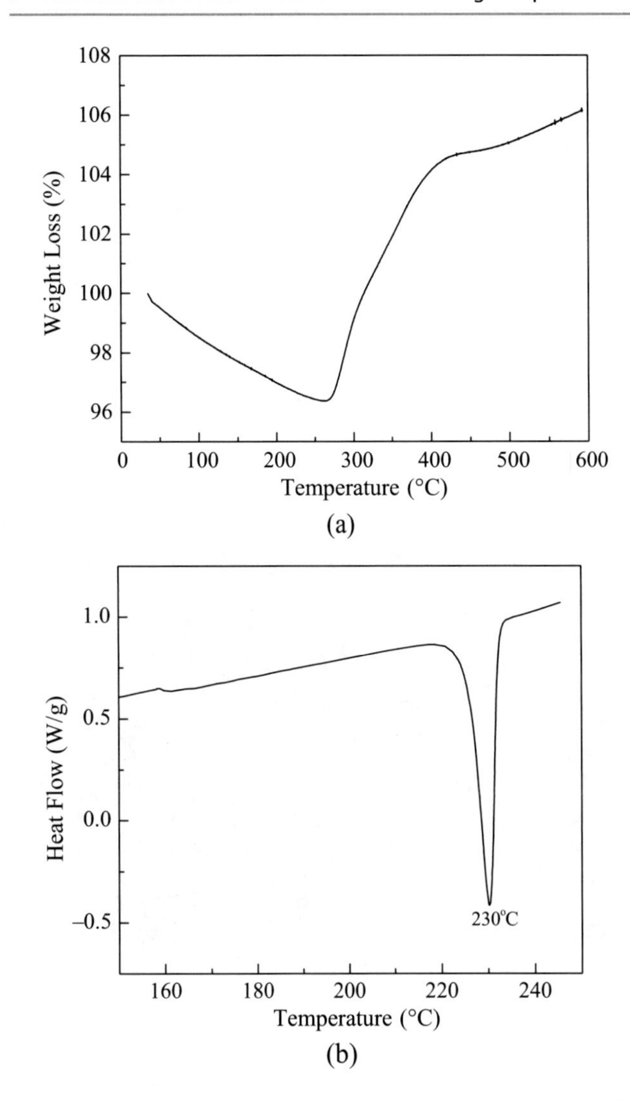

Fig. 5.4 TGA (**a**) and DSC (**b**) curves of the as-synthesized Sn nanoparticles

tetragonal structure of Sn. In the XRD pattern (Fig. 5.3b), all the peaks can be indexed to a tetragonal cell of Sn with $a = 0.582$ and $c = 0.317$ nm. The relative intensity of the peaks was consistent with that of the Sn nanoparticles reported elsewhere [27]. No obvious peaks at 20.6°, 33.8°, and 51.8° were found, which match the crystal planes of SnO2: (100), (101), and (211) [28]. However, the obvious oxidation peaks of SnO2 were observed from the X-ray diffraction (XRD) pattern of the Sn nanoparticles without surfactants. The XRD and high-resolution transmission electron microscopy (HRTEM) results showed that the Sn nanoparticles synthesized with surfactants in this study were nearly oxide-free.

Figure 5.4a shows the TGA curve of the as-synthesized Sn nanoparticles. The weight loss below 275 °C may be due to the evaporation of absorbed moisture and the decomposition of the surfactants. Above 275 °C, the weight gain was observed. This may be attributed to thermal oxidation of the pure Sn nanoparticles. The TGA results showed that the Sn nanoparticles were covered with the surfactants by ~4 wt% to the particle weight. And these surfactants can protect the Sn nanoparticles from oxidation. Figure 5.4b shows the thermal profiles of the as-synthesized Sn nanoparticles obtained from DSC where the melting point (T_m) was observed at 230 °C. Compared to the T_m of micron-sized Sn particles, the Sn nanoparticles exhibited the T_m depression by 2–3 °C.

The oxide-free Sn nanoparticles were obtained by using the different molar ratios between a precursor and a surfactant. Thus, different-sized Sn nanoparticles can be obtained.

Figure 5.5a, b shows the TEM image and DSC profile of the Sn nanoparticles which were synthesized by using 4.2×10^{-4} mol tin (II) acetate as a precursor in the presence of 0.045 mol surfactants. The average particle size was around 52 nm. The melting point of the Sn nanoparticles was around 228.0 °C, which was 4 °C lower than that of micron-sized Sn particles. Figure 5.5c, d shows the TEM image and DSC profile of the Sn nanoparticles which were synthesized by using 1.1×10^{-3} mol tin (II) acetate as a precursor in the presence of 0.045 mol surfactants. The average particle size was around 85 nm. The melting point of the as-synthesized Sn nanoparticles was 231.8 °C, which was still lower than the melting point of micron-sized Sn particles. The TEM image of Sn nanoparticles which were synthesized by using 1.75×10^{-4} mol tin (II) acetate as a precursor in the presence of 0.045 mol surfactants is shown in Fig. 5.5e. The average particle size was around 26 nm. The melting point of these particles was around 214.9 °C, which was 17.7 °C lower than that of micron-sized Sn particles. The melting transition of this sample took place over a temperature range of about 60 °C, which was much wider than the other three samples. This phenomenon can be attributed to a broadening of the phase transition due to the finite size effect [29] and wide size distribution of the Sn nanoparticles. Schmidt et al. also found a melting temperature range of 60 K for cluster size with the number of atoms between 70 and 200 [8].

From Table 5.2, it can be seen that the larger molar ratios between the surfactants and the precursors were used, the smaller the particle size was. This is because a larger amount of surfactants can restrict the growth of Sn nanoparticles. The surfactant molecules coordinate with the nanoclusters, resulting in the capping effect to restrict the particle growth. This was also found by Pal et al. on their gold nanoparticle synthesis that increasing the concentrations of surfactants would limit the particle size through the restriction of particle growth [30]. The DSC results showed size-dependent melting depression behavior and size-dependent latent heat of fusion.

Figure 5.6 shows the size dependence of the melting points of the synthesized Sn nanoparticle powders, which was compared with Lai et al.'s model [3]. The solid line

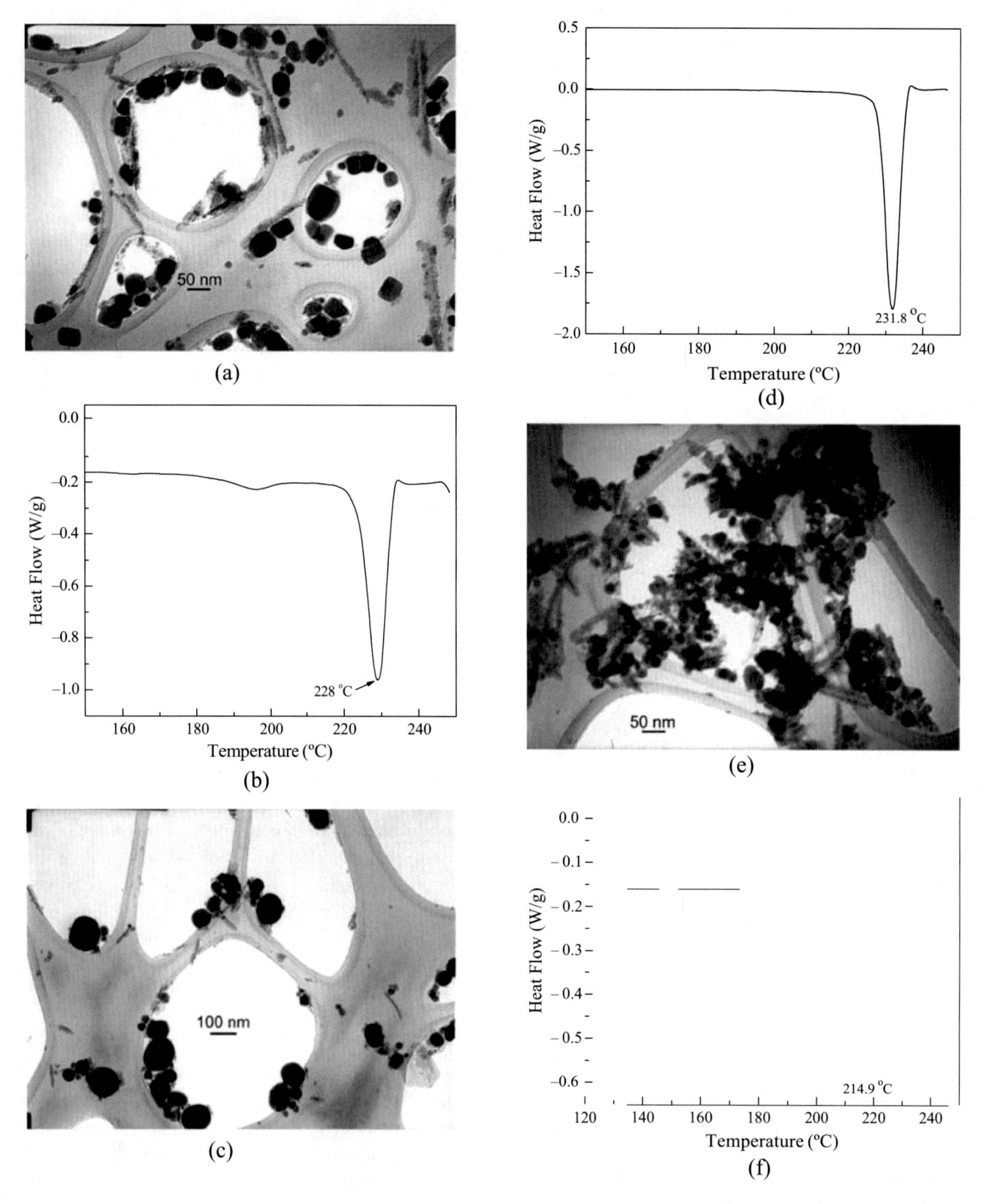

Fig. 5.5 TEM image and thermal behavior of Sn nanoparticles which were synthesized by using 4.2×10^{-4} mol (**a**) and (**b**), 1.1×10^{-3} mol (**c**) and (**d**), 1.75×10^{-4} mol (**e**) and (**f**) tin (II) acetate as a precursor in the presence of 0.045 mol surfactants

was calculated from Eq. (5.1). Lai et al. obtained this equation based on the model of Hanszen [31] in which it was assumed that the solid particle was embedded in a thin liquid overlayer and the melting temperature was taken to be the temperature of equilibrium between the solid sphere core and the liquid overlayer of a given critical thickness t_0.

Table 5.2 Melting points and heat of fusion of different-sized Sn nanoparticles

Samples	Surfactants/precursor	Average size (nm)	Melting point (°C)	*f..H* (J/g)
1	N/A	Micron size	232.6	60.0
2	$0.045/1.1 \times 10^{-3}$	85 ± 10	231.8	32.1
3	$0.045/2.1 \times 10^{-4}$	61 ± 10	230.0	24.5
4	$0.045/4.2 \times 10^{-4}$	52 ± 8	228.0	28.7
5	$0.045/1.75 \times 10^{-4}$	26 ± 10	214.9	27.4

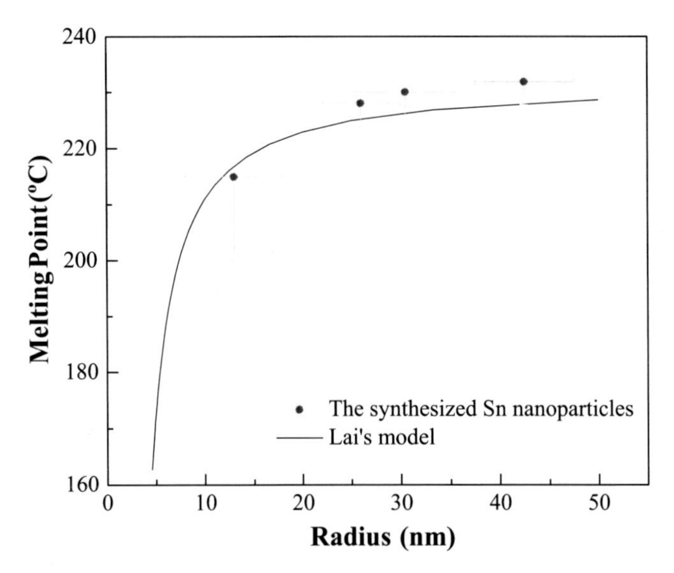

Fig. 5.6 Size dependence of the melting points of Sn nanoparticles (Y *error bars* stand for the melting temperature from the onset point to the peak point of DSC curves). The *solid line* is calculated from Eq. (5.1) by Lai et al. [3]

$$T_r = 232 - 782\left[\frac{\sigma_{sl}}{15.8(r - t_0)} - \frac{1}{r}\right] \tag{5.3}$$

where T_r (°C) is the melting point of particles with a radius of r (Å), t_0 (Å) is the critical thickness of the liquid layer, and σ_{sl} is the interfacial surface tension between the solid and the liquid and the liquid and its vapor. From their experiment, σ_{sl} is determined to be 48 ± 8 mN/m and the best fit for t_0 is 18 Å.

It is found that our experimental results are in reasonable agreement with the Lai et al.'s model and the melting point depends nonlinearly on the cluster radius.

From Table 5.1, it could also be found that the latent heat of fusion of the different-sized Sn nanoparticles was smaller than that of micron-sized Sn powders. Ercolessi et al. have found that *f..H*m of gold nanoparticles decreases steadily from 114 meV/atom (bulk) to 23 meV/atom ($N = 879$) and 10 meV/atom ($N = 477$) by molecular dynamic simulation [32]. It is also found by experiments that the normalized heat of fusion of Sn nanoparticles decreases markedly from the bulk value (58.9 J/g) by as much as 70% when the particle size is reduced, which can be interpreted as a solid core melting following the gradual surface melting for small particles [3]. We tried to compare the heat of fusion (*f..H*m) of our synthesized Sn particles with Lai et al.'s model as well [3]. It was found that the *f..H*m values of our Sn particles were smaller than those of their models. This might be due to the existence of the surfactants, solvents, and some small amounts of oxides which might develop from the heating process of DSC test.

5.2 Size-Dependent Melting of Tin/Silver Alloy Nanoparticles

Figure 5.7a shows the TEM image of the SnAg alloy nanoparticles synthesized by using 7.4×10^{-4} mol tin (II) 2-ethylhexanoate and 3.0×10^{-5} mol silver nitrate as precursors in the presence of 5.6×10^{-4} mol surfactants at 0 °C. The average diameter of the particles was ~24.0 nm. The XRD patterns of the as-synthesized SnAg alloy nanoparticles are shown in Fig. 5.7b. In addition to the peaks indexed to a tetragonal cell of Sn with $a = 0.582$ and $c = 0.317$ nm, the Ag3Sn phase (~39.6°) was found in the XRD patterns, indicating the successful alloying of Sn and Ag after the reduction process [26, 33]. No prominent oxide peak was observed from the XRD patterns. This indicates that the surfactants could help to protect the synthesized SnAg alloy nanoparticles from oxidation [34, 35].

Figure 5.8 shows the HRTEM image of the as-synthesized SnAg alloy nanoparticles, where the particles showed core–shell structures. The dark core and the brighter shell correspond to the crystalline metal and the amorphous organic surfactants, respectively. The surfactant shells on the particle surface protected the SnAg alloy nanoparticles from oxidation.

Figure 5.9a shows the TGA curve of the dried SnAg alloy nanoparticles in a nitrogen atmosphere. The weight loss below 180 °C might be due to the evaporation of a small amount of absorbed moisture and surfactants. Above 180 °C, the weight gain was observed, which was attributed to thermal oxidation of the SnAg alloy nanoparticles. Figure 5.9b displays the thermal profile of the dried SnAg alloy nanoparticles. The melting point of the SnAg nanoparticles was found at ~209.5 °C, about 13 °C lower than that of the

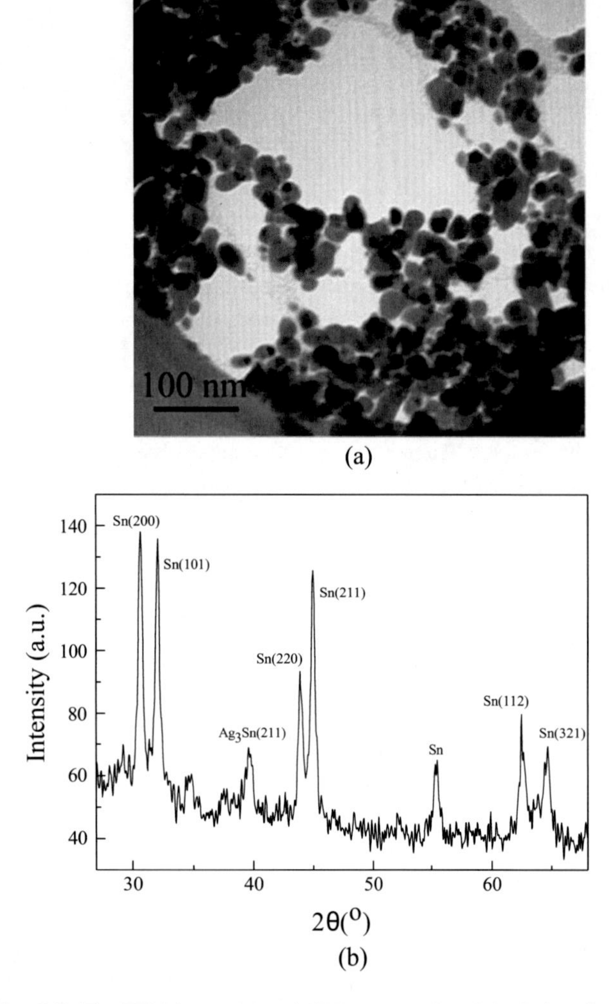

Fig. 5.7 The TEM image (**a**) and XRD patterns (**b**) of the SnAg alloy nanoparticles which were synthesized by using 7.4 × 10^{-4} mol tin (II) 2-ethylhexanoate and 3.0 × 10^{-5} mol silver nitrate as precursors in the presence of 5.6 × 10^{-4} mol surfactants

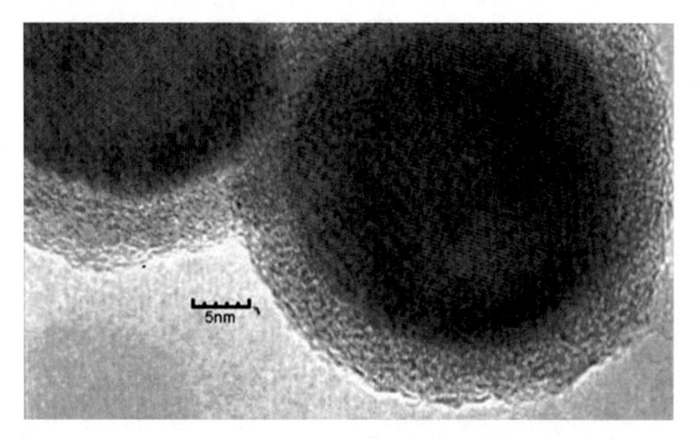

Fig. 5.8 HRTEM image of SnAg alloy nanoparticles which were synthesized by using 7.4 × 10^{-4} mol tin (II) 2-ethylhexanoate and 3.0 × 10^{-5} mol silver nitrate as precursors in the presence of 5.6 × 10^{-4} mol surfactants

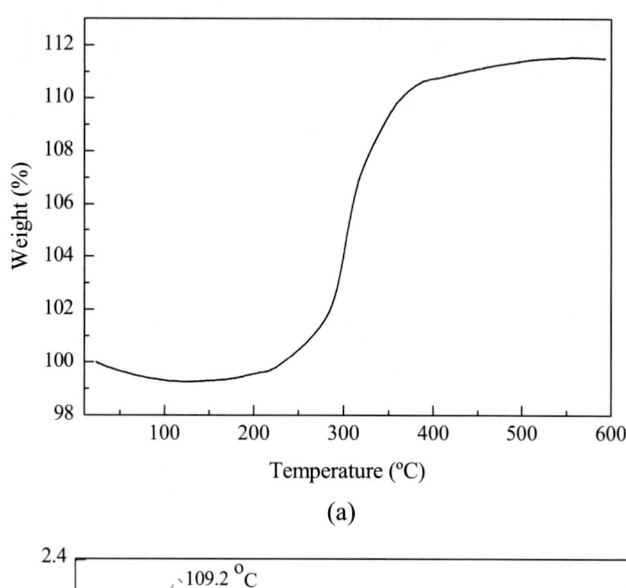

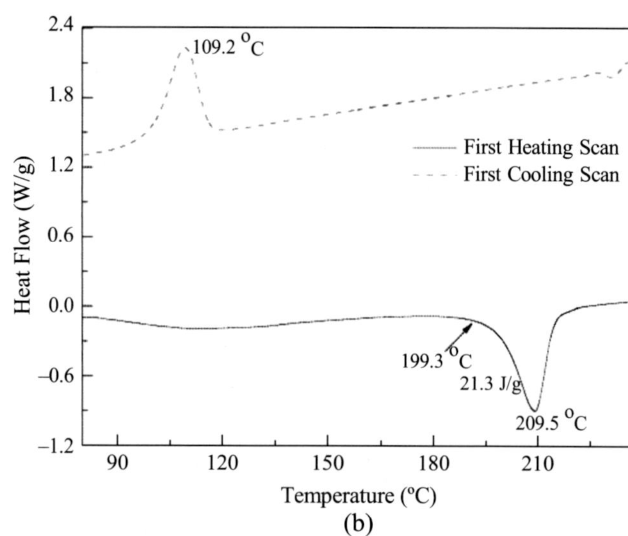

Fig. 5.9 The TGA (**a**) and DSC (**b**) curves of the SnAg alloy nanoparticles which were synthesized by using 7.4 × 10^{-4} mol tin (II) 2-ethylhexanoate and 3.0 × 10^{-5} mol silver nitrate as precursors in the presence of 5.6 × 10^{-4} mol surfactants

micron-sized 96.5Sn3.5Ag particles (222.6 °C). This was an obvious size-dependent melting point depression. The latent heat of fusion (*f..H*m) of the SnAg alloy nanoparticles (24.2 J/g) was smaller than that of micron-sized 96.5Sn3.5Ag powders (68.6 J/g). It has already been observed by experiments that the normalized heat of fusion of Sn nanoparticles decreases markedly from the bulk value by as much as 70% when the particle size is reduced. This can be interpreted as a solid core melting following the gradual surface melting for small particles [3]. The recrystallization peak of the as-synthesized SnAg alloy nanoparticles was found at 109.2 °C, which was 91.2 °C lower than that of the micron-sized 96.5Sn3.5Ag particles (200.4 °C). Such a supercooling effect in the recrystallization of the melted Sn nanoparticles has already been observed [36], which can be explained by the critical-sized stable grain that has to form for solidification to take place [37]. Solidification of the melted nanoparticles can only occur once the temperature is low enough so that the critical-sized solidification grain can be accommodated in a small volume.

The TEM image and the corresponding DSC curves of SnAg alloy nanoparticles which were synthesized by using 7.4×10^{-4} mol tin (II) 2-ethylhexanoate and 3.0×10^{-5} mol silver nitrate as precursors in the presence of 1.1×10^{-3} mol surfactants at;20 °C are shown in Fig. 5.10a, b. The average diameter of the as-synthesized nanoparticles was ~10 nm (Fig. 5.10a). From the DSC studies, the first peak melting temperature was 194.3 °C, which was about 28.3 °C lower than that of micron-sized 96.5Sn3.5Ag particles. The onset peak temperature was 172.6 °C, which was 46.9 °C lower than that of micron-sized particles (219.5 °C). The melting transition of this sample took place over a temperature range of about 21.7 °C, much wider than the micron-sized particles (3.1 °C). This phenomenon can be attributed to a broadening of the phase transition due to the finite size effect [29]. A small peak at 226.0 °C was also observed in the first heating scan, which might have come from the melting of a very small amount of larger-sized particles existing in this sample.

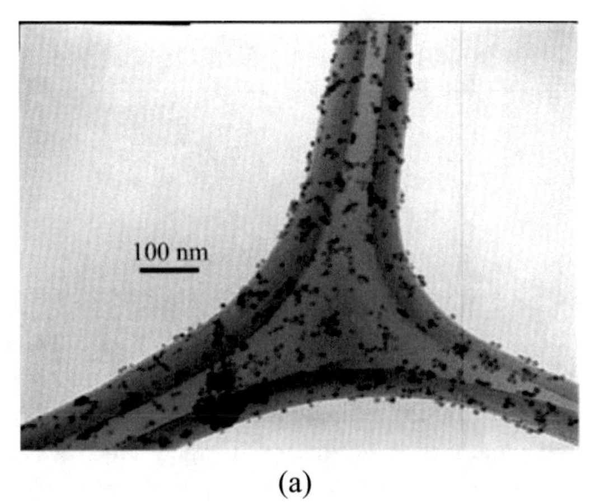

(a)

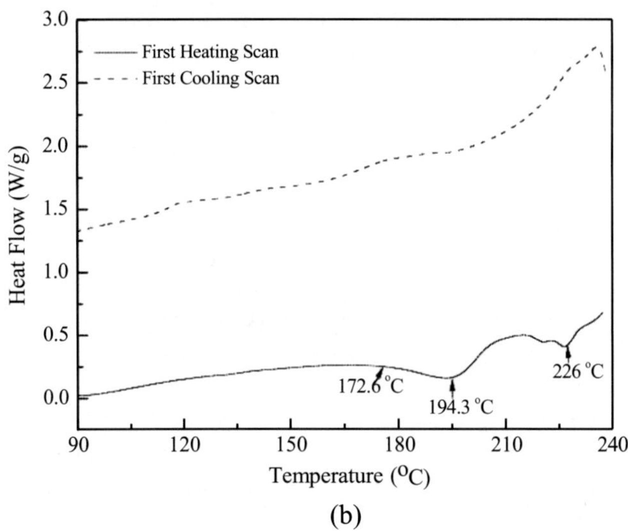

(b)

Fig. 5.10 The TEM image (**a**) and DSC curves (**b**) of the SnAg alloy nanoparticles which were synthesized by using 7.4×10^{-4} mol tin (II) 2-ethylhexanoate and 3.0×10^{-5} mol silver nitrate as precursors in the presence of 1.1×10^{-3} mol surfactants

The melting point and latent heat of fusion for micron-sized 96.5Sn3.5Ag are 222.6 °C and 68.5 J/g, respectively.

Table 5.3 shows the melting point and latent heat of fusion of different-sized SnAg alloy nanoparticles. We plotted the melting point and *f..H*m vs. the corresponding particle radius in Fig. 5.11. Both the particle size-dependent melting point depression and latent heat of fusion have been observed. This is due to the surface premelting of nanoparticles. It has already been found that surface melting of small particles occurs in a continuous manner over a broad temperature range whereas the homogeneous melting of the solid core occurs abruptly at the critical temperature T_m [38, 39]. For smaller-sized metal nanoparticles, the surface melting is strongly enhanced by curvature effects. Therefore, with the decreasing of particle size, both the melting point and latent heat of fusion will decrease too. Among all the synthesized particles in Table 5.3, the 10 nm (average diameter) SnAg alloy nanoparticles have a melting point at ~194.3 °C, which will be a good candidate for the low melting temperature lead-free solders and can solve the issues from the high-temperature reflow for the micron-sized lead-free particles.

5.3 Size-Dependent Melting of Tin/Silver/Copper Alloy Nanoparticles

SnAgCu alloy nanoparticles were synthesized by a chemical reduction method. Figure 5.12 shows the TEM image of the as-synthesized SnAgCu alloy nanoparticles. It can be calculated that the average diameter of the particles is around 22 nm.

The XRD patterns of the as-synthesized SnAgCu alloy nanoparticles are shown in Fig. 5.13. In addition to the peaks

Table 5.3 The melting points and heat of fusion of the as-synthesized different-sized SnAg alloy nanoparticles

No.	Surfactants (mol)	Reaction temperature (°C)	Average diameter (nm)	Melting point (°C)	*f..H*m (J/g)
1	5.6×10^{-4}	25	64	220.0	44.7
2	5.6×10^{-4}	0	24	209.5	24.2
3	8.0×10^{-4}	10	17	206.0	15.1
4	1.1×10^{-3}	20	10	194.3	9.95

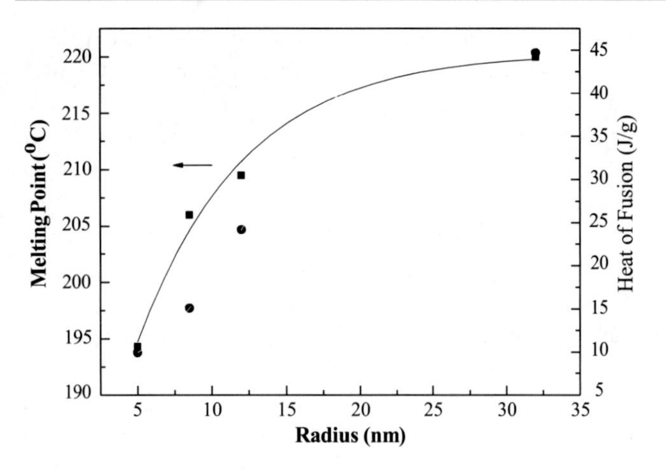

Fig. 5.11 The relationship between the radius of the as-synthesized SnAg alloy nanoparticles and their corresponding melting points and heat of fusion

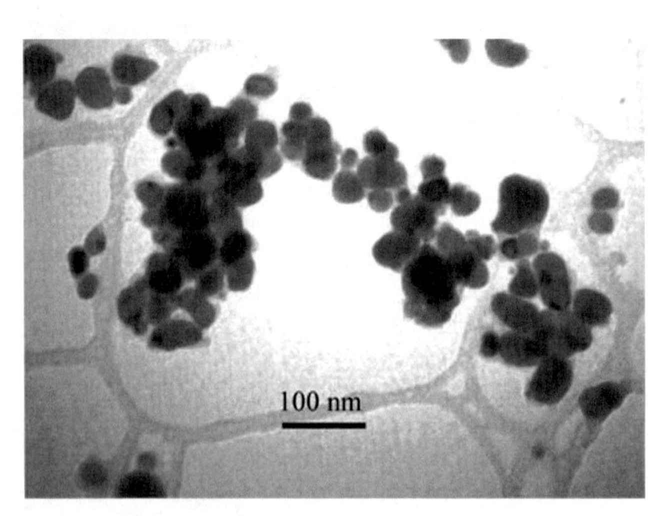

Fig. 5.12 The TEM image of ~22 nm SnAgCu alloy nanoparticles

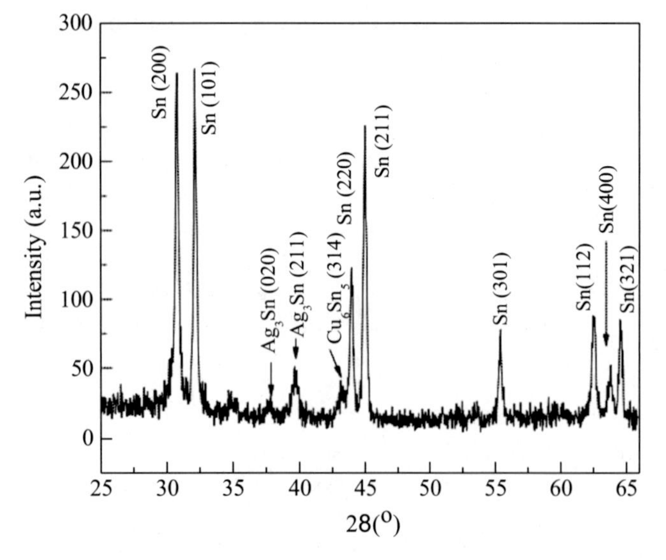

Fig. 5.13 The XRD patterns of ~22 nm SnAgCu alloy nanoparticles

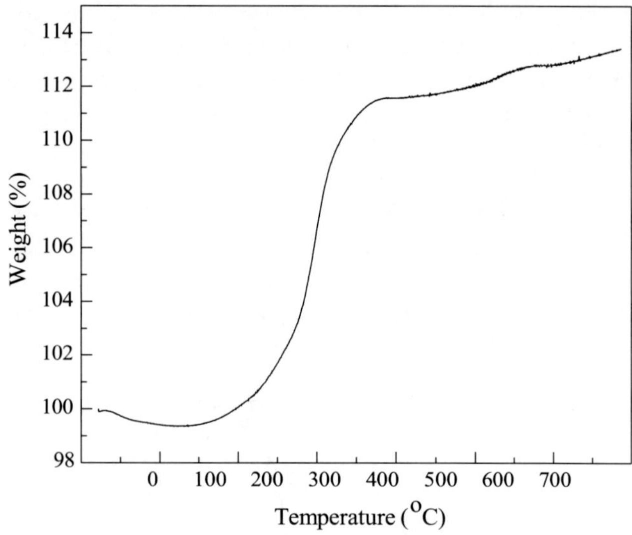

Fig. 5.14 The TGA curve of ~22 nm SnAgCu alloy nanoparticles

indexed to a tetragonal cell of Sn with $a = 0.582$ and $c = 0.317$ nm, the Ag3Sn phase (39.6°) was found in the XRD patterns, indicating the successful alloying of Sn and Ag after the reduction process [26, 33, 40].

At the same time, Cu6Sn5 was formed which was due to the alloying of Sn and Cu [26]. No prominent oxide peaks were observed from the XRD patterns. This indicates that surfactants can help protect the synthesized SnAgCu alloy nanoparticles from oxidation [34, 40]. The HRTEM characterizations already showed that the surfactants covered the particle surface and formed a core–shell structure [40]. The core was from the crystalline metal particles and the shell was from the amorphous surfactants. The amorphous surfactant shells on the particle surface helped prevent the diffusion of oxygen to SnAgCu alloy nanoparticles.

Figure 5.14 shows the TGA curve of the as-synthesized dried SnAgCu alloy nanoparticles in a nitrogen atmosphere. The weight loss below 180 °C might be due to the evaporation or decomposition of a small amount of absorbed moisture and surfactants. Above 180 °C, the weight gain was observed, which can be attributed to thermal oxidation of the SnAgCu alloy nanoparticles.

The thermal properties of the as-synthesized SnAgCu alloy nanoparticles were studied by a differential scanning calorimeter (Fig. 5.15). In the first heating scan of the DSC curve, an endothermic peak point at ~207 °C was obtained, which is around 10–12 °C lower than the melting point of micron-sized SnAgCu (217–219 °C) alloy particles. In the first cooling scan, the supercooling of the nanoparticles with a crystallization peak at 113.3 °C was observed, lower than that of micron-sized SnAgCu alloy nanoparticles (145 °C). Such a supercooling effect in the recrystallization of the melted Sn and SnAg alloy nanoparticles has already been

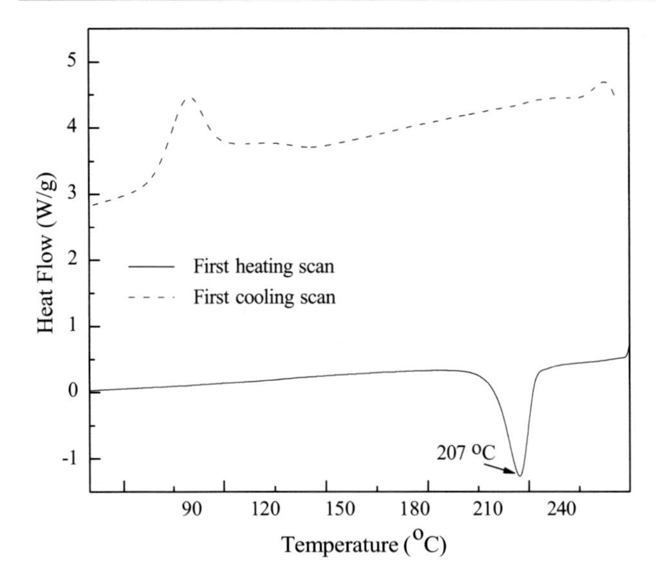

Fig. 5.15 The DSC curves of ~22 nm SnAgCu alloy nanoparticles

observed [36, 40], which can be explained by the critical-sized stable grain that needs to form in order for solidification to take place [37]. Solidification of the melted nanoparticles can only occur once the temperature is low enough so that the critical-sized solidification grain can be accommodated in a small volume.

Figure 5.16a shows the TEM image of 10–13 nm SnAgCu alloy nanoparticles which were synthesized by the chemical reduction method at a different reaction temperature. From DSC studies (Fig. 5.16b), the peak melting temperature was 199 °C, which was around 20 °C lower than that of micron-sized 96.5Sn3.0Ag0.5Cu particles. The onset peak temperature is 177 °C, 37.5 °C lower than micron-sized particles (214.5 °C). The melting transition of this sample took place over a temperature range of about 22 °C. This phenomenon can be attributed to broadening of the phase transition due to the finite size effect [29].

Table 5.4 shows the melting point, latent heat of fusion, and recrystallization temperature of different-sized SnAgCu alloy nanoparticles. Both size-dependent melting point depression and latent heat of fusion have been observed. It has already been found that surface melting of small particles occurs in a continuous manner over a broad temperature range whereas homogeneous melting of the solid core occurs abruptly at the critical temperature T_m [39]. For smaller-sized metal nanoparticles, surface melting is strongly enhanced by the curvature effects. Therefore, with the decrease of particle size, both the melting point and latent heat of fusion decrease as well. Among all the synthesized particles in Table 5.4, the 10–13 nm (average diameter) SnAgCu alloy nanoparticles have the lowest melting point at ~199 °C, indicating it to be a good candidate for lead-free solders as its melting point is comparable to the reflow temperature of the conventional eutectic micron-sized SnPb alloy particles.

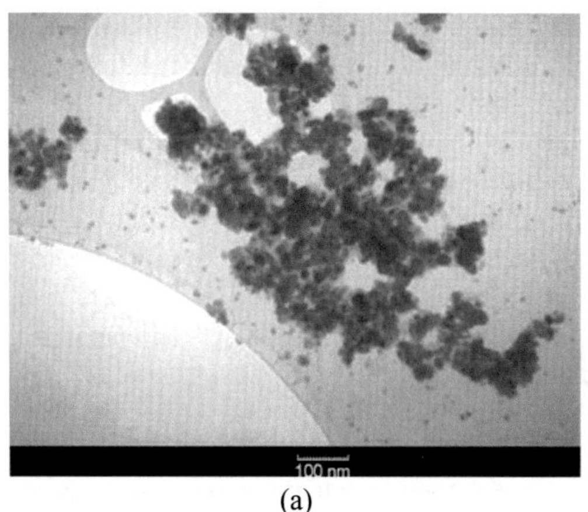

(a)

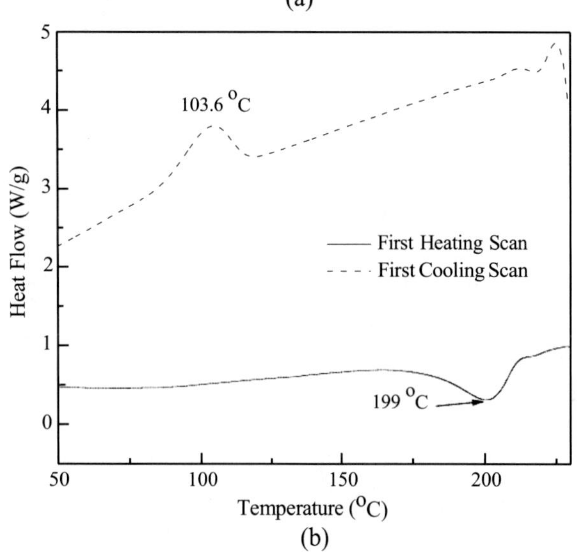

(b)

Fig. 5.16 The TEM image (**a**) and DSC curves (**b**) of 10–13 nm SnAgCu alloy nanoparticles

Table 5.4 The melting and recrystallization points, heat of fusion of different-sized SnAgCu alloy nanoparticles

Size	Melting point (°C)	f..H (J/g)	Recrystallization (°C)
Micron	217	72.2	145
~28 nm	210.4	42.7	112.3
~22 nm	207.3	31.0	109.9
~18 nm	206.1	24.7	108.7
~10–13 nm	199	15.9	103.6

5.4 Wetting Properties of Tin/Silver and Tin/Silver/Copper Alloy Nanoparticle Pastes

The solder paste is a viable interconnecting material, providing electrical, thermal, and mechanical properties applicable to electronics assemblies. It is a homogeneous and kinetically stable mixture of solder alloy powder, flux, and vehicle, which is capable of forming metallurgical bonds at given soldering conditions (Fig. 5.17).

Solder alloy powders are usually tin/lead (SnPb), lead-free solder materials, such as tin/silver (SnAg), tin/silver/copper (SnAgCu), and tin/copper (SnCu). The flux can cause the alloy powders and the surfaces to be joined to maintain a clean and metallic state. The vehicle is a carrier for the solder powder, provides a desirable rheology, and protects the molten solder and the cleaned substrates from re-oxidation.

The elements used in solder alloys are tin, lead, silver, copper, bismuth, indium, antimony, and cadmium. Phase diagram can be used to determine the eutectic point and the corresponding compositions for the solid solution. Several methods can be used to make the solder alloy powders, such as chemical reduction, decomposition, mechanical processing of solid metal, and atomization of liquid metals. Chemical reduction method and decomposition can be used to prepare fine solder alloy particles. Mechanical processing of solder metal can be used to prepare flake-like particles, while powders to be used in a solder paste are mostly produced by atomization because of desirable inherent morphology and shape of resulting particles.

Melting range, flow rate, and particle morphologies are important parameters for solder powders. The melting range of solder powders is usually tested by differential scanning calorimetry (DSC). Table 5.1 shows the melting point of SnPb solders and lead-free solders. The particle size, size distribution, and shapes are usually tested by scanning electron microscopy (SEM), light scattering, and so on.

Some of the other physical properties, like viscosity, surface tension, thermal properties, electrical properties, and thermal expansion coefficient, are also very important for the solder materials. For the SnPb alloys, the surface tension and thermal and electrical conductivity decrease with increasing lead content, while viscosity and thermal expansion coefficient increase with increasing lead content.

The function of fluxes in a solder paste is to chemically clean the surfaces to be joined, to clean the surface of solder powder, and to maintain the cleanliness of both substrate surface and solder powder surface during reflow so that a metallic continuity at the interface and a complete coalescence of the solder powder during reflow can be achieved (Fig. 5.18).

There are several kinds of fluxes:

- Type R-rosin flux, the weakest, contains only rosin without the presence of activator.
- Type RMA (mildly activated rosin) is a system containing both rosin and activator.
- Type RA is a fully activated rosin and resin system, having a higher flux strength than the RMA type.
- Type OA is an organic acid flux and possessing a high fluxing activity and is generally considered corrosive.
- Type SA (synthetic activator) is designed to improve the fluxing activity on hard-to-solder surfaces. The SA flux displays better wetting ability than RA flux and is equivalent to OA flux.

The major components of unmodified rosin are abietic acid, isopimaric acid, neoabietic acid, pimaric acid, dihydroabietic acid, and dehydroabietic acid. The chemical structures of these acids are shown in Fig. 5.19.

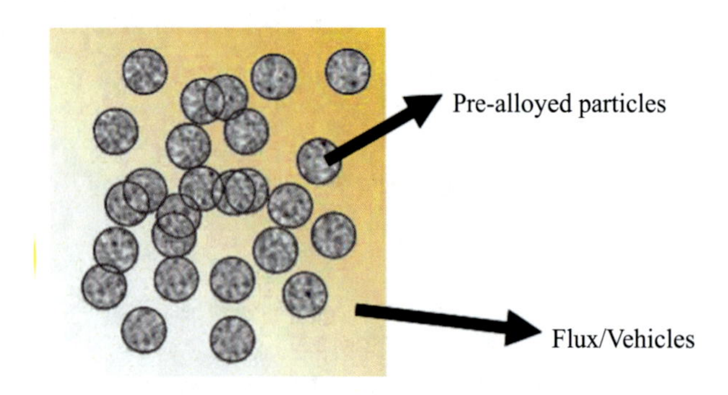

Fig. 5.17 The composition of solder pastes

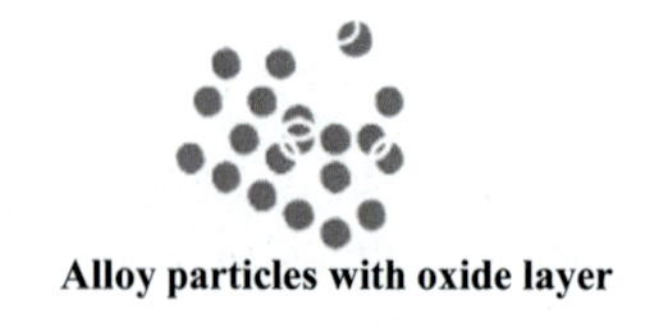

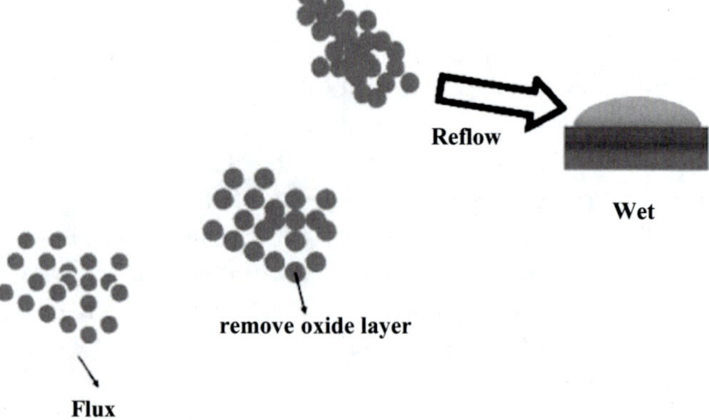

Fig. 5.18 The function of flux

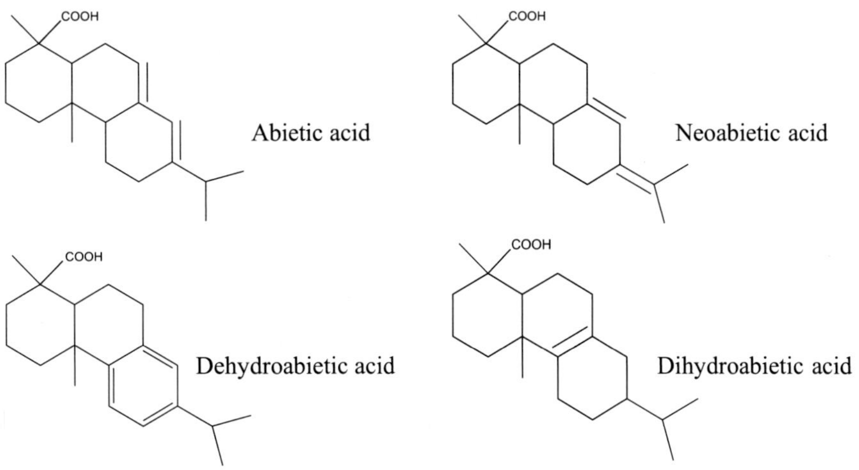

Fig. 5.19 The chemical structures of the acids in rosin

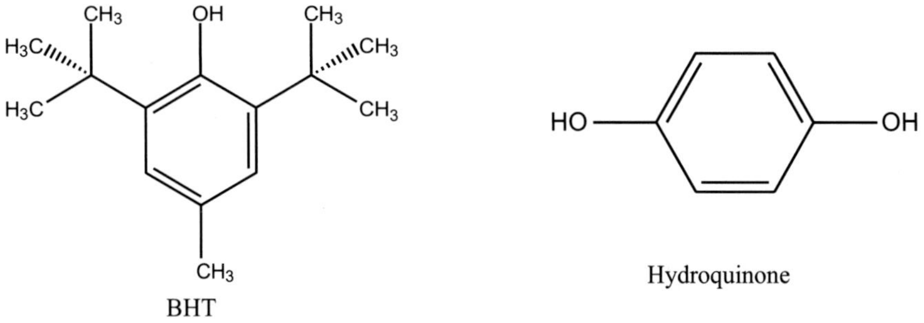

COOH Abietic acid
COOH Neoabietic acid
COOH Dehydroabietic acid
COOH Dihydroabietic acid

Fig. 5.20 The chemical structures of BHT and hydroquinone

CH_3 OH CH_3 H_3C CH_3 H_3C CH_3 CH_3
BHT

HO OH
Hydroquinone

The main compositions of vehicles we used in this section are tackifier, solvent, antioxidation agent, and surfactants.

The tackifier is typically a medium-to-high viscosity, high surface tension liquid serving to wet the printed circuit board and the component and retain the component in position during handling and reflow soldering. The tackifier usually comprises one or more alcohols, aromatic hydrocarbon solvents, aliphatic hydrocarbon solvents, or polymers.

A suitable solvent vehicle includes any solvent which is chemically inert with the other components in the flux. Three important parameters for solvents are boiling point, viscosity, and polarity. Alcohols are usually used as solvents. The more hydroxyl groups the solvents have, the better activity the flux will have.

Antioxidation agents are used to protect the Sn and its alloy nanoparticles from oxidation. Butylated hydroxytoluene (BHT) and hydroquinone (Fig. 5.20) are the typically used antioxidation agents. Oxygen reacts preferentially with BHT or hydroquinone rather than oxidizing Sn alloy, thereby protecting them from oxidation.

The solder surfactant is a compound which improves the solder wetting rate of a surface and enables better and more uniform spreading of molten solder across the surface to be soldered. Suitable surfactants include polybasic acids, e.g., polycarboxylic acids such as dicarboxylic and tricarboxylic acids. The dibasic acids typically have 4–10 carbon atoms. Suitable tricarboxylic acids typically comprise acids having 6–7 carbon atoms. Other suitable surfactants include hydroxyl-substituted polybasic acids, such as tartaric acid and citric acid. The selected surfactant is present in the flux mixture in an amount of at least 1 wt% of the resultant flux mixture.

The solderability, flow properties, wetting properties, and solidification are important physical properties for the solder pastes. Solderability is the ability to achieve a clean, metallic surface on solder powder and on substrates during the dynamic heating process so that a complete coalescence of solder powder particles and good wetting of molten solder on the surface of the substrates can be formed. Solderability

Table 5.5 The composition of flux and vehicle made by our group

Fluxing agent	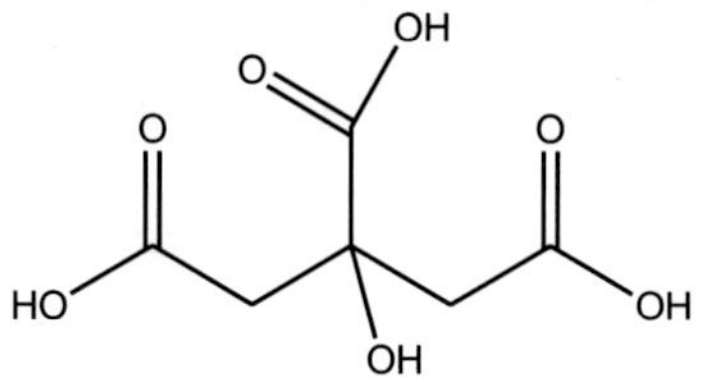
Tackifier	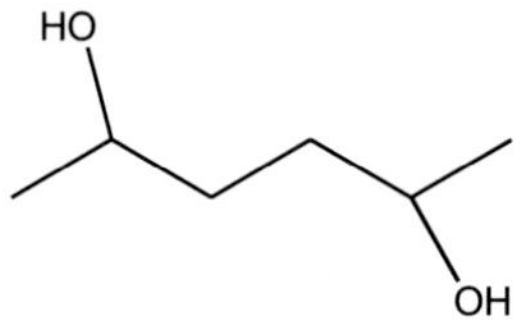
Surfactant	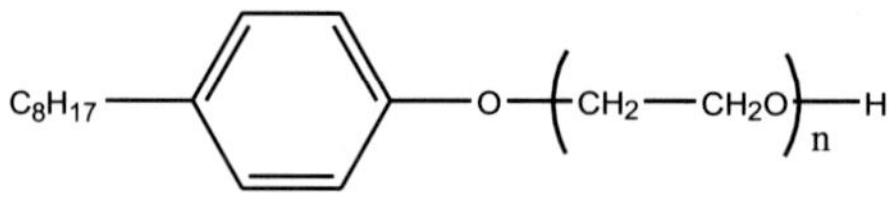 C_8H_{17}–C_6H_4–O–$(CH_2$–$CH_2O)_n$–H
Solvent	CH_3CH_2OH; 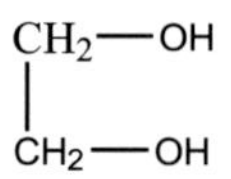 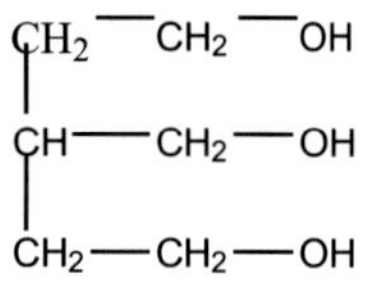 CH_2–OH / CH_2–OH; CH_2–CH_2–OH / CH–CH_2–OH / CH_2–CH_2–OH
Antioxidation	BHT

depends on fluxing efficiency by the solder pastes and the quality of surface of substrates.

When heat is applied to the paste through any means, the paste tends to spread or slump due to gravity and thermal energy generated. The surface energies of the liquid and the solid substrates are key factors in determining the spreading and wetting properties. For a system with liquid to wet the solid substrate, the spreading occurs only if the surface energy of the substrate to be wetted is higher than that of the liquid to be spread.

The solidification of liquid metal occurs by nucleation and crystal growth. Dendritic growth during solidification is a common phenomenon in pure metals and alloys. The heat flow during the solidification and the crystal structure of the alloy are crucial factors to the properties and structure of the solidified alloy.

Application techniques for the solder pastes usually include printing and paste dispensing. Printing is a viable method to economically produce accurate and reproducible transfer of paste onto the designed pattern, but only suitable for a flat surface. However, paste dispensing can be used for irregular surface or hard-to-reach area.

Conduction reflow is most suitable for the assemblies with flat surfaces, composed of thermal conductive materials as the substrates and with single-side component/device populations (fast heating rate and operational simplicity). Infrared reflow is a dynamic process. The precise temperature profile depends on solder alloy composition, properties of the pastes, and assembly involved. Some other reflow methods, such as vapor phase reflow, convection reflow, hot gas reflow, resistance reflow, laser reflow, and induction reflow, are also being used.

Residue is composed of polar organics, nonpolar organics, ionic salts, and metal salts of organics. The cleaning solvents include trichlorotrifluoroethane, 1,1,1-trichloroethane, acetone, methylene chloride, low carbon-chain alcohols, and water. Basic techniques include vapor degreasing, liquid spray, liquid immersion, high-pressure spray, and liquid immersion with ultrasonic acid.

A certain amount of fluxing agent, solvent, antioxidation agent, surfactants, and tackifier was mixed together to prepare the flux and vehicle for the SnAg or SnAgCu alloy nanoparticles. Table 5.5 shows the composition of the flux and vehicle we made and their corresponding chemical structures.

The SnAg alloy nanoparticles (sample 1 in Table 5.3) were mixed with an acidic-type flux to form the nanosolder pastes at room temperature. A copper foil was cleaned by hydrochloric acid to get rid of the oxide layer and then rinsed with DI water for four times. Thereafter, the nanosolder pastes placed on top of the copper foil were put into a 230 °C oven in an air atmosphere for 5 min. The cross section of the sample after reflow was shown in Fig. 5.21. It was observed that the SnAg alloy nanoparticles completely melted and wetted on the cleaned copper foil surface. The energy-dispersive spectroscopy (EDS) results revealed the

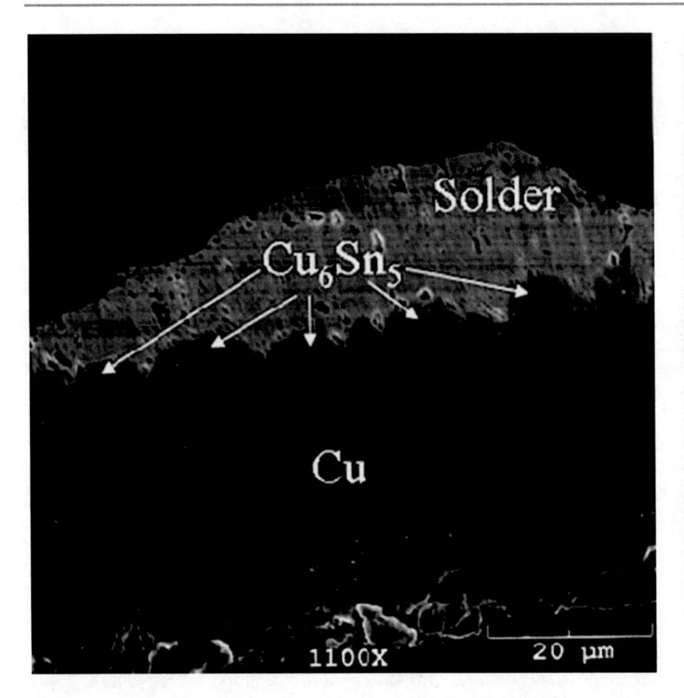

Fig. 5.21 SEM image of the cross section of the wetted SnAg alloy nanoparticles (sample 1 in Table 5.3) on the cleaned copper foil [46]

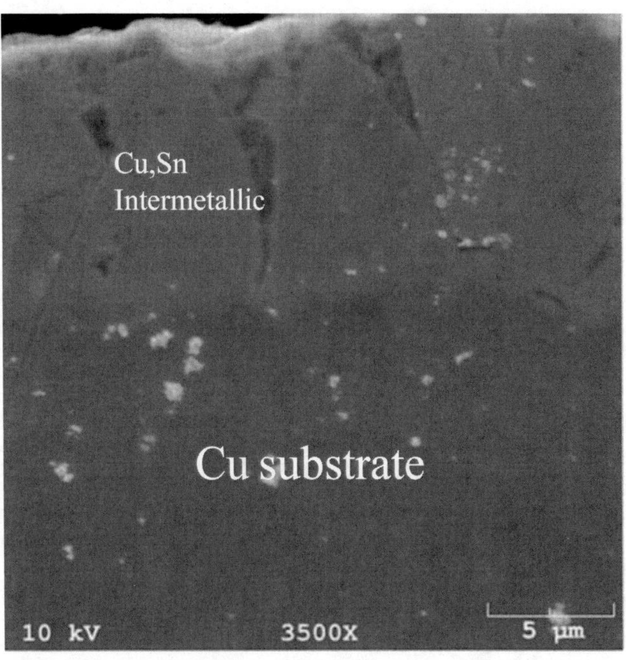

Fig. 5.22 SEM image of the cross section of the wetted 50 nm SnAgCu alloy nanoparticles on copper foil

formation of the intermetallic compound (IMC) of Cu6Sn5, which showed scallop-like morphologies in Fig. 5.21. The thickness of the IMC was approximately 4.0 µm. Further studies on the wetting properties of different-sized SnAg alloy nanoparticles at different reflow temperatures are still ongoing.

Nanolead-free solder pastes were formed by dispersing the synthesized SnAgCu alloy nanoparticles into the flux and vehicle system. Carboxylic acid was used as the fluxing agent. The vehicle system mainly consists of a tackifier, a surfactant, solvents, antioxidation agents, etc. The tackifier is typically a medium-to-high viscosity, high surface tension liquid serving to wet the printed circuit board and the component and retain the component in position during the handling and reflowing process. The surfactant is a compound which enables better and more uniform spreading of molten solder across the surface to be soldered. The antioxidation agents were used to protect the molten solder and substrates from oxidation.

A copper foil was cleaned by hydrochloric acid to get rid of the oxide layer and then rinsed with DI water for four times. Thereafter, the ~50 nm (average diameter) SnAgCu alloy nanoparticle pastes placed on top of the copper foil were put into a 220 °C oven in an air atmosphere for 5 min. The cross section of the sample after reflow was shown in Fig. 5.22. It was observed that the SnAgCu alloy nanoparticles completely melted and wetted on the cleaned copper foil surface. The energy-dispersive spectroscopy (EDS) results revealed the formation of the intermetallic compounds (Cu6Sn5), which showed scallop-like morphologies in Fig. 5.22. The thickness of intermetallic compounds was approximately 10.0 µm. The wetting properties of different-sized SnAgCu alloy nanoparticles at different reflow temperatures need to be further studied.

5.5 Conclusion

This chapter addresses the synthesis routes for lead-free nanoparticles including a tin single element and alloys and their structures and thermal behavior based on melting point depression. During this study, we observed obvious melting point depression behavior of synthesized nanoparticles and learned, most of all, that the surface oxidation of the non-noble nanoparticle is a critical obstacle in the case where melting of the particles is needed. Therefore, we found the surface capping by functionalization was imperative and when and how the surface can be capped were very important factors to obtain oxide-free non-noble metal nanoparticles, together with the careful selection of the surfactant. Finally in order to make nanosolder pastes, the appropriate flux vehicle and the reflow profile optimization were very important, which are also the lessons we learned.

The aim of this study is the reduction of the reflow temperature in microelectronic packaging processes which may be a representative case study to employ cutting-edge nanoscience and technology into the real-world research and development arena.

References

1. Abtew, M., Selvaduray, G.: Lead-free solders in microelectronics. Mater. Sci. Eng. R. Rep. **27**, 95 (2000)
2. Wronski, C.R.M.: Size dependence of melting point of small particles of tin. Br. J. Appl. Phys. **18**, 1731 (1967)
3. Lai, S.L., Guo, J.Y., Petrova, V., Ramanath, G., Allen, L.H.: Size-dependent melting properties of small tin particles: nanocalorimetric measurements. Phys. Rev. Lett. **77**(1), 99–102 (1996)
4. Bachels, T., Guntherodt, H.J., Schafer, R.: Melting of isolated tin nanoparticles. Phys. Rev. Lett. **85**(6), 1250–1253 (2000)
5. Zhao, S.J., Wang, S.Q., Cheng, D.Y., Ye, H.Q.: Three distinctive melting mechanisms in isolated nanoparticles. J. Phys. Chem. B. **105**(51), 2857–12860 (2001)
6. Shvartsburg, A., Jarrold, M.F.: Solid clusters above the bulk melting point. Phys. Rev. Lett. **85**, 2530 (2000)
7. Cleveland, L., Luedtke, W.D., Landman, U.: Melting of gold clusters: icosahedral precursors. Phys. Rev. Lett. **81**, 2036 (1998)
8. Schmidt, M., Kusche, R., Issendroff, B., Haberland, H.: Irregular variations in the melting point of size-selected atomic clusters. Nature. **393**(6682), 238–240 (1998)
9. Lewis, L.J., Jensen, P., Barrat, J.L.: Melting, freezing, and coalescence of gold nanoclusters. Phys. Rev. B. **56**, 2248 (1997)
10. Cleveland, C.L., Landman, U., Luedtke, W.D.: Phase coexistence in clusters. J. Phys. Chem. **98**, 6272 (1994)
11. Shi, F.G.: Size-dependent thermal vibrations and melting in nanocrystals. J. Mater. Res. **9**, 1307 (1994)
12. Jiang, Q., Shi, F.G.: Entropy for solid-liquid transition in nanocrystals. Mater. Lett. **37**, 79 (1998)
13. Allen, L., Bayles, R.A., Gile, W.W., Jesser, W.A.: Small particle melting of pure metals. Thin. Solid. Film. **144**, 297 (1986)
14. Buffat, P., Borel, J.P.: Size effect on melting temperature of gold particles. Phys. Rev. A. **13**, 2287 (1976)
15. Birringer, R., Gleiter, H., Klein, H.P., Marquart, P.: Nanocrystalline materials an approach to a novel solid structure with gas-like disorder? Phys. Lett. **102A**, 365 (1984)
16. Lee, B.I., Pope, E.J.A.: Chemical Processing of Ceramics. Marcel Dekker, New York (1994)
17. Raabe, O.G.: In: Liu, B.Y.H. (ed.) Fine Particles, p. 60. Academic, New York (1975)
18. J. P. Wilcoxon, A. Martino, R. L. Baughmann, E. Klavetter, A. P. Sylwester, "Synthesis of transition metal clusters and their catalytic and optical properties", in "Nanophase and Nanocomposite Materials" S. Komarneni, J. C. Parker and G. J. Thomas, MRS, Pittsburgh, 1993: p. 131
19. Thomas, J.: Preparation and magnetic properties of colloidal cobalt particles. J. Appl. Phys. **37**, 2914 (1966)
20. Rochfort, G.L., Rieke, R.D.: Preparation, characterization, and chemistry of activated cobalt. Inorg. Chem. **25**, 348 (1986)
21. Koch, C.C.: Materials synthesis by mechanical alloying. Ann. Rev. Mater. Sci. **19**, 121 (1989)
22. Klabunde, K., Li, Y., Tan, B.: Solvated metal atom dispersed catalysts. Chem. Mater. **3**, 30 (1991)
23. Mafune, F., Kohno, J.Y., Takeda, Y., Kondow, T.: Dissociation and aggregation of gold nanoparticles under laser irradiation. J. Phys. Chem. B. **105**, 9050 (2001)
24. Zhao, Y.B., Zhang, Z.J., Dang, H.X.: Preparation of tin nanoparticles by solution dispersion. Mater. Sci. Eng. **A359**, 405 (2003)
25. Zhao, Y.B., Zhang, Z.J., Dang, H.X.: Synthesis of In-Sn alloy nanoparticles by a solution dispersion method. J. Mater. Chem. **14**, 299 (2004)
26. Hsiao, L.Y., Duh, J.G.: Synthesis and characterization of lead-free solders with Sn-3.5Ag- xCu (x=0.2, 0.5, 1.0) alloy nanoparticles by the chemical reduction method. J. Electrochem. Soc. **152**(9), J105–J109 (2005)
27. Kwon, Y., Kim, M.G., Kim, Y., Lee, Y., Cho, J.: Effect of capping agents in tin nanoparticles on electrochemical cycling. Electrochem. Solid-State Lett. **9**, A34 (2006)
28. Wang, Y., Lee, J.Y., Deivaraj, T.C.: Controlled synthesis of V-shaped SnO2 nanorods. J. Phys. Chem. B. **108**, 13589 (2004)
29. Imry, Y., Bergman, D.: Critical points and scaling laws for finite systems. Phys. Rev. A. **3**(4), 1416 (1971)
30. Mandal, M., Ghosh, S.K., Kundu, S., Esumi, K., Pal, T.: UV photoactivation for size and shape controlled synthesis and coalescence of gold nanoparticles in micelles. Langmuir. **18**, 7792 (2002)
31. Hanszen, K.J.: Theoretische untersuchungen uber den schmelzpunkt kleiner kugelchen – Ein beitrag zur thermod ynamik der grenzflachen. Z. Phys. **157**, 523–553 (1960)
32. Ercolessi, F., Andreoni, W., Tosatti, E.: Melting of small gold particles – mechanism and size effects. Phys. Rev. Lett. **66**(7), 911–914 (1991)
33. Lai, H.L., Duh, J.G.: Lead-free Sn-Ag and Sn-Ag-Bi solder powders prepared by mechanical alloying. J. Electron. Mater. **32**(4), 215–220 (2003)
34. Jiang, H.J., Moon, K., Dong, H., Hua, F., Wong, C.P.: Size-dependent melting properties of tin nanoparticles. Chem. Phys. Lett. **429**, 492–496 (2006)
35. Balan, L., Schneider, R., Billaud, D., Ghanbaja, J.: A new organometallic synthesis of size- controlled tin(0) nanoparticles. Nanotechnol. **16**(8), 1153–1158 (2005)
36. Banhart, F., Hernandez, E., Terrones, M.: Extreme superheating and supercooling of encapsulated metals in fullerenelike shells. Phys. Rev. Lett. **20**(18), 185502 (2003)
37. Christenson, H.K.: Confinement effects on freezing and melting. J. Phys. Condens. Matter. **13**(11), R95–R133 (2001)
38. Garrigos, R., Cheyssac, P., Kofman, R.: Melting of lead particles of small sizes – influence of surface phenomena. Z. Phys. D. **12**, 497 (1989)
39. Hu, W.Y., Xiao, S.G., Yang, J.Y., Zhang, Z.: Melting evolution and diffusion behavior of vanadium nanoparticles. Eur. Phys. J. B. **45**(4), 547–554 (2005)
40. Jiang, H., Moon, K., Hua, F., Wong, C.P.: Synthesis and thermal and wetting properties of tin/silver alloy nanoparticles for low-melting point lead-free solders. Chem. Mater. **9**(8), 4482–4485 (2007)
41. Jiang, H., Moon, K., Sun, Y., Wong, C.P., Hua, F., Pal, T., Pal, A.: Tin/indium nanobundle formation from aggregation or growth of nanoparticles. J. Naonopart. Res. **10**(1), 41–46 (2008)
42. Jiang, H., Zhu, L., Moon, K., Wong, C.P.: The preparation of stable metal nanoparticles on carbon nanotubes whose surface were modified during production. Carbon. **45**(3), 655–661 (2007)
43. Jiang, H., Moon, K., Zhang, Z., Pothukuchi, S., Wong, C.P.: Variable frequency microwave synthesis of silver nanoparticles. J. Nanopart. Res. **8**(1), 117–124 (2006)
44. Jiang, H., Moon, K., Li, Y., Wong, C.P.: Surface functionalized silver nanoparticles for ultra- highly conductive polymer composites. Chem. Mater. **18**(13), 2969–2973 (2006)
45. Jiang, H., Moon, K., Lu, J., Wong, C.P.: Conductivity enhancement of nano Ag filled conductive adhesives by particle surface functionalization. J. Electron. Mater. **34**(11), 1432–1439 (2005)
46. Jiang, H., Moon, K., Wong, C.P.: Tin/silver/copper alloy nanoparticles for low temperature sol- der pastes interconnect. In: IEEE 58th Electronic Components & Technology Conference, May 27–30, 2008, pp. 1400–1404 (2008)

Introduction to Nanoparticle-Based Integrated Passives

6

Ranjith John and Ajay P. Malshe

6.1 Introduction and Background

Faster, denser, cheaper: this has been the battle cry of the electronic age and nanotechnology even though in its early stage it has shown considerable promise in taking us further along the road. The dawn of the electronic age was marked by the invention of vacuum tubes. Vacuum tubes (VTs) slowly but surely drove the commercialization of radios, television, radar, and digital computers. But with the invention of transistors, VTs lost their appeal as transistors offered a greater reduction in size which made electronic devices more appealing to consumers. Figure 6.1 shows the difference in size between a VT and a transistor. Additionally, transistors offered increased reliability in comparison to VT.

Transistors started replacing VTs effectively in electronic devices, but when engineers tried building complex devices such as computers, they quickly realized that transistors had their limitations. For instance, assembly workers had to manually solder each component to the circuit using long metal wires. This resulted in reducing the effectiveness of the computer and other electronic devices. So they ran into the problem of building complex circuitry using transistors.

In the summer of 1958, Jack Kilby a Texas Instruments engineer invented the integrated circuit (also known as IC or silicon chip) which revolutionized the electronic industry. He built and tested the first IC which had a transistor and other components built together [2]. Figure 6.2 shows the first working IC built by Jack Kilby. After some skeptical analysis, ICs have successfully replaced discrete transistors in radios, television sets, and many other applications. In fact we can even say that without the invention of ICs, none of the modern technology could be possible.

An IC is a miniaturized electronic circuit consisting of both active components (which have a gain, such as transistors, logic gates) and passive components (which do not have gain, such as resistors, capacitors, inductors). Due to IC's mass production capability, reliability, and building block approach to circuit design, it has made a rapid transition in replacing discrete transistors. Even though an early attraction of the IC technology was a dramatic reduction in size due to the reduction in the number of components in electronic systems, it never came to be a reality. Engineers have worked on incorporating more features and functionality into ICs. The past four decades have seen tremendous progress in miniaturization and integration of ICs in the field of electronics. For instance, we have moved from few hundred transistors to a billion transistors on an IC [3]. This progress has brought electronic packaging technology (the craft of miniaturizing and interconnecting complex circuitry while maximizing the functionality of the circuit) to the forefront of the semiconductor industry. The evolution of the electronic packaging technology started during the early 1980s when through-hole packages such as dual-in-line packages (DIPs) were being replaced by surface mount packages such as quad flat packages (QFP), thin quad flat packages (TQFP). The 1990s saw the emergence of packaging technology called chip scale packaging (CSP) technology (packages are of the size 1.2× a die area) which offered smaller and lighter package. The most recent technology in packaging is the wafer-level chip scale package (WLCSP) which involves the blending of both the front-end and the back-end processes of packaging. There is strong research in academia and industry toward the development of a system-level integrated package. For instance, the NSF Packaging Research Center (PRC) of Georgia Institute of Technology has proposed a system-on-package which is shown in Fig. 6.3.

Even though the development of ICs has advanced by leaps and bounds, passive components are still lagging and have evolved at a slower pace with respect to their size and density. This continuous need for miniaturization of

R. John · A. P. Malshe (✉)
College of Engineering, University of Arkansas, Fayetteville, AR, USA
e-mail: rjohn@uark.edu; apm2@uark.edu

C. P. Wong et al. (eds.), *Nano-Bio- Electronic, Photonic and MEMS Packaging*, https://doi.org/10.1007/978-3-030-49991-4_6

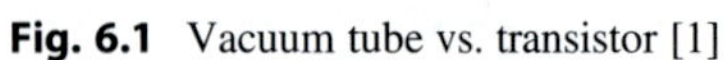

Fig. 6.1 Vacuum tube vs. transistor [1]

Fig. 6.2 Jack Kilby's first working IC [2]. (Photo courtesy of Texas Instruments)

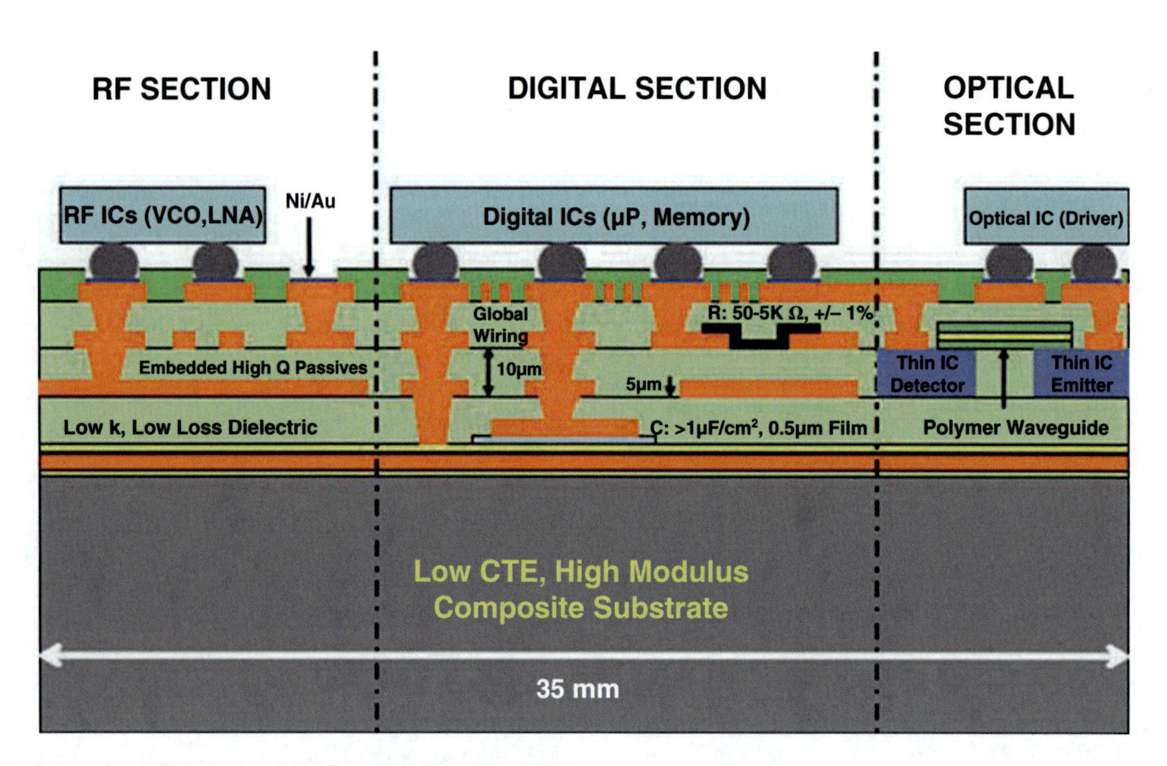

Fig. 6.3 Schematic of SOP structure [4]. (Copyright © 2004 by IEEE)

electronic devices is the driving force behind integrated passives (IPs). The rest of this chapter will address the background of passives, their classification, the need for nanoparticle (NP)-based IPs, the advantages and challenges of NP-based IPs, and future work.

6.2 History of Passive Technology

"Passives" usually refer to resistors, capacitors, and inductors but could include any device which does not have a gain. The "passive" industry has grown into a multibillion dollar business which supports electronic products in automotive, telecommunications, computer, consumer, and aerospace industries, for digital and analog–digital applications. Passives are generally used for connecting the complex network of ICs. They primarily serve as decoupling capacitors, bypass filters, and other purposes by which the functionality of the ICs is preserved.

Passives can be classified into two basic types, namely, discrete passives (DPs) and integrated passives (IPs). DPs encompass single passive devices which have their own leaded or surface mount technology (SMT) package (technology in which electronic components are mounted directly on the surface of printed circuit boards (PCB)). Today's handheld devices have greater functionality, and the complexities of these devices have greatly increased the number of discrete components. Figure 6.4 shows a circuit with discrete passives, occupying a majority of the space on a mother board.

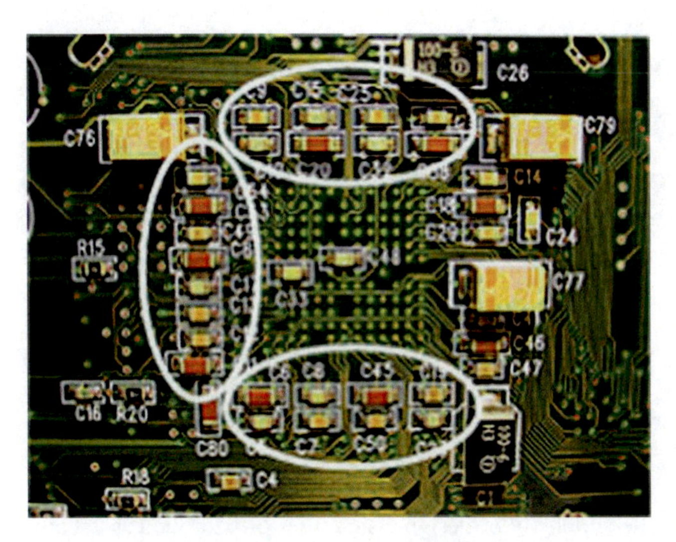

Fig. 6.4 Surface mount DPs [5]. (Photo courtesy of nGimat)

Analysts believed that DPs would be integrated away into ICs; however, exactly the opposite has occurred [6]. In 1984, passives accounted for 25% of all components on a PCB; by 1998 this grew to about 90% [7], and with the incessant growth toward complexity, this figure is expected to go up further. Additionally a typical handheld device today averages about 15–20 ICs and 300–400 passives. This results in a 20:1 ratio between passive and active devices [8]. Even though DPs are readily available and cheap, the challenges posed by them indirectly to cost and reliability are vital to the designer. For example, studies show that even though the cost of individual components averages only a few cents, the cost of assembly involving DPs is great [9]. Additionally, the reliability cost is another component of indirect cost. For instance, published studies show that the solder joint failure is one of the most important reasons for product failure [10] which leads to hidden cost. So relative to their integrated circuit (IC) counterparts which cannot give up the space (due to increasing functionality of products), passives are being closely investigated to provide the necessary reduction in size and cost while maintaining performance.

The idea of "integrated passives" includes the manufacturing of passives in groups or on a common substrate rather than in their own individual packages. They can be further classified as "embedded passives" and "passive arrays" depending on the packaging technology provided. Embedded passive (EP) technology involves packaging of passive devices inside the primary interconnect substrate rather than being on the surface. Passive arrays are those in which an array of passive devices are packaged in a single SMT case and mounted on the primary substrate. Figure 6.5 shows the hierarchy of IPs [12].

IPs are acknowledged to be the most likely alternative to DPs. Since, they promise to offer better reliability and more area saving on a PCB, there is widespread research being done to successfully incorporate them as a part of everyday semiconductor technology. Figure 6.6 shows the size reduction of IPs in comparison to DPs.

Fig. 6.5 Hierarchy of IPs [11]. (Copyright © 1996 by IMAPS)

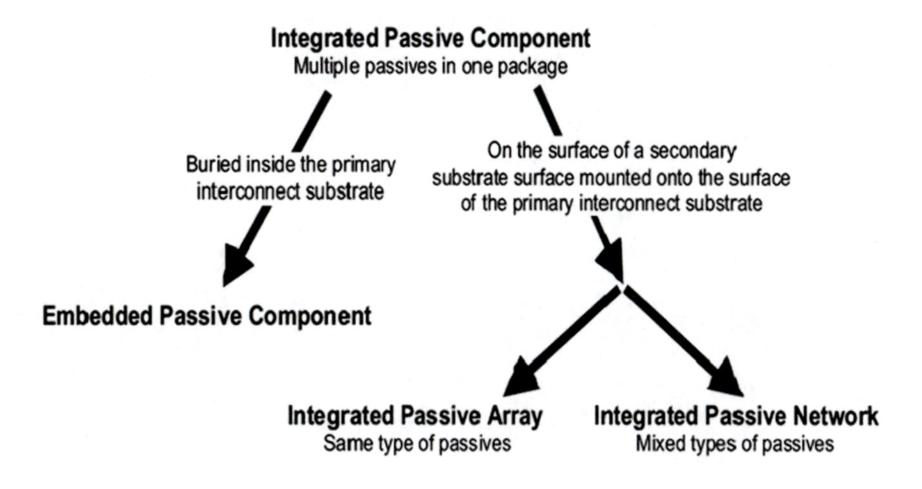

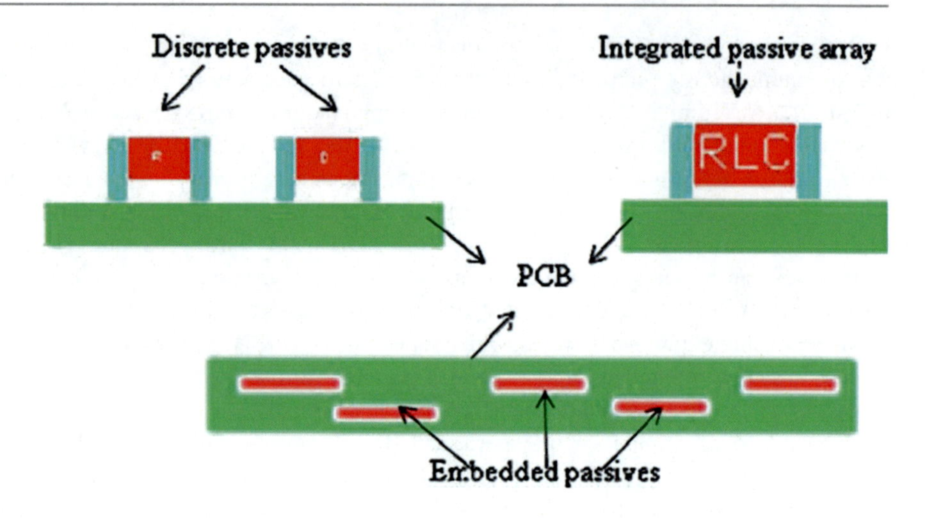

Fig. 6.6 Simple schematic of size reduction offered by IPs in comparison with DPs

Table 6.1 Projected needs for the portable emulator [11]

First year of significant production	Metric	2003	2005	2007	2013
Board assembly cost	c per I/O	0.5	0.45	0.4	0.3
Package I/O pitch	mm	0.5	0.5	0.5	0.5
Substrate lines and spaces	Microns	75	65	65	35
Substrate pad diameter	Microns	225	200	175	125
Max components per cm^2	$\#/cm^2$	50	55	60	25
Frequency on board	MHz	150	250	300	400
Solder	Composition	Lead/lead-free	Lead/lead-free	Lead-free	Lead-free
RF component thickness	mm	2.5	1.5	1.5, MEMS	1.5, MEMS
Passives		0201	0201	EPs	EPs
Product introduction cycle time, platform	Months	10	9	8	6
Product introduction cycle time, spin	Months	4	3	2	1

6.2.1 Next-Generation IP Needs

In 2000 the US market share for passives was estimated to be $18 billion. This huge multibillion dollar industry is expected to grow by about 3.7% per year [13]. Table 6.1 shows the trend in parameter of semiconductor technology which forecasts a paradigm shift in the passive industry.

As per the forecast of NEMI to meet the ever-increasing need for miniaturization and reduction of price, while maintaining or increasing the performance of commercial electronic devices (such as cell phones, IPODS, handheld cameras), the passive industry will have to adapt and make the transition from DPs to IPs.

IPs have several benefits over DPs, namely:

1. Overall system mass reduction
2. Better electrical performance and reliability
3. Design flexibility
4. Reduced unit cost

As indicated in the Introduction, DPs occupy a major portion of the substrate due to their bulky structure and can be replaced by EPs and IP arrays which can either be placed inside a layer of the substrate or can be arrayed into multiple passives which are then mounted on the surface of the substrate, thereby freeing up the board space and providing a major reduction in footprint. Also, IPs provide better electrical performance than their discrete counterparts because of the simplicity of design. The issues of parasitic capacitance and resistance which are present in DPs can be reduced greatly with the use of IPs. For instance, compared to their discrete counterparts, IP arrays do not require as many leads for electrical connection, thereby reducing the effect of parasitic, and they can be reduced further by the application of EPs which are fabricated under the surface of the board. This goes a long way in reducing the failure of product by reducing the most common failure seen in the industry – failure of solder joints. IPs also provide the designer with the design flexibility by giving options as to where and how to address the placement of passives. Additionally, since IPs can be formed simultaneously with active devices, the extra steps in manufacturing and assembly are eliminated.

6.2.2 Advantages of EPs

By definition, EP technology involves the processes of burying passive devices inside the substrate (such as ceramic, FR4, or silicon). EPs have been recognized as the most promising technology to replace DPs as they provide several advantages over DPs. As seen in Fig. 6.7, EPs offer a sizable advantage in size reduction, but apart from that they also completely dissolve the need for solder joints, thereby eradicating failure of solder joints [14]. Additionally, due to the fact that they are embedded in the substrate, they provide better electrical performance as they have shorter interconnect paths which reduce the signal delay and parasitic effects. Also, they provide increased silicon device efficiency and eliminate the need for separate packaging process and provide efficient circuit design [15–22]. Finally, placement time for discrete components is one of the major factors that determine the cost of passive components; by using EPs which are mass produced during the fabrication of substrate, there is a substantial reduction in assembly time and cost.

So if EPs offer all these features which will take us to our destination of smaller, cheaper, and better products, then why are they only used for niche applications? The answer to this question deals with multiple issues such as materials, processes, design tools, tolerance issues, and lack of a clear understanding of cost models. There are scores of research projects which have proven that EPs are attainable in almost any substrate, but there is still the need for identifying new materials and processes before EPs are standardized. The call for product scale-up and qualification is still in the developmental stages. For example, manufacturers who previously made only boards have to expand their business if they are to incorporate EPs. They will need to develop the background not just in processing but in issues such as performance, sizing, tolerance, parasitic, and reliability which were previously dealt with by a separate entity. Additionally the placement of DPs is one of the last steps to be performed on a PCB prior to its inclusion in the system, but with EPs which are fabricated during the process of board manufacturing and portions of board formed earlier, they should be able to withstand the thermal, chemical, and mechanical stress during the fabrication process. Also, the issues with rework due to yield problems will be compounded and might result in the scrapping of an entire board due to problem with one bad component. Finally, there is great demand for design tools to be established to effectively incorporate IP design into a system. All these aspects of the IP technology are being inspected in order to keep up with the demands of modern electronics.

6.2.3 Applications of IPs

Passives have different roles in the three main sectors of electronics – RF and microwave, digital, and analog–mixed-signal applications. The emergence of the wireless telecommunications network and the demand for high-frequency devices required by the military have led to the demand for integration of passives, and interestingly enough, the passive integration technology in this field (RF and

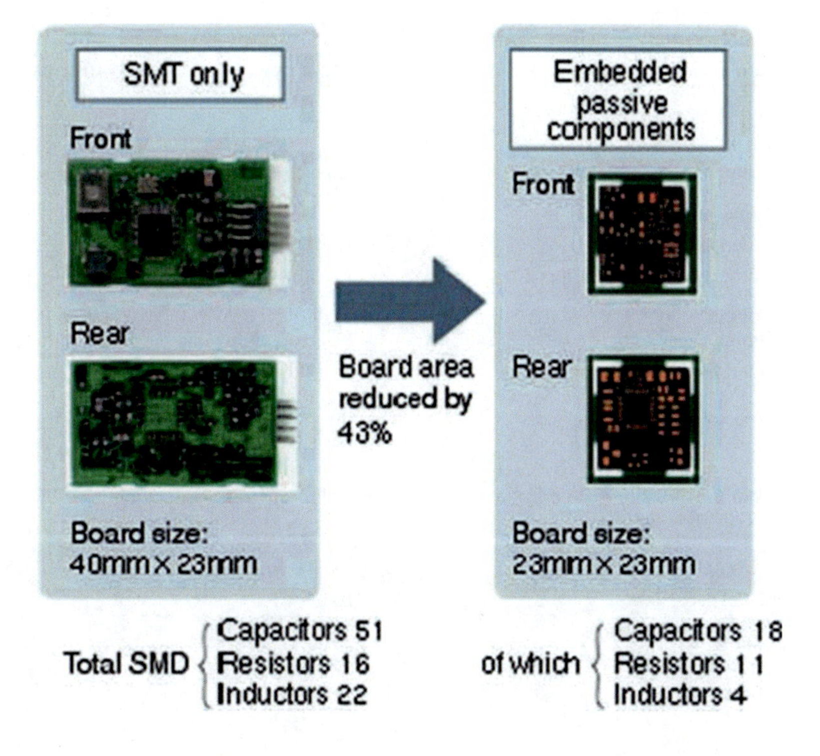

Fig. 6.7 Real estate saving and size reduction using EPs [14]. (Photo courtesy of TechOn)

microwave) is one which has been in practice for several decades. The main proponent of IPs in RF applications is the necessity to reduce the effects of parasitics. So there is a great demand to produce high operating frequency components, which have good electrical characteristics. Table 6.2 addresses the application range of passives in RF.

In digital applications ICs are accompanied by usually one and sometimes more than one decoupling capacitor whose main requirement is to be large in value so their tolerance is usually not an issue. Also, resistors are used as pull-up and pull- down devices between a signal line and either a power or ground supply terminal. Additionally, resistors are employed in terminating interconnections in high-speed digital circuits. This is not a comprehensive list of application for passive but does state some of the most important applications of passives in digital circuits. Table 6.3 lists the applications of passives in digital circuits.

In analog and mixed-signal applications, decoupling capacitors are found to accompany ICs. Since these circuits are not responsive to high frequencies, it is not important to have low series inductance capacitors. Capacitors and inductors are used to isolate analog parts of a circuit from digital parts. Table 6.4 shows the requirements of passives in analog and mixed-signal applications.

To summarize, passives find a wide range of applications and have a wide range of values in each area of application, placing severe demands on the need to inspect new materials, processes, and design tools for successfully integrating passives for future needs, and the general expectation is that nanotechnology will provide solutions for these needs making EPs the staple of next-generation devices.

6.3 Nanotechnology and Nanoparticles

Nanotechnology can be defined as the science and engineering of designing physical, chemical, and related functional properties of nanomaterials and nanostructures (<100 nm) and their fabrication, integration, and manufacturing to realize advanced integrated systems, many times across the scale boundaries. It is a highly diverse field which shows promise in the development of new materials and devices which are functional in the nanometer scale. It has wide ranging applications in the field of medicine, electronics, and energy. Even though nanotechnology is new, history shows us that research in the nanometer (10^{-9} m) scale is not new. For instance, gold nanoparticles have been used as an inorganic dye by the Chinese for more than a thousand years

Table 6.2 Passive component requirements for RF and microwave applications [9]

Application	Value range	Tolerance (%)	Requirement
Signal inductors	1–20 nH	1–10	High Q and self-resonant frequency
Signal capacitors	1–20 pF	5–10	High Q and self-resonant frequency
Decoupling capacitors	0.01–0.1 μF	10–20	Low series inductance
Choke inductors	1–10 μH	10–20	High Q and self-resonant frequency
Signal resistors	10–100 Ω	1–10	Tightly matched ratios
Terminating resistors	20–100 Ω	1–10	

Table 6.3 Passive component requirement for digital circuits [9]

Application	Value range	Tolerance (%)	Requirement
Decoupling capacitor	0.01–0.1 μF	10–20	Low series inductance
Pull-up/pull-down resistors	1–30 KΩ	10–20	
Terminating resistors	2–100 Ω	1–10	
Timing capacitors	10–100 pF	10–20	
Filter resistors	1–10 MΩ	20	

Table 6.4 Passive component requirements for analog and mixed-signal circuits [9]

Application	Value range	Tolerance (%)	Requirement
Resistors	10–100 MΩ	1–10	Tightly matched
Signal capacitors	10 pF–10 nF	5–10	Tightly matched
Decoupling capacitors	0.01–0.1 μF	10–20	
EMI filter capacitors	1–10 nF	10–20	
Choke inductors	1–10 μH	10–20	

[23, 24]. The use of colloidal gold is seen in treating arthritis and the diagnosis of a number of diseases by their interaction with spinal fluids of patients [25]. The semiconductor industry remaining true to Moore's law [26] combined with the recent explosion in our ability to image, engineer, and manipulate in the nanometer scale has brought about the drive toward the era of nanotechnology.

Over the last decade, nanoparticles have been of great scientific interest. They range from several nanometers to several tens of nanometers and act as a bridge between the bulk materials and atomic structures. Properties of materials change as they approach the nanometer scale, and these nanoparticles exhibit special properties relative to bulk materials [4, 27–30]. Nanoparticles can be used as agents of change for different processes or as building blocks of new materials or devices. They show a strong dependence on size, shape, and surface structure. Over the last couple of years, various commercial suppliers (NNI, Nanoscale, NanoSonic, etc.) have started making unique nanomaterials. These nanomaterials are unique in size, shape, and functionality due to surface chemistry. They have exceptional functional properties which are not possible in microsized particles. Further, they provide scientists and engineers the opportunity to assemble these particles in 1D, 2D, and 3D configurations through self-assembly and directed assembly processes. Due to these remarkable qualities, NPs have already found widespread use in application such as coatings, dispersions, and biological and environmental systems. In the semiconductor industry, scientists and engineers aim to take advantage of the new physical, chemical, and functional properties caused by this size scaling to reach new heights in the passive technology.

6.3.1 Synthesis of Nanoparticles

The science of synthesis of nanoparticles has gained a lot of attention over the past two decades and has seen enormous improvement. In fact the synthesis of NPs is a vast topic which cannot be thoroughly explained as a section of a chapter. There are countless articles and several books which deal exclusively with the different practices involved in the fabrication of nanoparticles. So in this section we will briefly discuss a few common methods used in the synthesis of nanoparticles. The most common methods used in the synthesis of nanoparticles are (i) ball milling, (ii) chemical reduction, (iii) laser ablation, and (iv) sol–gel method.

- Ball milling is a top-down approach where nanoparticles are produced from microsized particles. In this process microsized particles are loaded into vials and pounded by steel balls for a long duration. This grinding of microsized particles results in fine-sized nanoparticles whose size depends on the initial particle size and the number of hours spent in the ball mill. The possible drawback of this process is the presence of impurities in the nanoparticles contributed by the steel balls [31].
- The chemical reduction process is generally used in the reduction of metal compounds into elementary nanosized metal particles. Generally, a solution of the metallic compound is mixed with an alcohol (such as ethanol). Once it is well mixed, an alkaline solution such as NaOH/NaHCO3/NaBethyl solution is added. The mixture is heated to a given temperature under nitrogen gas, and the resulting liquid contains the reduced metal nanoparticles. For example, a reaction equation can be written as (M:metal, Et:Ethyl) [32]

$$\mathrm{MCl}x + x\mathrm{NaBet_3H} \rightarrow \mathrm{M} + x\mathrm{NaCl} + x\mathrm{Bet_3} + \frac{1}{2}x\mathrm{H_2}$$

- Laser ablation is the process where solid material is reduced to nanoparticles by a laser beam. This is very similar to the cutting of either metal or semiconductor materials. The raw material is the bulk form of the nanomaterial which is subjected to a focused laser beam, resulting in the production of fine nanoparticles. For instance, Lee et al. [33] used Lumonics KrF excimer laser of wavelength 248 nm and pulse-width of 12 ns in the production of well-dispersed 60–100 nm permalloy nanoparticles. The size of the nanoparticles can be controlled by process parameters, such as laser fluence. Nanoparticles produced in this method have the distinction of being very pure, spherical in shape, and not agglomerated.
- Sol–gel method is a wet chemical process which results in the synthesis of colloidal dispersions of organic, inorganic, and hybrid materials. Generally, sol–gel processing involves hydrolysis and condensation of both organic or inorganic salts and metal alkoxides. Organic or aqueous solvents are used to dissolve the above materials, and some catalysts are usually added to promote the reaction. One of the best known examples of sol–gel process is probably the production of SiO2. The hydrolysis and subsequent condensation reaction of Si–O methyl compound result in the formation of SiO2 gel or SiO2 nanocomposite (which is a gel-like material with nanoparticles). Below is an example of the chemical reaction in a sol–gel process [34].

Hydrolysis:

$$\mathrm{M(OEt)4} + x\mathrm{H2O} \Leftrightarrow \mathrm{M(OEt)4} - x\mathrm{(OH)}x + x\mathrm{EtOH}$$

Condensation:

$$\mathrm{M(OEt)}4 - x + \mathrm{(OH)}x + \mathrm{M(OEt)}4 - x + \mathrm{(OH)}x \Leftrightarrow \mathrm{(OEt)}4 - x\mathrm{(OH)}x - 1\mathrm{MOM(OEt)}4 - x\mathrm{(OH)}x - 1 + \mathrm{H2O}$$

6.3.2 Nanocomposites

Over the last decade, nanocomposites have garnered enormous interest due to the potential in building large area, flexible electronic devices [34–38]. Scientists and engineers have been intrigued by the remarkable physical and electrical properties of this unique material. For instance, upon optical pumping, the addition of ZnO nanoparticles into a polymer matrix results in laser-like behavior [33]. Addition of silicate nanoparticles into gel electrolytes results in high conductance [36]. Additionally, the application of nanocomposites to meet the continuous demand for miniaturization, high density, high speed, and flexible microelectronic devices is being investigated in the area of embedded passives.

6.3.2.1 Nanocomposite-Embedded Capacitors

In today's electronic devices, capacitors occupy about 50–60% of overall passive components [39]. So, the replacement of discrete capacitors with embedded capacitors offers the best potential in the drive toward miniaturization and high- density electronic packaging. They can be integrated into the substrate to provide decoupling, bypass, termination, and frequency-determining functions [39]. The packaging technologies available for integrating capacitors into a substrate are of three kinds: (i) MCM-C (packaging technology in which the substrate is ceramic), MCM-D (packaging technology in which the substrate is glass or silicon), and MCM-L (which uses PWBs as the substrate). The pros and cons of each packaging technology have been discussed in detail in several books and publications [19, 40]. Of the three, MCM-L technology is considered to be the best option for embedded capacitors due to the fact that PWBs offer the possibility of processing large number of MCMs on a single substrate. The fabrication of embedded capacitors has been demonstrated through several approaches, namely, sputtering, chemical vapor deposition, anodization, and ceramic green-sheet processes for the above packaging technologies. Nevertheless, these processes are too expensive or not process compatible with PCB fabrication. Nanoparticles dispersed in a polymer offer the best of both worlds; they offer the low-temperature processing which is compatible with PCB manufacturing as well as the high dielectric properties of the nanoparticle fillers which are essential for integration of capacitors into substrate. Some of the advantages offered by these nanocomposites are (i) good process compatibility with PCB; (ii) good CTE match with PWB; (iii) high temperature stability; (iv) low moisture absorption; (v) good adhesion to copper, interlayer-dielectrics, and substrates; (vi) ability to withstand high pH electroless plating bath; and (vii) good mechanical properties [19]. There are several different factors that are critical to the integration of nanoparticle-based embedded capacitors, namely, (i) materials, (ii) fabrication, and (iii) characterization.

Materials

There is wide range of nanoparticles which are being currently investigated to increase the dielectric constant, capacitance density, and streamlining of NP-based embedded capacitors. The most common nanoparticles which are being investigated for the dielectric properties fall under two categories: (i) ceramic-in-polymer nanocomposites and (ii) metal-in-polymer nanocomposites. For instance, Biswas et al. [41] have investigated the use of a ceramic-in-polymer approach where barium titanate nanoparticles were deposited in a polar polyaniline–polyurethane polymer composite to achieve a capacitance density of 15 nF/cm^2 for radio-frequency (RF) applications. Table 6.5 is a list of material choices being explored by various academic research groups. Bhattacharya et al. [47] have investigated the properties of lead magnesium niobate–lead titanate (PMN-PT) ceramic powder in epoxy composites for the fabrication of parallel plate capacitors. A capacitance density of 10 nF/cm^2 was achieved in epoxy nanocomposites. Windlass et al. [48] have investigated the effects of both lead magnesium niobate–lead titanate and barium titanate as ceramic fillers in epoxy composites with capacitance densities of 35 nF/cm^2 with film thickness of 2–3 μm. Additionally, Das et al. [45] have also investigated the effects of barium titanate nanoparticles in epoxy polymer composites with resulting capacitance densities of 10–100 nF/in.^{2} at 1 MHz. There are others who have employed metal nanoparticles to enhance the dielectric properties of polymer composites in achieving high capacitance densities for embedded capacitors. For example, Lu et al. [43, 44] achieved high dielectric constant and low dissipation factor for nanoparticle-based embedded capacitors by using in situ formed silver nanoparticles in carbon black/polymer composites. Additionally, Xu et al. [42] have investigated the effects of aluminum nanoparticles in polymer composites for the fabrication of embedded capacitors.

Table 6.5 Material choices in nanoparticle-based embedded passives

Material	References
Polymer/Al nanocomposite	Xu et al. [42]
Silver/carbon black/epoxy nanocomposite	Lu et al. [43, 44]
Barium titanate/epoxy nanocomposites	Das et al. [45]
Resin-coated copper nanocomposites	Das et al. [46]

Fabrication

There are several approaches involved in the fabrication of nanoparticle-based composites for embedded capacitors. One of the basic steps involved in the fabrication is the dispersion of nanoparticles. The dispersion and the size of nanoparticles play a very important role in determining the quality and capacitance density of nanoparticle-based embedded capacitors. According to the dispersant first rule of Cannon et al. [49], the efficient de-agglomeration which is essential for the fabrication quality of capacitors can be achieved in shorter times by dispersing powder in the solvent before adding the polymer. The most prevalent technique in dispersion is to mix the nanoparticles in a solvent such as 95% ethanol or propylene glycol methyl ether and then mix the epoxy polymer to the above solvent with the help of an ultrasonicator or a high-shear homogenizer to achieve a homogenous mixture of the nanocomposite. The nanocomposite is then oven-dried at the chemical-specific temperature and then either laminated or deposited on the substrate by either spin coating or meniscus coating. For example, Bhattacharya et al. [47] blended epoxy and ceramic powder using Pro Scientific high-shear homogenizer (Model PRO200) and roll milled the mixture 3–5 days before fabrication to prevent agglomeration and settling. The mixture was then spin coated for small area copper clad FR4 substrate deposition and meniscus coated for substrates of size 300 $\times$ 300 mm glass and copper clad PCB panels. The samples then underwent photolithography processes before the top copper electrode was deposited and processed. Additionally, Xu et al. have demonstrated the importance of surface treatment of nanoparticles before dispersion in epoxy. Windlass et al. [48] have stated in his paper that good dispersion of filler results in a homogeneous packaging, resulting in high reliable, uniform properties of film which result in high dielectric constants. Das et al. have demonstrated the fabrication of multilayer embedded capacitors on large coatings using the resin-coated copper capacitive (RC3) lamination process [46] where barium titanate nanoparticles and epoxy were dispersed in organic solvents. A thin layer of this nanocomposite was then deposited on copper substrate and dried to produce free standing RC3 which can be directly laminated onto a substrate. Das et al. [50] have also demonstrated the use of laser micromachining in the fabrication of barium titanate nanocomposite-based flexible embedded capacitors. In the laser micromachining process, the barium titanate/epoxy nanocomposite is deposited onto a substrate and dried to remove the organic solvent and then micro-machined using Nd:Yag laser. This process resulted in high performance, large area, and flexible capacitors. The schematic and conditions of the laser micromachining process are as shown in Figs. 6.8 and 6.9, respectively.

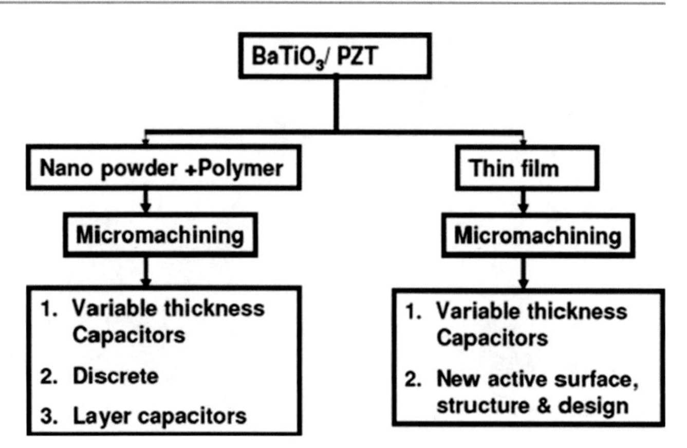

Fig. 6.8 Process flow chart [50]. (Copyright © 2007 by IEEE)

Additionally, Lu et al. [43, 44] have investigated the fabrication of embedded capacitors using an in situ processing of silver/epoxy nanocomposite, where silver nanoparticles were in situ synthesized in epoxy resin through the chemical reduction of silver nitrate. This was mixed with the carbon black/epoxy composite via ultrasonication to have a uniform, homogeneous mixture which resulted in a dielectric constant of $\sim$2200 at 10 kHz.

In summation, the general steps in fabrication of nanoparticle-based embedded capacitors are as follows:

1. Dispersion of nanoparticles in chemical specific solvent via ultrasonication.
2. Mixing of epoxy polymer in chemical specific solvent via ultrasonication.
3. Mixing of nanoparticle/solvent mixture and epoxy/solvent mixture via mixing and ultrasonicator.
4. Roll milling or ultrasonication of the nanoparticle/epoxy/solvent mixture to reduce settling and agglomeration.
5. The nanoparticle/epoxy/solvent mixture is deposited on substrate via spin coating or meniscus coating.
6. Then in case of photo-definable nanocomposite, mixture is patterned and oven-baked, whereas non-photo-definable nanocomposites are oven-baked to remove all organic solvents.
7. The top electrode is then deposited, patterned, and etched to form capacitors which can be laminated onto a substrate.

Characterization

Capacitance values are determined by the thickness, quality of the dielectric film, and the feature size. The characterization of embedded capacitors can be classified broadly into three different categories, namely, (i) characterization of nanocomposite, (ii) electrical characterization of the dielectric, and (iii) the reliability testing of the fabricated capacitors. So the first step in the characterization process is to ascertain

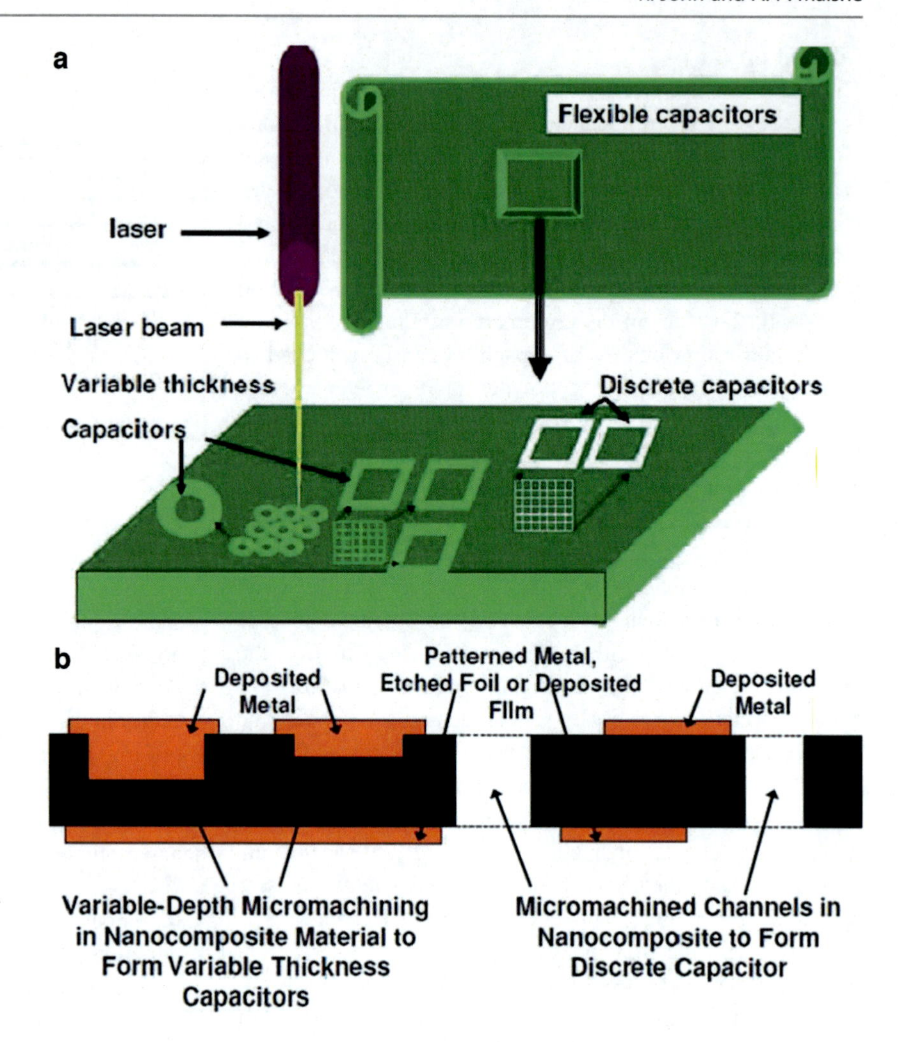

Fig. 6.9 Top (**a**) and cross-sectional (**b**) views of capacitors [50]. (Copyright © 2007 by IEEE)

the quality of the nanocomposite. This is done using techniques such as transmission electron microscopy (TEM), scanning electron microscopy (SEM), energy dispersive X-ray spectroscopy (EDS), Fourier transformed infrared spectroscopy (FTIR), X-ray diffraction (XRD), and stress rheometer. Once the quality of the nanocomposite has been characterized, the next step is to complete the electrical characterization of the dielectric which involves analyzing the thickness, dielectric constant, dissipation factor, frequency response, and capacitance. The final step in the characterization process is the reliability testing which is done by thermal shock, pressure cooker test (PCT), and IR reflow.

The challenge with nanoparticle/epoxy composites is the incompatibility between the hydrophilic nature of nanoparticles and the hydrophobic nature of epoxy. This opposing character of nanoparticles and epoxy results in agglomeration which results in poor quality of capacitors due to the non-uniformity of the dielectric thickness. So in the first step, characterization of the nanocomposite is performed using transmission electron microscopy (TEM) to verify the size, shape of the nanoparticle, and the dispersion of the nanoparticles in epoxy polymer matrix. For example, Xu et al. [42] claim that silane functionalization of aluminum nanoparticles helped in forming a uniform mixture of nanoaluminum/epoxy composite which was verified by the TEM images, which are shown in Fig. 6.10.

Second, it is important to identify the interlinking of the coupling agent or epoxy to the surface-treated nanoparticle. In the work done by Xu et al. [42], this verification was done via Fourier transformed infrared spectroscopy (FTIR). The surface chemistry of the aluminum nanoparticles was verified by oven-drying the aluminum nanocomposite and analyzing the dried KBr pellets prepared with the dried aluminum under the FTIR. The results obtained indicated the presence of Si and the absence of hydroxyl groups, indicating that there is good interaction between nanoaluminum and silane-coupling agent. There are other techniques such as X-ray diffraction (XRD) which can be used to determine the size of the nanoparticle in a given nanocomposite, and SEM and EDS have been utilized for several years to identify the presence of different chemicals in a given compound. For example, Lu et al. [43, 44] have employed TEM, SEM, EDS, and XRD to determine the size, size distribution, and verification of silver nanoparticles in the nanocomposite. The TEM pictures

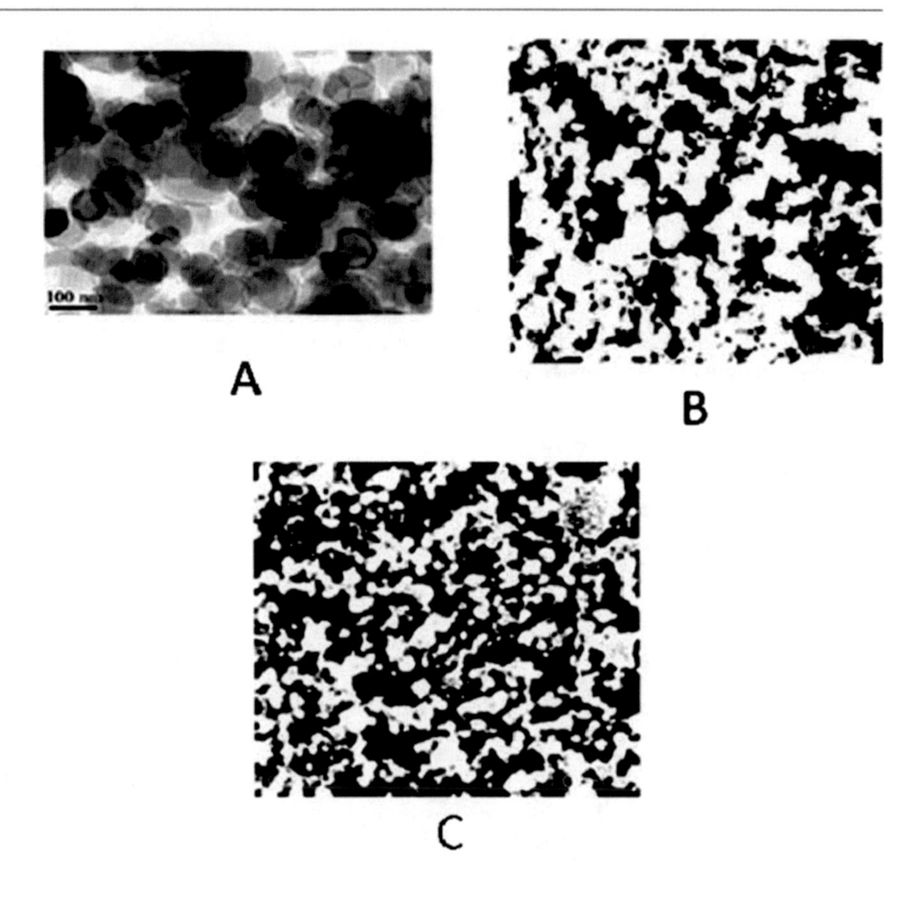

Fig. 6.10 (**a**) Nanoaluminum, (**b**) untreated nanoaluminum composite, and (**c**) silane-functionalized nanoaluminum composite [42]. (Copyright © (2005) by IEEE)

indicated the presence of capping agent and the effect of capping agent in inhibiting the growth and agglomeration of particles. SEM images verified the size and dispersion of the nanoparticles. EDS results confirmed the presence of silver nanoparticles in the polymer matrix. Finally XRD was used to verify the effects of the mixture on the size of silver nanoparticles in the epoxy polymer matrix. Additional tests are conducted to determine the rheology properties of the nanocomposite which is essential for the uniform coating of the substrate during the spin cycle (when the nanocomposite is deposited on a substrate at some predetermined conditions via spin coater or meniscus coater). The rheology properties of nanocomposites have been reported in several papers. For instance, Xu et al. [42] report that untreated nanoaluminum composites were more viscous than silane-treated composites. This indicates the importance of functionalization or presence of a coupling agent. Additionally, it also shows the effects of agglomeration of nanoparticles and how it adversely affects the dielectric properties. The next step involves the electrical characterization of the dielectric which accounts for the quality and capacitance density of the fabricated devices. The thickness of the dielectric plays a vital role in determining the capacitance. Capacitance is given as

$$C = \frac{\varepsilon_0 \varepsilon_r A}{d}$$

where C is the capacitance in farads, εo is the relative permittivity of free space, εr is the relative permittivity of the dielectric material, A is the area of each plate in square meters, and d is the distance between top and bottom plates.

The thickness of the nanocomposite film plays a very vital role in determining the capacitance and can be determined using a DEKTAK profilometer. Bhattacharya et al. [51] have reported the use of a DEKTAK profilometer as an electrical characterization tool by measuring the dielectric thickness. Additionally, the quality of the dielectric is imperative to the quality and capacitance density. In a two-phase composite system, the dielectric constant of the base polymer is vital in achieving higher permittivity, thereby higher capacitance density in the composite system [52]. The dielectric constant of the nanocomposite film can be measured via a precision LCZ meter, which measures the capacitance density and the dielectric constant as a function of temperature. There is a plethora of information that is provided by several authors who have used the LCZ meter to examine the capacitance and dielectric constant of capacitors. These measurements are at a certain frequency which also provides information upon the quality of the capacitors at different frequencies. For instance, Das et al. [46] report that the capacitance density of the RC3 nanocomposite-embedded capacitors decreased as the frequency was increased from 10 kHz to 1 MHz. The authors also indicate that the capacitance density depends on the loading of the nanomaterial by giving us evidence that the measured capacitance density of 33% loading nanomaterial has shown high capacitance value of 515 nF, which is six times of nanomaterial at 25% loading. Further electrical characterization can be done by obtaining the capacitance and dissipation factor of the nanocomposite film using an impedance/gain analyzer, and the frequency response of the nanocomposite film is obtained using a DEA test. The final stage in the characterization of embedded capacitors is the reliability testing. Embedded capacitors eliminate the failure due to solder joints, but they enhance the potential for defects due to material mismatch which leads to failure due to cracks and delamination. The reliability of nanoparticle-based embedded capacitors is examined by the tests such as IR reflow, pressure cooker test (PCT), and solder shock tests. These tests will accentuate the defects due to delamination or cracking. Additionally, thermal aging tests are run on these new-generation passives to make sure that they are at least as reliable as their surface mount counterparts.

In conclusion, the steps necessary to ascertain the quality of passives are in place and have been developed and implemented in universities and research labs. It is a matter of time before we figure out how to efficiently scale up the process and take advantage of this new line of passives which are on track to revolutionize the electronic industry.

6.3.2.2 Nanocomposite-Embedded Inductors

There is a great need for research and development in the field of embedded magnetic components which exhibit good performance characteristics. Magnetic components need compatible fabrication sequence in integrating them with embedded capacitors. A logical approach is to integrate these passive components using the appropriate MCM technology; the challenge is to make sure there is process compatibility while the quality is maintained.

In general, embedded inductors suffer from poor performance due to various core losses and low Q factor (which is the ratio of energy stored to energy lost). The losses in an inductor can be divided into three main categories: (i) eddy current loss, (ii) hysteresis loss, and (iii) residual loss. The ratios of these losses vary depending on the test and operating conditions such as frequency, magnetic flux density, and temperature. For instance, at high frequencies we see that the eddy current losses increase, but this loss can be controlled by increasing the core resistance and by the effect of grain boundaries which result in an increase in the electrical resistance at the interface. So it is important to have high core resistivity and narrow hysteresis to limit losses and improve performance. The integration of embedded inductors falls in the same categories as capacitors: (i) materials, (ii) fabrication, and (iii) characterization.

Materials

The first step in the integration of embedded inductors is the availability of materials. Due to the fact that high electrical resistivity is a need in producing inductors, ferrite materials play a vital role in the fabrication of inductors. The electromagnetic characteristics of these inductors depend on their composition and their structure. Since their discovery, ferrites have been involved in the fabrication of inductors but have been based on ceramic processing which involves high-temperature fabrication steps such as firing of ferrite-based paste. This cannot be integrated into mainstream electronics due to the fact that most mainstream electronic components use low-cost polymer substrates which cannot tolerate high temperatures. So, researchers have been working on developing ferrite-based polymer composites which can be used in the temperature-compatible fabrication of inductors. In fact there are several researchers who have developed ferrite particle-filled polymer composite inductors which can be used with organic substrates [53–59]. Additionally, there are several ferrite-filled polymer composites which are available commercially and are used in the fabrication of embedded capacitors. The shortcomings of these micron-sized ferrite particles are size limitations in the fabrication of thin film inductors and the presence of wide hysteresis loops which result in high hysteresis losses. These can be overcome by the use of nanoparticle-sized ferrite fillers which exhibit properties not seen in their bulk counterparts.

Nanoparticles of ferrites can be synthesized using anyone of the different techniques explained in earlier section, but one of the most common methods used is the method of chemical reduction. This method has been used significantly by researchers in the low-temperature synthesis. In this method the water-based reagents such as Co(NO3)2·6H2O and FeCl2 are mixed along with SDS which acts as a catalyst and forms a micellar solution. This solution is then heated to some predetermined temperature. This process results in the formation of nanophase CoFe2O4 which can be separated by centrifuge, and ferrite nanoparticles are obtained after oven drying for a predetermined time. Detailed procedure for the synthesis is documented elsewhere [57, 62]. The general procedure followed during the synthesis of ferrite nanoparticles is as follows. First, a salt solution that contains Fe compounds is prepared by dissolving Fe-linked chemicals in a predetermined ratio and then adding a hydroxyl group such as NH4OH solution into the ferrite solution. The hydroxyl group is added for pH adjustment of the salt solution. This results in the formation of Fe–O powder which can be reduced to nanophase ferrite solution at low temperature. The ferrite nanoparticles are then obtained by centrifuge and oven drying at certain predetermined level.

Fabrication

The fabrication of nanoparticle-based embedded inductors comprises the use of the binding properties of the polymer matrix to provide uniform adhesion to the magnetic nanoparticles. Commercially available epoxy resins are used as the base resin to which the nanoparticles are added. Then a commercial hardener is added to the mixture, and the composite is then cured for a predetermined time at a set temperature. The time and temperature for curing of the nanocomposite are dependent on the commercial epoxy and hardener that are used. For instance, Dong et al. [58] used EPON 8281 bisphenol A-type epoxy as the base resin and 4-methylhexahydrophthalic anhydride as a hardener and a methylimidazole as catalyst and added the ferrite nanoparticles to form the magnetic nanocomposite which was cured at 120 °C for 30 min and 150 °C for 2 h. In his work, Raj et al. [59] synthesized a nickel–zinc–ferrite nanoparticle of size 5–15 nm and dispersed them in a commercially available epoxy to form a thick paste. The paste was then screen printed onto an FR4 board and cured to a set temperature and fabricated the desired nanoparticle-based inductor to specific dimensions. Figure 6.11 shows the fabrication steps of a silicon-based inductor, and Fig. 6.12 shows the spiral inductor designed at the University of Toulouse. In conclusion the most common method of producing nanocomposites is to disperse synthesized nanoparticles into an epoxy/hardener mixture and cured to manufacturer specifications which are then characterized to determine their quality.

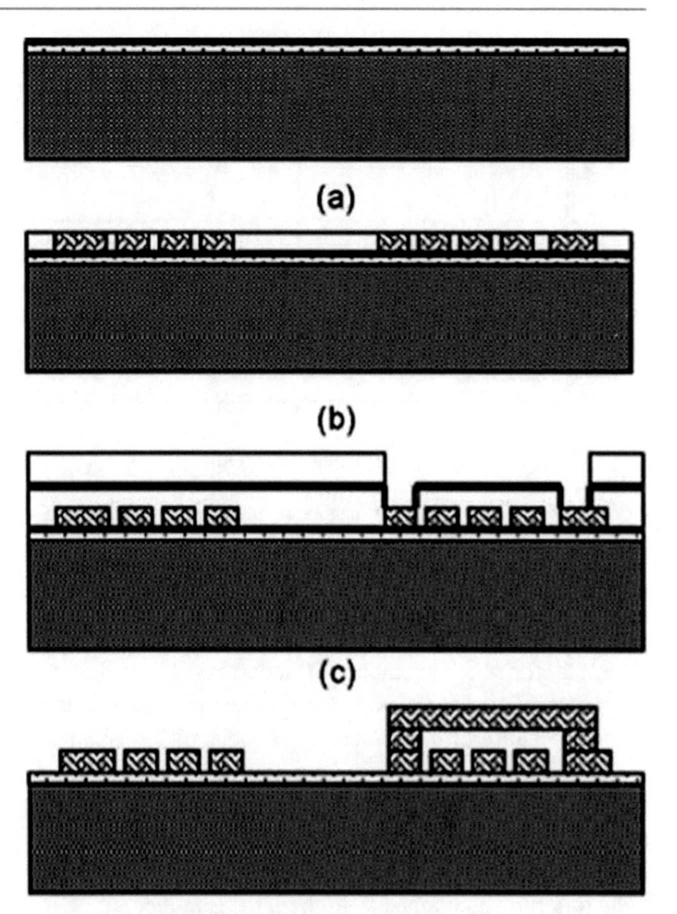

Fig. 6.11 Scheme of the inductor fabrication processes (**a**) 0.7 gm SiO2 and Ti/CU/Ti layer deposition. (**b**) First copper plating for the coil part of the inductor. (**c**) Air-bridge pattern and Cu/Ti layer deposition. (**d**) Air-bridge plating and seed layer removal [60]. (Copyright © (2006) by IEEE)

Characterization

The characterization process can be separated into different phases similar to the characterization of nanoparticle-based capacitors. The first step is to characterize the nanoparticles to verify the size and crystalline properties. The XRD is generally used to verify the crystalline structure of the nanoparticles. The nanoparticles are examined to see if their XRD curves match that of their bulk samples. Additionally, TEM or HRTEM is used to verify the size of the nanoparticles. The final and most important characterization of the nanoparticles is to analyze their magnetic property. This measurement is done using a SQUID (superconducting quantum interference device) magnetometer. The hysteresis response of the ferrite nanoparticles is plotted, and if the resulting loop is narrow, then the quality of the nanoparticle is considered to be very desirable for inductor application as it shows that there is almost no hysteresis loss. An example of a desirable hysteresis loop which will result in low loss and good quality inductor is shown in Fig. 6.13.

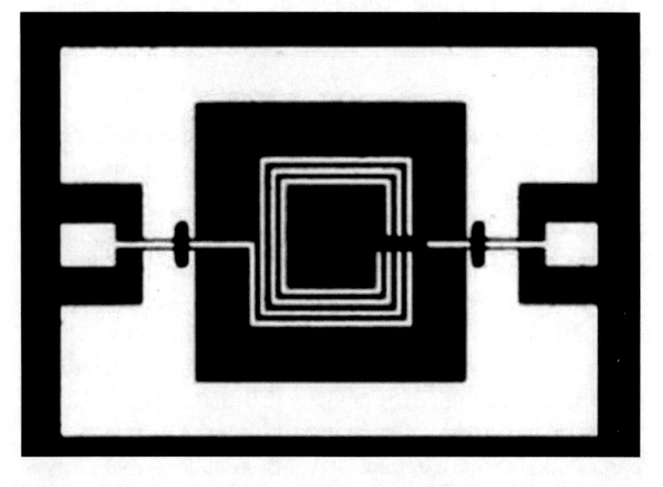

Fig. 6.12 Silicon-integrated inductors [61]. (Copyright © (2007) by IEEE)

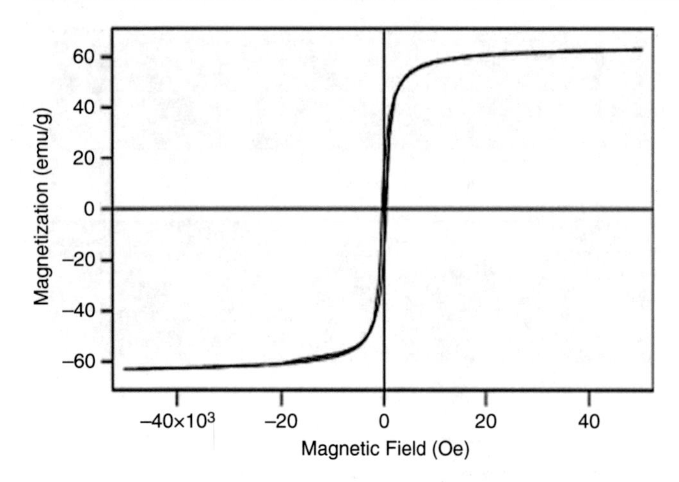

Fig. 6.13 Hysteresis curve of magnetic nanocomposite [55]. (Copyright © (2004) by IEEE)

Second, the electrical properties of the nanocomposite are tested to verify that the quality of the nanocomposite is desirable for the fabrication of inductors. The dielectric properties of the nanocomposite play an important role in the performance of the inductor. The dielectric constant and loss at low frequencies are measured using a precision LCR meter. The high-frequency measurement of the permittivity of the nanocomposites is conducted using an s-parameter Network Analyzer. The impedance of the fabricated inductors can be obtained from these measurements.

In conclusion, there is a lot of room for research and development in the area of nanoparticle-based embedded inductors. We still need to develop universally standard synthesis and fabrication techniques which when followed will result in good quality inductors. Additionally, there is room for exploration in the area of materials available for the fabrication of inductors. Finally there is a great need to develop standard methodology before nanoparticle-based embedded passives can be an integral part of mainstream electronics.

6.3.2.3 Nanoparticle-Based Embedded Resistors

The behavior of elemental metals is fairly predictable and has been well documented over the years. In order for nanoparticle-based embedded resistors to gain acceptance into mainstream electronics, it has to overcome several challenges. A few issues that need to be solved are tight tolerance, high yield, and reliability and availability of wide range of resistances. In fact iNEMI (International Electronics Manufacturing Initiative) predicts that maximum resistance of 100 kΩ/sq. is essential for embedded resistors to become a part of mainstream electronics. Another challenge is that even though solder joints are eliminated, better optimization of process flow is necessary as rework is not an option.

In embedded resistor technology, resistors are patterned as strips of the resistive material. The ends of the strip have pads for interconnection to other components on the board. The resistance is therefore dependent on the property of the material and the dimensions of the strip. The general formula for resistance is given below

$$R = \frac{\rho L}{A}$$

where R is the resistance (ohms), ρ is the resistivity of the material (Ω cm), L is the length of the strip (cm), and A is the cross-sectional area (cm^2). For majority of the applications, resistance that ranges from 10 Ω to 1 MΩ is a requirement [57]. The challenge in integrating resistors into the board deals with finding materials which are process compatible while giving the reliable performance.

Material Choices

The industry has used metal paste in the fabrication of resistors for a very long time. The paste is a mixture of micron-sized metal particles and some organic resin. This is applied onto a substrate and allowed to cure; after curing, the paste is fired, and the metal particles melt and fuse, forming a conductive film. This technology requires the substrate to withstand temperatures ranging between 550 °C and higher.

For embedded technology to become an everyday practice in industry, the fabrication of embedded resistors needs to be done at low temperatures. Organic boards can generally withstand a temperature in the order of few hundred degrees Celsius.

The key to nanoparticle-based embedded resistors is to use the natural property of nanoparticles. Nanoparticles have a tendency to agglomerate due to their surface energy, and this property of nanoparticles can be used effectively in the fabrication of nanoparticle-based embedded resistors. Metal nanoparticles can be mass produced by solid-state thermal decomposition process which is done at relatively low cost and at low temperatures. For instance, silver nanoparticles can be manufactured by heated decomposition of solid fatty acid silver at 200 °C for 5 h [57].

The synthesis of metal alloys for use as nanoparticles in the fabrication of embedded resistors is being actively pursued by several researchers in the industry and at educational institutions. Silver–palladium alloys and alloys of copper nanoparticles are being investigated in the fabrication of embedded resistors. For example, Nakamoto [57] has investigated the synthesis of silver–palladium alloys as nanoparticles for the fabrication of resistors. He claimed that silver–palladium nanoparticles which were synthesized have superior migration resistance which is a very important issue in the electronic industry. The fabrication of resistors using these silver–palladium alloys has been done at 300 °C for 30 min. The resistivity of these nanoparticle-based components was calculated to be 10 ohm-cm which is favorable for use with polyimide substrates. Water drop method evaluation showed that the migration resistance was higher than in pure nanoparticles. This results in the fabrication of narrow electronic structures with a very high reliability. Thus low-cost mass production of high-quality metal nanoparticles is feasible for volume production of nanoparticle-based embedded resistors.

Fabrication

Resistors are an integral part of every electronic circuit, so the fabrication of nanoparticle-based resistors is a vital component in realizing embedded circuits. There are several methods of fabricating embedded resistors which have been demonstrated over the years (embedded resistor fabrication). The most promising method for the fabrication of nanoparticle-based embedded resistors is the technology of inkjet printing.

Inkjet printing is the method of dispensing liquids and pastes at certain specific regions of a circuit which can be fired to form the required component. In one of the designs, this technology is based on the volumetric change in the fluid by displacement of piezoelectric material that is coupled to the fluid (a method of characteristics model of a DOD inkjet device using an integral method drop formation model). The volumetric change causes the conductive region to be fabricated without the need for photolithography and other practices which are common in the fabrication of passives in the semiconductor industry. Inkjet printing technology is considered to exhibit considerable promise in making embedded passive technology a reality and the backbone to meet the demands of next-generation electronic devices. There are two specific types of modes used in the dispensing of conductive solutions, namely, demand mode and continuous mode. In the demand mode, the conductive paste or fluid is dispensed at only a particular region, whereas in continuous mode the paste is dispensed in a continuous fashion to form passives over a specified area and can be used in the fabrication of capacitors and other devices. Shah et al. [60] reported the use of inkjet technology which has been employed in the fabrication of passives using droplets of polymers, solder, organometallic, and metal nanoparticle solutions of sizes 25–125 μm in diameter at rates up to 4000 droplets/s. Figures 6.14 and 6.15 show the embedded resistors printed using inkjet technology by Shah et al. at MicroFab Technologies.

The fabrication of embedded resistor technology has been demonstrated successfully by researchers at the National Institute of Standards and Technology (NIST) [58]. They have demonstrated the fabrication of embedded resistor technology by fabricating resistors ranging from 100 Ω to several mega ohms. Printed resistors have ranged from a few micrometers to several millimeters long. Thus there is ongoing research which has given us glimpses that the embedded resistor technology can become a reality and that nanoparticles will become an integral part in the drive toward miniaturized, reliable, electronic devices.

Fig. 6.14 Inkjet printed embedded resistor using conductive polymer (100 Ω/sq. resistivity) [63]. Copyright (c) 1999 by IMAPS

Characterization

The characterization techniques for nanoparticle-embedded resistors involve the analysis of the quality of the nanoparticles and quality and reliability of the fabricated resistors. The first step is to characterize the metal nanoparticles, and this can be done using TEM, SEM, and other analysis techniques described in earlier sections. These analysis tools will give a good understanding of the size, shape, and other qualities of the nanoparticles. The second step is to verify that the fabricated resistors give the correct resistance values for which they were designed. This can be accomplished using a two-probe method or four-probe methods which are well-known measurement techniques.

Additionally, reliability tests need to be completed on embedded resistors as it becomes difficult to rework or replace embedded resistors after they are integrated into the substrate. So it is imperative to complete reliability test before the embedded resistor technology can be called a success. The reliability of nanoparticle-based embedded resistors should be subjected to the same tests that embedded resistors

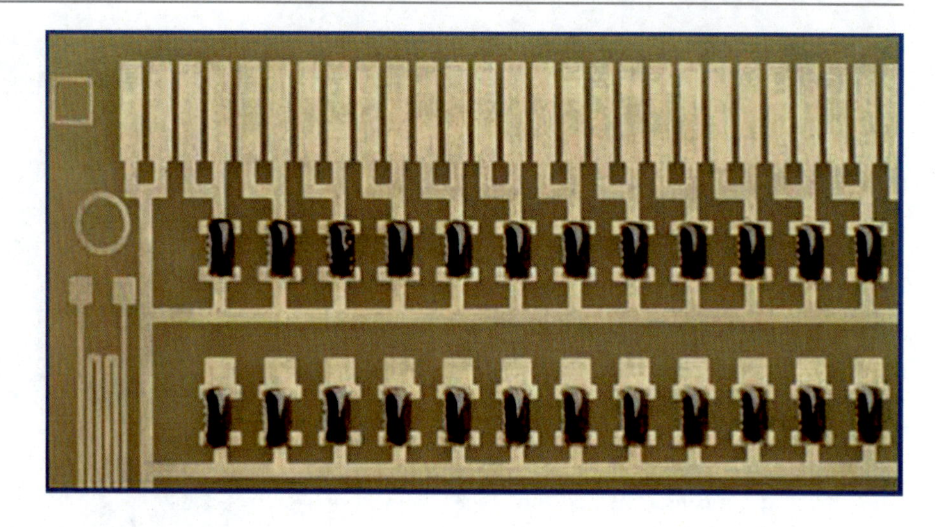

Fig. 6.15 Inkjet printed carbon nanotube filled epoxy resistor [60]. Copyright (c) 2002 by IMAPS

are subjected to. For instance, embedded passives are subjected to temperature shock test, high accelerated stress test (HAST), electrostatic charge discharge (ESD), thermal cycling, etc. Min et al. [31] describe in his work that embedded resistors fabricated during his collaborative effort with NIST have passed the reliability tests conducted.

In conclusion the synthesis, fabrication, and characterization of nanoparticle- based embedded resistors have been explored at universities' labs and are being currently investigated by the R&D departments at industries. The success of nanoparticle-based embedded passives depends on integrating all the different synthesis, fabrication, and characterization techniques into scalable production. Additionally, the rapidly evolving microelectronic technology coupled with the swift ascent of nanotechnology is leading us toward the goal of miniaturization. The unique properties found in nanoparticle which cannot be realized in microparticles and bulk material give us the opportunity to design devices with functional density which cannot be realized using other technologies. Thus it is not a question of whether nanoparticle-based embedded resistors will be realized but how soon will they be realized.

6.4 Summary

The advent of nanotechnology has provided glimpses of potential in solving some of the problems associated with embedded passives. The nanosize and the unique properties associated with this small size have brought into focus a new area for research and development. Literature, conferences, and workshops have started to focus on using the potential shown by nanoparticles in making nanoparticle-based embedded passives. Over the years, NEMI has clearly shown that the lack of infrastructure is one of the major problems associated with embedded passives, which also is needed for nanoengineered passives. In order for nanoparticle-based embedded passives to become a reality, there is a need for more research and development in materials, processes, designs, and proven scale-up and cost models along with environmental and health safety (EHS) data. This area is in the early stages of fundamental understanding and development. Depending upon the breakthroughs and systematic developments, the next decade could see integration of nanoparticle-based integrated passives.

Acknowledgments The authors would like to acknowledge the financial support of the Office of Naval Research (ONR). Also, one of the authors (APM) acknowledges the support of National Science Foundation (NSF; grant # 0501597).

References

1. http://nobelprize.org/educational_games/physics/integrated_circuit/history/(downloaded 10/3/08) Nobelprize.org. The history of integrated circuits 2005
2. http://www.ti.com/corp/graphics/press/image/on_line/co1034.jpg(downloaded 10/9/08) The Chip that Jack Built, (c. 2008), (HTML), Texas Instruments
3. Clarke, P.: Intel enters the billion-transistor processor era. EE Times 14
4. Sundaram, V., Tummala, R., Bhattacharya, S., Pulugurtha, R.M.: System-on-a-Package (SOP) Substrate and Module with Digital, RF and Optical Integration. Electronic Components and Technology Conference 54th Proceedings. **1**, 17–23 (2004)
5. http://www.ngimat.com/electronics/passives.html (downloaded 11/12/08) nGimat.com
6. Sandborn, P., Etienne, B., Subramanian, G.: Application-specific economic analysis of integral passives. IEEE Trans. Electron. Packag. Manufactur. **24**, 203–213 (2001)
7. Lasky, R.: Growth continues for passive components. Electron. Packag. Prod. **38**, 77–78 (1998)
8. Rector Jr., J., Doughtery, J., Brown, V., Galvagni, J., Pyrmak, J.: Integrated and Integral Passive Components: A Technology Roadmap. Electronic Components & Technology Conference. 713–723 (1997)

9. Frye, R.: Passive components in electronic applications: requirements and prospects for integration. Int. J. Microcircuit. Electron. Packag. **19**, 483–490 (1996)
10. Oberschmidt, J., Humenik, J.A.: Low Inductance Capacitor Technology. Electronic Components & Technology Conference 40th Proceedings. **1**, 284–288 (1990)
11. Dougherty, J.P., Galvani, J., Marcanti, L., Sheffield, R., Sandborn, P., Ulrich, R.K.: The NEMI roadmap: Integrated Passives Technology and Economics. Proceeding of the Capacitor and Resistor Technology. 1–11 (2003)
12. Ulrich, R.K., Schaper, L.W.: Integrated Passive Component Technology, pp. 19–22. Wiley-IEEE Press, New York (2003)
13. Electronic Industry Market Research and Knowledge Networks. Passive Components – Private Companies Report April 2001
14. http://techon.nikkeibp.co.jp/english/img2/nea0305manu1fig 1.jpg (downloaded 7/07/09) TechOn. Cover Story: Motorola Ships Passive-Embedded PCB for Mobile Phones May 2003
15. Bhattacharya, S.K., Tummala, R.R.: Epoxy nanocomposite capacitors for applications as MCM-L compatible integral passives. J. Electron. Packag. **124**, 1–6 (2002)
16. Tummala, R., Chahal, P., Bhattacharya, S.K.: Recent Advances in Integral Passives at PRC. Proceedings of the IMAPS 35th Nordic Conference. (1998)
17. Chahal, P., Tummala, R.R., Allen, M., Swaminathan, M.: A novel integrated decoupling capacitor for MCM-L technology. IEEE Trans. Compon. Packag. Manuf. Technol. **21**, 184–193 (1998)
18. Power, C., Realff, M., Bhattacharya, S.K.: A Decision Tool for Design of Manufacturing Systems for Integrated Passive Substrates. IMAPS Advanced Technology Workshop. (1999)
19. Bhattacharya, S.K., Tummala, R.R.: Next generation integral passives: materials, processes, and integration of resistors and capacitors on PWB substrates. J. Mater. Sci. **11**, 253–268 (2000)
20. Yoon, H., Hou, J., Bhattacharya, S.K., Chatterjee, A., Swaminathan, M.: Fault detection and automated fault diagnosis for embedded integrated passives. J. VLSI Sig. Process. **21**, 265–276 (1999)
21. Chahal, P., Tummala, R., Allen, M., White, G.: Electro-less Ni–P–W–P thin film resistors for MCM-L based technologies. Electronic Components and Technology Conference 48th Proceedings. 232–239 (1998)
22. Agarwal, V., Chahal, P., Tummala, R., Allen, M.: Improvements and Recent Advances in Nanocomposites Capacitors Using a Colloidal Technique. In: Proceedings of 48th Electronic Components and Technology Conference, pp. 165–170 (1998)
23. Camusso, L., Bortone, S.: Ceramics of the World: From 4000BC to the Present, p. 284. Harry N. Abrams, New York (1992)
24. Zhao, H., Ning, Y.: Techniques used for Preparation and Application of Gold powder in Ancient China. Gold Bull. **33**, 103 (2000)
25. Turkevich, J.: Colloidal gold: part I – historical and preparative aspects morphology and structure. Gold Bull. **18**, 86 (1985)
26. Moore, G.E.: Cramming more components onto integrated circuits. Proc. IEEE. **86**, 82–85 (1998)
27. Carpenter, E.E.: Iron nanoparticles as potential magnetic carriers. Sci. Direct. **255**, 17–20 (2001)
28. Papp, S., Szucs, A., Dekany, I.: Preparation of PdO nanoparticles stabilized by polymers and layered silicate. Appl. Clay Sci. **19**, 155–172 (2001)
29. Rataboul, F., Nayaral, C., Casanove, M.J., Maisonnat, A.: Synthesis and characterization of monodisperse zinc and zinc oxide nanoparticles from the organo-metallic precursor [Zn(C6H11) 2]. J. Organomet. Chem. **643–644**, 307–312 (2002)
30. Kruis, F.E., Fissan, H., Peled, A.: Synthesis of nanoparticles in the gas phase for electronic, optical and magnetic applications—a review. J. Aerosol Sci. **29**, 511–535 (1998)
31. Min, G.: Embedded passive resistors: challenges and opportunities for conducting polymers. In: Proceedings of the International Conference on Science and Technology of Synthetic Metals, pp. 49–52 (2005)
32. Xu, J.: Dielectric Nanocomposite for High Performance Embedded Capacitors in Organic Printed Circuit Boards. PhD Dissertation Georgia Institute of Technology. **29**, (2006)
33. Lee, J., Becker, F.M., Brock, R.J., Keto, W.J., Walser, M.R.: Permalloy nanoparticles generated by laser ablation. IEEE Trans. Magnet. **32**, 4484–4486 (1996)
34. Anglos, D., Stassinopoulos, A., Das, R.N., Zacharakis, G., Psyllaki, M., Anastasiadis, S.H., Vaia, R.A., Giannelis, E.P.: Random laser action in organic/inorganic nanocomposites. J. Opt. Soc. Am. B. **21**, 208–213 (2004)
35. Lappas, A., Zorko, A., Wortham, E., Das, R.N., Giannelis, E.P., Cevc, P., Arcon, D.: Low-energy magnetic excitations and morphology in layered hybrid perovskite- poly(dimethylsiloxane) nanocomposites. Chem. Mater. **17**, 1199–1207 (2005)
36. Scully, S.R., Lloyd, M.T., Herrera, R., Giannelis, E.P., Maliaras, G. G.: Dye-sensitized solar cells employing a highly conducting and mechanically robust nanocomposite gel-electrolyte. Synth. Met. **144**, 291–296 (2004)
37. Schmidt, D., Saha, D., Giannelis, E.P.: New advances in polymer/ layered silicate nanocomposites. Curr. Opinion Solid State Mater. Sci. **6**, 205–212 (2002)
38. Vaia, R.A., Giannelis, E.P.: Polymer nanocomposites. Status Opportunities MRS Bull. **26**, 394–401 (2001)
39. Herndon V.A.: Passive Components Technology Roadmap, National Electronics Manufacturing Technology Roadmaps, International Electronic Manufacturing Initiative (www.inemi.org), December 1998
40. Brown, W.D.: Advanced Electronic Packaging: With Emphasis on Multichip Modules, pp. 24–28. IEEE Press, Piscataway (1999)
41. Biswas, A.: Nanostructured Barium titanate composites for embedded radio frequency applications. Appl. Phys. Lett. **91**, 1–3 (2007)
42. Xu, J., Wong, C.P.: High-K Nanocomposites with Core-Shell Structured Nanoparticles for Decoupling Applications. Electronic Components and Technology Conference 55th Proceedings. 1234–1240 (2005)
43. Lu, J., Moon, K.S., Xu, J., Wong, C.P.: Synthesis and dielectric properties of novel high-K polymer composites containing in-situ formed silver nanoparticles for embedded capacitor applications. J. Mater. Chem. **16**, 1543–1548 (2006)
44. Lu, J., Wong, C.P.: Dielectric Performance Enhancement of Nanoparticle-Based Materials for Embedded Passive Applications. Electronic Components and Technology Conference 58th Proceedings. 736–741 (2008)
45. Das, R.N., Poliks, M.D., Lauffer, J.M., Markovich, V.R.: High Capacitance, Large Area, Thin Film, and Nanocomposite Based Embedded Capacitors. Electronic Components and Technology Conference 56th Proceedings. 1510–1515 (2006)
46. Das, R.N.: Resin Coated Copper Capacitive (RC3) Nanocomposites for Multilayer Embedded Capacitors. Electronic Components and Technology Conference 58th Proceedings. 729–735 (2008)
47. Bhattacharya, S.K.: Epoxy nanocomposite capacitors for applications as MCM-L compatible integral passives. J. Electron. Packag. **124**, 1–6 (2002)
48. Windlass, H., Raj, M., Balaraman, D., Bhattacharya, S.K., Tummala, R.R.: Colloidal processing of polymer ceramic nanocomposite integral capacitors. IEEE Trans. Electron. Packag. Manufactur. **26**, 100–105 (2003)
49. Cannon, W.R.: Dispersants for non-aqueous tape casting. Adv. Ceramics. **9**, 164–183 (1984)
50. Das, R.N.: Laser Micromachining of Nanocomposite-Based Flexible Embedded Capacitors. Electronics Components and Technology Conference 57th Proceedings. 435–441 (2007)
51. Bhattacharya, S.K., Tummala, R.R.: Integral passives for next generation of electronic packaging: application of epoxy/ceramic

nanocomposites as integral capacitors. Microelectron. J. **32**, 11–19 (2001)
52. Jayasundere, N., Smith, B.: Dielectric constant for binary piezoelectric 0-3 composites. J. Appl. Phys. **73**, 2462–2466 (1993)
53. Tang, S.C., Hui, S.Y.R., Chung, H.S.: A low-profile power converter using printed-circuit board (PCB) power transformer with ferrite polymer composite. IEEE Trans. Power Electron. **16**, 493–498 (2001)
54. Brandon, E.J., Wesseling, E.E., Vincent, C., Kuhn, W.B.: Printed micro-inductors on flexible substrates for power applications. IEEE Trans. Comp. Packag. Technol. **26**, 517–523 (2003)
55. Park, J.Y., Allen, M.G.: Low Temperature Fabrication and Characterization of Integrated Packaging-Compatible, Ferrite Core Magnetic Devices. Twelfth Annual APEC '97 Conference Proceedings. **1**, 361–367 (1997)
56. Park, J.Y., Allen, M.G.: Packaging-composite high Q microinductors and micro filters for wireless applications. IEEE Trans. Adv. Packag. **22**, 207–213 (1999)
57. Rondinone, A.J., Samia, A.C.S., Zhang, Z.J.: A chemo metric approach for predicting the size of magnetic spinel ferrite nanoparticles from the synthesis conditions. J. Phys. Chem. B. **104**, 7919–7922 (2000)
58. Dong, H., Liu, F., Wong, C.P.: Magnetic Nanocomposite for High Q Embedded Inductor. In: 9th International Symposium on Advanced Packaging Materials, pp. 171–174 (2004)
59. Raj, M., Muthana, P., Xiao, T.D., Wan, L., Balaraman, D., Abothu, I.R., Bhattacharya, S.K., Swaminathan, M., Tummala, R.R.: Magnetic Nanocomposites for Organic Compatible Miniaturized Antennas and Inductors. In: 10th International Symposium on Advanced Packaging Materials: Processes, Properties and Interfaces, pp. 272–275 (2005)
60. Shah, V.G., Hayes, D.J.: Fabrication of passive elements using ink-jet technology. Printed Circuit Fabrication. **25**, 36–38 (2002)
61. Nakamoto, M.: Fine electronic circuit pattern formation on plastic substrates by metal nano particle pastes. Weld. Int. **21**, 831–835 (2007)
62. Chao, T.Y., Cheng, Y.T.: Synthesis and Characterization of Cu/CoFe2O4 Magnetic Nanocomposite for RFIC Application. IEEE Trans. 810–813 (2006)
63. Hayes, D.J., Wallace, D.B., Cox, W.R.: Microjet Printing of Solders and Polymers for Multichip Modules and Chip Scale Packages. Proceedings, IMAPS International Conference on High Density Packaging and MCM's. 1–6 (1999)

Thermally Conductive Nanocomposites

7

Jan Felba

Miniaturization is a steady process, and the real progress of microelectronic circuits is influenced by practical state of technological level as scaling parameters, impurity, or contamination of the materials and processes, as well as the technical level of technological equipment. The heat dissipation problem is becoming a crucial barrier in miniaturization process, and new materials for this purpose are elaborated. The problem of heat transport and heat dissipation, especially in the first and second packaging level, is solved usually by conductive adhesives. At least, there are two main requirements for such material, namely, a sufficient mechanical strength of joined components and high thermal conductivity.

Generally, the thermally conductive adhesives can be treated as thermal interface materials (TIMs). Materials of this type play a key role for the heat dissipation at all levels within microsystems and are used for filling the gaps between surfaces to enhance the heat conduction in the package.

It means that the good adhesion to the surfaces cannot be always needed, while the requirement of high thermal conductivity remains crucial to its efficiency. Thermal interface materials can be classified into a few general categories [1, 2] such as thermal greases, phase-change materials, or filled polymers. It is worth to note that low melting point metal solder alloys commonly used for microelectronic packaging are also thermal interface materials and can be classified as phase-change materials.

In this chapter, only filled polymers will be discussed. These composites consist of the polymer-based material matrix and thermally conducting filler. All polymeric materials, epoxy or other types, thermoset or thermoplastic, have the thermal conductivity in the range of 0.2–0.3 W/m $\times$ K. Practically, only the filler material is responsible for heat transport. Conduction is provided by conductive additives, as high conductivity requires high filler content, considerably above the percolation threshold. It is believed that at this concentration, all conductive particles contact each other and form a three-dimensional network. The size, shape, and content of the filler particles change significantly the viscosity and rheology of the composite in comparison with the pure polymer. The expected conductivity value is obtained after the curing process due to better contacts between filler particles resulting from shrinkage of the polymer matrix. The thermal conductivity is limited by the so-called thermal contact resistance between filler particles. The contact thermal resistance depends on both properties of the materials and geometric parameters of the contact areas between the particles. The geometric parameters are related to the contact pressure within the contact area.

In modern microelectronics, gaps for thermal interface materials are usually not higher than tens of micrometers. This needs filler particles with much smaller dimensions—with size in micro- or even nanoscale. Today's technology makes it possible to produce metal particles with dimension of a few nanometers, which theoretically can be used as conductive fillers (mostly silver and gold). Also one of the most promising thermally conductive fillers—carbon nanotubes—has a diameter range of single nanometers or less [3]. This allows us to give such a material a name, *composites filled with nanosized particles* or simply *nanocomposites*. The important question which arises is the "nanoeffect" of nanoparticle or fiber inclusion relative to their larger-scale counterparts. In fact there are changes of glass transition temperatures for some polymer matrix with nanofiller incorporation [4]. Nevertheless, many nanocomposite systems can be modeled using continuum models where the absolute size is not important since only shape and volume fraction loading are necessary to predict their thermal properties.

J. Felba (✉)
Institute of Microsystem Technology, Wrocław University of Technology, Wrocław, Poland
e-mail: jan.felba@pwr.wroc.pl

C. P. Wong et al. (eds.), *Nano-Bio- Electronic, Photonic and MEMS Packaging*, https://doi.org/10.1007/978-3-030-49991-4_7

7.1 Model of Heat Transport

The general transient heat transfer equation can be written as follows:

$$\lambda \nabla^2 T - c\rho \frac{\partial T}{\partial t} + q_v = 0 \qquad (7.1)$$

where T is the temperature, λ the coefficient of thermal conductivity, c the specific heat, and ρ the material density, while q_v stands for other inside and outside heat sources or phenomena. The above equation consists of three components, of which:

- The first one is responsible for heat conductance
- The second one for heat accumulation
- While the third one for the internal heat generation or outside heat dissipation

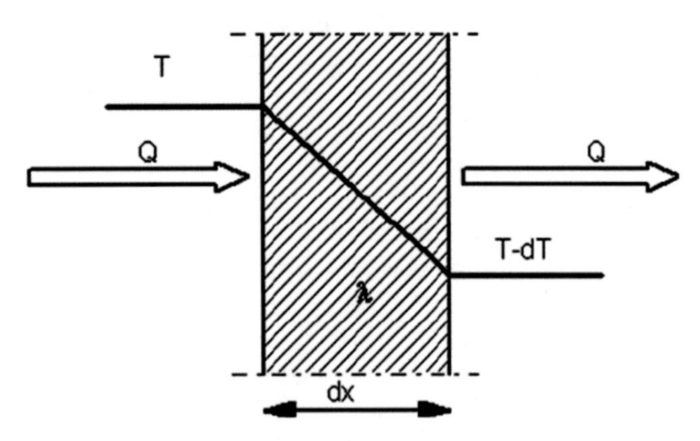

Fig. 7.1 Heat transfer through a layer

The basic principle of heat conductance through the thermally conductive adhesive layer dx (Fig. 7.1) is that heat is transferred in proportion to the gradient of temperature, which can be rewritten as follows:

$$\frac{dQ}{dt} = -\lambda A \frac{dT}{dx} \qquad (7.2)$$

where Q is the quantity of heat energy (J), t the time (s), λ the thermal conductivity (W/m × K), dT/dx the temperature gradient in the heat flow direction (K/m), x the direction of heat flow (m), and A the area of the cross section (m^2). Thermal conductivity is given by

$$\lambda = \frac{1}{3}\left(c_e v_e L_e + c_{ph} v_{ph} L_{ph}\right) = \lambda_e + \lambda_{ph} \qquad (7.3)$$

where c_e and c_{ph} are the heat capacities per unit volume (J/m^{3}K) of electrons and phonons, respectively, v_e and v_{ph} are their root-mean-square velocities, and L_e, L_{ph} are their mean free paths.

For metals, the λ_e is the dominant part of the thermal conductivity. Thermal conductivity depends on temperature, and for temperatures higher than room temperature, λ of pure metals remains essentially independent on temperature or decreases with increasing temperature in the case of metals of high conductivity (Fig. 7.2) [5].

The transport of heat in nonmetals occurs mainly by phonons. The thermal conductivity is restricted by various types of phonon scattering process. In order to maximize the thermal conductivity, these phonon scattering processes have

Fig. 7.2 Dependence of thermal conductivity on temperature for copper [5]

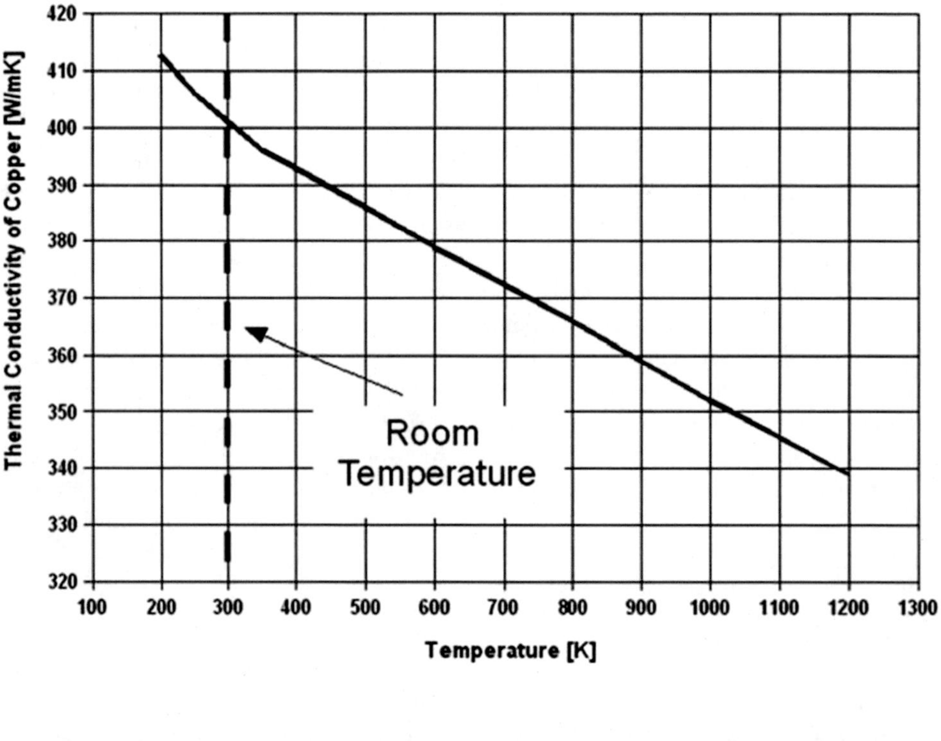

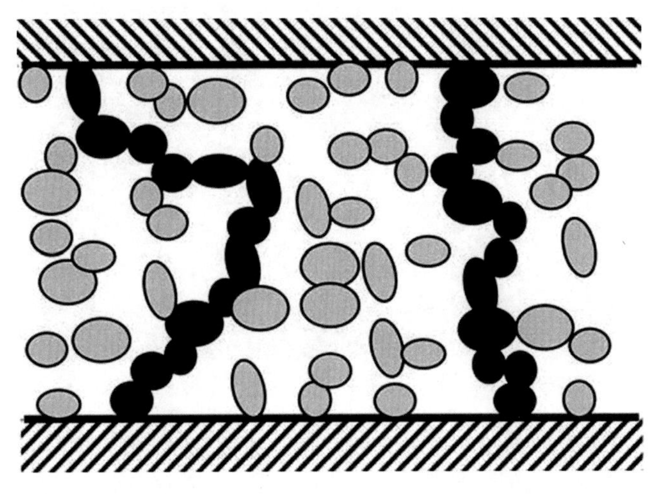

Fig. 7.3 The phenomenon of percolation occurring with the formation of a continuous chain (*dark balls*) between the two surfaces by the conducting particles

to be minimized. For insulators thermal conductivity linearly increases with temperature.

It is assumed that in macroscale the formulation of thermally conductive composites, which consist of the polymer-based material matrix, filler, and some special additives, is uniform and isotropic, although slight heterogeneity was also measured [6]. With enough high content of the filler, when the conductive paths appeared (Fig. 7.3), the heat conduction may depend on bulk thermal conductivity of a filler material. For the single particle of filler, its bulk conductivity is determined by the material used. From this point of view, the diamond with conductivity of 2000 W/m × K is a better material than alumina (slightly better than 30 W/m × K). In the case of composites formulated such as electrically conductive adhesives, which contain silver flakes with average particle dimensions of several micrometers, and thermosetting epoxy resin as polymer matrix, the thermal conductivity value does not exceed 3 W/m × K [7]. This enormous difference between thermal conductivity of pure silver (about 420 W/m × K) and thermally conductive adhesive with this filler is due to the constraint by the existing so-called thermal contact resistance between filler particles. It is the main "bottleneck" for heat transport inside composite formulation.

There are many analytical models to predict the heat transport between particles, which may be used as filler in thermally conductive composite. According to literature review concerning thermal interface materials [8, 9], the most common and important analytical models mostly taking into account spherical particles can be listed as follows:

- The Maxwell–Garnett effective medium model [9–12] can be considered as a good model for low-volume fraction, up to 40%. Therefore it is not used in modeling of the heat transport for most common composites in which conductive particle content exceeds 60% [8].
- The Bruggeman symmetric model for high-volume fraction. One of the biggest drawbacks of this model is that it does not include the interface resistance between the particle and the matrix [8].
- The Bruggeman asymmetric model. After modification, it includes thermal contact resistance between the particle and the matrix. This model is capable of predicting thermal conductivity of spherical particles for larger-volume fractions.
- Percolation models. Percolation is a geometrical phenomenon, which means that above a particular volume fraction (called the percolation threshold), there is a continuous path for heat conduction through the particles because the conducting particles start to touch each other as shown in Fig. 7.3.

The classic percolation theory assumes a statistical distribution of the filler particles in a matrix. Nevertheless, the distribution of nanoparticles in a polymer matrix cannot be completely random because of an aggregation of particles. In the case of nanosilver with epoxy resin, it was observed that a conductive network can be formed even if the content of particles is lower than the percolation threshold estimated conventionally [13]. With increasing temperature, self-organization of particles helps in the formation of the conducting path throughout the matrix.

7.2 Thermal Contact Resistance

There are two basic materials formulating thermally conductive composites— polymer matrix and thermally conductive filler. As polymeric materials are about 2000 times less conductive than metals with high thermal conductivity, such as silver or copper, and several thousand times less than carbon allotropes (e.g., carbon nanotubes or graphenes), the heat transport is possible by conducting pathways (Fig. 7.3). Theoretically, the effectiveness of heat transfer depends on conductivity of a filler bulk material and thermal contact resistance between filler particles.

Generally, the thermal resistance of both bulk material and contacts can be expressed as thermal conductors. This is allowed because of the analogy between thermal and electrical resistance defined by Ohm's law, where instead of the potential difference across the object and the current passing through the object, the temperature difference occurs provoking a heat flow. Due to this analogy, the concept of a thermal resistance can be introduced

$$\Theta_\lambda = \frac{l}{\lambda A} \tag{7.4}$$

where Θ_λ is the thermal resistance of a thermal conductor (K/W), λ the thermal conductivity, l the length of thermal conductor (m), and A the area of the conductor cross section (m^2). The total thermal resistance in "a three-dimensional network" of conducting filler particles can be summed using the concepts of serial and parallel resistances as in an electrical circuit. The thermal resistance Θ_p of the singular contact path in the space between two surfaces (filled by composite) in the steady-state conditions can be expressed in a simple equation:

$$\Theta_p = \Theta_{\lambda 1} + \Theta_{\lambda 2} + \cdots + \Theta_{\lambda n} + \Theta_{TC1} + \Theta_{TC2} + \cdots + \Theta_{TC(n-1)} \quad (7.5)$$

where Θ_λ is the thermal resistance of the filler particle (bulk resistance) and $\Theta_{TC}$the thermal contact resistance between particles forming a chain in the direction of the temperature gradient. The total thermal resistance of composite Θ_{TA} is

$$\Theta_{TA} = \left(\frac{1}{\Theta_{p1}} + \frac{1}{\Theta_{p2}} + \cdots + \frac{1}{\Theta_{pn}} \right)^{-1} \quad (7.6)$$

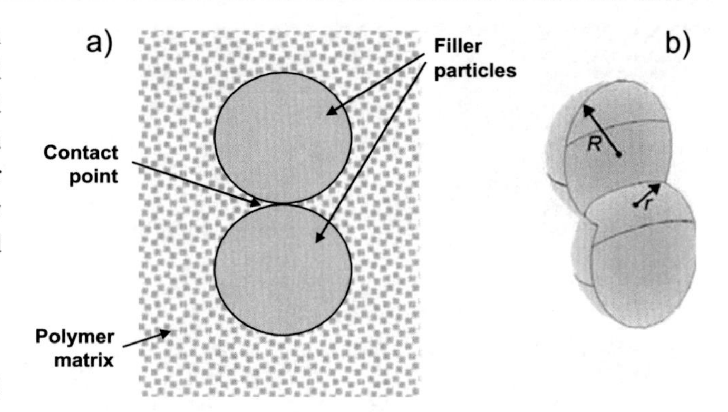

Fig. 7.4 Two-dimensional model of ideal balls (**a**) and ball-shape particle after deformation (**b**) filled thermally conductive polymer

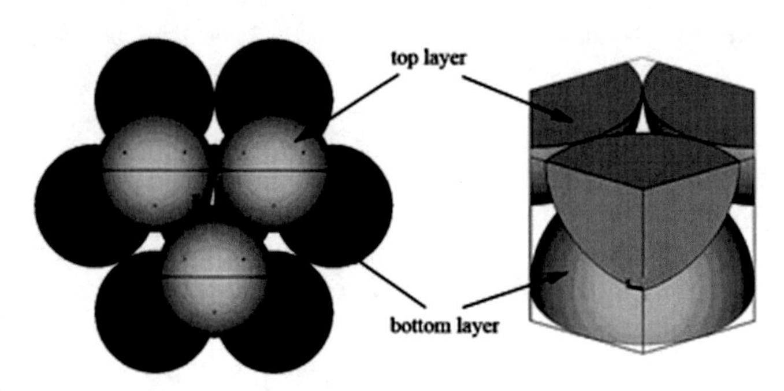

Fig. 7.5 Close-packed structure (*left*) and analyzed unit cell (*right*)

Table 7.1 Numerical simulation results

r/R	λ (W/m × K)
0.05	49.12
0.02	26.46

If contact members are in the form of perfect, hard balls, then they touch each other in one point (contact point "A" in Fig. 7.4a). In fact, this point turns into a small area of radius r of the filler particles contact area since contact materials are deformable (Fig. 7.4b).

The degree of this deformation influences the contact areas and consequently, the thermal contact resistance. To answer the question on how the contact state between particles affects the thermal conductivity of a polymer, the numerical modeling of silver-filled formulation was used [14, 15]. The numerical model was simplified by some assumptions: the identical-sized spherical-shaped silver particles were used as a filler and the unit cell has a maximal possible packing efficiency ($V_{filler}/V_{total} = 74.05\%$) in 3D spherical particles, i.e., hexagonal close-packed (HCP) or face-centered cubic (FCC) crystal packing structure (see Fig. 7.5). Additionally it was assumed that particles contact each other in the ideal pure metallic contact area (the impurity, roughness, etc. were neglected).

The unit cell shown in Fig. 7.5 was analyzed in ANSYS 8.0 commercial finite element method software. The two different contact area radii were taken into account represented by the relation to particle radius r/R where r is contact area radius and R is particle radius, as shown in Fig. 7.4b. The simulation results were gathered in Table 7.1.

As shown in Table 7.1, the contact area between filler particles has a strong influence on thermal conductivity of the adhesive. This area depends on stresses between particles, which occur due to shrinkage reaction of resin during curing. The role of polymer matrix for this seems to be crucial. The stress forces may decrease thermal contact resistance of adjoined particles of the filler, but the calculation of these stresses is quite complex, because of its relaxation in time due to viscoelastic behavior of the resin [16–18].

Instead of identical, close-packed, spherical-shaped filler particles, the heat flow can also be analyzed when particles touch each other by flat surfaces. If the contact areas are nominally flat and hard, then they touch each other maximum in three points (Fig. 7.6). In fact, these points become the small areas (a-spots) since contact member materials are deformable [19]. If the contact surfaces are perfectly clean, then the total metallic contact area A_c is the sum of all a-spot areas.

The steady-state heat flow from the top of the two-particle unit of higher temperature (T_1) to the bottom of lower temperature (T_2) is constricted through small conducting a-spots. Therefore, in the case of adhesives, when polymer matrix fills the contact areas (except A_c surface), the total thermal resistance of such units is the sum of the bulk thermal resistances $\Theta_{\lambda 1}$ and $\Theta_{\lambda 2}$ of both contact members, the constriction resistances Θ_{c1} and Θ_{c2} of contact members, and real thermal resistance of a-spot bump Θ_b (if a-spot is treated as three-

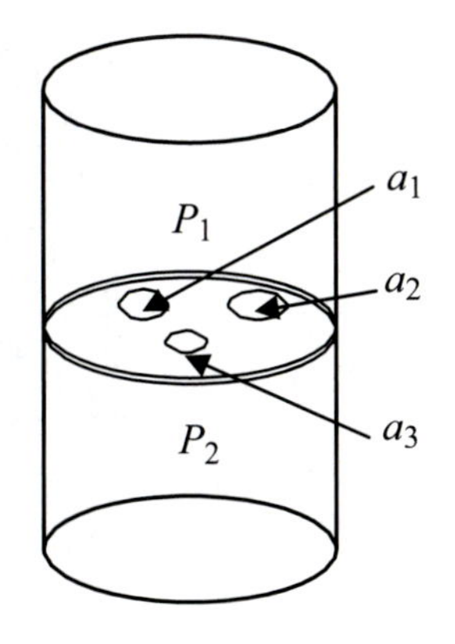

Fig. 7.6 Two cylindrical filler particles P_1 and P_2 forming the apparent contact area $A_c = a_1 + a_2 + a_3$

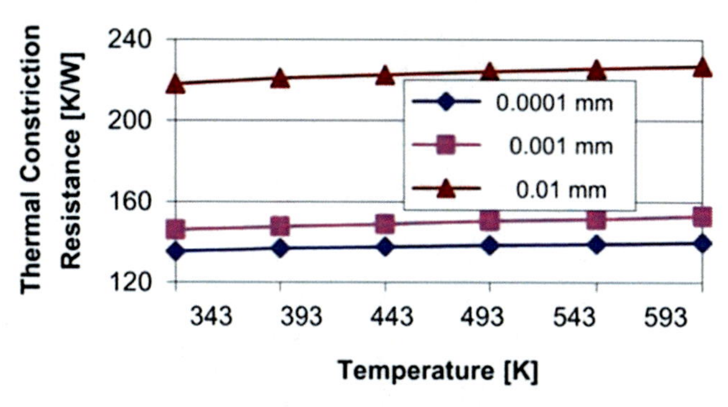

Fig. 7.7 The thermal constriction resistance of two-particle unit vs. temperature of heat source, with different gap distance between filler particles; the radius of a-spot is 0.01 mm [20]

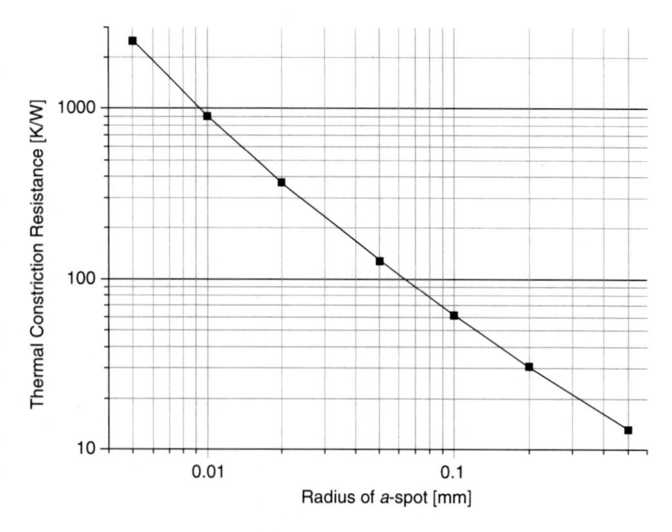

Fig. 7.8 The thermal constriction resistance of two-particle unit vs. contact area with different heat source temperature; the gap distance between filler particles is 10 μm [20]

dimensional structure). This effect of heat flow constriction through the a-spot is identified with thermal constriction resistance.

The thermal constriction resistances of two cylindrical copper particles (Fig. 7.6) were calculated using the finite element method [20]. Due to symmetry, only quarter of the system was analyzed. The computer calculations were performed assuming that the thermal conductivity of polymer matrix can be neglected, the contact surface is perfectly clean, without oxides and alien films, the a-spot is positioned on the axis of the contact system, and the radius a of this metallic contact area is equal to $0.0005R$ ... $0.05R$, where $R = 10$ mm is the radius of the contact members as well as of apparent contact area. The length of each contact member is 40 mm. The gap between both contact members ranges from 0.1 up to 10 μm. These values were chosen in such a way that the thermal constriction resistance could be regarded as "long resistance."

The result of calculation shows (Figs. 7.7 and 7.8) that both the gap between filler particles and the contact area influence thermal constriction resistance significantly. In fact, this resistance can be treated as the thermal contact resistance Θ_{TC}. For analyzed two-particle unit in temperature of 393 K (for copper $\lambda = 364.1$), the bulk resistance of both filler particles is about 0.7 K/W, while the thermal contact resistance $\Theta_{TC} = 221$ K/W for both the radius of a-spot and the gap is equal to 10 μm (from Fig. 7.7). It means that the total resistance of thermally conductive composite depends mainly on the values of thermal contact resistance and numbers of contacts between filler particles, making a chain in the direction of temperature gradient (Θ_{TC} and n in Eq. 7.5). The bulk resistance plays a minimal role.

Thermal conductive components usually are applied between two surfaces with different temperatures, e.g., between heat-generating component and the eventual heat sink. Some of those materials consist of permanent bonds, such as adhesive, but often these are only bolted between other elements. In such cases, the calculation of heat transfer is very difficult, as the real contact area is uncontrolled, a small fraction of the nominal or apparent area. This is mainly due to nonflat shape of contacting surfaces, surface roughness, deformation possibility of contact materials under pressure, additional materials between contacting surfaces, etc. In such a case of heat transport, the thermal resistance Θ_T can be expressed by the formula

$$\Theta_T = \Theta_{IR'} + \Theta_{TA} + \Theta_{IR''} \tag{7.7}$$

where $\Theta_{IR'}$ and $\Theta_{IR''}$ are interfacial thermal resistance values between a composite and surface elements. When thermally conductive adhesives are used for bonding of two elements in the way of heat transport, the thermal resistance Θ_T of a joint can be defined as

$$\Theta_T = \Theta_{BR'} + \Theta_{TA} + \Theta_{BR''} \tag{7.8}$$

where $\Theta_{BR'}$ and $\Theta_{BR''}$ stand for the bond thermal resistance of contacts between adhesive layer and joined elements, while Θ_{TA} represents the total thermal resistance of the adhesive

(Eq. 7.6). In such a case, an analysis of bond thermal resistance influence on Θ_T can be done [21–23].

7.3 Thermal Conductivity Measurements

Thermally conductive composites reported in the literature are characterized by thermal conductivity not higher than a few W/m × K for formulations with both commonly used filler materials (like silver [7, 24]) and special materials (diamond powder [19] or carbon fiber [25]). But there are also reports about polymer matrix and silver-filled adhesives with a few times higher thermal conductivity [26–28]. Probably various methods of thermal conductivity measurement and their errors as well as low accuracy cause significant differences in measured thermal conductivity of similar adhesive formulations. To analyze thermal data of adhesives, the knowledge about used measurement method is necessary.

The history of heat transfer measurement began in the second decade of the last century [29]. Currently there are two main categories of techniques to measure thermal conductivity, steady-state techniques, and transient techniques. The heat flow method and the guarded hot plate method are the best examples of steady-state techniques. Hot wire and laser flash can be listed as the most popular measurement methods of the transient technique, but also the 3ω, photoacoustic method, the pulsed photothermal displacement technique, and the thermal wave technique are in use [30].

The steady-state methods are based on establishing a steady temperature gradient over a known thickness of a sample and on controlling the heat flow from one side to the other. The determination of the thermal conductivity of the insulation follows from basic law of heat flow (Eq. 7.2). The principle of heat flow method is shown in Fig. 7.9.

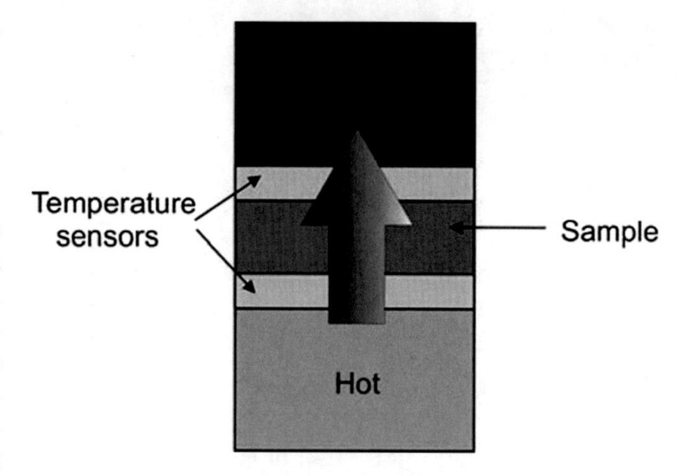

Fig. 7.9 The principle of heat flow method

Among the steady-state methods, the guarded hot plate is the most common technique for measuring the thermal conductivity of insulation or poor thermal conductor materials. Its operation is based on establishing a steady temperature gradient over a known thickness of a sample and on controlling the heat flow from one side to the other. The determination of the thermal conductivity of the insulation follows from the applicable approximation heat flow law in one dimension. The specimen is placed between a heat source and a "cold plate" or heat sink, both of which contain temperature sensors very close to their surfaces. The sensors measure the temperature drop across the sample after a steady-state heat flow has been established. The heat source is surrounded by a guard heater, maintained at the heat source temperature that minimizes heat losses out of the edges. In this method, the thermal resistance at the sample interfaces must be minimized and well controlled for consistent results. The most commonly employed instruments create a steady-state temperature gradient across a specimen of the material by sandwiching it between isothermal hot and cold plates (Fig. 7.10).

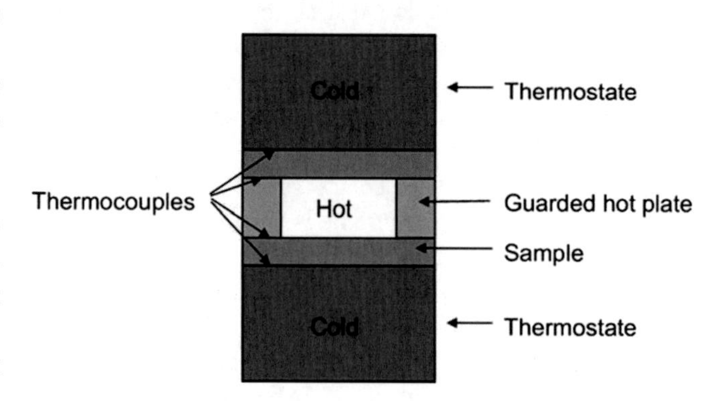

Fig. 7.10 The principle of the guarded hot plate method measurement

There are a few modifications of the basic guarded hot plate method. In contrast to a conventional system that used uniformly distributed heaters, the line heat source was applied, which improved measurement accuracy by determining the errors resulting from heat gains or losses at the edges of the specimens [31]. The method's error was diminished by more stable operation at liquid nitrogen temperatures [32] and measurement under vacuum conditions [33]. But generally, the heat flow method is based on American ASTM C 518-98 and European ISO 8301 standards, while guarded hot plate method on American ASTM C 177-97 and European ISO 8302 standards [34].

As the main disadvantage of the steady-state methods, the long time required to obtain a single measurement is mentioned. But it is worth to know that there are many sources of errors, which can influence dramatically the results of thermal conductivity measurement, especially when materials with low λ (like thermally conductive polymer-based materials) are tested. The high accuracy of measurement ought to meet at least the following requirements:

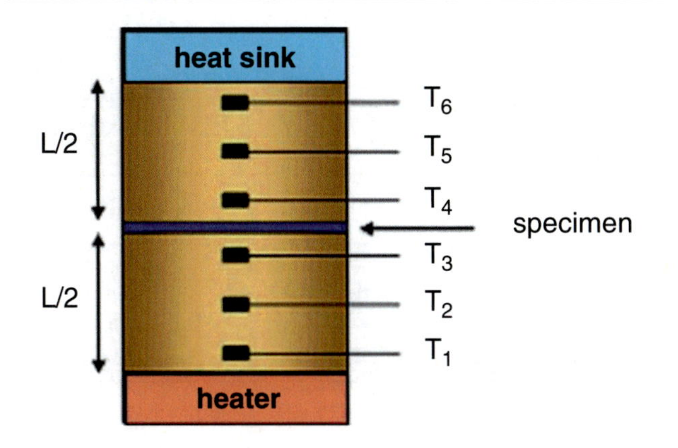

Fig. 7.11 An experimental setup (from [7])

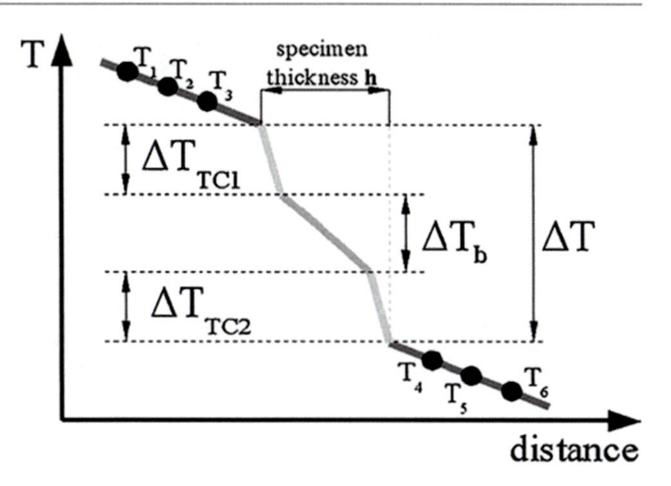

Fig. 7.12 Thermal distribution in steady state; T_1–T_6 measured temperature; $\Delta T_{TC1,2}$, temperature gradients in contact areas (unknown); ΔT_b, temperature gradient on measured sample (unknown); ΔT, the calculated temperature drop between a pair of iron contact members

1. Heat is transferred only by conduction in a measurement setup.
2. The heat flow is simply measured (e.g., by temperature drop).
3. The thermal resistances are taken into consideration.
4. Data of measured as well as reference materials are independent from temperature; or thermal conductivity changes vs. temperature (see Fig. 7.2) are taken into consideration.

An experimental setup which fulfills most of the above listed requirements is shown in Fig. 7.11. The system that is based on ASTM E 1530 Standard (Standard Test Method for Evaluating the Resistance to Thermal Transmission of Materials by the Guarded Heat Flow Meter Technique) consists of heater, heat sink, a pair of contact members (cylinders with a known thermal conductivity λ), as well as temperature sensors [35]. The whole setup is placed in high vacuum environment in order to minimize the effect of heat dissipation through convection. The radiation is out of the temperature work range and is therefore neglected, and the heat conduction is a dominating factor in transport heat energy. When the experimental setup reaches the steady state and then by extrapolating the readings of the temperature sensors, it is possible to find out the temperature drop ΔT at the contact interface [36], which is shown in Fig. 7.12.

The experimental setup [7] consists of a pair of iron contact members ($\lambda_{Fe} = 80.4$ W/m × K) with a diameter of 60 mm with six thermal sensors (T_1–T_6), a bulb halogen heater immersed in a bottom cylinder, and a water cooling system on the top. In the vacuum conditions, the readings of the temperature sensors were taken into calculation when the experimental setup reached the steady state (Fig. 7.13), and then by extrapolating, it is possible to find out the temperature drop ΔT (Fig. 7.12) at the tested specimen.

To eliminate the influence of interfacial thermal resistances ($\Theta_{IR'}$ and $\Theta_{IR''}$ in Eq. 7), it is assumed that both those values are equal and it is necessary to perform at least two measurements with different thickness of specimens. Changing thickness of a sample from h to $h + \Delta h$ causes the increase of thermal resistance of a material, but the thermal contact resistances remain unchanged. On the basis of Eqs. 7.2 and 7.6, it is possible to extract the thermal conductivity of the material tested [23]:

$$\lambda_b = \frac{\Delta h}{A} \cdot \left(\frac{\Delta T_2}{q_2} - \frac{\Delta T_1}{q_1} \right)^{-1} \quad (7.9)$$

where q_1 and q_2 are heat flows evaluated based on Eq. 7.2, taking into account temperature gradients and thermal conductivity λ of the contact members for both samples, and A is the contact area between the contact member and a measured sample. The measuring set was used successfully for thermal conductivity measurement of pure materials (e.g., polytetrafluoroethylene (PTFE)—0.246 W/m × K) as well as polymer composites filled with silver flakes (below 3 W/m × K).

As mentioned, it is very important to know uncertainty of results, especially when thermally conductive polymer-based materials are measured. The limiting error of the above-described measurement method is given by [37]

$$\Delta_{le}\lambda_b = \sum C_i \Delta_{le} X_i \quad (7.10)$$

where $\Delta_{le}\lambda_b$ is the limiting error of thermal conductivity of tested material, c_i is the partial differential, and $\Delta_{le}X_i$ is the limiting error of each variable. Uncertainty U of λ_b measurement with 95% probability is equal to:

$$U(\lambda_b) = 1.15\sqrt{\sum (c_i \cdot \Delta_{le} X_i)^2} \quad (7.11)$$

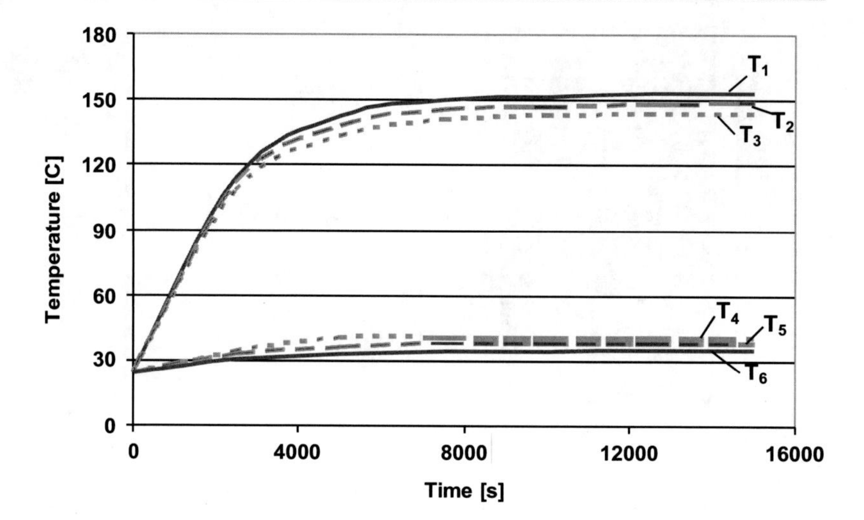

Fig. 7.13 The temperature distribution inside the contact members

In the case of polymer-based material filled with carbon nanotubes, its thermal conductivity λ_b equals 1.43 ± 1.1 W/m $\times$ K with 95% probability [37] was measured.

The transient methods measure a response as a signal is sent out to create heat in the sample. During the time of measurement, the change in temperature is monitored. For measurement of thermal conductivity of composites (also nanocomposites filled with carbon nanotubes [38]) the transient hot-wire technique can be used. This method is a transient dynamic technique based on a linear heat source (Fig. 7.14) of infinite length and infinitesimal diameter and on the measurement of the temperature rise at a defined distance from the linear heat source embedded in the test material. As an electric current of fixed intensity flows through the wire, the thermal conductivity can be derived from the resulting temperature change over a known time interval.

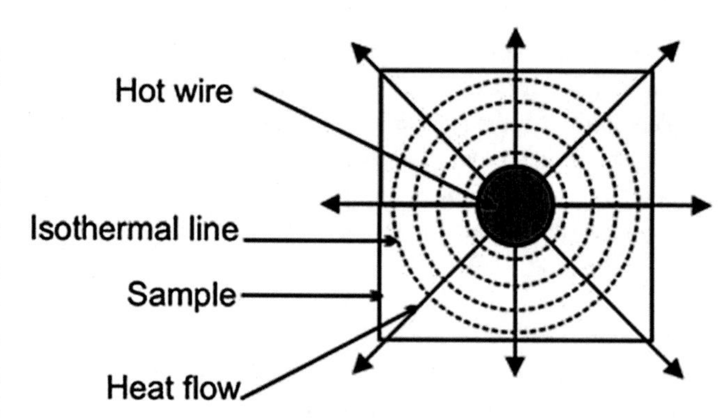

Fig. 7.14 The radial heat flow model of the hot-wire measurement method

If the temperature of a hot wire on the boundary surface between a reference medium and a medium of unknown thermal conductivity, λ, is measured by a thermocouple at times, t_1 and t_2, respectively, the thermal conductivity of the unknown medium can be obtained from the following equation [34]:

$$\lambda = K_1 \frac{P}{[T(t_2) - T(t_1)]} \ln\left(\frac{t_2}{t_1}\right) - H \tag{7.12}$$

where K_1 is a constant depending on the electric power supplied to the heating wire (P), H is a constant term that depends on the insulating material of the probe, and T is the temperature.

In this technique, small-diameter wires are immersed in the fluid and used simultaneously as electrical resistance heaters and as resistance thermometers to measure the resulting temperature rise due to the resistance heating. Two hot wires of differing length are operated in a differential mode to eliminate axial conduction effects due to the large diameter leads attached to the ends of each hot wire. Based on the transient line source model, the thermal conductivity can be found from the slope of the measured linear temperature rise as a function of elapsed time, while the thermal diffusivity can be found from the intercept of this same linear temperature rise curve [39].

The transient hot-wire technique has been modified, and nowadays four variations of the method are in use [40]: hot-wire standard technique, hot-wire resistance technique, two-thermocouple technique, and the hot-wire parallel technique. The theoretical model is the same, and the basic difference among these variations lies in the temperature measurement procedure. It makes possible to determine the thermal conductivity of a wide range of materials, including cured thermally conductive composites.

There are also a number of presently existing transient methods of measuring indirectly thermal conductivity—by measurement of material's thermal diffusivity. The relation between the thermal conductivity λ (W/m $\times$ K] and the thermal diffusivity α (m^2/s) is given by

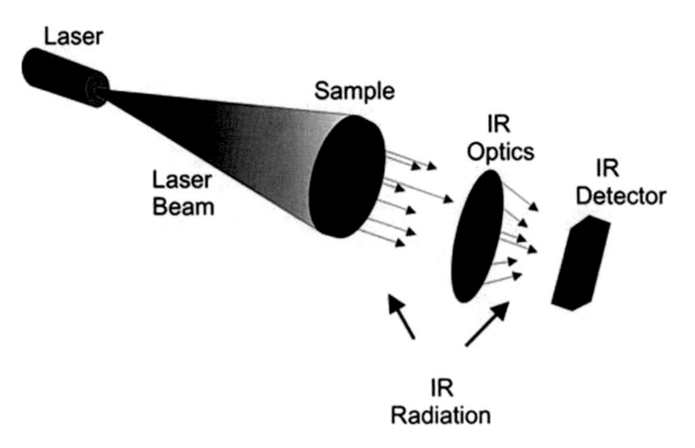

Fig. 7.15 The principle of thermal diffusivity measurement by using a laser flash method

$$\lambda = \alpha \cdot \rho \cdot c_p \tag{7.13}$$

where ρ (kg/m^3) is the density and c_p (J/kg × K) is the specific heat of the tested composite.

Among the transient methods, the so-called flash method is the most popular in determining the thermal diffusivity. The method was described in 1960 [41] and was developed during the next years [42–45]. In this method, a pulse of energy is absorbed instantaneously on the front face of a small disk-shaped specimen, and the subsequent temperature change of the rear face is recorded. The front surface of the sample must be uniformly irradiated at a short time compared to the rise time of the back surface temperature. Figure 7.15 illustrates the principle of flash method with a laser beam as an energy source.

In the determination of a thermal diffusivity of tested material, the magnitude of the temperature rise recorded on the back side of the sample and the amount of energy are not required. Only the shape of the time–temperature curve is used in further analysis [46]. The thermal diffusivity can be directly calculated from the curve and the sample thickness. If assumed that the sample is perfectly insulated from the environment during the test (there is no heat exchange) and the whole energy is absorbed instantaneously (zero pulse width) in a very thin layer of the sample material, then the thermal diffusivity can be evaluated by using the simple relation [41]:

$$\alpha = 0.1388 \frac{d^2}{t_{50}} \tag{7.14}$$

where d (m) is a sample thickness and t_{50} (s) is the time required for the back face of the sample to reach half of the peak value of temperature (Fig. 7.16). This method of calculation of the thermal diffusivity describes the ideal case, but because of the heat loss through radiation (especially when the measurements are done at high temperature) and nonzero pulse width, the other types of analyses are also used [42, 47].

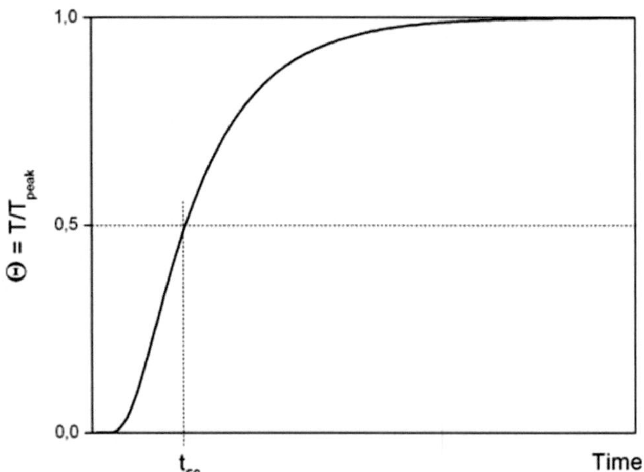

Fig. 7.16 Dimensionless plot of rear surface temperature history

The flash methods are distinguished in view of the source of different heat pulse, but today laser beam is the most popular as a heat source. The measured sample may be in the form of a disk with a diameter of a dozen or so millimeters and a few millimeters in thickness. The laser beam energy absorption depends on the sample surface. To minimize the mirror effect, the side on which the laser beam would hit may be coated with carbon. Under the carbon layer, the sample can be coated with gold. This layer is used to enhance the thermal contact and to prevent direct transmission of the laser beam through the specimen made with transparent polymer [48]. In spite of such procedure, the accuracy of the laser flash method cannot be too high. There are a lot of errors in the method caused by uncontrolled thermal conditions of the heated sample side, heat loss because of convection (immersing sample in air), contact with thermal sensor on the sample rear side, and measuring the specific heat of tested composite (necessary for thermal conductivity calculation—Eq. 7.13).

To avoid uncontrolled thermal conditions caused by laser beam reflection from a polymer composite surface, the electron beam as a more flexible heat source can be used. The absorptivity of an electron beam is extremely high in comparison to a laser beam, and almost the whole energy may be absorbed on sample surface. Additionally this heat source enables exact temperature control of heated sample's surface. For this purpose, the specially designed buffer of high conductivity material (e.g., pure copper) ought to be used to provide the heat pulse on a sample surface and protect it against being destroyed by high-power electron beam. The main problem with using an electron (e-) beam in the flash method is that it cannot be directly used on the surface of non conducting materials such as polymers. The measuring process is conducted in vacuum conditions, and heat is transferred only by conduction in the measuring setup. The possible setup for testing thermally conductive adhesives is presented in Fig. 7.17 [49]. The top surface of the

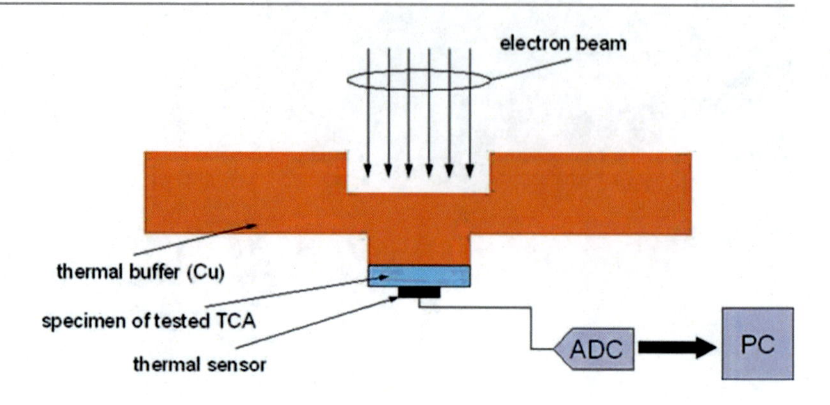

Fig. 7.17 Experimental setup designed for thermal diffusivity measurements of thermally conductive adhesives (TCA) using electron beam

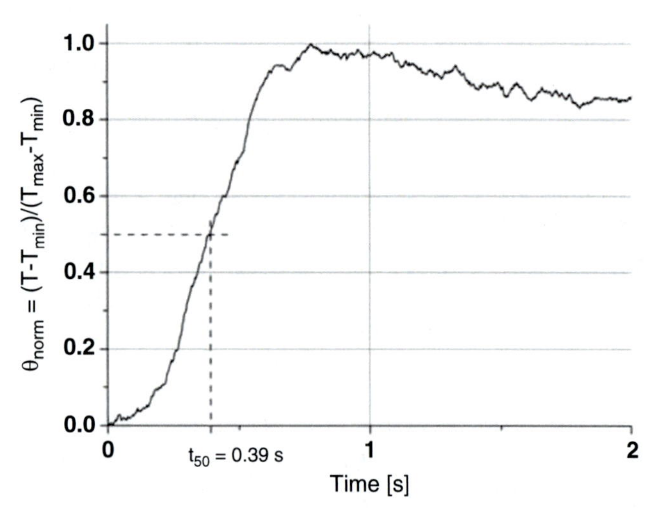

Fig. 7.18 The experimental result (normalized temperature measured on the rear face of sample vs. time) obtained for 0.75-mm-thick adhesive sample containing 70 wt% of Al_2O_3 as a filler

experimental setup was irradiated during test by pulse of defocused and deflected electron beam to obtain uniform energy distribution on the heated surface. The electron beam pulse shape and length was controlled by using a timer. The temperature was monitored on the back surface of a tested material specimen, which was located at the bottom side of the buffer.

Figure 7.18 presents the experimental results (normalized temperature on the rear face of sample vs. time) [49]. Based on Eq. (7.14) and data from Fig. 7.18, the thermal diffusivity for tested sample was calculated as 2.00×10^{-7} m^2/s.

The photoacoustic technique is another transient method for thermal conductivity of thin layer measurement. In this technique, a heating source (normally a laser beam) is periodically irradiated on a sample surface. The acoustic response of the gas above the sample is measured and related to the thermal properties of the sample. The method for single layer on a substrate was explained by Rosencwaig and Gersho [50], and also multilayered materials were measured [51].

A schematic diagram of the measuring setup is shown in Fig. 7.19 [52]. The laser beam is sinusoidally modulated by an acoustic-optical modulator (AOM) driven by a function generator. After being reflected and focused, the laser beam is directed onto the sample mounted at the bottom of the photoacoustic (PA) cell. This cell is pressurized by flowing compressed gas (He). A microphone, which is built into the PA cell, senses the acoustic signal and transfers it to a lock-in amplifier, where the amplitude and phase of the acoustic signal are measured (instead of a microphone, a piezoelectric transducer can also be used [53]). A personal computer is used for data acquisition and control of the experiment.

A tested sample can consist of any arbitrary number of layers and a backing material. Absorption of the laser beam is allowed in any layer and in more than one layer. The transient temperature field in the multilayer sample and gas can be derived by solving a set of one-dimensional heat conduction equations, and the transient temperature in the gas is related to the pressure, which is measured experimentally. Because the transient temperature in the gas is related to the thermal properties of the sample, measuring the pressure allows determination of the thermal quantities, usually the interfacial thermal resistance Θ_{IR} and thermal diffusivity.

Figure 7.19 presents equipment that was used for measurement of thermal conductance across multiwalled carbon nanotube array interfaces, but also liquid samples [54, 55], transparent materials [56], and other materials [57] including thin films [58] can be measured.

For thermal properties, estimation of dielectric composites and the three omega (3ω) thermal conductivity measurement method may be used. The method employs a metallic strip [59] or wire [60] in intimate contact with the specimen surface. An *AC* electrical current modulated at regular frequency ω is induced to flow in the strip causing heat generation in the strip. The heating has both a *DC* component which changes the average temperature of the specimen and an *AC* component at 2ω, which generates thermal waves in the specimen. Because the electrical resistance of the strip depends on the temperature, the resistance will be modulated at 2ω as well. Therefore, there will be an *AC* voltage drop across the ends of the strip at 3ω, $V_{3\omega}$, which is proportional to the *AC* temperature variation of the strip at 2ω, $T_{2\omega}$. $T_{2\omega}$ depends on thermal conductivity λ of the measured material.

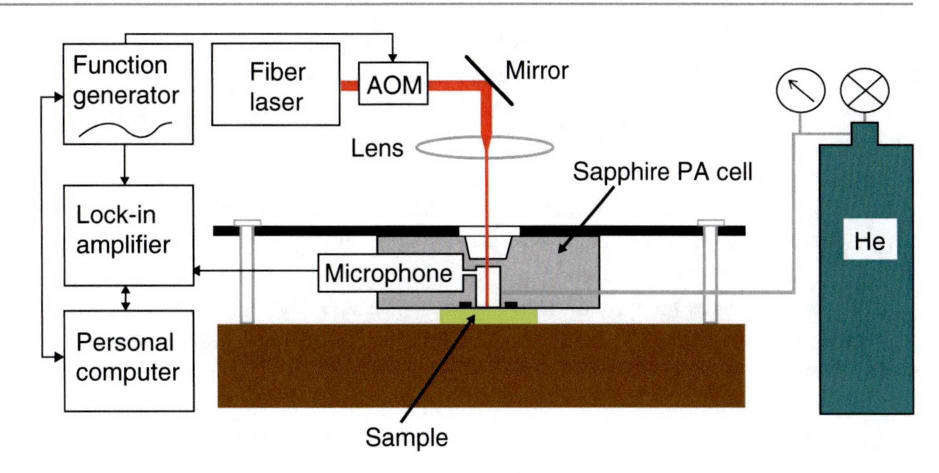

Fig. 7.19 The schematic diagram of the system for photoacoustic method [52]. (Courtesy of American Institute of Physics)

Table 7.2 Summary of thermal conductivity standard series ASTM and their applicability [64]

Standard	Conductivity range	Conductance range	Min ΔT	Remarks C177[a]	< 16 W/m^2 K	10 K	Thermally homogeneous
C518[b]	< 10 W/m^2 K	10 K	Thermally homogeneous				
C1114[c]		None	Thermally homogeneous, for specimens with moisture				
E1225[d]	0.2–200 W/m K	90–1300 K	Thermally homogeneous, opaque				
E1530[e]	25–500 W/m^2 K	150–600 K	Thermally homogeneous				
E1952[f]	0.1–1.0 W/m K	None	Thermally homogeneous				

[a]ASTM C177—Standard Test Method for Steady-State Heat Flux Measurements and Thermal Transmission Properties by Means of the Guarded-Hot-Plate Apparatus
[b]ASTM C518—Standard Test Method for Steady-State Thermal Transmission Properties by Means of the Heat Flow Meter Apparatus
[c]ASTM C1114—Standard Test Method for Steady-State Thermal Transmission Properties by Means of the Thin-Heater Apparatus
[d]ASTM E1225—Standard Test Method for Thermal Conductivity of Solids by Means of the Guarded-Comparative-Longitudinal Heat Flow Technique
[e]ASTM E 1530—Standard Test Method for Evaluating the Resistance to Thermal Transmission of Materials by the Guarded Heat Flow Meter Technique
[f]ASTM E1952—Standard Test Method for Thermal Conductivity and Thermal Diffusivity by Modulated Temperature Differential Scanning Calorimetry

Thus, it is possible to extract λ from a measurement of $V_{3\omega}$ versus ω [61].

The 3ω method can also be used for measuring thermal conductivity of quite different materials such as electrically conducting liquids [62] or the phase-change material [63].

The above-described transient method of thermal property measurement do not exhaust all possibilities of measurement. But other methods either are modifications of the presented ones (by source of heat or signal sensors) or cannot be used for measurement of thermal conductivity of nanocomposites or more general thermal interface materials.

For commercially available composites and quick comparison of the products, uniform methods for determining the thermal conductivity are proposed. The National Fenestration Rating Council [64] recommends ASTM standards. In accordance with these standards, at least three samples shall be measured and the mean value reported. The test shall be done in accordance with ASTM C177, C518, C1114, E1225, E1530, or E1952, as applicable for the specific material. The test equipment shall be calibrated at least once each year as per the recommendations in the appropriate ASTM documents. The choice of standard depends on the possibility of the heat transfer by measured sample (Table 7.2).

7.4 Composites with Metallic Fillers

Since the 1990s, isotropically conductive adhesives have been widely used in electronic packaging. Such composites consist of a polymer resin and electrically conductive fillers, mostly silver, although gold, nickel, and copper are also used. Silver is unique among all cost-effective metals by nature of its oxide, which unlike other metal oxides, is a good conductor. Silver with thermal conductivity 420 W/m × K has been considered to be the most suitable material and also as the thermally conductive filler. In fact, the typically formulated electrically conductive adhesives with silver filler in the form of micronized flakes can also be used as thermally conductive adhesive.

Equation 7.5 points out that the less the contact between particles, the greater the thermal conductivity of the composite. This requires bigger-sized fillers. It was found that at the alumina filler with a fixed loading level of 50 wt%, the thermal conductivity increases about 1.7 times when fillers of <10 μm average particle size were replaced by much bigger ones—not higher than 149 μm (−100 mesh) [65]. This way of composite thermal conductivity improvement is not suitable as the miniaturized packaging scale and technology of composite dispensing need dimension of filler particles much less than a few micrometers. According to Eq. 7.6, it is natural that the more the conducting path, the lower the thermal resistance of the composite. This can be provided by higher filler loadings. In Fig. 7.20, the thermal conductivity of the epoxy adhesives as a function of the volume loading of AlN filler is presented. The nonlinear conductivity increasing with an increase of the filler content suggests that additional paths for heat transport are being formed not only by new chains but also by multiplication of thermal contacts between filler particles in the whole network of particles. This way of conductivity improvement is also limited because of a need for the proper viscosity of formulation and mechanical strength of adhesion in the case of adhesives.

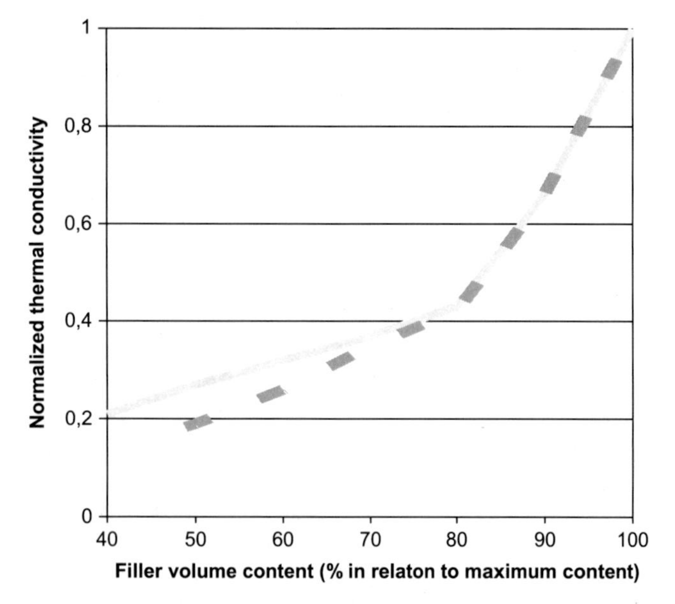

Fig. 7.20 Thermal conductivity of adhesive with electrically isolated filler vs. filler volume content; continuous line, mix of AlN whiskers and AlN particles with maximum content of 60 vol% (according to [48]); dotted line, AlN particles with maximum content of 62 vol% (according to [66])

As was stated earlier, bigger particles of the filler cause higher thermal conductivity of the composite. It means that with the use of nanosized silver particles, the thermal contact resistance (Θ_{TC}) between filler particles increases significantly, and the thermal resistance (Θ_p) of the singular contact path between two surfaces also increases.

Nanosilver is used in today's miniaturized electronics mostly for conductive microstructures and contacts with dimensions in the range of tens of micrometers. Such lines or patterns can be made by inkjet technology using a liquid-containing nanosized particles of silver [67]. There are many methods of producing nanosilver particles, even in the size below 8 nm [68–73]. During the production process, all particles are uniformly dispersed without possibility of agglomeration. For this, each particle is protected by a special kind of layer, usually an organic component, and single particles remain separated (Fig. 7.21a). The protection material occurs in the range of several percent of total metal mass (in case of Ag particles with average diameter of 6 nm—not higher than 4% [71]). As a result, this protective layer on single nanoparticles makes both *DC* electrical conduction impossible (at relatively low voltage) and heat transfer difficult. To obtain a good conductivity, each type of initial product needs additional energy—mostly thermal by heating.

Figure 7.21b presents the microstructure of a nanosized Ag sintered at 280 °C [74]. It can be seen that the structure changed into a dense network of Ag with micropores. Such

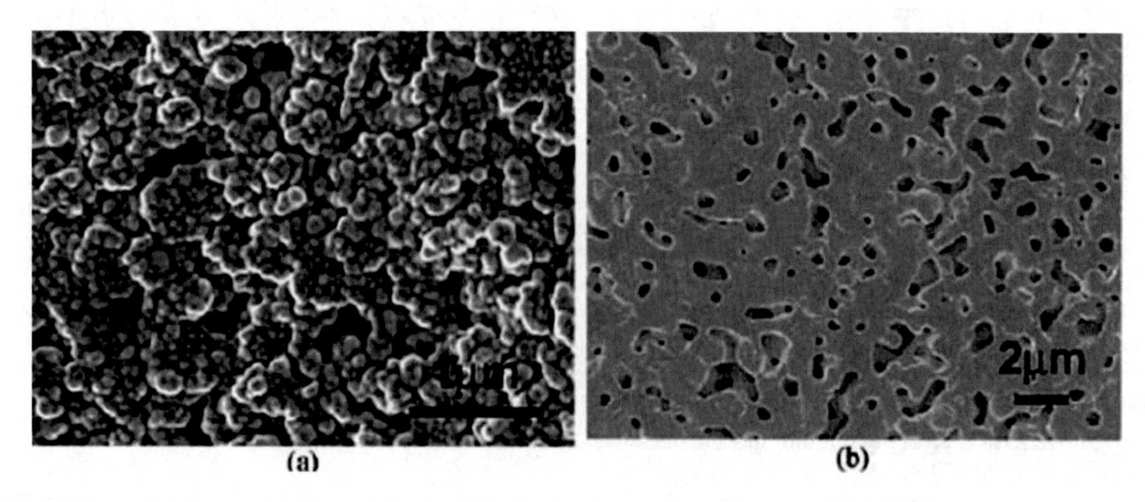

Fig. 7.21 SEM images of the nanosized Ag (**a**) before and (**b**) after sintering on silicon substrate at 280 °C [74]

structure surely conducts heat much better than before thermal process. Energy dispersion X-ray (EDX) analysis shows that almost all the organic components have burned off after the thermal process [75].

In case of thermally conductive composites, nanosized silver is added to a microcomposite rather than used as only a filler. It was stated that adding of silver nanoparticles to formulation with microsized silver particles may improve the composite thermal conductivity, even 2.55 times [26]. The conventional formulation with silver flake was tested as reference. It contains silver particles of 1–10 μm size (88 wt %), epoxy resin (9 wt%), and solvent (3 wt%). The hybrid silver formulation contains micro- and nanosized silver—together with 90 wt% and 10 wt% of phenol resin and a solvent. The size of silver microparticles is 3–5 μm, while silver nanoparticles with a coating material are 3–7 nm in diameter. Adding the silver nanoparticles makes the contact resistance between basic particles lower owing to the fusion process. The mechanism of the curing process of hybrid silver formulation can be explained by the following steps. In the initial stage, the formulation contains resin, solvent, silver particles (microsized), and silver nanoparticles with a coating material. During the first step of the curing process, the solvent starts to vaporize by heating. Next, the coating material begins to decompose and vaporize. Without the coating material, nanoparticles at high temperature (200 °C) fuse together simultaneously. The nanoparticles have grown from their initial sizes of 3–7 nm to 20–50 nm. However, what is crucial for heat flow, the shapes of new particles depend on the surface of their large-sized neighbors. The SEM image (Fig. 7.22) shows that new contacts between particles have extremely large areas, resulting in low thermal contact resistance Θ_{λ}.

There are also experiments with doping silver nanowires to enhance the electrical conductance of composites containing microsized silver particles [76, 77]. In the system composed of silver particles and nanowires, these slender wires are excellent materials to contact the particles together. From Fig. 7.23a, it is apparent that the nanowires are spanning over the gap between two flake silver particles and forming a conductive pathway. Unfortunately, for the heat transport, the role of such a single pathway is not important. In the experiment [76], wires of 10–50 μm long with an average diameter of about 100 nm were used. It is easy to calculate from Eq. 7.4 that the thermal bulk resistance Θ_{λ} of such structure is relatively high (for silver nanowires with diameters about 27 nm and lengths more than 70 μm [78], the resistance would be at least 20 times higher). Additionally, the contact area of the wire with the surface of the particle probably is in the range not higher than dozens of square nanometers, which make the value of thermal contact resistance (Θ_{TC} in Eq. 7.5) also high. But the conductive pathways could be increased in number (Fig. 7.23b) which, according

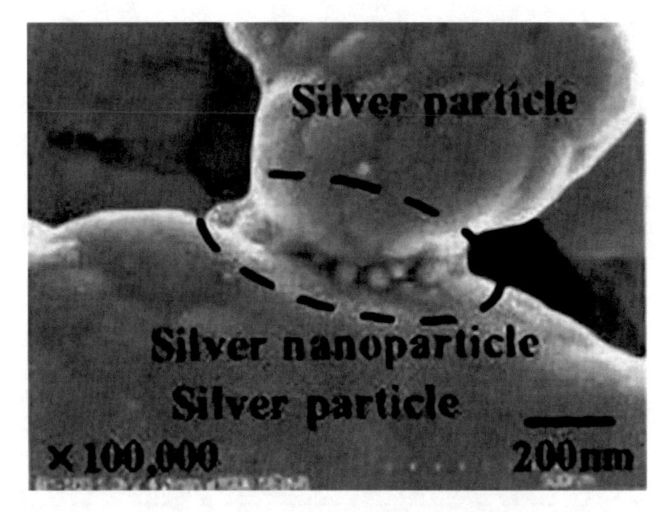

Fig. 7.22 SEM image of adhesive containing silver nanopaste [26]

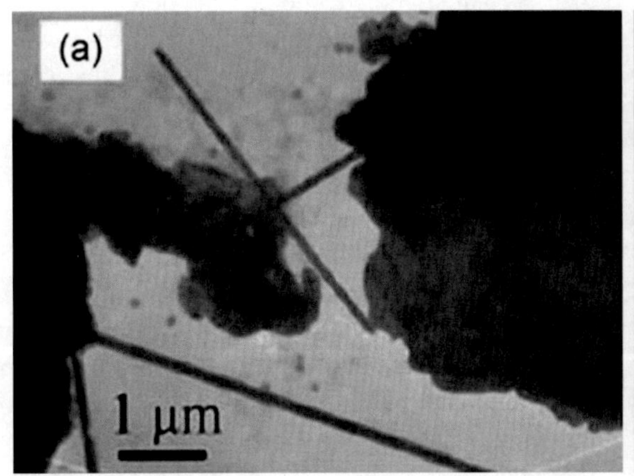

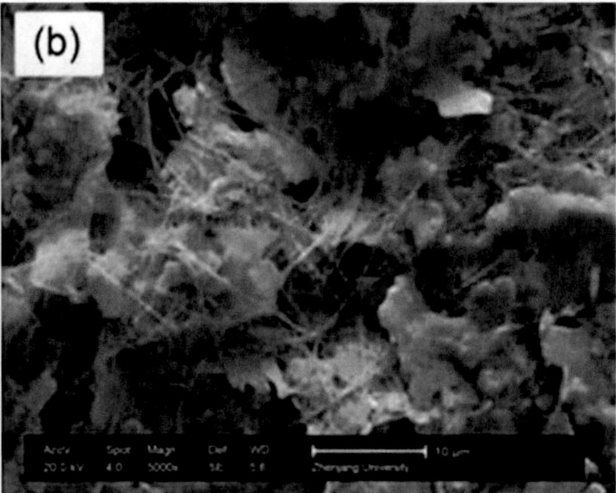

Fig. 7.23 Transmission electron microscopic (**a**) and scanning electron microscopic (**b**) images of local conductive pathways of flake silver particles doping nanowires [76]

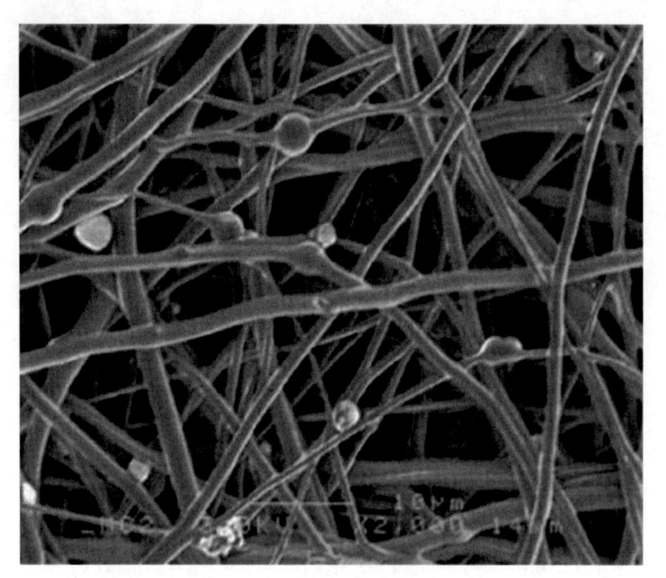

Fig. 7.24 Scanning electron microscopy image of polymer fibers embedded with nanosilver particles [80]. (Courtesy of IEEE)

to Eq. 7.6, causes lower value of total thermal resistance of composite Θ_{TA} and better heat transfer.

There are also investigations of composite in which nanosilver particles saturate polymer nanofibers [79, 80]. The polymer nanofibers were made in the electrospinning process, and nanoparticles of silver were added during this process as a thermal conductivity promoter. It was found that the presence of nanosilver slightly improves the thermal conductivity of the composite, especially in the range of 90–100 °C. The mechanism of better heat transport is unknown, as the SEM picture (Fig. 7.24) reveals only separate particles of silver attached to the fiber structure.

7.5 Composites with Carbon Allotropes Fillers

Thermally conductive composites filled with the carbon allotropes, including diamond, graphite, the buckminsterfullerene, carbon fibers, and carbon nanotube, seem to be very promising because of the high bulk thermal conductivity of such materials.

Unlike most electrical insulators, genuine diamonds are a very good thermal conductor because of the strong covalent bonding within the crystal. Specially purified synthetic diamonds could have the highest thermal conductivity of 2000–2500 W/m × K, which are five times more than any known solid element at room temperature. Unfortunately, the thermal measurement for micron-sized synthetic diamond powder as filler showed relatively a low thermal conductivity in comparison with micron-sized silver. This is believed to be due to the presence of nitrogen and other impurities in synthetic diamond [24]. On the other hand, thermal conductivity of diamond is strongly dependent on the microstructure of the material [81] and additionally is correlated to the grain size even for the bulk material [82]. Probably there are effects of structural defects such as stacking faults, twins, or dislocations from synthetic processes. Typically, where the grain size is in excess of 30–50 μm, the effects of grain boundaries on thermal conductivity become insignificant, whereas submicron grain sizes can reduce the thermal conductivity by more than two orders of magnitude. Because of this, the use of diamond with very low dimensional crystals (nanosized) seems to be unfounded. Such conclusions were confirmed by experiments [83] in which thermal conductivity of composites was measured in a steady heat flow in the temperature range of 50–200 °C in a vacuum. The results are presented in Fig. 7.25 as the function of different diamond grain content in the matrix of the natural single-crystal diamond powder with the fractal size of 10–14 μm.

From the histogram, the thermal conductivity of composites produced from 5 to 7 μm crystallites (region 4) is lower than that of composites synthesized from 10 to 14 μm crystallites (region 6) with the maximum thermal conductivity of about 500 W/m K. Adding nanodiamonds decreases component conductivity up to less than 10 W/m KK (region 1). This is likely to be due to the presence of a large number of grain boundaries, at which phonons are scattered.

Similar results were observed by testing of composites consisting of ultrafine (≈ 6 nm) diamond crystallites embedded into graphitic carbon matrix [84]. In such composites, thermal conductivity may increase by more than five times from a structure containing a small amount of carbon matrices (Fig. 7.26) up to components with 50% graphite.

Without a carbon matrix, the thermal contact between diamond nanoparticles is assumed to be very poor, so the very low thermal conductivity obtained for this specimen seems logical (Eq. 7.5). With the increase of the graphite matrix volume, more carbon fill the pores between diamond grains, so the composite conductivity increases with decreasing size of the remaining pores. It shows that graphite with the value of thermal conductivity smaller by at least one order of magnitude plays a much more important role in heat transport than nanosized diamond.

Figure 7.25 presents thermal conductivity of composites with adding of fullerene (region 5). Fullerenes and nanodiamonds added in small amounts (<1% mass) to powders of 10–14 μm diamond crystallites do not change markedly the thermal conductivity values, even though that fullerene can be characterized by so high thermal conductivity as single-walled carbon nanotubes [85]. It is also reported [86] that the conductivity of hard carbon prepared from fullerene C_{60} does not exceed a few W/m × K at room temperature.

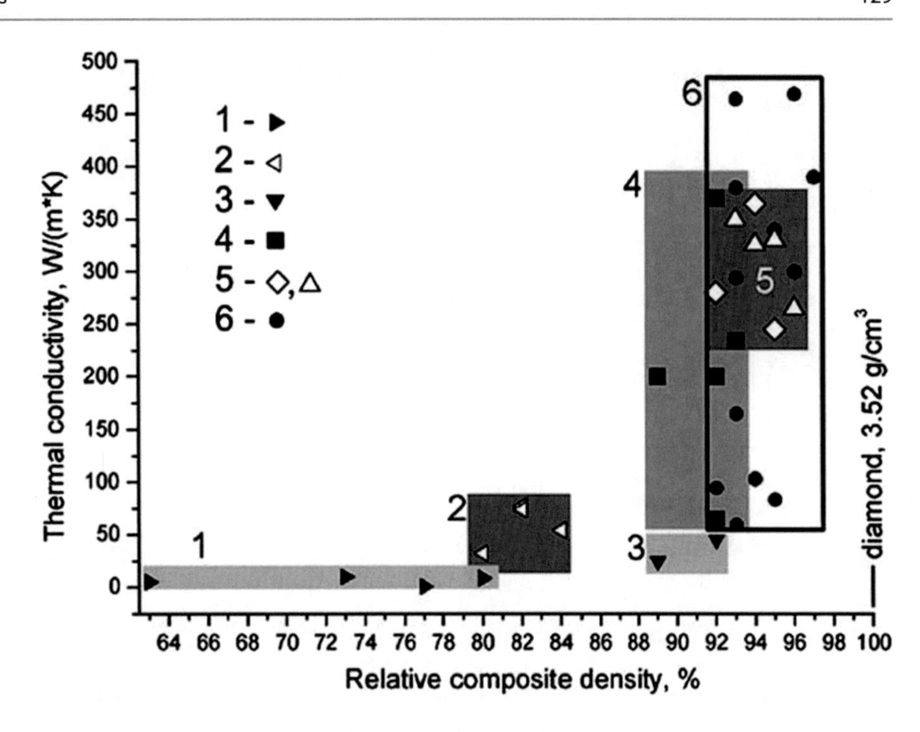

Fig. 7.25 The dependence of the thermal conductivity of composites sintered from micro- and nanodiamond powders on the density normalized to that of single-crystal diamond (3.52 g/cm^3). Regions showing the type of the sample composition: (*1*), 100% nanodiamonds; (*2*), 50% 10–14 μm diamonds and 50% nanodiamonds; (*3*), 85% 10–14 μm diamonds and 15% nanodiamonds; (*4*), 100% 5–7 μm diamonds; (*5*), 99.5% 10–14 μm diamonds and 0.5% fullerenes combined with data on 99% 10–14 μm diamonds and 1% nanodiamonds; (*6*), 100% 10–14 μm diamonds [83]

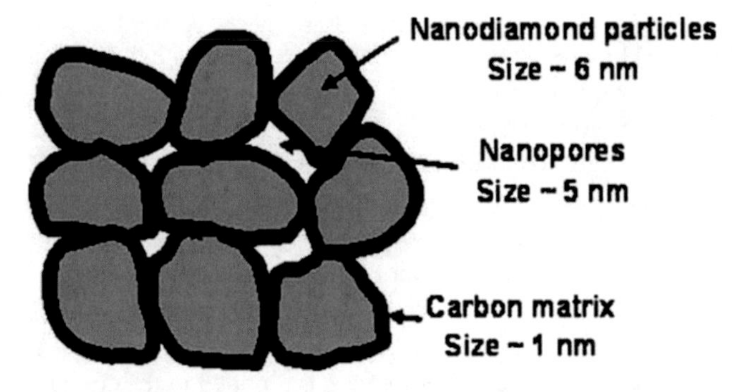

Fig. 7.26 Schematic presentation of composite structure with nanodiamond crystals

The use of carbon nanotubes (CNTs) as high-performance nanofillers to implement a new generation of high thermally conductive composites seems promising. Carbon nanotubes were discovered in 1991 by Sumio Iijima [3], first as multiwalled structures with outer diameters of 4–30 nm and a length of up to 1 μm, which consisted of two or more seamless graphene cylinders concentrically arranged. Single-walled carbon nanotubes, which are seamless cylinders each made of a single graphene sheet, were first reported 2 years later. Their diameters range from 0.4 to 2 to 3 nm, and their length is usually of the micrometer order, i.e., with a very high aspect ratio.

Numerous strategies for the synthesis of carbon nanotubes and more control over their characteristics (number of walls, diameter, length), while improving samples and reproducibility, were undertaken [87–89]. Basically, there are "high temperature" methods, such as the electric arc discharge and laser ablation, and "low temperature" ones, such as the chemical vapor deposition, with many technical variants for each one.

The high value of thermal conductivity of CNTs is the most desirable feature, which can improve the heat transport in composites. The thermal conductivity of individual multiwalled carbon nanotube is estimated to fall in the range of 2000– 3000 W/m × K [90–92]. As the measurements of thermal properties of a single nanotube are extremely difficult and conventional methods cannot be used for this, the values of λ presented in literature are different. Additionally, experimental results show [92] that the thermal conductivity of a carbon nanotube depends on both temperature and tube diameter. From Fig. 7.27, it can be seen that the λ value increases with decreasing tube diameter and increases with an increase in temperature and appears to have an asymptote near 320 K. The thermal conductivity values of one order of magnitude smaller were measured for bulk samples where the nanotubes have been aligned by filtration in a magnetic field. The measurement was taken in the direction parallel to the tubes [93].

The first composites with carbon nanotubes were proposed in the last decade of the twentieth century. Polymers such as epoxy, thermoplastics, gels, as well as poly(methyl methacrylate) have been used as the matrix [94]. The dispersion process requires a crucial technology to develop a nanocomposite because the specific surface area of carbon nanotubes is too large to well disperse in the matrix resin. According to theoretical calculations [95], the specific surface area of single-walled nanotube may reach a value of 1315 m^2/g. For multiwalled tube, the value is lower, but even for higher tube the diameter amounts to a few hundreds of

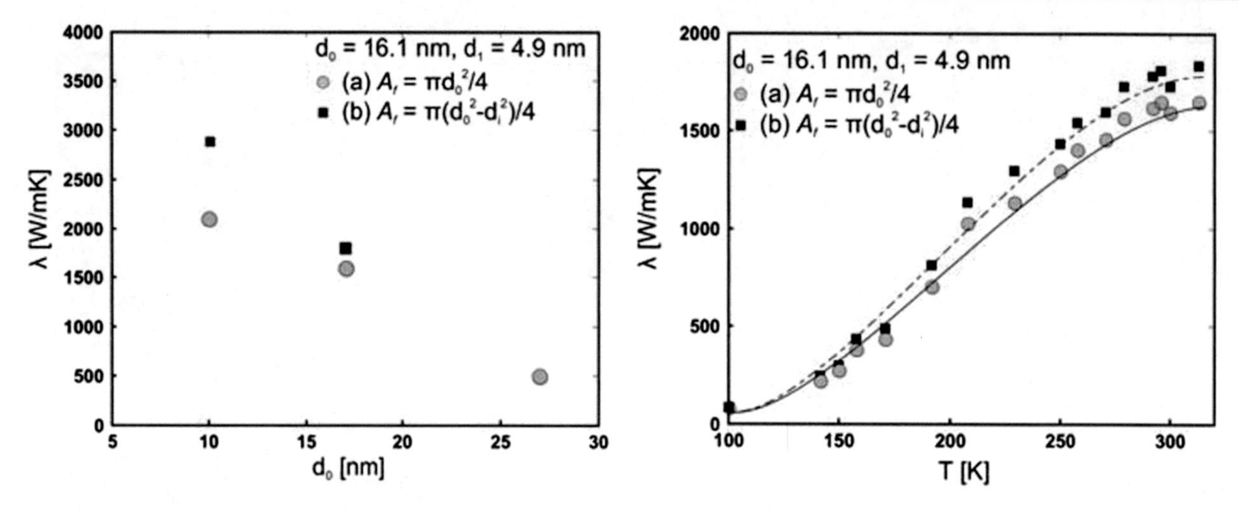

Fig. 7.27 Thermal conductivity of a single carbon nanotube at different diameters at room temperature (*left*) and different temperature (*right*) [92]

m^2/g. If the filler is not properly dispersed, the thermal property of composite cannot be significantly increased. Several processing methods are used for this reason, many of them based on improving nanotubes/matrix interactions [96–98], such as an ultrasonic bath, melt-mixing, in situ polymerization, or solution processing. To prepare carbon nanotubes for the proper dispersion on a macroscopic scale, some preprocessing is required. It can be purified to eliminate non-nanotube material followed by deagglomeration for dispersing individual nanotubes or chemical functionalization. However, the intensive chemical treatment may introduce more defects and reduce the intrinsic thermal conductivity of carbon nanotubes. The chemical treatment usually consists of many steps. For example, Fig. 7.28 presents the overall procedure modification of carbon nanotubes by an oxidation agent for improving the disparity between matrix and the filler.

In fact, even using complicated processes for improving nanotubes–matrix interactions, it is impossible to incorporate the filler content higher than a few weight percent. It is reported that the dispersion of 5 wt% CNTs in an epoxy resin was obtained by an ultrasonic treatment. Although the carbon nanotubes were well separated, they remained poorly distributed [99]. In such CNT content, the viscosity of the composite is more than ten times higher in comparison to pure matrix [94]. Generally, sonication method is only suitable for very low viscous matrix materials and small volumes because ultrasonic devices have a high impact of energy. A proper way to apply the sonication technique to produce nanocomposites with carbon nanotubes is to disperse them in an appropriate solvent (i.e., acetone, ethanol etc.), which first allows the agglomerates to be separated due to vibrational energy. The suspension can later be mixed with the polymer, and the solvent can simply be evaporated by heating.

For electrical current transport, the content level of the filler in which percolation threshold is achieved plays a crucial role. A review and analysis of electrical percolation in carbon nanotube polymer composites from 2009 [100] brings more information concerning the percolation threshold, but in most cases, the current is measured with the carbon nanotube content in the level of 0.01–0.4 vol% [101, 102] or 0.01–1.5 wt% [103, 104].

There are no observed correlations between "electrical" percolation threshold and heat transport. The experimental results of heat transport show that a random dispersion of multiwalled nanotubes in an organic fluid increases its thermal conductivity by more than 2.5 times at approximately 1% volume fraction of nanotubes [105]. For the same value of nanotube loading, a 125% increase is reported in the conductivity of industrial use epoxy due to a random dispersion of single-walled nanotubes [106]. Nevertheless, the thermal conductivity of the composites filled with single- or multiwalled carbon nanotube is lower than for composites with microsized fillers. When proper methods of thermal conductivity measurement were used, the λ value reached only 1 W/m K or less [107–110]. This is lower than calculated value based on the model of the heat transport with neglecting all interactions between the nanotubes [111]. For contents of "average in shape" multiwalled nanotubes in a single weight percent, the model predicts the thermal conductivity of a composite on the level of 1.8 W/m × K. It was calculated that effective conductivity increases linearly with increasing the nanotube loading, the conductivity is not very sensitive to the nanotube length (increases marginally with the increase in length), and the conductivity can change drastically with a change in the diameter and the smaller diameter nanotubes can significantly increase the overall conductivity.

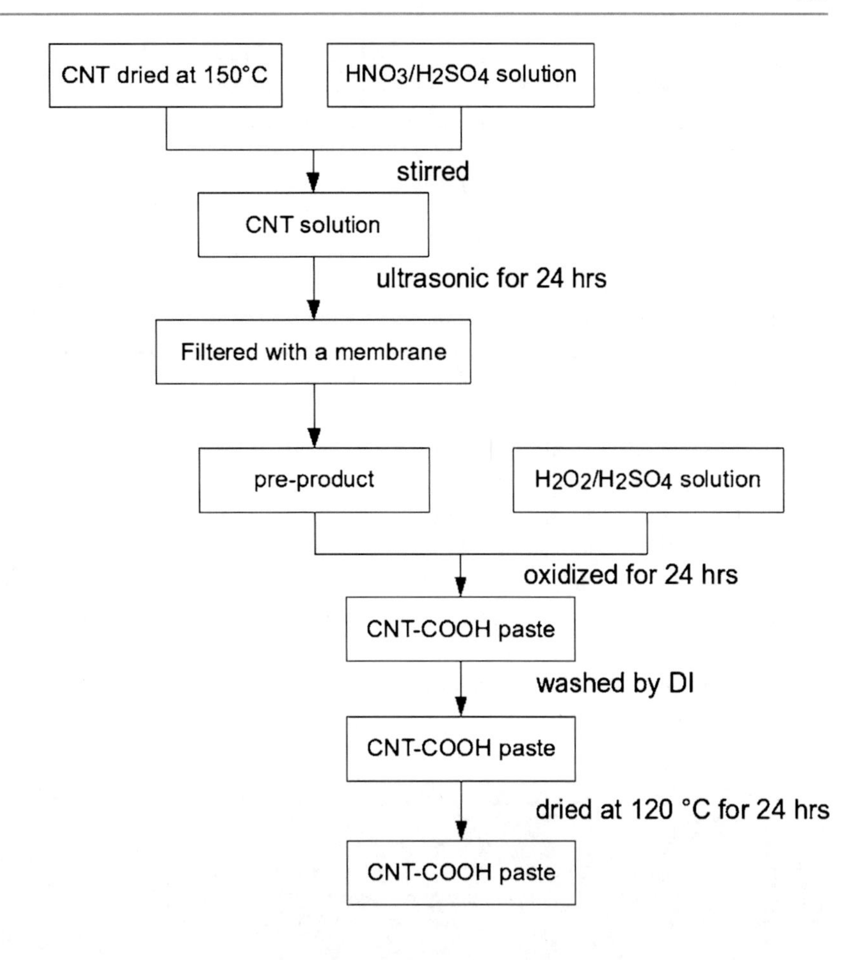

Fig. 7.28 Experimental procedure for acid-modified carbon nanotubes (CNT) [96]

7.6 Conclusive Remarks and Prospects

Temperature control is critical for both the operating performance and long-term reliability of single chip packages. The high junction-to-ambient thermal resistance resulting from an air-cooled heat sink provides inadequate heat removal capability at the necessary junction temperatures. The International Technology Roadmap for Semiconductors in 2007 Edition [112] projects power density and junction-to- ambient thermal resistance for high-performance chips at the nanometer generation on the level of >1 W/mm^2 and < 0.2 °C/W, respectively. The main bottlenecks in reducing the junction-to-ambient thermal resistance are the thermal resistances of the thermal interface material and the heat sink. There is a need for TIMs that provide the highest possible thermal conductivity, are mechanically stable during chip operation, have good adhesion, and conform to fill the gaps between two rough surfaces. The report [112] predicts that to address these needs, new TIMs containing carbon nanotubes will be elaborated.

In fact, carbon nanotubes due to their outstanding thermal properties have been the best candidate for a heat transport application. Unfortunately, the thermally conductive composites with CNTs are characterized by thermal conductivity not higher than a few W/m × K. Theoretically, according to Eqs. 7.5 and 7.6, the improvement of thermal conductivity of nanocomposites saturated by filler with very low thermal resistance Θ_λ can be done by:

- Decreasing the thermal contact resistance between particles forming a chain for heat transport in the direction of temperature gradient (Θ_{TC})
- Lessening the numbers contact resistance between particles responsible for heat transport
- Increasing the number parallel contact path in the direction of temperature gradient

All these requirements would be fulfilled with high content of carbon nanotube in the polymer matrix. Unfortunately, the results presented in the literature show that it is impossible to incorporate the CNT in composite higher than a few weight percent. In such cases there is not too many contact points between singular carbon nanotube, and even applying a polymer matrix with high cure shrinkage would not bring about a considerable improvement of a thermal conductivity of the TIM.

The diminishing of contacts can be achieved by a stable, conductive network formed by the CNT fillers. It seems to be necessary to develop a systematic method of producing various multiple junctions of nanotubes to establish a three-dimensional network instead of the set of singular structures. There is information about experiments aimed at establishing a junction between crossed nanotubes; the first attempt at doing such web-like CNTs has been achieved with electron beam, ion beam, laser, or other techniques of nanowelding [113].

As an example, Fig. 7.29 presents the single-walled carbon nanotubes joined by electron beam welding to form molecular junctions. In experiments, single-walled nanotubes were dispersed ultrasonically in ethanol and deposited onto porous carbon grids for transmission electron microscopy (TEM) observations. TEM observations were performed under high voltage of 1.25 MeV at specimen temperatures of 800 °C. From the random crisscrossing distribution of individual nanotubes and nanotube bundles on the specimen grid, several contact points could be identified where tubes were crossing and touching each other. After a few minutes of irradiating and annealing, different shapes of junctions were established, even the tube diameters within the junction were not the same [114].

Using empirical potential molecular dynamics, it was demonstrated that also ion irradiation should result in welding of crossed nanotubes. A simulation of the irradiation of crossed single-walled carbon nanotubes was carried out, both suspended and deposited on substrates, with 0.4–1 keV argon ions. The optimum Ar ion energies should be in the range of 0.4–0.6 keV, whereas the optimum irradiation dose should be about 10^{15} cm^{-2} [115]. In practice, the junctions were obtained as the result of carbon ion irradiation. In the experiment [116], the multiwalled carbon nanotubes were transferred onto the porous carbon micro-grid for transmission electron microscopy examination with an accelerating voltage of 160 kV. Second, this porous carbon micro-grid filled with unirradiated nanotubes was managed to be irradiated perpendicularly by carbon ion beam with an energy of 50 keV in a vacuum chamber. The current and dose of carbon ion beam were about 10 μA and 1×10^{17} cm^{-2}, respectively. The results were presented in Fig. 7.30. Figure 7.30a shows the image of a typical web-like structure which contains four junctions (indicated by A, B, C, and D). High-resolution transmission electronic microscope images

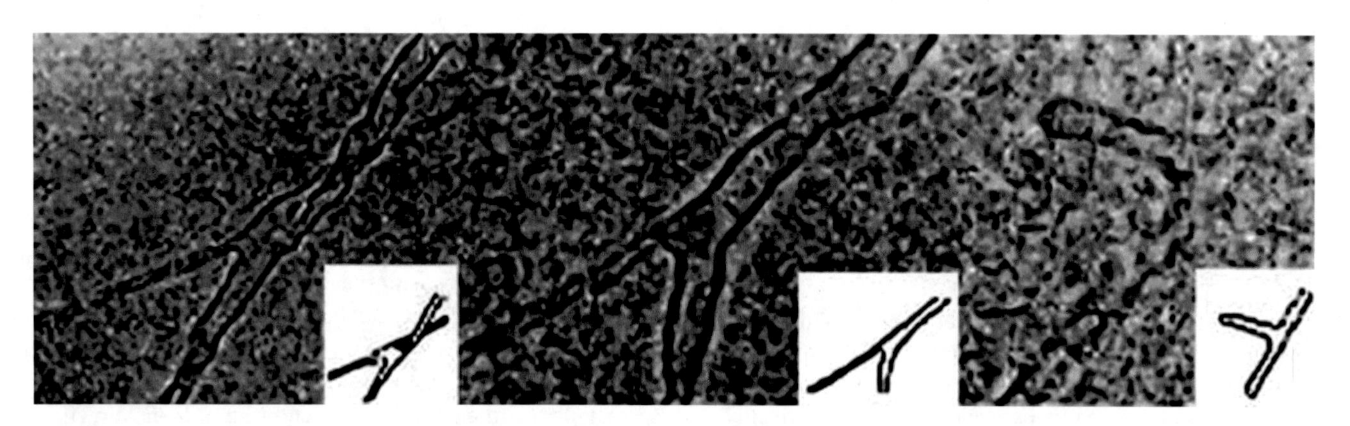

Fig. 7.29 Electron microscopy images of single-walled nanotube welding under intense electron beam [85]

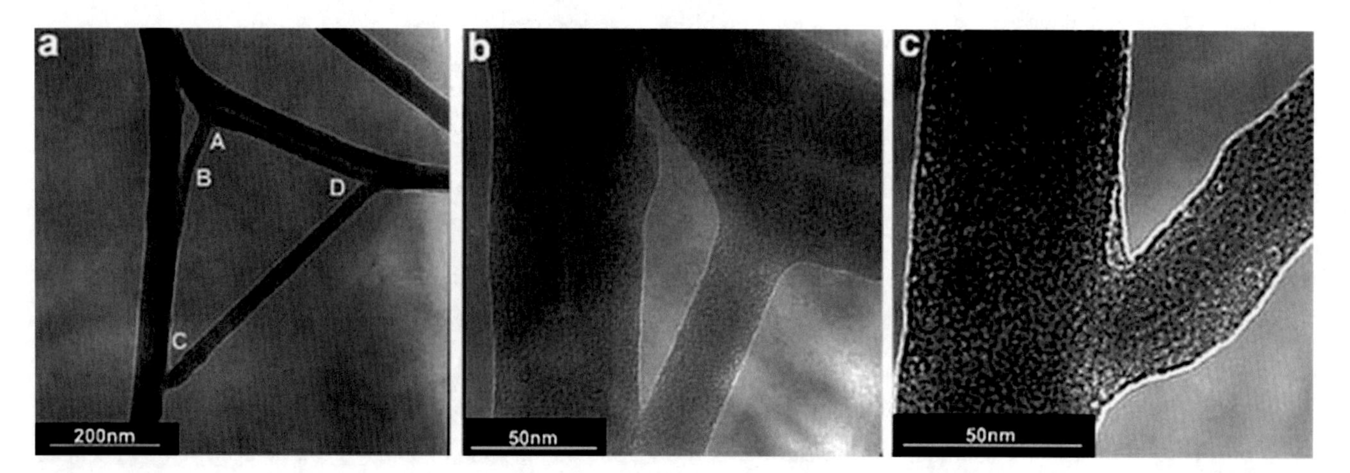

Fig. 7.30 (**a**) Web-like structure with junctions of amorphous carbon nanowires (indicated by A, B, C, and D). (**b**), (**c**) High-resolution transmission electron microscopy images for junctions at A and B showed in (**a**) [116]

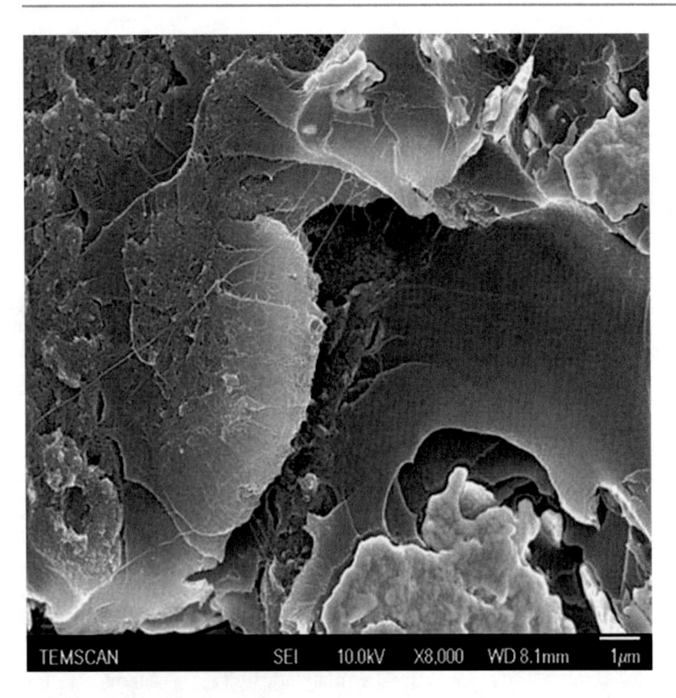

Fig. 7.31 Scanning electron microscopy image of composite containing silver flakes and carbon nanotubes as the filler. (Courtesy of CANOPY Project partners [117])

taken at the regions of A ("T" shape), B, and C ("Y" shape) junctions were shown in Fig. 7.30b, c. The merging region between two carbon nanowires was a uniform amorphous solid structure without any evident sign of the interface presented in an atomic scale.

The increasing of number parallel contact path in direction of temperature gradient can be achieved by different types of microfillers or nanofillers. Figure 7.31 presents an example of mixed fillers (microsized silver flakes and CNT). The low content of nanotubes (about 0.5 wt%) weakly influenced the total thermal conductivity of the composite yet, and efforts of research labs are aimed at achieving the significant increase of the CNT content [117].

Also adding of nanofillers can improve the thermal conductivity. It was observed that the presence of nanosized carbon black may improve the thermal conductivity of epoxy-based composite with 2 wt% carbon nanotubes [97] by ca. 20%. The carbon black with particle diameters of 22–60 nm was added to the mixture after it was sonicated in acetone. The prepared mixture was put into a vacuum chamber at 100 °C for 4–5 h to remove the acetone and the trapped air.

References

1. Liu, J., Wang, T., Carlberg, B., Inoue, M.: Recent Progress of Thermal Interface Materials. In: 2nd Electronics System integration Technology Conference, Greenwich, p. 351 (2008)
2. Sarvar, F., Whalley, D.C., Conway, P.P.: Thermal Interface Materials – A review of the State of Art. In: 1st Electronics System integration Technology Conference, Dresden, p. 1292 (2006)
3. Iijima, S.: Carbon nanotubes: past, present, and future. Physica B. **323**, 1 (2002)
4. Paul, D.R., Robeson, L.M.: Polymer nanotechnology: nanocomposites. Polymer. **49**, 3187 (2008)
5. Efunda – engineering fundamentals; http://www.efunda.com.
6. Damasceni, A., Dei, L., Guasti, F.: Thermal behaviour of silver-filled epoxy adhesives; technological implications in microelectronics. J. Therm. Anal. Calorim. **66**, 223 (2001)
7. Falat, T., Felba, J., Wymyslowski, A.: Improved Method for Thermal Conductivity Measurement of Polymer based Materials for Electronic Packaging. In: 28th International Conference of International Microelectronics and Packaging Society – Poland Chapter, Wroclaw, p. 219 (2004)
8. Prasher, R.: Thermal interface materials: historical perspective, status, and future directions. Proc. IEEE. **94**(8), 1571 (2006)
9. Nana, C.-W., Birringer, R., Clarke, D.R., Gleiter, H.: Effective thermal conductivity of particulate composites with interfacial thermal resistance. J. Appl. Phys. **81**(10), 6692 (1997)
10. Koledintseva, M.Y., DuBroff, R.E., Schwartz, R.W., Maxwell, A.: Garnett model for dielectric mixtures containing conducting particles at optical frequencies. Prog. Electromagn. Res. **63**, 223–242 (2006)
11. Mallet, P., Guérin, C.A., Sentenac, A.: Maxwell-Garnett mixing rule in the presence of multiple scattering: derivation and accuracy. Phys. Rev. B. **72**, 14205 (2005)
12. Levy, O., Stroud, D.: Maxwell Garnett theory for mixtures of anisotropic inclusions: application to conducting polymers. Phys. Rev. B. **56**(13), 8035 (1997)
13. Rong, M., Zhang, M., Liu, H., Zeng, H.: Synthesis of silver nanoparticles and their self- organization behavior in epoxy resin. Polymer. **40**(22), 6169 (1999)
14. Falat, T.: Heat Transfer Analysis in Composite Materials Filled with Micro and Nano-Sized Particles, Doctoral Thesis (in Polish). Faculty of Microsystem Electronics and Photonics, Wroclaw University of Technology, Wroclaw (2006)
15. Felba, J., Falat, T.: Thermally Conductive Adhesives for Microelectronics – Barriers of Heat Transport. In: 6th International IEEE Conference on Polymers and Adhesives in Microelectronics and Photonics, Polytronic, p. 228 (2007)
16. Falat, T., Felba, J.: Numerical Prediction of Influence Matrix and Filler Properties on Thermal Conductivity of Copper Filled TCA. In: 6th International IEEE Conference on Polymers and Adhesives in Microelectronics and Photonics, Polytronic, p. 114 (2007)
17. Felba, J., Falat, T., Wymyslowski, A.: Influence of thermo-mechanical properties of polymer matrices on the thermal conductivity of adhesives for microelectronic packaging. Mater. Sci.-Pol. **25**(1), 45 (2007)
18. Falat, T., Felba, J., Wymyslowski, A., Jansen, K.M.B., Nakka, J.S.: Viscoelastic Characterization of Polymer Matrix of Thermally Conductive Adhesives. In: 1st Electronics System integration Technology Conference, Dresden, p. 773 (2006)
19. Holm, R.: Electric Contacts – Theory and Application. Springer, Berlin (1967)
20. Wymyslowski, A., Friedel, K., Felba, J., Falat, T.: An Experimental-Numerical Approach to Thermal Contact Resistance. In: 9th International Workshop on Thermal Investigations of ICs and Systems, Aix-en-Provence, p. 161 (2003)
21. Madhusudana, C.V.: Thermal Contact Conductance. Springer, New York (1996)
22. Hasselman, D.P.H., Donaldson, K.Y., Barlow, F.D., Elshabini, A. A., Schiroky, G.H., Yaskoff, J.P., Dietz, R.L.: Interfacial thermal resistance and temperature dependence of three adhesive for electronic packaging. IEEE Trans. Compon. Packag. Technol. **23**(4), 633 (2000)
23. Wymyslowski, A., Falat, T., Friedel, K., Felba, J.: Numerical Simulation and Experiment Verification of the Thermal Contact Properties of the Polymer Bonds. In: 5th International Conference

on Thermal and Mechanical Simulation and Experiments in Microelectronics and Microsystems, Brussels, p. 177 (2004)
24. Bolger, J.C.: Prediction and Measurement of Thermal Conductivity of Diamond Filled Adhesives. In: 42nd Electronic Components and Technology Conference, p. 219 (1992)
25. Li, H., Jacob, K.I., Wong, C.P.: An improvement of thermal conductivity of underfill materials for flip-chip packages. IEEE Trans. Adv. Packag. **26**(1), 25 (2003)
26. Ukita, Y., Tateyama, K., Segawa, M., Tojo, Y., Gotoh, H., Oosako, K.: Lead Free Mount Adhesive using Silver Nanoparticles Applied to Power Discrete Package. In: International Symposium on Microelectronics IMAPS, Long Beach (2004)., session WA7
27. Macek, S., Rocak, D., Sebo, P., Stefanik, P.: The use of Polymeric Adhesives in Bonding Power Hybrid Circuits to Heat Sinks. In: 23rd International Spring Seminar on Electronics Technology ISSE, Balatonfured, p. 185 (2000)
28. Dietz, R., Robinson, P., Bartholomew, M., Firmstone, M.: High Power Application with Advanced High k Thermoplastic Adhesives. In: 11th European Microelectronics Conference, Venice, p. 486 (1997)
29. Zarr, R.R.: A history of testing heat insulators at the national institute standards and technology. ASHRAE Trans. **107**, Pt 2 (2001)
30. He, Y.: Rapid thermal conductivity measurement with a hot disk sensor. Part 1. Theoretical considerations. Thermochim. Acta. **436**, 122–129 (2005)
31. Tye, R.P.: Thermal conductivity. Nature. **205**, 636 (1964)
32. Smith, D.R., Hust, J.G., Van Poolen, L.J.: A Guarded-Hot-Plate Apparatus for Measuring Effective Thermal Conductivity of Insulations Between 80 K and 360 K. US National Bureau of Standards. (1982) NBSIR 81-1657
33. Stacey, C.: NPL Vacuum Guarded Hot-Plate for Measuring Thermal Conductivity and Total Hemispherical Emittance of Insulation Materials. Insulation Materials: Testing and Applications, 4. ASTM STP. **1426**, (2002)
34. Franco, A.: An apparatus for the routine measurement of thermal conductivity of materials for building application based on a transient hot-wire method. Appl. Therm. Eng. **27**, 2495–2504 (2007)
35. Salgon, J.J., Robbe-Valloire, F., Blouet, J., Bransier, J.: A mechanical and geometrical approach to thermal contact resistance. Int. J. Heat Mass Transfer. **40**(5), 1121 (1997)
36. Yeh, C.L., Wen, C.Y., Chen, Y.F., Yeh, S.H., Wu, C.H.: An experimental investigation of thermal contact conductance across bolted joints. Exp. Thermal Fluid Sci. **25**, 349 (2001)
37. Falat, T., Platek, B., Tesarski, S., Felba, J.: An Approach to Measurement and Evaluation of the Thermal Conductivity of the Thermal Adhesives in Electronic Packaging. In: 32nd International Spring Seminar on Electronics Technology, Brno (2009)
38. Assael, M.J., Antoniadis, K.D., Tzetzis, D.: The use of the transient hot-wire technique for measurement of the thermal conductivity of an epoxy-resin reinforced with glass fibres and/or carbon multiwalled nanotubes. Compos. Sci. Technol. **68**, 3178–3183 (2008)
39. Roder, H.M., Perkins, R.A., Laesecke, A.: Absolute steady-state thermal conductivity measurements by use of a transient hot-wire system. J. Res. Natl. Inst. Stand. Technol. **105**(2), 221 (2000)
40. dos Santos, W.N.: Advances on the hot wire technique. J. Eur. Ceram. Soc. **28**, 15 (2008)
41. Parker, W.J., Jenkins, R.J., Butler, C.P., Abbott, G.L.: Flash method of determining thermal diffusivity, heat capacity, and thermal conductivity. J. Appl. Phys. **32**, 1679 (1961)
42. Cowan, R.D.: Pulse method of measuring thermal diffusivity at high temperatures. J. Appl. Phys. **34**, 926 (1963)
43. Taylor, R.E., Cape, J.A.: Finite pulse-time effects in the flash diffusivity technique. Appl. Phys. Lett. **5**, 212 (1964)
44. Watt, D.A.: Theory of thermal diffusivity by pulse technique. Br. J. Appl. Phys. **17**, 231 (1966)
45. Walter, A.J., Dell, R.M., Burgess, P.C.: The measurement of thermal diffusivities using a pulsed electron beam. Rev. Int. Hautes Tempér. Réfract. **7**, 271 (1970)
46. Touloukian, Y.S., Powell, R.W., Ho, C.Y., Nicolaou, M.C.: Thermal Diffusivity. Thermophysical Properties of Matter, p. 10. IFI/Plenum, New York/Washington, DC (1973)
47. Clark, I.I.I.L.M., Taylor, R.E.: Radiation loss in the flash method for thermal diffusivity. J. Appl. Phys. **46**, 714 (1975)
48. Xu, Y., Chung, D.D.L., Mroz, C.: Thermally conducting aluminium nitride polymer-matrix composites. Composites A. **32**, 1749 (2001)
49. Falat, T., Felba, J.: Electron beam as a heat source in thermal diffusivity measurement of thermally conductive adhesives. Electrotech. Electron. **5–6**, 189 (2006)
50. Rosencwaig, A., Gersho, A.: Theory of the photoacoustic effect with solids. J. Appl. Phys. **47**, 64 (1976)
51. Hu, H., Wang, X., Xu, X.: Generalized theory of the photoacoustic effect in a multilayer material. J. Appl. Phys. **86**, 3953 (1999)
52. Cola, B.A., Xu, J., Cheng, C., Xu, X., Fisher, T.S., Hu, H.: Photoacoustic characterization of carbon nanotube array thermal interfaces. J. Appl. Phys. **101**, 054313 (2007)
53. Sun, L., Zhang, S.-Y., Zhao, Y.-Z., Li, Z.-Q., Cheng, L.-P.: Thermal diffusivity of composites determined by photoacoustic piezoelectric technique. Rev. Sci. Instrum. **74**, 834 (2003)
54. Balderas-López, J.A.: Thermal effusivity measurements for liquids: a self-consistent photoacoustic methodology. Rev. Sci. Instrum. **78**, 064901 (2007)
55. Leite, N.F., Miranda, L.C.M.: Thermal property measurements of liquid samples using photoacoustic detection. Rev. Sci. Instrum. **63**, 4398 (1992)
56. Balderas-López, J.A.: Self-normalized photoacoustic technique for thermal diffusivity measurements of transparent materials. Rev. Sci. Instrum. **79**, 024901 (2008)
57. Philip, J.: A photoacoustic scanning method to determine thermal effusivity of solid samples. Rev. Sci. Instrum. **67**, 3621 (1996)
58. Govorkov, S., Ruderman, W., Horn, M.W., Goodman, R.B., Rothschild, M.: A new method for measuring thermal conductivity of thin films. Rev. Sci. Instrum. **68**, 3828 (1997)
59. Kim, J.H., Feldman, A., Novotny, D.: Application of the three omega thermal conductivity measurement method to a film on a substrate of finite thickness. J. Appl. Phys. **86**(7), 3959 (1999)
60. Dames, C., Chen, G.: 1ω, 2ω, and 3ω methods for measurements of thermal properties. Rev. Sci. Instrum. **76**, 124902 (2005)
61. Cahill, D.G., Pohl, R.O.: Thermal conductivity of amorphous solids above the plateau. Phys. Rev. B. **35**(8), 4067–4073 (1987)
62. Choi, S.R., Kim, J., Kim, D.: 3ω method to measure thermal properties of electrically conducting small-volume liquid. Rev. Sci. Instrum. **78**, 084902 (2007)
63. Risk, W.P., Rettner, C.T., Raoux, S.: In situ 3ω techniques for measuring thermal conductivity of phase-change materials. Rev. Sci. Instrum. **79**, 026108 (2008)
64. National Fenestration Rating Council, Procedure for Determining Thermo-Physical Properties of Materials, www.nfrc.org
65. Fan, L., Su, B., Qu, J., Wong, C.P.: Effects of Nano-Sized Particles on Electrical and Thermal Conductivities of Polymer Composites. In: 9th International Symposium on Advanced Packaging Materials; Processes, Properties and Interfaces, p. 193 (2004)
66. Bujard, P., Ansermet, J.P.: Thermally Conductive Aluminium Nitride-Filled Epoxy Resin. In: 5th Semiconductor Thermal and Temperature Measurement Symposium, p. 126 (1989)
67. Felba, J., Schaefer, H.: Materials and technology for conductive microstructures. In: Morris, J.E. (ed.) Nanopackaging: Nanotechnologies and Electronics Packaging. Springer (2008)
68. Tilaki, R.M., Iraji Zad, A., Mahdavi, S.M.: Stability, size and optical properties of silver nanoparticles prepared by laser ablation in different carrier media. Appl. Phys. A. **84**, 215 (2006)

69. Saito, H., Matsuba, Y.: Liquid Wiring Technology by Ink-jet Printing Using NanoPaste. In: 35th International Symposium on Microelectronics IMAPS, San Diego (2006)
70. http://www.harima.co.jp/products/electronics
71. Moscicki, A., Felba, J., Sobierajski, T., Kudzia, J., Arp, A., Meyer, W.: Electrically Conductive Formulations Filled Nano Size Silver Filler for Ink-Jet Technology. In: 5th International IEEE Conference on Polymers and Adhesives in Microelectronic and Photonics, Wroclaw, p. 40 (2005)
72. Moscicki, A., Felba, J., Dudzinski, W.: Conductive Microstructures and Connections for Microelectronics Made by Ink-Jet Technology. In: 1st Electronics System integration Technology Conference, Dresden, p. 511 (2006)
73. Nakamoto, M., Yamamoto, M., Kashiwagi, Y., Kakiuchi, H., Tsujimoto, T., Yoshida, Y.A.: Variety of Silver Nanoparticle Pastes for Fine Electronic Circuit Pattern Formation. In: 6th International IEEE Conference on Polymers and Adhesives in Microelectronic and Photonics, Tokyo, p. 105 (2007)
74. Bai, J.G., Zhang, Z.Z., Calata, J.N., Lu, G.-Q.: Low-temperature sintered nanoscale silver as a novel semiconductor device-metallized substrate interconnect material. IEEE Trans. Compon. Packag. Technol. **29**(3), 589 (2006)
75. Bai, J.G., Calata, J.N., Lei, T.G., Lu, G.Q., Creehan, K.D.: Lead-free Die-attachment with High-temperature Capability by Low-temperature Nanosilver Paste Sintering. In: 35th International Symposium on Microelectronics IMAPS, San Diego (2006)
76. Chen, C., Wang, L., Li, R., Jiang, G., Yu, H., Chen, T.: Effect of silver nanowires on electrical conductance of system composed of silver particles. J. Mater. Sci. **42**(9), 3172 (2007)
77. Wu, H.P., Liu, J.F., Wu, X.J., Ge, M.Y., Wang, Y.W., Zhang, G. Q., Jiang, J.Z.: High conductivity of isotropic conductive adhesives filled with silver nanowires. Int. J. Adhes. Adhes. **26**, 617 (2006)
78. Graff, A., Wagner, D., Ditlbacher, H., Kreibig, U.: Silver nanowires. Eur. Phys. J. D. **34**, 263 (2005)
79. Liu, J., Olugbenga Olorunyomi, M., Lu, X., Wang, W.X., Aronsson, T., Shangguan, D.: New Nano-Thermal Interface Material for Heat Removal in Electronic Packaging. In: 1st Electronics System Integration Technology Conference, Dresden, p. 1 (2006)
80. Liu, J., Olugbenga Olorunyomi, M., Li, X., Shangguan, D.: Manufacturing and Characterization of Nano Silver Particles Based Thermal Interface Material. In: 57th Electronic Components and Technology Conference, p. 475 (2007)
81. Graebner, J.E., Jin, S.: Chemical vapor deposited diamond for thermal management. JOM. **50**(6), 52 (1998)
82. Brierley, C.J.: Thermal Management with Diamond. IEE Colloquium on Diamond in Electronics and Optics. (1993)
83. Kidalov, S.V., Shakhov, F.M., Ya, A.: Vul'. Thermal conductivity of nanocomposites based on diamonds and nanodiamonds. Diamond Related Mater. **16**(12), 2063 (2007)
84. Vlasov, A., Ralchenko, V., Gordeev, S., Zakharov, D., Vlasov, I., Karabutov, A., Belobrov, P.: Thermal properties of diamond/carbon composites. Diam. Relat. Mater. **9**(3–6), 1104 (2000)
85. Kawamura, T., Kangawa, Y., Kakimoto, K.: Investigation of the thermal conductivity of a fullerene peapod by molecular dynamics simulation. J. Cryst. Growth. **310**, 2301 (2008)
86. Smontara, A., Saint Paul, M., Lasjaunias, J.C., Bilusic, A., Kitamura, N.: Thermal and acoustic transport properties of hard carbon formed from C60 fullerene. Physica B. **316–317**, 250 (2002)
87. Terrones, M.: Carbon nanotubes: synthesis and properties, electronic devices and other emerging applications. Int. Mater. Rev. **49** (6), 325 (2004)
88. Popov, V.N.: Carbon nanotubes: properties and application. Mater. Sci. Eng. R. **43**, 61 (2004)
89. Yadav, Y., Kunduru, V., Prasad, S.: Carbon nanotubes: synthesis and characterization. In: Morris, J.E. (ed.) Nanopackaging: Nanotechnologies and Electronics Packaging. Springer (2008)
90. Kim, P., Shi, L., Majumdar, A., McEuen, P.L.: Mesoscopic thermal transport and energy dissipation in carbon nanotubes. Physica B. **323**, 67 (2002)
91. Pecht, M., Agarwal, R., McCluskey, P., Dishongh, T., Javadpour, S., Mahajan, R.: Electronic Packaging Materials and their Properties. CRC Press LLC, Washington, DC (1999)
92. Fujii, M., Zhang, X., Xie, H., Ago, H., Takahashi, K., Ikuta, T., Abe, H., Shimizu, T.: Measuring the thermal conductivity of a single carbon nanotube. Phys. Rev. Lett. **95**, 065502 (2005)
93. Hone, J., Liaguno, M.C., Biercuk, M.J., Johnson, A.T., Batlogg, B., Benes, Z., Fischer, J.E.: Thermal properties of carbon nanotubes and nanotube-based materials. Appl. Phys. A. **74**, 339 (2002)
94. Bal, S., Samal, S.S.: Carbon nanotube reinforced polymer composites–a state of the art. Bull. Mater. Sci. **30**(4), 379–386 (2007)
95. Peigney, A., Laurent, C., Flahaut, E., Bacsa, R.R., Rousset, A.: Specific surface area of carbon nanotubes and bundles of carbon nanotubes. Carbon. **39**, 507 (2001)
96. Lee, T.-M., Chiou, K.-C., Tseng, F.-P., Huang, C.-C.: High Thermal Efficiency Carbon Nanotube-Resin Matrix for Thermal Interface Materials. In: 55th Electronic Component & Technology Conference, Lake Buena Vista, p. 55 (2005)
97. Zhang, K., Xiao, G.-W., Wong, C.K.Y., Gu, H.-W., Yuen, M.M. F., Chan, P.C.H., Xu, B.: Study on Thermal Interface Material with Carbon Nanotubes and Carbon Black in High-Brightness LED Packaging with Flip-Chip Technology. In: 55th Electronic Component & Technology Conference, Lake Buena Vista, p. 60 (2005)
98. Titus, E., Ali, N., Cabral, G., Gracio, J., Ramesh Babu, P., Jackson, M.J.: Chemically functionalized carbon nanotubes and their characterization using thermogravimetric analysis, Fourier transform infrared, and Raman spectroscopy. J. Mater. Eng. Perform. **15**(2), 182 (2006)
99. Schadler, L.S., Giannaris, S.C., Ajayan, P.M.: Load transfer in carbon nanotube epoxy composites. Appl. Phys. Lett. **73**, 3842 (1998)
100. Bauhofer, W., Kovacs, J.Z.: A review and analysis of electrical percolation in carbon nanotube polymer composites. Compos. Sci. Technol. **69**, 1486–1498 (2009)
101. Chang, L., Friedrich, K., Ye, L., Toro, P.: Evaluation and visualization of the percolating networks in multi-wall carbon nanotube/epoxy composites. J. Mater. Sci. **44**, 4003–4012 (2009)
102. Wescott, J., Kung, P., Mati, A.: Conductivity of carbon nanotube polymer composites. Appl. Phys. Lett. **90**, 033116 (2007)
103. Kim, H.-S., Park, B.H., Yoon, J.-S., Jin, H.-J.: Thermal and electrical properties of poly (L-lactide)-graft-multiwalled carbon nanotube composites. Eur. Polym. J. **43**, 1729–1735 (2007)
104. Kovacs, J.Z., Velagala, B.S., Schulte, K., Bauhofer, W.: Two percolation thresholds in carbon nanotube epoxy composites. Compos. Sci. Technol. **67**, 922–928 (2007)
105. Choi, S.U.S., Zhang, Z.G., Yu, W., Lockwood, F.E., Grulke, E.A.: Anomalous thermal conductivity enhancement in nanotube suspensions. Appl. Phys. Lett. **79**(14), 2252 (2001)
106. Biercuk, M.J., Llaguno, M.C., Radosavljevic, M., Hyun, J.K., Johnson, A.T., Fischer, J.E.: Carbon nanotubes for thermal management. Appl. Phys. Lett. **80**, 2767 (2002)
107. Hong, W.-T., Tai, N.-H.: Investigations on the thermal conductivity of composites reinforced with carbon nanotubes. Diamond Relat. Mater. **17**, 1577 (2008)
108. Thostenson, E.T., Chou, T.-W.: Processing-structure-multi-functional property relationship in carbon nanotube/epoxy composites. Carbon. **44**, 3022 (2006)

109. Gojny, F.H., Wichmann, M.H.G., Fiedler, B., Kinloch, I.A., Bauhofer, W., Windle, A.H., Schulte, K.: Evaluation and identification of electrical and thermal conduction mechanisms in carbon nanotube/epoxy composites. Polymer. **47**, 2036 (2006)
110. Song, Y.S., Youn, J.R.: Evaluation of effective thermal conductivity for carbon nanotube/polymer composites using control volume finite element method. Carbon. **44**, 710–717 (2006)
111. Bagchi, A., Nomura, S.: On the effective thermal conductivity of carbon nanotube reinforced polymer composites. Compos. Sci. Technol. **66**, 1703 (2006)
112. International Technology Roadmap for Semiconductor.: http://www.itrs.net/Links/ 2007ITRS/Home2007.htm
113. Sahin, S., Yavuz, M., Zhou, Y.N.: Introduction to nanojoining. In: Zhou, Y.N. (ed.) Microjoining and Nanojoining. Woodhead Publishing Limited (2008)
114. Charlier, J.-C., Terrones, M., Banhart, F., Grobert, N., Terrones, H., Ajayan, P.M.: Experimental observation and quantum modeling of electron irradiation on single-wall carbon nanotubes. IEEE Trans. Nanotechnol. **2**(4), 349 (2003)
115. Krasheninnikov, A.V., Nordlund, K., Keinonen, J., Banhart, F.: Making junctions between carbon nanotubes using an ion beam. Nucl. Inst. Methods Phys. Res. B. **202**, 224 (2003)
116. Wang, Z., Yu, L., Zhang, W., Ding, Y., Li, Y., Han, J., Zhu, Z., Xu, H., He, G., Chen, Y., Hu, G.: Amorphous molecular junctions produced by ion irradiation on carbon nanotubes. Phys. Lett. A. **324**, 321 (2004)
117. EUREKA/EURIPIDES project "Carbon Nanotubes/epoxy composites" (2007–2010), acronym CANOPY

8 Physical Properties and Mechanical Behavior of Carbon Nano-tubes (CNTs) and Carbon Nano-fibers (CNFs) as Thermal Interface Materials (TIMs) for High-Power Integrated Circuit (IC) Packages: Review and Extension

Yi Zhang, Ephraim Suhir, and Claire Gu

In this chapter we address several major experimental techniques that have been developed during the last decade and are being used to characterize the physical/mechanical properties of the individual CNTs/CNTs and CNT/CNF arrays and to understand their physical (mechanical and thermal) behaviors and performance. We also discuss some results that have been obtained using these techniques. In the extension part of this chapter, we show how analytical stress modeling can be used to describe the mechanical behavior of CNTs/CNFs embedded into low-modulus elastic media. This is done to improve both the thermal and the mechanical performance of the CNT/CNF arrays. The developed models can be employed for the analysis and rational physical design of the structures in question.

8.1 Introduction

Since their discovery in 1991 [1], carbon nano-tubes (CNTs) have developed into a distinct branch of nano-technology, nano-material, and nano-mechanical engineering with numerous unique and useful applications. Carbon nano-tubes (CNTs) are unique tubular structures of nanometer diameter (say, <1–200 nm) and large length/diameter ratio (say, 500–1000 or so).

Y. Zhang (✉) · C. Gu
Department of Electrical Engineering, University of California, Santa Cruz, CA, USA
e-mail: claire@soe.ucsc.edu

E. Suhir
Bell Laboratories, Physical Sciences and Engineering Research Division, Murray Hill, NJ, USA

Department of Mechanical Engineering, University of Maryland, College Park, MD, USA

ERS Co., Los Altos, CA, USA

The following two major types of CNTs are distinguished:

1. *Single-wall* CNTs (SWCNTs) that consist of just one layer of tubular graphene.
2. *Multiwall* CNTs (MWCNTs) that consist of two or more concentric shells of carbon and a hollow inner capillary. The separation between the adjacent shells in MWCNTs is 0.34 nm.

There is also a third type of MWCNTs, in which there is no hollow inner capillary in the middle. Instead, the inner few layers of shells are linked together and form a "bamboo-like" structure. This type of CNTs is usually referred to as *carbon nano-fibers* (CNFs).

Whether it is a SWCNT, a MWCNT, or a CNF, the unique quasi-one-dimensional structure and the graphene/graphite arrangement of carbon atoms in the carbon shells in these structural elements result in some extraordinary properties of the CNT/CNF material: exceptionally high Young's modulus, significant high tensile strength, and extraordinarily high thermal conductivity. These unique properties lead to many attractive applications of CNTs and CNFs.

Thermal interface material (TIM) is one of the most attractive CNT/CNF applications [2–5]. Theoretically thermal conductivity of ideally structured SWCNTs can be as high as 6000 W/mK. This is even higher than that of the diamond [6]. Vertically aligned CNT/CNF arrays produced by using one of the *chemical vapor deposition* (CVD) techniques are viewed today as the most promising nanostructures and the most attractive fabrication technology, primarily because this technique enables one to produce consistent arrays with stable geometry [2].

Many theoretical and experimental studies have been conducted to evaluate the *thermal performance* of CNTs and CNFs. It has been established [7–9] that a single CNT exhibits thermal conductivities ranging from 1600 to 6000 W/mK. It is the common perception today that the predicted high thermal conductivity of single CNTs and

C. P. Wong et al. (eds.), *Nano-Bio- Electronic, Photonic and MEMS Packaging*, https://doi.org/10.1007/978-3-030-49991-4_8

CNFs might be retained in CNT/CNF arrays as well. This circumstance will lead to many breakthroughs in thermal management of IC and photonic devices. Particularly, it is expected that CNT/CNF-based systems might replace in the near future the existing conventional TIMs.

It should be pointed out, however, that in order to use CNTs or CNFs as effective and robust TIMs, these structural elements must possess, in addition to high thermal conductivity, also adequate mechanical properties; good "anchoring" strength (i.e., good "adhesion" to their substrates); ease in manufacturing; low enough cost; good wettability by the low-modulus embedding material, if any; etc.

There are several significant *challenges* one needs to overcome to make CNT/CNF systems suitable for TIM applications in electronic and photonic engineering. This is, first of all, the development and implementation of the well-controlled synthetic process of nano-materials and nano-structures. Second of all, it is the development of simple and reliable characterization techniques for both individual CNTs/CNFs and their "collective" properties, when arrays of CNT/CNFs are fabricated and used. Finally, one has to integrate these novel materials into the existing products, designs, and manufacturing technologies, or, if the advantages of the nano-technology-based TIMs are proven to be overwhelming, one has to be able to replace the existing designs, in terms of both better performance and lower cost. It is not easy to overcome the conservatism of the industry.

The ability to measure the physical (mechanical, thermal) properties of CNT/CNF systems (*metrobility* of nano-systems) is of an obvious practical importance as well. It is also an obvious challenge. The major reason for such a challenge is due, of course, to the ultra-small size of these structural elements. Specialized techniques and tools are needed even to pick up and install individual CNTs/CNFs into a metrological tool. In practical applications billions of CNTs/CNFs have to be grown, and CNT-based films (arrays) have to be fabricated to provide transfer of an appreciable amount of heat. There is a crucial need therefore for the appropriate characterization methods and adequate equipment that would enable one to carry out, with satisfactory accuracy and a timely fashion, thermal and mechanical measurements of the "collective properties" of CNT films and arrays.

In the review that follows, we address some existing *measurement techniques*, instrumentation, and equipment that were developed during the recent years to characterize the mechanical properties of the individual CNTs/CNFs as well as the physical (mechanical and thermal) properties of CNT/CNF arrays. We report also on some practically useful results obtained using these techniques. In the extension part of this chapter, we show how analytical *stress modeling* can be used to describe the mechanical behavior of CNTs/CNFs embedded into low-modulus elastic media. The developed models can be employed for the analysis and rational physical design of the structures in question.

8.2 Young's Modulus of Individual CNTs/CNFs

Young's modulus (YM) is an important material characteristic of any material, and therefore substantial effort has been dedicated to the evaluation [10–24] of the YM of CNTs/CNFs. It can be evaluated particularly by generating thermal vibrations of CNTs/CNFs and using *high-resolution tunneling electron microscope* (HRTEM) [10–14] to record the vibration characteristics. The YM can be then evaluated from the measured vibration frequency and the CNT/CNF geometry. The mechanical behavior of individual CNTs/CNFs was investigated also using the nano-probe manipulation method [21–23]. It has been found that the YM of a CNT/CNF might be diameter dependent. Based on some other literature, primarily theoretical, data, the YM of individual CNTs/CNFs was found to be in a rather wide range from 2 TPa to 200 GPa, when the CNT diameter changes from 5 to 40 nm. This information agrees satisfactorily with the experimental data obtained using HRTEM and assisted by an external oscillated electrical field [14]. It has been found also that the YM of the CNTs decreases exponentially with an increase in the CNT diameter [14].

CNFs produced by *plasma-enhanced chemical vapor deposition* (PECVD) are characterized by much larger diameters (about 100 nm) than CNTs. As is known, PECVD is a process mainly to grow CNTs or CNFs from a gas state (vapor) to a solid state on some substrates. In the case of CNT/CNF growth, three metals are used as catalysts: cobalt, nickel, or iron. Because of the large diameter and also because of the far-from-ideal "bamboo-like" morphology of the CNFs, their measured effective YM is considerably lower, by orders of magnitude, than the CNTs.

8.3 Tunneling Electron Microscopy (TEM)

Tunneling electron microscopy (TEM) is perhaps the most powerful tool currently available to observe the structure of nano-scale materials. This tool is able to provide real space resolution better than 0.1 nm, as well as atomic-scale structural information. Compared with other techniques, TEM is viewed as one of the best candidates to characterize the mechanical behavior of individual CNTs/CNFs.

Different approaches have been developed today to integrate the nano-structured materials (nano-tubes, nano-fibers, nano-wires) into the TEM high vacuum chamber with other additional specialized fixtures to obtain the in situ measurements of the CNT/CNF properties.

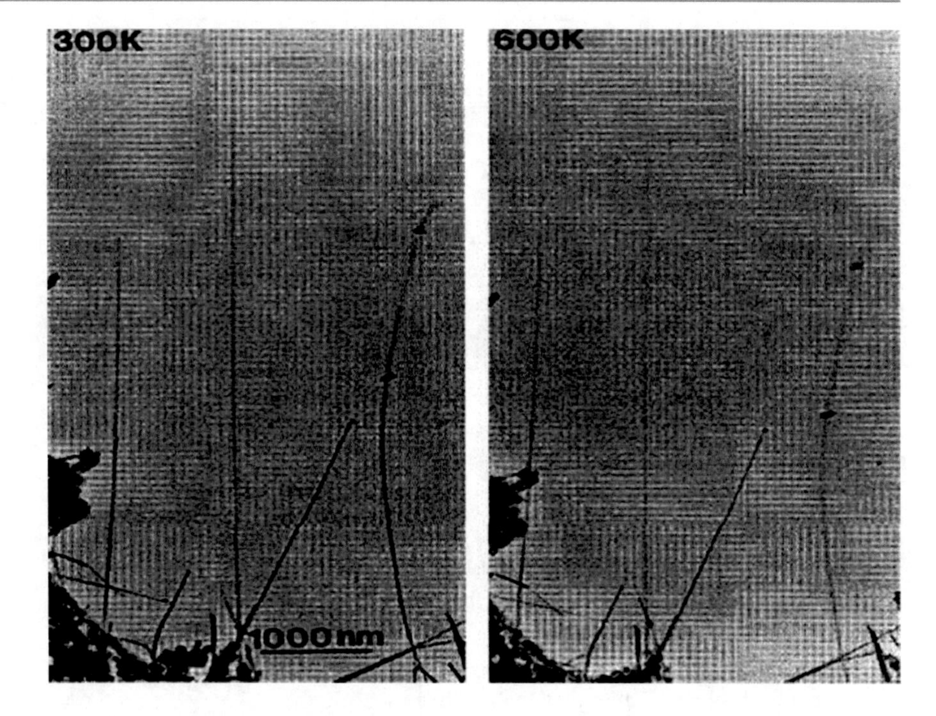

Fig. 8.1 TEM images show that the freestanding CNT vibrates at different temperatures (300 and 600 K) due to thermal effects [11]

8.4 Thermal Vibration Method

The first measurements conducted to evaluate the YM of individual CNTs were carried out by *thermal vibration method* in the TEM chamber by Treacy et al. back in 1996 [11]. The CNT used was synthesized by an arc-charged method. This method generally provides the CNT with the best structural quality. The CNT was mounted onto a specially designed holder and put into the vacuum chamber of the TEM. The high resolution of TEM enabled one to clearly see the structure and the dimensions of the CNTs. At room temperature (300 K), the image was blurred at the free end of the CNT; however, the supporting end had good focus. After the temperature was increased to 600 K, one could clearly observe that the tip became more blurry. It was concluded that the CNT was vibrating due to thermally induced vibrations (*Brownian motion*).

It has been found that the thermal vibration amplitude depends on the temperature: it has been 30 and 40 nm at 300 and 600 K, respectively. Detailed measurements of the change in the vibration amplitude with temperature are shown in Fig. 8.1. These measurements confirm the thermal origin of vibrations because of the Brownian motion.

When computing the YM of an individual CNT based on the thermal vibration phenomenon, the CNT was assumed to be a homogeneous cylindrical cantilever of certain length and diameter. It could be shown that the vibration amplitude, u, at the CNT tip is related to the CNT Young's modulus (YM) and the vibration energy W as $W = (1/2)\ cu^2$ (see [11]). Here c is the effective spring constant, which is

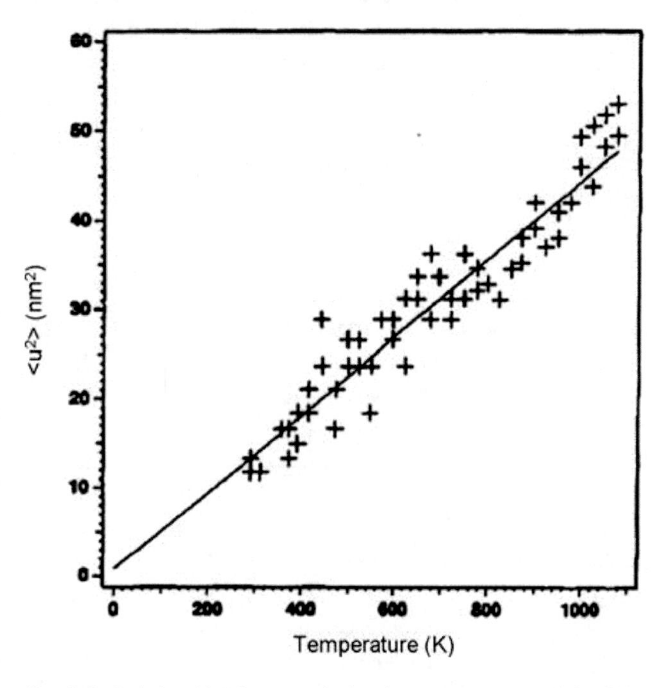

Fig. 8.2 Relationships between the heating temperature and the CNT thermal vibration amplitudes. The CNT was heated up to 800 °C with 25 °C as step. A linear relationship can be observed. This gives a YM value around 3.7 TPa for these particular CNTs [11]

proportional to the YM and is affected by the CNT structure and geometry.

The vibration amplitude could be obtained from Fig. 8.2 for different temperatures. As to the additional vibration energy, it could be obtained from the analysis of the Brownian motion as an average $= kT$. Here k is Boltzmann's constant and T is the absolute temperature. Then one could easily calculate the YM of an individual CNT at different vibration modes. CNTs of different lengths and different inner and outer diameters were measured, and Young's moduli were calculated. The calculated results indicated that the YM of CNT had a variation between 1.0 and 3.7 TPa, i.e., much higher than in the case of a typical CNF. The variations in the measured YM could be attributed to the uncertainties in the measurements of the CNT dimensions. Nonetheless, these data agree satisfactorily with the theoretical predictions.

8.5 Method Based on External Electric Field-Induced Vibrations

Poncharal et al. [15] developed a methodology for determining the YM of individual CNTs based on the observed, using TEM, electromechanical resonances. The CNTs were synthesized by an arc-charged method. The CNT was placed on a specially designed fixture that enabled one to apply the voltage across the CNT. Because of the small diameter of the CNT, the charges were accumulated on the CNT tip.

The CNTs experienced bending due to the electrostatic forces. By oscillating the voltage, one does so with the charges as well. The force direction also varies. When the applied frequency becomes close to the intrinsic (free) resonance frequency of the CNT, the CNT becomes subjected to elevated amplitudes of mechanical resonant vibrations.

For the first harmonic resonance, one can observe vibrations with the largest amplitude, as shown in Fig. 8.3. In order to calculate the YM of the CNT, an appropriate theoretical model has to be developed and applied. Based on the size of the CNT, one needs to decide whether the continuous elasticity theory is applicable to the case in question or a molecular dynamic approach should be considered. The authors calculated the node length of the second harmonic resonance, the frequency ratio, and the shape (deflected curve) of the oscillating CNT. They did that based on the elastic theory and compared the calculated data with the actually observed data. They found excellent agreement between the computed and the measured data. It has been concluded therefore that the model based on the continuous elastic approximation is still valid for the CNT data analysis. In this connection it should be pointed out that many well-established methods and approaches of engineering structural analysis and theory of elasticity that have been developed for continuous elastic media seem still be applicable to nano-structures and nano-materials despite their next-to-molecular size. It would be advisable, of course, to compare, whenever possible, the results based on engineering and applied mechanics with the results based on applied physics

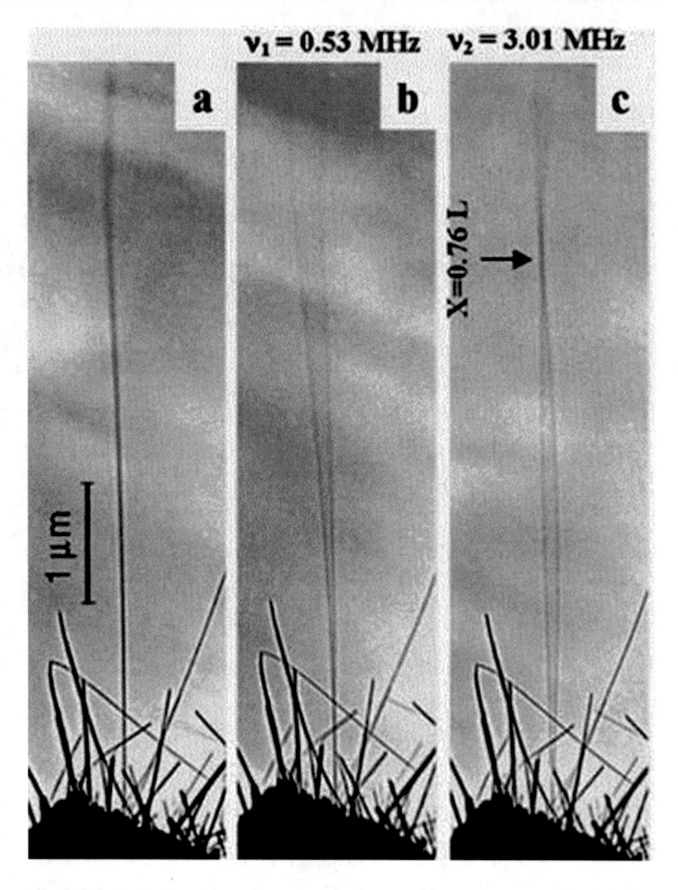

Fig. 8.3 A selected carbon nano-tube at (**a**) stationary, (**b**) the first harmonic resonance ($n1 = 1.21$ MHz), and (**c**) the second harmonic resonance ($n2 = 5.06$ MHz). The vibration of the CNT is caused by the alternating external electrical field [15]

and quantum mechanics for the given nano-structure and nano-material.

Assuming that a CNT can be idealized as a continuous hollow cantilever with one end fixed on the substrate, the resonance frequency can be found as [15].

$$\nu = \frac{\beta^2}{8\pi}\frac{1}{L^2}\sqrt{\frac{(D^2 + D_i^2)E}{\rho}}$$

Here D is the diameter of the CNT, L is the CNT length, ρ is the material's density, and E is the bending (Young's) modulus, which depends on the shape of the CNT. To apply the frequency of the excitations that matches the intrinsic resonance frequency of the CNT is not easy: one needs to omit the nonlinear term in the theoretical model as introduced in [15]. This is due to the fact that nonlinear frequencies are affected by the excitation force and, for this reason, are not characteristics of the structure and the material only. On the other hand, one needs to observe the largest oscillation amplitude in order to determine the intrinsic resonance frequency. As shown in Fig. 8.4, the oscillation amplitude is

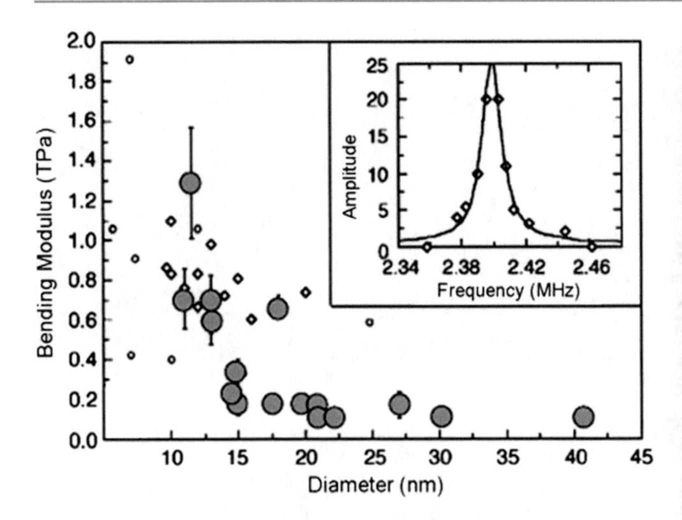

Fig. 8.4 Bending modulus of the multiwalled carbon nano-tube produced by arc discharge as a function of the outer diameter of the nano-tube. The larger CNT typically possesses a smaller YM compared with the thinner CNT. The inset shows the CNT vibration amplitude around intrinsic vibration frequency [15]

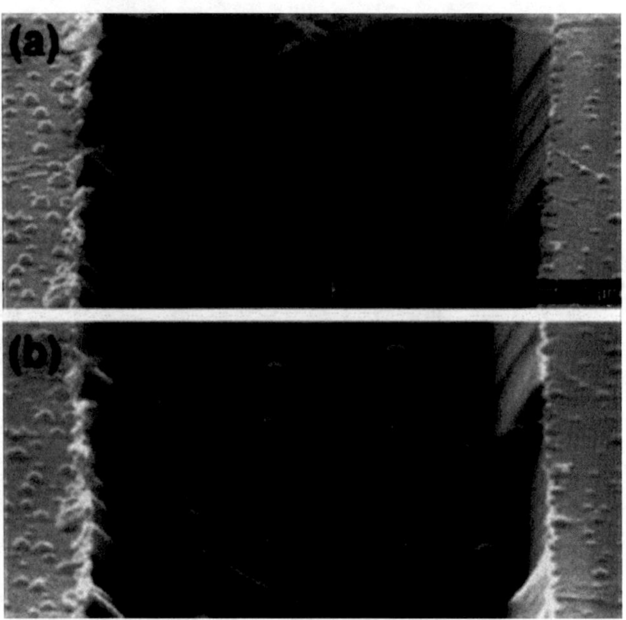

Fig. 8.5 Ropes of SWCNT suspended on the designed Si fixture; the CNT before and after sideway pulling were shown in (**a**) and (**b**) separately [26]

symmetric with respect to the central intrinsic frequency, with the integral 0.12 MHz.

In the CNT case, the bending of a CNT is determined not only by the YM of the material but also by the wall thickness and tube diameter. The in situ TEM experiments measured the bending modulus. Figure 8.4 shows a group of experimentally measured bending moduli of MWCNTs. It is apparent that the smaller-size nano-tubes have a bending modulus (i.e., YM computed from the bending tests) approaching the YM obtained for tensile deformations, while the difference in these moduli for the larger-size CNTs turned out to be much smaller. This size-dependent property could be attributed to the possible inelastic deformations of the CNTs.

8.6 Measurements Using Scanning Probe Microscope (SPM)

The YM of an individual CNT can also be evaluated by using *scanning probe microscope* (SPM), i.e., the SPM method [22–25]. The mechanical behavior of individual CNTs was investigated using the *nano-probe manipulation* method. The observed CNT structural failure was due to buckling of graphene layers. In the conducted experiments, the SWNT bundle rope was suspended on a specialized designed fixture, and the *atomic force microscope* (AFM) tip was launched onto it, thereby applying sideward pulling forces.

SWCNT "ropes" suspended on the designed Si fixture are shown in Fig. 8.5a, b. It was reported that the CNT was able to sustain reversibly many cycles of elastic elongation up to 6%. By assuming that the elongation is transmitted directly to the individual CNTs, the corresponding tensile strength was calculated to be above 45 GPa. This number is in satisfactory agreement with that for multiwalled tubes, although the details of the strain distribution could not be revealed in this experiment [26].

A direct tensile, rather than sideways, pull of a multiwall tube or a rope (i.e., bending deformations) has a clear advantage due to simpler load distribution and a straightforward way of measuring the YM. An important step in this direction has been recently reported by M. Yu et al. [22]. In this work, the CNT was attached between two atomic force microscope (AFM) tips, and the tensile pulling force was applied directly onto the CNT with the displacement and force relationship recorded. The direct tensile experiments for MWNTs report tensile strengths in the range of 11–63 GPa. The measured data did not show clear dependence of this strength on the outer diameter of the CNT shell. The nano-tube broke in the outermost layer, and the analysis of the stress–strain curves indicated that the YM for this layer was between 270 and 950 GPa. The measured normal strain at failure can be as high as 12%.

8.7 Method Using CNT Buckling

In addition to the deformations due to vibrations, deflections, and bending discussed above, the deformation associated with *buckling* (loss of elastic stability under the action of compressive loading) is another effective method to determine the YM of CNT/CNF materials and structures. This method was used to characterize the CNT by integrating it with the SPM tip [25].

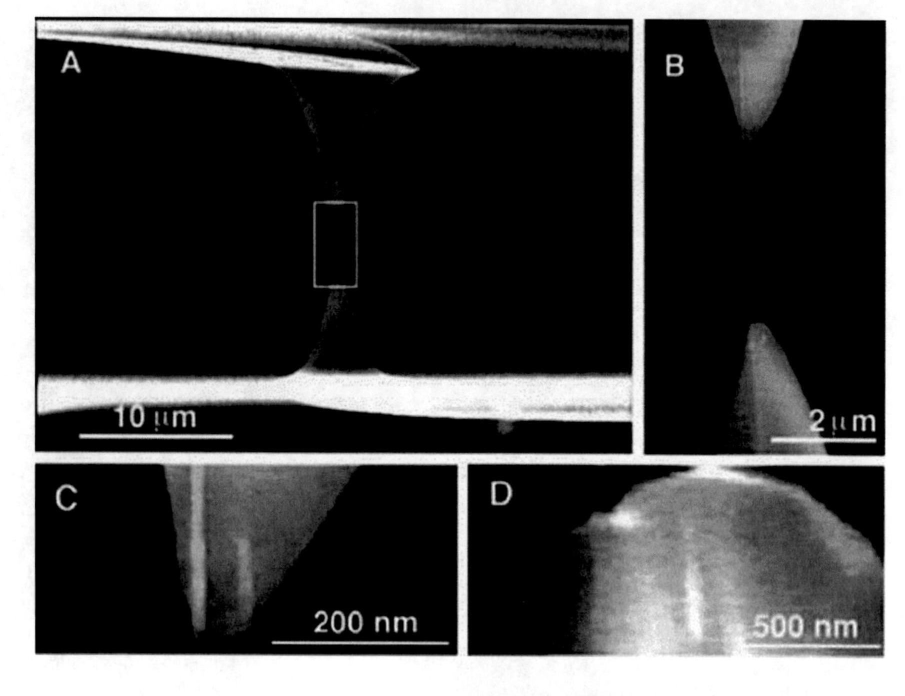

Fig. 8.6 SEM image of two oppositely aligned AFM tips holding a MWCNT attached at both ends on the AFM silicon tip surface by electron beam deposition of carbonaceous material. The lower AFM tip in the image is on a soft cantilever whose deflection is used to determine the applied force on the MWCNT. **B–D**: Large magnification SEM image of the indicated [22]

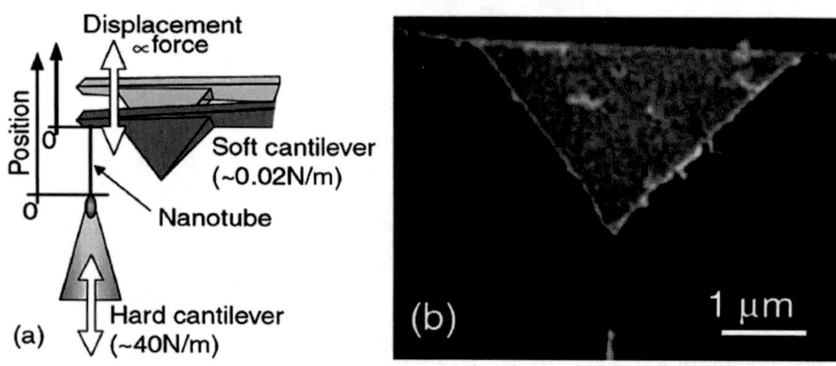

Fig. 8.7 Experimental setup schematic for using AFM to measure the direct tensile strength of individual CNT, which is attached onto the AFM tip; the actual SEM image for the testing sample is shown in (**b**) [25]

The concept is more or less straightforward: one needs to measure the buckling forces of the CNT and calculate the YM based on the Euler formula. The preparation of test specimens is not trivial, however, to say the least. The experiment was conducted in the SEM chamber, so one could record the displacement of the CNT by comparing the images. The forces were measured by the cantilever force sensor (Figs. 8.6 and 8.7).

The measured force vs. displacement relationship is shown in Fig. 8.8. The measured forces are proportional to the displacements. At the positions indicated by the thick arrows, the force abruptly decreases. This decrease should be interpreted as CNT buckling. The strain energy of the nano-tubes is released at this point. Young's moduli of the 20 shells of MWCNT and the processed nano-tubes, from which 14 inner shells were extracted, are estimated to be 0.77 and 0.80 TPa, respectively. These data follow from Euler's buckling model. The results imply that this model is applicable to the analysis of the axial bucking behaviors of the CNTs. It could also be applicable also for the characterization of mechanical properties of other types of nano-wires.

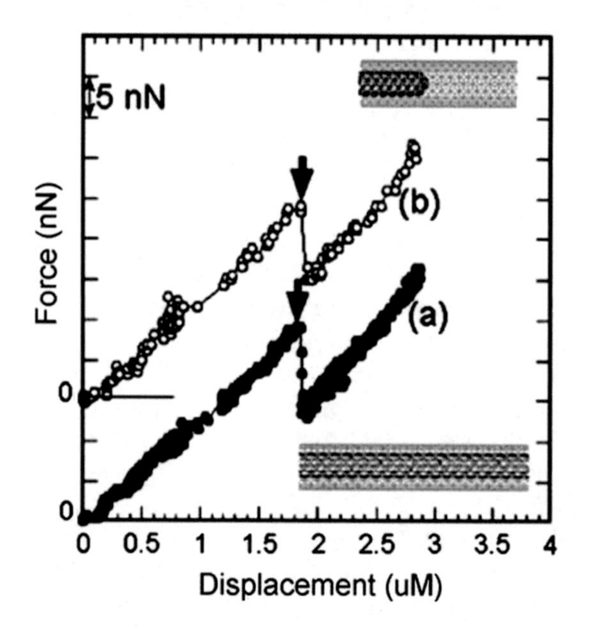

Fig. 8.8 The measured relationship between pulling force and the atomic force microscope (AFM) tip displacement, which is the same as the CNT extension length under tensile stress [25]

8.8 Buckling of CNT Shells

Shell buckling strength (critical loading) is another important mechanical property of the CNT/CNF material and structure. Ultrahigh-resolution AFM can be employed to characterize CNT shell buckling. M. Yu et al. [22] developed a method for measuring the shell bucking strength by applying the nano-indenter instead of the more conventional nano-manipulation technique.

With a specially designed CNT array, which was synthesized on an aluminum substrate, one can obtain well-oriented and vertically aligned CNT array with an aluminum matrix as filling materials to provide structure rigidity [22]. The synthesized CNTs had a relative large diameter (20 ~ 200 nm) with well-secured space in between (100 nm). With CNTs exposed out from the Al surface for a very short distance, one could make sure that the measured results are from the shell buckling ("local buckling") rather than of CNT "global" buckling or bending. The size of the nano-indenter tip is much bigger than the AFM tip and the CNT, so that when it was launched onto the CNT matrix surface, tens of CNTs were making direct contact. So, the measured results show the group appearance for the CNT, and one can extract the actual shell buckling strength (critical load) for each one with appropriate model. The compressed CNTs are shown in Fig. 8.9.

The load–displacement curves obtained in three sets of experiments are shown in Fig. 8.10. Each plot represents a

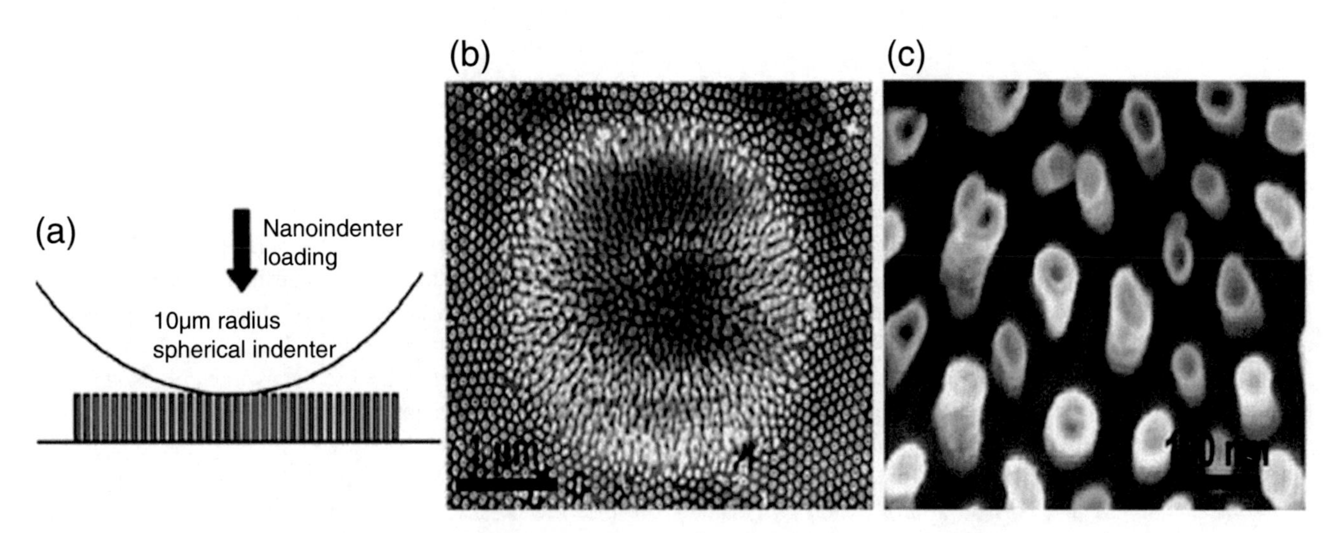

Fig. 8.9 Nano-indentation of individual carbon nano-tubes. (**a**) Schematic illustration of the experiment, in which a Berkovich indenter with 100 nm tip radius vertically compresses a multiwalled carbon nano-tube. (**b**) Schematic illustration of shell buckling in nano-tubes. (**c**) An in situ scanning image of the sample surface obtained immediately after compressing a nano-tube [22]

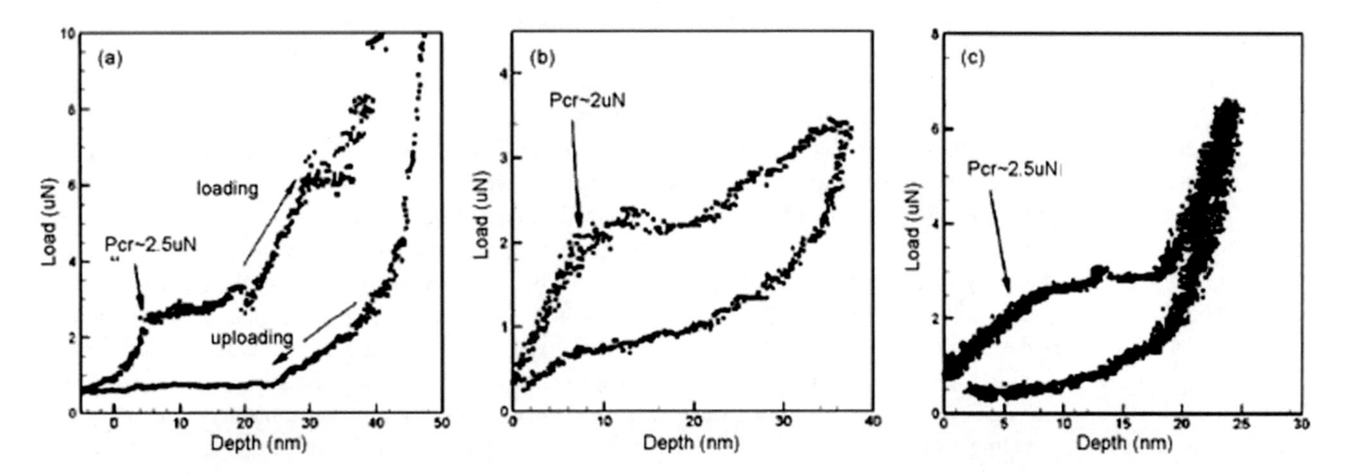

Fig. 8.10 Representative load–displacement data from a loading–unloading cycle. (**a**) Displacement-controlled indent on a 100-nm-long nano-tube, (**b**) load-controlled indent on a 100-nm-long nano-tube, and (**c**) load-controlled indent on a 50-nm-long nano-tube. In all cases, buckling, evident by a distinct drop slope, occurs between 2 and 2.5 μN. The critical buckling load was repeatable to within ±0.25 μN between experiments [22]

loading–unloading cycle. The loading portion consists of three stages: (8.1) an initial linear increase, followed by a (8.2) sudden drop in the slope and the curve becoming flat, and a (8.3) third stage comprising an increasing load. The sudden decrease in the slope is the signature of the shell buckling, which indicates the collapse process (Fig. 8.9). The critical buckling load has been consistently measured to be between 2 and 2.5 μN from multiple experiments. After buckling, the adjacent nano-tubes get into contact with the indenter tip. This results in an increase in the load, as shown in Fig. 8.10 (third stage). In each plot, the position of zero displacement corresponds to a nonzero load because a small preload was used to detect the location of the sample surface.

8.9 Effective Young's Modulus of CNT/CNF Arrays

8.9.1 Stress–Strain Relationship of PECVD-Synthesized CNT Film

In addition to the methods that have been developed to characterize the YM of individual CNT, there has been significant interest of utilizing CNT as TIMs or compressive foams [27, 28]. Within these applications, one is interested in studying mechanical properties of the CNT systems as quasi-bulk materials. The CNT/CNF array (CNFA) "brush" is treated [29], when such an approach is used, as a continuous elastic layer. As is known, the elastic response of a continuous, homogeneous, and isotropic elastic medium depends on two constants – YM and Poisson's ratio (or on two Lame constants, which are derivative of the above two). While Poisson's ratio changes in relatively narrow limits, the YM might be quite different for CNT/CNF materials obtained by using different synthesis methods. On the other hand, it is the YM that plays the most important role in the mechanical behavior and performance of materials. Although many phenomena of the CNT/CNF materials' behavior can be adequately described and explained only on the basis of quantum mechanics, it is often assumed [30, 31] that methods of the theory of elasticity and methods of engineering mechanics can be successfully employed to evaluate the YM of the CNT/CNF materials.

The main objective of the study, whose major findings are described below, was to develop a practical methodology for the evaluation of the effective Young's modulus (EYM) of vertically aligned PECVD-synthesized CNFA. This could be done on the basis of the measured compressive force vs. axial displacement in the post-buckling mode.

There are several reasons why the axial displacements should be large enough to ensure that the CNFA behaves in the post-buckling mode conditions. First, the force–displacement relationships in the post-buckling mode can be obtained in a very wide range of measured forces and displacements. Second, exact solutions exist (Euler's "elastica") for the evaluation of the highly geometrically nonlinear bending deformations of flexible rods (wires). Treating CNFs as nonlinear wires, one can use the "elastica" solutions to predict the EYM from the experimentally obtained force–displacement relationship. If the material exhibits nonlinear stress–strain relationship, i.e., if the EYM is stress dependent, then this dependency could be obtained as a suitable correction to the YM value obtained on the basis of the "elastica" solution. Finally, it is important that the interfacial pressure does not change significantly with the change in the axial displacements of the CNFA. This indeed takes place, when the compressed CNFA is operated in the post-buckling mode not very much remote from the buckling condition. Since in many practical applications CNFs are intended to perform in the post-buckling mode, the evaluation of their behavior in such a mode enables one to obtain valuable information for these structural elements in the use conditions.

The samples used in our experiments were synthesized using plasma-enhanced chemical vapor deposition (PECVD). The fabrication procedures of CNFs in an embedded vertical array were described in detail in [32, 33]. Briefly, a 30-nm-thick layer of Ti was deposited on a Si substrate, followed by a 35-nm-thick layer of Ni catalyst, both deposited by ion beam sputtering. The CNF array was grown using a PECVD process with C2H2 as the hydrocarbon source and NH3 as a diluent/etchant gas [32]. The table version of the Instron tester (Model 5542) was used in our analysis to measure the compressive force. The CNFA was placed on the lower anvil of the Instron machine, while the displacements of the upper anvil were automatically and thoroughly controlled, particularly, when approaching the CNFA surface with a speed of about 5 μm/min. The maximum capacity of the Instron and the maximum force applied during the experimentation was about 0.5 N ($\sim$500 gf), and the resolution of the measured force was about 0.01 N. This provided sufficient accuracy to distinguish the changes in the compressive strain.

The axial displacements of the CNFA were on the micrometric level. The regular Instron equipment loses its accuracy, when measuring such small axial displacements. This is because the measured displacements become comparable with the displacements of the beam in Instron's load cell. For this reason, the *Capacitec equipment* was employed in our experimental setup to measure the relative displacements of the Instron anvils and, hence, the axial displacements of the CNFA. The Capacitec equipment uses a non-contact way of determining the displacement by measuring, with a rather high accuracy, the change in the electrical capacity of the air gap between the Instron anvils. The anvils play in this situation the role of two electrodes. The strain signal is imported into the Instron device for

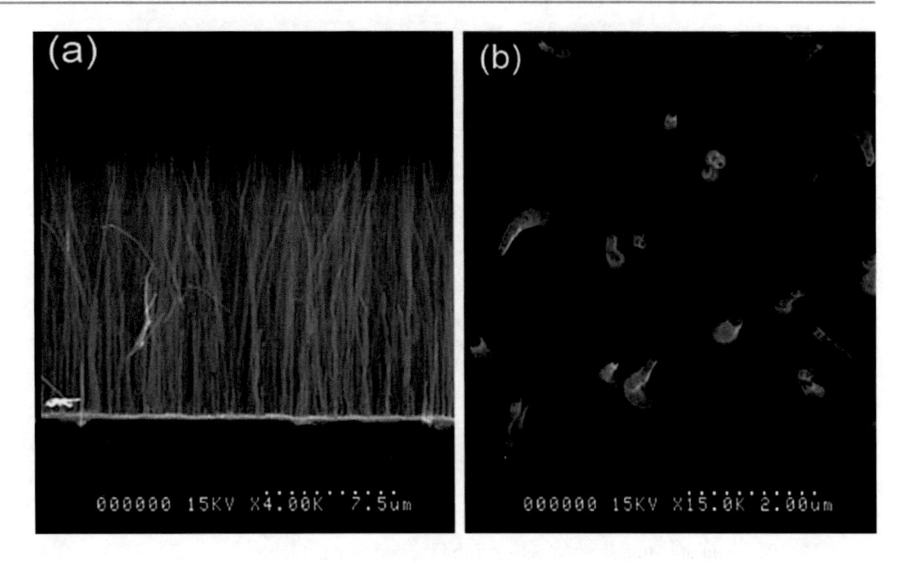

Fig. 8.11 Microscopic picture for a typical CNT/CNF samples used in the experiment. (**a**) SEM *cross-section* view of the CNT sample. (**b**) SEM *top* view of the CNT sample shows the density of the CNT generated. The scale bar of these images is 7.5 and 2 μm

calibration. By initializing the beginning strain point to correspond to zero displacement, one is able to measure the strain signal accurately and to transfer this signal into the displacement data, which has a higher resolution at ~0.5 μm. By using a combination of Instron and Capacitec devices, we were able to measure the force vs. displacement with a rather high resolution and, owing to that, to calculate the EYM with a high accuracy.

The quality and the structure of the CNFs obtained, as has been indicated, by using PECVD, were observed under the scanning electron microscope (SEM). An example of the SEM image is shown in Fig. 8.11. The length (thickness) of the CNFA was about 15 μm with about 2 μm variation. The diameter of the CNTs was about 80 ~ 100 nm. The samples had a high CNF coverage density (~10^8 CNF/cm^2), and the space between the two adjacent CNFs was about 1 μm. CNFA sample with lower density was used to obtain such a relatively larger space. The properties of the CNFA could be controlled by varying the synthesis conditions. The size of the CNFA in our test samples was 2 × 2 (length×width) cm^2.

The measured signals were the compressive force and the axial displacement. Based on the developed predictive model for a flexible rod (Euler's "elastica"), one is able to obtain the relationship between the T/Te and δ/l ratios, where T is the compressive force, Te is the buckling force [31, 34] (that could be determined assuming that an individual CNF could be treated as a cantilever rod clamped at the CNF substrate), δ is the axial displacement of an individual CNF, and l is its length. As a result of measuring the force–displacement relationship using the Instron–Capacitec device, one could obtain the relationship between the applied force, T, and the relative axial displacement, δ/l. Based on this relationship, if one assumes the YM ($E0$) at low strain levels of the CNF array to be constant, then it could be directly calculated from [31, 34].

The actual measurements indicated that the experimentally obtained T/Te vs. δ/l relationships deviate from the predictions based on the "elastica" solution, i.e., on the solution obtained for strain-independent YM. This deviation should be attributed to the nonlinear stress–strain relationship for the CNFA material. To account for such nonlinearity, we use the following approximation for the "nonlinear" EYM of the CNFA:

$$E \cong E_0 + \gamma\sigma$$

Here, σ is the compressive stress and γ is the factor of nonlinearity. Both the stress and the factor of nonlinearity could be obtained from the compressive stress vs. displacement relationship. In order to determine the best fit γ value, a series of experimental stress vs. displacement curves was generated (Fig. 8.2). These curves enable one to approximate ("fit") the experimental results by theoretical curves and to obtain the appropriate γ values that result in the best agreement between the experimental and the theoretical data. The γ values obtained in such a fashion could be then used to predict the EYM of the CNFA.

In Fig. 8.12, some simulation results are shown for different values of the parameter γ, along with the experimentally obtained curves for a particular specimen. As evident from the processed data, the theoretical curve with the parameter of nonlinearity of $\gamma = 20$ satisfactorily matches the experimentally obtained curve. Hence, it is this γ value that could be used in the case in question to evaluate the EYM for CNFA of a particular batch ("run").

Five samples from the same wafer ("batch") were prepared. Because of the same synthesis conditions, these samples were assumed to be identical. SEM images were taken for all the samples. Figure 8.1 contains, as an example, the SEM image for one of the samples. During the

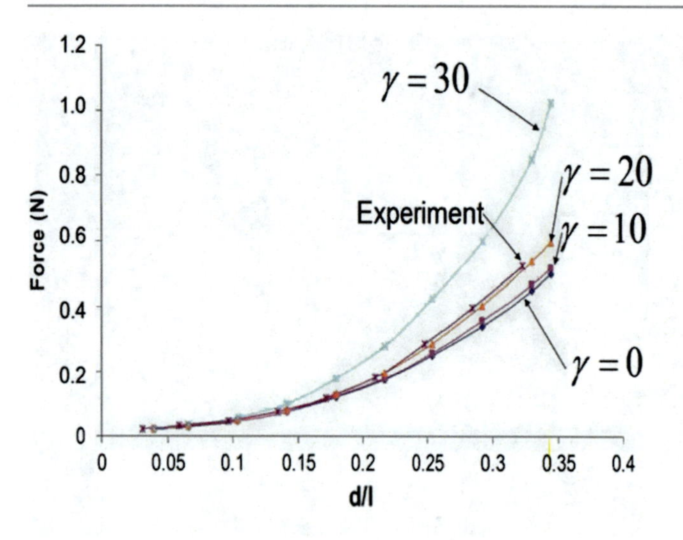

Fig. 8.12 The simulated relationship between the applied force, T, and the displacement-to-length ratio δ/l corresponding to different nonlinear coefficient. The nonlinear coefficient $\gamma = 20$ fits best (for the given specimen) the experimental curve

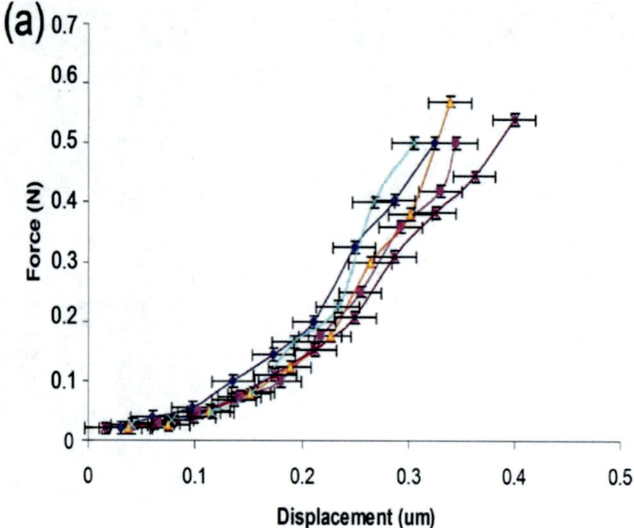

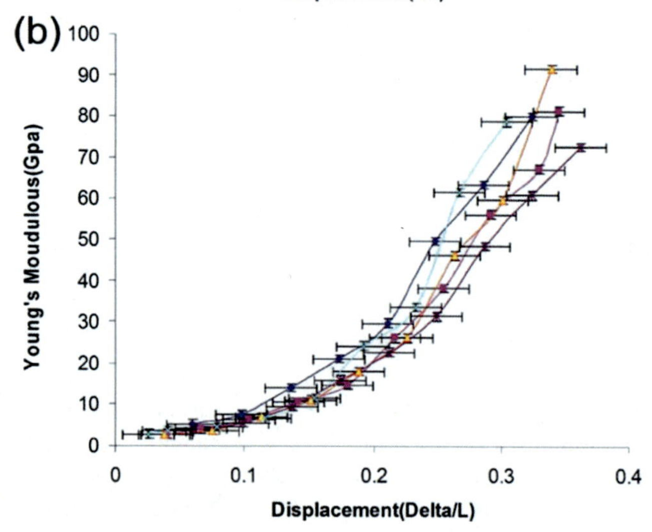

Fig. 8.13 (**a**) Measured compressive force as a function of the δ/l ratio for five samples taken from the same wafer (run). (**b**) Calculated YM of these samples based on the measured compressive force in (**a**); it reaches 90 GPa corresponding to a 30% lateral displacement of the CNT array

experiment, all the test parameters were kept the same for all the samples. The relationship between the compressive forces and the displacement was recorded for each sample and introduced into the calculation method (Fig. 8.3a). As a result, one could obtain the relationship between the calculated YM and the δ/l ratio for the tested samples. The EYM of the CNFA at 30% deflection is about 90 GPa and exhibits clearly pronounced nonlinear behavior.

Some scattering of the experimental data should be attributed not only to some differences in the test samples. The error bar in Fig. 8.13 characterizes the system measurement error. For some samples, the measured data still fall outside of the error bar. Thus the samples of the same run ("batch") possess some minor sample-to-sample variations. But one could nevertheless assume that these samples possess similar mechanical properties.

CNFs in our tests were synthesized at five different deposition conditions. The catalyst density was kept the same in order to obtain the same CNFA density for all the different synthesis conditions. The SEM images of all the samples were taken. The diameter and density of the CNFAs remained the same for all the samples. There was, however, about $2 \sim 3$ μm variation in the length of the CNFs in the CNFA. The growth time was controlled to be the same, in order to make sure that this parameter does not affect the CNF characteristics. Figure 8.4 shows the EYM for the five tested samples. The computed EYM varies from 5 to 35 GPa for different samples. The CNFs synthesized at optimized growth condition typically possess better atomic structure than those synthesized at other conditions. This would make CNFs synthesized at optimized condition "stiffer," i.e., their YM would be higher, when subjected to a compressive force.

From Fig. 8.14, one could observe that under small strain, like 10% in compression, the measured YM of the CNFs appears to be very similar to each other. It is only when larger compression is achieved that the nonlinearity of the measured YM of samples begins to get manifested clearly. For the samples of the same batch, we used, as a suitable characteristic of the material's nonlinearity, the parameter γ averaged over the CNF length. This parameter is certainly different for samples taken from different batches and is determined by both the quality of the CNFs and the density distribution.

It is noteworthy that the developed experimental methodology along with the predictive stress models provides a rather high resolution for distinguishing between the mechanical properties of CNFA synthesized under different conditions. Despite some unavoidable "noises" in our experiments, the developed methodology for the EYM

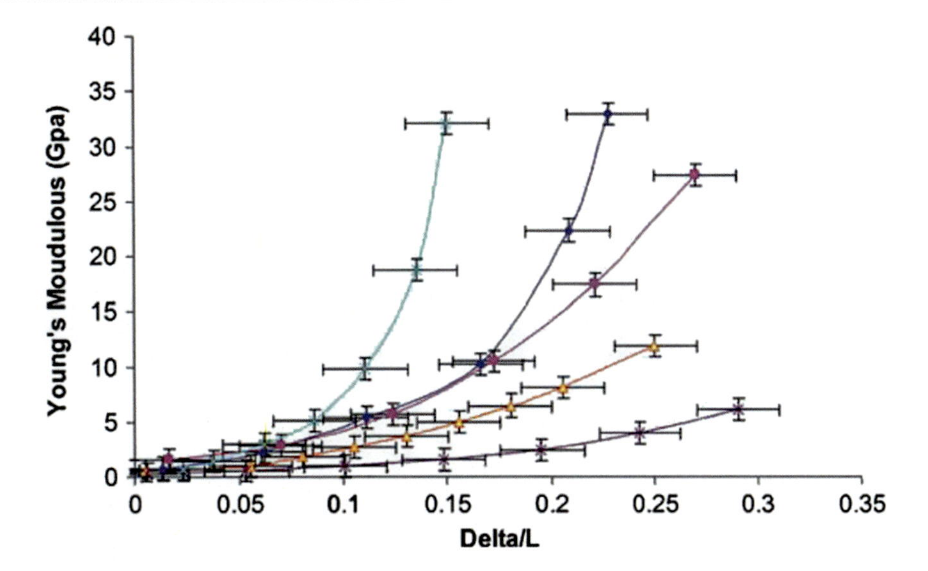

Fig. 8.14 Calculated Young's modulus vs. δ/l ratios for samples taken from different "runs" (different growth conditions). This data indicate that different growth conditions result, in effect, in considerably different materials

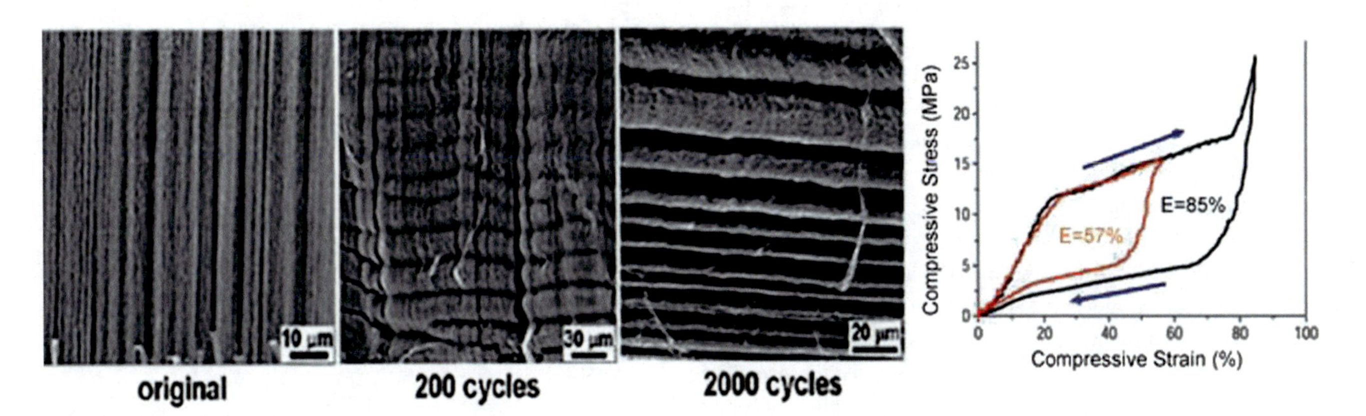

Fig. 8.15 SEM images for the CNT array under different cycles of compressive load and the measured stress–strain curve [19]

evaluation could be used, in particular, to establish which method for monitoring the CNF growth has the highest potential to obtain the CNFs of the highest quality possible.

8.9.2 Stress–Strain Relationship of TECVD-Synthesized CNT Film

Compared with the well-aligned CNT synthesized by PECVD, the TECVD-synthesized CNT is more entangled and appears as matrix. So the mechanical properties of these two types of film should behave differently. Previously, millimeter-long carbon nano-tubes synthesized by TECVD have been used as super-compressive foam [19]. The experiment was demonstrated on the same Instron equipment used by us. The CNT array was placed between the anvils. With the loading compressive pressure, the CNT film was compressed. It was discovered that the CNT film could perform like foam, which means the CNT array would bound back to its original length or maybe reasonably shorter. In addition, the CNT array was also compressed repetitively by a few thousand times under the same condition, and they buckled like the regular spring. Figure 8.15 shows the SEM images of the CNT array under different cycles of compressive load. The CNTs perform like spring. Also the force vs. displacement curve was presented as well. Based on the measured stress– strain relationship, one is allowed to calculate the effective YM of CNT film similar to the method introduced before.

8.9.3 Effective Young's Modulus Has a Certain Limit

We have been able to build a model to calculate the effective Young's modulus (YM) of the CNT array from the force vs. displacement provided by Instron and Capacitec. Here with the same methodology, we calculated the effective

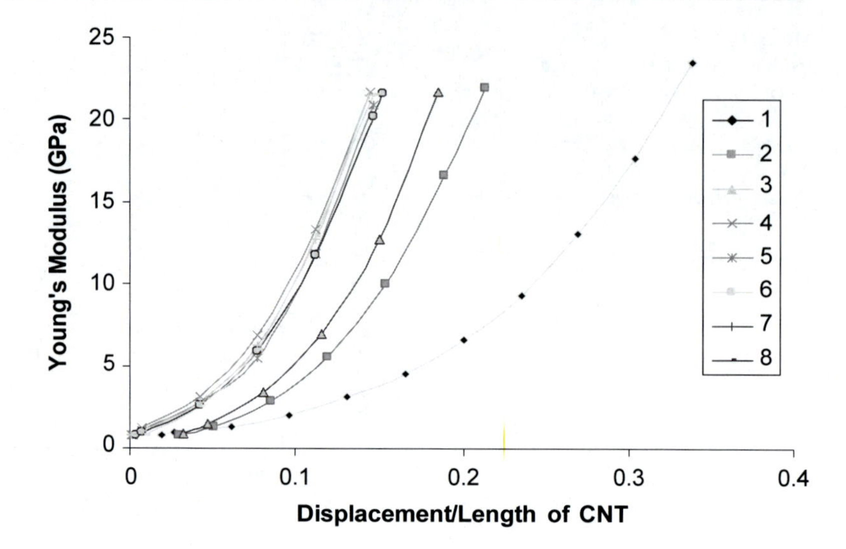

Fig. 8.16 Young's modulus converges to a certain limit when the sample is compressed with 50 gf for a few times

YM of the CNT array from the relationship between force and displacement of CNT. The load cell was launched onto the CNT array as described before, and the maximum force applied on the CNT array is about 50 gf. With this amount of force, the CNT array will not be damaged. It is confirmed by the scanning electronic microscope images, with the same sample placed in the same location. Instead of applying the force on the sample one time as before, same force is applied to the sample a few times, and the YM is calculated accordingly. It is interesting to observe that the effective YM of the CNT reaches a limit when the sample is pressed by the load cell for a few times (Fig. 8.16). In other words, the effective YM of the CNT array was increasing and finally saturated with certain amount of compressions. Although the amount of forces applied here is relatively small, however, it is already large to cause the CNT film to deflect. The measurement technique is accurate, since one can clearly observe the difference in mechanical properties at each time of compression.

More interestingly, it was discovered that the effective YM of the PECVD-synthesized CNT tends to increase to certain limit under the repetitive load. By setting different strain point (displacement/length), the final YM of the CNT tends to increase to a certain limit and gets saturated. However, compared with the previous publication in [35], it was discovered that the YM of the CNT array tends to decrease with the repetitive load. The exact reason that caused that difference is still unknown. However, it could possibly attribute to the mechanical properties difference and different buckling mechanisms (Fig. 8.17).

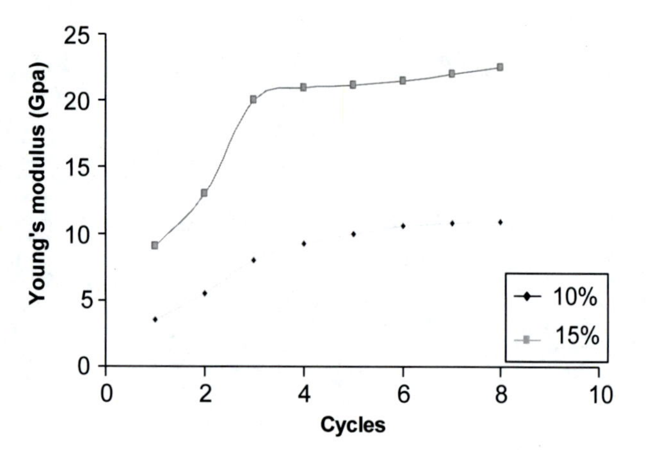

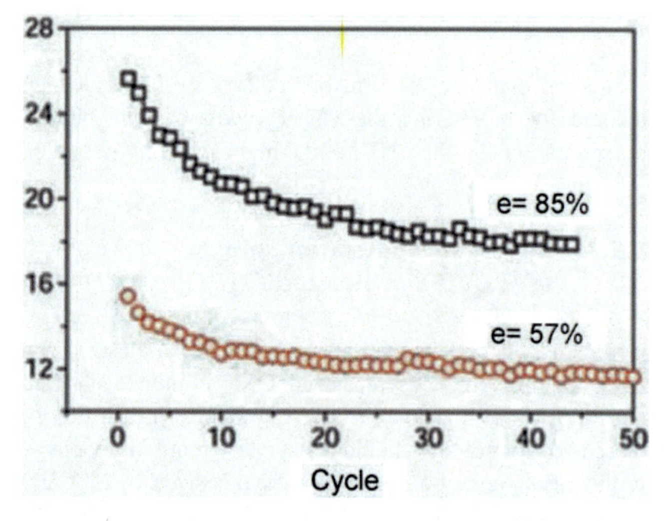

Fig. 8.17 At different displacement positions, both plots show that Young's modulus of CNF array and of the individual CNFs begin to increase for PECVD-synthesized CNT; however, it shows the opposite for TECVD-synthesized CNT [19]

8.10 CNT/CNF-Based TIMS: Requirements for Physical (Mechanical) Properties

8.10.1 CNT/CNF Compliance

We have indicated above that the CNTs are good candidates for the *TIM applications* mainly because of their good thermal properties. It is equally important that the employed CNT/CNF arrays possess adequate physical/mechanical properties, especially those that directly affect their performance as TIMs.

The reason for bringing in a TIM is to fill the gaps between two rough solid surfaces (bond lines), thereby reducing the thermal resistance of the heat-removing device. CNT/CNF arrays can be grown on copper (Cu) substrates or on Cu films. They can be grown, of course, on other materials as well. Cu has an excellent thermal conductivity and is generally used as heat sink; however, due to the rigidity of Cu, the surface roughness cannot been filled. That is why one needs a material with good compliance (softness) to compensate for the "hardness" of Cu.

Thermal resistance of CNFs as TIMs includes the bulk resistance of CNFs and contact resistances between the CNF tips and the hot contact surface. Due to the good thermal properties of CNFs, the physical nature of the contact resistance tends to take precedence. CNFs also need to possess good compliance properties as conventional *phase changing materials* (PCMs) to change the orientation of the CNF array according to the contacting surface roughness and configuration. The nano-scale dimension nature of CNFs allows them to fit well with the surface roughness and compliance nature of CNFs and make good contact with the hot surface under certain pressures. The CNT array synthesized by thermal enhanced chemical vapor deposition (ECVD) has also been demonstrated to have super foam-like nature and could be compressed by up to 85% of its original length, when it is experiencing a 13 MPa compressive pressure [35].

In the section, we elaborate on the compliance properties of CNT/CNF arrays and their effect on the thermal properties of the CNF arrays [27, 28], as well as on CNF arrays filled with Cu as TIMs. The CNF–Cu composites are fabricated by *electrochemical deposition* (ECD), where by varying the duration time of the ECD process, amount of filling Cu, and the numbers/lengths of undeposited CNFs can be controlled. Briefly, a thin layer of titanium (30 nm) was deposited on a Cu substrate, followed by a 35-nm-thick layer of nickel catalyst, both deposited by ion beam sputtering. The CNF array was grown using a PECVD process with acetylene (C2H2) as the hydrocarbon source and NH3 as a diluent/etchant gas. The quality and the structure of the CNF could be observed under the *scanning electron microscope* (SEM) and *high-resolution transmission electron microscope* (HRTEM). The length of the CNF arrays was about 20 μm with a 3 $\sim$ 5 μm variation (see Fig. 8.18). The assessed diameter of the CNFs was about 100 nm (see Fig. 8.18b). The sample had a high CNF coverage density ($\sim 10^9$ CNF/cm^2), as one could see from Fig. 8.18b.

Sibling samples synthesized with the same conditions as in Fig. 8.18a, b were used to fabricate the CNF–Cu composites. They were prepared in this work by Cu ECD using a three-electrode configuration in a Cu-containing electrolyte bath [36]. The titanium–copper (Ti–Cu) substrate, on which the CNF array is grown, acts as the working electrode for Cu deposition. Titanium is chosen due to its favorable electrochemical properties that lead to good filling characteristics compared to other metals, such as chromium. The potential and current control of the electrolyte bath with respect to the working electrode allows us to control deposition rate and amount of Cu deposited.

By controlling the deposition time, the thickness of Cu layer filled into the gap between the CNFs, amount of filling Cu, and the numbers/lengths of undeposited CNFs can be controlled. For 20 min soaking time of CNF array in the

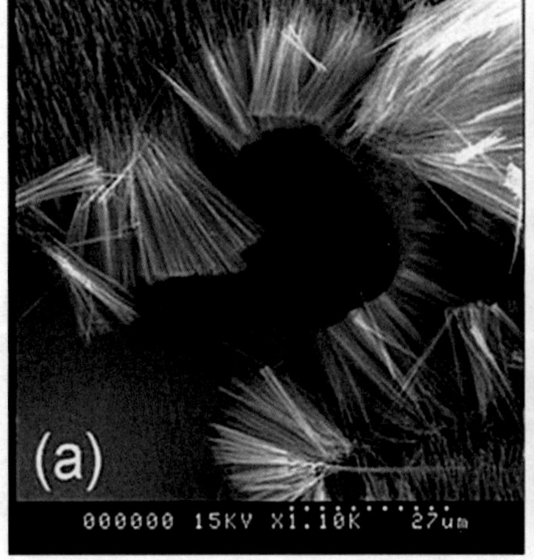

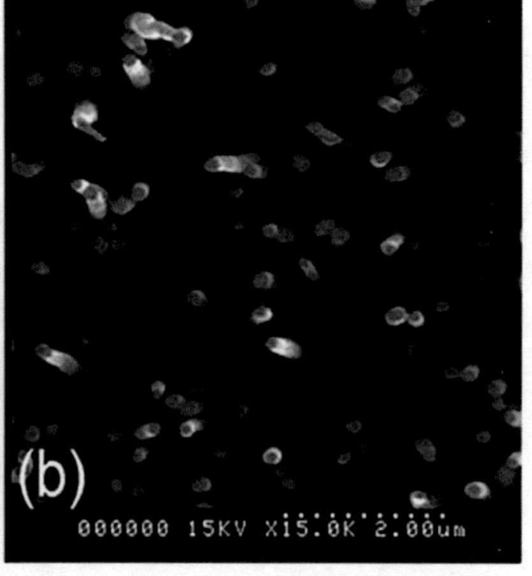

Fig. 8.18 *Tilt* view and *top* view of the CNT samples synthesized on Cu substrate

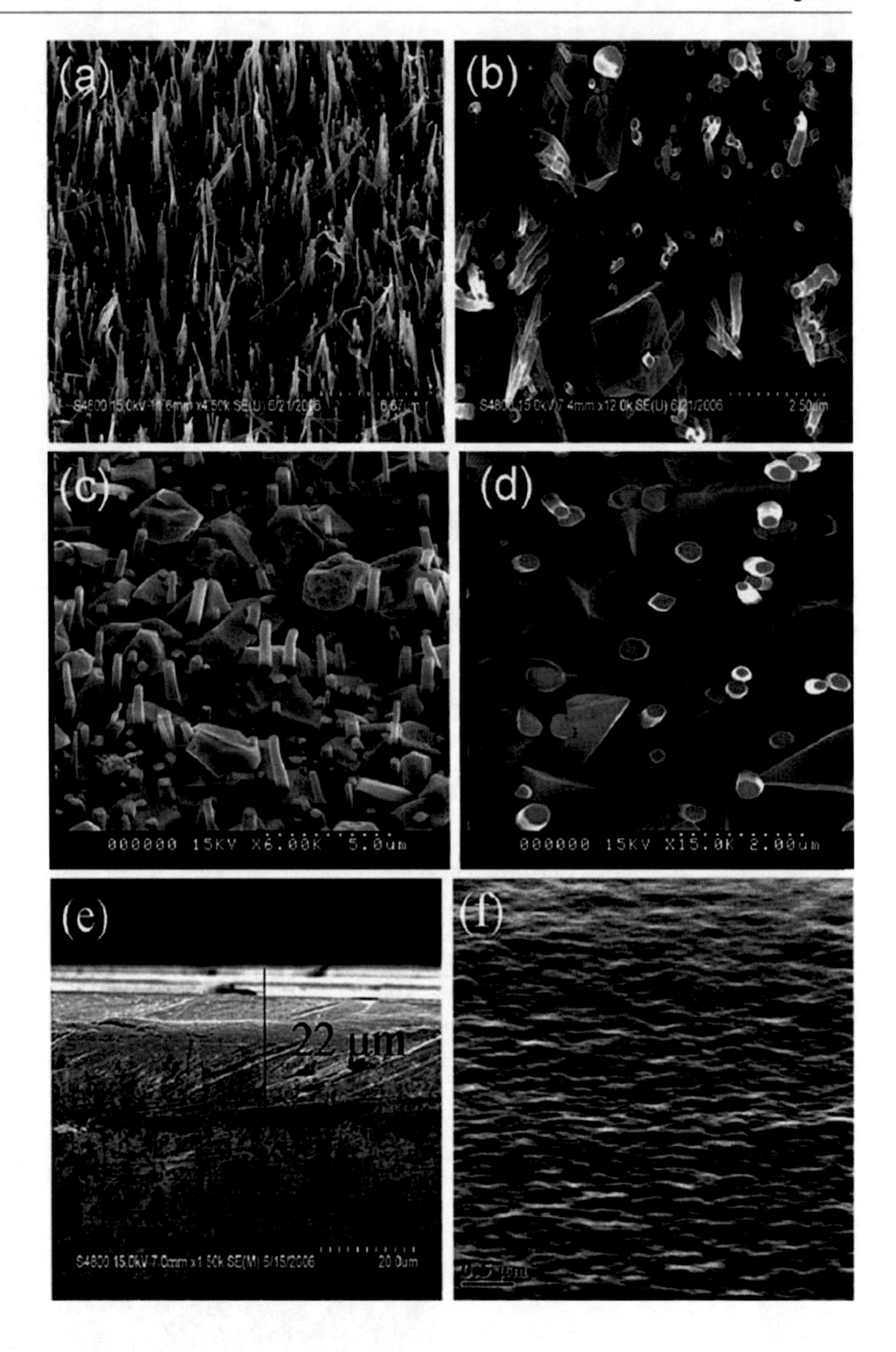

Fig. 8.19 SEM images for the ECD-prepared CNT samples. The sibling CNTs possess the sample length; however, after the ECD bath for different period of time, the gap between the CNTs was filled with Cu for different thickness. So the amount of exposed CNT appears differently, and also the rigidity of the CNT sample has an effect on the CNT thermal performance

electrolyte bath, the 30° tilt view indicates the CNF protruding from the Cu matrix is about 4 ~ 6 μm (Fig. 8.19a) and the density of the CNF array decreased to approximately 5 μm^{-2} (Fig. 8.19b). It is interesting to observe that after the ECD process and the soak of CNF in Cu-containing solution, the CNF array still possesses the vertical aligned nature. Furthermore, another sibling sample was soaked in the solution for 25 min. In this case, a relative thicker ECD Cu layer is obtained, the SEM images indicate that the length of the CNF protruding out is about 2 ~ 3 μm, there are a few CNFs relatively longer, but they are not the majority part of the CNFs. One could clearly observe the good "rock" like crystallized Cu. The CNFs are nicely surrounded by the ECD Cu. Visually there is no significant amount of voids introduced by the ECD process. If one continues to do the ECD deposition, eventually, all the CNT will be buried, and it ends up with a relative smooth Cu surface (Fig. 8.19e, f.)

Usually, there would be a following electro-polish step, which helps one obtain the smooth CNF–Cu surface and leave exposed nano-fibers protruding from the surrounding Cu matrix. However, we observed that with our CNF array synthesized on Cu substrate, after ECD process for 30 min, all the CNFs are covered by the ECD Cu, and we surprisingly observe an excellent flat Cu surface (Fig. 8.19f), and the surface roughness is measured to be ~100 nm. The SEM image of the side view of the sample indicates that the ECD Cu layer is about 22 μm thick, which matches the length of the bare CNF array in Fig. 8.18a.

After the CNF array and CNF–Cu composites are prepared, in order to study the compliance property of CNF array and its effect on the thermal properties of CNFs as TIMs, the total thermal resistances for each group of samples are measured based on the recently developed photo-thermal method. The experimental setup that used to measure the effective thermal resistance of CNT array has been described in detail earlier. The phase-shift signals measured for CNF array and CNF–Cu composites with different copper thicknesses (ECD treatment times) are shown in Fig. 8.20. The calibrated phase shifts of the samples are obtained by subtracting the system reference phase shift from the phase shift of the thermal radiation from the Ni film. The calibrated phase shift signal characterizes the thermal properties of the CNFs and CNF composite samples. In the experiment, the phase shift is measured over a wide-frequency range. Different values of the unknown properties are used to calculate the theoretical phase shift and compare with the experimental result. The properties giving the best fit of the experimental data are taken as the properties of the CNF composite sample.

Based on the calibrated phase shift fitting and the employed physical model, the effective total thermal resistances of CNF and CNF–Cu with 20, 25, and 30 min deposition are measured to be 0.13 ± 0.01, 0.18 ± 0.01, 0.25 ± 0.01, and 0.32 ± 0.01 cm^2 K/W, respectively. The total thermal resistances of the sample were calculated by considering the actual heating radius of the laser spot

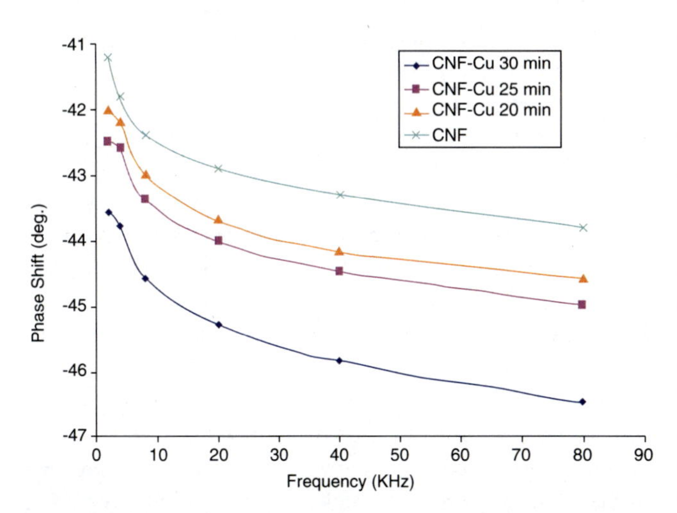

Fig. 8.20 Phase shift measurement results for CNT–Cu sample under different time period of Cu bath

(A = 1 mm [2]). From the measured thermal resistance, one can observe that the bare CNF array sample has the smallest thermal resistance, which means the CNFs are making good contact with the Ni hot surface by filling the air gaps and transfer the heat efficiently away. By depositing Cu into the gap between CNFs, the number of CNF protruding out of the Cu matrix is reduced, the length of the CNFs array are getting shorter, in which case, the compliance nature of CNF is strongly restricted but not able to make good contact with the contacting surface. Furthermore, from Fig. 8.13, one could observe the rough ECD Cu surface, which is also counted for the worse thermal properties of the composite.

However, although there is a relative flat ECD Cu surface after 30 min deposition, the measured total thermal resistance for the CNF–Cu with 30 min deposition possessed the largest thermal resistance (0.32 ± 0.01 cm^2 K/W). Clearly, there are no phase changing properties for Cu; although the thermal conductivity of Cu is excellent, it does not make good contact with the hot surface compared with CNF. The phase shift would still allow us to calculate the total thermal resistance of the CNF–Cu composite. However, major part is dominated by the interface resistance. In this case, it would not be very helpful to distinguish the interface resistance with the bulk resistance of the CNF–Cu composite.

The relationship between the applied forces and CNF–Cu thermal performance as TIMs is also investigated. In our experiment, a load cell is placed against a Cu substrate to measure the forces applied on the CNF composite samples. The applied pressure is increased gradually, and the total effective thermal resistance is calculated corresponding to phase delay of the CNF composite sample (Fig. 8.20). For each point in this figure, the phase delay signal is acquired for the entire frequency region, and the corresponding thermal resistance is calculated. One can see that the total effective thermal resistance including the interface resistance is reduced and reached its steady-state value at a pressure of 80 psi. With high pressure applied on the CNF and CNF composites, it ensures that better contact between CNF arrays, CNF filling materials, and the Ni film make the bulk resistance to be dominating the total thermal resistance. As a result, the heat transfer from the Ni film to the Cu substrate is mostly affected by the CNF array composite and makes it possible to measure the thermo-physical properties of CNF array composite. If no pressure or low pressure is applied on the ZnSe, the CNF composite will have poor contact with the Ni film due to the surface roughness, i.e., less number of CNF and filling materials would be in contact with Ni film, leading to a large contact resistance between CNF arrays and the Ni film. Thus the compliance properties of CNF array are very important for utilizing it as TIMs.

8.10.2 Bonding Strength of CNTs to Its Substrate

It goes without saying that satisfactory *adhesion* of the CNF array (CNFA) to its substrate is critical for making the CNF-based technology practical. Accordingly, the analysis that follows is aimed at emphasizing importance of CNF adhesion to its substrate as TIM and also the easy-to-use and effective experimental method for the evaluation of such an adhesion. Note that the adhesion strength of a single CNF to its substrate was addressed qualitatively, apparently for the first time, by Cui et al. [37] and by Chen et al. [38]. This was done in connection with the use of PECVD-synthesized CNFs as probe tips in the atomic force microscopy imaging equipment. In the experiments described by these investigators, individual CNFs were directly grown on tipless cantilevers. In the reported observation, the CNF probe was operated on a continuous scan mode for 8 h, and no degradation in image resolution has been observed. The CNFA in our tests, however, consisted of billions of CNFs. The obtained information characterizes therefore the performance of an ensemble of a plurality of CNFs. In order to translate the obtained experimental data into the corresponding shearing stresses, we developed a simple analytical stress model that enables one to calculate the magnitude and the distribution of the interfacial shearing stress from the measured (given) external force. Testing has been conducted using specially designed shear-off test specimens.

In the analysis we determine the maximum effective *shear stress-at-failure* for a CNFA fabricated on a thick Cu substrate. We use the term "effective" to emphasize that we address a plurality of CNFs and treat the CNFA "brush" as a sort of a continuous bonding layer. This approach enabled us to use an analytical stress model that we developed for the evaluation of the interfacial shearing stress in an assembly with a continuous bonding layer. The model used in our analysis is a modification of the models that were suggested earlier for the evaluation of the interfacial thermally induced stresses in thermally mismatched assemblies [39–42]. The samples in experiments were also synthesized using plasma-enhanced chemical vapor deposition (PECVD).

A 20 × 20 × 1 mm copper plate was used in the experimental specimens as a central block. A 20 × 10 mm copper foil was placed on and then pressed onto the central block, so that the edges of the foil had the same boundary as the central copper block. Two identical CNTA samples grown on 20 × 20 × 1 mm copper substrates were prepared and pressed against the copper foil. Only a portion of the surface of the CNFA was actually in contact with the copper foil and was able to provide adhesion. The test specimen was placed on Instron's lower anvil. Then the upper anvil with the load cell was lowered to approach the copper central block as shown in Fig. 8.21. Once the load cell got into contact with

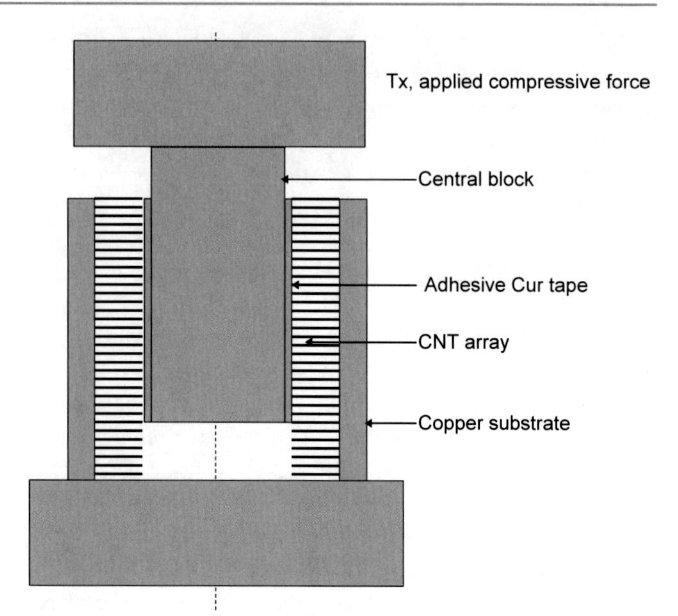

Fig. 8.21 Schematics of the test specimen assembly for the shear stress test

the central block, the Instron tester, with the Capacitec equipment mounted on it, started to record the forces and the corresponding displacements.

The copper foil used in our experiments was strong enough to hold in place the top surface of the CNFA. The thickness of the CNFA "brush" (the lengths of the CNFs) was about 30 μm, i.e., considerably thicker/longer than the thickness of the copper foil, so that the CNFs were able to deform in a beam-like fashion, while the CNFA "brush" as a whole experienced shear deformations.

The experimental data obtained for five CNTF test specimens are shown in Fig. 8.19. The force applied to the sample gradually increases until the interfacial failure occurs, and the CNF is sheared off from the substrate. If the test specimen were ideally well assembled, both the CNFA samples would have been simultaneously separated from the central block. In reality, however, only one of the CNFA interfaces was typically broken.

The force-at-failure varied in our tests appreciably from specimen to specimen. The most important source of uncertainty in our tests is the actual percentage of the contact area between the CNFs in the copper foil. If the CNFA is not attached well to the copper foil, the CNFs in the CNFA will not be deformed and would not be sheared off from the substrate (Fig. 8.23). Clearly, with more CNFs in contact with the copper foil, the force holding the CNFA on its substrate would be larger, and, as a result, the induced shear-off stress would be higher. As one could see from Fig. 8.22, the maximum force-at-failure varied in our experiments from 2 to 5 kg, depending on the CNFA bonding strength and on how many CNFs were in contact with the copper foil. The failure site was observed under a SEM, after the test samples were taken out from the Instron tester. One could clearly see that about 80% of CNFs were in contact with the copper foil. This case corresponded to the 5 kg force value. Unfortunately, there was no good way to quantitatively correlate the percentage of the CNTs in contact with the copper foil and the corresponding shear-off stress. This stress was calculated from the measured shearing force on the basis of the formula (see Appendix):

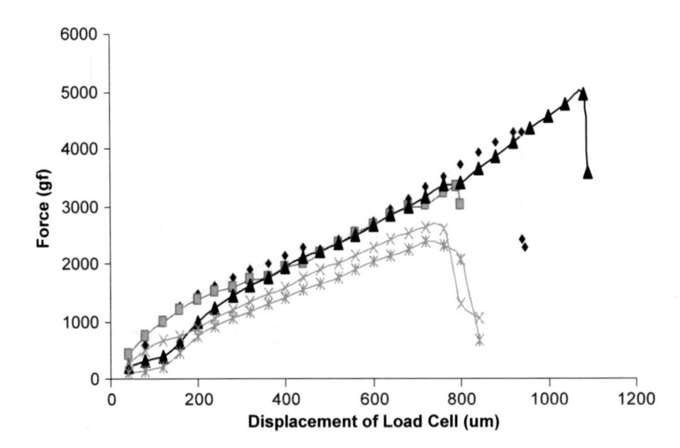

Fig. 8.22 The relationship between the forces applied on the surface of the vertical aligned CNF array. With different amount of CNFs in contact with the copper tape, the maximum breaking forces for the samples are different. The displacement is actually the displacement of the load cell pushed up by the CNFs

$$\tau(x) = \tau_{\max} e^{-kx}$$

where

$$\tau_{\max} = kT$$

is the maximum shearing stress, T is the measured force, and k is the factor of the interfacial shearing stress [39–42]. The origin of the coordinate, x, is at the assembly end, at which the external force was applied. The calculated shear-off stress was 300 psi for the case of a 5 kg force-at-failure. Note that only about half of the sample was in contact with the copper foil. SEM images were subsequently taken for the tested samples. Figure 8.23a shows the top view for the CNF samples prior to the shear-off testing. The density of the CNFA was about $10^9/\mathrm{cm}^2$, the average diameter of the CNFs was about 80 ~ 100 nm, and the lengths of the particular CNFs in the CNTA had a variation of about 3 μm. This would cause a small percentage of CNFs that would not be in contact with the copper foil. After the shear-off testing, part of the CNFs was sheared off completely from the substrate. With a close look at the substrate, one could observe a ring-shaped bottom part of the remaining CNFs. Only a few very short CNFs, which did not have contact with the copper foil, still remained on the substrate. The tilt view at the CNF roots of the CNTs remained on the substrate (Fig. 8.23c) which shows that the roots of the CNFs have graphitic walls with hollow cores at the center.

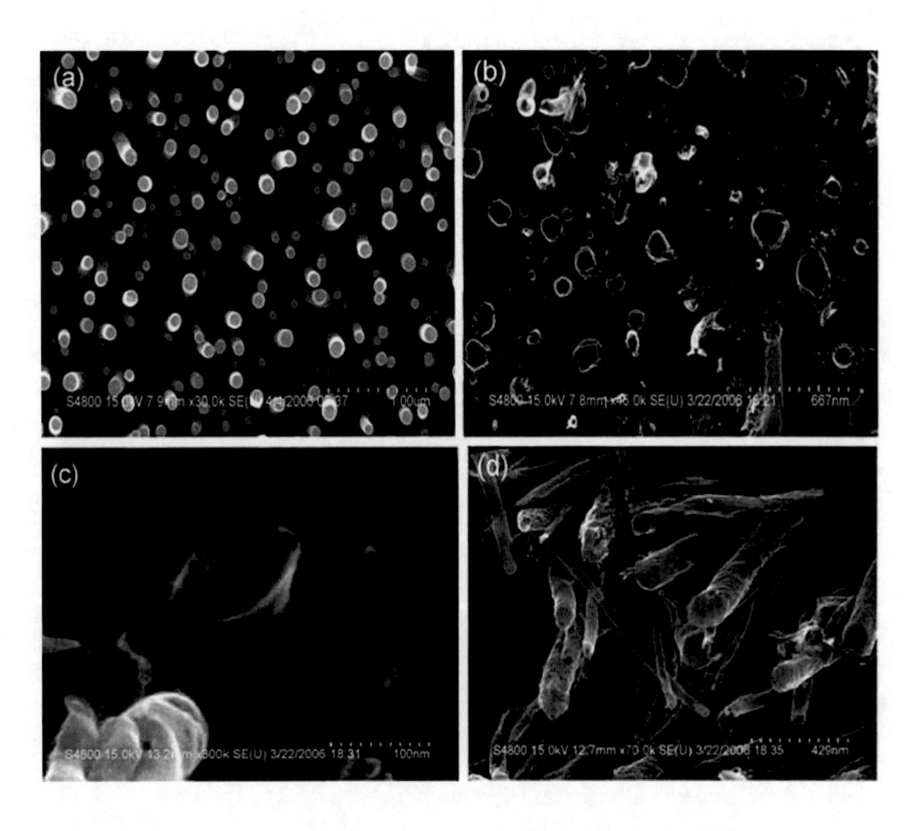

Fig. 8.23 SEM images reveal the *top* view of the CNF array before (**a**) and after (**b**) the shear-off test. Most of the CNFs are sheared off from the substrate. (**c**) is the *tilt* view of the CNF left on the original substrate; it reveals that our CNFs have the real hollow center and graphitic outside walls. (**d**) shows the CNFs sheared off from the substrate and stay on the copper tape

Although the PECVD-synthesized CNFs have a typical bamboo structure, they still have a partially hollow core structure, as CNFs synthesized by other methods do. In Fig. 8.23d, one can see sheared-off CNFs on the copper foil. The close end of these CNTA has a bamboo-like structure, which resembles the roots showed in Fig. 8.23c. The most likely breaking point takes place at the connection part of the bamboo-like CNFs.

8.11 Conclusions

We have addressed some of the existing methods that have been developed and applied for the characterization of the physical/mechanical properties of CNTs and CNFs. We have particularly addressed those methods and properties that have to do with the thermal performance of these structural elements as parts of the new generation of TIMs.

Appendix

Calculated Interfacial Shearing Stress from the Measured Shearing Force in a Bi-Material Assembly

Let a bi-material assembly be subjected to a shearing force, $\widehat{T}$. The objective of the analysis that follows is to evaluate the magnitude and the distribution of the interfacial shearing stress, τ (x), from the measured force, $\widehat{T}$. We proceed from the following approximated formulas for the interfacial longitudinal displacements for the assembly components #1 and #2:

$$u_1(x) = -\lambda_1 \int_0^x T(\xi)d\xi + \kappa_1\tau(x)$$
$$u_2(x) = \lambda_2 \int_0^x T(\xi)d\xi + \kappa_2\tau(x) \quad (8.1)$$

where

$$T(x) = \int_{-l}^{x} \tau(\xi)d\xi \quad (8.2)$$

is the force acting in the cross-section, x, of the assembly, τ (x) is the shearing stress acting in the same cross-section,

$$\lambda_1 = \frac{1-\nu_1}{E_1h_1}, \lambda_2 = \frac{1-\nu_2}{E_2h_2} \quad (8.3)$$

are the axial compliances of the assembly components, $h1$ and $h1$ are their thicknesses, $E1$ and $E2$ are Young's moduli of the component materials, $\nu1$ and $\nu2$ are Poisson's ratios,

$$\kappa_1 = \frac{h_1}{3G_1}, \kappa_2 = \frac{h_2}{3G_2} \quad (8.4)$$

are the interfacial compliances of the assembly components,

$$G_1 = \frac{E_1}{2(1+\nu_1)}, G_2 = \frac{E_2}{2(1+\nu_2)} \quad (8.5)$$

are the shear moduli of the component materials, and l is the assembly length. The origin of the coordinate x is at the cross-section where the force $\widehat{T}$ is applied. The first terms in the right parts of the formulas (8.1) are due to the forces $T(x)$ and are evaluated based on Hooke's law. The second terms account for the fact that the interfacial shearing displacements are somewhat larger than the displacements of the inner points of the given cross-section.

The compatibility condition for the displacements (8.1) can be written as

$$u_1(x) = u_2(x) - \kappa_0\tau(x) \quad (8.6)$$

where

$$\kappa_0 = \frac{h_0}{\varepsilon_0} \quad (8.7)$$

is the interfacial compliances of the bonding layer,

$$G_0 = \frac{E_0}{2(1+\nu_0)} \quad (8.8)$$

is the shear modulus of the bonding material, and $E0$ and $\nu0$ are the elastic constants of this material.

Introducing the formulas (8.1) into the displacement compatibility conditions (8.6), we have

$$-\lambda \int_0^x T(\xi)d\xi + \kappa\tau(x) = 0 \quad (8.9)$$

where the following notation is used:

$$\lambda = \lambda_1 + \lambda_2, \kappa = \kappa_0 + \kappa_1 + \kappa_2 \quad (8.10)$$

Here λ is the total axial compliance of the assembly, and k is its total interfacial compliance. From (8.9) we find, by differentiation:

$$\tau'(x) - k^2 T(x) = 0 \tag{8.11}$$

where

$$k = \sqrt{\frac{\lambda}{\kappa}} \tag{8.12}$$

is the parameter of the interfacial shearing stress.

The obvious boundary conditions for the force $T(x)$ are

$$T(0) = \widehat{T}, T(l) = 0 \tag{8.13}$$

Then the equation (8.11) results in the following boundary conditions for the interfacial shearing stress:

$$\tau'(0) = k^2\widehat{T}, \tau'(l) = 0 \tag{8.14}$$

The next differentiation of the equation (8.11), considering the formula (8.2), yields

$$\tau''(x) - k^2\tau(x) = 0 \tag{8.15}$$

This equation has the following solution:

$$\tau(x) = C_1 \sinh kx + C_2 \cosh kx \tag{8.16}$$

Here $C1$ and $C2$ are constants of integration. Introducing the sought solution (8.16) into the boundary conditions (8.14), we conclude that these constants are expressed as

$$C_1 = k\widehat{T}, C_2 = -k\widehat{T}\coth kl \tag{8.17}$$

so that the solution (8.16) leads to the following simple formula for the distributed shearing stress:

$$\tau(x) = k\widehat{T}\frac{\cosh[k(l-x)]}{\sinh kl} \tag{8.18}$$

After introducing this formula into the formula (8.2), we obtain the following expression for the distributed force, $T(x)$:

$$T(x) = \widehat{T}\frac{\tanh[k(l-x)]}{\sinh kl} \tag{8.19}$$

One could easily check that the conditions (8.13) are fulfilled for this force. The solution (8.18) yields

$$\tau(0) = -k\widehat{T}\coth kl, \tau(l) = -\frac{k\widehat{T}}{\sinh kl} \tag{8.20}$$

These formulas indicate that for sufficiently long (large l value) and/or stiff (large k value) assembles, when the product kl is, say, $kl = 2.5$, the stress τ (l) is next to zero, and the stress τ (0) at the virgin is

$$\tau(0) = \tau_{\max} = -k\widehat{T} \tag{8.21}$$

For large enough kl values, the formula (8.18) can be simplified:

$$\tau(x) = -k\widehat{T}e^{-kx} \tag{8.22}$$

This result indicates that the interfacial shearing stress is the largest at the edge, where the force is applied, and decreases exponentially with an increase in the distance x from this edge. The sign "–" in front of the formulas (8.18) and (8.22) indicates that the direction of the interfacial shearing stress is opposite to the direction of the external shearing force ^.

References

1. Iijima, S.: Nature. **354**, 56 (1991)
2. Ngo, Q., Cruden, B.A., Cassell, A.M., Sims, G., Meyyappan, M., Li, J., Yang, C.Y.: Nano Lett. **4**, 2403 (2004)
3. Xu, J., Fisher, T.S.: Int. J. Heat Mass Transf. **49**, 1658–1666 (2006)
4. Wang, X.W., Zhong, Z., Xu, J.: J. Appl. Phys. **97**(064302), (2005)
5. Xu, Y., Zhang, Y., Wang, X.W.: J. Appl. Phys. **100**(074302), (2006)
6. Ruoff, R.S., Lorents, D.C.: Carbon. **33**(925), (1995)
7. Berber, S., Kwon, Y., Tománek, D.: Phys. Rev. Lett. **84**, 4613 (2000)
8. Maruyama, S.: Microscale Thermophys. Eng. **7**, 41 (2003)
9. Osman, M.A., Srivastava, D.: Nanotechnology. **12**, 21 (2001)
10. Lourie, O., Wagner, H.D.: JMR. **13**(9), 2418–2422 (1998)
11. Treacy, M.M.J., Ebbesen, T.W., Gibson, J.M.: Nature. **381**(678), (1996)
12. Yao, N., Lordi, V.: Young's Modulus of single walled carbon nanotubes. J. Appl. Phys. **84**, 1939 (1998)
13. Hernandez, E., Gose, C., Bernier, P., Rubio, A.: Elastic properties of c and bxcynz composite nanotubes. Phys. Rev. Lett. **80**, 4502 (1998)
14. Krishnan, A., Dujardin, E., Ebbesen, T.W., Yianilos, P.N., Treacy, M.M.J.: Young's Modulus of single walled nanotubes. Phys. Rev. B. **58**, 14013 (1998)
15. Poncharal, P., Wang, Z.L., Ugarte, D., de Heer, W.A.: Science. **283**, (1513, 1999)
16. Gaillard, J., Skove, M., Rao, A.M.: Mechanical properties of chemical vapor deposition- grown multiwalled carbon nanotubes. Appl. Phys. Lett. **86**, 233109 (2005)
17. Wei, C., Srivastava, D.: Nanomechanics of carbon Nanofibers: structural and elastic properties. Appl. Phys. Lett. **85**, 2208 (2004)
18. Salvetat, J.P., et al.: Elastic Modulus of ordered and disordered multi-walled carbon nanotubes. Adv. Mater. **11**, 161 (1999)
19. Cao, P.D., Sawyer, W.G., Ghasemi-Nejhad, M., Ajayan, P.: Super compressible foam like carbon nanotube film. Science. **310**(1307), (2005)
20. Zhou, X., Zhou, J.J., Ou-Yang, Z.C.: Strain energy and Young's Modulus of Single–Wall carbon nanotubes calculated from electronic energy-band theory. Phys. Rev. B. **62**, 13692 (2000)

21. Ru, C.Q.: Effective bending stiffness of carbon nanotubes. Phys. Rev. B. **62**, 9973 (2000)
22. Yu, M.F., Lourie, O., Dyer, M.J., Moloni, K., Kelly, T.F., Ruoff, R. S.: Science. **287**(637), (2000)
23. Kuzumaki, T., Mistuda, Y.: Appl. Phys. Lett. **85**(1250), (2004)
24. Hishio, M., Akita, S., Nakayama, Y.: Jpn. J. Appl. Phys. **44**, L1097 (2005)
25. Wagner, H.D., Lourie, O., Feldman, Y., Tenne, R.: Appl. Phys. Lett. **72**, 188 (1998)
26. Walters, D.A., Ericson, L.M., Casavant, M.J., Liu, J., Colbert, D.T., Smith, K.A., Smalley, R.E.: Appl. Phys. Lett. **74**, 3803 (1999)
27. Zhang, Y., Suhir, E., Xu, Y., Gu, C.: JMR. **21**(11), (2006)
28. Y. Zhang, Y Xu, C Gu, and E Suhir, "Predicted shear-off stress in bonded assemblies: review and extension", IEEE COMT ASTR Workshop, 2006, San Francisco
29. Li, J., Papadopoulos, C., Xu, J.M., Moskovits, M.: Appl. Phys. Lett. **75**(367), (1999)
30. Harik, V.M.: Solid State Comm. **120**, 331 (2001)
31. Timoshenko, S.P., Gere, J.: Theory of Elastic Stability. McGraw-Hill, New York (1988)
32. Cruden, B.A., Cassell, A.M., Ye, Q., Meyyappan, M.: J. Appl. Phys. **94**, 4070 (2003)
33. Li, J., Steven, R., Delzeit, L., Ng, H.T., Cassell, A., Han, J., Meyyappan, M.: Appl. Phys. Lett. **81**(910), (2002)
34. Suhir, E.: Structural Analysis in Microelectronic and Fiber Optic Systems: Basic Principles of Engineering Elasticity and Fundamentals of Structural Analysis. Van Nostrand Reinhold, New York (1991)
35. Zhang, Y., Xu, Y., Suhir, E.: J. Phys. D. Appl. Phys. **39**, 4878 (2006)
36. Zhang, Y., Xu, Y., Suhir, E.: JMR. **21**(11), (2006)
37. Cui, H., Kalinin, S.V., Yang, X., Lowndes, D.H.: Nano Lett. **4**, 2157 (2004)
38. Chen, I.C., Chen, L.H., Ye, X.R., Daraio, C., Jin, S., Orme, C.A., Quist, A., Lal, R.: Appl. Phys. Lett. **88**, 153102 (2006)
39. Suhir, E.: Interfacial stresses in bi-metal thermostats. ASME J. Appl. Mech. **56**, 595 (1989)
40. Suhir, E.: Thermomechanical stress modeling in microelectronics and photonics. Electron. Cooling. **7**(4), 2001
41. Suhir, E.: Thermal stress in an adhesively bonded joint with a low modulus adhesive layer at the ends. J. Appl. Phys. **93**(3657), (2003)
42. Suhir, E.: Analysis of interfacial thermal stresses in a tri-material assembly. J. Appl. Phys. **89**, 3685 (2001)

On-Chip Thermal Management and Hot-Spot Remediation

9

Avram Bar-Cohen and Peng Wang

9.1 Introduction

The Moore's law progression in semiconductor technology, leading to shrinking feature size, increasing transistor density, faster circuit speeds, and higher chip performance, continues unabated. As shown in Fig. 9.1, the 2005 International Technology Roadmap for Semiconductors (ITRS) [1] predicts a continuous decrease in transistor size, which can be expected to lead to functioning 20 nm devices at more than a billion transistors per square centimeter (10^9/cm^2) by 2010 and to continue down to 10 nm, along with a rise in transistor density toward 10 billion transistors/cm^2 (10^{10}/cm^2), by 2018. These changes in semiconductor technology can be expected to lead to ever faster and more computationally complex chips. Moreover, despite the extensive efforts that have been made to reduce both capacitance and operating voltage on the chip, the chip power dissipation – at the very cutting edge of the technology – continues to rise. In the absence of more aggressive thermal control techniques, the elevated temperatures resulting from higher-power dissipation can be expected to decrease CMOS transistor switching speeds and accelerate some key failure mechanisms, thus compromising the performance and reliability of such advanced semiconductor devices.

The Moore's law-driven transition to the widespread use of nanoscaled electronics poses three specific thermal management challenges, which have motivated significant recent activity in microprocessor cooling [2–7]. Foremost among these is the drive to improve component speed, which has motivated circuit designers to compress the "core" of the microprocessor to an ever smaller size. Along with the reduced "time-of-flight" between transistors, this spatial compression leads disproportionately to high heat flux in the "core" areas of the silicon chip. Second, higher interconnect current densities, due to higher chip power, and the extreme interconnect aspect ratios, resulting from the high transistor densities, result in rapidly increasing interconnect temperatures in today's IC technology. The problem is aggravated by the trend to replace the SiO_2 interconnect passivation layer with lower dielectric constant materials, such as novel organic and porous dielectrics, which also possess lower thermal conductivity and greatly impede the conduction of heat away from the interconnect and transistor. Third, the use of novel nanoscaled electronics technologies, e.g., narrow channels, further aggravates the local temperature rise around individual transistors. Figure 9.2 shows the temperature contour around a single nanoscaled transistor and indicates there is nanometer scale hot spot in the transistor drain region [8].

Under the combined influence of these trends, chip power dissipation and heat flux are expected to increase further over the next decade. According to the International Electronics Manufacturing Initiative Technology Roadmap (iNEMI 2006) [9], the maximum chip power dissipation is projected to be 510 W and the maximum chip heat flux to be 300 W/cm^2 for the high-power automotive devices in 2015, as indicated in Fig. 9.3.

Moreover, as seen in Fig. 9.4, these advanced nanoelectronic chips are characterized by substantial nonuniformities in power dissipation, resulting in localized, high heat flux "hot spots" with a large on-chip temperature gradient [3]. Because chip thermal management must ensure that all junction temperatures in the microprocessor do not exceed an application-driven maximum temperature, typically in the range of 90–110 °C, it is often these hot spots, not the entire chip power dissipation, which drive the thermal design. This leads to two undesirable consequences: (9.1) nonuniform heat generation limits the total heat dissipation that can be managed by a conventional thermal solution, and, thus, a much more aggressive thermal solution than would be required for uniform heating, is necessary and (9.2) the focus on controlling the temperature of the hot spot can lead to

A. Bar-Cohen (✉) · P. Wang
Department of Mechanical Engineering, University of Maryland, College Park, MD, USA
e-mail: abc@umd.edu; wangp@umd.edu

C. P. Wong et al. (eds.), *Nano-Bio- Electronic, Photonic and MEMS Packaging*, https://doi.org/10.1007/978-3-030-49991-4_9

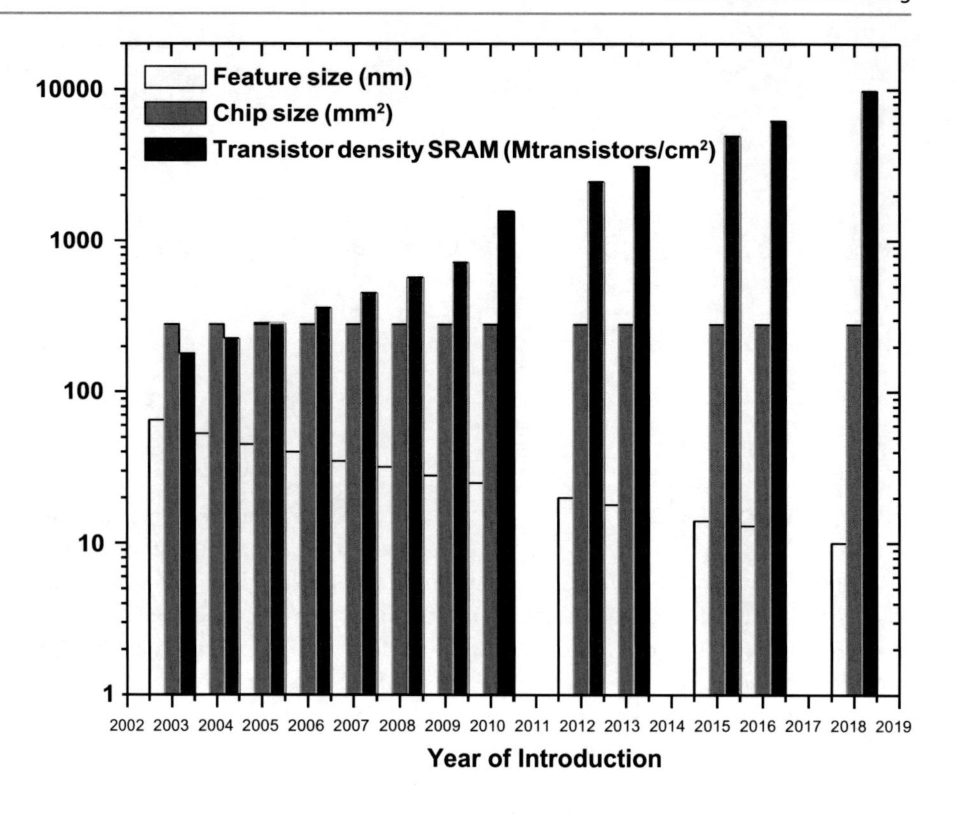

Fig. 9.1 The 2005 ITRS predictions of feature size, chip size, and transistor density for high-performance microprocessor chips [1]

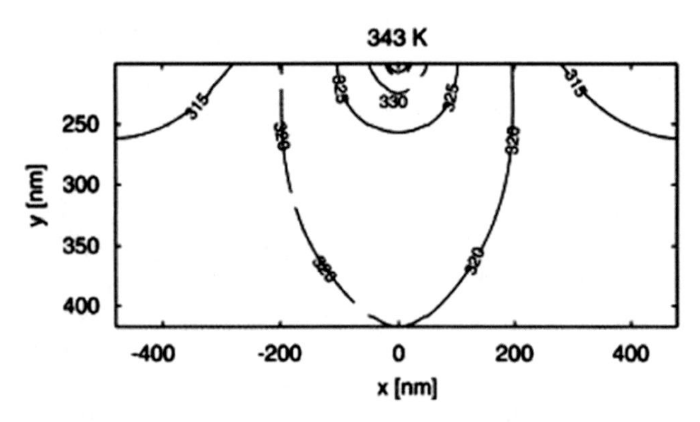

Fig. 9.2 The temperature field from a device-like hot spot in bulk 90 nm silicon transistor [8]

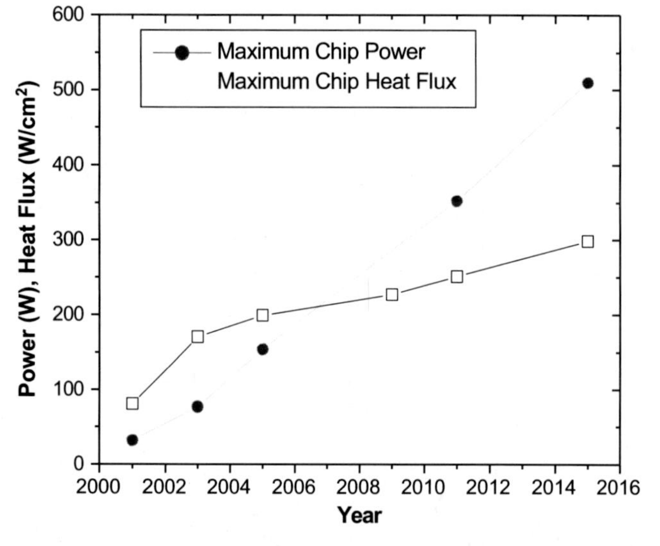

Fig. 9.3 The 2006 iNEMI prediction of chip power dissipation and heat flux for the highest-power automotive devices [9]

overdesign of the microprocessor cooling solution. As shown in Fig. 9.5, the 2006 iNEMI Roadmap predicted that desktop PC second-level thermal solution requirements for a uniformly powered chip will become more demanding, pushing the limits of current air-cooled systems. Once the nonuniform distribution of power across the chip is considered, air-cooling system will have to be replaced with more aggressive cooling solutions. Due to its complexity, on-chip hot-spot cooling has become one of the most active and challenging research areas in thermal management of electronic devices and packages.

9.1.1 Potential Hot-Spot Cooling Solutions

Thermal management for high flux electronic silicon chips can be classified into two strategies: passive cooling and active cooling, both of which have continued to be extensively studied during the past few years.

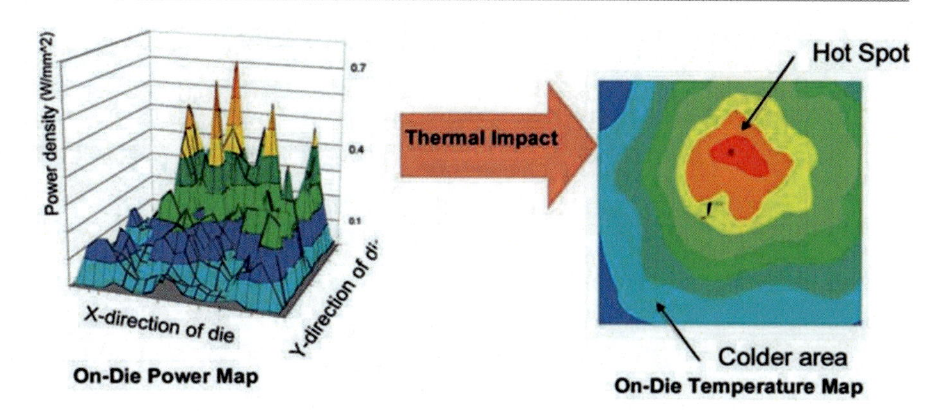

Fig. 9.4 Schematic illustrating typical die power map

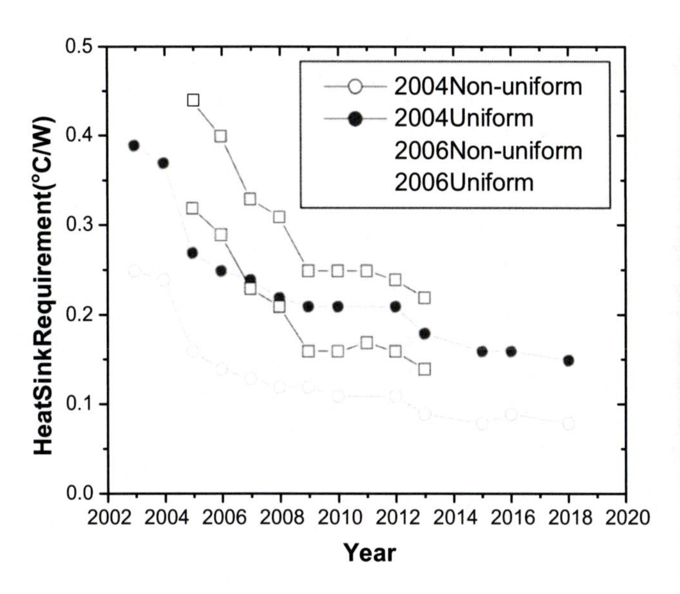

Fig. 9.5 The 2006 iNEMI predictions of desktop PC second-level thermal solution requirement

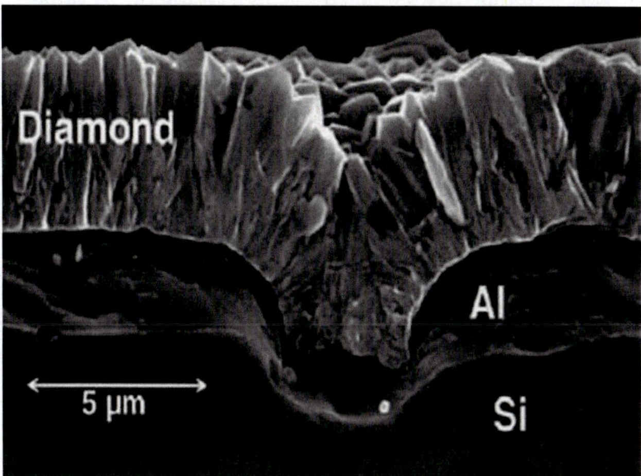

Fig. 9.6 Diamond deposited on aluminum metallization in a silicon substrate using a microwave plasma technique [12]

9.1.1.1 Passive Cooling Solutions

Passive cooling solutions are those that do not have moving parts and generally require no external electric power to activate or assist thermal transport. These techniques mainly rely on heat spreading in high-conductivity materials, such as bonded spreaders or diamond coatings on the silicon chip, and/or on vapor transport along with evaporation and condensation in tubes and channels, to transport the heat from the high-flux regions to the areas of lower heat flux. Due to the high thermal conductivity of silicon (~150 W/m K), only modest spreading improvements can be obtained from the use of traditional spreading materials, such as copper (390 W/m K), beryllia (250 W/m K), aluminum nitride (220 W/m K), or other composites [10]. Alternatively, diamond is a very attractive material for passive cooling of high-flux regions on a chip because it has the highest thermal conductivity of any known materials and also has a very high electric resistance (~10^8 Ωm). The thermal conductivity is in the range of 1500–2100 W/m K for single-crystal diamond fabricated by the high-pressure synthesis method and 500–1300 W/m K for polycrystalline diamond fabricated by CVD low-pressure synthesis [11]. Deposition of diamond on substrates, such as silicon, is a reasonably mature technology, and there are now multiple techniques that provide high-quality single-crystal or polycrystalline diamond films. The fabrication of a diamond heat spreading layer on silicon's active region includes direct growth of diamond on the silicon substrate or bonding of a polished diamond film onto the silicon substrate. Figure 9.6 shows an example of diamond deposited directly on the aluminum metallization layers of a silicon chip. However, the contamination with impurities that may occur in the silicon wafer during diamond deposition, associated with the carbon, nitrogen, oxygen, hydrogen, and other elements diffusing into the device wafer from reactive gases, has – thus far – kept diamond deposition from becoming the technique of choice. To avoid these difficulties, diamond is often bonded to silicon using a gold-tin eutectic alloy solder film. However, the thermal contact resistance at the silicon substrate/diamond interface and metallization layers/diamond interface and poor adhesion of diamond to the metallization layer, aggravated by cyclic thermally induced

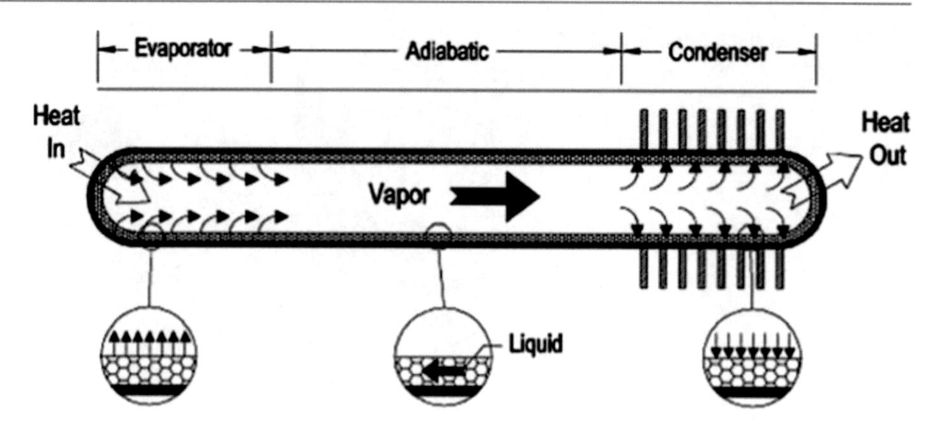

Fig. 9.7 Structure of typical wicked heat pipe

stress variations, has restricted the application of diamond as a reliable heat spreader.

Since the thermal conductivity of silicon is relatively high, while the heat transfer resistance through the thickness is relatively small, a heat pipe micromachined in the silicon substrate or a flat-plate heat pipe attached to the silicon chip is advantageous in providing in-plane spreading of the heat released by localized regions of high heat flux [13, 14]. A heat pipe is a passive heat transfer device with an extremely high effective thermal conductivity, resulting from the evaporation and condensation of a suitable working fluid within an evacuated, hermetically sealed enclosure. Figure 9.7 is a side view of a heat pipe showing the wick and the vapor/liquid flow characteristics.

In the past 10 years, extensive research has demonstrated that heat spreading performance comparable to that of diamond substrates can be obtained [15–17]. In addition, the factors that affect the performance of micro or miniature heat pipes such as the frictional vapor flow, the vapor space, the shape and dimensions of the microchannels, the operating temperature, the disjoining pressure, and the evaporating heat resistance have been systematically investigated. These investigations have provided insight into the design of highly efficient heat pipes, which can meet the new requirements of high heat flux cooling. Adkins at Sandia National Laboratories, Albuquerque, NM, investigated a heat pipe heat spreader embedded in a silicon substrate as an alternative to the conductive cooling of integrated circuits using diamond films. In their design miniaturized heat pipes, created by 35 μm grooves in the silicon wafer, function as highly efficient heat spreaders, collecting heat from the localized hot spots and dissipating the heat over the entire chip surface, with an effective thermal conductivity of at least 800 W/m K [18, 19].

Recently research has been focused on achieving further increases in performance by using thermally driven pulsating two-phase flows [20, 21], new capillary structures [22], and MEMS-based heat pipes [23, 24]. Plesch et al. reported that flat miniature heat pipes with axial grooves, using water as the working fluid, were capable of sustaining heat fluxes on the order of 40 W/cm^2 [25]. Ma et al. fabricated the heat pipes with microscaled sintered powder wicks, which can remove heat fluxes of up to 80 W/cm^2 without any sign of evaporator dry out [26]. Gillot fabricated flat miniature heat pipes with microcapillary grooves inside a silicon substrate, and the heat removal capability was reported to be 110 W/cm^2 [27]. Using miniaturized heat pipes, Lin et al. reported that a cooling heat flux of 140 W/cm^2 was achieved [28]. While heat pipes possess certain inherent advantages and current research may raise the observed thermal performance limits, heat removal rates of the currently available heat pipes are generally still not suitable for controlling the temperature of the most severe, high-flux hot spots on the chip.

9.1.1.2 Active Cooling Solutions

Active cooling solutions usually involve moving parts and require the input of electric energy for their operation. The most common active cooling solution is air-cooled, forced convection heat sinks, which have been long used for a wide range of electronic equipment, including office and desktop computers. However, this conventional active cooling method has very limited capability for dealing with high-flux zones on microelectronic chips, due to its inherently low heat transfer coefficients. Compared to air cooling, the use of liquid coolants has many advantages such as high thermal conductivity, high specific heat, low viscosity, and high latent heat of evaporation for two-phase application. As a result, different active liquid-cooling technologies, such as microchannel heat sinks and direct jet impingement, have been developed due to the higher heat transfer coefficient and high cooling flux achieved as compared to air-cooling heat sink.

With microfabrication techniques developed by the electronics industry, it is possible to manufacture microscaled three-dimensional structures, typically consisting of closely spaced parallel channels with rectangular, trapezoidal, or triangular cross sections and hydraulic diameters ranging from 100 to 1000 μm (see Fig. 9.8). Such microchannel "compact heat exchangers" may be fabricated in the chip itself or in the heat sink to which a chip or array of chips is

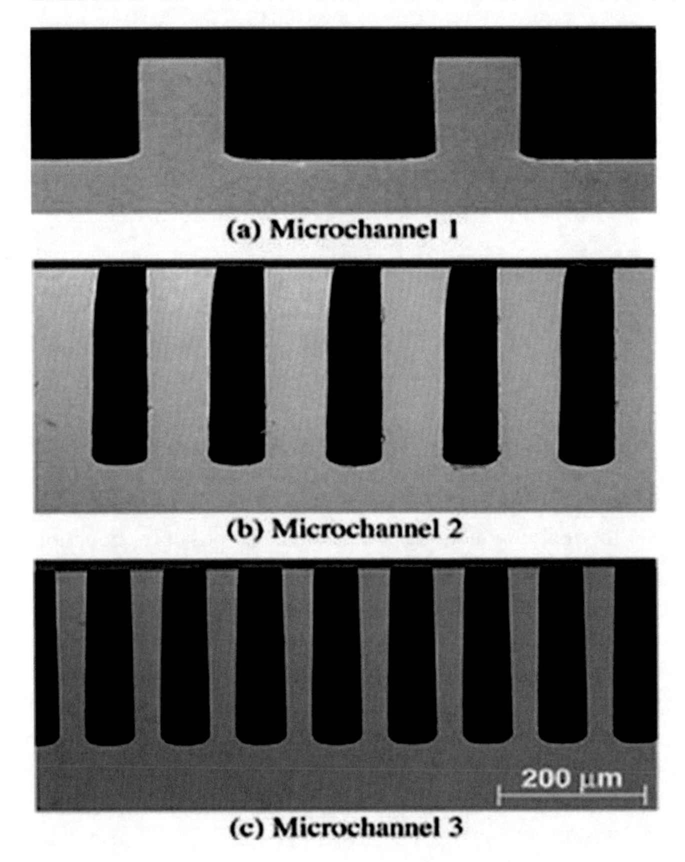

Fig. 9.8 SEM photos of cross sections of three different microchannels [30]

attached. Microchannel heat sinks can be used either with single-phase flow, where heat is transferred from the electronic chip via sensible heat gain to the coolant, or with two-phase flow, which also utilizes the latent heat of the coolant during liquid/vapor phase change. In both thermal transport modes, microchannel coolers are capable of significantly increasing the wetted surface area and of thinning the thermal boundary layers, thus achieving very high heat transfer rates. Single-phase liquid-cooling systems utilize a pump to actively circulate the liquid to the microchannels and have been studied for many years. In their 1981 pioneering work, Tuckerman and Pease used single-phase convection with water flowing through a microchanneled silicon chip to demonstrate heat removal of 790 W/cm^2 with a flow rate of 1×10^{-5} m^3/s and the pressure drop below 3.45×10^5 Pa [29]. Although the high heat flux capability is promising, and considerable pressure reductions and improvements in flow distribution have been achieved in recent years, the integration of such a microchannel cooler into a closed-loop plumbing/pumping system and justifying the required pumping power has proven challenging. Moreover, the industry's resistance to flowing liquids directly through active microchanneled chips has necessitated the use of thermal interface materials for the attachment of the microchannel coolers and considerably raised the thermal resistance of this thermal management approach.

Two-phase microchannel cooling, exploiting the latent heat of the coolant to reduce the flow rates and pumping power requirements and the high heat transfer coefficients associated with boiling, has received growing attention, focusing on reductions in pumping power and dealing with the performance limitations posed by flow instabilities and maldistribution. Bowers et al. showed that two-phase microchannels, operating with water flow rates of less than 1.08×10^{-6} m^3/s and pressure drops of 3.45×10^5 Pa could remove more than 200 W/cm^2 [31]. Mudawar also demonstrated a heat removal rate of 361 W/cm^2 with two-phase forced convective cooling on an enhanced surface with FC-72 as the dielectric coolant [32]. However, in a recent study of water-cooled two-phase microchannel coolers, Prasher found the biggest challenge to the use of two-phase microchannel coolers in hot-spot cooling is the nonuniformity in flow distribution and the resulting large temperature nonuniformities [30]. The testing results show that the worst-case temperature fluctuations are on the order of 30 °C for 0.6 W 400×400 μm hot-spot heating condition, on the order of 20 °C for 0.4 W 400×400 μm hot-spot heating condition, and on the order of 15 °C for the uniform heating condition, showing that temperature fluctuations depend on the power being dissipated from localized hot spots. In particular, poor flow distribution in two-phase microchannels might lead to less flow in the high-flux regions, leading to localized dry out on the hot spot, which will result in large and rapid increase in the hot-spot temperature.

Jet impingement cooling, with high-velocity liquid streams directly impinging onto the hot surface, is an alternative to microchannel coolers. This method offers several potential advantages, such as high heat transfer coefficients associated with the evaporation of thin liquid films and the ability to form patterns during the liquid distribution through an array of jets. The jets can be fabricated by circular or slot-shaped orifices or nozzles of various cross sections and can flow into a gaseous medium (vapor or air) forming the so-called free jets or into a liquid medium, leading to a "submerged jet." As a final distinction, jet impingement cooling of electronic components may involve forced convection alone or localized flow boiling, with or without net vapor generation [33–35]. Although electronic cooling applications will require the use of dielectric liquids, much of the available data is for jet impingement of water, including the work by Zhang demonstrating that two-phase jet impingement is capable of removing a heat flux of more than 100 W/cm^2 at water flow rates below 2.5×10^{-7} m^3/s [36], use of boiling macrojets for removal of a heat flux over 400 W/cm^2 [37], and the results by Kiper, using microscaled direct water impingement from an orifice plate to cool VLSI circuits dissipating more than 500 W/cm^2 [38]. The need to achieve this level of performance with dielectric coolants and concerns over the reliability, complexity, volume, weight, and cost of such jet impingement systems have posed significant barriers to the successful commercial implementation of these approaches.

9.1.1.3 Solid-State Cooling Solutions

In recent years, there has been increased interest in the application of solid-state thermoelectric coolers for high-flux thermal management because of their compact structure, fast response, high-flux spot cooling capability, and high reliability and their absence of moving parts [39–46]. Another major advantage of a thermoelectric cooler is that it can be miniaturized and integrated into the chip package. A thermoelectric cooler is based on the Peltier effect and consists of n-type and p-type thermoelectric elements, as shown in Fig. 9.9, where the n-type and p-type thermoelectric elements are joined by metallic connectors at the top and bottom. When a DC current goes through these thermoelectric element/metal contacts, heat is either released or absorbed in the contact region depending on the direction of the current. Attaching the TEC cold junction to a working device makes it possible to lower its temperature below its surroundings and possibly even below the ambient temperature.

In the conventional thermoelectric cooler (TEC) design, the maximum achievable temperature reduction across the thermoelectric cooler can be determined by:

$$\Delta T_{\max} = \frac{S^2 T_c^2}{2\rho k} \tag{9.1}$$

and the maximum achievable cooling flux on the cold side of thermoelectric cooler by Eq. (9.2)

$$q''_{\max} = \frac{S^2 T_c^2}{2\rho d} \tag{9.2}$$

where S is the Seebeck coefficient, k is the thermal conductivity, ρ is the electrical conductivity, T_c is the absolute temperature at the cold side, and d is the thickness of thermoelectric elements [47–49]. Thus, the largest temperature reduction is attained for thermoelectric materials when the Seebeck coefficient is large, and the electrical resistivity and the thermal conductivity are as small as possible. Equation (9.2) indicates that the maximum cooling flux is inversely proportional to thermoelectric element thickness, and thus the main advantage of going to thin-film thermoelectric coolers (TFTECs) is the dramatic increase in cooling heat flux. As shown in Fig. 9.10, Fleurial estimated that the heat flux of several hundred Watt per square centimeter could be removed with thin-film thermoelectric coolers when thermoelectric element thickness is on the order of 20 μm [50].

Recent attempts to improve the cooling performance of TECs have focused on low-dimensional nanostructured superlattices which are capable of suppressing the thermal conductivity through phonon trapping and improving the TEC's figure of merit Z ($Z = S^2/k\rho$). Venkatasubramanian reported thin-film superlattice Bi_2Te_3/Sb_2Te_3 coolers capable of providing up to 32 °C net cooling measured on the cold side of the coolers and a maximum estimated cooling heat

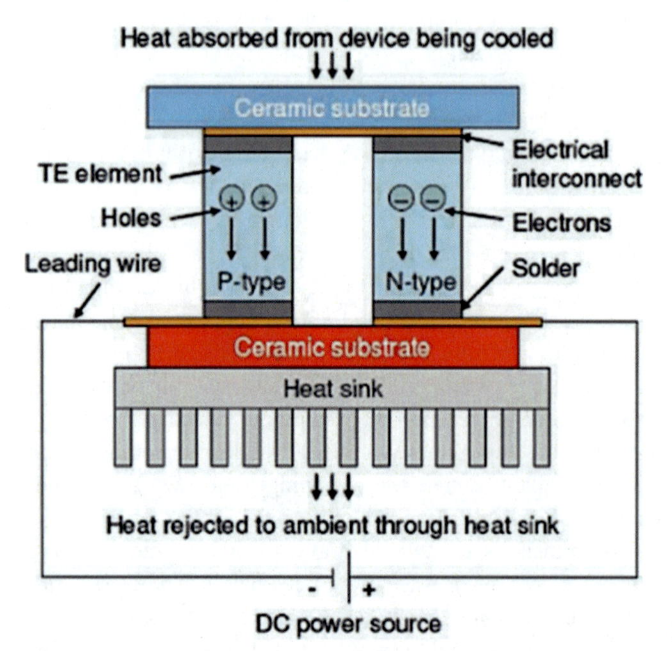

Fig. 9.9 Schematic diagram of a thermoelectric cooler

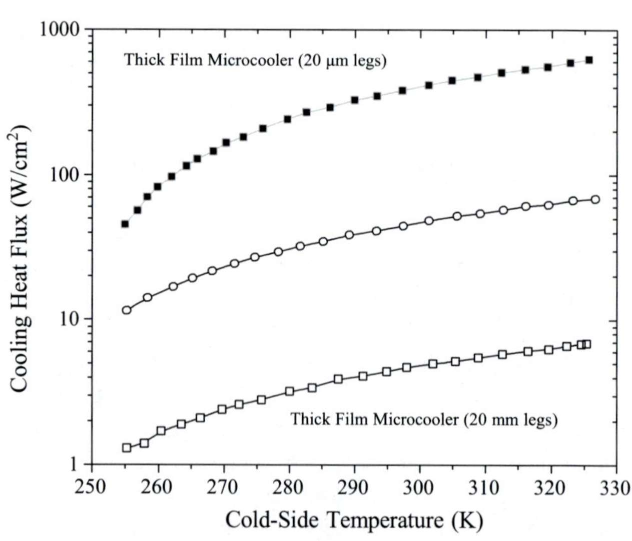

Fig. 9.10 Cooling heat flux as a function of thermoelectric leg thickness when diamond is used as the TEC substrate and bulk Bi_2Te_3 as thermoelectric material ($ZT = 0.9$)

flux of 700 W/cm^2 at room temperature [51]. Harman demonstrated a thin-film cooler based on quantum dot n-type PbSeTe/PbTe superlattice structure, which provided 43.7 °C net cooling at room temperature [52]. Fan and Shakouri demonstrated net cooling of up to 2.5 °C at room temperature and 7 °C at 100 °C ambient temperature and a maximum cooling heat flux as high as 680 W/cm^2 on the surface of a microcooler for p-type thin-film superlattice SiGeC/Si microcoolers TEC device miniaturization [53], to extract greater performance from existing bulk thermoelectric materials, is the other approach to TEC improvement currently receiving attention. Using ceramic thinning technology, Semenyuk developed and commercialized a 130 μm-thick miniatured Bi_2Te_3 thermoelectric cooler which can provide more than 100 W/cm^2 cooling heat flux at the cold side junction [54–57]. These reported results show solid-state thermoelectric coolers that provide a powerful alternative to traditional high-flux coolers and offer great promise for reducing the severity of on-chip hot spots.

Succeeding sections of this chapter will address the possible application of these proposed approaches to the thermal management of nanoelectronic hot spots. Attention will be focused on silicon thermoelectric coolers, mini-contact, miniaturized TECs that overcome the present heat flux limitations of conventional Bi_2Te_3 devices, two-phase micro-gap coolers that can directly cool high heat flux chips without the deleterious effects of contact resistance, and the use of anisotropic thermal interface materials (TIM) to enhance the effectiveness of both passive and active cooling, in suppressing the temperature rise in high-flux regions of the chip.

9.2 On-Chip Hot-Spot Cooling Using Thermoelectric Microcoolers

Since the discovery of the Peltier effect in 1834, extensive research has been done to develop solid-state refrigeration devices and solid-state energy generators based on thermoelectric effects. However, only limited applications of thermoelectric cooling existed until the middle 1950s, when it was discovered that doped semiconductors could achieve better thermoelectric properties than metallic materials. After several decades of research, the efficiency of thermoelectric cooling devices still reached only about 10% of Carnot efficiency in comparison with the 30% efficiency typical of vapor compression refrigerators. The progress in thermoelectrics research declined until the early 1990s, when theoretical predications indicated that low-dimensional structures, such as two-dimensional superlattices, could be used to produce high-performance thermoelectric materials. Since then extensive studies, theoretical as well as experimental, have been conducted to explore new materials and new designs for fabrication of high-performance thermoelectric coolers. Since 2000, the demand for hot-spot cooling in microprocessors, as well as in power electronic, RF, and laser components, has generated renewed interest in exploring novel thermoelectric microcoolers, which can provide localized, high-flux cooling capability.

9.2.1 Principle of Conventional Thermoelectric Cooler (TEC)

The principle of conventional thermoelectric coolers was developed 50 years ago by Ioffe and coworkers [58]. A typical TEC consists of an array of n-type and p-type thermoelectric elements, two ceramic substrates that provide mechanical support for the TEC, electric conductors that provide a serial electric connection for the thermoelectric elements and electric contacts to lead wires, solders that join the thermoelectric elements, and lead wires that are connected to the ending conductors and deliver power from a DC electrical source. When semiconductor TEC devices are assembled, the array of heavily doped p-type and n-type semiconductor elements is soldered to ceramic substrates so that they are connected electrically in series and thermally in parallel. As is known, electrons can move freely in electrical conductors but less in semiconductors. When the electrons leave an electric conductor and enter the p-type semiconductor, they drop down to a lower energy level and release heat at the interface. However, as the electrons move across the bonded interface from the p-type semiconductor into the electric conductor, they transition to a higher energy level and absorb heat. When these electrons then move into the n-type semiconductor, they rise further in energy level so that additional heat is absorbed. When the electrons leave the n-type semiconductor and enter the conductor, they drop down to a lower energy level and release heat during the process. Thus, in this thermoelectric circuit, heat is always absorbed when electrons enter n-type semiconductor or leave p-type semiconductor, and heat is always released when electrons enter p-type semiconductor or leave n-type semiconductor. The precise heat pumping capacity of a TEC is proportional to the current and is dependent on the element geometry, number of couples, and thermoelectric properties of the materials.

Figure 9.11 shows the basic configuration of a thermoelectric cooler with one p-type element and one n-type element. For a single-stage TEC, as shown in Fig. 9.11, the amount of heat that can be pumped at the cold side of a TEC is the net of three contributions. If we assume to have perfect thermal interfaces at the cold side and the hot side and neglect the Thomson effect, the net cooling power on the cold side of the TEC can be expressed by:

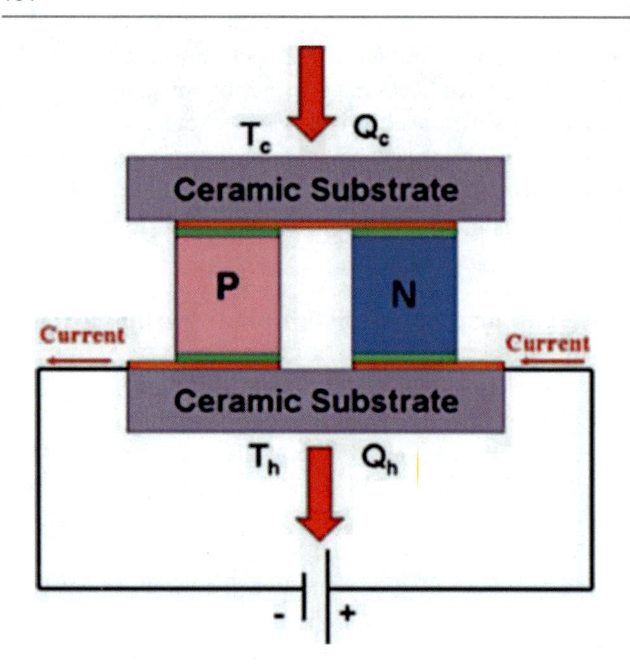

Fig. 9.11 Sketch of a thermoelectric cooler

$$q_c = (S_p - S_n)IT_c - K(T_h - T_c) - \frac{1}{2}I^2R_e \qquad (9.3)$$

where R is the overall electrical resistance of the TEC:

$$R_e = \rho_p \frac{L_p}{A_p} + \rho_n \frac{L_n}{A_n} \qquad (9.4)$$

K is the overall thermal conductance of the TEC

$$K = k_p \frac{L_p}{A_p} + k_n \frac{L_n}{A_n} \qquad (9.5)$$

and S, k, ρ, A, and L represent the Seebeck coefficient, thermal conductivity, electrical resistivity, the cross-section area, and the thickness of thermoelectric element, respectively. The p and n denote p-type and n-type thermoelectric materials, while Th and Tc are the temperatures at the cold side and the hot side of the TEC, respectively. The overall cooling rate is driven by the Peltier cooling (the first term) and reduced by the heat flowing back from the hot side to the cold side of the TEC (the second term) and Joule heating in the element (the third term).

When a current is applied to the TEC, a voltage drop is generated, which includes the resistive voltages and the Seebeck voltages across the thermoelectric elements and is given by:

$$V = (S_p - S_n)(T_h - T_c) + IR \qquad (9.6)$$

Therefore, the electric power consumption of the TEC system is equal to:

$$W = (S_p - S_n)(T_h - T_c)I + I^2R \qquad (9.7)$$

The coefficient of performance (COP) is used to describe the cooling efficiency of the TEC and is the ratio of the net cooling power at the cold side and the power consumption of the system and is given by:

$$\text{COP} = \frac{q_c}{W} \frac{(S_p - S_n)IT_c - K(T_h - T_c) - \frac{1}{2}I^2R_e}{(S_p - S_n)(T_h - T_c)I + I^2R} \qquad (9.8)$$

The maximum cooling rate that can be achieved by this TEC device can be determined by differentiating the cooling rate given in Eq. (9.3) to find the optimal current I_{opt}:

$$\left(\frac{dq_c}{dI}\right)_{opt} = 0 \Rightarrow I_{opt} = \frac{(S_p - S_n)T_c}{R_e} \qquad (9.9)$$

When $I = I_{opt}$, the heat removal rate attains its maximum, value given by:

$$q_{max} = \frac{1}{2} \frac{(S_p - S_n)^2 T_c^2}{R_e} - K(T_h - T_c) \qquad (9.10)$$

To determine the largest temperature reduction, i.e., the deepest cooling, that the TEC can achieve, the heat removed from the cold side is set equal to zero, $q_{max} = 0$, yielding the maximum temperature difference across the TEC, i.e., from the cold side to the hot side of the thermoelectric cooler:

$$\Delta T_{max} = \frac{(S_p - S_n)^2 T_c^2}{2KR_e} = \frac{ZT_c^2}{2} \qquad (9.11)$$

where Z is known as the TEC figure of merit, given by:

$$Z = \left[\frac{(S_p - S_n)}{(k_p\rho_p)^{0.5} + (k_n\rho_n)^{0.5}}\right]^2 \qquad (9.12)$$

For simplification, it is frequently assumed that the semiconductor Seebeck coefficients of the two materials are equal but opposite in sign, i.e., $S_p = -S_n = S$, and that the thermal conductivity, electrical resistivity, and dimensions are equal for the two elements, i.e., $\rho_p = \rho_n = \rho$, $k_p = k_n = k$, $A_p = A_n = A$, $L_p = L_n = L$, so Eq. (9.12) can be simplified as:

$$Z = \frac{S^2}{k\rho} \qquad (9.13)$$

Then the maximum achievable temperature difference at the optimized current, $I = I_{opt}$, when the heat flow is zero, can be calculated as:

$$\Delta T_{max} = \frac{S^2 T_c^2}{2k\rho} = \frac{1}{2} Z T_c^2 \quad (9.14)$$

Similarly, the maximum cooling rate at the optimized current ($I = I_{opt}$), when the hot side and cold side are equal in temperature can be calculated as:

$$q_{max} = \frac{S^2 T_c^2}{2R_e} = \frac{S^2 T_c^2 A}{2d\rho} \quad (9.15)$$

The corresponding cooling heat flux at the cold side of the TEC can be calculated as:

$$q''_{max} = \frac{q_{max}}{2A} = \frac{S^2 T_c^2}{2d\rho} \quad (9.16)$$

It is interesting to note that the maximum achievable cooling ΔT_{max} only depends on the figure of merit Z and the temperature of the cold side of the TEC, but it does not change with the cross-sectional area or thickness of the TEC elements. To attain the lowest temperature, it is desirable to have a thermoelectric material with a high value of Z. Alternatively, when thermal conduction is not an important parasitic effect, the power factor *P*, defined by Eq. (9.17), is often used to characterize the thermoelectric properties of a given material:

$$P = \frac{S^2}{\rho} \quad (9.17)$$

The maximum COP can be found by differentiating the COP given in Eq. (9.8) to find the COP-optimal current $I_{COP,opt}$:

$$\left(\frac{dCOP}{dI}\right)_{opt} = 0 \Rightarrow I_{COP,opt} = \frac{(S_p - S_n)(T_h - T_c)}{R_e(1 + ZT_{ave})^{0.5} - 1} \quad (9.18)$$

When $I = I_{COP,opt}$, the maximum value of COP can be calculated:

$$COP_{max} = \frac{T_c}{T_h - T_c} \frac{(1 + ZT_{ave})^{0.5} - T_h/T_c}{(1 + ZT_{ave})^{0.5} + 1} \quad (9.19)$$

where $T_{ave} = (T_h + T_c)/2$ is the mean temperature of the TEC.

9.2.2 Thermoelectric Cooling Materials and Devices

As described in the last section, for conventional thermoelectric applications, the best thermoelectric materials are those providing the highest *Z* values, usually associated with high Seebeck coefficients, high electrical conductivities, and low thermal conductivities. So far, the best thermoelectric material properties are found in heavily doped semiconductors. In semiconductors, the thermal conductivity is established by the flow of both electrons and phonons but with much of the thermal transport ascribed to phonons. The phonon thermal conductivity can be reduced without causing significant reduction in the electrical conductivity. A common approach to reduce the phonon thermal conductivity is through alloying or doping because the mass difference scattering in alloys or doped semiconductors reduces the lattice thermal conductivity significantly without much degradation in the electrical conductivity.

Bismuth telluride-based compounds, using alloys of Bi_2Te_3 with Sb_2Te_3 (p-type) and Bi_2Te_3 with Bi_2Se_3 (n-type), are the best commercial state-of-the-art materials for thermoelectric cooling, with the highest values of the figure of merit *ZT*. In bulk materials, a *ZT* of 0.75 for p-type $(BiSb)_2Te_3$ at room temperature was reported about 40 years ago. Since the 1960s, much effort has been made to raise the *ZT* of bulk materials based on bismuth telluride by doping or alloying other elements in various fabrication processes. Recently, a *ZT* of 1.14 at 300 K has been reported for the p-type $(Bi_{0.25}Sb_{0.75})_2(Te_{0.97}Se_{0.03})_3$ alloy [59]. By annealing the ingots prepared by the Bridgman method, Yamashita et al. have most recently achieved a significant increase in the *ZT* value to 1.19 at 298 K for the *n*-type $Bi_2(Te_{0.94}Se_{0.06})_3$ and 1.41 at 308 K for the p-type $(Bi_{0.25}Sb_{0.75})_2Te_3$ alloy, so that both *ZT* values exceed 1 [60–62]. In addition to alloying, several other approaches have been proposed to enhance the *ZT* value, through either improving electrical conductivity or reducing thermal conductivity. In this respect, low-dimensional materials, such as quantum wells, superlattices, quantum wires, and quantum dots, offer new ways to manipulate the electron and phonon properties of a given material. Some experiments have demonstrated that superlattice thermoelectric materials can achieve *ZT* values greater than 2.0. These inspiring results show the feasibility of applying thermoelectric materials to high-flux cooling applications.

Thermoelectric coolers (TECs) have traditionally been fabricated using bulk bismuth telluride materials and traditional processing techniques such as hot pressing sintering and extrusion. Commercial bulk thermoelectric coolers are made from such thermoelectric elements, typically several millimeters in thickness and combined into arrays that span 1.8 × 3.4 mm (and 2.4 mm thick) to 62 × 62 mm (and

5.8 mm thick). For such commercial TE modules, the maximum cooling at room temperature is about 70 °C, with relatively low cooling heat flux of 5–10 W/cm^2, which makes it impossible to use such TECs for high-flux hot-spot cooling application [63]. Because the maximum cooling heat flux of a TEC is inversely proportional to the thickness of its elements, there have been extensive studies focusing on microscaled thin-film TECs, miniatured TECs, and nanostructured superlattice TECs, which could realize high-flux cooling requirement for on-chip hot-spot reduction. In recent years, significant progresses have been reported in making microscaled thermoelectric coolers, as described in the three sections that follow.

9.2.2.1 Thin-Film Thermoelectric Coolers (TFTECs)

It is widely accepted that thin-film thermoelectric coolers (TFTECs) have great potential for high-flux cooling because of the dramatic theoretical increase in cooling heat flux made possible by decreasing thermoelectric element thickness as given by Eq. (9.16). Among the available deposition methods, electrochemical deposition is very attractive from an application perspective, due to its ability to deposit thin films at high deposition rates of tens of microns per hour and at the much lower batch processing cost when compared to other state-of-the-art thin-film fabrication processes [64]. Snyder et al. used electrochemical-MEMS technique to fabricate thin-film thermoelectric microcoolers, which contains 63 n-type Bi_2Te_3 elements and 63 p-type $Bi_{2-x}Sb_xTe_3$ elements, each element being 20 μm in thickness and 60 μm in diameter with bridging metal interconnects, as shown in Fig. 9.12 [65]. Unfortunately, the defect structure in this MEMS TEC produced a high concentration of low-mobility carriers, yielding an estimated Z value of 3.2 × 10–5 (1/K) and producing a maximum cooling of 2 °C and a maximum cooling heat flux of 7 W/cm^2.

Using co-evaporation as the deposition method, da Silva et al. fabricated thin-film thermoelectric microcoolers, which provided 60 n-type and p-type thermoelectric element pairs, with the thickness and width of the elements approximately 4.5 and 40 μm, respectively, as shown in Fig. 9.13. In da

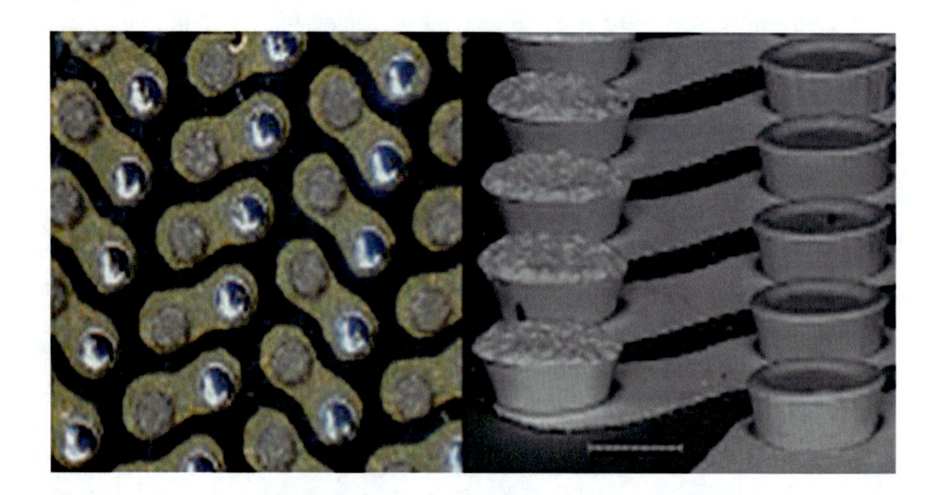

Fig. 9.12 Thin-film thermoelectric microcooler fabricated by MEMS [65]

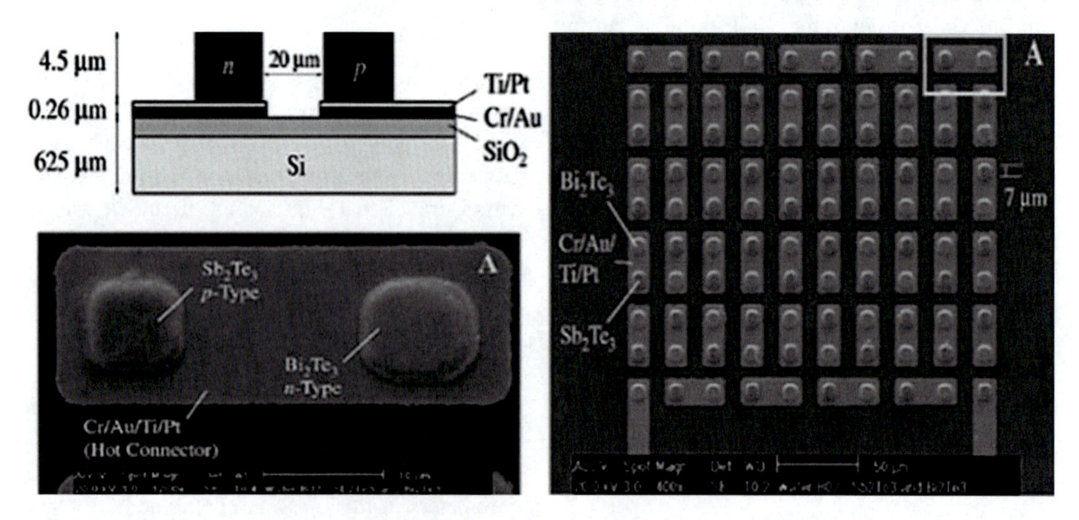

Fig. 9.13 Bi_2Te_3 and Sb_2Te_3 films deposited on Cr/Au/Ti/Pt bottom connectors [66]

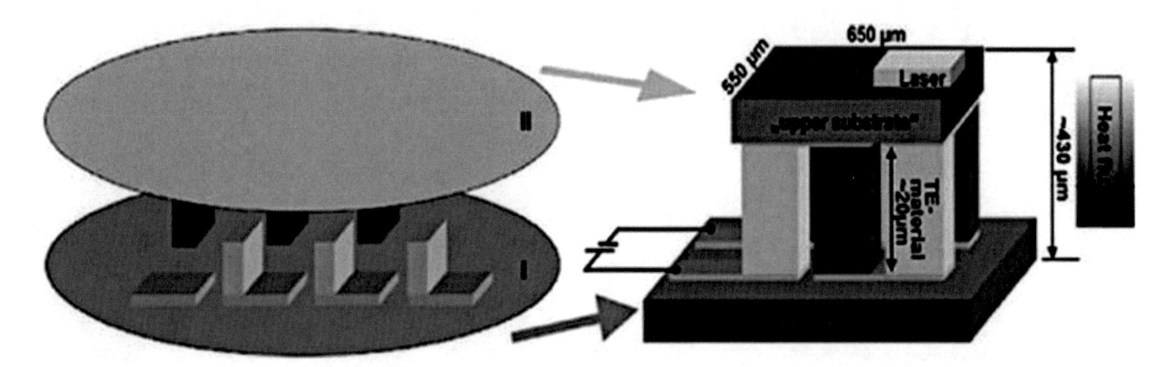

Fig. 9.14 Micro-Peltier cooler. *Left*: Schematic drawing of the developed two wafer (I,II) concept. *Right*: Schematic drawing of the thermoelectric cooler used for telecommunication device [67]

Silva's most recent work, the Seebeck coefficient, electrical resistivity, and power factor values of 149 μV/K, 1.25×10^{-5} Ω m, and 1.78 mW/K^{2}m, respectively, were achieved for p-type Sb_2Te_3 thin films at the optimized deposition temperature [66]. However, the overall thermoelectric cooling performance achieved with such improved thin films has not been reported.

Böttner et al. developed a two-wafer process to fabricate thin-film thermoelectric coolers. Figure 9.14 depicts a schematic drawing of the two-wafer process on the left and a schematic drawing of the resulting device on the right. The polycrystalline n-type $Bi_2(Se,Te)_3$ and p-type $(Bi,Sb)_2Te_3$ materials were deposited by co-sputtering from 99.995% element targets (Bi, Sb, Te). However, these alloys were not grown very well in thin-film form due to delivery problems of the Se-target suppliers. For these coolers, the thickness of n-type Bi_2Te_3 elements and p-type $(Bi,Sb)_2Te_3$ elements is about 20 μm as shown in Fig. 9.14 [67]. While he initially reported a net cooling of 11 °C at 60 °C ambient, more recently, Böttner et al. reported achieving a maximum temperature differences of nearly 48 °C at an applied current of 2.1 A and a maximum cooling flux of ~100 W/cm^2 for the complete device [68].

Zou et al. found that direct vapor deposition of bismuth telluride compounds is made difficult by the large difference in the vapor pressure between antimony, bismuth, and tellurium, which could result in noncongruence and in a lack of stoichiometry. In his work, Sb_2Te_3 films were deposited by co-evaporation of antimony and tellurium and Bi_2Te_3 thin films by co-evaporation of bismuth and tellurium onto heated, clean glass substrates. The figure of merit *Z* for the p-type Sb_2Te_3 film and n-type Bi_2Te_3 film was calculated and found to be approximately 1.04×10^{-3} at room temperature, corresponding to *ZT* of 0.32. The maximum temperature difference measured between the hot and cold ends was 15.5 °C at a current of 55 mA, showing a promising procedure for fabricating thermoelectric microcooler [69].

In using the state-of-the-art deposition techniques to develop thermoelectric thin films, a primary difficulty is to maintain the stoichiometry of the bismuth telluride compounds. For example, the problem of resputtering during the film growth is present in sputter deposition, while differences in volatility of the component elements pose difficulty in vacuum evaporation. A large deviation from stoichiometry arises in vapor deposition because the constituent elements in the target exhibit dissimilar sticking coefficients on the substrate. In addition, there is a tendency for reevaporation of certain elements from the deposited thin films because of their higher vapor pressure. Therefore, the thermoelectric properties of these thin films reported in the above publications vary widely, and the figure of merit (*ZT*) is always much small than ~1.0 for bulk bismuth telluride material. As shown in Table 9.1, to date, thin-film thermoelectric coolers are still not well developed, and complete characterization of the material properties of the various TEC thin films is lacking.

9.2.2.2 Bulk Miniaturized Thermoelectric Coolers

Although thin-film deposition technology has an advantage for mass production, currently, it appears not to provide thermoelectric cooling performance comparable to that available in bulk TECs, due to the difficulty in controlling thin-film growth conditions to obtain the desired stoichiometry and a defect-free microstructure. Alternatively, attention can be turned to miniaturized thermoelectric coolers, based on thinning of bulk materials, which can reduce thermoelectric element thickness down to tens of microns and, at the same time, maintain the excellent thermoelectric properties of the bulk materials. Table 9.2 shows the progress in developing bulk miniaturized TECs since the 1960s [70].

In 1967 Semenyuk began developing bulk bismuth telluride material, which could be used to form very short thermoelectric elements [71]. Following years of development, in 2006, Semenyuk thinned the thermoelectric element down to 130 μm, as shown in Fig. 9.15. The extruded p-type and n-type bismuth telluride thermoelectric materials were in the form of rods with *Z* values of 3.02×10^{-3} K^{-1}, corresponding to *ZT* value of 0.9. The 200 μm thick p-type

Table 9.1 Summary of cooling performance of Bi_2Te_3-based thin-film TEC

Author (year)	Growth method	ΔTmax (°C)	q''max (W/cm^2)	TE properties
Snyder (2002)	Electrochemical deposition	2@80 °C	7	$S = 60–100\ \mu V/K$
				$ZT = 0.01$ (estimated)
Zuo (2002)	Co-sputtering	15.5@25 °C	N/A	p-type:
				$S = 160\ \mu V/K$
				$\rho = 3.12 \times 10^{-5}\ \Omega m$
				n-type:
				$S = -200\ \mu V/K$
				$\rho = 1.29 \times 10^{-5}\ \Omega\ m$
da Silva (2005)	Co-evaporation	1.0@25 °C	N/A	p-type:
				$S = 228\ \mu V/K$
				$\rho = 2.83 \times 10^{-5}\ \Omega\ m$
				n-type:
				$S = -149\ \mu V/K$,
				$\rho = 1.25 \times 10^{-5}\ \Omega\ m$
Böttner (2005)	Co-sputtering	48@25 °C	100@25 °C	p-type:
				$S = 180\ \mu V/K$
				$\rho = 1.30 \times 10^{-5}\ \Omega\ m$
				n-type:
				$S = -175\ \mu V/K$
				$\rho = 1.95 \times 10^{-5}\ \Omega\ m$

Table 9.2 Thermoelectric cooling of bulk Bi_2Te_3-based miniaturized TEC

		TE element			
Year	TEC configuration	Thickness (μm)	TE properties	ΔT_{max} (°C)	q''_{max} (W/cm^2)
1967	Single TE couple	130	$ZT = 0.54$	38 @30 °C	95
1994	20 couple TECs	100	$ZT = 0.78$	50 @30 °C	100
1997	120 couple TECs	200	$ZT = 0.78$	67 @30 °C	65
2002	18 couple	200	$ZT = 0.90$	70.6 @30 °C	80
	TECs			91.8 @85 °C	98
2006	18 couple	130	$ZT = 0.90$	64.2 @30 °C	110
	TECs			83.5 @85 °C	132

Fig. 9.15 Thermion TECs with 130 μm thick TE elements

and n-type slices were initially cut from the rods using electroerosion process. Then the slices were lapped to the final thickness of 130 μm, etched electrochemically, and nickel plated.

AlN ceramic substrates were used with metal patterns obtained by standard microelectronics processing, including vacuum deposition of thin-film adhesive layers followed by electrochemical growth of thick copper films through photoresist processing, nickel plating, and finally Tin solder electrodeposition. The modules were found to provide a maximum cooling of 64.2 and 83.5 °C and a maximum cooling flux of 110 and 132 W/cm^2 when operated at 30 °C and 85 °C, respectively, offering somewhat lower temperature reductions than achieved by the miniaturized TEC with 200 μm thick elements but improving the cooling heat flux by 30% under both conditions.

9.2.2.3 Nanostructured Thermoelectric Cooler

Nanostructured low-dimensional materials, such as superlattices, quantum wells, quantum wires, and quantum dots, offer opportunities to manipulate the electron and phonon properties of a given material leading to new ways to increase *ZT*. Ina low-dimensional n-type material, the Fermi level is lower, and the Seebeck coefficient is higher than that for corresponding bulk semiconductors with the same electron concentration, enhancing the value of the product of S^2/ρ, the power factor. Dresselhaus and coworkers theoretically predicted that the use of quantum well nanostructures could increase the power factor via quantum size effects, which improve the electron performance by taking advantage of sharp features in the electron density of states and result in *ZT* values in the range of 2–3 [72]. In addition, thermal conductivity can be reduced due to significantly modified phonon dispersion and enhanced phonon scattering mechanisms using the short period superlattice to impede phonon transport without excessively restricting the carrier flow [73, 74]. Experimental studies have demonstrated significant thermal conductivity reduction in a wide variety of nanostructured superlattices [75], leading to significant enhancements of the thermoelectric figure of merit for Bi_2Te_3/Sb_2Se_3 and PbTe/PbTeSe superlattice nanomaterials [76, 77]. Table 9.3 compares the reported power factor, *ZT*, and thermal conductivity of these nanostructured materials with that of their corresponding bulk materials at room temperature. It is clear that thermal conductivity reduction plays a significant role in the reported *ZT* enhancement, and consequently, there is only a small improvement in the power factor. It is interesting to note that these nanostructured materials can also be used to achieve multilayer thermionic (TI) cooling, [78, 79] which allows for reduced parasitic Joule heating, since transport through the thin barriers is largely ballistic.

Venkatasubramanian used metal-organic chemical vapor deposition (MOCVD) to epitaxially grow a 5 μm thick Bi_2Te_3/Sb_2Te_3 nanostructured superlattice on GaAs substrates in 2001 [81]. These are phonon-blocking/electron-transmitting superlattices, which are produced by alternately depositing thin (1–4 nm) films of Bi_2Te_3 and Sb_2Te_3. *ZT* was reported to be 2.4 for p-type Bi_2Te_3/Sb_2Te_3 superlattices and 1.4 for n-type $Bi_2Te_3/Bi_2Te_{2.83}Se_{0.17}$ at room temperature. This high *ZT* was explained by a reduction of the lattice thermal conductivity due to scattering of the phonons at the superlattice interfaces. The maximum cooling of 32.2 and 40 °C was measured using an infrared camera, and the maximum cooling flux of 585 and 700 W/cm^2 was estimated for p-type Bi_2Te_3/Sb_2Te_3 superlattice at the temperatures of 25 °C and 80 °C, respectively, with a response time of only about 5 μs.

More recently, PbSeTe-based quantum dot superlattice grown by molecular beam epitaxy (MBE) was reported by Herman's group for thermoelectric cooling applications [39]. The superlattice thin film with a thickness of approximately 100 μm was grown on BaF_2 substrates. The developed superlattice thin-film n-type PbSeTe/PbTe has a *ZT* of 1.6–2.0 at room temperature and achieved a maximum cooling of 43.7 °C at 700 mA, under vacuum conditions, as shown in Fig. 9.16. It should be noted that theoretically increasing the *ZT* value will increase the maximum temperature differential of the TEC through $\Delta T_{max} = 0.5ZT_c^2$. However, for on-chip hot-spot cooling, *ZT* does not appear to be the relevant figure of merit. It was found that the materials with the same *ZT* will not necessarily provide equal degrees of hot-spot cooling, and, among three material parameters which determine *ZT* value, increase of Seeback coefficient is most effective to improve hot-spot cooling than decrease of electrical resistivity and thermal conductivity of the thermoelectric materials [82].

Shakouri and coworkers fabricated thin-film SiGe/Si and SiGeC/Si thermoelectric microcoolers based on superlattice nanostructures using molecular beam epitaxy (MBE) [83–87]. As the SiGe/Si superlattice microcoolers can be monolithically integrated with microelectronic components to achieve localized cooling and temperature control, such devices could provide advantages for on-chip hot-spot cooling. The microcooler structure is based on cross-plane electrical transport, and the main part of the microcooler is a

Table 9.3 Thermoelectric properties of nanostructured materials with high *ZT* [80]

Thermoelectric properties	PbTe–PbSeTe quantum dot	PbTe–PbSe	Bi_2Te_3–Sb_2Te_3	Bi_2Te_3–Sb_2Te_3
At 25 °C	Superlattices	Bulk alloy	Superlattices	Bulk alloy
Figure of merit *ZT*	1.6	0.35	2.4	1.0
Thermal conductivity	0.6	2.5	0.5	1.45
(W/m K) power factor (mW/K^{2}m)	3.2	2.8	4.0	5.0

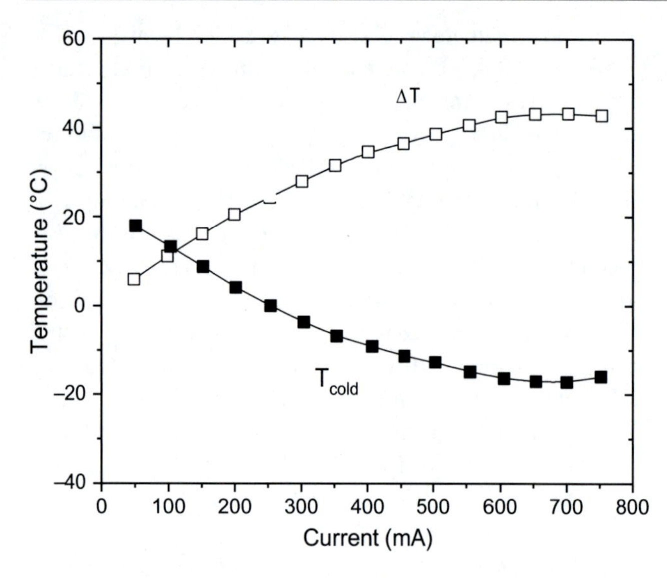

Fig. 9.16 Thermoelectric cooling characteristics of one-leg device made from n-type PbSeTe/PbTe superlattice thermoelectric cooler. ΔT represents measured data points of temperature differential between the hot junction temperature T_{hot} and T_{cold} cold junction temperature

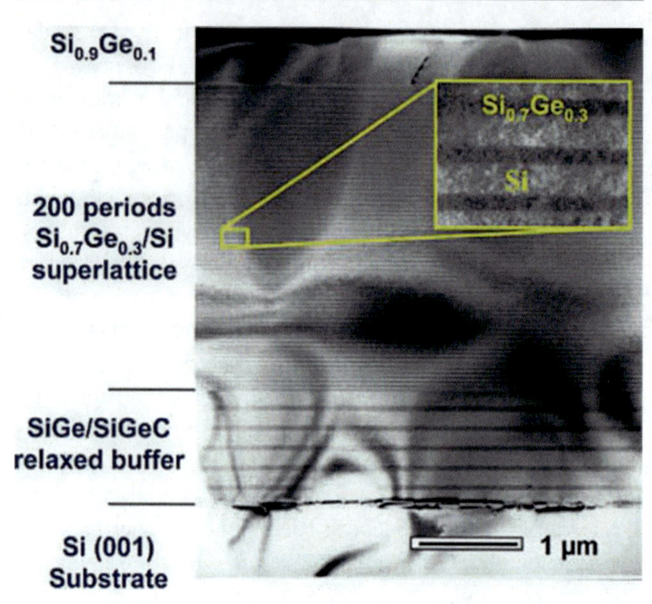

Fig. 9.17 Transmission electron micrograph of 3 μm thick $Si_{0.7}Ge_{0.3}$/Si nanostructured superlattice grown on a buffer layer

Table 9.4 Summary of cooling performance of nanostructured superlattice TEC

Year	Superlattice	ΔTmax (K)	$q$max (W/cm^2)	Properties
LaBounty (2000)	InGaAs/InGaAsP	1.2@25 °C 2.3@90 °C	N/A	N/A
Fan (2002)	SiGe/Si SiGeC/Si	4.5@25 °C 7.0@100 °C 14.0@250 °C	680@25 °C	$S = 200$ μV/K $k = 6.8$–8.7 W/m K $P = 2.2$ mW/K^{2}m $ZT = 0.085$
Venkatasubramanian (2001)	Bi2Te3/Sb2Te3	32.2@25 °C 40@80 °C	585@25 °C 700@80 °C	$ZT = 2.4$ $P = 4.0$ mW/K^{2}m
Herman (2002)	PbSeTe/PbTe	43.7 @25 °C	N/A	$ZT = 1.6$ $P = 3.2$ mW/K^{2}m
Zhang (2003)	AlGaAs/GaAs	0.8@25 °C 2.0@100 °C	N/A	N/A

3 μm thick strain-compensated SiGe/Si superlattice, as shown in Fig. 9.17. It consists of 200 periods of 12 nm $Si_{0.9}Ge_{0.1}$/3 nm Si, doped with boron to about 6×10^{19} cm^{-3}. A maximum cooling of 4.5 °C at 25 °C, 7 °C at 100 °C, and 14 °C at 250 °C was demonstrated. The maximum cooling heat flux increases with decreasing microcooler size, increasing from 120 to 680 W/cm^2 when the microcooler sizes reduce from 100 × 100 μm to 60 × 60 μm. Table 9.4 is the summary of cooling performance of thin-film nanstructured superlattice microcoolers developed since 2000.

9.2.2.4 Silicon Thermoelectric Materials and Microcoolers

While single-crystal silicon has been the key semiconductor material for much of the microelectronics era, silicon's thermoelectric potential has been largely ignored because of its high thermal conductivity and thus low value of the traditional TEC figure of merit, ZT (≈ 0.017) [88, 89]. However, silicon thermoelectric microcoolers, when formed on the back of the silicon chip for hot-spot cooling, provide unique advantages over TFTECs. As may be seen in Table 9.5, which provides the thermal and electrical properties for three conventional thermoelectric materials – bulk Bi_2Te_3 alloy bulk SiGe alloy and single-crystal silicon – at room temperature, single-crystal silicon appears to offer the highest power factor of the materials shown, due to its high Seebeck coefficient and low electrical resistivity, and thus constitutes a very viable candidate for high-flux on-chip cooling.

Zhang and Shakouri developed a silicon thermoelectric microcooler using bulk silicon which is p-type boron doped at a doping concentration of around 10^{19} cm^{-3} [91]. The device structure, which was fabricated with standard microfabrication techniques: dry etch, lithography, metal

Table 9.5 Typical values on the thermoelectric properties for Bi_2Te_3, SiGe, and single-crystal silicon at room temperature [90]

Material	Seebeck coefficient, S (μV/K)	Electrical resistivity, ρ (μΩ m)	Thermal conductivity, k (W/m K)	Figure of merit	Power factor P (mW/K^{2}m)
Bi2Te3 (n-type)	−240	10	2.02	$Z = 2.85 \times 10^{-3}$ $ZT = 0.86$	5.76
Bi2Te3 (p-type)	162	5.5	2.06	$Z = 2.32 \times 10^{-3}$ $ZT = 0.70$	4.77
SiGe (n-type)	−136	10.1	4.45	$Z = 0.328 \times 10^{-3}$ $ZT = 0.1$	1.83
SiGe (p-type)	144	13.2	4.80	$Z = 0.413 \times 10^{-3}$ $ZT = 0.12$	1.57
Silicon (p-type)	450	35	150	$Z = 0.039 \times 10^{-3}$ $ZT = 0.012$	5.79

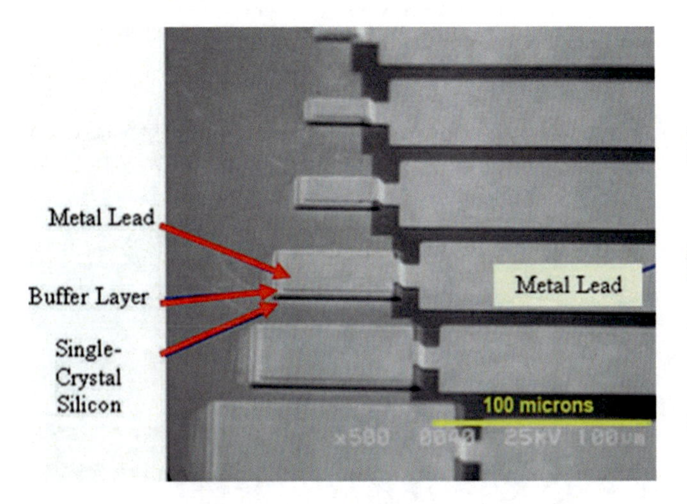

Fig. 9.18 A SEM photo of silicon microcooler [91]

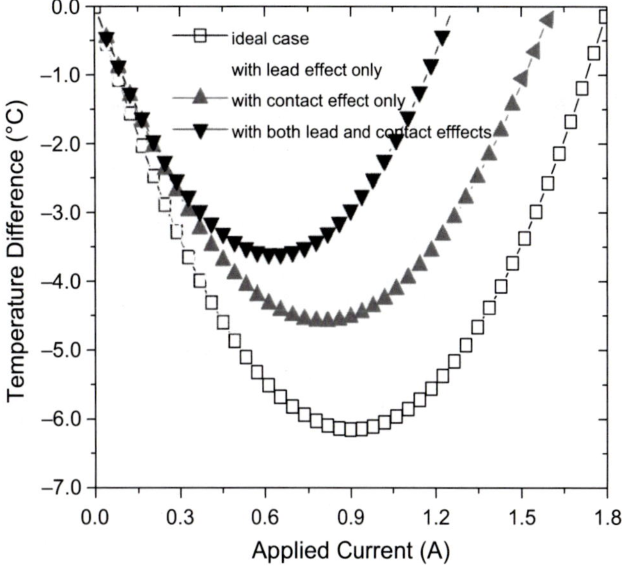

Fig. 9.19 Variation of silicon microcooler performance with applied current

evaporation, etc., is illustrated in Fig. 9.18. They experimentally demonstrated the ability of such silicon thermoelectric microcoolers to achieve a surface temperature reduction of 1.2 °C and a cooling flux of 580 W/cm^2 for a 40 μm × 40 μm microcooler operating at room temperature at an optimized current of 0.1 A. In anticipation of the application of these microcoolers to the thermal management of microprocessor hot spots, and using an analytical model – with the embedded temperature dependence of electrical resistivity, thermal conductivity, and Seebeck coefficient based on values reported for single-crystal silicon [92, 93], Bar-Cohen et al. predicted the achievable temperature reduction, cooling heat flux, and parametric sensitivities of such thermoelectric microcoolers at 100 °C [94]. The results displayed in Fig. 9.19 reveal that, in the absence of parasitic effects – Joule heating from electric contact resistance and heat conduction from metal lead, the silicon microcooler with the described configuration could achieve a maximum temperature reduction of 6.2 °C on the microcooler at the optimum current of 0.9 A.

Figure 9.20 shows the maximum attainable temperature reduction on the microcooler, at an operating temperature of 100 °C, for various doping concentrations with the microcooler size ranging from 20 × 20 μm to 100 × 100 μm. The highest maximum temperature reduction of 6.2 °C is achieved at a doping concentration of 2.5 × 10^{19} cm^{-3} and is independent of microcooler size. However, as shown in Fig. 9.20b, smaller microcoolers do achieve the optimal performance at lower currents. In Fig. 9.20a, the maximum average temperature reduction over the entire microcooler surface (average cooling) is also included for comparison. It is found that the maximum average cooling is approximately 30% lower than the maximum peak cooling.

One of the main advantages of silicon microcoolers is the very high cooling heat flux made possible by the high power factor for silicon. As with any thermoelectric cooler, the maximum cooling flux is achieved at a negligibly small temperature reduction, while the greatest temperature reduction is achieved with negligibly small heat flux. For the present microcooler configuration, Fig. 9.21 shows that the maximum cooling heat flux attains a predicted maximum value of 1 kW/cm^2 for 100 × 100 μm microcooler and

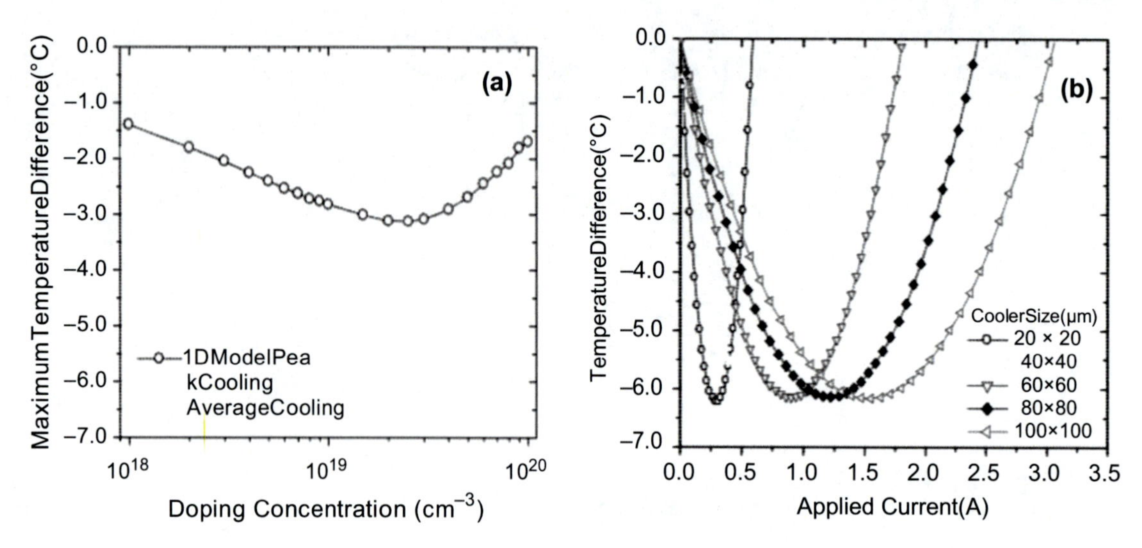

Fig. 9.20 (**a**) Variation of maximum temperature difference with the doping concentration for the ideal case and (**b**) Dependence of temperature difference on the applied current for different microcooler sizes at 100 °C

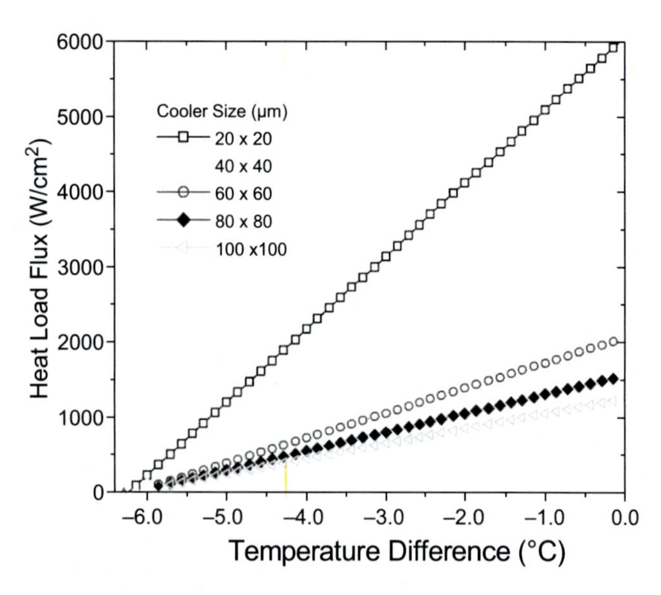

Fig. 9.21 Variation of heat load flux with temperature difference on the microcooler at 100 °C

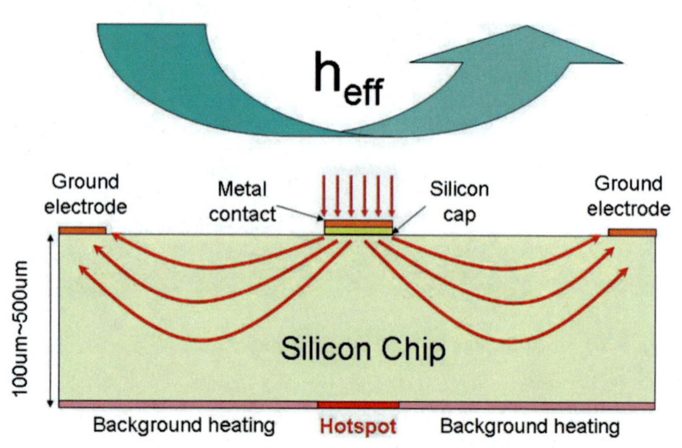

Fig. 9.22 Silicon thermoelectric microcooler for on-chip hot-spot cooling (The *arrows* indicate the direction for electric current)

6 kW/cm^2 for 20 × 20 μm microcooler. These results support the expectation that silicon microcoolers provide a very promising approach to high-heat flux spot cooling in silicon microprocessors.

9.2.3 Hot-Spot Cooling Using Silicon Thermoelectric Microcooler

The concept of silicon thermoelectric microcooler for on-chip hotspot cooling, fabricated on the back of the silicon chip, is illustrated in Fig. 9.22, which displays a single microcooler, activated by an electric current entering the silicon chip through the metal contact and the silicon cap, flowing laterally through the chip, and exiting at the ground electrode located on the periphery of the chip. In a thermoelectric circuit, the flow of electrons across the interface between dissimilar materials, each with a distinct Seebeck coefficient, induces the Peltier effect, providing localized cooling when the direction of the current flow is from the low Seebeck coefficient to the high Seebeck coefficient material. The flow of electric current also serves to transport the absorbed heat away from that junction and to deposit that heat at a secondary interface where the electric current flows from the high Seebeck coefficient to the low Seebeck coefficient material. Joule heating, associated with the resistance to electric current in the thermoelectric circuit, and heat conduction from the hot junction to the cold junction of the thermoelectric circuit limit the thermoelectric cooling that can be achieved. The possible use of silicon thermoelectric microcoolers for the remediation of on-chip hot spots is facilitated by the use of well-established metal-on-silicon fabrication techniques, yielding a very low thermal contact resistance between the electrodes and the chip. In addition, incorporation of the silicon chip into the thermoelectric circuit makes it possible to transfer the absorbed energy via the electric current to the edge of the chip, far from the location of the hot spot, thus substantially reducing the detrimental effect of thermoelectric heating on the temperature of the active circuitry.

Referring to the structure of the on-chip silicon microcooler depicted in Fig. 9.22, it may be seen that Peltier cooling occurs at the junction between the metal contact and the silicon cap which is highly doped silicon with a doping concentration of more than 1×10^{20} cm^{-3} and again at the silicon cap/silicon chip interface and that Peltier heating is encountered at the silicon chip/ground electrode interface, located on the periphery of the chip, where the electrons must shed some of their energy in entering the highly conductive metal. The overall Peltier heat transfer (cooling) rate of the silicon microcooler can be expressed as:

$$q_{\mathrm{TE \times c}} = -S_{\mathrm{Si}} T_{\mathrm{c}} I \tag{9.20}$$

where Tc is the absolute temperature at the microcooler, SSi the Seebeck coefficient of the silicon chip, and I the applied current. Similarly, the Peltier heating rate at the silicon chip/ground electrode interface can be represented as:

$$q_{\mathrm{TE,h}} = S_{\mathrm{Si}} T_{\mathrm{ed}} I \tag{9.21}$$

where T_{ed} is the absolute temperature at the ground electrode. In addition to volumetric Joule heating inside the silicon chip, the silicon cap, and the metal contact, these parasitic effects also arise at both the metal contact/silicon cap interface and the silicon chip/ground electrode interface. The interfacial Joule heating at the metal contact/silicon cap interface is given by:

$$q_{\mathrm{contact}} = I^2 R_{\mathrm{cont}} = I^2 \rho_{\mathrm{c}} / A_{\mathrm{cont}} \tag{9.22}$$

where R_{cont} is the electric contact resistance, A_{cont} the cross-sectional area of metal contact, and ρ_{c} the specific electric contact resistance at this interface. Equation (9.22) applies as well at the peripheral ground electrode/silicon chip interface, with the appropriately adjusted contact area and the specific electric contact resistance.

In the work of Bar-Cohen [95], a test vehicle was used, consisting of a 12 × 12 mm silicon chip with 70 W/cm^2 background heat flux and 70 × 70 μm hotspot, with a heat flux of 680 W/cm^2, located at the center of the active side of the chip, to explore the hot-spot cooling potential of this approach. In their extensive modeling studies, the back of the chip was cooled with a high-performance 25 °C air-cooled heat sink, capable of producing an effective heat-transfer coefficient of 8700 W/m^2 K representing the combined effect of the heat sink, heat spreader, and thermal interface materials used for electronic packages [96, 97]. The thermal conductivity of the silicon chip is assumed to be 110 W/m K, appropriate for 100 °C operating temperature [98].

The prediction of hot-spot cooling achievable with on-chip silicon microcooler, as described in Fig. 9.23, requires the solution of the three-dimensional Poisson's equation for the temperature distribution in a volume subjected to nonuniform heat generation, associated with the Joule heating in the silicon chip, heating and cooling boundary conditions, associated with Peltier cooling and Peltier heating on the back surface, along with the microprocessor heat generation on the front surface (active circuitry), i.e.,

$$\frac{\delta^2 T}{\delta x^2} + \frac{\delta^2 T}{\delta y^2} + \frac{\delta^2 T}{\delta z^2} + \frac{q'''_{\mathrm{Si}}(x, y, z)}{k_{\mathrm{Si}}} = 0 \tag{9.23}$$

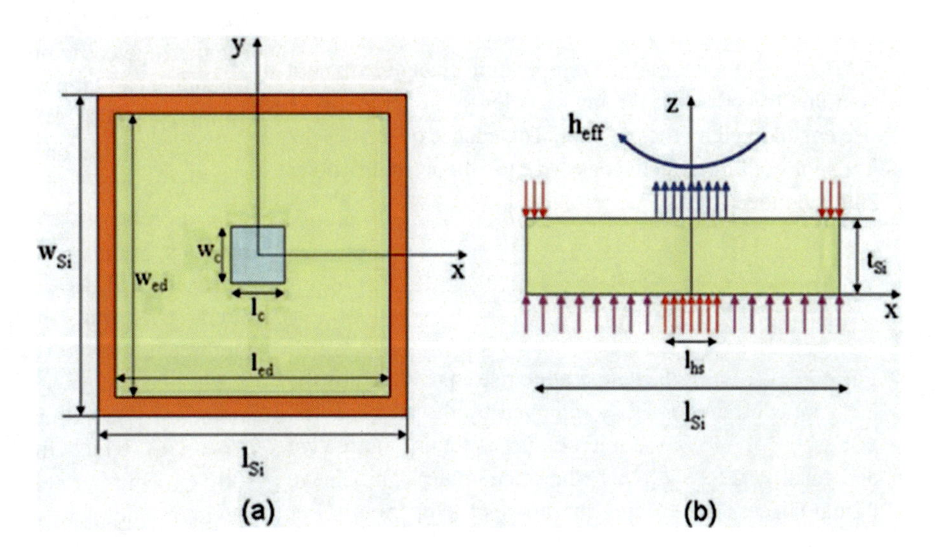

Fig. 9.23 (a) Coordinate system and (b) boundary conditions in the analytical model for the silicon chip integrated with silicon thermoelectric microcooler

where $q'''_{Si}(x,y,z)$ is the nonuniform volumetric heat generation due to the silicon Joule heating, and k_{Si} is the thermal conductivity of the silicon chip. Unfortunately, solution of Eq. (9.23) requires detailed knowledge of the internal heat generation function, $q'''_{Si}(x,y,z)$. Determination of this function requires a parallel solution of the Laplace's equation for the electric potential field, which will vary significantly with the geometries of the silicon chip and the silicon microcooler and the placement of the ground electrode. The resulting highly nonuniform heat generation function can be expected to render Eq. (9.23) analytically unsolvable for all but the simplest approximations of $q'''_{Si}(x,y,z)$.

Alternatively, considering the common use of "allocation factors" in determining the performance of one-dimensional thermoelectric devices [99] and the successful application of this approach to silicon thermoelectric microcoolers in an earlier publication by the authors, it is possible to reformulate Eq. (9.23) in the Laplace's form by allocating an appropriate fraction of the Joule heating to the microcooler (α) and the hot spot (β), respectively. With this approach, the volumetric silicon Joule heating is replaced with modified boundary conditions at the microcooler and the hot spot, respectively, and the Poisson's equation can then be transformed into the Laplace's equation for this same domain, which can be solved analytically.

In subsequent sections, the hot-spot remediation capability of silicon microcoolers will be characterized by three distinct metrics, including:

1. ΔT – the temperature reduction anywhere in the studied domain that is achievable by activating the microcooler. This metric characterizes the intrinsic thermoelectric cooling capability of the silicon microcooler. It is generally applied to the hot spot or the microcooler in this study and given by:

$$\Delta T = T_{\text{cooler,on}} - T_{\text{cooler,off}} \tag{9.24}$$

2. $\Delta T_{\text{hot spot}*}$ – the ratio of the temperature change at the hot spot due to activating the microcooler to the temperature rise engendered by the hot spot. This metric quantifies the hot-spot cooling effectiveness of the silicon microcooler and is defined as:

$$\Delta T^*_{\text{hot spot}} = \frac{T_{\text{hot spot on,cooler off}} - T_{\text{hot spot on,cooler on}}}{T_{\text{hot spot on,cooler off}} - T_{\text{hot spot off,cooler off}}} \tag{9.25}$$

For $\Delta T_{\text{hot spot}*} = 1$, the temperature rise engendered by the hot spot can be completely removed by the microcooler. For $\Delta T_{\text{hot spot}*} = 0$, the microcooler is totally ineffective and for $0 < \Delta T_{\text{hot spot}*} < 1$, the microcooler can achieve partial success in reducing the hot-spot temperature. For $\Delta T_{\text{hot spot}*} > 1$, the micro-cooler is capable of "overcooling" the hot spot relative to the base temperature of the silicon chip.

3. π – the thermal impact factor, which provides a measure of the power needed, P_{in}, to achieve a specified temperature reduction at the hot spot, $\Delta T_{\text{hot spot}*}$. This dimensional metric (K/W$_{\text{elec}}$) can be expressed as:

$$\pi = \frac{-\Delta T_{\text{hot spot}}}{q_{\text{in}}} \tag{9.26}$$

Clearly, as π increases less electric power is required in order to achieve a specific temperature reduction at the hot spot.

Following the transformation above, Bar-Cohen and coworkers developed a three-dimensional analytical thermal model of on-chip hot-spot cooling to investigate the effectiveness of such silicon thermoelectric microcoolers for a wide range of hot-spot sizes and heat fluxes, microcooler sizes, silicon chip thicknesses, doping concentrations, and electric contact resistances. The analytical solution yields the temperature distribution in the silicon chip, under the influence of hot-spot heating and background heating from related circuitry on the active surface, Peltier cooling, Peltier heating and conductive/convective cooling on the opposite surface, volumetric Joule heating inside the silicon chip, and interfacial Joule heating at the electric contact created by the silicon microcooler. The analytical solution employs numerically derived allocation factors to redistribute the Joule heating inside the chip to the hot spot and the microcooler. Results obtained from a three-dimensional electrothermal finite-element numerical simulation were used to validate and calibrate the analytical model. The parametric trends revealed by use of this analytical model are presented and discussed in subsequent sections.

9.2.3.1 Doping Concentration Effect

The thermoelectric properties of semiconductors are strongly dependent on doping concentration but modestly on the doping type [100–103]. Figure 9.24 shows that the electrical resistivity of silicon decreases with increasing doping concentration, while the Seebeck coefficient also displays an inverse relationship with doping concentration. Thus, increasing doping concentration results in lower electrical resistivity and, as a consequence, less Joule heating in the silicon chip, but, the associated decrease in the Seebeck coefficient leads to reduced thermoelectric cooling power.

The largest possible thermoelectric cooling power is attained by maximizing the thermoelectric power factor P $(= S^2\rho)$, which for boron-doped single-crystal silicon at 100 °C occurs at about 2.5×10^{19} cm^{-3}. The variation of maximum hot-spot cooling with doping concentration for

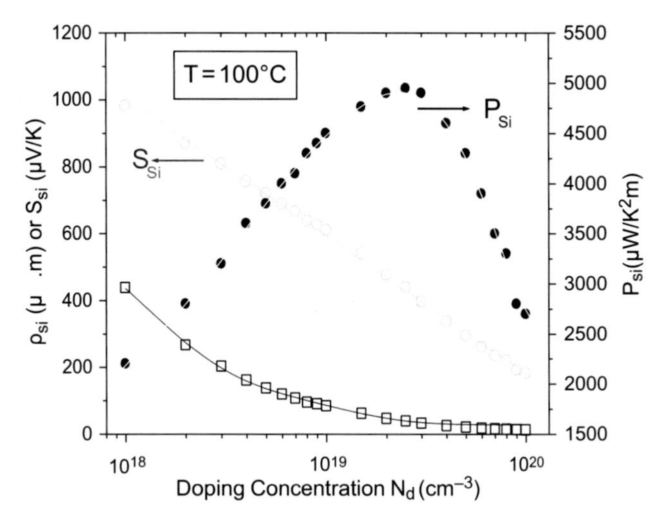

Fig. 9.24 Electrical resistivity (ρ), Seebeck coefficient (S), and power factor (P) as a function of boron doping concentration (N_d) in single-crystal silicon [104–106]

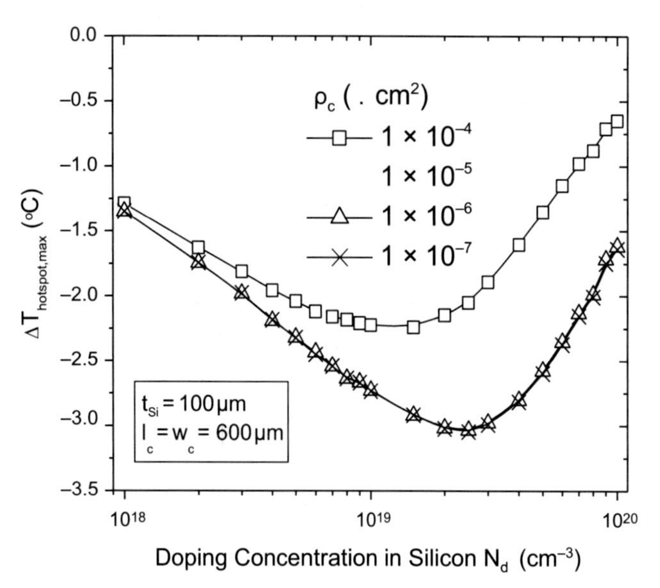

Fig. 9.25 Hot-spot cooling as a function of boron doping concentration for various specific electric contact resistances. The hot spot is 70 × 70 μm with a heat flux of 680 W/cm² and the microcooler size is 600 × 600 μm. The microcooler size is 600 × 600 μm

various microcooler sizes is presented in Fig. 9.24 for 100 μm thick chip and the specific electric contact resistance ranging from 1×10^{-7} to 1×10^{-4} Ω cm², revealing – as expected – that across the range of microcooler sizes studied, with increasing doping concentration, the hot-spot cooling increases until reaching a maximum value and then decreases with further increases in the doping concentration.

It is interesting to find that, despite the three-dimensional characteristic of heat spreading and electrical current spreading in the silicon chip surrounding the micro-cooler, for small electric contact resistance, e.g., $\rho_c < 1 \times 10^{-5}$ Ω cm², the optimum doping concentration is nearly equal to 2.5×10^{19} cm^{-3}, which yields the maximum power factor shown in Fig. 9.24. However, it is to be noted that the parasitic effect from larger electric contact resistance does have an influence on the optimum doping concentration, yielding a lower optimized doping concentration of 1.5×10^{19} cm^{-3} for a 600 × 600 μm microcooler with the specific electric contact resistance of 1.0×10^{-4} Ω cm². It has been found that this trend becomes more pronounced as the microcooler size gets smaller (Fig. 9.25).

It should be noted that the optimum doping level for a silicon thermoelectric microcooler is, thus, likely to be substantially higher than commonly used in semiconductor silicon chips. However, as is almost always the case for chip thermal management, the present analysis assumes that the back of the chip is used for cooling while the front is used for the active circuitry. Consequently, the doping concentration on the back side of the chip need not equal the more common doping concentration in the active semiconductor regions at the front of the chip, e.g., 1×10^{16} cm^{-3}. Due to the far higher electrical resistivity in the active silicon layer, the electric current that is used to activate thermoelectric cooling is not expected to penetrate into this region.

9.2.3.2 Microcooler Size Effect

The effect of microcooler size on cooling performance involves the interplay of thermoelectric cooling by the microcooler and thermal diffusion from the hot spot to the microcooler. With decreasing microcooler size, the effective cooling flux and thus the temperature reduction at the microcooler increases, while the thermal resistance between the hot spot and the microcooler also increases. Consequently, this larger cooling flux at smaller microcoolers cannot effectively translate into larger temperature reduction at the hot spot. On the other hand, with smaller thermal resistances between the hotspot and the microcooler, the more modest cooling flux on larger microcoolers can be projected effectively onto the hot spot, narrowing the temperature difference between the hot spot and the microcooler. However, the modest cooling flux achievable on the larger microcoolers reduces the beneficial temperature reduction at both the hot spot and the microcooler. The competition between these two effects results in an optimum microcooler size.

Figure 9.26 displays this behavior and shows the temperature reductions at the hot spot and the microcooler for a wide range of microcooler sizes for 100 μm thick chip operating under the background and hot-spot heat fluxes of 70 W/cm² and 680 W/cm², respectively. For each microcooler, the applied current is carefully optimized in order to achieve the maximum temperature reductions at the hot spot. It is

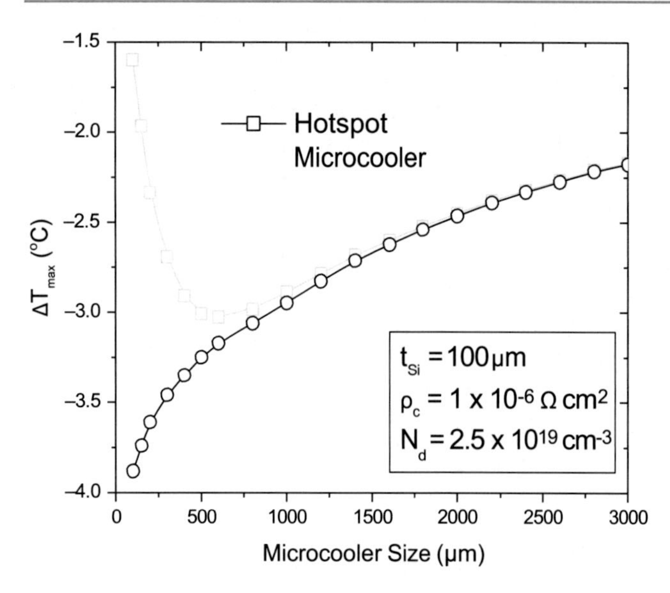

Fig. 9.26 Variation of temperature reductions at the hot spot and the microcooler with microcooler size. The hotspot is 70 × 70 µm with a heat flux of 680 W/cm^2

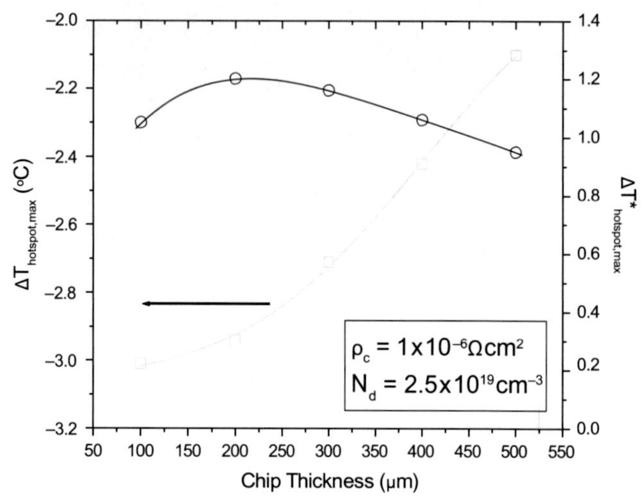

Fig. 9.27 Hot-spot cooling and hot-spot cooling effectiveness as a function of chip thickness. The hot spot is 70 × 70 µm with a heat flux of 680 W/cm^2

seen that the temperature reduction at the hot spot first increases with microcooler size and, after reaching the maximum value of 3.0 °C for 600 × 600 µm microcooler, decreases with a further increase in microcooler size. Interestingly, the temperature reduction at the microcooler varies monotonically with microcooler size, yielding progressively larger temperature reductions, to as much as 3.9 °C, as the microcooler dimension shrinks to 100 × 100 µm, which, however, only provides 1.6 °C temperature reduction at the hot spot.

9.2.3.3 Chip Thickness Effect

In the application of silicon microcoolers to hot-spot remediation, the silicon chip plays multiple roles, functioning as a thermoelectric material, to provide on-chip cooling and, at the same time, as an electrical conductor to transfer electrons from the ground electrode to the microcooler, and as a thermal conductor to provide a diffusion path for the heat generated in the chip to the ambient. Therefore, the chip thickness influences Joule heating distribution inside the chip, heat spreading from the hot spot, heat diffusion from the hot spot to the microcooler, and heat diffusion from the ground electrode, where Peltier heating occurs, to the hot spot. As the chip becomes thinner, the thermal resistance between the microcooler and the hot spot decreases, allowing the microcooler to achieve greater hot-spot temperature reductions, e.g., 2.05–3.03 °C as the chip thickness decreases from 500 to 100 µm, for the conditions of Fig. 9.28. However, due to the smaller heat spreading effect in thinner chips, the temperature rise engendered by the hot spot is also higher for thinner chips and increases with decreasing chip thickness from 2.2 °C for a 500 µm thick chip to 2.9 °C for 100 µm thick chip. These two trends compete with each other, yielding the maximum hot-spot cooling effectiveness at the chip thickness of 200 µm, with $\Delta T_{\text{hot spot}*} = 1.2$ as shown in Fig. 9.27. At this chip thickness, the silicon microcooler is, thus, capable of reducing the hot-spot temperature below the baseline temperature of the chip by approximately 0.5 °C. Moreover, for the present 70 × 70 µm hot spot with a heat flux of 680 W/cm^2, the silicon microcooler is capable of completely suppressing or overcooling the hot spot, with $\Delta T_{\text{hot spot}*} \geq 1$, for the chip thicknesses between 100 and 475 µm.

It is interesting to find that the chip thickness also influences the optimum micro-cooler size. As shown in Figs. 9.28 and 9.29, the thicker the silicon chip, the larger the optimized microcooler size. For 100 µm thick chip, the maximum hot-spot cooling of 3.0 °C, for the conditions studied, is achieved with a 600 × 600 µm microcooler, while for 500 µm thick chip, 2500 × 2500 µm silicon microcooler is required in order to attain the maximum hot-spot cooling of 2.1 °C. It should be noted that the optimum ratio of microcooler size to chip thickness is, thus, approximately 5.5, with a modest sensitivity to chip thickness, reaching 6.0 for 100 µm, 200 µm and 300 µm thicknesses, dropping to 5.5 for the 400 µm thick chip, and to 5.0 for the 500 µm chip. This decreasing ratio can be related to the growing contributions of silicon Joule heating and Peltier heating to the hot-spot temperature as the optimized current – necessitated by the larger microcooler – increases.

Figure 9.30 shows the dependence of the thermal impact factor, π, of the silicon microcooler, on the microcooler size for different chip thicknesses, revealing that this factor and the relative benefit of input power decrease steeply with the microcooler size but more gently with the chip thickness. For

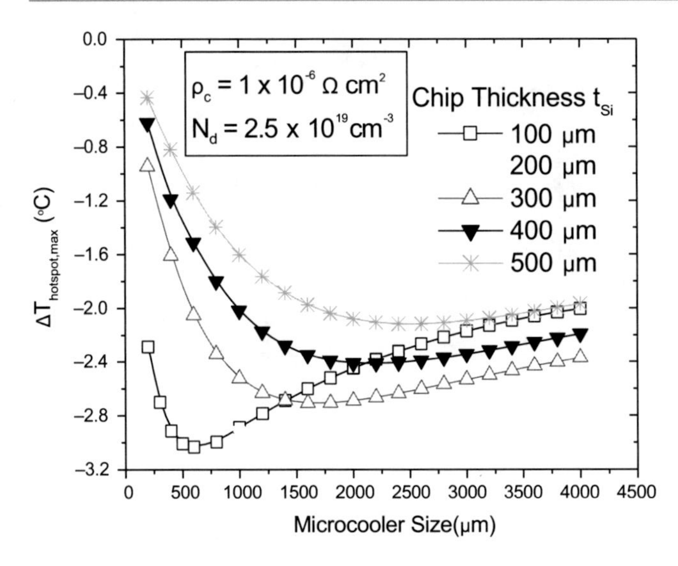

Fig. 9.28 Hot-spot cooling as a function of microcooler size for various chip thicknesses. The hot spot is 70 × 70 μm with a heat flux of 680 W/cm²

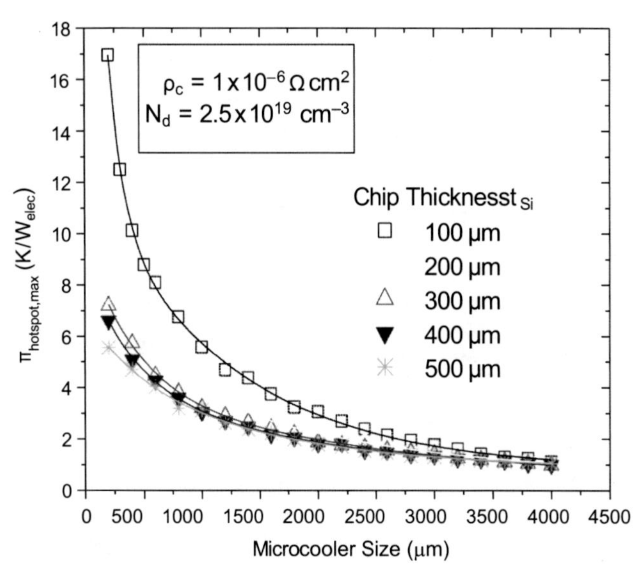

Fig. 9.30 Thermal impact factor as a function of microcooler size for various chip thicknesses. The hot spot is 70 × 70 μm with a heat flux of 680 W/cm²

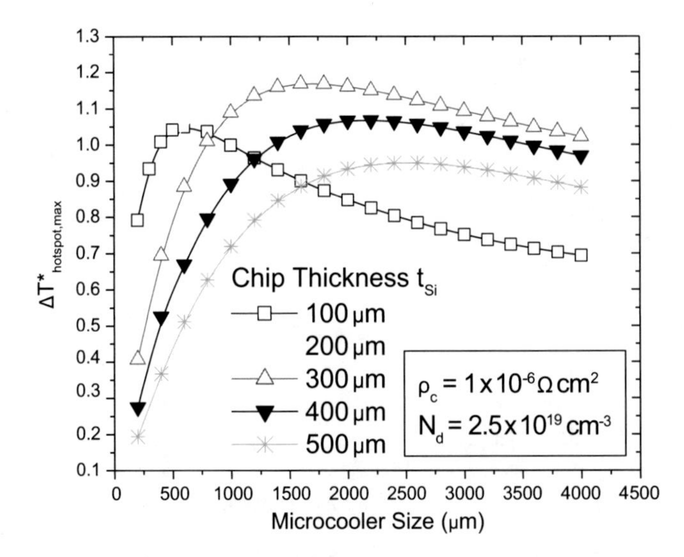

Fig. 9.29 Hot-spot cooling effectiveness as a function of microcooler size for various chip thicknesses. The hotspot is 70 × 70 μm with a heat flux of 680 W/cm²

example, 200 × 200 μm microcooler can achieve a π of 17.0 in comparison with 1.2 for 4000 × 4000 μm microcooler on 100 μm thick chip. With an increase of micro-cooler size, the effect of chip thickness on the thermal impact factor becomes less important. Consequently, this example for the specific parameters of the "test vehicle" used to explore the cooling potential of the silicon microcoolers suggests that more generally the largest π values and the best returns on invested energy are attained when smaller microcoolers are used to remediate hot spots on thinner chips.

9.2.3.4 Electric Contact Resistance Effect

The miniaturization of thermoelectric coolers tends to exacerbate the deleterious effects of the electric contact resistance, which is expected to occur at the interface between the metal contact and the silicon cap. The theoretical value of the specific electric contact resistance between highly doped silicon and a metal contact is in the range of 1×10^{-9} Ω cm^2 at room temperature or above [104]. However, due to process-related limitations, the typical specific electric contact resistance at such an interface usually ranges from 1×10^{-7} to 1×10^{-5} Ω cm^2, with significant batch to batch variations. Figure 9.31 shows the impact of the electric contact resistance on hot-spot cooling for different microcooler sizes on 100 μm thick chip. It should be noted that for a typical state-of-the-art thin-film process, which yields an average specific electric contact resistance of approximately 1×10^{-6} Ω cm^2 [105], the results displayed in Fig. 9.31 reveal that the electric contact resistance induced degradation in hot-spot cooling can be neglected.

More generally, as the specific electric contact resistance increases, hot-spot cooling performance is degraded, but the electric contact resistance has a larger impact for smaller microcooler sizes because the contact resistance is inversely proportional to the microcooler area. For an increase in the specific electric contact resistance from 1×10^{-9} to 1×10^{-4} Ω cm^2, hot-spot cooling will be degraded by a factor of 6.5 for 100 × 100 μm microcooler but only by 5% for 3000 × 3000 μm microcooler.

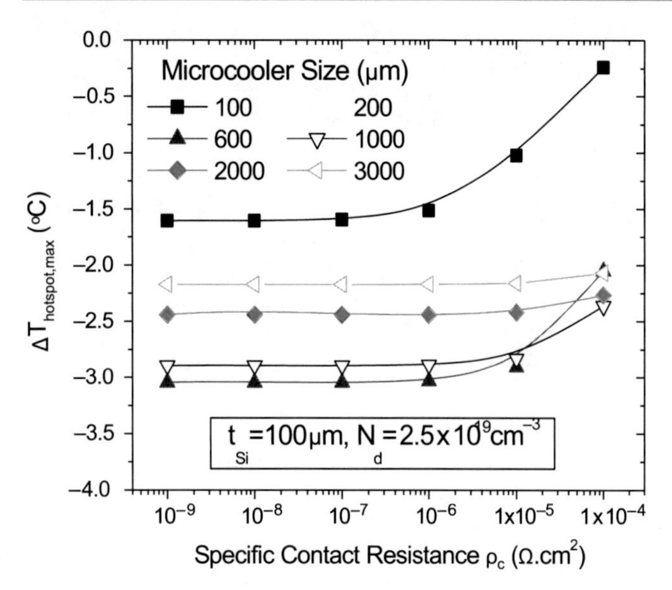

Fig. 9.31 Hot-spot cooling as a function of specific electric contact resistance for various micro-cooler sizes. The hot spot is 70 × 70 μm with a heat flux of 680 W/cm^2

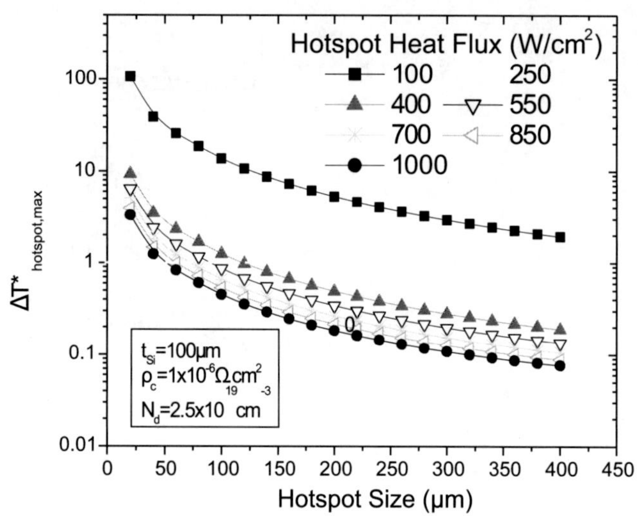

Fig. 9.33 Hot-spot cooling effectiveness as a function of hot-spot size and hot-spot heat flux

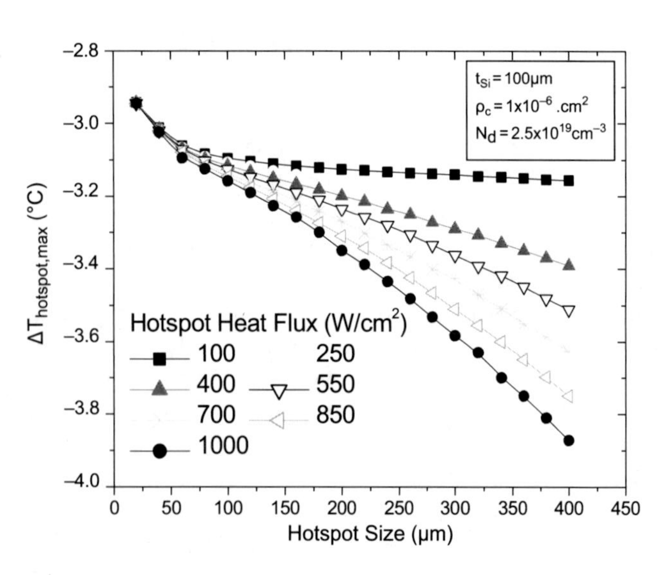

Fig. 9.32 Hot-spot temperature reduction as a function of hot-spot size and hot-spot heat flux

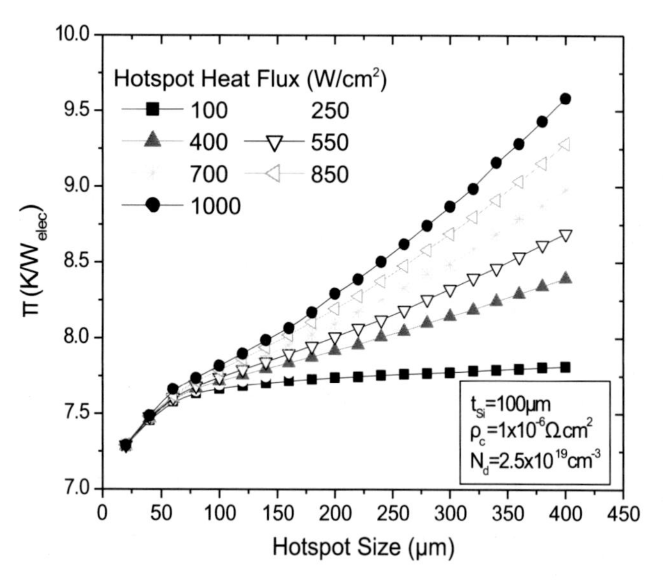

Fig. 9.34 Thermal impact factor as a function of hot-spot size and hot-spot heat flux. The microcooler size is 600 × 600 μm

9.2.3.5 Hot-Spot Parameter Effect

Finally, attention is turned to the effects of the hot-spot parameters – namely, hot-spot size and hot-spot heat flux – on cooling performance, as evaluated by the three proposed metrics – ΔT, ΔT^*, and π. For each hot-spot size and hot-spot heat flux, the applied current, the microcooler size, and the doping concentration have been optimized in order to achieve the maximum hot-spot temperature reduction, while the specific electric contact resistance is fixed at 1×10^{-6} Ωcm^2. It was found that the optimized current and thus the optimized input power increase slightly with hot-spot size and hot-spot heat flux if the chip thickness and the doping concentration in the silicon chip remain constant. As may be seen in Figs. 9.32, 9.33, and 9.34 for 100 μm thick chip, the efficacy of the silicon microcooler varies with these hot-spot parameters in a complex manner. For example, the maximum temperature reduction at the hot spot, shown in Fig. 9.32, increases from 3.0 °C for 70 × 70 μm hot spot with 680 W/cm^2 heat flux to 3.90 °C for 400 × 400 μm hot spot with 1000 W/cm^2 heat flux, primarily because of the effect of the higher chip temperature (105 vs. 150 °C) on the Peltier cooling rate. However, as seen in Fig. 9.33, the maximum cooling effectiveness decreases steeply with hot-spot size and hot-spot heat flux, from 1.05 for 70 × 70 μm hot-spot with 680 W/cm^2 heat flux to 0.08 for 400 × 400 μm hot-spot with

1000 W/cm^2 heat flux. Interestingly, since as the hot-spot size and the hot-spot heat flux increases, the maximum hot-spot temperature reduction increases, while the optimized input power remains nearly constant, π, the thermal impact factor, increases with the hot-spot size and the heat flux, as shown in Fig. 9.34. It should, thus, be understood that the silicon microcoolers can produce the largest cooling effect for constant input power when encountering large, high heat flux hot spots.

9.2.4 Mini-Contact-Enhanced TEC for Hot-spot Cooling

Solid-state thermoelectric coolers (TECs), which are highly reliable, can be locally applied for spot cooling, and can be integrated with IC processing, have been proposed for hot-spot thermal management. However, the relatively low cooling heat flux, 5–10 W/cm^2, of conventional TEC modules severely limits the application of these devices to hot-spot remediation. Recently, a novel use of a mini-contact pad, which connects the TEC and the silicon chip, thus concentrating the thermoelectric cooling power on the top of the silicon chip to significantly improve hot-spot cooling performance, was proposed and investigated [106]. The physical phenomenon underpinning the use of mini-contact-enhanced thermoelectric coolers is displayed in Fig. 9.35, where the mini-contact is seen to concentrate the thermoelectric cooling power on the reduced cross-sectional area of the mini-contact tip. It can be expected that, to a first approximation, the smaller the cross-sectional area of the mini-contact tip, the larger the cooling flux on the top of the silicon chip. Moreover, the localization of the TEC cooling flux to the region most affected by the hot spot reduces the overall cooling requirements and the input power needed to effectively utilize the TEC.

To analyze and optimize the on-chip hot-spot cooling performance of the mini-contact-enhanced TEC, a three-dimensional numerical thermal model was developed using the commercial finite element software ANSYSTM and applied to a typical chip package with a mini-contact-enhanced TEC, as shown in Fig. 9.36. This package consists of a silicon chip, thermal interface materials (TIM), a copper integrated heat spreader (IHS), an air-cooled aluminum heat sink, and a miniaturized bismuth telluride-based TEC. The TEC is integrated with a mini-contact pad, attached on the top of the silicon chip, and then embedded inside the thermal interface materials, TIM1 and TIM2. These could be different thermal interface materials such as solder or thermal grease. The TEC consists of a 4 × 4 array of 400 × 400 × 20 μm thermoelectric elements that are sandwiched between two 50 μm thick ceramic substrates and is 120 μm in overall height [107]. The copper mini-contact pad features a 2.4 mm × 2.4 mm × 100 μm base to facilitate heat spreading and a 2.4 mm × 2.4 mm × 50 μm tip to concentrate the thermoelectric cooling capability. For the purposes of this study, the thickness of TIM1 was held constant at 300 μm so that the mini-contact-enhanced TEC and the TIM1 layer could be accommodated within the height of the TIM1 thermal interface layer.

When a mini-contact-enhanced TEC is integrated into the chip package, it introduces several thermal interfaces. In consideration of possible assembly procedures for such an enhanced TEC, it is assumed that the two most important thermal interfaces occur between the top ceramic substrate and the TIM2 and between the mini-contact tip and the silicon die, as indicated in Fig. 9.36. R_{c1} and R_{c2} are used to represent these two thermal contact resistances, with each varying from 1×10^{-7} to 1×10^{-4} m^2 K/W, the typical range reported for electronic package application [108, 109]. In the simulation, thermoelectric cooling rate, $Q_{\mathrm{TE\ cooling}}$, is determined as the product of the Seebeck coefficient, temperature, and current, ST_cI, where S is the Seebeck coefficient of the thermoelectric material, T_c the cold-side junction temperature at the TEC, and I the applied electrical current on the TEC. Similarly, the Peltier heating

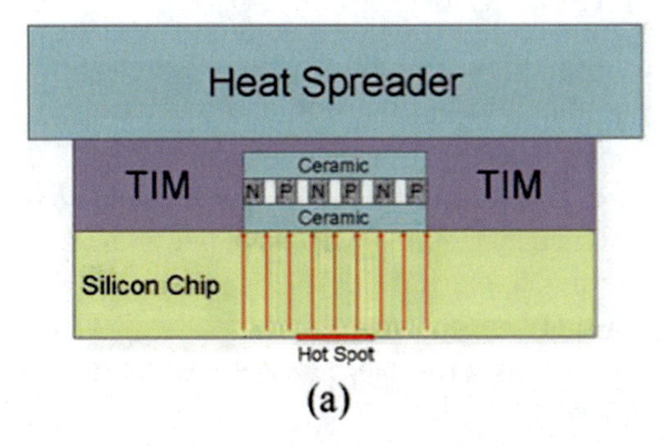

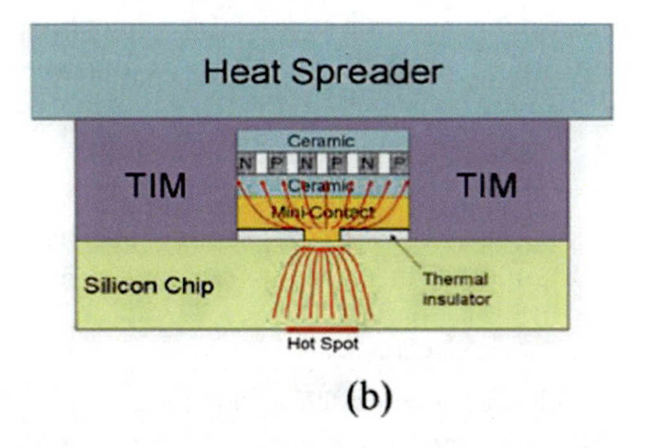

Fig. 9.35 Schematic of a TEC (consisting of numerous N-type and P-type TE elements sandwiched between two ceramic substrates) attached on silicon chip and embedded inside thermal interface material (TIM) of the chip package: (**a**) TEC without a mini-contact and (**b**) TEC with an integrated mini-contact. The arrows indicate the heat flow pattern in silicon chip and mini-contact. The detailed chip package is not shown in this figure

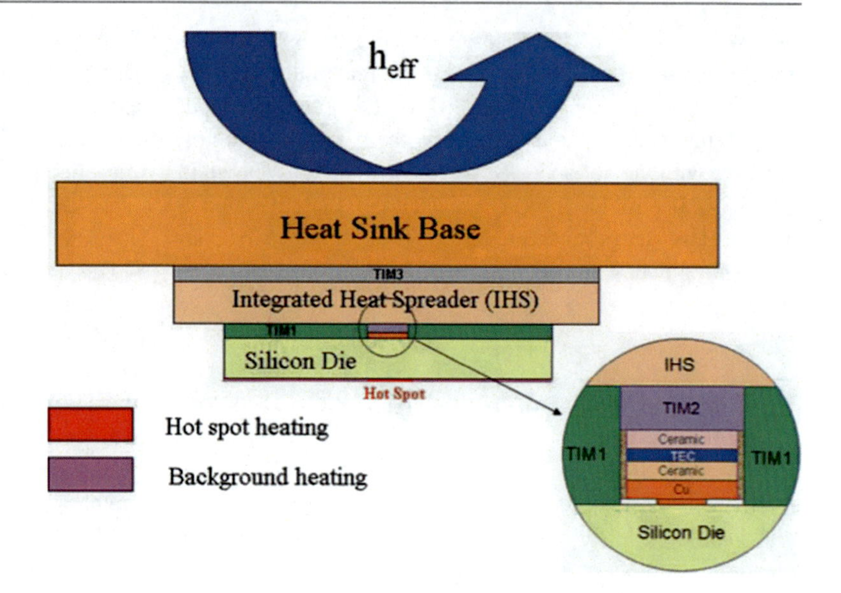

Fig. 9.36 Schematic of a typical chip package with a mini-contact-enhanced TEC

rate at the heat rejection side of the device, $Q_{\mathrm{TE\ heating}}$, is given by the product $ST_{\mathrm{h}}I$, where T_{h} is the hot side junction temperature at the TEC. Thermoelectric cooling and heating effects are represented as heat flux boundary conditions in the numerical simulation and directly added to the cold and hot sides of the TEC, respectively, while Joule heating is modeled as uniform volumetric heat generation inside the bismuth telluride elements. Joule heating from the electrical contact resistance is modeled as a surface boundary condition at the two TEC junctions [110]. Since the hot-spot cooling performance is strongly dependent on the input power supplied to the TEC, in the course of this simulation, various electric currents are applied to the TEC until the lowest hot-spot temperature is achieved, and usually it takes about 10 min to complete a simulation using Pentium IV processor.

9.2.4.1 Effect of Input Power on TEC

Thermoelectric cooling performance is dependent on the applied current or input power to the TEC in a nonlinear manner, as Peltier cooling has a favorable linear dependence on electric current, while the parasitic Joule heating effect has a quadratic dependence on electric current. The competition of these two opposite contributions leads to an optimum current or input power at which the maximum hot-spot cooling, or lowest hot-spot temperature, can be achieved. Figure 9.37 demonstrates the variation of the hot-spot temperature with the input power to the TEC, for a TE element thickness of 20 μm and the mini-contact tip size of 1250 × 1250 μm. The thermal contact resistance is chosen to be 1×10^{-7} Km^2/W at both the mini-contact tip/silicon chip interface and the ceramic/TIM2 interfaces. For a 400 × 400 μm hot spot with a heat flux of 1250 W/cm^2, if there is no TEC, the peak hot-spot temperature is found to reach 137.0 °C. However, if the TEC is activated, the hot-spot temperature decreases steeply as the power increases,

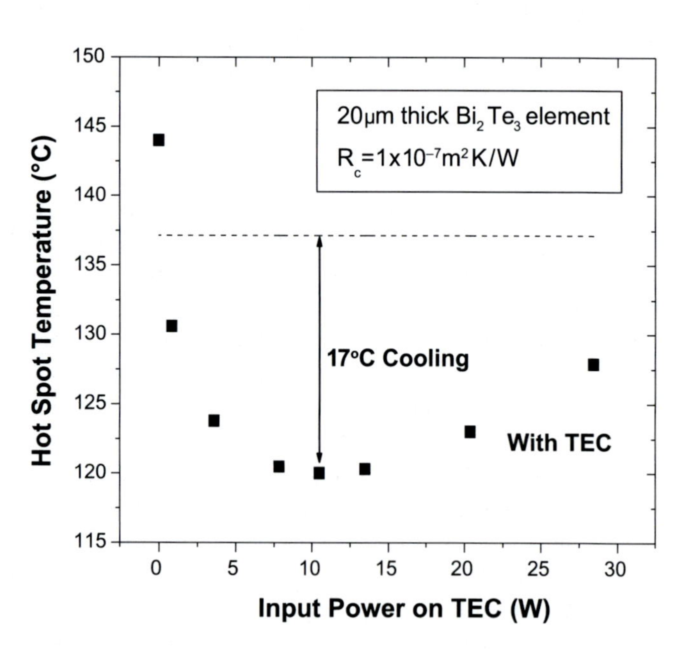

Fig. 9.37 Variation of typical hot-spot temperature with the input power on TEC (Thermoelectric element thickness = 20 μm, copper mini-contact tip size = 1250 × 1250 μm)

reaching a minimum of 120 °C at approximately 10 W, which corresponds to a temperature reduction of 17.0 °C at the hot spot compared to the temperatures encountered without the TEC and then rises slowly as the power increases further. It is worth noting that if the TEC is present but not activated, the hot-spot temperature will increase by 7 °C, due to the additional thermal resistance to heat flow created by the presence of the TEC.

9.2.4.2 Effect of Mini-Contact Size

The mini-contact pad, sandwiched between the TEC and the silicon chip, is used to concentrate the thermoelectric cooling

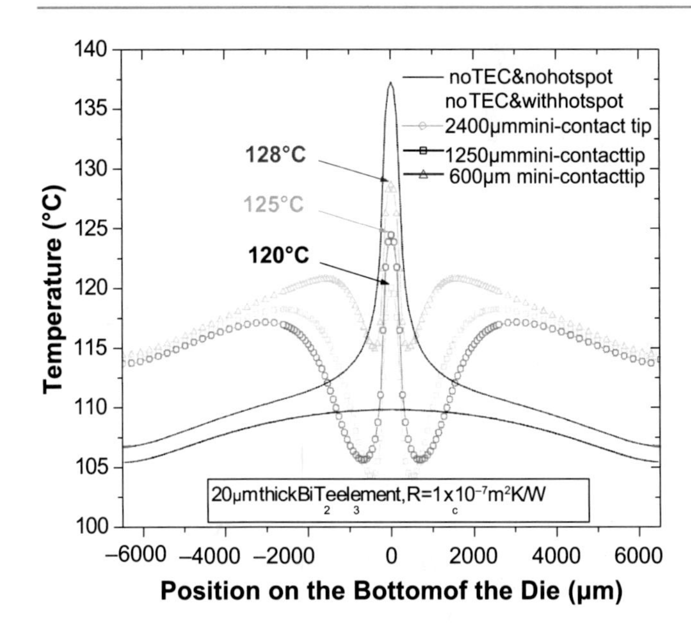

Fig. 9.38 Effect of mini-contact size on TEC-induced temperature profile (Thermoelectric element thickness = 20 μm, input power = 10 W)

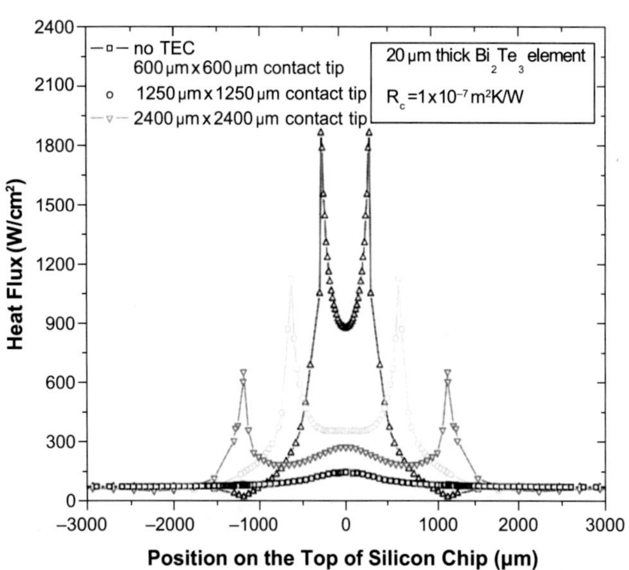

Fig. 9.39 Effect of mini-contact size on heat flux on the top of silicon chip (Thermoelectric element thickness = 20 μm, input power = 10 W)

rate on the top of the silicon chip. Its beneficial effect on hot-spot cooling could be limited by the heat spreading resistance inside the mini-contact pad as well as inside the silicon chip. Consequently, care should be taken to optimize the geometric configuration to achieve the maximum hot-spot cooling performance. Figure 9.38 shows the temperature profiles achieved along a line bisecting the bottom of the silicon chip, with and without an embedded TEC, and revealing the characteristic "W"-shaped temperature profile created by the mini-contact-enhanced TEC cooler. It may be seen that if there is no hot spot on the chip and no TEC embedded inside the package, the peak chip temperature is about 109 °C. However, if there is a 400 × 400 μm hot spot with a heat flux of 1250 W/cm^2, the peak chip temperature will increase to 137 °C. Therefore, the concentrated heat flux leads to about a 28 °C peak temperature rise on the chip. If the TEC with a thermoelectric element thickness of 20 μm is activated with a 10 W input and enhanced with a 600 × 600 μm mini-contact pad, the temperature profile created around the high-flux region shows a 9 °C reduction in the hot-spot temperature (to 128 °C), a low-temperature ring with temperatures below the uniform heat dissipation values, and a warm outer ring with slowly decaying temperatures in the radial direction. If the mini-contact tip grows to 1250 × 1250 μm, the hot-spot temperature will reduce further, down to 120 °C, resulting in 17 °C maximum cooling at the hot spot. However, if we expand the mini-contact size further, to 2400 × 2400 μm, the hot-spot cooling is limited to just 12 °C. Obviously there exists an optimum mini-contact size for each configuration. The observed temperature increases in the outer ring or "side lobes" of the profile, as well as a very modest increase in the average chip temperature, is due to the additional power dissipation associated with the operation of the TEC device.

Figure 9.39 shows the heat flux distribution along the line bisecting the top surface of the silicon chip, just below the TEC enhanced with three different sizes of copper mini-contacts. On the top of the silicon chip and far away from the contact zone, the heat flux is approximately 70 W/cm^2, the same as the background heat flux on the bottom of the silicon chip. However, at the interface between the mini-contact and the silicon chip, the heat flux increases significantly, indicating a strong cooling effect from the activated TEC. The heat flux averaged over the entire surface of the mini-contact/silicon chip interface increases from 250 to 640 W/cm^2, and then to 1600 W/cm^2 when the copper mini-contact size decreases from 2400 × 2400 μm to 1250 × 1250 μm, and then to 600 × 600 μm, suggesting that reduction of the mini-contact sizes can significantly increase the local cooling heat flux.

It is also interesting to note that the heat flux is highly nonuniform at the mini-contact/silicon chip interface, with the highest value occurring at the corner and the lowest value at the center. Although the cooling heat flux continues to increase with decreasing mini-contact size, the combined effect of the thermal resistance between the hot spot and the mini-contact area and this cooling flux results in the optimum mini-contact size previously shown in Fig. 9.38.

9.2.4.3 Effect of Thermoelectric Element Height

Thermoelectric element height is a key parameter for improving the hot-spot cooling performance as the maximum achievable cooling flux of the TEC is inversely proportional

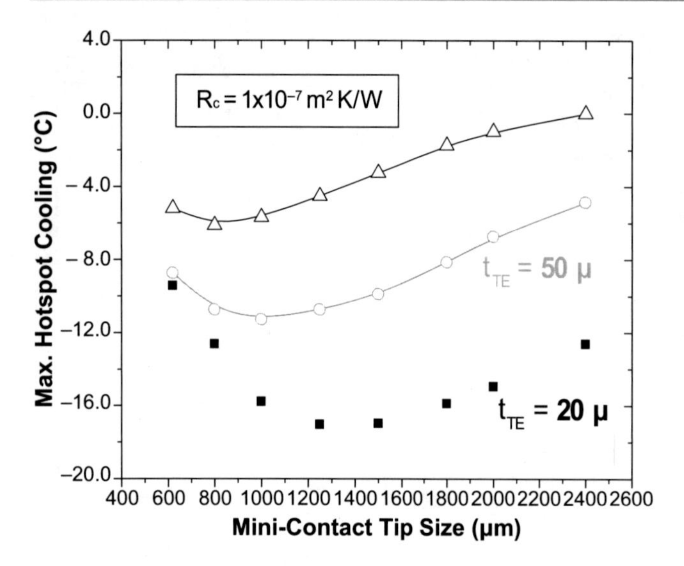

Fig. 9.40 Effect of thermoelectric element thickness *t*TE on the maximum hot-spot cooling performance. The optimized input power of 10, 7.5, and 6.1 W is applied on the TEC with the TE element thickness of 20, 50, and 100 μm, respectively

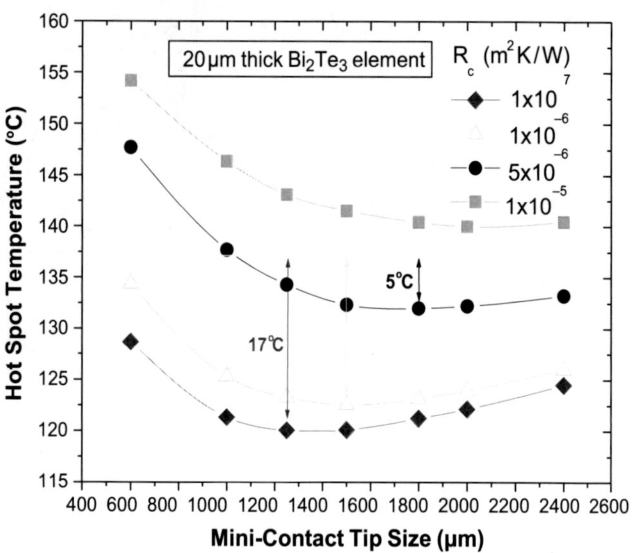

Fig. 9.41 Influence of thermal contact resistance on hot-spot cooling performance

to the thermoelectric element thickness. Figure 9.40 illustrates the variation of hot-spot cooling with the mini-contact size for three different thermoelectric element heights, under optimized input power on the TEC. As is expected, shorter thermoelectric elements allow the TEC to achieve better hot-spot temperature reductions, e.g., 6 °C to 11.2 °C and to 17.0 °C as the thermoelectric element decreases from 100 to 50 μm and to 20 μm in thickness, using the optimum mini-contact tip size. Even when the mini-contact tip size is kept constant, shorter thermoelectric elements always yield better hot-spot cooling than longer elements due to the higher fluxes extracted by the cold side of the TEC. However, it is interesting to find that for fixed contact resistances the optimum mini-contact tip size increases with decreasing element height, from 800 × 800 μm for a 100 μm thick element to 1000 × 1000 μm for a 50 μm thick element, and to 1250 × 1250 μm for a 20 μm high element.

As may be seen in Fig. 9.40, the TEC with 20 μm high elements has more dependence on the mini-contact size than the TECs with 50 μm or 100 μm high elements. It should be noted that the improvement provided by use of the mini-contact pad is larger with taller thermoelectric elements than with shorter elements. For the TEC with 20 μm thick elements, the addition of an optimally sized mini-contact pad improves the cooling by 4.3 °C, from hot-spot cooling of 12.7 °C with no mini-contact to 17 °C with a 1250 × 1250 μm mini-contact. However, for the TEC with 100 μm thick elements, the addition of an optimally sized mini-contact pad reduces the hot-spot temperature by an additional 6 °C, from hot-spot cooling of 0.1 °C with no mini-contact to 6.1 °C with a 800 × 800 μm mini-contact. Interestingly, the deterioration in performance with suboptimum contact pads displays the reverse trend, with the hot-spot temperature rising steeply with reduced mini-contact area for the 20 μm thick TEC but more gradually for the 100 μm thick TEC.

9.2.4.4 Effect of Thermal Contact Resistance

Low thermal resistance interfaces are critical to mini-contact-enhanced hot-spot cooling, since a high thermal resistance at the mini-contact/chip interface – where the cooling flux is highest – will significantly reduce the effectiveness of the mini-contact enhancement. Moreover, a bad thermal interface between the TEC and the TIM2 will impede the conduction of thermoelectric and Joule heat into the heat spreader and then into the heat sink and the ambient. Figure 9.41 displays the interplay between the thermal contact resistance and the achievable hot-spot cooling, with the assumption of equal thermal contact resistance at the two interfaces (e.g., $R_{c1} = R_{c2} = R_c$) and reveals that with increasing thermal contact resistance at both interfaces, the net cooling achievable on the hot spot diminishes. It may be seen that if the thermal contact resistance is 1×10^{-5} Km^2/W or higher, the hot-spot temperature will exceed 140 °C, and the embedded TEC will raise rather than lower the hot-spot temperature.

The thermal contact resistance also has an impact on optimized mini-contact size. As shown in Fig. 9.41, with increasing thermal contact resistance, the optimized mini-contact increases from 1250 × 1250 μm for the thermal contact resistance of 1×10^{-7} Km^2/W, representing a nearly perfect interface, to 2000 × 2000 μm for a thermal contact resistance of 1×10^{-5} Km^2/W, typical of thermal grease interfaces. It should be noted that if the thermal contact

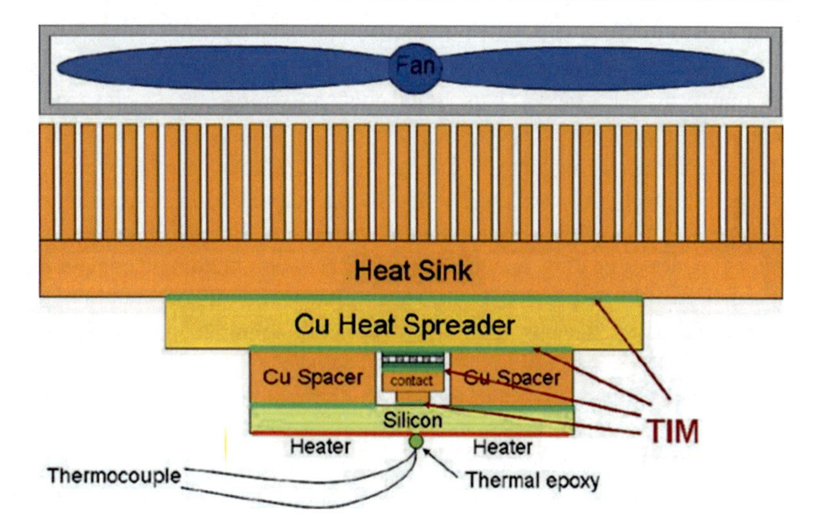

Fig. 9.42 Schematic of test vehicle for mini-contact-enhanced TEC for spot cooling

resistance is 5×10^{-5} Km^2/W or lower, the use of an optimized mini-contact tip always provides a lower hot-spot temperature than achieved without the mini-contact. However, the mini-contact is seen to provide diminishing returns as the contact resistances increase and to elevate the hot-spot temperatures for thermal contact resistances of 1×10^{-5} Km^2/W or higher.

9.2.4.5 Experimental Demonstration

Thermal measurements were performed on the chip package test vehicle shown in Fig. 9.42 to quantify the spot cooling improvement provided by the mini-contact pad and to determine its relationship to the TEC input power and power dissipation on the silicon chip. In this "proof-of-concept" experiment, there are no micro-scaled hot spots and temperature sensors on the chip. Instead, a uniform heat flux is imposed on the bottom of the chip, by attaching four, thin-film heaters, and the mini-contact-enhanced TEC is used to locally cool that chip below the temperature of the surrounding silicon. The schematic of the experimental structure is shown in Fig. 9.42, displaying a 2.5 mm × 2.5 mm × 500 μm silicon chip attached to four, thin-film heaters used to simulate chip power dissipation. The copper mini-contact pad with various mini-contact tip sizes, varying from 0.8 × 0.8 mm to 3.6 × 3.6 mm, was bonded onto the silicon chip using an indium-based solder. To facilitate the soldering process, a 200 nm thick Ni thin-film adhesion layer deposited on the silicon and then a 200 nm thick Au layer deposited onto the Ni layer by an e-beam process were used [111]. The copper mini-contact pad was then soldered onto the silicon by indium solder, which reacted with the Au thin film at around 160 °C to form an AuIn2 intermetallic compound.

Miniaturized TECs from Thermion (model number: 1MC04-018-02-2200D) [112] with a dimension of 3.6 × 3.6 × 1.6 mm and a total of 36 diced p-type and n-type 200 μm thick bismuth telluride thermoelectric elements, were used in these experiments. The thermal conductivity of bismuth telluride in the Thermion TEC is reported to be 1.3–1.4 W/m K, the Seebeck coefficient 200 μV/K, and the electrical resistivity 10 μΩ m, with a figure of merit value Z of 3×10^{-3} K^{-1} [113]. The two ceramic substrates are made of AlN, each with a thickness of 635 μm, and indium-tin solder was pre-tinned to the end faces to facilitate solder connections. In the present experiment, the TEC was attached to the mini-contact using thermal grease. The copper mini-contact pad and the TEC are accommodated inside the copper spacer. Above the copper spacer and the TEC, the copper heat spreader was attached using thermal grease. The heat spreader was then attached to the air-cooled copper heat sink.

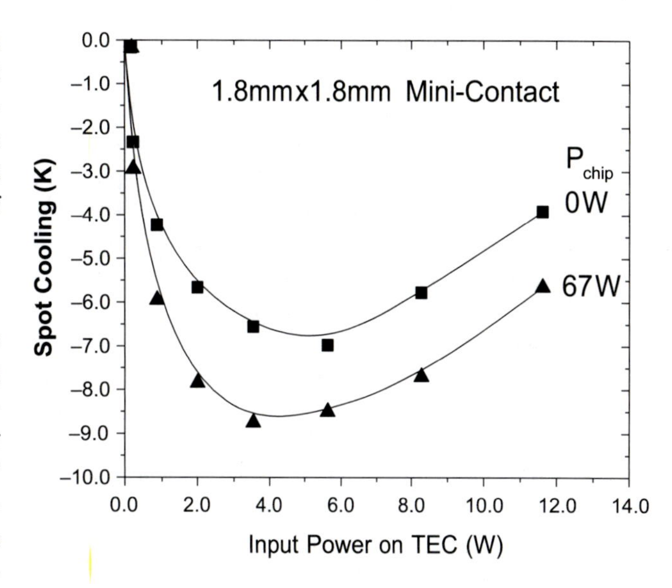

Fig. 9.43 Variation of measured spot cooling with TEC input power

Figure 9.43 shows the experimentally determined dependence of spot cooling on the TEC input power for a 500 μm thick silicon chip, with the mini-contact tip size kept at 1.8 × 1.8 mm and the power dissipation on the silicon chip varying from 0 to 67 W. It is found that the temperature reduction at the targeted spot (the center of the silicon chip) varies parabolically with the TEC input power, reflecting the competing mechanisms of rapidly improving Peltier cooling at lower input powers (or current) and progressively more damaging Joule heating, as well as reverse heat conduction, at the higher input powers. In this test vehicle, the silicon chip, thus, experiences its largest value of spot cooling at an input power of 3.5–5.5 W.

It is interesting to note that the power dissipation on the silicon chip has some effect on spot-cooling performance and that higher chip power dissipation leads to greater achievable spot cooling on the silicon chip. For example, if there is no power dissipation on the silicon chip, a maximum cooling of about 7.0 °C can be achieved.

However, if the power dissipation on the chip is increased to 67 W, the maximum spot cooling increases to around 9 °C. This improvement in cooling performance is related primarily to the increase in the cold side junction temperature of the TEC, which raises the Peltier cooling rate. Interestingly, this Peltier cooling improvement also leads to lower values of the optimum TEC input power, since at the higher temperature, more effective cooling can be achieved at lower current or power. This trend is illustrated in Fig. 9.8, where an increase of the power dissipation on the silicon wafer from 0 to 70 W, is seen to produce a decrease in the optimum TEC input power from 5.7 to 3.6 W.

The experimentally observed effect of the mini-contact tip size on the temperature reduction at the targeted spot is displayed in Fig. 9.44. For the three different power dissipations and a 500 μm thick chip, the maximum spot cooling is seen to display a parabolic dependence on the mini-contact tip size, showing very favorable improvements as the mini-contact tip size decreases in area from the "full coverage" limit, but ultimately reversing direction as the tip size shrinks below an optimum value and approaches point contact. The presence of an optimum tip size reflects the competing effects of the favorable cooling flux concentration and the parasitic spreading resistance in the mini-contact tip. As shown in Fig. 9.44, for the case of no power dissipation on the silicon chip, if the mini-contact is of the same size as the TEC base, the measured maximum spot cooling is about 3.3 °C. However, if a 1.8 × 1.8 mm copper mini-contact is integrated onto the TEC, 7.1 °C maximum spot cooling can be obtained, which results in 115% improvement on spot-cooling performance. Similarly, spot-cooling performance can be improved by 100 and 80% if the power dissipation of the silicon chip is 30 W and 67 W, respectively. It is interesting to note that the power dissipation on the silicon chip has an impact on the optimized mini-contact size, and the larger the power dissipation on the silicon chip, the smaller the optimized mini-contact size. As clearly illustrated in Fig. 9.44, as the power dissipation on the silicon chip increases from 0 to 67 W, the optimum mini-contact size decreases from 1.8 × 1.8 mm to 1.3 × 1.3 mm.

9.2.5 Applications in Biomedical Systems

Temperature control is one of the most essential operations in many biomedical systems [114]. Active temperature control is capable of accurately and rapidly bringing the intended

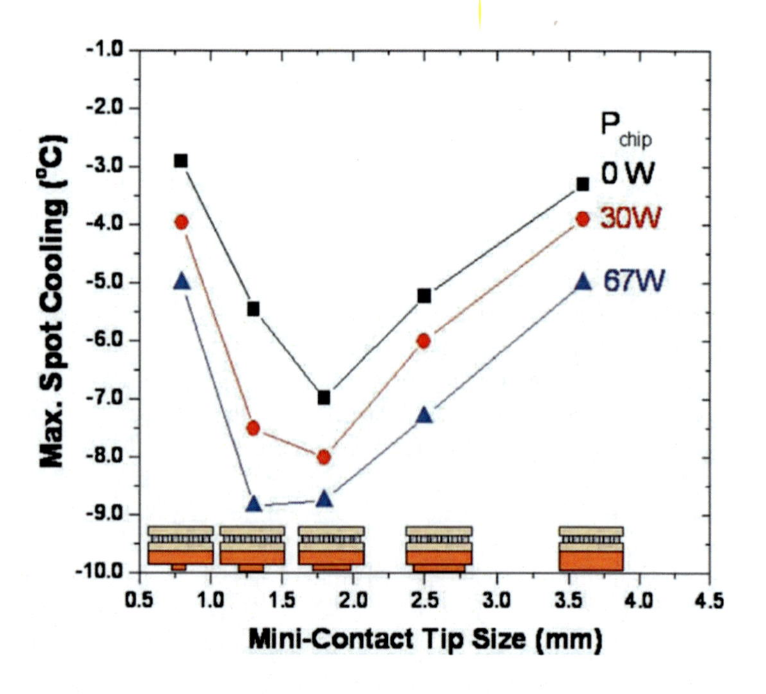

Fig. 9.44 Variation of measured maximum spot cooling on copper mini-contact size

temperature above, below, or in between preset limits. In particular, in biomedicine, thermal control systems are encountered frequently, for example, in electrophysiology [115], radiopharmaceutical synthesis [116], microbial studies [117], rapid thermal cycling of cells [118], and DNA sequencing [119]. Wijngaards et al. developed a concept of an active micro-thermostat system using an integrated thin-film poly-SiGe thermoelectric cooler, capable of stabilizing the temperature of a suspended structure at ambient temperature and above. This opens up the way to a large number of applications that need to be thermally controllable in the range of 10–50 °C, especially in the field of biomedicine.

Precise fluid temperature control in microfluidic channels is a requirement for many lab-on-a-chip and microreactor devices, especially in biotechnology where most processes are highly temperature sensitive. Microheaters integrated with microfluidic channels have been proposed but are of limited use for temperature control, as they can only be used for raising the fluid temperature. In addition, they are limited in creating large thermal gradients, in the order of 10 K/mm, due to thermal conduction in the substrate. One feasible method that scales well on the microscale is thermoelectric cooling. Maltezos demonstrated the concept of a Bi_2Te_3-based thermoelectric microcooler integrated into a microfluidic channel in order to give rapid and localized fluid cooling [120]. They reported an on-chip thermoelectric refrigerator and heat exchanger for microfluidic devices. The microfluidic chamber was cycled between −3 and over 120 °C, thus spanning water freezing and boiling, and the entire PCR temperature range. For smaller chambers, it was shown that it is possible to cool reagent from room temperature to freezing within 10–20 s and to obtain relatively good temperature stability ($<\pm 0.2$ °C) over long periods of time. The ability to localize heating and cooling in microfluidic chambers and channels enable massive parallelization of chemical reactions in which the temperature of each reaction vessel can be independently controlled. Thus, these thermal management systems enable the fabrication of complex chip based chemical and biochemical reaction systems in which the temperature of many processes can be controlled independently.

9.3 On-Chip Hot-Spot Cooling Using Anisotropic Heat Spreader

Anisotropic heat spreaders, with a high in-plane thermal conductivity, provide a very promising passive approach for hot-spot remediation and constitute a very attractive component for active cooling of nanoelectronic chips. This anisotropy can be used to laterally spread the hot spot heat to cooler regions of the chip and thus can significantly reduce the hot-spot temperature. Recently, Bar-Cohen et al. used analytical and numerical models to investigate the feasibility of hot-spot remediation using the lateral heat spreading capability of a thermal interface material (TIM) with orthotropic thermal conductivity [121]. It was found that when used together with existing cooling solutions, such materials, bonded directly to the silicon chip as shown in Fig. 9.45, can substantially reduce the temperature rise associated with a severe flux spot and more uniformly distribute the heat at the interface between the TIM and the next element in the heat transfer path.

In their work, the bilayer slab shown in Fig. 9.46 was used to investigate the hot-spot remediation provided by an orthotropic spreader attached directly to the back of a square chip with a single, centrally located, square flux spot. All external surfaces are assumed adiabatic, except for a heat flux boundary condition at the flux spot and a convective boundary condition on the back of the orthotropic spreader. Background heating on the active side of the chip is forgone because only the hot-spot temperature rise is sought. The boundary conditions are such that inclusion of background heating would simply elevate the entire temperature field. This effect becomes nontrivial if temperature-dependent material properties and heat transfer coefficients are employed, but all properties and heat transfer coefficients are taken to be constant in this analysis. The convective boundary condition represents the influence of the "global" cooling scheme and is modeled by a uniform heat transfer coefficient on the back of the spreader.

Following Muzychka' work [122], an analytical solution was developed using the separation of variables method, and

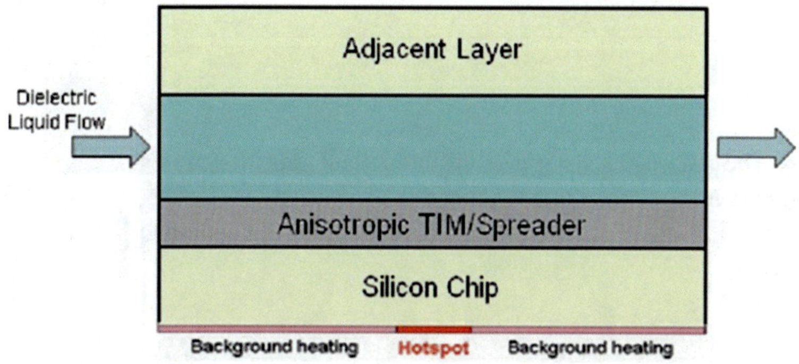

Fig. 9.45 Implementation of an anisotropic TIM/Spreader in a three-dimensional stack up

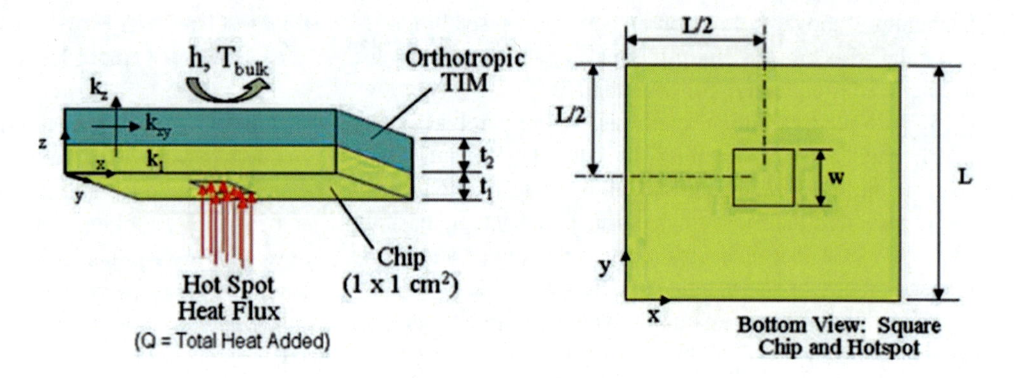

Fig. 9.46 Schematic of compound chip/spreader system

the excess temperature on the active side of the chip ($z = 0$) is given by the following equation:

$$\Delta T_{\text{bulk}}(x, y, z = 0) = A_0 + \sum_{m=1}^{\infty} A_m \cos(\lambda_m x) + \sum_{n=1}^{\infty} A_n \cos(\delta_n y) \cdots + \sum_{m=1}^{\infty}\sum_{n=1}^{\infty} A_{mn} \cos(\lambda_m x)\cos(\delta_n y) \quad (9.27)$$

The eigenvalues in Eq. (9.27) are given by $\lambda_m = \frac{m\pi}{L}$, $\delta_n = \frac{n\pi}{L}$ and $\beta_{mn} = \sqrt{\lambda_m^2 + \delta_n^2}$ the Fourier coefficients, A_o, A_n, A_m, and A_{mn}, are found from the application of the boundary conditions in the z-direction. Imposing the boundary conditions at $z = 0$ results in expressions for the "A" coefficients as follows:

$$A_0 = \frac{Q}{L^2}\left(\frac{t_1}{k_1} + \frac{t_2}{k_2} + \frac{1}{h}\right) \quad (9.28)$$

$$A_m = \frac{2Q\left[\sin\left(\frac{L+w}{2}\lambda_m\right) - \sin\left(\frac{L-w}{2}\lambda_m\right)\right]}{L^2 w k_1 \lambda_m^2 \phi(\lambda_m)}$$

$$A_n = \frac{2Q\left[\sin\left(\frac{L+w}{2}\delta_n\right) - \sin\left(\frac{L-w}{2}\delta_n\right)\right]}{L^2 w k_1 \delta_n^2 \phi(\delta_n)}$$

$$A_{mn} = \frac{16Q\cos\left(\frac{\lambda L}{2}\right)\sin\left(\frac{\lambda_m w}{2}\right)\cos\left(\frac{\delta_n L}{2}\right)\sin\left(\frac{\delta_n w}{2}\right)}{L^2 w^2 k_1 \beta_{mn} \lambda_m \delta_n \phi(\beta_{mn})} \quad (9.29)$$

The parameter, ϕ, appearing throughout Eq. (9.29), is a spreading parameter that is a function of a dummy variable ζ, with $\phi(\zeta)$ given by: On-Chip Thermal Management and Hot-Spot

$$\varphi(\zeta) = \frac{\left(\alpha e^{4\zeta t_1} - e^{2\zeta t_2}\right) + \psi\left[e^{2\zeta(2t_1+t_2)} - \alpha e^{2\zeta(t_1+t_2)}\right]}{\left(\alpha e^{4\zeta t_1} + e^{2\zeta t_2}\right) + \psi\left[e^{2\zeta(2t_1+t_2)} - \alpha e^{2\zeta(t_1+t_2)}\right]} \quad (9.30)$$

and ζ is replaced by λ_m, δ_n, or β_{mn} in Eq. (9.30) as appropriate. The new parameters in Eq. (9.30) are to be evaluated as $\psi = \frac{\zeta + h/k_2}{\zeta - h/k_2}$ and $\alpha = \frac{1-k_2/k_1}{1+k_2/k_1}$ where ζ is again replaced by λ_m, δ_n, or β_{mn} as appropriate. The conductivity of the second layer, k_2, is assumed to be isotropic in the above expressions. However, as suggested in the literature, if either of the layers exhibits orthotropic conductivity, the solution for purely isotropic layers can be used when the following length scale and conductivity transformations are employed in the subject orthotropic layer:

$$k \to k_{\text{eq}} = \sqrt{k_{xy}k_z} \qquad t \to t_{\text{eq}} = \frac{t}{\sqrt{k_z/k_{xy}}} \quad (9.31)$$

Thus, k_2 and t_2 in Eqs. (9.28), (9.29), and (9.30) can be replaced by the transformations in Eq. (9.31) to account for the anisotropicity in the spreader.

Equations (9.27), (9.28), (9.29), (9.30), and (9.31) provide a full solution for the excess temperature on the active side of the compound structure shown in Fig. 9.46. The overall resistance to heat transmission for the system shown in Fig. 9.46 is comprised of (9.1) the resistance to one-dimensional heat flow and (9.2) the spreading resistance. The first term in Eq. (9.27) is the Fourier coefficient, A_0, which is given by Eq. (9.28) and is attributable to uniform one-dimensional conduction through the compound bilayer slab. The three remaining terms are related to thermal spreading and thus vanish as the hot-spot size approaches the chip size. The total thermal resistance can be related to the average excess temperature at the hot spot through the following definition:

$$R_{\text{T}} = \frac{\overline{\Delta T_{\text{bulk}}}}{Q} \quad (9.32)$$

where R_{T} is the total thermal resistance, including thermal transport by both conduction and convection. The term $\overline{\Delta T_{\text{bulk}}}$ in Eq. (9.32) is found by integrating Eq. (9.27) over the flux spot region and dividing by the flux spot area, or, expressed mathematically:

$$\overline{\Delta T_{\text{bulk}}} = \frac{1}{A_S} \iint_{A_S} \Delta T_{\text{bulk}}(x, y, 0) dA_s \tag{9.33}$$

As noted, the total thermal resistance is also the sum of the one-dimensional resistance and the spreading resistance as follows:

$$R_T = R_{1D} + R_S \tag{9.34}$$

The one-dimensional resistance to heat conduction is easily found to be:

$$R_{1D} = \frac{t_1}{k_1 L^2 +} \frac{t_2}{k_z L^2} + \frac{1}{hL^2} \tag{9.35}$$

Meanwhile, the spreading resistance, R_s, can be found by substituting Eqs. (9.32), (9.33), and (9.35) into Eq. (9.34):

$$R_S = \frac{1}{A_S Q} \iint_{A_S} \Delta T_{\text{bulk}}(x, y, 0) dA_S - \left(\frac{t_1}{k_1 L^2} + \frac{t_2}{k_z L^2} + \frac{1}{hL^2} \right) \tag{9.36}$$

The integration in Eq. (9.36) yields the following expression for the spreading resistance in a bi-layer structure with a centrally located flux spot:

$$R_S = \frac{1}{2(w/2)^2 (L/2)^2 k_1} \left[\sum_{m=1}^{\infty} \frac{\sin^2(w\delta_m/2)}{\delta_m^3 \varphi(\delta_m)} + \sum_{n=1}^{\infty} \frac{\sin^2(w\lambda_n/2)}{\lambda_n^3 \varphi(\lambda_n)} \right] \cdots + \frac{1}{(w/2)^4 (L/2)^2 k_1} \sum_{m=1}^{\infty} \sum_{n=1}^{\infty} \frac{\sin^2(w\delta_m/2) \sin^2(w\lambda_n/2)}{\delta_m^2 \lambda_n^2 \beta_{mn} \varphi(\beta_{mn})} \tag{9.37}$$

where the eigenvalues are the same as those for Eq. (9.27) and the parameter φ is given by Eq. (9.30).

9.3.1 Effect of In-Plane Spreader Thermal Conductivity

Successful hot-spot remediation via an orthotropic spreader depends on the ability of the spreader to conduct heat laterally from local regions of high heat flux to other parts of the chip with lower thermal loads and is most directly influenced by its in plane thermal conductivity, k_{xy}. This conclusion is supported by the thickness and conductivity transformations in Eq. (9.31), which clearly show that any increase in k_{xy} for fixed values of k_z and t will result in larger values k_{eq} and t_{eq}, through the indicated square root dependence. Therefore, an increase in the conductivity ratio, k_{xy}/k_z, leads to an attendant decrease in the overall thermal resistance, and thus a decrease in the average hot-spot temperature.

Table 9.6 Parameter settings for k_{xy} variation

Parameter	Description	Value	Units
t_1	Chip thickness	250	μm
t_2	Spreader thickness	500	μm
k_1	Isotropic chip conductivity	163	W/m K
k_z	Thru-plane spreader conductivity	5	W/m K
k_{xy}	In-plane spreader conductivity	Varies	W/m K
q''	Hot-spot heat flux	1.4	kW/cm^2
h	Effective heat transfer coefficient	10,000	W/m^2 K
T_{bulk}	Ambient temperature for convective transfer	25	C
L	Chip size, square	1	cm
W	Hot-spot size, square	500	μm

A representative chip/spreader system with the parameter settings listed in Table 9.6 was used to determine the magnitude of the benefits of increasing k_{xy}. Please note that, unless otherwise specified, the parameters in Table 9.6 will be used throughout this section. The thru-plane conductivity of the spreader, k_z, was chosen to be 5 W/m K because this is a nominal thru-plane conductivity for some natural graphite materials as well as pyrolytic graphite [123]. A heat transfer coefficient of 10,000 W/m^2 K was applied to represent the presence of an aggressive cooling approach (e.g., pool boiling or a microchannel cold plate) [124]. The in-plane conductivity was varied between 5 and 1800 W/m K in order to determine the effect that the degree of anisotropy had on hot-spot remediation.

The excess temperature profiles on the active side of the chip, subjected to a 1.4 kW/cm^2, 0.5 mm^2 flux spot, are shown for several different in-plane conductivities in Fig. 9.47. The Gaussian-like temperature profile for the isotropic spreader ($k_z = k_{xy} = 5$ W/m K) is seen to yield a hot-spot temperature that is nearly 47.5 K above ambient. The use of the orthotropic spreaders is seen to produce similar profiles but to considerably reduce the peak temperature at the center of the hot spot and over the adjacent silicon while modestly elevating the temperature (~1 °C), in the edge regions of the chip. As expected, increasing the in-plane conductivity, k_{xy}, decreases the excess temperature. Thus, for $k_z = 5$ and $k_{xy} = 350$, which is representative of natural graphite sheets, the hot-spot temperature is ~9.3 °C below that obtained through the use of the isotropic spreader. If the in-plane conductivity is further increased to 1800 W/m K, a hot-spot suppression of ~14.3 °C is attained.

The variation of hot-spot remediation with the in-plane thermal conductivity of an orthotropic heat spreader, relative

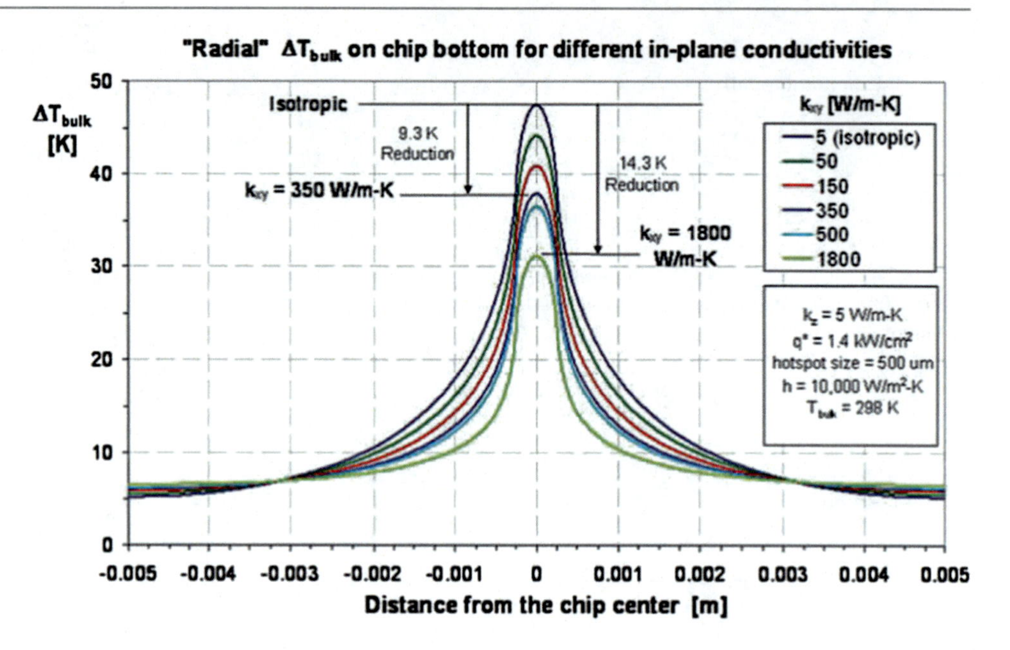

Fig. 9.47 Excess temperature profiles taken through the middle of the active chip surface for various k_{xy}

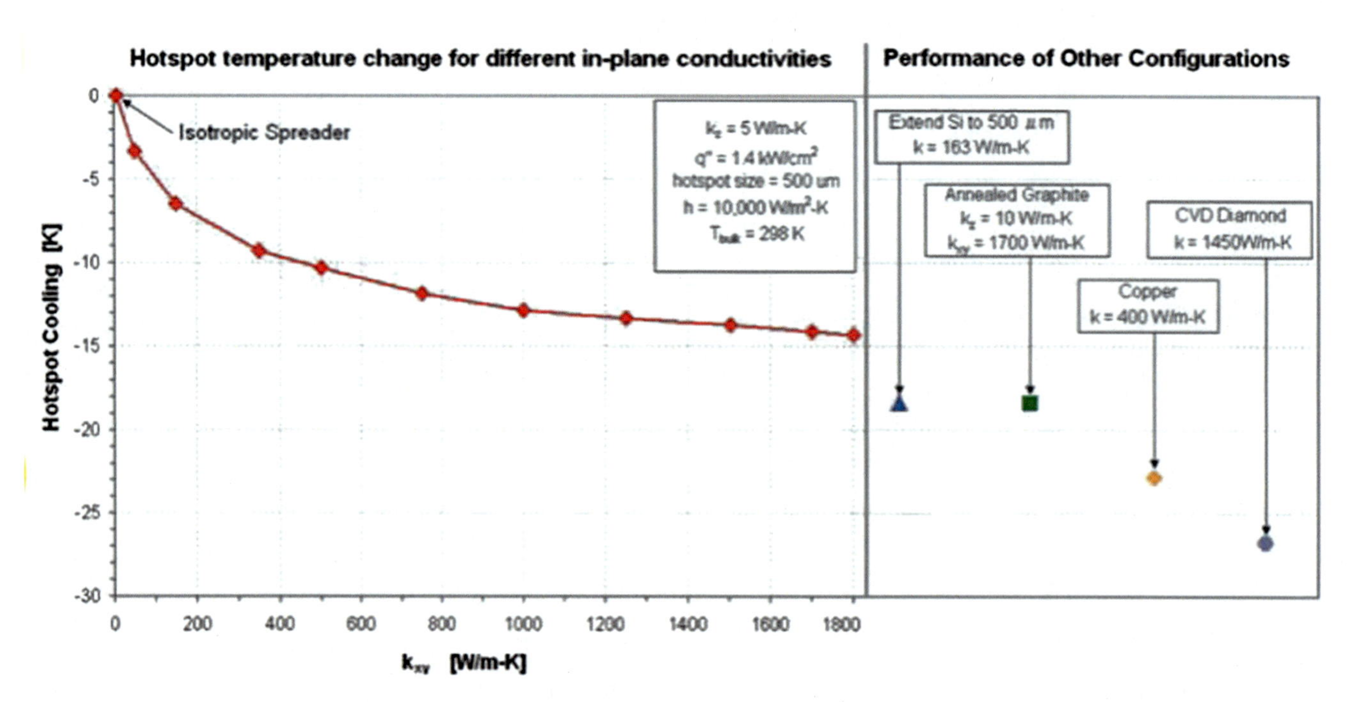

Fig. 9.48 Variation of hot-spot cooling with in-plane thermal conductivity in an orthotropic heat spreader

to a low-conductivity isotropic spreader, is shown in Fig. 9.48. The peak temperature reduction is seen to asymptotically approach 15 °C as the in-plane conductivity increases toward 1800 W/m K. In order to put these results into context, the performance of several spreaders with varying conductivities are evaluated for the same geometric parameters and thermal boundary conditions, as listed in Table 9.6. With the isotropic spreader ($k = 5$ W/m K) as a baseline, the hot-spot cooling achieved by each alternative spreader is shown on the right side of Fig. 9.48. The first data point corresponds to an isotropic spreader with conductivity of 160 W/m K, representing the use of a thick silicon chip. It is found that increasing the thickness of the silicon provides approximately 18.3 °C of cooling, which exceeds by 4 °C the best hot-spot suppression achieved. However, it can be seen in Fig. 9.48 that an orthotropic spreader with $k_z = 10$ W/m K and $k_{xy} = 1700$ W/m K – which is characteristic of annealed pyrolytic graphite (APG) [125] can provide the same hot-spot suppression as the thicker silicon chip.

Implementation of a copper spreader with isotropic thermal conductivity of 400 W/m K provides further hot-spot suppression, reaching a temperature reduction of ~23 °C. Of the alternative spreaders considered, the best hot-spot suppression of ~26.6 °C was provided by isotropic CVD diamond film with a thermal conductivity of 1450 W/m K in all three directions.

The above results show that increasing the in-plane thermal conductivity of an orthotropic spreader can provide substantial hotspot temperature reduction. For the parameters considered, the cooling performance of a highly orthotropic APG spreader matches the cooling provided by an equal thickness of pure silicon. This is an important result given the general reluctance of chip manufacturers to allocate valuable semiconductor-grade silicon for thermal management functions. Despite the good performance of the best highly orthotropic spreaders, an equally sized copper spreader provides about 4 °C better hot-spot remediation for the conditions examined. However, natural graphite orthotropic spreaders can be made extremely pliable and offer a weight advantage over copper spreaders of equal size since the density of natural graphite is approximately just 25% that of copper. This weight difference could be significant in weight-constrained mobile applications. Furthermore, highly orthotropic spreaders with low thru-plane conductivity have the ability to reduce hot-spot temperatures while simultaneously insulating adjacent layers of a chip stack.

Figure 9.49 shows the temperature rise of the silicon, copper, and APG spreaders as a function of location on the back of the spreader. The excess temperature displayed in Fig. 9.49 is the difference between the local temperature and the edge temperature; thus, the value of ΔT vanishes for all profiles as the edge of the spreader is approached. For the parameters considered, it can be seen that the silicon and copper spreaders allow a maximum temperature variation of 8 °C and 4.3 °C on the back of the spreader, respectively, compared to a maximum variation of 0.05 °C for the APG spreader. Consequently, in hybrid cooling systems – using two-phase cooling with an anisotropic spreader attached to the chip – an APG spreader would offer only a modest temperature rise on the rear of the spreader and yield a system that is less susceptible to local dryout or critical heat flux.

9.3.2 Variation of Spreader Thickness

The thickness of the orthotropic TIM/spreader will not only establish the magnitude of hot-spot cooling but will also determine the viability of this thermal solution for volume-constrained three-dimensional stacks of hot-spot-laden chips. It is, therefore, important to explore the tradeoffs between spreader thickness and cooling performance when assessing the merits of an anisotropic spreader. The impact of spreader thickness on hot-spot remediation is best understood in terms of the overall thermal resistance of the system, R_T, which is the sum of the spreading resistance, R_s, and the resistance to one-dimensional conduction and convection, R_{1D}. MATLAB codes were developed to aid in the evaluation of R_{1D} (Eq. 9.35) and R_s (Eq. 9.37) for various spreader thicknesses.

Using the model in Fig. 9.46 and the parameter settings in Table 9.6, the thickness of each of the spreaders represented in Fig. 9.48 was varied to determine the effect on R_{1D}, R_s, and hence R_T. Typical trends seen during this analysis are depicted in Fig. 9.50 where it is clear that the total thermal resistance generally experiences a minimum for some critical value of the spreader thickness. Near a spreader thickness of zero, the total thermal resistance of the system approaches that of a single layer of silicon. But, as the thickness increases, the one-dimensional resistance, R_{1D}, grows, and

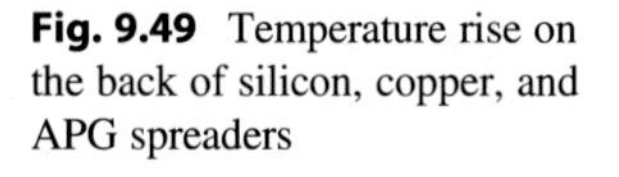

Fig. 9.49 Temperature rise on the back of silicon, copper, and APG spreaders

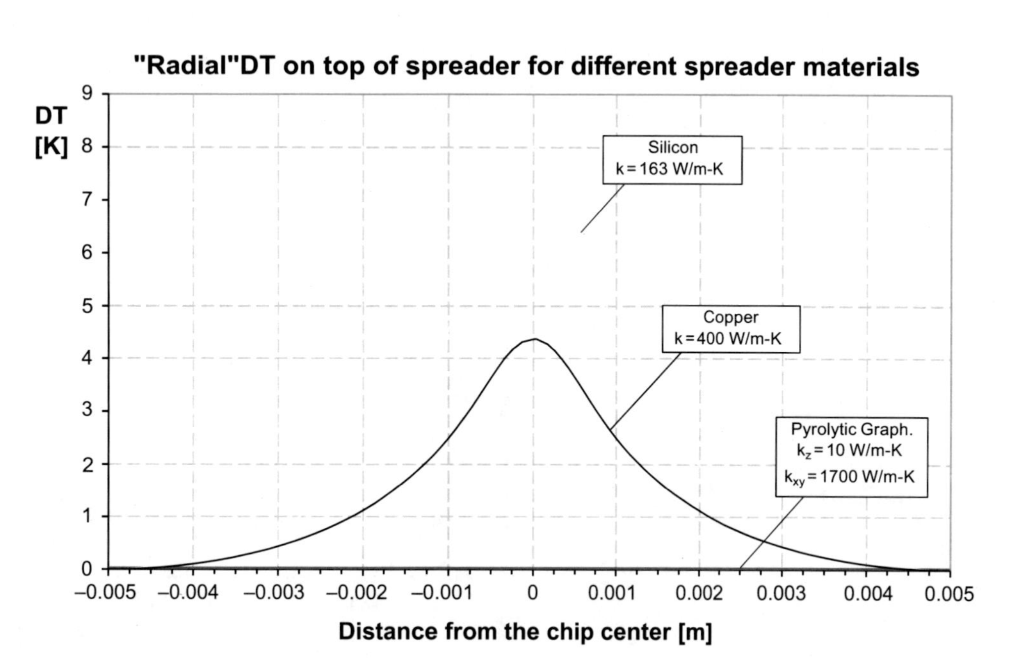

the spreading resistance, R_s, falls leading to a nonmonotonic variation in the overall thermal resistance. When the negative slope of R_s equals the positive slope of R_{1D} the minimum thermal resistance, and hence minimum average hot-spot temperature, is attained. For any increase in spreader thicknesses beyond this optimum, the linear rise in R_{1D} is greater than the decrease in R_s, and the total thermal resistance increases.

The resulting variation in overall thermal resistance for a spreader with a thru-thickness conductivity of 5 W/m K is shown in Fig. 9.50 for a range of in-plane conductivities. As anticipated, the total thermal resistance for each of the k_{xy} curves approaches the resistance of a directly cooled chip, i.e., $R_T \rightarrow 10.84$ K/W, as spreader thickness approaches zero. For increasing values of the in-plane conductivity (above 5 W/m K), the total thermal resistance and average hot-spot excess temperature generally decrease and display an optimum value. The specific optimum thickness varies with in-plane conductivity as shown in Fig. 9.51, exhibiting a peak optimum thickness of approximately 160 μm in the vicinity of $k_{xy} = 200$ W/m K and of approximately 100 μm at a thermal conductivity of 1800 W/m K. However, as seen

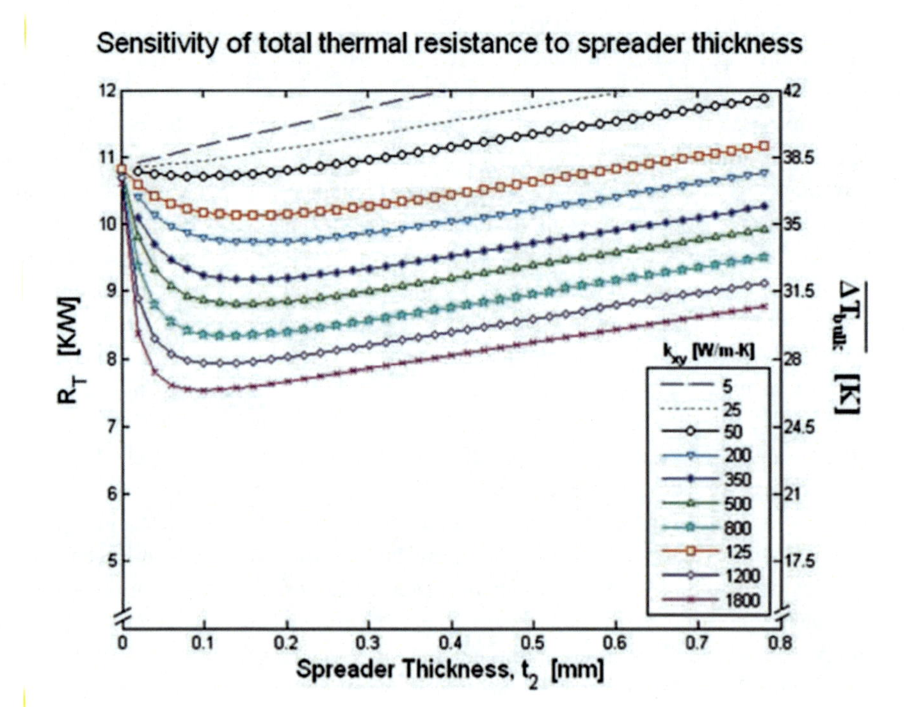

Fig. 9.50 Variation of R_T and average hot-spot excess temperature for increasing spreader thickness ($k_z = 5$ W/m K)

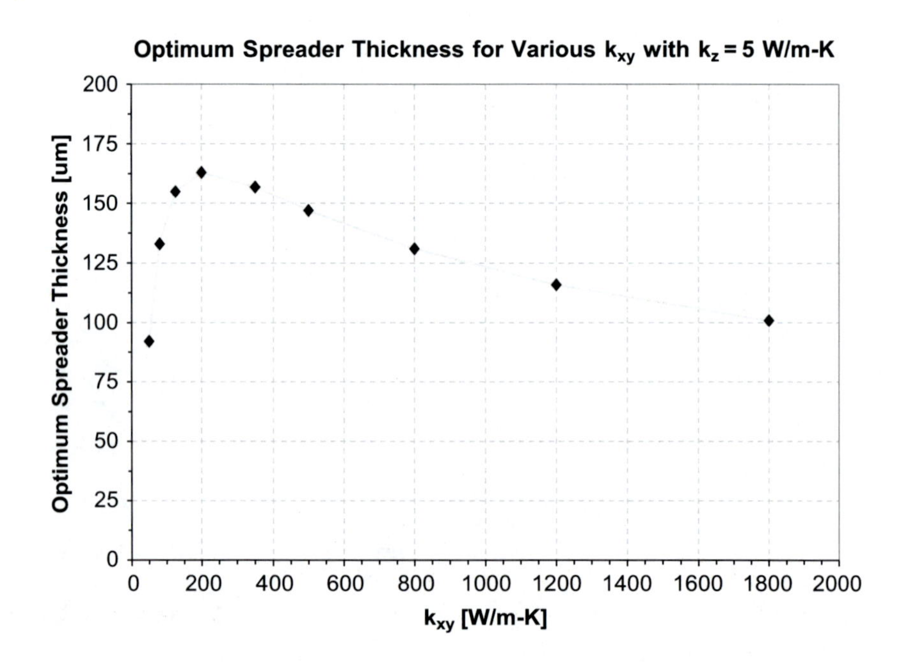

Fig. 9.51 Optimum spreader thickness varies with in-plane thermal conductivity

for the dashed lines where $k_{xy} = 5$ and 25 W/m K, for low values of the in-plane conductivity, there is no optimum thickness and R_T continuously increases. For the parameters considered, it was found that the development of a monotonically increasing total thermal resistance occurs for $k_{xy} \leq 14.5$ W/m K.

The variation of R_T with the TIM/spreader thickness, t_2, for each of these alternatives can be seen in Fig. 9.52. It is found that the highly orthotropic APG exhibits similar behavior to that shown in Fig. 9.50, with a distinct minimum occurring at a spreader thickness of 157 μm, for the stated conditions. However, the silicon, copper, and diamond spreaders all exhibit a broad "plateau" for which the thermal resistance remains relatively constant with thickness (these spreaders do indeed have minimum values of R_T, but the minima occur beyond the 2 mm thickness at which plotting was stopped in Fig. 9.52).

Figures 9.51 and 9.52 reveal that highly orthotropic graphite TIMs/spreaders yield optimum hot-spot cooling performance at relatively low thicknesses – under 165 μm – for the conditions examined. Furthermore, it is interesting to note in Fig. 9.52 that the orthotropic APG spreader yields lower average hot-spot temperatures than copper for thicknesses up to ~200 μm and lower temperatures than silicon up to ~500 μm. Also, the minimum average hot-spot excess temperature for APG is 24.4 K at 157 μm, which is only 5.0 °C and 1.3 °C hotter than that provided by nine times the thickness of copper and silicon (~1.4 mm), respectively. The exceptional performance of highly orthotropic TIMs/spreaders at low thickness may lead them to be favored over conventional heat spreading materials in space constrained 3D chip stacks.

Up to this point, the thickness of the silicon chip in Fig. 9.46 has been fixed at 250 μm for all cases. The variation of R_T with spreader thickness for a spreader with $k_z = 5$ W/m K and $k_{xy} = 350$ W/m K, where each of the plotted line represents a different chip thickness (all other parameters remain unchanged; see Table 9.6) is shown in Fig. 9.53. The total thermal resistance is seen to decrease with increasing chip thickness. However, it is clear that the additional hot-spot cooling provided by an optimally thick spreader becomes less dramatic for greater chip thicknesses. This is more clearly shown in Fig. 9.54 where the R_T data for each plotted line in Fig. 9.53 has been normalized by the total thermal resistance that would exist if the spreader were removed, and the bare chip were cooled directly (this resistance is called $R_{T,bare}$).

It is seen that an optimally thick spreader reduces $R_{T,bare}$ by ~43% when the chip is 75 μm thick but reduces $R_{T,bare}$ by only ~7% when the chip is 400 μm thick. The reduction in spreader effectiveness for increasing chip size is the result of more effective heat spreading in the thicker silicon, which reduces the role played by the orthotropic spreaders. Alternatively, as chip thicknesses shrink – the more likely scenario as chip manufacturers strive to more efficiently utilize valuable silicon ingots and package electronics in thinner packages – the orthotropic TIM/spreaders can compensate for the loss of inherent spreading in these thinner silicon chips of the future.

Figure 9.54 also reveals that increasing chip thicknesses are accompanied by a steady decrease in the optimum

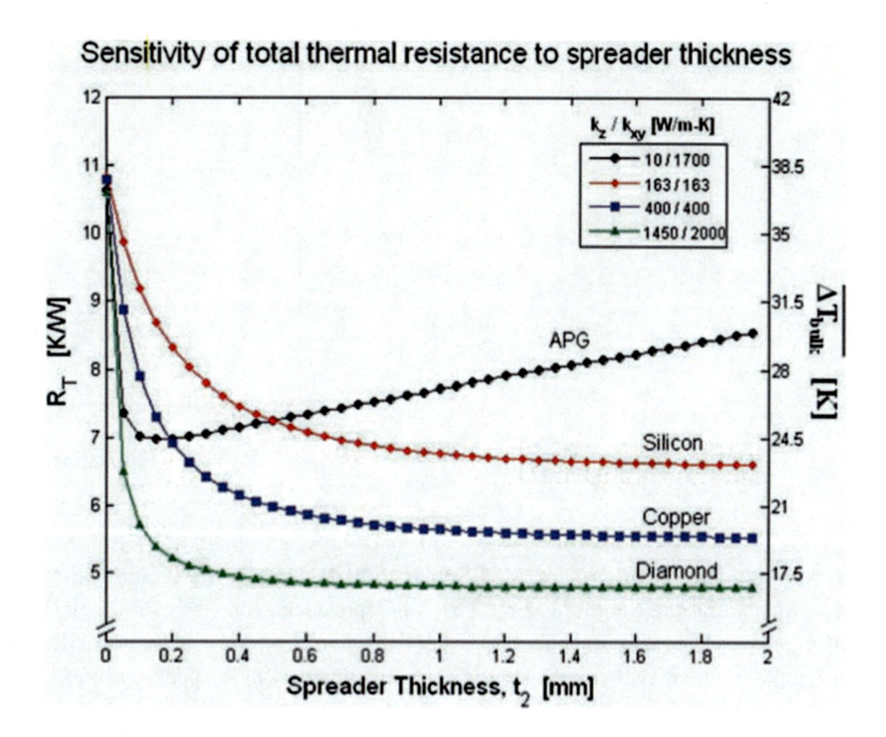

Fig. 9.52 Variation of R_T for increasing spreader thickness for alternative spreaders

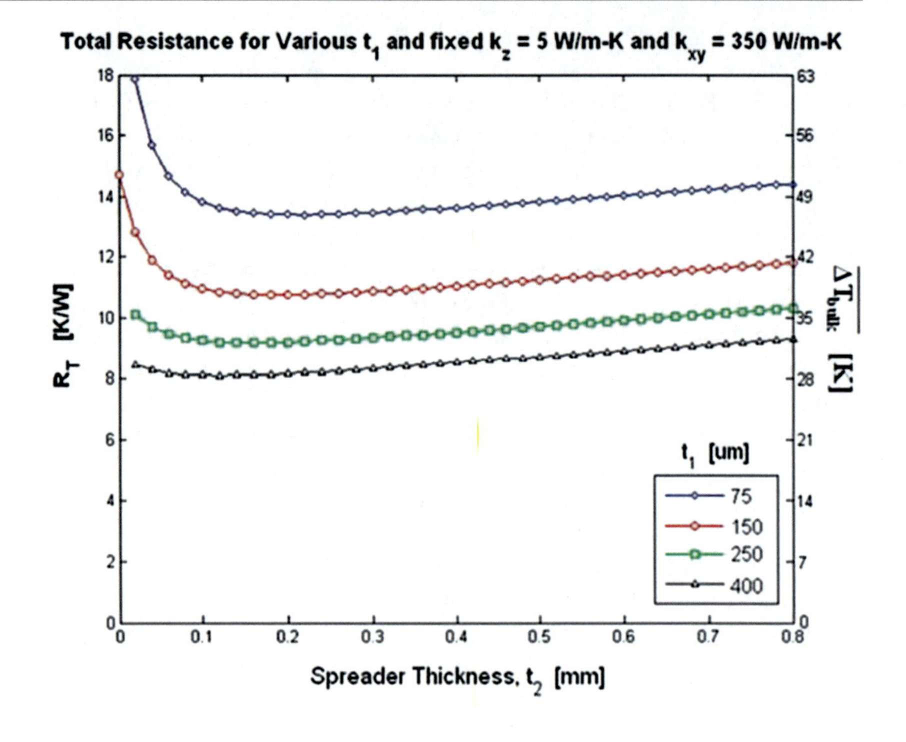

Fig. 9.53 Effect of chip thickness of spreader thickness variation for an orthotropic spreader with $k_z = 5$ W/m K and $k_{xy} = 350$ W/m K

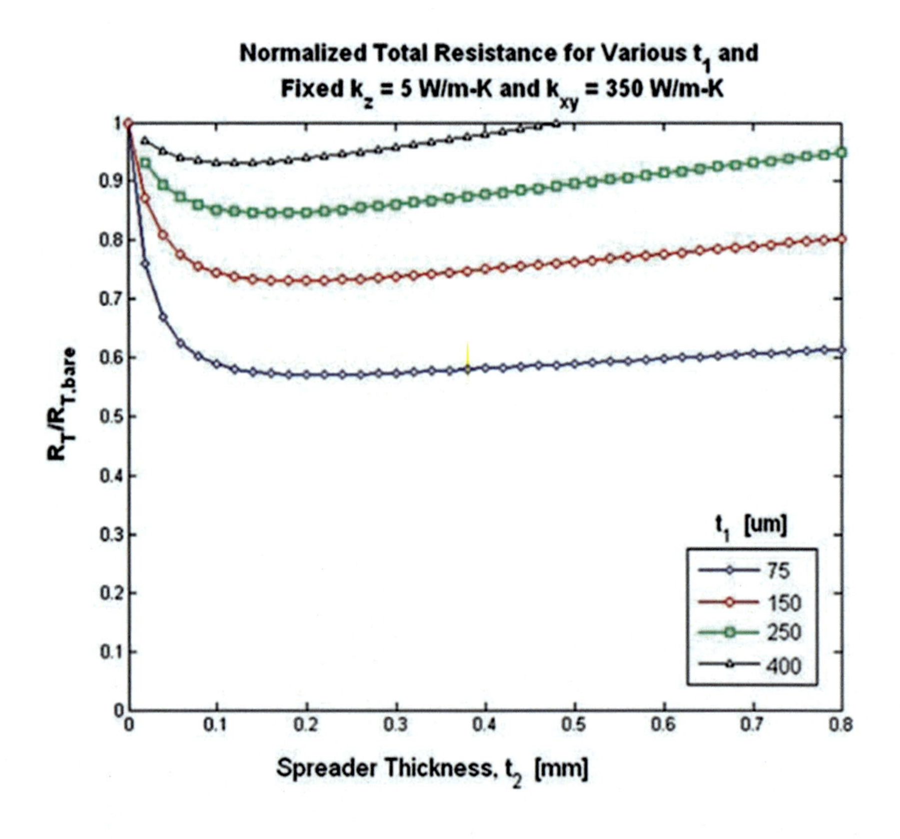

Fig. 9.54 Normalized total resistance for different chip thicknesses and an orthotropic spreader with $k_z = 5$ W/m K and $k_{xy} = 350$ W/m K

spreader thickness for a given k_{xy}. In order to better understand this variation, the plot in Fig. 9.51 was reproduced for different values of chip thickness, t_1. The results can be seen in Fig. 9.55, and it is clear that smaller chip thicknesses yield a larger optimum spreader thickness for a given k_{xy}. Also, thicker chips yield lower sensitivity of optimum thickness to in-plane conductivity, as evidenced by the suppression of the peak optimum thickness in Fig. 9.55 for larger values of t_1.

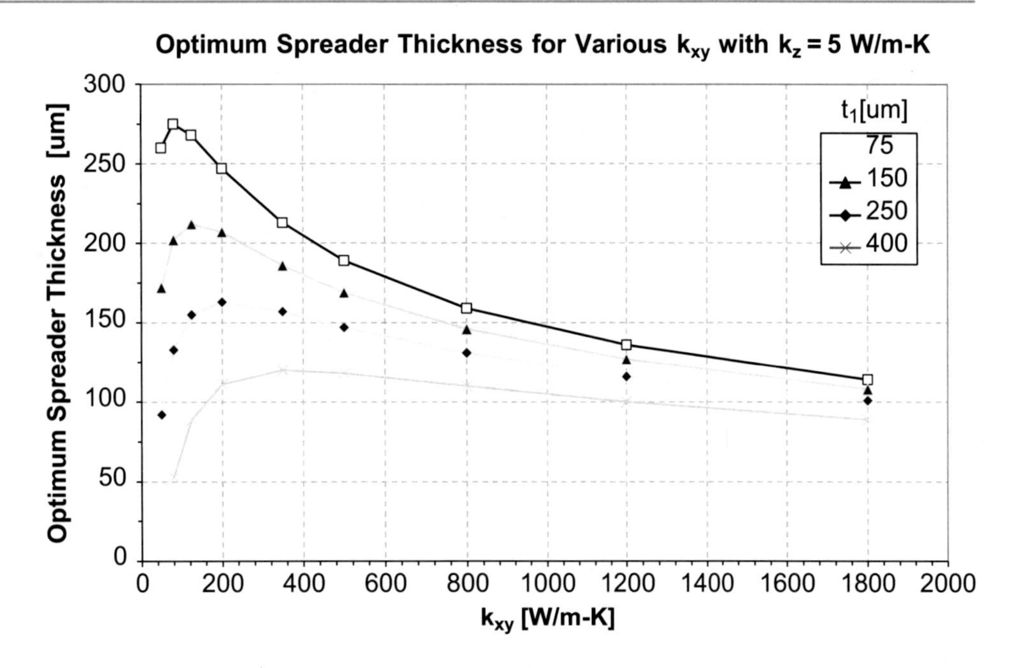

Fig. 9.55 Change in optimum spreader thickness for various k_{xy} and t_1

9.3.3 Numerical Simulations and Contact Resistance Variation

The parametric results above indicate that the use of an orthotropic spreader is a promising approach to reducing hot-spot temperatures. However, it may be anticipated that the physical attachment of the orthotropic material to the back of the chip, as depicted in Fig. 9.46, may well result in the creation of a potentially significant and deleterious thermal contact resistance. Typical contact resistances for electronic packaging applications are reported to be in the range of 10^{-7}–10^{-4} K m^2/W. The thermal contact resistances of $\sim 10^{-7}$ K m^2/W are representative of an excellent interface achieved by monolithic growth or eutectic interface attachment, while resistances of $\sim 10^{-4}$ K m^2/W represent a relatively poor thermal interface achieved through the use of phase change materials and elastomeric pads [126]. While it is, thus, desirable to assess the impact of thermal contact resistance at the chip/spreader interface on hot-spot remediation, to the authors' knowledge, there is no analytical solution that explicitly accounts for this effect in a layered structure. Consequently, the parametric sensitivity of hot-spot temperature to contact resistance is explored through numerical simulations with ANSYS for contact resistances varying from 0 to 10^{-4} K m^2/W.

Figure 9.56 shows the resulting heat flux vectors near the flux spot in the silicon chip for each contact resistance, with the results for perfect thermal contact on the top and poor thermal contact on the bottom, for the previously described conditions. Comparing the two heat flux plots reveals the subtle influence of contact resistance on the flow of heat in the system. In the case of perfect thermal contact, the heat flux vectors display a large thru-plane component, reflecting the relative ease with which heat can flow across the interface and into the spreader. Alternatively, the heat flux vectors for poor thermal contact exhibit a larger in-plane component, reflecting the additional resistance to heat flow across the interface. The contact resistance thus acts to more evenly distribute the heat flux imposed on the spreader, thereby reducing the spreader's effectiveness. Ultimately, the presence of the contact resistance results in larger peak and average temperature rises at the flux spot (these peak and average are 61.9 and 57.7 °C for the perfect interface, respectively, with the poor thermal interface resulting in 96.2 and 91.6 °C peak and average temperatures, respectively).

With the expectation that hot-spot temperatures should increase for escalating contact resistance, a total of 78 ANSYS simulations were run for contact resistances ranging from 0 to 10^{-4} K m^2/W, with model parameters defined in Table 9.6. Figure 9.57 depicts the increase in hot-spot temperature over the stated contact resistance range for differing degrees of spreader anisotropy. The reader is reminded that 10^{-7} K m^2/W is a very low contact resistance that may be achieved by monolithic growth on the back of the chip or through the use of a soldered interface. Alternatively, a poor interface, such as a lightly loaded interface with a phase change material or elastomeric pad, is represented by a contact resistance near 10^{-4} K m^2/W.

For the conditions studied, it is found that the contact resistance has a significant effect on hot-spot temperature, particularly when extreme anisotropy is present in the spreader. In order for a spreader with $k_z = 5$ W/m K and $k_{xy} = 1800$ W/m K to provide at least 10 °C better hot-spot cooling than the isotropic spreader, the contact resistance

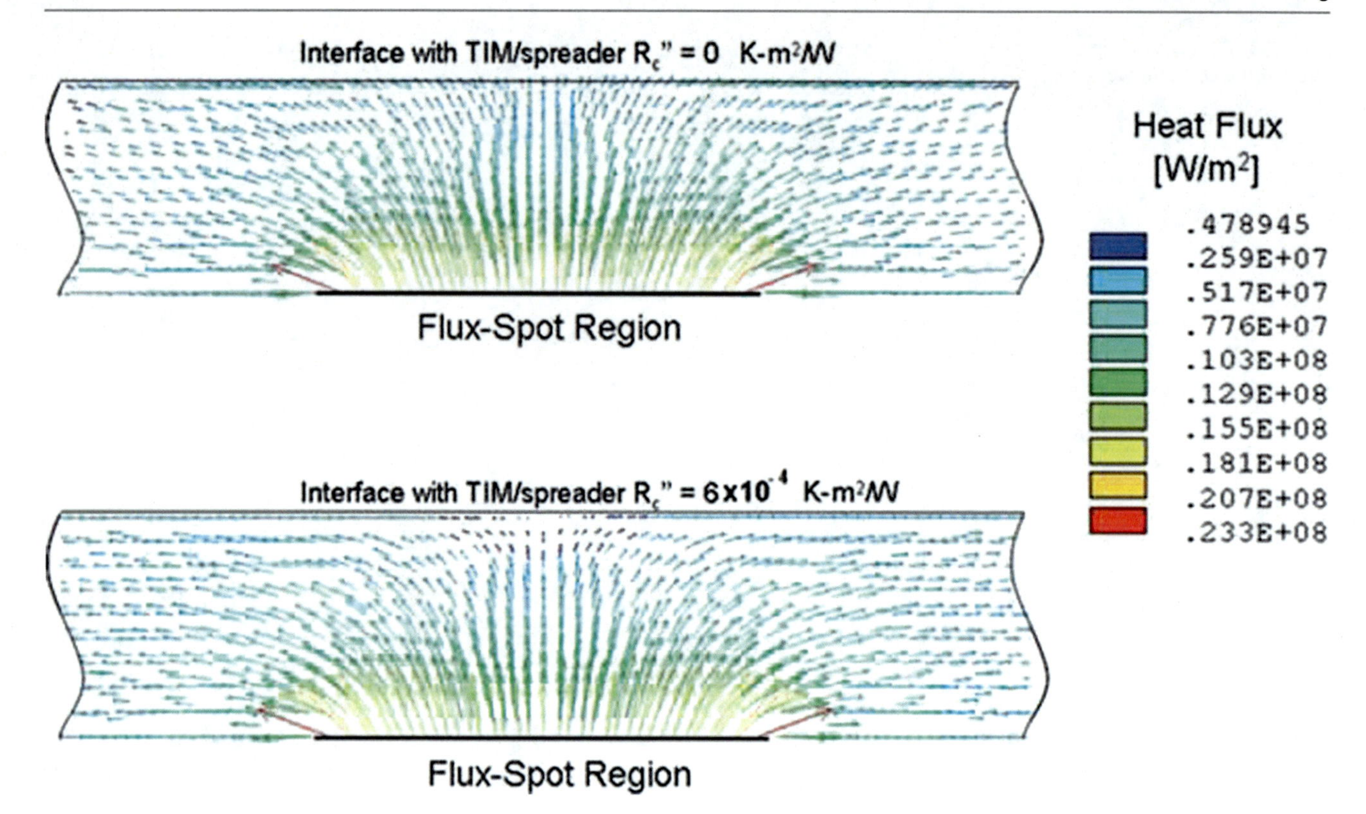

Fig. 9.56 Heat flux plots in the silicon chip for perfect and poor thermal contact

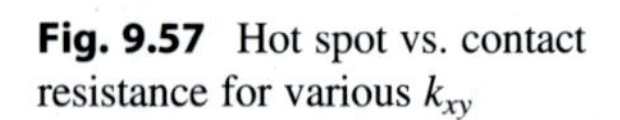
Fig. 9.57 Hot spot vs. contact resistance for various k_{xy}

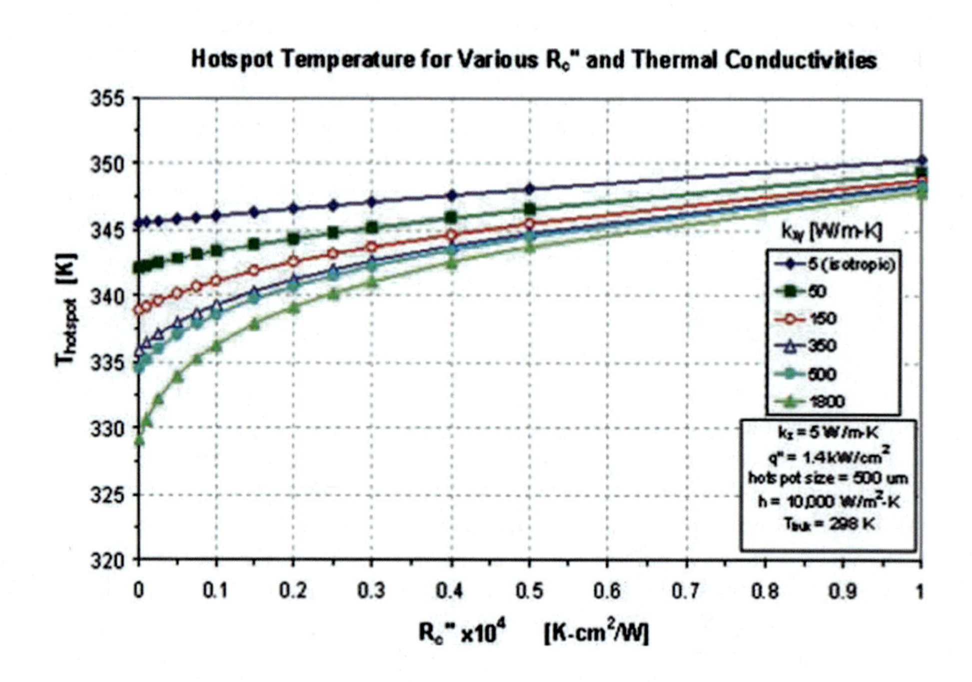

must be maintained below 0.1×10^{-4} K m²/W. However, even for a contact resistance of 0.5×10^{-4} K m²/W, a nearly 5 °C temperature reduction can be achieved by the best orthotropic material and 3 or 4 °C for more commonly available graphite TIM/spreaders.

9.3.4 Experimental Demonstration

Graphite has an anisotropic crystal structure that results in different properties in different directions. Its in-plane thermal conductivity can range from 140 to 1650 W/m K, while its thru-thickness thermal conductivity is much lower and ranges between 2 and 10 W/m K. Based on the analytical and numerical results described in the previous section, it would appear that such anisotropy can be used advantageously to both reduce the severity of a hot spot and to spread the heat in the TIM, thus shielding a surface adjacent to the heat source from a high heat flux. Recently, Xiong et al. experimentally demonstrated that a thin graphite heat spreader can reduce the hot spot significantly [127]. In his experiment, two graphite heat spreaders with different in-plane thermal conductivities and thicknesses were tested and compared. The first graphite material had an in-plane thermal conductivity of 425 W/m K, a thru-thickness thermal conductivity of 3.2 W/m K, and a thickness of 110 μm. The second material was an experimental grade of graphite with an in-plane thermal conductivity estimated to be greater than 1000 W/m K, a thru-thickness thermal conductivity estimated between 5 and 6 W/m K, and a thickness of 30 μm. A 50 × 70 × 1 mm acrylonitrile butadiene styrene (ABS) plate was used as the substrate. A thin graphite heat spreader was adhesively bonded to the ABS substrate, and a constant heat flux was applied over a limited area on one side, while the other side of the ABS substrate was exposed to the ambient for natural convection cooling. Three cases were investigated in their work:

Case 1: without a heat spreader and with a power of 0.75 W on the hot spot.
Case 2: with a 425 W/m K, 110 μm-thick graphite heat spreader, and with a power of 3.0 W on the hot spot.
Case 3: with a 1000 W/m K, 30 μm-thick graphite heat spreader, and with a power of 3.0 W on the hot spot.

Figure 9.58 shows the IR thermal image of the ABS substrate without an attached heat spreader, at a power dissipation of only 0.75 W on the hot spot. Because of the low thermal conductivity of the ABS (~0.25 W/m K), the heat from the hot spot did not spread well, and thus, there is a localized hot spot immediately above the heat source, and the temperature drops dramatically a short distance away from the heat source. The hot-spot temperature is 90 °C, and most of the spreader is at a temperature less than 30 °C.

Figure 9.59 is a thermal image of the ABS with an attached 425 W/m K, 110 μm-thick graphite heat spreader, while Fig. 9.66 is a thermal image of the ABS plate with a 1000 W/m K, 30 μm-thick graphite heat spreader. Note that in both cases, the excellent spreading effect of graphite spreader allowed the hot spot power to be increased to 3.0 W. These results demonstrate the significant heat-spreading effect of graphite heat spreaders. As shown in Figs. 9.59 and 9.60, with much higher power applied on the hot spot, the hot-spot temperatures are only 64 and 69 °C, respectively, much lower than that of the ABS without the spreader as shown in Fig. 9.58. The temperature gradients across the surfaces are also much lower with the entire ABS plate now above 47 °C. These results demonstrate that a very thin spreader made of graphite, with an in-plane thermal

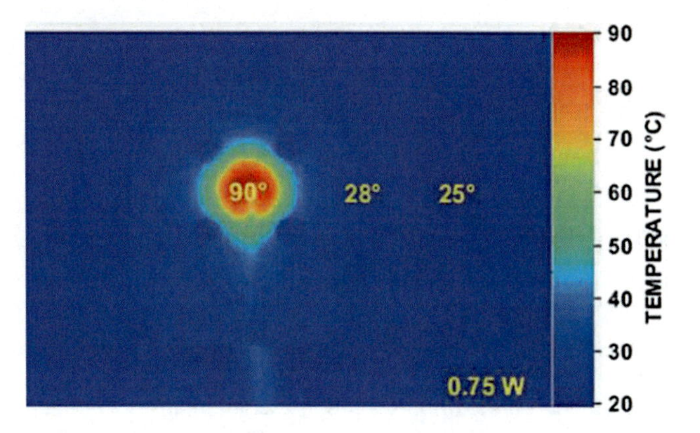

Fig. 9.58 Surface temperature distribution on the ABS substrate without heat spreader (0.75 W was applied on the hot spot)

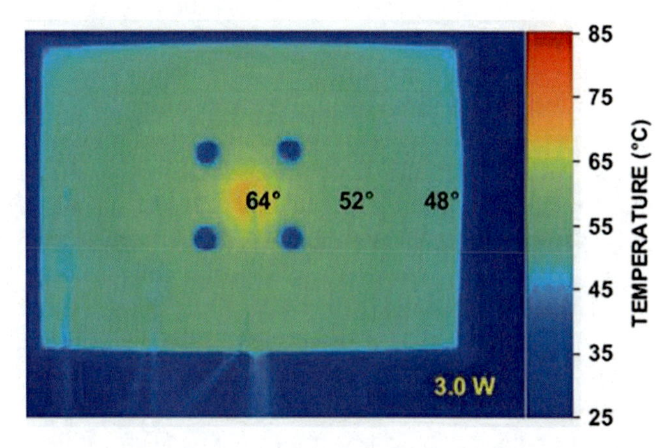

Fig. 9.59 Surface temperature distribution on an ABS substrate with an attached 425 W/m K graphite heat spreader (3.0 W was applied on the hot spot)

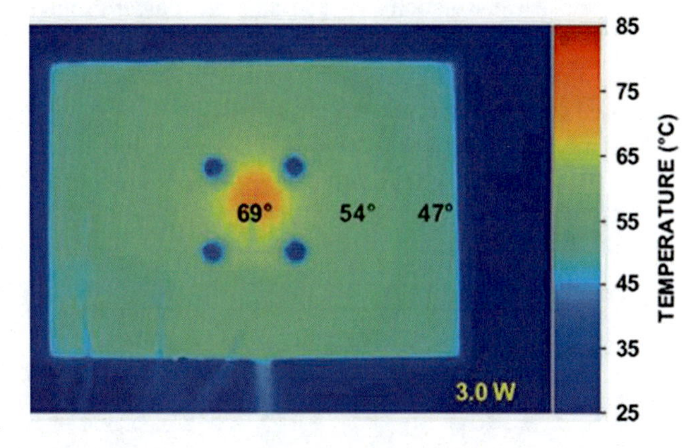

Fig. 9.60 Surface temperature distribution on an ABS substrate with an attached 1000 W/m K graphite heat spreader (3.0 W was applied on the hot spot)

conductivity on the order of 500–1000 W/m K, or other highly orthotropic materials, is a practical thermal solutions for hot-spot remediation in both a passive cooling mode and an active cooling mode.

9.4 On-Chip Hot-Spot Cooling Using Micro-Gap Cooler

As is made abundantly clear in the previous sections of this chapter, the thermal interface resistance between the "cooling solution" and the chip poses a considerable barrier to effective remediation of on-chip hot spots. Direct liquid cooling techniques, which allow for direct contact between an inert dielectric liquid and the surface of the chip and eliminate the TIM (thermal interface material), hold great promise for hot-spot-driven thermal management of ICs. Moreover, the use of phase-change processes, including pool boiling, gas-assisted evaporative cooling, jet impingement, and spray cooling, exploits the latent heat of these liquids to reduce the required mass flow rates and can provide the added advantage of inherently high heat transfer coefficients.

However, direct cooling of microelectronic components imposes stringent chemical, electrical, and thermal requirements on the liquids to be used in this thermal control mode. Direct liquid cooling of microelectronic components requires compatibility between the liquid coolant and a system-specific combination of the chip, chip package, substrate, and printed circuit board materials, e.g., silicon, silicon dioxide, silicon nitride, alumina, o-rings, plastic encapsulants, solder, gold, and epoxy glass. In addition, a liquid coolant must possess the dielectric strength needed to provide electrical isolation between adjacent power/ground conductors and signal lines. Fortunately, 3 M's family of perfluorinated liquids possess the required attributes and has been used extensively in the electronic industry [128, 129].

Recently, Bar-Cohen et al. proposed the use of micro-gap coolers to achieve a volume-efficient application of direct liquid cooling while providing high heat transfer coefficients – in the range needed to control the temperature of on-chip hot spots – on the back of the chip [130]. As shown in Fig. 9.61, in a micro-gap cooler a narrow submillimeter channel is created above the chip or substrate and liquid is pumped through the channel, thus removing the heat dissipated from the chip. Compared with a more conventional microchannel cooler, which needs to be attached with a TIM to the chip/substrate, micro-gap coolers require no attachment and no micromachining and could be a very attractive cooling approach for on-chip hot-spot remediation. Subsequent subsections deal with the empirical validation of this novel thermal management approach.

9.4.1 Single-Phase Experiments

Following single-phase heat transfer experiments, with water and FC-72 as the working fluid, and successful comparison to the predicted heat transfer coefficients obtained with established correlations and CFD simulation using IcePak [131], the researchers turned their attention to two-phase thermal transport in the micro-gap channel. The test results for the 210 μm gap two-phase heat transfer experiment are shown in Fig. 9.62. In these experiments, the mass flux was varied from 130 to 660 kg/m^2 s in steps of 130 kg/m^2 s, and the corresponding inlet fluid velocity was 0.0794, 0.159, 0.238, 0.317, 0.397 m/s. The average heat-transfer coefficient was found to generally display a parabolic variation with the heat flux, increasing toward a peak value as the channel condition changed from subcooled to saturated flow boiling and then decreasing with further increases of heat flux.

The 110 μm gap channel shows some similarities to the heat transfer coefficient variation observed in the 210 μm gap but – as shown in Fig. 9.63 – it appears to possess only the downsloping branch of the previously observed parabolic trend. The peak heat transfer coefficients attained in the 110 μm gap channel are higher than the 210 μm gap or the

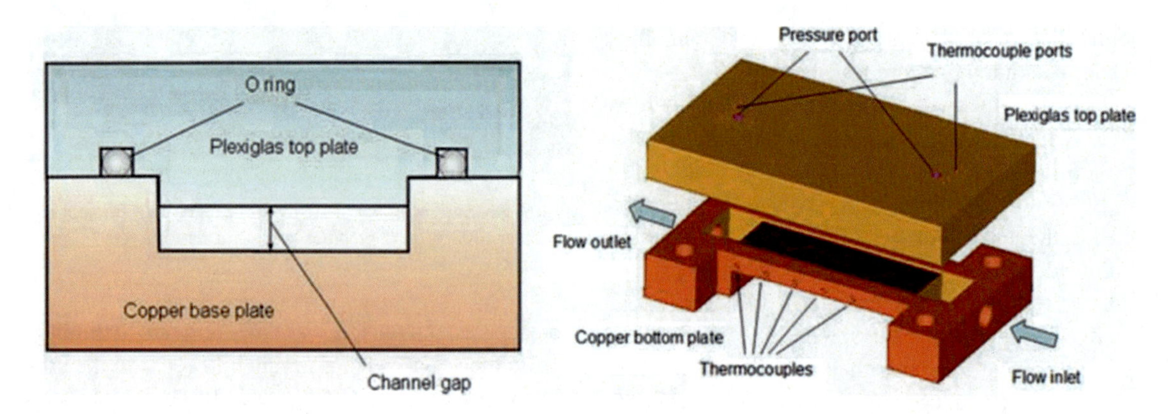

Fig. 9.61 Schematic diagram of micro-gap cooler

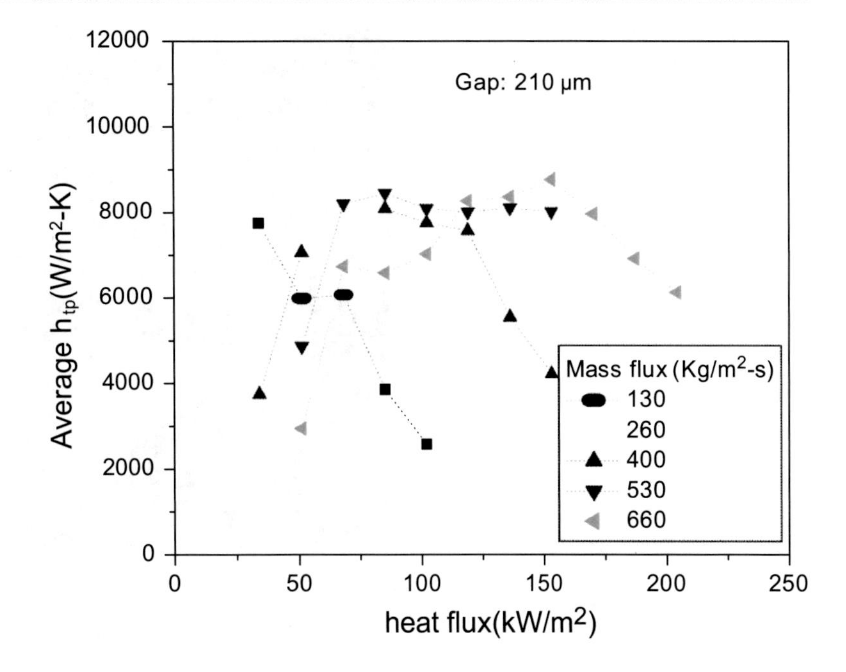

Fig. 9.62 Average heat transfer coefficient vs. heat flux of 210 μm-gap cooler

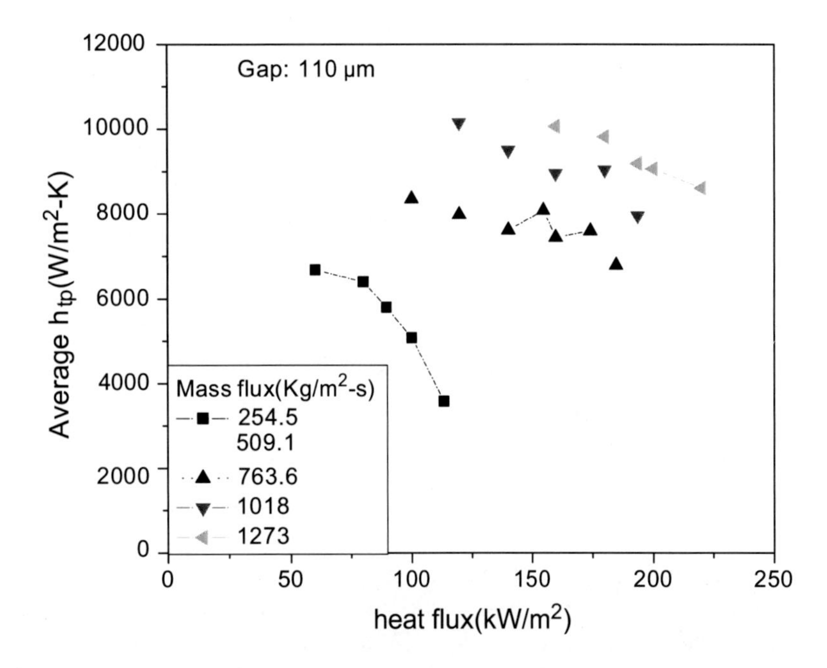

Fig. 9.63 Average heat transfer coefficient vs. heat flux for 110 μm-gap cooler

500 μm gap channels, and the downward trend with heat flux moderates substantially as the mass flux increases.

The above and other reported results show that the area-averaged heat transfer coefficients for FC-72, of between 7.5 and 15.5 kW/m^2 K, can be attained in micro-gap channels of 110–500 μm. These values are significantly higher than the single phase FC-72 values, showing that two-phase micro-gap cooler can provide the high cooling capability needed for hot spot remediation.

9.4.2 Application to Hot-Spot Remediation

To assess the efficacy of micro-gap cooler for thermal management of hot spots, it is instructive to simulate the thermal performance of a notional advanced semiconductor chip cooled by the micro-gap coolers. A 10 × 10 mm silicon chip, 500 μm thick, dissipating a uniform heat flux of 100 W/cm^2 across nearly all the active chip area serves as the test vehicle for this simulation. In keeping with the theme

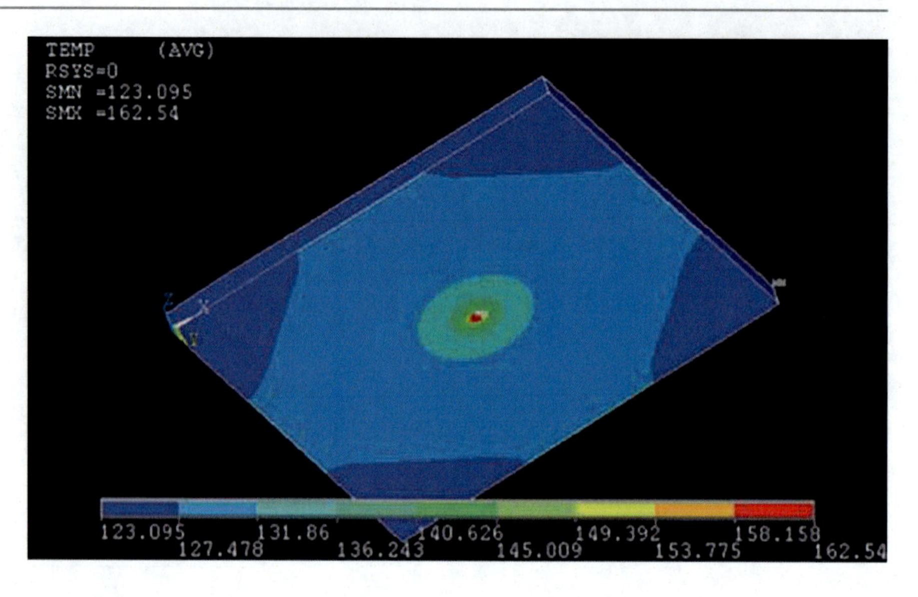

Fig. 9.64 Three-dimensional temperature profile for direct liquid cooling of advanced semiconductor chip [10 × 10 × 0.5 mm, $q'' = 100$ W/cm^2, $q''_{hs} = 2$ kW/cm^2, $d_{hs} = 400$ μm, $h = 10$ kW/m^2 K]

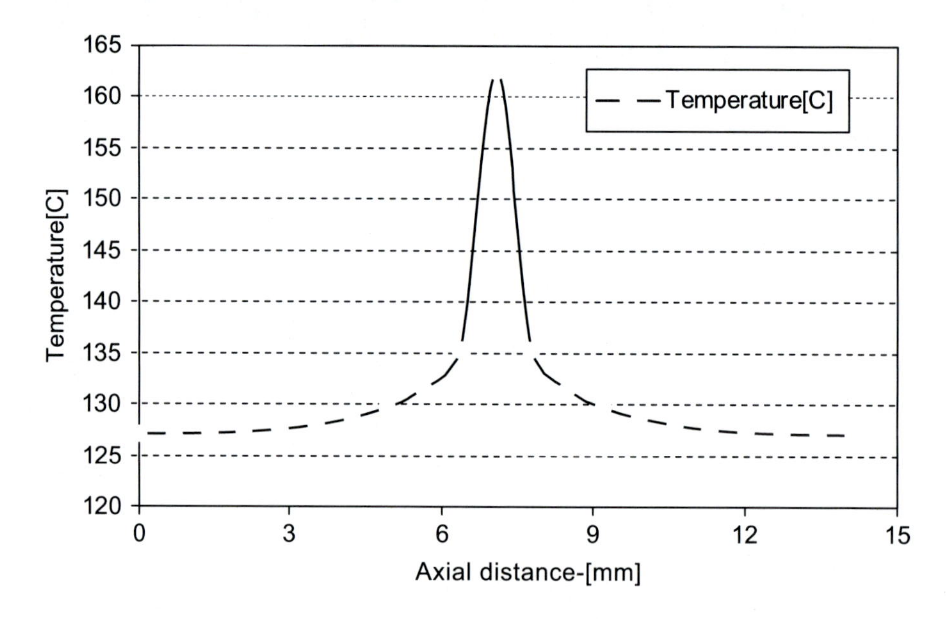

Fig. 9.65 Temperature distribution across the diagonal of the chip [$q_{chip} = 100$ W/cm^2, $q_{spot} = 2000$ W/cm^2, $h = 10{,}000$ W/m^2 K, $T_c = 22$ °C]

of this chapter, the chip is assumed to possess a central, circular hot spot, varying from 100 to 400 μm in diameter and dissipating between 1 and 2 kW/cm^2. It is further assumed that the thermal conductivity of the silicon chip is invariant at 125 W/m K, that it is cooled from the back surface (opposite to that of the active circuitry) with heat transfer coefficients that can vary from 5 to 20 kW/m^2 K, reflective of the values that can be potentially achieved with micro-gap coolers, and that the liquid temperature is 22 °C.

Figure 9.64 presents the three-dimensional temperature profile while, Fig. 9.65 depicts the temperature along a diagonal, on the active face of the silicon chip for a baseline microgap-cooled chip configuration with a 400 μm hot spot, generating a 2 kW/cm^2 heat flux. As observed in Figs. 9.64 and 9.65, when this notional baseline chip, with a very severe hot spot, is cooled by a micro-gap cooler with an h equal to 10 kW/m^2 K, it experiences an elevated average temperature of approximately 130 °C and a significant hot spot with a maximum temperature of 163 °C, or some 33 °C above the average chip temperature. The average and peak temperatures for various other combinations of the specified parameters are shown in Tables 9.7 and 9.8.

Tables 9.7 and 9.8 present the results for a hot-spot diameter of 100 μm and 400 μm, respectively, with various hot-spot heat fluxes and a range of heat transfer coefficients associated with the micro-gap coolers. In these tables, the first and second columns present the hot-spot flux and convective coefficients, respectively, while the next three columns provide the average temperature on the active side of the chip, the average temperature on the cooled (back) side

Table 9.7 Temperatures for 100 μm hot-spot diameter for various heat flux and cooling conditions

q''_{spot} [W/cm^2]	h [W/m^2 K]	T_{chip} [°C]	T_{conv} [°C]	T_{spot} [°C]	T_{spot_max} [°C]
1000	20,000	76.2	74.2	78.9	79.7
1000	10,000	126.2	124.2	129.0	129.7
1000	5000	226.3	224.3	229.1	229.8
2000	20,000	76.4	74.4	82.2	83.8
2000	10,000	126.5	124.6	132.3	134.0
2000	5000	226.7	224.7	232.5	234.1

Table 9.8 Temperatures for 400 μm hot-spot diameter for various heat flux and cooling conditions

q''_{spot} [W/cm^2]	h [W/m^2 K]	T_{chip} [°C]	T_{conv} [°C]	T_{spot} [°C]	T_{spot_max} [°C]
1000	20,000	7.75	75.6	89.1	92.2
1000	10,000	128.2	126.4	140.2	143.3
1000	5000	229.4	227.6	241.6	244.7
2000	20,000	79.2	77.5	103.7	110.3
2000	10,000	130.7	129.0	155.9	162.5
2000	5000	233.2	231.6	259.0	265.6

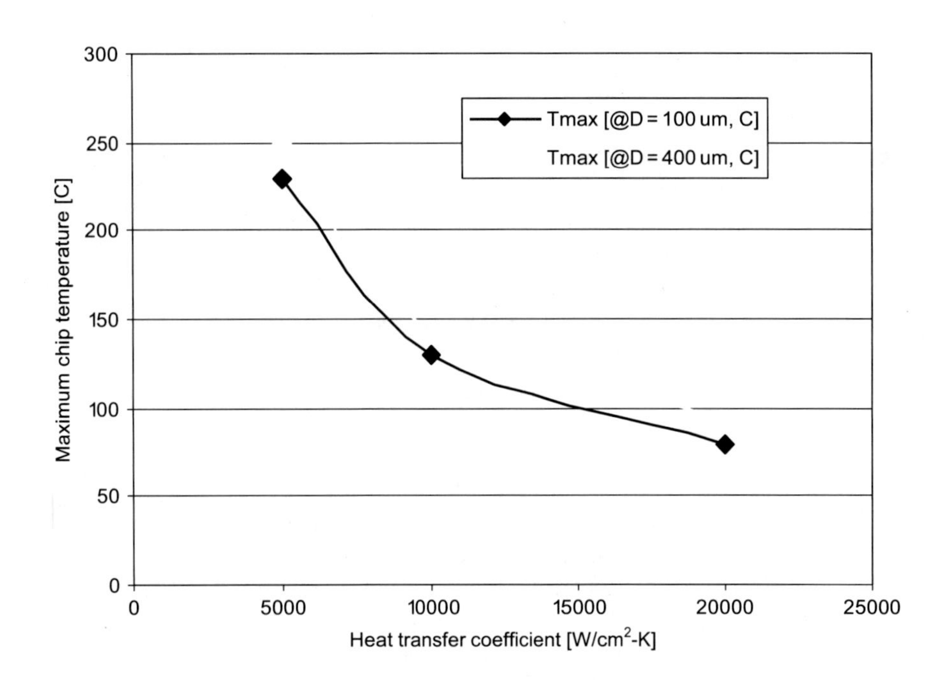

Fig. 9.66 Effect of convective coefficient on the hot-spot temperature for $q''_{hs} = 1000$ W/cm^2

of the chip, and the average hot-spot temperatures. The last column in Tables 9.7 and 9.8 presents the maximum temperature on the active side of the silicon chip. Not surprisingly, the average chip temperature (on both the active and wetted surfaces) is seen to vary directly with the heat transfer coefficient, while the on-chip hot-spot temperature rise is conduction limited and – for the fixed chip geometry and thermal conductivity – is driven by the heat flux and size of the hot spot. Thus, as seen in Table 9.8, while raising the heat transfer coefficient to 20 kW/m^2 K lowers the average chip temperature to 77 °C for a chip heat flux of 100 W/cm^2, the on-chip temperature rise for a 1 kW/cm^2, 400 μm hot spot, remains at approximately 15 °C across the range of heat transfer coefficients from 5 kW to 20 kW/cm^2 K. However, since the peak chip temperature is established by the superposition of these two effects, any reduction in the average chip temperature has a salutary effect on the peak chip temperature, as well.

Thus, as also revealed in Fig. 9.66, micro-gap coolers, along with effective thermal spreading in the chip, appear to offer the potential for successfully limiting the chip and hot-spot temperature rise to acceptable levels for a wide range of operating conditions. Most significantly, a heat transfer coefficient of 20 kW/m^2 K, which is thought to be attainable in a micro-gap cooler [132] could be used to effectively cool a most challenging 2 kW/cm^2, 400 μm hot spot, along with maintaining an acceptable average temperature, for a 100 W/cm^2 chip.

Interestingly, the on-chip temperature rise – of the hot-spot center relative to the chip average – can be seen to vary almost directly with the product of the heat flux and diameter, yielding a ninefold increase from 3.6 °C for a 100 μm, 1 kW/cm^2 hot spot to 32.4 °C for a 400 μm, 2 kW/cm^2 hot spot. Due to the superposition of the convective and conductive effects, it may also be noted that while a large change (~50%) in the maximum excess temperature results from increasing the micro-gap heat transfer coefficient from 5 to 10 kW/m^2 K, further increases to 20 kW/m^2 K only reduce the maximum temperature rise by approximately 30%.

9.5 Conclusions

The preceding chapter addresses on-chip, hot-spot cooling, which has become one of the most active and challenging domains in the thermal management of nanoelectronic devices and packages. Following a brief discussion of several passive and active high heat flux thermal management techniques, attention was turned to the physical phenomena underpinning the most promising on-chip thermal management approaches, including thin-film and miniaturized thermoelectric coolers, orthotropic TIMs/heat spreaders, and phase-change micro-gap coolers and their use for remediation of these hot spots. It was shown that, with proper thermal optimization, mini-contact-enhanced miniaturized "bulk" thermoelectric coolers can yield hot-spot temperature reductions in excess of 15 °C for near millimeter-sized hot spots with kW/cm^2-level heat fluxes but to be vulnerable to the deleterious effects of thermal contact resistance. Micro-scaled thin-film silicon thermoelectric coolers, monolithically grown, or fashioned, on the back of silicon chips, were similarly found to provide effective thermal management of high heat flux spots, and to be capable of neutralizing the local temperature rise for a large variety of submillimeter hot spots. Orthotropic TIM/spreaders with high in-plane thermal conductivities were also shown to offer significant temperature reduction capability even for large, high-flux spots, when used in conjunction with a more conventional, very high heat transfer coefficient thermal management approaches. Initial research results for micro-gap coolers, relying on the boiling and evaporation of a dielectric liquid flowing in a miniature gap at the back of the chip, strongly suggest that this cooling technique could provide both the local and global heat transfer coefficients needed to meet many of the most demanding nanoelectronic cooling challenges, including severe on-chip hot spots.

References

1. ITRS, The International Technology Roadmap for Semiconductors, 2004, Semiconductor Industry Association, http://www.ITRSnemi.org
2. Shelling, P., Li, S., Goodson, K.E.: Managing heat for electronics. Mater. Today. **8**, 30–35 (2005)
3. Mahajan, R., Chiu, C., Chrysler, G.: Cooling a microprocessor chip. Proc. IEEE. **94**, 1476–1486 (2006)
4. Pedram, M., Nazarian, S.: Thermal modeling, analysis, and management in VLSI circuits: principles and methods. Proc. IEEE. **9**, 1487–1501 (2006)
5. Bar-Cohen, A., Arik, M., Ohadi, M.: Direct liquid cooling of high flux micro and nano electronic components. Proc. IEEE. **94**, 1549–1570 (2006)
6. Mudawar, I.: Assessment of high-heat-flux thermal management schemes. IEEE Trans. Compon. Packag. Technol. Part A Packag. Technol. **24**, 122–141 (2001)
7. Garimella, S.V.: Advances in mesoscale thermal management technologies for microelectronics. Microelectron. J. **37**, 1165–1185 (2006)
8. Sinha, S., Goodson, K.E.: Thermal conduction in sub-100 nm transistors. Microelectron. J. **37**, 1148–1157 (2006)
9. NEMI, Electronics Manufacturing Initiative Technology Roadmap, 2006, http://www.nemi.org
10. Jannadham, K., Watkins, T.R., Dinwiddie, R.B.: Novel heat spreader coatings for high power electronic devices. J. Mater. Sci. **37**, 1363–1376 (2002)
11. Dahlgren, S.: High-pressure polycrystalline diamond as a cost effective heat spreader. Proc. Therm. Thermomech. Phenomena Electron. Syst. (ITHERM 2000). **1**, 23–26 (2000)
12. Goodson, K.E., Ju, Y.S.: Heat conduction in novel electronic films. Annu. Rev. Mater. Sci. **29**, 261–293 (1999)
13. Le Berre, M., Pandraud, G., Morfouli, P., Lallemand, M.: The performance of micro heat pipes measured by integrated sensors. J. Micromech. Microeng. **16**, 1047–1050 (2006)
14. Peterson, G.P., Duncan, A.B., Weichold, M.H.: Experimental investigation of micro heat pipes fabricated in silicon wafers. J. Heat Transf. **115**, 750–756 (1993)
15. Peterson, G.P., Duncan, A.B., Weichold, M.H.: Experimental investigation of micro heat pipes fabricated in silicon wafers. ASME J. Heat Trans. **115**, 751–756 (1993)
16. Mallik, A.K., Peterson, G.P., Weichold, M.H.: Fabrication of vapor deposited micro heat pipes arrays as an integral part of semiconductor devices. IEEE J. Microelectromechan. Syst. **4**, 119–131 (1995)
17. Karimi, G., Culham, J.R.: Review and assessment of pulsating heat pipe mechanism for high heat flux electronic cooling. Proc. ITHERM. **2**, 52–59 (2004)
18. Benson, D.A., Adkins, D.R., Peterson, G.P., Mitchell, R.T., Tuck, M.R., Palmer, D.W.: Turning silicon substrates into diamond: micro machining heat pipes. Proc. Adv. Design Mater. Process Apr. 19–21 (1996)
19. Suman, B.: Modeling, experiment, and fabrication of micro-grooved heat pipes: an update. Appl. Mech. Rev. **60**, 107–119 (2007)
20. Akachi, H.; Structure of a heat pipe, U.S. Patent #4,921,041 (1990)
21. Zuo, Z.J., North, M.T., Wert, K.L.: High heat flux heat pipe mechanism for cooling of electronics. IEEE Trans. Compon. Packag. Technol. **24**, 220–225 (2001)
22. Lin, C., Ponnappan, R., Leland, J.: High performance miniature heat pipe. Int. J. Heat Mass Transf. **45**, 3131–3142 (2002)
23. Lee, M., Wong, M., Zohar, Y.: Integrated micro-heat-pipe fabrication technology. J. Microelectromech. Syst. **12**, 138–146 (2003)

24. Lee, M., Wong, M., Zohar, Y.: Characterization of an integrated micro heat pipe. J. Micromech. Microeng. **13**, 58–64 (2003)
25. Khrustalev, D., Faghri, A.: Thermal characteristics of conventional and flat miniature axially-grooved heat pipes. J. Heat Transf. **117**, 740–747 (1995)
26. Ma, H.B., Lofgreen, K.P., Peterson, G.P.: An experimental investigation of a high flux heat pipe heat sink. J. Electron. Packag. **128**, 18–22 (2006)
27. Gillot, G., Avenas, Y., Cezac, N., Poupon, G., Schaeffer, C., Fournier, E.: Silicon heat pipes used as thermal spreaders. IEEE Trans. Compon. Packag. Technol. **26**, 332–339 (2003)
28. Lin, C., Ponnappan, R., Leland, J.: High performance miniature heat pipe. Int. J. Heat Mass Transf. **45**, 3131–3142 (2002)
29. Tuckerman, D.B., Pease, R.F.W.: High-performance heat sinking for VLSI. IEEE Electron Device Lett. **EDL-2**, 143–150 (1981)
30. Prasher, R., Chang, J., Sauciuc, I., Narasimhan, S., Chau, D., Chrysler, G., Myers, A., Prstic, A., Hu, C.: Nano and micro technology-based next-generation package level cooling solutions. Int. J. Technol. **9**, 285–296 (2005)
31. Bowers, M., Mudawar, I.: High-flux boiling in low-flow rate, low-pressure drop mini-channel and microchannel heat sinks. Int. J. Heat Mass Transf. **37**, 321–332 (1994)
32. Mudawar, I., Maddox, D.E.: Enhancement of critical heat flux from high power micro-electronic heat sources in a flow channel. J. Electron. Packag. **112**, 241–248 (1990)
33. Bar-Cohen, A., Arik, M., Ohadi, M.: Direct liquid cooling of high flux micro and nano electronic components. Proc. IEEE. **94**, 1549–1570 (2006)
34. Garimella, S.V., Singhal, V., Liu, D.: On-chip thermal management with microchannel heat sinks and integrated micropumps. Proc. IEEE. **94**, 1534–1548 (2006)
35. Prasher, R.: Thermal interface materials: historical perspective, status and future directions. Proc. IEEE. **94**, 1571–1586 (2006)
36. Zhang, L., Wang, E.N., Koo, J.M., Goodson, K.E., Santiago, J.G., Kenny, T.W.; Microscale liquid jet impingement, Proceedings of AMSE IMECE; Vol. 2: Paper No. MEME–23820 (2001)
37. Wolf, D.H., Incropera, F.P., Viskanta, R.: Local jet impingement boiling heat transfer. Int. J. Heat Mass Transf. **39**, 1395–1406 (1996)
38. Kiper, A.M.: Impinging water jet cooling of VLSI circuits. Int. Commun. Heat Mass Transf. **11**, 126–129 (1984)
39. Harman, T.C., Taylor, P.J., Walsh, M.P., LaForge, B.E.: Quantum dot superlattice thermoelectric materials and devices. Science. **297**, 2229–2232 (2002)
40. Venkatasubramanian, R., Siivola, E., Colpitts, T., O'Quinn, B.: Thin-film thermoelectric devices with high room-temperature figures of merit. Nature (London). **413**, 597–602 (2001)
41. Fan, X., Zeng, G., LaBounty, C., Bowers, J., Croke, E., Ahn, C., Huxtable, S., Majumdar, A.: SiGeC/Si superlattice micro-coolers. Appl. Phys. Lett. **78**, 1580–1600 (2001)
42. Chen, C., Yang, B., Liu, W.L.: Engineering nanostructures for energy conversion. In: Faghri, M., Sunden, B. (eds.) Heat Transfer and Fluid Flow in Microscale and Nanoscale Structures. WIT Press, Southampton (2004)
43. Yang, B., Liu, W.L., Wang, K.L., Chen, G.: Simultaneous measurements of Seebeck coefficient and thermal conductivity across superlattice. Appl. Phys. Lett. **80**, 1758–1760 (2002)
44. Shakouri, A., Zhang, Y.: On-chip solid-state cooling for integrated circuits using thin-film microrefrigerators. IEEE Trans. Compon. Packag Technol. **28**, 65–69 (2005)
45. Zhang, Y., Zeng, G.H., Piprek, J., Bar-Cohen, A., Shakouri, A.: Superlattice microrefrigerators fusion bonded with optoelectronic devices. IEEE Trans. Compon. Packag Technol. **28**, 658–666 (2005)
46. Simons, R.E., Ellsworth, M.J., Chu, R.C.: An assessment of module cooling enhancement with thermoelectric coolers. J. Heat Transf. **127**, 76–84 (2005)
47. Yeh, L., Chu, C.: Thermal Management of Microelectronic Equipment. ASME Press, New York (2002)
48. Kraus, A.D., Bar-Cohen, A.: Thermal Analysis and Control of Electronic Equipment. Hemisphere Publishing Corporation, New York (1983)
49. Fan, X.; Ph. D. Thesis, University of California at Santa Barbara, March 2002
50. Fleurial, J.-P., Borshchevsky, A., Ryan, M.A., Phillips, W., Kolawa, E., Kacisch, K., Ewell, R.: Thermoelectric microcoolers for thermal management applications. In: Proceedings of 16th International Conference on Thermoelectrics, pp. 641–645 (1997)
51. Venkatasubramanian, R., Siivola, E., Colpitts, T., O'Quinn, B.: Thin-film thermoelectric devices with high room-temperature figures of merit. Nature (London). **413**, 597–602 (2001)
52. Harman, T.C., Taylor, P.J., Walsh, M.P., LaForge, B.E.: Quantum dot Superlattice thermoelectric materials and devices. Science. **297**, 2229–2232 (2002)
53. Fan, X., Zeng, G., Croke, E., LaBounty, C., Shakouri, A., Bowers, J.E.: Integrated SiGeC/Si micro cooler. Appl. Phys. Lett. **78**(11), 1580–1582 (2001)
54. Semenyuk, V.: Thermoelectric micro modules for spot cooling of high density heat sources. In: Proceedings of the 20th International Conference on Thermoelectrics, pp. 391–396 (2001)
55. Semenyuk, V.: Cascade thermoelectric micro modules for spot cooling high power electronic components. In: Proceedings of the 21st International Conference on Thermoelectrics, pp. 531–534 (2002)
56. Semenyuk, V.: Thermoelectric cooling of electro-optic components. In: Rowe, D.M. (ed.) Thermoelectrics Handbook: Macro to Nano. CRC Press, Boca Raton (2006)
57. www.thermion–company.com
58. Ioffe, A.F.: Semiconductor Thermoelements and Thermoelectric Cooling. Infosearch Ltd, London (1957)
59. Ettenberg, M.H., Jesser, M.A., Rosi, E.D.: A new n-type and improved p-type pseudo-ternary (Bi2Te3)(Sb2Te3)(Sb2Se3) alloy for Peltier cooling. In: Proceedings of the 15th International Conference on Thermoelectrics, pp. 52–56 (1996)
60. Yamashita, O., Tomiyoshi, S.: Effect of annealing on thermoelectric properties of bismuth telluride compounds. Jpn. J. Appl. Phys. **42**, 492–500 (2003)
61. Yamashita, O., Tomiyoshi, S., Makita, K.: Bismuth telluride compounds with high thermoelectric figures of merit. J. Appl. Phys. **93**, 368–374 (2003)
62. Yamashita, O., Tomiyoshi, S.: High performance n-type bismuth telluride with highly stable thermoelectric figure of merit. J. Appl. Phys. **95**, 6277–6283 (2004)
63. Shakouri, A., Zhang, Y.: On-chip solid-state cooling for integrated circuits using thin-film microrefrigerators. IEEE Trans. Compon. Packag Technol. **28**, 65–69 (2005)
64. Pandey, R.K., Sahu, S.N., Chandra, S.: Handbook of Semiconductor Deposition. Marcel Dekker, New York (1996)
65. Snyder, G.J., Lim, J.R., Huang, C., Fleurial, J.-P.: Thermoelectric microdevice fabricated by a MEMS-like electrochemical process. Nat. Mater. **2**, 528–531 (2003)
66. da Silva, L.W., Kaviany, M., Uher, C.: Thermoelectric performance of films in the bismuth-tellurium and antimony-tellurium systems. J. Appl. Phys. **97**, 114903 (2005)
67. Böttner, H., Nurnus, J., Gavrikov, A., Kühner, G., Jägle, M., Künzel, C., Eberhard, D., Plescher, G., Schubert, A., Schlereth, K.: New thermoelectric components using microsystem technologies. J. Microelectromech. Syst. **13**, 414–420 (2004)

68. Bottner, H.: Micropelt miniaturized thermoelectric devices: small size, high cooling power densities, short response time. In: Proceedings of the 24th International Conference on Thermoelectrics, pp. 1–8 (2005)
69. Zhou, H., Rowe, D.M., Williams, S.: Peltier effect in a co-evaporated Sb2Te3(P)– Bi2Te3(N) thin film thermocouple. Thin Solid Films. **408**, 270–274 (2002)
70. Semenyuk, V.: Miniature thermoelectric modules with increased cooling power. In: Proceedings of the 25th International Conference on Thermoelectrics, pp. 322–326 (2006)
71. Semenyuk, V., Dissertation, P.D.: Odessa Technological Institute of Food and Refrigeratinbg Engineering. USSR, Odessa (1967). (in Russian)
72. Hicks, L.D., Dresselhaus, M.S.: Thermoelectric figure of merit of a one-dimensional conductor. Phys. Rev. B. **47**, 16631–16634 (1993)
73. Balandin, A., Wang, K.L.: Effect of phonon confinement on the thermoelectric figure of merit of quantum wells. J. Appl. Phys. **84**, 6149–6153 (1998)
74. Balandin, A., Lazarenkova, O.L.: Mechanism for thermoelectric figure-of-merit enhancement in regimented quantum dot superlattices. Appl. Phys. Lett. **82**, 415–417 (2003)
75. Yang, B., Chen, G.: In: Tritt, T.M. (ed.) Thermal Conductivity: Theory, Properties and Application. Kluwer Press, New York (2005)
76. Harman, T.C., Taylor, P.J., Walsh, M.P., LaForge, B.E.: Quantum dot superlattice thermoelectric materials and devices. Science. **297**, 2229–2232 (2002)
77. Venkatasubramanian, R., Silvona, E., Colpitts, T., O'Quinn, B.: Thin-film thermoelectric devices with high room-temperature figures of merit. Nature. **413**, 597–602 (2001)
78. Mahan, G.D., Woods, L.M.: Multilayer thermionic refrigeration. Phys. Rev. Lett. **80**, 4016–4019 (1998)
79. Shakouri, A., LaBounty, C., Piprek, J., Abraham, P., Bowers, J.E.: Thermionic emission cooling in single barrier heterostructures. Appl. Phys. Lett. **74**, 88–89 (1999)
80. Rowe, D.M.: Thermoelectrics Handbook Macro to Nano. CRC Press, Boca Raton (2005)
81. Venkatasubramanian, R., Colpitts, T., Liu, S., El-Masry, N., Lamvik, M.: Low-temperature organometallic epitaxy and its application to superlattice structures in thermoelectrics. Appl. Phys. Lett. **75**, 1104–1106 (1999)
82. Zhang, Y., Zeng, G., Bar-Cohen, A., Shakouri, A.; Is ZT the main performance factor for hot spot cooling using 3D microrefrigerators?, IMAPS on Thermal Management, 2005, Oct. 26th – 28th, Palo Alto, CA
83. Zeng, G., Shakouri, A., LaBounty, C., Robinson, G., Croke, E., Abraham, P., Fan, X., Reese, H., Bowers, J.E.: SiGe micro-cooler. Electron. Lett. **35**, 2146–2147 (1999)
84. Zeng, G., Fan, X., LaBounty, C., Croke, E., Zhang, Y., Christofferson, J., Vashaee, D., Shakouri, A., Bowers, J.E.: Cooling power density of SiGe/Si superlattice micro refrigerators. Proc. Thermoelectr. Mater. Res. Appl. **793**, 43–49 (2003)
85. Fan, X., Zeng, G., LaBounty, C., Vashaee, D., Christofferson, J., Shakouri, A., Bowers, J.E.: Integrated cooling for Si-based microelectronics. In: Proceedings of 20th International Conference on Thermoelectrics, pp. 405–408 (2001)
86. Fan, X.: SiGeC/Si superlattice microcoolers. Appl. Phys. Lett. **78**, 1580–1582 (2001)
87. Fan, X., Zeng, G., LaBounty, C., Croke, E., Vashaee, D., Shakouri, A., Ahn, C., Bowers, J.E.: High cooling power density SiGe/Si micro coolers. Electron. Lett. **37**, 126–127 (2001)
88. Herwaarden, A.W., Sarro, P.M.: Thermal sensors based on the Seebeck effect. Sensors Actuators. **10**, 321–346 (1986)
89. Geballe, T.H., Hull, G.W.: Seebeck effect in silicon. Phys. Rev. **98**, 940–970 (1955)
90. Rowe, D.M.: CRC Handbook of Thermoelectrics. CRC Press, Roca Raton (1995)
91. Zhang, Y., Shakouri, A., Zeng, G.: High-power-density spot cooling using bulk thermoelectrics. Appl. Phys. Lett. **85**, 2977–2979 (2004)
92. Nieveld, G.D.: Thermopiles fabricated using silicon planar technology. Sensors Actuators. **3**, 179–183 (1983)
93. Chapman, P.W., Tfte, O.N., Zook, J.D., Long, D.: Electrical properties of heavily doped silicon. J. Appl. Phys. **34**, 3291–3295 (1963)
94. Wang, P., Bar-Cohen, A., Yang, B.: Analytical modeling of silicon thermoelectric microcooler. J. Appl. Phys. **100**, 14501 (2006)
95. Wang, P., Bar-Cohen, A.: On-chip hot spot cooling using silicon-based thermoelectric microcooler. J. Appl. Phys. **102**, 034503 (2007)
96. Solbrekken, G.L., Zhang, Y., Bar-Cohen, A., Shakouri, A.: Use of superlattice thermionic emission for "Hotspot" reduction in convectively-cooled chip. In: Proceedings of 9th ITHERM's;04, pp. 610–616 (2004)
97. Wang, P., Bar-Cohen, A., Yang, B., Solbrekken, G.L., Zhang, Y., Shakouri, A.; Thermoelectric microcooler for hotspot thermal management, Proceedings of Inter PACK's;05, 2005; Paper No: 2005-7324
98. Lide, D.R.: CRC Handbook of Chemistry and Physics, 75th edn. CRC Press, Boca Raton (1994)
99. Kraus, A.D., Bar-Cohen, A.: Thermal Analysis and Control of Electronic Equipment. Hemisphere Publishing Corporation, New York (1983)
100. Geballe, T.H., Hull, G.W.: Seebeck effect in silicon. Phys. Rev. **98**, 940–970 (1955)
101. Herwaarden, A.W., Sarro, P.M.: Thermal sensors based on the Seebeck effect. Sensors Actuators. **10**, 321–346 (1986)
102. Horn, F.H.: Densitometric and electrical investigation of boron in silicon. Phys. Rev. **97**, 1521–1525 (1955)
103. Fritzsche, H.: A general expression for the thermoelectric power. Solid State Commun. **9**, 1813–1815 (1971)
104. Chang, C.Y., Fang, Y.K., Sze, S.M.: Specific contact resistance of metal-semiconductor barriers. Solid State Electron. **14**, 541–550 (1971)
105. Fan, X., Silicon, M.; Ph. D. Thesis, University of California at Santa Barbara (2002)
106. Yang, B., Wang, P., Bar-Cohen, A.: Mini-contact enhanced thermoelectric cooling of hot spot in high power devices. IEEE Trans. Compon. Packag Technol., Part A. **30**, 432–438 (2007)
107. Narasimhan, S., Lofgreen, K., Chau, D., Chrysler, G.; Thin film thermoelectric cooler thermal validation and product thermal performance estimation, Proceedings of 10th Intersociety Conference on Thermal and Thermo-mechanical Phenomena in Electronics Systems, San Diego, CA, May 30–June 2 (2006)
108. Gwinn, J.P., Webb, R.L.: Performance and testing of thermal interface materials. Microelectron. J. **34**, 215–222 (2003)
109. Singhal, V., Siegmund, T., Garimella, S.V.: Optimization of thermal interface materials for electronics cooling applications. IEEE Trans. Compon. Packag Technol. **27**, 244–252 (2004)
110. Labudovic, M., Li, J.: Modeling of TE cooling of pump lasers. IEEE Trans. Compon. Packag Technol. **27**, 724–730 (2004)
111. So, W.W., Lee, C.C.: Fluxless process of fabricating In-Au joints on copper substrates. IEEE Trans. Compon. Packag Technol. **23**, 377–382 (2000)
112. www.thermion–company.com
113. Semenyuk, V.: Thermoelectric cooling of electro-optic components. In: Rowe, D.M. (ed.) Thermoelectrics Handbook: Macro to Nano. CRC Press, Boca Raton (2006)
114. Wijngaards, D.D.L., de Graaf, G., Wolffenbuttel, R.F.: Single-chip micro-thermostat applying both active heating and active cooling. Sensors Actuators A. **110**, 187–195 (2004)

115. Corrèges, P., Bugnard, E., Millerin, C., Masiero, A., Andrivet, J.P., Bloc, A., Dunant, Y.: A simple, low-cost and fast Peltier thermoregulation set-up for electrophysiology. J. Neurosci. Methods. **83**, 177–184 (1998)
116. McKinney, C.J., Nader, M.W.: A Peltier thermal cycling unit for radiopharmaceutical synthesis. Appl. Radiat. Isot. **54**, 97–100 (2001)
117. Elsgaard, L., Jørgensen, L.W.: A sandwich-designed temperature gradient incubator for studies of microbial temperature responses. J. Microbiol. Methods. **49**, 19–29 (2002)
118. Reid, G., Amuzescu, B., Zech, E., Flonta, M.L.: A system for applying rapid warming or cooling stimuli to cells during patch clamp recording or ion imaging. J. Neurosci. Methods. **111**, 1–8 (2001)
119. Hodgson, J.: Gene sequencing's industrial revolution. IEEE Spectr. **37**, 36–42 (2000)
120. Maltezos, G., Johnston, M., Scherer, A.: Thermal management in microfluidics using micro-Peltier junctions. Appl. Phys. Lett. **87**, 154105 (2005)
121. Bachmann, C., Bar-Cohen, A.: Hotspot remediation with anisotropic thermal interface materials. Proc. ITHERM. **2008**, 238–247 (2008)
122. Muzychka, Y.S., Culham, J.R., Yovanovich, M.M.: Thermal spreading resistance of eccentric heat sources on rectangular flux channels. J. Electron. Packag. **125**, 178–185 (2003)
123. Smalc, M.; Thermal performance of natural graphite heat spreaders, Proceedings IPACK2005, San Francisco, California, USA, July 17–22, PaperNumber: IPACK2005-73073
124. Bar-Cohen, A., Arik, M., Ohadi, M.: Direct liquid cooling of high flux micro and nano electronic components. Proc. IEEE. **94**, 1549–1570 (2006)
125. Montesano, M.: Annealed pyrolytic graphite. Adv. Mater. Process. 1–3 (2006, June)
126. Viswanath, R., Wakharkar, V., Watwe, A., Lebonheur, V.: Thermal performance challenges from silicon to systems. Int. Technol. J. Quart. **3**, 1–16 (2000)
127. Xong, Y., Smalc, M., et al.: Thermal tests and analysis of thin graphite heat spreader for hot spot reduction in handheld devices. Proc. ITHERM. **2008**, 583–590 (2008)
128. Bergles, A.E., Bar-Cohen, A.: Immersion cooling of digital computers. In: Kakac, S., Yuncu, H., Hijikata, K. (eds.) Cooling of Electronic Systems, pp. 539–621. Kluwer Academic Publishers, Boston (1994)
129. Bergles, A.E., Bar-Cohen, A.: Direct liquid cooling of microelectronic components. In: Bar-Cohen, A., Kraus, D. (eds.) Advances in Thermal Modeling of Electronic Components and Systems, vol. 2, pp. 241–250. ASME, New York (1990)
130. Kim, D., Rahim, E., Bar-Cohen, A., Han, B.: Thermofluid characteristics of two-phase flow in micro-gap channels. Proc. ITHERM. **2008**, 979–992 (2008)
131. Kim, D.W.; Ph.D. Thesis, University of Maryland (2007)
132. Bar-Cohen, A., Rahim, E.: Modeling and prediction of two-phase microgap channel heat transfer characteristics. Heat Transf. Eng. **30**, 601–625 (2009)

Some Aspects of Microchannel Heat Transfer

10

Y. Joshi, X. Wei, B. Dang, and K. Kota

10.1 Fundamentals of Microchannel Pressure Drop and Heat Transfer

Microchannels may utilize single-phase flow or phase change. It is now generally believed that for microchannels of few microns or larger in hydraulic diameter involving single-phase flow, the fluid flow and heat transport are adequately characterized by the same pressure drop and heat-transfer correlations as larger sized channels. On the other hand, two-phase flows are significantly affected by microchannel sizes. We summarize the key aspects of pressure drop and heat transfer in single-phase and two-phase heat transfer.

10.1.1 Single-Phase Flows

If the length of the microchannel is significantly longer than the hydraulic entrance length, $X_{\mathrm{fd,h}}$, and thermal entrance length, $X_{\mathrm{fd,th}}$, as defined in Eqs. (10.1) and (10.2), respectively, the flow and heat transfer are considered fully developed. In Eqs. (10.1) and (10.2), Re and Pr are the Reynolds number and the Prandtl number for the fluid, respectively, and d_{h} is the hydraulic diameter:

$$X_{\mathrm{fd,h}} \approx 0.05\,\mathrm{Re} \cdot d_{\mathrm{h}} \tag{10.1}$$

$$X_{\mathrm{fd,th}} \approx 0.05\,\mathrm{Re} \cdot d_{\mathrm{h}} \cdot \mathrm{Pr} \tag{10.2}$$

As a result, the nondimensional Nusselt number, Nu, is a constant for particular channel geometry and boundary conditions. Thus, the heat-transfer coefficient, h, is inversely proportional to the hydraulic diameter, as shown in Eq. (10.3), where k is the thermal conductivity of the fluid:

$$Nu = \frac{h \cdot d_{\mathrm{h}}}{k} = \mathrm{const} \Rightarrow h \propto \frac{1}{d_{\mathrm{h}}} \tag{10.3}$$

Following this equation, heat-transfer coefficient in microchannels is orders of magnitude higher when compared to channels of macroscale. This benefit is demonstrated in Tuckerman and Pease (1981), who reported a microchannel heat sink with thermal resistance 0.09 K cm^2/W. This is three times or more lower than conventional heat sinks. Typical microchannel heat sinks as described in this work consist of parallel microflow passages 50 μm wide and 300 μm deep.

Following Tuckerman and Pease [1], several early studies [2–10] indicated that the flow and heat-transfer parameters deviate from the classical theory developed for macrosize channels and the transition from laminar to turbulent flow occurs at a considerably smaller critical Reynolds number. However, there is no consensus on the trend of this deviation among different researchers. Some of the reported deviations may be attributed to experiment conditions such as surface roughness and flow maldistribution. In contrast, results from more recent studies [11–16] appear to indicate that the pressure drop and transport characteristics for macroscale channels also hold for the microchannels for various applications with channel hydraulic diameters of 10 μm or larger, with gases or liquids, such as in microelectronics cooling.

10.1.1.1 Simplified Model for Single-Phase Microchannel Heat Sink

A simplified model of a microchannel heat sink incorporating a stack of microchannels seen in Fig. 10.1 [17] is described in this section. At each layer, a number of parallel

Y. Joshi (✉) · X. Wei · B. Dang · K. Kota
School of Mechanical Engineering, Georgia Institute of Technology, Atlanta, GA, USA
e-mail: yogendra.joshi@me.gatech.edu

C. P. Wong et al. (eds.), *Nano-Bio- Electronic, Photonic and MEMS Packaging*, https://doi.org/10.1007/978-3-030-49991-4_10

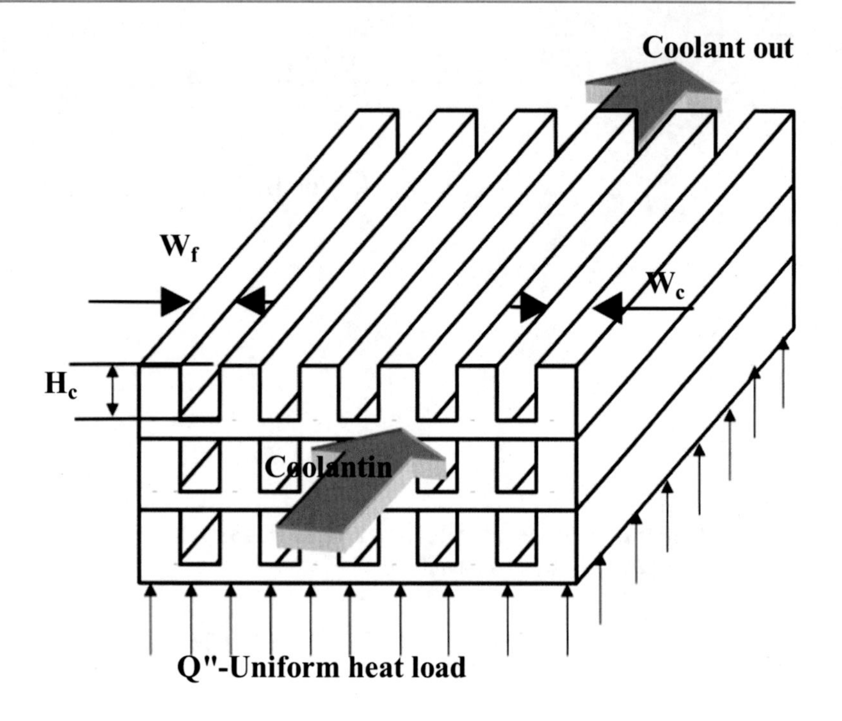

Fig. 10.1 Three-dimensional stack of microchannels. (Reprinted with permission from Wei and Joshi [17])

microchannels are machined in the surface of a substrate, e.g., copper, silicon, or diamond. These layers are then stacked and bonded to form a micro-heat sink. The benefits of using stacked microchannels are the significant reduction in pressure drop and temperature gradient, when compared with a single-layered microchannel heat sink. It also allows tailored design of the fluid passage for potential hot-spot cooling for highly nonuniform heat density applications. Although the focus of this section is on stacked microchannel, the same preliminary modeling procedure can be followed for a single-layered microchannel heat sink, since it is simply a special case of the stacked microchannel heat sink.

10.1.1.2 Correlations for Friction Factor and Nusselt Number (Wei and Joshi 2003)

The pressure drops associated with the stack structure include the contraction and expansion pressure drops in the inlet and outlet, respectively, the pressure drop due to 90° bend and the flow friction loss. As pointed out by many researchers [1, 18], the friction losses dominate the pressure drop in laminar flow in rectangular duct. Therefore, only the friction loss is included here.

To consider the friction factor (f_{app}) for developing laminar flow in a rectangular duct, a Churchill–Usagi asymptotic type of model shown in Eqs. (10.4), (10.5), and (10.6) is used [19].

$$f_{\mathrm{app}}\,\mathrm{Re}_{\sqrt{A_c}} = \left[\left(\frac{3.44}{\sqrt{y^+}}\right)^2 + \left(8\sqrt{\pi}G(\alpha)\right)^2\right]^{0.5} \tag{10.4}$$

$$G(\alpha) = \left[1.086957^{1-\alpha}\left(\sqrt{\alpha} - \alpha^{1.5}\right) + \alpha\right]^{-1} \tag{10.5}$$

$$y^+ = \frac{y}{\mathrm{Re}_{\sqrt{A_c}}\sqrt{A_c}} \tag{10.6}$$

In this equation, y^+ is the nondimensional length scale for friction factor calculation; α is the aspect ratio or the inverse of the aspect ratio of the microchannel that is always smaller than 1. It is noted here that the length scale for the Reynolds number (Re) and the dimensionless length is the square root of the channel cross-sectional area (A_c). The above equations allow the calculation of pressure drop across a microchannel.

Similarly, the Nusselt number for thermally developing condition in a rectangular duct is given as [20]

$$Nu_{\sqrt{A_c}}(y*) = \left[\left(C1\cdot C2\left(\frac{8\sqrt{\pi}G(\alpha)}{y*}\right)^{\frac{1}{3}}\right)^5 + \left(C3\left(\frac{G(\alpha)}{\alpha^\gamma}\right)\right)^5\right]^{0.2} \tag{10.7}$$

$$y* = \frac{y}{\mathrm{Re}_{\sqrt{A_c}}\cdot \mathrm{Pr}\cdot\sqrt{A_c}} \tag{10.8}$$

$C1 = 1$ for isoflux condition, $C2 = 0.501$, $C3 = 3.66$, $\gamma = 0.1$, y^* is the nondimensional length scale for Nusselt number calculation, and $G(\alpha)$ is the aspect ratio function. The heat-transfer coefficient can be calculated from the Nusselt number relations.

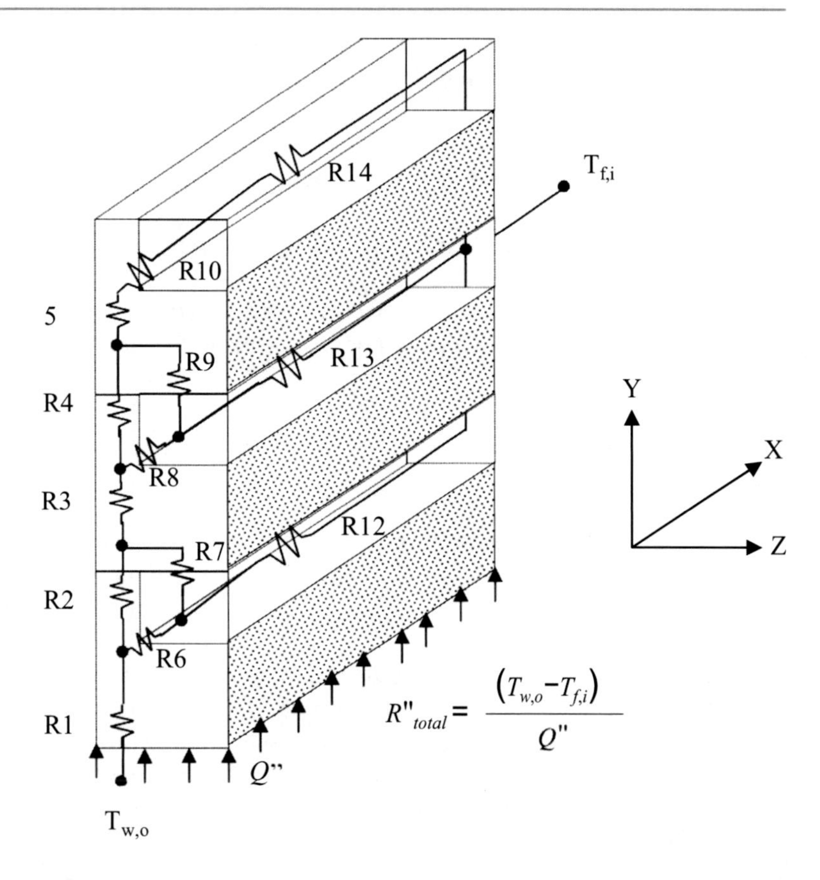

Fig. 10.2 Thermal resistance network for a three-layered microchannel stack. (Reprinted with permission from Wei and Joshi [17])

Table 10.1 Component resistances for the network

$R_1 = R_{cond} + R_{cont}$	$R_6 = R_{conv,fin}$	$R_7 = R_{conv,c}$	$R_{12} = R_{bulk}$
$R_3 = R_{cond} + R_{cont} + R_{spread}$	$R_8 = R_{conv,fin}$	$R_9 = R_{conv,c}$	$R_{13} = R_{bulk}$
$R_5 = R_{cond} + Rcont + R_{spread}$	$R_{10} = R_{conv,fin}$		$R_{14} = R_{bulk}$

10.1.1.3 Thermal Resistance Network Analysis

A thermal resistance analysis based on existing correlations for heat-transfer coefficients and friction factor in laminar channel flow [17] is described below. Heat spreading and constriction resistances for small-sized heat sources on larger substrates, adapted from [18], are also incorporated into the resistance network.

The thermal resistance network for a three-layered microchannel stack is illustrated in Fig. 10.2. All component resistances except R_2 and R_4 are listed in Table 10.1. To calculate these resistances, one-dimensional conduction is assumed. This assumption is reasonable as long as the aspect ratio is sufficiently large. In the base area, however, the aspect ratio is close to one. Two-dimensional heat conduction in that region is taken into account by considering constriction resistance for area contraction and spreading resistance for area expansion.

$$R_{\text{spread}} = \frac{L.n\left[1/\sin\left(\frac{\pi}{2}\frac{1}{1+W_c/W_f}\right)\right]}{k_S WL} \quad (10.11)$$

$$R_{\text{Conv,fin}} = \frac{1}{h \cdot n \cdot L \cdot W_c + 2 \cdot h \cdot \eta_f \cdot n \cdot L \cdot H_c} \quad (10.12)$$

$$R_{\text{Conv,c}} = \frac{1}{h \cdot n \cdot L \cdot W_c} \quad (10.13)$$

$$R_{\text{bulk}} = \frac{1}{\rho \cdot C_p \cdot v_m \cdot n \cdot H_c \cdot W_c} \quad (10.14)$$

where t is the base thickness for microchannel, k_s the solid thermal conductivity, k_f the fluid thermal conductivity, W_c the channel width, W_f the fin width, $\underline{L}$ the channel length, n the number of channels in each layer, h the heat-transfer coefficient, H_c the channel depth, η_f the fin efficiency, ρ the fluid density, C_p the specific heat of the fluid, and v_m the mean fluid velocity.

The fin equation shown in Eq. (10.15) can be solved to get the two conduction resistances R_2 and R_4:

$$\frac{d^2\theta}{dx^2} = m^2 \cdot \theta, 0 < x < H_c \quad (10.15)$$

where $\theta = T - T_f, m = \sqrt{2 \cdot h / k_S . W_f}$.

Since the boundary conditions for the fin equation are part of the solution, the solution procedure is iterative in nature. The simple model described in this section can be readily applied in an optimization algorithm to achieve optimum designs of microchannel heat sinks [17].

10.1.2 Two-Phase (Liquid–Vapor) Flows

In general, flow boiling regime maps and heat-transfer and pressure drop correlations developed for macroscale channels may show significant errors when the channel sizes go down below 1–2 mm. Within the past 5 years, flow regime maps for microchannel flows have been characterized. Evidence for the transition from macroscale to microscale in the form of flow regime changes and the associated parameters is described below.

10.1.2.1 Two-Phase Flow Regimes for Microchannels

Flow Regimes

Revellin and Thome [21] proposed a diabatic flow regime map for microchannels based on isolated bubble (IB) flow, coalescing bubble (CB) flow, annular (A) flow, and dryout (DO) flow. The transition from IB to CB was found to be governed by the boiling number (Bo_i), the liquid Reynolds number (Re_L), and the vapor Weber number (We_G). The following correlation predicted the transition vapor quality:

$$x_{IB/CB} = 0.763\,(\mathrm{Re}_L Bo_i/We_G)^{0.41} = 0.763\,(qrGs/u_L h_{LG} m^2)^{0.41} \tag{10.16}$$

where

$$\mathrm{Re}_L = GD/\mu_L \tag{10.17}$$

$$Bo_i = q/(Gh_{LG}) \tag{10.18}$$

$$We_G = G^2 D/(\rho_G \sigma) \tag{10.19}$$

Only two channel diameters were employed in this correlation, and hence, no diameter dependence was discerned.

The CB/A transition occurs when the bubble frequency reaches zero or at the end of distinct vapor bubbles and liquid slugs. This is predicted by

$$x_{CB/A} = 1.4 \times 10^{-3}\,\mathrm{Re}_L^{1.47} We_L^{-1.23} \tag{10.20}$$

where the liquid Weber number is given by

$$We_L = G^2 D/(\rho_L \sigma) \tag{10.21}$$

The A/DO correlation was obtained by combining the critical heat flux correlation of Wojtan, Revillin, and Thome [22] and an expression for critical vapor quality, x_{crit}. The resulting expression is

$$x_{crit} \frac{q_c}{G(h_{LV} + h_{sub})} \frac{4L_h}{D} \tag{10.22}$$

with G as the mass flux and L as the channel length, D is the diameter, h_{LV} is the latent heat of vaporization, h_{sub} is the subcooling enthalpy, q is the heat flux, σ is the surface tension, μ is the kinematic viscosity, and q_c is the critical heat flux with subscripts L and G signifying liquid and vapor phases, respectively.

Flow Transitions

In microscale two-phase horizontal flows, the effects of gravity are insignificant, resulting in the disappearance of stratified wavy and fully stratified flows, found in larger diameter channels. Microscale flows provide a uniform film thickness in annular and slug flows due to the small diameters. In vertical tube flows for diameters below 2 mm, the observed flow patterns in R-134a were found to be very different from macroscale flows [23]. Void fractions also seem to show deviations from homogeneous void fraction laws, as sizes are reduced. This has been observed for nitrogen in water, ethanol, and water–ethanol mixtures in channels from 0.25 to 0.05 mm. Thus, size effects in two-phase flows are seen for considerably larger sizes (up to ~2 mm) than in single-phase flows.

Transition Criteria

Kew and Cornwall [24] found that as the passage diameter is reduced, the influence of gravity is suppressed in comparison to surface tension for boiling number, $Bo_i < 4$. This results in a threshold diameter d_{th} below in which the macroscale methods are not suitable for predicting heat-transfer coefficients or flow patterns. The threshold to confined bubble flow to the channel diameter is covered by the equation, where Co is the confinement number (>0.5) and σ, ρ_L, ρ_G, and g are the surface tension, densities of liquid and vapor phases, and gravitational acceleration, respectively:

$$d_{th} = \frac{\sqrt{\sigma/(\rho_L - \rho_G)g}}{Co} \tag{10.23}$$

This ignores flow forces, i.e., no superficial velocities and shear effects on bubbles and annular films. Further refinements have introduced additional dimensional groups to define the threshold [25].

Jacobi and Thome [26] assumed that no stratification exists at the microscale. During bubble growth, the bubble diameter reaches tube diameter, prior to departure. The subsequent growth then takes place only along the tube length in the flow direction. Bubble growth relations from nucleate pool boiling have typically been used.

Li and Wang [28] performed an analysis to determine the transition from symmetric to asymmetric flow during condensation in tubes and is described in a later section on condensation in microchannels. These criteria could also be applied to adiabatic flows and flow boiling as well.

Computations of slug flows of elongated bubbles have been carried out for R134a at a saturation pressure of 7.74 bar ($T_{sat} = 30$ °C). The computations have compared the liquid film thickness at the top of the bubble to that at the bottom of the bubble, as a function of the channel diameter and six nondimensional parameters. The transition criteria suggest that the transition is in the range of 0.1–0.5 mm channel sizes, comparable to Li and Wang [28].

10.1.2.2 Pressure Drop Correlations and Models

The available evidence for pressure drop seems to be counterintuitive, suggesting that in the range of 0.1–3 mm diameter, the pressure drops estimated from several macroscale models, including homogeneous flow, predicted microscale data quite well, even for flows that would fall in slug, and annular regimes that provided the flow patterns are properly matched.

The pressure drop is calculated using the following equation for two-phase flow in microchannels:

$$\frac{dp_{tp}}{dz} = \frac{dp_f}{dz} + \frac{dp_m}{dz} + \frac{dp_S}{dz} \tag{10.24}$$

where the terms on the right-hand side of the equation represent the losses due to friction, acceleration of the fluid, and the static head due to gravitational forces (which is usually negligible).

Though a generic model is not available to predict pressure drops in two-phase flows, most of the proposed pressure drop correlations in the literature are based on one of the two classical two-phase flow models. For modeling two-phase pressure drops, the two commonly used models are the homogeneous model and the separated flow model. In the former, the two-phase fluid is assumed to behave as a single-phase fluid with pseudo-properties. The effective density, viscosity, and specific heat are evaluated based on vapor quality. In the latter approach, the two phases are segregated into separate streams, which interact through an interface.

Homogeneous Model

In the homogeneous model, the two phases are assumed to be constituents of a single fluid, and the properties of the phases are calculated based on the vapor mass quality (x). Both fluids are assumed to have the same velocity with a slip ratio of 1. The density of the mixture is based on the mass quality and is defined as in Eq. (10.25):

$$\rho_m = \left(\frac{x}{\rho_G} + \frac{1-x}{\rho_L}\right)^{-1} \tag{10.25}$$

Different empirical correlations are proposed in literature for the effective viscosity of the mixture. Few of the references are McAdams et al. [28], Cicchitti et al. [29], Dukler et al. [30], and Awad and Muzychka [31]. The most common definition of viscosity is by McAdams et al. [28] as shown in Eq. (10.26):

$$\mu_m = \left(\frac{x}{\mu_G} + \frac{1-x}{\mu_L}\right)^{-1} \tag{10.26}$$

Several researchers have proposed new definitions for viscosity and have compared their predictions with the published data for frictional pressure drop based on the homogeneous model. The homogeneous model is found to be more appropriate to predict pressure drops of bubbly flows.

Separated Flow Model

In the separated flow model, the frictional pressure drop is expressed in terms of frictional multipliers. It is defined as the ratio of the frictional pressure gradient to that for one or the other phase flow alone or for a flow at the same total mass flux having gas or liquid properties:

$$\frac{dp_f}{dz} = \phi_L^2 \frac{dp_L}{dz} = \phi_G^2 \frac{dp_G}{dz} \tag{10.27}$$

where Φ_L and Φ_G are two-phase multipliers.

The pressure gradient due to acceleration or deceleration effects is given by

$$\frac{dp_m}{dz} = G^2 \frac{d}{dz}\left[\frac{(1-x)^2}{(1-\alpha)\rho_L} + \frac{x^2}{\alpha \rho_G}\right] \tag{10.28}$$

where G is the mass flux, z is the length variable in flow direction, and α is the void fraction defined as the ratio of gas flow cross-sectional area to the total cross-sectional area. This model is more suitable for predictions in slug flows.

10.1.2.3 Heat-Transfer Correlations and Models

Heat-Transfer Coefficient

Some of the experimental evidence for R134a suggests an increase in the average heat-transfer coefficient in transitioning from macroscale to microscale (~2 mm).

Other studies indicate a weaker or no dependence.

Agostini and Thome [32] compared flow boiling heat-transfer coefficient data in microchannels from 13 studies. Wide variations were found in the level and trends due to differences in the conditions of the experiments. For example, the flow maldistribution resulting in mass flow rate variations within a multichannel evaporator fed from an inlet plenum could be addressed by having orifices at the inlet of each channel. The following general conclusions could be drawn from the studies.

At very low vapor qualities ($x < 0.05$), the heat-transfer coefficient increases with heat flux and either increases with or is insensitive to vapor quality.

At low-to-medium quality ($0.05 < x < 0.5$), the heat-transfer coefficient increases with heat flux and decreases with or is relatively insensitive to vapor quality.

For $x > 0.5$, the heat-transfer coefficient decreases sharply with vapor quality and does not depend on heat flux or mass velocity.

The heat-transfer mechanisms depend on the flow regime during boiling. In bubbly flow nucleate boiling, liquid convection can be assumed to be dominant. In slug flow, thin-film evaporation is the dominant mode. Liquid convection and vapor convection may also be important, the latter if there is a dry zone. In annular flow, thin liquid film evaporation is the dominant mode. In mist flow, vapor heat-transfer processes along with droplet impingement are important.

Kandilkar and coworkers (e.g., Kandlikar and Balasubramanian [33] and Peters and Kandlikar [34]) have proposed the determination of two-phase heat-transfer coefficient for the nucleate boiling dominant (NBD) and convective boiling dominant (CBD) regions:

$$h_{\text{tp,NBD}} = 0.6683 Co^{-0.2}(1-x)^{0.8} f_2(Fr_{\text{lo}}) h_{\text{lo}} + 1058 Bo_i^{0.7}(1-x)^{0.8} F_{fl} h_{\text{lo}} \tag{10.29}$$

$$h_{\text{tp,NBD}} = 1.136 Co^{-0.9}(1-x)^{0.8} f_2(Fr_{\text{lo}}) h_{\text{lo}} + 667.2 Bo_i^{0.7}(1-x)^{0.8} F_{\text{fl}} h_{\text{lo}} \tag{10.30}$$

The first term in each equation is referred to as the convective boiling term and the second as the nucleate boiling term. The authors propose the use of these two equations for calculating the heat-transfer coefficient during flow boiling, based on the flow regime and microchannel configuration. The larger of the two is to be used as the two-phase heat-transfer coefficient, h_{tp}. The h_{lo} in the above equations is calculated as a single-phase heat-transfer coefficient appropriate for the flow condition, assuming that the entire mass flux is liquid. Suggested correlations are as follows.

Turbulent Flow ($\text{Re}_{\text{lo}} \geq 3000$):
$3000 \leq \text{Re}_{\text{lo}} \leq 10^4$: Gnielinski correlation

$$h_{\text{lo}} = \frac{(\text{Re}_{lo} - 1000)\text{Pr}_l (f/2)(K_1/D)}{1 + 12.7\left(\text{Pr}_l^{2/3} - 1\right)(f/2)^{0.5}} \tag{10.31}$$

$10^4 \leq \text{Re}_{\text{lo}} \leq 5 \times 10^6$: Petukhov and Popov correlation

$$h_{\text{lo}} = \frac{(\text{Re}_{lo} - 1000)\text{Pr}_l (f/2)(K_1/D)}{1 + 12.7\left(\text{Pr}_l^{2/3} - 1\right)(f/2)^{0.5}} \tag{10.32}$$

$$\text{where } f = [1.58 \ln(\text{Re}_{\text{lo}}) - 3.28]^{-2} \tag{10.33}$$

Laminar Flow ($\text{Re}_{\text{lo}} < 1600$)

$$h_{\text{lo}} = \frac{Nu.k}{D_{\text{h}}} \tag{10.34}$$

where Nu is chosen based on the channel geometry and the wall boundary condition.

Transition Flow

h_{lo} is taken as a linear interpolation between the laminar value at $\text{Re}_{\text{lo}} = 1600$ and Gnielinski correlation value at $\text{Re}_{\text{lo}} = 3000$.

Peters and Kandlikar [34] propose the following modifications for $\text{Re}_{\text{lo}} < 410$: For $100 < \text{Re}_{\text{lo}} < 410$:

$$h_{\text{tp}} = h_{\text{tp,NBD}} = 0.6683 Co^{-0.2}(1-x)^{0.8} h_{\text{lo}} + 1058 Bo_i^{0.7}(1-x)^{0.8} F_{\text{fl}} h_{\text{lo}} \tag{10.35}$$

For $\text{Re}_{\text{lo}} < 100$, the convective boiling contribution is negligible, resulting in

$$h_{\text{tp}} = 1058 Bo_l^{0.7}(1-x)^{0.8} F_{\text{fl}} h_{\text{lo}} \tag{10.36}$$

where the parameter F_{fl} accounts for the effects of the fluid–surface interactions in nucleate boiling. For copper and brass surface, $F_{\text{fl}} = 1$ for water, and $F_{\text{fl}} = 1.63$ for R-134a.

Jacobi and Thome [26] proposed an elongated bubble (slug) flow boiling model to predict the heat-transfer coefficient. They accounted for the evaporation of the thin liquid film trapped between the bubble and the channel wall, as well as the convection by the liquid slugs between bubbles. The former was significantly larger than the latter. Their thin film evaporation model predicted $h_{\text{tp}} \sim q^n$, where the exponent n depends on elongated bubble frequency and initial liquid film thickness resulting from the passing bubble.

Critical Heat Flux (*CHF*)

Critical heat flux is an extremely important operational limit in two-phase flow-based cooling devices. It signifies the highest heat flux, which can be sustained for a particular set of conditions. Exceeding this value results in a dryout at the wall and an associated large increase in the wall temperature. Qu and Mudawar [35] reported water CHF data in microtubes from 1 to 3 mm diameter. These show mean absolute error of ~17% and an error band of ±40% compared to Katto and Ohno correlation for macrotubes and channels.

Bar-Cohen and coworkers [36] have studied the effect of flow patterns on CHF by considering over 4000 data points from the literature. The data have been categorized in terms of the Taitel and Duckler [37] flow regime maps, in terms of annular, intermittent, and bubbly flows. They suggest that a similar CHF mechanism as reported by Qu and Mudawar [35] would apply within any given flow regime.

Qu and Mudawar [35] considered a multi-microchannel heat sink made up of 21 parallel microchannels, 0.215 mm wide and 0.821 mm high. Instabilities in the form of vapor backflow into the inlet plenum were observed as the CHF was approached. The backflow resulted in negating any advantage of inlet subcooling on CHF. They recommend the following CHF correlation for saturated boiling. This correlation predicted their data for water and R-113 within 4%:

$$q_{\mathrm{CHF}} = 33.43\left[h_{\mathrm{LG}}(\rho_{\mathrm{G}}/\rho_{\mathrm{L}})^{1.11}G\right]/\left[We^{0.21}(L/d_{\mathrm{h}})^{0.36}\right] \quad (10.37)$$

Wojtan, Revillin, and Thome [22] investigated CHF in stainless steel tubes of diameter 0.509 and 0.790 mm for refrigerants R-134a and R-245fa. A mass velocity range of 400–1600 kg/m^2 s was examined, as well as four heated lengths (between 20 and 70 mm), two saturation temperatures (30–35 °C), and subcooling in the range 2–15 K. The following correlation was proposed:

$$q_{\mathrm{cr}} = 0.437(\rho_{\mathrm{V}}/\rho_{\mathrm{L}})^{0.073} \cdot We^{-0.24}(L_{\mathrm{h}}/D)^{-0.72} \cdot Gh_{\mathrm{LV}} \quad (10.38)$$

More recent data by Agostini et al. [38] support this correlation. The same research group estimates that a 2 × 20 mm silicon chip may dissipate 112–250 W/cm^2 in flow boiling, based on this correlation.

Based on the hypothesis that CHF at a microchannel location is achieved when during evaporation in annular flow, the height of the interfacial waves reaches the mean thickness of the annular film, Revellin and Thome [39] proposed a CHF model. The governing equations for mass, momentum, and energy assuming annular flow, ignoring interfacial wave formation, were first solved to determine the annular liquid film thickness along the channel length. The wave height was modeled using the Kelvin–Helmholtz critical wavelength criterion and the slip ratio. They compared their predicted results with several data sets, both from their group and from others, with a resulting agreement within ±20% for over 90% of the data.

Revillin et al. [40] extended the calculations for a local hot spot on a microprocessor chip. The question investigated was whether the presence of the hot spot may trigger CHF during forced convection boiling. A 0.4 mm hot spot along the periphery of the entire channel of 0.5 mm diameter and 20 mm length was considered for R-134a. The saturation temperature was 30 °C and the mass flux 500 kg/m^2 s, with no inlet subcooling. It was found that at a uniform heat flux of 218 kW/m^2, the hot spot should have a heat flux of 3000 kW/m^2 for CHF to occur. The value of CHF without the hot spot for the same conditions was 396 kW/m^2. The location of hot spots near the exit of the channel was found to increase the likelihood of getting CHF.

10.1.2.4 Condensation in Microchannels

Two-phase thermal management devices most commonly utilize a closed-loop arrangement, where the working fluid collects the heat in the evaporator and rejects it to the ambient or another working fluid in a condenser. As such, the performance of miniature condensation devices is of great importance. Most of the prior work on condensation flow regime maps was for tube diameters greater than 5 mm (Breber et al. [41] and Soliman [42]). The major nondimensional parameters considered in these studies are the liquid–vapor volume ratio, Froude number, and Weber number. These maps primarily focused on the hydraulic diameters between ~5 and ~25 mm and are therefore meant for determining the transitions between gravity-dominated stratified wavy flows and shear-dominated annular flows. For small diameter tubes, the tube diameter affects the transition boundaries significantly, and the traditional flow regime maps fail to predict the flow behavior accurately. This is because of the increased influence of surface tension relative to gravity. This effect particularly in noncircular channels is usually termed as "Gregorig effect" and was found to occur with hydraulic diameters around 1 mm [43]. Because of this, the condensate surface curvature changes, which in conjunction with surface tension leads to pressure gradients in the condensate film and enhanced heat transfer. Some of the prior efforts that focused on small-sized channels include Coleman and Garimella [44] and Yang and Shieh [45]. Dimensional parameters such as liquid and vapor superficial velocities or mass flux and quality were used to characterize the different flow regimes.

Coleman and Garimella [44] observed annular, wavy, intermittent, and dispersed flows. In addition, they found that high-pressure refrigerant flow regimes and transitions are significantly different from those for air–water mixtures.

Hence, flow regime maps for air–water data cannot be directly applied to refrigerants. In agreement with Coleman and Garimella [44], Chen and Cheng [46] reported results of visualization studies of condensation of steam in trapezoidal silicon microchannels with hydraulic diameters of 75 μm and 80 mm. They observed that droplet condensation took place near the inlet of the microchannels, while an intermittent flow of vapor and condensate was observed downstream of the channels. The traditional annular flow, wavy flow, and dispersed flow were not observed in microchannels.

Bandhauer et al. [47] and Nema [48] derived a set of nondimensional parameters to characterize flow regimes for condensation in practical microchannels. They found that the important flow transitions include annular or wavy regime to intermittent, discrete to disperse wave, intermittent or wavy to annular film flow, intermittent to dispersed flow, and transition to mist flow. They modeled the transition to the intermittent regime (consisting of slug and plug flow patterns) from the wavy or annular regimes using the vapor-phase Weber number (We_G) and the Martinelli parameter (X_{tt}). A critical Bond number ($Bo_{Critical}$) determined whether the transition occurs from the wavy or the annular regime:

$$Bo_{Critical} = \frac{(\rho_L - \rho_G) D_h^2 g}{\sigma} = \frac{1}{\left(\frac{\rho_L}{(\rho_L - \rho_G)} - \frac{\pi}{4}\right)} \quad (10.39)$$

Below the critical Bond number, they found that annular flow transitions directly to intermittent flow, suppressing the wavy flow regime. This is because of the increased prominence of surface tension over gravity forces as the Bond number or tube diameter is decreased. This transition criterion predicted 85% of data in the correct regimes. The corresponding transition liquid volume fraction was modeled as follows:

$$X_{tt_{slug}} = X_{tt_0} \text{ if } Bo \leq Bo_{Critical}, \text{ where } X_{tt_0} = 0.3521 \quad (10.40)$$

$$X_{tt_{slug}} = X_{tt_0} + \frac{X_{tt_1}(Bo - Bo_{Critical})}{Bo - Bo_{Critical} + e} \text{ if } Bo > Bo_{Critical} \quad (10.41)$$

where $Xtt1 = 1.6 – X_{tto}$ and $e = 5.5$.

They also found that a constant value of the modified Froude number could be used to model this transition and this criterion predicted 93% of the data correctly. The annular flow data were divided into three categories: annular film, annular film with mist, and mist flow. To model the transitions, annular film and annular film with mist were considered as one region (annular film), while mist flow was treated as the other region. The transition to the annular film regime from intermittent flow was established based on the vapor-phase Weber number and Martinelli parameter as described above. It was also found that most of the mist flow data existed only for liquid volume fractions (represented by the Martinelli parameter) less than a critical value. This transition criterion using constant vapor-phase Weber number and Martinelli parameter predicted 67% of the data accurately.

The sustenance of annular flow was modeled using the vapor-phase Weber number and Bond number as follows:

For $Bo \leq Bo_{Critical}$, annular film flow exists if $We_G > 6$.
For $Bo > Bo_{Critical}$, annular film flow exists if $We_G > 6 + c(Bo - Bo_{Critical})^d$, where the constants c and d depend on the size of the flow channel.

Based on the Young–Laplace equation, Li and Wang [27] propose the following critical and threshold values of tube diameter:

$$d_{cr} = 0.224\sqrt{\sigma/(\rho_L - \rho_G)g} \quad (10.42)$$

$$d_{th} = 7.81\, d_{cr} \quad (10.43)$$

The following regimes for condensation were defined based on these values:

(a) When $d_{cr} \leq d \leq d_{th}$, the flow condensation will be asymmetrical, and the corresponding flow regimes would be annular, asymmetrical annular, lengthened bubble, and bubble flow.
(b) When $d \leq d_{cr}$, the flow patterns will be annular, lengthened bubble, and bubble flow.
(c) When $d \geq d_{th}$, the flow regimes will be similar to those for macro-sized channels/tubes, except for the decreasing effect of surface tension with increasing tube size.

In accordance with lack of sufficient and reasonably accurate flow regime maps for condensation in microchannels, there is a dearth of definitive correlations for predicting the pressure drop and heat-transfer performance. Wang and Rose [43, 49] developed a theoretical model for heat transfer during condensation in square and triangular section horizontal microchannels including the effects of interfacial shear stress, gravity, and transverse surface tension force on the motion of the condensate film. Results are presented for fluids of R134a, R22, and R410A in square and triangular section channels of sides in the range 0.5–5 mm and for mass velocities in the range 100–1300 kg/m^2 s. The results demonstrated significant heat-transfer enhancement by surface tension toward the channel entrance. For smaller channels, the initial enhancement is higher but was found to fall off after a shorter distance along the channel; higher mass velocities were found to result in increased length of enhanced heat transfer.

As the channel sizes become small, different mixture viscosity models as against conventional ones need to be used to make homogeneous flow models work. This is attributed to the lower mixing losses due to the weak momentum coupling between the phases in the smaller channels. Garimella et al. [50] represented the total pressure drop as the summation of the frictional pressure drop in the slug and bubble regions and the pressure drop associated with the transitions between these regions.

Baird et al. [51] conducted an experimental investigation to determine the local heat-transfer coefficient during condensation in 0.92 and 1.95 mm internal diameter tubes. The data showed a strong influence of mass flux and local quality on the heat-transfer coefficient and a relatively weaker influence of system pressure. The observed heat-transfer coefficient was found to increase with mass flux. To predict the heat-transfer coefficients, they proposed an approach similar to that developed by Moser et al. [52] based on a core annular shear-driven gas–liquid flow in which the gas–liquid interface is assumed to be smooth and the liquid film is turbulent, with modifications to the film thickness parameter.

Bandhauer et al. (2006) [53] developed a model based on boundary layer analyses including the shear–stress estimation from the pressure drop models of Garimella et al. [54] developed specifically for microchannels. Their model also indirectly accounts for surface tension through a surface tension parameter in the pressure drop used for the shear–stress calculation.

Agarwal [55] found that both pressure drop and heat transfer increased with increasing vapor quality, increasing mass flux, and decreasing saturation temperature. The flow velocities and other parameters determined for the pressure drop model were used as inputs for the heat-transfer models. The slug and bubble regions were again analyzed separately to determine the slug and film heat-transfer coefficients. A time-averaged refrigerant heat-transfer coefficient was determined by combining the slug and film heat-transfer coefficients according to their transit times through the channel. The proposed pressure drop (Eq. $^{10.44}$) and heat-transfer models (Eq. 10.45) predict 95% and 94% of the data, respectively, within ±25%. As the channel hydraulic diameter decreases and the flow velocity increases, the pressure drop and heat-transfer coefficients were found to increase due to a decrease in film thickness and channel diameter. As the aspect ratio increases, the pressure drop and heat-transfer coefficients increased due to an increased occurrence of slugs.

The total pressure drop according to their model is given by

$$\Delta P_{\text{fric,mod el}} = N_{UC} \cdot \Delta P_{\text{transition}} + \Delta P_{\text{fric,only}} \quad (10.44)$$

where N_{UC} depends on the microchannel dimensions and aspect ratio, Reynolds number, flow quality, and density of phases, and $\Delta P_{\text{transition}}$ and $\Delta P_{\text{fric,only}}$ (which mainly depends on the lengths of slug and bubble flow regimes) are given in Agarwal [55].

On a similar note, they predicted the heat-transfer coefficient for a refrigerant as follows:

$$h_{\text{refg,model}} = h_{\text{slug}} \cdot \left(\frac{l_{\text{slug}}}{l_{\text{slug}} + l_{\text{bubble}}} \right) + h_{\text{film}} \cdot \left(1 - \frac{l_{\text{slug}}}{l_{\text{slug}} + l_{\text{bubble}}} \right) \quad (10.45)$$

where the method to obtain h_{slug} and h_{film} is detailed in Agarwal (2006) and l_{slug} and l_{bubble} are the slug and bubbly flow regime lengths, respectively. The equation shows that the heat-transfer coefficient also depends on the lengths of the slug and bubbly flow regimes.

10.1.3 Rotating Flows

Rotating flows in microchannels can occur in microfluidics applications for flow control, like in electro-osmosis and mixing as in a lab-on-a-chip system. Owing to the nature of the applications, mostly single-phase flow characteristics were focused upon in literature on radially rotating microchannels with centrifugal force-driven flow. Conventional pressure drop estimations seem to work well for rotating microchannels also for slow rotational speeds. But at high flow and rotational Reynolds numbers, Coriolis force dominates the centrifugal force component and needs to be included in estimating the frictional pressure drop [56]. This is found to be true also in the case of microchannels rotating about their own axis [57].

10.2 Numerical Techniques

For numerical modeling of flow and heat transfer in microchannels for hydraulic diameters of 1 μm and beyond, continuum-based mass, momentum, and energy transport equations are usually sufficient. For nanoscale channels, molecular techniques have been employed in modeling of transport phenomena. The following section briefly describes some of these techniques.

10.2.1 Continuum Models

Single-phase liquid flows are usually modeled using continuum forms of the momentum and energy conservation equations, with appropriate boundary conditions. These can include various boundary conditions and fluid enhancements such as microsized particle suspension flows [58–61]. These equations for fluids are simultaneously coupled with heat conduction in microchannel walls. Conventional computational methods to predict the flow and temperature fields provide reasonable accuracy when compared with experiments.

For modeling two-phase flows, flow boiling models are usually preferred, which must take into account the incipience of boiling, predict heat-transfer coefficient and pressure drop, and identify the critical heat flux. Modeling in general involves finding the point of incipience of boiling and then proceeding with using the conservation equations depending on the model employed. Boiling heat-transfer models are mostly based on specific flow pattern observations from experiments. For example, in the microlayer model of Jacobi and Thome [26], the critical nucleation radius is experimentally estimated and is used to solve the coupled algebraic differential equation set based on momentum and energy balance. The annular model of Qu and Mudawar [62] treats the liquid and vapor zones as separate, and conservation equations for individual phases are solved. Sarangi et al. [63] used both the homogeneous model and the annular model and used a known exit pressure condition to march backward in the microchannel rather than assuming the pressure at the boiling incipience point based on the assumed inlet conditions. Their results were found to either lie above or below (based on the model type) the experimental results of Qu and Mudawar [62]. The major drawback of most of the two-phase flow models is their applicability to specific flow conditions.

10.2.2 Molecular Models

For conditions in microchannels, where continuum approaches fail (Knudsen number >0.1) or in flows where capturing of particle–particle and particle–fluid interactions is of significance (e.g., estimating thermophysical properties such as thermal conductivity of nanoparticle suspensions), molecular approaches such as molecular dynamics simulations (MDS) and direct simulation Monte Carlo (DSMC) are used. This subsection provides the basics of lattice Boltzmann method (LBM), a macromolecular model applied to suspension flows.

The lattice Boltzmann model most widely uses the lattice Bhatnagar–Gross– Krook (LBGK) single-relaxation-time model. This model is based on a discrete Boltzmann equation that is usually solved on a regular lattice. Rather than considering a large number of individual molecules like in the molecular dynamics approach, LBM method considers a much smaller number of fluid "particles" (a large group of molecules) which is much larger than a molecule but still considerably smaller than the smallest length scale of the simulation. This reduces the amount of data which need to be stored for simulation. The particles are allowed to interact through hardsphere, inelastic, and frictional collisions. The dynamics of the method is given by the LB equation:

$$f_i\left(\vec{r}, \vec{c}_i, t+1\right) = f_i\left(\vec{r}, t\right) + \frac{1}{\tau} \times \left[f_i^{\text{eq}}\left(\vec{r}, t\right) - f_i\left(\vec{r}, t\right)\right] \tag{10.46}$$

in which f_i is the distribution function of the fluid particles moving in the $\vec{c}_i$ direction, and $f_i^{\text{eq}}(\vec{r}, t)$ is the equilibrium distribution toward which the distribution functions are relaxed. The density (ρ) and velocity ($\vec{u}$) of the fluid are determined at each lattice node from the first and second velocity moments of the distribution function f_i, respectively. One of the advantages of this method is the easy implementation of the fluid–solid no-slip boundary condition by using the heuristic bounce-back rule. Ladd [64, 65] generalized the bounce-back rule for moving boundaries.

$$f_i\left(\vec{r}, t+1\right) = f_{-i}\left(\vec{r}, t\right) + \frac{2\rho t_i}{c_s^2}\left(\vec{u}_{\text{w}} \cdot \vec{c}_i\right) \tag{10.47}$$

Here, $-i$ denotes the link opposite to i in the lattice, $\vec{u}_{\text{w}}$ is the local velocity of the moving boundary, t_i is a weight factor related to i direction, and c_s is the speed of sound. This generalization makes the simulation of moving particles straightforward. It does not only describe the action of the particles on the fluid but can also be used to compute the hydrodynamic forces and torques action on the particles at a very small computational cost. When Eq. (10.46) is solved, the velocities, angular velocities, and positions of the particles can be computed by using methods similar to molecular dynamics. This simulation will provide the time-averaged particle concentration, fluid and particle velocities, fluid–particle interaction force, shear and normal stresses due to particle–particle interaction, and fluid-phase viscous stresses across a microchannel. From this simulation, a rheological constitutive model can be obtained that describes shear and normal stresses in neutrally buoyant suspensions as functions of apparent shear rate and volume fraction of solids. LBM has also been applied for simulating multiphase flows (e.g., Chang [66]).

10.3 Experimental Techniques

Characterizing the transport phenomena inside microchannels is of vital importance to the design of these devices. Existing research has often focused on measuring bulk transport characteristics such as friction factors or mean thermal resistances for idealized geometries of microtubes (Judy et al. [67]), rectangular microchannels (Harns et al. (1999) [68]), and trapezoidal microchannels (Wu and Cheng [69]). Such bulk transport characteristics are often influenced by factors, which are otherwise negligible for macroscale channels such as flow rate distribution, surface roughness, and flow inlets and outlets. A direct measurement of the fluid flow and heat transfer inside the microchannel becomes necessary. To this end, this section discusses some nonintrusive optical measurement techniques for flow and/or temperature field inside microchannels. Most of the techniques involve some types of seeding particles that are used to trace the flow and temperature field through tracking the position and intensity, respectively. Section 10.3.1 gives a detailed introduction to the microparticle image velocimetry (μPIV) system for velocity measurement. Section 10.3.2 discusses various temperature measurement techniques including μPIV using Brownian motion, two-color laser-induced fluorescence (LIF) for fluid temperature, single quantum dot tracking for near-wall temperature, and velocity measurement.

10.3.1 Measuring Full-Field Flow Inside Microchannels Using Micro-PIV

Particle image velocimetry (PIV) is a well-established technique for measuring flow in macroscale channels. In a typical PIV implementation [70], the flow field to be measured is seeded with flow-tracing particles, which are illuminated with either a pulsed light source or a continuous light source gated by the camera shutter. The images of the particles are taken at two known instants, and the displacement of the particle images is then statistically determined by correlation techniques. Significant success has been achieved in developing microresolution PIV (micro-PIV) systems. The first implementation of micro-PIV was reported in Santiago et al. [71], where an epifluorescent microscope was coupled with a continuous Hg-arc lamp and a CCD camera to record the flow around a nominally 30×10^{-6} m elliptical cylinder in a Hele–Shaw flow cell. The resolution of this system was reported to be 6.9×10^{-6} m $\times$ 6.9×10^{-6} m $\times$ 1.5×10^{-6} m. As the time delay between image exposures was controlled through the camera shutter, application of such a system is limited to very slow flow. A micro-PIV system capable of measuring over a broader range of velocity was described in Meinhart et al. [72]. Velocity distribution in a glass microchannel, 30×10^{-6} m deep and 300×10^{-6} m wide, was measured. A 5×10^{-9} s (5 ns) pulsed Nd:YAG laser was used as the light source to illuminate the 200×10^{-9} m diameter fluorescent particles. The images of the flow field were recorded using a 1300 × 1030 × 12-bit CCD camera. The streamwise velocity profile estimated from the PIV measurements on one plane agreed within 2% with analytical results.

Recently, the effects of out-of-focus particles were examined in [73–76] considering the effects of out-of-plane motions and Brownian motions. These works laid the foundation for applying the micro-PIV technique to characterize flow inside microchannels [77]. Park et al. [78] demonstrated that confocal laser scanning microscopy (CLSM) can be used to improve the out-of-plane resolution. Mielnik and Sætran [79] showed that the out-of-focus resolution can be further improved using a selective-seeding technique to reduce the background noise caused by out-of-focus particles associated with volume seeding technique.

In a recent work by Wei and Joshi [16], velocity distributions inside slightly tapered microchannels were directly measured using the nonintrusive micro-PIV technique. An image parity exchange technique developed by Tsuei and Savas [80] for macroscale PIV was adapted to improve the accuracy in velocity measurement near the wall–fluid interface. The measured results agree well with theoretical predictions. It was found that the location of the maximum velocity deviates significantly from the midplane in the depth direction for the slightly tapered microchannel.

10.3.1.1 Typical Micro-PIV System for Fluid Flow

A typical micro-PIV system is illustrated in Fig. 10.3. The imaging system includes an epifluorescent inverted microscope (Nikon Eclipse TE2000-S), dual Nd:YAG lasers (Gemini PIV 200), and a cooled interline CCD camera. The epifluorescent filter cube consists of an excitation filter, a dichromatic mirror, and an emission filter. The pulsed green light from the dual lasers expands through a beam expander and passes through the diffuser, the excitation filter, and the dichromatic mirror to the microscope objectives. Fluorescent particles (e.g., Nile Red, Molecular Probe) excited by this incident green light emit red light peaked around 560×10^{-9} m (560 nm). The emitted red light passes through the long pass emission filter to reach the CCD camera, which operates in the double-frame single-exposure mode. Scattered green light from the background and ambient is blocked at the emission filter.

For a particular optical system, a scale factor needs to be determined as the ratio of the known physical size of the channel width to the image size in the CCD camera array. This scale factor is used to obtain the actual displacement between two laser pulses with known time intervals. The fine focus knob of the microscope objectives can be used to traverse the measurement focus plane; two-dimensional velocity measurement is thus obtained at different depths of the channel.

It is noted here that the displacement of the objectives (L_o) is different from the displacement of the objective planes. If an air-immersion lens is used to record the image and the

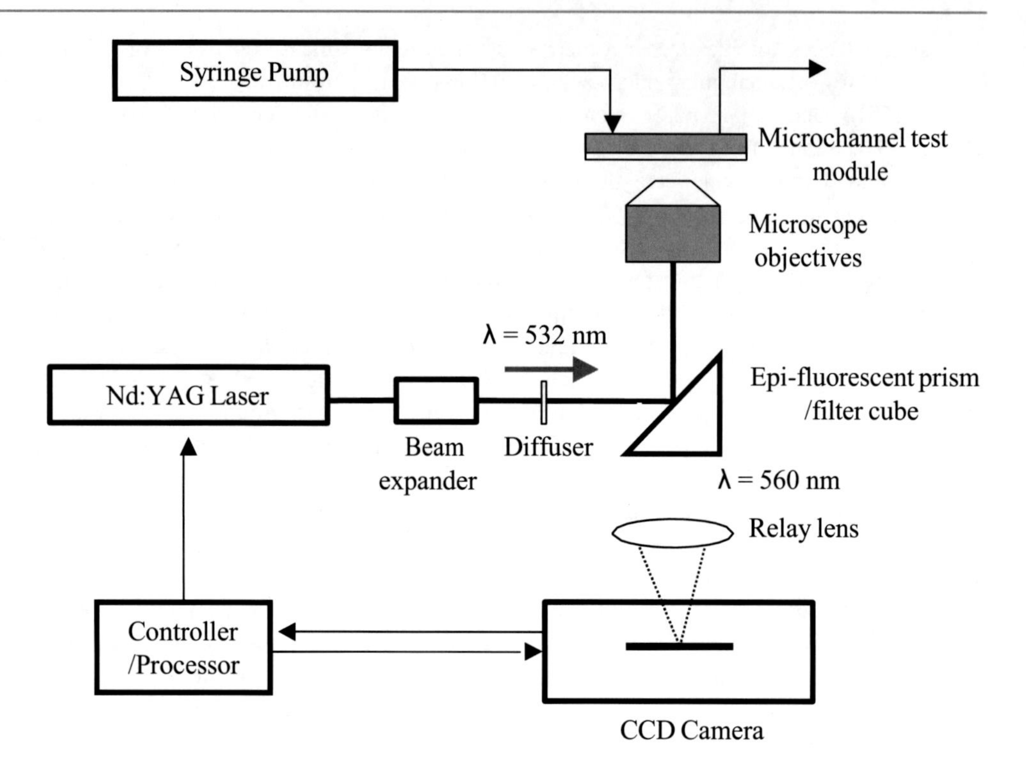

Fig. 10.3 Schematic of a micro-PIV test setup

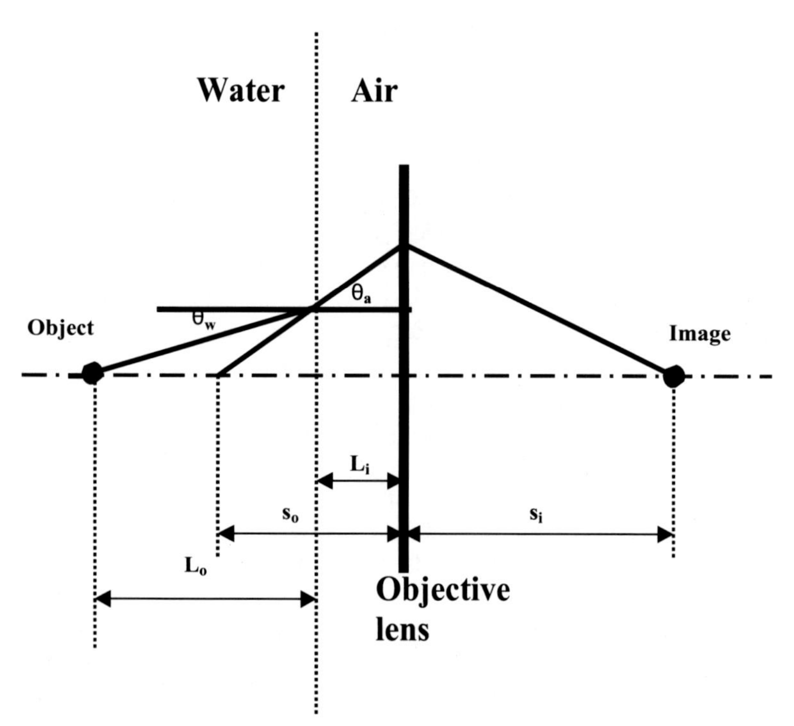

Fig. 10.4 Effects of the refractive index on the object plane displacement

objective plane is located in water, which has a different refractive index, from Fig. 10.4, Eqs. (10.48) and (10.49) can be derived based on the geometry and the Snell's law of refraction. The ratio of the measurement plane displacement to that of objective lens displacement is equal to 1.333, which is the ratio of the refractive index of water (n_w) to air (n_a). A near optical axis condition is assumed in deriving Eqs. (10.48) and (10.49):

$$L_o = (s_o - L_i)\frac{n_W}{n_a} \tag{10.48}$$

$$\Delta L_o = -\Delta L_i \frac{n_w}{n_a} \tag{10.49}$$

One important parameter for PIV measurement is the time interval between the two laser pulses. The displacement between the two interrogation pairs is the product of the time interval and the velocity. An insufficient time interval will make it difficult to detect a slow velocity, whereas an excessive time interval will cause severe loss of pairs as a lot of particles will travel out of the second interrogation area during the interval. As a general rule, the time interval for cross-correlation is selected so that the particles will travel no larger than one-fourth of the interrogation window width. For typical experimental conditions, the time interval in the range 5–100 μs can be used as a starting point.

Compared to a traditional PIV, micro-PIV is different in that volume illumination is used instead of the light sheet method. In addition, due to the small channel size, it is important to carefully select the seeding particles and apply proper image processing techniques to improve the signal-to-noise ratio (SNR).

10.3.1.2 Seeding Particles

In selecting particle sizes, consideration should be given to tracking fidelity, visibility, spatial resolution, and Brownian motion effects. A convenient measure of particle tracking fidelity is the relaxation time, which basically describes how long it takes for the particle to attain velocity equilibrium with the fluid. For Stokes flow, the relaxation time is proportional to the square of the particle diameter. Small particles should be used in order to track the fluid flow faithfully. For the particles used in the current study, the relaxation time is in the order of 10^{-8} s. However, particles should also be large enough to be visible by the imaging optics.

Another issue closely related to the particle size is the Brownian motion effects. Brownian motion results from the random collisions between the fluid molecules and the suspended microparticles. The relative error in measuring velocity caused by the Brownian motion is estimated to be

$$\varepsilon_{\mathrm{B}} = \frac{1}{u}\sqrt{\frac{2D}{\delta t}} \tag{10.50}$$

where u is the fluid velocity, δt is the time interval, and D is the Brownian motion diffusion coefficient given by

$$D = \frac{\kappa T}{3\pi \mu d_{\mathrm{p}}} \tag{10.51}$$

In this equation κ is the Boltzman's constant, T is absolute temperature, μ is the dynamic viscosity, and d_{p} is the particle diameter. Clearly, smaller particles will cause larger uncertainties in the velocity measurement due to Brownian motion. For this study, the relative error due to Brownian motion is typically less than 2%. The error due to Brownian motion is unbiased and can be further reduced by ensemble averaging over repeated measurements. This diffusive error is inversely proportional to the square root of the size of the ensemble.

Particle concentration is another important parameter to be considered in a micro-PIV experiment. Too many particles will increase the background glow and thus reduce the signal-to-noise ratio (SNR). On the other hand, insufficient seeding will result in a weak signal. In general at least 2–3 particles are required for each interrogation area. The SNR can be enhanced by the ensemble averaging technique as described in Meinhart et al. [81]. It was shown that ensemble averaging of the correlation function before the peak detection is superior to averaging the image or the vector map to enhancing the SNR.

For typical applications, the volumetric concentration of the fluorescent particles should be maintained in the range 0.05–0.07%. The size of the ensemble is typically around 40 to sufficiently enhance the SNR and reduce the uncertainties caused by Brownian motion.

10.3.1.3 Spatial Resolution and Depth of Measurement

The spatial resolution of any PIV measurement is defined by the size of the interrogation region and ultimately limited by the effective particle image size. The image of a finite-diameter particle is the convolution of the diffraction limited point response function with the geometric image of the particle. In a study performed at Georgia Tech, a long working distance objective with magnification $M = 60$ and numerical aperture NA $= 0.7$ was used to resolve the 0.5×10^{-6} m particles. The effective image diameter was about 52×10^{-6} m, and it corresponds to 0.87×10^{-6} m when projected back to the flow field. The interrogation area used was 64 pixels in streamwise direction and 32 pixels in the transverse direction. This corresponds to spatial resolution of $7.3 \times 10^{-6} \times 3.6 \times 10^{-6}$ m. When 50% overlap was used for the interrogation, about 20 vectors were measured across the channel width of 34×10^{-6} m.

One important parameter in interpreting the velocity data is the depth of measurement, since the out-of-focus particles also contribute to the correlation function. Olsen and Adrian [75] derived the depth of correlation based on the assumption that the particle image intensity follows Gaussian distribution and the illumination over the interrogation area is uniform. The depth of correlation is given as

$$\delta z_{\mathrm{corr}} = 2\left[\frac{(1-\sqrt{\varepsilon})}{\sqrt{\varepsilon}}\left(f^{\#2}d_p^2 + \frac{5.95(M+1)^2\lambda^2 f^{\#4}}{M^2}\right)\right]^{1/2} \tag{10.52}$$

where ε is a threshold defined as the ratio of the weighting function for the out-of-focus particles to that for the in-focus particles, *f#* is the *f*-number of the optics, λ is the recording wavelength, d_p is the particle diameter, and M is the magnification. Typically, ε is chosen as 0.01 [75]. For the current imaging system settings, Eq. (10.52) gives a measurement depth of about 2.65 μm.

10.3.1.4 Image Processing and Correlation Analysis

In PIV, the measured velocity is defined by the particle image displacement divided by the product of the magnification and the time intervals. The measured velocity is essentially an average over the interrogation region and is assigned to the centroid of the interrogation area. Effectively, the velocity is thus given by [75]

$$u_{\mathrm{m}} = \frac{\int u(X,t)\,W(X)\mathrm{d}X}{\int W(X)\,\mathrm{d}X} \tag{10.53}$$

where $W(X)$ is the weighting function, which determines the relative contribution to the correlation function at different positions of the interrogation area. If the velocity is fully developed, it is only dependent on the y-coordinates as shown in Fig. 10.5. The velocity within the interrogation area can be expanded into Taylor series around the centroid of the interrogation area as

$$u(y) = u_0 + \frac{\partial u}{\partial y}|_{y_0}(y - y_0) + \frac{1}{2}\frac{\partial^2 u}{\partial y^2}|_{y_0}(y - y_0)^2 + \mathrm{O}(y - y_0)^3 \tag{10.54}$$

where u_0 is the true velocity at the centroid. Clearly, the velocity is biased when there is a velocity gradient within the interrogation area. If a uniform weighting function is used, the measurement bias is the second order as the first-order term is terminated during the integration across the interrogation area.

If the interrogation area extends beyond the wall boundary, however, additional large bias in the velocities will occur. The largest bias occurs when the centroid of the interrogation area is located on the solid–liquid interface. As the velocity should be zero for the no-slip boundary condition, a nonzero velocity will be measured asonly half of the interrogation area contributes to the correlation function. This issue could be addressed if the velocity field can be extended to the wall area according to Eq. (10.49), which is proposed by Tsuei and Savas [80] for macroscale PIV:

$$u_{\mathrm{wall}}(y) = 2u(y_i) - u(2y_i - y) \tag{10.55}$$

where y is the coordinate orthogonal to the flow direction, as shown in Fig. 10.5, and y_i is the location of the interface. A Taylor series expansion around the centroid reveals that the extension in Eq. (10.51) is accurate to the first order; consequently using Eq. (10.51) improves the accuracy near the boundary region to the same degree for the bulk fluid region. Implementation of Eq. (10.51) is straightforward. With reference to Fig. 10.5, point y in the wall region and point $(2y_i–y)$ inside the channel are symmetric about the interface located at y_i. Since the interface velocity $u(y_i)$ is zero for the no-slip boundary condition, Eq. (10.51) indicates a reflection and a reverse of the velocity field inside the **channel** about the interface. This is achieved through the image process steps illustrated in Fig. 10.6 such that images for both frames are reflected about the interface. The wall region of the second frame is replaced with the reflected image of the fluid region in the first frame and vice versa. This process is referred to as image parity exchange (IPX) in Tsuei and Savas [80] and will be used hereafter. In micro-PIV measurement, the wall boundary is usually hard to define as it blurs as a result of defocusing. In the present study, a MATLAB (commercial technical computing software) function which searches for

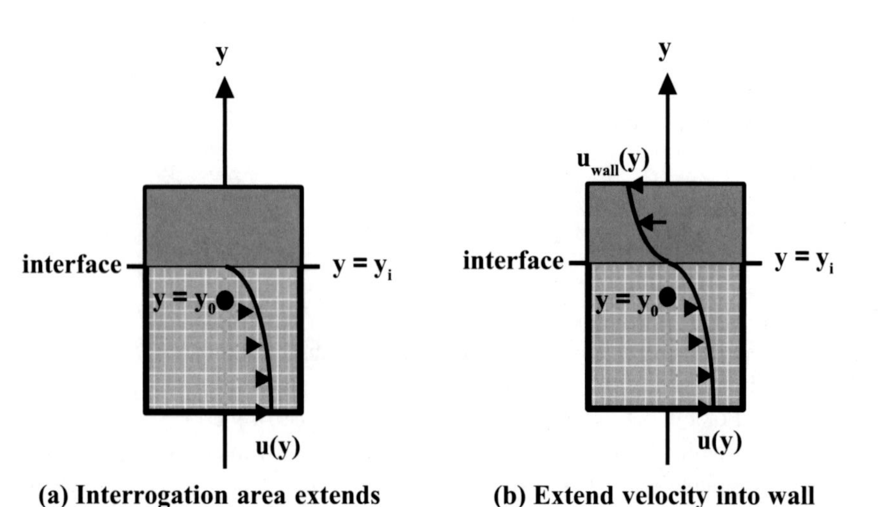

Fig. 10.5 Effects of the presence of wall boundary inside interrogation area on the velocity measurement

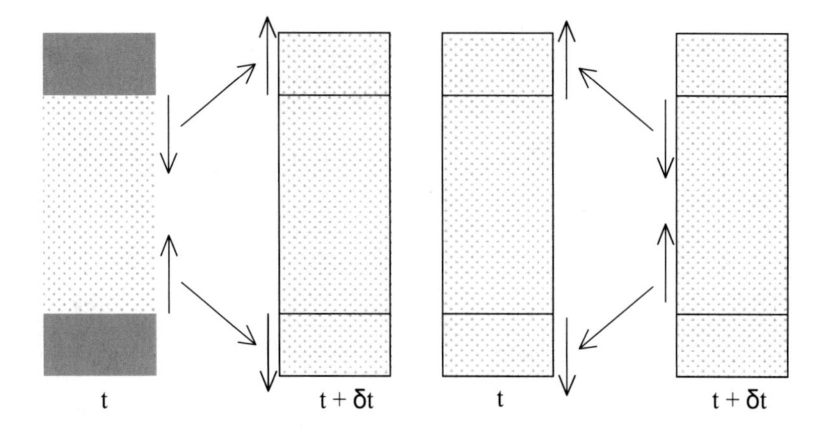

Fig. 10.6 Implementation of Eq. (10.55) to extend the velocity field to the wall region

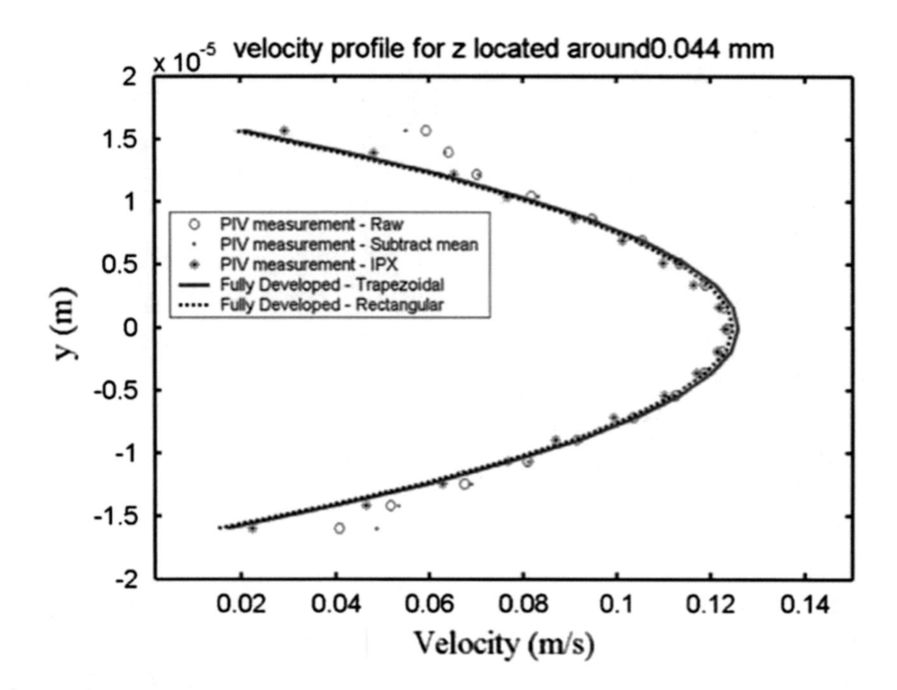

Fig. 10.7 Velocity profile using different image processing techniques

the maximum image intensity gradient is used to locate the channel boundary.

It is noted here that before the IPX process, the mean pixel intensity value is subtracted from the raw image maps to reduce the background noise. To identify the effects of IPX, correlation analysis is also performed on the image maps where only mean intensity is subtracted. These results are compared with predicted velocity profiles from the series solution for rectangular channel [82] and numerical solution for trapezoidal channel. As can be seen from Fig. 10.7, applying IPX before the correlation significantly improves the accuracy of the velocity profile prediction near the wall. This is particularly useful for measuring velocity for very narrow channels where boundary effects are significant such as the channel sizes in the current study.

To determine the velocity vectors, adaptive correlation can be first conducted to the image maps. The average velocity map obtained can then be used as the offset for the ensemble average correlation. This offset is applied to reduce the loss-of-pair issue successfully.

10.3.1.5 Velocity Profiles for Fully Developed Flow

Figure 10.8 plots the velocity profile for a channel of 34 μm wide and 118 deep μm at different depth positions for the flow rate of 2.78×10^{-10} m^3/s (1 ml/h).

The predicted results based on analytical solutions for an effective rectangular channel and numerical simulations for a trapezoidal channel are plotted for comparison. As can be clearly seen, the measured velocity profile matches very well with results predicted from laminar theory for all the planes measured. Even though the actual channel is not perfectly trapezoidal, the trapezoidal model results generally agree better with experimental measurements than the rectangular model, particularly for the measurement plane located away

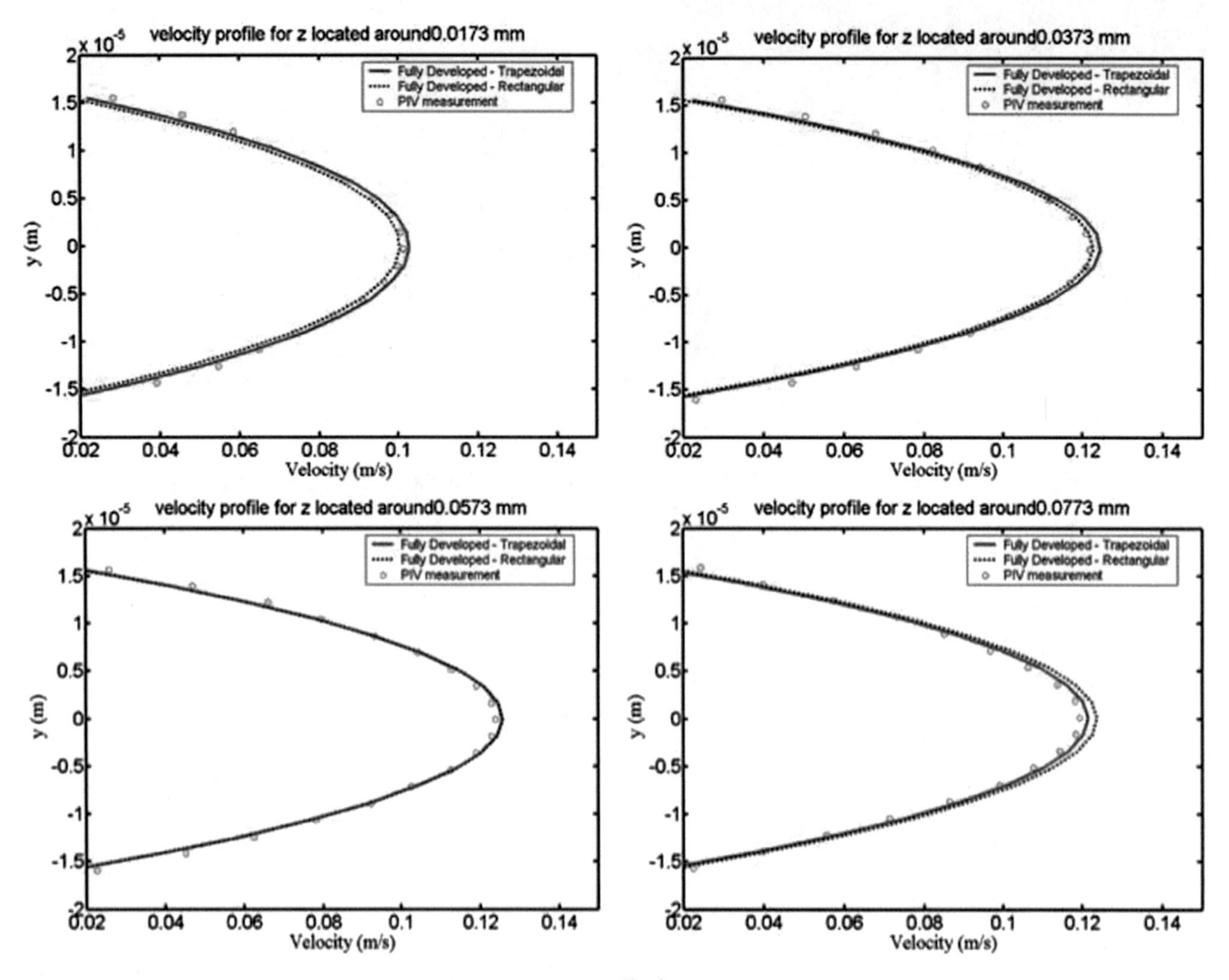

Fig. 10.8 Velocity profile at different focus planes (flow rate 2.78×10^{-10} m^3/s or mean velocity 0.07 m/s)

from the geometric symmetry plane. These comparisons clearly prove the validity of the laminar theory for microchannels. These results also confirm the feasibility of using micro-PIV as a diagnostic tool for microfluidics for channels of these sizes.

10.3.2 Measuring Fluid Temperature Inside Microchannels

Measuring temperature field inside microchannels is very challenging due to the small length scale involved. Optical nonintrusive techniques are probably the only feasible way for this type of work. This section discusses various temperature measurement techniques including micro-PIV using Brownian motion, two-color laser-induced fluorescence (LIF) for fluid temperature, single quantum dot tracking for near-wall temperature, and velocity measurement.

10.3.2.1 Temperature Measurement Using Micro-PIV

As discussed in the previous section, velocity measurement using micro-PIV is often affected by the Brownian motion of the seeding particles. Derived from the balance of the Stokes drag force and the diffusion force, the diffusion coefficient is a function of the absolute temperature and the viscosity as indicated in Eq. (10.51), which is rewritten here for convenience:

$$D = \frac{\kappa T}{3\pi \mu d_\mathrm{p}} \tag{10.56}$$

where κ is the Boltzmann constant, μ is the viscosity of the fluid, and d_p is the particle size, and T is the absolute temperature.

The RMS displacement x_rms by the particles within the time window of Δt is given by

$$x_{rms} = \sqrt{2D\Delta t} \quad (10.57)$$

Clearly, this displacement is proportional to the square root of the ratio of temperature to viscosity. In the presence of the Brownian motion, according to Hohreiter et al. [83], the probability function for the particles follows a Gaussian distribution, which is shown here in Eq. (10.58):

$$f = (4\pi D\Delta t)^{-3/2} EXP\left(-(x - x_0 - u\Delta t)^2/4D\Delta t\right) \quad (10.58)$$

where x is the position of the particle, x_0 is the original position of the particle, and u is the mean velocity. The net effects are the broadening of the peak and decrease in the peak height in cross-correlation. The broadening in the peak width can be quantified by the difference in the correlation peak areas between the autocorrelation and the cross-correlations, as given by [83]

$$\Delta A = 4M^2\kappa/3d_p(T/\mu)\Delta t \quad (10.59)$$

where M is the magnification of the objective lens.

This technique is implemented in Hohreiter et al. [83] to measure the water temperature inside a microchannel with no net flow. Images were taken sequentially at a fixed time interval. The peak width area of the autocorrelation of the first frame is used as the reference. The first frame is then correlated with the second and subsequent frames. As shown in Eq. (10.59), the increase in the peak area is linearly proportional to time. In the plot of the peak area and time, the slope represents the temperature multiplied by a constant. This relationship holds true as long as the heat transfer is in steady state such that the temperature and viscosity does not change with time. It is noted that the viscosity and particle size are needed as the input to calculate the temperature.

The measurement uncertainty of the temperature field depends on the accurate account of the peak area, particle diameter, and time interval. Due to the temperature dependency of the viscosity, the measurement uncertainty is also depending on the temperature. For the system described, the uncertainty is ±3 °C.

Simultaneous measurement of the temperature and the velocity is limited by the proper time intervals between the image frames. On the one hand, the time interval should be sufficiently long to allow measurable peak area increase. On the other hand, the time delay has to be short to avoid losing pairs between the cross-correlation pairs. Hohreiter et al. [83] reported that the measurable velocity for water with temperature higher than 20 °C is less than 8 mm/s for their system. Such difficulties stem from the fact that the same tracer particles are used for both Brownian motion and bulk flow. The following section introduces a technique with dual tracing particles where the temperature measurement and velocity can be separated.

10.3.2.2 Two-Color Laser-Induced Fluorescence (LIF) for Temperature Measurement

Temperature-sensitive fluorescent dyes have been used for temperature measurement in fluid systems [84]. The fluorescent intensity absorbed by the imaging camera is proportional to the quantum efficiency, incident light intensity, dye concentration, and absorptivity of the dye. The fluid field temperature can be mapped if the incident light intensity is uniform and constant and the dye concentration is constant. However, the requirement for uniform and constant incident light is very difficult to achieve, and the imaging system itself adds uncertain noises to the signal. One elegant way to address this is to introduce a second dye, which is nonsensitive to the temperature. The ratio of the intensity between the two should help to reduce the dependency on the incident light and the noises by the cameras and optics. This is shown in Eq. (10.60) for the system of Rhodamine-B and Rhodamine-110 by Kim et al. [85]. The ratio of the absorption spectral intensity $\frac{I_{0rhb}}{I_{0rh110}}$ is fixed for single light source. The fluorescent intensity ratio $\frac{I_{rhb}}{I_{rh110}}$ is only dependent on the quantum efficiency ratio $\frac{Q_{rhb}}{Q_{rh110}}$ since the absorptivity ratio $\frac{\varepsilon_{rhb}}{\varepsilon_{rh110}}$ and the concentration ratio $\frac{C_{rhb}}{C_{rh110}}$ can be fixed:

$$\frac{I_{rhb}}{I_{rh110}} = \frac{I_{0rhb}\varepsilon_{rhb}C_{rhb}Q_{rhb}}{I_{0rh110}\varepsilon_{rh110}C_{rh110}Q_{rh110}} \quad (10.60)$$

If the difference in the quantum efficiency between the two dyes are significant enough, the temperature can be mapped based on the intensity ratios extracted from the images. The dual dye system of Rhodamine-B and Rhodamine-110 [85] fits this requirement. Rhodamine-B has a temperature sensitivity of 2% per K, while Rhodamine-110 has a temperature sensitivity of 0.13% per K.

A typical optical system [86] is illustrated in Fig. 10.9, which includes an Ar-ion laser (wavelength = 488 nm). A scanning mirror sweeps vertically to form the light sheet to illuminate the test field of interest. The fluorescent light emitted from the particles was splitted by a beam splitter, which transmits the wavelengths $\lambda_T \leq 520$ nm and reflects wavelengths $\lambda_R \geq 570$ nm. As can be seen from Table 10.2, this splitter transmits light emission from Rhodamine-110, but reflects majority of Rhodamine-B. Since the background light has wavelength of 488 nm, another filter rejects wavelengths below 495 nm for Rhodamine-110. An optional filter helps eliminate wavelengths higher than 570 nm for Rhodamine-B. The CCD cameras are arranged to detect Rhodamine-110 and Rhodamine-B, respectively.

To accurately calculate the intensity ratios at every location, it is very important to align the two cameras with each

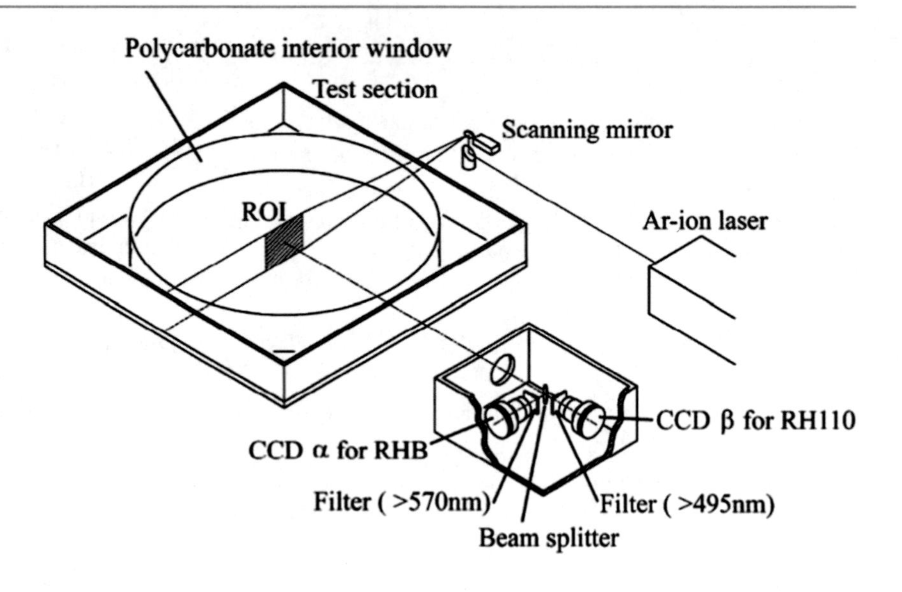

Fig. 10.9 Measurement of temperature field of a Rayleigh–Benard convection using two-color laser-induced fluorescence. (Reprinted with permission from Sakakibara and Adrian [86])

Table 10.2 Characteristics of Rhodamine-B and Rhodamine-110

Dye	Molecular weight	Peak absorption wavelength	Peak emission wavelength	Quantum yield	Absorptivity at 488 nm
Rhodamine-B	479.02	554	575	0.31	4.4
Rhodamine-110	366.8	496	520	0.8	34

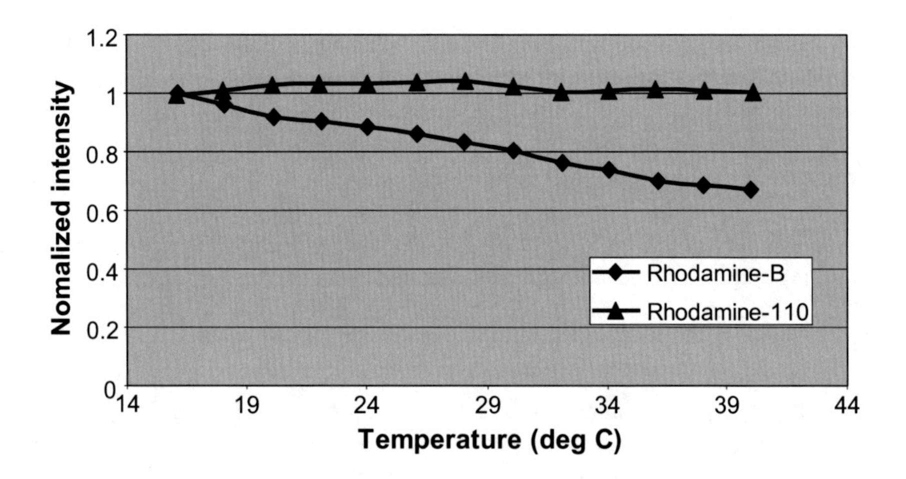

Fig. 10.10 Comparison of fluorescence intensities of Rhodamine-B and Rhodamine-110, normalized by the corresponding intensities at 14 °C. (Reprinted with permission from Kim et al. [85])

other. A calibration plate with equally spaced grid pattern can be used to calibrate the image coordinates against the physical coordinates. A simple correlation can be performed to give the coefficients for the transfer function from the physical coordinates to the image coordinates. The dual camera system allows the simultaneous measurement of the image intensity for both dyes. However, it is found in Sakakibara and Adrian [86] that intensity ratio measured by the two cameras is spatially dependent even when the temperature field is uniform. This may have been caused by the differences in object lens and the astigmatic aberrations by the inclined beam splitter. To overcome this problem, a location by location correction has to be performed through Fourier transformation [86].

A simplified version of the test system is used in Kim et al. [85], where the beam splitter is removed. To split between the Rhodamine-B and Rhodamine-110, an alternating filter is used before the emission from the particle reaches the single camera. Since only one camera is used, there is no need to do a pixel-by-pixel correction of the image coordinates.

Before actual temperature field measurement, the system is calibrated for a constant fluid temperature case. The intensity ratios are correlated with the known temperature. Such an example is given in Fig. 10.10. As can be clearly seen, the temperature sensitivity for Rhodamine-B is significantly higher than that of Rhodamine-110. The calibrated system can then be used to measure the actual field temperature. Kim et al. [85] reported that the uncertainty is 2.3 °C for spatial resolution of 19 μm.

One limitation of the system reported here is that a light sheet has to be employed. Similar to micro-PIV techniques, a volumetric illumination system can also be implemented where the emissions from the dual dyes are received by a single camera sequentially through an alternating camera. However, such a system is subject to the optical influences of the out-of-focus particles much like the micro-PIV. Furthermore, since the temperature measurement is based on the image intensity, the background noise by the out-of-focus particles is even more unpredictable. To address this issue, a total internal reflection fluorescence (TIRF) technique can be used [87].

10.3.2.3 TIRF for Simultaneous Measurement of the Temperature and Velocity for Near Regions

Total reflection happens when the incident light angle is larger than the critical angle determined by the refractive index for both media. An evanescent wave is generated when total reflection happens. The evanescent electromagnetic field decays exponentially from the interface as shown in Eq. (10.61):

$$J(z) = J_0 \exp\left(-z/d\right) \tag{10.61}$$

where $J(z)$ is the intensity distribution, J_0 is the wall intensity, z is the distance from the wall, and d is the penetration depth. Equation (10.62) is used to determine the penetration depth:

$$d = \frac{\lambda_0}{4\pi}\left[n_1^2 \sin^2\theta - n_2^2\right]^{-1/2} \tag{10.62}$$

where λ_0 and θ are the incident wavelength and angle, respectively, and n_1 and n_2 are the refractive index for the medium, respectively.

Typical penetration depth is about 100–200 nm into the fluid field and thus allows selective illumination of the fluid field. Unlike the flood illumination used in micro-PIV, contributions from the bulk of the fluid are minimized.

Using this technique, Guasto and Breuer [88] conducted simultaneous measurement of temperature and velocity with single quantum dot tracking. Quantum dots (QDs) are nanometer-sized tracer particles made out of semiconductor nanocrystals. These QDs are single fluorophores with significant quantum efficiency and resistance to photo-bleaching. The small size (5–25 nm in diameter) allows QDs to be used for true nanoscale measurements. They are typically coated with surface coatings to achieve solubility in either organic or inorganic solvents. In the system reported in Guasto and Breuer [88], CdSe/ZnS QDs are used together with the temperature-insensitive Rhodamine-110 in a borate buffer. QDs exhibit significant temperature-dependent quantum efficiency. Similar to LIF, Rhodamine-110 is used as the background dye for normalization purposes since its quantum efficiency is insensitive to the temperature. A schematic illustration of the optical system used in Guasto et al. [87] is reproduced in Fig. 10.11 The PDMS channel dimension is 340 × 30 × 10 mm. Dual wavelength images produced by the fluorescence emission of the QDs and Rhodamine-110 are separated by wavelength using an image splitter through a dichronic mirror with cutoff wavelength at 560 nm. The images are filtered before recombination and projection onto the ICCD. A silicon reference grid is used for image coordinates calibration purposes. The images of the QDs on the camera are defined by the diffraction limited spot since the diameter is significantly smaller than the light wavelength. QDs are detected by thresholding the maximum pixel value of an image until a possible particle is found. A set of the brightest particles are taken to achieve sufficient signal-to-noise ratio (SNR). Roughly, 25 QDs are detected for an effective area of 22 × 22 μm.

To measure the temperature, the peak intensity of the single QD is normalized by the single pixel intensity from the Rhodamine-110 reference images. To account for the temporal fluctuations, the Rhodamine-110 reference intensity is averaged over time.

To determine the temperature sensitivity of the QDs and Rhodamine-110, verification experiments are done first using bulk solutions with flood illumination. The result indicates significantly higher temperature sensitivity for QDs than for Rhodamine-110. The calibrated curve for intensity and temperature is then used for temperature measurement of the microchannel flow shown in the figure. The temperature of the flow is controlled by the controller attached to the microscope stage. Although the measured intensity does not follow a perfect Gaussian distribution, the mean value when normalized by the reference intensity shows good agreement with the temperature measured independently. The uncertainty is within 5°C although better accuracy is expected since some transient effects may have not been completely removed in the experiments.

After the QDs locations are detected in the image, the displacement of the QDs can be used to determine the stream and cross-stream velocity. Due to the extremely small diameter of the QDs, the Brownian motion effects are significant when measuring the velocity. The mean displacement is inversely proportional to the square root of the diameter. The net effect of the Brownian motion is the broadening of the velocity distribution. To minimize this effect, ensemble average is performed over 2000 samples, which reduced the standard deviation significantly.

It is noted here that QDs are typically subject to bleach and photostability issues. Although the shelf life is typically 6 months, QDs degrade rapidly once diluted. It is therefore

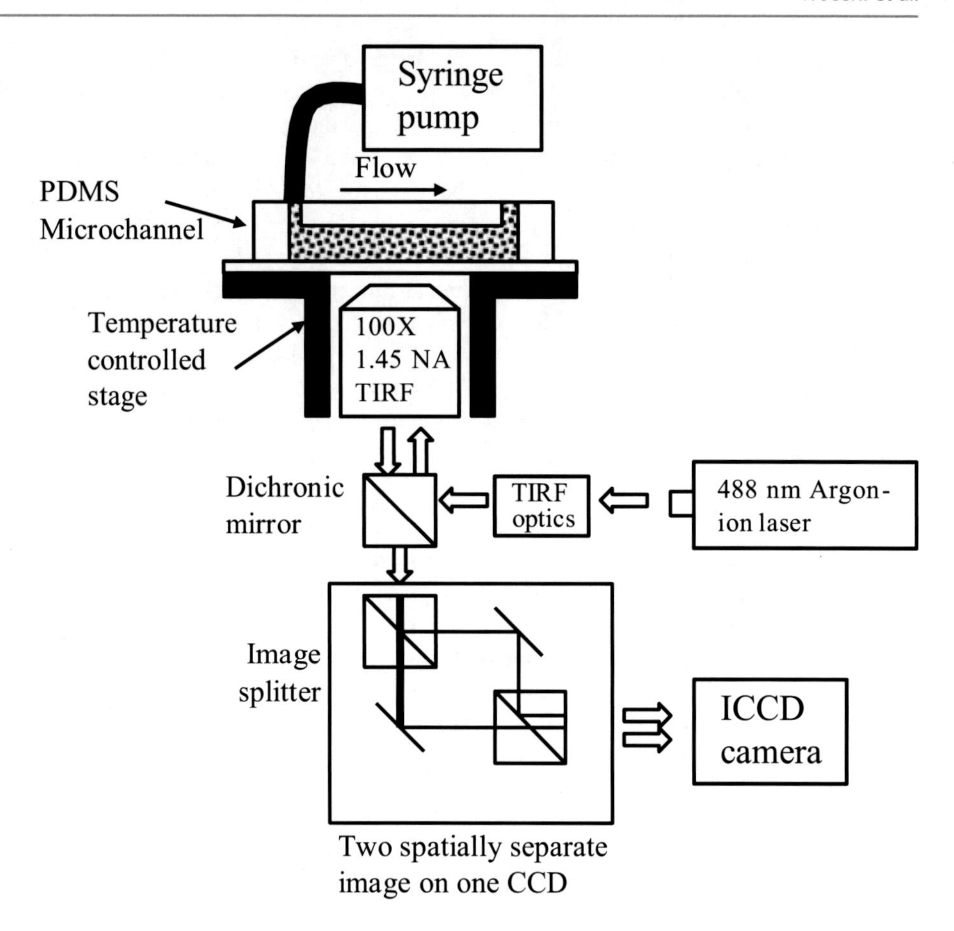

Fig. 10.11 Setup for simultaneous imaging of solutions of Rhodamine-110 and QDs using TIRF. (Reprinted with permission from Guasto et al. [87])

important to carry out experiments within a short period of time (hours). QD delays due to QD blinking phenomena could further impose illumination time limits on the test.

In summary, this section provided introduction of the measurement techniques inside microchannel for velocity and temperature. Details of the micro-PIV technique for velocity measurement are provided. While velocity measurement techniques have seen significant progress in micro-PIV, temperature measurement has remained a challenge. A brief introduction is given to three representative techniques in micro-PIV with Brownian motion, two-color LIF, and TIRF using QDs. Significant development is needed to improve the measurement accuracy in the future.

10.4 Microfabrication and Assembly

Currently, most microchannels used in applications fall into the range of 30–300 μm [89]. Microchannels can be fabricated using micromachining, molding, or embossing in a variety of materials, such as Si, glass, metals, or polymers [90–92]. The most important feature of the microchannels is their large surface-to-volume ratio, which leads to a high rate of heat transfer (or heat-transfer coefficient). In addition, microchannels can be integrated close to the Si chips to enable compact heat sinks.

Microchannel heat sinking was first proposed for very large-scale integrated (VLSI) electronic components in the 1980s. As shown in Fig. 10.12, Tuckerman and Pease [1] studied chip cooling with an on-chip microfluidic heat sink and demonstrated that a heat flux of ~790 W/cm^2 can be removed with a temperature rise of 71 °C. This translates into a thermal resistance as low as 0.09 °C/W. The deep microtrenches were first formed through micromachining in the backside of a Si substrate with integrated circuits. A cover plate was then bonded on top of the microtrenches to form enclosed microchannels. Following this landmark work, numerous studies have been carried out to implement on-chip microchannel cooling [93–95]. The most common bonding technique for high-quality bonding of Si wafers is field-assist bonding (anodic bonding) process. Such bonding technique usually requires ultra-smooth/flat/clean bonding interfaces, as well as high pressure, high temperature (up to ~1000 °C), and high voltage (500–1000 V). These processing conditions are not very friendly to back-end-of-the-line (BEOL) wafers, where CMOS devices and on-chip interconnects are completed. Therefore, CMOS-compatible

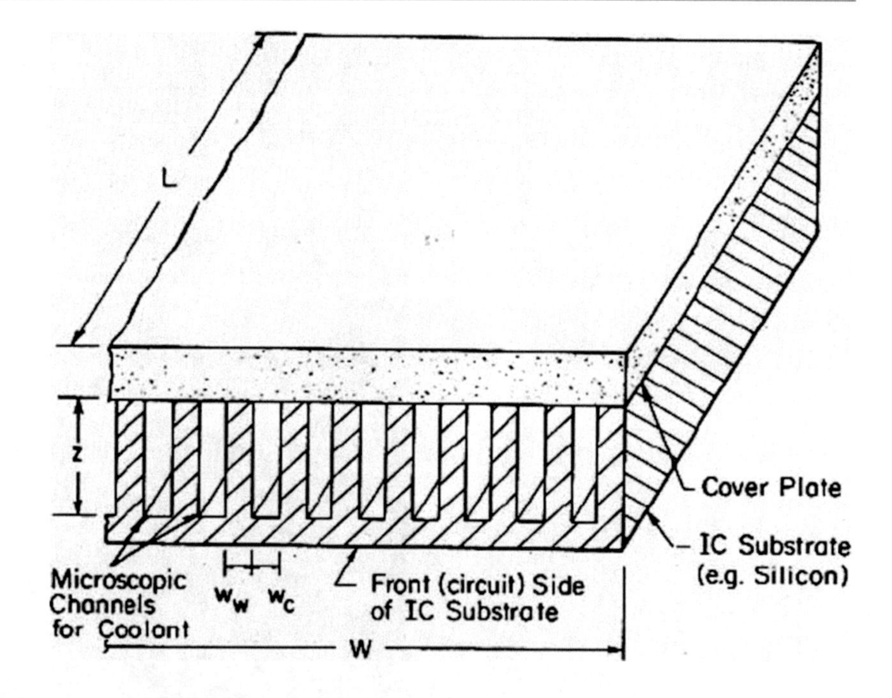

Fig. 10.12 Microchannel heat sink integrated in IC substrate

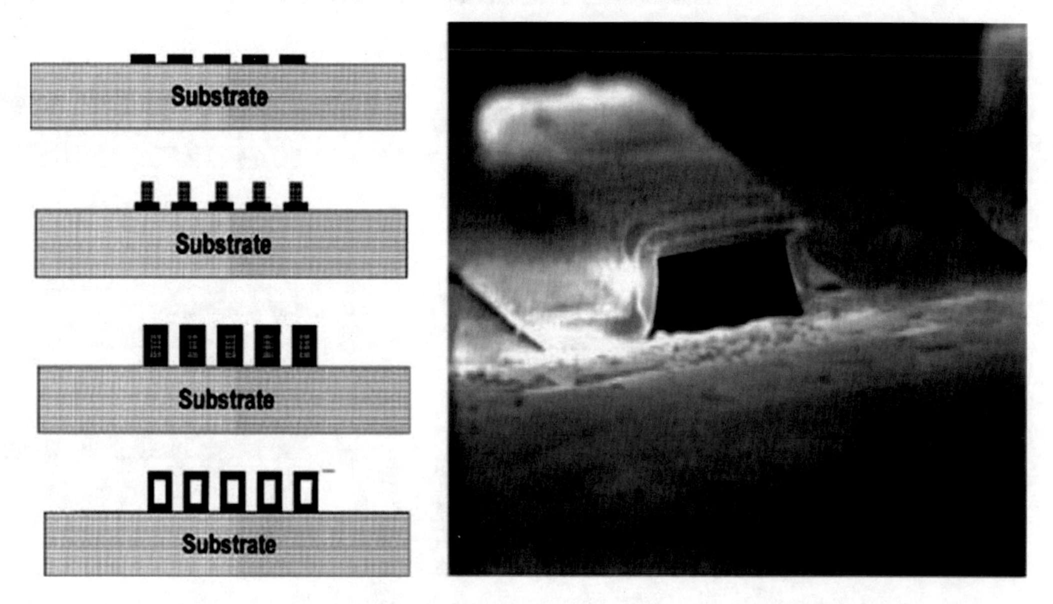

Fig. 10.13 Metallic microchannels formed by electroplating around sacrificial photoresist lines. (Papautsky et al. [92])

processes are necessary for the fabrication of on-chip microchannels following BEOL processing.

Several low-temperature CMOS-compatible microchannel fabrication approaches have been studied in the past decades. Electroplating method to form metallic microchannels is shown in Fig. 10.13 [92]. In this approach, a photoresist was used to define the microchannels and metallic shells were then deposited through electroplating. After the photoresist is dissolved in acetone bath, hollow microchannels can be formed. As shown in Fig. 10.14, another example is the buried microchannels fabricated in Si substrate through two-step electrochemical process, which combines formation of porous Si and electropolishing [93].

Dang et al. [97, 98] also studied a monolithic method to fabricate polymer-capped microchannels in Si substrate using a sacrificial polymer at low temperature, as illustrated in Fig. 10.15. Instead of wafer bonding, the deep trenches are first filled with a sacrificial polymer by spin coating. After surface planarization through mechanical polishing, the trenches are overcoated with a layered structure. The sacrificial polymer is then decomposed by heat treatment. The layered overcoat consists of a SiO_2 thin film (1–2 μm)

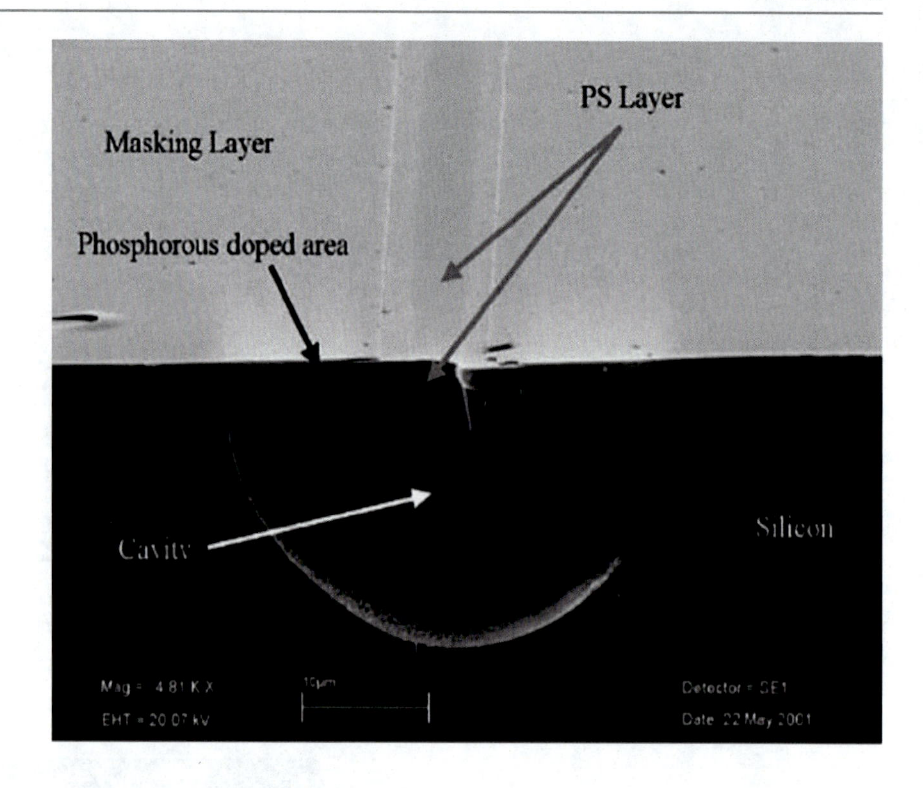

Fig. 10.14 Planar buried microchannels in Si substrate. (Kaltsas et al. [96])

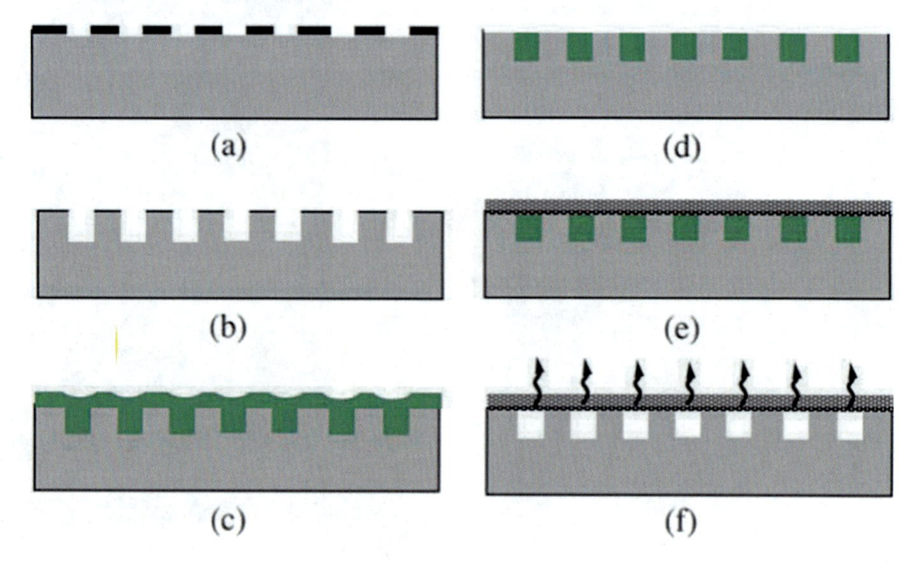

Fig. 10.15 Schematic illustration of low-temperature monolithic process of microchannel fabrication in Si wafer through a sacrificial polymer and an overcoat. (**a**) Trench patterning, (**b**) trench etching and surface cleaning, (**c**) trench refilling by spin coat, (**d**) polishing and descum, (**e**) overcoating with a layered structure (a porous SiO_2 layer and a polymer layer), and (**f**) sacrificial polymer decomposition and overcoat curing [101]

deposited with PECVD at low temperature (~150 °C) and a thicker layer of Avatrel 2195P polymer that is spin coated and cured as the structural polymer (~25 μm). Since the SiO_2 film is porous, it allows permeation of the gaseous products from decomposition of the sacrificial polymer. The SiO_2 also plays an important role in maintaining the structural integrity of the overcoat by preventing "sagging" of the above structural polymer. In fact, decomposition of the sacrificial polymer and curing of the structural polymer occur simultaneously during a heat treatment in a program-controlled oven. To further increase the mechanical strength and hermeticity of the microchannels, an additional overcoat layer can be easily applied, as shown in Fig. 10.16 (a, b). In addition to the dielectric polymer, many other materials may also be used as water-proof coating, such as metal layers deposited by DC sputtering and electro/electroless plating or spin-on glass.

High-pressure fluidic testing confirmed good adhesion, hermeticity, and high mechanical strength of the polymer overcoat. The polymer overcoat of only 30 μm in thickness can withstand a DI water pressure of 2.5 atm (35 psi) during the continuous testing. No delamination or cracking was observed. In addition, the adhesion between the overcoat polymer and the Si substrate did not degrade, and no cracking

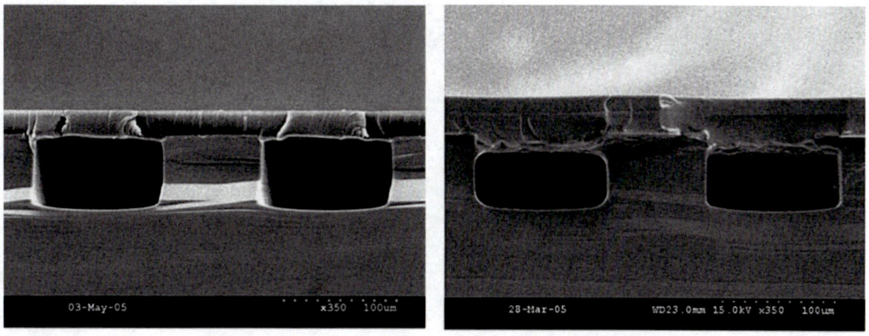

(a) (b)

Fig. 10.16 Cross-sectional SEM photographs of completed microfluidic channels enclosed by a layered overcoat after decomposition and curing. (**a**) Single overcoat layer and (**b**) double overcoat layer [101]

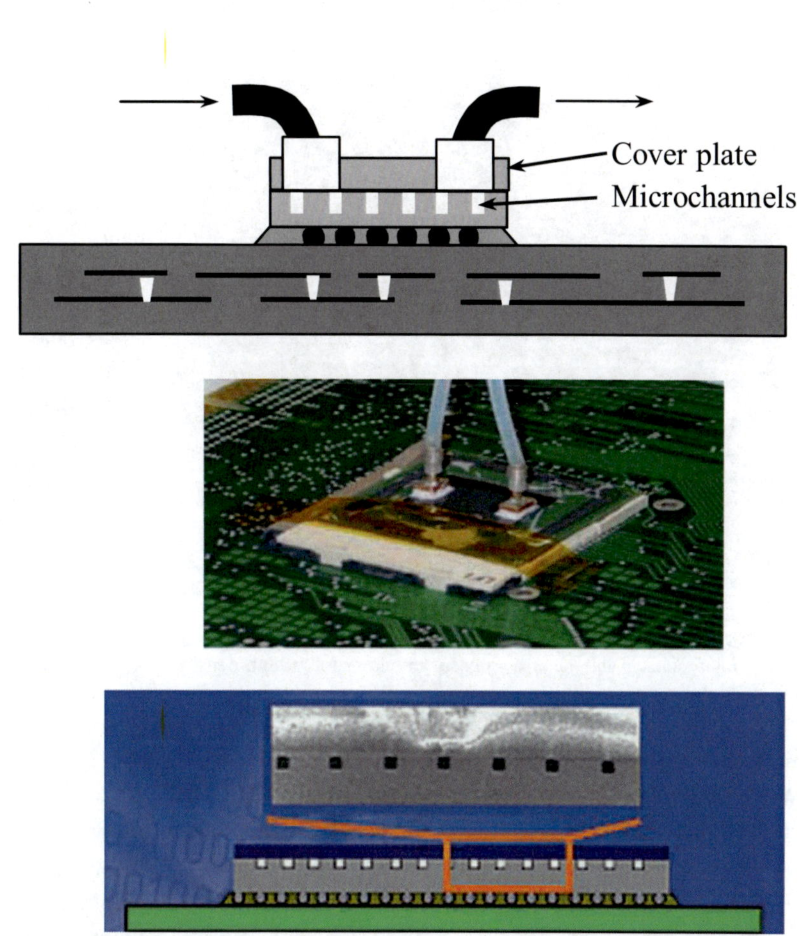

Fig. 10.17 Schematic of typical microchannel heat sinks in literature (the microchannel heat sink is integrated as part of a Si chip). [99, 100]

or delamination occurred during thermal cycling test based on JEDEC Standard (Condition J of JESD 22-A104-B, 0–100 °C, 3 cycles per hour).

In addition to microchannel fabrication, fluidic interconnection and assembly raise another challenge in on-chip microfluidic cooling system. Conventional microchannel interconnection method is illustrated in Fig. 10.17 [99, 100]. Inlet and outlet of the microchannel heat sink are located at the back side of a chip and are connected to the external fluidic circulation system.

Dang et al. [98] proposed a compact microelectronic system with board-level fluidic delivery for on-chip microchannel heat sink using thermal–fluidic I/O interconnection. The basic concept of the thermal–fluidic I/O assembly is illustrated in Fig. 10.18 [101]. Micropipes are fabricated and integrated with C4 (Controlled Collapse Chip Connection) solder bumps to provide simultaneous electrical and thermal–fluidic I/Os, as shown in Fig. 10.19.

As shown in Fig. 10.20, with a precision flip-chip bonder, a compression force is applied after the chip and the package substrate are aligned. Since the polymer pipes are usually

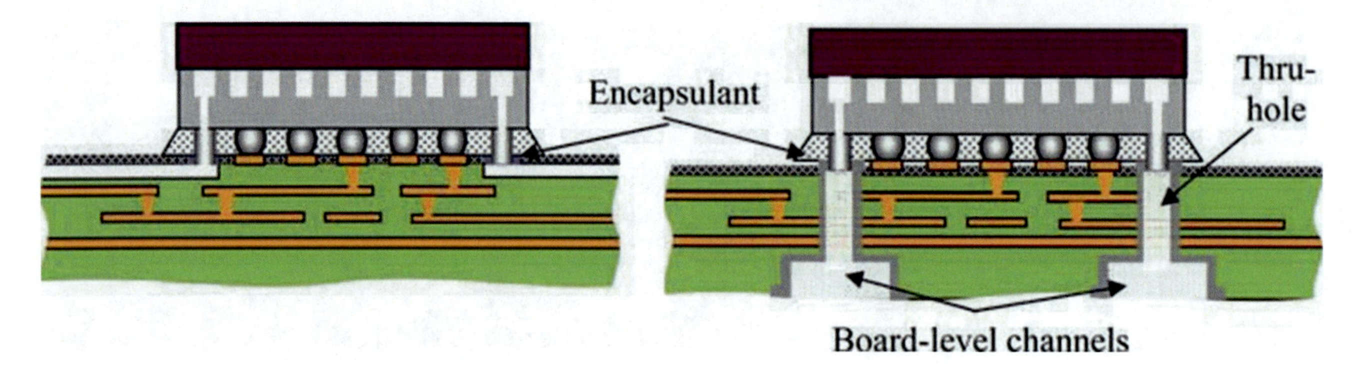

Fig. 10.18 Integrated thermal–fluidic I/O interconnection configurations [101]

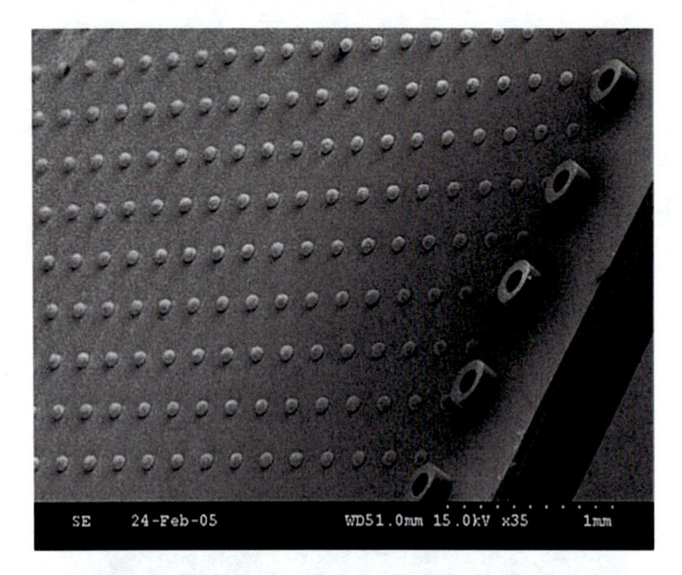

Fig. 10.19 SEM photographs of a row of micro-polymer pipes as the fluidic interconnects and the area array solder bumps for as high-density electrical interconnects [98]

taller than the solder bumps, a portion of the polymer pipes will be inserted into the orifices, when the solder balls come into contact with the metal pad on the substrate. Once reflowed, metallurgical joining will be formed by the solder bumps, and a chip is then mechanically bonded with the board substrate.

The micropipes are designed to have the same diameter as that of the orifices. To ensure hermetic sealing of the micropipes, a void-free polymer should be applied at the chip-edge after assembly, which is the same as the conventional underfilling process.

Figure 10.21 shows a "microfluidic flip chip" assembled onto a substrate with fluidic path for DI water circulation. For the purpose of simplicity, the microchannels on the test thermal chip were enclosed by glue bonding of a Si cover plate. The height of the microchannels is approximately 200 μm, and their width is around 100 μm. After underfilling and curing, both fluidic continuity and electrical continuity were achieved successfully.

10.5 Application in Electronics Thermal Management

Electronics are finding newer applications in the areas of nano- and biotechnologies, which lead to high-power densities. Microchannel cooling is very attractive owing to its simplicity in the implementation and facilitation for both single-phase and two-phase cooling with high heat flux removal capabilities (usually, 100–1000 W/cm^2). With microchannels (and their progressive variations) proven over the last two decades to be suitable for high heat flux removal with narrow range of operating temperatures, they can be used to provide highly effective and competitive cooling solutions and as high-precision temperature control systems for nano- and bioelectronic devices in future.

Different variations of microchannels were successfully implemented in recent years for cooling of electronics. Stacked microchannels [17] and manifold microchannels [102] are two major examples. While stacked microchannels were discussed before in this chapter, manifold microchannels (MMCs) are primarily designed to significantly reduce the pressure drop by having a normal flow. Pressure drop can be a major impediment to implementing the microchannel technology for electronics especially for sub-hundred microchannel sizes, and tailored stacked and manifold microchannels offer an attractive solution under such scenarios. An array of microchannels having a fraction of length of the total heat sink constitutes an MMC heat sink and has been demonstrated successfully [103].

In addition to variations in microchannel designs, efforts have also been made to improve thermophysical properties of the working fluids in microchannels. Examples include nanoparticle-loaded fluids [104], micro-encapsulated and nano-encapsulated solid–liquid-phase change material slurries [61], and emulsions, which ameliorate thermal conductivity and/or heat capacity of the working fluid for better heat penetration and absorption compared to pure fluids. Studies found an optimum loading of thermophysical property boosting media, inlet temperature, and fluid and loaded

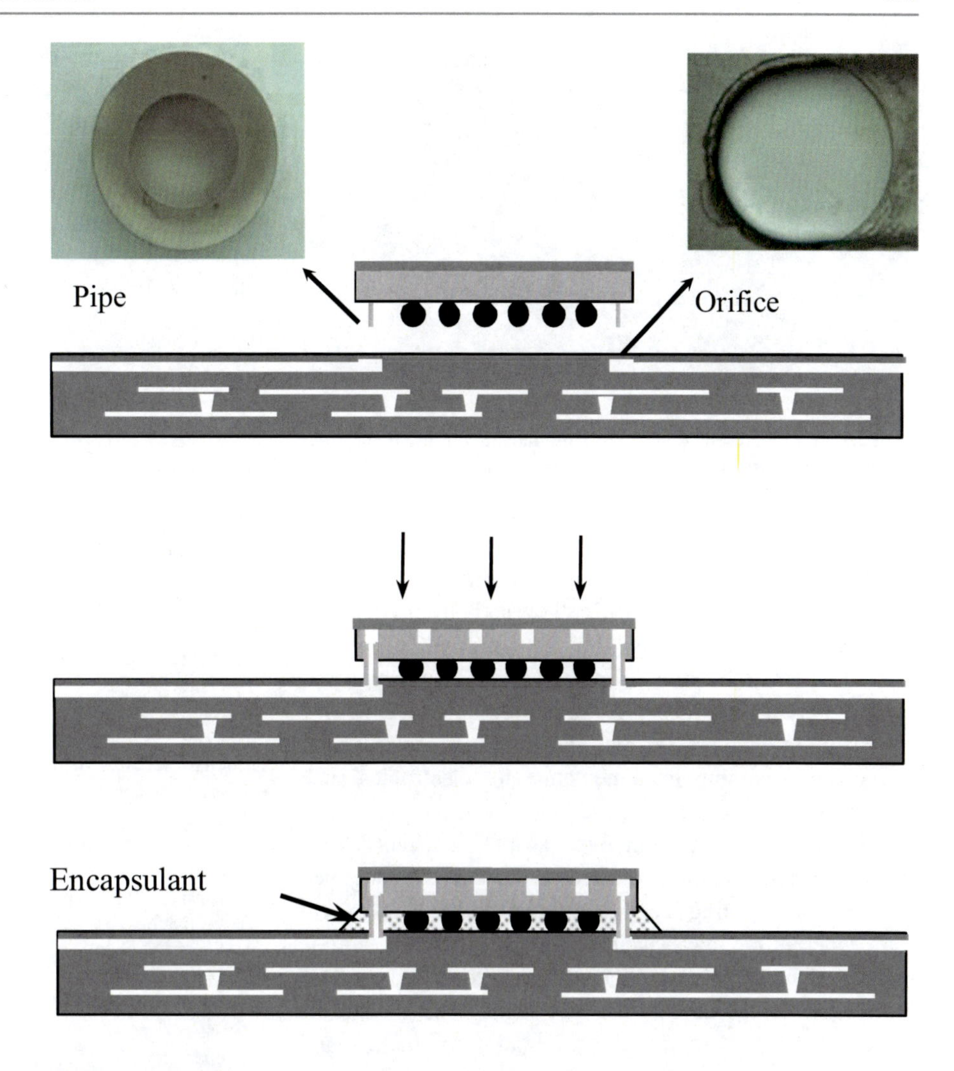

Fig. 10.20 Schematic of the assembly process of a "microfluidic flip chip" (**a**) alignment, (**b**) bonding through compression and heating, and (**c**) encapsulation with underfill [101]

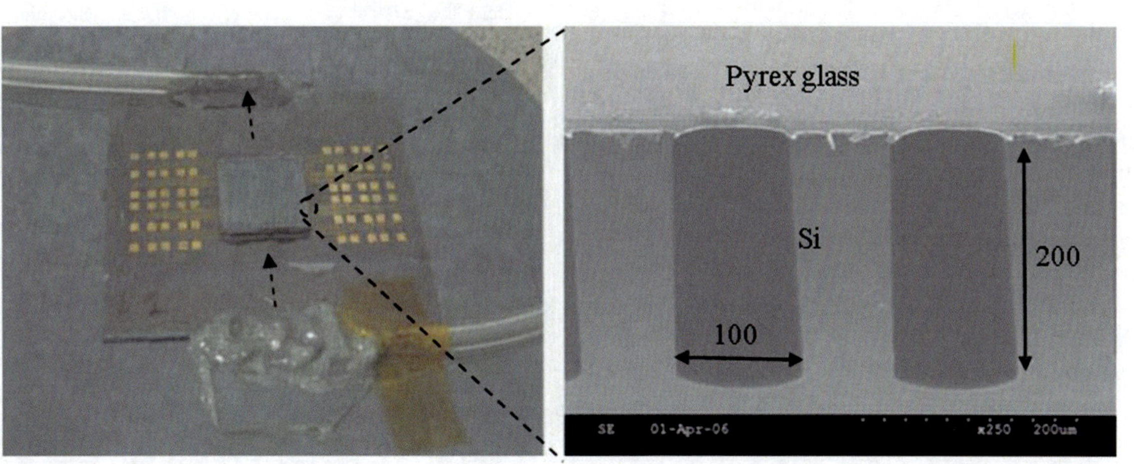

Fig. 10.21 A test thermal chip bonded on a substrate with through-vias connecting with fluidic tubes (51 microchannels distributed across a chip area of 1 cm^2) [101]

particle type are all important to achieve the best thermal performance. The advantage of high latent heat associated with liquid–vapor phase change of the working fluids has also been studied in microchannels and has been reviewed under Sect. 10.1 of this chapter.

Recent efforts [105–107] studied the effectiveness of passive heat-transfer enhancement surface features like etched dimples/concavities to improve heat transfer between the working fluid and the microchannel walls for cooling microelectronics. They found that such surface features help in

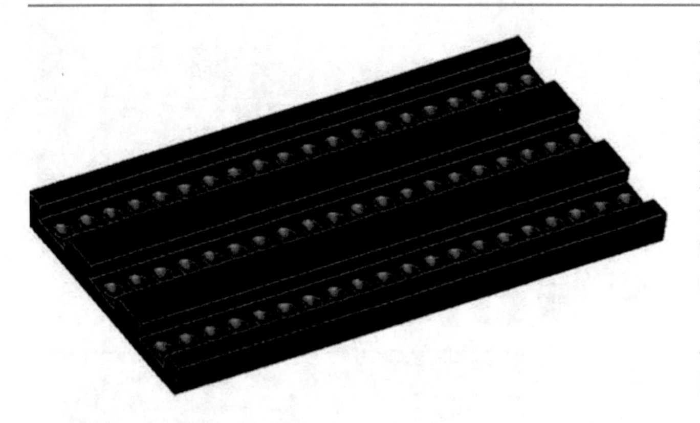

Fig. 10.22 3D representation of bottom-dimpled wall microchannels. (Reprinted with permission from Wei et al. [105])

increasing the heat transfer by even up to ~11% in the laminar flow regime, which is the case with most microchannel flows. Figure 10.22 shows a bottom wall dimpled microchannel geometry [105] as an example.

While most of the prior efforts by various research groups focused on reducing the fluid or solid–fluid interface thermal resistance, the largest thermal resistance in most typical microchannel cooling implementation for electronics such as microprocessors occurs at the junction between the heat sink and the electronic component, which is usually filled with thermal conductive grease. Hence, integrated microchannel cooling has gained prominence in recent years, wherein the microchannels are formed directly in one or more of the components that constitute the package. This technology is particularly useful when the electronic structures are stacked to achieve small form factors and high-power density.

Sekar et al. [108] presented a technology that involves 3D devices with integrated microchannel cooling. They propose the idea of fabricating through-hole silicon vias (TSVs) on a wafer using a two-step lithography process involving chip fabrication and etching to form the fluidic TSVs and microchannels. The device is then spin coated and polished with a sacrificial polymer material, which is spun-on, patterned, and cured to form a cover for the TSVs and microchannels. Cooling fluid was suggested to be delivered to the 3D stack either using tubes on the back side of the 3D stack or using fluidic channels on the substrate. Using this technology, their study showed a junction-to-ambient thermal resistance of 0.24 °C/W for a two-chip 3D stack.

Thermal management is a crucial facilitating technology in the development of modern electronics. As the power trends in electronics increase, the problem of thermal control becomes critical. This has led to the development and implementation of advanced thermal management solutions that are reliable, simple, and effective. Of them, microchannel cooling is established as one of the forefront technologies that can enable many of the Moore's law advances in micro-, nano-, and bioelectron- ics in future. With recent developments in microfabrication methods and materials, microchannels with advanced technologies such as using tailored single-phase and two-phase fluids, active/passive heat-transfer-enhancement structures, and substrate integration could be realized and will play a prominent role in thermal management of future high-power density nano- and bioelectronics.

References

1. Tuckerman, D.B., Pease, R.F.W.: High-performance heat sinking for VLSI. IEEE Electron. Device Lett. **2**, 126–129 (1981)
2. Wu, P., Little, W.A.: Measurement of friction factors for the flow of gases in very fine channels used for microminiature joule–Thompson refrigerators. Cryogenics. **23**, 273–277 (1983)
3. Pfahler, J., Harley, J., Bau, H.H., Zemel, J.: Gas and Liquid Flow in Small Channels. Micromechanical Sensors, Actuators and Systems. ASME, New York (1991)
4. Peng, X.F., Peterson, G.P., Wang, B.X.: Frictional flow characteristics of water flowing through rectangular microchannels. Exp. Heat Transf. **7**, 249–264 (1994)
5. Peng, X.F., Peterson, G.P., Wang, B.X.: Heat transfer characteristics of water flowing through rectangular microchannels. Exp. Heat Transf. **7**, 265–283 (1994)
6. Papautsky, I., Brazzle, J., Ammel, T., Frazier, A.B.: Laminar fluid behavior in microchannels using micropolar fluid theory. Sensors Actuators. **73**, 101–108 (1999)
7. Harms, T.M., Kazmierczak, M.J., Gerner, F.M.: Developing convective heat transfer in deep rectangular microchannels. Int. J. Heat Fluid Flow. **210**, 149–157 (1999)
8. Mala, G.M., Li, D.: Flow characteristics of water in microtubes. Int. J. Heat Fluid Flow. **20**, 142–148 (1999)
9. Mala, G.M., Li, D., Werner, C., Jacobasch, H.J., Ning, Y.B.: Flow characteristics of water through microchannels between two parallel plates with electrokinetic effects. Int. J. Heat Fluid Flow. **18**, 489–496 (1997)
10. Tso, C.P., Mahulikar, S.P.: The use of the Brinkman number for single phase forced convective heat transfer in microchannels. Int. J. Heat Mass Transf. **41**, 1759–1769 (1998)
11. Xu, B., Ooi, K.T., Wong, N.T.: Experimental investigation of flow friction for liquid flow in microchannels. Int. Commun. Heat Mass Transf. **27**, 1165–1176 (2000)
12. Qu, W., Mudawar, I.: Flow boiling heat transfer in two-phase micro-channel heat sinks –II. Annular two-phase flow model. Int. J. Heat Mass Transf. **46**, 2773–2784 (2002)
13. Liu, D., Garimella, S.V.: Investigation of liquid flow in microchannels. In: Eighth AIAA/ASME Joint Thermophysics and Heat Transfer Conference (2002)
14. Lee, P., Garimella, S.V.: Experimental investigation of heat transfer in microchannels. In: ASME Summer Heat Transfer Conference (2003)
15. Kohl, M.J., Abdel-Khalik, S.I., Jeter, S.M., Sadowski, D.L.: An experimental investigation of microchannel flow with internal pressure measurements. Int. J. Heat Mass Transf. **48**, 15518–11533 (2005)
16. Wei, X., Joshi, Y.: Experimental and numerical study of sidewall profile effects on flow and heat transfer inside microchannels. Int. J. Heat Mass Transf. **50**, 4640–4651 (2007)
17. Wei, X., Joshi, Y.: Optimization study of stacked micro-channel heat sinks for micro-electronic cooling. IEEE Trans. Compon Packag Technol. **26**, 55–61 (2003)

18. Phillips, R.J.: Micro-Channel Heat Sinks: Advances in Thermal Modeling of Electronic Components and Systems. ASME, New York (1990)
19. Muzychka, Y.S., Yovanovich, M.M.: Modeling friction factors in non-circular ducts for developing laminar flow. In: 2nd AIAA Theoretical Fluid Mechanics Meeting (1998)
20. Muzychka, Y.S., Yovanovich, M.M.: Modeling Nusselt numbers for thermally developing laminar flow in non-circular ducts. In: Seventh AIAA/ASME Joint Thermophysics and Heat Transfer Conference (1998)
21. Revellin, R., Thome, J.R.: A new type of diabatic flow pattern map for boiling heat transfer in microchannels. J. Micromech. Microeng. **17**, 788–796 (2007)
22. Wojtan, L., Revellin, R., Thome, J.R.: Investigation of critical heat flux in single, uniformly heated microchannels. Exp. Thermal Fluid Sci. **30**, 765–774 (2007)
23. Chen, L., Tian, Y.S., Karayiannis, T.G.: The effect of tube diameter on vertical two-phase flow regimes in small tubes. Int. J. Heat Mass Transf. **49**, 4220–4230 (2006)
24. Kew, P., Cornwall, K.: Correlations for the prediction of boiling heat transfer in small diameter channels. Appl. Therm. Eng. **17**, 705–715 (1997)
25. Kawaji, M., Chung, M.Y.. Unique characteristics of adiabatic gas–liquid flows in microchannels: diameter and shape effects on flow patterns, void fraction and pressure drop. In: First ASME International Conference on Microchannels and Minichannels (2003)
26. Jacobi, A., Thome, J.R.: Heat transfer model for evaporation of elongated bubble flows in microchannels. J. Heat Transf. **124**, 1131–1136 (2002)
27. Li, J., Wang, B.: Size effect on two-phase regime for condensation in micro/mini tubes. Heat Transf. Asian Res. **32**, 65–71 (2003)
28. McAdams, W.H., Woods, W.K., Heroman, L.C.: Vaporization inside horizontal tubes II-benzene-oil mixtures. Trans. ASME. **64**, 193–200 (1942)
29. Cicchitti, A., Lombaradi, C., Silversti, M., Soldaini, G., Zavattarlli, R.: Two-phase cooling experiments – pressure drop heat transfer burnout measurements. Energia Nucleare. **7**, 407–425 (1960)
30. Dukler, A.E., Moye, W., Cleveland, R.G.: Frictional pressure drop in two-phase flow. Part A: a comparison of existing correlations for pressure loss and holdup, and part B: an approach through similarity analysis. AICHE J. **10**, 38–51 (1964)
31. Awad, M.M., Muzychka, Y.S.: Effective property models for homogeneous two-phase flows. Exp. Thermal Fluid Sci. **33**, 106–113 (2008)
32. Agostini, B., Thome, J.R.: Comparison of an extended database of flow boiling heat transfer coefficient in multi-microchannel elements with the three-zone model. In: ECI International Conference on Heat Transfer and Fluid Flow in Microscale (2005)
33. Kandlikar, S.G., Balasubramanian, P.: Extending the applicability of the flow boiling correlation at low Reynolds number flows in microchannels. In: First ASME International Conference on Microchannels and Minichannels (2003)
34. Peters, J.V.S., Kandlikar, S.G.: Further evaluation of a flow boiling correlation for microchannels and minichannels. In: Proceedings of the Fifth International Conference on Nanochannels, Microchannels and Minichannels (2007)
35. Qu, W., Mudawar, I.: Measurement and correlation of critical heat flux in two-phase microchannel heat sinks. Int. J. Heat Mass Transf. **47**, 2045–2059 (2004)
36. Pribyl, D.J., Bar-Cohen, A., Bergles, A.E.: An investigation of critical heat flux and two-phase flow regimes for upward steam and water flow. In: Proceedings of the Fifth International Conference on Boiling Heat Transfer (2003)
37. Taitel, T., Dukler, A.E.: A model for flow regime transition in horizontal and near horizontal gas–liquid flows. AICHE J. **19**, 47–55 (1975)
38. Agostini, B., Thome, J.R., Fabbri, M., Michel, B., Calmi, D., Kloter, U.: High heat flux flow boiling in silicon multi-microchannels – part II: heat transfer characteristics of refrigerant R245fa. Int. J. Heat Mass Transf. **51**, 5415–5425 (2008)
39. Revellin, R., Thome, J.R.: A theoretical model for the prediction of the critical heat flux in heated microchannels. Int. J. Heat Mass Transf. **51**, 1216–1225 (2008)
40. Revellin, R., Quiben, J.M., Bonjour, J., Thome, J.R.: Effect of local hot spots on the maximum dissipation rates during flow boiling in a microchannel. IEEE Trans. Compon. Packag. Technol. **31**, 407–416 (2008)
41. Breber, G., Palen, J.W., Taborek, J.: Prediction of horizontal tubeside condensation of pure components using flow regime criteria. J. Heat Transf. **102**, 471–476 (1980)
42. Soliman, H.M.: Correlation of mist-to-annular transition during condensation. Can. J. Chem. Eng. **61**, 178–182 (1983)
43. Wang, H.S., Rosė, J.W.: A theory of film condensation in horizontal noncircular section microchannels. J. Heat Transf. **127**, 1096–1105 (2005)
44. Coleman, J.W., Garimella, S.: Two-phase flow regimes in round, square and rectangular tubes during condensation of refrigerant R134a. Int. J. Refrig. **26**, 117–128 (2003)
45. Yang, C.-Y., Shieh, C.-C.: Flow pattern of air-water and two-phase R-134a in small circular tubes. Int. J. Multiphase Flow. **27**, 1163–1177 (2001)
46. Chen, Y.P., Cheng, P.: Condensation of steam in silicon microchannels. Int. Commun. Heat Mass Transf. **32**, 175–183 (2005)
47. Bandhauer, T.M., Agarwal, A., Garimella, S.: Condensation pressure drop in circular microchannels. Heat Transf. Eng. **26**, 28–35 (2005)
48. Nema, G.: Flow Regime Transitions during Condensation in Microchannels. The George W. Woodruff School of Mechanical Engineering. Georgia Institute of Technology, Atlanta (2008)
49. Wang, H.S., Rose, J.W.: Film condensation in horizontal microchannels: effect of channel shape. Int. J. Therm. Sci. **45**, 1205–1212 (2006)
50. Garimella, S., Killion, J.D., Coleman, J.W.: An experimentally validated model for two-phase pressure drop in the intermittent flow regime for circular microchannels. J. Fluids Eng. **124**, 205–214 (2002)
51. Baird, J.R., Fletcher, D.F., Haynes, B.S.: Local condensation heat transfer rates in fine passages. Int. J. Heat Mass Transf. **46**, 4453–4466 (2003)
52. Moser, K.W., Webb, R.L., Na, B.: A new equivalent Reynolds number model for condensation in smooth tubes. J. Heat Transf. **120**, 410–417 (1998)
53. Bandhauer, T.M., Agarwal, A., Garimella, S.: Measurement and modeling of condensation heat transfer coefficients in circular microchannels. J. Heat Transf. **128**, 1050–1059 (2006)
54. Garimella, S., Agarwal, A., Killion, J.D.: Condensation pressure drop in circular microchannels. Heat Transf. Eng. **26**, 28–35 (2005)
55. Agarwal, A.: Heat Transfer and Pressure Drop during Condensation of Refrigerants in Microchannels. The George W. Woodruff School of Mechanical Engineering. Georgia Institute of Technology, Atlanta (2006)
56. Maruyama, T., Maeuchi, T.: Centrifugal-force driven flow in cylindrical micro-channel. Chem. Eng. Sci. **63**, 153–156 (2008)
57. Ducree, J., Haeberle, S., Brenner, T., Glatzel, T., Zengerle, R.: Patterning of flow and mixing in rotating radial microchannels. Microfluid. Nanofluid. **2**, 97–105 (2006)
58. Hao, Y.L., Tao, Y.-X.: A numerical model for phase-change suspension flow in microchannels. Numer. Heat Transf. A. **46**, 44–77 (2004)

59. Xing, K.Q., Tao, Y.X., Hao, Y.L.: Performance evaluation of liquid flow with PCM particles in microchannels. J. Heat Transf. **127**, 931–940 (2005)
60. Sabbah, R., Farid, M.M., Al-Hallaj, S.: Micro-channel heat sink with slurry of water with micro-encapsulated phase change material: 3D-numerical study. Appl. Therm. Eng. **29**, 445–454 (2009)
61. Kuravi, S., Kota, K., Du, J., Chow, L.: Numerical investigation of flow and heat transfer performance of nano-encapsulated phase change material slurry in microchannels. J. Heat Transf. **131**, 062901, 1–10 (2009)
62. Qu, W., Mudawar, I.: Experimental and numerical study of pressure drop and heat transfer in a single-phase micro-channel heat sink. Int. J. Heat Mass Transf. **45**, 2549–2565 (2002)
63. Sarangi, R.K., Bhattacharya, A., Prasher, R.S.: Numerical modeling of boiling heat transfer in microchannels. Appl. Therm. Eng. **29**, 300–309 (2009)
64. Ladd, A.J.C.: Numerical simulation of particulate suspensions via a discretized Boltzmann equation, part I. theoretical foundations. J. Fluid Mech. **271**, 285–309 (1994)
65. Ladd, A.J.C.: Numerical simulation of particulate suspensions via a discretized Boltzmann equation, part II. Numerical results. J. Fluid Mech. **271**, 311–339 (1994)
66. Chang, Q.: Lattice Boltzmann Method for Thermal Multiphase Fluid Dynamics. Mechanical and Aerospace Engineering/Case Western Reserve University, Cleveland (2005)
67. Judy, J., Maynes, D., Webb, B.W.: Characterization of frictional pressure drop for liquid flows through microchannels. Int. J. Heat Mass Transf. **45**, 3477–3489 (2002)
68. Harms, T.M., Kazmierczak, M.J., Gerner, F.M.: Developing convective heat transfer in deep rectangular microchannels. Int. J. Heat Fluid Flow. **20**, 149–157 (1999)
69. Wu, H.Y., Cheng, P.: Friction factors in smooth trapezoidal silicon microchannels with different aspect ratios. Int. J. Heat Mass Transf. **46**, 2519–2525 (2003)
70. Adrian, R.J.: Particle imaging techniques for experimental fluid mechanics. Annu. Rev. Fluid Mech. **23**, 261–304 (1991)
71. Santiago, J.G., Wereley, S.T., Meinhart, C.D., Beebe, D.J., Adrian, R.J.: A particle image velocimetry system for microfluidics. Exp. Fluids. **25**, 316–319 (1998)
72. Meinhart, C.D., Wereley, S.T., Santiago, J.G.: PIV measurements of a microchannel flow. Exp. Fluids. **27**, 414–419 (1999)
73. Meinhart, C.D., Wereley, S.T., Gray, M.H.B.: Volume illumination for two-dimensional particle image velocimetry. Meas. Sci. Technol. **11**, 809–814 (2000)
74. Olsen, M.G., Adrian, R.J.: Brownian motion and correlation in particle image velocimetry. Opt. Laser Technol. **32**, 621–627 (2000)
75. Olsen, M.G., Adrian, R.J.: Out-of-focus effects on particle image visibility and correlation in microscopic particle image velocimetry. Exp. Fluids. **29**, 166–174 (2000)
76. Olsen, M.G., Bourdon, C.J.: Out-of-plane motion effects in microscopic particle image velocimetry. J. Fluids Eng. **125**, 895–901 (2003)
77. Liu, D., Garimella, S.V., Wereley, S.V.: Infrared micro-particle image velocimetry of fluid flow in silicon-based microdevices. In: ASME Heat Transfer/Fluids Engineering Summer Conference (2004)
78. Park, J.S., Choi, C.K., Kihm, K.D.: Optically sliced micro-PIV using confocal laser scanning microscopy (CLSM). Exp. Fluids. **37**, 105–119 (2004)
79. Mielnik, M.M., Saetran, L.R.: Improvement of micro-PIV resolution by selective seeding. In: Proceedings of 4th International Conference on Nanochannels, Microchannels and Minichannels (2006)
80. Tsuei, L., Savas, O.: Treatment of interfaces in particle image velocimetry. Exp. Fluids. **29**, 203–214 (2000)
81. Meinhart, C.D., Wereley, S.T., Santiago, J.G.: A PIV algorithm for estimating time-averaged velocity fields. J. Fluids Eng. **122**, 285–289 (2000)
82. Shah, R.K., London, A.L.: Laminar Flow Forced Convection in Ducts. Academic Press, New York (1978)
83. Hohreiter, V., Wereley, S.T., Olsen, M.G., Chung, J.N.: Cross-correlation analysis for temperature measurement. Meas. Sci. Technol. **13**, 1072–1078 (2003)
84. Sakakibara, J., Hishida, K., Maeda, M.: Measurements of thermally stratified pipe flow using image processing techniques. Exp. Fluids. **16**, 82–96 (1993)
85. Kim, H.J., Kihm, K.D., Allen, J.S.: Examination of ratiometric laser induced fluorescence thermometry for microscale spatial measurement resolution. Int. J. Heat Mass Transf. **46**, 3967–3974 (2003)
86. Sakakibara, J., Adrian, R.J.: Measurement of temperature field of a Rayleigh–Benard convection using two-color laser induced fluorescence. Exp. Fluids. **37**, 331–340 (2004)
87. Guasto, J.S., Huang, P., Breuer, K.S.: Statistical particle tracking velocimetry using molecular and quantum dot tracer particles. Exp. Fluids. **41**, 869–880 (2006)
88. Guasto, J.S., Breuer, K.S.: Simultaneous, ensemble-average measurement of near-wall temperature and velocity in steady state micro-flows using single quantum dot tracking. Exp. Fluids. **45**, 157–166 (2008)
89. Gad-el-Hak, M.: MEMS Handbook. CRC/Taylor & Francis, Florida (2006)
90. de Boer, M.J., Tjerkstra, R.W., Berenschot, J.W., Jansen, H.V., Burger, G.J., Gardeniers, J.G.E., Elwenspoek, M., van den Berg, A.: Micromachining of buried micro channels in silicon. J. Microelectromech. Syst. **9**, 94–103 (2000)
91. Jayachandran, J.P., Reed, H.A., Zhen, H., Rhodes, L.F., Henderson, C.L., Allen, S.A.B., Kohl, P.A.: Air-channel fabrication for microelectromechanical systems via sacrificial photosensitive polycarbonates. J. Microelectromech. Syst. **12**, 147–159 (2003)
92. Papautsky, I., Brazzle, J., Swerdlow, H., Frazier, A.B.: A low-temperature IC-compatible process for fabricating surface-micromachined metallic microchannels. J. Microelectromech. Syst. **7**, 267–273 (1998)
93. Pijnenburg, R.H.W., Dekker, R., Nicole, C.C.S., Aubry, A., Eummelen, E.H.E.C.: Integrated micro-channel cooling in silicon. In: Proceedings of the 34th European Solid-State Device Research Conference (2004)
94. Oprins, H., Nicole, C.C.S., Baret, J.C., Van der Veken, G., Lasance, C., Baelmans, M.: On-chip liquid cooling with integrated pump technology. In: Proceedings of 21st Annual IEEE Semiconductor Thermal Measurement and Management Symposium (2005)
95. Chang, J.Y., Prasher, R.S., Chau, D., Myers, A., Dirner, J., Prstic, S., He, D.: Convective performance of package based single phase microchannel heat exchanger. In: Proceedings of ASME InterPACK (2005)
96. Kaltsas, G., Pagonis, D.N., Nassiopoulou, A.G.: Planar CMOS compatible process for the fabrication of buried microchannels in silicon, using porous-silicon technology. J. Microelectromech. Syst. **12**, 863–872 (2003)
97. Dang, B., Joseph, P.J., Bakir, M.S., Spencer, T., Kohl, P., Meindl, J.: Wafer-level microfluidic cooling interconnects for GSI. In: Proceedings of IEEE International Interconnect Technology Conference (2005)
98. Dang, B., Bakir, M., Meindl, J.: Integrated thermal-fluidic I/O interconnects for an on-chip microchannel heat sink. IEEE Electron. Device Lett. **27**, 117–119 (2006)
99. Kandlikar, S.G., Upadhye, H.R.: Extending the heat flux limit with enhanced microchannels in direct single-phase cooling of computer

chips. In: Proceedings of 21st Annual IEEE Semiconductor Thermal Measurement and Management Symposium (2005)
100. Pal, A., Joshi, Y.K., Beitelmal, M.H., Patel, C.D., Wenger, T.M.: Design and performance evaluation of a compact thermosyphon. IEEE Trans. Compon. Packag. Technol. **25**, 601–607 (2002)
101. Dang B.: Integrated Input/Output Interconnection and Packaging for GSI, Ph.D. Thesis, The School of Electrical and Computer Engineering, Georgia Institute of Technology, Atlanta (2006)
102. Copeland, D., Behnia, M., Nakayama, W.: Manifold microchannel heat sinks: isothermal analysis. IEEE Trans. Compon. Packag. Manuf. Technol. Part A. **20**, 96–102 (1997)
103. Kim, Y., Chun, W., Kim, J., Pak, B., Baek, B.: Forced air cooling by using manifold microchannel heat sinks. J. Mech. Sci. Technol. **12**, 709–718 (1998)
104. Xu, D., Pan, L.: Numerical study of nanofluid flow and heat transfer in microchannels. Int. J. Nanosci. **5**, 747–752 (2006)
105. Wei, X.J., Joshi, Y.K., Ligrani, P.M.: Numerical simulation of laminar flow and heat transfer inside a microchannel with one dimpled surface. J. Electron. Packag. **129**, 63–70 (2007)
106. Silva, C., Marotta, E., Fletcher, L.: Flow structure and enhanced heat transfer in channel flow with dimpled surfaces: application to heat sinks in microelectronic cooling. J. Electron. Packag. **129**, 157–166 (2007)
107. Silva, C., Park, D., Marotta, E., Fletcher, L.: Optimization of fin performance in a laminar channel flow through dimpled surfaces. J. Heat Transf. **131**, 021702 (1–9, 2009)
108. Sekar, D., King, C., Dang, B., Spencer, T., Thacker, H., Joseph, P., Bakir, M., Meindl, J.: A 3D-IC technology with integrated microchannel cooling. In: Proceedings of IEEE International Interconnect Technology Conference (2008)

Part II

BioMEMs Packaging

Nanoprobes for Live-Cell Gene Detection

11

Gang Bao, Won Jong Rhee, and Andrew Tsourkas

11.1 Introduction

The ability to image specific ribonucleic acid (RNA) in living cells in real time can provide essential information on RNA synthesis, processing, transport, and localization and on the dynamics of RNA expression and localization in response to external stimuli; it will offer unprecedented opportunities for advancement in molecular biology, disease pathophysiology, drug discovery, and medical diagnostics. Over the last decade or so, there is increasing evidence to suggest that RNA molecules have a wide range of functions in living cells, from physically conveying and interpreting genetic information, to essential catalytic roles, to providing structural support for molecular machines, and to gene silencing. These functions are realized through control of the expression level and stability, both temporally and spatially, of specific RNAs in a cell. Therefore, determining the dynamics and localization of RNA molecules in living cells will significantly impact on the molecular biology and medicine.

Of particular interest is the fluorescence imaging of specific messenger RNAs (mRNAs), both their expression level and subcellular localization, in living cells. As shown schematically in Box 11.1, for eukaryotic cells a pre-mRNA molecule is synthesized in cell nucleus. After processing (including splicing and polyadenylation), the mature mRNAs are transported from cell nucleus to cytoplasm and often localized at specific sites. The mRNAs are then being translated by ribosome to make specific proteins and degraded by RNases after a certain amount of time. The limited lifetime of mRNA enables a cell to alter protein synthesis rapidly in response to its changing needs. During the entire life cycle of an mRNA, it is always complexed with RNA-binding proteins to form a ribonucleoprotein (RNP). This has significant implications to the live-cell imaging of mRNAs.

Messenger RNA (mRNA) that encodes the chemical "blueprint" for a protein is synthesized (transcribed) from a DNA template, and the pre-mRNA is processed (spliced) to have mature mRNA, which is transported to specific locations in the cell cytoplasm. The coding information carried by mRNA is being used by ribosomes to produce proteins (translation). After a certain amount of time, the message is degraded. mRNAs are almost always complexed with RNA-binding proteins to form ribonucleoprotein (RNP) molecules.

Many in vitro methods have been developed to provide a relative (mostly semiquantitative) measure of gene expression level within a cell population using purified DNA or RNA obtained from cell lysate. These methods include polymerase chain reaction (PCR) [1], Northern hybridization (or Northern blotting) [2], expressed sequence tag (EST) [3], serial analysis of gene expression (SAGE) [4], differential display [5], and DNA microarrays [6]. These technologies, combined with the rapidly increasing availability of genomic data for numerous biological entities, present exciting possibilities for understanding human health and disease. For example, pathogenic and carcinogenic sequences are increasingly being used as clinical markers for diseased states. However, using in vitro methods to detect and identify foreign or mutated nucleic acids is often difficult in a clinical setting due to the low abundance of diseased cells in blood, sputum, and stool samples. Further, these methods cannot reveal the spatial and temporal variation of RNA within a single cell.

Labeled linear oligonucleotide (ODN) probes have been used to study intracellular mRNA via in situ hybridization (ISH) [7] in which cells are fixed and permeabilized to increase the probe delivery efficiency. Unbound probes are removed by washing to reduce background and achieve specificity [8]. To enhance the signal level, multiple probes targeting the same mRNA can be used [7]. However, fixation

G. Bao (✉) · W. J. Rhee · A. Tsourkas
Department of Biomedical Engineering, Georgia Institute of Technology and Emory University, Atlanta, GA, USA
e-mail: gang.bao@bme.gatech.edu

C. P. Wong et al. (eds.), *Nano-Bio- Electronic, Photonic and MEMS Packaging*, https://doi.org/10.1007/978-3-030-49991-4_11

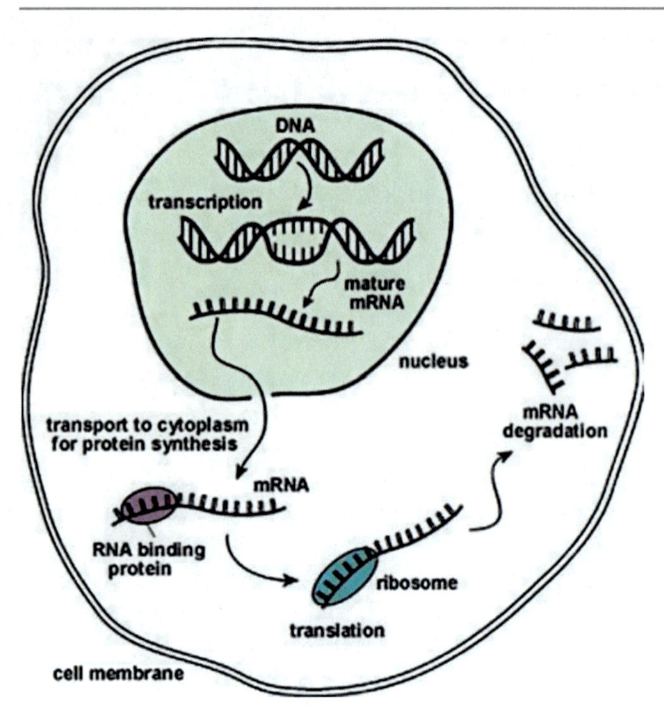

Box 11.1 mRNA life cycle

agents and other supporting chemicals can have considerable effect on signal level [9] and possibly on the integrity of certain organelles such as mitochondria. Thus, fixation of cells, by either cross-linking or denaturing agents, and the use of proteases in ISH assays may prevent from obtaining an accurate description of intracellular mRNA localization. It is also difficult to obtain a dynamic picture of gene expression in cells using ISH methods.

In order to detect RNA molecules in living cells with high specificity, sensitivity, and signal-to-background ratio, especially for low-abundance genes and clinical samples containing a small number of diseased cells, the probes need to recognize RNA targets with high specificity, convert target recognition *directly* into a measurable signal, and differentiate between true- and false-positive signals. It is important for the probes to quantify low gene expression levels with high accuracy and have fast kinetics in tracking alterations in gene expression in real time. For detecting genetic alterations such as mutations, insertions, and deletions, the ability to recognize single-nucleotide polymorphisms (SNPs) is essential. To achieve this optimal performance, it is necessary to have a good understanding of the structure–function relationship of the probes, probe stability, and RNA target accessibility in living cells. It is also necessary to achieve efficient cellular delivery of probes with minimal probe degradation.

In the remaining sections, we review commonly used fluorescent probes for RNA detection, discuss the critical issues in living cell RNA detection, including probe design, target accessibility, cellular delivery of probes, and detection sensitivity, specificity, and signal-to-background ratio. Emphasis is placed on the design and application of molecular beacons, although some of the issues are common to other oligonucleotide probes.

11.2 Fluorescent Probes for Live-Cell RNA Detection

Several classes of molecular probes have been developed for RNA detection in living cells, including (1) tagged linear oligonucleotide (ODN) probes; (2) oligonucleotide hairpin probes; and (3) probes using fluorescent proteins as reporter. Although probes composed of full-length RNAs (mRNA or nuclear RNA) tagged with a fluorescent or radioactive reporter have been used to study the intracellular localization of RNA [10–12], these probes are not discussed here since they cannot be used to measure the expression level of specific RNAs in living cells.

11.2.1 Tagged Linear ODN Probes

Single fluorescently labeled linear oligonucleotide probes have been developed for RNA tracking and localization studies in living cells [13–15]. In many cases, the ODN probe has a DNA backbone. A DNA molecule is a long polymer made from repeating nucleotides including adenine (A), quinine (G), cytosine (C), and thymine (T). For RNA, thymine is replaced by uracil (U). Typically G pairs with C, and A pairs with T (or U), according to Watson–Crick base pairing. Although these probes may recognize specific endogenous RNA transcripts in living cells via Watson–Crick base pairing and reveal subcellular RNA localization, this approach lacks the ability to distinguish background from true signal, since both bound probes (i.e., probes hybridized to RNA target) and unbound probes give fluorescence signal. It may also lack detection specificity since partial match between the probe and target sequences could induce probe hybridization to RNA molecules of multiple genes. A novel way to increase signal-to-noise ratio and improve detection specificity is to use two linear probes with a fluorescence resonance energy transfer (FRET) pair of (donor and acceptor) fluorophores [13]. However, the dual linear probe approach may still have a high background signal due to direct excitation of the acceptor and emission detection of the donor fluorescence. Further, it is difficult for linear probes to distinguish targets that differ by a few bases since the difference in free energy of the two hybrids (with and without mismatch) is typically rather small. This limits the application of linear ODN probes in biological and disease studies.

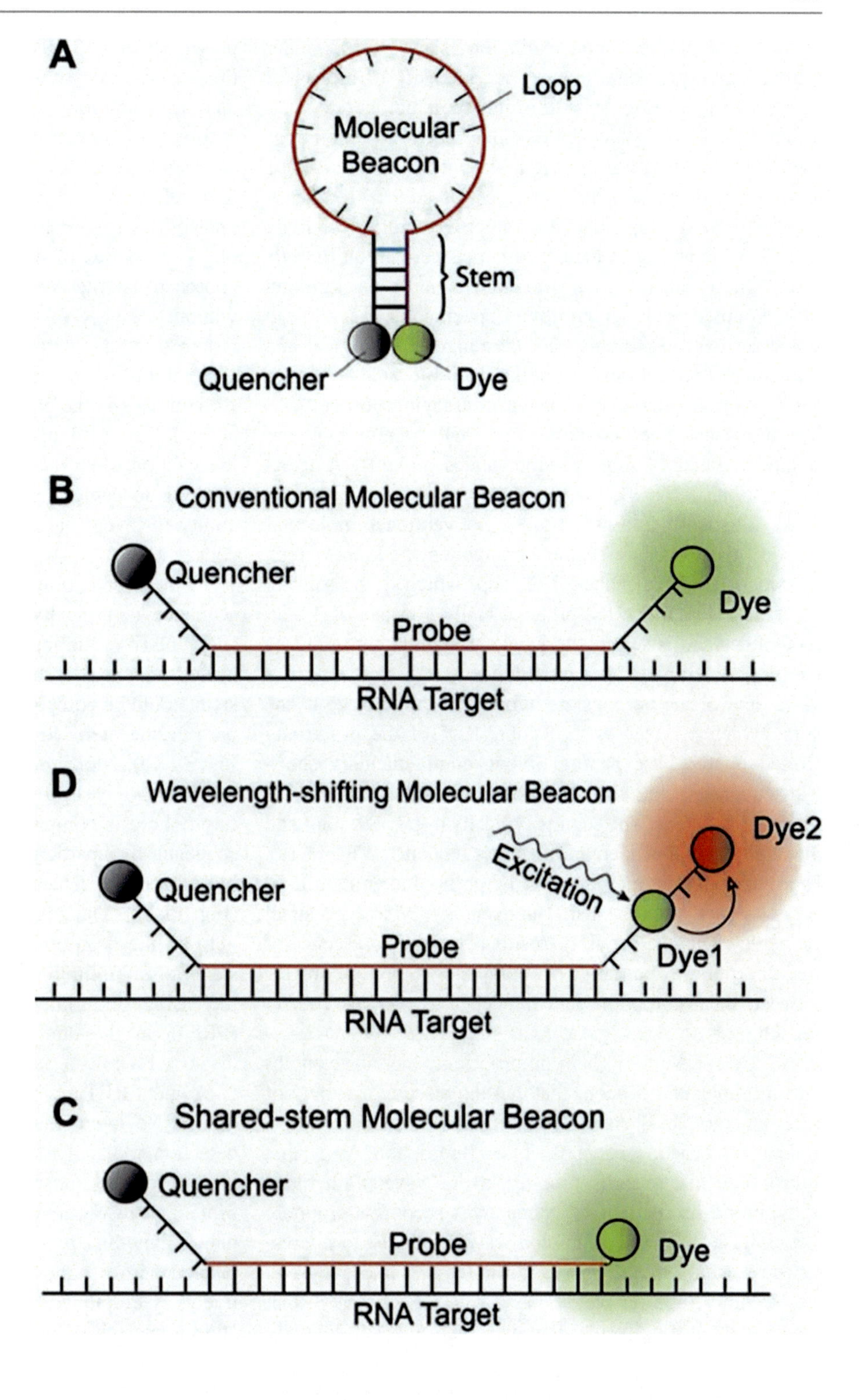

Fig. 11.1 Illustrations of molecular beacons. (**a**) Molecular beacons are stem-loop hairpin oligonucleotide probes labeled with a reporter fluorophore at one end and a quencher molecule at the other end. (**b**) Conventional molecular beacons are designed such that the short complementary arms of the stem are independent of the target sequence. (**c**) Wavelength-shifting molecular beacons containing two fluorophores, one absorbs in the wavelength range of the monochromatic light source and the other emits light at the desired emission wavelength due to FRET. (**d**) Shared-stem molecular beacons are designed such that one arm of the stem participates in both stem formation and target hybridization

11.2.2 ODN Hairpin Probes

Hairpin nucleic acid probes have the potential to be highly sensitive and specific in live-cell RNA detection. As shown in Fig. 11.1a, b, one class of such probes is molecular beacons, which are dual-labeled oligonucleotide probes with a fluorophore at one end and a quencher at the other end [16]. They are designed to form a stem-loop hairpin structure in the absence of a complementary target so that fluorescence of the fluorophore is quenched by the quencher molecule (typically through FRET between fluorophore and quencher and contact quenching). Hybridization with the target nucleic acid opens the hairpin and physically separates the fluorophore from quencher, allowing a fluorescence signal to be emitted upon excitation (Fig. 11.1b). Under optimal conditions, the fluorescence intensity of molecular beacons can increase by >200-fold upon binding to their targets [16]. This enables a molecular beacon to function as a sensitive probe with a high signal-to-background ratio. The stem-loop hairpin structure provides an adjustable energy penalty for hairpin opening which improves probe specificity [17, 18]. The ability to transduce target recognition *directly* into a fluorescence signal with high signal-to-background ratio, coupled with an improved specificity, has allowed molecular beacons to enjoy a wide range of biological and biomedical applications, including multiple analyte detection, real-time enzymatic cleavage assaying, cancer cell detection, real-time monitoring of PCR, genotyping and mutation detection, viral infection studies, and mRNA detection in living cells [14, 19–32].

As illustrated in Fig. 11.1a, a conventional molecular beacon has four essential components: loop, stem, fluorophore, and quencher. The loop, which is the sensing (targeting) domain of the probe, usually consists of 15–25 nucleotides and is selected to have a unique target sequence and proper melting temperature (defined as the temperature at which half of the beacons are hybridized to the target). The stem, which services as the "controller" of the probe and formed by the base pairing of two complementary short-arm sequences, is to close the loop domain therefore having a hairpin structure. The stem is typically 4–6 bases long and chosen to be independent of the target sequence (Fig. 11.1a). The fluorophore is a reporter of the probe that emits a fluorescent signal when excited. The quencher, typically a small molecule such as Dabcyl, is to suppress the fluorescence of the fluorophore partially or completely through FRET between the fluorophore and quencher as well as contact quenching.

A novel design of hairpin probes is the wavelength-shifting molecular beacons that can fluoresce in a variety of different colors [33]. As shown in Fig. 11.1c, in this design, a molecular beacon contains two fluorophores: a first fluorophore that absorbs strongly in the wavelength range of the monochromatic light source and a second fluorophore that emits at the desired emission wavelength due to fluorescence resonance energy transfer from the first fluorophore to the second fluorophore. It has been demonstrated that wavelength-shifting molecular beacons are substantially brighter than conventional molecular beacons that contain a fluorophore that cannot efficiently absorb energy from the available monochromatic light source.

One major advantage of the stem-loop hairpin probes is that they can recognize their targets with higher specificity than linear ODN probes. Solution studies suggested that [17, 18], using molecular beacons, it is possible to discriminate between targets that differ by a single nucleotide. In contrast to current techniques for detecting single nucleotide polymorphism (SNP), which are often labor-intensive and time-consuming, molecular beacons may provide a simple and promising tool for detecting SNPs in disease diagnosis.

Box 11.2 compares the basic features of molecular beacon versus fluorescence in situ hybridization (FISH). Specifically, molecular beacons are dual-labeled hairpin probes of 15–25 nucleotides (nt), while FISH probes are dye-labeled linear oligonucleotides of 40–50 nt. The molecular beacon-based approach has the advantage of detecting RNA in live cells without the need of cell fixation and washing. However, it requires cellular delivery of the probes and has low target accessibility (discussed in later sections). The advantage of FISH assays is the ease of probe design due to better target accessibility. Although FISH assays can be used to image the localization of mRNA in fixed cells, they rely on stringent washing to achieve signal specificity and do not have the ability to image the dynamics of gene expression in living cells.

In the conventional molecular beacon design, the stem sequence is typically independent of the target sequence (Fig. 11.1b), although sometimes two end bases of the probe sequence, each adjacent to one arm sequence of the stem, could be complementary with each other, thus forming part of the stem (light blue base of the stem shown in Fig. 11.1a). Molecular beacons can also be designed such that all the bases of one arm of the stem (to which a fluorophore is conjugated) are complementary to the target sequence, thus participating in both stem formation and target hybridization (shared-stem molecular beacons) [34] (Fig. 11.1d). The advantage of this shared-stem design is to help fix the position of the fluorophore that attached to the stem arm, limiting its degree-of-freedom of motion, and increasing the fluorescence resonance energy transfer (FRET) in the dual FRET molecular beacon design, as discussed below.

A dual FRET molecular beacons approach was developed [26–28] to overcome the difficulty that, in live-cell RNA detection, molecular beacons are often degraded by nucleases or open due to nonspecific interaction with hairpin-binding proteins, causing a significant amount of false-positive signal. In this dual probe design, a pair of molecular beacons labeled with a donor and an acceptor fluorophore, respectively, is employed (Fig. 11.2). The probe sequences are chosen such that this pair of molecular beacons hybridizes to adjacent regions on a single RNA target (Fig. 11.2). Since FRET is very sensitive to the distance between donor and acceptor fluorophores and typically occurs when the donor and acceptor fluorophores are within ~10 nm, FRET signal is generated by the donor and acceptor beacons only if both probes are bound to the same RNA target. Thus, the sensitized emission of the acceptor fluorophore upon donor excitation serves as a positive signal in the FRET-based detection assay, which is differentiable from non-FRET

Box 11.2 Comparison of molecular beacon and FISH approaches

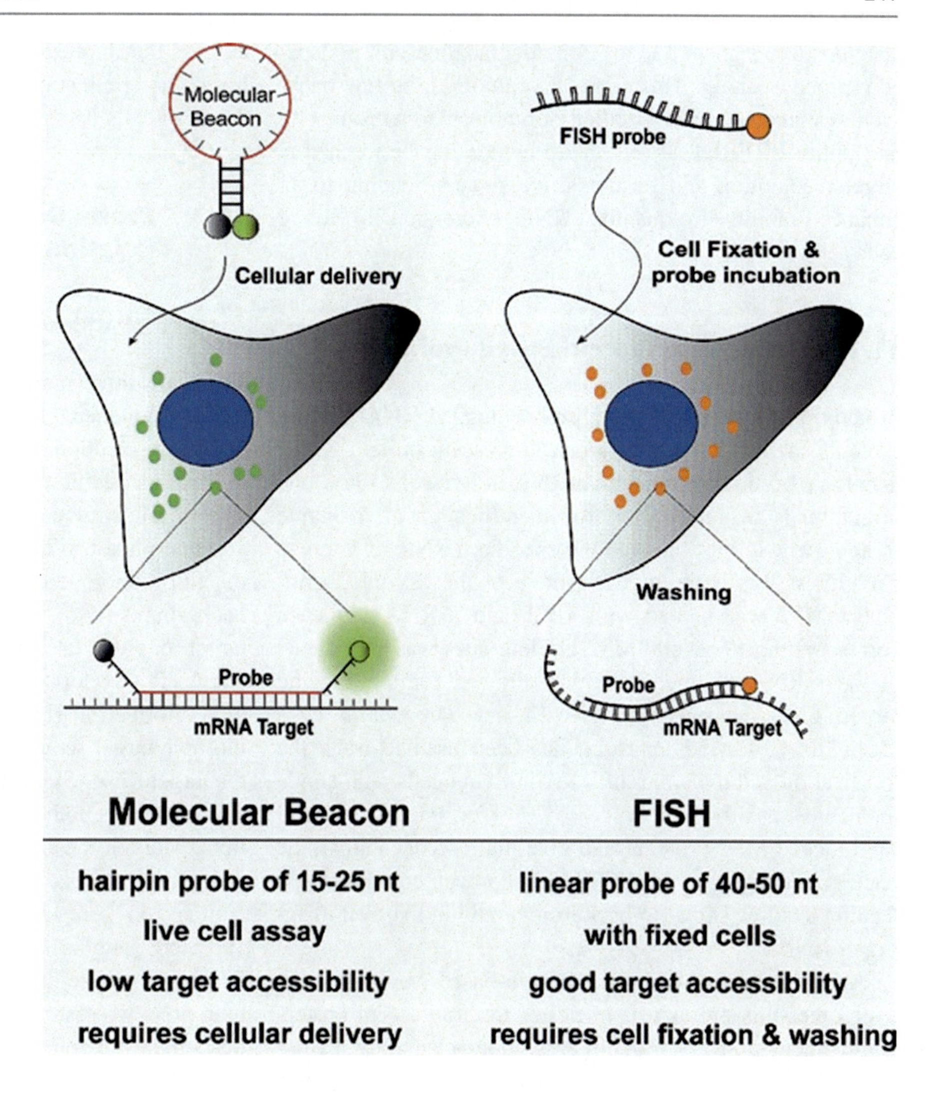

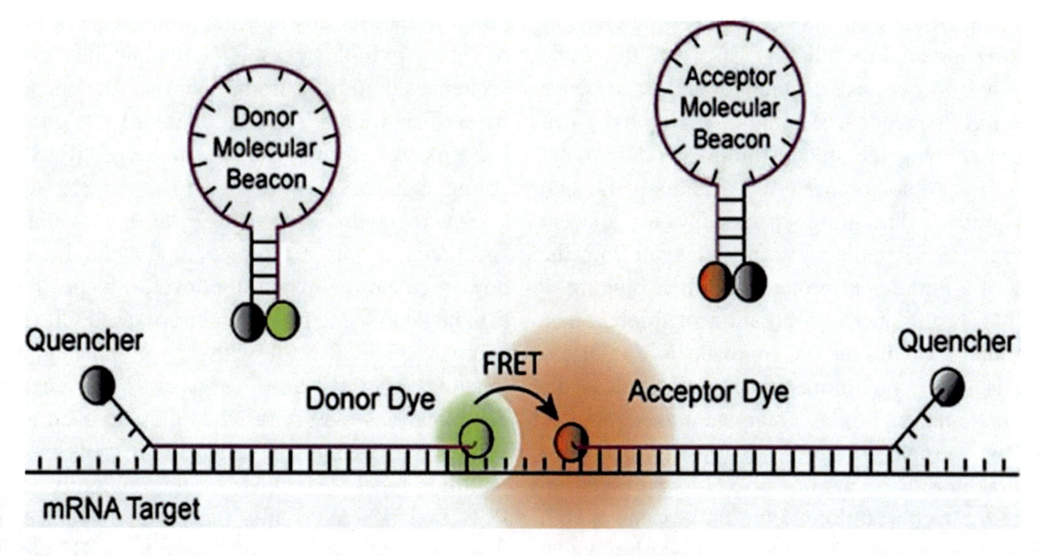

Fig. 11.2 A schematic showing the concept of dual FRET molecular beacons. Hybridization of donor and acceptor molecular beacons to adjacent regions on the same mRNA target results in FRET between donor and acceptor fluorophores upon donor excitation. By detecting FRET signal, fluorescence signals due to probe–target binding can be readily distinguished from that due to molecular beacon degradation and nonspecific interactions

false-positive signals due to probe degradation and nonspecific probe opening. This approach combines the low background signal and high specificity of molecular beacons with the ability of FRET assays in differentiating between true target recognition and false-positive signals, leading to an enhanced ability to quantify RNA expression in living cells [28].

11.2.3 Fluorescent Protein-Based Probes

In addition to oligonucleotide probes, tagged RNA-binding proteins such as those with green fluorescent protein (GFP) tags have been used to detect mRNA in live cells [35]. One limitation is that it requires the identification of a unique protein, which only binds to the specific mRNA of interest. To address this issue, a coat protein of the RNA bacteriophage MS2 was tagged with GFP, and a RNA sequence corresponding to several MS2-binding sites was introduced to the mRNA of interest, which allowed for the specific targeting of the nanos mRNA in live *Drosophila* eggs [36]. The GFP-MS2 approach has been used to track the localization and dynamics of RNA in living cells with single molecule sensitivity [37, 38]. However, since unbound GFP-tagged MS2 proteins also give fluorescence signal, the background signal in the GFP-MS2 approach could be high, leading to a low signal-to-background ratio in live-cell imaging of RNA.

An interesting fluorescent-protein-based approach that overcomes this problem is to utilize the fluorescent protein complementation [39, 40]. In this approach (split-GFP), a RNA-binding protein is dissected into two fragments, which are respectively fused to the split fragments of a fluorescent protein. Binding of the two tagged fragments of the RNA-binding protein to adjacent sites on the same mRNA molecule (or two parts of an aptamer sequence inserted to the mRNA sequence) brings the two halves of the fluorescent protein together, reconstituting the fluorescent protein and restoring fluorescence [40]. Alternatively, two RNA-binding proteins that bind specifically to adjacent sites on the same mRNA molecule can be tagged with the split fragments of a fluorescent protein, and their binding to the target mRNA results in the restoration of fluorescence [39]. The advantage of this novel approach is that background signal is low – no fluorescence signal unless the RNA-binding proteins (or protein fragments) are bound to the target mRNA. The split-GFP method, however, may have difficulties in tracking the dynamics of RNA expression in real time, since the reconstitution of the fluorescent protein from the split fragments typically takes 2–4 h, during which the RNA expression level may change. Transfection efficiency could also be a major concern in the GFP-based approaches in that usually only a few percent of the cells express the fluorescent proteins following transfection. This limits the application of the split-GFP methods in detecting diseased cells using mRNA as a biomarker for the disease.

11.3 Probe Design and Structure–Function Relations

11.3.1 Target Specificity

There are three major design issues of molecular beacons: probe sequence, hair- pin structure, and fluorophore–quencher selection. In general, the probe sequence is selected to ensure specificity and to have good target accessibility. The hairpin structure as well as the probe and stem sequences are determined to have the proper melting temperature, and the fluorophore–quencher pair should give high signal-to-background ratio. To ensure specificity, for each gene to target, one can use the Basic Local Alignment Search Tool (BLAST) developed by the National Center for Biotechnology Information (NCBI) [41] or similar software to select multiple target sequences of 15–25 bases that are unique for the target RNA. Since the melting temperature of a molecular beacon affects both the signal-to-background ratio and detection specificity, especially for mutation detection, it is often necessary to select the target sequence with a balanced G–C content (the percentage of G–C pairs) and to adjust the loop and stem lengths and the stem sequence of the molecular beacon to realize the optimal melting temperature. In particular, it is necessary to understand the effect of molecular beacon design on melting temperature so that, at 37 °C, single-base mismatches in target mRNAs can be differentiated. This is also a general issue for detection specificity in that, for any specific probe sequence selected, there might be multiple genes in the mammalian genome that have sequences that differ from the probe sequence by only a few bases. Therefore, it is important to design the molecular beacons so that only the specific target RNA would give a strong signal.

Several approaches can be taken to validate the signal specificity. For example, one could upregulate or downregulate the expression level of a specific RNA, quantify the level using reverse transcriptase PCR (RT-PCR), and compare the PCR result with that of molecular beacon-based imaging of the same RNA in living cells. However, complications may arise when the approach used to change the RNA expression level in living cells has an effect on multiple genes, leading to some ambiguity even when the PCR and beacon results match. Perhaps the best way to downregulate the level of a specific mRNA in live cells is to use small interfering RNA (siRNA) treatment, which typically leads to >80% reduction of the specific mRNA level. Since the effect of siRNA treatment varies depending on the specific siRNA probes used, the siRNA delivery method, and the cell type, optimization of the protocol (probe design and delivery method/conditions) is often needed.

11.3.2 Molecular Beacon Structure–Function Relations

The loop, stem lengths, and sequences are critical design parameters for molecular beacons, since at any given temperature they largely control the fraction of molecular beacons that are bound to the target [17, 18]. In many applications, the choices of the probe sequence are limited by target-specific considerations, such as the sequence surrounding a single-nucleotide polymorphism (SNP) of interest. However, the probe and stem lengths, and stem sequence, can be adjusted to optimize the performance (i.e., specificity, hybridization rate, and signal-to-background ratio) of a molecular beacon for a specific application [17, 34].

To demonstrate the effect of molecular beacon structure on its melting behavior, the melting temperature for molecular beacons with various stem-loop structures is displayed in Fig. 11.3a. In general, it was found that the melting temperature increased with probe length but appeared to plateau at a length of ~20 nucleotides. It was also found that the stem length of the molecular beacon could strongly influence the melting temperature of molecular beacon–target duplexes. If the melting temperature (T_m) is too high, the beacon becomes very stable, resulting in very slow hybridization kinetics and loss of the ability to discriminate mismatches between probe and target at 37 °C. On the other hand, if the T_m is too low, molecular beacons are too each to open, thus increasing the background signal. Therefore, the melting temperature of molecular beacons is an important parameter that has a large influence on signal-to-background ratio, hybridization kinetics, and detection specificity.

While both the stability of the hairpin probe and its ability to discriminate targets over a wider range of temperatures increase with increasing stem length, it is accompanied by a decrease in hybridization on-rate constant, as shown in Fig. 11.3b. For example, molecular beacons with a 4-base stem had an on-rate constant up to 100 times greater than molecular beacons with a 6-base stem. Changing the probe length of a molecular beacon may also influence the rate of hybridization, as demonstrated by Fig. 11.3b.

From the thermodynamic and kinetic studies, it was found that, if the stem length is too large, it will be difficult for the beacon to open upon hybridization. On the other hand, if the stem length is too small, a large fraction of beacons may open due to the thermal force. Similarly, relative to the stem length, a longer probe may lead to a lower dissociation constant; however, it may also reduce the specificity, since the relative free energy change due to one base mismatch would be

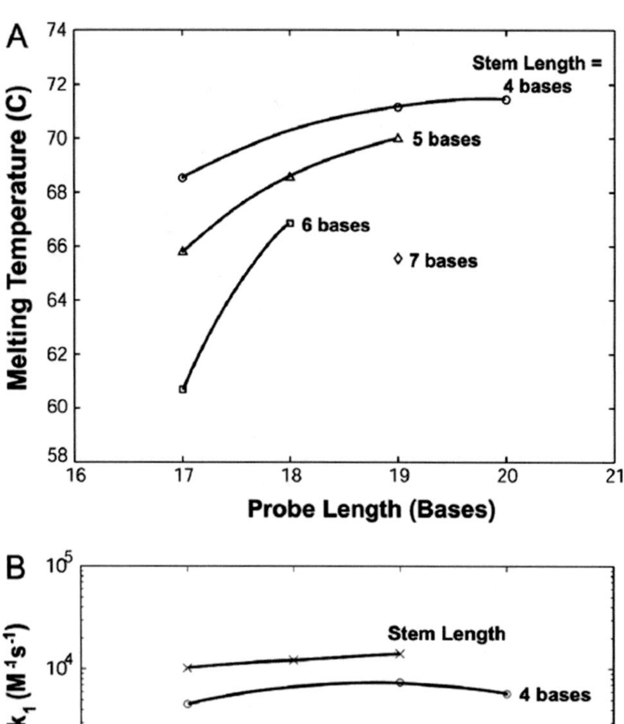

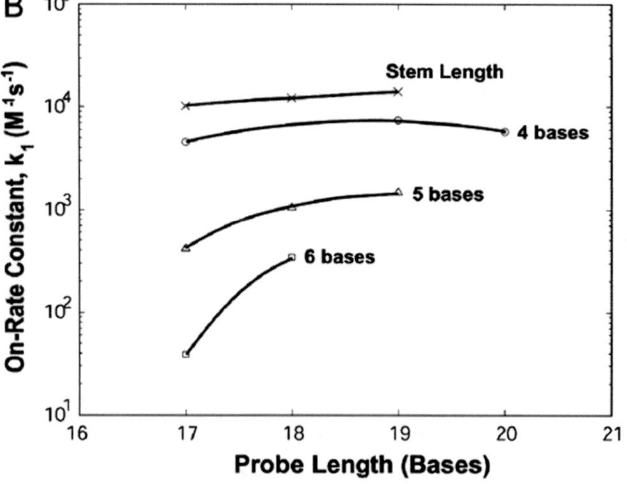

Fig. 11.3 Structure–function relations of molecular beacons. (**a**) Melting temperatures for molecular beacons with different structures in the presence of target. (**b**) The on-rate constant of hybridization k_l for molecular beacons with various probe and stem lengths hybridized to their complementary targets

smaller. A long probe length may also lead to coiled conformations of the beacons, resulting in reduced kinetic rates. Therefore, the stem and probe lengths need be carefully chosen in order to optimize both hybridization kinetics and MB specificity [17, 34]. In general, it has been found that molecular beacons with longer stem lengths have an improved ability to discriminate between wild-type and mutant targets in solution over a broader range of temperatures. This can be attributed to the enhanced stability of the molecular beacon stem-loop structure and the resulting smaller free energy difference between closed (unbound) molecular beacons and molecular beacon–target duplexes, which generates a condition where a single-base mismatch reduces the energetic preference of probe–target binding. Longer stem lengths, however, are accompanied by a decreased probe–target hybridization kinetic rate. Similarly,

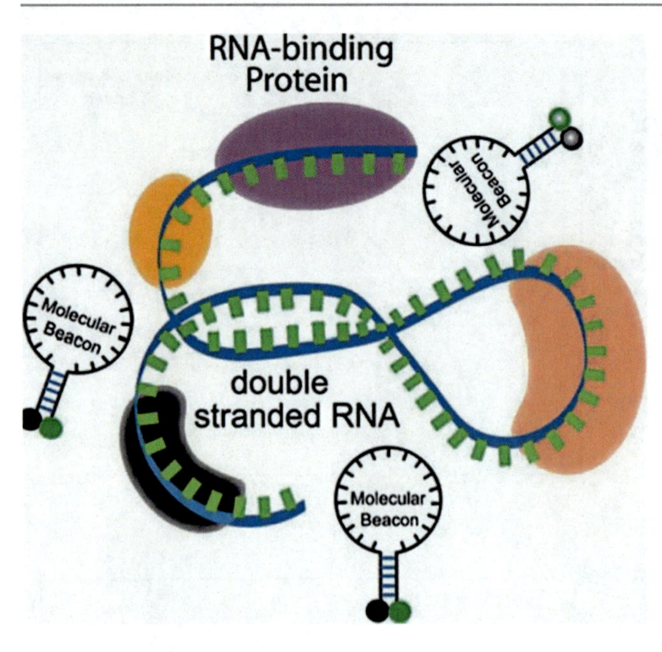

Fig. 11.4 A schematic illustration of a segment of the target mRNA with a double-stranded portion and RNA-binding proteins. A molecular beacon has to compete off an mRNA strand or RNA-binding protein (s) in order to hybridize to the target

molecular beacons with short stems have faster hybridization kinetics but suffer from lower signal-to-background ratios compared with molecular beacons with longer stems.

11.3.3 Target Accessibility

A critical issue in molecular beacon design is target accessibility, as is the case for most oligonucleotide probes for live-cell RNA detection. It is well known that a functional mRNA molecule in a living cell always has RNA-binding proteins on it, forming a ribonucleoprotein (RNP). Further, an mRNA molecule often has double stranded portions and forms secondary (folded) structures (Fig. 11.4). Therefore, in designing a molecular beacon, it is necessary to avoid targeting mRNA sequences that are double-stranded, or occupied by RNA-binding proteins, for otherwise the probe has to compete off the RNA strand or the RNA-binding protein in order to hybridize to the target. Indeed, molecular beacons designed for targeting a specific mRNA often show no signal when delivered to living cells. One difficulty in the molecular beacon design is that, although predictions of mRNA secondary structure can be made using software such as *Beacon Designer* (www.premierbiosoft.com) and *mfold* (http://www.bioinfo.rpi.edu/applications/mfold/old/dna/), they may be inaccurate due to limitations of the biophysical models used and the limited understanding of protein–RNA interaction. Therefore, for each gene to target, it may be necessary to select multiple unique sequences along the target RNA and have corresponding molecular beacons designed, synthesized, and tested in living cells to select the best target sequence.

To uncover the possible molecular beacon design rules, the accessibility of BMP- 4 mRNA was studied using different beacon designs [42]. Specifically, molecular beacons were designed to target the start codon and termination codon regions, the siRNA and antisense oligonucleotide probe sites identified previously, and sites that were randomly chosen. All the target sequences are unique to BMP-4 mRNA. Of the eight molecular beacons designed to target BMP-4 mRNA, it was found that only two beacons gave strong signal, one targets the start codon region, and the other targets the termination codon region. It was also found that, even for a molecular beacon that works well, shifting its targeting sequence by just a few bases toward the 3^t or 5^t ends would significantly reduce the fluorescence signal from beacons in a live-cell assay, indicating that the target accessibility is quite sensitive to the location of the targeting sequence. These results, together with molecular beacons validated previously, suggest that the start codon and termination codon regions and the exon–exon junctions are more accessible than other locations in an mRNA.

11.3.4 Fluorophores and Quenchers

With proper backbone synthesis and fluorophore–quencher conjugation, in theory, a molecular beacon can be labeled with any desired reporter–quencher pair. However, proper selection of the reporter and quencher could improve the signal-to-background ratio and multiplexing capabilities. Selecting a fluorophore label for a molecular beacon as the reporter is usually not as critical as the hairpin probe design since many conventional dyes can yield satisfactory results. However, proper selection could yield additional benefits such as an improved signal-to-background ratio and multiplexing capabilities. Since each molecular beacon utilizes only one fluorophore, it is possible to use multiple molecular beacons in the same assay, assuming that the fluorophores are chosen with minimal emission overlap [19]. Molecular beacons can even be labeled simultaneously with two fluorophores, i.e., "wavelength shifting" reporter dyes (Fig. 11.1c), allowing multiple reporter dye sets to be excited by the same monochromatic light source yet fluorescing in a variety of colors [33]. Clearly, multicolor fluorescence detection of different beacon/target duplexes can become a powerful tool for the simultaneous detection of multiple genes.

For dual FRET (fluorescence resonance energy transfer) molecular beacons (Fig. 11.2), the donor fluorophores typically emit at shorter wavelengths compared with that of acceptor. Energy transfer occurs as a result of long-range dipole–dipole interactions between the donor and the acceptor. The efficiency of energy transfer depends upon the extent of the spectral overlap of the emission spectrum of the donor with the absorption spectrum of the acceptor, the quantum yield of the donor, the relative orientation of the donor and acceptor transition dipoles [43], and the distance between the donor and acceptor molecules (usually 4–5 bases). In selecting donor and acceptor fluorophores, in order to have high signal-to-background ratio, it is important to optimize the above parameters and to avoid direct excitation of the acceptor fluorophore at the donor excitation wavelength, as well as minimizing donor emission detection at the acceptor emission detection wavelength. Examples of FRET dye pairs include Cy3 (donor) and Cy5 (acceptor), TMR (donor) and Texas Red (acceptor), fluorescein (FAM) (donor), and Cy3 (acceptor).

It is relatively straightforward to select the quencher molecules. Organic quencher molecules such as Dabcyl, BHQ2 (Black Hole Quencher II) (Biosearch Tech), BHQ3 (Biosearch Tech), and Iowa Black (IDT) can all effectively quench a wide range of fluorophores by both fluorescence resonance energy transfer (FRET) and the formation of an exciton complex between the fluorophore and the quencher [44].

11.4 Cellular Delivery of Nanoprobes

One of the most critical aspects of measuring the intracellular level of RNA molecules using synthetic probes is the ability to deliver these probes into cells through the plasma membrane, which is quite lipophilic and restricts the transport of large and charged molecules. Therefore, it is a very robust barrier to polyanionic molecules such as hairpin oligonucleotides. Further, even if the probes enter the cells successfully, the efficiency of delivery in an imaging assay should be defined not only by how many probes enter the cell or how many cells have probes internalized but also how many probes remain functioning inside cells. This is different from both antisense and gene delivery applications where the reduction in level of protein expression is the final metric used to define efficiency or success. For measuring RNA molecules (including mRNA and rRNA) in the cytoplasm, a large amount of probes should remain in the cytoplasm.

Existing cellular delivery techniques can be divided into two categories: endocytic and nonendocytic methods. Endocytic delivery typically employs cationic and polycationic molecules such as liposomes and dendrimers, while nonendocytic methods include microinjection and the use of cell-penetrating peptides (CPP) or streptolysin *O* (SLO). Probe delivery via the endocytic pathway typically takes 2–4 h. It has been reported that ODN probes internalized via endocytosis are predominately trapped inside endosomes and often lysosomes and being degraded there due to cytoplasmic nucleases [45]. Consequently, only 0.01–10% of the probes remain functioning after escaping from endosomes and lysosomes [46].

Oligonucleotide probes (including molecular beacons) have been delivered into cells via microinjection [47]. In most of the cases, the ODNs exhibited a fast accumulation in the cell nucleus, preventing the probes from targeting mRNAs in the cell cytoplasm. Depletion of intracellular ATP or lowering the temperature from 37 to 4 °C did not have a significant effect on ODN nuclear accumulation, ruling out active, motor-protein-driven transport [47]. It is unclear if the rapid transport of ODN probes to nucleus is due to electrostatic interaction, or driven by microinjection-induced flow, or the triggering of some signaling pathway. There is no fundamental biological reason why ODN probes should accumulate in the cell nucleus. To prevent nuclear accumulation, streptavidin (60 kDa) molecules were conjugated to linear ODN probes via biotin [13]. After microinjected into cells, dual FRET linear probes could hybridize to the same mRNA target in the cytoplasm, resulting in a FRET signal. More recently, it was demonstrated that when tRNA transcripts are attached to molecular beacons with 2^{t}-*O*-methyl back-bone and injected into the nucleus of HeLa cells, the probes are exported into the cytoplasm. When these constructs are introduced into the cytoplasm, they remain cytoplasmic [48]. However, even without the problem of unwanted nuclear accumulation, microinjection is inefficient in delivering probes into a large number of cells.

Another nonendocytic delivery method is toxin-based cell membrane permeabilization. For example, streptolysin O (SLO) is a pore-forming bacterial toxin that has been used as a simple and rapid means of introducing oligonucleotides into eukaryotic cells [49–51]. SLO binds as a monomer to cholesterol and oligomerizes into a ring-shaped structure to form pores of approximately 25–30 nm in diameter, allowing the influx of both ions and macromolecules. It was found that SLO-based permeabilization could achieve an intracellular concentration of ODNs of approximately 10 times that of electroporation and liposomal-based delivery. Since cholesterol composition varies with cell types, the permeabilization protocol needs to be optimized for each cell type by varying temperature, incubation time, cell number, and SLO concentration. An essential feature of this technique is that the toxin-based permeabilization is reversible. This can be achieved by introducing oligonucleotides with SLO under serum-free conditions and then removing the mixture and adding normal media with serum [50, 52].

Cell penetrating peptides (CPP) have been used to introduce proteins, nucleic acids, and other biomolecules into living cells [53–55]. Among the family of peptides with membrane translocating activity are antennapedia, HSV-1

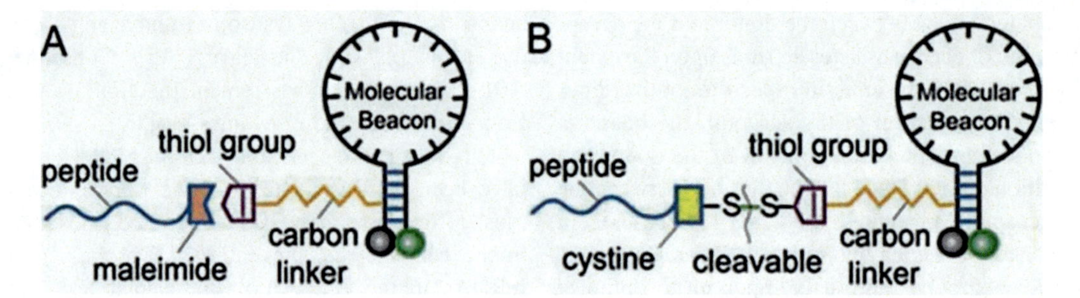

Fig. 11.5 A schematic of peptide-linked molecular beacons. (**a**) A peptide-linked molecular beacon using the thiol-maleimide linkage in which the quencher arm of the molecular beacon stem is modified by adding a thiol group, which can react with a maleimide group placed to the C terminus of the peptide to form a direct, stable linkage. (**b**) A peptide-linked molecular beacon with a cleavable disulfide bridge in which the peptide is modified by adding a cysteine residue at the C terminus, which forms a disulfide bridge with the thiol-modified molecular beacon. This disulfide bridge design allows the peptide to be cleaved from the molecular beacon by the reducing environment of the cytoplasm

VP22, and the HIV-1 Tat peptide. To date the most widely used peptide are HIV-1 Tat peptide and its derivatives due to their small size and high delivery efficiency. The Tat peptide is rich in cationic amino acids especially arginines, which is very common in many of the cell-penetrating peptides. However, the exact mechanism for CPP-induced membrane translocation remains elusive.

A wide variety of cargos have been delivered to living cells both in cell culture and in tissue using cell-penetrating peptides [56, 57]. For example, Allinquant et al. [58] linked antennapedia peptide to the 5^t end of DNA oligonucleotides (with biotin on the 3^t end) and incubated both peptide-linked ODNs and ODNs alone with cells. By detecting biotin using streptavidin-alkaline phosphatase amplification, it was found that the peptide-linked ODNs were internalized very efficiently into all cell compartments compared with control ODNs. No indication of endocytosis was found. Similar results were obtained by Troy et al. [59] with a 100-fold increase in antisense delivery efficiency when ODNs were linked to antennapedia peptides. Recently, Tat peptides were conjugated to molecular beacons using different linkages (Fig. 11.5); the resulting peptide-linked molecular beacons were delivered into living cells to target GAPDH and survivin mRNAs [29]. It was demonstrated that, at relatively low concentrations, peptide-linked molecular beacons were internalized into living cells within 30 min with nearly 100% efficiency. Further, peptide-based delivery did not interfere with either specific targeting by or hybridization-induced florescence of the probes, and the peptide-linked molecular beacons could have self-delivery, targeting, and reporting functions. In contrast, liposome-based (Oligofectamine) or dendrimer-based (SuperFect) delivery of molecular beacons required 3–4 h and resulted in a punctate fluorescence signal in the cytoplasmic vesicles and a high background in both cytoplasm and nucleus of cells [29]. It was clearly demonstrated that cellular delivery of molecular beacons using the peptide-based approach has far better performance compared with conventional transfection methods.

11.5 Living Cell RNA Detection Using Molecular Beacons

Sensitive gene detection in living cells presents a significant challenge. In addition to issues with target accessibility, detection specificity, and probe delivery as discussed above, achieving high detection sensitivity and signal-to-background ratio requires not only careful design of the probes and advanced fluorescence microscopy imaging but also a better understanding of RNA biology and probe–target interactions. It is likely that different applications have different requirements on the properties of probes. For example, rapid determination of RNA expression level and localization requires fast probe–target hybridization kinetics, while long-time monitoring of gene expression dynamics requires probes having high intracellular stability.

To demonstrate the capability of molecular beacons in sensitive detection of specific endogenous mRNAs in living cells, dual FRET molecular beacons were designed to detect K-ras and survivin mRNAs in HDF and MIAPaCa-2 cells, respectively [28]. K-ras is one of the most frequently mutated genes in human cancers [60]. A member of the *G*-protein family, K-ras is involved in transducing growth-promoting signals from the cell surface. Survivin, one of the inhibitor of apoptosis proteins (IAPs), is normally expressed during fetal development but not in most normal adult tissues [61], thus can be used as a tumor biomarker for several types of cancers. Each FRET probe pair consisted of two molecular beacons, one labeled with a donor fluorophore (Cy3, donor beacon) and a second labeled with an acceptor fluorophore (Cy5, acceptor beacon). These molecular beacons were designed to hybridize to adjacent regions on an mRNA target so that the two fluorophores lie within the FRET range (~6 nm) when probe–target hybridization occurs for both beacons. BHQ-2 and BHQ-3 were used as quenchers for the donor and acceptor molecular beacons, respectively. One pair of molecular beacons targets a segment of the wild-type K-ras gene whose codon 12 mutations are involved in the pathogenesis of many cancers. A negative control dual FRET molecular beacon pair was also designed ("random beacon pair") whose specific 16-base target sequence was selected using random walk, thus having no exact match in the mammalian genome. It was found that detection of the FRET signal significantly reduced false positives, leading to sensitive imaging of K-ras and survivin mRNAs in live HDF and MIAPaCa-2 cells. For example, FRET detection gave a ratio of 2.25 of K-ras mRNA expression in stimulated versus unstimulated HDF, comparable to the ratio of 1.95 using RT-PCR, and in contrast to single-beacon result of 1.2. The detection of survivin mRNA also indicated that, compared with the single-beacon approach, dual FRET molecular beacons gave lower background signal, thus having a higher signal-to-background ratio [28].

11.5.1 Biological Significance

An intriguing discovery in detecting K-ras and survivin mRNAs using dual FRET molecular beacons is the clear and detailed mRNA localization in living cells [28]. To demonstrate, in Fig. 11.6a, a fluorescence image of K-ras mRNA in stimulated HDF cells is shown, indicating an intriguing filamentous localization pattern. The localization pattern of K-ras mRNA was further studied and found to be colocalized with mitochondria inside live HDF cells [62]. Since K-ras proteins interact with proteins such as Bcl-2 in mitochondria to mediate both antiapoptotic and proapoptotic pathways, it seems that cells localize certain mRNAs where the corresponding proteins can easily bind to their partners.

The survivin mRNA, however, is localized in MIAPaCa-2 cell very differently. As shown in Fig. 11.6b in which the fluorescence image was superimposed with a white-light image of the cells, survivin mRNAs seemed to localize in a nonsymmetrical pattern within MIAPaCa-2 cells, often to one side of the nucleus of the cell. These mRNA localization patterns raise many interesting biological questions. For example, how mRNAs are transported to their destination and how the destination is recognized? To what subcellular organelle might the mRNAs be colocalized? What is the biological implication of mRNA localization? Although mRNA localization in living cells is believed to be closely related to post-transcriptional regulation of gene expression, much remains to be seen if such localization indeed targets a protein to its site of function by producing the protein "right on the spot."

The transport and localization of oskar mRNA in *Drosophila melanogaster* oocytes have also been visualized [26]. In this work, molecular beacons with 2^{t}-*O*-methyl backbone were delivered into cells using microinjection, and the migration of oskar mRNAs was tracked in real time, from the nurse cells where it is produced to the posterior cortex of the oocyte where it is localized. Clearly, the direct visualization of specific mRNAs in living cells with molecular beacons will provide important insight into the intracellular trafficking and localization of RNA molecules.

As another example of targeting specific genes in living cells, molecular beacons were used to detect the viral genome and characterize the spreading of bovine respiratory syncytial virus (bRSV) in living cells [63]. It was found that molecular beacon signal could be detected in single living cells infected by RSV with high detection sensitivity, and the signal revealed a connected, highly three-dimensional, amorphous inclusion body structure not seen in fixed cells. Figure 11.7 shows the molecular beacon signal indicating the spreading of viral infection at days 1, 3, 5, and 7 post-infection (PI), which demonstrates the ability of molecular beacons to monitor and quantify in real time the viral infection process. Molecular beacons were further used to image the viral genomic RNA (vRNA) of human RSV (hRSV) in live Vero cells, revealing the dynamics of filamentous virion egress and

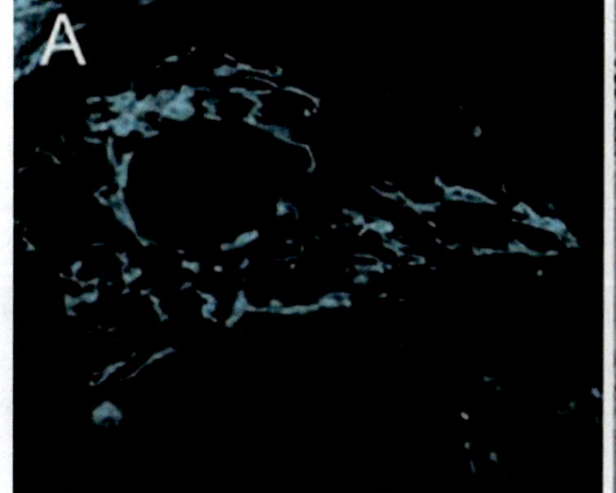

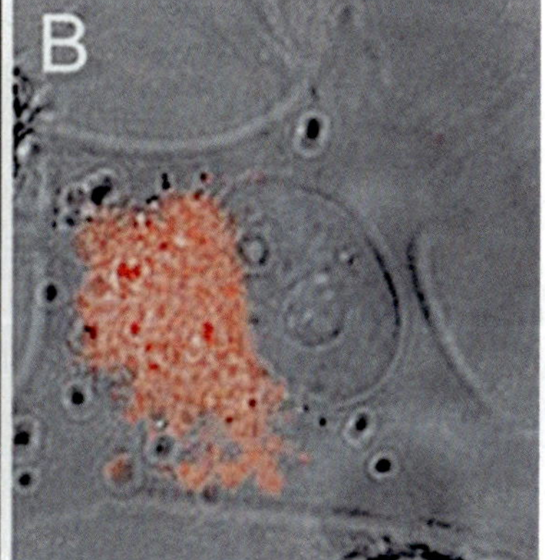

Fig. 11.6 mRNA localization in HDF and MIAPaCa-2 cells. (**a**) Fluorescence images of K-ras mRNA in stimulated HDF cells. Note the filamentous K-ras mRNA localization pattern. (**b**)A fluorescence image of survivin mRNA localization in MIAPaCa-2 cells. Note that survivin mRNAs often localized to one side of the nucleus of the MIAPaCa-2 cells

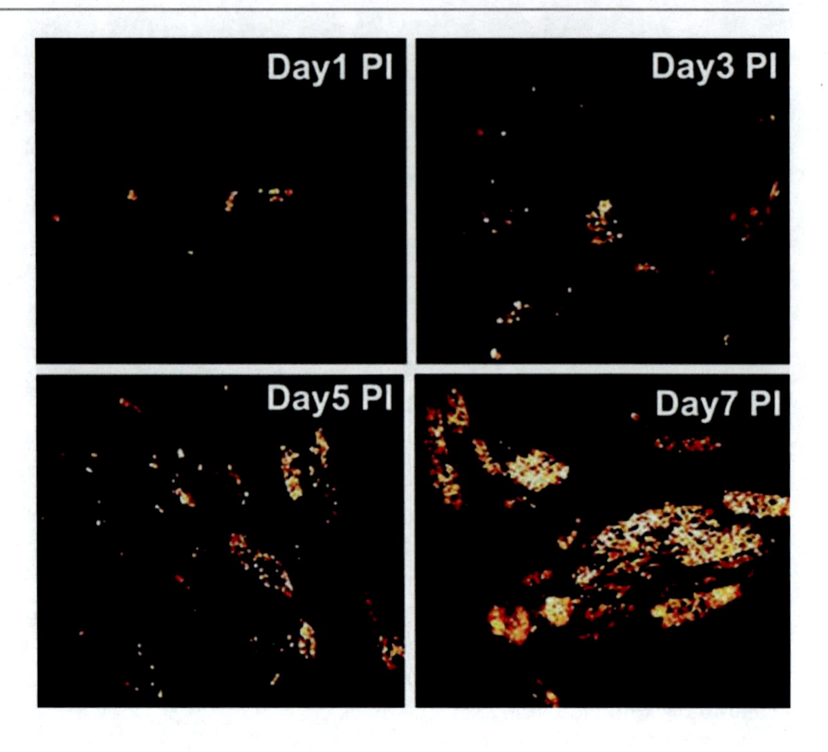

Fig. 11.7 Live-cell fluorescence imaging of the genome of bovine respiratory syncytial virus (bRSV) using molecular beacons shows the spreading of infection in host cells at days 1, 3, 5, and 7 post-infection (PI). Primary bovine turbinate cells were infected by a clinical isolate of bRSV, CA-1, with a viral titer of $2 \times 10^{3.6}$ $TCID_{50}$/ml. Molecular beacons were designed to target several repeated sequences of the gene-end-intergenic-gene-start signal within the bRSV genome, with a signal-to-noise ratio of 50–200

providing insight into how viral filaments bud from the plasma membrane of the host cell [64].

11.6 Engineering Challenges

Nanostructured molecular probes such as molecular beacons have the potential to enjoy a wide range of applications that require sensitive detection of genomic sequences. For example, molecular beacons are used as a tool for the detection of single-stranded nucleic acids in homogeneous in vitro assays [65, 66]. Surface-immobilized molecular beacons used in microarray assays allow for the high-throughput parallel detection of nucleic acid targets while avoiding the difficulties associated with PCR-based labeling [65, 67]. Another novel application of molecular beacons is the detection of double-stranded DNA targets using PNA "openers" that form triplexes with the DNA strands [68]. Further, proteins can be detected by synthesizing "aptamer molecular beacon" [69, 70] which, upon binding to a protein, undergoes a conformational change that results in the restoration of fluorescence.

The most exciting application of nanostructured oligonucleotide probes, however, is living cell gene detection. As demonstrated, molecular beacons can detect endogenous mRNA in living cells with high specificity, sensitivity, and signal-to- background ratio, thus having the potential to provide a powerful tool for laboratory and clinical studies of gene expression in vivo. For example, molecular beacons can be used in high-throughput cell-based assays to quantify and monitor the dose-dependent changes of specific mRNA expression in response to different drug leads. The ability of molecular beacons to detect and quantify the expression of specific genes in living cells will also facilitate disease studies, such as viral infection detection and cancer diagnosis.

There are a number of challenges in detecting and quantifying RNA expression in living cells. In addition to issues of probe design and target accessibility, quantifying gene expression in living cells in terms of mRNA copy number per cell poses a significant challenge. For instance, it is necessary to distinguish true and background signals, determine the fraction of mRNA molecules hybridized with probes, and quantify the possible self-quenching effect of the reporter, especially when mRNA is highly localized. Since the fluorescence intensity of the reporter may be altered by the intracellular environment, it is also necessary to create an internal control by, for example, injecting fluorescently labeled oligonucleotides with known quantity into the same cells and obtaining the corresponding fluorescence intensity. Further, unlike in RT-PCR studies where the mRNA expression is averaged over a large number of cells (usually over 1 million), in optical imaging of mRNA expression in living cells, only a relatively small number of cells (typically less than 1000) are observed. Therefore, the average copy number per cell may change with the total number of cells observed due to the (often large) cell-to-cell variation of mRNA expression.

Another issue in living cell gene detection using hairpin ODN probes is the possible effect of probes on normal cell function, including protein expression. As revealed in the antisense therapy research, complementary pairing of a short segment of an exogenous oligonucleotide to mRNA can have a profound impact on protein expression levels and even cell fate. For example, tight binding of the probe to the translation start site may block mRNA translation. Binding of a DNA probe to mRNA can also trigger RNase H-mediated mRNA degradation. However, the probability of eliciting antisense effects with hairpin probes may be very low when low concentrations of probes (<200 nM) are used for mRNA detection, in contrast to the high concentrations (typically 20 μM; [51]) employed in antisense experiments. Further, it generally takes 4 h before any noticeable antisense effect occurs, whereas visualization of mRNA with hairpin probes requires less than 2 h after delivery. However, it is important to carry out a systematic study of the possible antisense effects, especially for molecular beacons with 2^{t}-*O*-methyl backbone, which may also trigger unwanted RNA interference.

As a new approach for in vivo gene detection, the nanostructured probes can be further developed to have enhanced sensitivity and a wider range of applications. For example, it is likely that hairpin ODN probes with quantum dot as the fluorophore will have a better ability to track the transport of individual mRNAs from the cell nucleus to the cytoplasm. Hairpin ODN probes with near-infrared (NIR) dye as the reporter, combined with peptide-based delivery, have the potential to detect specific RNAs in tissue samples, animals, or even humans. It is also possible to use lanthanide chelate as the donor in a dual FRET probe assay and perform time-resolved measurements to dramatically increase the signal-to-noise ratio, thus achieving high sensitivity in detecting low-abundance genes. Although very challenging, the development of these and other nanostructured ODN probes will significantly enhance our ability to image, track, and quantify gene expression in vivo and provide a powerful tool for basic and clinical studies of human health and disease.

There are many possibilities for nanostructured ODN probes to become clinical tools for disease detection and diagnosis. For example, molecular beacon could be used to perform cell-based early cancer detection using clinical samples including blood, saliva, and other bodily fluid. In this case, cells in the clinical sample are separated, and molecular beacons designed to target specific cancer genes are delivered to the cytoplasm for detecting mRNAs of the cancer biomarker genes. Cancer cells having a high level of the target mRNAs (such as survivin) or mRNAs with specific mutations that cause cancer (such as K-ras codon 12 mutations) would show high level of fluorescence signal, while normal cells would show just low background signal.

Table 11.1 Descriptions of key biological and technical terminologies

Deoxyribonucleic acid (DNA)	A DNA molecule is a nucleic acid that contains the genetic instructions used in the development and functioning of all known living organisms and some viruses. It is a long polymer made from repeating nucleotides including adenine (A), quanine (G), cytosine (C), and thymine (T). Typically G pairs with C and A pairs with T, according to Watson–Crick base pairing
Ribonucleic acid (RNA)	RNA is a biologically important type of molecule that consists of a long chain of nucleotide units. Each nucleotide contains a ribose sugar, with carbons numbered 1^{t} through 5^{t}. A base is attached to the 1^{t} position, generally adenine (A), cytosine (C), quanine (G), or uracil (U). Typically G pairs with C and A pairs with U
Oligonucleotide (ODN)	An oligonucleotide is a short nucleic acid polymer, typically with 20 or fewer bases. Because oligonucleotides readily bind to their respective complementary nucleotides, they are often used as probes for detecting DNA or RNA
Polymerase chain reaction (PCR)	PCR is a technique to amplify a single or few copies of a piece of DNA across several orders of magnitude, generating thousands to millions of copies of a particular DNA sequence. Almost all PCR applications employ a heat-stable DNA polymerase which enzymatically assembles a new DNA strand from DNA building blocks, the nucleotides, by using single-stranded DNA as a template and DNA oligonucleotides (DNA primers) which are required for initiation of DNA synthesis
Fluorescence in situ hybridization (FISH)	FISH is a cytogenetic technique used to detect and localize the presence or absence of specific DNA sequences on chromosomes. FISH uses fluorescent probes that bind to only those parts of the DNA that show a high degree of sequence similarity. FISH is often used for finding specific features in DNA for use in genetic counseling, medicine, and species identification. FISH can also be used to detect and localize specific mRNAs within tissue samples
Ribonucleoprotein (RNP)	RNP is a nucleoprotein that contains RNA and protein, i.e., it is an association that combines ribonucleic acid and protein together.
Fluorescence resonance energy transfer (FRET)	FRET is the non-radiative transfer of energy from one molecule to another based on the proximity of the two molecules and their spectral overlap. It is a widely applied tool for optical sensing and studies of biomolecular interaction
Single-nucleotide polymorphism (SNP)	A single-nucleotide polymorphism (SNP) is a DNA sequence variation occurring when a single nucleotide—A, T, C or G,—in the genome (or other shared sequence) differs between members of a species (or between paired chromosomes in an individual)
Fluorophore	A fluorophore, in analogy to a chromophore, is a component of a molecule which causes a molecule to be fluorescent. It is typically a functional group in a molecule which will absorb energy of a specific wavelength and re-emit energy at a different (but equally specific) wavelength
Quencher	A quencher is a substance that absorbs excitation energy from a fluorophore; a dark quencher dissipates the energy as heat thus eliminates a fluorophore's ability to fluoresce
Codon	Codon is a sequence of three adjacent nucleotides constituting the genetic code that specifies the insertion of an amino acid in a specific structural position in a polypeptide chain during the synthesis of proteins

In this approach, the target mRNAs would not be diluted compared with the approaches using cell lysate. Thus, molecular beacon-based assay has the potential to positively identify cancer cells in a clinical sample with high specificity and sensitivity. It may also be possible to detect cancer cells in vivo by using NIR-dye-labeled molecular beacons in combination with endoscopy. Although there remain significant challenges, imaging methods using nanostructured probes have a great potential in becoming a powerful clinical tool for disease detection and diagnosis. To help readers in engineering, key biological and technical terminologies used are summarized in Table 11.1.

Acknowledgments This work was supported by the National Heart, Lung, and Blood Institute of the NIH as a Program of Excellence in Nanotechnology (HL80711), by the National Cancer Institute of the NIH as a Center of Cancer Nanotechnology Excellence (CA119338), and by the NIH Roadmap Initiative in Nanomedicine through a Nanomedicine Development Center award (PN2EY018244).

References

1. Saiki, R.K., Scharf, S., Faloona, F., Mullis, K.B., Horn, G.T., Erlich, H.A., Arnheim, N.: Enzymatic amplification of beta-globin genomic sequences and restriction site analysis for diagnosis of sickle cell anemia. Science. **230**, 1350–1354 (1985)
2. Alwine, J.C., Kemp, D.J., Parker, B.A., Reiser, J., Renart, J., Stark, G.R., Wahl, G.M.: Detection of specific RNAs or specific fragments of DNA by fractionation in gels and transfer to diazobenzyloxymethyl paper. Methods Enzymol. **68**, 220–242 (1979)
3. Adams, M.D., Dubnick, M., Kerlavage, A.R., Moreno, R., Kelley, J. M., Utterback, T.R., Nagle, J.W., Fields, C., Venter, J.C.: Sequence identification of 2,375 human brain genes. Nature. **355**, 632–634 (1992)
4. Velculescu, V.E., Zhang, L., Vogelstein, B., Kinzler, K.W.: Serial analysis of gene expression. Science. **270**, 484–487 (1995)
5. Liang, P., Pardee, A.B.: Differential display of eukaryotic messenger RNA by means of the polymerase chain reaction. Science. **257**, 967–971 (1992)
6. Schena, M., Shalon, D., Davis, R.W., Brown, P.O.: Quantitative monitoring of gene expression patterns with a complementary DNA microarray. Science. **270**, 467–470 (1995)
7. Bassell, G.J., Powers, C.M., Taneja, K.L., Singer, R.H.: Single mRNAs visualized by ultrastructural in situ hybridization are principally localized at actin filament intersections in fibroblasts. J. Cell Biol. **126**, 863–876 (1994)
8. Buongiorno-Nardelli, M., Amaldi, F.: Autoradiographic detection of molecular hybrids between RNA and DNA in tissue sections. Nature. **225**, 946–948 (1970)
9. Behrens, S., Fuchs, B.M., Mueller, F., Amann, R.: Is the in situ accessibility of the 16S rRNA of *Escherichia coli* for Cy3-labeled oligonucleotide probes predicted by a three-dimensional structure model of the 30S ribosomal subunit? Appl. Environ. Microbiol. **69**, 4935–4941 (2003)
10. Huang, Q., Pederson, T.: A human U2 RNA mutant stalled in 3ᵗ end processing is impaired in nuclear import. Nucleic Acids Res. **27**, 1025–1031 (1999)
11. Glotzer, J.B., Saffrich, R., Glotzer, M., Ephrussi, A.: Cytoplasmic flows localize injected oskar RNA in Drosophila oocytes. Curr. Biol. **7**, 326–337 (1997)
12. Jacobson, M.R., Pederson, T.: Localization of signal recognition particle RNA in the nucleolus of mammalian cells. Proc. Natl. Acad. Sci. U. S. A. **95**, 7981–7986 (1998)
13. Tsuji, A., Koshimoto, H., Sato, Y., Hirano, M., Sei-Iida, Y., Kondo, S., Ishibashi, K.: Direct observation of specific messenger RNA in a single living cell under a fluorescence microscope. Biophys. J. **78**, 3260–3274 (2000)
14. Dirks, R.W., Molenaar, C., Tanke, H.J.: Methods for visualizing RNA processing and transport pathways in living cells. Histochem. Cell Biol. **115**, 3–11 (2001)
15. Molenaar, C., Abdulle, A., Gena, A., Tanke, H.J., Dirks, R.W.: Poly (A)+ RNAs roam the cell nucleus and pass through speckle domains in transcriptionally active and inactive cells. J. Cell Biol. **165**, 191–202 (2004)
16. Tyagi, S., Kramer, F.R.: Molecular beacons: probes that fluoresce upon hybridization. Nat. Biotechnol. **14**, 303–308 (1996)
17. Tsourkas, A., Behlke, M.A., Rose, S.D., Bao, G.: Hybridization kinetics and thermodynamics of molecular beacons. Nucleic Acids Res. **31**, 1319–1330 (2003)
18. Bonnet, G., Tyagi, S., Libchaber, A., Kramer, F.R.: Thermodynamic basis of the enhanced specificity of structured DNA probes. Proc. Natl. Acad. Sci. U. S. A. **96**, 6171–6176 (1999)
19. Tyagi, S., Bratu, D.P., Kramer, F.R.: Multicolor molecular beacons for allele discrimination. Nat. Biotechnol. **16**, 49–53 (1998)
20. Li, J.J., Geyer, R., Tan, W.: Using molecular beacons as a sensitive fluorescence assay for enzymatic cleavage of single-stranded DNA. Nucleic Acids Res. **28**, E52 (2000)
21. Molenaar, C., Marras, S.A., Slats, J.C., Truffert, J.C., Lemaitre, M., Raap, A.K., Dirks, R.W., Tanke, H.J.: Linear 2ᵗ O-Methyl RNA probes for the visualization of RNA in living cells. Nucleic Acids Res. **29**, E89 (2001)
22. Sokol, D.L., Zhang, X., Lu, P., Gewirtz, A.M.: Real time detection of DNA. RNA hybridization in living cells. Proc. Natl. Acad. Sci. U. S. A. **95**, 11538–11543 (1998)
23. Vet, J.A., Majithia, A.R., Marras, S.A., Tyagi, S., Dube, S., Poiesz, B.J., Kramer, F.R.: Multiplex detection of four pathogenic retroviruses using molecular beacons. Proc. Natl. Acad. Sci. U. S. A. **96**, 6394–6399 (1999)
24. Kostrikis, L.G., Tyagi, S., Mhlanga, M.M., Ho, D.D., Kramer, F.R.: Spectral genotyping of human alleles. Science. **279**, 1228–1229 (1998)
25. Piatek, A.S., Tyagi, S., Pol, A.C., Telenti, A., Miller, L.P., Kramer, F.R., Alland, D.: Molecular beacon sequence analysis for detecting drug resistance in *Mycobacterium tuberculosis*. Nat. Biotechnol. **16**, 359–363 (1998)
26. Bratu, D.P., Cha, B.J., Mhlanga, M.M., Kramer, F.R., Tyagi, S.: Visualizing the distribution and transport of mRNAs in living cells. Proc. Natl. Acad. Sci. U. S. A. **100**, 13308–13313 (2003)
27. Tsourkas, A., Behlke, M.A., Xu, Y., Bao, G.: Spectroscopic features of dual fluorescence/luminescence resonance energy-transfer molecular beacons. Anal. Chem. **75**, 3697–3703 (2003)
28. Santangelo, P.J., Nix, B., Tsourkas, A., Bao, G.: Dual FRET molecular beacons for mRNA detection in living cells. Nucleic Acids Res. **32**, e57 (2004)
29. Nitin, N., Santangelo, P.J., Kim, G., Nie, S., Bao, G.: Peptide-linked molecular beacons for efficient delivery and rapid mRNA detection in living cells. Nucleic Acids Res. **32**, e58 (2004)
30. Tyagi, S., Alsmadi, O.: Imaging native beta-actin mRNA in motile fibroblasts. Biophys. J. **87**, 4153–4162 (2004)
31. Peng, X.H., Cao, Z.H., Xia, J.T., Carlson, G.W., Lewis, M.M., Wood, W.C., Yang, L.: Real-time detection of gene expression in cancer cells using molecular beacon imaging: new strategies for cancer research. Cancer Res. **65**, 1909–1917 (2005)
32. Medley, C.D., Drake, T.J., Tomasini, J.M., Rogers, R.J., Tan, W.: Simultaneous monitoring of the expression of multiple genes inside of single breast carcinoma cells. Anal. Chem. **77**, 4713–4718 (2005)

33. Tyagi, S., Marras, S.A., Kramer, F.R.: Wavelength-shifting molecular beacons. Nat. Biotechnol. **18**, 1191–1196 (2000)
34. Tsourkas, A., Behlke, M.A., Bao, G.: Structure–function relationships of shared- stem and conventional molecular beacons. Nucleic Acids Res. **30**, 4208–4215 (2002)
35. Brodsky, A.S., Silver, P.A.: Identifying proteins that affect mRNA localization in living cells. Methods. **26**, 151–155 (2002)
36. Forrest, K.M., Gavis, E.R.: Live imaging of endogenous RNA reveals a diffusion and entrapment mechanism for nanos mRNA localization in Drosophila. Curr. Biol. **13**, 1159–1168 (2003)
37. Shav-Tal, Y., Darzacq, X., Shenoy, S.M., Fusco, D., Janicki, S.M., Spector, D.L., Singer, R.H.: Dynamics of single mRNPs in nuclei of living cells. Science. **304**, 1797–1800 (2004)
38. Haim, L., Zipor, G., Aronov, S., Gerst, J.E.: A genomic integration method to visualize localization of endogenous mRNAs in living yeast. Nat. Methods. **4**, 409–412 (2007)
39. Ozawa, T., Natori, Y., Sato, M., Umezawa, Y.: Imaging dynamics of endogenous mitochondrial RNA in single living cells. Nat. Methods. **4**, 413–419 (2007)
40. Valencia-Burton, M., McCullough, R.M., Cantor, C.R., Broude, N. E.: RNA visualization in live bacterial cells using fluorescent protein complementation. Nat. Methods. **4**, 421–427 (2007)
41. States, D.J., Gish, W., Altschul, S.F.: Improved sensitivity of nucleic acid database searches using application-specific scoring matrices. Methods. **3**, 66–70 (1991)
42. Rhee, W.J., Santangelo, P.J., Jo, H., Bao, G.: Target accessibility and signal specificity in live-cell detection of BMP-4 mRNA using molecular beacons. Nucleic Acids Res. (2007) submitted
43. Lakowicz, J.R.: Principles of Fluorescence Spectroscopy, 2nd edn. Springer, New York (1999)
44. Marras, S.A., Kramer, F.R., Tyagi, S.: Efficiencies of fluorescence resonance energy transfer and contact-mediated quenching in oligonucleotide probes. Nucleic Acids Res. **30**, e122 (2002)
45. Price, N.C., Stevens, L.: Fundamentals of Enzymology: the Cell and Molecular Biology of Catalytic Proteins, 3rd edn. Oxford University Press, New York (1999)
46. Dokka, S., Rojanasakul, Y.: Novel non-endocytic delivery of antisense oligonucleotides. Adv. Drug Deliv. Rev. **44**, 35–49 (2000)
47. Leonetti, J.P., Mechti, N., Degols, G., Gagnor, C., Lebleu, B.: Intracellular distribution of microinjected antisense oligonucleotides. Proc. Natl. Acad. Sci. U. S. A. **88**, 2702–2706 (1991)
48. Mhlanga, M.M., Vargas, D.Y., Fung, C.W., Kramer, F.R., Tyagi, S.: tRNA-linked molecular beacons for imaging mRNAs in the cytoplasm of living cells. Nucleic Acids Res. **33**, 1902–1912 (2005)
49. Giles, R.V., Ruddell, C.J., Spiller, D.G., Green, J.A., Tidd, D.M.: Single base dis- crimination for ribonuclease H-dependent antisense effects within intact human leukaemia cells. Nucleic Acids Res. **23**, 954–961 (1995)
50. Barry, M.A., Eastman, A.: Identification of deoxyribonuclease II as an endonuclease involved in apoptosis. Arch. Biochem. Biophys. **300**, 440–450 (1993)
51. Giles, R.V., Spiller, D.G., Grzybowski, J., Clark, R.E., Nicklin, P., Tidd, D.M.: Selecting optimal oligonucleotide composition for maximal antisense effect following streptolysin O-mediated delivery into human leukaemia cells. Nucleic Acids Res. **26**, 1567–1575 (1998)
52. Walev, I., Bhakdi, S.C., Hofmann, F., Djonder, N., Valeva, A., Aktories, K., Bhakdi, S.: Delivery of proteins into living cells by reversible membrane permeabilization with streptolysin-O. Proc. Natl. Acad. Sci. U. S. A. **98**, 3185–3190 (2001)
53. Snyder, E.L., Dowdy, S.F.: Protein/peptide transduction domains: potential to deliver large DNA molecules into cells. Curr. Opin. Mol. Ther. **3**, 147–152 (2001)
54. Wadia, J.S., Dowdy, S.F.: Protein transduction technology. Curr. Opin. Biotechnol. **13**, 52–56 (2002)
55. Becker-Hapak, M., McAllister, S.S., Dowdy, S.F.: TAT-mediated protein transduction into mammalian cells. Methods. **24**, 247–256 (2001)
56. Wadia, J.S., Dowdy, S.F.: Transmembrane delivery of protein and peptide drugs by TAT-mediated transduction in the treatment of cancer. Adv. Drug Deliv. Rev. **57**, 579–596 (2005)
57. Brooks, H., Lebleu, B., Vives, E.: Tat peptide-mediated cellular delivery: back to basics. Adv. Drug Deliv. Rev. **57**, 559–577 (2005)
58. Allinquant, B., Hantraye, P., Mailleux, P., Moya, K., Bouillot, C., Prochiantz, A.: Downregulation of amyloid precursor protein inhibits neurite outgrowth in vitro. J. Cell Biol. **128**, 919–927 (1995)
59. Troy, C.M., Derossi, D., Prochiantz, A., Greene, L.A., Shelanski, M. L.: Downregulation of Cu/Zn superoxide dismutase leads to cell death via the nitric oxide- peroxynitrite pathway. J. Neurosci. **16**, 253–261 (1996)
60. Minamoto, T., Mai, M., Ronai, Z.: K-ras mutation: early detection in molecular diagnosis and risk assessment of colorectal, pancreas, and lung cancers–a review. Cancer Detect. Prev. **24**, 1–12 (2000)
61. Altieri, D.C., Marchisio, P.C.: Survivin apoptosis: an interloper between cell death and cell proliferation in cancer. Lab. Investig. **79**, 1327–1333 (1999)
62. Santangelo, P.J., Nitin, N., Bao, G.: Direct visualization of mRNA colocalization with mitochondria in living cells using molecular beacons. J. Biomed. Opt. **10**, 44025 (2005)
63. Santangelo, P., Nitin, N., LaConte, L., Woolums, A., Bao, G.: Live-cell characterization and analysis of a clinical isolate of bovine respiratory syncytial virus, using molecular beacons. J. Virol. **80**, 682–688 (2006)
64. Santangelo, P.J., Bao, G.: Dynamics of filamentous viral RNPs prior to egress. Nucleic Acids Res. **35**, 3602–3611 (2007)
65. Liu, X., Tan, W.: A fiber-optic evanescent wave DNA biosensor based on novel molecular beacons. Anal. Chem. **71**, 5054–5059 (1999)
66. Kambhampati, D., Nielsen, P.E., Knoll, W.: Investigating the kinetics of DNA– DNA and PNA–DNA interactions using surface plasmon resonance-enhanced fluorescence spectroscopy. Biosens. Bioelectron. **16**, 1109–1118 (2001)
67. Steemers, F.J., Ferguson, J.A., Walt, D.R.: Screening unlabeled DNA targets with randomly ordered fiber-optic gene arrays. Nat. Biotechnol. **18**, 91–94 (2000)
68. Kuhn, H., Demidov, V.V., Coull, J.M., Fiandaca, M.J., Gildea, B. D., Frank-Kamenetskii, M.D.: Hybridization of DNA and PNA molecular beacons to single-stranded and double-stranded DNA targets. J. Am. Chem. Soc. **124**, 1097–1103 (2002)
69. Hamaguchi, N., Ellington, A., Stanton, M.: Aptamer beacons for the direct detection of proteins. Anal. Biochem. **294**, 126–131 (2001)
70. Yamamoto, R., Baba, T., Kumar, P.K.: Molecular beacon aptamer fluoresces in the presence of Tat protein of HIV-1. Genes Cells. **5**, 389–396 (2000)

Packaging for Bio-micro-electro-mechanical Systems (BioMEMS) and Microfluidic Chips

12

Edward S. Park, Jan Krajniak, and Hang Lu

12.1 Introduction

In recent years, applications of lab-on-a-chip, microfluidics, and bioMEMS devices have ranged from basic research on macromolecules (e.g., DNAs and proteins), cells, tissues, and organisms to clinical applications (e.g., blood analysis, disease diagnosis, and drug delivery). The device design and the final packaging of the devices demand two very important aspects of compatibility. First, the fabrication processes and sequences of all components (including device or reagents) need to be compatible with each other. Second, the devices and their packaging need to be biocompatible, whether it is referring to minimizing sample contamination, optimizing sample interaction with device/packaging materials, or eliminating immune responses to implanted devices. Therefore, in designing these biochips, one has to consider not only the functionalities but also the packaging scheme with the final product and the environment in which it will be used.

This chapter is organized to address packaging issues in a variety of applications, packaging schemes that are commonly used, the types of interfaces one needs to think about (e.g., interface with the outside environment and interface between devices), and the materials of choice and their modifications. In designing a good packaging scheme, one needs to strategize in terms of constraints and a series of trade-offs. No packaging scheme is universal, and there exist many options. Depending on how the devices are designed and how they are intended to be used, the packaging scheme can be very different, and aspects that have smaller design space (miniaturization) need to be considered first. For example, if an implantable device needs to be packaged, the most important aspect is the biocompatibility; next, one needs to consider where the device is to be implanted and then match the mechanical properties of the tissues and consider all the geometrical constraints. More than often, the infrastructure that is available to manufacture the devices and the packaging will also put constraints on selecting the processes. For industrial applications, overall cost is likely an important parameter.

This chapter aims to point out some standard solutions and research opportunities in bioMEMS packaging and will refer to many primary research articles for detailed information on existing technologies.

12.2 Packaging Schemes Based on Application

12.2.1 Portable and Point-of-Care (POC) Diagnostics and Analysis

Owing to their small size, bioMEMS are ideal platforms upon which to design portable and point-of-care (POC) diagnostics and analytical systems. Portability is an essential consideration in the design and packaging of systems intended for use in resource-poor or similarly challenging environments. These environments lack the infrastructure of typical "developed-world" clinical laboratories, which provide stable electrical power, cleanliness, refrigeration, and highly trained personnel. Interest in developing bioMEMS-based portable analytical systems originated in the 1990s from the US Defense Advanced Research Projects Agency (DARPA), seeking to enhance chemical and biological weapons detection and to improve battlefield medical care. More recently, interest from both governmental and charitable organizations has led to significant investments in developing portable diagnostics to avert epidemics in poor and remote geographies [1, 2].

The design and packaging of portable diagnostic systems must satisfy a challenging set of requirements. Such requirements include (i) small size and weight, (ii) full

E. S. Park · J. Krajniak · H. Lu (✉)
School of Chemical Biomolecular Engineering, Georgia Institute of Technology, Atlanta, GA, USA
e-mail: edward.park@chbe.gatech.edu; jan.krajniak@chbe.gatech.edu; hang.lu@chbe.gatech.edu

C. P. Wong et al. (eds.), *Nano-Bio- Electronic, Photonic and MEMS Packaging*, https://doi.org/10.1007/978-3-030-49991-4_12

integration of necessary functions (unit operations), (iii) low power consumption, (iv) easy operation by potentially non-skilled personnel, (v) functionality in wide temperature ranges, (vi) ruggedness and mechanical protection, (vii) protection from dust or other contaminants, and (viii) low cost.

The packaging schemes that have emerged to address the above requirements can be grouped into two broad categories. The first category focuses upon integrating as many functions as possible onto a single chip in order to maximize portability. The second category consists of systems that are partitioned into two components: (i) a disposable chip (or cartridge) and (ii) a readout machine (or scanner). The cartridge is inserted into the scanner, read and analyzed, and then discarded. Both the integrated and partitioned categories are discussed in the following sections.

12.2.1.1 Integrated BioMEMS Packaging Schemes

Integrating functions onto a single bioMEMS device are the central aim for packaging diagnostic systems when size and weight are the most important criteria. Basic functions in a typical diagnostic system include fluid handling (i.e., pumping and mixing), reagent storage, species separation and concentration, and detection. Today, diagnostic chips that integrate all necessary functions are rare. In most cases, the devices are designed with a specific function in mind.

The major challenges in portable chip-scale integration involve fluid handling, detection, and electrical power. Fluid pumping for analytical purposes has traditionally been carried out using macroscale machines (i.e., peristaltic, syringe pump). Likewise, analyte detection is normally accomplished using complex optical or spectroscopic techniques that require bench-top equipment, which are usually expensive and bulky. In such forms, both functions require power sources that exceed the latest capabilities of portable batteries. To circumvent these challenges, integrated biochips must often rely upon passive fluid transport mechanisms (e.g., capillary action) and simplified "yes/no" detection readouts for portable diagnostics.

Immunochromatographic Test Strips The classic example of integrated, portable diagnostic devices is immunochromatographic strip (ICS) tests. ICS tests are packaged with a strip of fibrous material that is divided into a sample pad, conjugation/labeling region, and detection zone. The strip is contained in a plastic casing with access windows over the sample pad and detection zone. Fluid samples are transported along the strip via capillary action through the pores and channels in the fibrous material and are bound to an antigen (or by an antibody) as they pass through the labeling region. Typically, a dye is labeled to the analyte, allowing for easy visual readout by the operator. ICS tests have been successfully deployed "in the field" for over 10 years to diagnose infectious agents like diphtheria toxin [3] and various sexually transmitted diseases [4–6]. The ICS packaging scheme is advantageous because it does not include any active power source. In addition, materials and reagents are off-the-shelf, inexpensive, and compatible with high-volume production methods.

The primary drawback of ICS design is its relatively crude "yes/no" readout mechanism (although visual readouts can also be an advantage if the operator is non-skilled). Paper-based microfluidic systems are being developed that provide colorimetric readout of digital images [7–9]. Such systems are comprised of a paper sheet that is embedded with a photoresist pattern that guides sample fluid into different detection zones. Each zone contains different labels for parallel diagnoses. Paper-based systems exhibit the same packaging advantages as ICS systems. However, the image acquisition is performed using a digital camera, which reduces the degree of integration but increases the precision of the measurement due its semi-quantitative nature. For both ICS and paper-based microfluidics, this simple packaging scheme is compatible with the economic requirement of the applications.

Electrical Detection In order for portable diagnostics to provide fully quantitative measurements, researchers are investigating ways to miniaturize sophisticated detection components and package them on-chip. The most actively researched techniques are electrical, mechanical, and optical. To achieve electrical detection on-chip, conductive electrodes are integrated into the flow path of the sample fluid. The electrodes are typically metallic and are sometimes functionalized with molecular probes or modified with a selective membrane. The simplest packaging configuration for electrical detection is to pattern electrodes on a planar substrate, which is interfaced with another substrate that contains the sample flow path (Fig. 12.1). Several examples of this configuration exist, most notably for glucose and urea detections [10, 11], where a current or a potential is measured as a result of redox (reduction and oxidation) reactions on the substrate. A similar packaging configuration enables ICS test strips to give fully quantitative outputs using screen-printed, disposable electrodes that couple to quantum dot labels [12]. Also, highly sensitive and reversible electrical detection of antibodies has been demonstrated using silicon nanowires grown on an oxidized silicon surface and coupled to a polydimethylsiloxane (PDMS) flow channel [13, 14]. The challenge is that often sensitive off-chip electronics are needed to detect minute changes in current or potential.

Mechanical Detection On-chip mechanical detection has been achieved using a two-substrate packaging scheme similar to electrical detection. Mechanical detection has been demonstrated using micro- and nanoscale cantilever beams

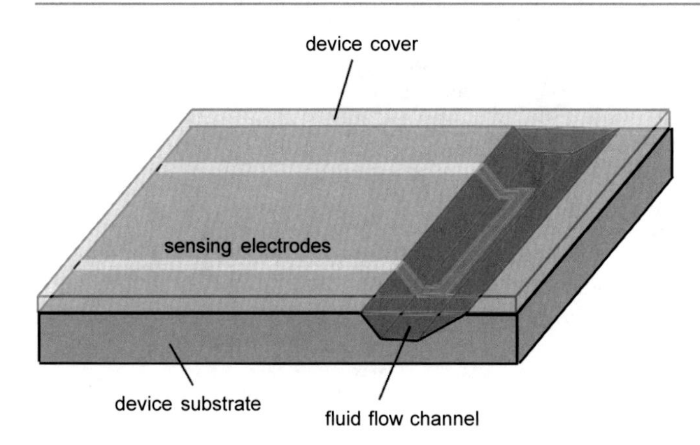

Fig. 12.1 Electrodes used for sensing are deposited on the surface of the device, with the sensing portion protruding into the fluidic components. Packaging materials compatible with the manufacturing requirements have to be chosen. For example, if the device is to be polymer based, deposition of metal electrodes will be complicated and will require modification of the polymer

that are machined from bulk silicon using MEMS fabrication techniques and functionalized with the appropriate capture probes [15, 16]. A change in the surface stress or resonance frequency of the cantilever indicates binding of biological species. The cantilever's mechanical output is then measured in two ways: (i) optically, by tracking the change in location of a laser beam reflected off the cantilever or (ii) electrically, by coupling a piezoresistive element to the edge of the beam. As a result, cantilever detectors must be packaged in a manner that can interface to optical components or electrical leads, respectively. Specifically, optical detection would require a viewing aperture with a transparent substrate (and laser beam alignment and position detection), while electrical detection would require attachment of leads to bond pad arrays (Fig. 12.2).

Because of such additional components, portable and fully integrated cantilever- based diagnostic systems have yet to be realized. However, a major advantage of cantilever-based detection, which is the ability to detect molecules without attaching optical (e.g., fluorescent) labels to the molecules themselves, motivates continuing research. Label-free, stress-based detection has been demonstrated with DNA and proteins [17, 18, 21, 22]. In addition, resonance-based detection has been demonstrated with cells [20] and viruses [19]. A particularly novel resonator is configured so that the sample flows through a microchannel in a hollow cantilever, which enables sub-femtogram measurement resolution [23, 24]. High sensitivity is made possible by packaging the cantilever and drives electrode under high vacuum in a hermetically sealed cartridge. Such innovation in both fabrication and packaging is a good example of collaboration between academia (MIT) and industry (Innovative Micro Technology, Inc. and Affinity Biosensors). Although some progress has been made in this area in miniaturizing and integrating the detection modules, most mechanical detection chips still require some off-chip components (i.e., optics). This area therefore remains an active research area for miniaturization.

Optical Detection The miniaturization and integration of optical detection on-chip is also an active area of research. Packaging schemes vary dramatically, depending on which optical components are deemed critical for integration. The most basic approach is to attach a separate module containing a miniaturized excitation source and photodetector directly to a chip containing the appropriate flow network. Although it is not true chip-scale integration, the modular approach confers advantages with respect to flexibility and convenience, while maintaining acceptable (though not ideal) size and weight. Such an approach has been successfully employed to couple a glass capillary electrophoresis chip to a light-induced fluorescence (LIF) module to detect protein biotoxins (i.e., ricin, botulinum toxin) without preconcentration [25].

A similar modular approach assembles a miniature diode laser, sample well plate, and detection chip. The use of a self-contained detection chip in lieu of a full-size camera makes the assembly better suited for field use, and diffractive optics split the excitation beam into 16 channels, resulting in parallel analysis of 16 corresponding wells. The chip-to-plate approach is shown to detect low concentrations of *Bacillus globigii*, a surrogate marker for *Bacillus anthracis*, utilizing ELISA signal amplification to compensate for the detection chip's limited sensitivity [26].

True on-chip integration of optical components has been achieved on a limited basis. The concept of a "microscope-on-a-chip," integrating miniature lenses and complementary metal oxide semiconductor (CMOS) detectors on-chip, has been investigated [27]. Optical fibers micromachined with angled grooves have been embedded underneath microfluidic chips to guide excitation beams into multiple flow channels simultaneously [28]. Similarly, optical fibers have also been directly inserted into waveguides patterned in photoresist (SU-8, a multifunctional epoxy negative photoresist, manufactured by MicroChem Corp.) [29].

Using soft lithography, the fabrication of several optical components, such as beam splitters, lenses, and prisms, has been shown [30]. Moreover, integrated circuit (IC) fabrication techniques have been utilized to produce monolithic light- emitted diodes (LEDs), photodetectors, and optical fibers [31]. From a packaging standpoint, this represents a high degree of integration, uniting the excitation source, waveguides, and detectors onto a single silicon chip that is attached to a PDMS microfluidic flow channel. In a similar display of high integration, optical waveguides in photoresist, an optically pumped dye laser, and photodiodes embedded in silicon have been integrated on-chip

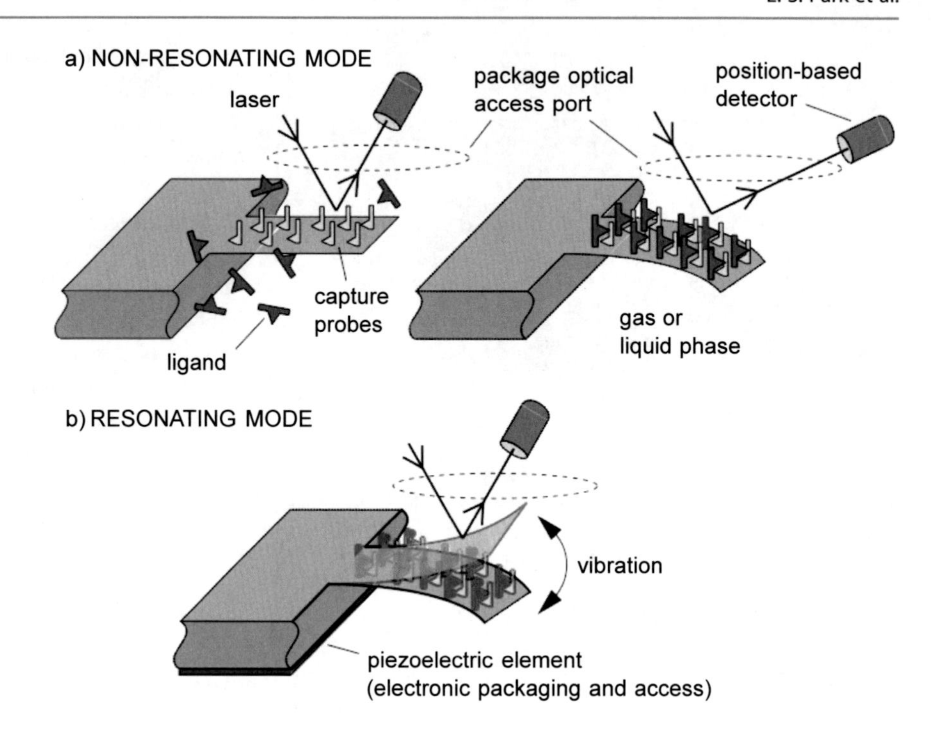

Fig. 12.2 Mechanically based bio-detection via cantilever beams requires optical and electrical access in packaging. (**a**) A stress-based cantilever functionalized with capture probes is exposed to fluid containing a ligand of interest; attachment of the ligand causes the beam to deflect, changing the angle of reflected light. DNA hybridization and protein–protein interactions have been reported using stress-based methods [17, 18]. (**b**) A resonance-based cantilever responds by changes in resonance frequency due to changes in mass as ligand attaches to capture probes [19, 20]; piezoelectric actuation requires electrical leads and interconnects in packaging.

[32]. On-chip optical integration is a large and diverse field [12, 33]. Nevertheless, the field is still in its infancy, providing few portable packages for diagnostic applications today.

Other Components The portability of diagnostic devices has been improved by the integration of other components, such as vapor barriers to prevent the evaporation of fluids from microfluidic systems constructed from gas-permeable elastomers [34]. Also, on-chip diaphragm-based pumps and valves, which rely upon thermal actuation from resistive losses from adjacent electrodes, have been integrated into a portable polymerase chain reaction-capillary electrophoresis (PCR- CE) system [35]. Furthermore, the storage and release of reagents on-chip has been addressed by the integration of torque-actuated screw valves, which eliminates the necessity for electrical power when closing or opening microfluidic flow paths [36].

Integration Outlook Although the full integration of all essential functions onto a single chip has not been realized, some devices have come very close to being total analysis system "on a chip" [37–39]. As previously discussed, fluid handling, detection (i.e., electrical, mechanical, optical), and electrical power are the major challenges in developing a fully integrated device. In addition, stored reagents could also add to the bulk of a diagnostic chip. The trade-off one makes as more functions are integrated on-chip is that detection sensitivity typically decreases. In addition, the cost per chip increases. Because of these limitations, the idea of inserting a simple and disposable diagnostic chip into a separate scanner box could provide a workable compromise between portability and performance.

12.2.1.2 Portable BioMEMS Chip in Tandem with a Readout Box

By partitioning a portable diagnostic system into a disposable chip and a scanner, not only can high performance be provided, but also cleaning and per-test costs are reduced. From the design standpoint, a major task is to decide what mix of functions will be located on the disposable chip versus on the reader. Careful consideration will result in functions that are located in a manner that maximizes the combined benefit of detection sensitivity, ease of use (during testing and maintenance), and low cost. From there, packaging can be developed that properly interfaces the chip to the reader and the reader to the operator. Figure 12.3 illustrates the concept of the partitioned approach.

A champion of the partitioned design concept is the Yager Research Group at the University of Washington. Its development of a disposable diagnostic card for saliva testing is an insightful case study [2, 40]. Most active functions, such as fluid pumping and surface plasmon resonance [41] excitation and detection, are located on the scanner. Functions located on the chip, which include sample preconditioning, mixing, and assaying, are specifically designed to be passive in nature. For instance, the mixing region relies upon chaotic mixing that is induced by herringbone structures as fluid flows through the channel. Also, the diffusion of small molecules across the interface of two laminar flow streams (buffer and sample) is utilized to precondition the sample before assay. Such a preconditioning approach cleverly

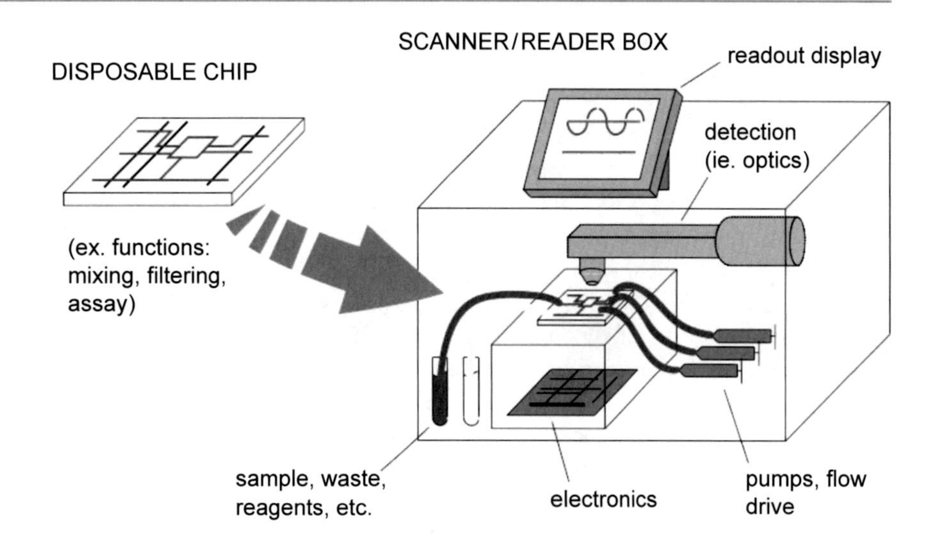

Fig. 12.3 A bioMEMS chip used in tandem with a scanner. Functions are partitioned between the disposable chip and scanner box, thereby optimizing critical factors (i.e., cost, detection sensitivity, portability) [2]. Interconnect between the two elements, as well as scanner maintenance, is an additional packaging consideration

removes the necessity for less portable preconditioning methods (e.g., centrifugation). The chip (roughly 4 inches by 2 inches in size) consists of a PDMS slab containing the flow channel network that is bonded to a glass slide, which is gold-coated and functionalized for antibody capture. Rubber gaskets (O-rings) are embedded into the fluidic inlet/outlet ports to facilitate connection with off-chip pumps and other active elements.

A similar partitioned package, consisting of a microfluidic chip whose channels are loaded with polyacrylamide gel, has been used to perform electrophoretic immunoassays to measure the concentration of the protein MMP-8 from a saliva sample [42]. The glass chip is inserted into a portable scanner containing a user interface, miniature high-voltage power supply, data acquisition, and software control. Although some sample preparation and fluorescence imaging was performed with standard laboratory equipment, the system proves the principle of a modular, point-of-care microfluidic assay for a clinically relevant disease marker. The choice of glass also minimizes work that has to be performed for surface modification and ensures optical compatibility for the readout.

The partitioned design strategy is being pursued in other contexts also [43]. A capillary electrophoresis (CE) chip for the separation and detection of organophosphate nerve agents has been demonstrated as a chemical weapon detector [44]. The glass chip contains a CE channel coupled to a screen-printed carbon electrode for amperometric detection. The scanner houses the electrochemical detector, reference/counter electrodes, and buffer/sample reservoirs. A portable, highly parallel SPR system has also been developed for chemical weapons detection [45]. The system is based on the Texas Instruments' Spreeta sensing chip, which contains three LEDs and a diode array detector. The sensing chips are inserted into a control box that houses the power supply, LCD touch screen display, custom electronic board, and fluidic elements, such as pumps and valves. Each chip is replaced via a snap-in mechanism that mates it to a silicone flow cell. The combination of eight chips, each having three channels, enables the simultaneous monitoring of up to 24 analytes.

Yet another embodiment of the tandem strategy utilizes a cartridge in which cells are cultured and stimulated for use as a cell-based biosensor [46]. The cartridge is made of a CMOS silicon chip bonded to a PDMS slab patterned with cell chambers. Circuitry on the CMOS chip enables temperature control and electrophysiological sensing with microelectrodes. The cartridge interfaces with a handheld electronic reader that monitors the action potential (AP) activity of cardiomyocytes within the cell chambers. The system has successfully monitored the AP response of cells to different biochemical stimulation under conditions outside the laboratory (desert field testing).

12.2.1.3 Outlook for Portable Diagnostics

Portable diagnostic and analysis systems have the potential to dramatically improve the health of people in the poorest, most remote areas in the world, especially with respect to infectious diseases. Moreover, the benefits of portability will undoubtedly contribute to the ease and decentralization of disease management in the developed world as well. One could imagine patients wearing portable diagnostic devices that monitor their response to drugs for screening or dose titration purposes, which would be a critical step toward realizing the concept of "personalized medicine" [47, 48].

The major challenges facing packaging, as previously discussed, are function integration and partitioning between disposable and non-disposable components. Another challenge is cost. Virtually all prototype devices today have been fabricated out of relatively expensive base materials, such as glass, silicon, or PDMS. Furthermore, their fabrication processes may not be ideal for mass production, unlike thermoplastics that can be hot-embossed or injection-molded, which makes them disposable. For example, a microfluidic immunosensor made of embossed plastic has recently been applied to monitor cardiac inflammation markers from human serum [49]. On the other hand, some applications have stringent materials requirements that disqualify some plastic materials; hence, glass and PDMS continue to serve in many areas as the materials of choice.

Besides the device components, the packaging of reagents presents an additional challenge to system portability. To reduce bulk and transportation costs, reagents would ideally be packaged in dry form and reconstituted into solution at the point of care. In situations where the reagent can neither be dried nor reconstituted with local solvents to maintain activity (e.g., antibodies), the reagent solution must typically be packaged off-chip due to inadequate space within the miniscule volume of the on-chip microfluidic network. To address this, the use of plastic clinical tubing as a reagent cartridge has been investigated [50]. Plugs of different reagents are sequentially loaded into a tube and spaced by air. Similar packaging may enable reliable transport of valuable, ready-to-use solutions in portable diagnostic systems.

12.2.2 Implantable Devices

In the recent years, MEMS technology has become popular in the medical field due to its many advantages over traditional technologies. BioMEMS chips are highly functional, versatile, and power efficient and can be fabricated much smaller than other implantable systems. Furthermore, since bioMEMS fabrication processes originate in the IC industry, the infrastructures of the fabrication are well established, and these techniques are reasonably well developed and standardized [51]. Therefore, it is no surprise that many implantable devices that primarily use IC fabrication technologies, such as wireless pressure sensors [52], subcutaneous drug delivery chips [53], and glucose sensors [54, 55], have been and are being developed and commercialized.

The major packaging challenges for implantable MEMS devices include but are not limited to biocompatibility and size considerations. Another consideration is matching the mechanical properties and sometimes optical properties of the surrounding tissue. Implantation is an invasive process and triggers a healing and immune response from the body. Not only do implants have to be designed to minimize the immediate and long-term immune response, but they are also required to be non-toxic to the surrounding tissue. For this purpose, various packaging methods, materials, and protective coatings are available.

These designs must be developed with the miniaturization requirement in mind. They need to add as little volume or length to the device as possible and have to be adaptable to the specific tissue environment and their applications. For example, ocular implants need to be small enough for implantation through hypodermic needles, flexible enough to not cause irritation to the patients, and to match the mechanical properties of the surrounding tissues, but mechanically sturdy enough to survive the implantation process. The latter is particularly critical because inadequate protection during implantation can cause a complete loss of device functionality [56]. This section will review some successful strategies for implantable devices with these considerations in mind.

12.2.2.1 Drug Delivery Devices

Packaging for drug delivery devices is highly dependent on the duration of the intended delivery applications. Short- or intermediate-term delivery can be achieved through oral or subcutaneous administration, whose packaging requirement is minimal, typically involving encapsulating drugs in microbeads or delivery through various types of microneedles [57–60]. However, for long-term controlled delivery, bioMEMS devices seem to be most suitable: these devices can be designed to be highly functional; in addition, if proper packaging is designed, the drug reservoirs for controlled delivery can be made refillable [61].

A highly efficient and reliable controlled-release biochip design for implantation was introduced by Santini et al. [62]. The release of drug occurs from an array of wells covered by a gold thin-film membrane, which is electrochemically dissolved during the time of release (Fig. 12.4a). The reservoirs are fabricated using standard wet-etch MEMS technology in silicon wafers. The gold membrane covering the wells serves simultaneously as an electrical component and as a protective layer for the content of the reservoirs. The gold membrane is chemically stable and mechanically able to resist pressures exerted by the tissue and interstitial fluids [62]. Applying proper voltage to the anodic membrane leads to the dissolution of the membrane. This process is dependent on the presence of chlorine ions (Cl^-) in bodily fluids.

Soluble salts of gold form due to the applied potential and the presence of Cl^-, and instead of which dissolve into the surrounding fluid. Santini et al. established that these salts are biocompatible and are present in very low concentrations after the dissolution of the gold metal [63]. Because silicon is the structural material for the chip, the device overall is suitable for implantation. To protect the metallic components of the chip further from the interstitial fluids, SiO2 is deposited as a protective coating for the device.

This technology is being further developed for commercial application by MicroCHIPS [64]. The aim is to couple the release of drug to sensor devices for improved release timing and schemes. As such, a microchip packaged with a sensor array and a power source would regulate the release of drugs from the reservoirs [53] (Fig. 12.4b). A modified version of the approach, which relies on re-sealable compartments, is being

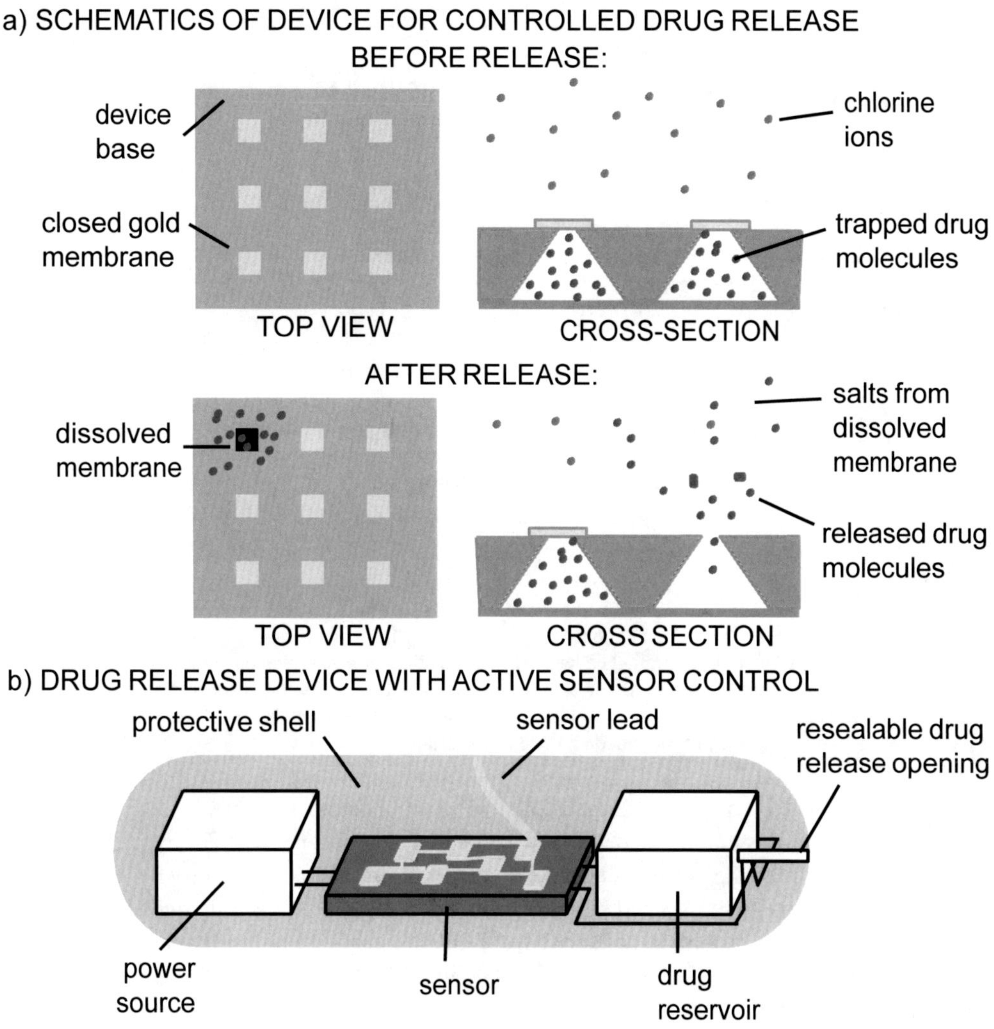

a) SCHEMATICS OF DEVICE FOR CONTROLLED DRUG RELEASE
BEFORE RELEASE:
device base
closed gold membrane
chlorine ions
trapped drug molecules
TOP VIEW
CROSS-SECTION
AFTER RELEASE:
dissolved membrane
salts from dissolved membrane
released drug molecules
TOP VIEW
CROSS SECTION
b) DRUG RELEASE DEVICE WITH ACTIVE SENSOR CONTROL
protective shell
sensor lead
resealable drug release opening
power source
sensor
drug reservoir

Fig. 12.4 Implantable drug release devices. (**a**) Basic design schematic of the device design by Santini et al. Prior to release, gold membranes form protective caps for drug compartments etched into the device. Once current is applied to the membrane, the gold forms soluble salts with Cl^- ions from the environment, dissolving the membrane and releasing the drug. Coating for protection and passivation can be applied. (**b**) Concept drawing of a drug release system combining active sensing technology and controlled drug release, contained within an implantable protective shell. Concept adapted from research by MicroCHIPS and ChipRx

investigated by ChipRx [64]. This technique could prove useful in limiting the size requirement for larger amounts of drugs through the use a single refillable compartment. While silicon is a chemically and mechanically robust material and is usually easy to be rendered biocompatible, one drawback of silicon-based devices is that these devices may not be suitable for other drug delivery formats that require soft and flexible surface materials, such as ocular drug delivery. New approaches and new materials need to be considered.

12.2.2.2 Ocular Implants

Li et al. [61, 65] developed a drug delivery device enclosed in silicone rubber for ocular drug delivery (Fig. 12.5a). The soft material protects a two-part chip – a reservoir and a delivery chip; the refillable drug reservoir is bound to the delivery chip during the packaging process. The packaging is comprised of a silicone rubber outer surface molded to fit the curvature of the eye, which is filled with paraffin wax and encloses the delivery chip, provided that the chip is small compared to the eye. The advantage of this approach is that the device is completely enclosed in the flexible and biocompatible material [61, 66]. Furthermore, since the enclosure is made of flexible material, the reservoir can be punctured by a needle and refilled with drug, after which the silicone rubber seals itself [61]. This already efficient packaging could be further improved through various surface modification technologies. Integration of traditional IC or MEMS components is also made possible with flexible polymeric packaging [67–69]. The chip device can thus be fabricated from the required or desired material and packaged for flexibility. Another option is to design a hard-shell packaging scheme, which must minimally interfere with the tissue in the eye [70, 71].

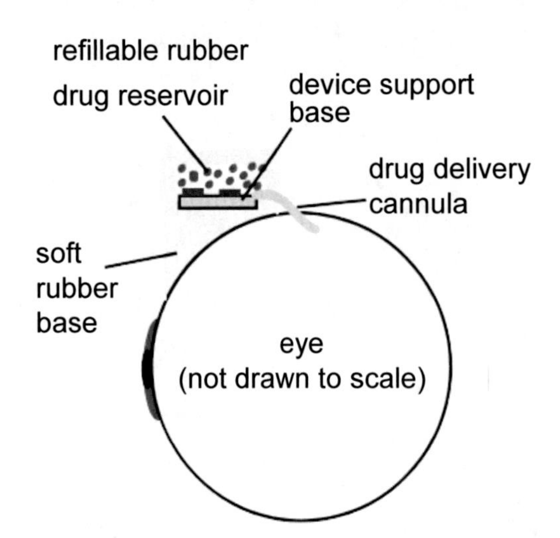

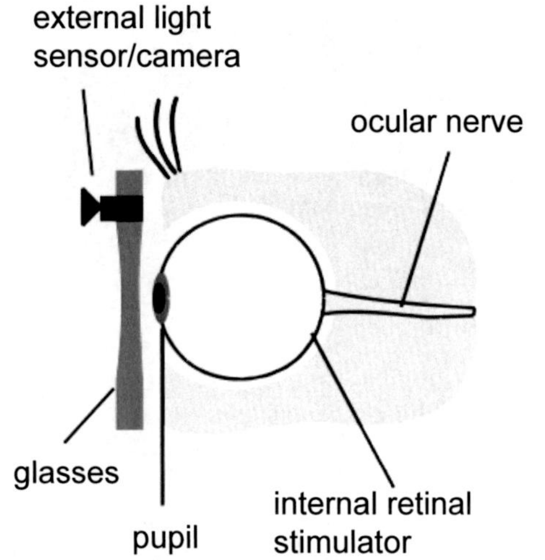

Fig. 12.5 Examples of devices for ocular implantation. (**a**) Example of an ocular drug delivery device fabricated with a soft silicone rubber shell for minimal eye irritation is designed by Li et al. [61, 65]. The base is molded to conform to the curvature of the eye. The reservoir can be refilled by puncturing the rubber reservoir cap, which re-seals itself after the procedure. The protective silicone rubber is a soft and biocompatible material and thus suitable for implantable devices. (**b**) The light sensor is separated from the retinal stimulator in this ocular implant. Only the stimulation chip is implanted in the eye, while the light sensor or camera is worn on the outside. This allows for minimal invasiveness, especially if the implanted chip is made from soft and flexible biocompatible materials

Another type of ocular implant is the retinal stimulation device (Fig. 12.5b). The purpose of this implant is to stimulate the ocular nerve in diseases caused by photoreceptor degeneration. These implants incorporate image sensors and electrode stimulators to facilitate image capture and nerve stimulation, which renders the chip more complex [69]. Because of the electrical components, CMOS fabrication technology is used to produce the device components. Typically all of the components are fabricated on a highly flexible silicon substrate. The silicon substrate between these components is then thinned, leading to an overall flexible device [69].

Monolithic integration of all parts for ocular implant devices can be advantageous in terms of simpler design because of the electrical connectivity and signal transduction among the sensing, stimulation, and electronics components. However, this monolithic scheme comes with the price of having larger devices. The alternative is then to reduce the number of components in the system to be implanted. In the case of ocular implants, image sensors/CMOS cameras can be worn on the outside, while only a signal receiver and the stimulation chip (with its electrodes) need to be implanted. In one packaging scheme, the implanted components are then coated with Parylene-C to form a biocompatible protective layer [67, 68] where the device is fabricated on a flexible (but not necessarily biocompatible) substrate, such as polyimide [68].

Hard-shell protection for the implanted device is also an option. Since the shell is not flexible, its shape and size should be minimally invasive to the eye. Tube-like capsules can be fabricated from glass, ceramics, metals, or silicon, all of which must be hermetically sealed to the device [71]. This method has been pursued by BION™ implants. These implants incorporate circuitry, a power source, and other functional parts within a hermetically sealed tantalum tube [70]. The rugged yet biocompatible tube is small enough to be implanted through a needle.

12.2.2.3 Neural Interface Implants

Some of the most common neural interfaces include pacemakers, neural probes, brain stimulators, and cochlear implants [72, 73]. The connection to the nervous tissue usually occurs through electrodes implanted into the brain tissue and in the vicinity of neurons and their processes. These electrodes are commonly organized into arrays such as the Michigan probe array and Utah array [74–76]. Both of these are variations of arrays of long silicon multi-electrode probes (Fig. 12.6) and are commonly coated with silicon nitride (SiN), silicon dioxide (SiO2), or gold [73] to improve biocompatibility.

Besides silicon, which is mechanically rigid and easy for insertion during surgery, other materials are used to create flexible arrays. Common choices include polyimide, benzocyclobutene, and Parylene. Some of these devices also include microfluidic channels for delivery of drugs to discourage scar tissue buildup for long-term recording and stimulation of the electrodes. In most arrays (stiff or soft), it is common to coat the surfaces with (i) poly(ethylene glycol) (PEG) films to minimize protein adsorption and (ii) drug eluting coatings [73]. The latter is achieved by using amphiphilic hydrogel particles that are capable of releasing drugs

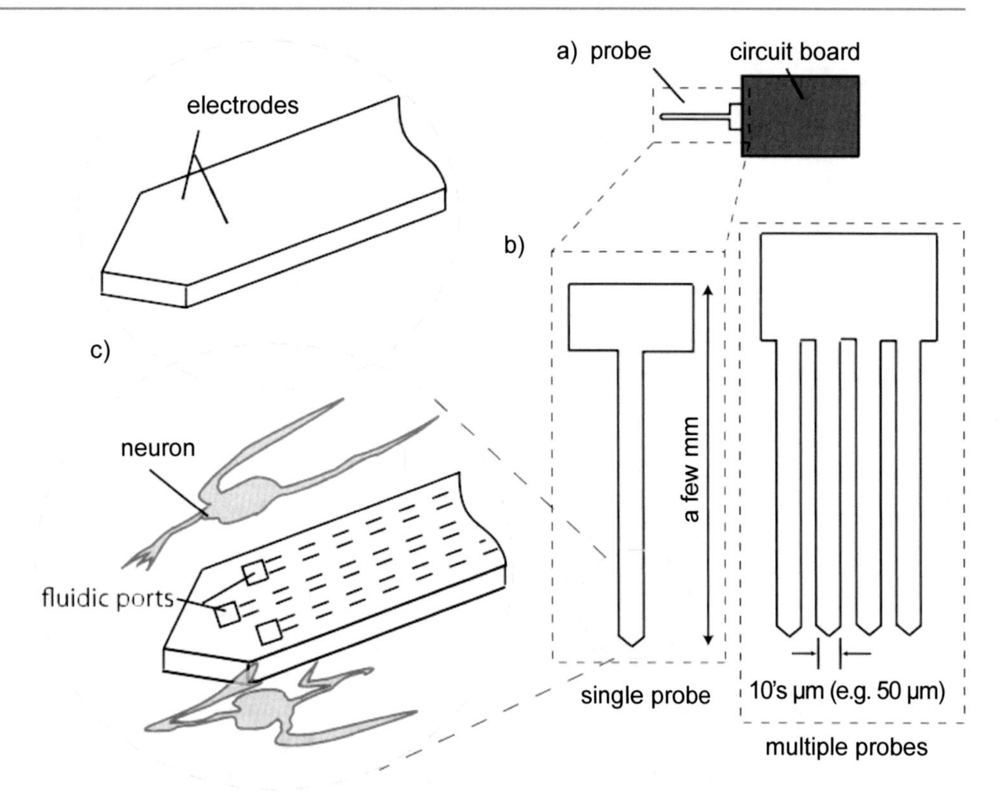

Fig. 12.6 Device details and packing of typical neural probes. (**a**) Microfabricated neural probes are usually mechanically mounted on and electrically connected to a circuit board. The probes are inserted into brain tissues for recording of neural activities. (**b**) These microfabricated probes can include either a single probe head or multiple parallel probe heads. (**c**) The tip of the probes can incorporate both metal electrode leads and fluid delivery ports. Materials to render the devices biocompatible or to include additional drug delivery capabilities (not drawn)

over time. One challenge of incorporating these particles into the coating is the choice of polymers, which need to be stable and biocompatible over time and also giving the desired drug release profile. Modification of the surfaces to increase porosity may also be used to reduce cell adhesion, which may be especially useful in preventing isolation and preserving functionality of probes [73].

Tokuda et al. [77] developed a general neural interface based on a multi-chip system similar to the segmented ocular implant described above. Several interconnected unit chips equipped with stimulating and recording electrodes can be implanted as a thin sheet. Due to the small size of each chip and flexible interconnects, the array can be bent and extended. However, for added mechanical protection, the chips are anchored on a polyimide film and covered with epoxy resin, which protects the body of the chip while leaving the electrodes exposed [77].

As was the case with ocular implants, hard-shell enclosures may be used with neural interface implants. Biocompatible materials such as titanium could be used to manufacture the shell. The shell has to be hermetically sealed, which can be achieved through the use of medical-grade epoxy, which also provides additional shock absorption for the circuitry [78].

12.2.2.4 Cardiovascular Implants

Stents Though stents are not typically considered bioMEMS devices, packaging technologies for stents can be applied to MEMS devices designed to interact with the cardiovascular system. The cardiovascular system is very sensitive to disturbances, and the presence of foreign objects can lead to problems such as intimal hyperplasia, or thickening of the blood vessel wall due to injury, and blood coagulation.

Stents often suffer from restenosis or the re-narrowing of the blood vessel wall, rendering them not functional. To prevent restenosis, various protective coatings have been developed, with the goal to provide a biocompatible interface, which passively or actively limits vessel wall thickening or fouling. TiNO*x* films can be deposited by physical vapor deposition (PVD); tuning the ratios of nitrogen to oxide allows for control of electrical properties of the film [79]. Polymeric coatings are also often used to provide a protective function. Drugs to prevent restenosis and hyperplasia are often contained in the polymer matrix and eluted from the matrix after the implantation [80, 81]. Although the type of drug may vary, the coating procedure generally relies on the application of a thin layer of polymer–drug mixture in solvents (e.g., by dip coating). Some of common polymers used for this purpose include methyacrylate and ethylene-based polymers [81].

Stents can also be implanted in combination with a pressure sensor to monitor the success of the implantation procedure [52]. CardioMEMS Inc. has developed a capacitance-based sensor enclosed in flexible polymers (Fig. 12.7). The capacitance change of the sensor can be correlated to deflection of the capacitor component from which pressure data can be calculated. The sensor is constructed from ultra- flexible alloys for additional flexibility. Pressure-related data are obtained wirelessly from an external readout machine. Communication between the devices is achieved through an energizing radio frequency (RF) signal sent by the readout machine. The signal is coupled to the sensor via magnetic coupling and induces a current in the sensor, which vibrates at a specific frequency depending on the applied pressure. The readout machine receives the response from the sensor through magnetic coupling and determines pressure based on the frequency of the sensor [82]. In this case, the packaging materials should not interfere with the RF signals.

Pacemakers Pacemakers are electrical devices that interact with the cardiovascular system and the nervous system. Although typically not considered as MEMS devices, they are sometimes interfaced with MEMS devices and have successful packaging schemes that implantable MEMS devices can borrow. Packaging schemes for different components of pacemaker devices face the challenge of different requirements. Typically the power source and control mechanism can be housed in traditional enclosures manufactured from titanium [83], PDMS, to a myriad of other materials. However, packaging of the pacemaker leads has to be treated separately.

Pacemakers are packaged to minimize the required pacing current and the polarization of the lead, to protect the lead from damage, and to reduce the inflammatory response around the lead. The packaging schemes are similar to that of stents; for example, protective polymeric sheaths, such as polyether polyurethane sheaths, are further covered by silicone rubber sheaths [84–86]. As was the case with stents, these sheaths may elute drugs, such as steroids, to prevent inflammatory response [84, 85].

12.2.2.5 Implantable Biosensors

Glucose Sensors Continuous glucose measurement is critical in monitoring the well-being of diabetes patients. A myriad of glucose measurement techniques is available for subcutaneously implantable glucose sensors. In these techniques, blood sugar can be measured through changes in fluorescence of a glucose sensitive system [87], output current in electrochemistry [54], or changes in viscosity [55]. Since each of these methods relies on different chemical and/or physical principles, packaging schemes must be fitted to reflect the different demands.

Encapsulation of fluorescently sensitive systems, whether they are housed on-chip or in microparticles, can be achieved through coating by PEG hydrogels [87]. To obtain a flexible sensor, which can be worn or implanted, biocompatible polymers can be used [54]. Flexible polymers are particularly useful if bendable electrodes are used in a current-sensing-based device. Bioinert polymers, such as PDMS and dimethylacrylamide (DMA), may be combined with other types of polymers for additional functionality, such as increased permeability of small molecules [54]. Biocompatible rigid sensor enclosures can also be manufactured from materials such as silicon, glass, or medical-grade epoxy [55] to reduce the risk of immune response.

Pressure Sensors MEMS-based minimally invasive pressure sensors can be used to monitor intraocular, intracranial, and cardiovascular pressures [52], each being of great interest to parts of the medical community [88]. Although the designs may differ, these sensors are generally fabricated using CMOS technology and enclosed in silicon [89–91]. The overall device may be further coated with deposited silicon nitride or silicon oxide layers.

As with other implantable devices, pressure sensors can also be coated with a polymer film for protection during the implantation process and normal operation. One such example is the use of Parylene-C as a protective membrane for ocular implants, which can be deposited through CVD processes at room temperature [92]. The polymer package may be further treated to reduce bubble formation on the surface through oxygen plasma roughening [92].

Communication between the sensors and the outside world must be considered in the packaging scheme. For example, ocular pressure sensors can be observed directly through the tissue of the eye, since this tissue is mostly transparent. Choosing transparent packaging such as biocompatible polymers will thus enable readouts by direct observation [93]. However, this option is not available for sensors implanted in non-transparent tissue, such as the CardioMEMS sensor described previously. For deep tissue implants, other readout methods are required, such as the RF-based method.

12.2.3 BioMEMS Packaging for Clinical Applications

BioMEMS technology is rapidly gaining attention in the clinical field due to its high versatility and broad range of functions. The most common bioMEMS uses can be divided into two broad categories: (i) analytical tools and (ii) components of medical devices, such as endoscopes or catheters. Section 12.2.2 reviews many implantable devices. This section will include additional devices used in medical diagnostics. Analytical devices used in the laboratory environment share designs common with point-of-care diagnostic devices, with some key differences in packaging schemes. The major difference is rooted in the availability of off-chip analytical technology and a clean and protected processing environment in the laboratory. Many of these devices are a direct miniaturization of existing conventional macroscopic assays. Even though all sample modification and processing occur on chip, analysis of

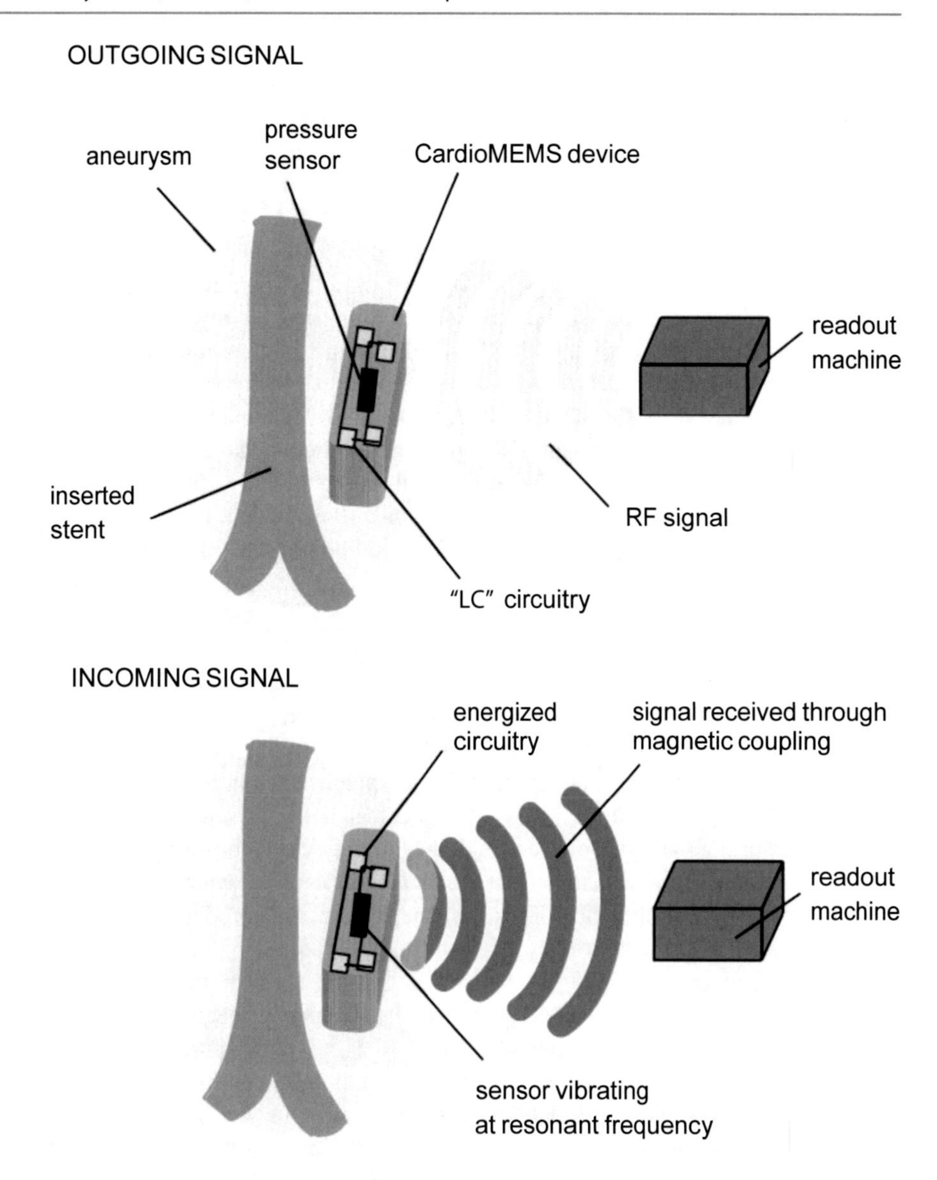

Fig. 12.7 CardioMEMS wireless pressure sensor. The sensor is implanted inside an aneurism along with the stent. Pressure measurements are obtained using an external readout module. This module sends out RF signals, which energize the LC circuit on the implanted device, including a pressure sensor. The sensors resonant frequency is determined through a series of RF signals with different frequencies; the module receives data through magnetic coupling. Once the resonant frequency has been determined, the pressure reading can be calculated. Packaging for this device not only needs to be biocompatible but also to pose minimal interference to the incoming and outgoing signals

the results can be performed off-chip [94] without exposure to an interfering environment. Hence, the challenge for packaging schemes lies in providing ease of interfacing to external instruments and primarily a high level of re-usability of bioMEMS devices.

Enzyme-linked immunosorbent assay (ELISA) devices are a prime example of the common device design. ELISA chips utilize antibody–antigen interactions to detect the presence of specific molecules in the analyte fluid. The sample is pumped through the chip resulting in (i) bonding of specific analyte to a localized protein or enzyme and (ii) changes in chemiluminescence, from which presence of the desired analyte can be inferred. The change is observed by an external camera [94] or on a microscope and processed on a computer. Analogous designs may include a sensor array and wireless transmitters to gather signals and relay them to the final analysis system [95]. Packaging schemes for these types of devices do not face biocompatibility issues but must allow for interfacing the chip to the outside world. This most commonly occurs in the fluidic, optical, and electrical domains, since analytes are introduced in solution and signal gathering occurs through imaging or microelectrode sensing.

Polymerase chain reaction (PCR) chips are also heavily used in the medical field when genetical analysis is required. PCR chips can be manufactured using typical MEMS materials such as silicon, glass, epoxy, SU-8 (an epoxy negative photoresist, MicroChem Inc.), PDMS, poly(methyl methacrylate) (PMMA), poly(ethylene terephthalate) (PET or PETE), and common fabrication processes [96, 97]. These devices operate on fluids, making use of standard microfluidic approaches for pumping, mixing, and flow control. As such, their packaging must allow for easy sample introduction and removal of product. However, reuse of PCR chips (or similar biological or medical analytics devices) is complicated due to risk of contamination from previous samples [96]. This risk is minimized through the use of various cleaning and packaging schemes. Prakash et al. use a removable silane coating in a glass-based device to prevent cross-contamination. A silane layer is first applied to improve the hydrophilicity of internal channels for PCR. After the reactions, the silane is stripped and reapplied for a new run [96]. This method or its analogs for different coatings may improve the re-usability of bioMEMS devices used in the clinical environment. PCR chip packaging is further discussed in Sect. 12.2.4.1.

Microfluidic-based bioMEMS chips are also used for analysis of human blood [98, 99]. The purpose of these devices is to separate specific molecules or cells from the sample for analysis or transfusion. For example, microfluidic devices can be used to remove leukocytes from blood before it is received by the patient to minimize the immune response. Micronics, Inc. has designed a microfluidic device to work in tandem with a readout machine for the purpose of blood (and even urine) analysis [99]. The analysis occurs on a microfluidic card designed to perform all functions related to sample manipulation. The card is placed into a manifold, which allows the card to interact with a computerized pumping and control mechanism. The system can then automatically perform an analysis of the sample included in the card. The packaging for this system must ensure that the card interacts with the control device. Thus, proper precautions have to be taken, especially at the card–manifold interface, where leaking can occur.

Unlike stand-alone bioMEMS devices, packaging schemes for bioMEMS components of medical tools, such as endoscopes [100], are heavily influenced by the tool design and its application. Packaging has to allow for incorporation into the overall apparatus without inhibiting function. Since there are no general device types, no general packaging schemes exist for this application. Rather, each device or chip must be integrated into the complete system with specific considerations for that system.

12.2.4 General Research for the Life Sciences

Whether a research laboratory is academic or industrial, it is undoubtedly in need of data of higher quality, greater throughput, and potentially newer forms. To address these unmet needs, bioMEMS-based research devices are being developed to offer (i) greater sensitivity and lower noise; (ii) higher throughput, automation, and standardization; and (iii) new functional capabilities.

The packaging schemes for research bioMEMS are diverse and non- standardized. This diversity spans both academic and industrial systems and is due in part to the relative youth of the bioMEMS field. The following subsections will highlight general packaging strategies and considerations for four of the most actively researched applications: PCR–CE analysis, microarrays, microfluidic large-scale integration (MLSI), and cell culture.

12.2.4.1 Genetic Analysis via PCR and CE

The analysis of DNA and RNA is fundamental to the life sciences. One of the most widely used methods for such analysis is the polymerase chain reaction (PCR). The PCR process amplifies minute amounts of nucleic acids by subjecting a sample solution to a sequence of temperature cycles. PCR is one of the first applications for which biochips have been designed and where integration of different functionalities is demonstrated and therefore is a good example for packaging considerations (Fig. 12.8). The advantages of miniaturized PCR include shorter cycle times, reduced reagent consumption, lower fabrication costs, and reduced contamination.

Materials One of the most important factors to consider in the packaging of PCR chips is material selection. First, the choice of material will dictate its thermal conductivity, which in turn affects the device's temperature cycling rate and overall throughput. Second, optical transparency must be considered, as less transparent materials may limit the device's utility for real-time optical detection of amplification products. Third, the binding affinity of sample molecules to the material will determine whether the inner surface of the reaction chamber requires a passivation coating. Finally, adequate chemical resistance and dimensional stability are required for cleaning steps and temperature cycles, respectively. One should bear in mind that the factors just discussed are by no means an exhaustive list. For example, cost and manufacturability may also be important.

PCR biochips have been manufactured in silicon, glass, and various polymers. A thorough review of PCR chip technology is by Zhang et al. [97], which among other things includes a detailed list of materials used in PCR chips. In recent years, polymers have gained the most interest due to the potential for very low cost for both raw materials and fabrication. Notable polymers are embossed polycarbonate [101, 102], embossed PMMA [103], and compression-molded poly(cyclic olefin) [104]. Also, hybrid PCR chips constructed out of two base materials have also been investigated. Silicon–glass [105, 106], polymer–silicon [107], and polymer–glass [108, 109] chips have also been realized for various applications.

Temperature Cycling and Heating Another factor in PCR chip packaging is the choice of heating elements. Thin-film electrodes have been integrated on-chip out of platinum [110], other metals, and doped polysilicon [111]. Indium tin oxide (ITO) thin film has also been used due to its optical

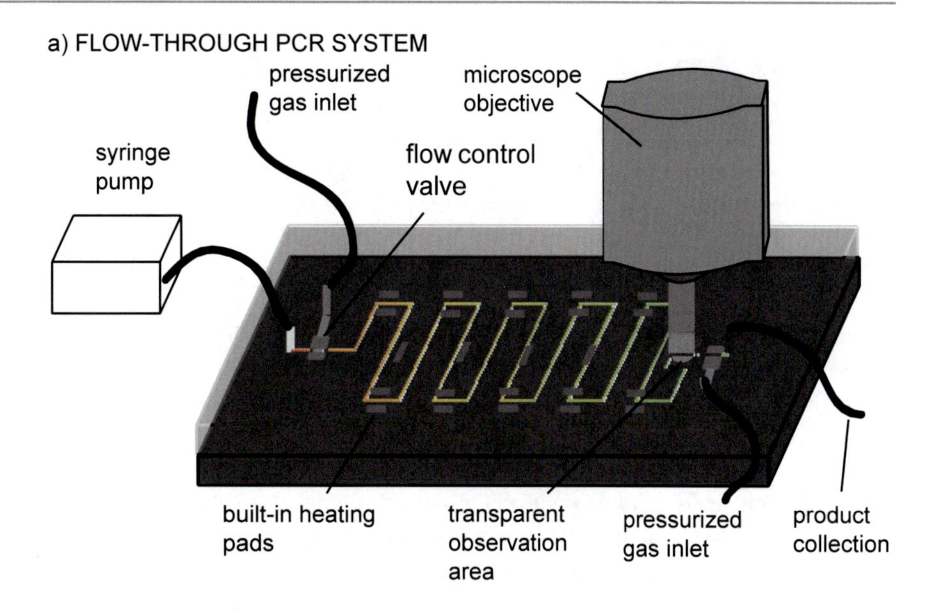

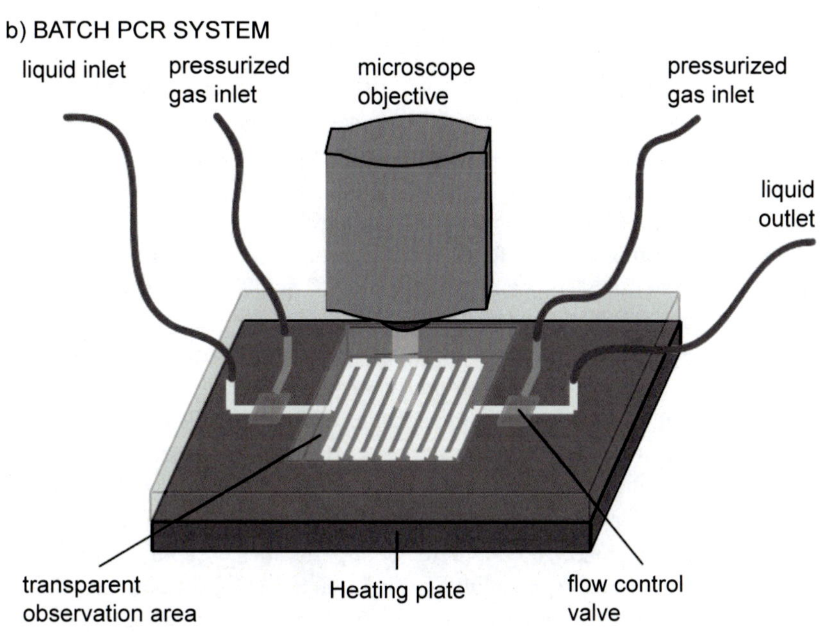

Fig. 12.8 Examples of different types of PCR systems. (**a**) In a flow-through system, a syringe pump/injector drives the fluid through the device. The fluid is heated in a cyclic fashion for the PCR to occur, and the result, such as an increase in fluorescence, is observed at the end. Packaging for flow-through systems has to allow for connecting the syringe pump and including inputs for valve control and a source of heat to run the PCR and a transparent area for microscopy. (**b**) Batch PCR systems do not require an input for a syringe pump, but all other components of the packaging scheme need to be present, such as a heating pad, flow control, and transparent packaging for imaging purposes

transparency [112]. The advantage of thin-film electrodes is the small thermal mass of the integrated chip and rapid heating. This comes at the cost of a more complex fabrication process and the risk of sample contamination due to electrode degradation.

External heating strategies have also been employed. For example, Peltier-based thermoelectric heating plates have been attached to PCR chips [113–115]. To achieve intimate thermal contact, a layer of material with high thermal conductivity, such as mineral oil [113, 116] or metal [117], is placed at the interface between the Peltier heater and PCR chip. Peltier heaters are a reliable and modular temperature cycling method. However, their large thermal mass makes it difficult for a single Peltier heater to achieve an adequate ramp rate; as a result, two or more heaters must sometimes be packaged above and below the chip [113, 116, 118]. Other external heaters, such as commercial thin-film resistors [104] and resistive heater coils [119, 120], have been utilized.

PCR heating is also achieved without direct physical contact between the chip and the heat source. This scheme makes the fabrication process simpler. Examples of non-contact heat sources are hot/cold air streams [121, 122], infrared (IR) radiation [123–125], and lasers [126]. When non-contact heating methods are employed, the package must provide good contact and thermal conductivity to the chip. In addition, thermal insulation is also important in IR heating, where heat should not be easily lost to the surroundings.

Flow Control In flow-through PCR systems, where the sample undergoes temperature cycles by flowing past a series of heating zones, a fluid pump must be included in the system. Syringe pumps [127] and peristaltic pumps [128] are examples of flow drivers. In such cases, the pump and PCR chip are separate modules within the entire package. In a few cases, the flow driver has been integrated into the chip itself. Miniaturized peristaltic pumps have been integrated [129]. Also, electrokinetic pumping driven by on-chip electrodes has been used [37]. In these flow delivery schemes, sealing of fluids and interfaces between the chip and the macro-world needs to be considered.

CE and Integration of Other Functions Capillary electrophoresis (CE) benefits from miniaturization through dramatically increased separation efficiency. One noteworthy innovation in CE packaging is a radial array CE system, which utilizes a laser-excited rotary confocal scanner with four color detection channels, enabling the simultaneous analysis of 96 samples [130]. Moreover, PCR and CE are commonly integrated onto a single chip [37, 43, 104, 108, 113, 116, 131, 132]. By packaging PCR and CE together on-chip, sample handling and risk of contamination are minimized, sensitivity increases, and large-scale parallelization becomes possible. In addition, fluids can be driven by electrokinetic means, thereby obviating the need for an external pump.

The integration of sample preparation and PCR–CE onto a single chip is a less developed endeavor. Sample preparation includes cell separation, isolation, washing, and lysis. Attempts at integrating sample preparation steps include on-chip cell capture and lysis using immunomagnetic beads [38], as well as cell separation using dielectrophoresis [100, 133], or DEP coupled with field flow fractionation (DEP-FFF) [134]. Despite such efforts, sample preparation largely remains an off-chip operation; therefore, major opportunities exist for higher-speed and lower-sample consumption if sample preparation is integrated into a single-chip package.

The integration of detection systems on-chip is in a similarly early stage of development. PCR products are typically detected optically using laser-induced or other fluorescence methods. Optical excitation sources and emission detectors (e.g., mercury lamps and CCD cameras, respectively) are sophisticated and bulky instruments usually located off-chip. Using these systems, an integrated PCR–CE chip must be optically transparent and have an unobstructed observation path. Efforts to integrate detection on-chip are limited. For instance, photodiodes have been coupled to PCR chambers via integrated optical fibers, with the excitation source remaining off-chip [135].

12.2.4.2 Microarrays

Microarray technology is considered one of the first realizations of truly high- throughput biological analysis [136]. A microarray is a grid-like arrangement of micrometer-scale spots on a planar substrate. Each spot is a surface-deposited cluster of molecules known as capture probes. The two most common types of microarrays are DNA and protein arrays. The capture probes on DNA microarrays are either single-stranded oligonucleotides or complementary DNA (cDNA), whereas the capture probes on protein microarrays could by definition be one of many different types of proteins, including enzymes, antibodies, peptides, or protein complexes. When a sample solution containing DNA, RNA, or proteins (a.k.a. the target) is incubated with the array, the target molecules bind to the capture probes of the spots on the array to varying degrees. Binding and capturing are facilitated (i) by hybridization between complementary DNA–DNA or DNA–RNA sequences or (ii) by protein–protein or protein–DNA interactions. Fluorescent labeling of the target probes enables the researcher to optically detect the presence and/or interactions of thousands of different target probes simultaneously on a single array.

The emergence of microarrays can be attributed to a serendipitous merger of precision robotics, microelectronics fabrication, biology, and chemistry over a decade ago. Today, a wide variety of microarrays is commercially available from several manufacturers (e.g., Affymetrix Inc. and Nanogen Inc.). Microarrays are used for both basic and applied purposes, including gene expression analysis, mutation analysis, protein function studies, drug development, diagnostics, and forensics. Recent reviews offer a detailed discussion of current and future applications [137–141].

General Packaging and Fabrication The typical packaging scheme for a commercial microarray consists of a glass or quartz substrate that is encased by a card-like plastic cartridge. The substrate acts as a rigid support upon which the capture probe array is patterned and subsequently read. To facilitate the attachment of capture probes, the surface of the substrate must be pre-treated or otherwise modified (not discussed here) [136]. Furthermore, the substrate in the optical path is required to be transparent and minimally autofluorescent so as not to interfere with the probes' fluorescence signals.

The capture probe array is patterned using one of many techniques, including an in situ photolithographic synthesis or an electric field-mediated attachment (for DNA arrays), as well as robotic spotting or microstamping (for protein arrays) [136, 142].

Cartridge The cartridge serves a number of purposes: (i) protecting the substrate from mechanical damage, (ii) forming an enclosed reaction chamber and flow path, and (iii) interfacing the substrate with external supporting equipment. An enclosed reaction chamber is beneficial because it lowers the usage of sample/wash buffers and prevents the evaporation to the external environment; the enclosure also permits straightforward connection to external fluid ports (which in turn link with pumps, sample, and waste) and eases handling by the operator. Rubber septa/gaskets are embedded at the inlets to the cartridge to interface with external fluid connections, forming a leak-free re-sealable fluid port. Also, the cartridge must have a viewing window for optical interrogation of the array in addition to a shape that facilitates easy loading and automated processing in a scanning system.

Supporting Infrastructure The microarray package is supported by a substantial collection of bench-scale equipment. This includes a fluidic module (i.e., pumps and reservoirs for samples, buffers, and waste), scanning system and housing, computer, and equipment for sample preparation (e.g., PCR, fluorescent labeling) [143]. The scanning system contains the excitation source, such as laser or mercury lamp, and the emission detector, such as photomultiplier tube (PMT) or charge-coupled device (CCD).

Integrated Packages The footprint, complexity, and sample usage of the total microarray system can be reduced by miniaturizing and packaging elements of the supporting infrastructure into the cartridge. In one instance, sample preparation, including PCR and target labeling, was integrated with the array on a single chip that is less than the size of a credit card [144]. The chip interfaced with 10 buffer connections, three Peltier heater–coolers, and over 100 gas pressure lines to actuate on-chip valves. Functionality of the chip was demonstrated by detecting mutations from a low copy sample of RNA. Another device was demonstrated to have even greater upstream integration by including cell capture and lysis, as well as PCR and DNA hybridization, on a single chip [38]. In this setup, no external pressure sources were required, as fluid pumping was provided by thermoelectrically actuated on-chip valves. In addition, the hybridization was detected using an integrated electrochemical sensor. Although the chip's resolution and sensitivity are less favorable in comparison with that of optical arrays, the simplicity to detect the presence of bacteria from a sample of whole blood was an advantage.

The integration of new functions can also make dramatic improvements in performance. A notable example is the desire to speed up hybridization. Under normal circumstances, hybridization is driven solely by molecular diffusion of targets to the capture probes, which could take many hours. In one device, the target solution was oscillated by an on-chip pump, thereby introducing convection to the fluid and increasing the hybridization rate many times over [145]. Cavitation microstreaming, which mixes fluids by vibration, has also been integrated on-chip and shows a fivefold increase in hybridization [38, 146]. Because DNA molecules are charged, it is also possible to use electrical field to increase the mass transfer of the target molecules to the surface. The packaging of any of these flow devices would require somewhat more complex schemes, where a pump, a piezo-actuator, or electrical leads need to be accommodated.

Integration Outlook Naturally, higher integration will lead to more interconnect-related challenges in the package, particularly for electrical and fluidic interconnects. In addition, the increasingly commercial nature of the microarray field will demand fabrication processes and raw materials of lower cost. The use of polymer substrates that are amenable to high-volume manufacturing (i.e., embossing, injection-molding) may bring the end results closer to both goals [147–149]. The integration of new functions and materials, while introducing short-term challenges, will likely lead to higher performance and more economical microarray packages [150].

12.2.4.3 Microfluidic Large-Scale Integration

Microfluidic large-scale integration (MLSI) is a term describing microchips that possess a dense arrangement of fully integrated valves, channels, and chambers [151]. Invented by researchers in Stephen Quake's group, the enabling technology for MLSI is a miniaturized, elastomeric valve that is fabricated monolithically into the microfluidic network of the device using soft lithography [152] and thermal bonding [153]. Such valves are used to perform fluid handling operations, such as pumping and mixing, and to isolate chambers (or sections) of a device from fluid flow.

Out of all bioMEMS platforms, the layout and packaging of MLSI devices are most analogous to that of microelectronic integrated circuits (ICs). In MLSI, multiple layers of channels, chambers, and valves are stacked upon each other in a grid-like configuration that is reminiscent of the multi-layer architecture comprised of metal lines, transistors, and gate electrodes found in ICs. Moreover, fluidic interconnections between MLSI inlets (or outlets) and external tubes are accomplished by rows of densely packed, hollow metal pins positioned in a manner that resemble the bond pads and vias at the edge of ICs (Fig. 12.9).

The applications of MLSI are diverse. The detection of mRNA from single cells [154], synthesis of precious reagents [155], and realization of a non-fouling bacterial chemostat [156] are a few questions that have been addressed by exploiting the highly integrated nature of MLSI. In addition, commercial MLSI systems are currently used as tools for protein crystallization screening and genetic analysis [157–160].

Packaging and Infrastructure Although MLSI systems are employed in diverse applications, their design layout and packaging are similar. All MLSI devices are fabricated out

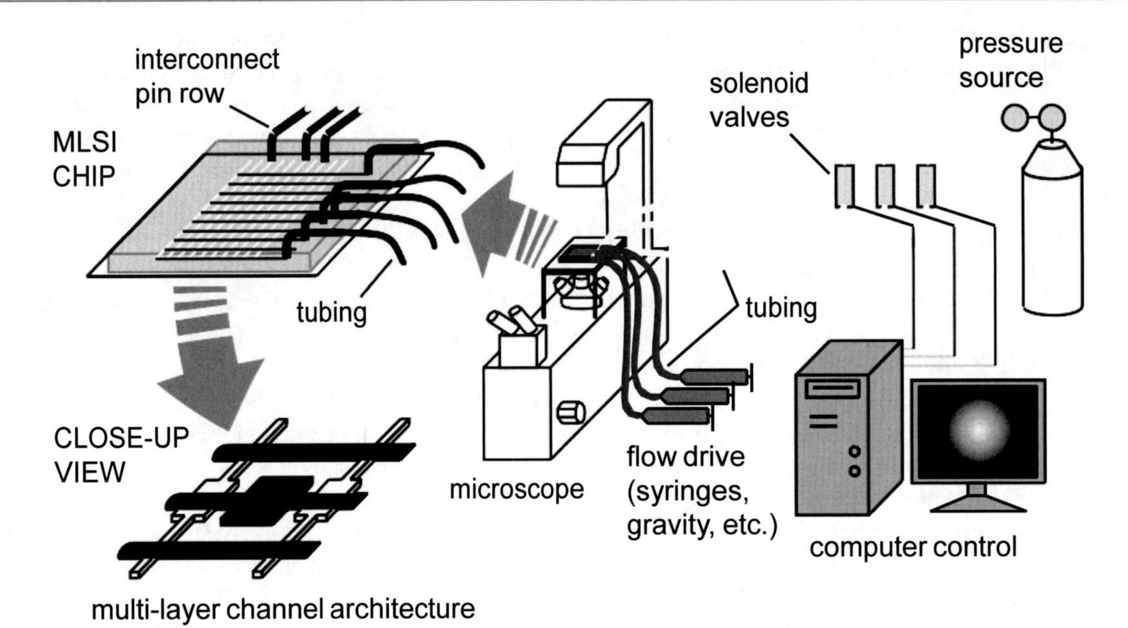

Fig. 12.9 An illustration of a MLSI system. The MLSI chip is mounted to a microscope stage for optical interrogation. The chip is comprised of an elastomeric block, patterned with a microfluidic network, which is bonded to a glass substrate. Rows of hollow L-shaped pins connect the on-chip network to external tubing. A magnified view shows a portion of the multilayer architecture that enables the construction of monolithic valves and dense arrays. MLSI systems typically require extensive off-chip infrastructure, including solenoid valves, a pressure source, and computer control

of the elastomeric polydimethylsiloxane (PDMS) or a fluorinated elastomer that behaves similarly to PDMS [161], which is both amenable to replica molding and exhibits plastic deformability that is essential for valve operation. The features in MLSI devices are typically configured in an array of modules that share common bussing channels, pumps, or specific unit operations. Also, MLSI systems are almost universally packaged to interface with external optical detection instruments. Therefore, glass slides are used as the device's support substrate, and inlets/outlets (with accompanying interconnect pins) are located near the edge of the chip to allow for unobstructed viewing.

The off-chip infrastructure consists of a computer-controlled battery of solenoid valves that stand between the chip and a pressure source (i.e., compressed gas tank). Selective actuation of solenoid valves actuates corresponding on-chip valves via pressurization or depressurization. Furthermore, reagents are commonly stored off- chip in vials or syringes.

Outlook The dense packing and computer control of fluidic components have led to unprecedented parallelization and automation, respectively. However, the extensive off-chip infrastructure and the lack of standardized fabrication and interconnect methods are aspects of ongoing research, which will potentially enable wide applications of MLSI. Strategies to minimize the number of off-chip solenoid valves for a given set of on-chip operations, as well as general design rules, are being advocated [146, 162]. The development of alternative valve actuation techniques could ease implementation by eliminating the need for solenoid valves and bulky compressed gas tanks. For example, bioMEMS chips have been mounted to Braille displays, whose piezoelectric pins slide up and down to actuate on-chip valves [163–165]. Similar to pneumatic monolithic valves, when Braille pins slide upward, they deform the PDMS membrane and seal the flow channel directly above them. The tandem package of a chip and Braille display could substantially improve MLSI device portability; however, it may restrict the optical access of the chip. In addition, standardized interconnect strategies are likely being pursued by commercial manufacturers on a proprietary basis [166, 167]. This could lead to the development and acceptance of industry-wide standards in the near future.

12.2.4.4 Cell Culture and Assay

One of the most promising applications of bioMEMS and microfluidics is the culture and assay of living cells. Microtechnologies possess unique capabilities that can dramatically increase the resolution and content of data from cellular experiments. In addition, experimental parallelization and throughput can be greatly improved [168–170].

The maintenance of living entities brings with it additional requirements, which in turn affect the packaging of devices.

One requirement is the control of temperature at physiological (or experimentally perturbed) levels. Also, the transport of nutrients to the cells must be provided on a continuous or semi-continuous basis to match the cells' metabolic rate. Added to that must be the means to monitor the levels of critical nutrients or other culture conditions, such as pH and dissolved gas concentration. Moreover, biocompatibility of the device interior must be assured through appropriate selection of base material, polymer coatings, or adsorption/functionalization with specific biomolecules.

Integration of Functions One way to accommodate for the additional requirements of cell-based systems is to integrate more functions on-chip. Specifically, many requirements are met by integrating electrical components into the device. Temperature control has been demonstrated by the use of microfabricated indium tin oxide (ITO) electrodes [171, 172]. In addition to being conductive, ITO has the added benefit of being transparent, making it an ideal choice when optical interrogation is required. Also, the on-chip electrochemical generation [173, 174] and measurement [65, 173, 175] of dissolved gas have been demonstrated. Similarly, integrated measurement of pH has been achieved [176]. Moreover, the manipulation and positioning of cells into specific locations within a device has been investigated using dielectrophoresis [177–179]. As a consequence of integrating electrical functions on-chip, the packaging of bioMEMS is complicated by the attachment (or bonding) of electrical leads to bond pads at the edge of the chip. External power supplies and driver electronics are also required.

Integration of other functions can be achieved through MLSI or arraying approaches. The culture of stem cells has been demonstrated in an array of cell chambers, each of which was semi-continuously perfused with the necessary growth medium [180]. Fluid handling within the device is performed by integrated valves and pumps, and the corresponding fluid interconnect is achieved via rows of densely packed hollow pins. The pins are in turn connected to a computer-controlled row of solenoid valves that lead to a compressed gas pressure source. Other cell-based bioMEMS with innovative linear [181, 182] and radial [183] arrays avoid reliance upon integrated pumps, which reduces the on-chip complexity. However, this necessitates the use of additional off-chip equipment, such as syringe pumps or rotary motors. These trade-offs need to be evaluated according to the specific needs of the applications.

Optical Interface One off-chip function that has a great impact upon cell-based device packaging is optical detection. Many assays require very high magnification microscope systems. For instance, the forces exerted by cells as they crawl along a substrate have been measured by observing minute changes in the location of dots embedded in an elastomeric substrate [184] and bending of elastomeric pillars [185]. To manipulate and position cells optically, high-power laser beams are necessary using optical tweezing techniques [186, 187]. BioMEMS for such applications must be fabricated of optically transparent materials, and just as importantly, their thickness must accommodate the short working distance of the high-magnification objectives (Fig. 12.10). Occasionally when there is thermal stress (induced by laser power adsorption by the medium or the cell), one must also provide effective means to dissipate the heat so as to reduce cellular damage.

12.3 BioMEMS Chip Interfacing

The previous section covers the specific strategies of packaging in many applications of microtechnologies in biological and medical research and practice. This section reviews two important interfaces in biochip designs: the interface between various components of a microsystem and the chip-to-world interface that is critical to operations and functions of each chip.

12.3.1 Interfacing On-Chip Components in BioMEMS

The integration of various types of components on a bioMEMS chip can prove challenging if the fabrication methods are incompatible. For example, bioMEMS chips often contain immobilized biomolecules, such as DNA or proteins. The high temperatures required for some of the traditional MEMS fabrication steps easily denature these molecules. For instance, although room temperature deposition has been achieved through ammonia catalysis [188], chemical vapor deposition of SiO2 and SiN commonly occurs at temperatures in excess of 250 °C [189]. Therefore, intermediate protective packaging steps need to be introduced to allow the integration and interfacing of seemingly incompatible components, such as the deposition of biomolecules and CMOS technology.

The order of manufacturing steps becomes part of the packaging scheme if protective packaging or modification of fabrication technology is unavailable or infeasible. The order is especially important in manufacturing of hybrid MEMS chips, such as CMOS/microfluidic chips. CMOS and microfluidic components will generally have to be integrated following separated fabrication steps. Integration of biological functionality such as adhesion of DNA, proteins, or enzymes will have to occur last and in already covered devices.

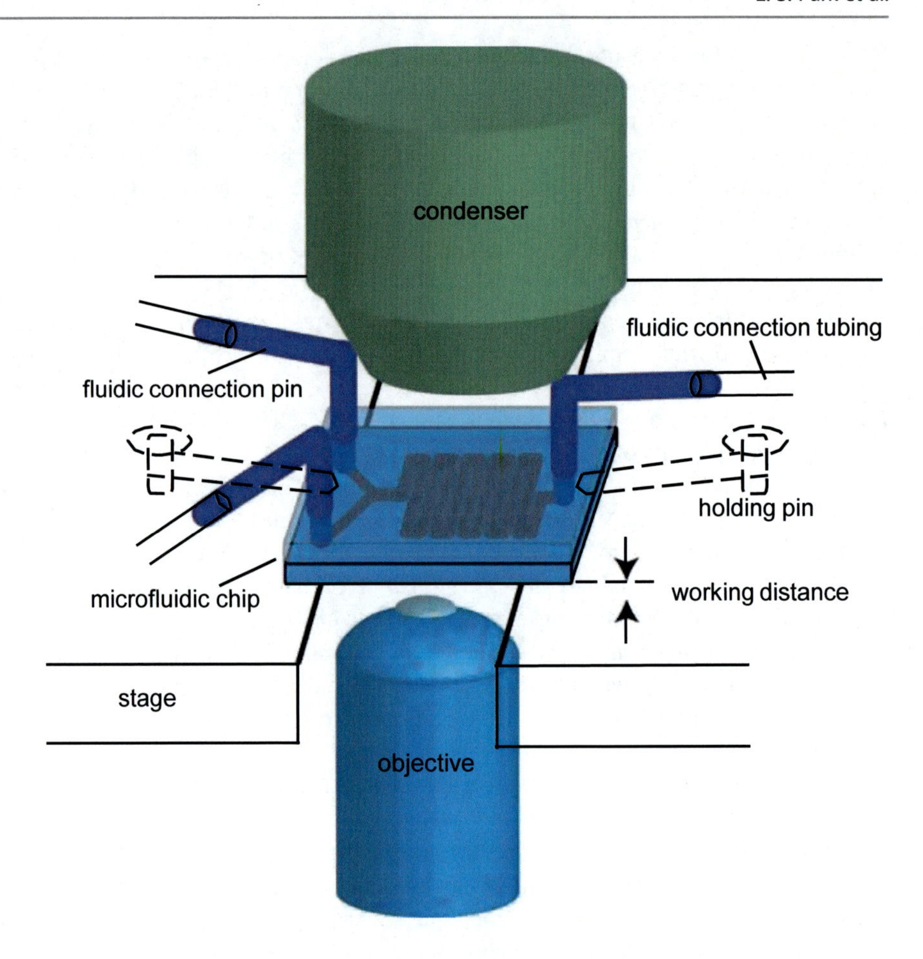

Fig. 12.10 Optical access to the chip, especially at high magnifications, strongly depends on the chip packing. In transmission mode, both the condenser and the objective need to have access to the chip without hindrance of the fluidic, electrical, and mechanical connectors. The working distance of the lenses is a critical parameter to consider. In fluorescent mode, the objective still needs access to the chip; since no condenser is necessary, the fluidic and electrical connections as well as the mechanical clamping mechanisms can be placed on the open side of the device

Lastly, interfacing of different components can be difficult when the materials used for their fabrication are incompatible. For example, common bioMEMS materials, such as PDMS, can be difficult to interface with CMOS technologies. PDMS is a poor substrate for the adhesion of deposited metals due to the low surface energy of PDMS [190, 191]. If PDMS packaging is to be used with CMOS technology, better adhesion between the metallic and polymeric layers must be achieved. Furthermore, due to the elevated temperatures during metal deposition, the PDMS surface can buckle upon cooling and cause uneven surfaces and cracks in the metallic layer. Therefore, sufficient cooling of the PDMS layer during metal deposition is required to minimize this effect [192].

12.3.1.1 Protecting Biomolecules with Intermediate Packaging Steps and Modification of Manufacturing Methods

Because some fabrication steps can be detrimental to proteins and DNAs, these biomolecules must be protected to allow for their introduction before the last steps in a fabrication process. Trau et al. [193] address this type of complications by passivating the DNA by a protective gold layer. The protective gold layer was deposited by chemical vapor deposition (CVD), while the wafer temperature was not allowed to exceed 45 °C, a temperature that is deemed benign to DNA molecules. Parts of the gold layer were etched away to expose desired areas on the wafer for traditional wet etching, metal deposition, and other common techniques. By applying this method, they were able to integrate the deposition process of DNA oligonucleotides and proteins during routine microfabrication. The biomolecules maintained functionality through both wet and dry etching processes [193]. After the microfabrication processes, the molecules were deprotected by a solution of potassium cyanide (KCN) in phosphate buffered saline solution (PBS) [193]. This may have been a successful example of protecting biomolecules; in general, however, it is nontrivial to maintain the bioactivities of molecules such as proteins under microfabrication conditions. Rigorous tests need to be performed to assess the bioactivities of protected and deprotected molecules before a method can be adopted. If the biomolecules can be applied after the completion of the microfabrication, it is perhaps the least complicated method for a reliable deposition/introduction of the biomolecules.

Construction of the final protective enclosure for a device can also cause damage to biomolecules, the choice of which also needs careful considerations. For instance, bonding of a protective casing or membrane to the device may require exposure to high temperatures over a long period of time.

Therefore the choice of materials is critical. Novel bonding methods performed at lower temperatures may be better alternatives. One example of such low temperature bonding process is through the use of a UV-curable adhesive as illustrated by Kentch et al. [194]. The adhesive is applied in ultra-thin-film form on the surface, followed by an alignment of surfaces and curing by UV light. Another formulation of the adhesive, Vitralit, allows for bonding of different combinations of materials, such as SU-8 with glass or cyclic olefin copolymer with poly(methyl methacrylate) (PMMA) [194]. Many other light-curable epoxies, two-part epoxies, and medical-grade epoxies (e.g., surgical glues) also exist for various applications.

Certain obstacles must be overcome before this type of packaging method can be universally applied. For applications where thicker films are required, care must be taken not to cause warping of substrates and thus incomplete adhesion in the longer term. In addition, in some coating applications, due to evaporation of solvent from the adhesive, the solvent may condense inside the structures of a device, such as the channels of a microfluidic device [194]. Therefore, great care must be taken in either designing a device system with high tolerance or selecting adhesives and solvents used. Furthermore, the strength of the material interface can be weakened after prolonged exposure to oxygen or corrosive media and buffer such as PBS [194]. This issue may be overcome by choosing an adhesive appropriate for each particular application.

12.3.1.2 Order of Manufacturing Steps

Hybrid bioMEMS devices more often require precise sequences of fabrication steps for different types of components. For example, CMOS components may need integration of microfluidic channels that require bulk and surface micromachining steps, which may not be directly and fully compatible with the CMOS processes [195]. Similarly, biomolecules may have to be deposited after the last micromachining step if no protective process is available during the deposition process [193, 196]. This type of requirements calls for an optimized order of fabrication steps and also new technologies that allow for monolithic, sequential (i.e., one component at a time), or a hybrid of two approaches.

Microfluidic channels can be formed on top of a CMOS chip through a direct- write fabrication process [195]. Channel features are patterned by the deposition of an organic ink – a sacrificial material. Following the ink deposition, uncured epoxy is deposited over the channels and then cured. Then the organic ink is removed to form the microfluidic network. The downside of this technique is that the channel dimensions are constrained by the dimensions of the ink deposition nozzle. Commercially available nozzles come with a diameter of 100 μm. While smaller nozzles and more precise manipulation stages are being developed, the designs have to contend with high nozzle fragility and surface tension due to viscosity [195]. Additionally, this process is not easily scaled up for batch processing. Ongoing work may explore other techniques (e.g., using photolithography on photopatternable materials as sacrificial materials) to overcome this drawback.

When interfacing components of different types, such as CMOS and microfluidic systems, interfaces and the different size scales must be considered [197]. CMOS chip sizes tend to be in the millimeter range, while microfluidic chips are generally larger, especially if various functionalities such as pumping and mixing are included on chip. Furthermore, inputs and outputs for fluidic systems require standard (macroscale)-sized pins and tubing, even though on device channels are much smaller. These interfaces with the macroworld usually take up large portions of the chip area.

The connection between the CMOS and microfluidic world may be achieved through effective layout design [197]. The CMOS chip can be selectively exposed to some functional areas on the microfluidic chip, such as a mixing or dilution chamber, or electrical or optical interconnects. All other operations on the fluid occur prior to entry in this area, and pins and tubing for inlets and outlets can remain the standard size. If a barrier between the two components is required for fluidic isolation or electrical insulation, it can be as simple as a layer of Pyrex® glass or a membrane made from one of structural materials, such as PDMS [197].

In some applications, it is also necessary to introduce electrical components to the outside of a chip. Some materials pose adhesion issues. For example, PDMS is commonly used in microfluidics and is the subject of numerous studies, but metal layers do not easily adhere using current processes [191]. This problem can be partially resolved by incorporating carbon black or other charged particles in the polymer matrix. In addition, PDMS-based composites such as Ag-PDMS, which includes silver microparticles, and C-PDMS, which includes carbon black nanoparticles, may be used to increase conductivity of the matrix such that electrical contacts can be made on the matrix directly [191].

In dealing with biomolecules, such as enzymes, the molecules are either protected [193] or deposited after microfabrication [196], as stated above. Novel methods are yet to be developed to achieve biomolecular functionalization for wafer-level fabrication and high-throughput processes. Zimmermann et al. [196] have achieved this by introducing enzymes in a polymer solution into target channels as the last step of the microfabrication process. The polymer was then crosslinked through UV exposure, and the uncrosslinked polymer was washed out. This procedure was performed after the final microfabrication step, after the wafer-level bonding was already completed. Similarly in Herr et al. [42], forming gels inside protein analysis chips were also done after the device fabrication was completed. Careful choices of sequences of UV crosslinking, washing, and fluid pumping allowed for a complete integration of sophisticated multistep protein analysis on-chip.

12.3.2 Interfacing BioMEMS with External Systems

The external interface of a bioMEMS or microfluidic device is where the "chip" ends and the "rest of the world" begins. In many applications, the "chip" is an assembly of two planar substrates, at least one of which containing microfabricated or replica-molded features, that are bonded together to seal a microfluidic flow network. One of the substrates is usually rigid in order to provide mechanical support. All off-chip equipment is characterized as "external" to the chip. The interface between the bioMEMS chip and external systems is also called the "world-to-chip" connections or "interconnects," which is the equivalent term from the microelectronics field.

Due to similarities in form and fabrication, biochips face similar interconnect issues as traditional ICs and MEMS. For instance, the reliability of bioMEMS is heavily dependent on proper interconnection to external systems; most failure mechanisms are associated with interconnect failure. Also, as chip size continues to decrease and feature density increases, space for interconnect junctions becomes more constrained. This motivates the development of techniques and platforms to smoothly transition from macro- to microscale. The microelectronics industry has well-developed approaches for scaling, such as flip-chip packaging and fully automated assembly. On the other hand, bioMEMS interconnect and scaling approaches are relatively unsophisticated. Possible contributing factors to the lack of sophistication include (i) the bioMEMS field being less well developed as compared to the IC industry, (ii) lack of interest (and funding) in packaging, as the majority of current work is in academic research, (iii) diversity in the applications and hence diversity in the needs of packaging, and (iv) traditional reliance on manual techniques that are prone to variability. As the field grows, more research has been devoted to biochip interconnect, as shown in recent reviews [41, 194, 198].

This section covers five areas: (i) fluidic, (ii) electrical, (iii) optical, (iv) thermal, and (v) mechanical. The following subsections will discuss the basic issues of each regime.

12.3.2.1 Fluid Interconnect

Fluid interconnect relates to the means by which fluids enter and exit from a biochip. In most cases, the fluids also carry biological entities ranging from proteins (often many kinds of proteins), to DNA and RNA, to cells and even organisms. It is perhaps the greatest differentiating issue between bioMEMS interconnect strategies and those of IC/MEMS. Proper fluid interconnect is critical to averting numerous failure modes, including leakage, infiltration (or nucleation) of gas bubbles, and introduction of contaminants. Achieving hermetic seal and having little or no dead volume are the ultimate goals of fluid interconnect. In addition, robust and rapid assembly of interconnects (i.e., alignment, attachment, and bonding) is an active field of research [199–202]. Fluid interconnect is most commonly achieved by the use of pins and ports; in practice, both custom and commercial techniques are employed.

Interconnect Pins Interconnect pins are hollow metal tubes, usually bent 90° into an "L" shape. One end of the pin is inserted vertically into an inlet port on the chip, while the other end points horizontally toward the exterior of the chip. Plastic tubing is attached to the horizontal end of the pin by friction fit. From the pin, the tubing extends to off-chip components, such as reagent sources, pumps, or waste reservoirs. The vertical end of the pin also fits by friction into the chip's inlet port. Consequently, the substrate into which the pin is inserted must be elastomeric or otherwise plastically deformable. The inlet port on the chip is slightly higher in gauge (smaller in diameter) to provide for a tight, leak-free fit when the pin is inserted.

Pins are a popular means for interconnect in biochips fabricated in PDMS, given its favorable elastomeric properties. The most striking demonstration of this is with MLSI devices. Rows of over 50 pins, spaced by approximately one millimeter, have been packed onto the edge of MLSI chips less than 3 in on a side [180]. At such densities, space becomes constrained, and pin placement becomes a critical factor in chip design. Another design constraint is substrate thickness. The need for a stable and secure fit between an inlet and pin requires that the thickness of the elastomer substrate be roughly a few millimeters. Therefore, thinner device designs may preclude pins.

Progress in pin technology will be hindered without the evolution of standardized pin sizes and placement templates. Also, current methods for pin insertion and hole punching (for inlet and outlet ports) are painstakingly labor intensive, causing high variation and low throughput. The development of automated assembly methods may improve reliability and allow for more ambitious chip designs. Such automation might already exist within companies that manufacture PDMS-based biochips.

Connector Ports Another fluid interconnect approach is to use connector ports. A connector port usually comes in the form of a small cylindrical assembly, where one end is bonded to the inlet of the chip and the other end acts as a fitting for external tubing. The bond between the substrate and connector port is achieved using epoxy or similar adhesive.

Connector ports are well suited for attachment to hard substrates, such as glass and silicon. Furthermore, the strong bond between the substrate and port makes it possible for the interconnection point to withstand pressures as high as 100 bars. In contrast, interconnect pins are designed for soft, elastomeric substrates, and the nature of their attachment to the substrate (i.e., friction fit) limits the operating pressure to less than 10 bars.

Connector ports are a flexible interconnect platform due in part to the variety of materials (polymers, metals, etc.) and commercial manufacturers from which they are available (e.g., Upchurch Scientific). However, the assembly process is complex. Of the major procedures, (i) holes must first be drilled (or etched) through a hard substrate, (ii) adhesive must be dispensed, (iii) the port must be carefully aligned and attached, and (iv) time must be allotted for the adhesive to cure.

Commercial and "Plug-and-Play" Interconnect The growth of commercial bioMEMS is giving rise to a new set of approaches for fluid interconnects. Commercial bioMEMS are typically packaged as a cartridge that is inserted into a bench-scale scanner box. Examples of such configurations are the Agilent Lab-on-a-Chip, Affymetrix GeneChip®, and Fluidigm Topaz® system. In each of these systems, rubber gaskets and septa around the chip's inlet holes are used to seal against a chuck (or frame) when the cartridge is pressed into the scanner box. This reduces or eliminates manual attachment of fluid interconnect leads, resembling a so-called plug-and-play approach.

Others are developing the plug-and-play concept as a means to build custom microfluidic systems [203–205]. Custom plug-and-play kits consist of a set of bioMEMS building blocks, each encasing an individual feature or unit operation, which can then be assembled to perform a desired set of functions. The plug-and-play approach will inevitably demand higher standardization, which must exist either for how the building blocks interface with each other or for how they interface with a generalized assembly template. One idea of an assembly template is the microfluidic "breadboard" [206]. The breadboard contains a pattern of bussing channels, which fluidically link different building blocks together based on where each is placed on the board (Fig. 12.11).

Although plug-and-play and breadboarding have been proven as concepts, the standardization of fluid interconnect is still in its infancy. Fundamental challenges to standardization persist, such as (i) the need to uniquely tailor surface chemistries for each application and (ii) the question of how to combine electrical and optical components into assembly templates (i.e., multifunctional "breadboards"). By addressing such challenges, the bioMEMS community will move closer to a level of accessibility and scalability comparable to that which is enjoyed by the microelectronics field today.

12.3.2.2 Electrical Interconnect

Alongside fluid interconnect, electrical interconnect is the most common assembly issue of bioMEMS. Fortunately, their form and fabrication makes them well suited to the same approaches used in IC/MEMS interconnect. Given the well-developed state of the microelectronics interconnect field, a thorough discourse will not be provided here, but instead only a few fundamental issues and noteworthy examples will be highlighted. For further information, the reader is encouraged to examine more general reviews [207–209].

In bioMEMS, electrical components are used for a wide variety of purposes, such as transductive readout (electrical or mechanical), resistive heating, electrochemical reactions (providing potential or current), and electrokinetically driven fluid flow. Some applications need simple electrodes, and metal wires can serve the purpose. For others, wire bonding to the chip to interface with the on-chip electrodes is used. Similar to ICs and MEMS, the electrodes for bioMEMS "fan out" and expand into bond pads at the edge of the chip. Electrical interconnect is an essential design factor in bioMEMS with a high density of electrical components, such as (i) the stimulation (and measurement) of action potentials in living neural circuits [210], (ii) the manipulation and sorting of cells by dielectrophoresis [100, 177], and the movement of droplets via electrowetting techniques in digital microfluidics [211]. Each application requires a complex array of electrodes that is coupled to a fluid channel or chamber.

Microelectronic assembly techniques and configurations have been applied to bioMEMS on a limited basis. For instance, flip-chip techniques have been utilized to unite microfluidic to electronic substrates [120, 212, 213]. However, the sensitivity of many bioMEMS to heat might limit the annealing temperature for solder bumps or other contact points. Additionally, some bioMEMS are directly attached to printed circuit boards (PCBs) [214]. The disadvantage of the PCB approach is that by mounting a PCB to the back surface of the chip, it precludes the use of transmissive optical interrogation.

Another noteworthy issue is corrosion. Although interconnect corrosion is a well-known concern for IC/MEMS, the problem is more pronounced in bioMEMS because of the "wetware." Buffers and reagents used in bioMEMS often contain salts. These fluids are prone to contaminate electrical leads and bond pads while preparing and running experiments. Furthermore, salt solutions commonly flow directly over electrodes within the device's fluid network.

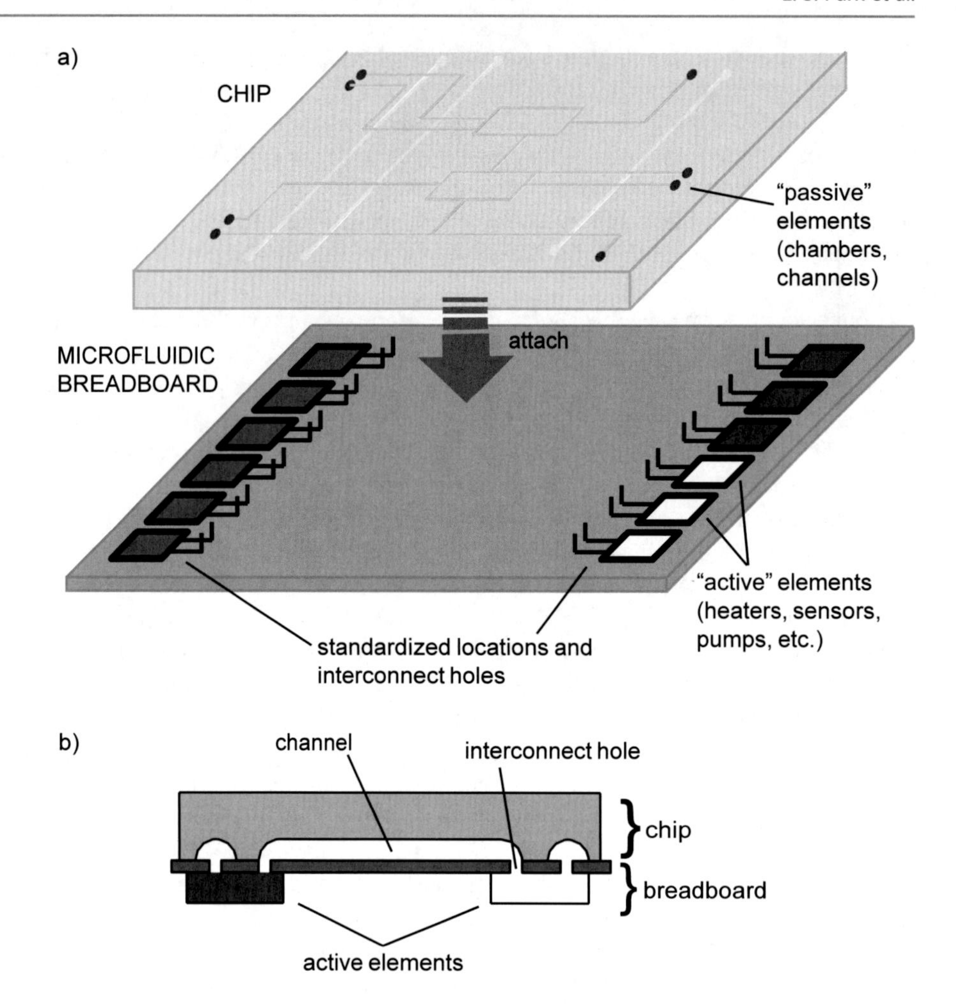

Fig. 12.11 Scheme of a microfluidic breadboard. (**a**) A microfluidic chip is attached to the breadboard, which possesses an array of active functional elements that are fabricated by standard methods (i.e., CMOS/IC processing) and placed in standardized locations [206]. The on-chip microfluidic network is customized to link with specific elements on the breadboard depending on application. (**b**) Side-view of a channel linking two active elements on a breadboard

Approaches to passivate electrical components are similar to those of IC/MEMS, although corrosion prevention will also depend heavily on better fluid interconnect technologies.

12.3.2.3 Optical Interconnect

Optical interconnect is typically achieved through non-contact methods, such as excitation by an off-chip laser or observation through a microscope objective. In such cases, the main concerns are (i) the transparency of the chip's substrate to the wavelengths of interest, (ii) providing an unobstructed path (or window) for excitation and emission (for fluorescence measurements), (iii) distortion of light as it travels through the chip, and (iv) autofluorescence of the substrate material.

The third and fourth concerns are particularly relevant to PDMS, one of the most popular base materials for biochips. Optical distortion may occur in zones of non- uniform crosslink density. This is in part caused by inadequate mixing of the PDMS prepolymer and crosslinking agent before it is cured. Distortion may also be caused by the insertion of fluidic interconnect pins, which generate non-uniform stress by compression in the polymer. To avoid such distortion, the observation zone must be located a safe distance from the stress fields. Moreover, some polymers (including PDMS and PMMA) are known to be autofluorescent [215], which may disqualify them from fluorescence applications that demand very little background signal.

Contact methods for optical interconnect also exist. For example, on-chip waveguides are often coupled to optical fibers that guide light to off-chip detectors [216]. Also, the concept of an optical backplane, analogous to electrical backplanes (or breadboards), has been investigated [217]. Care must be taken to align and secure the optical fibers to the on-chip components.

12.3.2.4 Thermal Interconnect

Thermal interconnect is an important issue for many bioMEMS applications, particularly PCR and cell culture. Not only could heat be generated from the application of electrical field (or magnetic field), which needs to be dissipated promptly, but maintaining temperature for biological entities (such as proteins and cells) is critical. Similar to optical interconnect, thermal interconnect can often be achieved by non-contact means. Non-contact heating mechanisms for PCR include infrared radiation [123–125] and convective heating by fans [121, 122]. For cell culture, convective heating via fans or environmental chambers is the common non-contact method.

Heating by contact can be achieved by using Peltier heaters (and coolers) and, if the device is mounted to a microscope, stage and objective heaters. To facilitate intimate thermal contact between a Peltier surface and a biochip, intermediate layers of mineral oil [113, 116] and thin metal [117] have been used.

An important factor that determines the success of the thermal interconnection is the thermal mass of the package. As the chip-heater package becomes smaller, heat can be added and removed more quickly. For applications that require high temporal resolution in heating cycles, packages with minimal thermal mass are the most viable options. For instance, rapid cycling times increase PCR throughput; hence, low-mass IR techniques have been developed [123, 125]. However, when the temperature must be sustained for long periods (i.e., for cell culture), thermal mass is not as important, and perhaps a larger thermal mass is preferred to dampen potential temperature fluctuation.

Other factors to consider for thermal interconnect are (i) material stability (dimensional and chemical) at elevated temperatures and (ii) differences in the coefficient of thermal expansion (CTE) between substrates in hybrid bioMEMS.

Heating via integrated electrodes deserves a brief mention. A key advantage of integrated electrodes is their ability to heat areas with high spatial resolution. This is particularly important for flow-through PCR, where hot and cold zones must be defined in close proximity [37]. Standard heaters cannot provide such precise local heating. An additional advantage is low thermal mass. The thin-film electrodes add no additional mass to the package, which is advantageous for high temporal resolution applications as well.

Although heating and maintaining temperature above room temperature is the main discussion here, integrated cooling systems and Peltier off-chip coolers are some technologies that are used to lower the temperature on-chip for some biological applications [218, 219]. The design considerations are similar to the discussion above on heating schemes.

12.3.2.5 Mechanical Interconnect

Mechanical interconnect deals with the manner in which a bioMEMS device relates to solid objects around it. First, bioMEMS must be protected from physical breakage. This is especially important for portable bioMEMS, which cannot be handled as carefully in the field as they are in the laboratory. Mechanical protection is provided either by packaging the chip within a protective enclosure or by fabricating it out of mechanically tough materials. Polymers are the materials of choice due to their resilience, as well as their low cost.

Second, bioMEMS must have the appropriate shape and form to fit into external systems. For instance, bioMEMS are often designed to be attached to standard-sized microscope slides (or cover slips). Therefore, they can be easily mounted to most microscope systems. The same consideration follows for other standardized chucks or frames found on other external systems.

12.4 Biocompatibility of BioMEMS

Besides the types of interconnects for biochips, an additional consideration is the biocompatibility of the chip and the packaging materials. The criteria for biocompatibility largely depend on the context and the applications the chips are intended for. The major concern for bioanalytical chips is to avoid contamination, so reducing foreign molecules that can get into the analytes or react with analytes is important. On the other hand, for cell culture chips and implant devices, avoiding non-physiological response from the cells or the body is important. To render devices biocompatible, however, one defines biocompatibility, the choices of materials and surface modifications are critical. This section aims to review these two important aspects of chip design and packaging schemes for a series of common materials and processes in bioMEMS.

12.4.1 Biocompatibility of Fabrication and Packaging Materials

The biocompatibility of bulk materials used in bioMEMS and their packaging is essential to clinical and commercial success. One such class of devices is implants and cell culture systems. For implants, the primary concerns are whether the material is toxic or elicits an immune response. In cell culture systems, toxicity is one concern, but additionally it is important to consider whether the material causes an unexpected physiological response that confounds the analysis of experimental data. The lack of biocompatibility with the bulk material could be driven by its inherent chemistry or by the presence of impurities that may leach out of the bulk (usually residues from the manufacturing process). A noteworthy point is that these issues could become exacerbated in bioMEMS cell culture systems because of a substantial increase in the surface-to-volume ratio (SVR) as size is reduced from the macro- to the microscale.

Much of what is known about material biocompatibility has been learned in the medical implant and drug delivery fields. Therefore, a large body of knowledge already exists for certain well-utilized metals, ceramics, glasses, and polymers, which could be applied here. Included below is a brief summary of the biocompatibility of some of the most popular bioMEMS materials.

Glass and Silicon Glass and silicon are the archetypical bulk materials for bioMEMS due to their compatibility with IC/MEMS-based microfabrication and micromachining processes. Although the biocompatibility of glass is well established, that of silicon is still being investigated. In a short-term (<21 day) study, a comparison of the inflammatory and wound healing responses to silicon implanted in rodents showed no difference between implanted and non-implanted cohorts [220]. Silicon has proven itself as a base material for selected bioMEMS applications, including cell culture with multiparametric electrical monitoring [221] and neural prosthetic implantation [222]. In the future, the long-term biocompatibility of silicon may be proven by clinical studies of implanted drug delivery devices [62, 223] and transdermal silicon microneedles [224].

PDMS and SU-8 PDMS is widely used in medical applications [225, 226], and it is emerging as one of the materials of choice for cell culture in bioMEMS. Cells cultured on various compositions/formulations of PDMS have shown growth rates similar to that of polystyrene culture dishes (with some dependency on cell types) [227]. However, the high SVR of bioMEMS culture chambers, combined with the possible leaching of residual molecules (i.e., uncrosslinked oligomer, low molecular unreactive dimethylsiloxane cyclics, solvents, and platinum catalyst) from PDMS, could significantly reduce the viability of mammalian cells in PDMS devices [228]. Fortunately, cell viability can be significantly improved when the devices are pre-washed with several cleaning solvents and sterilized [228, 229]. As for SU-8, its use as a bioMEMS material is far less developed than that of PDMS, although its biocompatibility has been demonstrated in short-term clinical studies [220].

12.4.2 Surface Modification

When there are limited choices of device materials, surface treatment is most commonly used to render materials more biocompatible and to minimize biofouling, for instance, on implants and in microfluidic channels. This type of surface modifications can be considered part of the packaging scheme since it is an integral part of the post-fabrication process to deliver the functionality and usability of the chips. Additionally, surface modification can serve to protect devices from the environment.

Implantation of foreign material could lead to an immune response from the body, the result of which includes, but is not limited to, inflammation, scarring, fibrous tissue buildup, and atypical cell growth. A properly passivated or functionalized surface can reduce or effectively eliminate unwanted immune response. Furthermore, surface modification can facilitate not only a passive response but also improved adaptation of the device to best suit the environment. For implants, this could lead to a controlled growth of cells around the implant leading to faster healing after insertion and normal tissue formation. Moreover, some proper functionality of the device can be maintained over a longer period of time. In addition, surface treatment can be a useful tool for protecting the device from its corrosive surroundings. Common issues such as fouling and device isolation can thus be addressed.

This following section will provide an overview of existing and emerging techniques which could eventually be used to mitigate the adverse effects of the interaction between the biochip and its environment and enhance the device functionality through active manipulation of that interaction.

12.4.2.1 Basic Principles of Biological Surface Chemistry – Biorecognition

Although modern bioMEMS devices can be fabricated with micrometer and nanometer features and can be manufactured small enough to fit into a hypodermic needle, the basic device–environment interaction occurs at the molecular level. Thus, for a biological environment, surface treatment schemes have to be designed with biorecognition in mind. Biorecognition is the highly developed ability of biological systems to recognize specially designed features on the molecular scale, whether it is through topographic architecture, chemical architecture, or dynamic properties [230]. This feature of biological systems can determine the rate of adsorption of proteins and other biomolecules onto a surface, the ability of cells to cover a specific surface, and the stability of the interactions.

Furthermore, biorecognition on the nanoscale can affect interactions on much larger length scales [230]. Due to mass transport properties such as diffusivity, the order of arrival of different types of molecules at the device surface is different. The first molecules to arrive are water molecules, followed by proteins and similar types of organic molecules. In the case of implants, the effect of the surface type, along with the arrival order, can influence the ability of cells to cover a device surface [230]. In particular, the initial water shell exhibits different properties based on the type of material with which it interacts. As such, water can form hydration shells, which through interaction with the hydration shells of biomolecules may determine whether proteins denature, how they are oriented, and how effective they can cover the surface [230]. Cells require an extracellular matrix of proteins (ECM) to successfully adhere to a surface and proliferate. If the ECM is altered due to interaction with the surface, the overall cell–surface interaction is also modified. Thus, awareness and understanding of biorecognition becomes imperative for successful design of surfaces.

12.4.2.2 Surface Treatments for Common BioMEMS Materials

BioMEMS fabrication techniques differ from traditional MEMS microfabrication techniques to various extents, but they share common elements, such as the most typical materials. One such material for IC and CMOS fabrication is silicon; silicon is generally regarded as non-toxic and inert and thus a good material for certain implant and cell culture applications.

Along with silicon, SiO2 and SiN are used in the fabrication process and often as a protective layer for the device. A layer of the oxide or nitride can be deposited by chemical vapor depositions [189]. These materials do not leech into the environment and under certain conditions are also deemed non-toxic and biocompatible. Other common, non-polymeric materials include metals such as platinum, titanium oxides (TiO2), gold, and alloys, as well as ceramics.

Polymer-based devices have become increasingly popular due to their simple and relatively inexpensive manufacturing processes. Polymers can be used not only as the structural material of the bioMEMS device but also as a protective material for silicon-based devices. PDMS, Parylene-C, and PMMA are commonly used polymers as described earlier. They are easy to process, have physical properties (e.g., stiffness) that can be controlled or modified, and are largely non-toxic. Polymer-based photoresists may also be used as structural materials or protective films, but often the concerns are costs and solvents that cannot be completely removed in the process. SU-8, an epoxy-based negative resist, is becoming increasingly popular due to its low toxicity.

It is important to note the limitations of materials' applicability. Many chips are manufactured using a combination of materials, each requiring a different treatment. For example, traditional CMOS components may be combined with polymer-based device components in a non-monolithic process. Therefore, it is important to understand the complexity of a device and choose compatible treatment techniques that will achieve the desired effect.

Optimizing the Surface of PDMS The surface of PDMS is naturally hydrophobic and thus may be unsuitable for certain biological applications. Furthermore, PDMS surfaces can adsorb certain biological molecules, but not in a completely controllable fashion, i.e., the surface property of PDMS may vary depending on the processing conditions. To minimize cell adhesion (e.g., for implant applications, for instance) or to better control surface properties, several techniques have been developed to make the surface hydrophilic or to limit adsorption of biomolecules, such as plasma processing, surfactants, polyelectrolyte multilayers, and graft polymerization through radiation exposure or cerium (IV) catalysis [231].

Exposing PDMS to plasma, discharge [231, 232] can alter its surface properties. Through the use of oxygen plasma, the surface of the PDMS in direct contact with the plasma becomes ionized and forms excited species such as ions and radicals. Although the reactions are complex and the mechanisms are not fully understood, the result of the plasma treatment is the formation of a transient silica-like hydrophilic layer of SiO_x with high oxygen content [231]. This renders the surface much more hydrophilic than the original surface.

Generating reactive elements on the surface of PDMS is also possible through exposure to radiation [233]. Free radical formation on the surface as a result of exposure to UV light provides the reactive component required to graft polymerize acrylic acid and other monomers, such as acrylamide (AM), hydroxyethyl acrylate (HEA), PEG, and DMA, onto the surface. A layer of polymer can thus be grafted in the exposed area leading to the desired functionality. Graft polymerization can also be achieved through cerium (IV) catalytic reactions, although the process condition can be more complex [231, 234].

Another simple surface modification to make it hydrophilic and charged is through exposure to surfactants below the critical micelle concentrations [231]. The concentration must be low enough to prevent micelle formation yet sufficiently high enough to ensure coating of the PDMS surface. The process facilitates the hydrophobic tail of the surfactant attaching to the surface, leaving the hydrophilic end exposed to the environment. Various surfactants have been used, such as sodium dodecyl sulfate [235] and 2-morpholinoethanesulfonic acid [236]. Because surfactants could be damaging to cell membranes, one needs to be judicious in choosing and using surfactants when the chips are used for cell-based experiments.

Exposing a PDMS surface to cationic and anionic polyelectrolyte solutions in an alternating fashion forms a polyelectrolyte multilayer [231]. Although time consuming (because many layers have to be deposited with many rinses in between each deposition), layer-by-layer assembly usually gives robust films. Deposition of a film of this type is also possible with small organic molecules, polymers, natural proteins, inorganic clusters, clay particles, and colloids [237]. The adsorption of polyelectrolytes to a solid surface occurs in solution, in which case substrate size and topology should not play a role in coating efficiency.

Although useful for short-term modifications of surface properties, some of these approaches do not always fully address the issue of long-term stable hydrophilicity and hydrophobic recovery [238]. For example, over time, the silica-like layer on PDMS that forms upon exposure to plasma or radiation may revert back to hydrophobic surfaces due to diffusion of the hydrophilic species from the surface into the bulk and potentially also crack and deform due to mismatch of physical properties (stiffness and flexibility) from the supporting PDMS layer. Long-term stable films may be achieved through electrostatic layer-by-layer self-assembly of polyelectrolytes with chemical crosslinking [238]. This method uses the same basic principle as the formation of polyelectrolyte multilayers but introduces crosslinking between the layers for additional stability. It is also possible to covalently bond PEG to the layer-by-layer modified surface for additional protein adsorption resistance [238].

It is important to note that all of the above techniques need to be incorporated in the appropriate order during the manufacturing process. For example, if a device contains biological molecules for analytical purposes, exposure to oxygen plasma without protection may result in unwanted reactions. Thus, the surface of the packaging should be modified either prior to loading with biological molecules or those areas of the chip need to be isolated from the plasma. Exposure to UV light and electrolyte solutions may also render certain components of a device non-functional, and thus they should only be used with proper protection for vulnerable elements.

Modification of Silicon Surfaces Using silicon in the biological environment presents several challenges. One of the major issues is biofouling, leading to interference and the suboptimal operation of the device [239]. Silicon surface is hydrophobic and hence potentially attractive to protein deposition. Surface modification provides a convenient way to limit or direct adsorption of biological molecules such as proteins on, and in effect the interactions of cells to, the silicon surface. Other than depositing or growing SiO2 and SiNx layers, the most common approach to treating silicon surfaces is to form protective films of other materials.

Plasma polymerization has been used to polymerize a layer of tetraethylene glycol dimethyl ether (tetraglyme) onto a silicon surface [240]. The resulting protective coating is a PEG-like polymer thin film; silicon surfaces thus treated exhibited less protein adsorption and less fouling. One possible drawback is that coating of high aspect ratio can suffer from non-uniformities [239] and can be as thin as 10 nm [240].

The formation of the protective coating can be achieved by exposing a cleaned silicon surface to tetraglyme vapor in a plasma chamber. The procedure is also applicable to SiO2 and SiN surfaces which may be desirable for electrical passivation and compatible with photoresist patterned surfaces and may conceivably be used on metallic surfaces [239]. In addition, the film does not interfere with electrical conductivity of metallic electrodes used for neural sensing. As such, it can serve as an optimal protective coating for electrodes used to monitor brain activity and many other applications. However, the formation process is not fully compatible with biological molecules and could destroy biologically active chip components. Hence, as a surface treatment technique, it should only be applied after the device has been completely sealed, preventing vapor access to destructible elements, or prior to the introduction of biologically functional components.

Acrylic acid can also be successfully grafted onto silicon through plasma polymerization [241]. Polymeric acrylic acid coatings lead to higher fluid velocity along the coated surface. Controlling flow and velocity is particularly useful for devices analyzing or interacting with biological fluids. Higher velocity leads to lower biomolecule adsorption, which in the case of blood can lead to less coagulation. Lowering adsorption on the surface can thus lead to less coagulation. Devices aimed at implantation in the cardiovascular system could greatly benefit from understanding this mechanism.

Facilitating controlled cell adhesion and growth is imperative for implantable devices, such as the multi-electrode arrays. Ultrananocrystalline diamond (UNCD) films are emerging as a new type of protective film. This type of films is chemically and electrically inert, has a low friction coefficient and high wear resistance, provides a good substrate for cell attachment, and can be used with a silicon surface as the coating base [242]. Furthermore, the film formation technique is compatible with some common MEMS fabrication processes. The deposition of the film on silicon wafers occurs via microwave plasma chemical vapor deposition, in an argon (99%) and methane (1%) atmosphere [242]. This process leads to C2 dimer formation, which supplies nucleation species for the entire film. The deposition occurs at a temperature of 400–800 °C and is then followed by several cleaning steps before the surface is ready to be dried and used. Compared to pure silicon and platinum surfaces, HeLa, PC12, and MC3T3 cells adhered much better to the UNCD surface. Furthermore, cells attached to this surface exhibited lowest cell rounding and highest cell spreading, which seems to indicate that the UNCD surface is the most biocompatible surface of the three [242].

The drawback of this technique is the complexity of the process and cost, making it unsuitable for single-use applications, for example. Additionally, the high temperature required to form the film and the aggressive cleaning solutions used in the process (e.g., piranha solution) may not be compatible with other processes. The temperature range at which the deposition occurs affects not only biological molecules or reagents but would also melt some metals commonly used in CMOS processes. Although it is possible to protect different components during the processing, it would lead to additional manufacturing steps and higher cost. Nevertheless, the UNCD approach is perhaps one of the more reliable ways to modify silicon surface and can be useful when processes are well designed.

Control over the thickness of protective films is a process feature desired for every manufacturing process. Atomic layer deposition (ALD) is a process capable of depositing ultra-thin conformal films with atomic-level thickness control [243]. Via this method, materials, such as Al2O3, SiC, and biocompatible materials (e.g., TiO2), may be deposited on devices to form protective films. Similar in principle to polyelectrolyte multilayers, the ALD technique relies on a series of reactions, each of which terminates after one step, between the gaseous precursors and the solid substrate. Once a precursor molecule reacts with the surface, the particular site is unavailable for further reaction. Furthermore, the molecule cannot react with itself, limiting the film thickness to just one layer. The next layer is applied to regenerate the ability of the desired molecule to adsorb. This cycle can be repeated to obtain the desired thickness [243]. The deposition of Al2O3 using this technique can be performed at 177 °C. Although this temperature is high enough to destroy most biomolecules, it is low enough for the process to be compatible with most CMOS processes. Thus, (i) if insulation of biological chip components can be achieved or (ii) if the process can be applied prior to the introduction of such components, this technique could be used to form protective films of controlled thickness on silicon devices. The advantages of this approach are a low temperatures required (compared to CVD), the ability to form several layers, unlike self-assembling monolayer techniques [243], and most importantly the high degree of conformity of the deposited films [243].

Modification of SU-8 Surfaces Due to its rigidity, chemical stability, low apparent toxicity, and transparency to visible light, SU-8 has been used as a structural component in biochip fabrication [244]. However, since the surface of SU-8 is hydrophobic, it faces similar challenges as other materials for controlling protein deposition and cell interactions.

Modification of SU-8 surfaces has been achieved through methods similar to PDMS surface modification. PEG graft polymerization has been achieved through exposure to radiation [245], and hydrophilicity has been achieved through treatment with ethanolamine [246] or grafting of amine groups to the surface via a CVD treatment [247]. It is also possible to apply the cerium (IV)-based surface graft polymerization process to modify SU-8 surfaces. This can be performed at room temperature and can be used to graft a variety of monomers onto the SU-8 [244]. The method relies on the opening of residual epoxide rings using a mixture of nitric acid and cerium (IV) ammonium nitrate, followed by an incubation of the surface in the same solution containing monomer, after which the grafting reaction is allowed to occur at room temperature. Using this method, polyacrylic acid can be grafted onto the surface to enhance cell attachment [244]. Although this methodology is simple and straightforward, care must be taken to protect vulnerable components of the bioMEMS chip from the nitric acid.

12.4.2.3 Modification of Surface Topography for Improved and Directed Cell Attachment and Growth

Although the mechanism is not fully understood yet, surface topography can affect cell attachment, growth, and spreading. Thus, topographic features (in conjunction with other surface treatment technologies) could be used to promote or limit cell growth on specific components of a bioMEMS chip, improve integration of implants, and even improve on the functionality of cell-based sensors.

To achieve this goal, a technique for selectively patterning ECM on a surface has been suggested [248]. Microcontact printing techniques based on self-assembling monolayer technology on silicon wafers have been used to fabricate substrates with well-defined regions of ECM [248]. Using this technique and by controlling geometry [248], it is possible to coordinate cell attachment density, at least in the initial phase. Cells, however, are able to migrate onto regions not covered by the ECM after a while, possibly due to protein adsorption from the solution or by secretion of ECM components by the cells themselves [248]. Therefore, it is imperative to passivate and limit the adsorption of protein to areas not covered by the patterned ECM.

Introducing various microtopography features such as ridges of different depth, spacing, and width can influence cell adhesion and spreading [249]. Through photo ablation with an excimer laser and the use of a projection lithography/etching technique, well-defined surface features down to submicron scale can be fabricated. The desired pattern is ablated from a polyethylene terephthalate (PET) surface on a fully automated, computer-controlled stage. Although the original experiment tested PET and polyvinyl alcohol (PVA), the methodology is also applicable to mineral or organic surfaces, since it eliminates the need for a photoresist [249]. The laser technology can also be used with temperature-sensitive materials, since heating occurs only in the area targeted by the laser and the process occurs rapidly – 1 s/cm^2 – thus minimizing heat flux to surround areas [250].

Although no theory exists that thoroughly describes the effects of topography on cell growth and orientation, empirical results can be used to determine the type of groove dimensions best suited for the surface treatment [249]. For example, smoother transitions in surface topography appear to promote cell growth [43, 249]. However, this effect was highly dependent on cell type, cell size, and the surface chemistry, such as formation of carbon-rich species at the ablated surface. Nevertheless, the technology provides a valuable blueprint for future research and may soon find practical use.

12.4.3 Other Considerations

It can be pointed out that miniaturization of devices in some sense may make implantable devices more biocompatible. For example, miniaturized implants can be inserted using needles and catheters instead of using traditional surgical methods [70]. Small wounds elicit smaller immune response. However, there are cases where miniaturization is not the solution. Szarowski et al. [222] tested the response of brain tissue to implanted micromachined silicon devices. Following implantation, cell density increased in the surrounding area, as many others have previously discovered with implanted devices not free to follow the movement of tissue [251–253]. Furthermore, it was determined that the long-term brain response to devices of various shapes and with different degrees of surface smoothness and corner roundedness was effectively the same and led to device encapsulation through formation of dense tissue around the device [222]. For this specific case, since the response was independent of size, techniques other than miniaturization will have to be utilized to improve the device. In general, one would need to consider multiple facets of a packaging scheme in order to determine the optimal strategy to render devices biocompatible.

12.5 Conclusion

There is no universal packaging scheme that suits all applications for biochips. Many aspects, including materials, geometry, and costs, have to be considered in order to design the best packaging schemes. Most likely, some trade-offs have to be made – sacrificing some flexibility in certain aspects to ensure functionality of the device as a whole. Moreover, continuing research from both academia and industries in the packaging area will further mature this field and advance lab-on-a-chip, microfluidics, and bioMEMS in the long run.

References

1. Chin, C.D., Linder, V., Sia, S.K.: Lab-on-a-chip devices for global health: past studies and future opportunities. Lab Chip. **7**(1), 41–57 (2007)
2. Yager, P., Edwards, T., Fu, E., Helton, K., Nelson, K., Tam, M.R., Weigl, B.H.: Microfluidic diagnostic technologies for global public health. Nature. **442**(7101), 412–418 (2006)
3. Engler, K.H., Efstratiou, A., Norn, D., Kozlov, R.S., Selga, I., Glushkevich, T.G., Tam, M., Melnikov, V.G., Mazurova, I.K., Kim, V.E., Tseneva, G.Y., Titov, L.P., George, R.C.: Immunochromatographic strip test for rapid detection of diphtheria toxin: description and multicenter evaluation in areas of low and high prevalence of diphtheria. J. Clin. Microbiol. **40**(1), 80–83 (2002)
4. Arai, H., Petchclai, B., Khupulsup, K., Kurimura, T., Takeda, K.: Evaluation of a rapid immunochromatographic test for detection of antibodies to human immunodeficiency virus. J. Clin. Microbiol. **37**(2), 367–370 (1999)
5. Patterson, K., Olsen, B., Thomas, C., Norn, D., Tam, M., Elkins, C.: Development of a rapid immunodiagnostic test for Haemophilus ducreyi. J. Clin. Microbiol. **40**(10), 3694–3702 (2002)
6. Zarakolu, P., Buchanan, I., Tam, M., Smith, K., Hook, E.W.: Preliminary evaluation of an immunochromatographic strip test for specific Treponema pallidum antibodies. J. Clin. Microbiol. **40**(8), 3064–3065 (2002)
7. Martinez, A.W., Phillips, S.T., Butte, M.J., Whitesides, G.M.: Patterned paper as a platform for inexpensive, low-volume, portable bioassays. Angew. Chem. Int. Ed. **46**(8), 1318–1320 (2007)
8. Martinez, A.W., Phillips, S.T., Carrilho, E., Thomas, S.W., Sindi, H., Whitesides, G.M.: Simple telemedicine for developing regions: camera phones and paper-based microfluidic devices for real-time, off-site diagnosis. Anal. Chem. **80**(10), 3699–3707 (2008)
9. Martinez, A.W., Phillips, S.T., Whitesides, G.M.: Three-dimensional microfluidic devices fabricated in layered paper and tape. Proc. Natl. Acad. Sci. **105**(50), 19606–19611 (2008)
10. Hintsche, R., Moller, B., Dransfeld, I., Wollenberger, U., Scheller, F., Hoffmann, B.: Chip biosensors on thin-film metal-electrodes. Sensors Actuators B Chem. **4**(3–4), 287–291 (1991)
11. Shulga, A.A., Soldatkin, A.P., Elskaya, A.V., Dzyadevich, S.V., Patskovsky, S.V., Strikha, V.I.: Thin-film conductometric biosensors for glucose and urea determination. Biosens. Bioelectron. **9**(3), 217–223 (1994)
12. Hunt, H.C., Wilkinson, J.S.: Optofluidic integration for microanalysis. Microfluid. Nanofluid. **4**(1–2), 53–79 (2008)
13. Cui, Y., Wei, Q.Q., Park, H.K., Lieber, C.M.: Nanowire nanosensors for highly sensitive and selective detection of biological and chemical species. Science. **293**(5533), 1289–1292 (2001)
14. Zheng, G.F., Patolsky, F., Cui, Y., Wang, W.U., Lieber, C.M.: Multiplexed electrical detection of cancer markers with nanowire sensor arrays. Nat. Biotechnol. **23**(10), 1294–1301 (2005)
15. Bashir, R.: BioMEMS: state-of-the-art in detection, opportunities and prospects. Adv. Drug Deliv. Rev. **56**(11), 1565–1586 (2004)
16. Waggoner, P.S., Craighead, H.G.: Micro- and nanomechanical sensors for environmental, chemical, and biological detection. Lab Chip. **7**(10), 1238–1255 (2007)
17. Fritz, J., Baller, M.K., Lang, H.P., Rothuizen, H., Vettiger, P., Meyer, E., Guntherodt, H.J., Gerber, C., Gimzewski, J.K.: Translating biomolecular recognition into nanomechanics. Science. **288** (5464), 316–318 (2000)
18. Hansen, K.M., Ji, H.F., Wu, G.H., Datar, R., Cote, R., Majumdar, A., Thundat, T.: Cantilever- based optical deflection assay for discrimination of DNA single-nucleotide mismatches. Anal. Chem. **73**(7), 1567–1571 (2001)
19. Gupta, A., Akin, D., Bashir, R.: Single virus particle mass detection using microresonators with nanoscale thickness. Appl. Phys. Lett. **84**(11), 1976–1978 (2004)
20. Ilic, B., Czaplewski, D., Craighead, H.G., Neuzil, P., Campagnolo, C., Batt, C.: Mechanical resonant immunospecific biological detector. Appl. Phys. Lett. **77**(3), 450–452 (2000)
21. McKendry, R., Zhang, J.Y., Arntz, Y., Strunz, T., Hegner, M., Lang, H.P., Baller, M.K., Certa, U., Meyer, E., Guntherodt, H.J.,

Gerber, C.: Multiple label-free biodetection and quantitative DNA-binding assays on a nanomechanical cantilever array. Proc. Natl. Acad. Sci. U. S. A. **99**(15), 9783–9788 (2002)
22. Wu, G.H., Datar, R.H., Hansen, K.M., Thundat, T., Cote, R.J., Majumdar, A.: Bioassay of prostate-specific antigen (PSA) using microcantilevers. Nat. Biotechnol. **19**(9), 856–860 (2001)
23. Burg, T.P., Godin, M., Knudsen, S.M., Shen, W., Carlson, G., Foster, J.S., Babcock, K., Manalis, S.R.: Weighing of biomolecules, single cells and single nanoparticles in fluid. Nature. **446**, 1066–1069 (2007)
24. Burg, T.P., Mirza, A.R., Milovic, N., Tsau, C.H., Popescu, G.A., Foster, J.S., Manalis, S.R.: Vacuum-packaged suspended microchannel resonant mass sensor for biomolecular detection. J. Microelectromech. Syst. **15**(6), 1466–1476 (2006)
25. Fruetel, J.A., Renzi, R.F., VanderNoot, V.A., Stamps, J., Horn, B. A., West, J.A.A., Ferko, S., Crocker, R., Bailey, C.G., Arnold, D., Wiedenman, B., Choi, W.Y., Yee, D., Shokair, I., Hasselbrink, E., Paul, P., Rakestraw, D., Padgen, D.: Microchip separations of protein biotoxins using an integrated hand-held device. Electrophoresis. **26**(6), 1144–1154 (2005)
26. Stratis-Cullum, D.N., Griffin, G.D., Mobley, J., Vass, A.A., Vo-Dinh, T.: A miniature biochip system for detection of aerosolized Bacillus globigii spores. Anal. Chem. **75**(2), 275–280 (2003)
27. Psaltis, D., Quake, S.R., Yang, C.H.: Developing optofluidic technology through the fusion of microfluidics and optics. Nature. **442** (7101), 381–386 (2006)
28. Irawan, R., Tjin, S.C., Fang, X.Q., Fu, C.Y.: Integration of optical fiber light guide, fluorescence detection system, and multichannel disposable microfluidic chip. Biomed. Microdevices. **9**(3), 413–419 (2007)
29. Lin, C.H., Lee, G.B., Chen, S.H., Chang, G.L.: Micro capillary electrophoresis chips integrated with buried SU-8/SOG optical waveguides for bio-analytical applications. Sensors Actuators A Phys. **107**(2), 125–131 (2003)
30. Kou, Q., Yesilyurt, I., Studer, V., Belotti, M., Cambril, E., Chen, Y.: On-chip optical components and microfluidic systems. Microelectron. Eng. **73–74**, 876–880 (2004)
31. Misiakos, K., Kakabakos, S.E., Petrou, P.S., Ruf, H.H.: A monolithic silicon optoelectronic transducer as a real-time affinity biosensor. Anal. Chem. **76**(5), 1366–1373 (2004)
32. Balslev, S., Jorgensen, A.M., Bilenberg, B., Mogensen, K.B., Snakenborg, D., Geschke, O., Kutter, J.P., Kristensen, A.: Lab-on-a-chip with integrated optical transducers. Lab Chip. **6**(2), 213–217 (2006)
33. Dandin, M., Abshire, P., Smela, E.: Optical filtering technologies for integrated fluorescence sensors. Lab Chip. **7**(8), 955–977 (2007)
34. Prakash, A.R., Adamia, S., Sieben, V., Pilarski, P., Pilarski, L.M., Backhouse, C.J.: Small volume PCR in PDMS biochips with integrated fluid control and vapour barrier. Sensors Actuators B Chem. **113**(1), 398–409 (2006)
35. Kaigala, G.V., Hoang, V.N., Stickel, A., Lauzon, J., Manage, D., Pilarski, L.M., Backhouse, C.J.: An inexpensive and portable microchip-based platform for integrated RT-PCR and capillary electrophoresis. Analyst. **133**(3), 331–338 (2008)
36. Weibel, D.B., Kruithof, M., Potenta, S., Sia, S.K., Lee, A., Whitesides, G.M.: Torque-actuated valves for microfluidics. Anal. Chem. **77**(15), 4726–4733 (2005)
37. Burns, M.A., Johnson, B.N., Brahmasandra, S.N., Handique, K., Webster, J.R., Krishnan, M., Sammarco, T.S., Man, P.M., Jones, D., Heldsinger, D., Mastrangelo, C.H., Burke, D.T.: An integrated nanoliter DNA analysis device. Science. **282**(5388), 484–487 (1998)
38. Liu, R.H., Yang, J.N., Lenigk, R., Bonanno, J., Grodzinski, P.: Self-contained, fully integrated biochip for sample preparation, polymerase chain reaction amplification, and DNA microarray detection. Anal. Chem. **76**(7), 1824–1831 (2004)
39. Pal, R., Yang, M., Lin, R., Johnson, B.N., Srivastava, N., Razzacki, S.Z., Chomistek, K.J., Heldsinger, D.C., Haque, R.M., Ugaz, V. M., Thwar, P.K., Chen, Z., Alfano, K., Yim, M.B., Krishnan, M., Fuller, A.O., Larson, R.G., Burke, D.T., Burns, M.A.: An integrated microfluidic device for influenza and other genetic analyses. Lab Chip. **5**(10), 1024–1032 (2005)
40. Fu, E., Chinowsky, T., Nelson, K., Johnston, K., Edwards, T., Helton, K., Grow, M., Miller, J.W., Yager, P.: SPR Imaging-Based Salivary Diagnostics System for the Detection of Small Molecule Analytes. Oral-Based Diagnostics, pp. 335–344. Blackwell Publishing, Oxford (2007)
41. Velten, T., Ruf, H.H., Barrow, D., Aspragathos, N., Lazarou, P., Jung, E., Malek, C.K., Richter, M., Kruckow, J.: Packaging of bio-MEMS: strategies, technologies, and applications. IEEE Trans. Adv. Packag. **28**(4), 533–546 (2005)
42. Herr, A.E., Hatch, A.V., Throckmorton, D.J., Tran, H.M., Brennan, J.S., Giannobile, W.V., Singh, A.K.: Microfluidic immunoassays as rapid saliva-based clinical diagnostics. Proc. Natl. Acad. Sci. U. S. A. **104**(13), 5268–5273 (2007)
43. Lagally, E.T., Scherer, J.R., Blazej, R.G., Toriello, N.M., Diep, B. A., Ramchandani, M., Sensabaugh, G.F., Riley, L.W., Mathies, R. A.: Integrated portable genetic analysis microsystem for pathogen/ infectious disease detection. Anal. Chem. **76**(11), 3162–3170 (2004)
44. Wang, J., Chatrathi, M.P., Mulchandani, A., Chen, W.: Capillary electrophoresis microchips for separation and detection of organophosphate nerve agents. Anal. Chem. **73**(8), 1804–1808 (2001)
45. Chinowsky, T.M., Soelberg, S.D., Baker, P., Swanson, N.R., Kauffman, P., Mactutis, A., Grow, M.S., Atmar, R., Yee, S.S., Furlong, C.E.: Portable 24-analyte surface plasmon resonance instruments for rapid, versatile biodetection. Biosens. Bioelectron. **22**(9–10), 2268–2275 (2007)
46. DeBusschere, B.D., Kovacs, G.T.A.: Portable cell-based biosensor system using integrated CMOS cell-cartridges. Biosens. Bioelectron. **16**(7–8), 543–556 (2001)
47. Hood, L., Heath, J.R., Phelps, M.E., Lin, B.Y.: Systems biology and new technologies enable predictive and preventative medicine. Science. **306**(5696), 640–643 (2004)
48. Weston, A.D., Hood, L.: Systems biology, proteomics, and the future of health care: toward predictive, preventative, and personalized medicine. J. Proteome Res. **3**(2), 179–196 (2004)
49. Bhattacharyya, A., Klapperich, C.M.: Design and testing of a disposable microfluidic chemiluminescent immunoassay for disease biomarkers in human serum samples. Biomed. Microdevices. **9**(2), 245–251 (2007)
50. Linder, V., Sia, S.K., Whitesides, G.M.: Reagent-loaded cartridges for valveless and automated fluid delivery in microfluidic devices. Anal. Chem. **77**(1), 64–71 (2005)
51. Grayson, A.C.R., Shawgo, R.S., Johnson, A.M., Flynn, N.T., Li, Y.W., Cima, M.J., Langer, R.: A BioMEMS review: MEMS technology for physiologically integrated devices. Proc. IEEE. **92**(1), 6–21 (2004)
52. Fonseca, M. A.M.D.S.J.W.J.K.; Cardiomems, Innc., assignee. Implantable Wireless Sensor for Pressure Measurement within the Heart. US patent 6855115. (2005 February 15)
53. Santini, J.T.M.J.C.R.S.L.; MIT, assignee. Microchip Drug Delivery Devices. US. (1998 August 25)
54. Kudo, H., Sawada, T., Kazawa, E., Yoshida, H., Iwasaki, Y., Mitsubayashi, K.: A flexible and wearable glucose sensor based on functional polymers with Soft-MEMS techniques. Biosens. Bioelectron. **22**(4), 558–562 (2006)
55. Zhao, Y.J., Li, S.Q., Davidson, A., Yang, B.Z., Wang, Q., Lin, Q.: A MEMS viscometric sensor for continuous glucose monitoring. J. Micromech. Microeng. **17**(12), 2528–2537 (2007)

56. Jauniaux, E., Watson, A., Ozturk, O., Quick, D., Burton, G.: In-vivo measurement of intrauterine gases and acid-base values early in human pregnancy. Hum. Reprod. **14**(11), 2901–2904 (1999)
57. Prausnitz, M.R.: Microneedles for transdermal drug delivery. Adv. Drug Deliv. Rev. **56**(5), 581–587 (2004)
58. Prausnitz, M.R., Allen, M.G., Gujral, I.J. Microneedle Drug Delivery Device. US Patent 7, 226, 439; (2007)
59. Gujral I.J., Allen, M.G., Prausnitz M.R. Microneedle Device for Extraction and Sensing of Bodily Fluids. US Patent 7,344,499; (2008)
60. McAllister, D.V., Wang, P.M., Davis, S.P., Park, J.H., Canatella, P.J., Allen, M.G., Prausnitz, M.R.: Microfabricated needles for transdermal delivery of macromolecules and nanoparticles: fabrication methods and transport studies. Proc. Natl. Acad. Sci. **100** (24), 13755–13760 (2003)
61. Li, P.Y., Shih, J., Lo, R., Saati, S., Agrawal, R., Humayun, M.S., Tai, Y.C., Meng, E.: An electrochemical intraocular drug delivery device. Sensors Actuators A Phys. **143**(1), 41–48 (2008)
62. Santini, J.T., Cima, M.J., Langer, R.: A controlled-release microchip. Nature. **397**(6717), 335–338 (1999)
63. Voskerician, G., Shawgo, R.S., Hiltner, P.A., Anderson, J.M., Cima, M.J., Langer, R.: In vivo inflammatory and wound healing effects of gold electrode voltammetry for MEMS micro-reservoir drug delivery device. IEEE Trans. Biomed. Eng. **51**(4), 627–635 (2004)
64. Razzacki, S.Z., Thwar, P.K., Yang, M., Ugaz, V.M., Burns, M.A.: Integrated microsystems for controlled drug delivery. Adv. Drug Deliv. Rev. **56**(2), 185–198 (2004)
65. Wu, C.C., Yasukawa, T., Shiku, H., Matsue, T.: Fabrication of miniature Clark oxygen sensor integrated with microstructure. Sensors Actuators B Chem. **110**(2), 342–349 (2005)
66. Wu, H.K., Huang, B., Zare, R.N.: Construction of microfluidic chips using polydimethylsiloxane for adhesive bonding. Lab Chip. **5**(12), 1393–1398 (2005)
67. Hungar, K., Gortz, M., Slavcheva, E., Spanier, G., Weidig, C., Mokwa, W.: Production processes for a flexible retina implant (Eurosensors XVIII, Session C6.6). Sensors Actuators A Phys. **123–24**, 172–178 (2005)
68. Schanze, T., Hesse, L., Lau, C., Greve, N., Haberer, W., Kammer, S., Doerge, T., Rentzos, A., Stieglitz, T.: An optically powered single-channel stimulation implant as test-system for chronic biocompatibility and biostability of miniaturized retinal vision prostheses. IEEE Trans. Biomed. Eng. **54**(6), 983–992 (2007)
69. Schwarz, M., Ewe, L., Hauschild, R., Hosticka, B.J., Huppertz, J., Kolnsberg, S., Mokwa, W., Trieu, H.K.: Single chip CMOS imagers and flexible microelectronic stimulators for a retina implant system. Sensors Actuators A Phys. **83**(1–3), 40–46 (2000)
70. Loeb, G.E., Peck, R.A., Moore, W.H., Hood, K.: BION (TM) system for distributed neural prosthetic interfaces. Med. Eng. Phys. **23**(1), 9–18 (2001)
71. Weiland, J.D., Liu, W.T., Humayun, M.S.: Retinal prosthesis. Annu. Rev. Biomed. Eng. **7**, 361–401 (2005)
72. Schwartz, A.B.: Cortical neural prosthetics. Annu. Rev. Neurosci. **27**(1), 487–507 (2004)
73. Cheung, K.C.: Implantable microscale neural interfaces. Biomed. Microdevices. **9**(6), 923–938 (2007)
74. Kipke, D.R., Vetter, R.J., Williams, J.C., Hetke, J.F.: Silicon-substrate intracortical microelectrode arrays for long-term recording of neuronal spike activity in cerebral cortex. IEEE Trans. Rehabil. Eng. **11**(2), 151–155 (2003)
75. Rutten, W.L.C.: Selective electrical interfaces with the nervous system. Annu. Rev. Biomed. Eng. **4**(1), 407–452 (2002)
76. Cogan, S.F.: Neural stimulation and recording electrodes. Annu. Rev. Biomed. Eng. **10**(1), 275–309 (2008)
77. Tokuda, T., Pan, Y.L., Uehara, A., Kagawa, K., Nunoshita, M., Ohta, J.: Flexible and extendible neural interface device based on cooperative multi-chip CMOS LSI architecture. Sensors Actuators A Phys. **122**(1), 88–98 (2005)
78. Smith, B., Tang, Z.N., Johnson, M.W., Pourmehdi, S., Gazdik, M. M., Buckett, J.R., Peckham, P.H.: An externally powered, multichannel, implantable stimulator-telemeter for control of paralyzed muscle. IEEE Trans. Biomed. Eng. **45**(4), 463–475 (1998)
79. Windecker, S.I., Mayer, G., de Pasquale, W., Maier, O., Dirsch, P., de Groot, Y.P., Wu, G., Noll, B., Leskosek, B., Meier, O.M., Hess, C., Working Grp Novel Surface: Stent coating with titanium-nitride-oxide for reduction of neointimal hyperplasia. Circulation. **104**(8), 928–933 (2001)
80. Grube, E., Gerckens, U., Rowold, S., Muller, R., Selbach, G., Stamm, J., Staberock, M.: Inhibition of in-stent restenosis by the Quanam drug eluting polymer stent; Two year follow-up. J. Am. Coll. Cardiol. **37**(2), 74A–74A (2001)
81. Hiatt, B.L., Ikeno, F., Yeung, A.C., Carter, A.J.: Drug-eluting stents for the prevention of restenosis: in quest for the holy grail. Catheter. Cardiovasc. Interv. **55**(3), 409–417 (2002)
82. Allen, M. G M.E.J.K.D.J.M.; CardioMEMS, Inc., assignee. Communication with an Implanted Wireless Sensor. US. (2007)
83. Klose, J., Rehtanz, E., Rothe, C., Eulitz, I., Guther, V., Beck, W.: Manufacture of titanium implants. Mater. Werkst. **39**(4–5), 304–308 (2008)
84. Wiegand, U.K.H., Potratz, J., Luninghake, F., Taubert, G., Brandes, A., Diederich, K.W.: Electrophysiological characteristics of bipolar membrane carbon leads with and without steroid elution compared with a conventional carbon and a steroid-eluting platinum lead. Pacing Clin. Electrophysiol. **19**(8), 1155–1161 (1996)
85. Wiegand, U.K.H., Zhdanov, A., Stammwitz, E., Crozier, I., Claessens, R.J.J., Meier, J., Bos, R.J., Bode, F., Potratz, J.: Electrophysiological performance of a bipolar membrane-coated titanium nitride electrode: a randomized comparison of steroid and nonsteroid lead designs. Pacing Clin. Electrophysiol. **22**(6), 935–941 (1999)
86. Wiggins, M.J., Wilkoff, B., Anderson, J.M., Hiltner, A.: Biodegradation of polyether polyurethane inner insulation in bipolar pacemaker leads. J. Biomed. Mater. Res. **58**(3), 302–307 (2001)
87. Russell, R.J., Pishko, M.V., Gefrides, C.C., McShane, M.J., Cote, G.L.: A fluorescence-based glucose biosensor using Concanavalin a and dextran encapsulated in a poly(ethylene glycol) hydrogel. Anal. Chem. **71**(15), 3126–3132 (1999)
88. Receveur, R.A.M., Lindemans, F.W., de Rooij, N.F.: Microsystem technologies for implantable applications. J. Micromech. Microeng. **17**(5), R50–R80 (2007)
89. Mokwa, W., Schnakenberg, U.: Micro-transponder systems for medical applications. IEEE Trans. Instrum. Meas. **50**(6), 1551–1555 (2001)
90. Flick, B.B., Orglmeister, R.: A portable microsystem-based telemetric pressure and temperature measurement unit. IEEE Trans. Biomed. Eng. **47**(1), 12–16 (2000)
91. Esashi, M., Sugiyama, S., Ikeda, K., Wang, Y.L., Miyashita, H.: Vacuum-sealed silicon micromachined pressure sensors. Proc. IEEE. **86**(8), 1627–1639 (1998)
92. Chen, P.J., Rodger, D.C., Agrawal, R., Saati, S., Meng, E., Varma, R., Humayun, M.S., Tai, Y.C.: Implantable micromechanical parylene-based pressure sensors for unpowered intraocular pressure sensing. J. Micromech. Microeng. **17**(10), 1931–1938 (2007)
93. Chen, L., Manz, A., Day, P.J.R.: Total nucleic acid analysis integrated on microfluidic devices. Lab Chip. **7**(11), 1413–1423 (2007)
94. Heyries, K.A., Loughran, M.G., Hoffmann, D., Homsy, A., Blum, L.J., Marquette, C.A.: Microfluidic biochip for chemiluminescent detection of allergen-specific antibodies. Biosens. Bioelectron. **23** (12), 1812–1818 (2008)

95. Isoda, T., Urushibara, I., Sato, M., Uemura, H., Sato, H., Yamauchi, N.: Development of a sensor-array chip with immobilized antibodies and the application of a wireless antigen-screening system. Sensors Actuators B Chem. **129**(2), 958–970 (2008)
96. Prakash, R., Kaler, K.: An integrated genetic analysis microfluidic platform with valves and a PCR chip reusability method to avoid contamination. Microfluid. Nanofluid. **3**(2), 177–187 (2007)
97. Zhang, C.S., Xu, J.L., Ma, W.L., Zheng, W.L.: PCR microfluidic devices for DNA amplification. Biotechnol. Adv. **24**(3), 243–284 (2006)
98. Sethu, P., Sin, A., Toner, M.: Microfluidic diffusive filter for apheresis (leukapheresis). Lab Chip. **6**(1), 83–89 (2006)
99. Battrell, C. F M.S.B.H.W.J.M.H.C.A.L.W.B.; Micronics, Inc., assignee. Method and System for Microfluidic Manipulation, Amplification and Analysis of Fluirds, for Example, Bacteria Assays and Antiglobulin Testing. US. (2004)
100. Chamot, S.R., Depeursinge, C.: MEMS for enhanced optical diagnostics in endoscopy. Minim. Invasive Ther. Allied Technol. **16**(2), 101–108 (2007)
101. Hupert, M.L., Witek, M.A., Wang, Y., Mitchell, M.W., Liu, Y., Bejat, Y., Nikitopoulos, D.E., Goettert, J., Murphy, M.C., Soper, S. A.: Polymer-based microfluidic devices for biomedical applications. Proc. SPIE. **4982**, 52–64 (2003)
102. Mitchell, M.W., Liu, X., Bejat, Y., Nikitopoulos, D.E., Soper, S. A., Murphy, M.C.: Modeling and validation of a molded polycarbonate continuous flow polymerase chain reaction device. Proc. SPIE. **4982**, 83–98 (2003)
103. Lee, D.S., Park, S.H., Yang, H.S., Chung, K.H., Yoon, T.H., Kim, S.J., Kim, K., Kim, Y.T.: Bulk- micromachined submicroliter-volume PCR chip with very rapid thermal response and low power consumption. Lab Chip. **4**(4), 401–407 (2004)
104. Koh, C.G., Tan, W., Zhao, M.Q., Ricco, A.J., Fan, Z.H.: Integrating polymerase chain reaction, valving, and electrophoresis in a plastic device for bacterial detection. Anal. Chem. **75**(17), 4591–4598 (2003)
105. Krishnan, M., Burke, D.T., Burns, M.A.: Polymerase chain reaction in high surface-to-volume ratio SiO2 microstructures. Anal. Chem. **76**(22), 6588–6593 (2004)
106. Woolley, A.T., Hadley, D., Landre, P., de Mello, A.J., Mathies, R. A., Northrup, M.A.: Functional integration of PCR amplification and capillary electrophoresis in a microfabricated DNA analysis device. Anal. Chem. **68**(23), 4081–4086 (1996)
107. West, J., Karamata, B., Lillis, B., Gleeson, J.P., Alderman, J., Collins, J.K., Lane, W., Mathewson, H.B.: Application of magnetohydrodynamic actuation to continuous flow chemistry. Lab Chip. **2**(4), 224–230 (2002)
108. Hong, J.W., Fujii, T., Seki, M., Yamamoto, T., Endo, I.: Integration of gene amplification and capillary gel electrophoresis on a polydimethylsiloxane-glass hybrid microchip. Electrophoresis. **22** (2), 328–333 (2001)
109. Shen, K.Y., Chen, X.F., Guo, M., Cheng, J.: A microchip-based PCR device using flexible printed circuit technology. Sensors Actuators B Chem. **105**(2), 251–258 (2005)
110. Daniel, J.H., Iqbal, S., Millington, R.B., Moore, D.F., Lowe, C.R., Leslie, D.L., Lee, M.A., Pearce, M.J.: Silicon microchambers for DNA amplification. Sensors Actuators A Phys. **71**(1–2), 81–88 (1998)
111. Northrup, M.A., Benett, B., Hadley, D., Landre, P., Lehew, S., Richards, J., Stratton, P.: A miniature analytical instrument for nucleic acids based on micromachined silicon reaction chambers. Anal. Chem. **70**(5), 918–922 (1998)
112. Sun, K., Yamaguchi, A., Ishida, Y., Matsuo, S., Misawa, H.: A heater-integrated transparent microchannel chip for continuous-flow PCR. Sensors Actuators B Chem. **84**(2–3), 283–289 (2002)
113. Khandurina, J., McKnight, T.E., Jacobson, S.C., Waters, L.C., Foote, R.S., Ramsey, J.M.: Integrated system for rapid PCR-based DNA analysis in microfluidic devices. Anal. Chem. **72**(13), 2995–3000 (2000)
114. Lin, Y.C., Huang, M.Y., Young, K.C., Chang, T.T., Wu, C.Y.: A rapid micro-polymerase chain reaction system for hepatitis C virus amplification. Sensors Actuators B Chem. **71**(1–2), 2–8 (2000)
115. Lin, Y.C., Yang, C.C., Huang, M.Y.: Simulation and experimental validation of micro polymerase chain reaction chips. Sensors Actuators B Chem. **71**(1–2), 127–133 (2000)
116. Zhou, Z.M., Liu, D.Y., Zhong, R.T., Dai, Z.P., Wu, D.P., Wang, H., Du, Y.G., Xia, Z.N., Zhang, L.P., Mei, X.D., Lin, B.C.: Determination of SARS-coronavirus by a microfluidic chip system. Electrophoresis. **25**(17), 3032–3039 (2004)
117. Gulliksen, A., Solli, L., Karlsen, F., Rogne, H., Hovig, E., Nordstrom, T., Sirevag, R.: Real- time nucleic acid sequence-based amplification in nanoliter volumes. Anal. Chem. **76**(1), 9–14 (2004)
118. Matsubara, Y., Kerman, K., Kobayashi, M., Yamamura, S., Morita, Y., Tamiya, E.: Microchamber array based DNA quantification and specific sequence detection from a single copy via PCR in nanoliter volumes. Biosens. Bioelectron. **20**(8), 1482–1490 (2005)
119. Curcio, M., Roeraade, J.: Continuous segmented-flow polymerase chain reaction for high- throughput miniaturized DNA amplification. Anal. Chem. **75**(1), 1–7 (2003)
120. Sethu, P., Mastrangelo, C.H.: Cast epoxy-based microfluidic systems and their application in biotechnology. Sensors Actuators B Chem. **98**(2–3), 337–346 (2004)
121. Swerdlow, H., Jones, B.J., Wittwer, C.T.: Fully automated DNA reaction and analysis in a fluidic capillary instrument. Anal. Chem. **69**(5), 848–855 (1997)
122. Zhang, N.Y., Yeung, E.S.: On-line coupling of polymerase chain reaction and capillary electrophoresis for automatic DNA typing and HIV-1 diagnosis. J. Chromatogr. B Analyt. Technol. Biomed. Life Sci. **714**(1), 3–11 (1998)
123. Ferrance, J.P., Wu, Q.R., Giordano, B., Hernandez, C., Kwok, Y., Snow, K., Thibodeau, S., Landers, J.P.: Developments toward a complete micro-total analysis system for Duchenne muscular dystrophy diagnosis. Anal. Chim. Acta. **500**(1–2), 223–236 (2003)
124. Huhmer, A.F.R., Landers, J.P.: Noncontact infrared-mediated thermocycling for effective polymerase chain reaction amplification of DNA in nanoliter volumes. Anal. Chem. **72**(21), 5507–5512 (2000)
125. Oda, R.P., Strausbauch, M.A., Huhmer, A.F.R., Borson, N., Jurrens, S.R., Craighead, J., Wettstein, P.J., Eckloff, B., Kline, B., Landers, J.P.: Infrared-mediated thermocycling for ultrafast polymerase chain reaction amplification of DNA. Anal. Chem. **70** (20), 4361–4368 (1998)
126. Tanaka, Y., Slyadnev, M.N., Hibara, A., Tokeshi, M., Kitamori, T.: Non-contact photothermal control of enzyme reactions on a microchip by using a compact diode laser. J. Chromatogr. A. **894**(1–2), 45–51 (2000)
127. Schneegass, I., Brautigam, R., Kohler, J.M.: Miniaturized flow-through PCR with different template types in a silicon chip thermocycler. Lab Chip. **1**(1), 42–49 (2001)
128. Chou, C.F., Changrani, R., Roberts, P., Sadler, D., Burdon, J., Zenhausern, F., Lin, S., Mulholland, N.S., Terbrueggen, R.: A miniaturized cyclic PCR device - modeling and experiments. Microelectron. Eng. **61–62**, 921–925 (2002)
129. Liu, J., Enzelberger, M., Quake, S.: A nanoliter rotary device for polymerase chain reaction. Electrophoresis. **23**(10), 1531–1536 (2002)
130. Shi, Y.N., Simpson, P.C., Scherer, J.R., Wexler, D., Skibola, C., Smith, M.T., Mathies, R.A.: Radial capillary array electrophoresis

microplate and scanner for high-performance nucleic acid analysis. Anal. Chem. **71**(23), 5354–5361 (1999)
131. Waters, L.C., Jacobson, S.C., Kroutchinina, N., Khandurina, J., Foote, R.S., Ramsey, J.M.: Multiple sample PCR amplification and electrophoretic analysis on a microchip. Anal. Chem. **70**(24), 5172–5176 (1998)
132. Waters, L.C., Jacobson, S.C., Kroutchinina, N., Khandurina, J., Foote, R.S., Ramsey, J.M.: Microchip device for cell lysis, multiplex PCR amplification, and electrophoretic sizing. Anal. Chem. **70**(1), 158–162 (1998)
133. Perch-Nielsen, I.R., Bang, D.D., Poulsen, C.R., El-Ali, J., Wolff, A.: Removal of PCR inhibitors using dielectrophoresis as a selective filter in a microsystem. Lab Chip. **3**(3), 212–216 (2003)
134. Gascoyne, P., Mahidol, C., Ruchirawat, M., Satayavivad, J., Watcharasit, P., Becker, F.F.: Microsample preparation by dielectrophoresis: isolation of malaria. Lab Chip. **2**(2), 70–75 (2002)
135. Namasivayam, V., Lin, R.S., Johnson, B., Brahmasandra, S., Razzacki, Z., Burke, D.T., Burns, M.A.: Advances in on-chip photodetection for applications in miniaturized genetic analysis systems. J. Micromech. Microeng. **14**(1), 81–90 (2004)
136. Kumar, A., Goel, G., Fehrenbach, E., Puniya, A.K., Singh, K.: Microarrays: the technology, analysis and application. Eng. Life Sci. **5**(3), 215–222 (2005)
137. Bulyk, M.L.: DNA microarray technologies for measuring protein-DNA interactions. Curr. Opin. Biotechnol. **17**(4), 422–430 (2006)
138. Cretich, M., Damin, F., Pirri, G., Chiari, M.: Protein and peptide arrays: recent trends and new directions. Biomol. Eng. **23**(2–3), 77–88 (2006)
139. Hoheisel, J.D.: Microarray technology: beyond transcript profiling and genotype analysis. Nat. Rev. Genet. **7**(3), 200–210 (2006)
140. Hultschig, C., Kreutzberger, J., Seitz, H., Konthur, Z., Bussow, K., Lehrach, H.: Recent advances of protein microarrays. Curr. Opin. Chem. Biol. **10**(1), 4–10 (2006)
141. Stoughton, R.B.: Applications of DNA microarrays in biology. Annu. Rev. Biochem. **74**, 53–82 (2005)
142. Barbulovic-Nad, I., Lucente, M., Sun, Y., Zhang, M.J., Wheeler, A.R., Bussmann, M.: Bio-microarray fabrication techniques - a review. Crit. Rev. Biotechnol. **26**(4), 237–259 (2006)
143. http://www.affymetrix.com
144. Anderson, R.C., Su, X., Bogdan, G.J., Fenton, J.: A miniature integrated device for automated multistep genetic assays. Nucleic Acids Res. **28**(12), e60i–e60vi (2000)
145. Lenigk, R., Liu, R.H., Athavale, M., Chen, Z.J., Ganser, D., Yang, J.N., Rauch, C., Liu, Y.J., Chan, B., Yu, H.N., Ray, M., Marrero, R., Grodzinski, P.: Plastic biochannel hybridization devices: a new concept for microfluidic DNA arrays. Anal. Biochem. **311**(1), 40–49 (2002)
146. Liu, J., Hansen, C., Quake, S.R.: Solving the "world-to-chip" interface problem with a microfluidic matrix. Anal. Chem. **75** (18), 4718–4723 (2003)
147. Martynova, L., Locascio, L.E., Gaitan, M., Kramer, G.W., Christensen, R.G., MacCrehan, W.A.: Fabrication of plastic microfluid channels by imprinting methods. Anal. Chem. **69**(23), 4783–4789 (1997)
148. Qi, S.Z., Liu, X.Z., Ford, S., Barrows, J., Thomas, G., Kelly, K., McCandless, A., Lian, K., Goettert, J., Soper, S.A.: Microfluidic devices fabricated in poly(methyl methacrylate) using hot-embossing with integrated sampling capillary and fiber optics for fluorescence detection. Lab Chip. **2**(2), 88–95 (2002)
149. Soper, S.A., Ford, S.M., Qi, S., McCarley, R.L., Kelly, K., Murphy, M.C.: Polymeric microelectromechanical systems. Anal. Chem. **72**(19), 642A–651A (2000)
150. Situma, C., Hashimoto, M., Soper, S.A.: Merging microfluidics with microarray-based bioassays. Biomol. Eng. **23**(5), 213–231 (2006)
151. Thorsen, T., Maerkl, S.J., Quake, S.R.: Microfluidic large-scale integration. Science. **298**(5593), 580–584 (2002)
152. Duffy, D.C., McDonald, J.C., Schueller, O.J.A., Whitesides, G.M.: Rapid prototyping of microfluidic systems in poly(dimethylsiloxane). Anal. Chem. **70**(23), 4974–4984 (1998)
153. Unger, M.A., Chou, H.P., Thorsen, T., Scherer, A., Quake, S.R.: Monolithic microfabricated valves and pumps by multilayer soft lithography. Science. **288**(5463), 113–116 (2000)
154. Marcus, J.S., Anderson, W.F., Quake, S.R.: Microfluidic single-cell mRNA isolation and analysis. Anal. Chem. **78**(9), 3084–3089 (2006)
155. Lee, C.C., Sui, G.D., Elizarov, A., Shu, C.Y.J., Shin, Y.S., Dooley, A.N., Huang, J., Daridon, A., Wyatt, P., Stout, D., Kolb, H.C., Witte, O.N., Satyamurthy, N., Heath, J.R., Phelps, M.E., Quake, S. R., Tseng, H.R.: Multistep synthesis of a radiolabeled imaging probe using integrated microfluidics. Science. **310**(5755), 1793–1796 (2005)
156. Balagadde, F.K., You, L.C., Hansen, C.L., Arnold, F.H., Quake, S. R.: Long-term monitoring of bacteria undergoing programmed population control in a microchemostat. Science. **309**(5731), 137–140 (2005)
157. Anderson, M.J., Hansen, C.L., Quake, S.R.: Phase knowledge enables rational screens for protein crystallization. Proc. Natl. Acad. Sci. U. S. A. **103**(45), 16746–16751 (2006)
158. Hansen, C.L., Classen, S., Berger, J.M., Quake, S.R.: A microfluidic device for kinetic optimization of protein crystallization and in situ structure determination. J. Am. Chem. Soc. **128** (10), 3142–3143 (2006)
159. Hansen, C.L., Sommer, M.O.A., Quake, S.R.: Systematic investigation of protein phase behavior with a microfluidic formulator. Proc. Natl. Acad. Sci. U. S. A. **101**(40), 14431–14436 (2004)
160. http://www.fluidigm.com
161. Schorzman, D.A., Desimone, J.M., Rolland, J.P., Quake, S.R., Van Dam, R.M.: Solvent- resistant Photocurable "liquid Teflon" for microfluidic device fabrication. J. Am. Chem. Soc. **126**(8), 2322–2323 (2004)
162. Melin, J., Quake, S.R.: Microfluidic large-scale integration: the evolution of design rules for biological automation. Annu. Rev. Biophys. Biomol. Struct. **36**, 213–231 (2007)
163. Kamotani, Y., Bersano-Begey, T., Kato, N., Tung, Y.C., Huh, D., Song, J.W., Takayama, S.: Individually programmable cell stretching microwell arrays actuated by a Braille display. Biomaterials. **29**(17), 2646–2655 (2008)
164. Song, J.W., Gu, W., Futai, N., Warner, K.A., Nor, J.E., Takayama, S.: Computer-controlled microcirculatory support system for endothelial cell culture and shearing. Anal. Chem. **77**(13), 3993–3999 (2005)
165. Gu, W., Zhu, X.Y., Futai, N., Cho, B.S., Takayama, S.: Computerized microfluidic cell culture using elastomeric channels and Braille displays. Proc. Natl. Acad. Sci. U. S. A. **101**(45), 15861–15866 (2004)
166. Enzelberger, M.M., Hansen, C.L., Liu, J., Quake, S.R., Ma, C. Nucleic acid amplification using microfluidic devices, WO/2002/081729. World Intellectual Property Organization; (2002)
167. Lee, C., Sui, G., Elizarov, A., Kolb, H.C., Huang, J., Heath, J.R., Phelps, M.E., Quake, S.R., Tseng, H., Wyatt, P. Microfluidic Devices with Chemical Reaction Circuits. EP Patent 1,838,431; (2007)
168. El-Ali, J., Sorger, P.K., Jensen, K.F.: Cells on chips. Nature. **442** (7101), 403–411 (2006)
169. Park, T.H., Shuler, M.L.: Integration of cell culture and microfabrication technology. Biotechnol. Prog. **19**(2), 243–253 (2003)
170. Sims, C.E., Allbritton, N.L.: Analysis of single mammalian cells on-chip. Lab Chip. **7**(4), 423–440 (2007)

171. Cheng, J.Y., Yen, M.H., Kuo, C.T., Young, T.H.: A transparent cell-culture micro chamber with a variably controlled concentration gradient generator and flow field rectifier. Biomicrofluidics. **2**(2), 12 (2008)
172. Petronis, S., Stangegaard, M., Christensen, C.B.V., Dufva, M.: Transparent polymeric cell culture chip with integrated temperature control and uniform media perfusion. BioTechniques. **40**(3), 368–376 (2006)
173. Park, J., Bansal, T., Pinelis, M., Maharbiz, M.M.: A microsystem for sensing and patterning oxidative microgradients during cell culture. Lab Chip. **6**(5), 611–622 (2006)
174. Maharbiz, M.M., Holtz, W.J., Sharifzadeh, S., Keasling, J.D., Howe, R.T.: A microfabricated electrochemical oxygen generator for high-density cell culture arrays. J. Microelectromech. Syst. **12** (5), 590–599 (2003)
175. Vollmer, A.P., Probstein, R.F., Gilbert, R., Thorsen, T.: Development of an integrated microfluidic platform for dynamic oxygen sensing and delivery in a flowing medium. Lab Chip. **5**(10), 1059–1066 (2005)
176. Ges, I.A., Ivanov, B.L., Schaffer, D.K., Lima, E.A., Werdich, A. A., Baudenbacher, F.J.: Thin-film IrOx pH microelectrode for microfluidic-based microsystems. Biosens. Bioelectron. **21**(2), 248–256 (2005)
177. Taff, B.M., Voldman, J.: A scalable addressable positive-dielectrophoretic cell-sorting array. Anal. Chem. **77**(24), 7976–7983 (2005)
178. Voldman, J., Gray, M.L., Toner, M., Schmidt, M.A.: A microfabrication-based dynamic array cytometer. Anal. Chem. **74** (16), 3984–3990 (2002)
179. Wang, X.B., Yang, J., Huang, Y., Vykoukal, J., Becker, F.F., Gascoyne, P.R.C.: Cell separation by dielectrophoretic field-flow-fractionation. Anal. Chem. **72**(4), 832–839 (2000)
180. Gomez-Sjoberg, R., Leyrat, A.A., Pirone, D.M., Chen, C.S., Quake, S.R.: Versatile, fully automated, microfluidic cell culture system. Anal. Chem. **79**(22), 8557–8563 (2007)
181. Hung, P.J., Lee, P.J., Sabounchi, P., Lin, R., Lee, L.P.: Continuous perfusion microfluidic cell culture array for high-throughput cell-based assays. Biotechnol. Bioeng. **89**(1), 1–8 (2005)
182. Lii, J., Hsu, W.J., Parsa, H., Das, A., Rouse, R., Sia, S.K.: Real-time microfluidic system for studying mammalian cells in 3D microenvironments. Anal. Chem. **80**(10), 3640–3647 (2008)
183. Madou, M., Zoval, J., Jia, G.Y., Kido, H., Kim, J., Kim, N.: Lab on a CD. Annu. Rev. Biomed. Eng. **8**, 601–628 (2006)
184. Balaban, N.Q., Schwarz, U.S., Riveline, D., Goichberg, P., Tzur, G., Sabanay, I., Mahalu, D., Safran, S., Bershadsky, A., Addadi, L., Geiger, B.: Force and focal adhesion assembly: a close relationship studied using elastic micropatterned substrates. Nat. Cell Biol. **3** (5), 466–472 (2001)
185. Tan, J.L., Tien, J., Pirone, D.M., Gray, D.S., Bhadriraju, K., Chen, C.S.: Cells lying on a bed of microneedles: an approach to isolate mechanical force. Proc. Natl. Acad. Sci. U. S. A. **100**(4), 1484–1489 (2003)
186. Hellmich, W., Pelargus, C., Leffhalm, K., Ros, A., Anselmetti, D.: Single cell manipulation, analytics, and label-free protein detection in microfluidic devices for systems nanobiology. Electrophoresis. **26**(19), 3689–3696 (2005)
187. Munce, N.R., Li, J.Z., Herman, P.R., Lilge, L.: Microfabricated system for parallel single-cell capillary electrophoresis. Anal. Chem. **76**(17), 4983–4989 (2004)
188. Klaus, J.W., George, S.M.: SiO2 chemical vapor deposition at room temperature using SiCl4 and H2O with an NH3 catalyst. J. Electrochem. Soc. **147**(7), 2658–2664 (2000)
189. Senturia, S.D.: Microsystem Design. Springer Science+Business Media, LLC, New York (2005)
190. Lim, K.S., Chang, W.J., Koo, Y.M., Bashir, R.: Reliable fabrication method of transferable micron scale metal pattern for poly (dimethylsiloxane) metallization. Lab Chip. **6**(4), 578–580 (2006)
191. Niu, X.Z., Peng, S.L., Liu, L.Y., Wen, W.J., Sheng, P.: Characterizing and patterning of PDMS- based conducting composites. Adv. Mater. **19**(18), 2682–2686 (2007)
192. Bowden, N., Brittain, S., Evans, A.G., Hutchinson, J.W., Whitesides, G.M.: Spontaneous formation of ordered structures in thin films of metals supported on an elastomeric polymer. Nature. **393**(6681), 146–149 (1998)
193. Trau, D., Jiang, J., Sucher, N.J.: Preservation of the biofunctionality of DNA and protein during microfabrication. Langmuir. **22**(3), 877–881 (2006)
194. Kentsch, J., Breisch, S., Stezle, M.: Low temperature adhesion bonding for BioMEMS. J. Micromech. Microeng. **16**(4), 802–807 (2006)
195. Ghafar-Zadeh, E., Sawan, M., Therriault, D.: Novel direct-write CMOS-based laboratory- on-chip: design, assembly and experimental results. Sensors Actuators A Phys. **134**(1), 27–36 (2007)
196. Zimmermann, S., Fienbork, D., Flounders, A.W., Liepmann, D.: In-device enzyme immobilization: wafer-level fabrication of an integrated glucose sensor. Sensors Actuators B Chem. **99**(1), 163–173 (2004)
197. Linder, V., Koster, S., Franks, W., Kraus, T., Verpoorte, E., Heer, F., Hierlemann, A., de Rooij, N.F.: Microfluidics/CMOS orthogonal capabilities for cell biology. Biomed. Microdevices. **8**(2), 159–166 (2006)
198. Pan, J.Y.: Reliability considerations for the BioMEMS designer. Proc. IEEE. **92**(1), 174–184 (2004)
199. Bhagat, A.A.S., Jothimuthu, P., Pais, A., Papautsky, I.: Re-usable quick-release interconnect for characterization of microfluidic systems. J. Micromech. Microeng. **17**(1), 42–49 (2007)
200. Christensen, A.M., Chang-Yen, D.A., Gale, B.K.: Characterization of interconnects used in PDMS microfluidic systems. J. Micromech. Microeng. **15**(5), 928–934 (2005)
201. Han, K.H., McConnell, R.D., Easley, C.J., Bienvenue, J.M., Ferrance, J.P., Landers, J.P., Frazier, A.B.: An active microfluidic system packaging technology. Sensors Actuators B Chem. **122**(1), 337–346 (2007)
202. Puntambekar, A., Ahn, C.H.: Self-aligning microfluidic interconnects for glass- and plastic-based microfluidic systems. J. Micromech. Microeng. **12**(1), 35–40 (2002)
203. Fujii, T., Sando, Y., Higashino, K., Fujii, Y.: A plug and play microfluidic device. Lab Chip. **3**(3), 193–197 (2003)
204. Igata, E., Arundell, M., Morgan, H., Cooper, J.M.: Interconnected reversible lab-on-a-chip technology. Lab Chip. **2**(2), 65–69 (2002)
205. Yuen, P.K.: SmartBuild–A truly plug-n-play modular microfluidic system. Lab Chip. **8**, 1374–1378 (2008)
206. Shaikh, K.A., Ryu, K.S., Goluch, E.D., Nam, J.M., Liu, J.W., Thaxton, S., Chiesl, T.N., Barron, A.E., Lu, Y., Mirkin, C.A., Liu, C.: A modular microfluidic architecture for integrated biochemical analysis. Proc. Natl. Acad. Sci. U. S. A. **102**(28), 9745–9750 (2005)
207. Ko, W.H.: Packaging of microfabricated devices and systems. Mater. Chem. Phys. **42**(3), 169–175 (1995)
208. Murarka, S.P.: Multilevel interconnections for ULSI and GSI era. Mater. Sci. Eng. R. Rep. **19**(3–4), 87–151 (1997)
209. Tong, H.M.: Microelectronics packaging – present and future. Mater. Chem. Phys. **40**(3), 147–161 (1995)
210. James, C.D., Spence, A.J.H., Dowell-Mesfin, N.M., Hussain, R.J., Smith, K.L., Craighead, H.G., Isaacson, M.S., Shain, W., Turner, J. N.: Extracellular recordings from patterned neuronal networks using planar microelectrode arrays. IEEE Trans. Biomed. Eng. **51** (9), 1640–1648 (2004)

211. Fair, R.B.: Digital microfluidics: is a true lab-on-a-chip possible? Microfluid. Nanofluid. **3**(3), 245–281 (2007)
212. Hartley, L., Kaler, K., Yadid-Pecht, O.: Hybrid integration of an active pixel sensor and microfluidics for cytometry on a chip. IEEE Trans. Circuits Syst. Regul. Pap. **54**(1), 99–110 (2007)
213. Huang, Y., Yang, J.M., Hopkins, P.J., Kassegne, S., Tirado, M., Forster, A.H., Reese, H.: Separation of simulants of biological warfare agents from blood by a miniaturized dielectrophoresis device. Biomed. Microdevices. **5**(3), 217–225 (2003)
214. Petrou, P.S., Moser, I., Jobst, G.: BioMEMS device with integrated microdialysis probe and biosensor array. Biosens. Bioelectron. **17** (10), 859–865 (2002)
215. Piruska, A., Nikcevic, I., Lee, S.H., Ahn, C., Heineman, W.R., Limbach, P.A., Seliskar, C.J.: The autofluorescence of plastic materials and chips measured under laser irradiation. Lab Chip. **5** (12), 1348–1354 (2005)
216. Bliss, C.L., McMullin, J.N., Backhouse, C.J.: Rapid fabrication of a microfluidic device with integrated optical waveguides for DNA fragment analysis. Lab Chip. **7**(10), 1280–1287 (2007)
217. Lee, K.S., Lee, H.L.T., Ram, R.J.: Polymer waveguide backplanes for optical sensor interfaces in microfluidics. Lab Chip. **7**(11), 1539–1545 (2007)
218. Chung, K., Crane, M.M., Lu, H.: Automated on-chip rapid microscopy, phenotyping and sorting of *C.elegans*. Nat. Methods. **5**(7), 637–643 (2008)
219. El-Ali, J., Gaudet, S., Gunther, A., Sorger, P.K., Jensen, K.F.: Cell stimulus and lysis in a microfluidic device with segmented gas-liquid flow. Anal. Chem. **77**(11), 3629–3636 (2005)
220. Voskerician, G., Shive, M.S., Shawgo, R.S., von Recum, H., Anderson, J.M., Cima, M.J., Langer, R.: Biocompatibility and biofouling of MEMS drug delivery devices. Biomaterials. **24**(11), 1959–1967 (2003)
221. Brischwein, M., Motrescu, E.R., Cabala, E., Otto, A.M., Grothe, H., Wolf, B.: Functional cellular assays with multiparametric silicon sensor chips. Lab Chip. **3**(4), 234–240 (2003)
222. Szarowski, D.H., Andersen, M.D., Retterer, S., Spence, A.J., Isaacson, M., Craighead, H.G., Turner, J.N., Shain, W.: Brain responses to micro-machined silicon devices. Brain Res. **983** (1–2), 23–35 (2003)
223. http://www.mchips.com
224. Shawgo, R.S., Grayson, A.C.R., Li, Y.W., Cima, M.J.: BioMEMS for drug delivery. Curr. Opin. Solid State Mater. Sci. **6**(4), 329–334 (2002)
225. Fallahi, D., Mirzadeh, H., Khorasani, M.T.: Physical, mechanical, and biocompatibility evaluation of three different types of silicone rubber. J. Appl. Polym. Sci. **88**(10), 2522–2529 (2003)
226. Mata, A., Fleischman, A.J., Roy, S.: Characterization of polydimethylsiloxane (PDMS) properties for biomedical micro/nanosystems. Biomed. Microdevices. **7**(4), 281–293 (2005)
227. Lee, J.N., Jiang, X., Ryan, D., Whitesides, G.M.: Compatibility of mammalian cells on surfaces of poly(dimethylsiloxane). Langmuir. **20**(26), 11684–11691 (2004)
228. Millet, L.J., Stewart, M.E., Sweedler, J.V., Nuzzo, R.G., Gillette, M.U.: Microfluidic devices for culturing primary mammalian neurons at low densities. Lab Chip. **7**(8), 987–994 (2007)
229. Kim, L., Toh, Y.C., Voldman, J., Yu, H.: A practical guide to microfluidic perfusion culture of adherent mammalian cells. Lab Chip. **7**(6), 681–694 (2007)
230. Kasemo, B.: Biological surface science. Surf. Sci. **500**(1–3), 656–677 (2002)
231. Makamba, H., Kim, J.H., Lim, K., Park, N., Hahn, J.H.: Surface modification of poly(dimethylsiloxane) microchannels. Electrophoresis. **24**(21), 3607–3619 (2003)
232. Fritz, J.L., Owen, M.J.: Hydrophobic recovery of plasma-treated polydimethylsiloxane. J. Adhes. **54**(1), 33–45 (1995)
233. Hu, S., Ren, X., Bachman, M., Sims, C.E., Li, G.P., Allbritton, N.: Surface modification of poly(dimethylsiloxane) microfluidic devices by ultraviolet polymer grafting. Anal. Chem. **74**(16), 4117–4123 (2002)
234. Slentz, B.E., Penner, N.A., Regnier, F.E.: Capillary electrochromatography of peptides on microfabricated poly(dimethylsiloxane) chips modified by cerium(IV)-catalyzed polymerization. J. Chromatogr. A. **948**(1–2), 225–233 (2002)
235. Ocvirk, G., Munroe, M., Tang, T., Oleschuk, R., Westra, K., Harrison, D.J.: Electrokinetic control of fluid flow in native poly (dimethylsiloxane) capillary electrophoresis devices. Electrophoresis. **21**(1), 107–115 (2000)
236. Dou, Y.H., Bao, N., Xu, J.J., Chen, H.Y.: A dynamically modified microfluidic poly(dimethylsiloxane) chip with electrochemical detection for biological analysis. Electrophoresis. **23**(20), 3558–3566 (2002)
237. Decher, G.: Fuzzy nanoassemblies: toward layered polymeric multicomposites. Science. **277**(5330), 1232–1237 (1997)
238. Sung, W.C., Chang, C.C., Makamba, H., Chen, S.H.: Long-term affinity modification on poly(dimethylsiloxane) substrate and its application for ELISA analysis. Anal. Chem. **80**(5), 1529–1535 (2008)
239. Hanein, Y., Pan, Y.V., Ratner, B.D., Denton, D.D., Bohringer, K. F.: Micromachining of non-fouling coatings for bio-MEMS applications. Sensors Actuators B Chem. **81**(1), 49–54 (2001)
240. Lopez, G.P., Ratner, B.D., Tidwell, C.D., Haycox, C.L., Rapoza, R.J., Horbett, T.A.: Glow- discharge plasma deposition of Tetraethylene glycol dimethyl ether for fouling-resistant biomaterial surfaces. J. Biomed. Mater. Res. **26**(4), 415–439 (1992)
241. Dhayal, M., Choi, J.S., So, C.H.: Biological fluid interaction with controlled surface properties of organic micro-fluidic devices. Vacuum. **80**(8), 876–879 (2006)
242. Bajaj, P., Akin, D., Gupta, A., Sherman, D., Shi, B., Auciello, O., Bashir, R.: Ultrananocrystalline diamond film as an optimal cell interface for biomedical applications. Biomed. Microdevices. **9**(6), 787–794 (2007)
243. Hoivik, N.D., Elam, J.W., Linderman, R.J., Bright, V.M., George, S.M., Lee, Y.C.: Atomic layer deposited protective coatings for micro-electromechanical systems. Sensors Actuators A Phys. **103** (1–2), 100–108 (2003)
244. Wang, Y.L., Pai, J.H., Lai, H.H., Sims, C.E., Bachman, M., Li, G. P., Allbritton, N.L.: Surface graft polymerization of SU-8 for bio-MEMS applications. J. Micromech. Microeng. **17**(7), 1371–1380 (2007)
245. Wang, Y.L., Bachman, M., Sims, C.E., Li, G.P., Allbritton, N.L.: Simple photografting method to chemically modify and micropattern the surface of SU-8 photoresist. Langmuir. **22**(6), 2719–2725 (2006)
246. Nordstrom, M., Marie, R., Calleja, M., Boisen, A.: Rendering SU-8 hydrophilic to facilitate use in micro channel fabrication. J. Micromech. Microeng. **14**(12), 1614–1617 (2004)
247. Joshi, M., Kale, N., Lal, R., Rao, V.R., Mukherji, S.: A novel dry method for surface modification of SU-8 for immobilization of biomolecules in Bio-MEMS. Biosens. Bioelectron. **22**(11), 2429–2435 (2007)
248. Chen, C.S., Mrksich, M., Huang, S., Whitesides, G.M., Ingber, D. E.: Micropatterned surfaces for control of cell shape, position, and function. Biotechnol. Prog. **14**(3), 356–363 (1998)
249. Duncan, A.C., Weisbuch, F., Rouais, F., Lazare, S., Baquey, C.: Laser microfabricated model surfaces for controlled cell growth. Biosens. Bioelectron. **17**(5), 413–426 (2002)

250. Duncan, A.C., Rouais, F., Lazare, S., Bordenave, L., Baquey, C.: Effect of laser modified surface microtopochemistry on endothelial cell growth. Colloids Surf. B. Biointerfaces. **54**(2), 150–159 (2007)
251. Edell, D.J., Toi, V.V., McNeil, V.M., Clark, L.D.: Factors influencing the biocompatibility of insertable silicon microshafts in cerebral-cortex. IEEE Trans. Biomed. Eng. **39**(6), 635–643 (1992)
252. Hoogerwerf, A.C., Wise, K.D.: A 3-dimensional microelectrode array for chronic neural recording. IEEE Trans. Biomed. Eng. **41** (12), 1136–1146 (1994)
253. Schmidt, S., Horch, K., Normann, R.: Biocompatibility of silicon-based electrode arrays implanted in feline cortical tissue. J. Biomed. Mater. Res. **27**(11), 1393–1399 (1993)

Packaging of Biomolecular and Chemical Microsensors

13

Peter J. Hesketh and Xiaohui Lin

Abbreviations

3D	Three dimensional
AFM	Atomic force microscope
ALD	Atomic layer deposition
BCB	Benzocyclobutene
CAAT	Silver-coated nickel particles in foil
CAD	Computer-aided design
CE	Capillary electrophoresis
CHEMFET	Chemically sensitive field effect transistor
COC	Cyclic olefin copolymer
CVD	Chemical vapor deposition
DIL	Dual in line
DMFC	Direct methanol fuel cell
DRIE	Deep reactive ion etching
FEA	Finite element analysis
FIB	Focused ion beam
ISFET	Ion selective field effect transistor
LPC	Liquid phase chromatography
MEMS	Microelectromechanical systems
MSI	Microfluidic system interface
MTS	Mercapto propyl trimethoxy silane
NEMS	Nano electromechanical systems
OD	Outside diameter
PCB	Printed circuit board
PCR	Polymerase chain reaction
PDMS	Polydimethylsiloxane or silicone rubber
PI	Polyimide
PMMA	Polymethylmethacrylate
PTFE	Polytetrafluoroethylene
RIE	Reactive ion etching
SAW	Surface acoustic wave
SEM	Scanning electron microscope
SLA	Stereolithographic assembly
SOI	Silicon on insulator
SPE	Solid-phase extraction
STL	Stereolithography interface format
SU-8	SU 8 epoxy resin
USB	Universal system bus
UV	Ultraviolet
VLSI	Very large scale integration
WLP	Wafer-level packaging

P. J. Hesketh (✉) · X. Lin
School of Mechanical Engineering, Georgia Institute of Technology, Atlanta, GA, USA
e-mail: peter.hesketh@me.gatech.edu; xlin3@mail.gatech.edu

Symbols

Ag	Silver
Au	Gold
Co	Cobalt
Cu	Copper
Fe	Iron
Ni	Nickel
Tg	Glass transition temperature
TiN	Titanium nitride
Pd	Palladium
Pt	Platinum

13.1 Introduction

The purpose of packaging a chemical sensor or biosensors is to expose a region to the analyte of interest, while providing mechanical support and both electrical and fluid connections for the sensor. The mechanical support is important, for example, if a membrane has to be held in place against the

C. P. Wong et al. (eds.), *Nano-Bio- Electronic, Photonic and MEMS Packaging*, https://doi.org/10.1007/978-3-030-49991-4_13

sensor. There are additional specific functions for a package. It can also include selective filter(s) to remove specific interfering species or a valve/pump to control the sample flow over the sensor or sensor array. The package also defines the volume of analyte in which a measurement is made, when configured in a microfluidic format. Finally, the thermal environment of the sensor is affected by the package, and it is important because the rate of reaction for chemical sensors is a function of temperature. In addition, many biological assays must be carried out at specific temperatures and have temperature limits to avoid denaturation of biological selective molecules.

Packages for electronic devices are well developed [1] and for electronic microsystems [2]. Commercially available semiconductor packages can be and often are adapted for sensor applications. For example, by mounting the die in the semiconductor package header, any customization can be accomplished with modifications to the cover. The sensor or sensor array will be exposed to the analytes, while electrical contacts are made through wire bonding. This approach has been very successful in many applications, particularly for gas sensors. There are few standard packages defined for sensors. Low-cost packaging for sensors is discussed by Tomescu [3] and in a recent book chapter by Brand [4]. Packaging for BioMEMS is discussed by Velten et al. [5], and some case studies for MEMS packaging are described by Lee et al. [6] and suggestions for future work along with some of the unsolved problems in packaging.

The functions of a package can be summarized as follows:

- Provide mechanical support and protection for sensor in the intended application
- Provide electrical connections to the sensor(s)
- Deliver the sample to the sensing interface
- Control thermal environment by creating regions with differing operating temperatures when required
- Protect any electronic sensor interface circuits from the environment
- Provide convenient integration at the system level
- Provide ease of calibration for sensor(s)
- Improve reliability
- Automate assembly
- Identify component part and date of manufacture

For nanotechnology the issue of packaging has not yet been developed. A future challenging issue for packaging is specifically how to take advantage of scaling for nanoscale sensors. In fact, both the alignment and electrical contacts to the nanowires have to be carefully considered. When the nanowires are grown in place with a catalytic seed, the placement is predetermined by the catalyst, however; otherwise, alignment has to be carried out with methods such as fluid flow, electric field, or AFM probes to direct the position of the nanowires.

When we consider the intended duration for use of the sensor, it has an important influence on the package type and material selection. When short-term use is required then stability and corrosion resistance are not as much a concern; however, in long-term use, these issues, in addition to hermeticity for protection of electronic interface circuit, also will be a requirement. For chemical and biomolecular sensors non-hermetic packaging is necessary because the sensor interface needs to be in contact with the environment to provide the sensing function. However, other MEMS devices such as microresonators require hermeticity in an inert environment or a vacuum to improve performance and stability.

Electrical connections are made through wire bonding or flip-chip bonding. Methods developed for semiconductor devices are routinely used for sensors. Packaging is a large topic and can hardly be covered in one chapter; therefore the scope here is limited to packaging for chemical and biomolecular sensors. Selected examples are used to illustrate the range of different approaches that have been developed, often for specific applications.

13.2 Chemical Sensors

A plethora of chemical sensing principles have been applied in a miniature format. Many excellent reviews of chemical sensors have been published, for example, the texts by Gardner [7], Peirce [8], Kovacs [9], and the seven volume series published by Wiley-VCH [10]. Since the first development of the ion selective field effect transistor (ISFETs) [11], chemically sensitive field effect transistor (CHEMFETs) [12], ion-controlled diodes [13], palladium gate gas sensor [14], and miniature Clark electrode [15], there has been a host of new development. Other miniature chemical sensors are based on miniaturization of larger-scale sensors: electrochemical sensor electrodes [16] or the pelistor or semiconductor metal oxides gas sensor for high temperatures and metal semiconductor (MOS) gas sensors.

13.2.1 ISFET Packaging

A recent review of ISFET devices is provided by Poghossian and Schoning [17]. Here we are concerned with the development of suitable packaging for the ISFET. Approaches include that of Smith et al. [18] which are based on precision hand assembly. The packaging of ISFETs has been challenging to avoid instability and drift that arise due to changes in impedance and leakage currents [19, 20] or aluminum

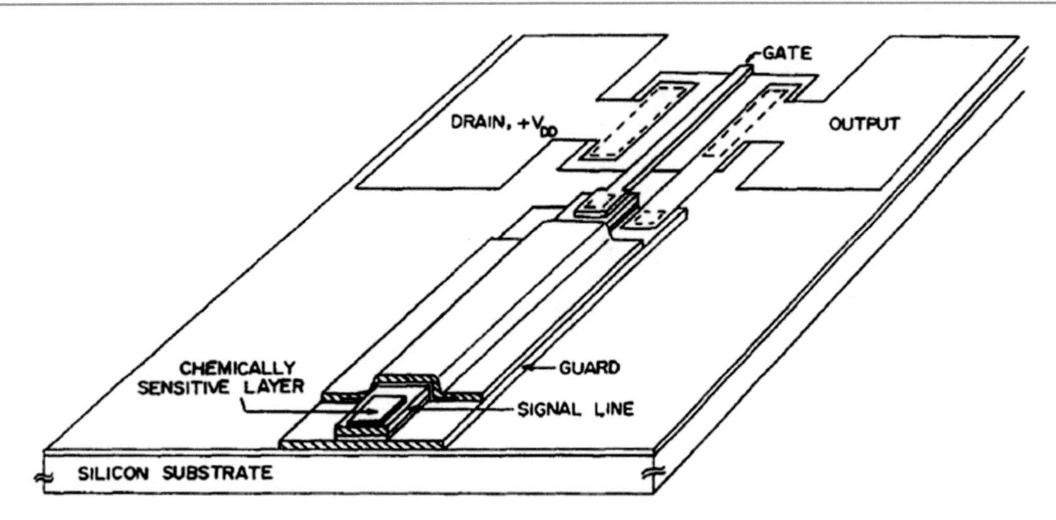

Fig. 13.1 Schematic diagram of extended gate field effect transistor. (Reprinted from Ref. [22] with permission from Elsevier © (1983))

electrodes neighboring the gate region to indicate degradation of the encapsulation [21]. The extended gate FET developed by van der Spiegel et al. [22, 23] located the chemically sensitive region a distance from the electronic device to reduce the requirements on encapsulation, as shown in Fig. 13.1. This method has been applied to tin oxide sensing by Chi et al. [24]. One of the key issues of using ISFETs is the necessity for a reference electrode, unless the device is operated as a Kelvin Probe [25]. This implies the system should include packaging of the miniature reference electrode to provide a functional measurement system. The advantages of miniaturization of FETs will be important for arrays where the sensors can share the same reference electrode. Differential FETs where one device is passivated or a device is functionalized for different analyte sensitivity provide a method of addressing the reference without a reference electrode. Back contact ISFETs were developed by van den Vlekkert et al. [26] with a novel micromachining process to keep the electrical contact isolated from the working fluid by placing them on the other side of the wafer; see Fig. 13.2. In addition, needle-type ISFETs have been demonstrated by Matsuo and Esashi [27], and the needle-type ISFETs extend the device into a probe which might be particularly useful for biomedical applications.

Photocurable polymer encapsulation was investigated by several groups [28–30]. A novel three-dimensional ISFET built on a mesa structure has been developed to facilitate packaging of the devices [31] as shown in Fig. 13.3. A Kapton (a flexible polyimide film, trademark of DuPont) foil is defined over the device which has electrical contacts integrated. Assembly which involves a manual alignment and bonding with a controlled collapse structure and heating to 300 °C with pressure is applied for 2 min. Epoxy resin is poured over the back of the Kapton tape to support the structure.

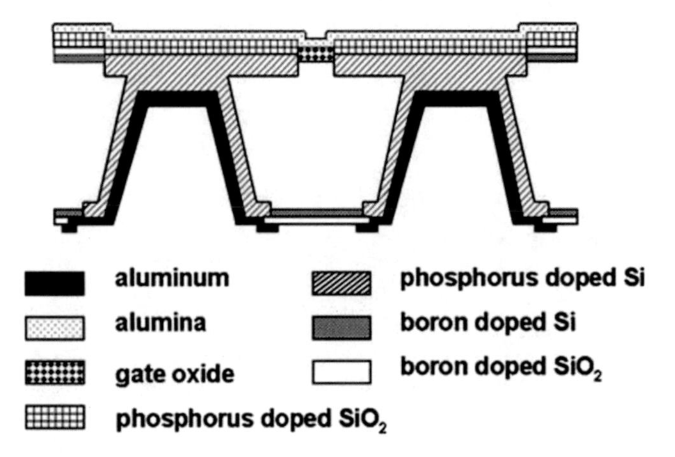

Fig. 13.2 Cross-sectional schematic diagram of a FET with back contact. (Reprinted from Ref. [26] with permission from Elsevier © (1988))

The photopolymerization of encapsulation resin provides encapsulation that is effective and has low leakage currents. The process involves first coating with an adhesion promoter, mercapto propyl trimethoxy silane (MTS), to ensure attachment of the resin to the silicon oxide/nitride layer of the ISFET followed by several coatings, each of 30 μm thickness, exposure to UV radiation in the areas where the polymer should be removed from the gate and contact pads, and finally development in isopropyl alcohol and curing. Figure 13.4 shows the summary of the process and the encapsulated sensor package.

A large array of ISFETs has been studied for extracellular recording by Milgrew et al. [32]. Figure 13.5 shows the array of 16 × 16 ISFETS along with signal conditioning and readout circuitry. The microfluidic chamber containing the cells is positioned over the array so that the cell culture can be grown over the array. A SU-8 photodefined channel is used to define the chamber, as indicated in the figure with a

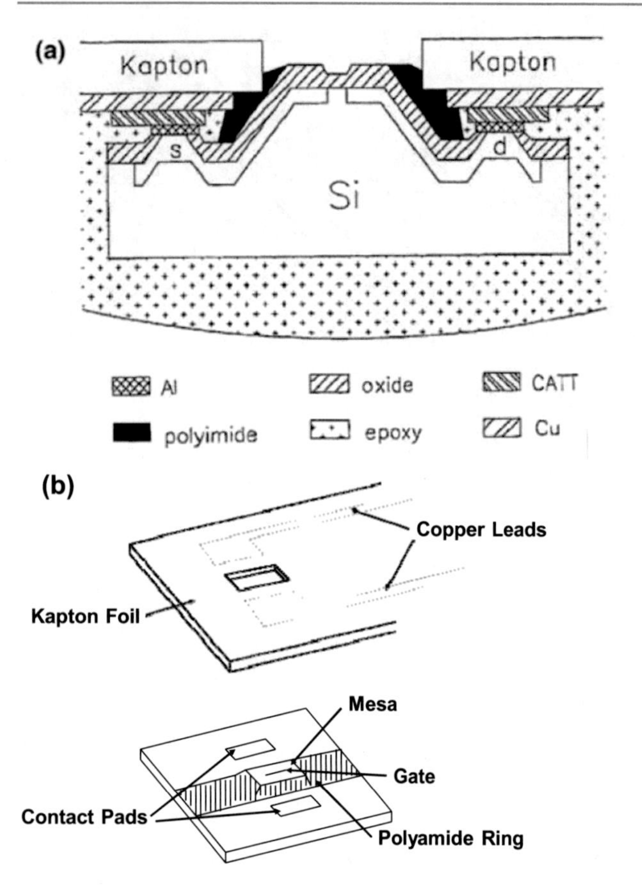

Fig. 13.3 (**a**) Cross-sectional view of a fully encapsulated mesa ISFET and (**b**) alignment of the Kapton foil over the mesa ISFET. (Reprinted from Ref. [31] with permission MYU publishing (Tokyo))

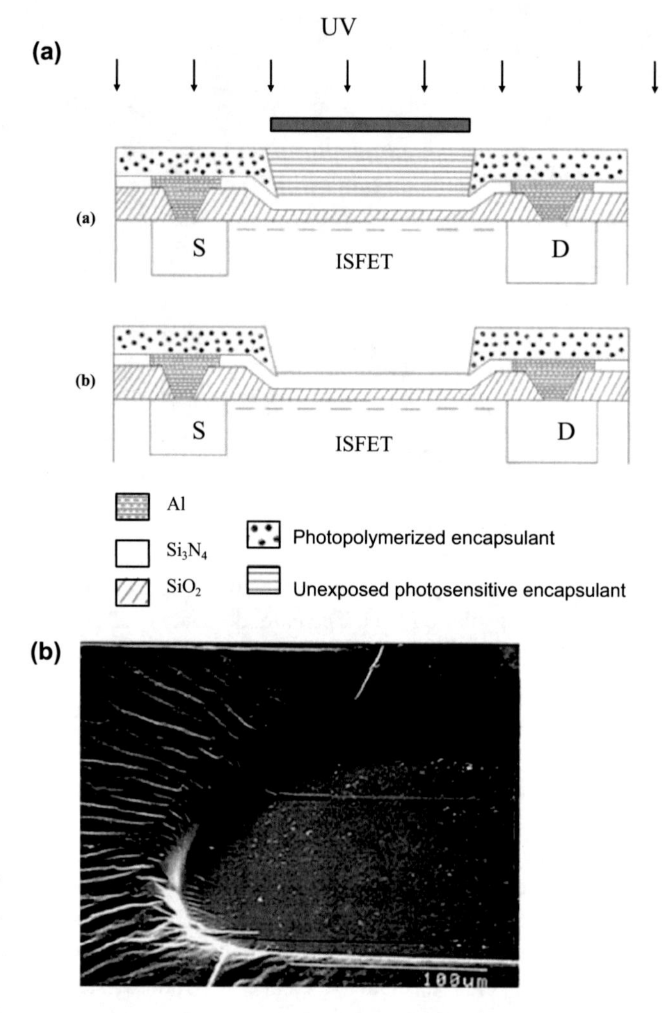

Fig. 13.4 (**a**) Schematic diagram of the direct photopolymerization of UV curable encapsulants on ISFET devices and (**b**) micrograph of gate-photocurable encapsulant for a specific formulation. (Reprinted from Ref. [29] with permission from Elsevier © (2005))

polydimethylsiloxane (PDMS)-molded structure defining the internal volume for the measurements. Electrical connections to the ISFETS are made with wire bonding and SU-8 (a photodefinable multifunction bis-A type epoxy) coating to provide electrical insulation on the PC board mounted die.

A comprehensive review of ISFET packaging is provided by Oelβner et al. [33]. Technology for encapsulation of ISFETs includes the use of molds developed at KSI Meinsberg (Germany). The molds are a methylpentene copolymer produced with an injection molding. They provide the shape for casting of the epoxy resin over the sensor body and integration of electrical wires. Figure 13.6 shows the ISFET-encapsulated sensor package, where the gate region is in a small circular region, or can be extended to include reference FET junctions. Probes can be placed in small fluid volumes and thereby allow study of corrosion, applications in food processes, or medical applications including ophthalmology, dermatology, gynecology, and dentistry. For example, Fig. 13.7 shows a needle-based encapsulation design with a preformed mold, as shown, for a single or multiple ISFET devices. Figure 13.8 shows the straightforward encapsulation technology processing steps that use a thermoplastic elastomeric mold and heat-cured epoxy resin injected into the mold.

13.2.2 Microhotplate Packaging

Metal oxide chemical sensors require controlled elevated temperature for their correct operation. The package must provide thermal control and isolation in addition to fluidic and electrical connections to the device. Early work on microhotplates was pioneered by Suehle et al. [34], thereafter further improvements in tin oxide (SnO2) film quality via nanostructured microspheres [35] and in sensing platform via nanostructured SnO2 grains [36] have been demonstrated. Excellent reviews are published by Simon et al. [37] and Semancik and Cavicchi [38].

Figure 13.9 shows a schematic diagram of a micromachined gas sensor developed by Motorola Inc. [39]. The tin oxide is formed by rheotaxial growth on a thin silicon nitride membrane to provide thermal isolation and hence reduce the power consumption in the range 30–150 mW. This allows the chip support to operate close to ambient temperature so that standard semiconductor

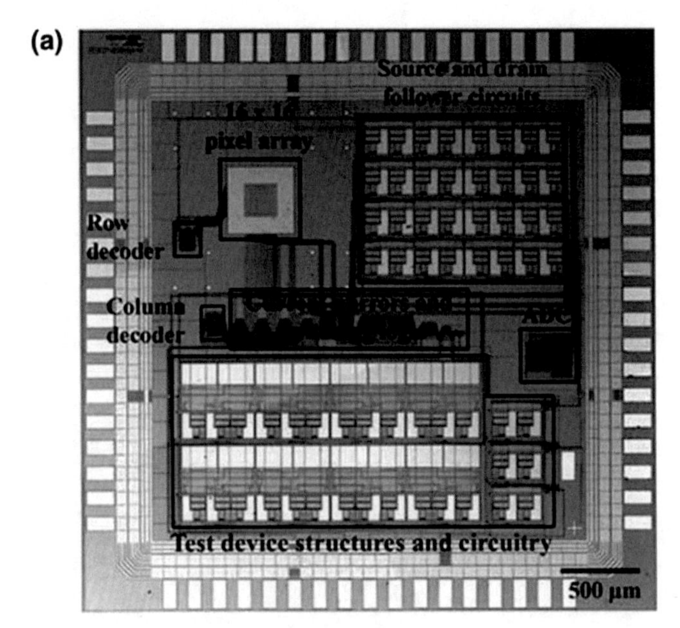

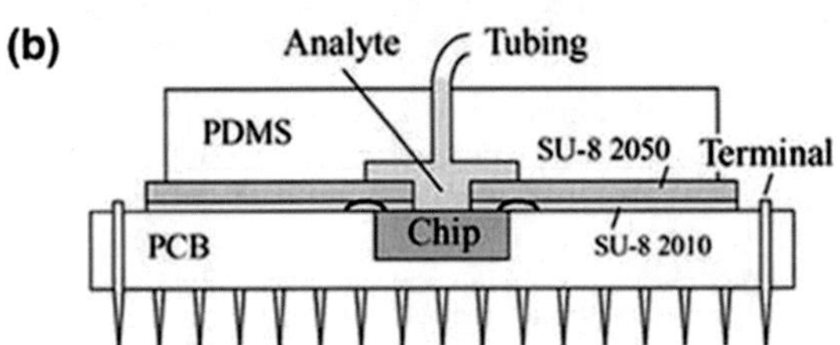

Fig. 13.5 (**a**) Micrograph of an integrated ISFET sensor array chip and (**b**) schematic diagram of the chip packaging process. (Reprinted from Ref. [32] with permission of the ECS)

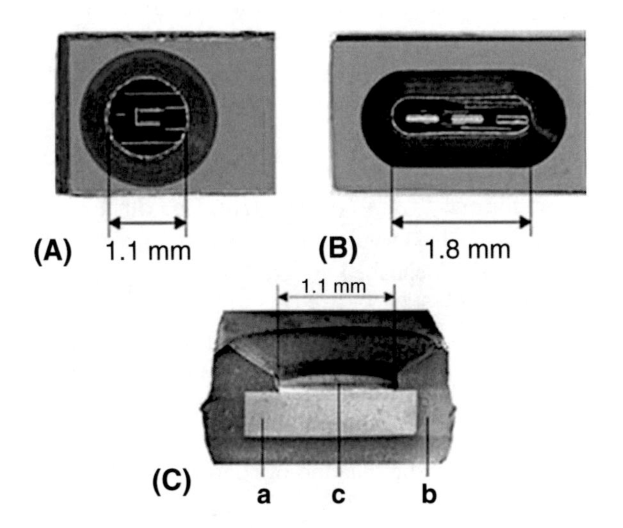

Fig. 13.6 (**a**) Circular shape and 1.1 mm diameter measuring cavity; (**b**) elliptical-shaped cavity with three sensors; and (**c**) micrograph of an ISFET sensor across the cavity: (**a**) ISFET chip, (**b**) encapsulation, (**c**) measuring cavity. (Reprinted from Ref. [33] with permission from Elsevier © (2005))

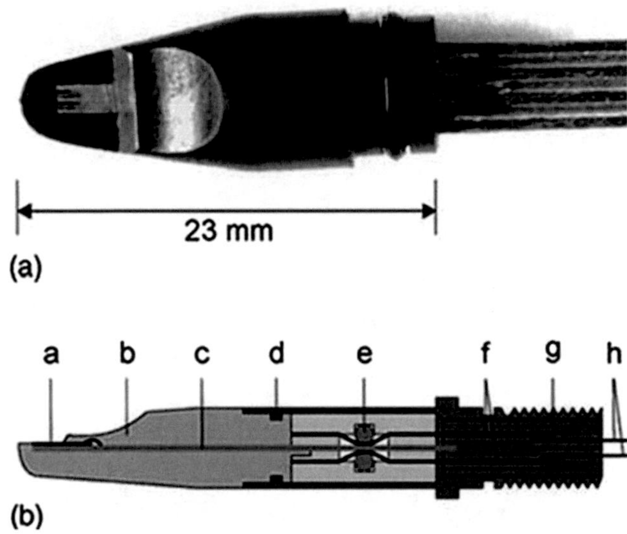

Fig. 13.7 (**a**) Photograph of an ISFET sensor element encapsulated according to molding process shown in Fig. 13.8 by KSI, Meinsberg, Germany. (**b**) Inserting and contacting the ISFET element of the sensor include: (**a**) ISFET chip, (**b**) epoxy encapsulation, (**c**) printed circuit board, (**d**) O-ring sealing, (**e**) elastic contacting elements, (**f**) flexible printed circuit boards, (**g**) sensor stainless steel shaft, and (**h**) connecting wires. (Reprinted from Ref. [33] with permission from Elsevier © (2005))

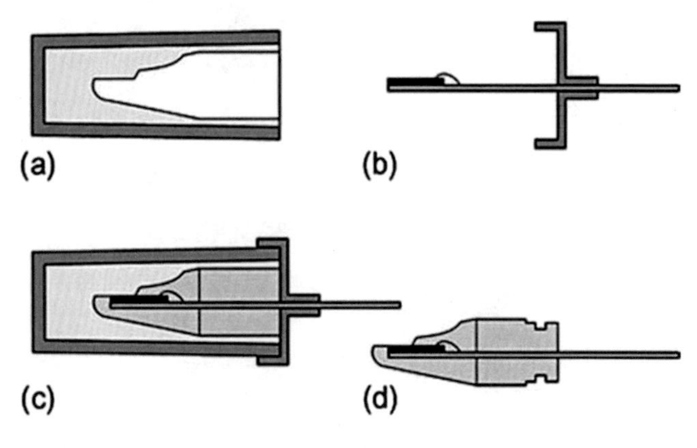

Fig. 13.8 Technology processing steps for heat-curing epoxy molding for the encapsulation of ISFET chips at the laboratory scale: (**a**) molds made from thermoplastic elastomer; (**b**) the chip is glued onto a circuit board and bonded and fixed to the cover of mold; (**c**) the chip carrier is positioned in the mold, cast, and heat cured with epoxy under specific conditions; and (**d**) only minor finishing work is necessary after the mold release. (Reprinted from Ref. [33] with permission from Elsevier © (2005))

package can be used as indicated in Fig. 13.9b. A charcoal filter is included to remove unwanted gases that might interfere with tin oxide response and a nylon cap with mesh opening for mechanical protection. The reliability of the thin silicon oxide–nitride membrane to thermal and

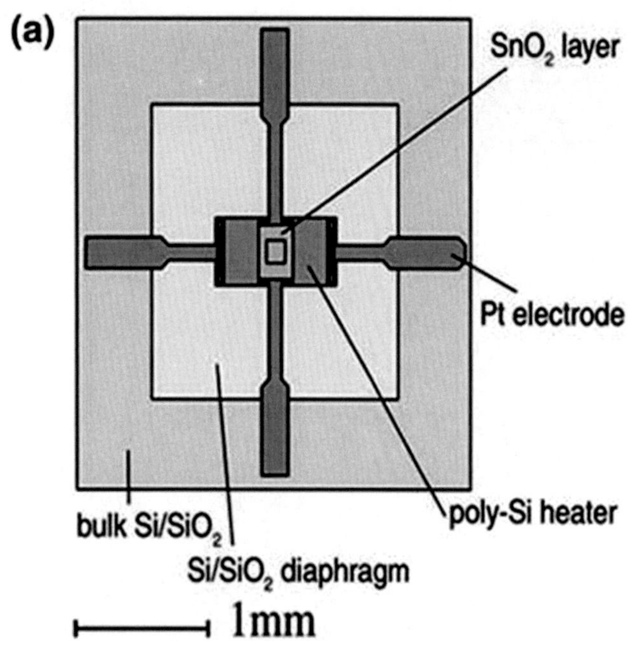

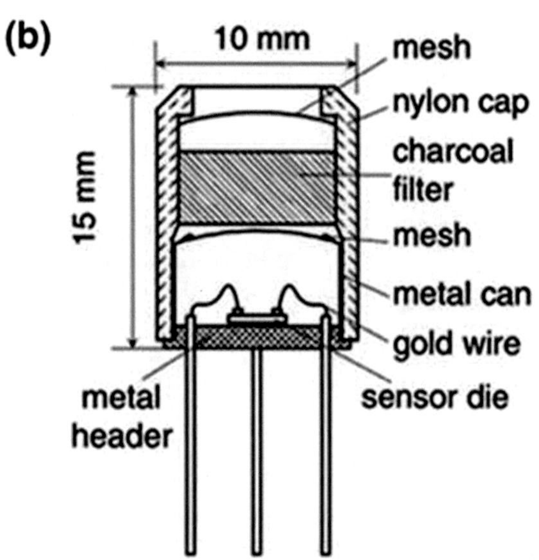

Fig. 13.9 Schematic diagram of the MGS 1100 sensor from Motorola Inc. The micromachined element on the *left* and the sensor housing on the *right*. The sensitive film was obtained by rheotaxial growth and thermal oxidation of tin layers deposited on to the silicon oxide–nitride membrane. (Reprinted from Ref. [37] with permission Elsevier © (2001))

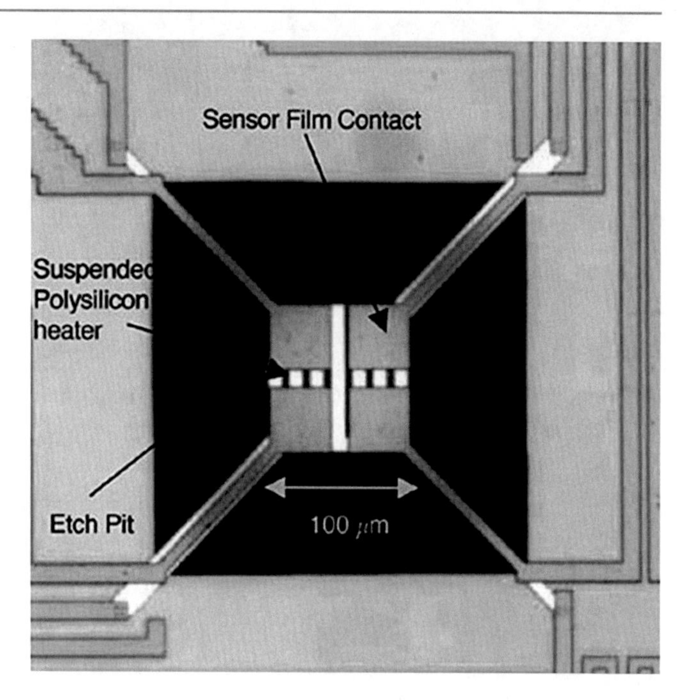

Fig. 13.10 Photograph of a typical four-element microhotplate to sense film deposition. The thin silicon dioxide structure 100 × 100 μm with a serpentine polysilicon heater is suspended with four silicon dioxide beams which have embedded polysilicon runners. The four TiN contact electrodes are defined on the top of the microhotplate. (Reprinted with permission from Ref. [41] © 2005 IEEE)

mechanical shock was investigated with accelerated testing to establish reliability for this device in real-world applications [39].

Figure 13.10 shows a microhotplate which utilizes a polysilicon heater and operates up to 500 °C with just 22 mW of electrical power. Femtomolar isothermal desorption has been carried out by Shirke et al. [40] with heating rates of up to 10^6 °C/s and minimal power consumption due to the small thermal mass of the microhotplates. The technique has been demonstrated for the isothermal desorption of benzoic acid from a single SnO2 covered microhotplate at 23–174 °C. In addition, the microhotplate has been demonstrated as a sensitive device for the detection of chemical warfare agents at 5 nmol/mol (ppb) concentration levels, and it shows stable operation for 14 h of use at elevated temperatures and with pulsed operation of 200 ms over the temperature range of room temperature to 480 °C [41].

Sensitive and selective response using microhotplates with ultrathin metal film coatings (Ag, Co, Cu, Fe, Ni) to volatile organics has been reported by Panchapakesan et al. [36]. The SnO2 is deposited by selective CVD by heating the microhotplates to 500 °C; see Fig. 13.11. The amount of material deposited was monitored by recording when the resistance reaches 2 k ohms. The resultant SnO2 grain size was in the range of 20–121 nm. Stable sensor operation for 200 h with methanol is reported at 80 ppm concentration levels. Finally, the combination of nanowires of SnO2 with the microhotplates has been demonstrated to produce enhanced performance at lower operating temperature, reduced power consumption, and good stability [42]. Figure 13.12 shows the hotplate with a single SnO2 nanowire integrated with FIB deposition necessary to make an electrical contact to the nanowire. The sensor response to reducing gases was observed, and due to the low thermal mass, a rapid

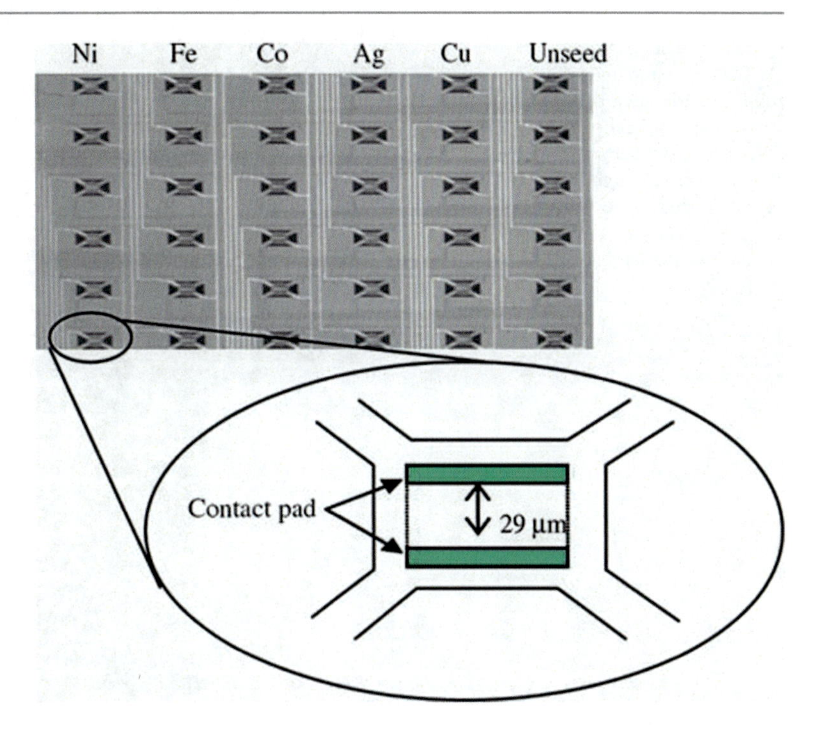

Fig. 13.11 Optical image and schematic illustration of the 36-element microhotplate array. Each hotplate has a different metal catalyst deposited onto it. The amount of material is monitored by measuring the conductivity between pair of contact pads. (Reprinted from Ref. [36] with permission of IOP Publishing)

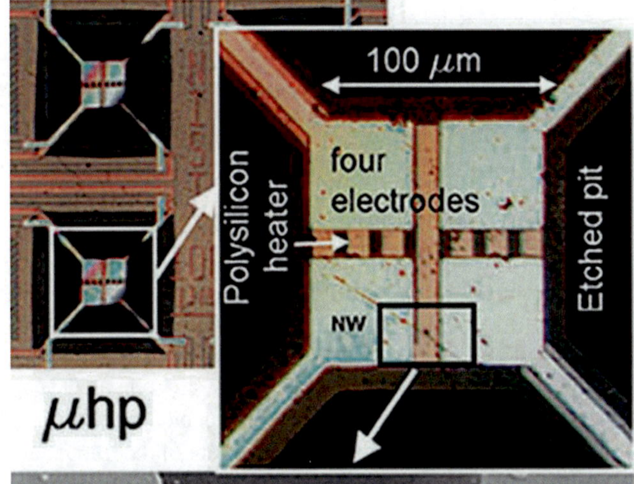

Fig. 13.12 (**a**) Optical image of a typical microhotplate and (**b**) SEM image of the individual tin oxide nanowire integrated as a chemiresistor using FIB technology. (Reprinted from Ref. [42] with permission of AIP)

examination of response as a function of temperature could be carried out to determine optimum operating conductions for highest gas sensitivity.

The KAMINA microarray tin oxide sensor developed by Goschnick [43] has 38 segments and operates at 1 W power consumption with a thin film of porous silica over the array with a gradient of thickness to produce differential response to different gases. The sensors operate at an elevated temperature, typically 300 °C, maintained by thin film Pt heaters on the back of the substrate, as indicated in the insert of Fig. 13.13a. The sensor is wire bonded into a ceramic board, shown in Fig. 13.13b. The surface temperature is a function of the gas flow in this package, which can vary by up to 10 °C; however, with the addition of a heat exchanger at the inlet, built-in stainless steel inlet, this variation is greatly reduced [44]. Finally, in the lid of the housing there is also included a temperature and humidity sensor (Fig. 13.13c).

Polyimide-based microhotplates offer low cost and low thermal conductivity substrate for tin oxide gas sensors, thereby lowering the power consumption compared to silicon-based sensors. The operating and fabrication temperatures are less than 450 °C [45]. There are several designs considered, as shown in Figs. 13.14 and 13.15. The hotplate is defined on polyimide sheets of Upliex-S (*Tg* ~500 °C, 50 μm thick, Hitachi Chemical Co.) with a spin on film of PI2731 (*Tg* ~350 °C from DuPont) and the spin-coated polyimide on a silicon substrate. The heating element is a thin Pt film photolithographically defined, and the tin oxide solution was deposited by dropping a microliter volume of a tin-containing solution onto the Pt conductivity sensor, followed by annealing at 450 °C for 10 min. The

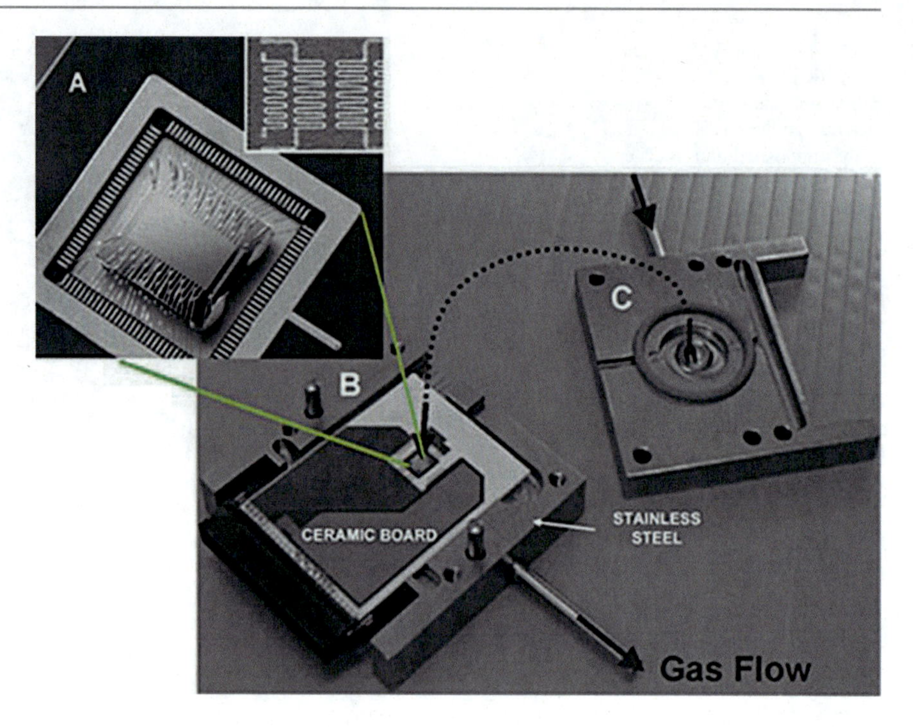

Fig. 13.13 The package of the KAMINA chip mounted with the ceramic board which is enclosed by two stainless steel parts providing a cavity above and below the chip through which the gas is delivered to the sensor. (Reprinted from Ref. [44] with permission Elsevier © (2007))

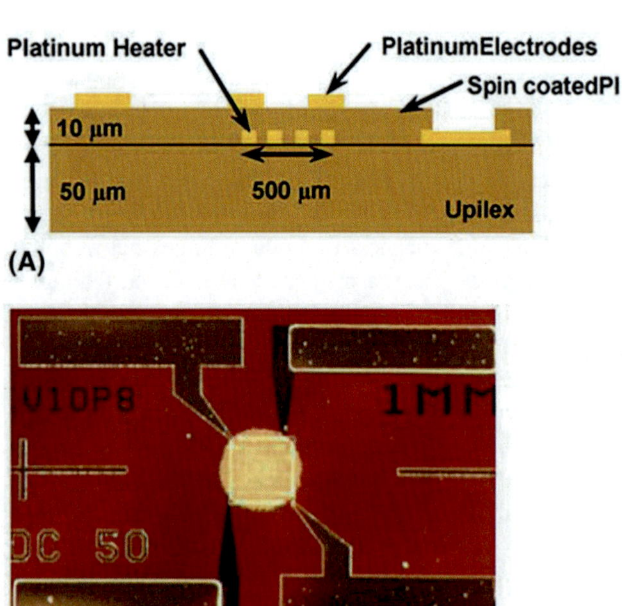

Fig. 13.14 Complete metal oxide microhotplate gas sensor made on a polyimide sheet, (**a**) cross-sectional diagram and (**b**) photograph of completed sensor, with 0.5 mm wide heating area. (Reprinted from Ref. [45] with permission Elsevier © (2008))

power input to achieve a temperature of 325 °C was in the order of 60 mW, and the calculated temperature distribution over the sensing area for the silicon-based device as shown in Fig. 13.15a is plotted in Fig. 13.15b. The polyimide layer was spin coated on to the oxidized silicon substrate and the removal of the bulk silicon under the heated platform carried out with deep RIE etching. Packaging was achieved by assembly of a gas permeable membrane to the sensors, placed into a TO-5 (transistor-outlined package-5) header with an epoxy, and wire bonding for electrical connections [46].

13.2.3 Microcantilever Packaging

Vacuum packaging of innovative hollow microcantilever resonators has been demonstrated by Burg et al. [47]. When the analyte flows through the inside of the cantilever resonator, a frequency shift is observed. They are operated at a reduced pressure or vacuum environment, to ensure a high Q, and therefore sensitive measurement of the frequency shifts due to selective adsorption inside the hollow cantilever. SEM micrographs of the cantilever are shown in Fig. 13.16. The electrical contacts are located on the glass lid, and fluid is injected into the hollow channel through pits defined through the thickness of the wafer. The sensors are attached to an adhesive backed gold-plated printed circuit board (PCB) and wire bonded. The assembly is clamped with Teflon (trademark from DuPont) against a PTFE manifold holding an array of 1/32^{tt} tubes through a polytetrafluoroethylene (PTFE) block as shown in Fig. 13.17, and the fluid connections have an "O" ring-sealed interface.

A research on CMOS-integrated microcantilever sensor arrays including their novel packaging has also been performed at ETH (Switzerland) [48]. Modified

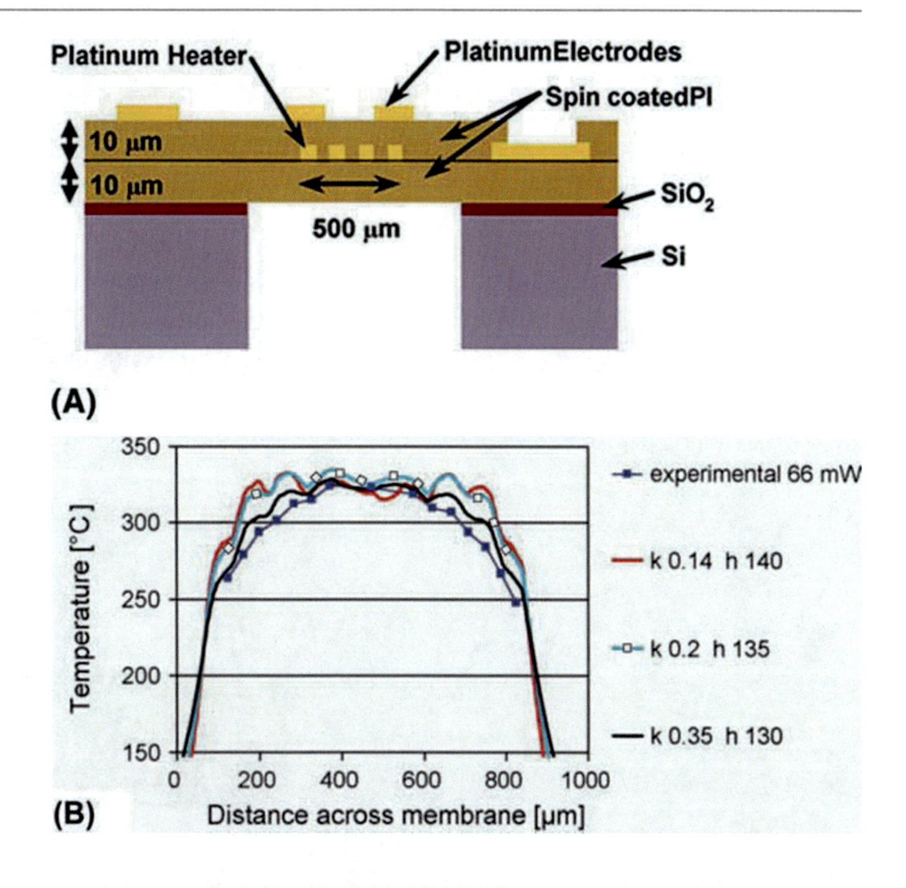

Fig. 13.15 (**a**) Schematic diagrams of the microhotplate gas sensor structure realized on a silicon substrate and (**b**) temperature distribution over the sensing area obtained from simulations, with thermal conductivity, k (W/m K) of Pt layer as a parameter and for several values of convection coefficient, h (W/m^2 K). (Reprinted from Ref. [46] with permission Elsevier © (2008))

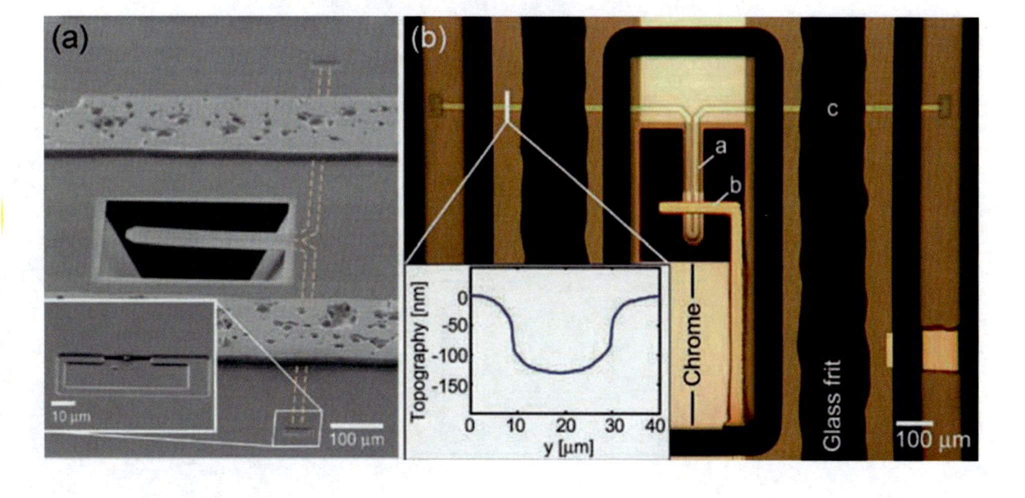

Fig. 13.16 (**a**) SEM micrograph of cantilever and glass frit seal after removal of the glass lid. The inset shows the channel inlet and (**b**) SEM micrograph of packaged cantilever resonator. The 300 μm long beam contains a 1 × 20 μm microfluidic channel; a, electrode on the glass surface above the cantilever enables electrostatic actuation; b, glass frit conforms to the surface topography (inset) and does not collapse the thin channel locations; and c, during bonding. (Reprinted with permission from Ref. [47] © 2006 IEEE)

semiconductor packages such as the dual-in-line (DIL) packages were wire bonded into customized packages with a ceramic support and a partial glob top to protect the electronic components while leaving the sensor area exposed. An array of piezoresistive microcantilevers with standard packaging is shown in Fig. 13.18. The piezoresistive detection method provides an attractive alternative to optical beam deflection methods normally used with atomic force microscope (AFM) probes. Here the probes have integrated silicon tips which are formed at the beginning of the cantilever micromachining process on a silicon on insulator (SOI) wafer [49]. A microheater is defined to allow actuation of the cantilever, with a piezoresistive sensor to detect deflection, thereby allowing feedback control of probe motion for an AFM imaging.

Packaging for resonant beam sensor arrays has also been developed [50]. The array of four resonant beam sensors with an integrated CMOS electronics were designed for measurement of volatile organics in a liquid water sample. Packaging is accomplished with a custom polymethyl

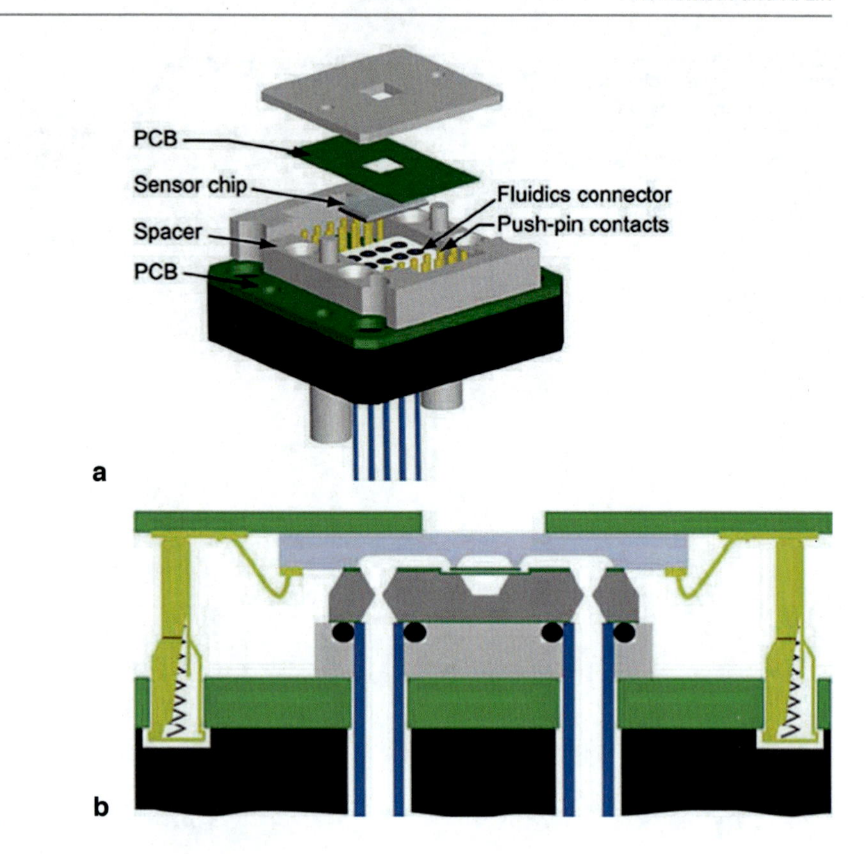

Fig. 13.17 (**a**) *Bottom* view of the sensor chip wire bonded to PCB. A colloidal silver paste is applied to the corner of the chip to connect the chromium layer on the resonator surface and a gold trace on the glass lid. (**b**) The sensor chip is clamped onto a Teflon manifold holding an array of standard 1/32'' OD Teflon tubes. Perfluoroelastomer O-rings seal the manifold to the chip, and an array of spring loaded contacts provides the electrical connections to the PCB. (Reprinted from with permission from Ref. [47] © 2006 IEEE)

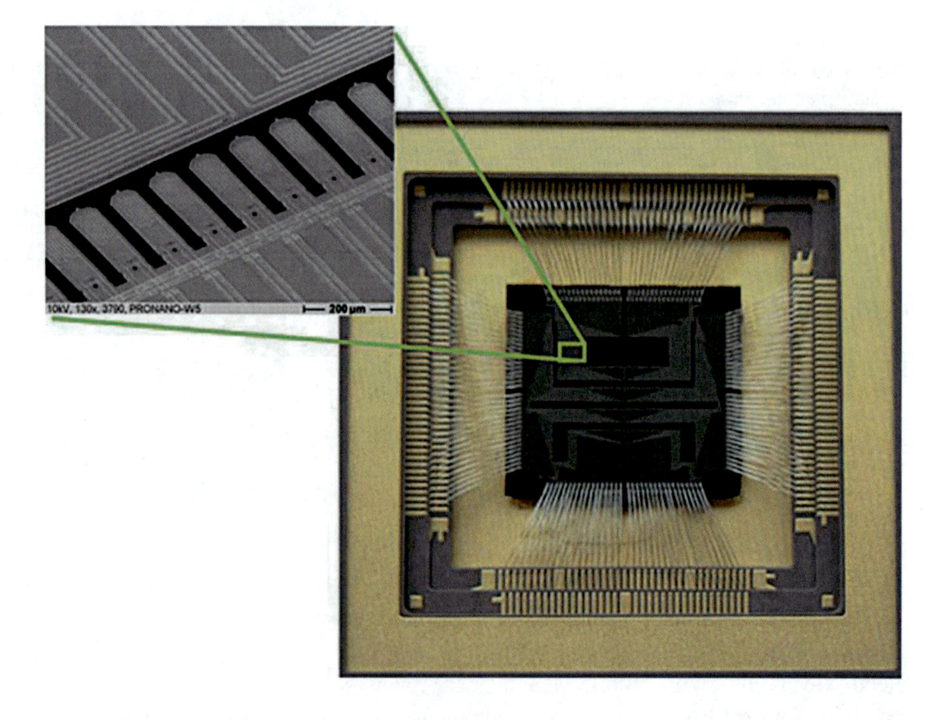

Fig. 13.18 A packaged fully addressable self-actuated piezoresistive cantilever array VLSI NEMS-chip incorporating 32 and 128 proximal probes and details of the array with the integrated silicon tips. (Reprinted from Ref. [49] with permission Elsevier © (2001))

methacrylate (PMMA) fluid housing fitted with threaded connectors for miniature liquid-phase chromatography (LPC) as shown in Fig. 13.19. The cantilever vibration is excited by an external magnetic field and based upon generation of a Lorenz force from the current flow in a conductor loop on the beam, I, as indicated in Fig. 13.19b. A miniature microcantilever sensor for electronic nose applications is packaged with SU-8 gasket and stainless steel housing by Pinnaduwage et al. [51]; see Fig. 13.20.

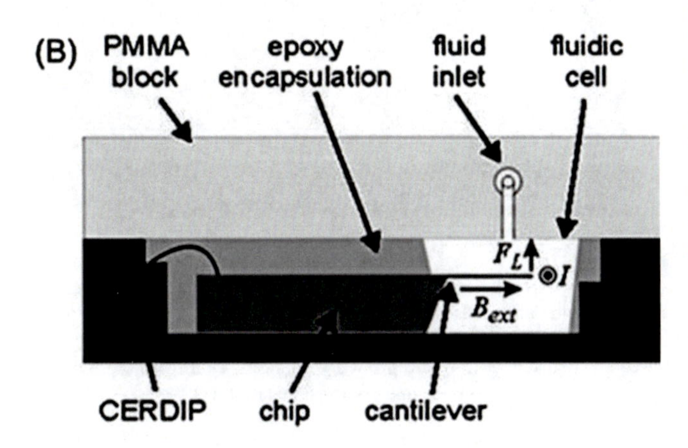

Fig. 13.19 (**a**) Photograph of the packaged single chip system with mounted PMMA block for fluid handing; (**b**) schematic cross-section of the packaged cantilever chip. The direction of current, *I*, at the cantilever tip indicates out of the plane [50]. (Reprinted with permission from the Analytical Chemistry of American Chemical Society (2007))

Customized three-dimensional packages can be made by stereolithographic assembly (SLA) which is discussed in detail in a later section of this chapter. Figure 13.21 shows a typical gas sensor package made using SLA. A cantilever gas sensor is placed in the chip housing with its cantilever array exposed to gas which flows through the channel. A ten microcantilever array with piezoresistive readout is built on a SOI wafer with standard micromachining processes [52]. The sensors are typically 450 μm in length and positioned on one side of the 6 × 13 mm die. A recessed region is defined in the SLA base to provide alignment of the die so that the cantilevers are positioned near the middle of the flow channel. Contact pads were then wire bonded to metal pins for signal output. Details of the SLA fabrication process are given later in this chapter in Sect. 13.3.4.

Fig. 13.20 Photograph of the sensor board with flow cell and interface electronics which allow the sensor to be connected to a computer USB port. (Reprinted from Ref. [51] with permission of AIP)

13.2.4 Metal Oxide High-Temperature Sensor Packaging

Higher temperature sensors have unique challenges and requirements for high-reliability applications. In particular for space applications, a number of sensors have been developed by Hunter et al. at NASA [53]. A careful study at elevated temperatures was carried out to determine the effect of alloy composition on the sensor response and stability. To this end standard packages have been modified very successfully to provide reliable working systems, as shown in Fig. 13.22. A series of other analytes, including hydrazine, hydrogen, and oxygen has also been investigated. Figure 13.22a shows the hydrogen sensor which operates at 425 °C [54]. The oxygen sensor, shown in Fig. 13.22b, utilizes conductivity measurements of a zirconia and operates at higher temperatures with a ceramic package.

13.2.5 Microfluidic Sensor Array Packaging

The function of a package is to direct fluid flow and provide mixing, flow splitting, and filtering functions, and expose the sensor array to the analyte under controlled conditions. The design must accommodate flow of sample to the sensor array

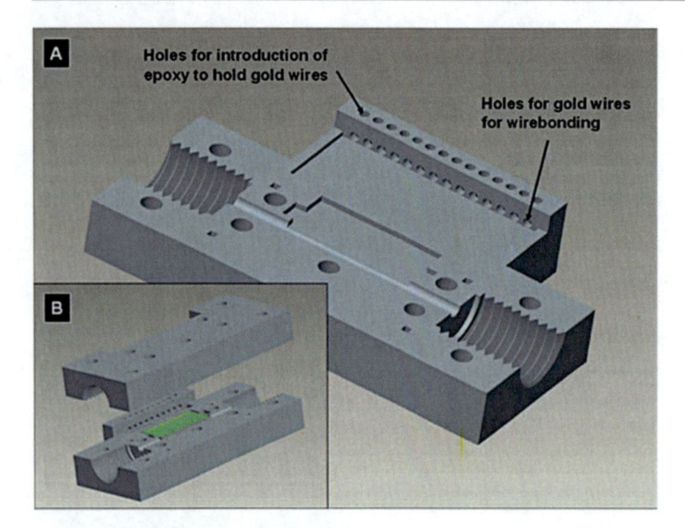

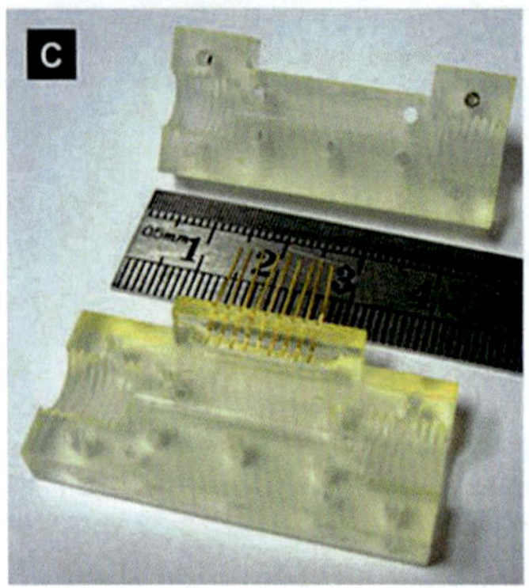

Fig. 13.21 Microcantilever packages built by stereolithography. (**a**) CAD image of the base of the flow cell threaded for connection to standard LPC connectors and indicating position for inserting gold wires for electrical connections. (**b**) CAD image of the two parts of flow cell where the sensor die is shown positioned next to the flow channel. (**c**) Photograph of the SLA package parts with the wire inserted [96]

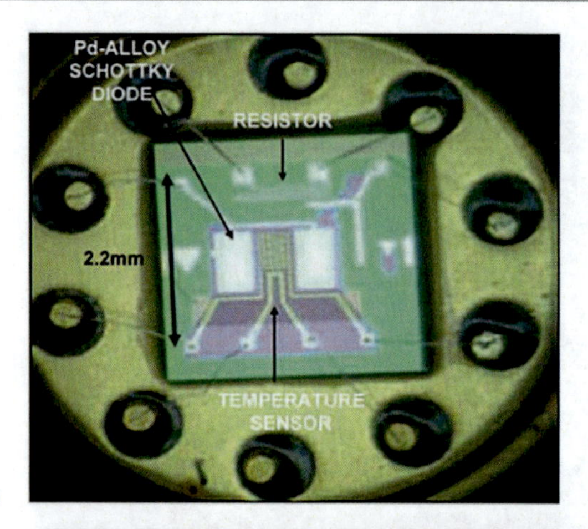

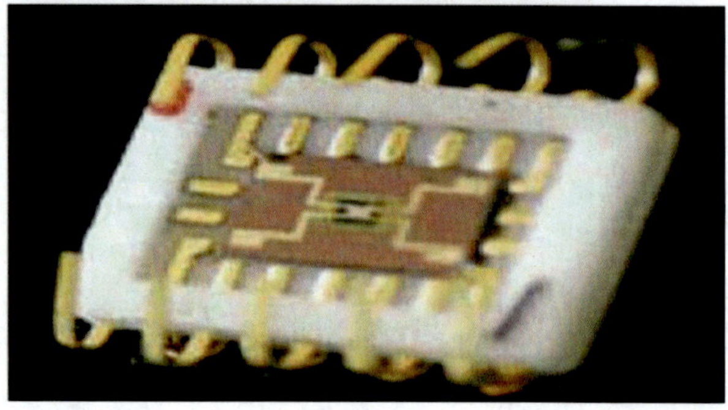

Fig. 13.22 (**a**) Picture of the packaged silicon-based hydrogen sensor. The Pd alloy Schottky diode (rectangular regions) resides symmetrically on either side of a heater and temperature sensor. The Pd alloy resistor is included for high concentration measurements. (**b**) Picture of zirconia-based oxygen sensor [54]. (Reprinted with permission from the American Institute of Aeronautics and Astronautics © (2005))

with necessary sampling volume, and there may be a distribution of concentration over the surface of the sensing interface depending upon specific flow conditions. Such deliberate modulation of the flow of analyte through the channel gives rise to a higher-order chemical sensing [55] in which the transient response is recorded as a function of flow, rather than the steady-state response. Finally, the thermal design for constant temperature operation may also be critical for correct sensor operation.

Novel packaging with SU-8 for optical sensors that provide both fluid channels and optical to fiber connectors is described by Ruano-Lopez et al. [56]. In addition optical sensors built with Mach-Zehnder interferometer on a CMOS chip are described by Blanco et al. [57]. Chang-Yen and Gale reported a microfluidic package for an optical biosensor array [58]. On this package the optical waveguide sensor was printed on a glass substrate, and optical fibers are aligned on the die with tapered SU-8 channels. Fluidic package was constructed of PDMS and is shown schematically in Fig. 13.23.

A variety of polymers, including SU8, had been applied to packaging of surface acoustic wave (SAW) sensors by Georgel et al. [59]. In this work the SAW device is flip-chip bonded to a PCB substrate providing electrical connections and a polymer sealing ring on the same substrate. The photosensitive SU-8 lithographically patterned was compared with a non-photosensitive benzocyclobutene (BCB, from Dow Chemical) for sealing the SAW sensors to the substrate. The SU-8 layer was 30 μm in thickness, and after bonding shear strength test was performed and results are plotted in Fig. 13.24. The stud bump formed of NiAu provides electrical connections from the SAW sensor to the PCB to facilitate testing of the sensor.

Miserendiono and Tai [60] describe a fully integrated microgaskets for MEMS sensors which make use of an SU-8 layer and a photosensitive silicone layer. The microgaskets fabrication process is shown schematically in Fig. 13.25. The microgasket is integrated with 3 μm thick Parylene film to form microfluidic channels defined with a

excitation from 450nm LED
liquid analyte
waste out
emission spectrometer
125um optical fiber
PDMS fluidic channel
backfilled core to seal fluidics to waveguide
dye/enzyme coated optical waveguide
130um waveguide core surrounded by 3 100um deep channels

a

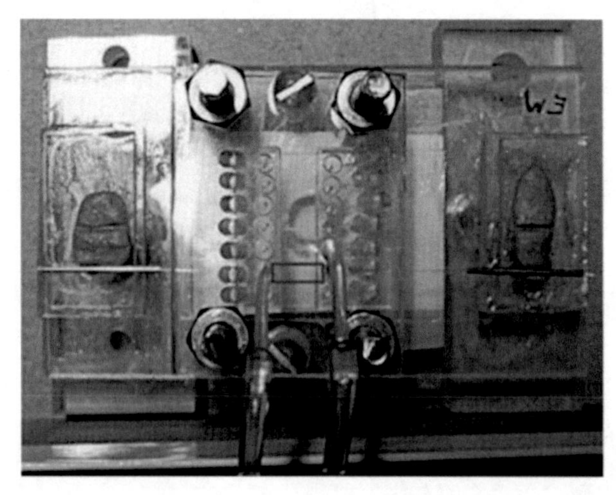

b

Fig. 13.23 (**a**) Diagram of the assembled waveguide sensor. The unusual cross-sectional area of the fluidic channel was created to limit the analyte exposure to the waveguide surfaces, allowing only one-dimensional analyte diffusion to the immobilized dye/enzyme layer [58]. (**b**) Photograph of assembled and packaged sensor. One of the seven channels is boxed in the center of the package [58]. (Reprinted with permission from the SPIE)

photoresist sacrificial layer. The microgaskets were capable of deforming approximately 25%, and this behavior was characterized through a novel capacitance method. The microgaskets are reusable and able to form leak-free seals operating at pressures up to 50 psi. The dead volume was estimated with finite element analysis (FEA) and found to be 9 nl per connection via; however, this can be reduced with a reduction in channel diameter.

Low-cost packaging has been manufactured by injection molding of cyclic olefin copolymer resin (COC) [61]. The COC polymer resin is heated to the liquid state and then injected into the mold. After the molding process, the mold is cooled and the package is formed. Figure 13.26 shows an example of modular package for fluidic components that can be snapped together to connect multiple sensors with standard diameter 1/8^{tt} OD tubing. The snap-in connector with standard tubing is combined with an electrode for an electrochemical-based chlorine sensor and a thin film resistor temperature sensor for an aqueous sample, as shown in Fig. 13.26c.

13.2.6 Biomedical Devices

Requirements for biosensors are application specific, and there are additional issues to consider including the biological fluid compatibility and the biocompatibility of materials which may be in contact with the body and need for sterilization. There are a host of requirements for implantable devices including chemical stability, corrosion resistance, and the effects on the host immune system. The harsh environment in the body is a result of the high salt concentrations and the foreign body reaction which takes place at the site of the implant. These issues are less stringent for semi-invasive devices, such as microneedles.

Microneedles and microdevices for drug delivery also require custom packages. A micropump integrated with microneedle array has been developed for painless drug delivery [62]. Figure 13.27a shows the needle array, and Fig. 13.27b shows an SEM micrograph of the microneedles. The needles are approximately 200 μm in length to provide penetration of the stratum corneum to allow drugs to be delivered subcutaneously, however with very little pain as the nerves are located deeper in the skin. The housing provides a volume of drug, and the piezoelectric pump, manufactured separately, provides the pressure to induce the follow of drug through the needles. Packaging of the pump includes gluing of the reservoir to the piezodisk and assembly to an outlet port. Packaging of the microneedle array is shown schematically in Fig. 13.27c. The silicon needle array is laminated to a polyimide support which has a hole cut out to allow drug flow into the needles. This assembly is then glued to the carrier base support piece with an epoxy.

Integration of a cannula, drug reservoir, and miniature pump on a single substrate has been examined by P.-Y. Li et al. [63] for the treatment of ocular diseases. The novel electrochemical pump generates pressure through electrolysis of water at low voltage and operates at low power. The gas pressure generated is sufficient to generate flow of drug at a controlled rate through the flexible Parylene cannula; see Fig. 13.28. The drug reservoir is made of PDMS and was molded with soft lithography mold [64]. The pumping rate was determined in the range of 2 μL/min suitable for the ocular drug delivery application. A range of flows could be generated from pL to μL making this a worthwhile approach; however, to further develop the full potential of this

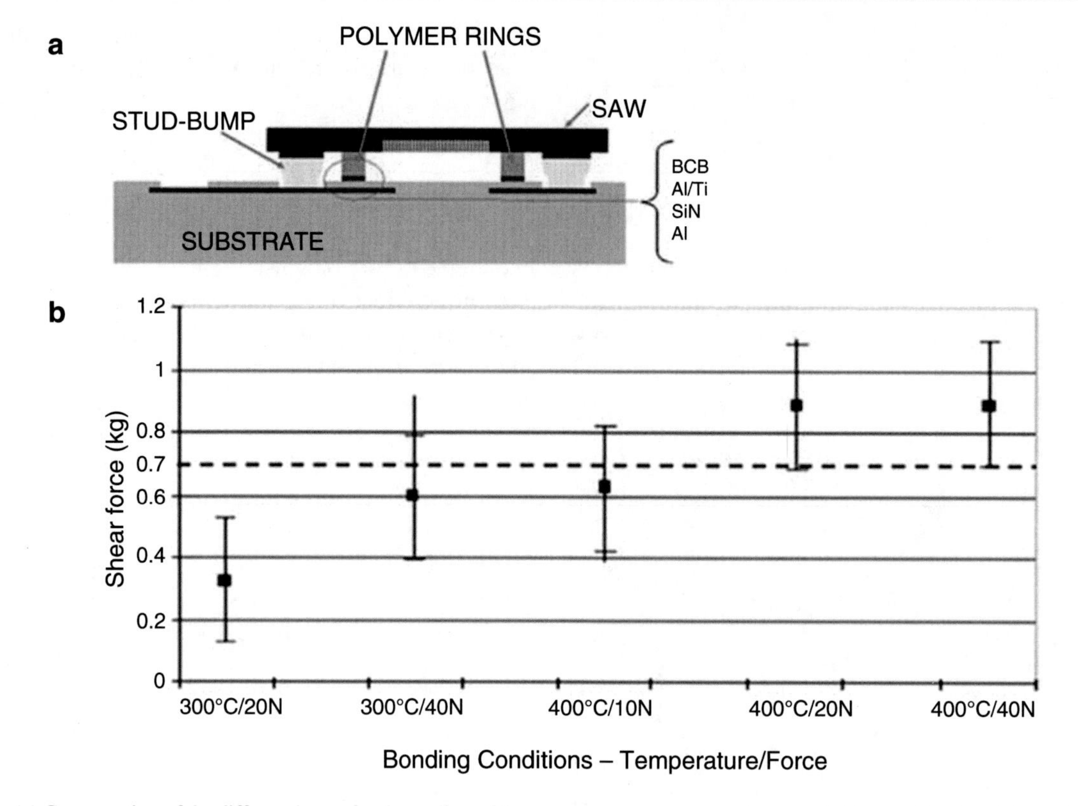

Fig. 13.24 (**a**) Cross section of the different layers for the sealing of the SAW sensor to the substrate, (**b**) arrangement for the shear test experiment, and (**c**) shear force at different bonding forces and process temperatures with SU-8 material bond. (Reprinted from Ref. [59] with permission Elsevier © (2008))

technology, extensive animal testing would be required. Other MEMS devices are finding increasing use in medicine, in particular for surgery [65] and drug delivery [66].

13.3 Packaging Technologies

An important part of packaging is the packaging technology including new materials and bonding of materials to form seals. A number of bonding methods have been developed; these include glass–silicon anodic bonding [67], silicon–silicon direct bonding [68], polymer bonding [69], solder bonding, and adhesive bonding. The temperature at which bonding takes place and any stresses generated by the bonded interface can have an influence on sensor performance.

13.3.1 Wafer-Level Bonding

With the development of wafer bonding technology, wafer-level packaging (WLP) is showing its potential advantage in system integration to reduce packaging cost as well as increasing the yield of devices after dicing and packaging; since WLP would allow a wafer-level burn-in and wafer-level testing as such, it solves the known good die (KGD) issue. A WLP has the advantage that multiple dies are packaged in a single step prior to dicing. This technology has been widely used in MEMS area such as the packaging of microscanning mirrors [70], RF MEMS switch [71], and other applications. First, it can serve as a functional mechanical protection for the chip. This cap wafer is often featured by vias filled with sputter-deposited metals that connect the pad on the chip to top solder bumpers. Yun et al. [72] demonstrated two approaches, so-called via-last and via-first method, to package micromachined accelerometers at the wafer level. The main character of this type of packaging is that the dicing procedure occurs after bonding which means a number of devices can be packaged at the same time. Second, the wafer-level package can be employed to realize the 3D structures. Takegawa et al. [73] presented a wafer-level packaging process using surface-activated bonding for silicon diaphragms. Diced chips were obtained with a high yield without damaging the interface and diaphragms. The structure also features sufficient bonding strength and low leak rate (1.2×10^{-12} Pa·m^3/s) which indicate its ability to satisfy the requirements for MEMS hermetic packaging. They also pointed out that this technology is very suitable for applications requiring low stress, stability, and hermeticity, such as bioimplants and mechanical sensors.

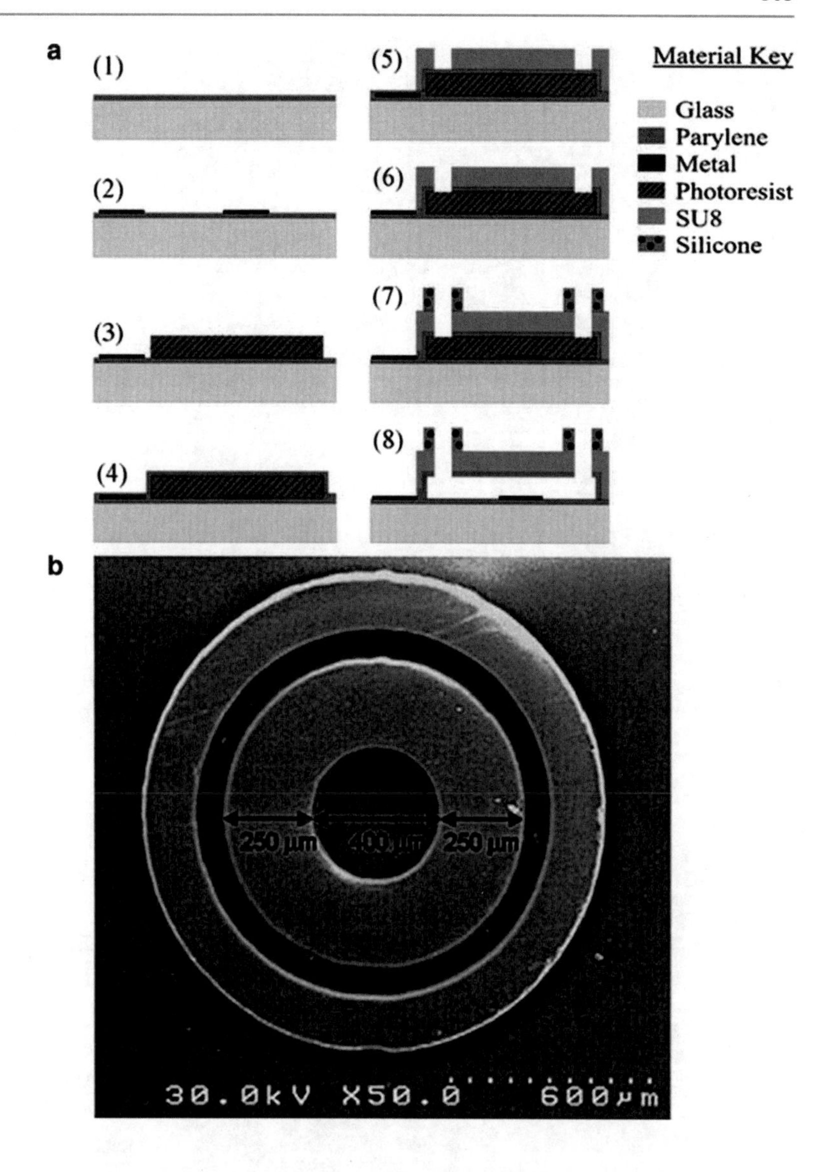

Fig. 13.25 (**a**) Cross-sectional diagram of device fabrication process: (1) Parylene-C deposited on substrate with adhesion promotor, (2) metal layer defined by lift-off, (3) thick photoresist layer defines microchannel, (4) 3-μm thickness layer of Parylene-C, (5) define fluidic vias 65-μm thick SU-8 layer, (6) oxygen plasma etching of Parylene, (7) photodefine silicone layer for gasket and descum with RIE, (8) dissolve photoresist in alcohol. (**b**) Scanning electron micrograph of a single photodefined silicone rubber MEMS O-ring, of thickness approximately 30 μm. (Reprinted from Ref. [60] with permission Elsevier © (2008))

When it comes to the packaging of chemical sensors, wafer-scale package has also been developed for gas sensors [74]. The sensor elements are fabricated by micromachining in the same manner as before, with the additional space for a Pyrex glass ring that is anodically bonded to the die. This supports a gas permeable membrane defined over the chemical sensor to allow analyte to diffuse to the sensor surface, while the bonding pads are not covered to facilitate electrical connections. Here the anodic bonding and membrane assembly are carried out at the wafer level, followed by dicing. The completed sensor is illustrated schematically in Fig. 13.29. This package has been further modified to be a smaller and a more cost-efficient design as shown in Fig. 13.29d in which the gas-sensitive layer was placed in cavities fabricated by deep reactive ion etching (DRIE) from the back and the whole wafer with these cavities is sealed with a gas permeable membrane. The modified concept makes use of the silicon as the supporting rim thus takes away the need for the Pyrex glass frame to realize the WLP of the sensors and simplify therefore their processing [75].

13.3.2 Localized Bonding

Localized anodic bonding of micromachined glass packaged is demonstrated by Zaire et al. [76] for encapsulation and protection of microelectronic circuits. Another example of localized bonding has been successfully demonstrated by Lin et al. [77, 78] in the form of bonding to form hermetic packages. Localization of heated regions can promote interdiffusion, the formation of intermetallics, and even localized chemical vapor deposition (CVD) to form new sealing materials at the interface. Polysilicon heaters were defined photolithographically to define local heated regions. The temperature in the microheater can increase to 1000 °C while in the neighboring areas less than several micrometer

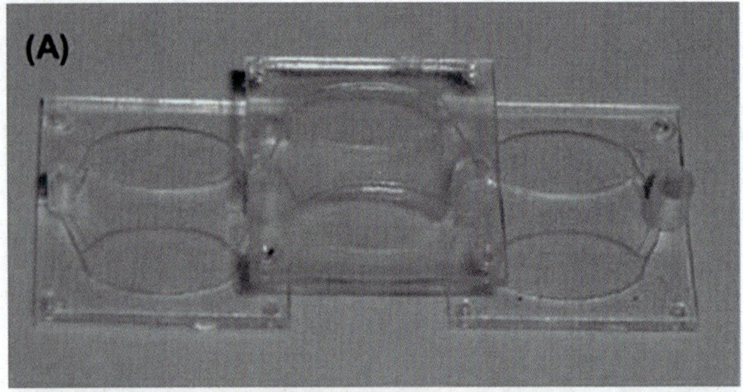

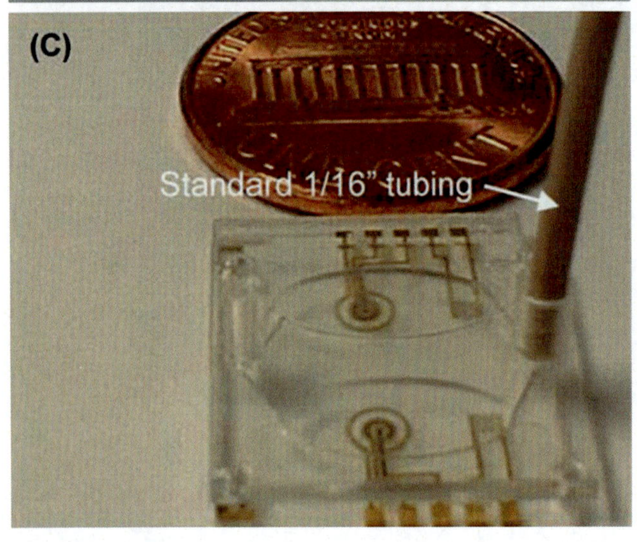

Fig. 13.26 (**a**) Fabricated interconnecting fluidic package showing package-to-package connection with mating inlet and outlet, (**b**) side view connectors, and (**c**) assembled sensor with microfluidic channels and snap-in connection. (Reprinted from Ref. [61] with permission Elsevier © (2007))

away are only at 100 °C. Building microheaters on the desired bonding area and applying pressure and electric current, the localized bonding process is activated, as shown in Fig. 13.30. Silicon-to-glass localized fusion bonding using polysilicon as microheater and silicon-to-gold eutectic bonding using gold as microheater are achieved with bonding strength higher than 10 MPa [77]. The local heating method has also been applied to bond solder and was able to overcome the surface roughness and create excellent step coverage by the reflow of the solder material. When applied as localized CVD bonding, this method is capable of activating the decomposition of silane locally by localized heating, thus connects the cap and device substrate firmly [78].

13.3.3 PDMS Fabrication

Polydimethylsiloxane (PDMS) or so-called silicon rubber is a material widely used in nowadays microdevice manufacturing and packaging. Here we are trying to reveal some of the functions of PDMS by describing its application in biosensors including gas sensor, MOSFET-type biosensor, optical biosensor array, and blood cell biochip.

13.3.3.1 Main Advantages and Disadvantages

PDMS has several attractive specific properties for the fabrication of microfluidic channels for use with aqueous systems of biochemicals and cells: it is transparent to visible and ultraviolet light, with good chemical and thermal stability, soft, flexible, electrically insulating, unreactive, and permeable to gases. Also, its surface can be rendered hydrophilic that allows it to adhere to other materials like polymers and glass easily. The main disadvantages of PDMS are that it absorbs a range of organic solvents and compounds (particularly alkyl and aryl amines) [79] and that its surface properties can be difficult to control [80].

13.3.3.2 Soft Lithography and Multilayer Devices

Soft lithography represents a non-photolithographic strategy based on self-assembly and replica molding for carrying out micro- and nanofabrication structures with feature sizes ranging from 30 nm to 100 μm. In soft lithography, PDMS has always been chosen as the material for fabricating elastomeric stamps with structures on its surface. The elastomeric stamp is prepared by cast molding from a master which has been lithographically defined, as shown in Fig. 13.31 [81]. The master is silanized by exposure to vapor for 30 min. The PDMS elastomer frequently used is Sylgard 184 (Dow Corning Inc.). It is supplied in a two part kit: a liquid silicon rubber base and a catalyst or curing agent. Once mixed and pored over the master and heated to elevated temperature, the liquid mixture becomes a cross-linked elastomer in a few hours. The cured PDMS is then peeled off the master so that each master can be used multiple times. In soft lithography, the featured PDMS stamp is often used as a master to further transport specific structures attached to it to another substrate. However, in some of the bio- or chemical application, the PDMS layers, flat or structured, serve directly as the body of the devices. In C.W. Cho's work, single-walled carbon nanotube network gas sensors were fabricated on PDMS [82].

Zhang et al. [83] present a PDMS microdevice with built-in optical biosensor array. Here the merging of different

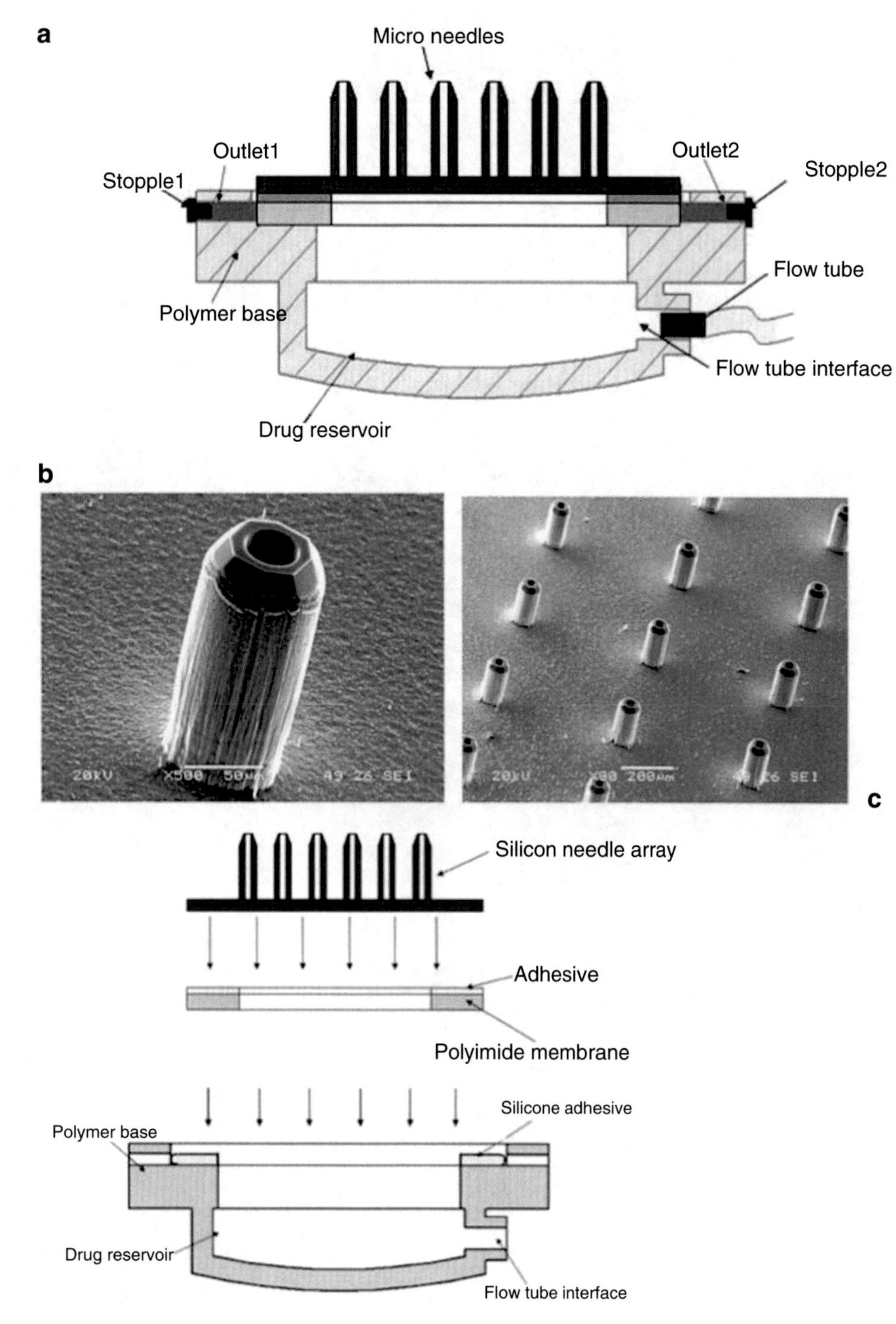

Fig. 13.27 (**a**) Packaged microneedle array assembled with drug reservoir; (**b**) SEM micrographs of silicon microneedles fabricated by micromachining, approximately 20 μm in length, inner diameter of 30 μm, and pitch 250 μm. (**c**) Packaging process for microneedle array includes silicon microneedle die to polyimide bonding, dispensing epoxy on the microneedle base, removal of excess epoxy, curing, and insertion of the tube. (Reprinted with permission from Ref. [62] © 2006 IEEE)

technologies including fluorescence-based optical sensing, layer-by-layer encapsulation, and PDMS-based microfluidics to produce a hybrid system which can provide on-site monitoring of the cellular microenvironment within the microchannels. The layout of their device is shown in Fig. 13.32. The top cover layer could be a thin PDMS slab or a microscope glass slide; the middle layer is the functional PDMS layer with fabricated microchannels and nanowells where both cell culture and in situ sensing take place. The bottom layer is fiber optics alignment layer which can also be made from PDMS. As mentioned above, PDMS material can be rendered hydrophilic thus here is applied as a pretreatment

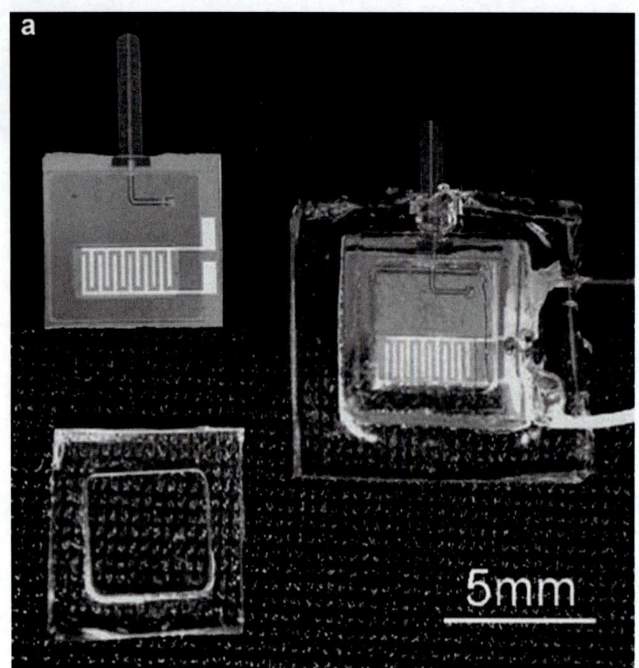

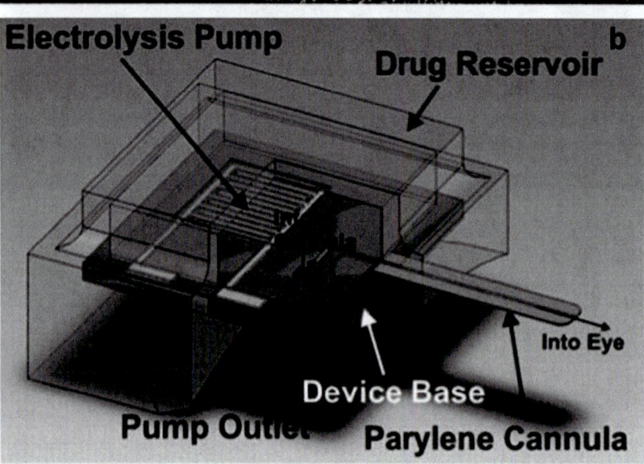

Fig. 13.28 (**a**) Pump and cannula chip (*upper left*), silicone drug reservoir (*lower left*), and packaged drug delivery device with attached wires for electrolysis actuation (*right*). (**b**) Schematic diagram illustrating the components of the drug delivery device with 1-mm wide cannula. The arrows follow the drug path from the reservoir through the channel of dimensions 100 × 25 × 5 μm in the Parylene cannula. (Reprinted from Ref. [63] with permission Elsevier © (2008))

to achieve water proof sealing between the top and middle layer with the aid of proper pressure. Figure 13.32 shows a molded PDMS device without the bottom layer. What should be mentioned here is that the PDMS is nonfluorescent and transparent down to a wavelength of 230 nm that satisfies the need of fluorescence measurement.

Another example comes from J.K. Shin's [84] work which reported a MOSFET-type biosensor assembled in a PDMS microfluidic channel package for the electrical detection of nanoscale biomolecules. As can be seen in Fig. 13.33, the upper PDMS layer with the microchannel was aligned and

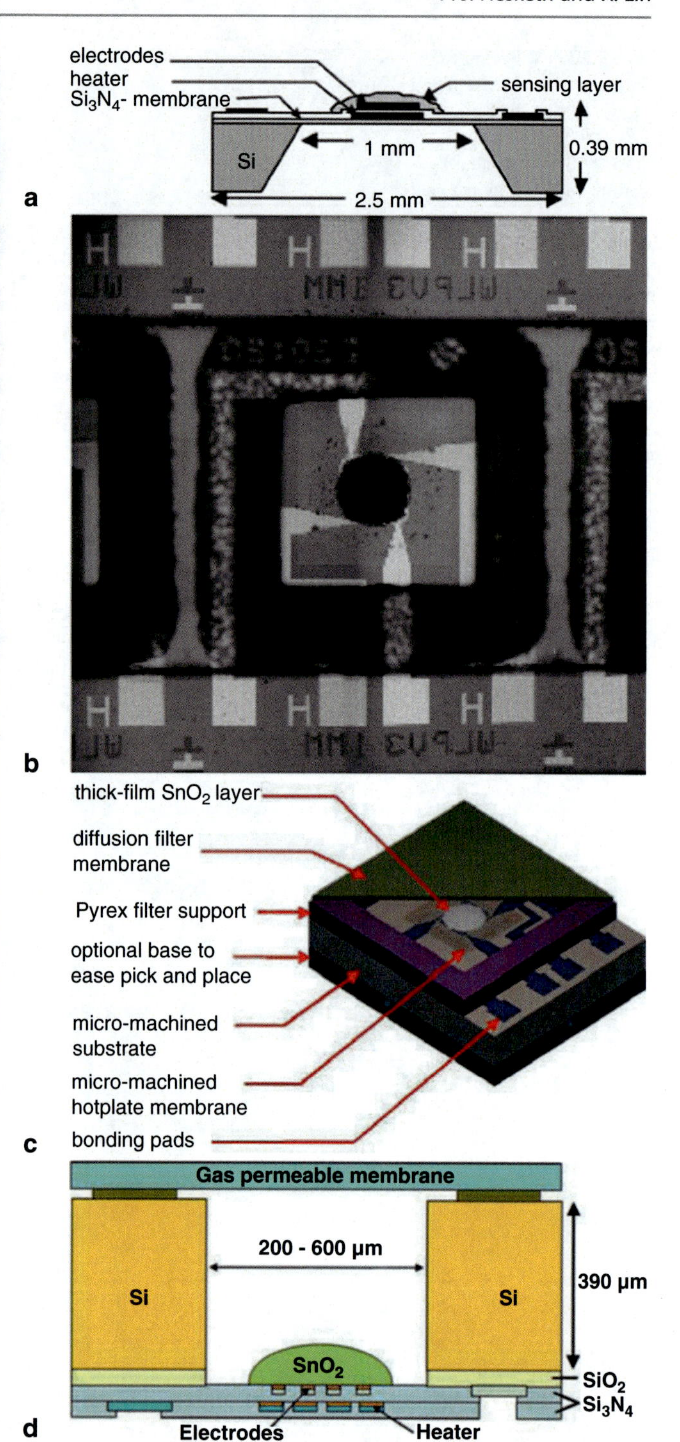

Fig. 13.29 (**a**) Schematic cross-sectional diagram of the tin oxide hotplate gas sensor, (**b**) image of the sensor after bonding to the Pyrex cover and before sealing and dicing, and (**c**) schematic drawing of the completed packaged gas sensor. Reprinted with permission from [74] © 2007 IEEE. (**d**) Schematic diagram of the miniaturized wafer-level packaged metal oxide gas sensors combining DRIE of silicon and drop coating of the gas-sensitive film [75]

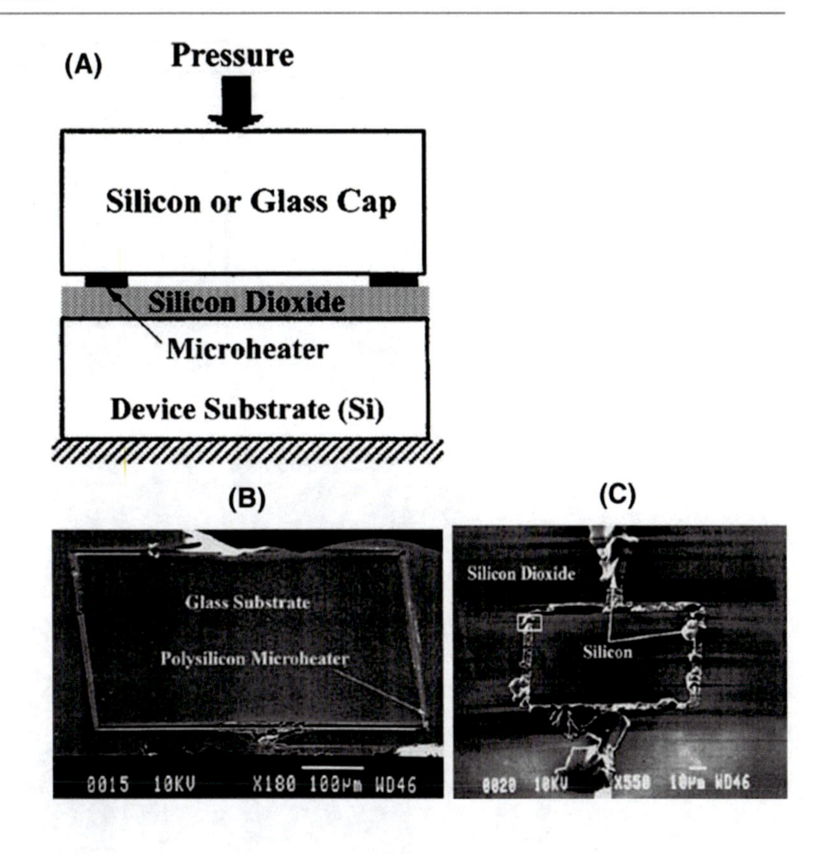

Fig. 13.30 (**a**) Schematic diagram showing the principle of localized bonding based upon microheater on insulating silicon dioxide film; (**b**) SEM micrograph showing the glass cap substrate is softened and has the shape of the polysilicon heater; (**c**) SEM micrograph shows localized silicon–gold eutectic formation. After the bond is forcefully broken, the silicon cap is torn apart and transferred to the device substrate. (Reprinted with permission from Ref. [77] © 2000 IEEE)

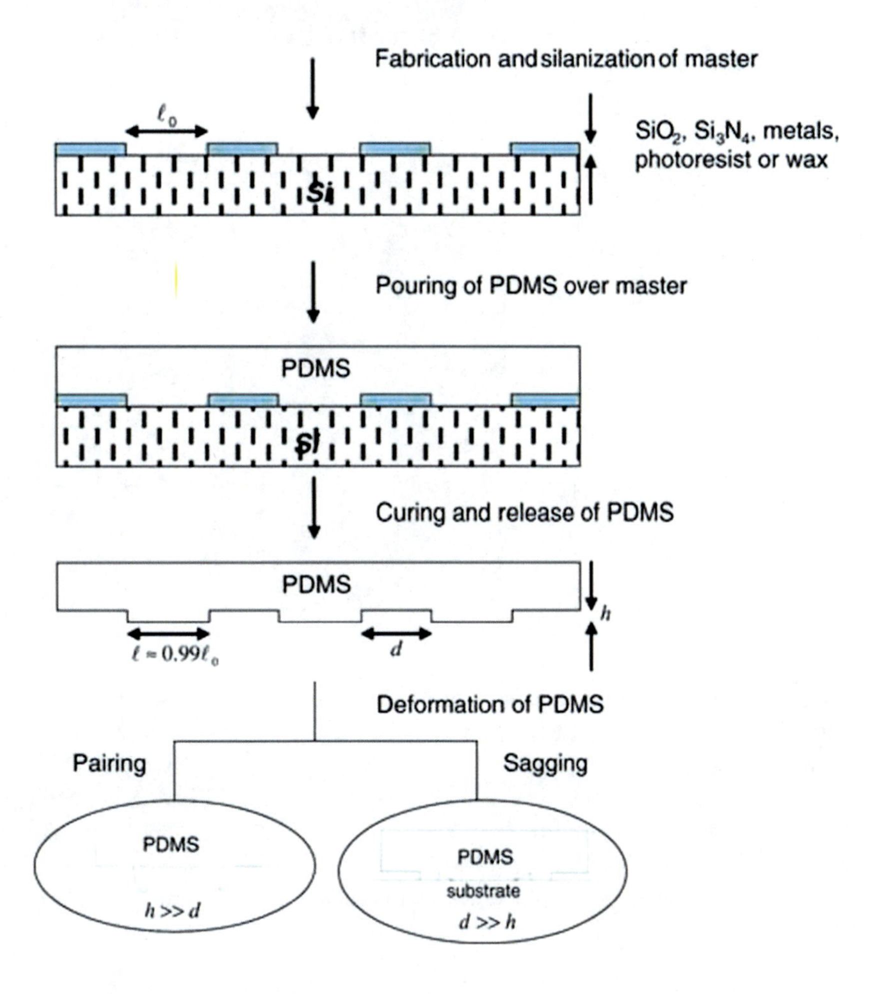

Fig. 13.31 Schematic illustration of the procedure for fabricating PDMS stamps from a master having relief structures on its surface [81]. (Reprinted with permission from Annual Reviews)

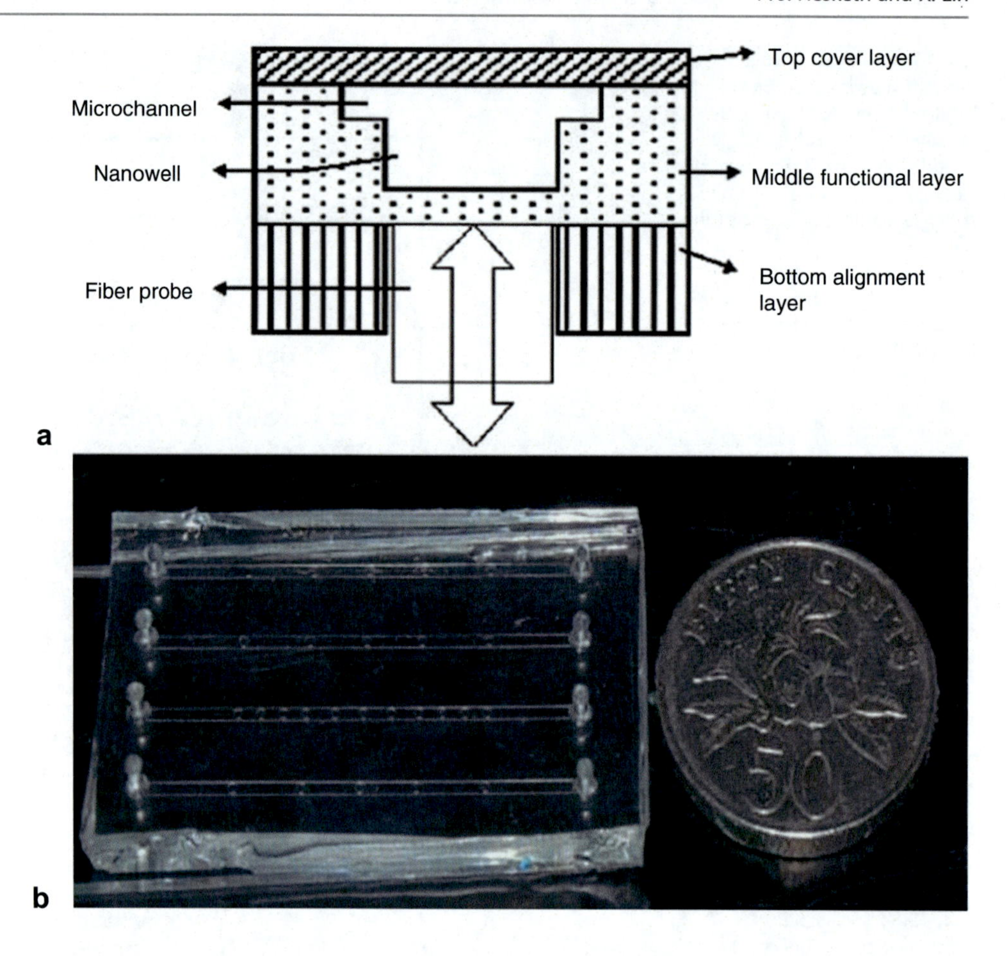

Fig. 13.32 (**a**) Cross section of the trilayer microdevice, where the *top layer* is glass, *middle layer* is PDMS, and *bottom layer* is alignment layer for the fiber optics. (**b**) Molded PDMS device (without the *bottom layer*) [83]. (Reprinted with permission from the SPIE)

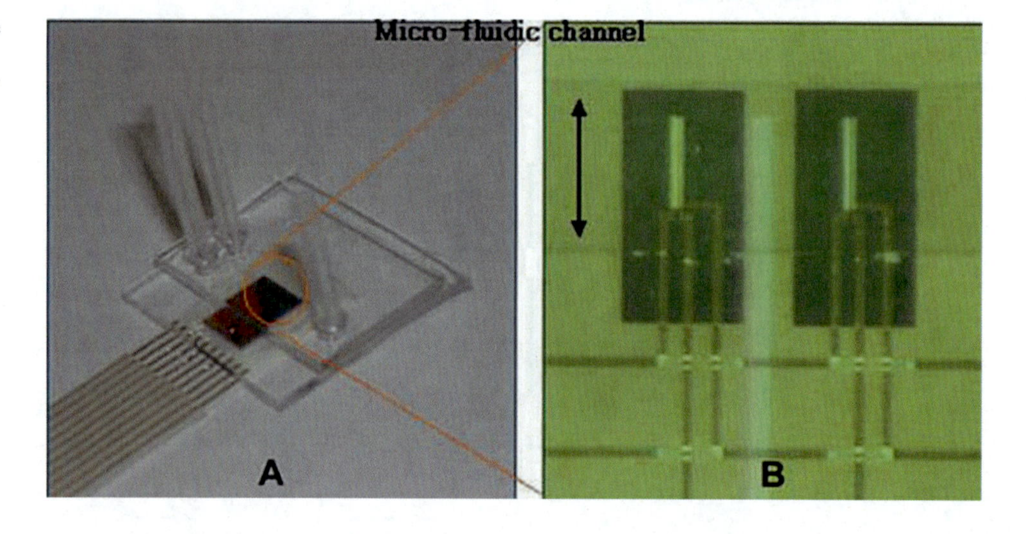

Fig. 13.33 (**a**) Photograph of the sensor device packaged with a metalized alumina header and (**b**) image of the MOSFET-type sensor and PDMS microfluidic channel formed on the device [84]. (Reprinted with permission from the SPIE)

bonded along with the sensing area of the dual-gated MOSFET device implanted in lower layer, which is also PDMS, except for bonding pad.

13.3.3.3 PDMS as a Sealing Material

Another important function for PDMS is serving as the sealing layer. Thanks to its properties of transparency and elasticity, PDMS layers are capable of performing both sealing and mechanical support function to the chip and device. Y. H. Cho [85] demonstrated the sealing function of PDMS in a MEMS-based biochip for the characterization of a single red blood cell. As shown in Fig. 13.34, PDMS is served as the sealing layer that can be flat or with a 3D structure. The 3D sealing layer required 3D master mold

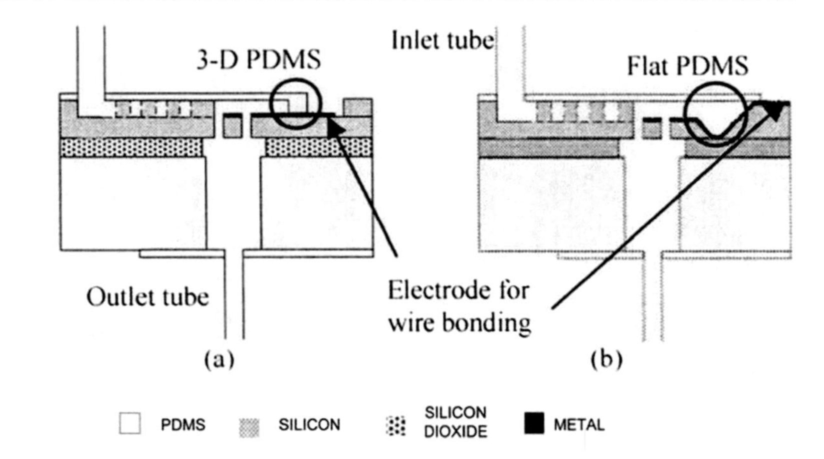

Fig. 13.34 Schematic view of the electrode sensor packaging, with use of (**a**) 3D-molded PDMS structure and (**b**) 2D flat PDMS structure. (Reprinted with permission from Ref. [85] © 2005 IEEE)

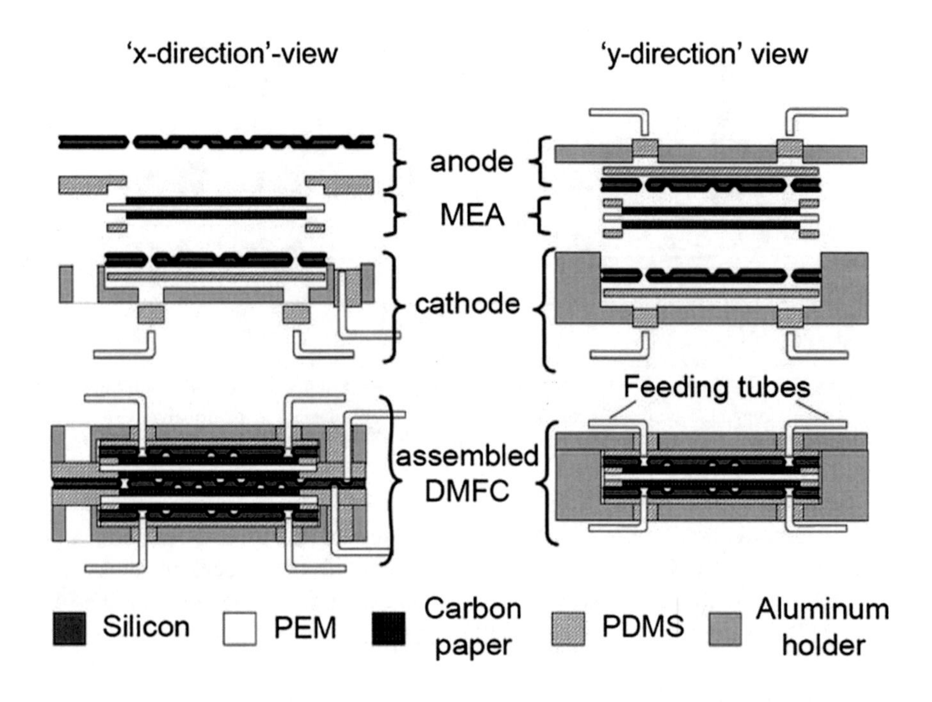

Fig. 13.35 Schematic cross sections in orthogonal planes, showing the components of the miniature direct methanol fuel cell stack. The prefabricated MEA and silicon anode with integrated electrodes are assembled with the molded PDMS membranes, approximate thickness 0.2–0.3 mm. The feed tubes and electrodes are connected to complete the assembly process. (Reprinted with permission from Ref. [86] © 2007 IEEE)

and higher requirement for alignment, while the flat one is easy to align. Another packaging sample showing the functions of PDMS can be found in Zhong's [86] work which presents a microdirect methanol fuel cell (μDMFC) stack implemented by microfabrication process and PDMS assembly method. The prefabricated PDMS pieces, approximately 200–300 μm in thickness, were used in the packaging for three purposes: (1) keeping the MEAs, leading channels and feeding holes from leakage by the PDMS sealing layers, (2) protecting the fragile silicon plates against mechanical stress with the PDMS buffering layers, and (3) fixing the feeding tubes with PDMS blocks. The assembly process of each parts of the μDMFC can be found in Fig. 13.35. The components are held together with an aluminum plate for testing the operation of the DMFC.

13.3.4 Stereolithography-Based Packaging

There are few reviews of earlier work on SLA for 3D MEMS [87] and custom packages [88]. A typical process for SLA manufacturing includes the following steps. First a 3D model should be built using any 3D modeling software which is able to convert its own format into STL file (3D Systems specific file format), such as Solidworks, Pro/Engineer, IDEAS, or AutoCAD. Then, with the help of "3D Lightyear SLA Platform Preparation Software," you can prepare STL three-dimensional model representations into build files that are to be built on an SLA solid imaging system. There are several things that need to be done in this preparation step, including verifying your 3D model, determining the build direction, editing the support layer, and slicing the part. Finally, at the

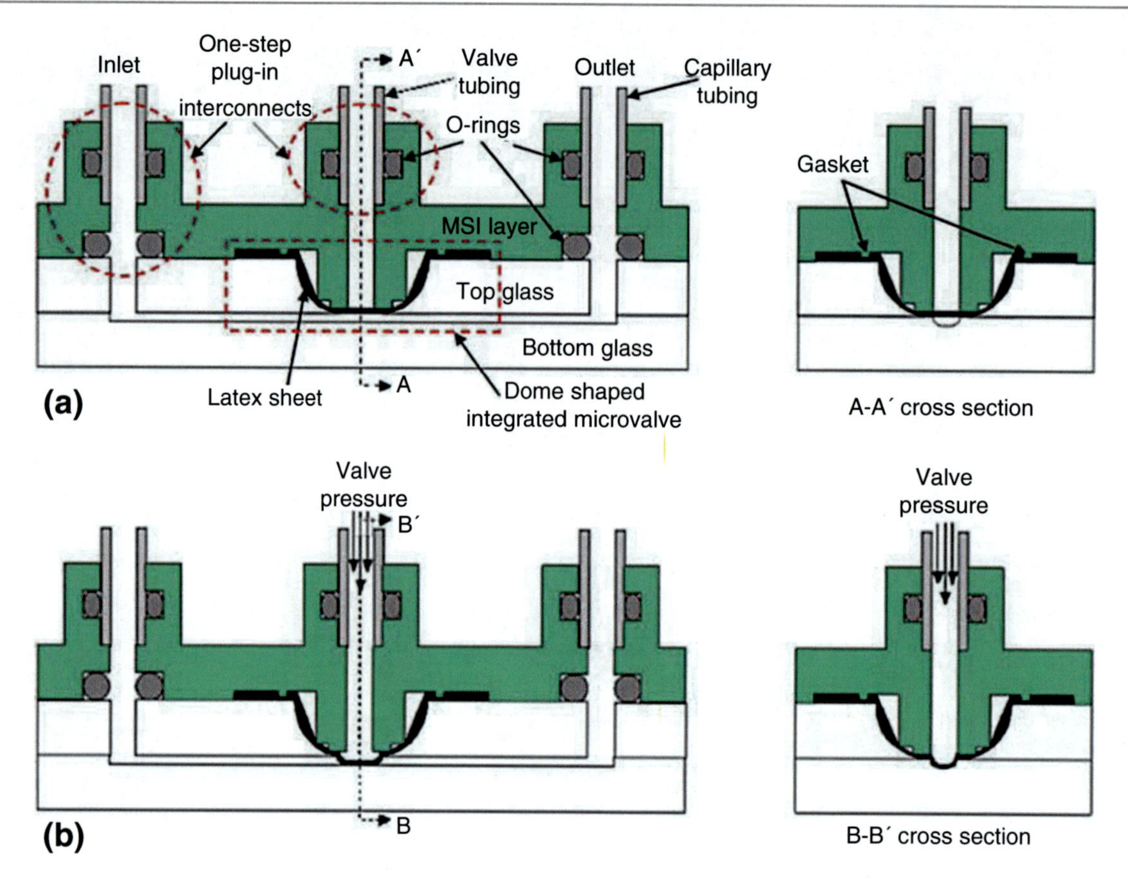

Fig. 13.36 Examples of the one-step plug-in microfluidic interconnects fabricated with SLA combined with two layers of glass that form microfluidic channels. The plug-in interconnects and dome-shaped monolithic microvalve are shown in cross section: (**a**) when the microvalve is open and (**b**) when the microvalve is closed. (Reprinted from [90] with permission Elsevier © (2007))

SLA station, the liquid resin is solidified using a computer-directed, ultraviolet (UV) laser beam. By exposing successive layers of photosensitive polymer resin, the CAD design is turned into a solid plastic object, which is rinsed in solvent and cured in a UV oven.

An alternative to SU-8 is the use of stereolithographic assembly (SLA) to define customized microfluidic channels and packages for sensor arrays. Han et al. [89, 90] describe microfluidic channels with integrated microvalves and plug-in connections with O-ring seals, as shown in Fig. 13.36. The valves are pneumatically actuated and have a dead volume of 24 nL. One-step plug-in microfluidic connections are made with an O-ring seal on each port. The lower O-ring is to prevent clogging when the microchannels are bonded with adhesive to the glass chip. One advantage of using this approach is that the glass microfluidic component can be fabricated independently from the microfluidic system interface (MSI) layer. SLA system enables the manufacture of complex microfluidic structures with very few design constraints and sufficient resolution (25 μm vertical, ~1 μm horizontal). Some of the important features should be considered and designed in the SLA parts to prevent leakage (space designed for O-ring or silicon gasket) as well as accomplishing their functions (microvalve). The MSI can be configured with integrated microvalves to facilitate fluid manipulation for chemical and bioassays.

Figure 13.37 shows an example of a customized package for a AFM microprobe that provides a fluid chamber attached to the probe [91]. The surface of SLA resin can be coated with Parylene to improve chemical resistance and potential for biocompatibility. Another example is a package that fits the chip containing a sensor array and then is inserted into a standard 1/8" diameter tube connector as shown in Fig. 13.38.

The advantage of SLA is that customized packaging geometries can be configured and the integration of microvalves can be easily accomplished to facilitate fluid manipulation, sample routing, and mixing, for example, for PCR analysis, as shown in Fig. 13.39. The MSI component was aligned and connected to other parts of the fluidic system made of glass with epoxy adhesive. Thermal management for PCR is important. Singh and Ekaputri [92] describe an ANSYS simulation of temperature profiles in a multilayer packaged system.

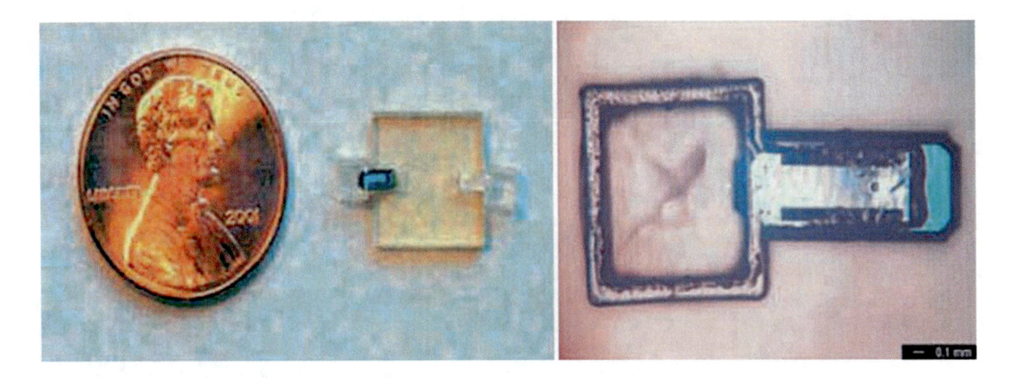

Fig. 13.37 SLA package for a AFM cantilever sensor that allows fluid introduction for the probe [91]

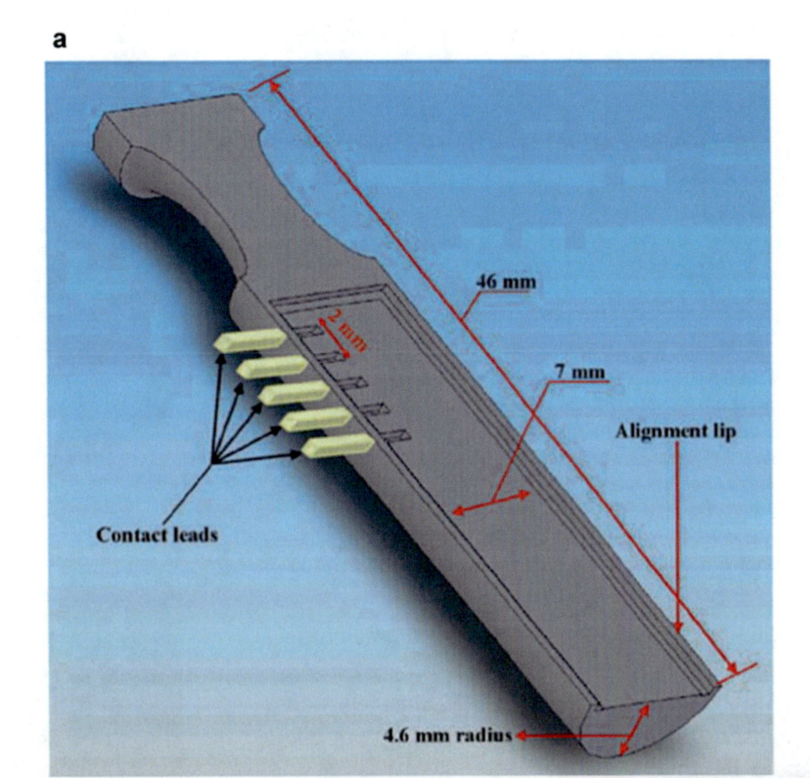

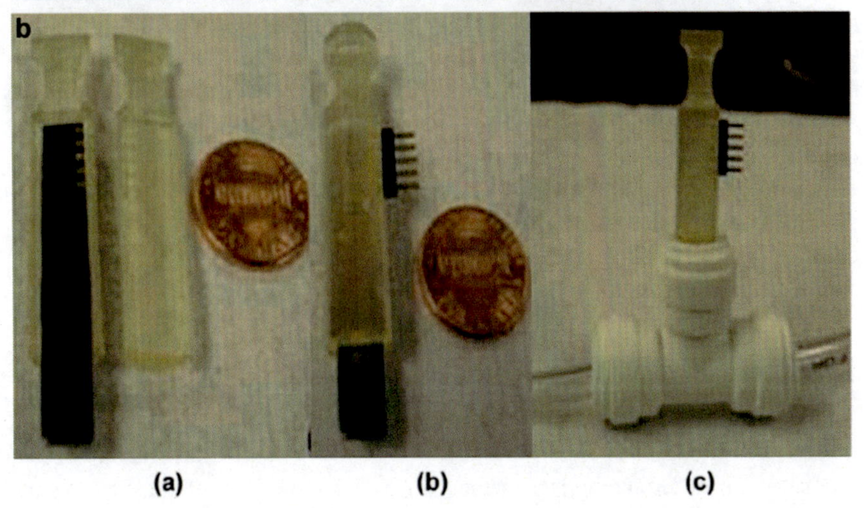

Fig. 13.38 (**a**) Solidworks 3D model of package for the sensor with clamping action and electrical contacts and (**b**) a snap-in connection of a standard tubing to the package for a sensor: (**a**) before assembly, (**b**) after assembly, and (**c**) in standard 3/8" diameter tube connector. (Reprinted from Ref. [61] with permission Elsevier © (2007))

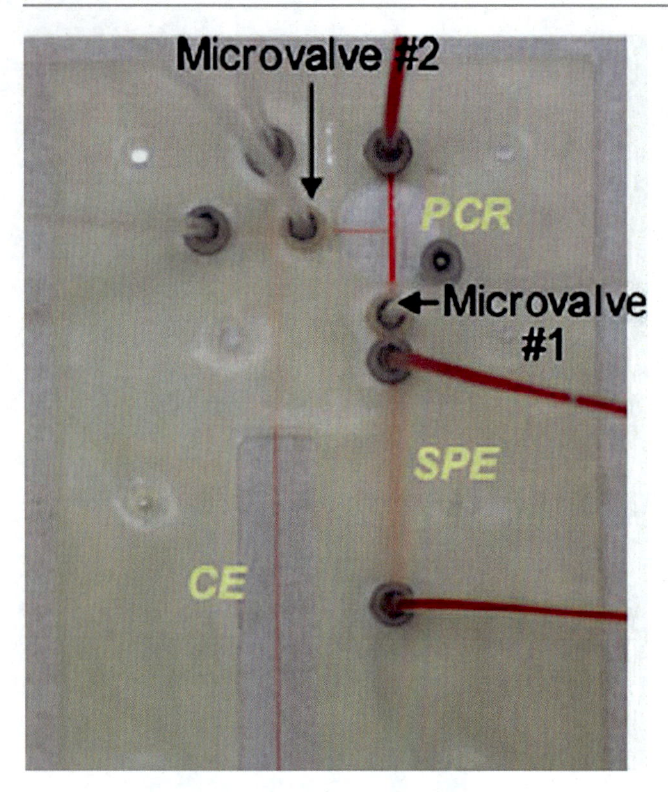

Fig. 13.39 Red dye flowing through the microfluidic compartment in the DNA separation system with microvalve #1 and #2 open. This simulates the sequence of events necessary for a PCR analysis. The integrated valves are operated at pressures of 400 kPa and when valve #1 and #2 were closed dye fills the SPE chamber. When valve #1 is open, sample is directed from SPE to PCR thermal cycling chamber, and finally when valve #2 is opened, sample is injected into CE separation column. (Reprinted from Ref. [90] with permission Elsevier © (2007))

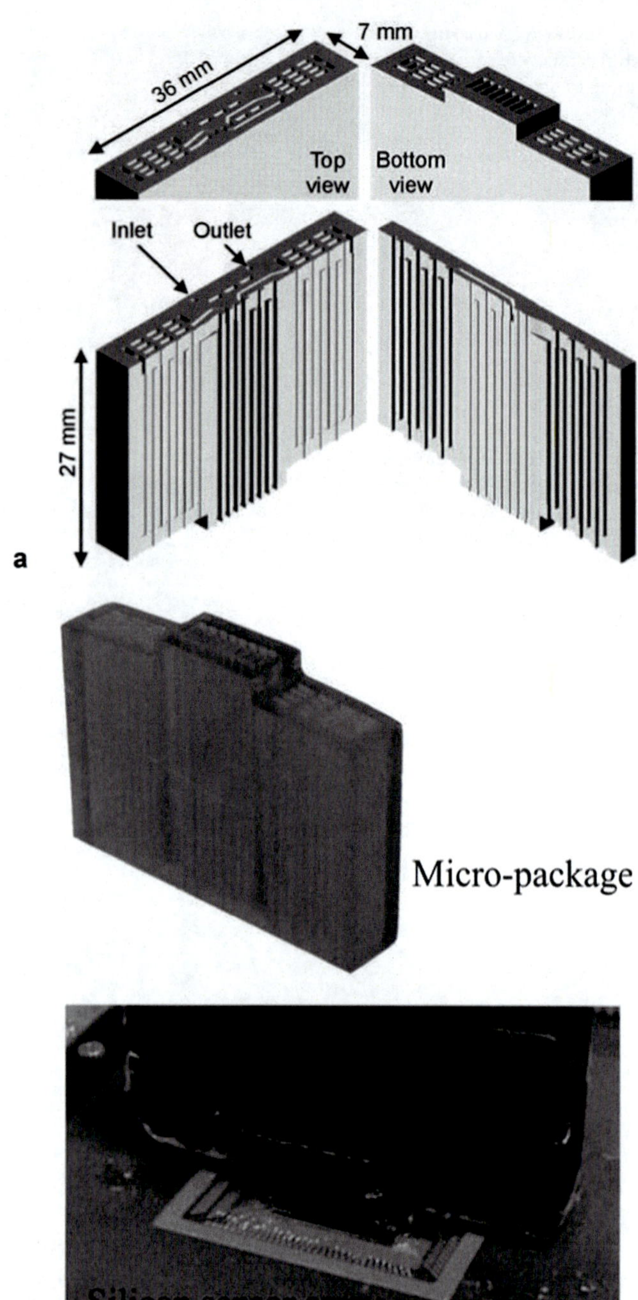

Fig. 13.40 (**a**) CAD drawing of MSL package, dimensions 36 × 27 × 7 mm, which was built by SLA and coated with Parylene-C and (**b**) photograph of the micropackage and final assembly to the sensor array providing channels for the gas separation [93]. (Reprinted with permission of IET Publishers)

The flexibility of SLA is to build custom fluidic channel designs as indicated by the example of mimicking the sense of olfaction [93]. Building a series of flow channels with a stationary phase of Parylene-C approximately 10 μm thick, analogous to the mucus coating of the nasal passages, is shown in the CAD drawing in Fig. 13.40a. The prototype olfactory mucosa is assembled onto a microsensor array, such that each sensor element is beneath one of the channels. Thus the sample is exposed to each sensor in series depending upon the flow and interaction with the stationary phase. Hence the temporal response of the sensors indicates some attributes of the gas species being detected. The SLA flow channels are attached to the die with additional uncured epoxy resin, as shown in Fig. 13.40b.

13.4 Discussion of Future Prospects and Conclusions

The selection of an appropriate method for packaging of a sensor or sensor array is an important and a complex issue that is often underestimated when designing systems. One must consider the mechanical design, thermal issues, chemical and biocompatibility, ease of manufacture, and cost. Based upon the intended application, material selection can have an influence on performance. For example, the influence of the stresses generated by package may act on the sensor to change performance or reliability and should be carefully considered. Important material properties include hermeticity, corrosion resistance, inertness of package, and possibility of absorption of chemicals and subsequent leaching from the package during the operational lifetime of sensor. Issue of reliability of packaging has been investigated by de Reus et al. [94], and more work is needed.

Many chemical sensor packages are hand assembled and require precision alignment under the microscope. Lower-cost packaging with injection molding of polymers has been successful for ISFET sensors in applications that require short-term use. The cost of materials and level of automation have an impact on the cost of assembly and manufacture.

The packaging requirement for nanosensors may not be that much different from microsensors. The size of the nanowires suggests that a smaller die could result and hence smaller-scale packages. One of the current issues for the further successful application of nanowire sensors is the ability to make reproducible and reliable electrical connections to the nanowires. However, packages suitable for MEMS chemical sensors [6] may also be useful because the ultimate function is identical.

Given the diversity of sensor platforms and the range of different applications, it appears that currently a package must be specifically designed for each individual application. This level of customization leads inevitably to increased cost. Until a generic packaging approach is developed for chemical and biomolecular sensors, the cost of the packaging will invariably exceed the cost of the sensor. New methods for packaging need to be developed that are amenable to automated assembly and mass production. An area where progress in wafer-level packaging has occurred includes new materials, such as polycrystalline diamond for integrated packaging [95]. The potential for further development in sensors with integrated packages is a fruitful one for study where perhaps progress in novel materials processes such as atomic layer deposition (ALD)/CVD and use of flexible substrates would be useful. Nevertheless, packaging remains a multidisciplinary area of study and needs development of generic methods in order to achieve reduced cost for chemical and biomolecular sensors.

References

1. Harper, C.A. (ed.): Electronic Packaging and Interconnection Handbook. New York, McGraw-Hill (2004)
2. Tummala, R. (ed.): Fundamentals of Microsystems Packaging. New York, McGraw-Hill (2001)
3. Tomescu, C.C.: Low-Cost Sensor Packaging. Delft University Press, Delft (2001)
4. Brand, O.: Packaging. In: Gianchandani, Y.B., Tabata, O., Zappe, H. (eds.) Comprehensive Microsystems, pp. 431–460. Elsevier, Amsterdam (2008)
5. Velten, T., Ruf, H.H., Barrow, D., Aspragathos, N., Lazarou, P., Jung, E., Malek, C.K., Richter, M., Kruckow, J., Wackerle, M.: Packaging of bio-MEMS: strategies, technologies, and applications. IEEE Trans. Adv. Packag. **28**, 533–546 (2005)
6. Lee, Y.C., Parviz, B.A., Chiou, J.A.: Packaging for microelectromechanical and nanomechanical systems. IEEE Trans. Adv. Packag. **26**, 217–226 (2003)
7. Gardner, J.W., Udrea, F.: Microsensors: Principles and Applications. Wiley, London (2009)
8. Pearce, T.C., Schiffman, S.S., Nagle, H.T., Gardner, J.W.: Handbook of Machine Olfaction: Electronic Nose Technology. Wiley-VCH, Weinheim (2003)
9. Kovacs, G.T.A.: Micromachined Transducer Sourcebook. McGraw Hill, New York (1998)
10. Göpel, W., Jones, T.A., Kleitz, M., Lundström, I., Seiyama, T.: Sensors: A Comprehensive Survey, Vol. 2, Pt. I, Chemical and Biochemical Sensors. Wiley-VCH, Weinheim (1997)
11. Bergveld, P.: Development of an ion-sensitive solid-state device for neurophysiological measurements. IEEE Trans. Biomed. Eng. **17**, 70–71 (1970)
12. Moss, S.D., Johnson, C.C., Janata, J.: Hydrogen, calcium, and potassium ion-sensitive FET transducers: a preliminary report. IEEE Trans. Biomed. Eng. **25**, 49–54 (1978)
13. Chern, G.-C., Zemel, J.N.: Temperature dependence of the gate-controlled portion of ion-controlled diodes. IEEE Trans. Electron Devices. **29**, 115–123 (1982)
14. Yamamoto, N., Tonomura, S., Matsuoka, T., Tsubomura, H.: A study on a palladium-titanium oxide Schottky diode as a detector for gaseous components. Surf. Sci. **92**, 400–406 (1980)
15. Koudelka, M.: Performance characteristics of a planar 'Clark-type' oxygen sensor. Sensors Actuators. **9**, 249–258 (1986)
16. Maclay, G.J., Buttner, W.J., Stetter, J.R.: Microfabricated amperometric gas sensors. IEEE Trans. Electron Devices. **35**, 793–799 (1988)
17. Poghossian, A., Schoning, M.J.: Chemical and biological field-effect sensors for liquids – a status report. In: Marks, R.S., Lowe, C.R., Cullen, D.C., Weetall, H.H., Karube, I. (eds.) Handbook of Biosensors and Biochips. Wiley, Chichester (2007)
18. Smith, R.L., Janata, J., Huber, R.J.: Transient phenomena in ion sensitive field effect transistors. J. Electrochem. Soc. **127**, 1599–1603 (1980)
19. Grisel, A., Francis, C., Verney, E., Mondin, G.: Packaging technologies for integrated electrochemical sensors. Sensors Actuators. **17**, 285–295 (1989)
20. Chovelon, J.M., Jaffrezic-Renault, N., Cros, Y., Fombon, J., Pedone, D.: Monitoring of ISFET encapsulation aging by impedance measurements. Sensors Actuators B. **3**, 43–50 (1991)
21. Gracia, I., Cane, C., Lozano, M., Esteve, J.: On-line determination of the degradation of ISFET chemical sensors. Sensors Actuators B. **8**, 218–222 (1993)
22. van der Spiegel, J., Lauks, I., Chan, P., Babic, D.: The extended gate chemically sensitive field effect transistor as multi-species microprobe. Sensors Actuators. **4**, 291–298 (1983)
23. Gracia, I., Cane, C., Lora-Tamayo, E.: Electrical characterization of the aging of sealing materials for ISFET chemical sensors. Sensors Actuators B. **24–25**, 206–210 (1995)
24. Chi, L.-L., Chou, J.-C., Chung, W.-Y., Sun, T.-P., Hsiung, S.-K.: Study on extended gate field effect transistor with tin oxide sensing membrane. Mater. Chem. Phys. **63**, 19–23 (2000)
25. Janata, J.: Thirty years of CHEMFETs – a personal view. Electroanalysis. **16**, 1831–1835 (2004)
26. van den Vlekkert, H.H., Kloeck, B., Prongue, D., Berthoud, J., Hu, B., de Rooij, N.F., Gilli, E., de Crousaz, P.H.: A pH-ISFET and an integrated pH-pressure sensor with back-side contacts. Sensors Actuators. **14**, 165–176 (1988)
27. Matsuo, T., Esashi, M.: Methods of ISFET fabrication. Sensors Actuators. **1**, 77–96 (1981)

28. Bratov, A., Mufioz, J., Dominguez, C., Bartroli, J.: Photocurable polymers applied as encapsulating materials for ISFET production. Sensors Actuators B. **24–25**, 823–825 (1995)
29. Munoz, J., Bratov, A., Mas, R., Abramova, N., Dominguez, C., Barfroli, J.: Planar compatible polymer technology for packaging of chemical microsensors. J. Electrochem. Soc. **143**, 2020–2025 (1996)
30. Munoz, J., Jimenez, C., Bratov, A., Bartroli, J., Alegret, S., Dominguez, C.: Photosensitive polyurethanes applied to the development of CHEMFET and ENFET devices for biomedical sensing. Biosens. Bioelectron. **12**, 577–585 (1997)
31. van Hal, R.E.G., Bergveld, P., Engbersen, J.F.J.: Fabrication and packaging of mesa ISFETs. Sensors Mater. **8**, 455–468 (1996)
32. Milgrewa, M.J., Riehle, M.O., Cumming, D.R.S.: A large transistor-based sensor array chip for direct extracellular imaging. Sensors Actuators B. **111–112**, 347–353 (2005)
33. Oelßner, W., Zosel, J., Guth, U., Pechstein, T., Babel, W., Connery, J.G., Demuth, C., Gansey, M.G., Verburg, J.B.: Encapsulation of ISFET sensor chips. Sensors Actuators B. **105**, 104–117 (2005)
34. Suehle, J.S., Cavicchi, R.E., Gaitan, M., Semancik, S.: Tin oxide gas sensor fabricated using CMOS micro-hotplates and in-situ processing. IEEE Electron Device Lett. **14**, 118–120 (1993)
35. Martinez, C.J., Hockey, B., Montgomery, C.B., Semancik, S.: Porous tin oxide nanostructured microspheres for sensor applications. Langmuir. **21**, 7937–7944 (2005)
36. Panchapakesan, B., Cavicchi, R., Semancik, S., DeVoe, D.L.: Sensitivity, selectivity and stability of tin oxide nanostructures on large area arrays of microhotplates. Nanotechnology. **17**, 415–425 (2006)
37. Simon, I., Barsan, N., Bauer, M., Weimar, U.: Micromachined metal oxide gas sensors: opportunities to improve sensor performance. Sensors Actuators B. **73**, 1–26 (2001)
38. Semancik, S., Cavicchi, R.: Kinetically controlled chemical sensing using micromachined structures. Acc. Chem. Res. **31**, 279–287 (1998)
39. Bosc, J.M., Odile, J.P.: Micromachined structure reliability testing specificity the Motorola MGSI100 gas sensor example. Microelectron. Reliab. **37**, 1791–1794 (1997)
40. Shirke, A.G., Cavicchi, R.E., Semancik, S., Jackson, R.H., Frederick, B.G., Wheeler, M.C.: Femtomolar isothermal desorption using microhotplate sensors. J. Vac. Sci. Technol. A. **25**, 514–526 (2007)
41. Meier, D.C., Taylor, C.J., Cavicchi, R.E., White, E., Ellzy, M.W., Sumpter, K.B., Semancik, S.: Chemical warfare agent detection using MEMS-compatible microsensor arrays. IEEE Sensors J. **5**, 712–725 (2005)
42. Meier, D.C., Semancik, S.: Coupling nanowire chemiresistors with MEMS microhotplate gas sensing platforms. Appl. Phys. Lett. **91**, 063118 (2007)
43. Goschnick, J.: An electronic nose for intelligent consumer products based on a gas analytical gradient microarray. Microelectron. Eng. **57–58**, 693–704 (2001)
44. Gmur, R., Goschnick, J., Hocker, T., Schwarzenbach, H., Sommer, M.: Impact of sensor packaging on analytical performance and power consumption of metal oxide based gas sensor microarrays. Sensors Actuators B. **127**, 107–111 (2007)
45. Briand, D., Colin, S., Gangadharaiah, A., Velab, E., Dubois, P., Thiery, L., de Rooij, N.F.: Micro-hotplates on polyimide for sensors and actuators. Sensors Actuators A. **132**, 317–324 (2006)
46. Briand, D., Colin, S., Courbat, J., Raible, S., Kappler, J., de Rooij, N.F.: Integration of MOX gas sensors on polyimide hotplates. Sensors Actuators B. **130**, 430–435 (2008)
47. Burg, T.P., Amir, R., Mirza, N.M., Tsau, C.H., Popescu, G.A., Foster, J., Manalis, S.R.: Vacuum-packaged suspended microchannel resonant mass sensor for biomolecular detection. J. Microelectromech. Syst. **15**, 1466–1476 (2006)
48. Song, W.H.: Packaging techniques for CMOS-based chemical and biochemical microsensors. Ph. D. Dissertation in Physical Electronics Laboratory, p. 104. Swiss Federal Institute of Technology, Zurich (2005)
49. Rangelow, I.W., Ivanov, T., Ivanova, K., Volland, B.E., Grabiec, P., Sarov, Y., Persaud, A., Gotszalk, P.Z., Zielony, M., Dontzov, D., Schmidt, B., Zier, M., Nikolov, N., Kostic, I., Engl, W., Sulzbach, T., Mielczarski, J., Kolb, S., Latimier, D.P., Pedreau, R., Djakov, V., Huq, S.E., Edinger, K., Fortagne, O., Almansa, A., Blom, H.O.: Piezoresistive and self-actuated 128-cantilever arrays for nanotechnology applications. Microelectron. Eng. **84**, 1260–1264 (2007)
50. Vancura, C., Li, Y., Lichtenberg, J., Kirstein, K.-U., Hierlemann, A.: Liquid-phase chemical and biochemical detection using fully integrated magnetically actuated complementary metal oxide semiconductor resonant cantilever sensor systems. Anal. Chem. **79**, 1646–1654 (2007)
51. Pinnaduwage, L.A., Gehl, A.C., Allman, S.L., Johansson, A., Boisen, A.: Miniature sensor suitable for electronic nose applications. Rev. Sci. Instrum. **78**, 55101–55104 (2007)
52. Choudhury, A., Hesketh, P.J., Hu, Z., Thundat, T.: Low noise chemical detection with a Piezoresistive Microcantilever. Trans. ECS. **3**, 473–482 (2006)
53. Hunter, G.W., Liu, C.C., Makel, D.: Microfabricated chemical sensors for aerospace applications. In: Gad-el-Hak, M. (ed.) MEMS Handbook: Design and Fabrication. CRC Press, Boca Raton, USA (2006)
54. Hunter, G.W., Xu, J.C., Neudeck, P.G., Makel, D.B., Ward, B., Liu, C.C.: Intelligent Chemical Sensor Systems for in-Space Safety Applications. First International Forum on Integrated System Health Engineering and Management in Aerospace. American Institute of Aeronautics and Astronautics, Napa (2005)
55. Hierlemann, A., Gutierrez-Osuna, R.: Higher-order chemical sensing. Chem. Rev. **108**, 563–613 (2008)
56. Ruano-Lopez, J.M., Aguirregabiria, M., Tijero, M., Arroyo, M.T., Elizalde, J., Berganzo, J., Aranburu, I., Blanco, F.J., Mayora, K.: A new SU-8 process to integrate buried waveguides and sealed microchannels for a Lab-on-a-Chip. Sensors Actuators B. **114**, 542–551 (2006)
57. Blanco, F.J., Agirregabiria, M., Berganzo, J., Mayora, K., Elizalde, J., Calle, A., Dominguez, C., Lechuga, L.M.: Microfluidic-optical integrated CMOS compatible devices for label-free biochemical sensing. J. Micromech. Microeng. **16**, 1006–1016 (2006)
58. Chang-Yen, D.A., Gale, B.K.: Design, fabrication, and packaging of a practical multianalyte-capable optical biosensor. J. Microlithogr. Microfabr. Microsyst. **5**, 21101–21107 (2006)
59. Georgel, V., Verjus, F., van Grunsven, E.C.E., Poulichet, P., Lissorgues, G., Chamaly, C.P.S., Bourouina, T.: A SAW filter integrated on a silicon passive substrate used for system in package. Sensors Actuators A. **142**, 185–191 (2008)
60. Miserendino, S., Tai, Y.-C.: Modular microfluidic interconnects using photodefinable silicone microgaskets and MEMS O-rings. Sensors Actuators A. **143**, 7–13 (2008)
61. Pepper, M., Palsandram, N.S., Zhang, P., Lee, M., Cho, H.-J.: Interconnecting fluidic packages and interfaces for micromachined sensors. Sensors Actuators A. **134**, 278–285 (2007)
62. Ma, B., Liu, S., Gan, Z., Liu, G., Cai, X., Zhang, H., Yang, Z.: A PZT insulin pump integrated with a silicon micro needle array for transdermal drug delivery. IEEE Electronic Compon. Technol. Conf. 677–681 (2006)
63. Li, P.-Y., Shih, J., Lo, R., Saati, S., Agrawal, R., Humayun, M.S., Tai, Y.-C., Meng, E.: An electrochemical intraocular drug delivery device. Sensors Actuators A. **143**, 41–48 (2008)
64. Unger, M.A., Chou, H.-P., Thorsen, T., Scherer, A., Quake, S.R.: Monolithic microfabricated valves and pumps by multilayer soft lithography. Science. **288**, 113 (2000)

65. Rebello, K.J.: Applications of MEMS in surgery. Proc. IEEE. **92**, 43–55 (2004)
66. Gilla, H.S., Prausnitz, M.R.: Pocketed microneedles for drug delivery to the skin. J. Phys. Chem. Solids. **69**, 1537–1541 (2008)
67. Knowles, K.M., van Helvoort, A.T.J.: Anodic bonding. Int. J. Mater. Rev. **51**, 273–311 (2006)
68. Alexe, M., Gosele, U., Gosle, U.: Wafer Bonding. Springer, New York (2004)
69. Kim, H., Najafi, K.: Characterization of low temperature wafer bonding using thin film parylene. J. Microelectromech. Syst. **14**, 1347–1355 (2005)
70. Oldsen M, Hofmann, U., Quenzer, H.J.; Janes, J.; Stolte, C.; Gruber, K.; Ites, M.; Sorensen, F.; Wagner, B. A novel fabrication technology for anti-reflex wafer-level vacuum packaged microscanning mirrors. Proceedings of the SPIE – The International Society for Optical Engineering. SPIE, 2008:688201–688211
71. Kim, J.-M., Lee, S., Baek, C.-W., Kwon, Y., Kim, Y.-K.: BCB-based wafer-level packaged single-crystal silicon multi-port RF MEMS switch. Electron. Lett. **44**, 118–119 (2008)
72. Yun C.H., Martin, J.R., Chen, T., Davis, D.: MEMS wafer-level packaging with conductive vias and wafer bonding. TRANSDUCERS'07 & Eurosensors XXI 2007 14th International Conference on Solid-State Sensors, Actuators and Microsystems, pp. 2091–2094 (2007)
73. Takegawa, Y., Baba, T., Okudo, T., Suzuki, Y.: Wafer-level packaging for micro-electromechanical systems using surface activated bonding. Jpn. J Appl Phys. **46**, 2768–2770 (2007)
74. Raible, S., Briand, D., Kappler, J., de Rooij, N.F.: Wafer level packaging of micromachined gas sensors. IEEE Sensors J. **6**, 1232–1235 (2006)
75. Briand D., Guillot, L., Raible, S., Kappler, J., de Rooij, N.F.: Highly integrated wafer level packaged MOX gas sensors. TRANSDUCERS'07 & Eurosensors XXI, 14th International Conference on Solid-State Sensors, Actuators and Microsystems, pp. 2401–2404 (2007)
76. Ziaie, B., Jeffrey, A., Von Arx, J.A., Dokmeci, M.R., Najafi, K.: A hermetic glass-silicon micropackage with high-density on-chip feedthroughs for sensors and actuators. J. Microelectromech. Syst. **5**, 166–179 (1996)
77. Lin, L.: MEMS post-packaging by localized heating and bonding. IEEE Trans. Adv. Packag. **23**, 608–616 (2000)
78. Cheng, Y.T., Lin, L., Najafi, K.: Localized silicon fusion and eutectic bonding for MEMS fabrication and packaging. J. Microelectromech. Syst. **9**, 3–5 (2000)
79. Weibel, D.B., Whitesides, G.M.: Applications of microfluidics in chemical biology. Curr. Opin. Chem. Biol. **10**, 584–591 (2006)
80. Lee, J.N., Park, C., Whitesides, G.M.: Solvent compatibility of poly (dimethylsiloxane)-based microfluidic devices. Anal. Chem. **75**, 6544–6554 (2003)
81. Xia, Y., Whitesides, G.M.: Soft lithography. Annu. Rev. Mater. Sci. **28**, 153–184 (1998)
82. Cho, C.-W., Lim, C.-H., Woo, C.-S., Jeon, H.-S., Park, B., Ju, H., Lee, C.-J., Maeng, S., Kim, K.-C., Kim, S.H., Lee, S.-B.: Highly flexible and transparent single wall carbon nanotube network gas sensors fabricated on PDMS substrates. Materials Research Society Symposium Proceedings, v 922, Organic and Inorganic Nanotubes: From Molecular to Submicron Structures. MRS. 95–101 (2006)
83. Zhang L., Cheung, K. Y., Chua, Y. R., Trau, D. PDMS microdevice with built-in optical biosensor array for oil-site monitoring of the microenvironment within microchannels. Progress in Biomedical Optics and Imaging – Proceedings of SPIE, v 6445, Optical Diagnostics and Sensing VII, p. 64450 W (2007)
84. Shin J-K, Kim, D.-S., Lim, G., Shoji, S. Electrical detection of biomolecules in a PDMS micro-fluidic channel using a MOSFET-type biosensor. The International Society for Optical Engineering, v 6416, Biomedical Applications of Micro- and Nanoengineering III, p. 641612 (2007)
85. Cho Y.H., N. Takama, T. Yamamoto, T. Fujii, B. J. Kim. MEMS based biochip for the characterization of single red blood cell. Proceedings of the 3rd Annual International IEEE EMBS Special Topic Conference on Microtechnologies in Medicine and Biology: Kahuku, Oahu, Hawaii, pp. 60–63 (2005)
86. Zhong, L., Wang, X., Jiang, Y., Qiu, X., Zhou, Y., Liu, L.: A silicon-based micro direct methanol fuel cell stack with compact structure and PDMS packaging. Micro Electro Mech. Syst. IEEE. 891–894 (2007)
87. Varadan, V., Jiang, X., Varadan, V.V.: Microstereolithography and other Fabrication Techniques for 3D MEMS. Wiley, New York (2001)
88. Rosen, D.W.: Stereolithography and rapid prototyping. In: Hesketh, P.J. (ed.) BioNanoFluidic MEMS, pp. 175–196. New York, Springer (2008)
89. Han, A., Wang, O., Graff, M., Mohanty, S.K., Edwards, T.L., Han, K.-H., Frazier, A.B.: Multi-layer plastic/glass microfluidic systems containing electrical and mechanical functionality. Lab Chip. **3**, 150–157 (2003)
90. Han, K.H., McConnell, R.D., Easley, C.J., Bienvenue, J.M., Ferrance, J.P., Landers, J.P., Frazier, B.: An active microfluidic system packaging technology. Sensors Actuators B. **122**, 337–346 (2007)
91. Tse, A.L., Hesketh, P.J., Gole, J.L., Rosen, D.W.: Rapid prototyping of chemical sensor packages with stereolithography. Microsyst. Technol. **9**, 319–323 (2003)
92. Singh, J., Ekaputri, M.: PCR thermal management in an integrated Lab on Chip. J. Phys. Conf. Ser. **34**, 222–227 (2006)
93. Covington, J.A., Gardner, J.W., Hamilton, A., Pearce, T.C., Tan, S. L.: Towards a truly biomimetic olfactory microsystem: an artificial olfactory mucosa. IET Nanobiotechnol. **1**, 15–21 (2007)
94. de Reus, R., Christensen, C., Weichel, S., Bouwstra, S., Janting, J., Eriksen, G.F., Dyrbye, K., Brown, T.R., Krog, J.P., Jensen, O.S., Gravesen, P.: Reliability of industrial packaging for microsystems. Microelectron. Reliab. **38**, 1251–1260 (1998)
95. Zhu, X., Aslam, D.M., Tang, Y., Stark, B.H., Najafi, K.: The fabrication of all-diamond packaging panels with built-in interconnects for wireless integrated microsystems. J. Microelectromech. Syst. **13**, 396–405 (2004)
96. Choudhury A.: A piezoresistive microcantilever array for chemical sensing applications. Ph. D. Dissertation in Mechanical Engineering. Georgia Institute of Technology: Atlanta, USA, p. 178 (2007)

Nanobiosensing Electronics and Nanochemistry for Biosensor Packaging 14

Dasharatham G. Janagama and Rao R. Tummala

Chapter Objectives

- Define bioelectronics, nanobioelectronic sensors, and their applications.
- Describe fundamentals of biology related to interfacing biological material with electronic systems.
- Present the elements or building blocks of nanobioelectronic sensors and fabrication methodologies.
- Describe functionality of nanobioelectronic sensors and nanopackaging.
- Identify the challenges in interface mechanisms between biosensor element and transducer, signal transduction, processing, and monitoring.

14.1 Introduction

14.1.1 Bioelectronics and Biosensors

Bioelectronics, including biosensors, is an emerging interdisciplinary field encompassing bioscience, chemistry, physics, materials science, electronics, and engineering. A key factor of bioelectronics is in understanding the mechanisms of interfaces between biological materials and electronics. The miniaturization of electronic circuits laid the foundation for building microelectronic devices and modules. Typical bioelectronic devices are biosensors, biochips, electronic pills, electronic eye, electronic nose, genetic toggle switches, biocomputing devices, and bioactuators (Fig. 14.1). Bioelectronics is revolutionizing the health-care industry, forensic medicine, food and drink industries, environmental protection, genome analysis of organisms, and communications.

Biosensor is an analytical device that converts a biological response into an electrical signal. It facilitates selective identification of biological materials such as peptide (two or more amino acids linked by the carboxyl group of one amino acid to the amino group of another), protein, deoxyribonucleic acid (DNA), and ribonucleic acid (RNA). The biochip is a collection of miniaturized test sites (microarrays), which perform thousands of biological reactions involving protein or DNA in a few seconds. The concept of electronic pills is the development of noninvasive microelectronics for monitoring physiological processes in situ. Electronic eye consists of a smart camera computer device, which facilitates blind people to perform visual tasks. Electronic nose device typically works as an animal nose. It recognizes any compound or combination of compounds. This device is very useful in the detection of spoilage of food stuff and pathogenic contamination, oil spills, and environmental pollutants. Genetic toggle switches are gene regulatory circuits with addressable cellular memory unit. They are used in biotechnology, biocomputing, and gene therapy applications. The strands of DNA (or RNA), the computer code of life, are used to create molecular-size circuits of self-assembly, which revolutionized the computing capabilities. It is known to surpass the capability of a super computer. Bioactuators are microsensor-based feedback devices, which are used, for example, in the detection of tumor microenvironment and biophysical cancer treatment. Biofuel cell is a device that converts the biochemical energy into electrical energy. They are used as small power sources for implanted medical sensors, actuators, and telemetry devices. Biofuel cell could replace rechargeable batteries in the near future.

14.1.2 Fundamentals of Biosensors and Microsystems/Nanosystems Packaging

Among the bioelectronic devices, biosensors came into prominence because of vast application potentials. A sensor

D. G. Janagama (✉) · R. R. Tummala
Package Research Center (PRC), Georgia Institute of Technology, Atlanta, GA, USA
e-mail: rao.tummala@ee.gatech.edu

C. P. Wong et al. (eds.), *Nano-Bio- Electronic, Photonic and MEMS Packaging*, https://doi.org/10.1007/978-3-030-49991-4_14

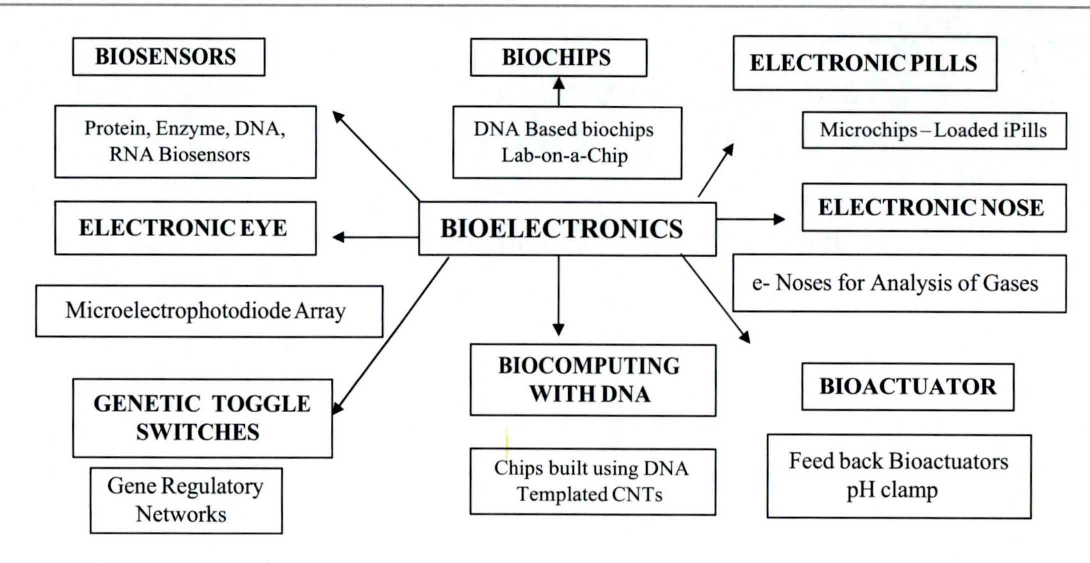

Fig. 14.1 Schematic of bioelectronics

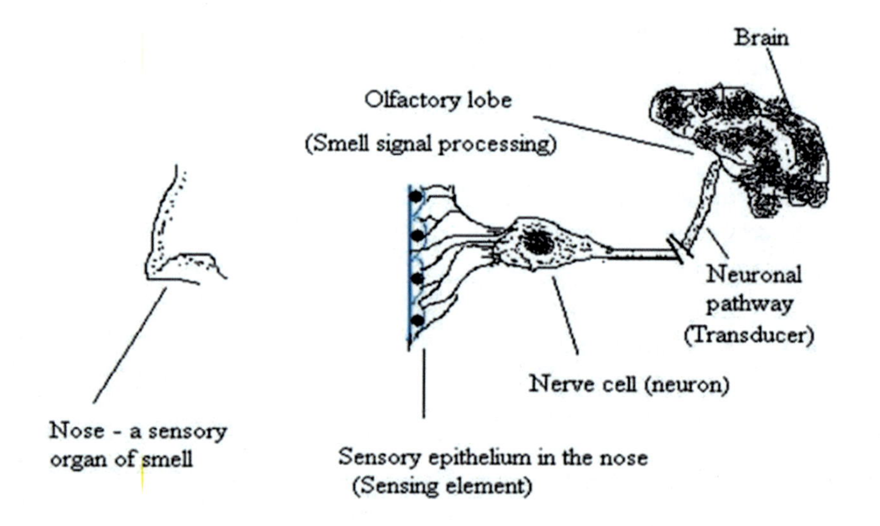

Fig. 14.2 Schematic of mechanism of sense of smell – a sensory system

is a device that senses or measures a real-world condition such as physical contact, motion, heat, or light and converts the impulses into analog or digital representation. The biosensors are used for detection, identification, quantification, and analysis of target molecules. A multifunctional sensory system, the nervous system, is evolved in the animals for sensing physical touch or contact, taste, smell, vision, and hearing. For example, the sense of smell is initiated by sensory receptors in the nasal epithelium (sensing element) upon inhaling certain molecules, carried through the neuronal pathway (transducer), and passed to the olfactory portion of the brain for processing the information as to the nature of gaseous molecules that come into contact with nasal epithelium in the nose (Fig. 14.2). These sensory organs in animals help in sensing the nature of surrounding environment, ultimately aimed at self-preservation and survival.

A typical biosensor is made up of three basic parts, namely, biosensing, transducer or signal conversion, and signal processing elements (Fig. 14.3). In a biosensor, the sensing or biorecognition element (biosensing) is a protein, peptide, deoxyribonucleic acid (DNA), ribonucleic acid (RNA), whole cell, or tissue, which interacts or hybridizes specifically with its complementary target or analyte molecules. It is the primary part of the sensor and determines the selectivity and sensitivity of the sensor. The transducer or signal conversion is a device such as electrode, piezoelectric crystal, photosensitive element, nanobelts/wire, carbon nanotubes (CNTs), or thermistor, which interfaces with the recognition element and converts the magnitude of the biological relevant information or response signal into an electronic signal. The vast gene diversity among animals and plants is manifested in the major biomolecules such as

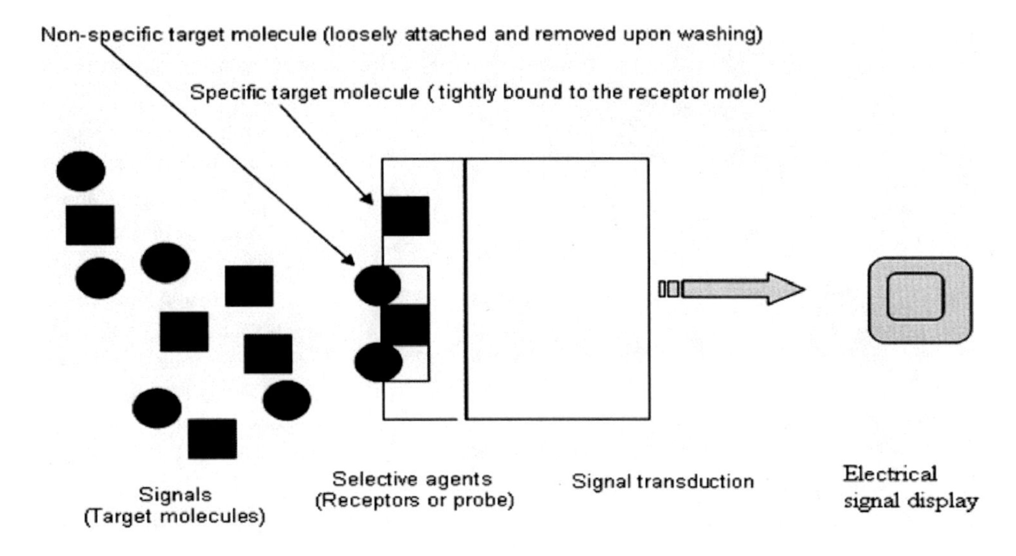

Fig. 14.3 Principle of a biosensor

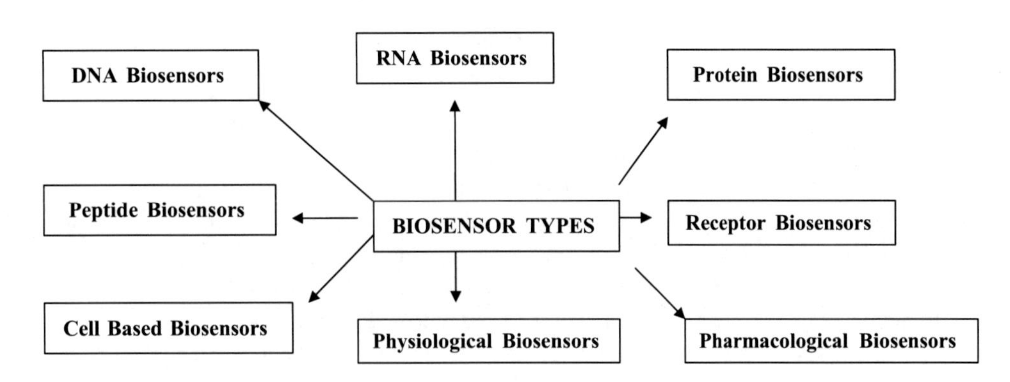

Fig. 14.4 Schematic of different types of biosensors

proteins, peptides, DNA, and RNA. Biosensors are used for detection, identification, quantification, and analysis of biological samples.

Biosensors are categorized into two main types based on the location of lodging of the device: in vivo or implantable biosensor, if inserted into the cell, tissue, or body for sensing the analyte, and in vitro biosensors, if placed in an artificial environment external to the cell, tissue, or body. The biosensor devices have been further categorized based on the types of molecules detected into nucleic acid sensors and DNA chips, immunosensors, enzyme-based biosensors, organism- and whole cell-based biosensors, and natural and synthetic receptor-based biosensors (Fig. 14.4). Innovative signal transduction technology, system integration, proteomics, bioelectronics, and nanoanalytical systems are essential for the development of smart biosensors.

The conventional detection system involves labeling of the probe molecules, a requirement of large amount of target molecules to enable detection and elaborate signal detection methods. Further, the conventional technologies based on the "top-down" approaches are difficult to continue to scale down due to limitations in lithography and interconnect schemes for design and fabrication of biosensors.

The novel properties or unusual combinations of properties possessed by nanostructured materials are utilized for fabrication of nanobiosensors. A "bottom-up" technology incorporating the integration of nanotechnology with biology and bioelectronics constitute the third generation of smart biosensors (Fig. 14.5).

Nanoscience and nanotechnology involves in manipulating the materials at the scale of 100 nm or less. The advancement in microfabrication/nanofabrication technologies enables fabrication and packaging of varieties of functional devices using nanomaterials for biological and nonbiological applications. The applications range across materials science, information technologies, environmental protection, and medicine. As many biological functions occur at microscale and nanoscale in the cell and subcellular systems, the nanodevices help in potential life science applications including improved drug and gene delivery, biocompatible materials for implants, and advanced sensors for early disease detection of diseases and therapies. These nanodevices made with smart nanomaterials are efficient, cost-effective, and easily implantable.

The biosensor packaging deals with integrating a number of functional active and passive components, such as

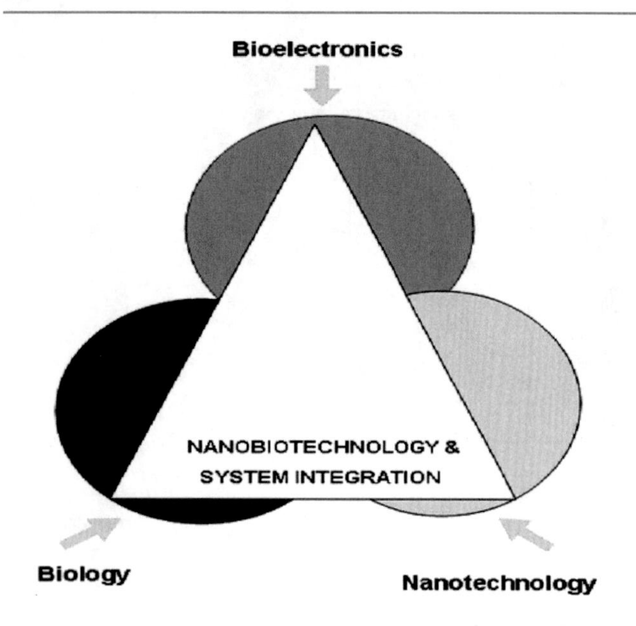

Fig. 14.5 Integration of nanotechnology with biology and bioelectronics to produce nanotechnology-based biosensors

biosensing elements, biosignal conversion elements, signal processing elements, microfluidic systems, and reagent amplification system into a single sensor chip and package. Lab-on-chip (LOC) concept refers to translating the entire laboratory testing processes and functions onto a tiny microelectronic device. The microfluidics incorporated into the package facilitates the flow of the samples and reagents in the chip. Biosensor detection systems applied to analyze DNA–DNA (deoxy ribonucleic acid), antibody–antigen, protein– protein, protein–nucleic acid, protein–lipid, protein–small molecule, and enzyme–substrate interactions. The biosensor packing includes surface modification chemistry, capture molecule attachment (biofunctionalization) with detection methods, and high throughput. However, for complete biosystem integration, the signal and power distribution, I/Os, and suitable interfaces are to be incorporated.

In this chapter of biosensors, we describe the application of biosensors in various walks of life, fundamentals of biology as applied to biosensors, biomechanisms, building blocks of biosensors, biological probe design and interfacing, various transducer mechanisms, challenges in integration of biomolecules with interface mechanisms and signal transduction, integration of nanobiosensors with electronics and fluidics using system-on-package (SOP) platform technologies, and future trends in biosensors. The following ten sections will briefly describe the above in a systematic way:

In Sect. 1, a brief introduction to bioelectronics and biosensors with emphasis on the interdisciplinary nature of the subject and historical account of the development of biosensors systems is presented.

In Sect. 2, the applications and importance of bioelectronics and biosensors in health care, food hygiene, environmental safety, industrial bioprocess control and monitoring, genome analysis, drug discovery, and drug abuse are described.

In Sect. 3, fundamentals of living cells as applied to different types of sensors are described citing the examples of prokaryotic cells and eukaryotic cells, DNA, RNA, proteins, and peptides.

In Sect. 4, building blocks of biosensors such as target molecules, probe molecules and their biomechanisms, interface mechanisms, and signal transduction are mentioned. The molecular aspects of biorecognition elements such as antibodies, nucleic acids (DNA/RNA), natural and synthetic receptors, enzymes, and their biomechanisms are discussed.

Detailed account of biosensing mechanisms is dealt in Sect. 5.

Section 6 deals with various transducer types and their mechanism of interfacing and signal transduction, and challenges in biofunctionalization of the devices and signal transduction are mentioned.

Nanostructures, biomolecules, and cells as components of biosensors are described in Sect. 7.

In Sect. 8, design and preparation of biological probes such protein and DNA for biofunctionalization of nanostructures, fabrication methods, and nanopackaging are described.

Section 9 deals with material characterization of biosensor structures.

Section 10 deals with the future trends in nanobiosensors. As anticipated, discovery of novel materials and innovative approaches may progress by leaps and bounds, as in the case of the computer industry after the development of integrated circuits. A brief summary of nanobiosensors is given.

14.2 Applications of Bioelectronics

The recent advances in miniaturization of electronic circuits provided the technological capability to utilize biological materials for fabricating sensors and circuitry. The microelectronics are the basis for biosensors; biochips; implantable medical devices; prosthetic devices; artificial organs, including electronic nose and electronic eye; electronic pills, surgical and medical devices, biofuel cells, molecular motors, molecular electronics, and biofabrication templates, which endow enormous power to the detection system.

14.2.1 Applications of Biosensors

The biosensors with electrical detection system are superior over the conventional system of immunohistochemical method of detection of biological samples. The importance of critical detection of hazardous pathogen and toxic molecules in the environment was realized by the events of September 11, 2001, attack on US soil, and the subsequent anthrax attacks and ricin (a potent poison from castor beans) attacks in the United Kingdom, which highlighted the value of using sensitive biosensors and bioelectronics all over the world. The important applications of biosensors are as follows:

(a) **Clinical Diagnostics**

The clinical biosensors offer sensitive and precision detection of physiological, prenatal, and postnatal diagnosis of genetic diseases and abnormalities. These biosensors are capable of detecting single-nucleotide polymorphisms, which are caused by gene mutations, including single-base substitution and small insertions or deletion bases in the genome. Rapid and high-throughput detection methods using biosensors are in dire need for early detection of breast cancer, prostate cancer, AIDS, genetic diseases, bacterial and viral infections, and bacterial and viral drug resistance. The other infectious and communicable disease detection include amebiasis; chickenpox; cholera; cyclosporiasis; diphtheria; encephalitis; hand, foot, and mouth disease; hepatitis A; malaria; measles; meningitis; rubella; and West Nile virus. The clinically relevant point mutations are detected by a novel piezoelectric biosensor [1].

(b) **Drug Discovery**

The cell-based biosensors (CBB) are engineered to optically or electrically report specific biological activity. The cellular biosensors are comprised of living cells and can be used in various applications, including screening chemical libraries for drug discovery. Panels of biosensors may also be useful for elucidating the function of novel genes.

(c) **Drug Abuse**

Nanoscale detection and quantification systems for various drug abuses are essential for law enforcement authorities to control the menace of drug abuse. Cocaine sensor has been developed for rapid detection and quantification of coca alkaloids [2]. DNA biosensors and DNA chips are the potential detection devices for investigation and evaluation of DNA–drug interaction mechanisms.

(d) **Genome Analysis**

The use of biosensors in basic research to measure biomolecular interactions in real time is increasing dramatically. Applications include protein–protein, protein– peptide, DNA–protein, DNA–DNA, and lipid–protein interactions. Such techniques have been applied to antibody–antigen, receptor–ligand, signal transduction, and nuclear receptor studies. Biosensors are also useful for the detection of genetic- engineered products, genome analysis [3], and detection of mutations in plants and animals by identifying DNA sequences. There is a great demand for cheap, portable, and easy-to-use devices for point-of-care analysis. These systems integrate on the same substrate the sites of the reaction, the sensors providing electric signals, and the circuits for signal conditioning and amplification.

(e) **Food Hygiene**

Food hygiene is of priority for any nation aspiring to be healthy and sound. The protection of dairy products from Botulinum toxins, foods from poisonous bacterial and fungal contamination and pathogens devastating animal livestock; the protection of meat from mad cow disease causing pathogens, castor bean ricin that are poisonous to people, animals, and insects are critically important. These biomolecules can be offered as the powerful weapons and "road map for terrorists" to destroy humans and livestock indiscriminately. *Salmonella* biosensor was developed, which detects the pathogen in food and water as low as 10^4 cell colony-forming unit/ml [4]. The timely detection of bird flu caused by the virus H5N1 will save the poultry industry and public health all over the world. Biosensors are also useful in monitoring the freshness of aquatic foods, including fish, and fermentation processes.

(f) **Industrial Bioprocess Control and Monitoring**

Those must be monitored and detected in time to prevent a national calamity. Bioprocess control and biosensors are important technological inputs in biotechnology industry. DNA chip array biosensors based on microelectrode arrays, which are fabricated with silicon technology and activated with oligonucleotide probes that bind to the target sequence, monitor the expression of selected genes. These devices are critical for production processes using microorganisms in bioreactors, particularly the protein products generated by recombinant DNA technology during microbial cell or mammalian cell fermentation. Further, they are used for controlling/monitoring extracellular parameters in a bioreactor. Thermal biosensors, for instance, are used for specific temperature monitoring and controlling of bioprocesses. Addition of a flow injection analysis (FIC) system to the biosensor facilitates the control delivery of all reagents to the bioreactor. We described electrical detection of enzymatic hydrolysis of starch using ZnO nanobelts/wire. The detection of the degree of hydrolysis of starch converting into glucose, cyclodextrins, and other monomer products is very critical in food and pharmaceutical industries. Further, real-time and online detection diction of starch hydrolysis events could be monitored through electronic detection system [5].

(g) **Environmental Safety**

Biosensors are widely used in protecting the environment by monitoring of pollutants in water, air, oils, and soil, toxic vapor, acid rain, and odor in the atmosphere. Analysis of explosive trinitrotoluene (TNT) in groundwater with detection limit of TNT of 20 μg/ml using portable fiber-optic biosensor was reported [6]. Pesticides in the water are detected using in situ optical biosensor with a detection limit of 2 ppm for atrazine and alachlor [7]. Ozone detection system was developed using fiber-optic near-infraviolet evanescent wave, which detects ozone over the range of 0.02–0.35 vol%, in the response time of about 1 min.

[8]. Sensitive and rapid biosensors are required for identification of pathogenic organisms in dealing with bioterrorism. Metro subways and other sensitive places need biosensors, which should detect toxins and explosive materials in trace amounts. The detection of industrial toxic effluents and human and animal waste into water bodies is an important aspect of environmental safety and protection.

14.2.1.1 Advantages of Biosensors over Conventional Detection Systems

The use of nanotechnology-based biosensors confers several advantages over the conventional detection systems. Magnetic materials show super magnetic behavior at nanoscale [9]. The semiconducting nanoparticles show resonance tunneling and Coulomb blockade effects [10]. Nanoparticles undergo widening their bandgap energy as the material dimensions reduce, resulting in blueshifts in optical properties caused by a quantum-confined effect. These unique properties do not exist in their bulk counterparts. Therefore, the functional nanostructures composed of metals, semiconductors, magnetic materials, quantum dots, molecular dyes, and other polymers are used in fabricating nanobiosensors, which show the potential advantages over the conventional types:

- In a biosensor, no involvement of exogenous molecules or labels such as conjugation with enzyme, radioactive fluorescence, or chemiluminescence molecules.
- The probe and target-binding events or enzyme and substrate reactions are recorded by the transducer in the absence of any label requirement.
- A polymerase chain reaction (PCR)-free detection systems for DNA identification.
- It is a nanoscale detection of the target molecules, a key factor in early detection of diseases such as breast cancer and AIDS.
- Rapid and high-throughput detection.
- Detection processes are simple, user-friendly, fast, and cost-effective.
- Reduced material requirement to fabricate the devices.
- Reduced power requirement and easier recycling.
- Novel properties and new capabilities.
- Repetitive, portability, and stability.

14.2.2 Applications of Other Bioelectronic Sensing Devices

(a) **Multielectrode Array (MEA)**

Microelectrode array is a conventional passive tool used to record electrical activity in an intact cellular environment of cell cultures, brain slices, and other tissues. It provides a simple interface for monitoring the electrical activity and impedance characteristics of populations of cultured cells over extended periods. The cell contact with electrode is created as cells attach directly to planar electrode structures.

It is used in drug screening, drug delivery, and monitoring of drug side effects. MEA technology is used in the study of neurite outgrowth, synaptogenesis, and regeneration and modulation of these processes by drug, neuronal stimulation, and recording. Figure 14.6 shows the layout of the electrodes fabricated on the cell culture dish (a), and cell culture setup (b) may be neuronal or any cell-type experimental cultures.

Microelectrode array (MEA) of microscale dimensions of Si-fabricated structures is advanced microdevices. They are used as retinal prosthetic devices to stimulate retinal nerve cells as a viable treatment for degenerative retinal diseases and several other applications. Figure 14.7 shows miniature array of microelectrodes. The array consists of a 4 mm × 4 mm base that contains 100 silicon spikes of 1–1.5 mm long.

(b) **Microphotodiode Array (MPDA)**

Microphotodiode array (MPDA) is a micromachined photovoltaic device, used as retinal prosthetics to replace degenerated photoreceptors of the retina of the eye.

MPDA in the subretinal space transforms light stimuli into electrical current for the stimulation of still healthy retinal neurons. They are fabricated as silicon-based multilayered microstructures.

It is made of silicon measuring 2 mm^2, implanted into the retina of the eye. As the silicon material is not biocompatible, other suitable material for implantation is highly sought after. The photodiodes are connected in series and can be independently stimulated. Figure 14.8 shows the schematic of MPDA implant in the retina of the eye.

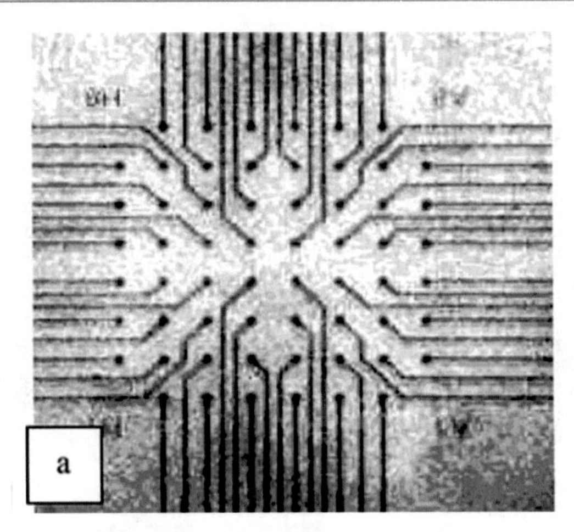

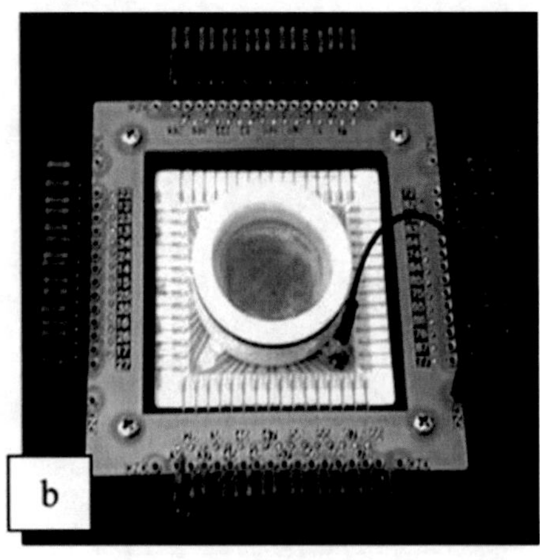

Fig. 14.6 Shows 2D multielectrode array (MEA): (**a**) layout of electrodes, (**b**) MEA assembly with cell culture system. (Source: Courtesy of Dr. Daniel A. Wagenaar, 2003)

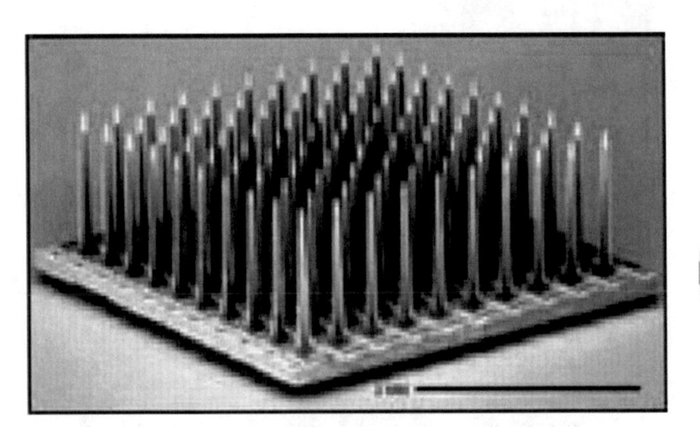

Fig. 14.7 Shows 3D MEA. (Source: Courtesy of Cyber Kinetics)

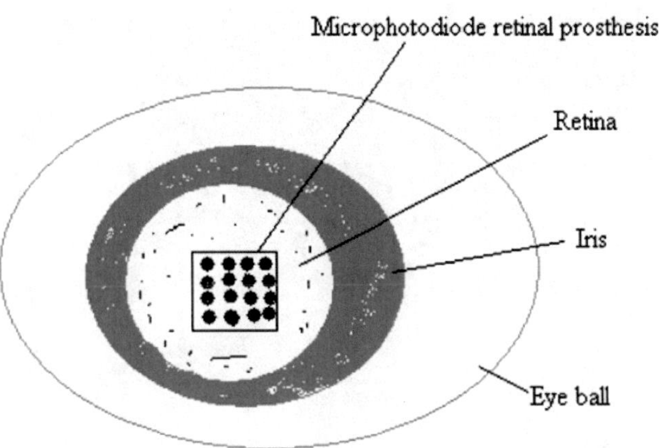

Fig. 14.8 Schematic of MPDA implant in the eye

14.3 Fundamentals of Biological Materials as Sensing Elements

The knowledge of fundamentals of living cells and biomolecules is a prerequisite for understanding the concepts of biosensors. All biological functions in the cell or organism such as digestion, excretion, respiration, reproduction, and sensing are carried out in the structures at nanoscale. The living cells in organisms have been categorized into prokaryotic cells with out-defined nucleus and eukaryotic cells with well-defined nucleus. The prokaryotes are less evolved unicellular organisms such as bacteria and virus with relatively simple subcellular structure and organization. Most cells are between 10 μm (micrometers) and 100 μm in diameter. The human cell is about 10 μm in diameter, and the bacterium is generally rod-shaped with dimensions 1 × 2 μm. The bacterial cell consists of a cell wall enclosing the cytoplasm, in which the genetic material (DNA/RNA), mesosomes, ribosomes, and other organelles are distributed (Fig. 14.9). The functional aspects of prokaryotic cells are related directly to the structure and organization of the macromolecules in their cell makeup, i.e., DNA, RNA, phospholipids, proteins, and polysaccharides, unlike in eukaryotes in which specific organelles are involved. The molecular diversity in the primary structure of these functional molecules is responsible for the diversity that exists among prokaryotes. The bacteria play a vital role in maintaining the Earth as a suitable place for inhabitation by other forms of life as they involve in putrefaction or decay and cleaning up of dead organisms. Many biotechnology products are obtained from useful bacteria, such as antibiotics, recombinant gene products, therapeutics, dairy products, and breweries.

A virus particle, also known as a virion, is essentially a nucleic acid (DNA or RNA) enclosed in a protein shell or coat (Fig. 14.10). Viruses are extremely small, approximately 15–25 nm in diameter.

The eukaryotic cells are represented in protozoa (single celled), mesozoa (few celled), and metazoan (multicelled) – the major branches of the animal kingdom. The eukaryotic cell, as in higher animals, has well-defined substructures or organelles with specific cellular functions. Unlike in bacteria and virus, the cell is covered with a double membrane, which is made up of mainly protein, lipid, and carbohydrate molecules.

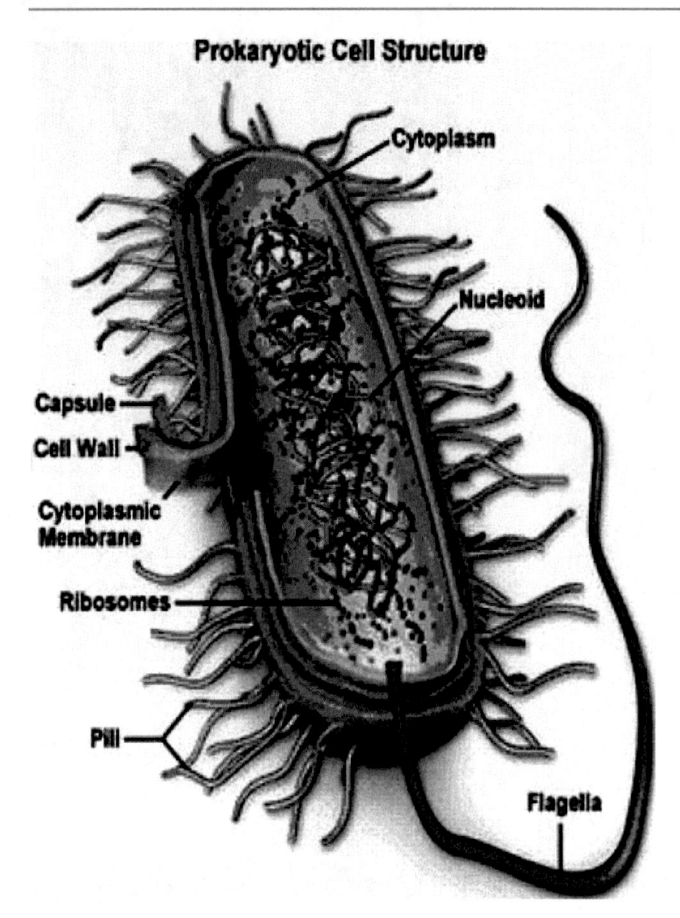

Fig. 14.9 Schematic diagram of a typical bacterial cell (note the absence of nucleus). (Source: http://www.williamsclass.com/ images/ JPGImages/BacteriaCell.jpg)

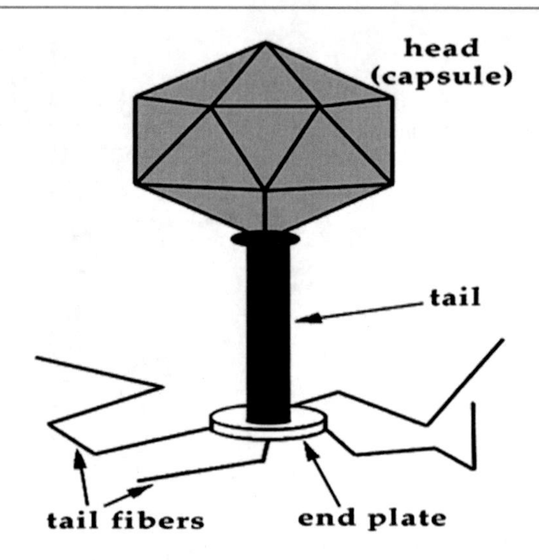

Fig. 14.10 Schematic diagram of a typical virus. (Source: Emiliani, C. 1993. Extinction and viruses)

DNA is a double helix of complementary strands, which contain the code for operation of all living organisms. As shown in Fig. 14.11, DNA is made up of four standard bases in long chains, known as adenosine (A), guanine (G), thymine (T), and cytosine (C). A–T and G–C are the complementary base pairs. Adenosine and guanine are purine bases, whereas cytosine, thymine, and uracil are pyrimidine bases. The sequences of nucleotides define "genes." The genes define RNA and proteins. Of the 22 amino acids commonly found in nature, a series of 3 DNA bases (triplet code) is enough to correspond one amino acid, which is called universal code, meaning the code applied to all organisms to follow ($4 \times 4 \times 4 = 64$ possible combination of four DNA bases).

DNA carries the genetic information of a cell and consists of thousands of genes. Each gene serves as a recipe on how to build a protein molecule. RNA is a single- strand molecule. As in DNA, RNA contains four nucleotides, except the base uracil instead of thymine (Fig. 14.12).

A crucial property of the purines and pyrimidines is their ability to form hydrogen-bonded pairs composed of one purine and one pyrimidine. The constant distance of 1.085 nm between the two helical strands, as rungs in a ladder, is due to similar base pairing pattern (G–C and A–T).

Amino acids are the building blocks (monomers) of proteins. Proteins and enzymes are essential structural and functional building blocks of all organisms. Generally, 22 different amino acids are used to synthesize proteins. The shape and other properties of each protein are dictated by the precise sequence of amino acids in it. By possible combinations of these amino acids, thousands of proteins are generated as thousands of words are formed using 26 alphabets in the English language.

Thus, variability in shape and size among different organisms is due to the vast diversity in their protein and enzyme structures.

Several amino acids linked through peptide bonds forming a functional gene product, namely, protein (Fig. 14.13). Proteins are synthesized in the cytoplasm of the cell under the direction of DNA. The chemical messages in DNA are transcribed into RNA (m-RNA) in the nucleus. The m-RNA move out of the nucleus through nuclear pores into the cytoplasm, with the help of ribosomes and transfer RNA (t-RNA), and other factors translate the m-RNA information into a specific long chain of amino acids called polypeptide. The polypeptide undergoes posttranslational modifications and forms 3D functional protein and enzyme molecules.

Proteins perform important tasks for the cell functions or serve as building blocks in the form of structural proteins and enzymes. Antigens are functional proteins on the cell surface as well as within the cell. Antibodies are particular proteins made by cells in response to the antigens.

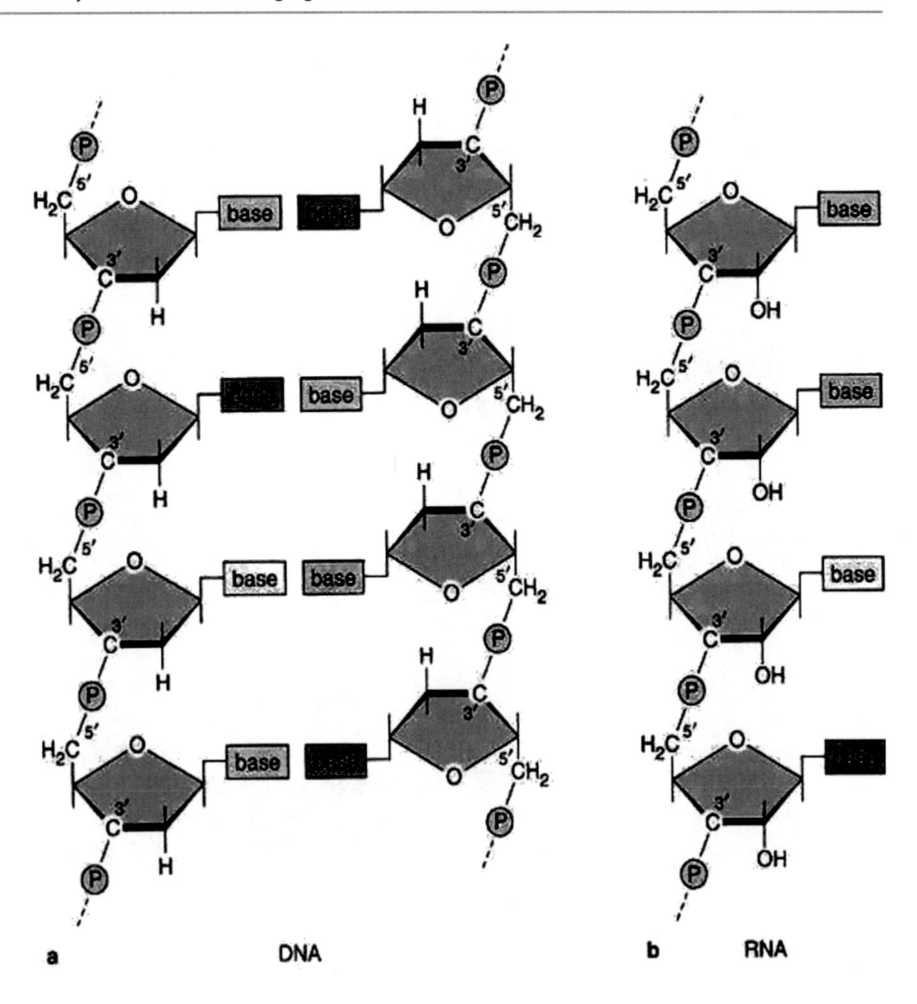

Fig. 14.11 Schematic of DNA and RNA structures. DNA double helix interconnected by bases (A–T, G–C). RNA is a single-stranded structure with uracil (U) base in place of thymine (T). (Source: http://faculty.uca.edu/johnc/DNA%20structure.jpgO)

Fig. 14.12 Purine and pyrimidine bases in nucleic acids

The purines

Adenosine

Guanine

The Pyrimidines

Cytosine

Thymine

Uracil

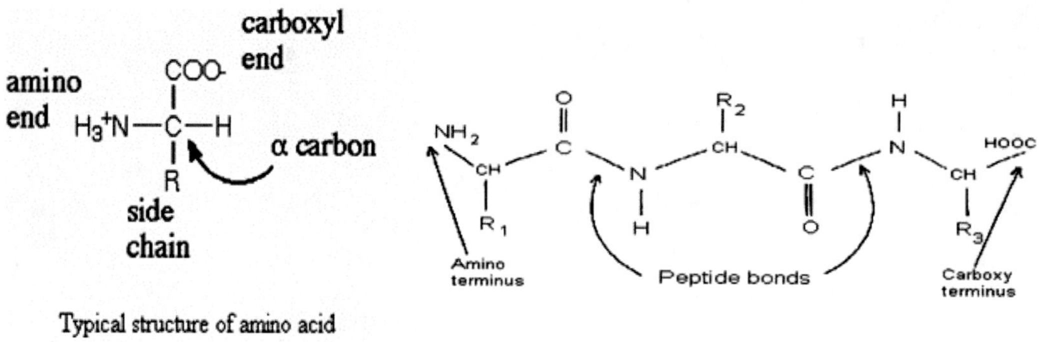

Typical structure of amino acid

Fig. 14.13 Structure of a typical amino acid with amide and carboxylic group at the end. Two amino acids linked by peptide bond

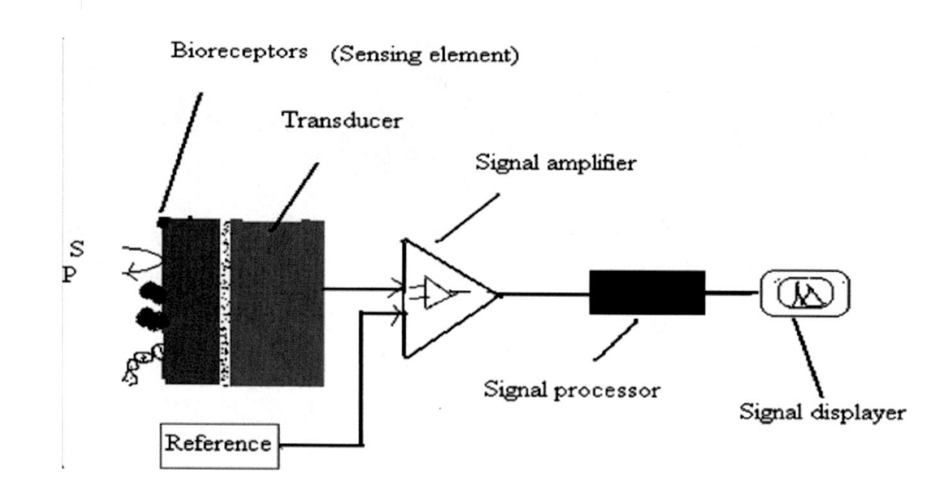

Fig. 14.14 Schematic of a typical biosensor: bioreceptors (note DNA, protein receptor molecules attached: *s* substrate, *p* products in an enzyme-based sensor), transducer, signal amplifier, signal processor, and signal displayer

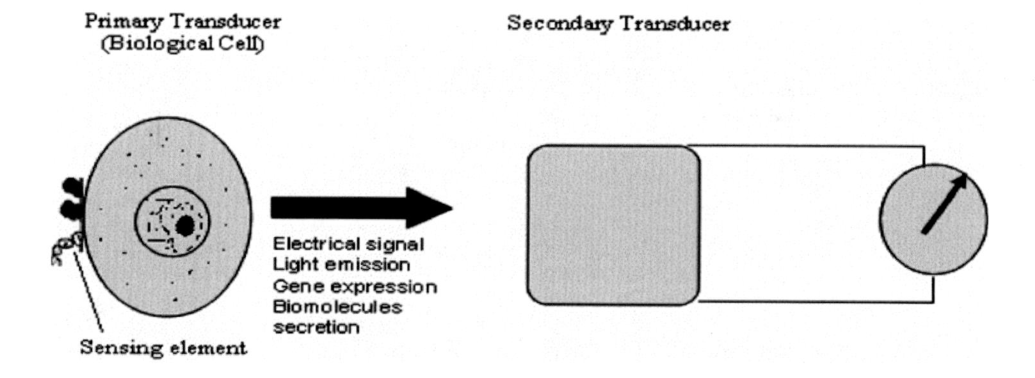

Fig. 14.15 Schematic of a typical cell-based biosensor showing primary and secondary components of the biosensor

14.4 Building Blocks of Biosensors

As shown in Fig. 14.14, the main building blocks of biosensors are biosensing element, signal conversion or transducer, and signal processing.

In cell-based biosensors (CBB), the cell with receptor molecules acts as primary transducer, and the secondary transducer [11] is typically an electrical device (Fig. 14.15).

The interface and transduction work by electrical, optical, acoustic, mechanical, thermal, chemical, and magnetic methods. As shown in Fig. 14.16, the bifunctional membrane is coated with probe molecules such as enzymes, antibodies, DNA, RNA, tissue, cells, organelles, and receptors (natural or synthetic). Upon hybridization or affinity binding with the target molecules, a signal is generated in the form of production of chemical substance, weight change, light, sound, heat, or electrical impulse, which ultimately converts into electrical signal.

The important biological component of a biosensor is a receptor, which recognizes the analyte or target molecules. These biorecognition elements are specific and sensitive for a

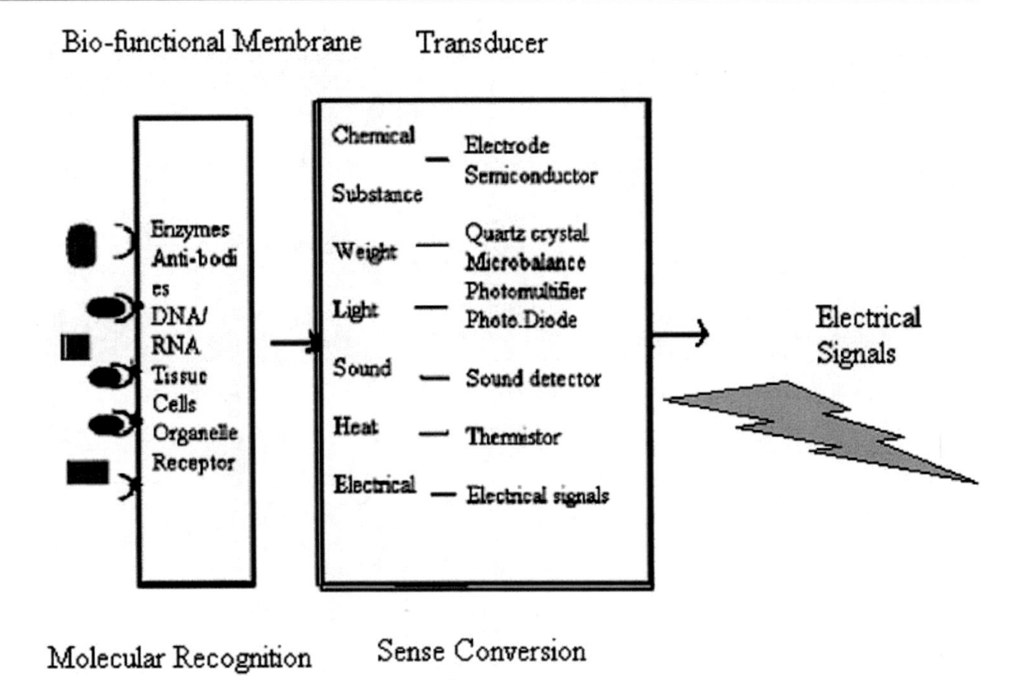

Fig. 14.16 Illustration of biomechanisms of biorecognition and signal transduction

particular substance but not others. There are four main groups of biological elements:

1. Antibodies (proteins)
2. Nucleic acids (DNA/RNA)
3. Natural and synthetic receptors
4. Enzymes (biocatalysts)

14.5 Biosensing Mechanisms

The biomechanisms for recognition of the important biomolecules such as proteins, DNA, and RNA lie in their molecular structure. They have a naturally evolved selectivity to biologically active analytes. The complementary structures of DNA and specific binding sites on antigen and antibody are utilized for the detection of these molecules for analytical purposes.

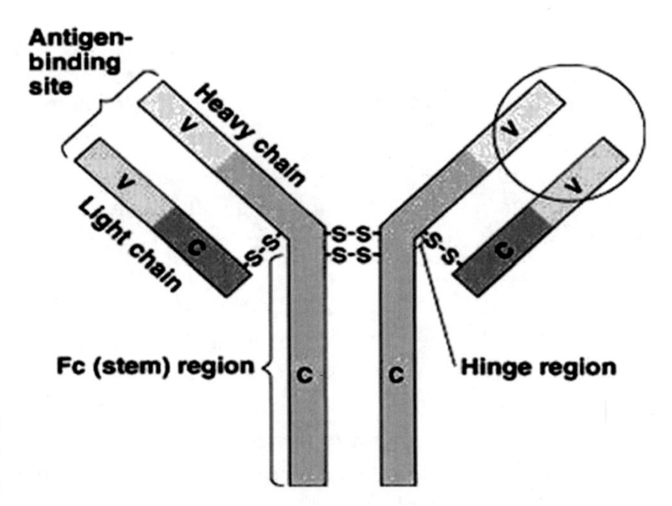

Fig. 14.17 Schematic diagram of immunoglobulin molecular structure. Binding sites (epitopes) of antigen and antibody reaction lead to the formation of covalently bonded immune complexes, which are neutralized proteins. (Courtesy of Indian River Community College, Health Science Center, USA)

14.5.1 Immune Complex Formation

Antigen–antibody reactions are the binding processes of an antigen molecule with its specific antibody. This reaction involves site or epitope recognition on the antigen by the antibodies and then binding (Fig. 14.17). Epitope is a portion of a molecule to which an antibody binds. It can be composed of sugars, lipids, or amino acids. The epitopes can be visualized by using monoclonal antibodies specific for that epitope. The affinity reactions are recognized by labeling the antibodies usually with radioisotopes, enzymes, fluorescent probes, chemiluminescent probes, or metal probes. Generally, it is visualized in gel, western blot, or labeling via immunofluorescence. It is a reversible chemical reaction with high specificity. The main forces that play a role in these reactions are hydrogen bonds, electrostatic forces, van der Waals bonds, and hydrophobic bonds. The stability of the antigen–antibody binding depends on the valency of antibodies, which are termed as avidity and affinity. Antibody upon cleavage with papain enzyme separates into two proteolytic fragments in the hinge region, namely, Fab and Fb. The Fab fragment retains the antigen-binding activity,

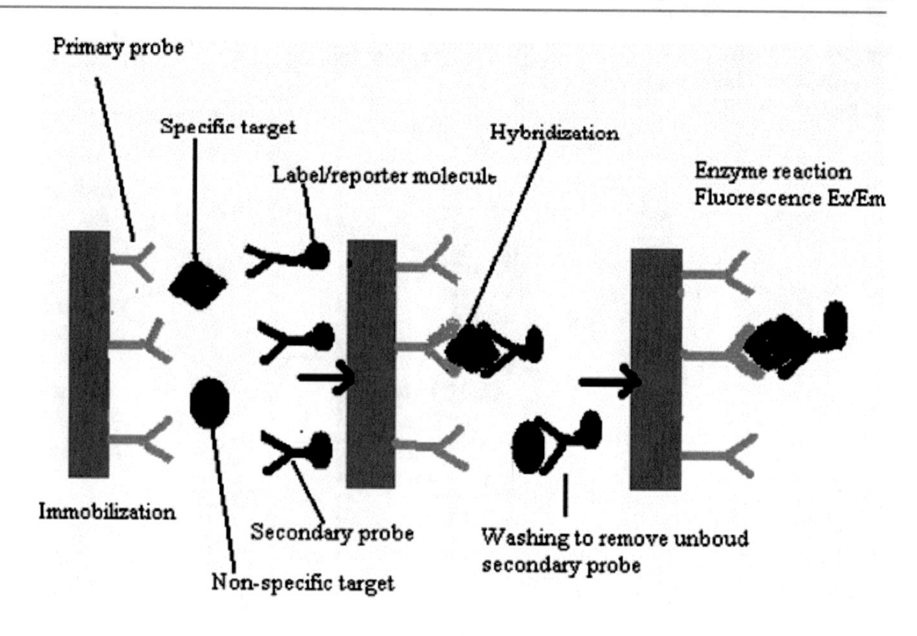

Fig. 14.18 Schematic of biosensing mechanism of antigen and antibody binding and immune complex formation

Watson-Crick base pairing: **A T T T C A T C G G (Target)**

T A A A G T A G C C (Probe) Complementary structure

Fig. 14.19 DNA and DNA molecules are hybridized with complementary DNA and RNA probes (Fig. 14.19b)

binding to a monovalent antigen with an affinity nearly as high as that of the entire antibody molecule. The Fc fragment mediates class-specific functions, such as complement fixation, exhibited by IgG and IgM antibodies, leading to lysis of the intruding cell. The in vitro antigen–antibody reactions (serological tests) are widely used in blood grouping, diagnosis of infectious diseases, and detection of specific protein of interest. These tests are mainly based on the mechanisms of agglutination, precipitation, complement fixation, and solid-phase immunoassays such as enzyme-linked immunosorbent assay (ELISA) and immunosensors.

In a conventional method of immunodetection of protein, the primary antibodies (probe molecules) are coated onto the microtiter plate, and allowed the hybridization to take place with labeled specific target molecules (antigen).The molecular labeling could be radioisotopes, enzyme, digoxigenin, fluorescent, or chemiluminescent. After hybridization with the target molecules, the plate is washed to remove the unbound nonspecific target molecules. Figure 14.18 shows only the specific target molecules hybridized with probe molecules, and the nonspecific molecules cannot hybridize with the receptor molecules.

14.5.2 DNA Hybridization

DNA and DNA molecules are hybridized with complementary DNA and RNA probes (Fig. 14.19).

The sequence-specific DNA and RNA hybridized molecules are detected in Southern and Northern blots, respectively. The nonspecific DNA sequences are removed by washing. In conventional detection system, nucleic acids require radioisotopes, enzyme, fluorescent, chemiluminescent, or metal labeling for the detection of hybridization signals.

14.5.3 Natural and Synthetic Receptors

Templating techniques, such as molecular imprinting (MIP) on polymers, are used to form natural or synthetic receptor sites on polymers [12]. The concept of cell imprinting is based on stamping of cells with minimum force into a polymerizing sensor layer. The patterned surface remained after the polymerization recognizes cells depending on geometrical fit and chemical interaction [13].

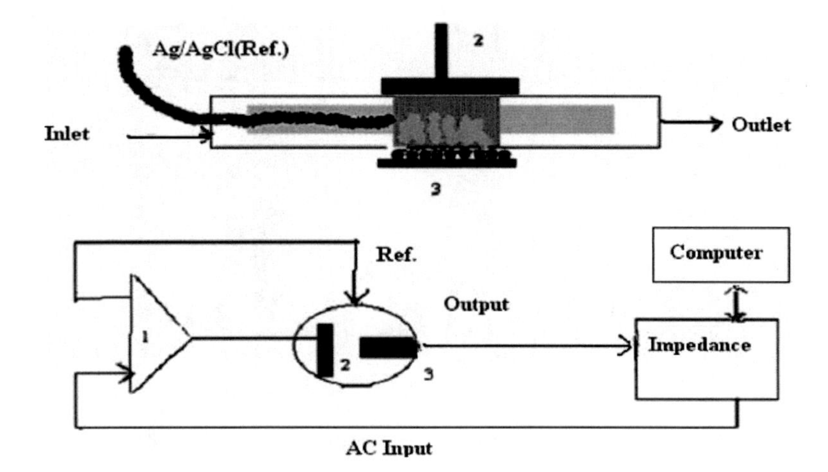

Fig. 14.20 Three-electrode chemical fluid cell biosensor: (1) potentiostat, (2) platinum foil counter electrode, (3) working electrode as DNA-modified silicon sample. (Source: Courtesy of Cai et al., 2003)

Yeast-imprinted polyurethane sensor layers in combination with microbalances were shown to detect yeast cells without cross-selectivity to bacteria [14]. Mass- sensitive detection of cells, viruses, and enzymes with artificial receptors was performed using lysozymes, tobacco mosaic virus (TMV), and erythrocyte imprinted polymers [15].

14.5.4 Biocatalysts

The catalytic biosensor is based on the recognition and binding of an analyte followed by catalytic conversion of the analyte. The reaction is detected by the transducer. Enzymes are the natural protein biocatalysts that specifically bind to a unique substrate or group of substrates. Enzyme stability is the key factor in determining the stability of enzyme biosensors. Biocatalysts include enzymes, organelles, whole cells, microorganisms, and tissues.

14.6 Functionality of Various Transducers and Challenges

The biosignals are converted into electrical, optical, and electromagnetic signals by a variety of signal conversion methods as described below:

- Electrochemical (conductimetric, amperometric, and potentiometric)
- Optical
- Acoustic wave
- Microelectromechanical (resonant)
- Magnetic
- Thermal

14.6.1 Electrochemical Method

Electrochemical fluid cell is used for the field-effect transistor (FET) detection of DNA or protein. It involves electrical impedance measurements of the probe and target molecules using a three-electrode electrochemical cell containing DNA or protein-modified Si sample as working electrode, Ag/AgCl reference electrode, and Pt counter electrode. A polydimethylsiloxane (PDMS) fluid cell (5 μl volume) is pressure sealed between DNA-modified Si sample (8 × 4 mm) and the Pt counter electrode (Fig. 14.39). Impedance is measured using a three-electrode potentiostat coupled to an impedance analyzer. All electrochemical potentials are considered with respect to the saturated Ag/AgCl reference electrode [16]. The multiwalled carbon nanotubes dispersed on the working electrode form the efficient working electrode (Fig. 14.20). The impedance is proportional to the target–probe molecular hybridization.

14.6.2 Optical Method

In optical signal conversion method, the bioprobe and target hybridization signals are converted into optical signals. For example, surface plasmon resonance (SPR) is an advanced optical signal conversion technique. The basic SPR configuration consists of a prism or glass slide coated with a thin metal film, usually silver or gold (Fig. 14.21). Bioprobe such as DNA or protein is immobilized on the thin gold film. When the light is passed through the prism and onto to the gold surface at angles and wavelengths near the surface plasmon resonance condition (the intensity of the reflected light is reduced at a specific incident angle producing a sharp shadow), the optical reflectivity of the gold changes very

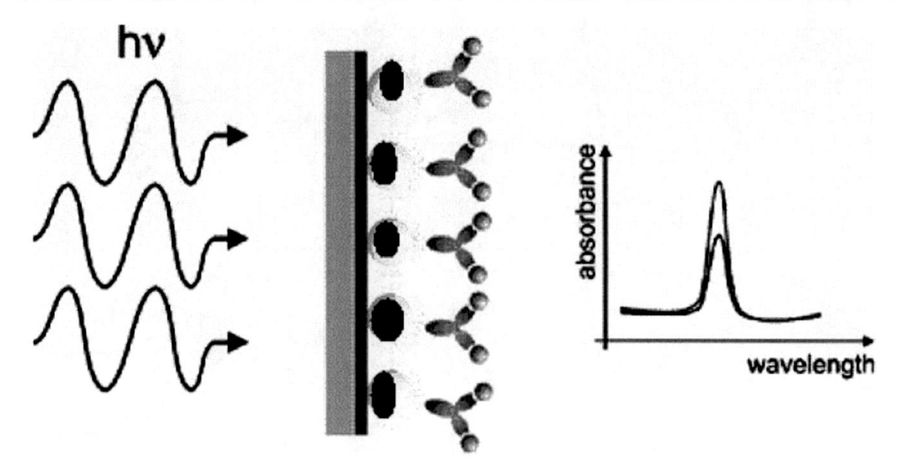

Fig. 14.21 Schematic of optical biosensor. Gold and silver nanoparticles are deposited on a quartz substrate. The substrate is coated with mercaptosilane adhesion layer and functionalized using gold or silver nanoparticles, self-assembled monolayers of functional thiols, antibodies, and antigen. The resulting absorption spectrum increases upon binding of analytes to the functional nanoparticles. (Courtesy of A. Campitelli, C. Van Hoof, IMEC)

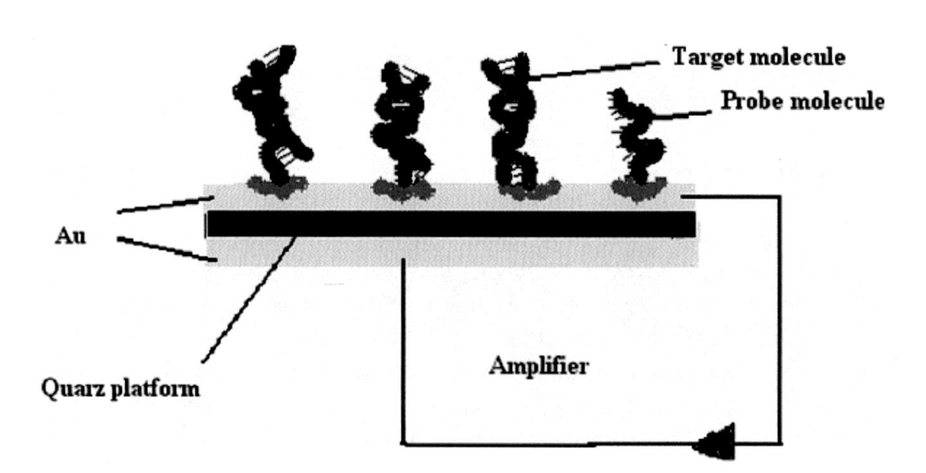

Fig. 14.22 Schematic of acoustic wave biosensor detects probe and target molecular binding on the surface of quartz platform caused difference in the voltage

sensitively in the presence of the gold surface. Similar effect is found using quartz substrate coated with gold and silver nanoparticles. The high sensitivity of the optical response is due to very efficient collective excitation of conduction electrons near the gold surface.

The optical components in an optical sensor could be fiber optics, wave guides, photodiode, spectroscopy, charge-coupled device (CCD), and interferometers. The changes in intensity, frequency, phase shift, and polarization are measured. The type of measurements involved is absorbance (oligo, 260 nm; peptide, 280 nm), fluorescence, refractive index, and light scattering. Surface plasmon resonance is particularly useful for the detection of biological molecules: (1) they can be extremely sensitive (nanomoles or less), and (2) they are nondestructive to the sample. Optical conversion technique offers an inexpensive system for the detection of heavy metals, particularly the high toxic arsenic in the environment [17]. In optical method, no reference electrode is needed, and there is no electrical interference.

14.6.3 Acoustic Wave Method

Acoustic wave is a kind of wave used in piezoelectric component, such as crystal, in circuits. The frequency characteristic in the crystal varies in response to the surrounding environment. In acoustic wave signal conversion method, an applied radio frequency produces mechanical stress in the crystal. As a result, surface acoustic wave (SAW) is induced. The SAW is received by the electrodes and is translated to voltage (Fig. 14.22). The probe and target molecular hybridization on the surface of quartz platform causes signal conversion change into the voltage. These are

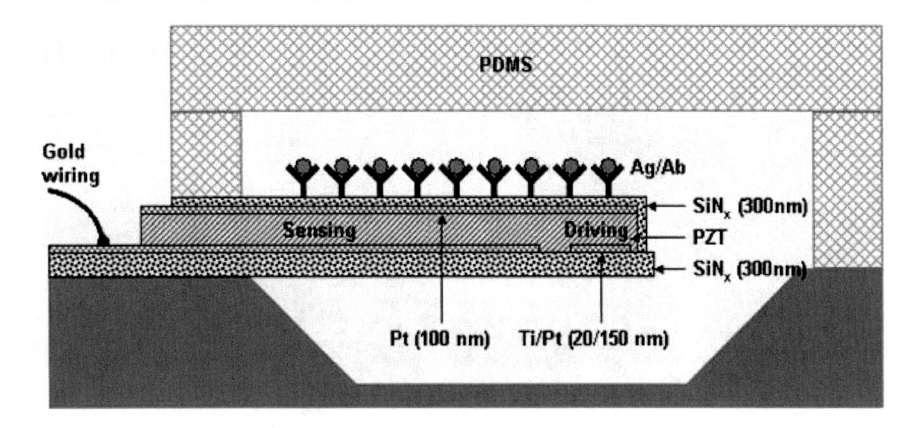

Fig. 14.23 Piezoelectric cantilever. (Source: Courtesy of Drs. Il-Han Hwang and Jong-Hyun Lee, 2006)

typically signal conversion components that detect the presence of low mass such as single molecule or cell.

14.6.4 Electromechanical Method

Nanocantilevers and quartz crystal microbalance sensors are extremely sensitive mass sensors, which are capable of measuring attogram (10^{-18} g) mass sensitivity. When target binding occurs between protein, DNA or RNA target molecules, and probe molecules on the cantilever beam, change of adsorption is directly related to the cantilever deflection. The molecular binding surface tension causes bending of cantilever beam. The deflection of the beam is measured optically (Fig. 14.23). The resonance frequency varies as a function of mass loading. Nanobelts, nanorings, and nanosprings with piezoelectric properties have been synthesized for use as nanoscale transducers, actuators, and sensors.

The cantilever biosensor can also be operated in a dynamic mode, where its frequency shift is measured before and after the binding of biomolecules on the cantilever surface [18].

Single-crystal zinc oxide (ZnO) and SnO2 nanobelts that spontaneously rolled them up for piezoelectricity into helical structures with widths in the range of 10–60 nm wide and 5–20 nm thick. These nanobelts/wires are quasi-one-dimensional semiconducting and piezoelectric nanostructures used to fabricate field-effect transistor (FET). These nanobelts induce a field effect in the presence of charged biological molecules such as proteins, peptides, DNA, and RNA. The ZnO and Si nanowires [19] and Si microresonators (Fig. 14.24) are in recent use for nanobiosensor device fabrication. They are synthesized by simply evaporating oxides of zinc, tin, indium, cadmium, and gallium, whose electrical performance is extremely sensitive to the surface-adsorbed molecules. The FET-based device offers as a potential technique for in situ, real-time, wireless, and implantable biosensors [20]. These devices facilitate detection of subnanomolar quantities of proteins, DNA, and RNA from anthrax, salmonella, smallpox, AIDS virus, and other pathogens and toxins, including ricin.

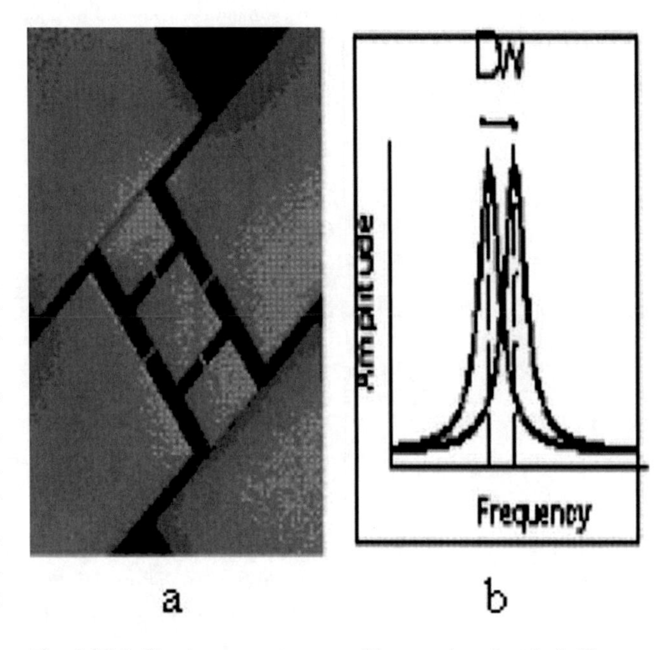

Fig. 14.24 Si microresonator, capacitive or piezoelectrical. (Source: Courtesy of Dr. Farrokh Ayazi, Georgia Tech)

14.6.5 Magnetic Method

Magnetic nanoparticles are also used in force amplified biological sensor (FABS). FABS use magnetic particles to label antibodies and sequence-specific DNA or RNA. Atomic force microscope (AFM) is employed to pull on and measure the strength of single DNA–DNA or antibody–antigen bonds [21]. Thus, the detection of binding and hybridization events of probe and target molecules is done using magnetic nanoparticles (Fig. 14.25).

The magnetic particles used for FABS must be smooth and spherical, so that the entire surface of each particle is

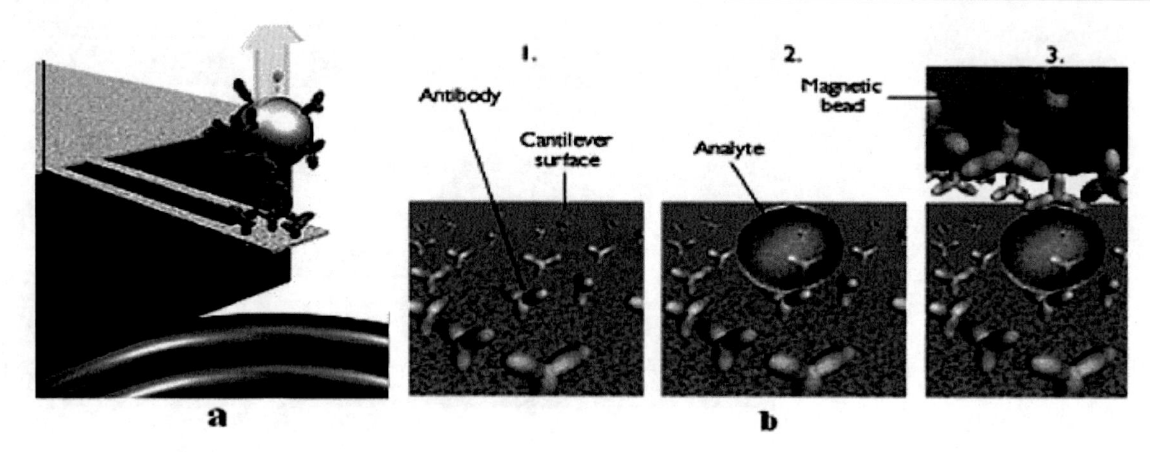

Fig. 14.25 Magnetic biosensor, in which the antibody-labeled magnetic particles are attracted in the magnetic field (Source: Courtesy of Drs. Baselt et al. 1998)

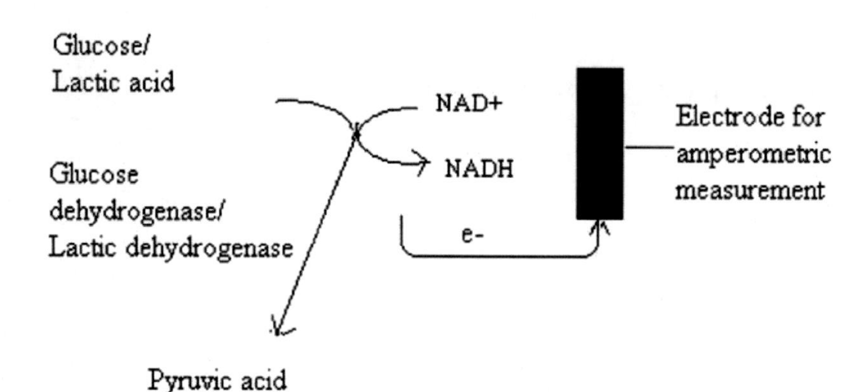

Fig. 14.26 Enzymes catalyzing oxidoreductions. The electrons produced in the reaction are a measure of glucose

available for binding to the flat cantilever surface. Irregular particles effectively have a reduced active area. The particles should exert a uniform force in a magnetic field and a high magnetic moment for maximum signal. They should not corrode when immersed in physiological salt solutions and should be low density and initially nonmagnetic to avoid coagulation of the solution.

14.6.6 Chemical and Enzymatic Method

Variety of glucose oxidoreductases and lactic dehydrogenases are used as a component for various enzyme sensors. Electron mediator-dependent glucose dehydrogenases (GDH) recognized as ideal constituents of mediator-type enzyme sensors. The dehydrogenases utilize the NAD+/NADH couple. The NADH generated (Fig. 14.26) is detected by the amperometric method.

In simple glucose biosensors, the enzymatic oxidation of glucose produces hydrogen peroxide, which in turn generates electrons by electrode reaction. The current density is used as a measure of glucose in the sample (Fig. 14.27).

14.6.7 Thermal Method

The thermal biosensor is an enzyme thermistor modified for split flow analysis. A miniaturized thermal biosensor involves in a flow injection analysis system for the determination of glucose in whole blood. The blood glucose is determined by measuring the heat evolved when samples containing glucose passed through a small column with immobilized glucose oxidase and catalase. A sample of whole blood as small as 1 μl can be measured directly using this thermal biosensor [22]. The detailed account of nanomaterials and fabrication methodologies of nanodevices are available [23, 24].

14.6.8 Challenges in Integration of Biomolecules with Interface Mechanisms and Signal Transduction

To make any biosensor device commercially viable, the following criteria should be met:

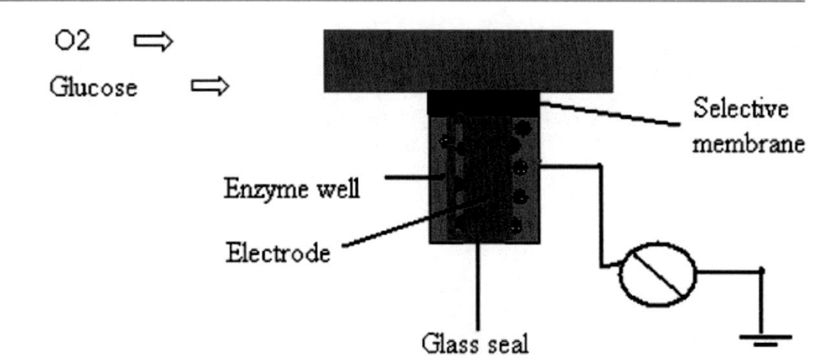

Fig. 14.27 The selective membrane in contact with glucose and the enzyme acts as a transducer

- Relevance of output signal to measurement environment
- Speed of response, accuracy, sensitivity, dynamic range, and repeatability
- Amenable to testing and calibration
- Reliability and self-checking capability
- Insensitivity to temperature (or temperature compensation)
- Insensitive to electrical and other environmental interference
- Biocompatible in vivo application
- Development of a new generation of biomimetic sensors for single-molecule detection

The technical issues that need to be addressed in integrating biomolecules with interface mechanisms and signal transduction include:

- Integration of different technologies from the interdisciplinary fields (electronics, chemistry, physics, and biology) for proper functioning of a biosensor. The interface between the transducer and the biological receptor may be made more exclusive, thus reducing interference and development of new receptors with improved or new affinities.
- The scaling issues involved in miniaturizing process without detriment to overall functionality of the biosensor.
- The control of size, shape, and magnetic properties of magnetic nanoparticles in fabricating nanoparticle-based biosensors.
- Proper retention of specificity and sensitivity of complex biological molecules that have their inherent instability. Necessary care should be taken to ascertain their viability on the attachment surfaces.
- Toxicity of nanoparticles for in vivo applications such as targeted drug delivery, MRI contrast agents, and pathogen cleansing.
- Magnetic "traps" or read/write architectures that is biocompatible and capable of manipulating single magnetic particles with nanoscale precision.
- Preventing the biosensor devices from biofouling, as protein buildup on the biological active interfaces.
- Finding efficient interfaces between biological materials and electronics.
- Evaluation of new manufacturing procedures for large-scale production.
- Optimized performance of the sensor is supported by associated electronics.
- Fluidics and integrating with system-on-package (SOP) platform.

14.7 Nanostructures, Biomolecules, and Cells as Components of Biosensors

The novel properties possessed by nanostructured materials open several possibilities toward fabricating advanced nanobiosensors with superior sensitivity and specificity. For instance, single-walled carbon nanotubes (SWNTs) and multiwalled carbon nanotubes (MWNTs) exhibit unique properties such as chemical stability, good mechanical properties, outstanding charge transport characteristics, and high surface area that makes them ideal for the design of nanobiosensors for detecting sub-femtogram quantities of target protein, DNA, and RNA. Among the approaches for exploiting nanotechnology in medicine, nanoparticles offer some unique advantages as sensing, image enhancement, and delivery agents [25].

The remarkable biorecognition capabilities of the biomolecules such as DNA, RNA, and proteins can be utilized for the fabrication of ultraminiature biological electronic sensing devices, by conjugating them with nanoscale solids (particles, dots, or wires). Varieties of nanoparticles are available commercially, including polymeric nanoparticles, metal nanoparticles, liposomes, micelles, quantum dots, dendrimers, and nanoassemblies. The gold nanoparticles of size 1.4 nm are used in the fabrication of biosensors. In the fabrication of immunosensors, the nanoparticles are conjugated with the protein probe, which involves in antigen–antibody immune complex formation. Nanoparticles are also used for conjugating DNA or RNA probe involving sequence-specific DNA or RNA hybridization detection, in the construction of nucleic acid probes. Nanoparticle-based

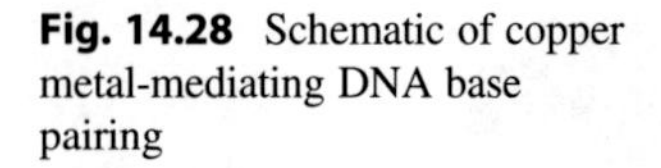

Fig. 14.28 Schematic of copper metal-mediating DNA base pairing

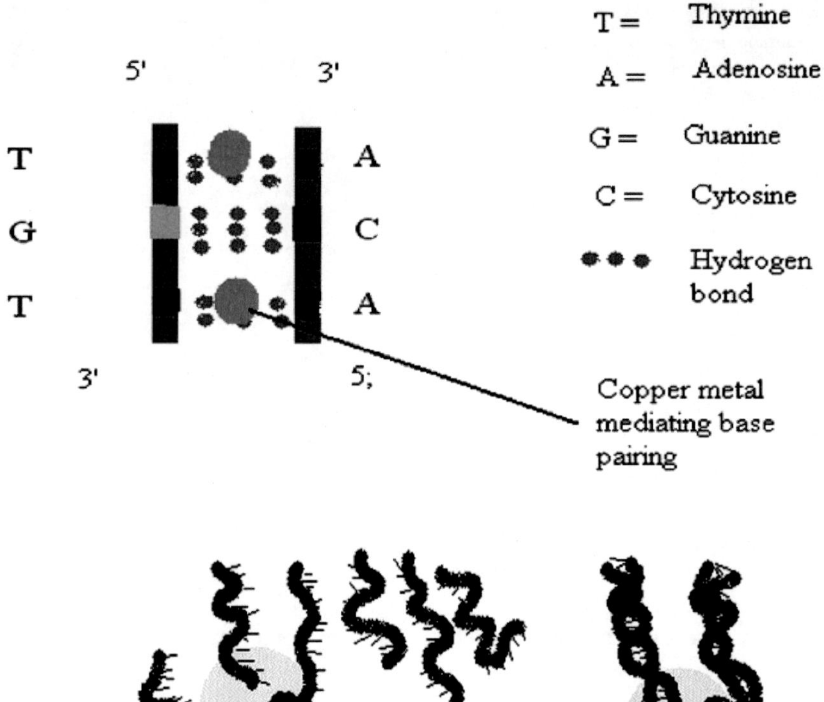

Fig. 14.29 Schematic diagram of hybridization of probe DNA on gold nanoparticles with sequence-specific target DNA

biosensors are sensitive, inexpensive, noninvasive detection systems for measuring biological molecules and are advanced over the existing technologies due to their nanosize and smart performance, constituting the third-generation biosensors. Aptamers are oligonucleotides (DNA and RNA) that can bind with high affinity and specificity to wide range of target molecules (proteins, peptides, drugs, vitamins, and other organic or inorganic compounds). The biosensors fabricated with resonant oscillating quartz crystal can detect minute changes in resonance frequency when probe ss-DNA hybridizes with target DNA. DNA-based supramolecular chemistry allowed the replacement of natural bases in the DNA molecule with artificial nucleotides or nucleotide mimics [26]. Further, new generation of such nucleotide mimics was reported in which the hydrogen bonding interactions were replaced by metal-mediating base pairs [27]. The metal incorporation into the interior of the DNA confers DNA metal-based functions such as thermal stability and magnetic properties (Fig. 14.28).

The cell-based biosensors are composed of two transducers, where the primary transducer is cellular and the secondary transducer is typically electrical. It is known that olfactory neurons respond to odorant molecules, and the retinal neurons in the eye are triggered by photons. These cells in the body act as primary transducers capturing the signals and in turn passed on to the secondary transducer.

14.7.1 The Gold Nanoparticles

Nanoparticles are used for fabrication of DNA-based biosensors [28, 29]. The hybridization or binding events of sequence-specific probe single-stranded DNA (ss-DNA) functionalized with gold particles and target molecules demonstrated changes in electrical conductivity. The target and probe-binding events localize gold nanoparticles in an electrode gap. The silver deposition facilitated by this gap leads to measurable conductivity changes. These nanoparticle-based devices could detect target DNA or RNA at concentration as low as 500 fmol. As shown in Fig. 14.29, sequence-specific DNA or RNA probe is attached to the gold nanoparticles and subjected to hybridization with target oligonucleotides.

The gold nanoparticles coated with probe protein or antibody is used for the detection of specific antigen in immunosensor system.

The enzyme electrodes are constructed by introducing nanoparticles into the critical sites of the enzyme, such as in

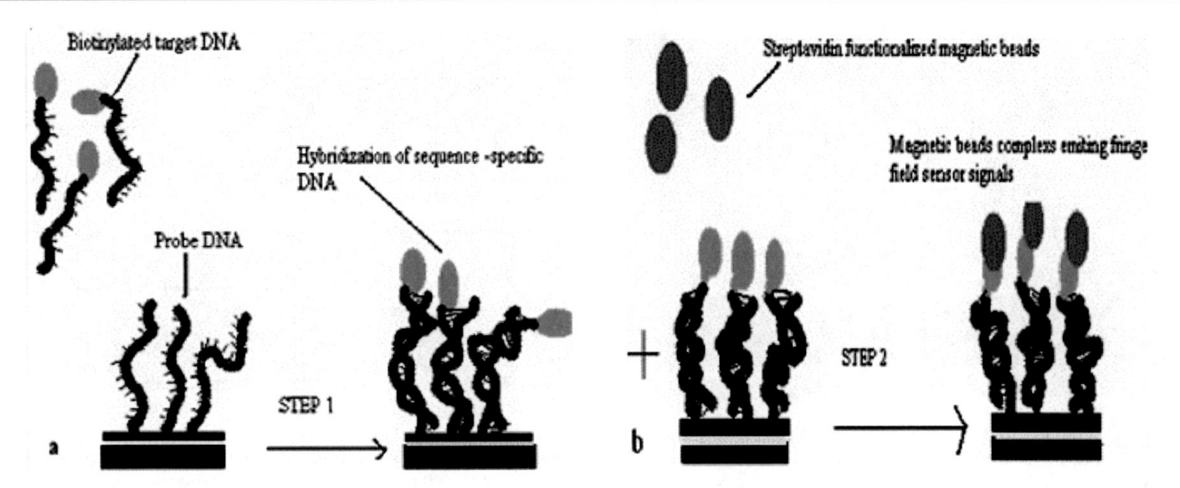

Fig. 14.30 Schematic diagram of hybridization of target DNA with probe DNA, followed by hybridization with streptavidin-functionalized magnetic beads

redox center of glucose-oxidizing enzymes. When the enzyme oxidizes glucose, electrons flow through the nanoparticle and into the electrode. The amount of current indicates the level of glucose present. These enzyme-conjugated electrodes are ultimately used in fabrication of implantable enzyme sensors inside the body for monitoring blood glucose.

Precipitation, physical vapor deposition (PVD), chemical vapor deposition (CVD), and organic encapsulations are the common methods of synthesizing nanoparticulate gold film on various supports.

14.7.2 Magnetic Nanoparticles

Magnetism-based biosensors use magnetic nanoparticles functionalized with biological molecules such as DNA, RNA, and protein for fabrication of variety of biosensors. Metals such as nickel (Ni), cobalt (Co), Fe, Fe3O4, and y-Fe2O3 and their oxides are used to synthesize magnetic beads or particles of length scales ranging from 1 to 100 nm. As shown in Fig. 14.30, the target DNA is biotinylated and hybridized with sequence-specific ss-DNA probe [30]. The streptavidin-functionalized magnetic beads interact with biotinylated hybridized DNA, and the hybridization event is detected by a change in resistance from a giant magnetoresistive (GMR).

Nanoparticles conjugated with biomolecules are used in the detection of antigens. Europium (III)-labeled nanoparticles coated with antibodies or streptavidin are used to detect prostate-specific antigen (PSA) in serum [31].

The important brain molecules, gamma-aminobutyric acid (GABA), are involved in Huntington's and Parkinson's disease, seizures, myoclonic discharges, and alcohol addiction. In real-time measurements of the molecules, ultraviolet (UV)–visible spectroscopic determination methods are insensitive. The nanoparticle- based neurotransmitter sensors fabricated using nanoparticles will facilitate accurate measurements of neurotransmitting molecules.

The magnetic nanoparticles with heterogeneous size from 200 to 400 nm are shown to be advantageous over magnetic microspheres. They exhibit high magnetization per unit weight, remain in suspension for longer periods of time without aggregating, and are faster – an important requirement for the more sensitive sensors and measurements.

14.7.3 Semiconductor Quantum Dots

The quantum dots are inorganic semiconductor nanocrystals. They are created by confining free electrons in a 3D semiconducting matrix. Those tiny islands or droplets of confined free electrons show many interesting electronic properties.

A key feature of these materials is that they can be modified with large number of small molecules and linker groups to optimize the functionality for particular applications. The quantum dots can be covalently linked with biorecognition molecules such as DNA, RNA, antibodies, peptides, and small molecule inhibitors for use as fluorescent probes, which exhibit improved brightness, photo stability, and broader excitation spectra (Fig. 14.31a, b) Thus, quantum dots are superior over the existing imaging techniques, which use natural molecules such as organic dyes and proteins (Fig. 14.31).

14.7.4 Nanowire

Semiconducting and piezoelectric nanostructures such as single-crystal zinc oxide (ZnO) nanobelts (Fig. 14.32) are

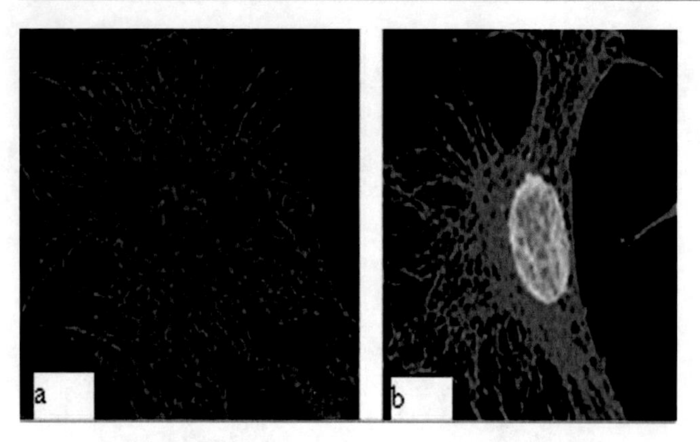

Fig. 14.31 (**a**) The microtubules of the cell were stained with 605 nm fluorescent quantum dot conjugate, and the nuclei were counterstained with Hoechst dye; (**b**) nucleus and microtubules were labeled with *red* and *green* quantum dot conjugates (Source: Quantum Dot Corp)

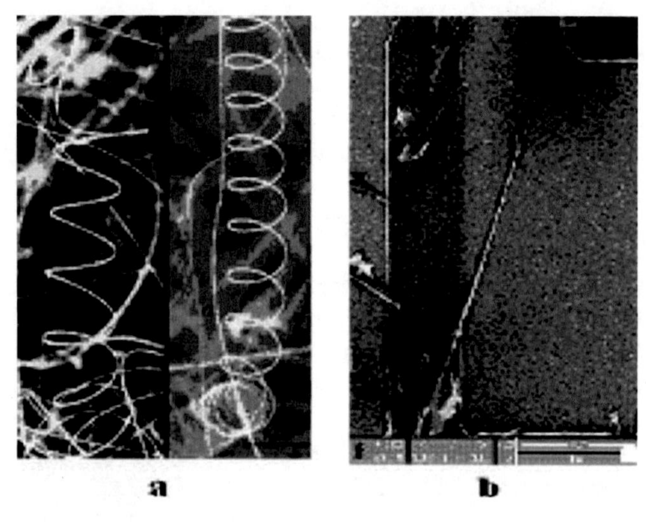

Fig. 14.32 (**a**) Single-crystal zinc oxide (ZnO) nanobelts, (**b**) ZnO nanowire mounted between the electrodes. (Source: Courtesy of Dr. ZL Wang, Georgia Institute of Technology)

used as sensitive signal conversion components, capable of measuring signal conversion from a single molecule or cell. ZnO nanobelts/wires (10–60 nm wide and 5–20 nm thick) induce a field effect in the presence of charged biological molecules such as proteins, peptides, DNA, and RNA [20]. ZnO and Si nanowires are used as efficient signal conversion components [19].

14.7.5 Carbon Nanotubes (CNTs)

The carbon nanotubes and silicon nanowires and actinyl peroxide self-assembled structures are the products of novel innovative technologies. Self-assembled carbon nanotubes produced in the arc discharge process are widely used in catalytic processes [32]. Based on the wall structure of the nanotubes, two types of tubes are categorized: single-walled and multiwalled. A single-walled carbon nanotube (SWNTs) consists of a single graphene cylinder, whereas a multiwalled carbon nanotube (MWNTs) comprises several concentric graphene cylinders. SWNTs can possess efficient electrical conductivity. The carbon atoms are strongly (covalently) bound to each other (carbon–carbon distance 0.14 nm). As shown in Fig. 14.33, rolling up the sheet along one of the symmetry axis gives either a zigzag ($m = 0$) tube or an armchair ($n = m$) tube.

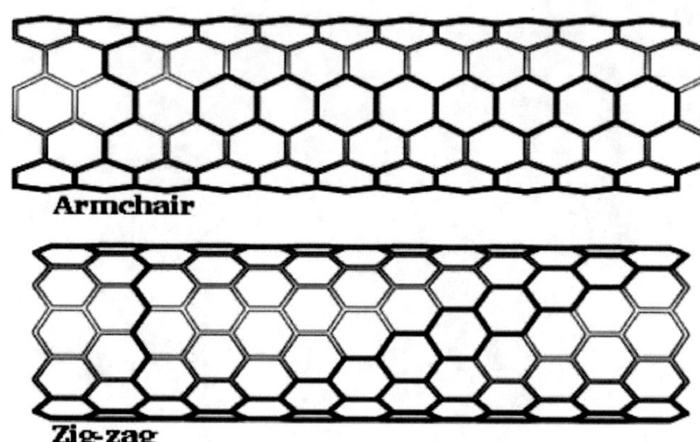

Fig. 14.33 Structure of carbon nanotubes. (Courtesy of Technion, Department of Physics, Israel University (http://phycomp.technion.ac.il/~talimu/structure.html)

The carbon nanotubes (CNTs) possess unique properties such as chemical stability, good mechanical properties, high surface area, and outstanding charge transport characteristics that make them ideal for fabrication of nanobiosensors.

14.7.6 Actinyl Peroxide Compounds

Actinyl peroxide compounds are known to self-assemble into nanosized hollow cages containing actinides of other elements such as potassium. Nanowire containing a step every five silicon atom rows and a row of gold atoms in the middle of the terrace is reported [33].

The fabrication of nanostructures from deoxyribonucleic acid (DNA) – DNA dendrimers, DNA nanotubes, and DNA diamond – adducts and assembles these nanostructures into large-scale functional constructs used in biosensor devices.

14.7.7 Conductive Polymer Membrane

Polymer conductive property has been utilized for fabricating conductive polymer-based biosensor array (electronic nose). When conductive polymers come into contact with analyte molecules, their conductivity changes, creating a current within the sensor that is proportional to the concentration of the analyte. These currents are picked up by signal processing circuits in the instrument and compiled into a response profile that characterizes the atmosphere around the sensor. The Langmuir–Blodgett film method is used for nanofilm fabrication in which monolayers of desired material efficiently are built one at a time. The electronic nose consists of two components: (1) an array of chemical sensors (usually gas sensors) and (2) pattern recognition algorithm. This sensor can be useful for detecting odor or smell emitting from biological organisms and chemical sources. The following are some of the examples of electron nose applications:

- Smelling foods for their freshness, including freshness of fish and other seafoods
- Bacterial growth on foods such as meat and fresh vegetables
- Process control of cheese, sausage, beer, and bread manufacture
- Identification of quality of coffee and monitoring of roasting process
- Rancidity measurements of oils
- Identification of spilled chemicals in the environment, particularly in gas transportation, and leakage of utility gases
- Diagnosis of pulmonary infections (e.g., TB or pneumonia)
- Diagnosis of ulcers in the lungs by breath tests

14.7.8 DNA, Proteins, and Cells

DNA, proteins, and cells are the basis for fabrication of biosensors and biodevices for various applications.

14.8 Biochemical Probe Design and Interfacing with Biosensors

The production, proper attachment, and optimization of probe biomolecules such as DNA, RNA, and antibodies on electrode surfaces are challenging tasks in fabricating a biosensor. Several critical steps are involved in immobilization and assessment of proper immobilization and viability.

14.8.1 Preparation of Protein Probe

The protein of interest to be used as protein probe is produced for biofunctionalization of biosensor devices. Proteins are produced mainly by three methods:

14.8.1.1 Proteins Produced from Cell and Tissue Cultures

The cells with the protein of interest are cultured using appropriate culture conditions. The cells are lysed in a cell lysis buffer with 0.1 mg/ml lysozyme, and proteins are extracted. A protein of interest is isolated and purified using standard protocols.

14.8.1.2 Proteins Produced by Gene Activation

As shown in Fig. 14.34, when a regulatory gene for the protein of interest is in "off" condition, the corresponding gene does not produce any protein in the bacteria. Addition of a new regulatory gene makes the gene in "on" position and facilitates the protein production.

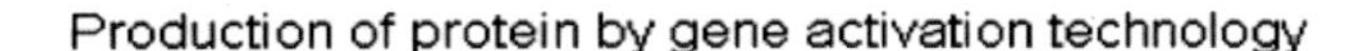

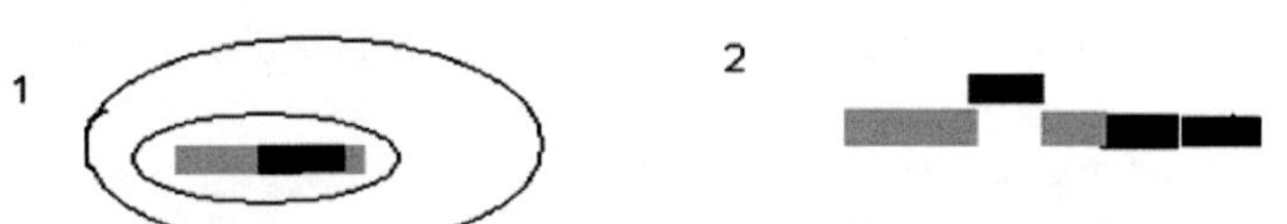

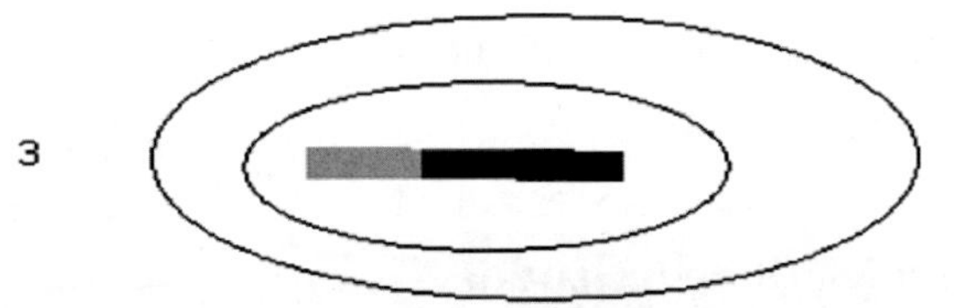

Fig. 14.34 Schematic of production of protein by gene activation

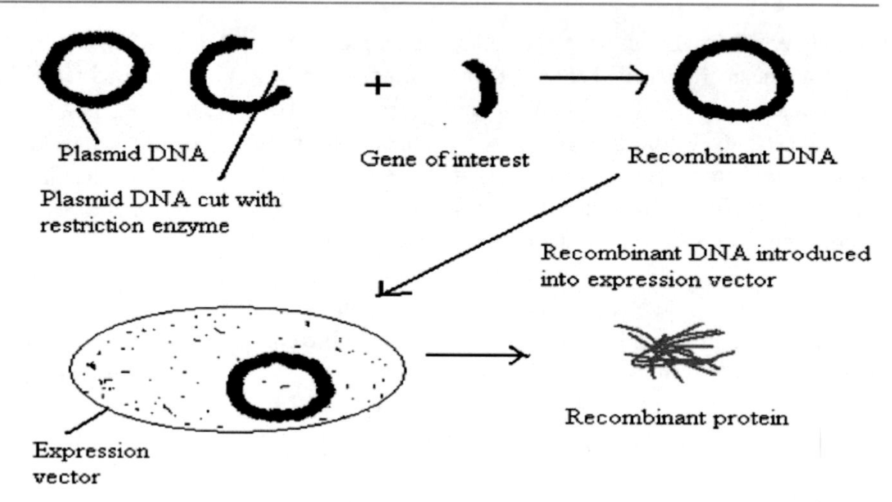

Fig. 14.35 Schematic of production of protein using recombinant gene technology

14.8.1.3 Recombinant Protein Production

Recombinant gene technology is used for the production of a specific single protein, unlike in other two methods where total proteins are in the pool. To overexpress the gene, the plasmid is transformed into the *E. coli* strain or any other expression vector. As shown in Fig. 14.35, the plasmid is cut open at an appropriate position using a restriction enzyme. The gene responsible for the protein of interest is inserted and attached to the cut ends of the plasmid using ligase enzyme. The resulting recombinant DNA is used for expression of the protein in an appropriate expression vector.

The proteins obtained by various methods are purified by standard protein purification methods.

14.8.1.4 Production of Polyclonal and Monoclonal Antibodies

The polyclonal antibodies are produced in vivo in response to immunization with different epitopes on the protein injected. Antiserum can be raised in mouse, rat, rabbit, sheep, goat, and multiple injections of the antigen along with the adjuvant (a nonspecific enhancer of the immune response) are possible. After repeated immunizations, the antibodies produced will be predominantly high-affinity IgG.

The monoclonal antibodies provide single epitope specificity. They are produced using hybridoma technology. The antibody-producing cells from the mouse's spleen and cancerous immune cells fuse to form hybridoma cells. Individual hybridoma cells are cloned and tested to find out the clones that produce the desired antibody.

14.8.2 Synthesis of DNA/RNA Oligonucleotide Probes

DNA can be synthesized using polymerase chain reaction (PCR). The appropriate primer, Taq polymerase, d-NTPs, and other ingredients in the reaction condition can generate desired oligo synthesis. The complementary oligonucleotides for a particular target gene are synthesized in commercially available DNA/RNA synthesizer. Figure 14.36 shows the rapid multiplication of the DNA molecules in PCR. Starting from single molecule, an astronomical figure of 68 billion molecules are obtained in 35 cycles of DNA synthesis.

All nucleotides are purified by gel filtration and reverse-phase chromatography using an Invitrogen RP-C18 column.

14.8.3 Electrically Conductive Contact Surface Materials

Nobel metals such as gold, platinum, and palladium are known to be biocompatible. Titanium is widely used as implants and surgical staples. These metals are also widely used as surface materials. The use of conductive polymers for surface coating with ligands offers great advantage over metals and inorganic semiconductors, including higher electrical conductivity than other conductive materials, plasticity, elasticity, low mass density, low coefficient of expansion, resistance to chemicals and corrosion, tunable optical properties, and electrochromism. The functionalization of their nanoparticles with FITC streptavidin is achieved by carbodiimide chemistry, similar to that of enzyme binding to nanoparticles.

14.8.4 The Process of Attachment or Immobilization of Biomolecules to the Surfaces

The attachment of probe biomolecules to the surface of electrodes is a critical step in the construction of a nucleic acid and protein-based biosensors. DNA, RNA, and proteins

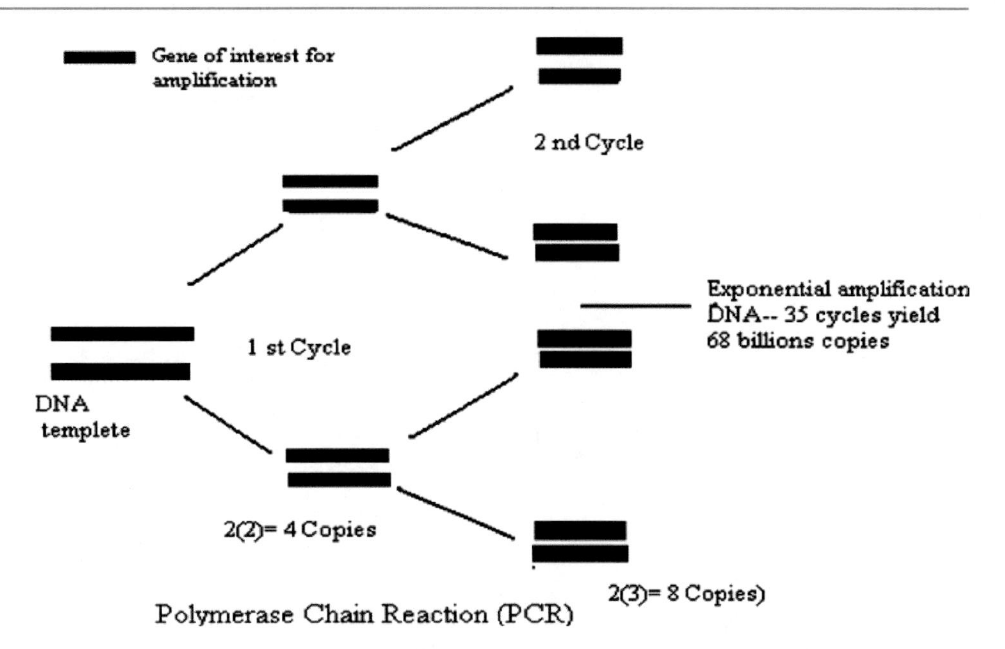

Fig. 14.36 Schematic diagrams of polymerase chain reaction (PCR). The reaction set 30–40 amplification cycles of 3 min each. Denaturation for 1 min heating at 94 °C, annealing for 45 s at 54 °C, and extension for 2 min at 72 °C; first produced two DNA molecules

may be passively immobilized through hydrophobic and ionic interactions. However, covalent immobilization is often necessary for proper adsorption, orientation, and conformation to the noncovalent surfaces. The immobilization process should occur rapidly and selectively in the presence of common functional groups, including amines (NH2), carboxyl (COOH), hydroxyl (OH), thiol (SH), and aldehyde (CHO). The surface density of the molecules should be optimized. Low-density surface coverage yields a correspondingly low frequency of binding sites. High density may result in inaccessible binding sites to the target molecules. The correct orientation of ligand molecules on the surface is required to facilitate the ligand available for the target. For efficient immobilization, the maximum biochemical activity and minimum nonspecific interactions are required. A brief description of attachment methods is as follows.

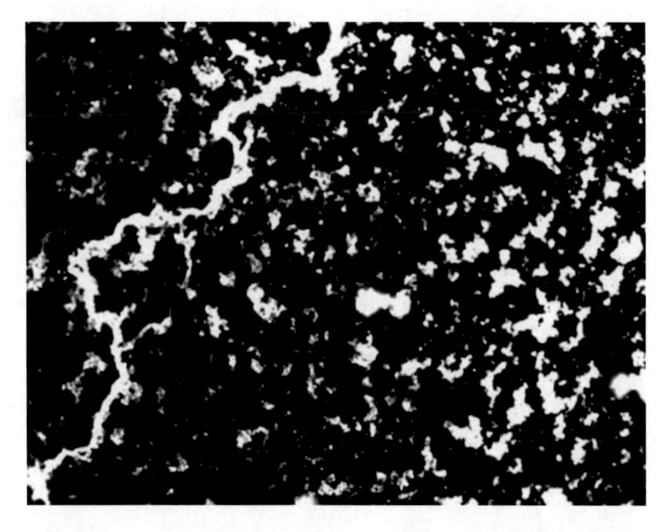

Fig. 14.37 Carbon nanotube (O.D. × wall thickness × length 20-–50 nm × 1–2 nm × 0.5–2 μm). Carbon nanotubes are biofunctionalized with protein and visualized using a confocal microscope

14.8.4.1 Surface Modification and Biofunctionalization of Carbon Nanotubes

The interface between biological molecules and novel nanomaterials is important in developing new types of miniature devices for biological applications. The carbon nanotubes are surface functionalized and biofunctionalized as shown in Fig. 14.37. Carbon nanotubes are surface modified for incorporating carboxylic group (-COOH) by refluxing in concentrated nitric acid for 4 h. The carboxylic groups formed on the surface of the nanotube giving a chemical handle for attaching carbon nanotubes to protein or DNA molecules. The modified carbon nanotubes are attached on glassy graphite electrode using *N,N*-dimethylformamide (DMF). The CNTs are biofunctionalized with antibodies or ss-DNA (probe) and hybridized with antigen or ss-DNA (analyte or target), respectively. Hybridization is detected with fluorescein isothiocyanate (FITC)-labeled IgG or cyanine (Cy3 or Cy5)-labeled target DNA using confocal microscope. In case of unlabeled probe molecules, the change in electrical measurement indicates probe–target molecular hybridization.

The streptavidin/biotin system is used to investigate the adsorption behavior of proteins on the sides of single-walled carbon nanotubes (SWNTs) [34]. Functionalization of SWNTs by coadsorption of a surfactant and poly(ethylene glycol) is found to be effective in resisting nonspecific adsorption of streptavidin. Specific binding of streptavidin

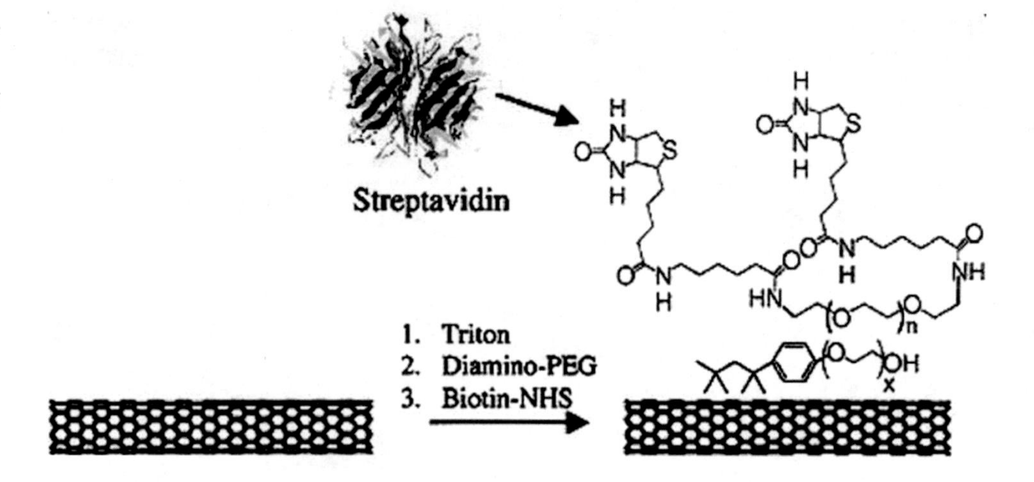

Fig. 14.38 Surface modification, biofunctionalization, and detection of biotin–streptavidin interactions. (Source: Courtesy of Moonsub Shim et al. 2002)

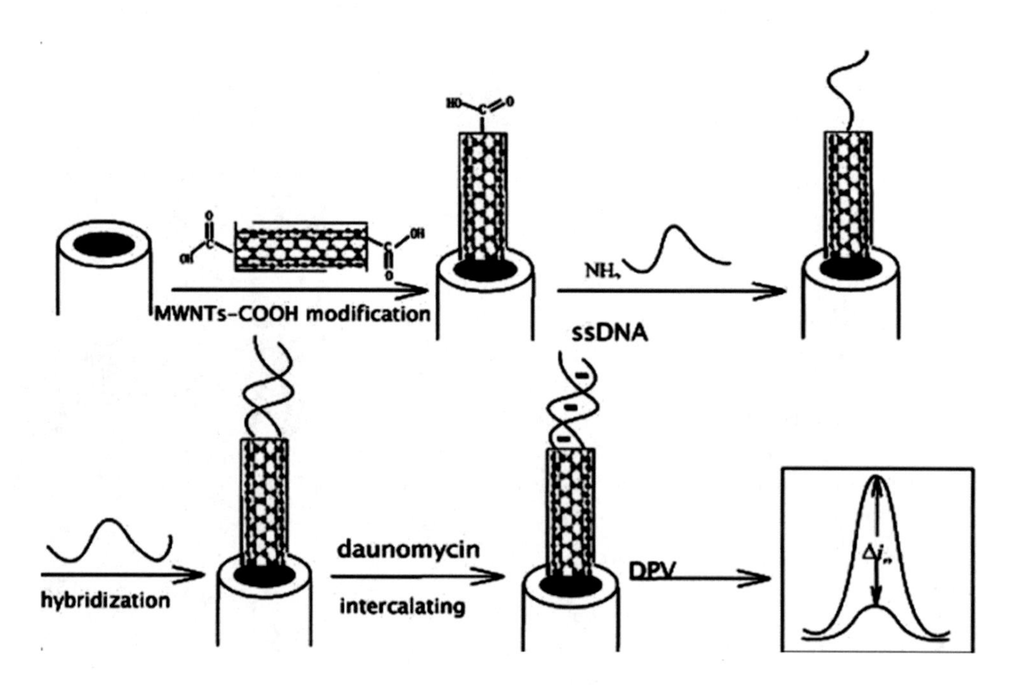

Fig. 14.39 Surface modification, biofunctionalization, and detection of DNA hybridization. (Source: Courtesy of Hong Cai et al. 2003)

onto SWNTs is achieved by cofunctionalization of nanotubes with biotin and protein-resistant polymers (Fig. 14.38).

Similarly, carbon nanotubes were functionalized with a carboxylic acid group (MWNTs–COOH) for covalent DNA immobilization and enhanced hybridization [35]. The MWNTs–COOH-modified glassy carbon electrode (GCE) was fabricated, and oligonucleotides with the 5^{t}-amino group were covalently bonded to the carboxyl group of carbon nanotubes. The hybridization reaction on the electrode was monitored by differential pulse voltammetry (DPV) analysis using an electroactive intercalator daunomycin as an indicator (Fig. 14.39).

14.8.4.2 Thiolated DNA for Self-Assembly on to Gold Transducers

Functionalized oligonucleotides are widely used in diagnostics, therapeutics, and genetic analysis. Thiolation of DNA involves condensation of DNA with thiolated polyethylenimine (PEI-SH), and the resulting nanoparticles are surface-coated using thiol-reactive poly[*N*-(2-hydroxypropyl) methacrylamide] (PHPMA) with 2-pyridyldisulfanyl or maleimide groups, forming reducible disulfide-linked or stable thioether-linked coatings, respectively.

14.8.4.3 Covalent Linkage to the Gold Surface

Self-assembling process of alkylthiols on gold is initiated by the strong chemical interactions between the sulfur and the gold surface. It involves chemisorption that forced a thiorate molecule to adsorb commensurate with a gold lattice. Then, the tail–tail interaction of the molecules created by lateral interchain nonbonded interactions, such as by van der Waals, steric, repulsive, and electrostatic forces, is strong enough to align the molecules parallel on the gold surface and create a

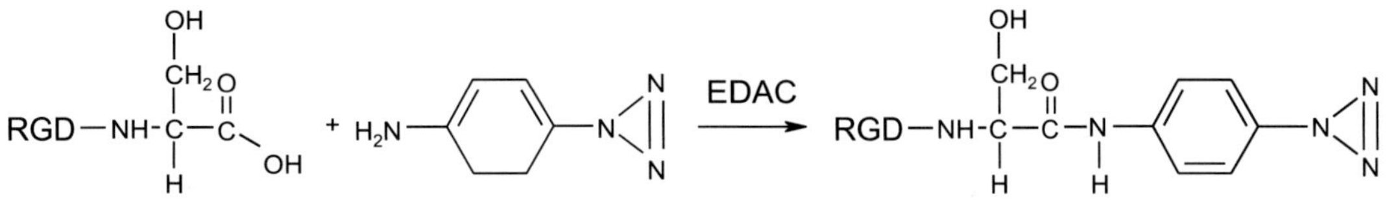

Fig. 14.40 Peptide cross-linkage to aniline in the presence of linker molecule EDAC

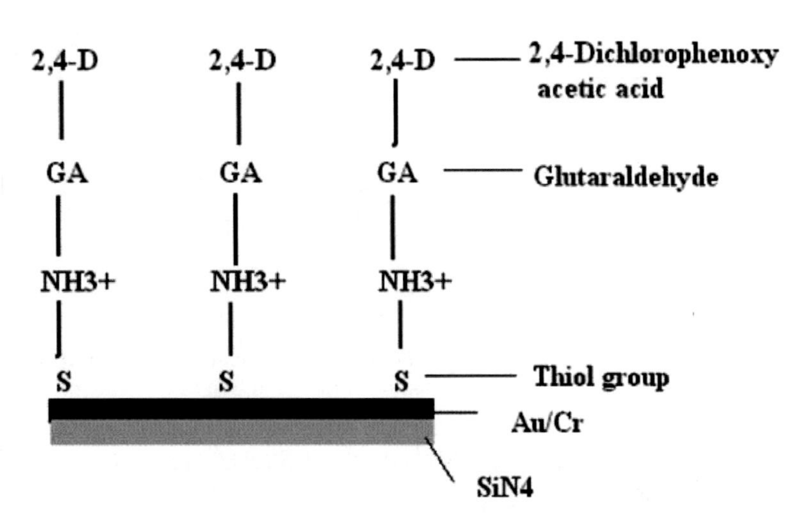

Fig. 14.41 Schematic of attachment of the probe molecules on gold layer coated onto silicon nitride electrode

crystalline film. In most cases, organic disulfides, thiols, and sulfides have been utilized for the preparation of stable SAMs on the gold surfaces. Methyl-terminated alkanethiol, biotin-terminated alkanethiol, COOH-terminated alkanethiol, CONHS-terminated alkanethiol, and NH2-terminated alkanethiol are main functional groups that establish covalent linkage. The photonic activation of disulfide bridges was shown to orient protein immobilization on biosensor surfaces [36].

14.8.4.4 Covalent (Carbodiimide) Coupling to Carbon Electrodes

The carbodiimides are hetero bifunctional coupling reagents that are utilized for coupling reactions through carboxyl and amino groups. The procedures generally used carbodiimide for coupling of peptides and proteins. Most peptides and proteins contain exposed carboxyl groups, so carbodiimides such as EDC are useful reagents for linking carboxyl groups to primary amine-containing matrices. The cross-linker, 1-ethyl-3-(3-dimethylaminopropyl) carbodiimide (EDAC), is widely used in immobilization of peptide, proteins, and sugars to various supports such as polystyrene plates, beads, gels, and biosensors. Cross-link is also used for structural studies, in terms of intramolecular and inter-subunit perspectives between proteins and DNA. The following is the typical RGD peptide (arginine–glycine–aspartic acid) cross-linking to aniline in the presence of EDAC (Fig. 14.40).

The contact surface of the electrode is washed thoroughly and coated with poly(terthiophene)-bearing carboxylic acid groups. This coated surface is used to attach DNA/RNA nucleotides. The conducting polymer poly(TTCA)-coated electrode is immersed in a 30 mM acetic acid/acetate buffer (pH 5.2) containing 4.0 mM EDAC for 1 h. The EDAC-attached conducting polymer-coated electrode is rinsed with a buffer solution and subsequently incubated in 11 μM probe DNA/RNA/acetate buffer solution for 6 h at 25 °C. In this process, NH2-linked C6-modified DNA/RNA is immobilized on the poly(TTCA) film by the formation of covalent bonds with carboxyl groups on the polymer.

Different cross-linkers such as cysteamine, glutaraldehyde, and bovine serum are used for attaching probe biomolecules on to gold layer on the silicon nitride electrodes. Figure 14.41 shows attachment of the herbicide protein, 2, 4-dichlorophenoxy acetic acid using glutaraldehyde cross-linker. The gold layer is about 40 nm in thickness, whereas the probe biomolecules along with linker molecules form 4 nm thickness.

14.8.4.5 Biotinylated DNA for Complex Formation with Avidin or Streptavidin

Single-stranded DNA (ss-DNA) is biotinylated on nylon membrane or electrode surfaces using UV light. The membrane or surface is subjected to a single wash step. The synthesized probe is completely free of unincorporated precursors. If biotinylation is performed under gentle conditions, the biological activity of the protein can be preserved.

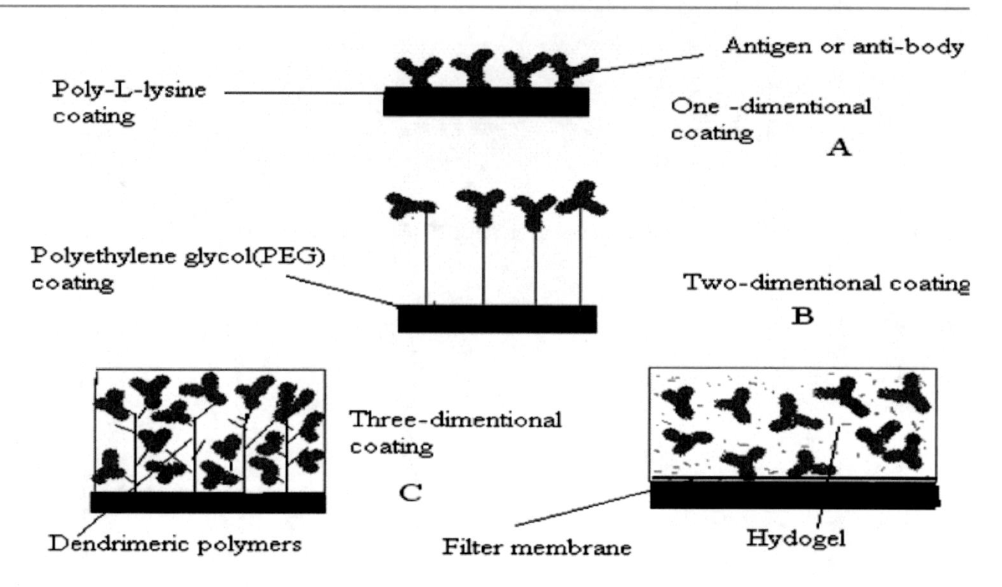

Fig. 14.42 Schematic of attachment of probe molecules. (**a**) One-dimensional coating, (**b**) two-dimensional coating, and (**c**) three-dimensional coating

Biotinylation of the target oligonucleotides can be achieved either enzymatically or chemically. Avidin is an egg white-derived glycoprotein with an extraordinarily high affinity (affinity constant >1015 M^{-1}) for biotin. Streptavidin is similar in properties to avidin but has a lower affinity for biotin. Many biotin molecules can be coupled to a protein, enabling the biotinylated protein to bind more than one molecule of avidin. By covalently linking avidin with different ligands such as fluorochromes, enzymes or electron microscopy (EM) markers can be utilized to study a wide variety of biological structures and processes. The extraordinarily high affinity between avidin or streptavidin and biotin results in rapidly formed and stable complex between the streptavidin conjugate and the biotin-labeled protein or DNA.

14.8.4.6 Polymers for Attachments of Probe Molecules

Hydrogels, filter membranes, dendrimeric polymers, poly-L-lysine, and polyethylene glycol (PEG) have been used as solid support for microarray assays (Fig. 14.42). In one-dimensional coating, poly-L-lysine is used for the development of antibody/antigen microarray assays [37]. Polyethylene glycol is used in two- dimensional coating. The PEG-treated surface provides large spacer molecule, which results in higher analyte capture capacity [38]. Dendrimeric polymers, filter membranes, and hydrogels facilitate the formation of monolayer of probe molecules [39]. Three-dimensional matrix structure reduces denaturation of protein because of the aqueous environment in the coating. The disadvantage with this coating is long incubation time required to probe–target molecular interactions.

14.8.4.7 Simple Adsorption to Carbon Surfaces

Self-assembled monolayer (SAM) coatings can be prepared at room temperature in the laboratory by simply dipping the desired substrate in the required millimolar solution for a specific period of time followed by rinsing with the same solution, water, ethanol and drying using a jet of dry nitrogen. The high affinity of gold for sulfur-containing molecules generates self-assembled monolayers. The quality of monolayers depends on several factors such as nature and roughness of the substrate, solvent used, temperature, and concentration of the adsorbate [40]. The polyamidoamine (PAMAM) dendrimer-modified magnetite nanoparticles are used to improve the efficiency of protein immobilization. First, aminopropyltrimethoxysilane (APTS) is immobilized on nanoparticles. Then, methylacrylate and ethylenediamine are added to form a dendritic structure on nanoparticle surface. It has been demonstrated that PAMAM dendrimer-coated magnetite nanoparticles are 3.9–7.7 times more efficient than aminosilane-modified magnetite nanoparticles [41].

14.8.5 Biosensor Device Fabrication Methods

The nanodevices are fabricated using one or more of the following common methods:

- Soft lithography, which is a new high-resolution patterning technique
- Thermal diffusion bonding of structural ceramics to superalloys
- UV laser micromachining for 2D and 3D medical devices
- Deep reactive ion etching in the substrate
- Electronic pick and place for heterogeneous integration
- Microfabricated fluidic nanotiterplate
- Micromechanical milling
- Microplastic injection molding
- Polymer embossing and polymer injection molding
- Micromanufacturing of cellular systems

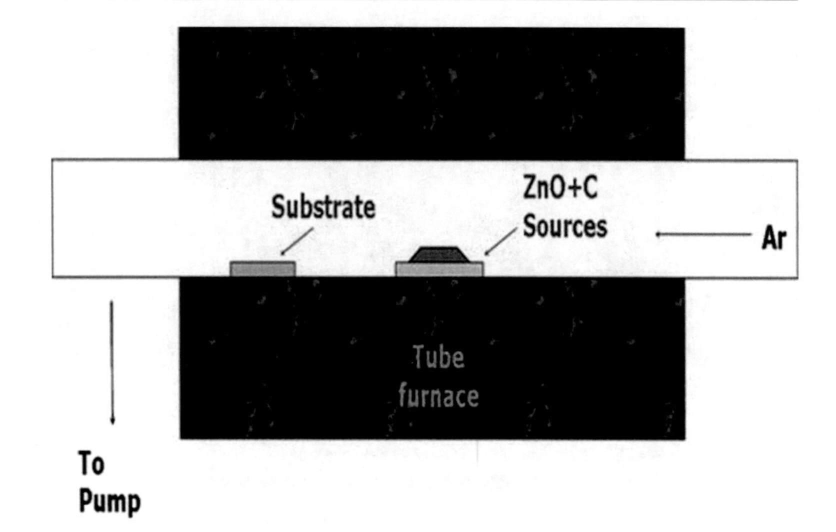

Fig. 14.43 The setup of the vapor–solid (VS) synthesis process. (Courtesy of PRC, Georgia Tech)

14.8.5.1 Fabrication Nanowire-Based Biosensors

The ZnO nanowire is synthesized through a high-temperature vapor–solid (VS) process that has been developed in Georgia Tech [42] and used for fabrication of ZnO nanowire-based biosensor system for detection of biological molecules and cancer cells [43]. The system setup is shown in Fig. 14.43.

The setup of the vapor–solid (VS) synthesis process consists of a horizontal high-temperature tube furnace with the length of ~50 cm, an alumina tube of ~75 cm, a rotary pump, and a gas-controlling system. About 2 g of the mixture of commercial ZnO powder (Alfa Aesar) and carbon powder is loaded into a polycrystalline Al2O3 boat, which is placed in the center of the alumina tube. The furnace is heated up to ~1400 °C in a ramp rate of 20 °C/min from room temperature. Gas Ar at a flow rate of 50 stand cubic centimeters per minute (sccm) is introduced into the tube as carrying gas. After holding at ~1400 °C for about 2 h, the furnace is slowly cooled down to room temperature. Polycrystalline Al2O3 substrates which are positioned ~11 cm from the ZnO source at the downstream side are used to collect the as-synthesized ZnO nanowires. The whole system is maintained at a pressure of ~200 mbar throughout. The as-synthesized sample is then inspected under an LEO 1530 field emission scanning electron microscope (SEM) operated at 5 kV.

14.8.5.2 Fabrication of Microelectrochemical Biosensors

As shown in Fig. 14.44, microelectrochemical sensor device is fabricated patterning the Pt electrodes using lift-off process. The patterning of the silver electrode is done by a semiadditive process (evaporate Ti/Ag, photoresist patterning, electroplating Ag, photoresist, and seed layer removal), chloridized in $FeCl_3$ to form Ag/AgCl [44].

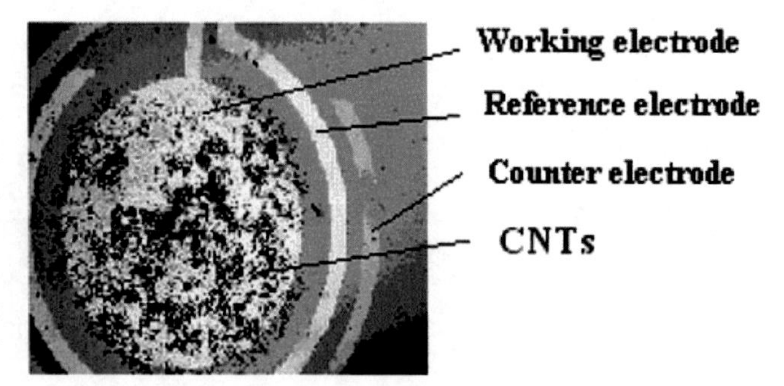

Fig. 14.44 Three-electrode structure of electrochemical biosensor. (Courtesy of PRC, Georgia Tech)

Sol–Gel System for Encapsulation of the Enzyme

Sol–gel chemistry is gaining importance in acting as an interface between biological materials and nonbiological components in biosensors. Zirconia/nafion sol–gel encapsulation of enzyme is used in fabrication of enzyme-based sensors. The zirconia sol–gel solution is prepared by mixing 8.0 ml of deionized water with 1.0 ml of 0.12 M ZrOpr dissolved in isopropanol and 0.25 ml of 11 M HCl solution. The zirconia/nafion composite is prepared by mixing zirconia solution with nafion. Glucose oxidase enzyme at 100 mg/ml is prepared in 0.05 M phosphate buffer at pH 7.0. The enzyme-encapsulating matrix is prepared by mixing 40 μl of zirconia/nafion stock solution and 40 μl of the enzyme solution, casted on the working electrode (Fig. 14.45). The advantage of the sol–gel encapsulation are as follows:

- It resists chemical attack and withstands high temperature.
- Nafion is highly conductive to cations, making it ideal for many membrane applications.
- Pure nafion films revealed large calcification and instability of the cast films.

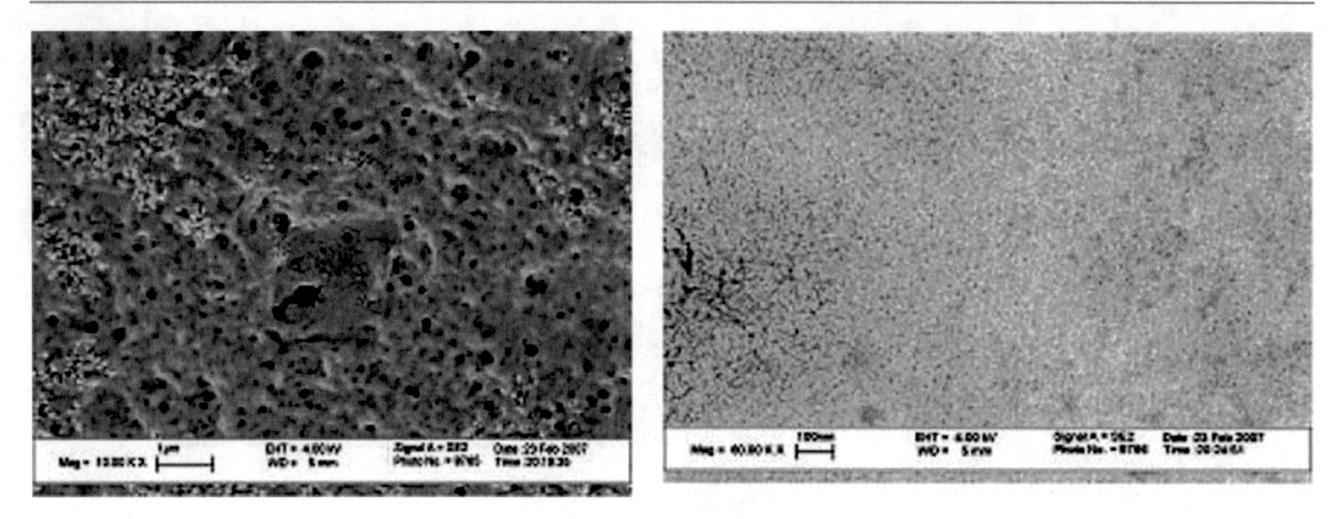

Fig. 14.45 FESEM micrographs of zirconia–nafion gels before (*top*) and after (*bottom*) enzyme entrapment. (Courtesy of PRC, Georgia Tech)

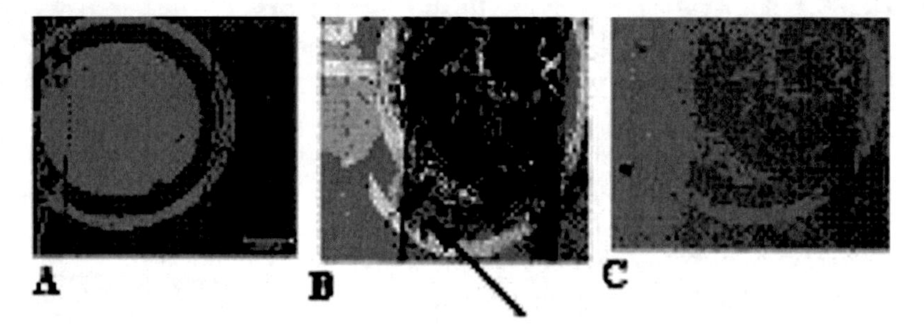

Fig. 14.46 Planar microelectrodes: (**a**) without CNTs and sol–gel enzyme; (**b**) CNTs and sol–gel enzyme, open microfluidic channel; (**c**) CNTs and sol–gel enzyme, sealed microfluidic channel

- It incorporates zirconia into anionic nafion assembly in order to prevent calcification.
- Encapsulation of the enzyme with the sol–gel protects from heat.

14.8.6 Microfluidic Channels

Microfluidics system is the manipulation of liquids in channels having cross- sectional dimensions on the order of 10–100 μm. Microfluidic channels are essential for guiding and helping biofluids carrying the target or the sample to be tested onto sensing element of the biosensor. As all biological molecular reaction takes place in liquid environment, microfluidic channels facilitate controlled flow of micro-quantities of fluids containing desired test molecules (Figs. 14.46 and 14.47). The common fluids used as target in microfluidic devices include whole blood samples, bacterial cell suspensions, protein or antibody solutions, and various buffers. Fluidics carries the biological molecules without involving mechanical moving parts that will wear out.

Materials for implantable devices and microfluidic channels need to be biocompatible, which is not producing a toxic, injurious, or host response in living tissue. Biocompatible materials such as polydimethylsiloxane (PDMS) and LCP materials are commonly used for fabrication of microfluidics. Further, laminated plastic microfluidic components made of polyimide, poly(methyl methacrylate) (PMMA), and polycarbonate materials with thickness between 25 and 125 μm have also been developed and used. These devices can be potentially manufactured in high volume with low unit cost because of advancements in the materials and chemical processes such as high-speed full-field excimer laser ablation process (Anvik Incorporated) to generate the fluidic channels. Low-cost PDMS adhesive-based bonding is used to seal the channels. Fabrication of both the submicron features of the photonic crystal sensor structure and the >10 μm features of a flow channel network in one step at room temperature on a plastic substrate was demonstrated [45, 46]. Polymer-based microfluidic devices can be created by combining lithographic and conventional manufacturing techniques (e.g., injection molding or embossing). Polymeric channels can also be made transparent in the UV, visible, and IR ranges. This property enables the use of generic off-chip detection systems, thereby reducing the cost and complexity of a microanalysis chip. As an

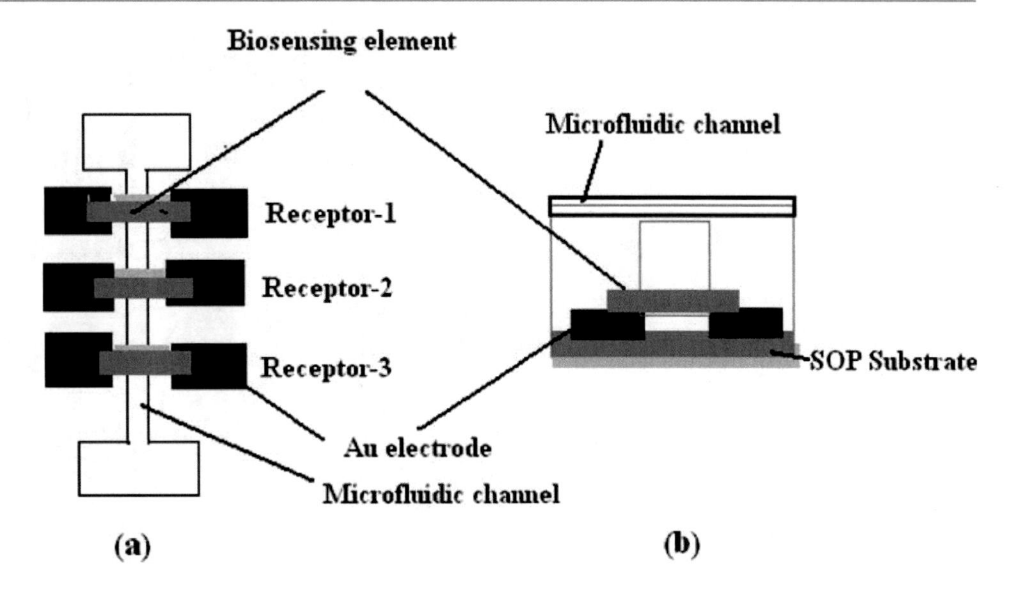

Fig. 14.47 (a) Schematic of microfluidics integrated, multiplexed biosensor system; (b) cross section of the device

example, laminated plastic microfluidic components of polyimide, poly(methyl methacrylate) (PMMA), and polycarbonate materials with thickness between 25 and 125 μm have been developed [47]. These devices can be potentially manufactured in high volume with low unit cost because of advances in the materials and processing tools made by the electronic industry.

14.8.7 Biosensor Packaging

The highly miniaturized electronic system technology mostly relies on integrated circuit (IC) integration (Moore's law) for performance improvement and cost reduction. System-on-package (SOP) technology paradigm (the second law of electronics) pioneered by Georgia Tech 3D Systems Packaging Research Center, since the early 1990s, provides system-level miniaturization in a package size that makes today's handheld devices into megafunctional systems, with applications ranging from computing, wireless communications, health care, and personal security. The SOP is a system miniaturization technology that ultimately integrates nanoscale thin film components for batteries, thermal structures, and active and passive components in low-cost organic packaging substrates, leading to microscale to nanoscale modules and systems. True miniaturization of products should take place not only at IC but also at system level, the latter made possible by thin film batteries, thermal structures, and embedded actives and passives in package-size boards. This is the fundamental basis for the SOP concept. The traditional packaging by which components are integrated into systems today presents several sets of barriers in cost, size, performance, and reliability. The SOP concept overcomes these barriers by the best of both the IC and the package integration at system level, i.e., the IC for transistor density, and the system package for component density of RF, optical, and digital functions and biofunctions. In addition to miniaturization, such a concept leads to lower cost, higher performance, and better reliability of all electronic systems including biosensor systems described in this chapter. The attributes of traditional electronic system are bulky, poor reliability, moderate performance, and medium cost compared to SOP-based systems, which show miniaturized, 3× improvement in reliability, high performance, and low cost. SOP-based biosensor systems can be greatly enhanced in terms of the functionalities in both clinical and nonclinical domains. As shown in Fig. 14.48, the energy source for many biomedical microsystems is through an external RF link. The system communicates with the outside world over an inductively coupled bidirectional wireless link [48].

14.9 Characterization of Functionalized Biosensor Structures

The functionalized electrode surface is characterized by transmission electron microscopy (TEM), Fourier transform infrared spectroscopy (FT-IR), fluorescent microscopy, X-ray photoelectron spectroscopy (XPS), scanning electron microscopy (SEM), and atomic force microscopy (AFM). Fourier transform infrared spectroscopy (FT-IR) is used to identify specific types of chemical bonds or functional groups based on their unique absorption signatures.

Transmission electron microscopy (TEM) is used to study the morphology and size of the particles.

Energy-dispersive X-ray spectrometer is used to confirm the composition of particles immobilized on gold surface. The ellipsometry is a real-time optical measurement technique that is used to study several material characteristics such as layer thickness, optical constants (refractive index

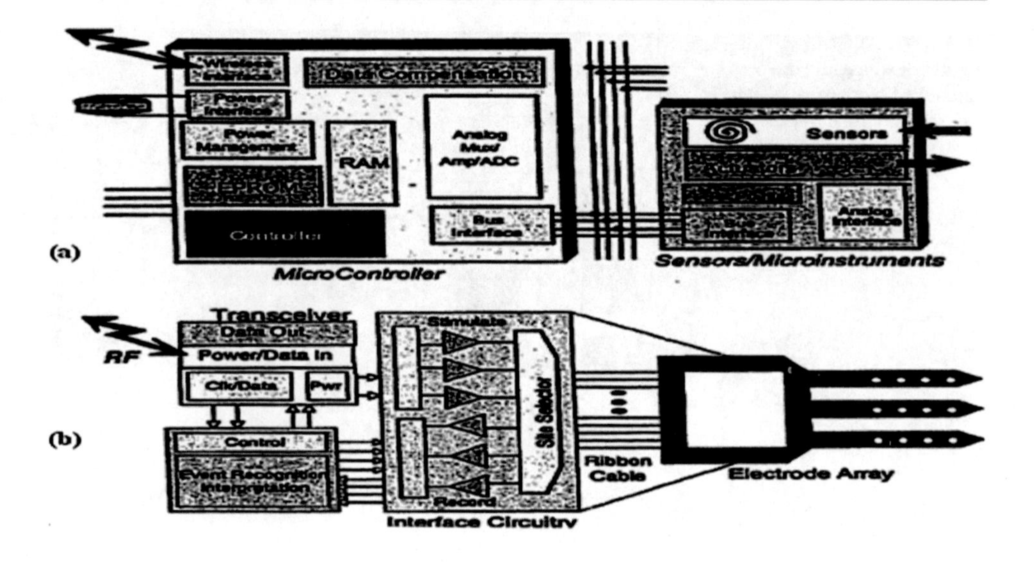

Fig. 14.48 (**a**) Schematic of a wireless integrated microsystem, (**b**) implantable version of the same microsystem. (Courtesy of Dr. Kensall D. Wise, University of Michigan)

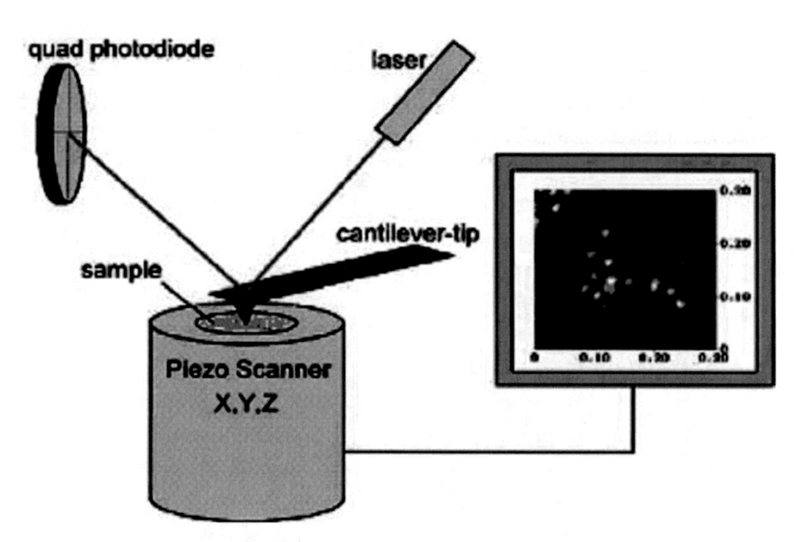

Fig. 14.49 Schematic of the atomic force microscope. (Source: Courtesy of Hafner et al., 2001)

and extinction coefficient), and optical anisotropy. It is used for quantification and visualization of the lateral thickness distribution of thin (0–30 nm) transparent layers on solid substrates. The thinness of the biofunctionalized molecules on a nanodevice surface is measured at the same time with a high lateral resolution.

Atomic force microscope (AFM) is used to take three-dimensional images of resolution on the order of 1–2 nm. It consists of cantilever tip assembly, piezo tube scanner, and laser deflection system (Fig. 14.49). The attachment of the biomolecules such as proteins and DNA and the molecular hybridization events on the biosensor devices can be characterized using AFM [49].

14.10 Future Trends and Summary

The tremendous advancements in the sensor technologies are due to the great technological demand for rapid, sensitive, and cost-effective biosensor systems in vital areas of human activity such as health care, genome analysis, food and drink, process industries, environmental monitoring, defense, and security. In fact, nanobioconjugates that consist of various functional nanoparticles or nanostructures linked to biological molecules revolutionized the biosensor systems. Biosensors have enormous potentials and prospects in wide range of applications. These nanobioelectronic devices will revolutionize the detection system and signal processing and enrich the quality of life.

At present, the nanotechnology-based biosensors are at the early stage of development. The vast applications of nanotechnology in such diverse fields as semiconductors, biological and medical devices, polymer composites, optical devices, dispersions, and coatings are amazing. With the discovery of novel materials and innovated approaches, it may progress by leaps and bounds, as in the case of the computer industry after the development of integrated circuits. As biosensors are essentially involved in revealing molecular information, they form a part of important information technology. The potentials and prospects of biosensors are effectively attracting myriad start-ups to the Fortune 500 companies to manufacture these devices.

In summary, the building blocks of biosensors, biomechanisms, biofunctionalization of nanodevices, search for suitable interfaces between biological material and electronics, intelligent signal processing of information transmitted by biosensors, and other related issues are discussed. The methods for protein and DNA and RNA molecular probe design and syntheses and microfabrication methods of the biosensor devices are enumerated. A detailed account on the application of nucleic acid sensors and DNA chips, immunosensors, enzyme-based biosensors, organism- and whole cell-based biosensors, ion sensors, natural and synthetic receptors for biosensors, new signal transduction technology, system integration, proteomics and single-cell analysis, bioelectronics, and nanoanalytical systems and the importance of nanoscale target molecular detection in cancer, AIDS and other diseases are discussed. The issues of technical challenges in designing and fabricating the smart biosensors and the integration of nanobiosensors with system on package (SOP) are discussed. Emphasis is made on nanochemistry applications for biosensor packaging. The potentials and prospects of biosensors and their commercial value are mentioned.

References

1. Dell'Atti, D., Tombelli, S., Minunni, M., Mascini, M.: Detection of clinically relevant point mutations by a novel piezoelectric biosensor. Biosens. Bioelectron. **27**(10), 1876–1879 (2006)
2. Toppozada, A.R., Wright, J., Eldefrawi, A.T., Eldefrawi, M.E., Johnson, E.L., Emche, S.D., Helling, C.S.: Evaluation of a fiber optic immunosensor for quantitating cocaine in coca leaf extracts. Biosens. Bioelectron. **12**(2), 113–124 (1997)
3. Nuwaysir, F., Huang, W., Albert, T.J., Jaz, S., Nuwaysir, K., Pitas, A., Richmond, T., Gorski, T., Berg, J.P., Ballin, J., McCormick, M., Norton, J., Pollock, T., Sumwalt, T., Butcher, L., Porter, D., Molla, M., Hall, C., Blattner, F., Sussman, M.R., Wallace, R.L., Cerrina, F., Green, R.D.: Gene expression analysis using oligonucleotide arrays produced by maskless photolithography. Genome Res. **12**(11), 1749–1755 (2002)
4. Zhou, C., Pivarnik, P., Rand, A.G., Letcher, S.V.: Acoustic standing-wave enhancement of a fiber-optic Salmonella biosensor. Biosens. Bioelectron. **13**(5), 495–500 (1998)
5. Liu, J., Janagama, G.D., Iyer, M.K., Tummala, R.R., Wang, Z.L.: ZnO Nanobelts/wire for electronic detection of enzymatic hydrolysis of starch International Symposium and Exhibition on Advanced Packaging Materials Processes, Properties and Interfaces. Georgia Tech, Atlanta, USA (2006)
6. Shriver-Lake, L.C., Donner, B.L., Ligler, F.S.: On-site detection of TNT with a portable fiber optic biosensor. Environ. Sci. Technol. **31**, 837–841 (1997)
7. Ragan, F., Meaney, M., Vos, J.G., MacCraith, B.D., Walsh, J.E.: Determination of pesticides in water using ATR-FTIR spectroscopy on PVC/chloroparaffin coatings. Anal. Chim. Acta. **334**, 85–92 (1996)
8. Potyrailo, R.A., Hobbs, S.E., Hieftje, G.M.: Near-ultraviolet evanescent-wave absorption sensor based on a multimode optical fiber. Anal. Chem. **70**, 1639–1645 (1998)
9. Woods, S.I., Kirtley, J.R., Sun, S., Koch, R.H.: Direct investigation of superparamagnetism in Co nanoparticle films. Phys. Rev. Lett. **87** (13), 137205 (2001)., 4 Pages
10. Kaplan, D.K., Sverdlov, V.A., Likharev, K.K.: Coulomb gap, coulomb blockade, and dynamic activation energy in frustrated single-electron arrays. Phys. Rev. B. **68**, 045321 (2003)., 4 pages
11. Pancrazio, J.J., Whelan, J.P., Borkholder, D.A., Ma, W., Stenger, D. A.: Development and application of cell-based biosensors. Ann. Biomed. Eng. **27**, 697–711 (1999)
12. Dickert, F.L., Thierer, S.: Molecularly imprinted polymers for optochemical sensors. Adv. Mater. **8**(12), 987–990 (2004)
13. Liebertzeit, P.A., Glanzning, G., Jenik, M., Gazda-Miarecka, S., Dickert, F.L., Leidl, A.: Soft lithography in chemical sensing—analytes from molecules to cells. Sensors. **5**, 509–518 (2005)
14. Hayden, O., Dickert, F.L.: Selective microorganism detection with cell surface imprinted polymers. Adv. Mater. **12**, 311–313 (2001)
15. Hayden, O., Bindeus, R., Haderspock, C., Mann, K., Wirl, B., Dickert, F.L.: Mass-sensitive detection of cells, viruses and enzymes with artificial receptors. Sensors Actuators B Chem. **91**, 316–319 (2003)
16. Cai, W., Peck, J.R., van der Weide, D.W., Hamers, R.J.: Direct electrical detection of hybridization at DNA-modified silicon surfaces. Biosens Bioelectron, Article in Press. **19**(91), 1013–1019 (2004)
17. Sahoo, S.K., Labhasetwar, V.: Nanotech approaches to drug delivery and imaging. Drug Disc Today. **8**, 1112–1120 (2003)
18. Hwang, I.-H., Lee, J.-H.: Self-actuating biosensor using a piezoelectric cantilever and its optimization. J Phys: Conference Series. **34**, 362–367 (2006)
19. Yi, C., Qingqiao, W., Hongkun, P., Lieber, C.M.: Nanowire nanosensors for highly sensitive and selective detection of biological and chemical species. Science. **293**, 1289–1292 (2001)
20. Wang Z L: Second International Workshop on nano-bio-electronics packaging held in Atlanta, (2005)
21. Baselt, D.R., G u, L., Natesan, M., Metzger, S.W., Sheehan, P.E., Colten, R.J.: A biosensor based on magnetoresistance technology. Biosens. Bioelectron. **13**, 731–739 (1998)
22. Harborn, U., Xie, B., Venkatesh, R., Danielsson, B.: Evaluation of a miniaturized thermal biosensor for the determination of glucose in whole blood. Clin. Chim. Acta. **267**(2), 225–237 (1997)
23. Wilson, M., Kannangara, K., Smith, G., Simmons, M., Raguse, B.: Nanotechnology, Basic Science and Emerging Technology. Chapman & Hall/CRC, New York (2002)
24. Goddard, W.A., Brenner, D.W., Lyshevski, S.E., Lafrate, G.J.: Hand Book of Nanoscience, Engineering and Technology. CRC Press, USA (2003)
25. Jones, E.F., He, J., VanBrocklin, H.F., Franc, B.L., Seo, Y.: Nanoprobes for medical diagnosis: current status of nanotechnology in molecular imaging. Curr. Nanosci. **4**(1), 17–29 (2008) (13)
26. Rist, M.J., Marino, J.P.: Fluorescent nucleotide base analogs as probes of nucleic acid structure, dynamics and interactions. Curr. Org. Chem. **6**(9), 775–793 (2002)
27. Kool, E.T.: Artificial DNA through metal-mediated base pairing: structural control and discrete metal assembly. Acc. Chem. Res. **35**, 936 (2002)
28. LaVan, D.A., Lynn, D.M., Langer, R.: Moving smaller in drug discovery and delivery. Nat. Rev. **1**, 77–84 (2002)
29. Niemeyer, C.M.: Self assembled nanostructures based on DNA: towards the development of nanobiotechnology. Curr. Opin. Chem. Biol. **4**, 609–618 (2000)
30. Langer, P.R., Waldrop, A.A., Ward, D.C.: Enzymatic synthesis of biotin-labeled polynucleotides: novel nucleic acid affinity probes. Proc. Natl. Acad. Sci. U. S. A. **78**, 6633–6637 (1981)
31. Huhtinen, P., Soukka, T., Lovgren, T., Harma, H.: Immunoassay of total prostate-specific antigen using europium (III) nanoparticle labels and streptavidin-biotin technology. J. Immunol. Methods. **294**(1–2), 111–122 (2004)

32. Ebessen, T.W. (ed.): Carbon Nanotubes: Preparation and Properties. CRC Press, Boca Raton (1997)
33. Riikonen, S., Sanchez-Portal, D.: Structural model for Si (553)- Au atomic chain reconstruction. Nanotechnology. **16**, 5218–5223 (2005)
34. Moonsub, S., Shi, K.N.W., Robert, C.J., Li, Y., Dai, H.: Functionalization of carbon nanotubes for biocompatibility and biomolecular recognition. Nano Lett. **2**(4), 285–288 (2002)
35. Hong, C., Xuni, C., Ying, J., He, P., Yuzhi, F., Hong, C., Xuni, C., Ying, J., He, P., Fang, Y.: Carbon nanotube-enhanced electrochemical DNA biosensor for DNA hybridization detection. Anal. Bioanal. Chem. **375**, 287–293 (2003)
36. Neves-Petersen, M.T., Snabe, T., Klitgaard, S., Duroux, M., Petersen, S.B.: Photonic activation of disulfide bridges achieves oriented protein immobilization on biosensor surfaces. Protein Sci. **15**(2), 343–351 (2003)
37. Langer, R., Tirrell, D.A.: Designing materials for biology and medicine. Nature. **428**, 487–492 (2004)
38. Kusnezow, W., Hoheisel, J.D.: Solid support for microarray immunoassays. J. Mol. Recognit. **16**, 165–176 (2003)
39. Beier, M., Hoheisel, J.D.: Versatile derivatisation of solid support media for covalent bonding on DNA-microchips. Nucl. Acids Res. **27**, 1970–1977 (1999)
40. Mourougou-Candoni, N., Naud, C., Thibaudau, F.: Adsorption of thiolated oligonucleotides on gold surfaces: an atomic force microscopy study. Langmuir. **19**(3), 682–686 (2003)
41. Pan, B.F., Gao, F., Gu, H.C.: Dendromer modified magnetic nanoparticles for protein immobilization. J. Colloid Interface Sci. **284**(1), 1–6 (2005)
42. Pan, Z.W., Dai, Z.R., Wang, Z.L.: Nanobelts of semiconducting oxides. Science. **291**, 1947 (2001)
43. Liu, J., Janagama, D.G., Iyer, M.K., Tummala, R.R., Wang, Z.L.: ZnO Nanobelts/wire for electronic detection of enzymatic hydrolysis of starch. International Symposium and Exhibition on Advanced Packaging Materials Processes, Properties and Interfaces, Georgia Tech., Atlanta, USA. (2006)
44. Janagama, D.G., Markondeya Raj, J.P., Iyer, M.K., Tummala, R.R.: Bio SOP with embedded Electrochemical Biosensors using MWCNT electrodes and Zirconia/Nafion based enzyme encapsulants. ECTC 57th Electronic components and Technology conference. (2007)
45. Choi, C.J., Cunningham, B.T.: Single-step fabrication and characterization of photonic crystal biosensors with polymer microfluidic channel. Lab Chip. **6**, 1373–1380 (2006)
46. Zaytseva, N.V., Goral, V.N., Montagna, R.A., Baeumner, J.: Development of a microfluidic biosensor module for pathogen detection. Lab Chip. **5**, 805–811 (2005)
47. Martin, P.M.: Laminated plastic microcomponents for biological and chemical systems. J. Vac. Sci. Technol. **17**(4), 2262–2269 (1999)
48. Wise, K.D.: Wireless integrated microsystems: coming breakthroughs in health care. Proceedings of International Electron Devices meeting. (2006)
49. Hafner, J.H., Cheung, C.L., Woolley, A.T., Lieber, C.M.: Structural and functional imaging with carbon nanotube AFM probes. Prog. Biophys. Mol. Biol. **77**(1), 73–110 (2001)

Biomimetic Lotus Effect Surfaces for Nanopackaging

15

Yonghao Xiu and C. P. (Ching-Ping) Wong

15.1 Introduction

Lotus effect or superhydrophobicity was first described by Barthlott in 1996 [1, 2]. When he studied lotus leaf surfaces, he found that the surface is rough instead of smooth. Combined with hydrophobic surface layers, a superhydrophobic state was observed. Water droplets falling on the surface bead up and roll off the surface instantaneously. Superhydrophobicity requires that both a surface hydrophobic layer and rough surface structures are present. This phenomenon is quite common in the biology arena, e.g., on some insects and plants leaves, shown in Fig. 15.1. Artificial superhydrophobic surfaces were prepared before the phenomenon was widely acknowledged. In 1986, Sacher et al. reported that very water-repellent films were obtained by depositing hexamethyldisiloxane on Si chips in a high-energy plasma at low temperature [3]. Drops of water had contact angles as high as 180° and rolled off the slightly inclined surfaces, which suggested a near-zero roll off angle. Morra et al. had prepared a superhydrophobic surface by O2 plasma etching on PTFE surfaces in 1989 [4]. Actually after 5 min of treatment, water drops rolled easily across the surface. Due to the absence of hysteresis, no obstacle to their movement occurred. The first super water- and oil-repellent surface was developed in Tsujii's group [5] on an anodically oxidized aluminum surface after a hydrophobic treatment.

The effect of roughness on surface wetting is scale-dependent. Besides the surface hydrophobicity, the surface roughness can be more important in achieving dewetting surfaces even without the most hydrophobic materials (e.g., fluorine containing polymers). On flat surfaces oil repellency is difficult to achieve. However, with appropriate surface roughness with special geometry designs, a superoleophobic surface may result. This signifies the importance of surface wetting/dewetting control via flexible, fine-tuned surface structure design at micro- and nanoscales.

The microelectronic industry is developing very fast. Microsystems include automotive, computer and business equipment, communications, consumer, industrial and medical, military, and aerospace. Packaging provides electrical connections for signal transmission, power input, and voltage control. It also provides for thermal dissipation and the physical protection required for reliability. Today assembly and packaging are a limiting factor in both cost and performance for electronic systems. Wireless and mixed signal devices, bio-chips, optoelectronics, and MEMS have placed new requirements on packaging and assembly. With the device size reduction, it poses even harsher requirement for the packaging which gives it protection not only for reliability but also for signal propagation (high speed), thermal dissipation, and power insulation.

There are several areas that require a water-/moisture-resistant surface layer. For polymer packaging, reduction in moisture absorption and diffusion is the most important in achieving system reliability. For MEMS, hermetic sealing is very important to prevent capillary force-induced failure. In addition, moisture diffusion can cause corrosion of the fine metal lines. Since package must function in humid environments, methods to reduce moisture absorption or diffusion would be essential. In recent years, the superhydrophobic surface is attracting intense research interest due to the promising applications in self-cleaning and moisture/water repellency, such as in anticorrosion, antistiction of MEMS device packaging, microfluidics, solar cell packaging, and bio-assay. These applications will be discussed in detail in Sect. 15.6.

Y. Xiu (✉) · C. P. Wong
School of Materials Science and Engineering, Georgia Institute of Technology, Atlanta, GA, USA
e-mail: yxiu@gatech.edu; cp.wong@mse.gatech.edu

C. P. Wong et al. (eds.), *Nano-Bio- Electronic, Photonic and MEMS Packaging*, https://doi.org/10.1007/978-3-030-49991-4_15

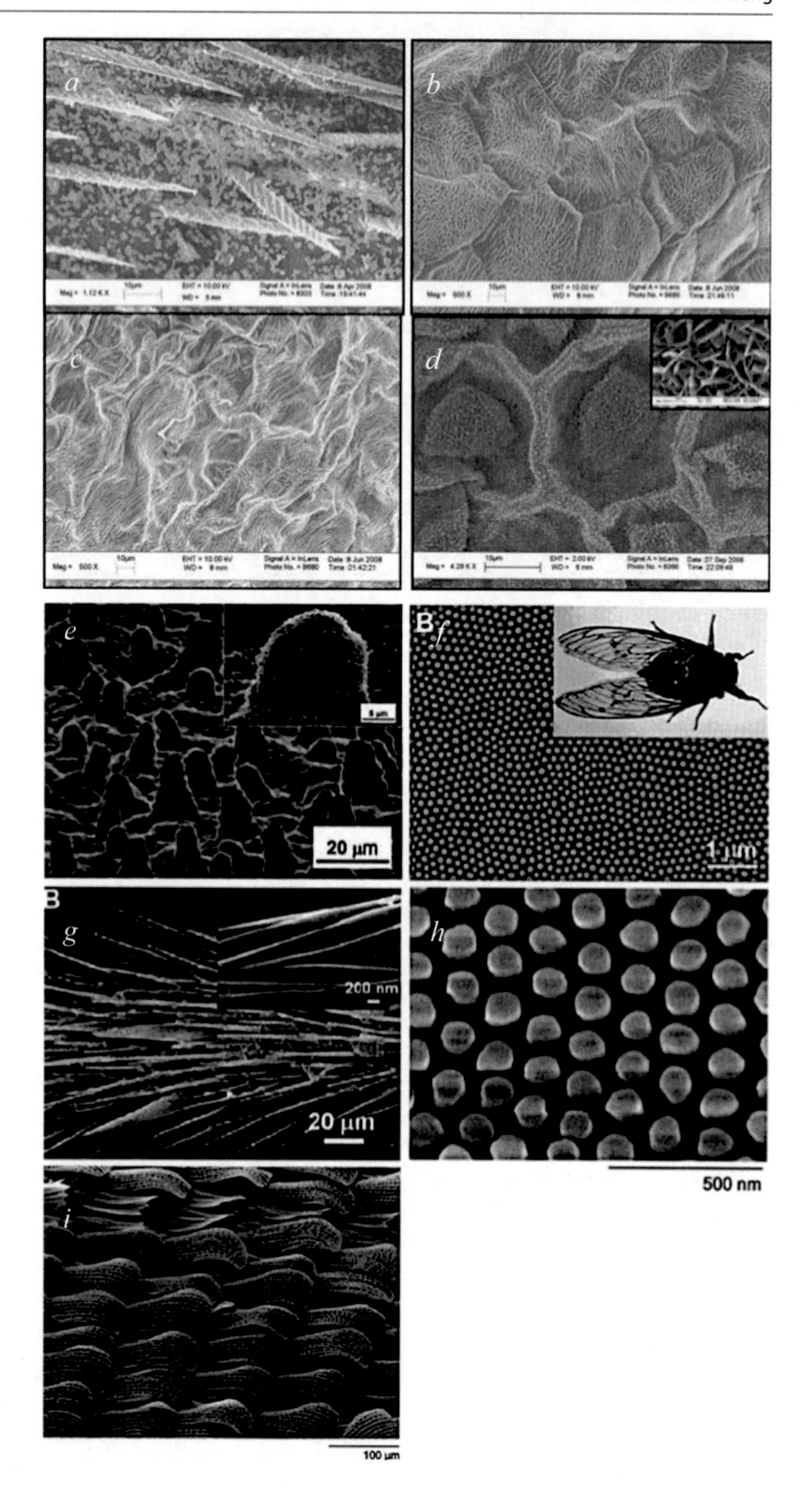

Fig. 15.1 SEM images of different biosurface structures: (**a**) wasp wing, (**b**) azaleas petal, (**c**) rose petal, (**d**) *Colocasia esculenta* leaf surface, (**e**) micro- and nanostructures on the lotus leaf (*Nelumbo nucifera*) [6], (**f**) cicada wing [7], (**g**) water strider [8], the leg with numerous oriented spindly microsetae, (inset) nanoscale groove structure on a seta, (**h**) surface of mosquito (*Culex pipiens*) eye [9], (**i**) the wings of *Papilio ulysses*; the way the tiles are displayed together with the detail of the texture confer anisotropy to the texture [10]

15.2 Thermodynamics Aspect of Superhydrophobic Surfaces

15.2.1 Contact Angle on a Rough Surface

On a flat surface, the change in surface free energy, ΔG, accompanying a small displacement of the liquid such that the change in area of solid covered, ΔA, is

$$\Delta G = \Delta A_{SL}(\gamma_{SL} - \gamma_{SV}) + \Delta A_{LV}\gamma_{LV} \quad (15.1)$$

$$\Delta G = \Delta A(\gamma_{SL} - \gamma_{SV}) + \Delta A\gamma_{LV}\cos(\theta - \Delta\theta) \quad (15.2)$$

At equilibrium,

$$\Delta = 0$$

And according to Young's equation

$$\gamma_{SL} - \gamma_{SV} + \gamma_{LV}\cos\theta = 0$$

or

$$\cos\theta = \frac{\gamma_{SV} - \gamma_{SL}}{\gamma_{LV}} \quad (15.3)$$

Here, γ_{SV} is the surface tension at solid/vapor interface, γ_{SL} is the interface tension at solid/liquid interface, and γ_{LV} is the surface tension at liquid/vapor interface.

On a smooth surface, Young's Eq. (15.3) determines contact angles of liquid droplet as shown in Fig. 15.2.

On a rough surface, however, Young's equation no longer holds for the apparent contact angles due to the introduction of a new factor: surface roughness. Water droplet/solid interface contact can be either in Wenzel state or in Cassie state (shown in Fig. 15.3), depending on the surface hydrophobicity and geometries of the surface structures.

Usually two equations were used for the calculation of the apparent contact angles corresponding to the two regimes. One is Wenzel equation as shown in Eq. (15.4):

$$\cos\theta_A = r\cos\theta_Y \quad (15.4)$$

This equation described the relation between the surface roughness r and the apparent contact angle, where r is determined by the ratio of total surface area to the projected surface area. This equation is valid for a complete solid/liquid contact interface.

For heterogeneous surfaces, Cassie equation can be used to describe the heterogeneity effect on the apparent contact angle [11],

$$\cos\theta_A = f\cos\theta + f - 1 \quad (15.5)$$

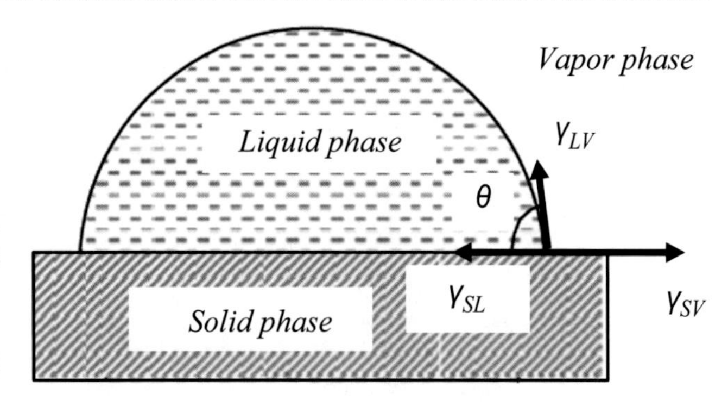

Fig. 15.2 The illustration of relationship between the surface and interface tensions according to Young's equation

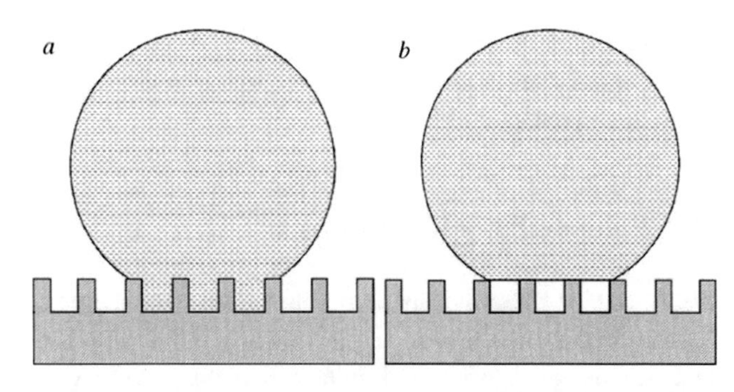

Fig. 15.3 The two contact regimes of water droplet with rough surfaces: (**a**) Wenzel contact regime, Cassie contact regime

where θ_A is the apparent contact angle on the rough surface, f is the projected surface fraction of solid, and θ Y is the contact angle on a flat surface as per Young's equation.

The surface roughness r can be described by Wenzel roughness factor:

$$r = \frac{A_{Actual}}{A_{projected}} \quad (15.6)$$

The development of Wenzel and Cassie equations is shown below. For water droplets on rough surfaces, according to Eq. (15.2),

$$\Delta G = \Delta Ar(\gamma_{SL} - \gamma_{SV}) + \Delta A\gamma_{LV}\cos(\theta - \Delta\theta)$$

And the equation reduced to

$$r(\gamma_{SL} - \gamma_{SV}) + \gamma_{LV}\cos\theta_A = 0 \quad (15.7)$$

Therefore,

$$\cos\theta_A = r\frac{\gamma_{SL} - \gamma_{SV}}{\gamma_{LV}} = r\cos\theta_Y$$

This is the Wenzel equation shown in Eq. (15.4) [12].

Similarly for Cassie equation, considering the fraction of solid/air and solid/liquid interface,

$$f_1(\gamma_{S1V} - \gamma_{S1L}) + f_2(\gamma_{S2V} - \gamma_{S2L}) - \gamma_{LV}\cos\theta_A = 0$$

or

$$\gamma_{LV}\cos\theta_A = f_1(\gamma_{S1V} - \gamma_{S1L}) + f_2(\gamma_{S2V} - \gamma_{S2L}) \quad (15.8)$$

When $S2$ is also air, γ S2V = 0 and $f1 + f2 = 1$, Eq. (15.8) reduced to Cassie equation (Eq. 15.5).

If the micropillar surface is composed of a second-scale roughness, then the Cassie equation can be written as [13, 14]

$$\cos\theta_A = rf\cos\theta_Y + f - 1 \quad (15.9)$$

where r is the Wenzel roughness factor of the structures at the solid–liquid contact interface.

15.2.2 Contact Angle Hysteresis

The roll-off feature of superhydrophobic surfaces can be characterized by the roll-off angle which is defined by the tilt angle at which a water droplet sitting on top starts to move; it can also be characterized by the contact angle hysteresis which is the difference between the advancing angle and the receding angle when the water droplet starts to move. They are illustrated in Fig. 15.4.

The relation between the contact angle hysteresis and the tilt angle can be illustrated in the following equation:

$$\sin\alpha = \frac{\gamma d(\cos\theta_R - \cos\theta_A)}{mg} \quad (10)$$

where d is the diameter of the moving droplet, m is the mass of the droplet.

To achieve roll-off superhydrophobicity, the contact angle hysteresis must be controlled. Only when the hysteresis is small can the surface be self-cleaning (usually <10°). If the surface contact angle hysteresis is high, the droplet will be difficult to roll off; instead, it will just stick to the rough surface. Therefore self-cleaning effect cannot be effectively achieved.

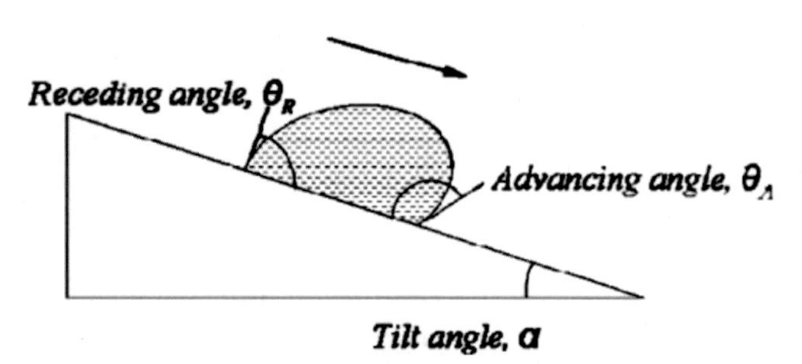

Fig. 15.4 The illustration of the tilt angle and the advancing and receding angles

15.3 Low Surface Energy Materials

A variety of materials have been used to prepare superhydrophobic surfaces including both organic and inorganic materials. For polymeric materials, which are generally inherently hydrophobic, fabrication of surface roughness is the primary focus. For inorganic materials, which are generally hydrophilic, a surface hydrophobic treatment is performed after the surface structures are fabricated.

Most polymeric materials can be used to fabricate superhydrophobic surfaces. For instance, plasma etching has been employed to generate roughness on polymer surfaces to achieve superhydrophobicity; polyethylene was etched by CF4/O2 plasmas [15] or direct catalytic polymerization process [16], etc. In addition, transparent and superhydrophobic surfaces can be achieved on poly(ethylene terephthalate) surfaces [17]. Examples of inorganic materials that have been used to create superhydrophobic surfaces are silica, alumina, hepatite, boehmite, titania, and copper which can be superhydrophobic if the roughened surfaces are treated appropriately with a hydrophobic material [18, 19]. In this sense, even for inorganic, the top surface can be covered with an organic layer or monolayer for the reduced surface energy.

Low-energy materials are particularly important in achieving stable superhydrophobic surfaces; among these materials, fluoropolymer/fluorocarbon is one of the lowest. The F–C bond is one of the strongest covalent bonds. In addition, the bulky F atoms cover well the C atoms beneath the surface (compared to H atoms), thus preventing chemical attack of the weaker C–C bonds. These facts lead to the significant potential of fluorocarbon layers for self-cleaning coatings. Indeed, Teflon has been roughened to produce superhydrophobic surfaces. Short-chain fluorocarbon silanes have been also investigated to form monolayers on a rough surface to achieve superhydrophobicity (10 Å thickness is enough to give a hydrophobic layer without any effect of the underlying substrates) [20]. Studies have shown that the surface energy with CF3 groups is lower than that of CF2 covered surfaces [21]. Another very important low surface energy material is silicone. The siloxane backbone shows very high bond energy, and the side CH3 gives the low surface energy.

15.4 Surface Structure Effect

For superhydrophobic surfaces, the surface structure/roughness is essential; the geometrical shape is also critical in achieving the desired surface superhydrophobicity. Patankar investigated the energy barrier between the Cassie and the Wenzel states for surface geometries of inclined structures. He demonstrated that an energy barrier to prevent the transition from a Cassie high energy state to a Wenzel low energy state as shown in Fig. 15.5 may not necessarily be present; that is, the transition may be spontaneous.

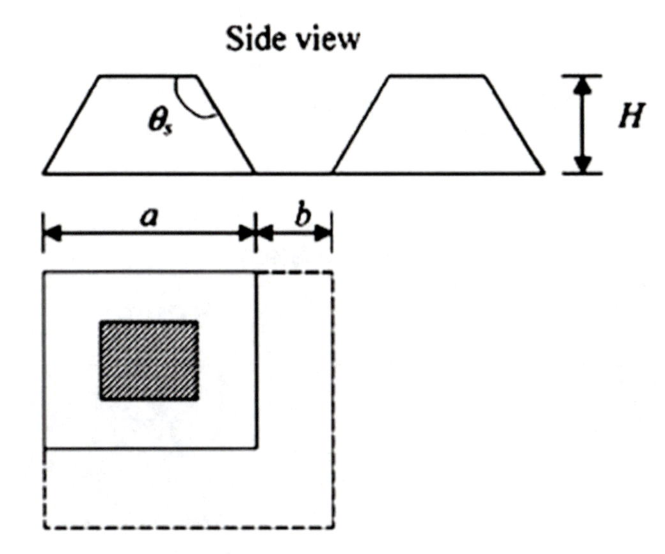

Fig. 15.5 Side and top views of geometrical roughness obtained by periodically placed pillars with inclined side walls [26]

A Cassie drop will be formed only if a liquid–vapor interface can connect from pillar to pillar without touching the valleys. Such an interface can establish equilibrium at the top corners of the pillars only if $\theta s \leq \theta e \leq 180°$. For other values of θe, a Cassie drop is not possible, and the valleys will always be filled with liquid.

A pyramidal Si surface represents a good model surface for the investigation of this structure inclination effect. It was shown mathematically that the inclination angle is very important in maintaining the Cassie state. As the inclination angle θs increases, the difficulty in achieving a Cassie superhydrophobic state increases. As predicted, the liquid–vapor interface on the composite surface cannot be maintained, and the droplet is always in the Wenzel state. However, by implementing a second scale of surface roughness on the pyramid surface, the liquid–vapor interface can be maintained on the pyramid top, and a Cassie state can be effectively achieved.

A superoleophobic surface has been also reported by using the structure shown in Fig. 15.6a. In order to achieve oil repellency, a reentrant (or overhang) structure design was employed. On this surface, a high contact angle can be achieved even though the surface is hydrophilic ($\theta < 90°$). The reentrant structure was also fabricated in electrospun polymer/polyhedral oligomeric silsesquioxane (POSS) composite fibers as shown in Figs. 15.6b and 15.7.

Similar results have been also reported by fabrication of a nanonail structure on Si surfaces using plasma etching techniques [23]. Instead of attempting to guarantee that the Cassie–Baxter state always remains the minimum-energy state, this approach created a system in which the height of the energy barrier that separates the metastable Cassie–Baxter state from the stable Wenzel state is designed to be

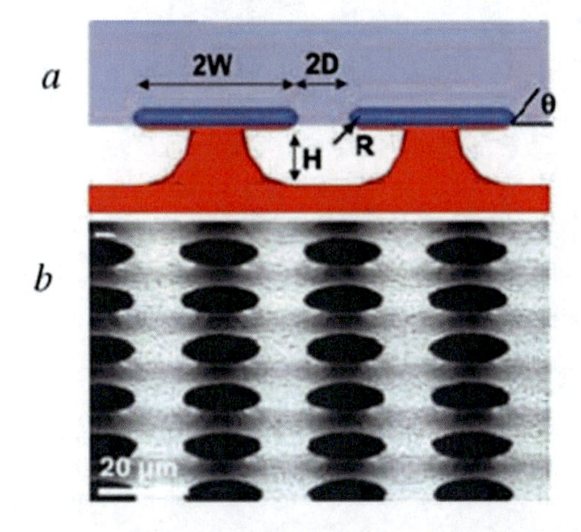

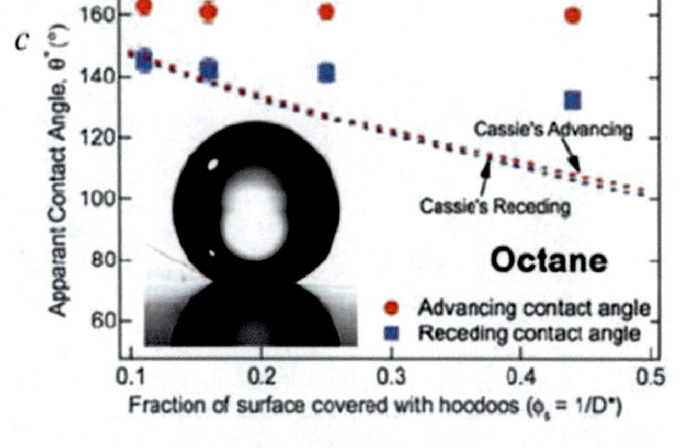

Fig. 15.6 (**a**) Cartoon highlighting the formation of a composite interface on surfaces with reentrant topography. The geometric parameters R, D, H, and W characterizing these surfaces are also shown. The *blue* surface is wetted, while the *red* surface remains nonwetted when in contact with a liquid whose equilibrium contact angle is θ Y (<90°); (**b**) SEM micrographs for so-called micro-hoodoo surfaces having circular flat caps. The sample is viewed from an oblique angle of 30°; contact angles for octane on silanized micro-hoodoos as a function of f (ϕs) [22]

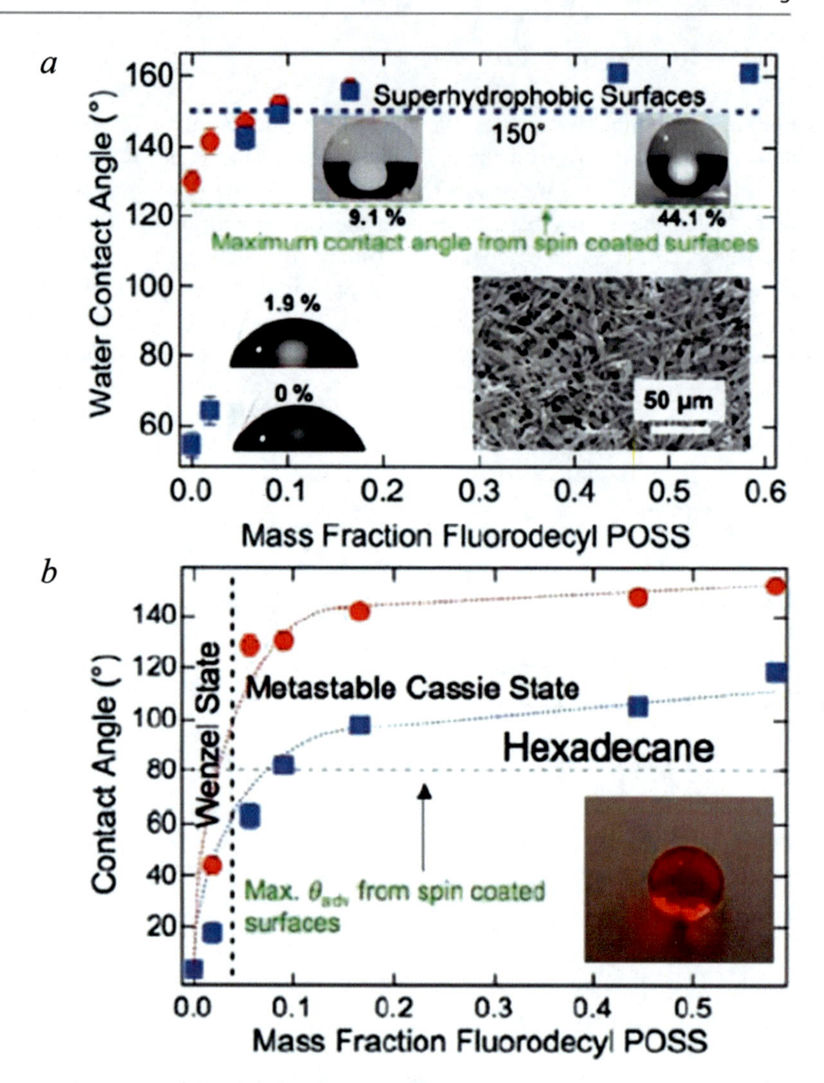

Fig. 15.7 (**a**) θadv (*dots*) and θrec (*squares*) for water on electrospun polymer surfaces. The *inset* shows a SEM micrograph for an electrospun surface containing 9.1 weight % POSS; (**b**) θadv (*dots*) and θrec (*squares*) for hexadecane on the electrospun surfaces, the *inset* of (**a**) shows a drop of hexadecane (dyed with oil red O) on a 44 weight % fluorodecyl POSS electrospun surface [22]

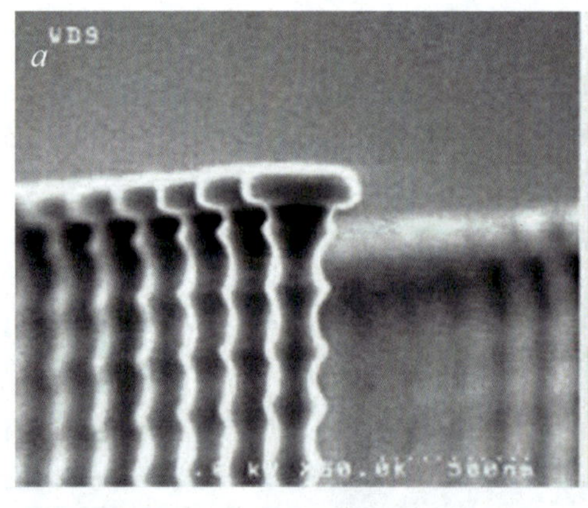

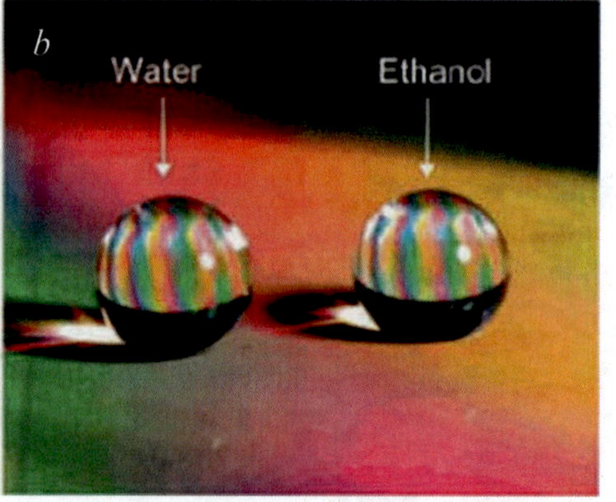

Fig. 15.8 (**a**) Scanning electron microscopy (SEM) image of 2-µm-pitch nanonails [23]. The nail head diameter D is ~405 nm, the nail head thickness h is ~125 nm, and the nanonail stem diameter d is ~280 nm; (**b**) a nanonail-covered substrate in action. Droplets of two liquids with the very different surface tensions, 72 mN/m (water) and 21.8 mN/m (ethanol), sit next to each other on the 2-µm-pitch nanonail substrate

sufficiently high to achieve effective pinning of the liquid in the desired nonwetting state. The idea is to create a special type of three-dimensional surface topography that normally inhibits transitions from the Cassie–Baxter state to the Wenzel state, but that would allow such transitions under the influence of external stimuli, such as electrical fields. A particular example of such topography is the "nanonail" structure in Fig. 15.8.

Porous Si surfaces from Au-assisted electroless etching process have shown the ability to achieve an oil-repellent surface [24, 25]. These structures show similar overhang character to that for nanonail structures as shown in Fig. 15.9; droplet contact angles of water and various organic solvents are shown in Fig. 15.10.

Achievement of high contact angles for both polar (water) and nonpolar droplets (e.g., hexadecane) requires the design of surface structures similar to those shown in Fig. 15.10. In order to achieve mechanical stability and optimized hierarchical surface structures for enhanced self-cleaning on superhydrophobic surfaces, it is critical to understand the fundamental surface physics, especially the surface–structure effect on superhydrophobicity to improve control over the surface structures (geometrical shapes, pitch, height, aspect ratio, hierarchical design, etc.). These design criteria can then be used to achieve the requirements necessary for various applications.

As described previously, for surface superhydrophobicity, there are two requirements. The first is high contact angle (>150°), which relates the contact angle to the surface structure (expressed by the Wenzel surface roughness factor *r*) and can be described by the improved Cassie–Baxter equation [13, 26]. The second parameter is the contact angle hysteresis, which is defined by the difference between the surface advancing contact angle and the surface receding contact angle.

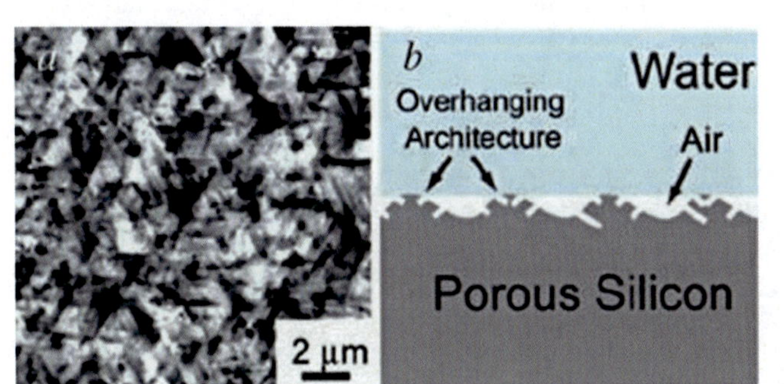

Fig. 15.9 Porous Si surface fabricated by Au-assisted electroless etching. (**a**) Top view SEM image and (**b**) schematic cross-sectional profile of water in contact with the porous Si surface [24]

15.5 Approaches to Preparing Superhydrophobic Surface Coatings

15.5.1 Plasma Etching Techniques

Plasma techniques are widely employed in microelectronics processing (Si surface pattern etching, cleaning, removal of photoresists, and dielectric film deposition such as inorganic or polymer materials) [27]. Plasma etching techniques are also capable of surface roughness generation and therefore offer a method to fabricate superhydrophobic surfaces.

15.5.1.1 Plasma Etching of Fluoropolymers

Using plasma etching techniques, polymer materials ranging from PTFE to silicones to polyethylene can be etched to produce a rough surface. To roughen PTFE, the technique of radiofrequency sputtering has been reported [4, 28]. For long sputtering times, a superhydrophobic Teflon surface was prepared by use of an O2 plasma; the resulting surface showed a water contact angle of 168° [29]. Plasma ion beam treatment using O2 and Ar has been also reported in preparation of superhydrophobic PTFE coatings [30] with a surface morphology as shown in Fig. 15.11.

By using a simple plasma-based technique that combines etching and plasma polymerization on Si substrates, superhydrophobic surfaces have been fabricated by tailoring their surface chemistry and surface topology [31]. This technique showed the capability of large area processibility and good reproducibility. Using SF6 as an etchant, surface roughness can be produced by over-etching the photoresist layer.

Fig. 15.10 Static contact angles of water, diethylene glycol, and hexadecane on flat silicon (Si), porous silicon (PSi) with tilted pores, flat Si coated with FTS (FTS-Si), and porous silicon with tilted pores coated with FTS (FTS-PS) [24]

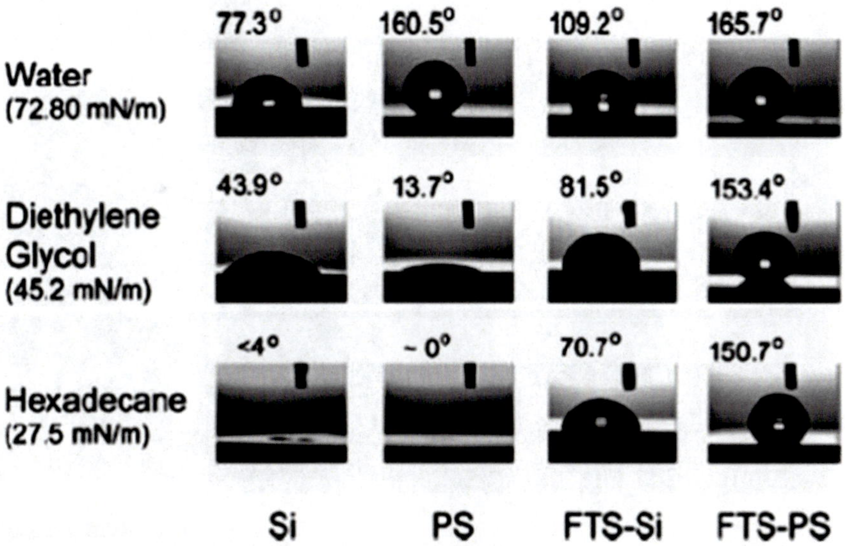

Overall, the method to generate superhydrophobic surfaces consisted of three steps. First, a 2.3-μm-thick photoresist polymer layer (microposit S1813 supplied by Shipley Company) is deposited by spin coating on a Si wafer. Then, photoresist layers are etched using an inductively coupled plasma with a magnetic pole-enhanced ICP (MaPEICP) source. Third, the CFx layer is deposited on etched photoresist/Si wafers from C4F8 using the same plasma reactor but with the 13.56 MHz r.f. capacitive plasma mode. The key step of this process is etching of the surface topography into the substrate to create high roughness before deposition of the fluorocarbon coating. Superhydrophobicity is achieved after the C4F8 plasma polymer deposition process.

Fig. 15.11 The SEM surface morphology of plasma-treated PTFE surface by Ar/O_2 (2 sccm/2 sccm) 1.5 KeV [30]

15.5.1.2 Plasma Etching of Si with Micromask

SF6 and CHF3 plasmas were used to investigate the surface roughness generation on Si surfaces for surface wettability modification [32]. Due to the nonuniform removal of photoresist (Microposit S1813, Shipley, spin coating at 2000 rpm), the residue that remains on the surface acts as localized masks to create roughness on the Si surface. After the surface roughness was formed as shown in Fig. 15.12, a CFx layer was deposited on top of the rough surface using plasma-enhanced chemical vapor deposition (PECVD). The surfaces treated by the CHF3 plasma are all superhydrophobic up to 10 min of etching, while SF6 etching will not produce superhydrophobic after 2.5 min etching; these results are due to differences in Si etch rate with the two fluorocarbon gases. After 5 min etching for SF6, the surface contact angle becomes 116°, which is approximately the contact angle on a flat surface for CFx coatings.

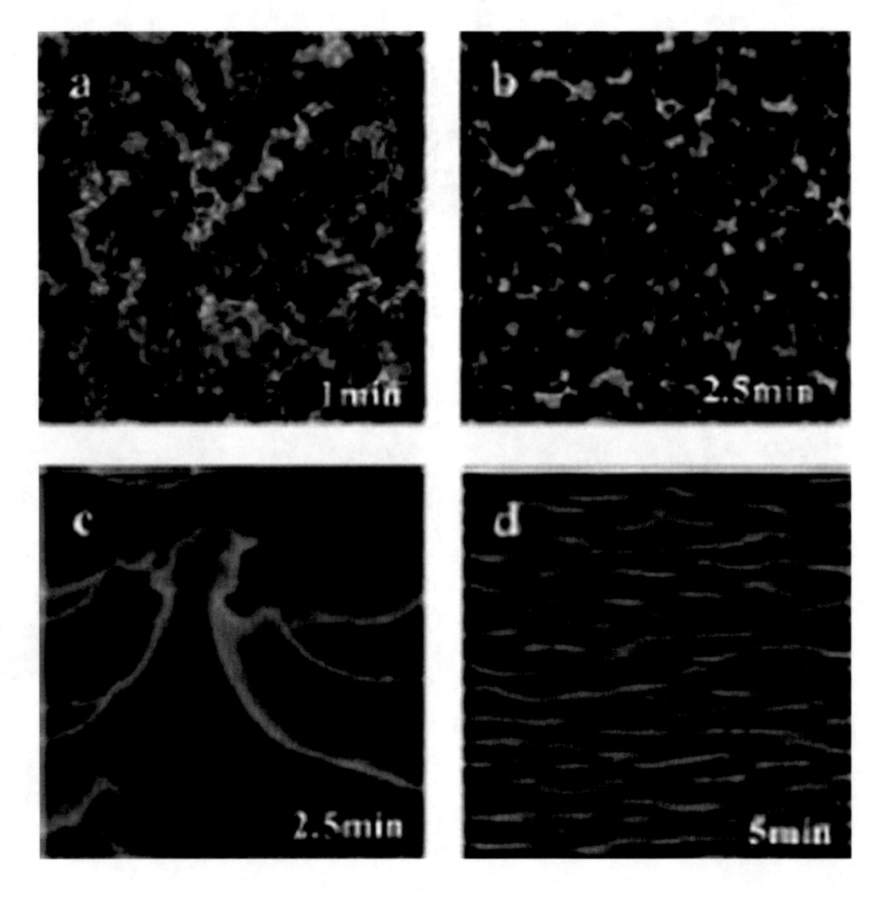

Fig. 15.12 SEM image of Si surface etched by SF6 for (**a**) 1, (**b**, **c**) 2.5, and (**d**) 5 min. Image (**c**) is a magnified view (with an observation angle of 45°) of one residual photoresist particle on a peaked feature. The areas in images (**a**), (**b**), and (**d**) are 20 × 20 μm; the area of image (**c**) is2 × 2 μm [32]

A detailed analysis of roughness generation was conducted in photoresist-masked isotropic Si plasma etch process performed in an inductively coupled plasma (ICP) etcher with SF6, by systematically changing the pressure (2.5–70 mTorr), SF6 flow rate (50–300 sccm), platen power (0–16 W), and ICP power (1000–3000 W) [33]. A high ICP power (3000 W) and a high flow rate (300 sccm) result in the lowest normalized bottom roughness. High pressures and low flow rates increase the roughness. Increasing the ion energy by applying a radiofrequency platen power of 16 W has a great impact on the roughness. Actually the bottom surface becomes completely smooth. The roughness morphology at these process conditions shows a crystallographic anisotropy, where (111) planes are revealed. When an oxide mask was used, the surface roughness was reduced compared to that for a photoresist mask. When there is no mask, the surface roughness is very low which suggests an effect due to the photoresist mask layer (which redeposits during etching) on the surface roughness.

15.5.1.3 Plasma Etching of Polydimethylsiloxane (PDMS)

Polydimethylsiloxane is a very low surface energy material with a surface energy of 21–22 (mJ/m^2), slightly higher than that of *poly(tetrafluoroethene)* (PTFE) (~18 mJ/m^2). However, after plasma treatment, the surface may show a much higher surface energy. For instance, superhydrophobic PDMS has been fabricated by an SF6 plasma treatment followed by fluorocarbon film deposition that generated hydrophobic high aspect ratio columnar-like nanostructures as shown in Fig. 15.13 [34].

In addition, under appropriate plasma processing conditions (SF6 plasma etching for a sufficient time to generate the appropriate surface roughness), superhydrophobic and simultaneously transparent surfaces can be produced [34], which is of great practical importance. In particular, nanostructured PDMS surfaces are visibly transparent for treatment times up to 2 min (height of column: 130 nm), although they become opaque for etching times longer than 4 min.

15.5.1.4 Plasma Etching of Polyethylene Terephthalate (PET) (Domain-Selective Plasma Etching) and Low-Density Polyethylene (LDPE, Different Etching Rate for Crystalline and Amorphous Phase)

To establish superhydrophobicity of PET, a method consisting of a two-step process comprised of nano-texturing by an oxygen plasma treatment and subsequent deposition of a hydrophobic coating by means of low-temperature chemical vapor deposition or plasma-enhanced chemical vapor deposition was employed to form ultra water-repellent polymer sheets [17, 35, 36]. After oxygen etching, many protrusions are observed on the polymer surface. In general, noncrystalline domains are more readily etched than the crystalline domains. PET contains crystalline and amorphous domains and the protrusions of plasma-etched PET were probably formed as a result of domain-selective plasma etching. This surface nano-texture remained after deposition of the hydrophobic coatings using organosilane precursors; surface topologies are shown in Fig. 15.14. The surface-modified substrate was transparent (the polymer is optically transparent) and ultra water-repellent, showing a water contact angle greater than 150°.

By selectively patterning a surface (deposited fluorocarbon polymer), a micro-patterned polymer substrate with superhydrophobic/superhydrophilic domains as shown in Fig. 15.15 can be prepared. Such surfaces are extremely useful in chemical and biological sciences to keep surfaces from retaining or being fouled by samples and reagents.

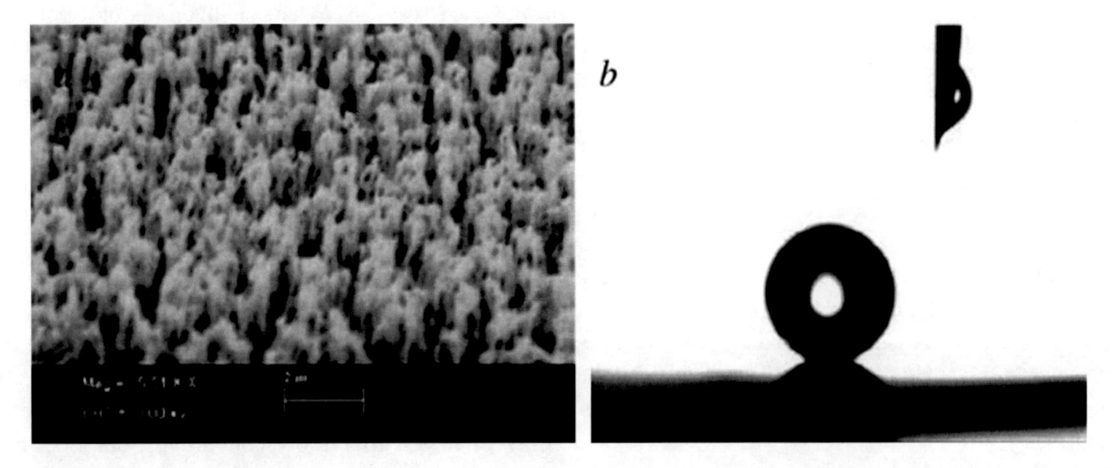

Fig. 15.13 SEM image of a PDMS elastomer (Sylgard 184) surface after a 6 min SF6 plasma treatment. 1.45 μm-high nano-columns are shown. (**b**) Image of a water droplet rolling off the surface in (**a**) after being conformally coated with a 20 nm-thick FC film [34]

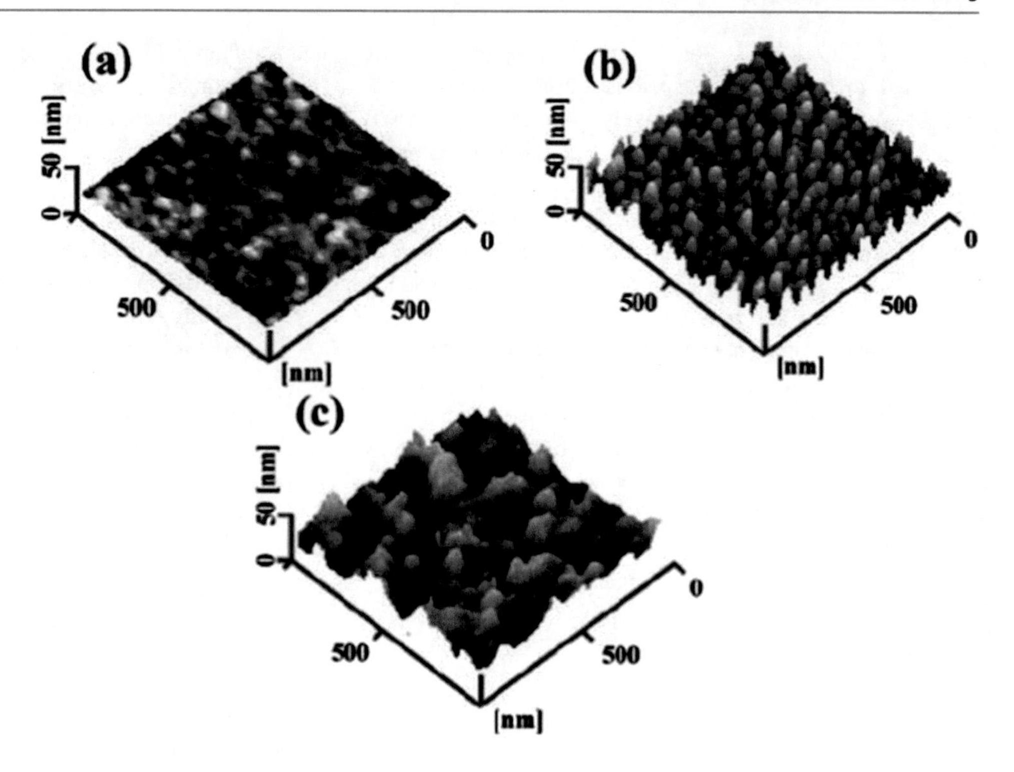

Fig. 15.14 AFM of PET (**a**) untreated, (**b**) O_2 plasma treatment, and (**c**) FAS coating after O_2 plasma [36]

Fig. 15.15 Environmental scanning electron microscope (ESEM) image of micro-patterned water droplets on a superhydrophobic/superhydrophilic surface prepared by a multistep drying process [37]

To achieve superhydrophobicity on LDPE surfaces, a similar process has been employed due to the fact that LDPE also shows noncrystalline and crystalline domains [38–40]. Two routes to achieve superhydrophobicity are possible. The first corresponds to a one-step synthesis (CF4 plasma modification of LDPE). The second route involves two steps (O2 plasma treatment followed by CF4 plasma deposition onto LDPE). However, these surfaces tend to lose their superhydrophobic behavior after water vapor condensation.

Superhydrophobicity can also be achieved by polystyrene (PS) etching in CF4– O2 [41]. By controlling the etching parameters, the surface wetting characteristics can be tailored from sticky to slippery superhydrophobicity. Using this process, the optical transparency can be maintained on the PS surfaces (Fig. 15.16).

Woodward et al. used a two-step CF4 plasma etching and fluorination process followed by curing to prepare a fluorinated rough polybutadiene surface (Fig. 15.17) [42]. This surface exhibited superhydrophobicity with a water contact angle of $\sim$175° and contact angle hysteresis of 0.4° [41].

Dorrer et al. reported a method to manipulate superhydrophobicity/superhydrophilicity by precise control of the surface chemistry using various polymer coatings on Si nanograss surfaces generated by anisotropic plasma etching (Fig. 15.18) [43]. When a hydrophobic polymer poly (heptadecafluorodecylacrylate) (PFA) was used to coat Si nanograss surfaces, superhydrophobicity (179°, no observable hysteresis) was achieved, and the surface showed "condensation resistance" to water vapor.

15.5.2 Chemical Vapor Deposition (CVD) Process

Silicone nanofilaments have been formed from trichloromethylsilane (TCMS) precursor via a gas-phase reaction route at room temperature and pressure without a carrier gas [44]. Silanes with hydrolysable groups (chloride) react with water to form silanols, which then couple to hydroxyl groups at the surfaces, and subsequently polymerize to a silicone

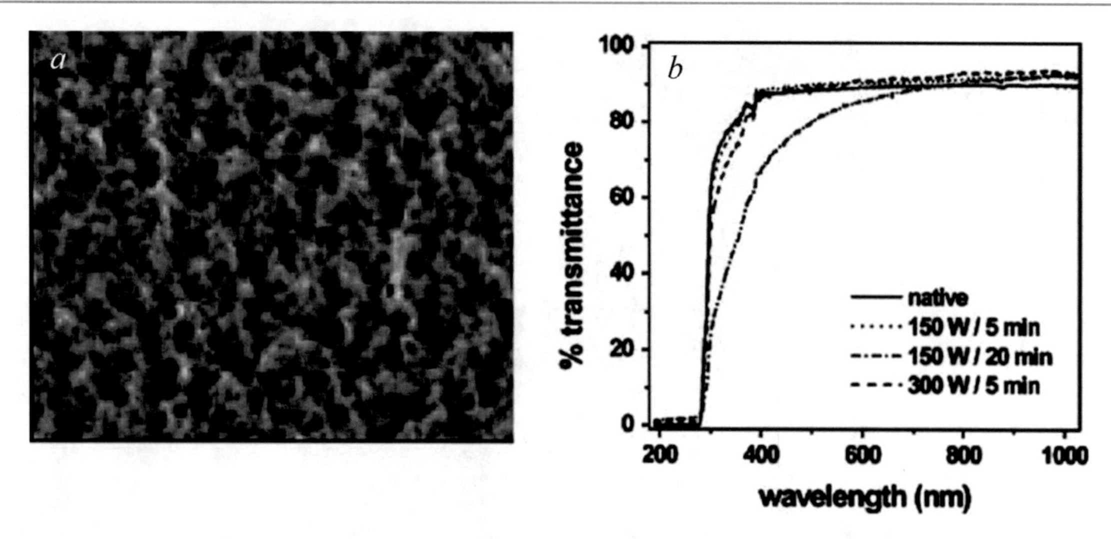

Fig. 15.16 (**a**) SEM image of PS samples treated with a CF4–O_2 discharge with 17% O_2 for 5 min [41]; 300 W 5-min plasma etching; (**b**) transmittance of the prepared surfaces

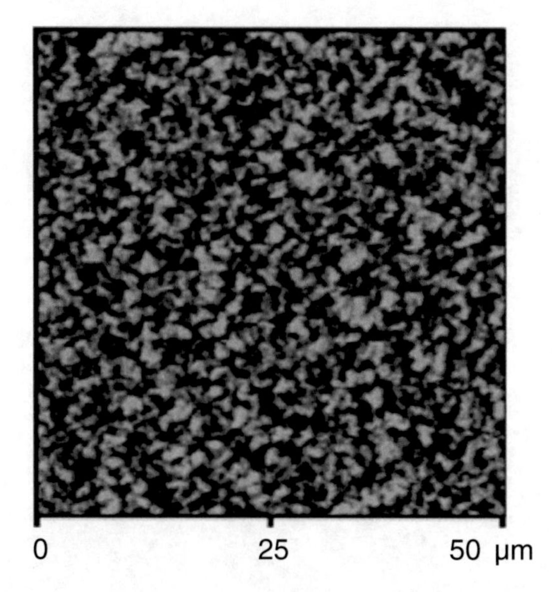

Fig. 15.17 AFM height image of polybutadiene in CF4 plasma (power 50 W) for 900 s [42]

layer, depending on the reaction conditions. When dry reaction conditions with only traces of water were maintained, monolayers or nearly monolayer thick coatings resulted. When equimolar amounts of liquid TCMS and water vapor were used in the reaction vessel, a superhydrophobic surface was achieved; the coating then consisted of polymethylsilsesquioxane nanofilaments. This technique can be used to coat a variety of surfaces such as cotton fiber, wood, polyethylene, ceramics, titanium, and aluminum. In addition, on glass surfaces, a transparent coating can be achieved as shown in Fig. 15.19. When the as-deposited surface was treated in an O2 plasma followed by a treatment with fluoroalkylsilane, oil-repellent surfaces resulted [45] as shown in Fig. 15.19d.

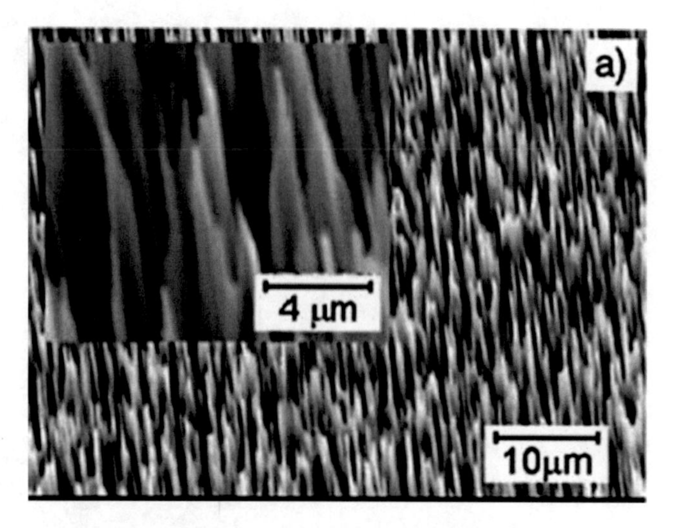

Fig. 15.18 Electron microscopy images of a nanograss surface coated with a PFA thin film [43]

Zhu et al. reported a method of superhydrophobic materials preparation by using two-tier CNT surfaces formed by growing patterned CNT arrays on CNT covered Si surfaces via CVD (Fig. 15.20a) [46]. Because of the presence of nanoroughness on the CNT film/array surface, a two-tier roughness can be achieved. Therefore, after coating the surfaces with a fluorocarbon film, superhydrophobicity was achieved with a contact angle of 167° and a hysteresis of 1°. A CNT nanoforest has been also prepared by PECVD (Fig. 15.20b) [47]. After the surface was coated with PTFE, superhydrophobicity was achieved down to the microscopic level. Even micrometer-sized water droplets can be suspended on top of the nanotube forest.

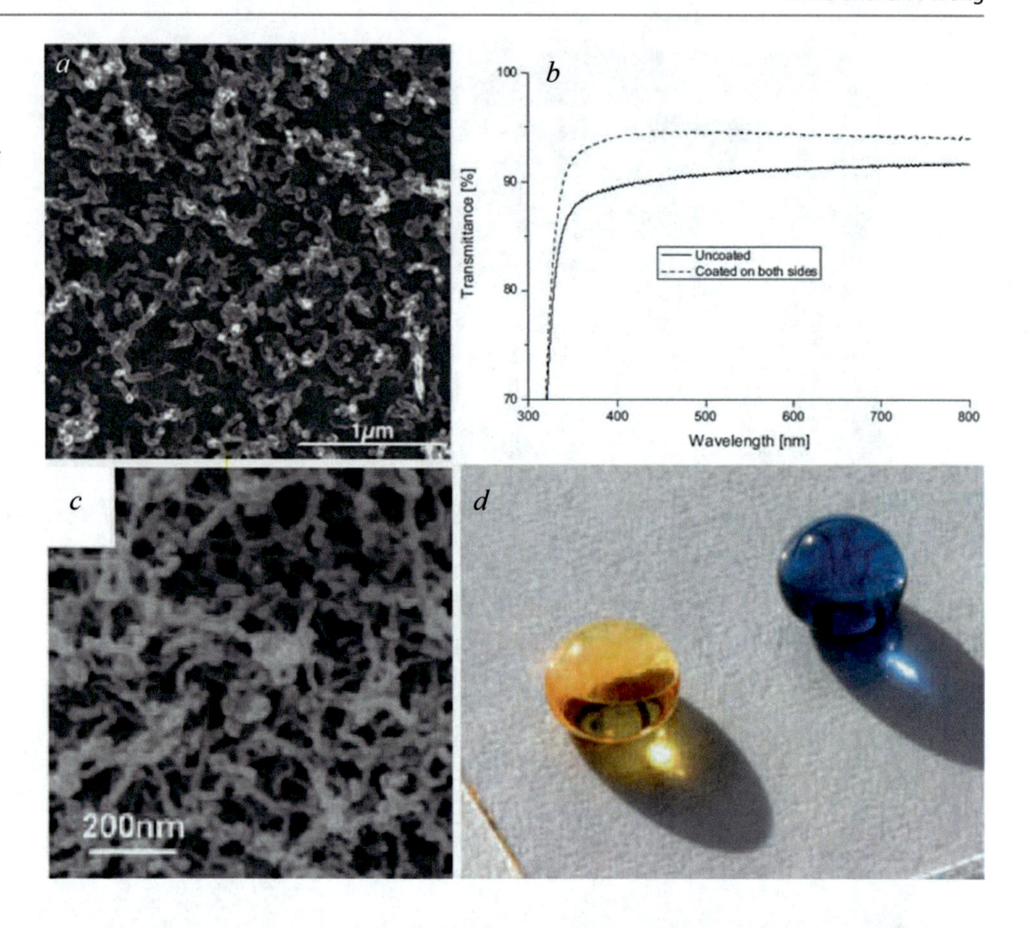

Fig. 15.19 Scanning electron microscopy images of silicone nanofilaments on Si. (**b**) UV–vis spectra of coated and uncoated glass slides [44]; (**d**) 10-ml drop of colored hexadecane (*yellow*) and colored water (*blue*) on a PFOTS-modified silicone nanofilament coating on glass [45]

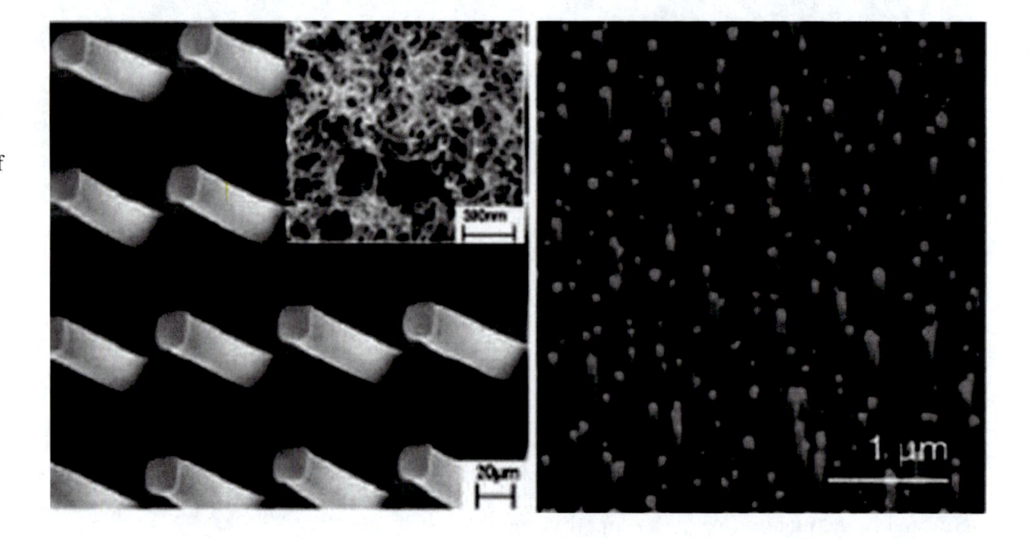

Fig. 15.20 (**a**) Typical SEM images of CNT arrays on Si substrates; (**b**) SEM images of carbon nanotube forests prepared by PECVD with nanotube diameter of 50 nm and a height of 2 μm [46, 47]

15.5.3 Sol–Gel Process

Sol–gel processes can also produce rough surfaces on a variety of oxides such as silica, alumina, and titania [48, 49]. This approach is a very versatile process for the preparation of superhydrophobic thin films or bulk materials. The sol-gel process can form a flat surface coating, xerogel coating, or aerogel coating depending on the process conditions; both xerogel and aerogel show rough or fractal surfaces.

Silica aerogels can contain up to 15 wt% of absorbed air and water vapor. In order to decrease the pore affinity to absorption and thus make aerogels less hydrophilic, a number of strategies have been used, such as avoiding the presence of terminal hydroxyl groups by co-gelling certain Si precursors containing at least one nonpolar chemical group, i.e., CH3–Si

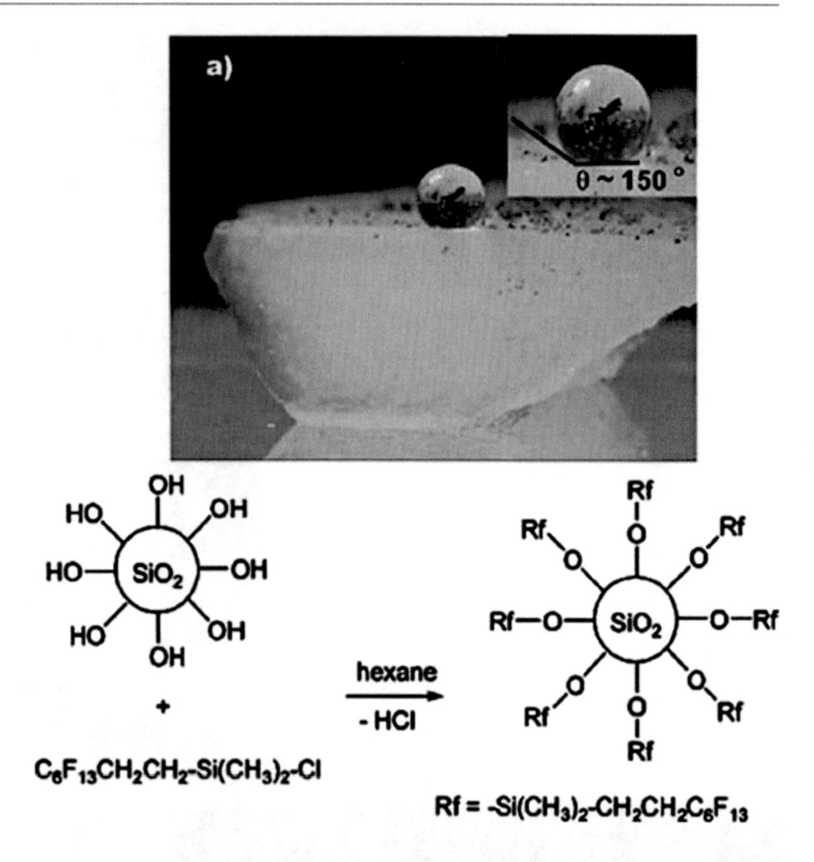

Fig. 15.21 Drop of water on top of fluorinated aerogel, inset shows a close-up [48]

[50]. Superhydrophobic silica aerogels can also be obtained by surface modification of standard silica gels with a heavily fluorinated silyl chloride followed by supercritical fluid solvent removal [48]. After either ambient drying or supercritical drying, both surfaces show superhydrophobic properties (Fig. 15.21).

Attempts have been made to synthesize highly flexible and superhydrophobic silica aerogels using methyltrimethoxysilane (MTMS) as a precursor by a two-step acid–base sol–gel process [51] (Figs. 15.22 and 15.23). The aerogels consist of crosslinked networks of silica polymer chains extended in three dimensions. Due to the presence of non-polar alkyl groups (i.e., methyl) attached to the silica polymer chains, the interchain cohesion is minimized resulting in an elastic and flexible three-dimensional network. Increased dilution of the MTMS precursor with methanol solvent yielded a silica network with a low degree of polymerization which exhibited higher flexibility. In contrast, for lower dilution of the MTMS precursor, an extensive polymerization resulted in dense and rigid structures. Because of the new property, i.e., flexibility, imparted to the aerogel, it can be bent into a variety of shapes and acts as a good shock absorber as well. A water droplet contact angle of 164° can be achieved on such aerogel surfaces.

Another method to produce superhydrophobic silica materials is the phase separation method in which a hardening process "freezes" a phase separation that occurs

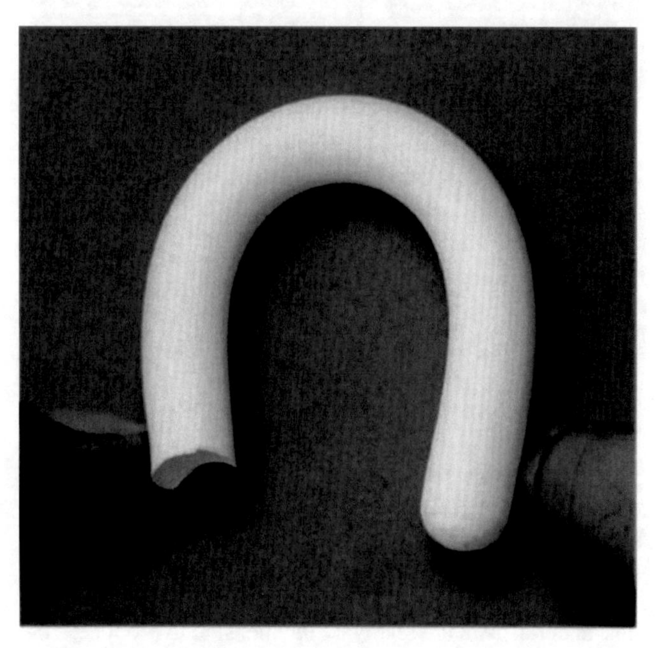

Fig. 15.22 Flexible silica aerogels prepared at two different MeOH/MTMS molar ratios of 35 [51]

concurrently with hardening of one of the phases [52]. This method produced co-continuous materials consisting of a solid phase and a liquid phase. When the liquid was removed, a porous structure remained. The materials system was comprised of condensed organotriethoxysilane in a mixture

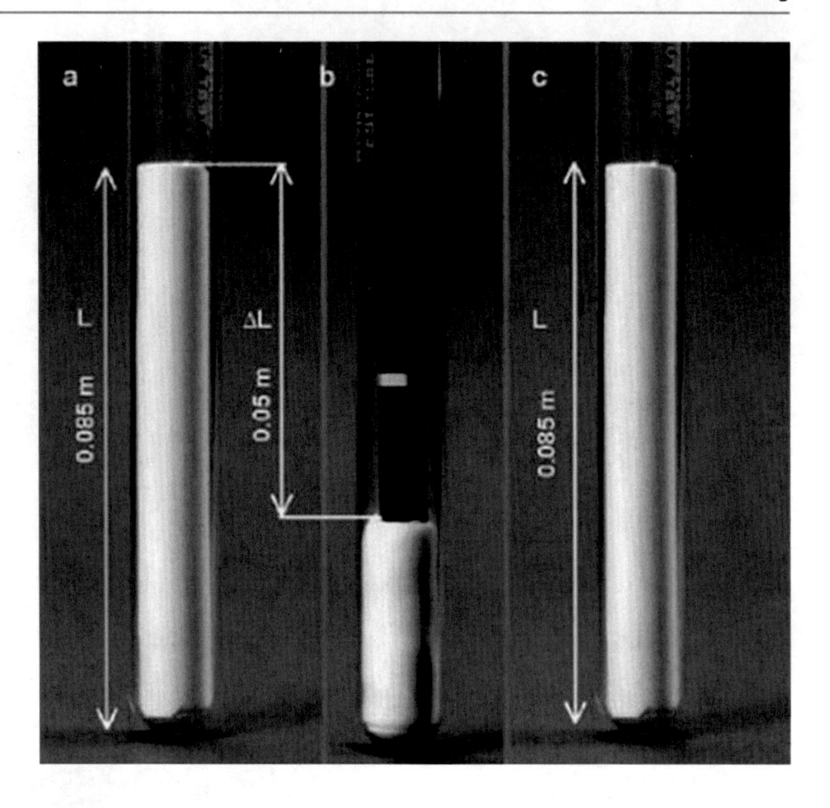

Fig. 15.23 Photograph of the three states of a flexible aerogel sample: (**a**) without stress, (**b**) with stress, and (**c**) after releasing the applied stress [51]

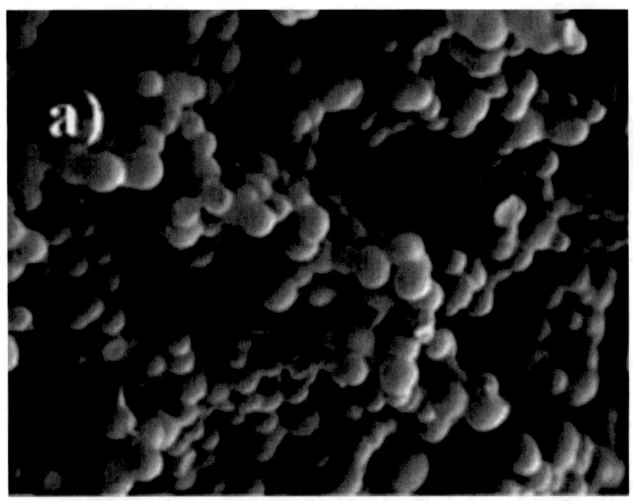

Fig. 15.24 Electron micrographs of gold-coated foams: MTEOS with 1.1 M ammonia, heated to 300 °C [52]

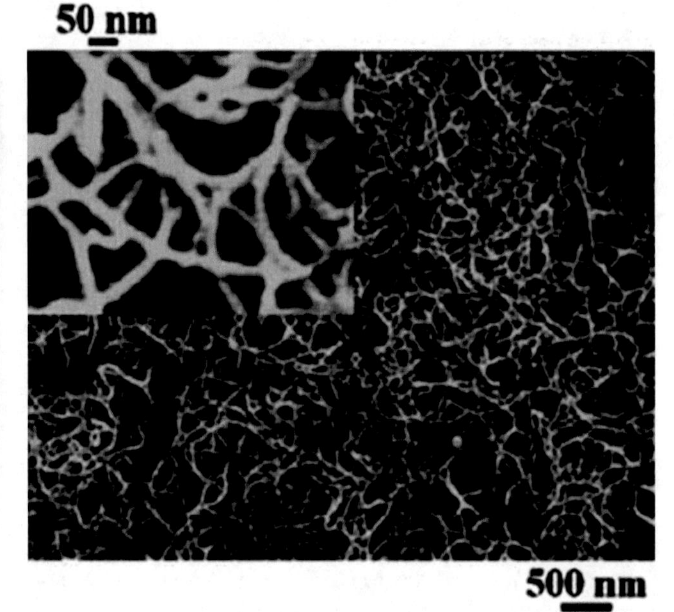

Fig. 15.25 SEM images of a superhydrophobic surface from CH_3SiCl_3 [53]

of organic solvent and water. Such sol-gel materials show a great promise as stronger superhydrophobic surfaces and as bulk materials and are relatively inexpensive to produce. The reaction occurs through hydrolysis of the ethoxy groups and polymerization of the resulting silanol groups. Polymerization causes a decrease in dipole moment, leading to hydrophobic phase separation. The dried material has the organic groups on its surface, causing the foams to be superhydrophobic (Fig. 15.24).

Despite abrasion of these sol-gel surfaces, the high water droplet contact angle can be maintained. This result is significant for applications wherein handling and mechanical abrasion are inevitable. Even at temperatures as high as 300 °C, no loss of superhydrophobicity is detected.

Gao et al. reported a method to fabricate a perfect superhydrophobic surface ($\theta A/\theta R = 180°/180°$) by immersion of a Si wafer into CH3SiCl/toluene solution; rinsing with toluene, ethanol, and water; and drying the surface coating [53]. The surface morphology generated is shown in Fig. 15.25.

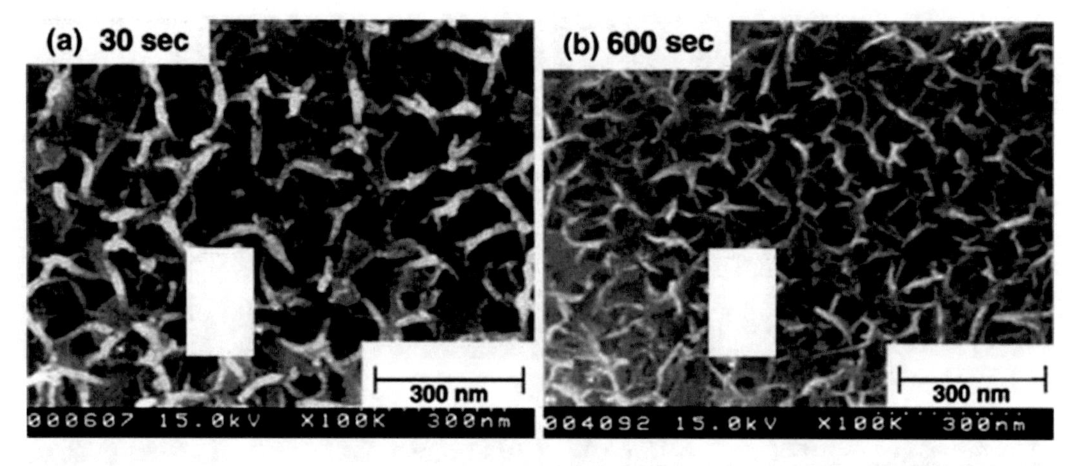

Fig. 15.26 FE-SEM photographs of the surface of the Al_2O_3 films immersed in boiling water for (**a**) 30 s and (**b**) 600 s [49]

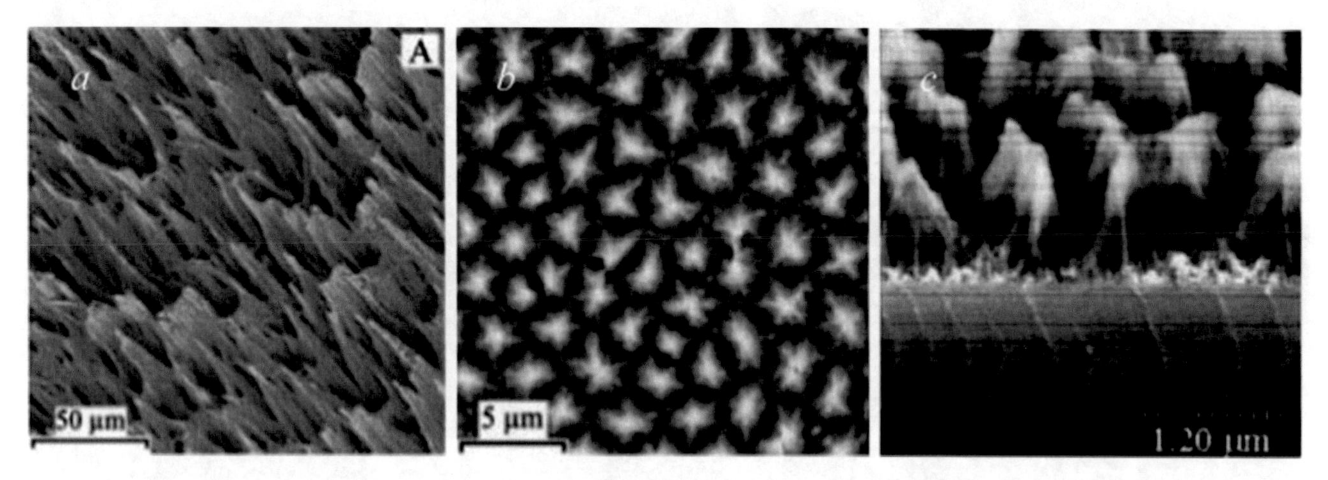

Fig. 15.27 SEM images of etched Si surfaces (**a**) from $AgNO_3$/HF etching, 0.02/5.0; (**b**) from $Ni(NO_3)_2$/HF etching 0.08/5.0 [56]; and (**c**) evaporated Pt on Si, etching in HF/H_2O_2 [54]

Alumina is a hard material that is abrasion resistant. Using a sol-gel process, nanoscale surface structures can be generated by various methods (Fig. 15.26). Tadanaga et al. showed that by immersing the porous alumina gel films prepared by the sol-gel method in boiling water, alumina thin films with a roughness of 20–50 nm were formed [49]. As a result, high optical transparency can be achieved. After a surface hydrophobic treatment, a superhydrophobic surface with a contact angle of 165° resulted.

15.5.4 Wet Chemical Etching

In addition to plasma etching, wet chemical etching has been also employed to generate surface roughness, frequently with using Si as a substrate. Wet chemical etching can be used for a variety of applications in both polishing and roughening. If the etch conditions are controlled properly, a rough surface can be prepared. For example, KOH etching of Si(100) can be controlled to achieve a pyramid-textured surface.

15.5.4.1 Metal-Assisted Etching

The metal-assisted etching process on Si surfaces involves (1) deposition of metal nanoparticles (Au, Pt, Ag, Pd, etc.) on the Si surface and (2) etching of the metal coated Si in HF/oxidant solutions [54]. During this process, Si is continuously etched away at the site of the metal/Si contacts and surface roughness formed by control of the metal particle density. Deposition methods including direct metal deposition by sputtering or electron beam evaporation and electrochemical deposition of metal particles from reactive metal precursors. Figure 15.27 shows the result of using different metals for assisted etching on Si surfaces. Metal has been deposited onto Si surfaces by reduction of metal precursors such as HAuCl4 [25], AgNO3 [55], and Ni(NO3) 2 [56]. E-beam evaporation or DC sputtering can also be used to deposit metal onto Si surfaces for this etching process [54].

Superhydrophobic surfaces with hierarchical structures on Si using metal-assisted etching of Si pyramid surfaces (Fig. 15.28) [57, 58] by Xiu et al. The surface can be made

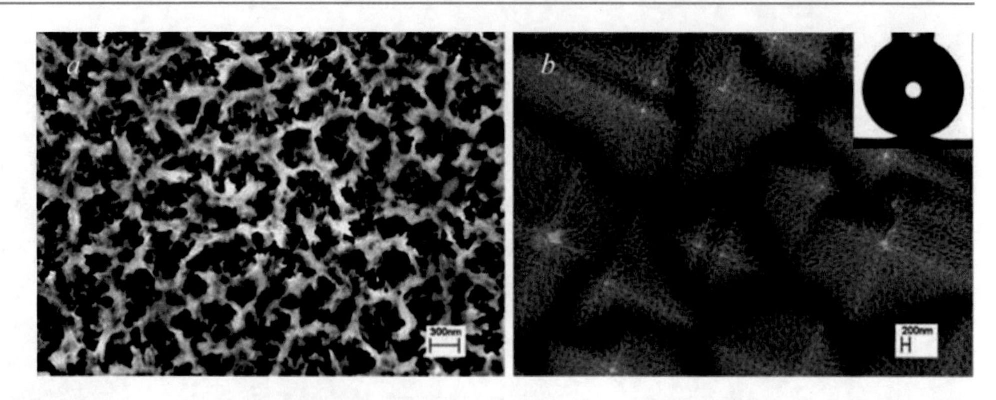

Fig. 15.28 (**a**) Si (100) surface etched by Au-assisted etching and (**b**) Si micro-pyramid surface etched by Au-assisted etching; inset: water droplet contact angle of ~165°

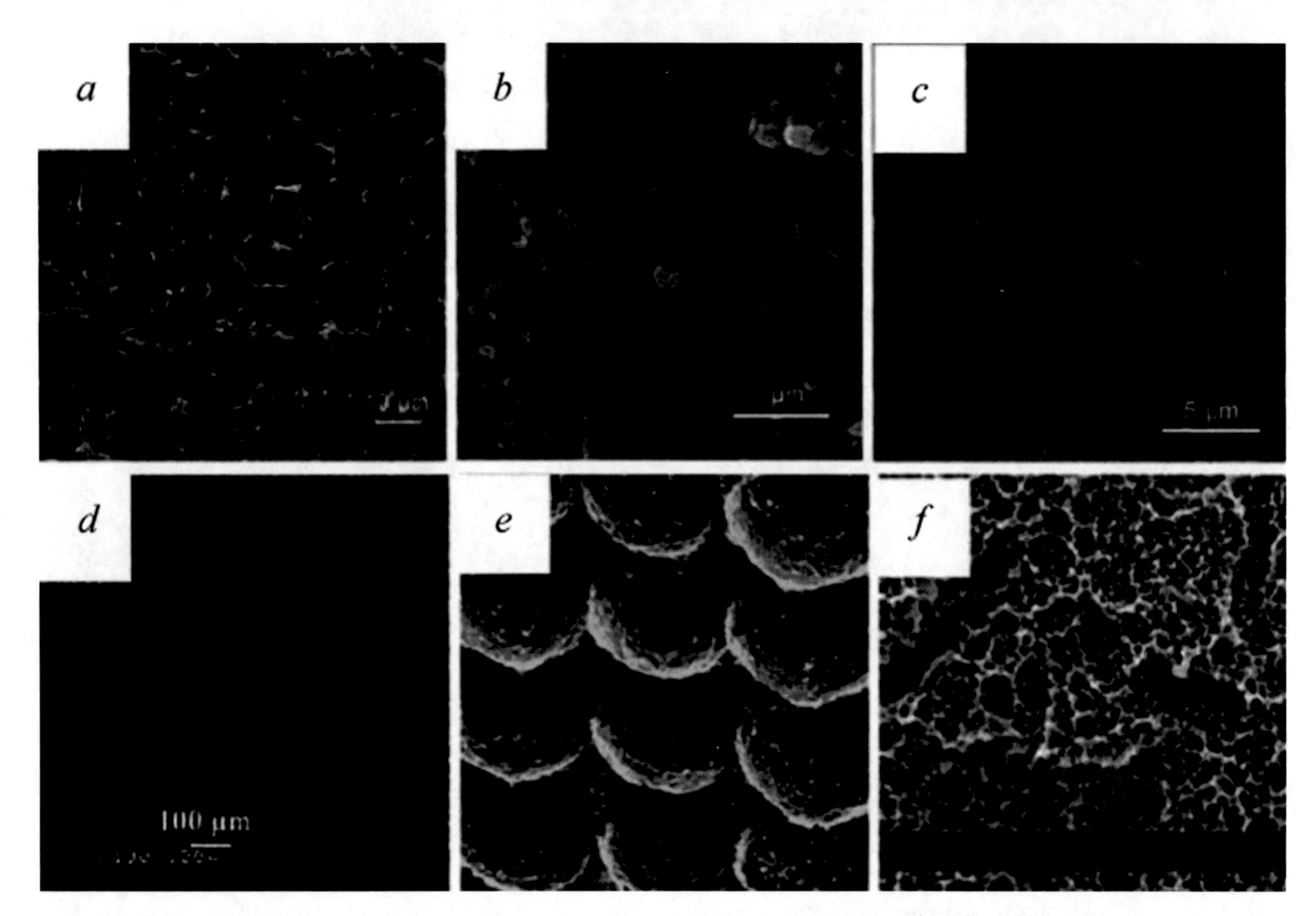

Fig. 15.29 SEM images of different etched surfaces: (**a**) zinc surface etched with 4.0 mol/L HCl solution for 90 s at room temperature [59]; (**b**) copper surface etched with a modified Livingston's dislocation etchant for 24 h at ambient temperature [59]; (**c**) Al etched by Beck's dislocation etchant for 10 s; (**d**) Al alloy (2024Al, 92.8 wt% Al, 5.5 wt% Cu) etching by NaOH [60]; (**e**) Cu surface [61], Cu etching was carried out in a potassium persulfate solution rough etched top and pits; (**f**) Ti surfaces from sandblasting and acid etching [62]

as not only superhydrophobic but also non-reflecting for light incident on to the surface (Fig. 15.29). Therefore, multifunctional surfaces are possible. This surface is promising to be employed in solar cell surface texturing and passivation when used with the appropriate chemical passivation on the surface.

15.5.4.2 Dislocation Selective Chemical Etching and Other Etching Methods

Dislocation selective chemical etching can be performed on the surface of polycrystalline metals such as aluminum, copper, and zinc to prepare rough surfaces as shown in Fig. 15.29 [59–62]. The key to this etching technique is the use of a

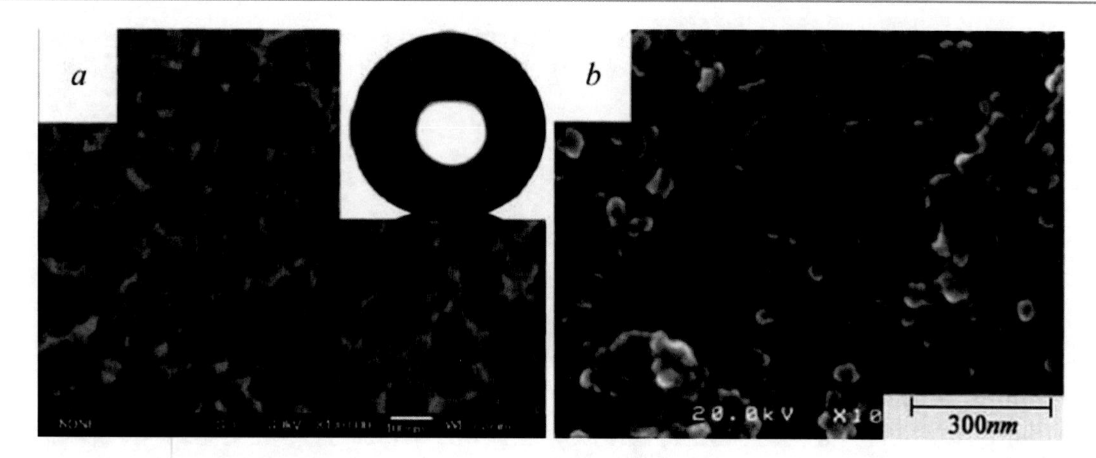

Fig. 15.30 Anodized surfaces: (**a**) Ti anodization, TiO_2 formed on surface [63]; (**b**) Al-anodized surface [5]

dislocation selective etchant that preferentially dissolves the dislocation sites in the grains. Water droplet contact angles higher than 150° and roll-off angles of less than 10° were obtained when surfaces were treated with fluoroalkylsilane.

Superhydrophobic surfaces can also be achieved by Al alloy etching using NaOH. Etching mainly occurred in the Al domain because of preferential dislocation etching; roughness is due to the remaining Cu and other elements on the surface. The as-prepared surface shows a contact angle of 154° and sliding angle of ~3° after hydrophobic treatment [60]. On Cu surfaces, the microstructures can be prepared by photolithography and subsequent etching using a potassium persulfate solution [61]. A Ti surface with low surface energy was prepared by sandblasting and acid etching (HCl/H2SO4) after a hydrophobic treatment [62] (Fig. 15.29f). The initial advancing contact angle on the surface is 139.9° without any hydrophobic treatment, which is the highest contact angle considering that there is no more hydrophobic treatment on the surface. However, the receding contact angle is <5°, and repeated measurement gave a contact angle of 0°.

Electrochemical anodization can also be used to generate surface roughness. For example, Ti was treated by electrochemical anodization to form a sponge-like nanostructured TiO_2 film on the surface [63] (Fig. 15.30a). After a hydrophobic surface treatment, the advancing angle on the surface was 160.1°, and the hysteresis was only 0.8°.

Al surfaces can also be anodized to prepare a fractal surface [5] (Fig. 15.30b). After a fluorinated monoalkyl phosphate (F-MAP) treatment, the surface showed a water droplet contact angle of 163° and rapeseed oil contact angle of 150°. The fluorination chain length was very important in this case, and the degree of oil repellency for the silane coupling agent was much less than that of F-MAP.

15.5.5 Other Methods

15.5.5.1 Monodisperse Nanoparticles

Colloidal crystals can be formed on a substrate from monodispersed nanospheres, e.g., SiO2 and polystyrene spheres, as a result of assembly into arrays of hexagonal close packed or FCC layers. In this scenario, the as-formed surface is rough due to the regular arrangement of the monodispersed spheres. Although this surface is not superhydrophobic even after a surface treatment with hydrophobic surface modification agents [64], superhydrophobicity can be achieved by imparting nanoscale structures on/in the surface to achieve multiscale roughness as shown in Fig. 15.31. Furthermore, this surface can serve as an ideal model surface for the investigation of the effect of surface roughness on superhydrophobicity. Monodispersed silica particles can be used to form a thin layer on a glass slide [14]. After deposition of a Au layer on top of the spheres followed by heat treatment to form Au nanoparticles, surface nanoroughness was generated. By controlling the Au deposition conditions, the necessary nanoroughness can be effectively designed into the surface. In addition, non-close-packed hexagonally ordered structures were also reported using a templating technique [65]. By using low reflective index and low surface energy materials such as fluorinated polymers, a self-cleaning and antireflection coating can be prepared. Such tunability of the structures suggests specific ways by which multifunctional surface properties such as superhydrophobicity, antireflection, mechanical stability, and durability can be achieved.

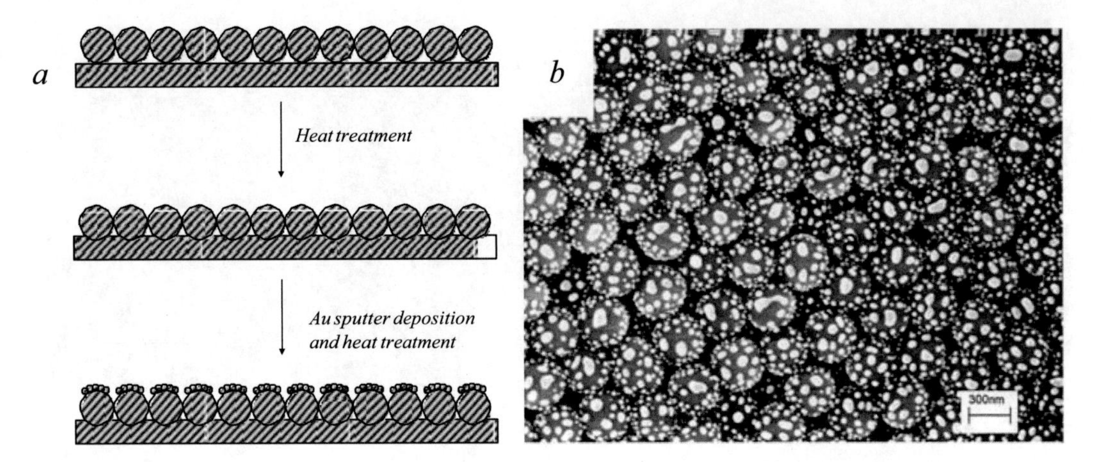

Fig. 15.31 (**a**) Illustration of the hierarchical roughness formation by monodisperse particle and Au deposition; (**b**) SEM image of the surface prepared from the process [14]

15.5.5.2 Layer-by-Layer Self-Assembly

Layer-by-layer (LbL) self-assembly, which is based on the alternating physisorption of oppositely charged building blocks, represents a method to immobilize polyelectrolytes, colloid nanoparticles, and biomacromolecules, such as enzymes, extracellular matrices (ECM), and deoxyribonucleic acid (DNA) [66]. The sensitivity of the LbL multilayer toward its environment (e.g., pH, ionic strength) further provides new approaches to adjust the layered nanostructure with a tailored composition and architecture. The use of the LbL technique for constructing superhydrophobic coatings from poly(acrylic acid) (PAA) and polyethyleneimine (PEI), especially using the silver ion (Ag^+) to enhance the exponential growth of the PEI/PAA multilayers, has been explored [67]. It was found that the addition of Ag^+ can significantly improve the surface roughness as shown in Fig. 15.32. On such a surface, a high contact angle of 172° can be achieved after surface fluorination by (tridecafluoroctyl)triethoxysilane.

Zhai et al. prepared a polyelectrolyte multilayer surface by LbL assembly and then overcoated the surface with silica nanoparticles to mimic the hierarchical scale that is present on lotus leaf surfaces [68] (Fig. 15.33). Superhydrophobicity was achieved after surface fluorination with a fluoroalkylsilane.

Honeycomb-like microporous polymer films with 500 nm to 50 μm pores (diameter) have been prepared by casting polymer solutions under humid conditions (Fig. 15.34). If a fluoropolymer was used, superhydrophobicity was achieved [69]. In this process, water droplets condensed on the solution film surface act as templates. Depending on the drying rate of solvents, the pore size can be controlled by controlling the water droplet size. This surface can also be optically transparent under the proper formation conditions.

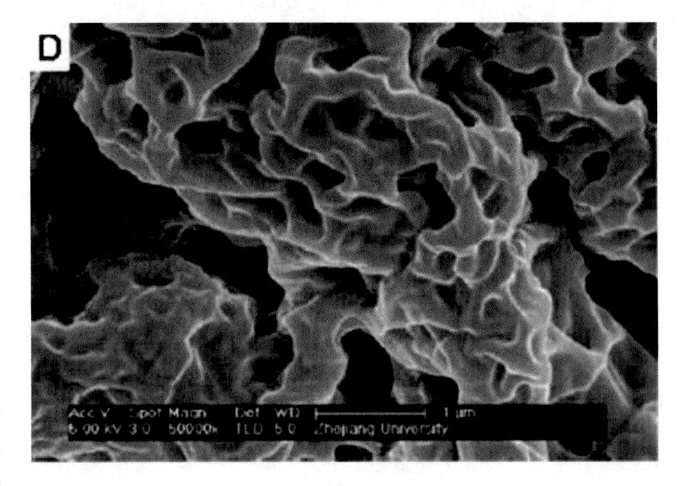

Fig. 15.32 SEM image of (PAA/PEI-Ag+) 7.5 film [67]

15.6 Applications

Superhydrophobicity is relatively a new research field in recent years describing the extreme water repellency of rough hydrophobic surfaces. However, the use of this technique is for a long time. Early products include breathable fabrics and some membrane filters, which use the same effect as superhydrophobic surfaces, but these are rarely referred to as being superhydrophobic. For example, Gore-Tex® is superhydrophobic which was patented in 1972.

Another important application for superhydrophobic technique may be in the field of oil/water separation [70]. Feng et al. demonstrated that a mesh coated with PTFE showed two extremities in wettability for water and oil. For water, it is extremely nonwetting surface (superhydrophobic), while for oil, it changed to complete wetting (superoleophilic). The

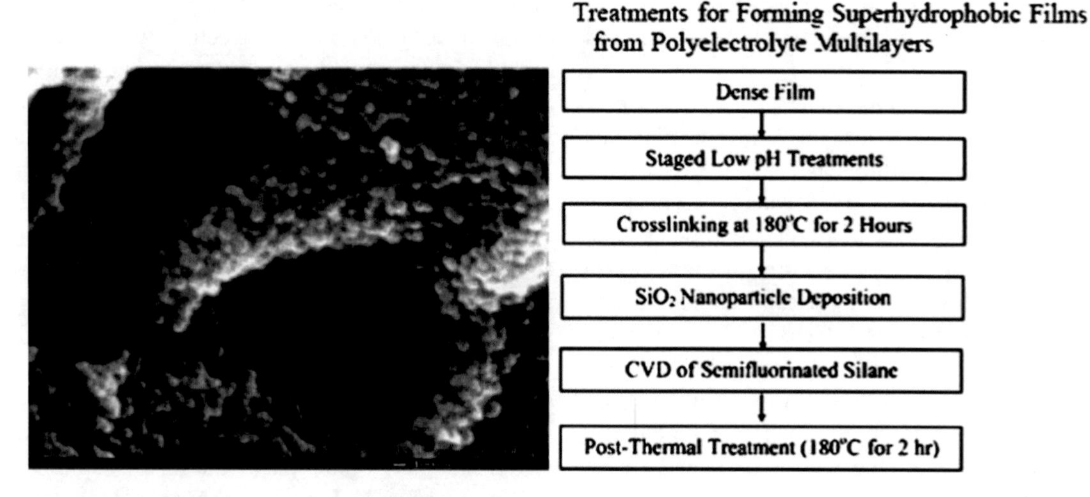

Fig. 15.33 SEM image of the fully treated polyelectrolyte structure (a 2 h immersion in a pH 2.7 solution followed by a 4 h immersion in a pH 2.3 solution, with no water rinse) with silica nanoparticles [68]

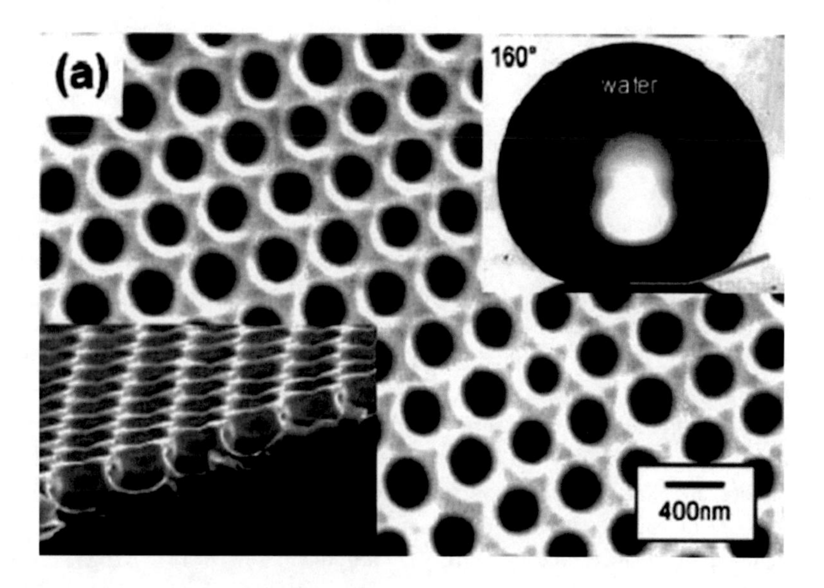

Fig. 15.34 Scanning electron micrograph of the 300-nm-sized honeycomb-patterned film (top image and cross section) and the water contact angle on this film [69]

extreme difference in wettability for water and oil showed the possibility in separation of oil and water. When a mixture of oil and water is put on the mesh film, water will remain on the upper part of the film, while the oil will penetrate through the mesh and be collected at the underneath.

Application of superhydrophobic coating in biomedical-related field is of great significance. One possible application may be in the self-cleaning coating of biomedical devices and equipment to prevent bacteria/virus growth and adhesion. The superhydrophobic surfaces have the great advantage of preventing water droplet staying on the surface and, at the same time, preventing water vapor from condensing onto the surface. This will effectively prevent bacteria growth on the surface, due to the fact that any organism growth requires sufficient water/moisture. In addition, the much reduced adhesion on the superhydrophobic surface will make bacteria hard to deposit/reside on the surface. Therefore, the propagation of the microorganisms is prohibited. Combination of these two effects, superhydrophobic surface coatings are very promising for the application in the biomedical device/equipment medical hygiene concerns. For the application in the biocompatible surfaces, the superhydrophobic surface is based on the bio-inert principle, similar to PTFE and silicone because of the very low surface energy. However, they can be much cheaper because the bulk materials selection is flexible and cheap while the surface will resemble PTFE or silicone for the reduced surface energy, therefore, adhesion. Superhydrophobic surface coatings have remarkable chemical inertness due to the much reduced surface energy, therefore, providing a minimized chronic foreign body reaction. This will help reduce the risk of pain-related complications for long-term patient comfort. In addition, this coating helps reduce the risk of infection by reducing potential bacteria colonization sites which is a problem for currently used polymer coatings.

The friction reduction in underwater coatings is a new field for the superhydrophobic coatings. Speedo™ is one proved example of using hydrophobic suit to reduce the friction when swimming in water. Regarding superhydrophobic surfaces, the drag reduction comes from (1) the hydrophobic surface (low surface energy) with reduced friction; (2) the rough surfaces can form air cushion when the surface is soaked in water, therefore the solid/water contact can be minimized, which will result in the much reduced friction originated from the water/solid contact. The application of superhydrophobic coatings onto the outside surface of submarine or ship will be promising in the reduction of the friction in motion. Therefore, energy consumption will be greatly reduced. This is beneficial not only in energy conservation but also in the control of CO2 reduction which is a critical issue in global warming.

15.6.1 Anticorrosion

Corrosion-induced failure plays an important role in microelectronic devices and package failures. Failures of microelectronic devices and packages not only cause the malfunction of the devices but also sometimes lead to catastrophic consequences for the whole systems. Corrosion in microelectronic packaging depends on the package type, electronic materials, fabrication and assembly processes, and environmental conditions such as moisture condensation, ionic or organic contaminants, temperature, residual and thermal stress and electrical bias, etc. With the ever-reducing feature sizes of microelectronic components and devices, they are more susceptible to corrosion-induced failures. The better performance and reliability requirements drive to improve corrosion resistance of packaging systems. There are basically three types of packaging for microelectronic components and devices. They are ceramic, metal, and plastic packaging. Ceramic and metal packages are hermetic packages mainly used in the military, aerospace, and automobile microelectronic devices for high reliability. Plastic packaging is non-hermetic package but become more and more widely used because of the low cost and better manufacturability. Compared to other package types, plastic-packaging systems have more corrosion-related problems because the polymeric materials used in plastic package systems are more vulnerable to the permeation of moisture into the inside dies, wires, bond pads, lead frames, and solder joints. Therefore, corrosion is becoming a more serious issue in modern microelectronic packages reliability.

Corrosion involves essentially electrochemical processes except for those oxidized at the elevated temperature under the dry environment. The basic requirements for electrochemical corrosion include electrically conductive anode, cathode, interconnecting electrolyte (humidity environment), and driving force (electrochemical potential). Impurity ions such as halides compounds which can be dissociated in water (electrolytes) are very dangerous resources in corrosion of electronics packaging impurity ions due to the formation of paths of electrochemical corrosion. Preventing corrosion is actually a matter of preventing moisture condensation on interest areas such as die interfaces, wires, bond pads, and solder joints. Superhydrophobic coatings, which are extremely water-repellent, show great potential for the anticorrosion applications in microelectronic devices.

Recent discoveries have linked the mechanism for the self-cleaning of a lotus plant to a microscopic morphology leading to superhydrophobic surfaces. This water-repellent surface suggests huge opportunities in the area of corrosion inhibition for metal components. A simple superhydrophobic-coating technique was developed by Barkhudarov et al. focusing on aluminum corrosion protection with a rough, highly porous organosilica aerogel-like film [71]. When silica-coated Al surfaces were submerged in 5 wt% NaCl in D2O, neutron reflection measurement was conducted to measure the increase of the oxidation layer and the decrease of Al layer. It was found that the superhydrophobic layer can effectively reduce the corrosion process as shown in Fig. 15.35.

15.6.2 Applications in MEMS Antistiction

Microelectromechanical systems (MEMS) have been used extensively to perform basic signal transduction operations in miniaturized sensors and actuators. However, autoadhesion, or spontaneous sticking (stiction) between MEMS structures, remains a major limitation in bringing this new class of engineering devices to the broader market. Free-standing mechanical structures fabricated from polycrystalline Si may strongly adhere to each other when brought into contact, due to the hydrogen bonding between surface hydroxyl groups [72]. In a high humidity ambient, this problem is exacerbated by adsorption of water and capillary condensation. Water capillaries or condensation can cause catastrophic failure of MEMS devices [73]. Low surface energy coatings for antistiction of MEMS are required for many practical MEMS devices [72–74].

Two primary methods have been invoked to prevent these failures. One is a surface treatment with low surface energy coatings, e.g., octadecyl trichlorosilane (OTS), to inhibit surface condensation of water [75]. The other method involves the fabrication of structured surfaces in order to reduce the contact area and thereby reduce the adhesion

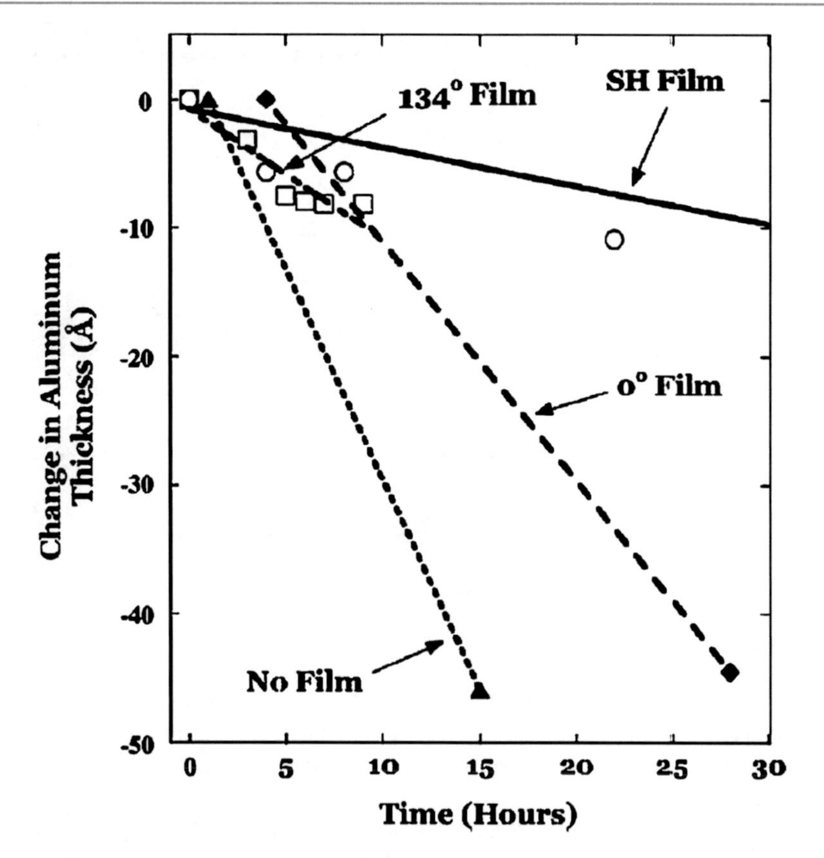

Fig. 15.35 The change in thickness of the aluminum layer vs. time for samples protected by films of varying contact angle and a sample with only native Al_2O_3 layer (without protective film). At each data point, the initial aluminum layer thickness for that sample was subtracted in order to leave only information about changes in thickness. The *solid* and *dashed lines* represent linear fits of the aluminum layer thickness decrease. Note: the figure does not show all data points used to obtain the linear fit for the SH film sample, which was measured up to 186 h (this is why the SH line does not appear to be an accurate fit for the points shown in this figure) [71]

Fig. 15.36 Surface–probe interactions as determined from AFM force measurements between a tipless Si_3N_4 probe and either a superhydrophobic film spin-casted from ethanol/sol ratio of 5:1 or a PFOS-treated glass slide

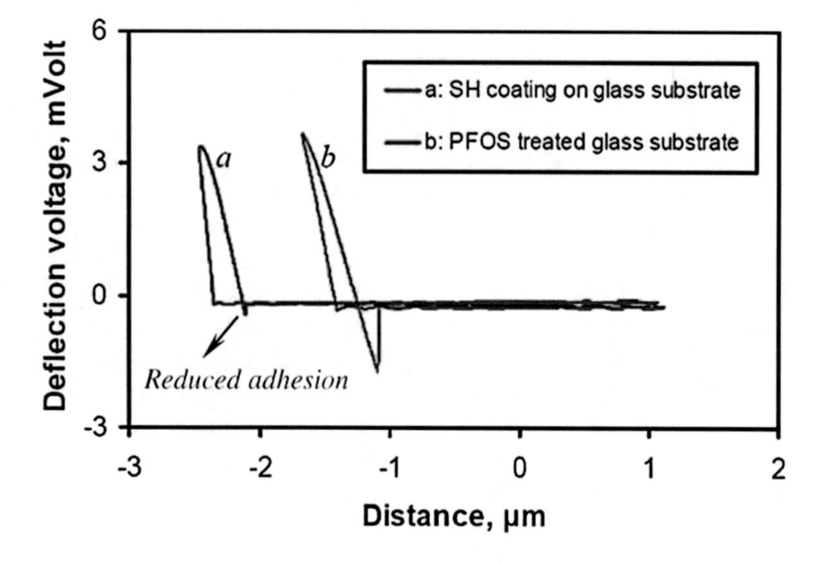

force between the free-standing structures and the substrate [76].

Superhydrophobic silica surfaces were proposed and demonstrated for the first time in MEMS antistiction by AFM adhesion tests using tipless probes as shown in Fig. 15.36. This approach greatly reduces the adhesion forces between the tipless probe and the superhydrophobicly coated substrate, which showed that it is promising to reduce stiction failures of MEMS devices. Water does not condense on this superhydrophobic surface thus eliminating capillary forces.

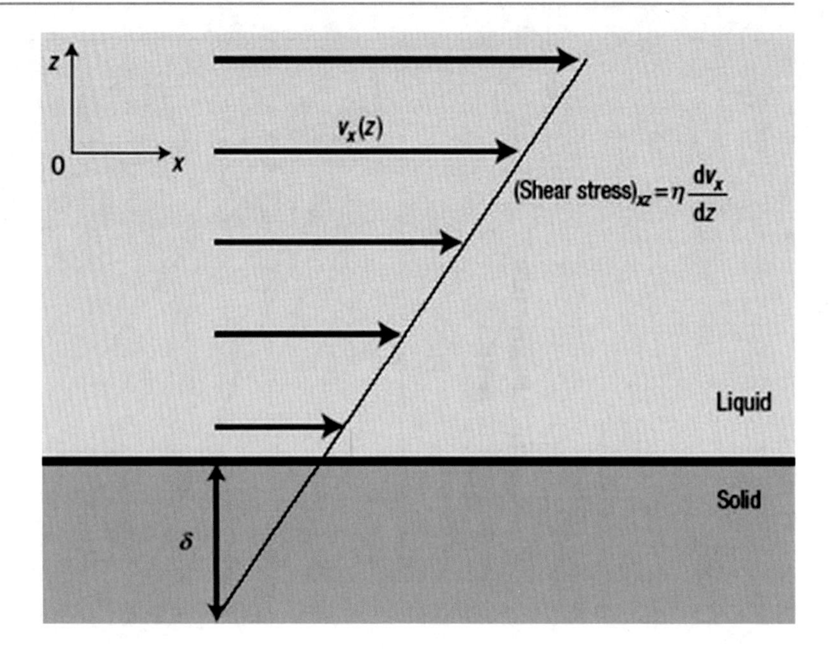

Fig. 15.37 Definition of the slip length. The linear velocity (*v*x) profile in the flowing, Newtonian fluid (characterized by a Newtonian viscosity η and a constant shear stress) does not vanish at the solid boundary. Extrapolation into the solid at a depth δ is necessary to obtain a vanishing velocity, as assumed by the macroscopic "no-slip" boundary condition [77]

Furthermore, the adhesion force between the free-standing structures and the substrate can also be effectively reduced by decreasing the contact area due to the presence of surface nanostructures.

15.6.3 Slip Flow in Microfluidics

In fluid mechanics, usually a no-slip boundary condition is assumed that the fluid velocity at a fluid–solid interface equals that of the solid surface as shown in Fig. 15.37. Recent development in microfluidics showed that the violation of the assumption may exist. For surfaces with large surface-to-volume ratio, surface properties may dramatically affect the flow resistance. The possibility of engineering slip properties in a controlled way is therefore crucial for microfluidic applications. Mechanisms behind the boundary slip include surface roughness and structural characteristics of roughness, roughness-induced dewetting on hydrophobic surfaces, dissolved gas, and bubbles on the surface. For hydrophilic rough surfaces, the no-slip boundaries can be achieved. However, for a hydrophobic surface, roughness may increase the slip due to the transition to a superhydrophobic state [77]. Liquid cannot enter the valleys of the rough surface; instead, water only stay at the top of them, with a minimum solid/liquid contact area. The as-formed air bubbles help lubricating the flow, reducing the friction of the flow process.

15.6.4 Solar Cells

Superhydrophobic surfaces have been prepared by using a variety of materials. In addition to the establishment of superhydrophobicity, these materials have been further functionalized to realize electroactive surfaces with controlled wetting properties [23], transparent surfaces [44], adhesive surfaces [78], photocatalytic surfaces [79], and photonic bandgap layers [80]. However, due to its widespread use in electronic and photovoltaic devices, incorporation of superhydrophobic properties into Si surfaces may extend and enhance the more traditional applications of these devices. Metal-assisted etching of Si surfaces has been employed to define specific structures on Si surfaces [54, 81]. After deposition of a thin layer of a metal such as Au, Pt, or Pt/Pd onto p- or n-type Si surfaces, nanostructures can be formed by immersing this substrate into various mixtures of hydrofluoric acid/oxidant/solvent. This approach has been used to prepare photoluminescent surfaces [82] and "black" anti-reflecting Si surfaces [54]. In fact, the reflectivity can be reduced to nearly zero if the appropriate structures are formed. The use of etched Si surfaces for photovoltaic applications has been also investigated, and a solar cell efficiency of 9.31% was reported [83]. On the hierarchically etched Si surfaces, the weighted light reflection loss can be reduced to below 4% without any other antireflection coatings as shown in Fig. 15.38. The etched surface showed a surface hierarchical structure shown in Fig. 15.39.

15.6.5 Bio-Assay

In biotechnology, one is interested in controlling droplets containing biologically relevant molecules (DNA and proteins), minimizing contamination. Already the construction of complementary DNA (cDNA)-microarrays prepared by spotting techniques requires the specific wetting

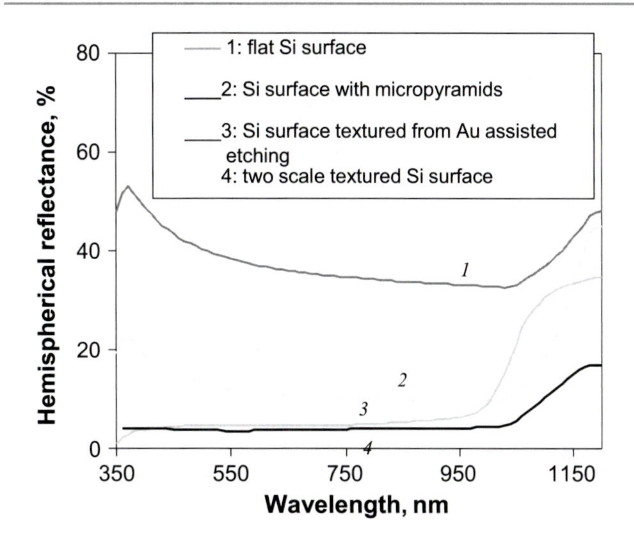

Fig. 15.38 Light reflection on Si nano-textured surfaces generated by Au-assisted etching. Au (5-nm thickness)-assisted etching in HF:H_2O_2:H_2O of 1:5:10 (v/v/v) for 2 min [58]

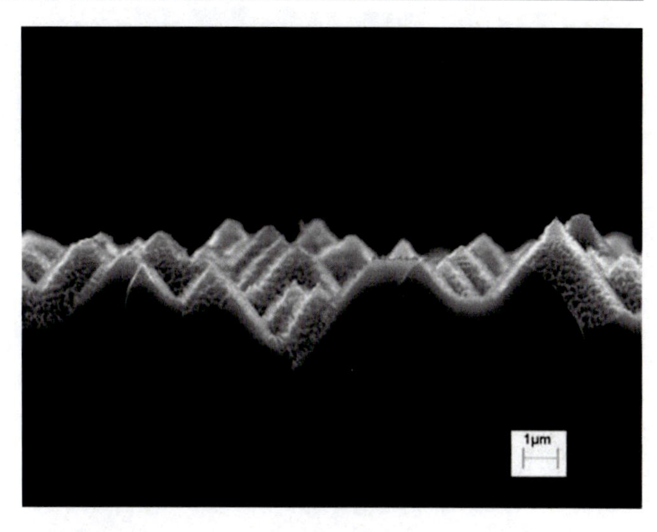

Fig. 15.39 Cross sections of Si pyramid surfaces etched by Au-assisted etching in HF/H_2O_2/H_2O (v/v/v 1:5:10) [58]

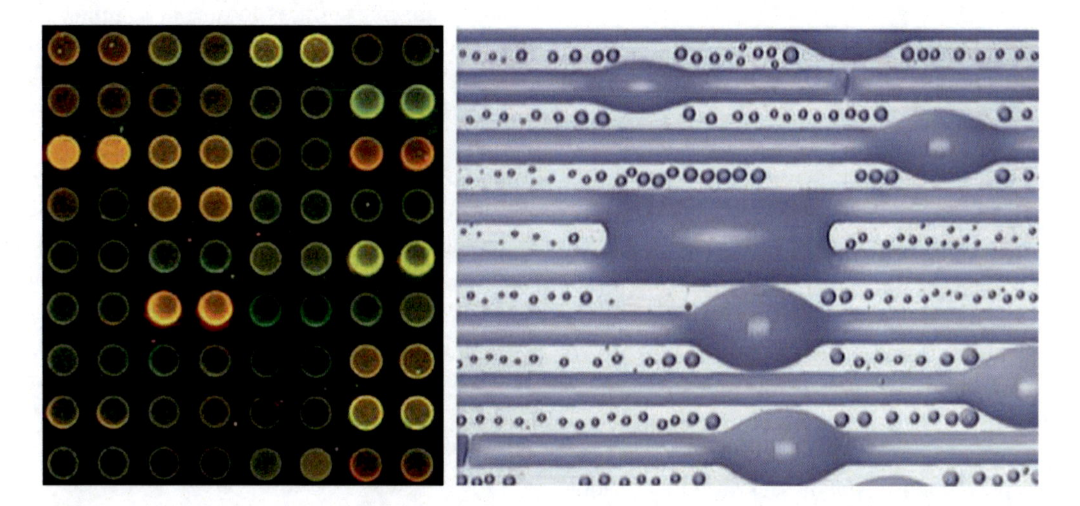

Fig. 15.40 (**a**) Complementary DNA microarray on a silanized glass plate. The doughnut-like shapes of the deposits are clearly visible; (**b**) liquid channels on a hydrophilically–hydrophobically patterned substrate: a possible pathway to surface-tension-controlled microfluidics [84]

properties of the substrates. Glass slides, the substrate most commonly used, are usually only mildly hydrophobized, such that the drops of drying cDNA solution produce unwanted ring-like structures, a nuisance known in the field as the "doughnut-effect" (see the following Fig. 15.40a). This effect is related to the well-studied "coffee-stain effect"; an evaporation-driven convection mechanism drives dissolved or dispersed particles inside the drop to the edge of the drop, where evaporation is fastest [84]; therefore, after drying, the deposited material is not a spot but a ring-like shape. Superhydrophobicity is advantageous in this application: the almost fully spherical drops on a superhydrophobic surface can shrink exactly the same as a drop in free air. In addition, the position and shape of spotted drops can be steered by combining hydrophilic and hydrophobic surface patterns. The usefulness of this idea is shown in a real biosystem: some desert beetles capture their drinking water by a hydrophobically–hydrophilically structured back. In the same way, by prepatterning of a substrate, hydrophilic prespotting (anchoring) on an otherwise hydrophobic or superhydrophobic surface results in new possibilities for improvements in spotting and analyzing DNA and proteins by avoiding wall contact.

Beyond the possibility of improving deposit shapes on substrates, the possibility of a guided motion of droplets on superhydrophobic surfaces offers the chance to develop a droplet-based microfluidics system, in contrast to the classical concept based on microfluidic channels. Driving the liquid along the channels and making them merged at predefined locations offers a novel way to mix reactants or steer biochemical reactions, defining the concept of a "liquid microchip" or "surface-tension confined microfluidics." One advantage of the open structures over capillaries, in addition to their ease of cleaning, is that blocking of the capillary by unforeseen chemical reactions cannot occur.

Microfluidics can also be based on droplets on superhydrophobic surfaces alone; because the drops have very low contact areas with the substrate, they are easy to move by external fields such as electrostatic forces or surface acoustic waves. Systems that make use of a droplet-based actuation mechanism are being developed by many researchers. The aim here is to control droplet positioning and motion on the substrates with as little surface contact as possible, and making the droplet-based system a programmable reactor, by which the liquid positions are prescribed and tuned externally.

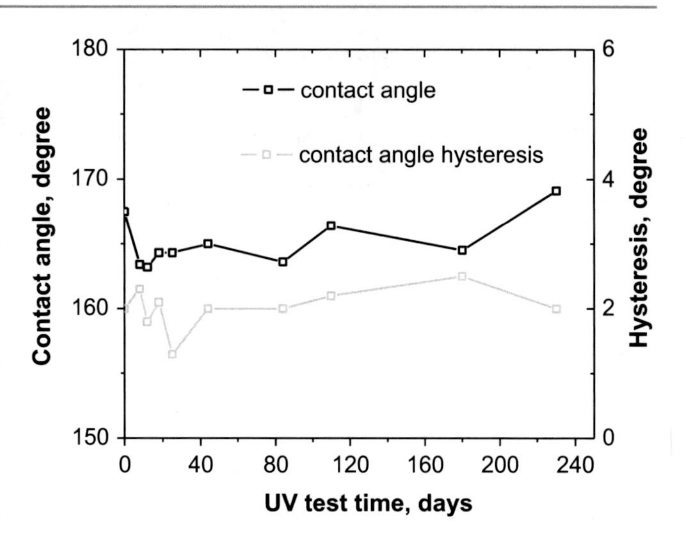

Fig. 15.41 Stability of a PFOS-treated rough silica thin film surface [85]

15.6.6 Superhydrophobic Surfaces for Outdoor Applications

For the self-cleaning applications, most involve the outdoor environment. This poses another engineering requirement: weathering resistance which mostly involves the UV exposure. Therefore a UV-resistant superhydrophobic surface is required.

For polymeric materials, UV degradation (unzipping of polymer chain and photooxidation) is inevitable due to the intrinsic defects in the polymer chain and impurities (acting as catalyst) inside the polymer. Formation of carbonyl or hydroxyl groups on the surface ensues. This happened first on the surface which means that before any mechanical property loss, surface properties changed first from hydrophobic to hydrophilic. Currently no efficient method has been reported for the prevention of polymer surface degradation under UV irradiation.

Xiu et al. reported a method using inorganic materials such as silica with the combination of fluoroalkylsilane treatment for the UV-resistant superhydrophobic surfaces [85]. The surface was tested for ~5500 h without obvious degradation in superhydrophobicity as shown in Fig. 15.41. This method has used small stable molecules such as fluoroalkylsilanes for the surface layer instead of polymers, which prevent the possible chain unzipping reactions and the subsequent loss of superhydrophobicity.

In summary, superhydrophobic surfaces have been successfully prepared by using a variety of materials, and the mechanism for self-cleaning and water repellency has been better understood. However, one of the biggest problems facing all self-cleaning or contaminant-free surface applications is degradation and failures under harsh environment. For biotechnology applications, this is not so relevant; the surfaces will often be used for analytic purposes and hence designed as disposables. If one thinks of contaminant-free surfaces for use in complicated conditions, the issues of robustness under mechanical abrasion (e.g., handing of the coated surfaces) has not yet been resolved. And yet a different matter is the use of self-cleaning surfaces in outdoor applications. In addition, stability under high humidity condition also needs improvement by designing small features with cost-effective approaches. The dew point may be reduced on the superhydrophobic surface with small features. Oil or surfactant repellent or superoleophobic surface may also be desirable which further add more difficulties to the design of superhydrophobicity. As coatings improve and prices fall down, the use of superhydrophobic surfaces is likely to increase, particularly as many of the drawbacks are now well-understood and some can be minimized by choosing specific coatings.

References

1. Wagner, T., Neinhuis, C., Barthlott, W.: Wettability and contaminability of insect wings as a function of their surface sculptures. Acta Zool. **77**, 213–215 (1996)
2. Neinhuis, C., Barthlott, W.: Characterization and distribution of water-repellent, self-cleaning plant surfaces. Ann. Bot. **79**, 667–677 (1997)
3. Sacher, E., Sapieha, J.K., Schrieber, H.P., Wertheimer, N.R., McIntyre, N.S.: Silanes, Surfaces, and Interfaces, vol. I. Taylor & Francis, London (1986)
4. Morra, M., Occhiello, E., Garbassi, F.: Contact-angle hysteresis in oxygen plasma treated poly(tetrafluoroethylene). Langmuir. **5**, 872–876 (1989)
5. Shibuichi, S., Yamamoto, T., Onda, T., Tsujii, K.: Super water- and oil-repellent surfaces resulting from fractal structure. J. Colloid Interface Sci. **208**, 287–294 (1998)

6. Nosonovsky, M., Bhushan, B.: Biologically inspired surfaces: broadening the scope of roughness. Adv. Funct. Mater. **18**, 843–855 (2008)
7. Sun, T.L., Feng, L., Gao, X.F.: Bioinspired surfaces with special wettability. Acc. Chem. Res. **38**, 644–652 (2005)
8. Gao, X.F., Jiang, L.: Water-repellent legs of water striders. Nature. **432**, 36 (2004)
9. Gao, X.F., Yan, X., Yao, X., Xu, L., Zhang, K., Zhang, J.H., Yang, B., Jiang, L.: The dry-style antifogging properties of mosquito compound eyes and artificial analogues prepared by soft lithography. Adv. Mater. **19**, 2213–2217 (2007)
10. Berthier, S.: Iridescences: the Physical Colors of Insects. Springer, New York (2007)
11. Cassie, A.B.D., Baxter, S.: Wettability of porous surfaces. Trans. Faraday Soc. **40**, 0546–0550 (1944)
12. Wenzel, R.N.: Resistance of solid surfaces to wetting by water. Ind. Eng. Chem. **28**, 988–994 (1936)
13. Marmur, A.: Wetting on hydrophobic rough surfaces: to be heterogeneous or not to be? Langmuir. **19**, 8343–8348 (2003)
14. Xiu, Y., Zhu, L., Hess, D.W., Wong, C.P.: Biomimetic creation of hierarchical surface structures by combining colloidal self-assembly and Au sputter deposition. Langmuir. **22**, 9676–9681 (2006)
15. Lu, X.Y., Zhang, J.L., Zhang, C.C., Han, Y.C.: Low-density polyethylene (LDPE) surface with a wettability gradient by tuning its microstructures. Macromol. Rapid Commun. **26**, 637–642 (2005)
16. Han, W., Wu, D., Ming, W.H., Niemantsverdriet, H., Thune, P.C.: Direct catalytic route to superhydrophobic polyethylene films. Langmuir. **22**, 7956–7959 (2006)
17. Teshima, K., Sugimura, H., Inoue, Y., Takai, O., Takano, A.: Ultra-water-repellent poly(ethylene terephthalate) substrates. Langmuir. **19**, 10624–10627 (2003)
18. Tadanaga, K., Morinaga, J., Matsuda, A., Minami, T.: Superhydrophobic-superhydrophilic micropatterning on flowerlike alumina coating film by the sol-gel method. Chem. Mater. **12**, 590–592 (2000)
19. Nakajima, A., Fujishima, A., Hashimoto, K., Watanabe, T.: Preparation of transparent superhydrophobic boehmite and silica films by sublimation of aluminum acetylacetonate. Adv. Mater. **11**, 1365–1368 (1999)
20. Israelachvili, J.: Intermolecular and Surface Forces, 2nd edn. Academic, New York (1991)
21. Nishino, T., Meguro, M., Nakamae, K., Matsushita, M., Ueda, Y.: The lowest surface free energy based on -CF3 alignment. Langmuir. **15**, 4321–4323 (1999)
22. Tuteja, A., Choi, W., Ma, M., Mabry, J.M., Mazzella, S.A., Rutledge, G.C., McKinley, G.H., Cohen, R.E.: Designing superoleophobic surfaces. Science. **318**, 1618–1622 (2007)
23. Ahuja, A., Taylor, J.A., Lifton, V., Sidorenko, A.A., Salamon, T.R., Lobaton, E.J., Kolodner, P., Krupenkin, T.N.: Nanonails: a simple geometrical approach to electrically tunable superlyophobic surfaces. Langmuir. **24**, 9–14 (2008)
24. Cao, L.L., Price, T.P., Weiss, M., Gao, D.: Super water- and oil-repellent surfaces on intrinsically hydrophilic and oleophilic porous silicon films. Langmuir. **24**, 1640–1643 (2008)
25. Cao, L.L., Hu, H.H., Gao, D.: Design and fabrication of micro-textures for inducing a superhydrophobic behavior on hydrophilic materials. Langmuir. **23**, 4310–4314 (2007)
26. Patankar, N.A.: Transition between superhydrophobic states on rough surfaces. Langmuir. **20**, 7097–7102 (2004)
27. Lieberman, M.A., Lichtenberg, A.J.: Principles of Plasma Discharges and Materials Processing, 2nd Translator. Wiley-Interscience, London (2005)
28. Stelmashuk, V., Biederman, H., Slavinska, D., Zemek, J., Trchova, M.: Plasma polymer films rf sputtered from PTFE under various argon pressures. Vacuum. **77**, 131–137 (2005)
29. Shiu, J.Y., Kuo, C.W., Chen, P.: Fabrication of tunable superhydrophobic surfaces. Proc. SPIE-Int. Soc. Optical Eng. **5648**, 325–332 (2005)
30. Byun, D., Lee, Y., Tran, S.B.Q., Nugyen, V.D., Kim, S., Park, B., Lee, S., Inamdar, N., Bau, H.H.: Electrospray on superhydrophobic nozzles treated with argon and oxygen plasma. Appl. Phys. Lett. **92**, 093507 (2008)
31. Larsen, K.P., Petersen, D.H., Hansen, O.: Study of the roughness in a photoresist masked, isotropic, SF6-based ICP silicon etch. J. Electrochem. Soc. **153**, G1051–G1058 (2006)
32. Lejeune, M., Lacroix, L.M., Bretagnol, F., Valsesia, A., Colpo, P., Rossi, F.: Plasma-based processes for surface wettability modification. Langmuir. **22**, 3057–3061 (2006)
33. Teshima, K., Sugimura, H., Inoue, Y., Takai, O., Takano, A.: Transparent ultra water-repellent poly(ethylene terephthalate) substrates fabricated by oxygen plasma treatment and subsequent hydrophobic coating. Appl. Surf. Sci. **244**, 619–622 (2005)
34. Tserepi, A.D., Vlachopoulou, M.E., Gogolides, E.: Nanotexturing of poly(dimethylsiloxane) in plasmas for creating robust super-hydrophobic surfaces. Nanotechnology. **17**, 3977–3983 (2006)
35. Teshima, K., Sugimura, H., Inoue, Y., Takai, O., Takano, A.: Wettability of poly(ethylene terephthalate) substrates modified by a two-step plasma process: ultra water-repellent surface fabrication. Chem. Vap. Depos. **10**, 295 (2004)
36. Teshima, K., Sugimura, H., Inoue, Y., Takai, O., Takano, A.: Transparent ultra water-repellent poly(ethylene terephthalate) substrates fabricated by oxygen plasma treatment and subsequent hydrophobic coating. Appl. Surf. Sci. **244**, 619–624 (2005)
37. Teshima, K., Sugimura, H., Takano, A., Inoue, Y., Takai, O.: Ultrahydrophobic/ultrahydrophilic micropatterning on a polymeric substrate. Chem. Vap. Depos. **11**, 347–352 (2005)
38. Fresnais, J., Chapel, J.P., Poncin-Epaillard, F.: Synthesis of transparent superhydrophobic polyethylene surfaces. Surf. Coat. Technol. **200**, 5296–5305 (2006)
39. Fresnais, J., Benyahia, L., Poncin-Epaillard, F.: Dynamic (de)-wetting properties of superhydrophobic plasma-treated polyethylene surfaces. Surf. Interface Anal. **38**, 144–149 (2006)
40. Fresnais, J., Benyahia, L., Chapel, J.P., Poncin-Epaillard, F.: Polyethylene ultrahydrophobic surface: synthesis and original properties. Eur. Phys. J. Appl. Phys. **26**, 209–214 (2004)
41. Di Mundo, R., Palumbo, F., d'Agostino, R.: Nanotexturing of polystyrene surface in fluorocarbon plasmas: from sticky to slippery superhydrophobicity. Langmuir. **24**, 5044–5051 (2008)
42. Woodward, I., Schofield, W.C.E., Roucoules, V., Badyal, J.P.S.: Super-hydrophobic surfaces produced by plasma fluorination of polybutadiene films. Langmuir. **19**, 3432–3438 (2003)
43. Dorrer, C., Ruhe, J.: Wetting of silicon nanograss: from superhydrophilic to superhydrophobic surfaces. Adv. Mater. **20**, 159–164 (2008)
44. Artus, G.R.J., Jung, S., Zimmermann, J., Gautschi, H.P., Marquardt, K., Seeger, S.: Silicone nanofilaments and their application as superhydrophobic coating. Adv. Mater. **18**, 2758–2762 (2006)
45. Zimmermann, J., Rabe, M., Artus, G.R.J., Seeger, S.: Patterned superfunctional surfaces based on a silicone nanofilament coating. Soft Matter. **4**, 450–452 (2008)
46. Zhu, L.B., Xiu, Y.H., Xu, J.W., Tamirisa, P.A., Hess, D.W., Wong, C.P.: Superhydrophobicity on two-tier rough surfaces fabricated by controlled growth of aligned carbon nanotube arrays coated with fluorocarbon. Langmuir. **21**, 11208–11212 (2005)
47. Lau, K.K.S., Bico, J., Teo, K.B.K., Chhowalla, M., Amaratunga, G.A.J., Milne, W.I., McKinley, G.H., Gleason, K.K.: Superhydrophobic carbon nanotube forests. Nano Lett. **3**, 1701–1705 (2003)
48. Roig, A., Molins, E., Rodriguez, E., Martinez, S., Moreno-Manas, M., Vallribera, A.: Superhydrophobic silica aerogels by fluorination at the gel stage. Chem. Commun. **20**, 2316–2317 (2004)

49. Tadanaga, K., Katata, N., Minami, T.: Formation process of super-water-repellent Al2O3 coating films with high transparency by the sol-gel method. J. Am. Ceram. Soc. **80**, 3213–3216 (1997)
50. Doshi, D.A., Shah, P.B., Singh, S., Branson, E.D., Malanoski, A.P., Watkins, E.B., Majewski, J., van Swol, F., Brinker, C.J.: Investigating the interface of superhydrophobic surfaces in contact with water. Langmuir. **21**, 7805–7811 (2005)
51. Rao, A.V., Bhagat, S.D., Hirashima, H., Pajonk, G.M.: Synthesis of flexible silica aerogels using methyltrimethoxysilane (MTMS) precursor. J. Colloid Interface Sci. **300**, 279–285 (2006)
52. Shirtcliffe, N.J., McHale, G., Newton, M.I., Perry, C.C.: Intrinsically superhydrophobic organosilica sol-gel foams. Langmuir. **19**, 5626–5631 (2003)
53. Gao, L.C., McCarthy, T.J.: A perfectly hydrophobic surface (theta (A)/theta(R)=180 degrees/180 degrees). J. Am. Chem. Soc. **128**, 9052 (2006)
54. Koynov, S., Brandt, M.S., Stutzmann, M.: Black nonreflecting silicon surfaces for solar cells. Appl. Phys. Lett. **88**, 203107/1 (2006)
55. Peng, K.Q., Wu, Y., Fang, H., Zhong, X.Y., Xu, Y., Zhu, J.: Uniform, axial-orientation alignment of one-dimensional single-crystal silicon nanostructure arrays. Angew. Chem. Int. Ed. **44**, 2737–2742 (2005)
56. Peng, K.Q., Yan, Y.J., Gao, S.P., Zhu, J.: Dendrite-assisted growth of silicon nanowires in electroless metal deposition. Adv. Funct. Mater. **13**, 127–132 (2003)
57. Xiu, Y., Zhu, L., Hess, D.W., Wong, C.P.: Hierarchical silicon etched structures for controlled hydrophobicity/superhydrophobicity. Nano Lett. **7**, 3388–3393 (2007)
58. Xiu, Y., Zhang, S., Yelundur, V., Rohatgi, A., Hess, D.W., Wong, C.P.: Superhydrophobic and low light reflectivity silicon surfaces fabricated by hierarchical etching. Langmuir. **24**, 10421–10426 (2008)
59. Qian, B.T., Shen, Z.Q.: Fabrication of superhydrophobic surfaces by dislocation-selective chemical etching on aluminum, copper, and zinc substrates. Langmuir. **21**, 9007–9009 (2005)
60. Guo, Z.G., Zhou, F., Hao, J.C., Liu, W.M.: Effects of system parameters on making aluminum alloy lotus. J. Colloid Interface Sci. **303**, 298–305 (2006)
61. Shirtcliffe, N.J., McHale, G., Newton, M.I., Perry, C.C.: Wetting and wetting transitions on copper-based super-hydrophobic surfaces. Langmuir. **21**, 937–943 (2005)
62. Rupp, F., Scheideler, L., Olshanska, N., de Wild, M., Wieland, M., Geis-Gerstorfer, J.: Enhancing surface free energy and hydrophilicity through chemical modification of microstructured titanium implant surfaces. J. Biomed. Mater. Res. A. **76A**, 323–334 (2006)
63. Lai, Y.K., Lin, C.J., Huang, J.Y., Zhuang, H.F., Sun, L., Nguyen, T.: Markedly controllable adhesion of superhydrophobic spongelike nanostructure TiO2 films. Langmuir. **24**, 3867 (2008)
64. Shiu, J.Y., Kuo, C.W., Chen, P.L., Mou, C.Y.: Fabrication of tunable superhydrophobic surfaces by nanosphere lithography. Chem. Mater. **16**, 561–564 (2004)
65. Sun, C.H., Jiang, P., Jiang, B.: Broadband moth-eye antireflection coatings on silicon. Appl. Phys. Lett. **92**, 051107 (2008)
66. Zheng, H., Okada, H., Nojima, S., Suye, S., Hori, T.: Layer-by-layer assembly of enzymes and polymerized mediator on electrode surface by electrostatic adsorption. Sci. Technol. Adv. Mater. **5**, 371–376 (2004)
67. Ji, J., Fu, J.H., Shen, J.C.: Fabrication of a superhydrophobic surface from the amplified exponential growth of a multilayer. Adv. Mater. **18**, 1441–1444 (2006)
68. Zhai, L., Cebeci, F.C., Cohen, R.E., Rubner, M.F.: Stable superhydrophobic coatings from polyelectrolyte multilayers. Nano Lett. **4**, 1349–1353 (2004)
69. Yabu, H., Shimomura, M.: Single-step fabrication of transparent superhydrophobic porous polymer films. Chem. Mater. **17**, 5231–5234 (2005)
70. Feng, L., Zhang, Z.Y., Mai, Z.H., Ma, Y.M., Liu, B.Q., Jiang, L., Zhu, D.B.: A super-hydrophobic and super-oleophilic coating mesh film for the separation of oil and water. Angew. Chem. Int. Ed. **43**, 2012–2014 (2004)
71. Barkhudarov, P.M., Shah, P.B., Watkins, E.B.: Corrosion inhibition using superhydrophobic films. Corros. Sci. **50**, 897–902 (2008)
72. Ashurst, W.R., Yau, C., Carraro, C., Lee, C., Kluth, G.J., Howe, R. T., Maboudian, R.: Alkene based monolayer films as anti-stiction coatings for polysilicon MEMS. Sensors Actuators A Phys. **91**, 239–248 (2001)
73. Mastrangelo, C.H., Hsu, C.H.: Mechanical stability and adhesion of microstructures under capillary forces: part I: basic theory. IEEE J. Microelectromech. Syst. **2**, 33–43 (1993)
74. Mastrangelo, C.H., Hsu, C.H.: Mechanical stability and adhesion of microstructures under capillary forces – part II: experiments. J. Microelectromech. Syst. **2**, 44–55 (1993)
75. Kulkarni, S.A., Mirji, S.A., Mandale, A.B.: Thermal stability of self-assembled octadecyltrichlorosilane monolayers on planar and curved silica surfaces. Thin Solid Films. **496**, 420–425 (2006)
76. Lee, C.C., Hsu, W.: Method on surface roughness modification to alleviate stiction of microstructures. J. Vac. Sci. Technol. B. **21**, 1505–1510 (2003)
77. Cottin-Bizonne, C., Barrat, J.L., Bocquet, L.: Low-friction flows of liquid at nanopatterned interfaces. Nat. Mater. **2**, 237–240 (2003)
78. Mahdavi, A., Ferreira, L., Sundback, C., Nichol, J.W., et al.: A biodegradable and biocompatible gecko-inspired tissue adhesive. Proc. Natl. Acad. Sci. **105**, 2307–2312 (2008)
79. Zhang, X.T., Jin, M., Liu, Z.Y., Nishimoto, S., Saito, H., Murakami, T., Fujishima, A.: Preparation and photocatalytic wettability conversion of TiO2-based superhydrophobic surfaces. Langmuir. **22**, 9477–9479 (2006)
80. Wang, J.X., Wen, Y.Q., Feng, X.J.: Control over the wettability of colloidal crystal films by assembly temperature. Macromol. Rapid Commun. **27**, 188–192 (2006)
81. Li, X., Bohn, P.W.: Metal-assisted chemical etching in HF/H2O2 produces porous silicon. Appl. Phys. Lett. **77**, 2572–2574 (2000)
82. Chattopadhyay, S., Li, X.L., Bohn, P.W.: In-plane control of morphology and tunable photoluminescence in porous silicon produced by metal-assisted electroless chemical etching. J. Appl. Phys. **91**, 6134–6140 (2002)
83. Peng, K.Q., Xu, Y., Wu, Y., Yan, Y.J., Lee, S.T., Zhu, J.: Aligned single-crystalline Si nanowire arrays for photovoltaic applications. Small. **1**, 1062–1067 (2005)
84. Deegan, R.D., Bakajin, O., Dupont, T.F., Huber, G., Nagel, S.R., Witten, T.A.: Capillary flow as the cause of ring stains from dried liquid drops. Nature. **389**, 827–829 (1997)
85. Xiu, Y.H., Hess, D.W., Wong, C.P.: UV and thermally stable superhydrophobic coatings from sol-gel processing. J. Colloid Interface Sci. **326**, 465–470 (2008)

Part III

Wearable Devices Packaging and Materials

Soft Material-Enabled Packaging for Stretchable and Flexible Hybrid Electronics 16

Herbert Robert and Woon-Hong Yeo

16.1 Introduction

Wearable medical electronics widely used today are typically rigid and bulky, which prevents long-term integration on the human body. Although these devices successfully achieve monitoring of various signals with the use of rigid electronics and packaging, new strategies are required to enable devices for continuous, mobile, and long-term applications. To achieve this new set of devices, designs and materials must allow integration onto the human skin without discomfort and allow endurance of typical strains during movement. Advancements in the field of soft materials and inorganic electronics have provided a foundation for creating stretchable and flexible hybrid electronics. The integration and packaging of the electronics rely on the application of soft materials to achieve seamless integration onto skin. Compared to conventional electronics, these wearable devices are thin and have low effective moduli, enabling conformal integration with skin and high mechanical stretchability and flexibility. In vivo demonstrations and clinical studies have begun to validate soft material-enabled electronics, which display a wide array of advantages over conventional devices. A number of considerations are required to achieve functional, high-performance wearable electronics. A comprehensive understanding of material selection for encapsulation and active layers in terms of mechanical design, electrical performance, and chemical properties is required when developing stretchable and flexible hybrid electronics. Fabrication strategies also play a significant role in manipulating and integrating materials for hybrid electronics. Recent advances in these areas have led to a variety of soft material-enabled packaging techniques that integrate high-performance electronics with soft materials in order to realize comfortable, wearable devices. This new class of electronics has enabled a wide range of applications, as illustrated in Fig. 16.1. Such devices continuously and wirelessly monitor physiological signals to improve diagnostics, real-time health tracking, rehabilitation, and human-machine interfaces. Examples include recording of acoustic speech signals, sweat content, respiratory data, pulse oximetry, and electromyography. Unobtrusive, conformal integration of electronic circuits onto skin enables long-term use without restricting patient movement or inducing discomfort. Highly stretchable and flexible sensors interfaced with these soft material packaged circuits enable high-performing, multiplex devices and provide a more comprehensive view of patient health. This additional data and broadened view, compared to the narrow window of observation in clinical settings, improves the monitoring of diseases and treatments. Collectively, this chapter provides a review of material considerations for both passive and active components when developing soft material-based stretchable and flexible hybrid electronics. Mechanical and electrical designs of recently developed concepts are reviewed. The significance of these characteristics is emphasized in a brief review of fabrication methods for wearable electronics. Integration and packaging concepts are then provided, fully incorporating material considerations and fabrication strategies to realize stretchable and flexible hybrid electronics that enable human-machine interfaces, recording of physiological signals, and more.

16.2 Materials and Designs for Stretchable and Flexible Hybrid Electronics

16.2.1 Soft Materials

To enable stretchable and flexible hybrid electronics, soft materials are often used to achieve desired mechanical characteristics and integrated with rigid components and

H. Robert · W.-H. Yeo (✉)
George W. Woodruff School of Mechanical Engineering, Wallace H. Coulter Department of Biomedical Engineering, Georgia Institute of Technology, Atlanta, GA, USA
e-mail: whyeo@gatech.edu

C. P. Wong et al. (eds.), *Nano-Bio- Electronic, Photonic and MEMS Packaging*, https://doi.org/10.1007/978-3-030-49991-4_16

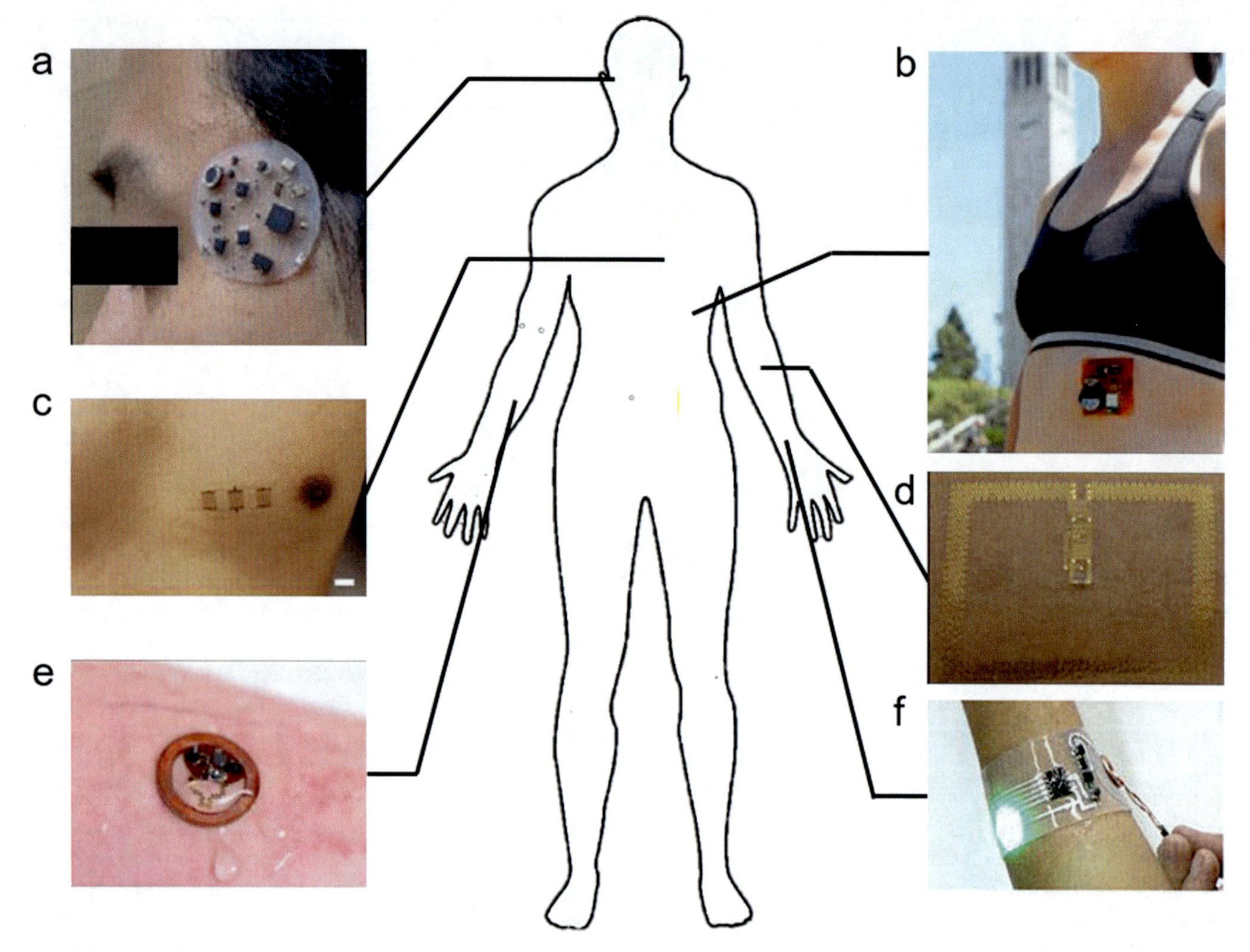

Fig. 16.1 Wearable electronics. (**a**) Wireless EOG monitor [70]. (**b**) ECG and body temperature sensor [163], Copyright 2016, John Wiley and Sons. (**c**) ECG electrodes [17], Copyright 2013, John Wiley and Sons. (**d**) RF power harvester [164]. (**e**) Wireless temperature and thermal properties of skin sensor [162], Copyright 2018, John Wiley and Sons. (**f**) Wirelessly powered circuit [159], Copyright 2018, John Wiley and Sons

other materials to achieve functional electronics. Soft materials are widely used for encapsulating layers in packaging schemes. In addition, composite materials based on elastomer matrix can create functional materials for interconnections, sensors, and other electronic components. A number of material properties are to be considered when designing wearable electronics to achieve desired performance criteria. Material modulus plays a significant role in determining overall stretchability and flexibility. Figure 16.2a displays the wide range of moduli for commonly used materials, along with the modulus of skin. Skin modulus ranges from 10 kPa up to a few hundred kPa, with elastic deformation up to 15% [1]. For optimal mechanical performance, it is important to mechanically match wearable electronics with skin to prevent device failure. Devices should apply minimal stress onto skin to ensure comfort. In order to achieve these mechanical characteristics, soft materials are often employed as substrates and encapsulation. As a substrate, the materials often act as a layer separating skin from more rigid components. Soft materials, such as silicone elastomers, have moduli similar to skin and thus are often used as device substrates and encapsulation layers to integrate more rigid materials onto skin. Unlike soft materials, metals and other active materials have moduli significantly higher than skin and are incompatible with direct integration onto skin. Commonly used substrates include silicone, such as PDMS and Ecoflex, which possess moduli similar to skin. Additional materials include Silbione, medical tapes, PVA, PET, PEN, and polyimide [2–4]. PLGA and silk fibroin are used as substrates for biodegradable devices [5, 6]. Selected substrates are often highly stretchable and flexible. Ecoflex 00-30, widely used in wearable applications, has a stretchability of 900% [7]. One demonstration of the advantage of a low moduli material is to develop skin-like electrodes [8]. Conventional rigid electrodes do not match well with skin, leading to increased

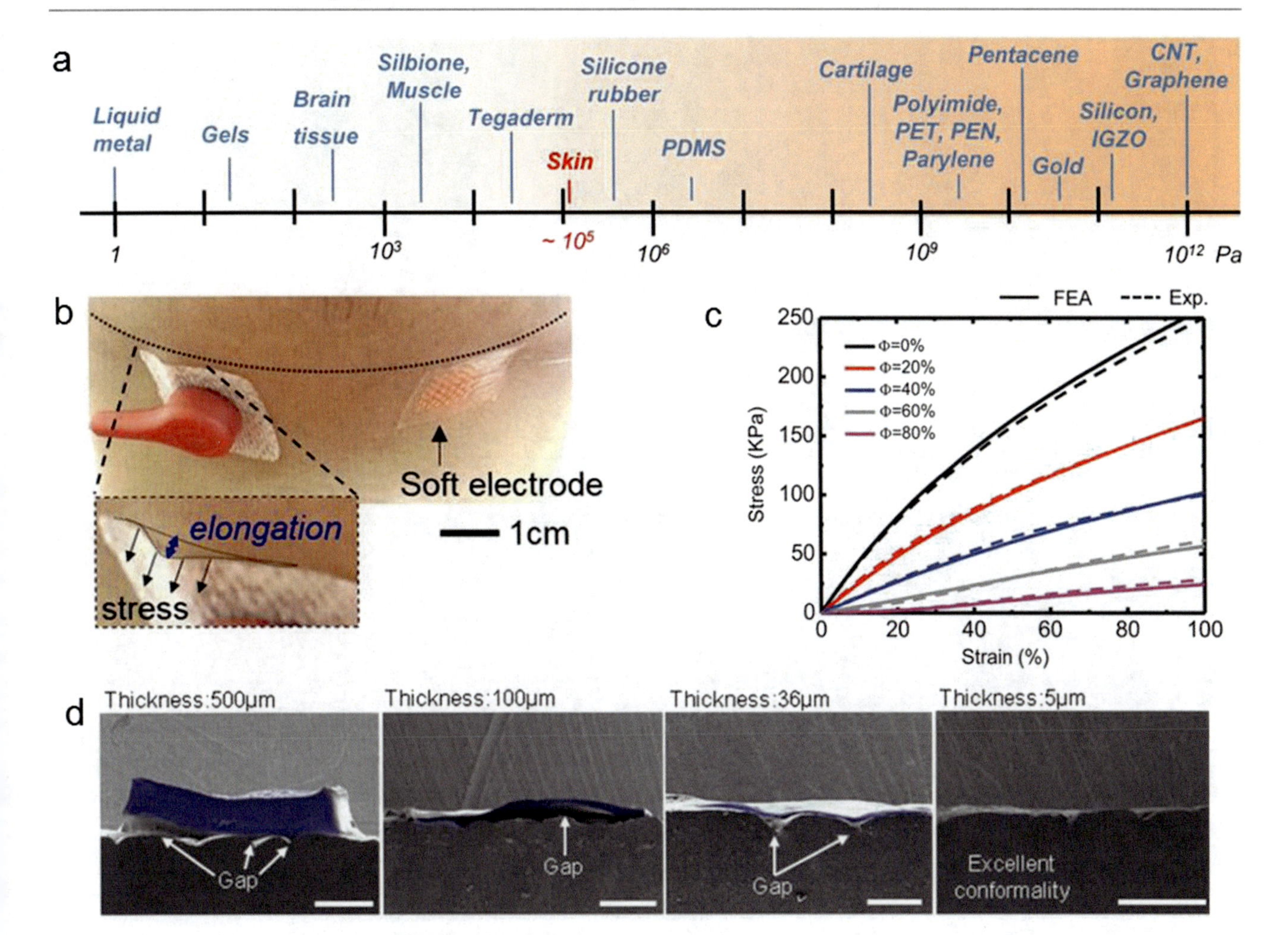

Fig. 16.2 Mechanical properties for stretchable and flexible electronics applied on the skin. (**a**) Moduli of soft materials and functional materials compared to human skin. Reprinted with permission from [4], Copyright 2017, American Chemical Society. (**b**) Soft, skin-like electrode minimizes stress on skin compared to conventional electrodes [8]. (**c**) Increasing the design porosity of a soft substrate decreases stress [9], Copyright 2017, John Wiley and Sons. (**d**) Experimental validation of the analytical model, where a lower thickness silicone layer improves conformality [11], Copyright 2013, John Wiley and Sons

stress on the skin (Fig. 16.2b). However, fabricating thin Au and PI electrodes on silicone elastomer significantly reduces stress as it seamlessly integrates onto skin. Mechanical designs may also be applied to further soften substrates and improve mechanical stability. Using a porous, open cellular design lowers the overall modulus, creating a more compliant structure and reducing stress (Fig. 16.2c) [9]. This porous design consists of a soft substrate with Au and PI films.

In addition to modulus, thickness determines flexibility of a structure. A thicker material shows a higher bending resistance and thus increased strain at a given bending radius. Flexibility is highly important to maintain conformal contact with skin. An analytical study of this interface indicated a method to identify requirements of conformal contact for different device characteristics [10]. For conformal adhesion to skin, the device is required to deform to the skin's surface waveform. In general, as material thickness increases, a corresponding increase in adhesion is required to maintain conformal contact. The filamentary serpentine epidermal electronic system (FS-EES) requires a lower adhesion value than the island-bridge system (IPS-EES) due to the former's lower metal and polymer thickness. Rougher skin, corresponding to higher-frequency or higher-amplitude sinusoid, requires a larger adhesion value to ensure seamless integration with skin. This calculation includes not only a soft substrate but also metals and polymers forming the electronics. A thicker layer, including elastomer, metal, or polymer, necessitates a more adhesive substrate. Figure 16.2d illustrates the improvement of conformality of silicone with skin by decreasing the material thickness [11]. Due to direct placement on skin, porosity and comfort are important properties for substrates, particularly for long-term applications. One wearable, intraoral device required high breathability of porous membranes [12]. An elastomeric spacer separated this porous substrate, which encapsulated the circuit and was in contact with surrounding tissue. The membrane using Soma Foama 15 (SF15) was superior in terms of porosity and permeability. Due to the increased

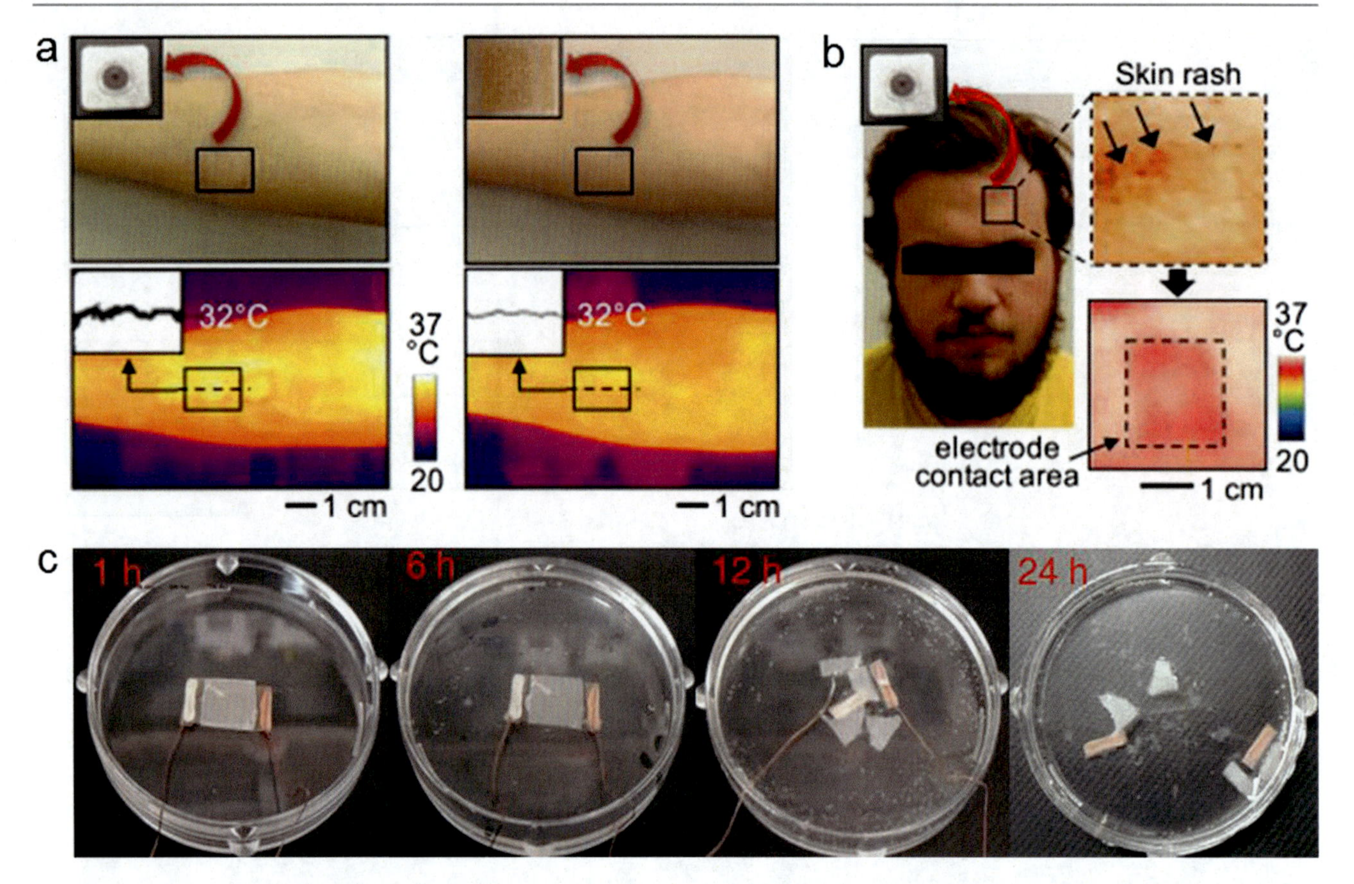

Fig. 16.3 Important substrate properties for wearable electronics. (**a**) Soft material-based electrode prevents damage to the skin, as indicated by no temperature increase compared to the conventional electrode. Reprinted from Ref. [13], Copyright 2017, with permission from Elsevier. (**b**) Skin irritation due to conventional electrodes. Reprinted from Ref. [13], Copyright 2017, with permission from Elsevier. (**c**) Degradation of a pressure sensor based on silk fibroin film [15], Copyright 2019, John Wiley and Sons

porosity, the modulus was lower than both Ecoflex 00-30 and PDMS. Moreover, SF15 showed the highest heat dissipation and lowered heat transfer to the surrounding tissues. Similar to the requirements for this intraoral device, wearable electronics integrated on the skin must minimize harm to the skin. One study compared the impact on skin from soft material-based electrodes with conventional rigid electrodes [13]. The rigid electrodes, used with electrolyte gels, caused erythema due to an increase in local temperature (Fig. 16.3a). However, over the same period of 10 h, the soft material-based electrodes with a fractal metal design caused no increase in temperature. Additionally, the conventional electrodes caused skin rash (Fig. 16.3b). Transparent encapsulation layers may be beneficial for select applications, such as for a sweat sensor that changes colors based on sweat contents [14]. The clear layers allow for optical monitoring. With application on human skin, biocompatibility is a significant consideration. Additionally, temporary electronics may be designed as biodegradable, as shown in Fig. 16.3c [15]. This capacitive pressure and strain sensor utilizes a silk fibroin film, which degrades over 24 h. Most common soft substrates discussed are considered biocompatible [2].

Along with mechanical characteristics, electrical performance is thoroughly considered to ensure wearable devices are reliable, high-performance electronics. Conformal contact significantly affects electrical performance, particularly for electrodes. More conformal contact reduces impedance, which improves signal quality and enables classifications that are more accurate. Material selection also affects signal-to-noise ratio (SNR). A high SNR is desired for high-quality signals and more accurate signal identifications. Figure 16.4a indicates that compared to a conventional needle electrode that penetrates the dermis, a soft material-enabled "biostamp" still maintains a functional signal-to-noise ratio for monitoring [16]. Figure 16.4b indicates that a fabricated electrode utilizing thin film metal on an elastomer membrane achieves a higher SNR than conventional electrodes, which may be caused by a difference in conformal skin contact [13]. Conformal contact, achieved with soft material packaging, reduces the impact of device movement on the recorded signals. These electrodes record high-quality signals of eye movements, as shown in Fig. 16.4c. Similarly, silicone-based electrodes achieve high-quality ECG signals similar to those recorded with conventional electrodes (Fig. 16.4d)

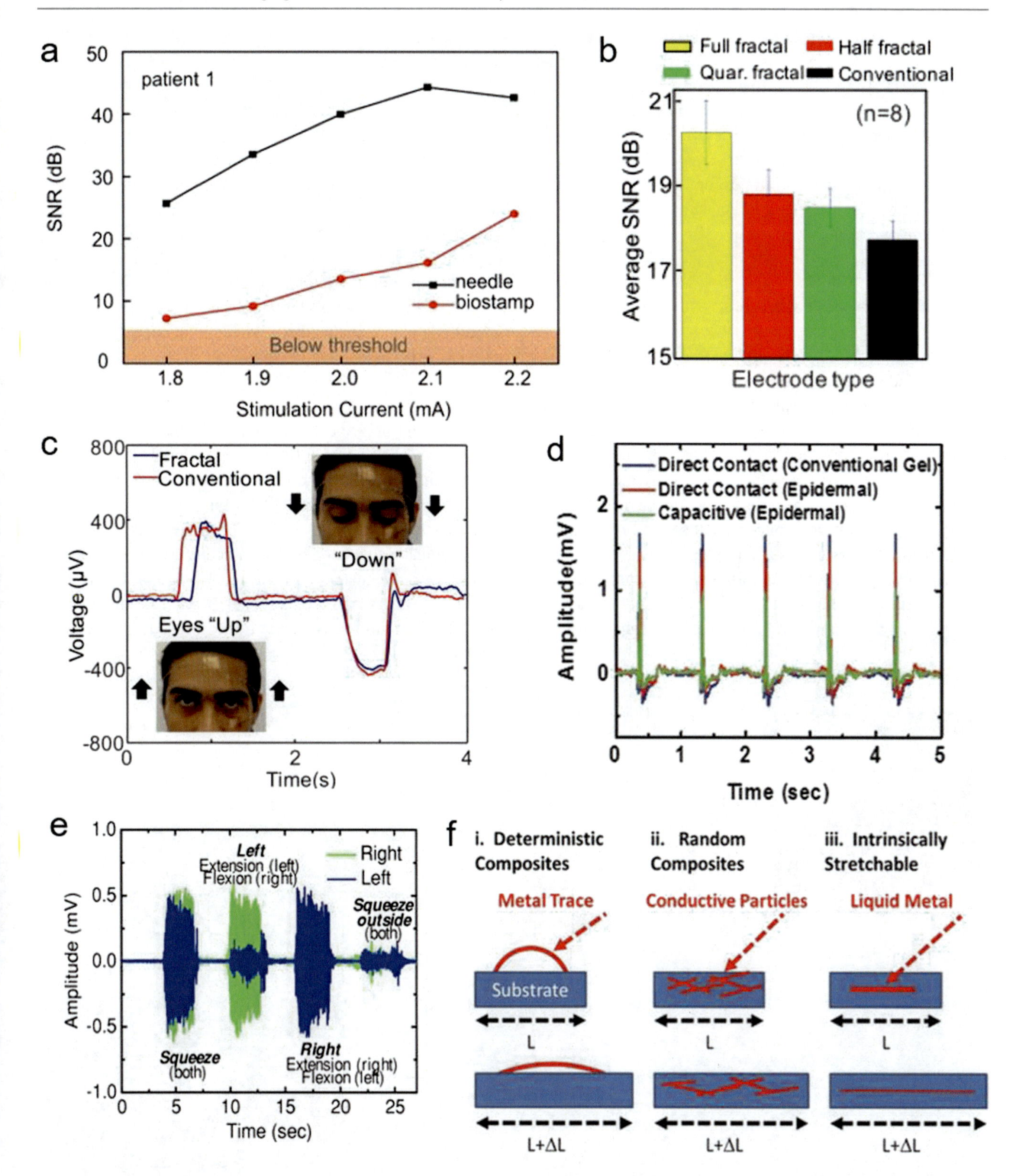

Fig. 16.4 Electrical signals and designs achieved with soft materials. (**a**) A hybrid electronic device achieves high signal quality, eliminating the need for invasive needle electrodes [16]. (**b**) An electrode consisting of thin metal film and elastomer membrane improves SNR. Reprinted from Ref. [13], Copyright 2017, with permission from Elsevier. (**c**) High-quality EOG signals recorded with soft electrode. Reprinted from Ref. [13], Copyright 2017, with permission from Elsevier. (**d**) ECG signals recorded with soft electrodes compared to conventional, rigid electrodes [17], Copyright 2013, John Wiley and Sons. (**e**) EMG signals recorded for human-machine interface [11], Copyright 2013, John Wiley and Sons. (**f**) Strategies for forming stretchable and flexible hybrid electronics with soft materials [85], Copyright 2017, John Wiley and Sons

[17]. Surface EMG signals have also been monitored for a human-machine interface using soft, fractal electrodes (Fig. 16.4e) [11]. To achieve higher-performing electronics, strategies have been developed for the integration of electronic components onto soft substrates. Soft materials may be used as a substrate and as the matrix material for composite structures. Figure 16.4f illustrates options for fabricating stretchable and flexible electronics, which are discussed in the following sections.

16.2.2 Functional Materials

While these organic, soft materials are often used as substrates and encapsulation layers due to their advantageous mechanical properties and biocompatibility, electronics require functional, or active, materials to form the electrical components. The previously discussed soft materials are generally poor electrical conductors and semiconductors. Thus, classes of active materials are utilized and may be achieved via two distinct processes: deterministic and intrinsic. Deterministic designs use inorganic, rigid materials, such as metals, in thin films. These thin films are highly flexible, and fabricating the films into various designs can achieve high stretchability. Intrinsic designs modify inherently soft materials with nanomaterials or other fillers to enhance electrical properties.

16.2.2.1 Deterministic Designs

Inorganic materials are generally used to achieve high-performance electronics due to their favorable properties as conductors and semiconductors. Inorganic semiconductors are the basis for existing conventional electronics [18]. These materials may be integrated with soft materials for functional, wearable devices. Thin films of these materials, including silicon, gold, copper, and silver, with optimized designs can allow the electronics to achieve high flexibility and stretchability when integrated with soft materials [4].

Thin Films

A vast amount of research has been performed to manipulate the mechanical structures of inorganic materials for wearable applications. Because bending strain is proportional to film thickness and bending radius, thin films of these materials are used to achieve flexibility [19]. The Euler-Bernoulli beam theory indicates that bending resistance is proportional to the thickness of the material to the third power [18]. This is demonstrated in Fig. 16.5a for silicon nanomembranes [20]. The bending resistance of the films decreases with

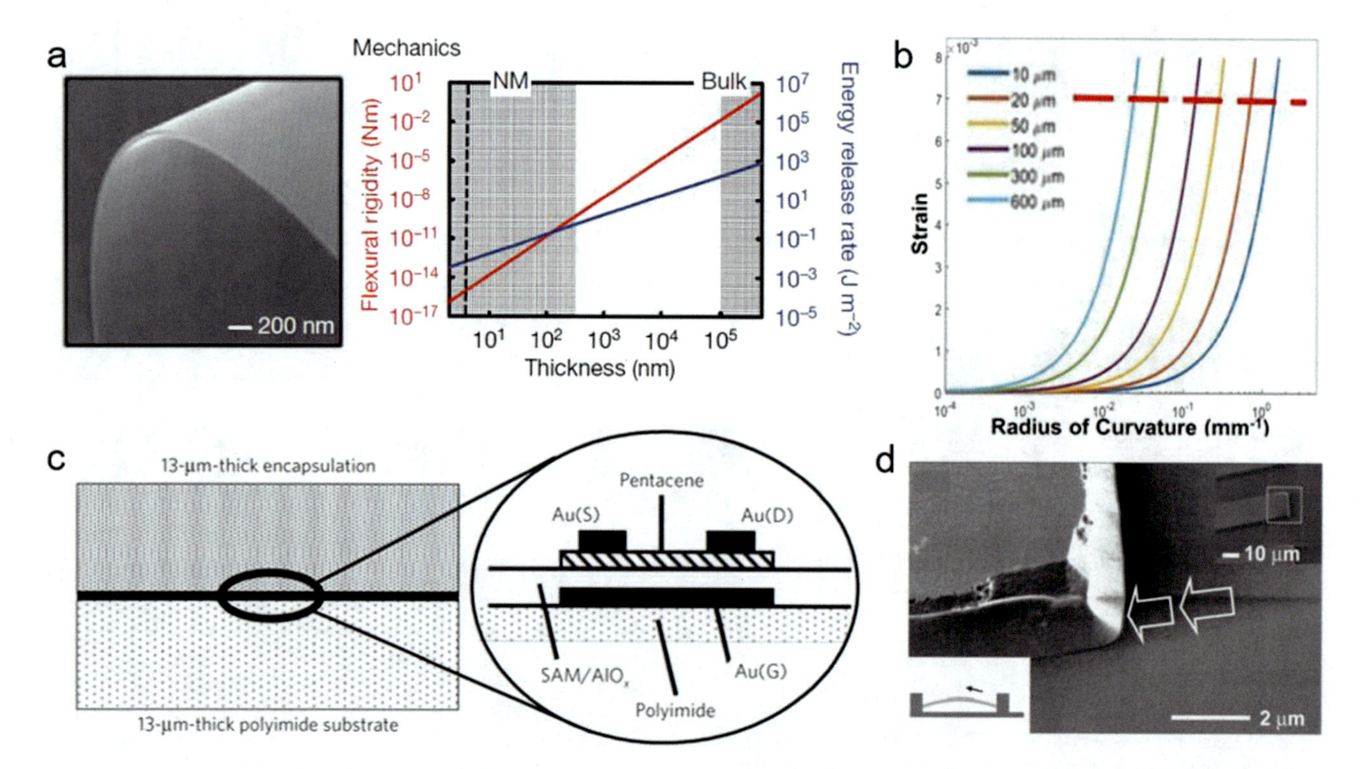

Fig. 16.5 Thin film flexibility and integration with soft materials. (**a**) Lower flexural rigidity of silicon nanomembrane by decreasing thickness. Reprinted by permission from Springer Nature: [20], Copyright 2011. (**b**) Strain induced by bending decreases for a thinner silicon chip [21]. (**c**) Placement of thin films and fragile components near the neutral plane of bending protects the components from fracture. Reprinted by permission from Springer Nature: [22], Copyright 2010. (**d**) Adhesion between thin films and soft substrates is important to prevent premature failure from slipping or delamination [23], Copyright 2008, John Wiley and Sons

thickness. Thus, for a given bending curvature, a thinner film will result in less strain. Figure 16.5b exemplifies the lowering of strain for a thinned silicon chip [21]. The dashed line indicates the strain of fracture. Due to this strain reduction, thin films are widely used in flexible electronics for mechanically stable configurations. To form such thin films, mechanical exfoliation, etching procedures, or epitaxial growth, or a combination of these, may be used [20]. The use of soft material layers can further reduce strain by positioning thin films at the neutral plane of bending, where compressive and tensile strains balance. This can be accomplished by using encapsulating materials or layers to position sensitive components at the neutral plane (Fig. 16.5c) [22]. By placing a transistor at the neutral plane, a bending radius of 100 μm was achieved before damage, compared to a bending radius of 3.5 mm when not located at the neutral axis. Adhesion between soft materials and thin films is also highly important for seamless integration. Weak adhesion between thin films and substrates can result in premature failure due to slipping or delamination from the substrate (Fig. 16.5d) [23]. Rather than cracking, the thin film may slip indicating that the interfacial shear stress between the substrate, or adhesive layer, and thin film can dictate bendability rather than fracture strain of the film. Delamination may occur because of interfacial normal stresses. The strain can be locally reduced and more uniformly distributed by increasing adhesion [4]. Poor adhesion causes increased strains where the thin film detaches from the substrate. Thin films can also be manipulated to alter thermal properties [20]. Adding an array of holes into a silicon nanomembrane can yield a thermal conductivity about 80 times smaller, without a change in electrical conductivity. However, wearable electronics also require high stretchability, which thin films do not readily achieve. Even though wearable applications can require 100% or more stretchability, many thin films fail with a strain near 1% [24]. Therefore, these thin films may be manipulated with the following fabrication techniques to achieve stretchable electronics.

Wrinkled Films

A number of methods for forming stretchable thin films have been researched and developed via analytical, computational, and experimental approaches. One design is to form the thin films in wavy, or wrinkled, structures on soft material substrates. A number of fabrication methods have been used to achieve this wavy structure, including mechanical straining, thermal straining, and ultraviolet/ozone radiation [25, 26]. Mechanical straining can achieve higher levels of prestrain compared to others [26]. Other methods may be applied to achieve a wrinkling design, such as via laser treatments and swelling [25]. The method of straining the substrate prior to integration of a thin film to create a wavy film structure is illustrated in Fig. 16.6a [27]. The substrate is mechanically strained prior to the thin film being transferred onto it [28]. Releasing the strain then causes the film to wrinkle due to translation of compressive strains into bending strain [29]. One such film is shown in Fig. 16.6b [30]. The level of prestrain is one method to control the resulting wavy design. This mechanical method can be further extended to 2D wavy structures, similar to that shown in Fig. 16.6c [31]. Here, the soft substrate is strained biaxially prior to the thin film placement. Releasing the strain forms the 2D wrinkled structure, which allows stretchability in two directions [31–34]. A common form is the herringbone as this possesses the lowest energy [31]. These wavy designs provide stress-strain curves resembling skin's curve (Fig. 16.6d) [28]. These designs show a bilinear shape, in which stress increases rapidly once a certain strain is reached. This fabrication method can also be tuned to achieve different wrinkled designs with the use of analytical models of buckling [31, 35]. Adding ridges to the substrate results in a non-uniform distribution of stress and yields a different wrinkled design [36]. Additionally, a more exact method of tuning wrinkled films is by patterning adhesion locations on a substrate. One method passed deep ultraviolet light through a photomask to convert the hydrophobic PDMS surface to an activated, highly polar surface. This activated surface is then able to form strong bonds with inorganic surfaces [30]. After transferring film ribbons to the substrate, the layers bond at the predetermined locations, yielding a more finely tuned film. Thus, geometry of the wavy patterns is more precisely controlled. Multiple types of patterns can be achieved with these processes and may be based on the buckling modes in the thin film [37–39]. The substrate thickness also impacts the buckling method and resulting design (Fig. 16.6e) [40]. The determination of global or local buckling is also based on modulus and the thickness ratio of the two materials. For PDMS and Si, a thickness ratio greater than 1000 leads to local buckling, which is the more commonly used method. Other parameters, such as wavelength and amplitude of the wavy pattern, are also dictated by mechanical properties and design [35, 41]. To enable design of such patterns, studies have introduced a variety of models to predict features [37, 42–44]. Materials used in these patterns include Si ribbons, metals, piezoelectrics, and nanomaterials [26]. This wavy pattern may also be applied for individual connections between rigid components, also termed as arc-shaped bridges [45, 46]. One study developed an array of solar cells with thin interconnects that was then transferred to a biaxially prestrained PDMS substrate [45]. This results in the interconnections buckling into premade trenches of the structured PDMS.

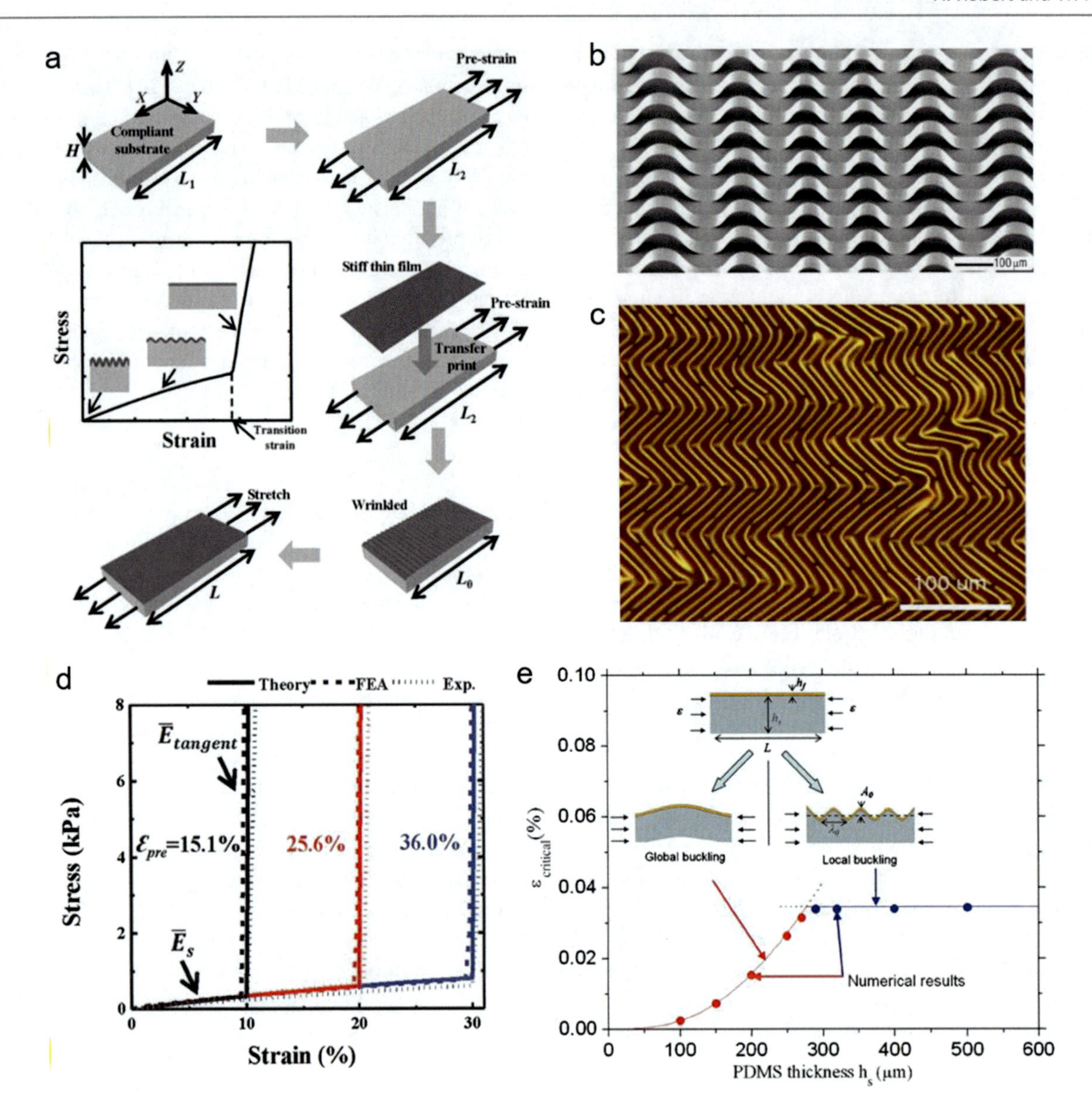

Fig. 16.6 Wrinkled and wavy thin films achieved with soft substrates. (**a**) Fabrication process of wrinkled thin films on soft substrate [27], Copyright 2016, John Wiley and Sons. (**b**) Wrinkled ribbons on soft material. Reprinted by permission from Springer Nature: [30], Copyright 2006. (**c**) 2D wrinkled film fabricated via biaxial prestrain. Reprinted from Ref. [31], with the permission of AIP publishing. (**d**) Bilinear stress-strain curve of wrinkled films for different prestrains [27], Copyright 2016, John Wiley and Sons. (**e**) Impact of substrate thickness on the resulting wrinkled film shape. Reprinted from Ref. [40], with the permission of AIP publishing

Serpentine Structures

In addition to wavy nonplanar structures, planar designs can be used to achieve high stretchability of thin films. The stretchable films or interconnects are formed from rigid materials on soft substrates by depositing materials in patterns. Numerous devices have utilized such designs, including wearable CMOS inverters, LED arrays, and wireless sweat sensors, among others [47–49]. A vast range of patterns have been developed and tested, such as horseshoe patterns, serpentines, and fractal-inspired designs. Unlike the wrinkled patterns described previously, these patterns are generally planar until stretched. One common design is a serpentine-shaped pattern, as shown in Fig. 16.7a, formed by repeating unit cells of arcs and lines [50, 51]. The pattern is generally formed for metal interconnections, which is sandwiched by PI layers for support (Fig. 16.7b). A number of methods may be used to fabricate these patterns, including lithography techniques and direct printing [48]. When this design is stretched, the pattern straightens and deforms in and out of plane. While these patterns unfold, the strain on the rigid material remains low, minimizing changes in resistance (Fig. 16.7c) [52]. Figure 16.7d shows a serpentine design being stretched on a soft substrate. A number of analytical and computational models have been developed for

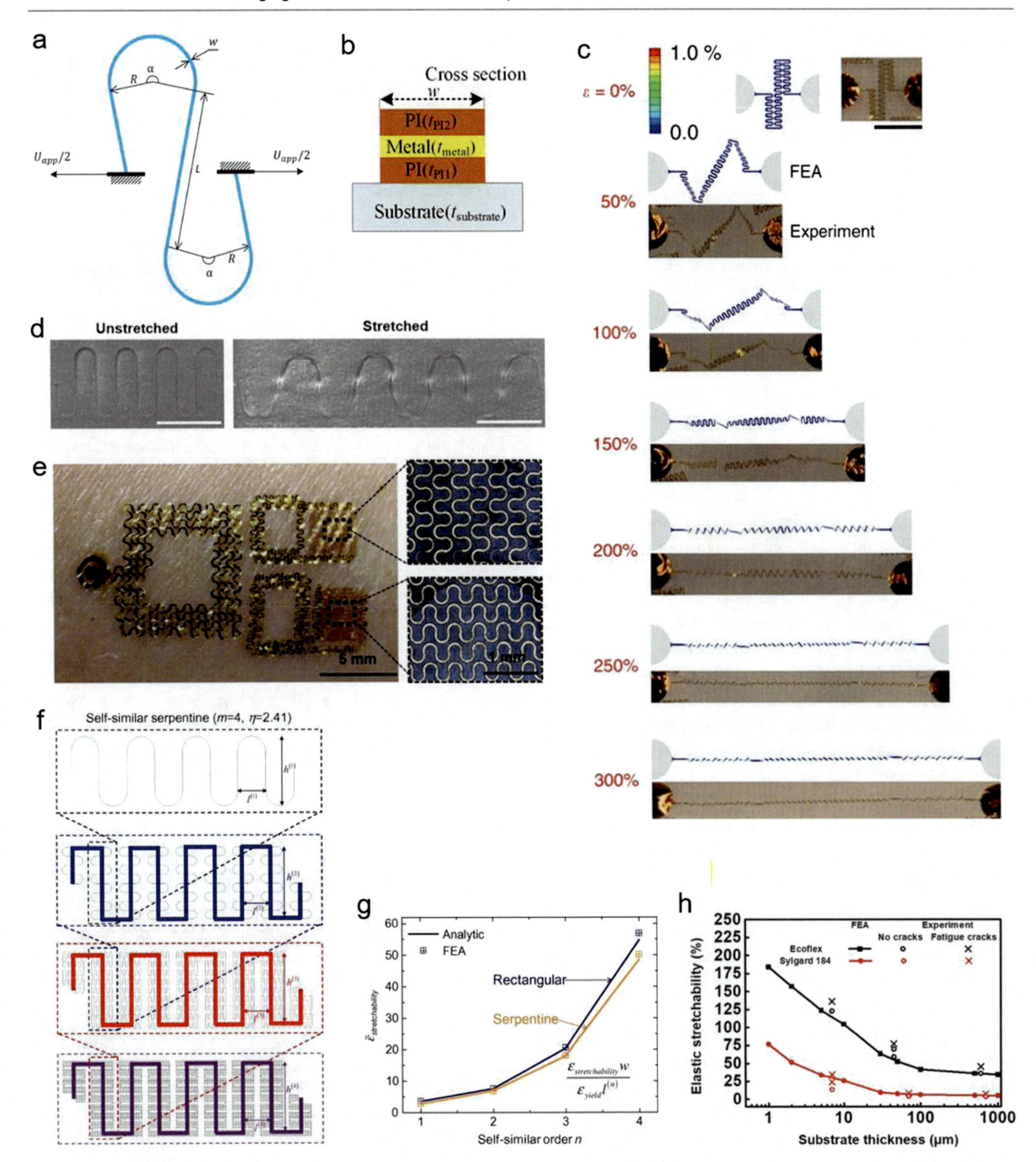

Fig. 16.7 Serpentine designs for stretchable electronics. (**a**) Unit cell of serpentine pattern. Reprinted from Ref. [50], Copyright 2016, with permission from Elsevier. (**b**) Typical fabrication layers consisting of PI layers supporting a thin metal film on a soft substrate. Reprinted by permission from Springer Nature: [54], Copyright 2017. (**c**) FEA of serpentine design unfolding when stretched. Reprinted by permission from Springer Nature: [52], Copyright 2013. (**d**) Stretching of serpentine pattern interconnect on soft substrate [65], Copyright 2017, John Wiley and Sons. (**e**) Wireless epidermal sensor utilizing serpentine geometries [57], Copyright 2014, John Wiley and Sons. (**f**) Illustration of self-similar serpentine pattern of orders 1–4. Reprinted from Ref. [51], Copyright 2013, with permission from Elsevier. (**g**) Stretchability of serpentines increases for higher-ordered patterns. Reprinted from Ref. [51], Copyright 2013, with permission from Elsevier. (**h**) Substrate thickness on the stretchability of serpentine patterns [65], Copyright 2017, John Wiley and Sons

comprehensive testing and optimization of these designs, such as the hierarchical computational model and scale law [50, 53, 54]. The serpentine design is identified by geometric parameters, such as line width, arc radius, arc angle, and straight line length [55]. Serpentines can also be in a noncoplanar form, allowing further stretching [19, 56]. These are fabricated with similar procedures as prestraining substrates for the wrinkled films. One device utilized these serpentine patterns to form a wireless epidermal sensor of strain and hydration (Fig. 16.7e) [57]. The patterns enable conformal contact and endure strain when on skin. This serpentine concept can be further extended to self-similar designs to achieve even higher stretchability. Here, the pattern is duplicated along the larger pattern, creating higher-ordered serpentine structures (Fig. 16.7f) [51]. Using a self-similar pattern and condensing the feature size, higher stretchability is achieved. The self-similar order, spacing ratio, number of unit cells, and shape combine to determine the overall stretchability of the design. Before the successive unraveling of each order, the highest order of the pattern unfolds first when strained since they easily deform (Fig. 16.7g) [58]. The higher-ordered structures reach higher stretchability with a compact design, but at the cost of increasing electrical path length and resistance. Fractal designs may also be used to enable higher flexibility and stretchability of devices [58, 59]. Such designs may be laminated onto freestanding or encapsulated substrates [60–62]. These freestanding designs can be achieved by bonding the ends of the serpentine connection, allowing the patterned film in between to delaminate [56]. Here, deformations occur both in plane and out of plane based on the thickness-to-width ratio [55, 63, 64]. For both deformation types, a variety of models predict stretchabilities of the freestanding and laminated serpentines in terms of thickness-to-width ratio, geometrical ratios, arc angles, and substrate properties [26]. For encapsulated serpentines, the patterns can be located closer to the neutral plane to lower bending strain. In addition to design of the rigid materials and integration onto substrate, the effect of substrate properties has also been studied [65]. One study determined that a critical substrate thickness exists, in which substrates below this thickness display a significant increase in stretchability by lowering the thickness (Fig. 16.7h). However, above the critical thickness, lowering and increasing the thickness shows minimal changes. The stretchability is enhanced due to the global buckling that occurs with thinner substrates. Another design variation of serpentine interconnects is using thick geometries instead of thin layers, which causes a scissor-like stretching mechanism [66]. By using the thicker design, electrical resistance can be lowered. These design strategies, including wrinkling and serpentine structures, also offer a unique property similar to skin and the wrinkled films [28]. These film designs exhibit curves similar to the J-shaped stress-strain curve of skin. Due to this skin-like property, layers with serpentine or fractal designs may be integrated between the skin and electronics to guard the electronics from high deformations [67]. This yields a strain-limiting layer beneficial for packaging to protect the electronics and skin.

Spirals and Helices

Another design for stretchable interconnects is a 2D spiral. The planar coil unwinds when stretched and can achieve higher stretchabilities per initial area than serpentines (Fig. 16.8a) [68]. Similar to serpentines, both in-plane and out-of-plane deformation occurs. Moreover, this spiral design can be integrated with serpentine structures to achieve even larger stretchabilities [69]. Modifying the spiral design in one study achieved a stretchability over 250%, with fracture occurring near 325%. The modified design utilized longer straight portions, as opposed to a continuous spiral, which enabled more out-of-plane deformation (Fig. 16.8b) [68]. In addition to planar spirals, 3D helices are another form of stretchable interconnects (Fig. 16.8c) [70]. 3D helical interconnects can offer spring-like structures. Printing techniques, manual wrapping, molding, and buckling methods exist to fabricate these structures [26]. Fabricating via buckling is accomplished similar to the wavy structures, as it is fabricated on a prestrained substrate with prescribed bonding locations [70, 71]. The bonding locations here were fabricated by depositing Ti/SiO2 onto a silicone elastomer. Then, transferring the 2D design onto the prestrained substrate and heating it formed strong siloxane bonds. The controlled buckling of PI and Au or Cu interconnects results in the helical configuration. Similar to other designs, this buckling method has associated theoretical models for designing the helix [72]. These 3D structures can more smoothly distribute stresses compared to 2D interconnects (Fig. 16.8d–f). These 3D interconnects are more independent of the substrate and allows for more uniform stress across the cross-section of the traces. Based on the substrate prestrain, the interconnect stretchability can exceed between 3 and 9.5 times that of 2D designs for prestrain between 50% and 300%.

16.2.2.2 Intrinsic Designs

Unlike the deterministic designs, intrinsic designs apply and modify intrinsically soft materials to improve electrical properties. Examples of intrinsic functional materials include hydrogels and liquid metals, which are inherently soft and may be manipulated to improve electrical functionality. Conductive polymers have also been developed and tested to achieve mechanical properties similar to polymers. Additionally, nanomaterials, such as CNTs and nanoparticles, are being applied to form composite structures, such as highly stretchable and conductive electrical interconnects.

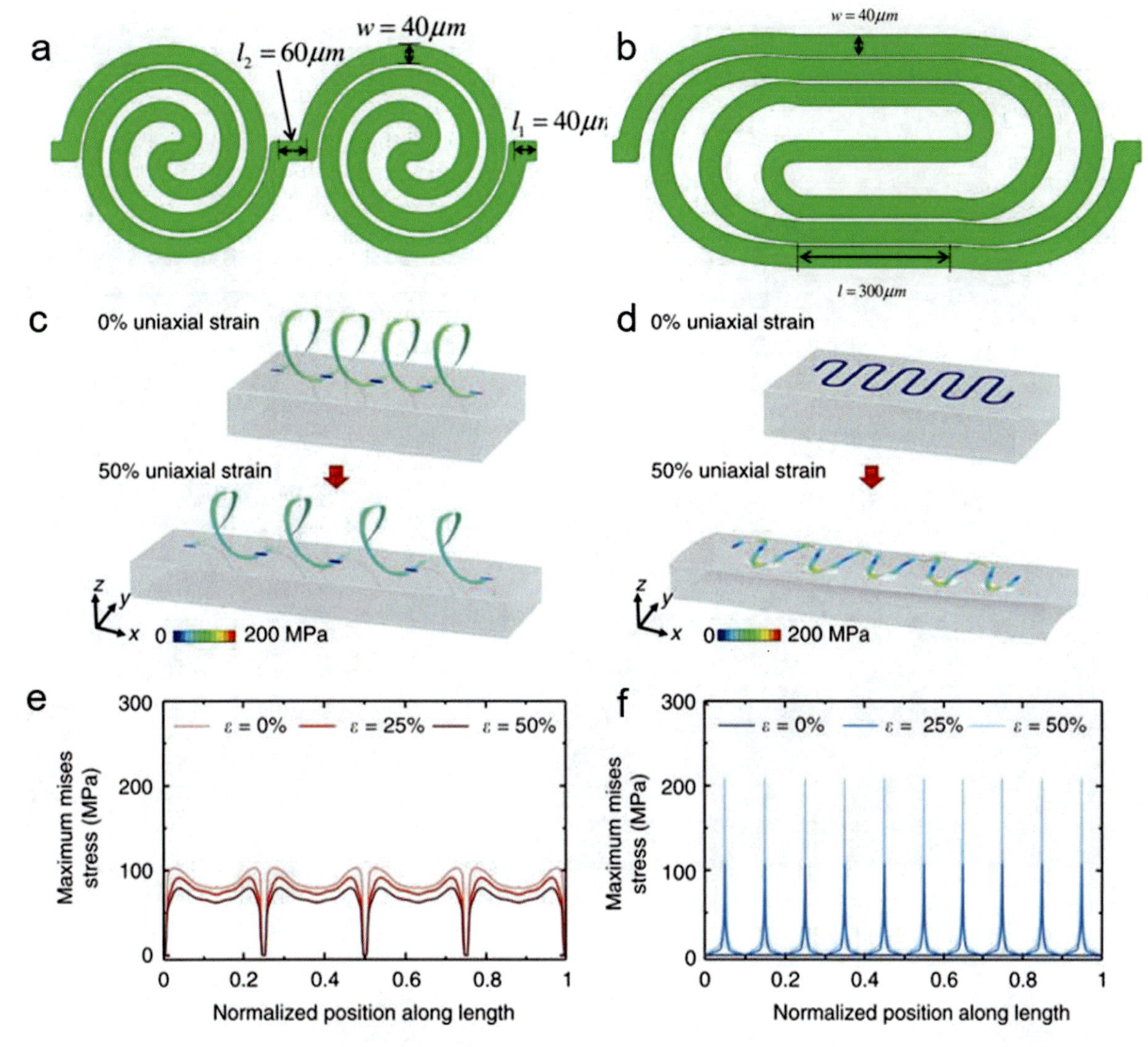

Fig. 16.8 Spiral and helical interconnects. (**a**) Planar coil interconnect that unwinds when stretched. Reprinted from Ref. [68], Copyright 2014, with permission from Elsevier. (**b**) Modified spiral coil to achieve higher stretchabilities. Reprinted from Ref. [68], Copyright 2014, with permission from Elsevier. (**c**) Helical coil interconnect compared to (**d**) serpentine interconnect [70]. (**e**) Helical interconnects distribute stress more uniformly than (**f**) spiral interconnects [70]

Hydrogels

One category of these soft materials is hydrogels, which are polymeric networks with 3D microstructures [73]. One study modified the synthesis of the hydrogel to achieve an increased conductivity of 0.11 Scm^{-1}. Inkjet printing and spray coating of this hydrogel was achieved to form supercapacitors and glucose oxidase sensors, indicating compatibility with existing processes. Hydrogels have high stretchability and self-healing capabilities [74]. A self-healing hydrogel, integrated with CNTs, graphene, and Ag NW, resulted in a piezoresistive sensor with a maximum strain of 1000% [75]. The self-healing feature of hydrogels can endure damage, which is important for long-term uses [76, 77]. To improve conductivity of hydrogels, nanomaterials can be mixed with it. Mechanical properties of these materials can also be tuned with nanocomposites, fiber reinforcement, and networks [78–80]. One study developed a hydrogel capable of 2000% stretching or 1700% stretching despite introducing a notch (Fig. 16.9a-b) [78]. Here, the alginate network unzips to dissipate strain energy, while the covalent bonds formed via the polyacrylamide network allow the hydrogel to recover. Another study formed a hydrogel to function as a substrate and encapsulation [81]. Electronic components were embedded into the hydrogel, and the hydrogel could deform without the components debonding. As a demonstration, an LED array was formed and stretched (Fig. 16.9c). Another example of a hydrogel substrate is where an adhesive hydrogel was used for transfer printing of prefabricated electronics [82]. One study developed a stretchable hydrogel circuit, where a

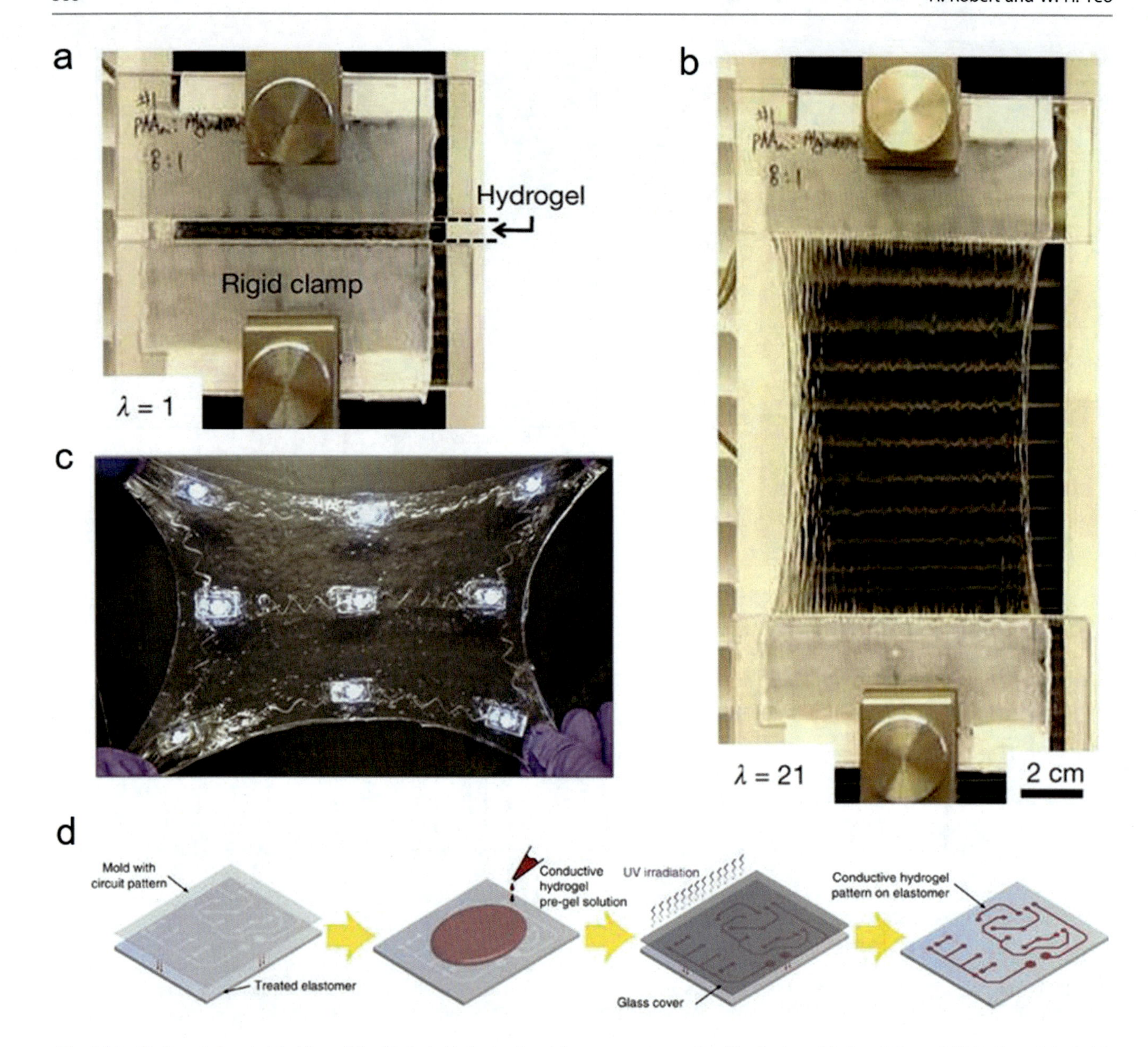

Fig. 16.9 Hydrogels for stretchable and flexible hybrid electronics. (**a**) Hydrogel capable of (**b**) 2000% stretching without fracture. Reprinted by permission from Springer Nature: [78], Copyright 2012. (**c**) LED array encapsulated by hydrogel being deformed [81], Copyright 2015, John Wiley and Sons. (**d**) Illustration of fabrication process for patterning conductive hydrogel on elastomer [83]

hydrogel was patterned onto elastomer and could endure large deformations, such as 100 cycles of 350% stretching [83]. Figure 16.9d illustrates the fabrication process of patterning hydrogel onto an elastomer substrate. Hydrogels are also generally biocompatible and can be biodegradable [2]. The water in hydrogels may be lost over time, resulting in stiff hydrogels, but one study replaced the water with an organic solvent to enable long-term uses [84].

Liquid Metals

Resistance of a material is highly important, particularly how it changes when strain is applied. Liquid metals show one of the highest conductivities at maximum strain, as shown in Fig. 16.10a [26]. Liquid metals are intrinsically conductive and deformable. Gallium alloys, specifically EGaIn, are commonly used liquid metals in wearable devices. Gallium possesses a range of favorable properties, including low viscosity, high electrical and thermal conductivity, and low toxicity [85]. One study coated Ag NP traces with the liquid metal EGaIn to enhance electrical conductivity and improve stretchability (Fig. 16.10b) [86]. Additionally, the liquid metal reacts with oxygen to form a thin oxide layer in ambient conditions [87]. Without affecting the electrical characteristics of the liquid metal, this oxide layer enables

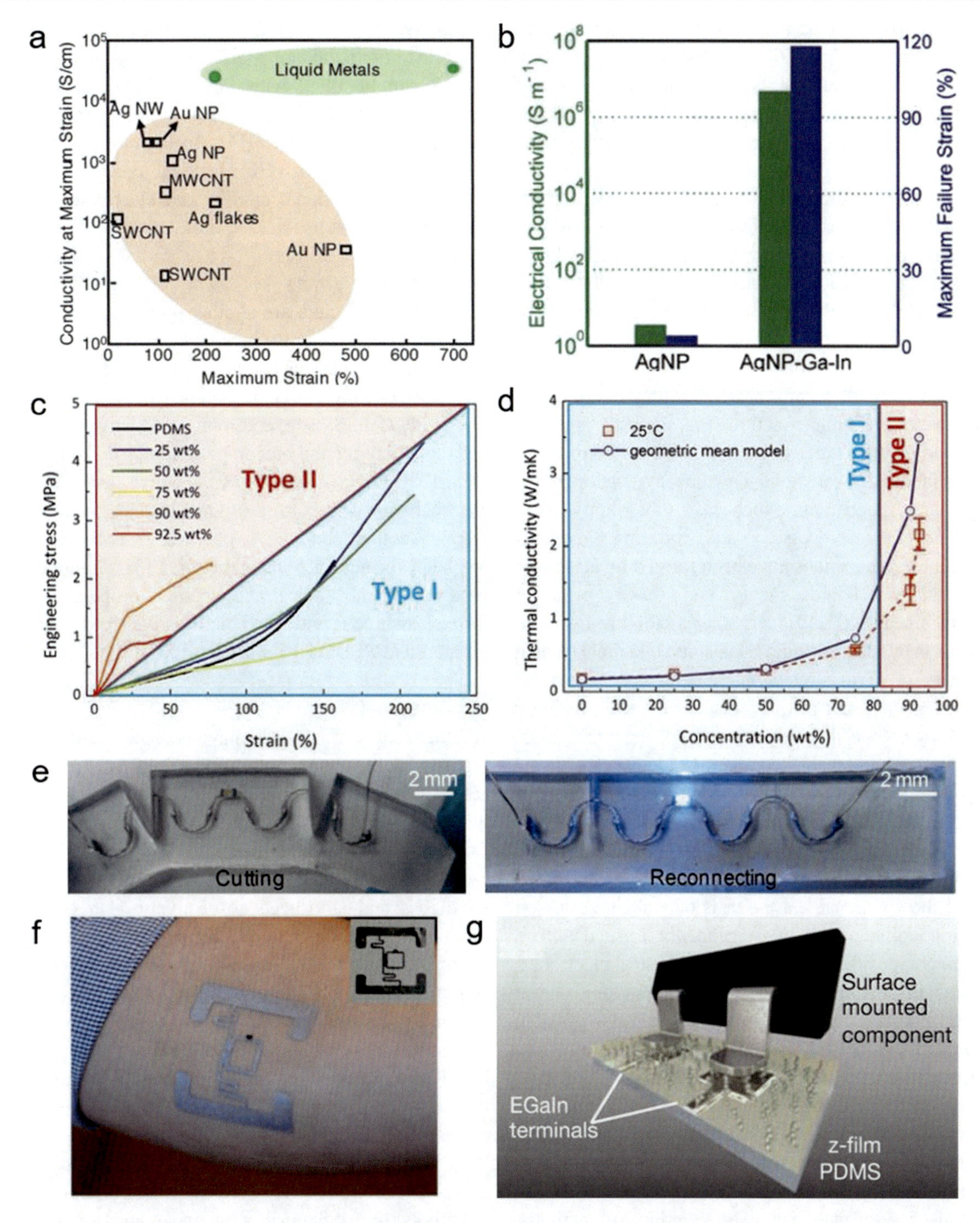

Fig. 16.10 Liquid metal and conductive polymers for wearable electronics. (**a**) Liquid metals provide a high electrical conductivity despite high strains [26], Copyright 2018, John Wiley and Sons. (**b**) Conductivity and maximum strain of Ag NP traces were improved by coating with liquid metal [86], Copyright 2018, John Wiley and Sons. (**c**) Stress-strain curve of PDMS modified with different fill fractions of liquid metal [89]. (**d**) Improvement of thermal conductivity of PDMS- liquid metal composite by increasing liquid metal concentration [89]. (**e**) Display of self-healing capability of liquid metal circuit after being separated. Reproduced from Ref. [95] with permission from the Royal Society of Chemistry. (**f**) Printed liquid metal on PDMS forming a highly conformal UHF RFID tag. Reproduced from Ref. [101] with permission from the Royal Society of Chemistry. (**g**) Electrical vias through a PDMS Ag-Ni microparticle composite for liquid metal circuits. Reprinted with permission from Ref. [104], Copyright 2015, American Chemical Society

the liquid metal to adhere to surfaces without limitations from surface tension. Biocompatibility has not been widely studied, but drug deliveries utilizing liquid metals have been used [88]. Thermal management is also critical for electronics, and wearable devices cannot use conventional cooling methods such as fans and heat sinks. Liquid metals address this due to their high thermal conductivity, enabling removal of heat [89, 90]. One work mixed a gallium-based liquid alloy into a PDMS elastomer matrix, maintaining a high stretchability while enhancing thermal conductivity [89]. A fill fraction greater than 75% resulted in a change of mechanical properties, as observed in Fig. 16.10c. In this range, a significant increase in thermal conductivity occurred, from 0.17 W/mK of PDMS to 0.58 W/mK at 75% filling fraction and 2.2 W/mK at 92.5% filling fraction (Fig. 16.10d). In order to apply liquid metals to stretchable and flexible electronics, a number of fabrication methods have been developed for patterning and depositing, such as lithography-based methods, injection processes, and printing methods [85]. Stretchable connections have been formed by elastomer fibers containing liquid metal and have achieved stretchabilities near 800% [91]. Another method to integrate liquid metals is by placing liquid metals into channels formed in elastomers to form stretchable, conductive elastomers [92]. Liquid metals can also be dispersed into elastomer prior to curing. Since the metals are fluid, the mechanical properties of the stretchable conductors are largely determined by the chosen elastomer [93]. Unlike rigid particles embedded in elastomers, liquid metals do not alter the composite's mechanical properties as significantly and show minimal hysteresis [93, 94]. The maximum strain and modulus of the composite elastomer is dependent solely on the inclusion volume fraction. A capacitor formed with a liquid metal composite dielectric and liquid metal electrodes, encapsulated in silicone, achieved 700% stretching before failure [93]. Similar to hydrogels, self-healing is also possible for liquid metal composites due to the oxide layer formed by the liquid metal [95, 96]. One study utilized an inkjet printing system to pattern a Galinstan circuit [95]. When joined back together after cutting and separating the circuit, the liquid metal self-heals, and the electrical pathway is reestablished (Fig. 16.10e). Due to the thin oxide layer formed with air, liquid metals can be patterned with a variety of methods, including soft lithography processes, printing, and injection [97–100]. One study used a stencil to print a liquid alloy onto semi-cured PDMS before electronic components were placed on top with a final top PDMS encapsulation [101]. This method was applied to form a UHF RFID tag, including both chip and antenna that conformally integrated onto skin (Fig. 16.10f). In addition to these dispersions and patterned depositions, liquid metal networks can be formed. One example of a network is using a 3D grid nanostructure of PDMS and then injecting liquid metal into the channels [102]. A conductivity over 24,100 Scm^{-1} was achieved at 220% strain. Similarly, a more random, sponge-like structure in PDMS can be filled with liquid metal [103]. One study developed vertical columns of Ag-Ni microparticles for electrical vias through PDMS (Fig. 16.10g) [104]. This conductive composite allows for electrical connections between embedded liquid metal circuits and electronic components mounted on the substrate.

Conductive Polymers

Conductive polymers are another option for combined conductivity and polymer properties. Conductive polymers are a set of organic materials with π-conjugated systems, whose conductivity may be controlled via material modifications [3]. PEDOT:PSS is a common conductive polymer with high conductivity and biocompatibility [105, 106]. For example, PEDOT:PSS was used as the active layer in an organic electrochemical transistor for monitoring glucose in saliva [107]. Another conductive polymer is polypyrrole, which has been formed into nanostructures [3]. Additives may be used to alter mechanical and electrical properties; one study formed solar cells and wearable resistive strain sensors based on modified PEDOT:PSS films [108].

Nanomaterials

A number of nanomaterials have become available for creating stretchable, composite conductors. The nanomaterials, acting as nanofillers, provide an electrical connection within the elastomer matrix. These nanomaterials are generally categorized as zero-dimensional, one-dimensional, and two-dimensional. This includes nanoparticles, nanowires or nanotubes, and nanosheets, respectively. Matrix materials are often a soft, elastomer matrix. One form of nanomaterials is nanoparticles, a zero-dimensional nanomaterial. Nanoparticles are widely used in forming thin films and conductive elastomers. One device applied Ag NPs in curved arrays through template-induced printing as a strain gauge for facial expression recognition [109]. Here, as strain is applied, the distance between NPs increased, resulting in a corresponding increase of resistance. Ag NPs have also been applied as temperature sensors [110]. Ag NPs were deposited in indium tin oxide nanomembranes to increase sensitivity to mechanical movement and temperature. Au NP-based strain gauges have shown significantly higher sensitivities compared to metal thin films [111]. In addition to metal NPs, oxide NPs have also been used. One type is ZnO NPs, where they were drop-casted on a flexible substrate to form a gas sensor highly sensitive to ethanol and can achieve 90° bending [112]. NPs can also be combined with other nanomaterials, such as CNTs [113]. The CNTs act as electrical bridges between ZnO NPs to improve conductivity, and the material acts as a nanogenerator. One-dimensional nanomaterials include nanowires and nanotubes. The piezoelectric property of ZnO NWs has been exploited for form wearable nanogenerators [114]. This nanogenerator, which had a total thickness of 16 um, had high conformability and was tested on an eyelid to detect movement of the eye. Polymer nanofibers, which are softer than metal nanowires, can also be used in similar applications [115]. One study deposited Pt on polymer fiber arrays and then embedded between two PDMS layers. External loads change the interaction of the fiber arrays, thus changing resistance. The highly sensitive structure could sense wrist pulses and the impact of individual water droplets. Due to Ag possessing the highest electrical conductivity among metals, Ag NWs are a widespread option for stretchable and flexible electronics [116]. Synthesizing Ag NWs can be accomplished via a number of methods [117]. To improve conductivity, longer NWs may be grown to enable longer percolation paths and limit NW junctions [118]. Similar to CNTs, Ag NWs have been embedded into elastomer to form a stretchable conductor [119]. Hysteresis may be reduced by using a sandwich structure of PDMS and Ag NWs [120]. This sandwich form has been applied as a wearable, stretchable heater using Ag NWs and SBS elastomer [121]. Another example is Cu NWs, which are typically less costly and more abundant than Ag NWs [122]. However, Cu NWs offer less favorable properties due to oxidation and poor solvent dispersion [123]. Thus, Au NWs are widely used due to their biocompatibility and high conductivity [124]. Initial methods provided only thick and short NWs, which resulted in mechanically and electrically poor characteristics, but ultrathin Au NWs were later developed [122]. Unlike other NWs, Au NWs are serpentine in shape and have an inherent degree of stretchability [125]. One study used Au NW films doped with polyaniline microparticles to enhance conductivity and sensitivity. This strain sensor was used to wirelessly control a robotic arm. A mesh-like film of Au NWs improves conductivity and has been applied for flexible touchscreens and strain sensors [126, 127]. Compared to a thin indium tin oxide film, the mesh Au NW film improves flexibility without increasing resistance (Fig. 16.11a) [126]. Using a sandwich structure of Au NWs and PDMS resulted in a highly sensitive pressure sensor [128]. Two PDMS sheets were integrated on either side, with one having interdigitated electrodes. Applying pressure changes the number of Au NWs electrically connecting the pairs of electrode fingers where pressure can be derived from a change in current. With a high sensitivity, the flexible sensor could record wrist pulses (Fig. 16.11b) [128]. CNTs can also be applied in multiple ways, including dispersing into elastomer for a composite material and depositing onto a substrate [129–131]. The amount of CNT mixed into PDMS controls both mechanical and electrical properties [132]. Adding CNT lowers stretchability while decreasing electrical resistance (Fig. 16.11c, d). One study used a mixture of CNT and adhesive PDMS to form an interfacial layer of an ECG electrode to achieve conformal contact with skin (Fig. 16.11e) [133]. A contact area of 99.7% was achieved, indicating exceptional conformal contact with skin and resulting in less noisy signals. Polymer nanofibers have also been used; electrospinning is one method to form the nanofibers [134]. One study used P3HT nanofibers and Au NPs to fabricate a nonvolatile flash transistor device on flexible substrate. Another study used the piezoelectric properties of PVDF to form flexible pressure sensors [135]. Polymer nanofibers have also been applied to develop a skin-like layer for use as a substrate (Fig. 16.11f) [136]. Here, the polymer fibers embedded in elastomer allows high stretchability at low strains but stiffens at higher strains. Such a substrate protects skin and electrical components from extreme strains. Another form of nanomaterials is nanosheets. These 2D nanomaterials can be fabricated and processed with many existing processes. A number of materials have been formed in nanosheets, including silicon. Silicon nanosheets have been incorporated into prosthetic skin to sense the environment, including strain, temperature, and humidity (Fig. 16.11g) [137]. MoS2 sheets have been applied in tactile sensors [138]. Similar to previously discussed composites, nanosheets may be integrated with nanoparticles and other structures. One study used graphene nanosheets with Au NPs to lower impedance and improve electrochemical sensing while maintaining a soft device (Fig. 16.11h) [139]. Another study used a nitrogen-doped graphene-MnO2 nanosheet in a flexible supercapacitor [140]. After 1500 bending cycles, the capacitor maintained over 90% of its initial capacitance (Fig. 16.11i).

16.3 Soft Material-Enabled Packaging Strategies

16.3.1 Packaging Strategies

For wearable devices, packaging of the electronics is required to maintain stretchability and flexibility for integration on skin. These packages should maintain conformal contact, be lightweight, and ensure comfort. Additionally, electronic components and interconnects need to be mechanically and electrically stable. A number of strategies have been developed to package electronics with the use of previously discussed designs and materials. The role of soft materials in these strategies is highly important for adhesive, conformal contact while maintaining high-performing electronic devices.

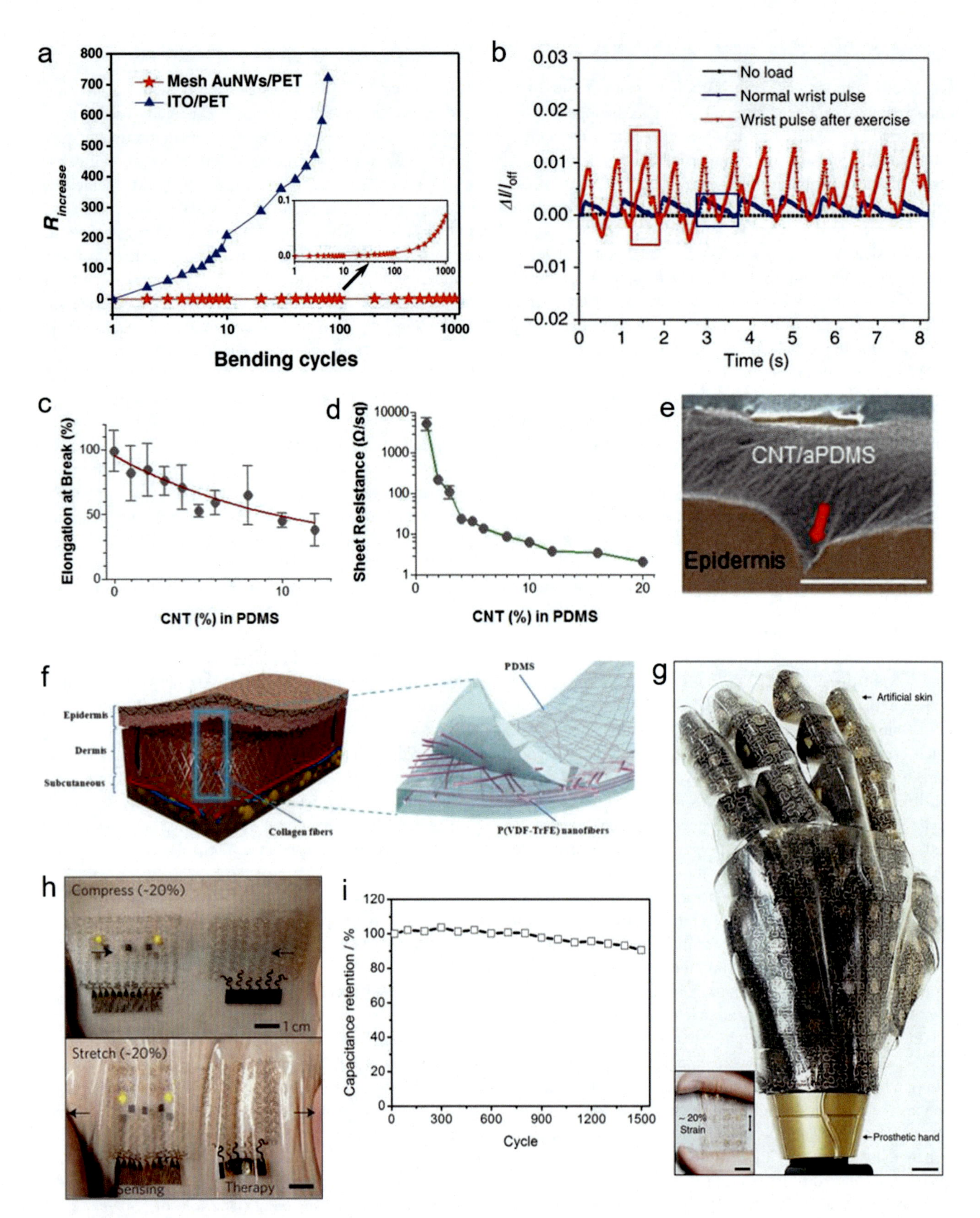

Fig. 16.11 Nanomaterials applied in wearable applications. (**a**) Au NW mesh film endures cycles of bending without significantly increasing resistance, unlike the solid ITO film [126], Copyright 2018, John Wiley and Sons. (**b**) Pressure sensor based on Au NWs recording wrist pulses. Reprinted by permission from Springer Nature: [128], Copyright 2014. (**c**) Change in stretchability and (**d**) reduction of electrical resistance of CNT-PDMS composite while increasing CNT content [132]. (**e**) CNT-PDMS composite conforming to sharp wrinkles in skin to decrease impedance [133]. (**f**) Illustration comparing skin to PDMS and polymer nanofiber composite as a strain-limiting layer. Reprinted with permission from [136], Copyright 2018, American Chemical Society. (**g**) Prosthetic hand with conformal sensors. Reprinted by permission from Springer Nature: [137], Copyright 2014. (**h**) Wearable electrochemical sensor based on graphene nanosheets and Au NPs. Reprinted by permission from Springer Nature: [139], Copyright 2016. (**i**) Flexible capacitor using graphene-MnO_2 nanosheets prevents decrease in capacitance during bending. Reprinted with permission from [140], Copyright 2016, American Chemical Society

16.3.1.1 Island-Bridge

One packaging strategy relies on an island-bridge configuration. Mesh layouts with rigid islands and stretchable connections provide high stretchability on the soft substrates [47, 141]. Figure 16.12a illustrates the components of this design, along with an example fabricated device [63]. Fabrication may involve transfer of circuits onto substrates, and prestraining the substrate can yield more stretchable circuits. The island-bridge structure is fabricated using lithography, deposition, and etching processes to produce the mesh layout with straight interconnects [47]. Then, the layout is transferred to a prestrained PDMS substrate. The rigid islands are coated with SiO2 to covalently bond with the treated PDMS surface. This allows the interconnects to delaminate when the strain is released, forming a wrinkled interconnect. Stretching this system then straightens the interconnects while

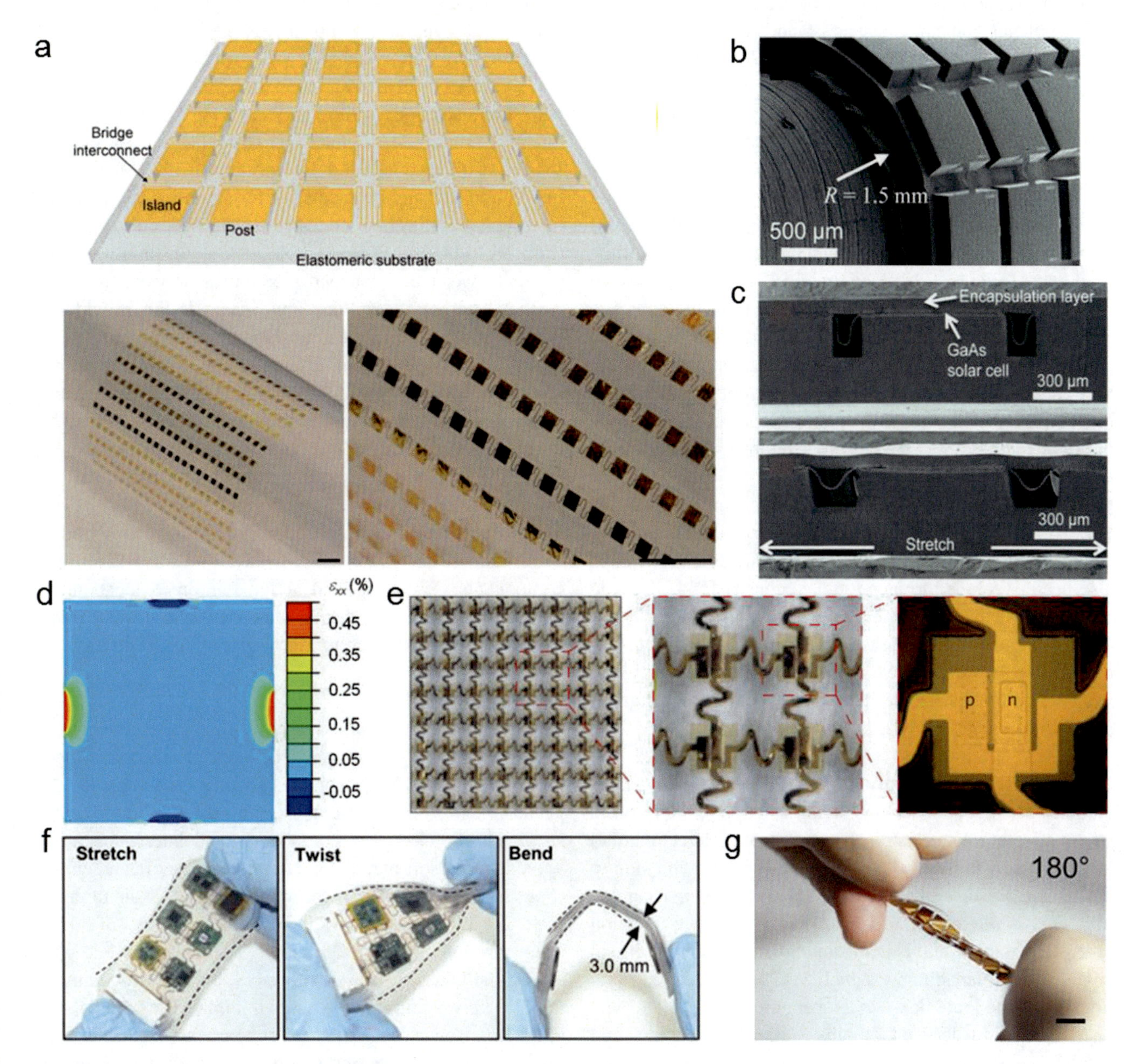

Fig. 16.12 Island-bridge structure for stretchable circuits. (**a**) Illustration of island-bridge components and images of fabricated device. Reproduced from Ref. [63] with permission from the Royal Society of Chemistry. (**b**) Suspended interconnects in trenches between rigid islands enables conformal contact around curve [46], Copyright 2011, John Wiley and Sons. (**c**) Stretching of suspended interconnects [46], Copyright 2011, John Wiley and Sons. (**d**) FEA showing stretchable interconnects limit the strain applied to rigid islands when device is stretched. Reprinted from Ref. [142], with the permission of AIP Publishing. (**e**) Array of diode sensors using island-bridge structure. Reprinted by permission from Springer Nature: [143], Copyright 2013. (**f**) Image of mechanical stretchability and flexibility enabled by organizing electronic components into islands [16]. (**g**) Display of high flexibility of a bilayer circuit [144], Copyright 2016, John Wiley and Sons

maintaining low strain on the rigid elements. One study developed a stretchable battery with a design based on island-bridge [52]. Stretchabilities up to 300% were achieved with self-similar serpentine interconnects. The study also included a stretchable wireless charging system for the battery using electromagnetic flux from a transmitting coil. It should be noted that the island-bridge design is often combined with other packaging schemes for enhanced control of strain distribution and protection of electronic components. Both coplanar and noncoplanar island-bridge designs have been developed, with the latter increasing stretchability. Straight interconnects may be bonded to a prestrained elastomer substrate to form a coplanar wavy design or unbounded to form noncoplanar interconnects. Serpentine and fractal interconnects may further increase stretchability and flexibility, in both coplanar and noncoplanar forms. Under applied strain, interconnects deform between the rigid islands and enable conformal contact throughout the circuit despite rigid areas (Fig. 16.12b, c). Figure 16.12c shows the use of suspended interconnects placed in trenches of the substrate [46]. The stretchable interconnects minimize strain applied to the fragile components on the islands, as confirmed by the FEA result in Fig. 16.12d [142]. It is critical that these interconnects are designed to endure cyclic strain. A variety of circuits and devices have been developed with this structure, including the diode sensor array shown in Fig. 16.12e [143]. The array of electronic component islands in Fig. 16.12f shows the highly stretchable and flexible form enabled by this packaging design [16]. Bilayer circuits have also been developed to increase area density [144]. Despite the increased thickness, the circuit maintains extreme flexibility due to the island-bridge design (Fig. 16.12g). A variety of analytical models have been developed to study the stretchability of different interconnects under a range of applied deformations [142, 145–147].

16.3.1.2 Strain Isolation

Another packaging scheme is strain isolation, which lowers strain affecting the electronic components. Here, a liquid layer or cavity is placed between the electronics and skin to mechanically isolate the two (Fig. 16.13a). The cavity in the elastomer is filled with an ionic liquid via syringe injection [148]. The ionic liquid is nonirritating and has low electrical conductivity to enable low signal loss. Tests indicated functionality of the ionic liquid at temperatures up to 180 °C and lasted through at least 11 months. Using water as the fluid resulted in evaporation and diffusion through the elastomer, limiting its lifespan. One potential problem with this design is the collapse of the cavity where the top and bottom elastomer layers of the cavity adhere. This negatively affects strain isolation, but an analytical study determined critical parameter relationships for both air- and liquid-filled cavities to understand cavity collapses. Additionally, parameters were optimized to minimize stress on the skin when subjected to stretching. Stresses below 2 kPa, the limit of sensation for extremely sensitive skin, were achieved despite 50% strains, well below a device without a cavity (Fig. 16.13b) [148]. This structure also minimizes the tendency for the device delaminating from the skin during use. Figure 16.13c shows results of testing of different cavity lengths. Demonstrations of this application included a mechano-acoustic sensor, with an accelerometer, to determine applied vibration frequency. Identified frequencies did not change for non-stretched devices without strain isolation, non-stretched devices with strain isolation, and 30% stretch with strain isolation. Additionally, an NFC die with temperature measurement was connected to a copper coil for wireless sensing. Collected data did not vary despite 40% stretching. In addition to using fluid cavities for strain isolation, strain-limiting layers may be used to protect the electronics and skin (Fig. 16.13d) [67]. One example of a strain-limiting layer, added between the skin and electronics, is using an open mesh design of polyimide. This mesh design provides a nonlinear stress-strain relationship, replicating the stress-strain curve shape of skin. Strains less than 20% result in a low modulus, but strains larger than 40% cause a sharp increase in modulus (Fig. 16.13e) [67]. This is a result of the mesh pattern straightening and bending, prior to the mesh material dominating the response. Another option is to use an elastic substrate with reinforced rigid islands [149]. This hybrid substrate of silicone and polymer islands minimized the strain at the location of these islands. Electronic components were integrated at the reinforced locations to minimize strains. This design was applied for soft contact lenses with wireless monitoring of glucose concentration.

16.3.1.3 Shell-Core

Another packaging method uses a shell-core structure to minimize strain applied to the fragile electronic components [16, 60, 67, 150, 151]. This design often accompanies the island-bridge concept. Here, a core material immediately surrounds the electronic components and interconnections, while another soft material encapsulates the core and circuit (Fig. 16.14a) [16]. Here, the core material is an ultrasoft material, such as Silbione RT Gel with a modulus of 5 kPa. The surrounding shell is another soft material, but with higher modulus, such as Ecoflex 00-30 with a modulus of 60 kPa. This configuration causes minimal stress on the skin and helps isolate the electronic components from the skin. Such devices may also be modified to replicate the human skin surface to further reduce stresses.

One device achieved a modulus of 22 kPa with electronics, well below that of skin, which is about 130 kPa [67]. However, the modulus using standard encapsulation resulted in a modulus of 2.8 MPa, well above that of the shell-core structure and that of skin. This decrease allows

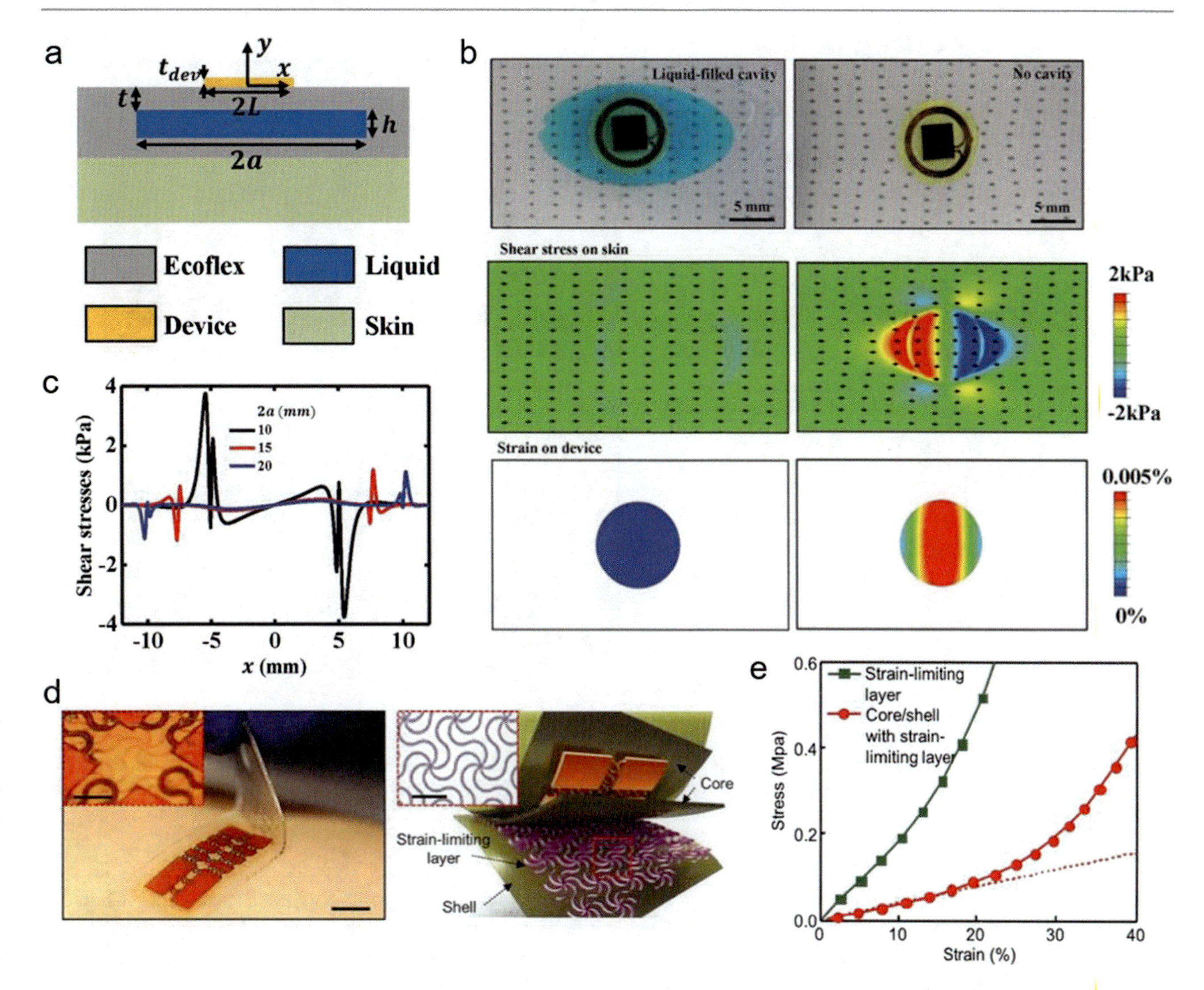

Fig. 16.13 Strain isolation and strain-limiting techniques for wearable electronics. (**a**) Diagram of strain isolation cavity [148], Copyright 2016, John Wiley and Sons. (**b**) FEA results showing the strain isolation cavity minimizes stress applied to both the (**c**) skin and device [148], Copyright 2016, John Wiley and Sons. (**d**) Strain-limiting layer formed with polyimide fractal design located between skin and circuit [67], Copyright 2015, John Wiley and Sons. (**e**) The strain-limiting layer significantly reduces strain applied to the device for a given stress [67], Copyright 2015, John Wiley and Sons

highly conformal integration with skin (Fig. 16.14b) [67]. Compared to only PDMS encapsulation, the shell-core structure prevents delamination of the edges of the device. Both computational and experimental analysis confirms that this package shows a significantly lower modulus, as shown in Fig. 16.14c [67]. The core thickness is the most sensitive parameter to achieve strain isolation, whereas shell thickness has minimal effects. Stresses on the skin decreased as the thickness of the core increased, but thicknesses over 300 μm lead to an asymptotic value of modulus. Changing shell thickness is significantly less important and is typically much thinner than the core. An example of this device is the integration of an accelerometer, temperature sensor, and Bluetooth communication connected via an island-bridge structure in the shell-core package. This wearable system resulted in a modulus 70 times smaller than standard packaging. Wireless, real-time tracking of exercises were demonstrated. Furthermore, to reduce stress concentrations near corners of the package, rounded corners and tapered edges were used and further reduced delamination.

This shell-core structure was also applied to develop a power management system with solar cells, rechargeable lithium-ion batteries, and integrated circuits [150]. The package here also enables a folded design, allowing the solar cells and batteries to be compact with minimal strain. The soft nature and low form factor allow for folding of the circuit. The substrate may also be prestrained to increase the areal density of components. The ultrasoft core material enables unconstrained motion of interconnects. This device could wrap around a human finger without failure. The modulus

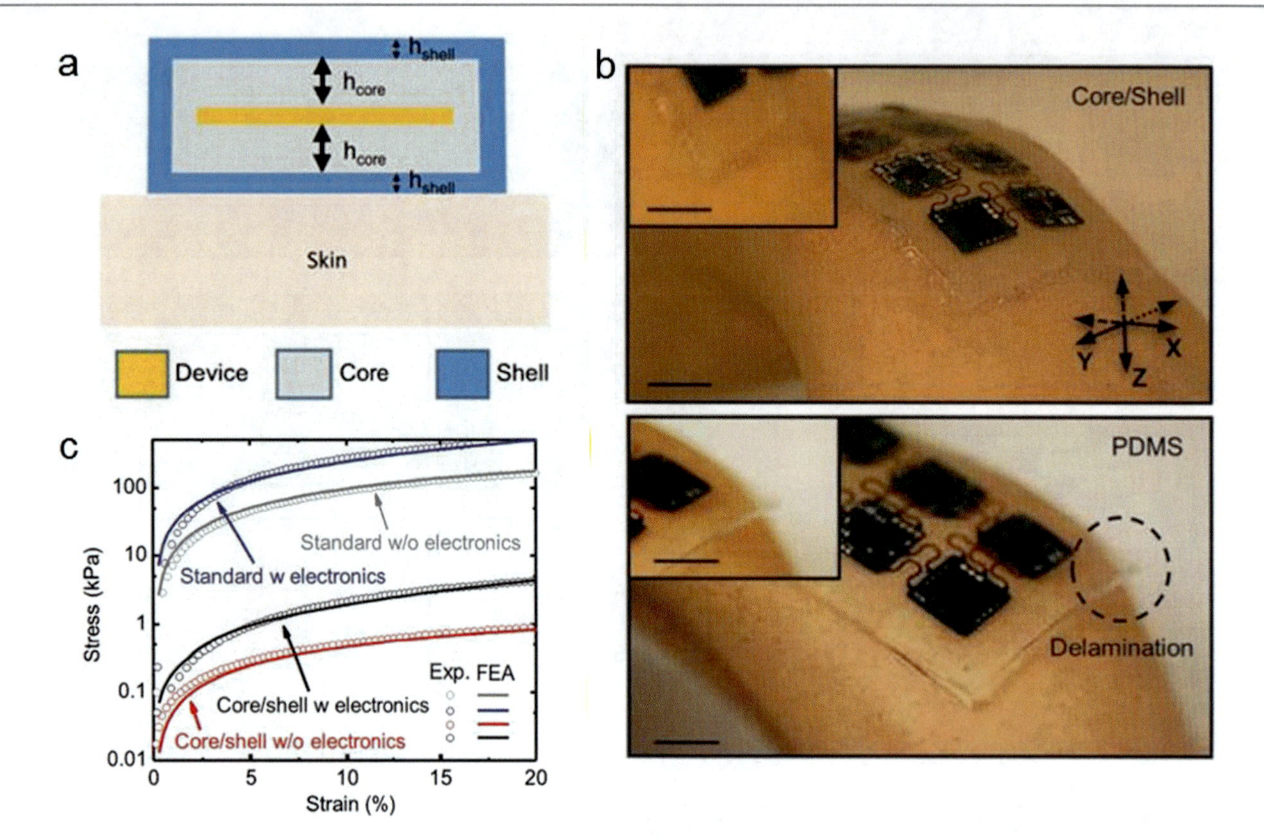

Fig. 16.14 Shell-core design of wearable devices. (**a**) Illustration of shell-core layers [16]. (**b**) The shell-core structure enables conformal contact, while encapsulation of PDMS results in delamination, particularly at edges of device [67], Copyright 2015, John Wiley and Sons. (**c**) Stress-strain curves for standard encapsulation and shell-core encapsulation [67], Copyright 2018, John Wiley and Sons

of the total system increases only 15% when electronic components are included, but removing the core causes a threefold increase in the modulus. Increasing the core thickness up to 200 μm causes an increase in stretchability before becoming more dependent on other material and design parameters. This trend agrees with the previously discussed work [67]. This study also determined that longer, more complex serpentines might not always increase stretchability due to higher strain concentrations. Despite the elastomers straining over 100%, strain in the circuit interconnects maintained below their yielding point. An array of solar cells achieved identical performance despite biaxial strains of 30%. This work also demonstrated the lamination of a solar module onto a battery module, creating an electrical connection that allows for charging of the batteries [150]. An LED module was also connected and tested via this method. The shell and core layers alleviate stress from the electronics at the folding location, as indicated in the figure. An NFC-based device for wireless temperature monitoring was demonstrated during exercise. Additionally, the resulting device is waterproof such that it was submerged in water for 8 min without failure. A similar design was used to develop mechano-acoustic sensing, which was applied to cardiovascular diagnostics and human-machine interfaces [151]. The copper traces forming interconnects are located at the neutral plane with polyimide insulation. Thoracic vibrations and ECGs were monitored. Speech recognition was achieved without ambient noise due to the conformal contact between the device and human skin.

Another study utilized the shell-core structure to create a "biostamp" for neurosurgical monitoring [16]. Rigid components are condensed into a set of islands connected with serpentine traces. The electrical components included Bluetooth, flash memory module, accelerometer, gyroscope, and wirelessly rechargeable battery for 16 h of monitoring. When strained to 20%, the stress on the skin was below 2 kPa, eliminating sensation in the skin. This device was clinically tested for continuous monitoring of muscle responses during surgeries of the peripheral nerve, spine, and cranium. Compared to the conventional monitoring methods, the biostamp offered multimodal sensing and a simplified operation due to the lack of wires and complex procedures. The soft material-enabled properties allow for a variety of uses in clinical applications.

An extension of the shell-core packaging structure is using fluid as the core material [60]. The components are rendered independent of the encapsulation and substrate layers, except for small sections of bonds between the substrate and selected components. To prevent adhesion to the substrate, an array of soft pyramidal features alleviated strong adhesion of the floating interconnects to the substrate. Barriers are also located on the substrate to prevent interconnects from entangling, and longer interconnects have midpoint bonding sites. The interconnects are transfer printed onto the substrate, while the chips are bonded via a low-temperature solder. A disadvantage of this structure is that the fluid may leak or evaporate over time. Secondly, the fluid is near electrical components and can cause damage. The fluid applied in the device must have a large resistance, high dielectric strength, thermal stability, and low-loss RF characteristics. Additionally, a transparent fluid enables easy inspection of the circuit. Due to the free-floating nature of the interconnects, the fluid allows for deformation with minimal constraint both in and out of plane. This design allows for increased stretchability, up to 20 times, compared to implementing on a substrate. This method achieved approximately 0.2% strain in interconnects when subjected to 50% stretching. The demonstrated device is capable of recording, filtering, amplifying, and wirelessly transmitting ECGs, EMGs, EOGs, and EEGs. External contact pads to epidermal electrodes are located on the edges of the microfluidic chamber. The device achieved 100% biaxial strain before failure and 75° twisting, while 49% strain did not cause plastic deformation of interconnects. The low stresses between the skin and device enable strong adhesion for recording of high-fidelity signals.

16.3.1.4 Thinned Chips

There is also interest in using ultrathin chips for flexible electronics [21]. A variety of methods have been applied for fabrication of ultrathin chips. Grinding of a wafer is one common method, resulting in a chip with 3 μm in thickness [152]. While this method more rapidly yields thinned chips, it may damage the material's crystal structure or while being handled. Various etching methods, including reactive ion etching and wet etching, have been used [153, 154]. To realize functional devices, these ultrathin chips are incorporated onto flexible substrates. Conventional wire bonding is not suitable for ultrathin chips as it can lead to cracking [21]. Flip chip assembly and laminating between layers are strategies for incorporating ultrathin chips on flexible substrates [155, 156]. Using a polyimide substrate, solder bumped die can be reflowed. By integrating the ultrathin chips between two flexible layers, the chip can be placed in the neutral plane of bending. Such a method has been used for wireless and wearable ECG monitoring [156, 157]. To print interconnections, the thin chips may be placed face-up on a flexible substrate. Screen-printing and inkjet printing can be applied to form the circuitry.

16.3.1.5 Additional Designs

In addition to these packaging strategies, a variety of other designs have been developed with the use of soft materials, deterministic designs, and intrinsic strategies. One circuit design acted as a wireless, intraoral system for continuous monitoring of sodium intake [12]. This design achieved a communication distance over 10 m, exceeding that of previous wearable, wireless electronics. The circuit is formed on rigid components on a serpentine mesh of interconnects. A dielectric membrane, for insulation and impedance matching, and stretchable ground plane complete the design before elastomer encapsulation. A 180° bending radius of 1.5 mm and 20% biaxial stretching was achieved without change in resistance. The resonant frequency shifted slightly, but sufficient power was still delivered. Another package technique is based on the previously discussed 3D helical interconnects (Fig. 16.15a) [70]. The study implemented about 250 3D interconnects with 500 bonding sites. The resulting device could wirelessly record health data. Electrical properties were considered heavily in this design, such as the length of the interconnect may cause resistive dissipation and electromagnetic noise. To minimize interference, the antenna and electrodes were distanced from the power lines. An ultrasoft modulus was used to encapsulate the interconnects and assembly, but the 3D network allowed the material to be introduced prior to releasing substrate prestrain. This difference allowed for improve stretchability of about 2.2 and 2.8 times of uniaxial and radial stretching. In addition, increasing the prestrain to 200% from 150% reduced the circuit area by 70%. These devices also experienced compressive forces during use, and the study determined that the helical interconnects could withstand compression ratios of 0.3 and pressures greater than 10 kPa. Demonstrated devices included a three-axis accelerometer and monitoring of ECG, EMG, EOG, and EEG. ECG data recorded with an undeformed device and a 50% radially stretched device were identical. Another packaging design can enable high areal density of components by using a stackable and foldable design [144]. Similar to the shell-core structure, this work used a silicone outer layer with a softer silicone in between components and the outer layers. The device was fabricated in planar form and then folded prior to encapsulation. Due to the high adhesion and low modulus of silicone, the two sides readily joined. Misalignment of the stacked layers was found to significantly modify the stretchability. A thicker core layer enabled higher stretchabilities when applied with a thin encapsulation layer. A thick encapsulation layer restricted out-of-plane deformation and thus lowers stretchability. The study also discovered that the elastic stretchability of the interconnects was dependent on both PI support thickness and interconnect design. For thicker PI of 7.5 μm, increasing

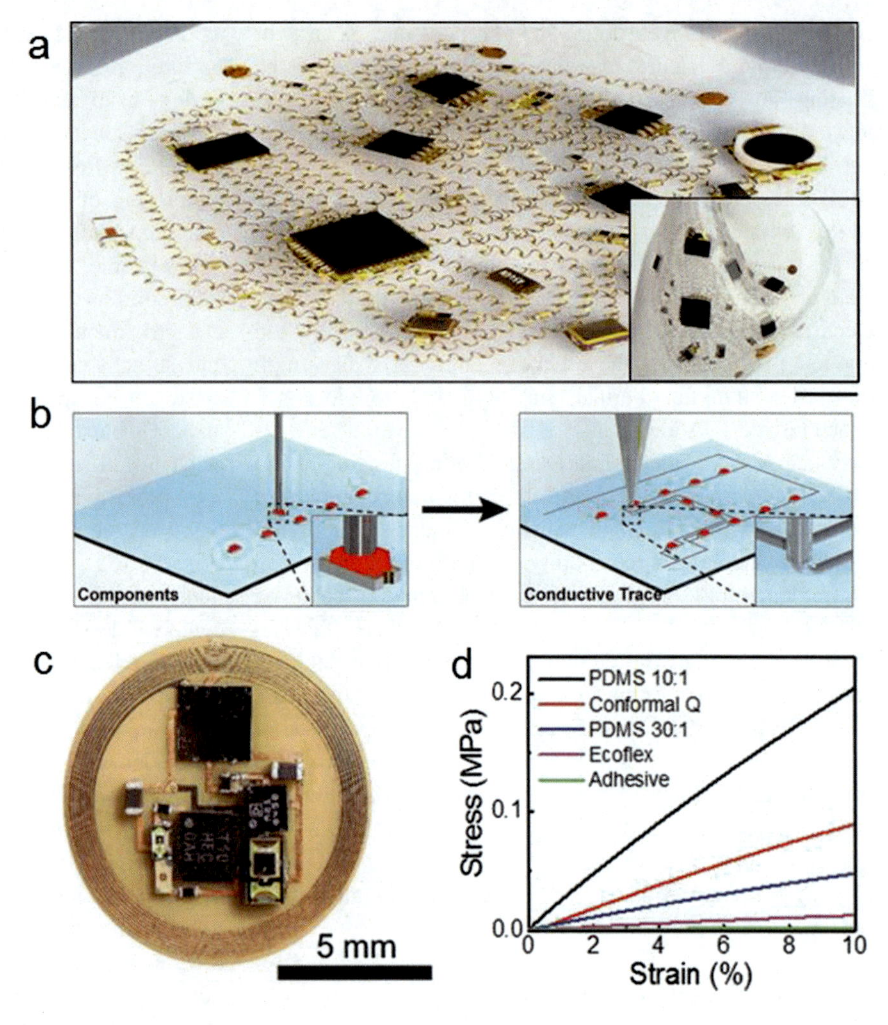

Fig. 16.15 Additional circuit design concepts for hybrid electronics. (**a**) 3D helical interconnects embedded in elastomer [70]. (**b**) Hybrid printing of conductive ink on elastomer to connect components [158], Copyright 2017, John Wiley and Sons. (**c**) Miniature, flexible device for wireless sensing of pulse oximetry [161], Copyright 2016, John Wiley and Sons. (**d**) Modification of device's stress-strain curve with different encapsulating materials [161], Copyright 2016, John Wiley and Sons

the interconnect amplitude lowered the stretchability from 180% to 140% in the tested range. However, thinner PI of 2.1um showed an increase of stretchability from 110% to 135%. A PI thickness of 4.8um yielded an increase of stretchability from 150% to 220% by increasing interconnect amplitude. This result is explained by buckling mechanics, which plays a significant role in determining circuit stretchability. Device demonstration used 36 functional components, with area coverage of 128% via folding. Another method of creating soft electronics is a hybrid 3D printing technique [158]. Here, electrical components can be placed onto a soft substrate in selected locations. Then, conductive traces between the placed components are printed to form interconnects (Fig. 16.15b). The conductive inks are based on biocompatible and stretchable thermoplastic polyurethane (TPU). Silver flakes are introduced to form the AgTPU ink. When stretched, the interconnects do not crack or delaminate, and low resistance is maintained despite 30% strain. The study tested a microcontroller device with a strain sensor to monitor joint angles. A similar packaging technique is based on printable Ni-GaIn [159]. The ink was directly printed on Ecoflex with a rolling brush and then encapsulated by Ecoflex. The study demonstrated monitoring of pulse waves and wireless powering of a circuit. Soft modular electronic blocks (SMEBs) can be used for customizable circuits based on body features and placement [160]. Individual blocks were designed using an island-bridge structure to ensure each met mechanical requirements for wearable applications. The following integration of the SMEBs used oxygen plasma treatment to allow for irreversible bonding between the SMEBs PEN layer and the base PDMS substrate. The rigid island in the SMEBs had PEN underneath to limit strain. Interconnect blocks were formed with an Ag NW and PDMS composite. The study demonstrated monitoring of joint flexion angle. This concept can enable rapid development of custom electrical circuits for wearable applications. Other packages emphasize miniaturization with thin flexible materials [161]. A wireless pulse oximetry sensor achieved a diameter of 10 mm with a thickness below 1 mm (Fig. 16.15c). Using a lower modulus encapsulation layer reduces stress for given strain and thus may improve conformality depending on location (Fig. 16.15d). A bilayer loop antenna is integrated with an NFC bare die chip for wireless data transmission, while an LED and photodetector are used to record pulse oximetry. The electronic components are mounted on the neutral plane of bending, while the encapsulating materials ensure conformal contact. Despite bending about a 5 mm radius, the RF properties do not fluctuate. The double-layer coil enhances the quality factor and allows for longer transmission distance. Moreover, the packages are washable and water resistant due to the polyimide and silicone encapsulation. A similar packaging technique allows for thermal characterization of skin with a device diameter of 16 mm [162]. The inductive planar coil harvests power for NFC-based communication. Such NFC designs and coils may also be directly integrated onto skin like a temporary tattoo [61]. Another wireless device used a passive LC circuit design to characterize sweat, avoiding the use of rigid electrical components [49].

16.4 Conclusions and Outlook

Recent developments and applications of soft materials for stretchable and flexible hybrid electronics have enabled high-performance, wearable devices for health monitoring and human-machine interfaces. Unlike conventional, rigid devices, these soft material-based systems enabled widespread, continuous monitoring with minimal impact on users. Soft substrates enable adhesive, conformal contact to skin without irritation. This improved, noninvasive contact enhances electrical performance, particularly the recording of electrophysiological signals. The design and fabrication of deterministic and intrinsic electronic materials further lower moduli of devices and enable more durable electronics. The mechanical and material designs have been manipulated to achieve a new class of packaging for wearable electronics. These packaging techniques rely on the use of soft materials, particularly as substrates, encapsulation, and strain isolation layers. Island-bridge designs of high-performance electronic components are combined with soft material packaging schemes, including shell-core structures and strain isolation layers, to realize comfortable, functional devices. Wireless capabilities continue to be improved, allowing for real-time, continuous monitoring without restricting patients. Further advancement of stretchable and flexible hybrid electronics will rely on soft material-based electronic components rather than the current rigid components. Self-healing materials, such as hydrogels and liquid metals, may be further explored for durable, long-term devices. Additional applications and high-performance epidermal electronics will continue to be realized with the continued development of soft material-based packaging techniques.

Acknowledgments This research was supported by the Imlay Innovation Fund and funds from the Marcus Foundation, the Georgia Research Alliance, and the Georgia Tech Foundation through their support of the Marcus Center for Therapeutic Cell Characterization and Manufacturing (MC3M) at Georgia Tech. This work was performed in part at the Institute for Electronics and Nanotechnology, a member of the National Nanotechnology Coordinated Infrastructure, which is supported by the National Science Foundation (Grant ECCS-1542174).

References

1. Liang, X., Boppart, S.A.: Biomechanical properties of in vivo human skin from dynamic optical coherence elastography. IEEE Trans. Biomed. Eng. **57**, 953–959 (2010)
2. Herbert, R., Kim, J.-H., Kim, Y., Lee, H., Yeo, W.-H.: Soft material-enabled, flexible hybrid electronics for medicine, healthcare, and human-machine interfaces. Materials. **11**, 187 (2018)
3. Xu, M., Obodo, D., Yadavalli, V.K.: The design, fabrication, and applications of flexible biosensing devices–a review. Biosens. Bioelectron. (2018)
4. Liu, Y., Pharr, M., Salvatore, G.A.: Lab-on-skin: a review of flexible and stretchable electronics for wearable health monitoring. ACS Nano. **11**, 9614–9635 (2017)
5. Yu, K.J., Kuzum, D., Hwang, S.-W., Kim, B.H., Juul, H., Kim, N. H., et al.: Bioresorbable silicon electronics for transient spatiotemporal mapping of electrical activity from the cerebral cortex. Nat. Mater. **15**, 782 (2016)
6. Wang, H., Zhu, B., Wang, H., Ma, X., Hao, Y., Chen, X.: Ultra-lightweight resistive switching memory devices based on silk fibroin. Small. **12**, 3360–3365 (2016)
7. Smooth-On. Ecoflex Series. Available: https://www.smooth-on.com/tb/files/ECOFLEX_SERIES_TB.pdf
8. Lee, Y., Nicholls, B., Lee, D.S., Chen, Y., Chun, Y., Ang, C.S., et al.: Soft electronics enabled ergonomic human-computer interaction for swallowing training. Sci. Rep. **7**, 46697 (2017)
9. Lee, Y.K., Jang, K.I., Ma, Y., Koh, A., Chen, H., Jung, H.N., et al.: Chemical sensing systems that utilize soft electronics on thin elastomeric substrates with open cellular designs. Adv. Funct. Mater. **27**, 1605476 (2017)
10. Wang, S., Li, M., Wu, J., Kim, D.-H., Lu, N., Su, Y., et al.: Mechanics of epidermal electronics. J. Appl. Mech. **79**, 031022 (2012)
11. Jeong, J.W., Yeo, W.H., Akhtar, A., Norton, J.J., Kwack, Y.J., Li, S., et al.: Materials and optimized designs for human-machine interfaces via epidermal electronics. Adv. Mater. **25**, 6839–6846 (2013)
12. Lee, Y., Howe, C., Mishra, S., Lee, D.S., Mahmood, M., Piper, M., et al.: Wireless, intraoral hybrid electronics for real-time quantification of sodium intake toward hypertension management. Proc. Natl. Acad. Sci. **115**, 5377–5382 (2018)
13. Mishra, S., Norton, J.J., Lee, Y., Lee, D.S., Agee, N., Chen, Y., et al.: Soft, conformal bioelectronics for a wireless human-wheelchair interface. Biosens. Bioelectron. **91**, 796–803 (2017)
14. Koh, A., Kang, D., Xue, Y., Lee, S., Pielak, R.M., Kim, J., et al.: A soft, wearable microfluidic device for the capture, storage, and colorimetric sensing of sweat. Sci. Transl. Med. **8**, 366ra165–366ra165 (2016)
15. Hou, C., Xu, Z., Qiu, W., Wu, R., Wang, Y., Xu, Q., et al.: A biodegradable and stretchable protein-based sensor as artificial electronic skin for human motion detection. Small. **15**, 1805084 (2019)
16. Liu, Y., Tian, L., Raj, M.S., Cotton, M., Ma, Y., Ma, S., et al.: Intraoperative monitoring of neuromuscular function with soft, skin-mounted wireless devices. npj Digital Med. **1**, 19 (2018)
17. Jeong, J.W., Kim, M.K., Cheng, H., Yeo, W.H., Huang, X., Liu, Y., et al.: Capacitive epidermal electronics for electrically safe,

long-term electrophysiological measurements. Adv. Healthc. Mater. **3**, 642–648 (2014)

18. Yu, K.J., Yan, Z., Han, M., Rogers, J.A.: Inorganic semiconducting materials for flexible and stretchable electronics. npj Flexible Electron. **1**, 4 (2017)
19. Kim, D.H., Xiao, J., Song, J., Huang, Y., Rogers, J.A.: Stretchable, curvilinear electronics based on inorganic materials. Adv. Mater. **22**, 2108–2124 (2010)
20. Rogers, J., Lagally, M., Nuzzo, R.: Synthesis, assembly and applications of semiconductor nanomembranes. Nature. **477**, 45 (2011)
21. Gupta, S., Navaraj, W.T., Lorenzelli, L., Dahiya, R.: Ultra-thin chips for high- performance flexible electronics. npj Flexible Electron. **2**, 8 (2018)
22. Sekitani, T., Zschieschang, U., Klauk, H., Someya, T.: Flexible organic transistors and circuits with extreme bending stability. Nat. Mater. **9**, 1015 (2010)
23. Park, S.I., Ahn, J.H., Feng, X., Wang, S., Huang, Y., Rogers, J.A.: Theoretical and experimental studies of bending of inorganic electronic materials on plastic substrates. Adv. Funct. Mater. **18**, 2673–2684 (2008)
24. Rogers, J.A.: Materials for semiconductor devices that can bend, fold, twist, and stretch. MRS Bull. **39**, 549–556 (2014)
25. Wang, Y., Li, Z., Xiao, J.: Stretchable thin film materials: fabrication, application, and mechanics. J. Electron. Packag. **138**, 020801 (2016)
26. Wang, C., Wang, C., Huang, Z., Xu, S.: Materials and structures toward soft electronics. Adv. Mater. **30**, 1801368 (2018)
27. Ma, Y., Jang, K.I., Wang, L., Jung, H.N., Kwak, J.W., Xue, Y., et al.: Design of strain- limiting substrate materials for stretchable and flexible electronics. Adv. Funct. Mater. **26**, 5345–5351 (2016)
28. Ma, Y., Feng, X., Rogers, J.A., Huang, Y., Zhang, Y.: Design and application of 'J- shaped' stress–strain behavior in stretchable electronics: a review. Lab Chip. **17**, 1689–1704 (2017)
29. Khang, D.-Y., Jiang, H., Huang, Y., Rogers, J.A.: A stretchable form of single-crystal silicon for high-performance electronics on rubber substrates. Science. **311**, 208–212 (2006)
30. Sun, Y., Choi, W.M., Jiang, H., Huang, Y.Y., Rogers, J.A.: Controlled buckling of semiconductor nanoribbons for stretchable electronics. Nat. Nanotechnol. **1**, 201 (2006)
31. Song, J., Jiang, H., Choi, W., Khang, D., Huang, Y., Rogers, J.: An analytical study of two-dimensional buckling of thin films on compliant substrates. J. Appl. Phys. **103**, 014303 (2008)
32. Huang, Z., Hong, W., Suo, Z.: Nonlinear analyses of wrinkles in a film bonded to a compliant substrate. J. Mech. Phys. Solids. **53**, 2101–2118 (2005)
33. Kim, P., Abkarian, M., Stone, H.A.: Hierarchical folding of elastic membranes under biaxial compressive stress. Nat. Mater. **10**, 952 (2011)
34. Chen, X., Hutchinson, J.W.: A family of herringbone patterns in thin films. Scr. Mater. **50**, 797–801 (2004)
35. Jiang, H., Khang, D.-Y., Fei, H., Kim, H., Huang, Y., Xiao, J., et al.: Finite width effect of thin-films buckling on compliant substrate: experimental and theoretical studies. J. Mech. Phys. Solids. **56**, 2585–2598 (2008)
36. Bae, H.J., Bae, S., Yoon, J., Park, C., Kim, K., Kwon, S., et al.: Self-organization of maze- like structures via guided wrinkling. Sci. Adv. **3**, e1700071 (2017)
37. Song, J., Jiang, H., Huang, Y., Rogers, J.: Mechanics of stretchable inorganic electronic materials. J. Vac. Sci. Technol. A. **27**, 1107–1125 (2009)
38. Ma, Y., Xue, Y., Jang, K.-I., Feng, X., Rogers, J.A., Huang, Y.: Wrinkling of a stiff thin film bonded to a pre-strained, compliant substrate with finite thickness. Proc. R. Soc. A Math. Phys. Eng. Sci. **472**, 20160339 (2016)
39. Wang, B., Bao, S., Vinnikova, S., Ghanta, P., Wang, S.: Buckling analysis in stretchable electronics. npj Flexible Electron. **1**, 5 (2017)
40. Wang, S., Song, J., Kim, D.-H., Huang, Y., Rogers, J.A.: Local versus global buckling of thin films on elastomeric substrates. Appl. Phys. Lett. **93**, 023126 (2008)
41. Li, B., Huang, S.-Q., Feng, X.-Q.: Buckling and postbuckling of a compressed thin film bonded on a soft elastic layer: a three-dimensional analysis. Arch. Appl. Mech. **80**, 175 (2010)
42. Khang, D.Y., Rogers, J.A., Lee, H.H.: Mechanical buckling: mechanics, metrology, and stretchable electronics. Adv. Funct. Mater. **19**, 1526–1536 (2009)
43. Liu, Y., Li, M., Liu, J., Chen, X.: Mechanism of surface wrinkle modulation for a stiff film on compliant substrate. J. Appl. Mech. **84**, 051011 (2017)
44. Chen, X., Hutchinson, J.W.: Herringbone buckling patterns of compressed thin films on compliant substrates. J. Appl. Mech. **71**, 597–603 (2004)
45. Shin, G., Jung, I., Malyarchuk, V., Song, J., Wang, S., Ko, H.C., et al.: Micromechanics and advanced designs for curved photodetector arrays in hemispherical electronic-eye cameras. Small. **6**, 851–856 (2010)
46. Lee, J., Wu, J., Shi, M., Yoon, J., Park, S.I., Li, M., et al.: Stretchable GaAs photovoltaics with designs that enable high areal coverage. Adv. Mater. **23**, 986–991 (2011)
47. Kim, D.-H., Song, J., Choi, W.M., Kim, H.-S., Kim, R.-H., Liu, Z., et al.: Materials and noncoplanar mesh designs for integrated circuits with linear elastic responses to extreme mechanical deformations. Proc. Natl. Acad. Sci. **105**, 18675–18680 (2008)
48. Hu, X., Krull, P., De Graff, B., Dowling, K., Rogers, J.A., Arora, W.J.: Stretchable inorganic-semiconductor electronic systems. Adv. Mater. **23**, 2933–2936 (2011)
49. Huang, X., Liu, Y., Chen, K., Shin, W.J., Lu, C.J., Kong, G.W., et al.: Stretchable, wireless sensors and functional substrates for epidermal characterization of sweat. Small. **10**, 3083–3090 (2014)
50. Fan, Z., Zhang, Y., Ma, Q., Zhang, F., Fu, H., Hwang, K.-C., et al.: A finite deformation model of planar serpentine interconnects for stretchable electronics. Int. J. Solids Struct. **91**, 46–54 (2016)
51. Zhang, Y., Fu, H., Su, Y., Xu, S., Cheng, H., Fan, J.A., et al.: Mechanics of ultra- stretchable self-similar serpentine interconnects. Acta Mater. **61**, 7816–7827 (2013)
52. Xu, S., Zhang, Y., Cho, J., Lee, J., Huang, X., Jia, L., et al.: Stretchable batteries with self-similar serpentine interconnects and integrated wireless recharging systems. Nat. Commun. **4**, 1543 (2013)
53. Zhang, Y., Fu, H., Xu, S., Fan, J.A., Hwang, K.-C., Jiang, J., et al.: A hierarchical computational model for stretchable interconnects with fractal-inspired designs. J. Mech. Phys. Solids. **72**, 115–130 (2014)
54. Dong, W., Zhu, C., Ye, D., Huang, Y.: Optimal design of self-similar serpentine interconnects embedded in stretchable electronics. Appl. Phys. A. **123**, 428 (2017)
55. Widlund, T., Yang, S., Hsu, Y.-Y., Lu, N.: Stretchability and compliance of freestanding serpentine-shaped ribbons. Int. J. Solids Struct. **51**, 4026–4037 (2014)
56. Kim, D.H., Liu, Z., Kim, Y.S., Wu, J., Song, J., Kim, H.S., et al.: Optimized structural designs for stretchable silicon integrated circuits. Small. **5**, 2841–2847 (2009)
57. Huang, X., Liu, Y., Cheng, H., Shin, W.J., Fan, J.A., Liu, Z., et al.: Materials and designs for wireless epidermal sensors of hydration and strain. Adv. Funct. Mater. **24**, 3846–3854 (2014)
58. Zhang, Y., Huang, Y., Rogers, J.A.: Mechanics of stretchable batteries and supercapacitors. Curr. Opinion Solid State Mater. Sci. **19**, 190–199 (2015)

59. Fan, J.A., Yeo, W.-H., Su, Y., Hattori, Y., Lee, W., Jung, S.-Y., et al.: Fractal design concepts for stretchable electronics. Nat. Commun. **5**, 3266 (2014)
60. Xu, S., Zhang, Y., Jia, L., Mathewson, K.E., Jang, K.-I., Kim, J., et al.: Soft microfluidic assemblies of sensors, circuits, and radios for the skin. Science. **344**, 70–74 (2014)
61. Kim, J., Banks, A., Cheng, H., Xie, Z., Xu, S., Jang, K.I., et al.: Epidermal electronics with advanced capabilities in near-field communication. Small. **11**, 906–912 (2015)
62. Jang, K.-I., Han, S.Y., Xu, S., Mathewson, K.E., Zhang, Y., Jeong, J.-W., et al.: Rugged and breathable forms of stretchable electronics with adherent composite substrates for transcutaneous monitoring. Nat. Commun. **5**, 4779 (2014)
63. Zhang, Y., Xu, S., Fu, H., Lee, J., Su, J., Hwang, K.-C., et al.: Buckling in serpentine microstructures and applications in elastomer-supported ultra-stretchable electronics with high areal coverage. Soft Matter. **9**, 8062–8070 (2013)
64. Li, T., Suo, Z., Lacour, S.P., Wagner, S.: Compliant thin film patterns of stiff materials as platforms for stretchable electronics. J. Mater. Res. **20**, 3274–3277 (2005)
65. Pan, T., Pharr, M., Ma, Y., Ning, R., Yan, Z., Xu, R., et al.: Experimental and theoretical studies of serpentine interconnects on ultrathin elastomers for stretchable electronics. Adv. Funct. Mater. **27**, 1702589 (2017)
66. Su, Y., Ping, X., Yu, K.J., Lee, J.W., Fan, J.A., Wang, B., et al.: In-plane deformation mechanics for highly stretchable electronics. Adv. Mater. **29**, 1604989 (2017)
67. Lee, C.H., Ma, Y., Jang, K.I., Banks, A., Pan, T., Feng, X., et al.: Soft core/shell packages for stretchable electronics. Adv. Funct. Mater. **25**, 3698–3704 (2015)
68. Lv, C., Yu, H., Jiang, H.: Archimedean spiral design for extremely stretchable interconnects. Extrem. Mech. Lett. **1**, 29–34 (2014)
69. Qaiser, N., Khan, S., Nour, M., Rehman, M., Rojas, J., Hussain, M. M.: Mechanical response of spiral interconnect arrays for highly stretchable electronics. Appl. Phys. Lett. **111**, 214102 (2017)
70. Jang, K.-I., Li, K., Chung, H.U., Xu, S., Jung, H.N., Yang, Y., et al.: Self-assembled three dimensional network designs for soft electronics. Nat. Commun. **8**, 15894 (2017)
71. Xu, S., Yan, Z., Jang, K.-I., Huang, W., Fu, H., Kim, J., et al.: Assembly of micro/nanomaterials into complex, three-dimensional architectures by compressive buckling. Science. **347**, 154–159 (2015)
72. Liu, Y., Yan, Z., Lin, Q., Guo, X., Han, M., Nan, K., et al.: Guided formation of 3D helical mesostructures by mechanical buckling: analytical modeling and experimental validation. Adv. Funct. Mater. **26**, 2909–2918 (2016)
73. Pan, L., Yu, G., Zhai, D., Lee, H.R., Zhao, W., Liu, N., et al.: Hierarchical nanostructured conducting polymer hydrogel with high electrochemical activity. Proc. Natl. Acad. Sci. **109**, 9287–9292 (2012)
74. Chen, S., Tang, F., Tang, L., Li, L.: Synthesis of Cu-nanoparticle hydrogel with self- healing and photothermal properties. ACS Appl. Mater. Interfaces. **9**, 20895–20903 (2017)
75. Cai, G., Wang, J., Qian, K., Chen, J., Li, S., Lee, P.S.: Extremely stretchable strain sensors based on conductive self-healing dynamic cross-links hydrogels for human- motion detection. Adv. Sci. **4**, 1600190 (2017)
76. Meng, H., Xiao, P., Gu, J., Wen, X., Xu, J., Zhao, C., et al.: Self-healable macro−/microscopic shape memory hydrogels based on supramolecular interactions. Chem. Commun. **50**, 12277–12280 (2014)
77. Xing, L., Li, Q., Zhang, G., Zhang, X., Liu, F., Liu, L., et al.: Self-healable polymer nanocomposites capable of simultaneously recovering multiple functionalities. Adv. Funct. Mater. **26**, 3524–3531 (2016)
78. Sun, J.-Y., Zhao, X., Illeperuma, W.R., Chaudhuri, O., Oh, K.H., Mooney, D.J., et al.: Highly stretchable and tough hydrogels. Nature. **489**, 133 (2012)
79. Cong, H.-P., Wang, P., Yu, S.-H.: Stretchable and self-healing graphene oxide–polymer composite hydrogels: a dual-network design. Chem. Mater. **25**, 3357–3362 (2013)
80. Imran, A.B., Esaki, K., Gotoh, H., Seki, T., Ito, K., Sakai, Y., et al.: Extremely stretchable thermosensitive hydrogels by introducing slide-ring polyrotaxane cross-linkers and ionic groups into the polymer network. Nat. Commun. **5**, 5124 (2014)
81. Lin, S., Yuk, H., Zhang, T., Parada, G.A., Koo, H., Yu, C., et al.: Stretchable hydrogel electronics and devices. Adv. Mater. **28**, 4497–4505 (2016)
82. Wu, H., Sariola, V., Zhu, C., Zhao, J., Sitti, M., Bettinger, C.J.: Transfer printing of metallic microstructures on adhesion-promoting hydrogel substrates. Adv. Mater. **27**, 3398–3404 (2015)
83. Yuk, H., Zhang, T., Parada, G.A., Liu, X., Zhao, X.: Skin-inspired hydrogel–elastomer hybrids with robust interfaces and functional microstructures. Nat. Commun. **7**, 12028 (2016)
84. Lee, Y.Y., Kang, H.Y., Gwon, S.H., Choi, G.M., Lim, S.M., Sun, J.Y., et al.: A strain-insensitive stretchable electronic conductor: PEDOT: PSS/acrylamide organogels. Adv. Mater. **28**, 1636–1643 (2016)
85. Dickey, M.D.: Stretchable and soft electronics using liquid metals. Adv. Mater. **29**, 1606425 (2017)
86. Tavakoli, M., Malakooti, M.H., Paisana, H., Ohm, Y., Green Marques, D., Alhais Lopes, P., et al.: EGaIn-assisted room-temperature sintering of silver nanoparticles for stretchable, inkjet-printed, thin-film electronics. Adv. Mater. **30**, 1801852 (2018)
87. Dickey, M.D., Chiechi, R.C., Larsen, R.J., Weiss, E.A., Weitz, D. A., Whitesides, G.M.: Eutectic gallium-indium (EGaIn): a liquid metal alloy for the formation of stable structures in microchannels at room temperature. Adv. Funct. Mater. **18**, 1097–1104 (2008)
88. Lu, Y., Lin, Y., Chen, Z., Hu, Q., Liu, Y., Yu, S., et al.: Enhanced endosomal escape by light-fueled liquid-metal transformer. Nano Lett. **17**, 2138–2145 (2017)
89. Jeong, S.H., Chen, S., Huo, J., Gamstedt, E.K., Liu, J., Zhang, S.-L., et al.: Mechanically stretchable and electrically insulating thermal elastomer composite by liquid alloy droplet embedment. Sci. Rep. **5**, 18257 (2015)
90. Zhu, J.Y., Tang, S.-Y., Khoshmanesh, K., Ghorbani, K.: An integrated liquid cooling system based on galinstan liquid metal droplets. ACS Appl. Mater. Interfaces. **8**, 2173–2180 (2016)
91. Zhu, S., So, J.H., Mays, R., Desai, S., Barnes, W.R., Pourdeyhimi, B., et al.: Ultrastretchable fibers with metallic conductivity using a liquid metal alloy core. Adv. Funct. Mater. **23**, 2308–2314 (2013)
92. Kramer, R.K., Majidi, C., Wood, R.J.: Masked deposition of gallium-indium alloys for liquid-embedded elastomer conductors. Adv. Funct. Mater. **23**, 5292–5296 (2013)
93. Bartlett, M.D., Fassler, A., Kazem, N., Markvicka, E.J., Mandal, P., Majidi, C.: Stretchable, high-k dielectric elastomers through liquid-metal inclusions. Adv. Mater. **28**, 3726–3731 (2016)
94. Kazem, N., Hellebrekers, T., Majidi, C.: Soft multifunctional composites and emulsions with liquid metals. Adv. Mater. **29**, 1605985 (2017)
95. Li, G., Wu, X., Lee, D.-W.: A galinstan-based inkjet printing system for highly stretchable electronics with self-healing capability. Lab Chip. **16**, 1366–1373 (2016)
96. Palleau, E., Reece, S., Desai, S.C., Smith, M.E., Dickey, M.D.: Self-healing stretchable wires for reconfigurable circuit wiring and 3D microfluidics. Adv. Mater. **25**, 1589–1592 (2013)
97. Gozen, B.A., Tabatabai, A., Ozdoganlar, O.B., Majidi, C.: High-density soft-matter electronics with micron-scale line width. Adv. Mater. **26**, 5211–5216 (2014)

98. Khoshmanesh, K., Tang, S.-Y., Zhu, J.Y., Schaefer, S., Mitchell, A., Kalantar-Zadeh, K., et al.: Liquid metal enabled microfluidics. Lab Chip. **17**, 974–993 (2017)
99. Yu, Y.-Z., Lu, J.-R., Liu, J.: 3D printing for functional electronics by injection and package of liquid metals into channels of mechanical structures. Mater. Des. **122**, 80–89 (2017)
100. Ladd, C., So, J.H., Muth, J., Dickey, M.D.: 3D printing of free standing liquid metal microstructures. Adv. Mater. **25**, 5081–5085 (2013)
101. Jeong, S.H., Hagman, A., Hjort, K., Jobs, M., Sundqvist, J., Wu, Z.: Liquid alloy printing of microfluidic stretchable electronics. Lab Chip. **12**, 4657–4664 (2012)
102. Park, J., Wang, S., Li, M., Ahn, C., Hyun, J.K., Kim, D.S., et al.: Three-dimensional nanonetworks for giant stretchability in dielectrics and conductors. Nat. Commun. **3**, 916 (2012)
103. Liang, S., Li, Y., Chen, Y., Yang, J., Zhu, T., Zhu, D., et al.: Liquid metal sponges for mechanically durable, all-soft, electrical conductors. J. Mater. Chem. C. **5**, 1586–1590 (2017)
104. Lu, T., Wissman, J., Majidi, C.: Soft anisotropic conductors as electric vias for ga- based liquid metal circuits. ACS Appl. Mater. Interfaces. **7**, 26923–26929 (2015)
105. Greco, F., Zucca, A., Taccola, S., Menciassi, A., Fujie, T., Haniuda, H., et al.: Ultra-thin conductive free-standing PEDOT/ PSS nanofilms. Soft Matter. **7**, 10642–10650 (2011)
106. Kim, N., Kee, S., Lee, S.H., Lee, B.H., Kahng, Y.H., Jo, Y.R., et al.: Highly conductive PEDOT: PSS nanofibrils induced by solution-processed crystallization. Adv. Mater. **26**, 2268–2272 (2014)
107. Liao, C., Mak, C., Zhang, M., Chan, H.L., Yan, F.: Flexible organic electrochemical transistors for highly selective enzyme biosensors and used for saliva testing. Adv. Mater. **27**, 676–681 (2015)
108. Savagatrup, S., Chan, E., Renteria-Garcia, S.M., Printz, A.D., Zaretski, A.V., O'Connor, T.F., et al.: Plasticization of PEDOT: PSS by common additives for mechanically robust organic solar cells and wearable sensors. Adv. Funct. Mater. **25**, 427–436 (2015)
109. Su, M., Li, F., Chen, S., Huang, Z., Qin, M., Li, W., et al.: Nanoparticle based curve arrays for multirecognition flexible electronics. Adv. Mater. **28**, 1369–1374 (2016)
110. Park, M., Do, K., Kim, J., Son, D., Koo, J.H., Park, J., et al.: Oxide nanomembrane hybrids with enhanced mechano-and thermo-sensitivity for semitransparent epidermal electronics. Adv. Healthc. Mater. **4**, 992–997 (2015)
111. Sangeetha, N.M., Decorde, N., Viallet, B., Viau, G., Ressier, L.: Nanoparticle-based strain gauges fabricated by convective self assembly: strain sensitivity and hysteresis with respect to nanoparticle sizes. J. Phys. Chem. C. **117**, 1935–1940 (2013)
112. Zheng, Z., Yao, J., Wang, B., Yang, G.: Light-controlling, flexible and transparent ethanol gas sensor based on ZnO nanoparticles for wearable devices. Sci. Rep. **5**, 11070 (2015)
113. Sun, H., Tian, H., Yang, Y., Xie, D., Zhang, Y.-C., Liu, X., et al.: A novel flexible nanogenerator made of ZnO nanoparticles and multiwall carbon nanotube. Nanoscale. **5**, 6117–6123 (2013)
114. Lee, S., Hinchet, R., Lee, Y., Yang, Y., Lin, Z.H., Ardila, G., et al.: Ultrathin nanogenerators as self-powered/active skin sensors for tracking eye ball motion. Adv. Funct. Mater. **24**, 1163–1168 (2014)
115. Pang, C., Lee, G.-Y., Kim, T.-i., Kim, S.M., Kim, H.N., Ahn, S.-H., et al.: A flexible and highly sensitive strain-gauge sensor using reversible interlocking of nanofibres. Nat. Mater. **11**, 795 (2012)
116. Yang, C., Gu, H., Lin, W., Yuen, M.M., Wong, C.P., Xiong, M., et al.: Silver nanowires: from scalable synthesis to recyclable foldable electronics. Adv. Mater. **23**, 3052–3056 (2011)
117. Langley, D., Giusti, G., Mayousse, C., Celle, C., Bellet, D., Simonato, J.-P.: Flexible transparent conductive materials based on silver nanowire networks: a review. Nanotechnology. **24**, 452001 (2013)
118. Lee, P., Lee, J., Lee, H., Yeo, J., Hong, S., Nam, K.H., et al.: Highly stretchable and highly conductive metal electrode by very long metal nanowire percolation network. Adv. Mater. **24**, 3326–3332 (2012)
119. Xu, F., Zhu, Y.: Highly conductive and stretchable silver nanowire conductors. Adv. Mater. **24**, 5117–5122 (2012)
120. Amjadi, M., Pichitpajongkit, A., Lee, S., Ryu, S., Park, I.: Highly stretchable and sensitive strain sensor based on silver nanowire–elastomer nanocomposite. ACS Nano. **8**, 5154–5163 (2014)
121. Choi, S., Park, J., Hyun, W., Kim, J., Kim, J., Lee, Y.B., et al.: Stretchable heater using ligand-exchanged silver nanowire nanocomposite for wearable articular thermotherapy. ACS Nano. **9**, 6626–6633 (2015)
122. Gong, S., Cheng, W.: One-dimensional nanomaterials for soft electronics. Adv. Electron. Mater. **3**, 1600314 (2017)
123. Tang, Y., Gong, S., Chen, Y., Yap, L.W., Cheng, W.: Manufacturable conducting rubber ambers and stretchable conductors from copper nanowire aerogel monoliths. ACS Nano. **8**, 5707–5714 (2014)
124. Kang, M., Lee, H., Kang, T., Kim, B.: Synthesis, properties, and biological application of perfect crystal gold nanowires: a review. J. Mater. Sci. Technol. **31**, 573–580 (2015)
125. Gong, S., Lai, D.T., Wang, Y., Yap, L.W., Si, K.J., Shi, Q., et al.: Tattoolike polyaniline microparticle-doped gold nanowire patches as highly durable wearable sensors. ACS Appl. Mater. Interfaces. **7**, 19700–19708 (2015)
126. Gong, S., Zhao, Y., Yap, L.W., Shi, Q., Wang, Y., Bay, J.A.P., et al.: Fabrication of highly transparent and flexible NanoMesh electrode via self-assembly of ultrathin gold nanowires. Adv. Electron. Mater. **2**, 1600121 (2016)
127. Gong, S., Lai, D.T., Su, B., Si, K.J., Ma, Z., Yap, L.W., et al.: Highly stretchy black gold E-skin nanopatches as highly sensitive wearable biomedical sensors. Adv. Electron. Mater. **1**, 1400063 (2015)
128. Gong, S., Schwalb, W., Wang, Y., Chen, Y., Tang, Y., Si, J., et al.: A wearable and highly sensitive pressure sensor with ultrathin gold nanowires. Nat. Commun. **5**, 3132 (2014)
129. Park, J., Lee, Y., Hong, J., Lee, Y., Ha, M., Jung, Y., et al.: Tactile-direction-sensitive and stretchable electronic skins based on human-skin-inspired interlocked microstructures. ACS Nano. **8**, 12020–12029 (2014)
130. Jung, S., Kim, J.H., Kim, J., Choi, S., Lee, J., Park, I., et al.: Reverse-micelle-induced porous pressure-sensitive rubber for wearable human–machine interfaces. Adv. Mater. **26**, 4825–4830 (2014)
131. Cai, L., Li, J., Luan, P., Dong, H., Zhao, D., Zhang, Q., et al.: Highly transparent and conductive stretchable conductors based on hierarchical reticulate single-walled carbon nanotube architecture. Adv. Funct. Mater. **22**, 5238–5244 (2012)
132. Kim, J.H., Hwang, J.-Y., Hwang, H.R., Kim, H.S., Lee, J.H., Seo, J.-W., et al.: Simple and cost-effective method of highly conductive and elastic carbon nanotube/polydimethylsiloxane composite for wearable electronics. Sci. Rep. **8**, 1375 (2018)
133. Lee, S.M., Byeon, H.J., Lee, J.H., Baek, D.H., Lee, K.H., Hong, J. S., et al.: Self- adhesive epidermal carbon nanotube electronics for tether-free long-term continuous recording of biosignals. Sci. Rep. **4**, 6074 (2014)
134. Chang, H.C., Liu, C.L., Chen, W.C.: Flexible nonvolatile transistor memory devices based on one-dimensional electrospun P3HT: Au hybrid nanofibers. Adv. Funct. Mater. **23**, 4960–4968 (2013)
135. Persano, L., Dagdeviren, C., Su, Y., Zhang, Y., Girardo, S., Pisignano, D., et al.: High performance piezoelectric devices based on aligned arrays of nanofibers of poly (vinylidenefluoride-co-trifluoroethylene). Nat. Commun. **4**, 1633 (2013)

136. Hanif, A., Trung, T.Q., Siddiqui, S., Toi, P.T., Lee, N.-E.: Stretchable, transparent, tough, ultrathin, and self-limiting skin-like substrate for stretchable electronics. ACS Appl. Mater. Interfaces. **10**, 27297–27307 (2018)
137. Kim, J., Lee, M., Shim, H.J., Ghaffari, R., Cho, H.R., Son, D., et al.: Stretchable silicon nanoribbon electronics for skin prosthesis. Nat. Commun. **5**, 5747 (2014)
138. Park, M., Park, Y.J., Chen, X., Park, Y.K., Kim, M.S., Ahn, J.H.: MoS2-based tactile sensor for electronic skin applications. Adv. Mater. **28**, 2556–2562 (2016)
139. Lee, H., Choi, T.K., Lee, Y.B., Cho, H.R., Ghaffari, R., Wang, L., et al.: A graphene- based electrochemical device with thermoresponsive microneedles for diabetes monitoring and therapy. Nat. Nanotechnol. **11**, 566 (2016)
140. Liu, Y., Miao, X., Fang, J., Zhang, X., Chen, S., Li, W., et al.: Layered-MnO2 nanosheet grown on nitrogen-doped graphene template as a composite cathode for flexible solid- state asymmetric supercapacitor. ACS Appl. Mater. Interfaces. **8**, 5251–5260 (2016)
141. Ko, H.C., Shin, G., Wang, S., Stoykovich, M.P., Lee, J.W., Kim, D.H., et al.: Curvilinear electronics formed using silicon membrane circuits and elastomeric transfer elements. Small. **5**, 2703–2709 (2009)
142. Song, J., Huang, Y., Xiao, J., Wang, S., Hwang, K., Ko, H., et al.: Mechanics of noncoplanar mesh design for stretchable electronic circuits. J. Appl. Phys. **105**, 123516 (2009)
143. Webb, R.C., Bonifas, A.P., Behnaz, A., Zhang, Y., Yu, K.J., Cheng, H., et al.: Ultrathin conformal devices for precise and continuous thermal characterization of human skin. Nat. Mater. **12**, 938 (2013)
144. Xu, R., Lee, J.W., Pan, T., Ma, S., Wang, J., Han, J.H., et al.: Designing thin, ultrastretchable electronics with stacked circuits and elastomeric encapsulation materials. Adv. Funct. Mater. **27**, 1604545 (2017)
145. Li, R., Li, M., Su, Y., Song, J., Ni, X.: An analytical mechanics model for the island- bridge structure of stretchable electronics. Soft Matter. **9**, 8476–8482 (2013)
146. Su, Y., Wu, J., Fan, Z., Hwang, K.-C., Song, J., Huang, Y., et al.: Postbuckling analysis and its application to stretchable electronics. J. Mech. Phys. Solids. **60**, 487–508 (2012)
147. Chen, C., Tao, W., Su, Y., Wu, J., Song, J.: Lateral buckling of interconnects in a noncoplanar mesh design for stretchable electronics. J. Appl. Mech. **80**, 041031 (2013)
148. Ma, Y., Pharr, M., Wang, L., Kim, J., Liu, Y., Xue, Y., et al.: Soft elastomers with ionic liquid-filled cavities as strain isolating substrates for wearable electronics. Small. **13**, 1602954 (2017)
149. Park, J., Kim, J., Kim, S.-Y., Cheong, W.H., Jang, J., Park, Y.-G., et al.: Soft, smart contact lenses with integrations of wireless circuits, glucose sensors, and displays. Sci. Adv. **4**, eaap9841 (2018)
150. Lee, J.W., Xu, R., Lee, S., Jang, K.-I., Yang, Y., Banks, A., et al.: Soft, thin skin-mounted power management systems and their use in wireless thermography. Proc. Natl. Acad. Sci. **113**, 6131–6136 (2016)
151. Liu, Y., Norton, J.J., Qazi, R., Zou, Z., Ammann, K.R., Liu, H., et al.: Epidermal mechano- acoustic sensing electronics for cardiovascular diagnostics and human-machine interfaces. Sci. Adv. **2**, e1601185 (2016)
152. Mizushima, Y., Kim, Y., Nakamura, T., Uedono, A., Ohba, T.: Behavior of copper contamination on backside damage for ultrathin silicon three dimensional stacking structure. Microelectron. Eng. **167**, 23–31 (2017)
153. Sevilla, G.A.T., Inayat, S.B., Rojas, J.P., Hussain, A.M., Hussain, M.M.: Flexible and semi-transparent thermoelectric energy harvesters from low cost bulk silicon (100). Small. **9**, 3916–3921 (2013)
154. Schander, A., Stemmann, H., Tolstosheeva, E., Roese, R., Biefeld, V., Kempen, L., et al.: Design and fabrication of novel multichannel floating neural probes for intracortical chronic recording. Sensors Actuators A Phys. **247**, 125–135 (2016)
155. Banda, C., Johnson, R.W., Zhang, T., Hou, Z., Charles, H.K.: Flip chip assembly of thinned silicon die on flex substrates. IEEE Trans. Electron. Packag. Manuf. **31**, 1–8 (2008)
156. Christiaens, W., Bosman, E., Vanfleteren, J.: UTCP: a novel polyimide-based ultra- thin chip packaging technology. IEEE Trans. Compon. Packag. Technol. **33**, 754–760 (2010)
157. Sterken, T., Vanfleteren, J., Torfs, T., de Beeck, M.O., Bossuyt, F., Van Hoof, C.: Ultra-Thin Chip Package (UTCP) and stretchable circuit technologies for wearable ECG system. In: 2011 Annual International Conference of the IEEE Engineering in Medicine and Biology Society, pp. 6886–6889 (2011)
158. Valentine, A.D., Busbee, T.A., Boley, J.W., Raney, J.R., Chortos, A., Kotikian, A., et al.: Hybrid 3D printing of soft electronics. Adv. Mater. **29**, 1703817 (2017)
159. Guo, R., Wang, X., Chang, H., Yu, W., Liang, S., Rao, W., et al.: Ni-GaIn amalgams enabled rapid and customizable fabrication of wearable and wireless healthcare electronics. Adv. Eng. Mater. **20**, 1800054 (2018)
160. Yoon, J., Joo, Y., Oh, E., Lee, B., Kim, D., Lee, S., et al.: Soft modular electronic blocks (SMEBs): a strategy for tailored wearable health-monitoring systems. Adv. Sci. 1801682 (2018)
161. Kim, J., Gutruf, P., Chiarelli, A.M., Heo, S.Y., Cho, K., Xie, Z., et al.: Miniaturized battery-free wireless systems for wearable pulse oximetry. Adv. Funct. Mater. **27**, 1604373 (2017)
162. Krishnan, S.R., Su, C.J., Xie, Z., Patel, M., Madhvapathy, S.R., Xu, Y., et al.: Wireless, battery-free epidermal electronics for continuous, quantitative, multimodal thermal characterization of skin. Small. **14**, 1803192 (2018)
163. Khan, Y., Garg, M., Gui, Q., Schadt, M., Gaikwad, A., Han, D., et al.: Flexible hybrid electronics: direct interfacing of soft and hard electronics for wearable health monitoring. Adv. Funct. Mater. **26**, 8764–8775 (2016)
164. Huang, X., Liu, Y., Kong, G.W., Seo, J.H., Ma, Y., Jang, K.-I., et al.: Epidermal radio frequency electronics for wireless power transfer. Microsyst. Nanoeng. **2**, 16052 (2016)

Part IV

Biomedical Devices Packaging

17 Evolution of Advanced Miniaturization for Active Implantable Medical Devices

Chunho Kim

17.1 Introduction

An active implantable medical devices (AIMD) is a powered device that is implanted into a patient's body through either a natural orifice or by surgical means and actively treat the targeted disease or monitor the targeted health conditions [1]. Such a device can be powered by a contained electrical power source (e.g., battery, energy harvest) or an external electrical power source that is transferred wirelessly to the device.

Miniaturization has been always an essential demand for AIMDs or wearable since miniaturized devices would provide better patient comfort, less invasive surgical procedures, lower infection risks, and shorter hospitalization. Over the last decades, such a continual demand has driven remarkable advances in medical device miniaturization for various applications, thanks to the technological advancements in integrated circuit (IC), discrete component, microelectromechanical systems (MEMS), materials, packaging, battery, telemetry communication, and manufacturing processes.

One noticeable trend enabled by the miniaturization of AIMDs is that their delivery methods are shifting from a traditional invasive surgical procedure to a less invasive delivery method, such as injection, insertion, and ingestion, and using a transcatheter [2–13], examples of which are illustrated in Fig. 17.1. All the new delivery methods are characterized by a smaller incision and a simpler delivery procedure in contrary to the traditional surgical methods requiring large incisions and relatively complex procedures. While such miniaturized devices occupy significantly smaller volume in common, the device form factor purposed for each of the invasive delivery methods is slightly different.

C. Kim (✉)
Medtronic plc, Phoenix, AZ, USA
e-mail: chunho.kim@medtronic.com

Injectable devices are intended to be injected by a hypodermic needle so the form factor is typically in cylindrical shape of small diameter and relatively long. Because of its tight size constraint, many of these devices are designed for wireless power and composed of simple electronics. Insertable devices are primarily subcutaneously inserted under the skin through a small incision of 5–15 mm in length, and the device form factor is typically of a thin brick shape. Such a brick shape may be preferred to a round shape for preventing a rotational movement under the skin, which may be an important consideration if a device functionality is sensitive to its rotational position. Ingestible devices are typically in a spherical or cylindrical shape with hemispherical ends, so that they are comfortably swallowed and easily pass through the esophagus and gastrointestinal (GI) track. The device has a diameter range of 9–12 mm, and it provides enough space to have active components, sensors, telemetry transmission, and a battery. Transcatheter-delivered devices are to be delivered by a catheter through veins. The diameter and length of a transcatheter-delivered device would be limited by the vein diameter and curvature in the delivery passage.

Figure 17.1 shows select miniaturized AIMD representatives of each of the four delivery methods, and Table 17.1 summarizes the devices' dimensional features, power sources, intended applications, and case materials.

This chapter reviews the historical evolution of AIMD miniaturization and the expected path toward the continual miniaturization. 3D electronics packaging technologies will be discussed as potential solutions to meet the future advanced miniaturization needs.

17.2 Evolution of AIMD Miniaturization

There is a broad range of diseases and conditions that are being addressed and potentially can be addressed by AIMDs, as shown in Fig. 17.2. The existing implantable medical devices have been miniaturized for patient benefits over last decades.

C. P. Wong et al. (eds.), *Nano-Bio- Electronic, Photonic and MEMS Packaging*, https://doi.org/10.1007/978-3-030-49991-4_17

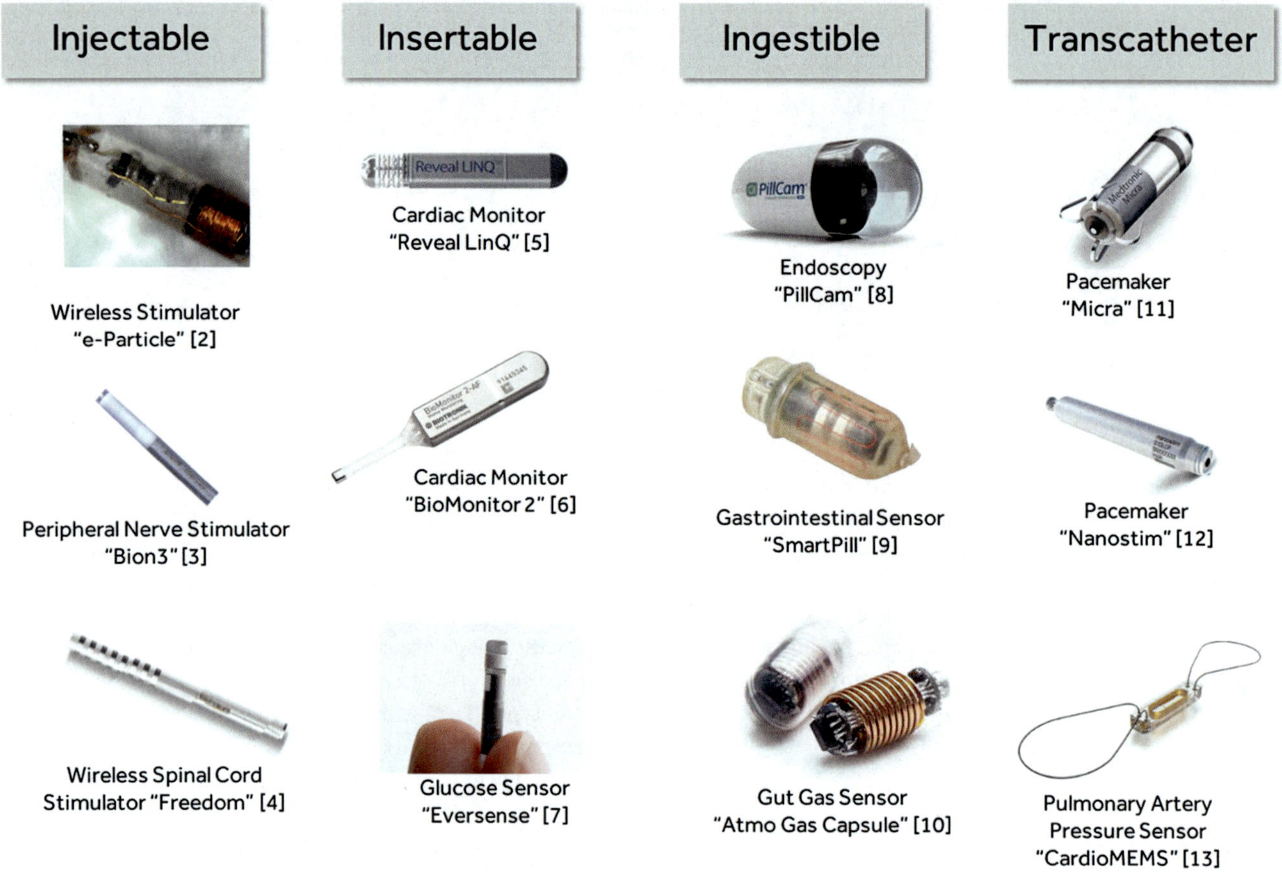

Injectable

Wireless Stimulator "e-Particle" [2]

Peripheral Nerve Stimulator "Bion3" [3]

Wireless Spinal Cord Stimulator "Freedom" [4]

Insertable

Cardiac Monitor "Reveal LinQ" [5]

Cardiac Monitor "BioMonitor 2" [6]

Glucose Sensor "Eversense" [7]

Ingestible

Endoscopy "PillCam" [8]

Gastrointestinal Sensor "SmartPill" [9]

Gut Gas Sensor "Atmo Gas Capsule" [10]

Transcatheter

Pacemaker "Micra" [11]

Pacemaker "Nanostim" [12]

Pulmonary Artery Pressure Sensor "CardioMEMS" [13]

Fig. 17.1 Select miniaturized AIMDs categorized by delivery methods

Table 17.1 Device features

Device name	Size	Application	Delivery type	Power source	Case materials
e-Particle	Ø0.5 mm × 2.3 mm	Neurostimulator	Injectable	Inductively powered	Epo-Tek 301
Bion3	Ø3.3 mm × 27 mm	Neurostimulator	Injectable	Rechargeable battery	Ceramic and titanium
Freedom	Ø1.3 mm × 120 mm	Neurostimulator	Injectable	RF powered	Polyurethane
Reveal LinQ	44.8 mm × 7.2 mm × 4.0 mm	Cardiac monitor	Insertable	Battery powered	Titanium and polyurethane
BioMonitor 2	88 mm × 15.2 mm × 6.2 mm	Cardiac monitor	Insertable	Battery powered	Titanium and silicon
Eversense	Ø3.5 mm × 18.3 mm	Glucose monitor	Insertable	RF powered	Polymethyl methacrylate
PillCam	Ø11 mm × 26 mm	Endoscopy monitor	Ingestible	Battery powered	Biocompatible plastic
SmartPill	Ø11.7 mm × 26.8 mm	Gastrointestinal monitor	Ingestible	Battery powered	Polyurethane
Atmo Gas Capsule	Ø9.8 mm × 26 mm	Gut gas monitor	Ingestible	Battery powered	Polyethylene
Micra	Ø6.7 mm × 25.9 mm	Pacemaker	Transcatheter	Battery powered	Titanium
Nanostim	Ø5.99 mm × 42 mm	Pacemaker	Transcatheter	Battery powered	Titanium
CardioMEMS	2 mm × 3.4 mm × 15 mm	Pulmonary artery pressure monitor	Transcatheter	RF powered	Fused silica

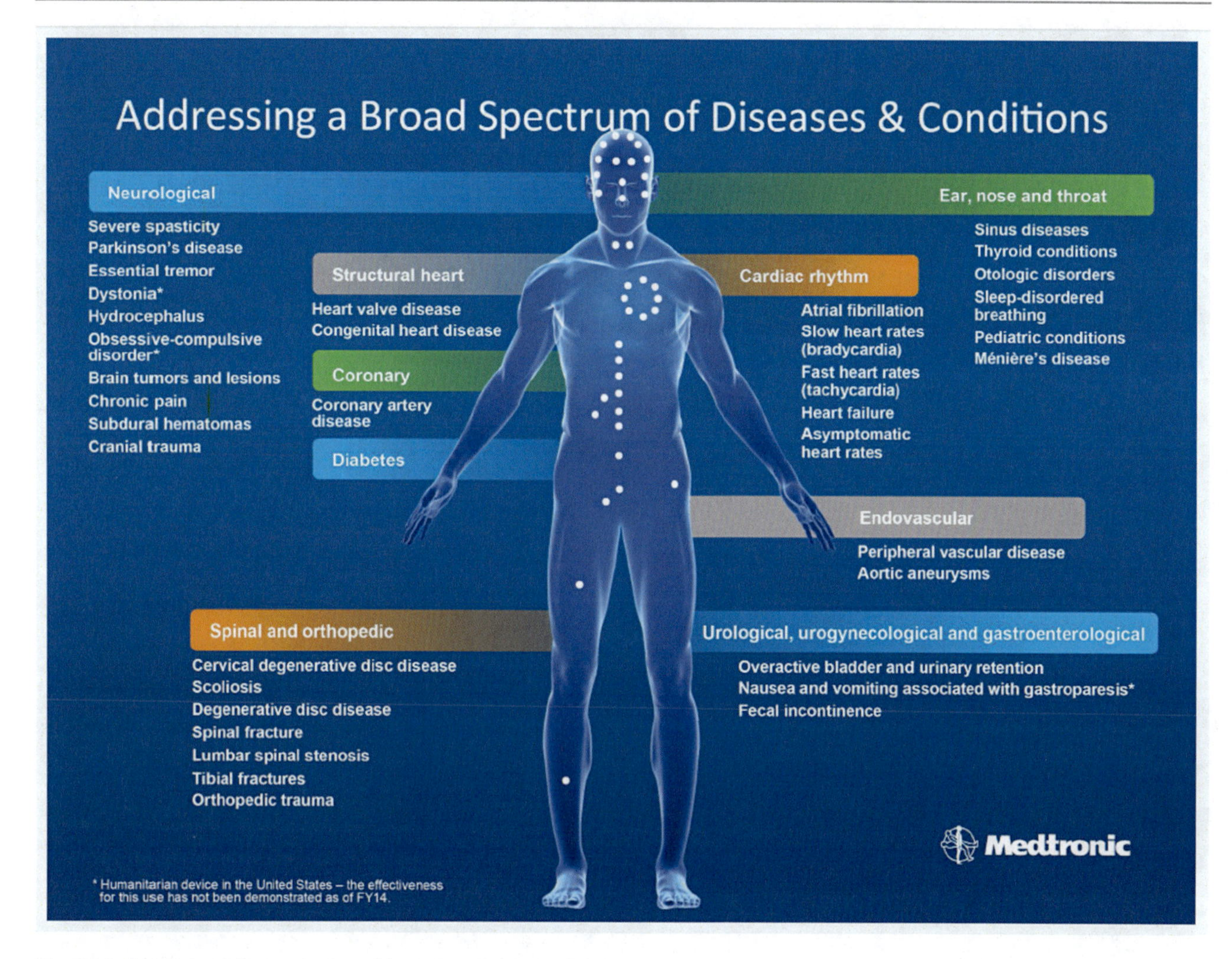

Fig. 17.2 Spectrum of diseases and conditions. (Reprinted from Ref. [14])

The historical evolution of the miniaturization of pacemakers, defibrillators, neurostimulators, and monitors will be reviewed.

17.2.1 Implantable Pacemaker Miniaturization

An implantable pacemaker is a surgically implanted electronic device composed of a generator and one to three leads to treat arrhythmias by sending electrical signals to the heart. The generator is composed of an electronic circuit and a battery. The pacemaker technology evolved toward miniaturization as shown in Fig. 17.3.

Historically, the Electrodyne PM-65 pacemaker in 1952 was an external device powered by an alternative current (AC) power from the wall outlet. It was bulky and heavy and was carried on a cart whose mobility was limited by the length of the power cable connecting the device to the wall outlet [16, 17]. More badly, it posed a danger of an unexpected power supply failure and the resulting disruption of the pacing of patients.

PM-65 was soon followed by a battery-powered pacemaker invented by Earl Bakken [18]. The smaller, lighter, and battery-powered device provided the patients with higher levels of safety, mobility, and efficiency. Despite these advancements, this device was still externally powered, and such an external power method could not be sustained for long because of the infection risk due to the wires passing through the skin incisions and the inconvenience of carrying the pacemaker externally.

To overcome these problems, Ake Senning and Rune Elmqvist invented the first fully implantable pacemaker in 1958. In their device, the two 60 mAh rechargeable batteries were encapsulated with the other electric components for pulse generation in epoxy resin, and two insulated, stainless steel leads came out through the epoxy encapsulation. Since then, the paradigm of full implantation continued, and the basic hardware configuration of using lead wires along with a generator and a battery have remained the same. In this paradigm, the pacemakers have had many complications related to the leads. Firstly, the malfunction of the leads

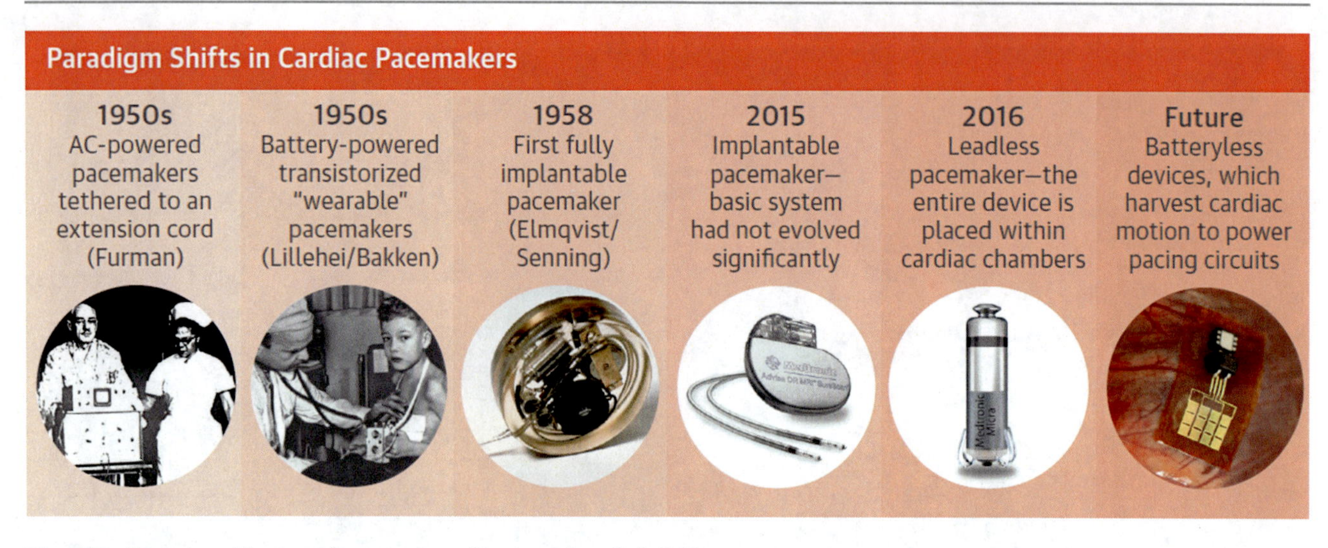

Fig. 17.3 Evolution of implantable pacemakers. (Reprinted from Ref. [15])

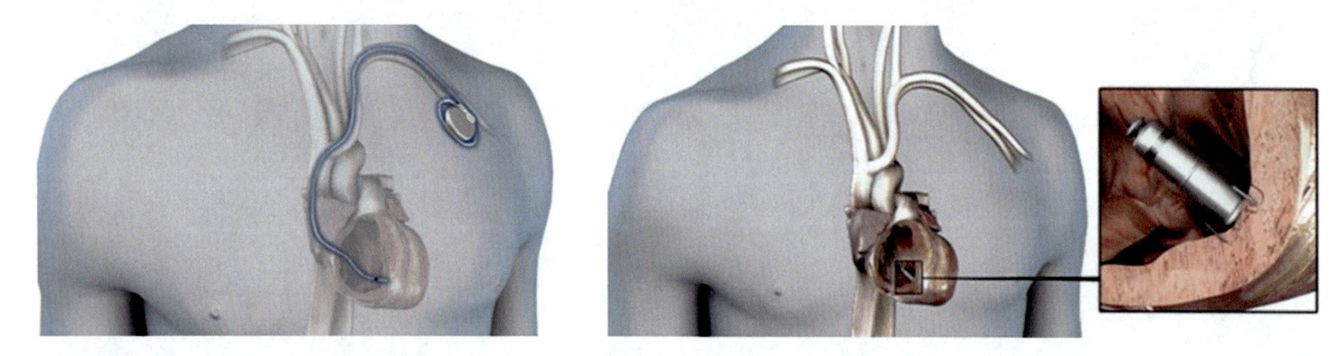

Fig. 17.4 Transvenous pacemaker implantation (left) and leadless pacemaker (Micra TPS) implantation (right). (*Courtesy:* Medtronic)

could occur because of the fracture or the breach of the insulation and the issues related to vascular access (such as use of veins needed for dialysis) or vein stenosis caused by the lead itself. Secondly, the pacemakers still require an incision below the clavicle to create a pocket for the generator, and such an implantation can induce discomfort and unwanted complications such as the erosion of the generator through the skin and infections of the wound and device [19, 20].

In the 2010s, a new paradigm came to birth that could obviate the aforementioned complications and was realized in clinical use by the leadless pacemaker, Micra™ Transcatheter Pacing System (TPS), manufactured by Medtronic (Fig. 17.4). The 25.9 mm long and 20 French diameter device is so small that the whole device is delivered directly to the right ventricle myocardium through the femoral vein by a catheter. This new paradigm would provide advantages not only avoiding the aforementioned complications associated with the traditional pacemakers but also making the delivery procedure less invasive, shortening the surgery time and the recovery time, eliminating visible lump and scar, reducing discomfort, and providing a better quality of life [20, 21]. Nanostim™ is another leadless pacemaker developed by Abbott [12]. However, despite all of its attractive advantages, the use of this novel technology is limited to the single-chamber ventricular pacing, and the end-of-life strategy is a challenge [19, 22–24]. To address those issues, new paradigms such as batteryless pacemaker [25] and wireless stimulation [26] could represent future opportunities.

17.2.2 Implantable Defibrillator Miniaturization

An implantable cardioverter defibrillator (ICD) is a surgically implanted electronic device composed of a generator and one to three leads to treat the life-threatening cases when the heart is beating chaotically and too fast by sending an electric shock to the heart. The generator is composed of an electronic circuit, a battery, and a high-capacity capacitor.

Figure 17.5 shows the progressive volume reduction of ICD devices over time. During the first 10 years of the ICD history, the device volume reduced rapidly, thanks to the advances in integrated circuit, battery capacity, capacitor

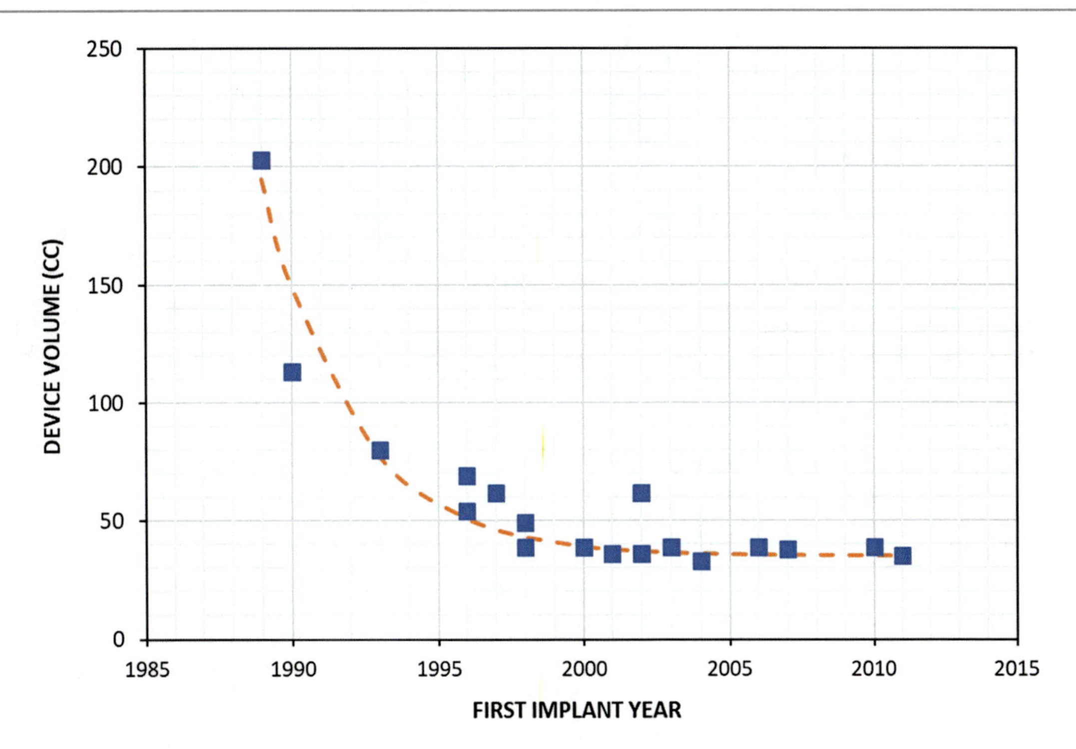

Fig. 17.5 Progressive volume reduction of ICDs

performance, and manufacturing technologies. However, since then, the device volume stopped reducing and became stagnant. Instead, the ICD performance was improved, and more functionalities were added gradually, while the volume remained constant.

Notably, ICDs have not seen disruptive miniaturization like the leadless pacemakers. An ICD is similar to a pacemaker in electrical configuration, but its size is larger to accommodate bigger and higher number of components (e.g., charging capacitors), so that the device can send a high-energy electrical shock to the heart when the heart stops pumping in a severe arrhythmia condition. Currently, making a leadless ICD small enough to be deliverable by a transcatheter method like the leadless pacemakers is challenging. Because the current ICDs are experiencing the same complications related to the leads mentioned before, a leadless ICD would be a significant beneficial innovation.

As an alternative to implanting the whole ICD device inside the heart, subcutaneous implantable cardioverter defibrillators (S-ICD) [27] and extravascular implantable cardioverter defibrillators (EV-ICD) [28] emerged. These devices can reduce the lead-related complications by having their leads placed outside of the heart and veins, i.e., underneath the skin and above the sternum (S-ICD) or under the sternum (EV-ICD). However, the pulse generators in S-ICD and EV-ICD retain similar size to those of the traditional transvenous ICDs and are still implanted in a pocket below the skin like the transvenous ICDs, and the leads are tunneled below the skin instead of through the veins and connected to the pulse generator. The visible lump and scar, discomfort and complications associated with the generator implantation and the leads remain unresolved.

17.2.3 Implantable Neurostimulator Miniaturization

An implantable neurostimulator is a surgically implanted device to deliver low electrical signals to muscles and nerves to disrupt the pain signals. Diverse implantable neurostimulators have been developed and are being used for clinical conditions as schematically shown in Fig. 17.6: deep brain stimulators (DBS) to cure disorders such as Parkinson disease, essential tremor, dystonia, and obsessive-compulsive disorder; motor cortex stimulators (MCS) to treat intractable partial onset epilepsy; spinal cord stimulators (SCS) to treat failed back surgery syndrome, complex regional pain syndrome, angina pectoris, ischemic limb pain, and abdominal pain; vagus nerve stimulators (VNS) to treat epilepsy and depression; and overactive bladder (OAB) stimulators to treat urinary incontinence, frequency, and incomplete bladder emptying.

Similar to the implantable pacemakers, the implantable neurostimulators consist of a pulse generator composed of an electrical circuit for controlling electrical signals and a battery and leads that deliver the electrical signals from the pulse generator to the target treated locations. On the other hand, the leads for the neurostimulators are different in the

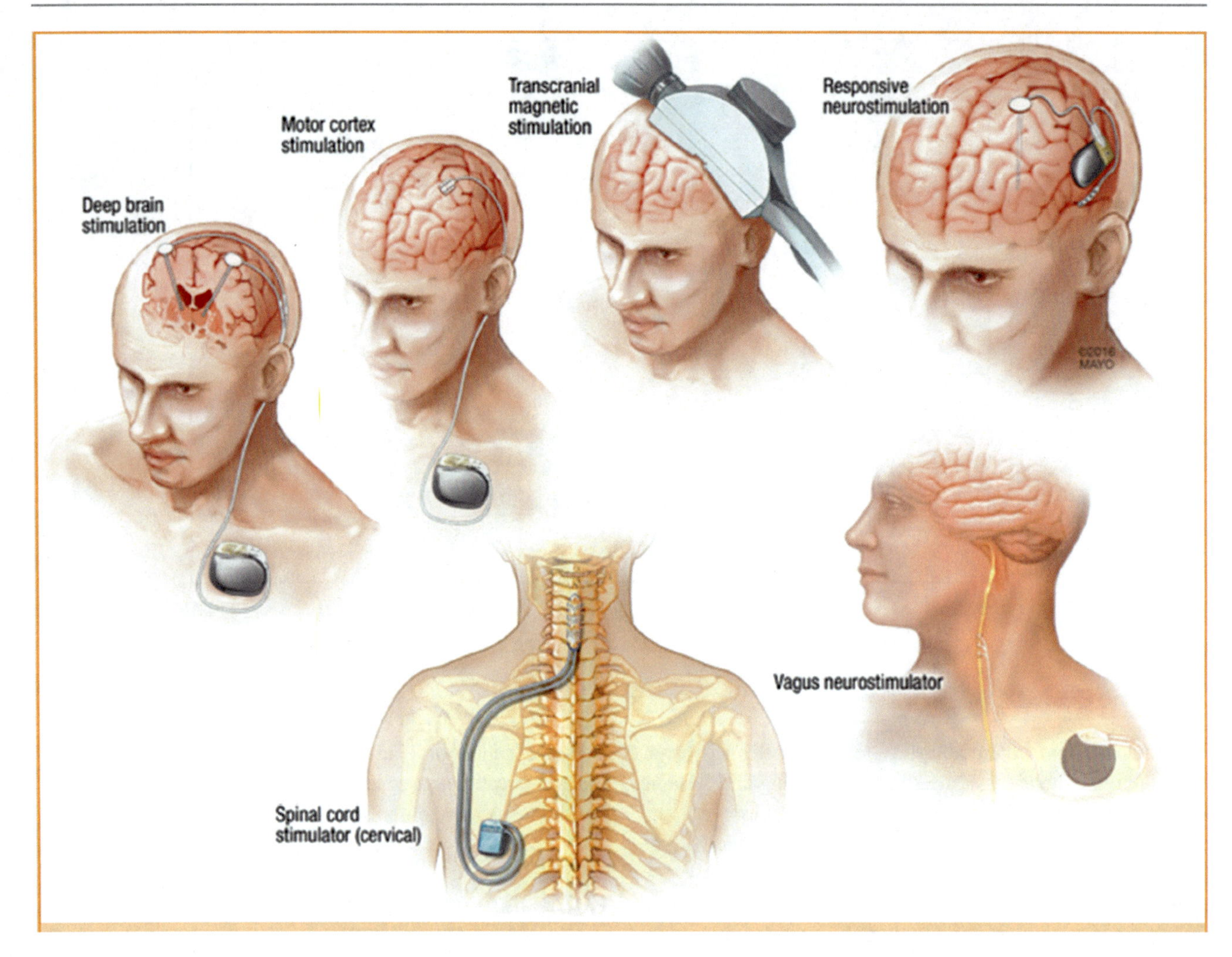

Fig. 17.6 Neurostimulators for various treatments. (Reprinted from Ref. [29])

number of electrodes from those for the pacemakers. Each lead has 4–16 electrodes depending on the condition being treated as well as the physician's preference, while a pacemaker's lead has one (unipolar) or two (bipolar) electrodes. In addition, the cross-sectional lead shape for neurostimulators can be either round (percutaneous deliverable) or flat (surgical deliverable), while a pacemaker lead shape is round only.

Recently, in the similar fashion and motivation to the leadless pacemakers, leadless neurostimulators have emerged [2–4], in which the electrodes are mounted directly onto the stimulators in order to avoid the complications associated with the traditional lead-based neurostimulators. Such leadless neurostimulators are even smaller than the leadless pacemakers (Medtronic Micra TPS, Abbott Nanostim), so they can be implanted by an injectable method using a needle. One way to make such a small neurostimulator is eliminating the battery from the implanted device and instead powering it wirelessly from an external wearable antenna assembly [2, 4]. The biggest advantage of this approach is such small devices could be applicable to many neurological treatment applications (e.g., DBS) in a minimally invasive delivery method [30]. In addition, the wirelessly powered implanted device could be used for much longer than the battery-powered implantable devices whose lifetime is limited by that of the primary or rechargeable battery. On the other hand, the obvious disadvantage is the need to constantly wear an external power transmitter to operate the device. The other way to make such a smaller neurostimulator is using a smaller rechargeable battery [3]. For a battery-powered stimulator, the battery drives a larger device size compared to the wirelessly powered stimulators, but the self-powered device would provide more comfort to the patients by not requiring to wear an external transmitter except to charge the battery. Considering continual advancement in the battery technology, the rechargeable option may become the major preference in size and patient comfort.

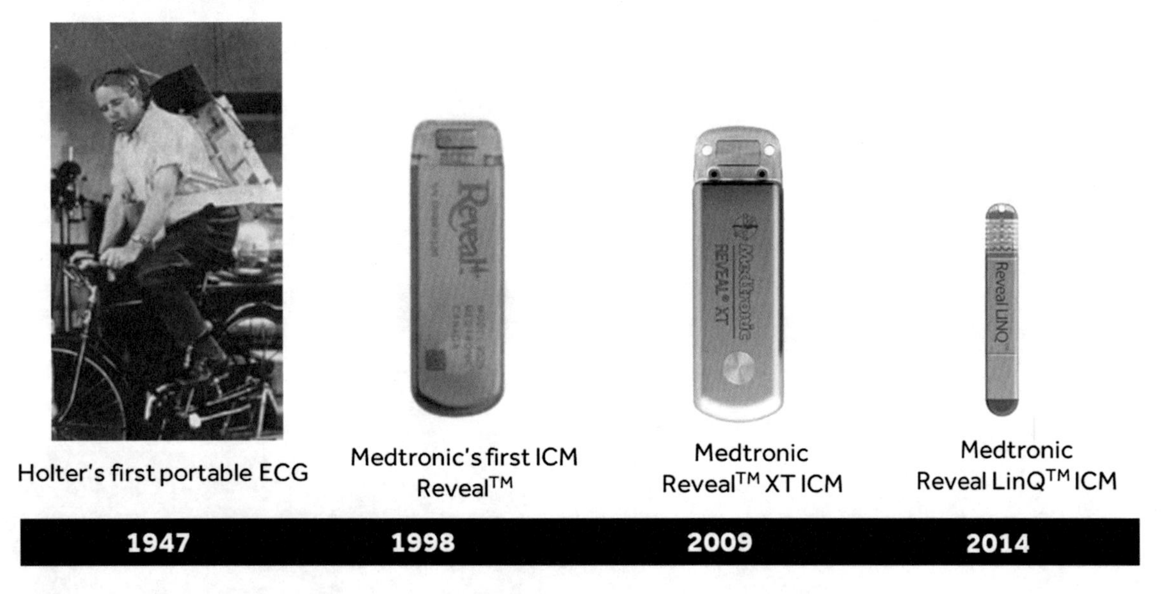

Fig. 17.7 Progressive miniaturization of cardiac monitor devices

17.2.4 Implantable Monitor Miniaturization

Jeff Holter invented the first portable electrocardiograph (ECG) in 1947. It was a 38 kg device wearable externally like a backpack as shown in Fig. 17.7 and could record several hours of data using a reel-to-reel technology [31]. Since then, the electronic technologies have advanced toward the development of the first implantable cardiac monitoring device (ICM), Reveal™ Insertable Loop Recorder (ILR) in 1998 by Medtronic with a 14-month battery life. Further technical advancement led to the more functional and longer life ICM, Reveal™ XT Insertable Cardiac Monitor (ICM), in a similar form factor in 2009, and more recently, a disruptively smaller ICM, Reveal LinQ™ Insertable Cardiac Monitor, was released in 2014 [4]. LinQ™ is small enough to be insertable under the skin with an incision less than 10 mm, and it can operate continuously for up to 3 years [32].

Another example of miniaturized implantable device is the CardioMEMS™ heart failure sensor designed to monitor the pulmonary artery (PA) pressure. This device worked in conjunction with an external electronics system, which reads the PA pressure from the sensor wirelessly [13]. CardioMEMS™ sensor is a simple passive electrical circuit composed of a three-dimensional coil and a pressure-sensitive capacitor, enclosed and sealed hermetically in a fused silica case. The case size is so small (2 mm × 3.4 mm × 15 mm) to be deliverable through the femoral vein using a 12 Fr sheath.

Continuous glucose monitoring (CGM) sensors have also evolved toward miniaturization. One example is the subcutaneously implantable CGM, "Eversense" [6]. This sensor is in a cylindrical shape of 3.5 mm in diameter and 18.3 mm long and invasively insertable in the upper arm via 5–8 mm incision. Like the wireless-powered neurostimulators, Eversense is powered wirelessly by a transmitter worn externally over the sensor. This sensor consists of core electronics and optics that are sealed in epoxy within a polymethyl methacrylate (PMMA) encasement. The sensor can last up to 90 days, longer than other implantable CGMs which typically last 7–14 days.

Ingestible sensing is another fast-emerging medical technology area, enabled by the advancements in electronics miniaturization. Ingestible sensing is a noninvasive and safe way to access the gut environment along the gastrointestinal track and collect various information such as endoscopic images, pH, temperature, pressure, gas, medication, ultrasound, etc. [33]. Among the ingestible sensors are the endoscopy sensor, "PillCam" [8], the gastrointestinal sensor "SmartPill" [9], and the gut gas sensor "Atmo Gas Capsule" [10]. Each of these ingestible devices consists of sensors, a wireless transmitter, an antenna, ICs, and a battery in a cylindrical capsule of a 9–12 mm diameter.

17.3 Electronic Packaging Solutions for Advanced Miniaturization

The electronics packaging technologies have been evolving rapidly over the last decades (Fig. 17.8). Thanks to the technology advancements, the modern consumer electronics offer smaller sizes, more functionality, faster speed, and lower-power consumption.

One noticeable trend is the transition from two-dimensional (2D) packaging to three-dimensional (3D) packaging in order to meet the continuing miniaturization demands, since 2D packaging has reached its limit of size reduction. 3D packaging allows the components to be

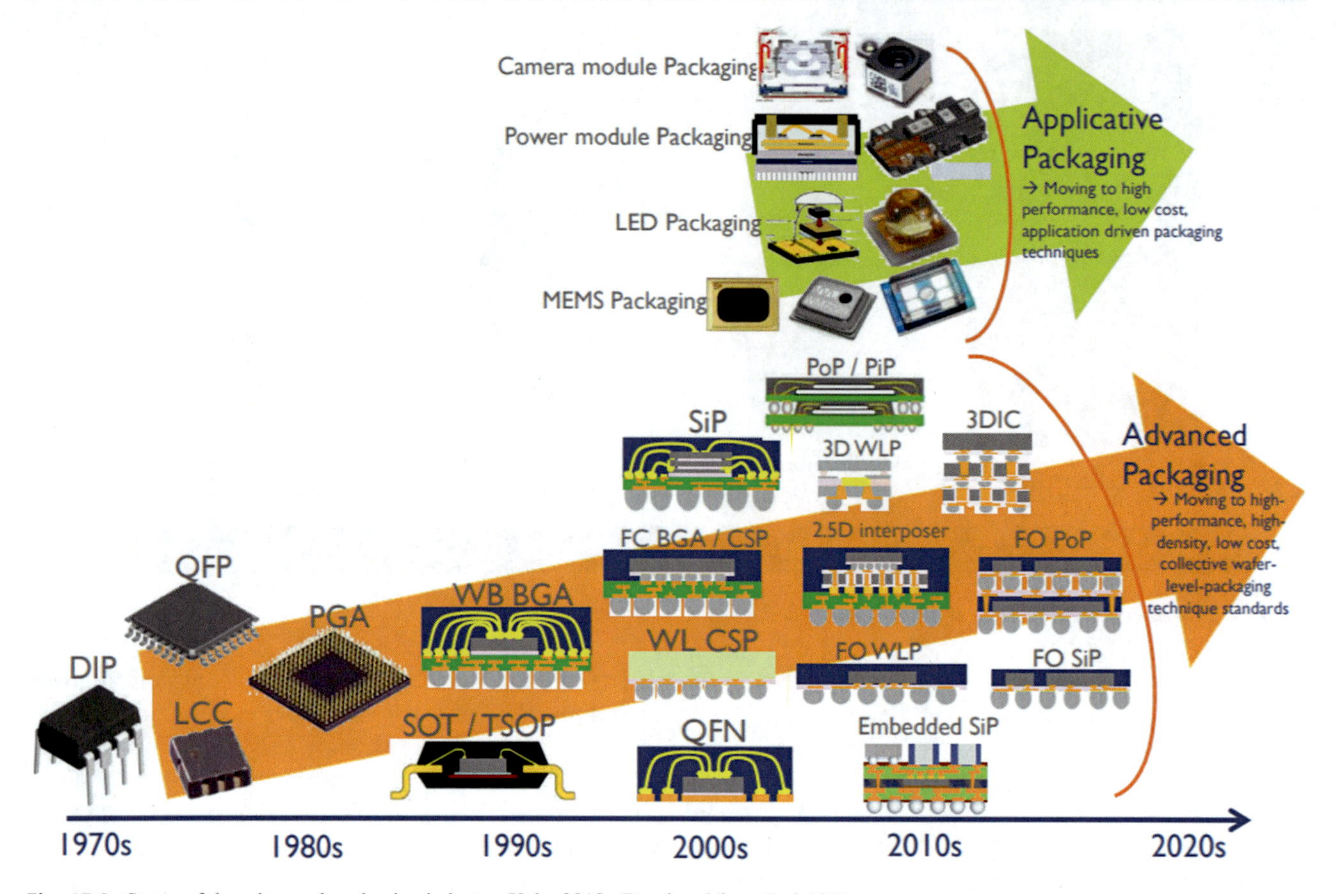

Fig. 17.8 Status of the advanced packaging industry, Yole, 2018. (Reprinted from Ref. [34])

stacked vertically as well as horizontally, while the 2D packaging allows only a horizontal component layout and requires a comparatively larger footprint. 3D packaging also facilitates a shorter interconnect path and thus a lower signal and power loss, a higher integration density, and the better utilization of space in three dimensions.

However, the apparent life cycles of medical devices are not as fast as the evolution of electronics packaging technology, and the 3D packaging technologies are not being used widely yet in medical devices. In order to continue driving the miniaturization of the medical devices, the 3D packaging technologies could be considered as viable solutions [35].

17.4 Concluding Remarks

The historical evolutions of AIMD miniaturization were reviewed. This continuous miniaturization has been driven to provide the patients with better comfort, less invasive surgical procedures, lower infection risks, and shorter hospitalization. Thanks to the continuous advancements of IC, passives, antenna, RF transmission, battery, materials, and manufacturing processes, the device miniaturization has been advancing incrementally but not disruptively over the last decades. However, the new recent paradigms such as leadless pacemakers, leadless neurostimulators, wireless powering, and wireless stimulation have emerged and led to disruptively smaller implantable devices that can be deliverable by injectable, insertable, ingestible, and transcatheter procedures. The disruptive miniaturization is expected to not only provide the patients with more benefits but also expand the range of treatable diseases and conditions.

An ideal AIMD should include a power source (primary battery, rechargeable battery or energy harvesting), antenna, sensors, stimulating electrodes, and electronic circuits, all integrated into a small case. 3D electronic packaging technologies could be potential viable solutions to integrate all the heterogeneous components. Even though the 3D electronic packaging technologies are being used widely in consumer products, they have yet seen limited applications in hearing aid devices. 3D packaging technologies along with continuous advancements of materials, battery, and manufacturing processes can help make ideal AIMDs possible to provide the maximal benefits to the patients for various medical applications.

References

1. Active Implantable Medical Device (AIMD): On the approximation of the laws of the Member States relating to active implantable medical devices. Directive 90/385/EEC (http://ec.europa.eu/growth/single-market/european-standards/harmonised-standards/implantable-medical-devices/)
2. Freeman, D.K., O'Bri, J.M., Kumar, P., Daniel, B., Irion, R.A., Shraytah, L., Ingersoll, B.K., Magyar, A.P., Czarneck, A., Wheeler, J., Coppeta, J.R., Abban, M.P., Gatzke, R., Fried, S.I., Lee, S.W., Duwel, A.E., Bernstein, J.J., Widge, A.S., Hernandez-Reynoso, A., Kanneganti, A., Romero-Ortega, M.I., Cogan, S.F.: A sub-millimeter, inductively powered neural stimulator. Front. Neurosci. **11**, Article 659 (2017)
3. Loeb, G.E., Richmond, F.J.R., Baker, L.L.: The BION devices: injectable interfaces with peripheral nerves and muscles. Neurosurg. Focus. **20**(5), E2 (2006)
4. Billet, B., Wynendaele, R., Vanquathem, N.E.: Wireless neuromodulation by a minimally invasive technique for chronic refractory pain. Report of preliminary observations. Med. Res. Archives. **5**(8) (2017)
5. Tomson, T.T., Passman, R.: The reveal LINQ insertable cardiac monitor. J. Expert. Rev. Med. Dev. **12**(1) (2015)
6. Reinsch, N., Ruprecht, U., Buchholz, J., Diehl, R.R., Kälsch, H., Neven, K.: The BioMonitor 2 insertable cardiac monitor: clinical experience with a novel implantable cardiac monitor. J. Electrocardiol. **51**(5), 751–755 (2018)
7. Christiansen, M.P., Klaff, L.J., Brazg, R., Chang, A.R., Levy, C.J., Lam, D., Denham, D.S., Atiee, G., Bode, B.W., Walters, S.J., Kelley, L., Bailey, T.S.: A prospective multicenter evaluation of the accuracy of a novel implanted continuous glucose sensor: PRECISE II. Diabetes Technol. Ther. **20**(3) (2018)
8. Van de Bruaene, C., De Looze, D., Hindryckx, P.: Small bowel capsule endoscopy: where are we after almost 15 years of use? World J. Gastrointest. Endosc. **7**(1), 13–36 (2015)
9. Saad, R.J., Hasler, W.L.: A technical review and clinical assessment of the wireless motility capsule. Gastroenterol Hepatol (NY). **7**(12), 795–804 (2011)
10. Kalantar-Zadeh, K., Berean, K.J., Ha, N., Chrimes, A.F., Xu, K., Grando, D., Ou, J.Z., Pillai, N., Campbell, J.L., Brkljača, R., Taylor, K.M., Burgell, R.E., Yao, C.K., Ward, S.A., McSweeney, C.S., Muir, J.G., Gibson, P.R.: A human pilot trial of ingestible electronic capsules capable of sensing different gases in the gut. Nat. Electron. **1**, 79–87 (2018)
11. Bernabei, M.A.: The Micra Transcatheter pacing study: the making of a revolution in pacemaking. J. Lancaster General Hospital. **9**(3) (2014)
12. Sperzel, J., Hamm, C., Hain, A.: Nanostim—leadless pacemaker. Herzschr Elektrophys. **29**, 327–333 (2018)
13. De Rosa, R., Piscione, F., Schranz, D., Citro, R., Iesu, S., Galasso, G.: Transcatheter implantable devices to monitoring of elevated left atrial pressures in patients with chronic heart failure. Transl. Med. UniSa. **17**, 19–24 (2017)
14. Medtronic: Integrated Performance Report, (2014)
15. Madhavan, M., Mulpuru, S.K., McLeod, C.J., Cha, Y.-M., Friedman, P.A.: Cardiac pacemakers: function, troubleshooting, and management. J. Am. Coll. Cardiol. **69**(2), 189–210 (2017)
16. Aquilina, O.: A brief history of cardiac pacing. Images Paediatr. Cardiol. **8**(2), 17–81 (2006)
17. Mittal, T.: Pacemakers—A journey through the years. Indian J. Thorac. Cardiovasc. Surg. **21**(3), 236–249 (2005)
18. Lillehei, C.W., Gott, V.L., Hodges Jr., P.C., Long, D.M., Bakken, E. E.: Transistor pacemaker for treatment of complete atrioventricular dissociation. J. Am. Med. Assoc. **172**, 2006–2010 (1960)
19. Bhatia, N., El-Chami, M.: Leadless pacemakers: a contemporary review. J. Geriatr. Cardiol. **15**, 249–253 (2018)
20. Sideris, S., Archontakis, S., Dilaveris, P., Gatzoulis, K.A., Trachanas, K., Sotiropoulos, I., Arsenos, P., Tousoulis, D., Kallikazaros, I.: Leadless cardiac pacemakers: current status of a modern approach in pacing. Hellenic J. Cardiol. **58**(6), 403–410 (2017)
21. Miller, M.A., Neuzil, P., Dukkipati, S.R., Reddy, V.Y.: Leadless cardiac pacemakers– Back to the future. JACC. **66**, 1179–1189 (2015)
22. Beurskens, N.E.G., Tjong, F.V.Y., Knops, R.E.: End-of-life management of leadless cardiac pacemaker therapy. Arrhythmia Electrophysiol. Rev. **6**(3) (2017)
23. Seriwala, H.M., Khan, M.S., Munir, M.B., Riaz, I.B., Riaz, H., Saba, S., Voigt, A.H.: Leadless pacemakers: a new era in cardiac pacing. J. Cardiol. **67**(1), 1–5 (2016)
24. Siderisa, S., Archontakisb, S., Dilaveris, P., Gatzoulis, K.A., Trachanas, K., Sotiropoulos, I., Arsenos, P., Tousoulis, D., Kallikazarosa, I.: Leadless cardiac pacemakers: current status of a modern approach in pacing. Hellenic Soc. Cardiol. **58**(6), 403–410 (2017)
25. Dagdeviren, C., Yang, B.D., Su, Y., Tran, P.L., Joe, P., Anderson, E., Xia, J., Doraiswamy, V., Dehdashti, B., Feng, X., Lu, B., Poston, R., Khalpey, Z., Ghaffari, R., Huang, Y., Slepian, M.J., Rogers, J. A.: Conformal piezoelectric energy harvesting and storage from motions of the heart, lung, and diaphragm. PNAS. **111**(5), 1927–1932 (2014)
26. Neuzil, P., Reddy, V.Y., Sedivy, A.B.: 17-01: Wireless LV endocardial stimulation for cardiac resynchronization: long-term (12 month) experience of clinical efficacy and clinical events from two centers (abstr). Heart Rhythm. **13**, S38 (2016) article title: AB17-01
27. De Maria, E., Olaru, A., Cappelli, S.: The entirely Subcutaneous Defibrillator (S-Icd): state of the art and selection of the ideal candidate. Curr. Cardiol. Rev. **11**(2), 180–186 (2015)
28. Chan, J.Y.S., Lelakowski, J., Murgatroyd, F.D., Boersma, L.V., Cao, J., Nikolski, V., Wouters, G., Hall, M.C.S.: Novel extravascular defibrillation configuration with a coil in the substernal space: the ASD clinical study. JACC: Clin. Electrophysiol. **3**(8) (2017)
29. Edwards, C.A., Kouzani, A., Lee, K.H., Ross, E.K.: Neurostimulation devices for the treatment of neurologic disorders. Mayo Clin. Proc. **92**(9), 1427–1444 (2017)
30. Maeng, L.Y., Murillo, M.F., Mu, M., Lo, M., De La Rosa, M., O'Brien, J.M., Freeman, D.K., Widge, A.S.: Behavioral validation of a wireless low-power neurostimulation technology in a conditioned place preference task. J. Neural Eng. **16** (2019) 026022 (14pp)
31. Ioannou, K., Ignaszewski, M., Macdonald, I.: Ambulatory electrocardiography: the contribution of Norman Jefferis Holter. BC Medical J. **56**(2) (2014)
32. Giancaterino, S., Lupercio, F., Nishimura, M., Hsu, J.C.: Current and future use of insertable cardiac monitors. JACC Clin. Electrophysiol. **4**(11), 1383–1396 (2018)
33. Kalantar-zadeh, K., Ha, N., Ou, J.Z., Berean, K.J.: Ingestible sensors. ACS Sens. **2**(4), 468–483 (2017)
34. Kumar, S.: Status of the Advanced Packaging Industry 2018, Yole (2018).
35. Yang, Y., Li, F.: Recent advances in 3D packaging for medical applications. In: 19th International Conference on Electronic Packaging Technology, 8–11 August 2018, Shanghai, China, pp 1193–1197 (2018)

Part V

Optical Device Packaging

18 An Introduction of the Phosphor-Converted White LED Packaging and Its Reliability

C. Qian, J. J. Fan, J. L. Huang, X. J. Fan, and G. Q. Zhang

18.1 Introduction

With ever-increasing competences in lumen efficacy, color consistence, and durability, phosphor-converted white LEDs (pc-white LEDs) have gained a wide range of applications including indoor lighting, commercial lighting, urban lighting, industrial lighting, automotive lighting, healthcare lighting, intelligent lighting, etc. [1]. Figure 18.1 illustrates a pc-white LED package that comprises of serval parts such as LED chip (also known as LED die), phosphors, encapsulant, bonding wire, die attach, lead frame, and housing. The LED chip is further divided by a stack of layers including the epi-layer, current spreading layer, chip substrate, chip electrodes, etc. The phosphors mixed with encapsulant undertake an important mission to convert the LED blue light into yellow and red spectrum. The housing acts as a solid shell to prevent mechanical damages to the pc-white LED. In order to enhance the light extraction efficiency (LEE) of the pc-white LED, the inner surface of the housing is always coated with reflecting materials, such as Ag. The lead frame is embedded inside of the housing to connect the LED chip with the external circuit via bonding wires for electricity conductivity. The die attach layer with solder paste is placed to fix the LED chip on the lead frame and also plays as a heat flow passage in between the LED chip and lead frame. Furthermore, different materials are employed in these different components, resulting in that the pc-white LED package is actually a complicated system.

For a pc-white LED package, the reliability issues are usually referred to three kinds of failures: catastrophic failure (i.e., sudden-death of the pc-white LED), luminous flux degradation (i.e., degradation of the luminous flux), and color shift (i.e., shift of the color coordinates) [2, 3]. The evolution of these failures is performed by a competition among the different failure modes existing in its different components [4]. In the early stage, failures of pc-white LEDs are mainly contributed by the failure of LED chips. Particularly, a majority of the catastrophic failures of pc-white LEDs occurred in the early period are because of the chip damages. Then, along with the increase of operational time, degradation in the phosphor/silicone composite takes more and more important effects on the failures (mostly the degradation failures) of the pc-white LEDs [4]. In this chapter, a couple of key individual components in a pc-white LED package and their reliability impact factors are discussed in Sects. 18.2, 18.3, 18.4, 18.5, and 18.6, respectively, and finally concluded in Sect. 18.7.

18.2 LED Chip

There are three types of LED chip structure, which are lateral, flip chip, and vertical LEDs as illustrated in Fig. 18.2a–c, respectively.

Figure 18.2a shows a lateral LED chip, considered as the most common LED architecture in the world. As shown in Fig. 18.2a, the p contact and n contact electrodes are placed on the same side, connecting to the electric circuit substrate via wire bonds. However, this brings out a problem that the horizontal current flow through the n-GaN layer creates a current crowding region where the hot spot will be generated. To minimize this problem, a transparent current spreading layer (CSL) created by is placed below the p contact electrode to homogenize the current density [5]. Note that the thickness

C. Qian
School of Reliability and Systems Engineering, Beihang University, Beijing, China

J. J. Fan
College of Mechanical and Electrical Engineering, Hohai University, Changzhou, China

J. L. Huang · G. Q. Zhang
EEMCS Faculty, Delft University of Technology, Delft, The Netherlands

X. J. Fan (✉)
Department of Mechanical Engineering, Lamar University, Beaumont, TX, USA
e-mail: xuejun.fan@lamar.edu

C. P. Wong et al. (eds.), *Nano-Bio- Electronic, Photonic and MEMS Packaging*, https://doi.org/10.1007/978-3-030-49991-4_18

of CSL needs to be carefully considered. It should be thick enough to uniformly distribute the current density, but not too thick to reduce the self-absorption of the light. The traditional material to produce the CSL is indium tin oxide (ITO). But lately the graphene-based CSL has gained more and more attention by its superior thermal and electrical conductivity and high transmittance [6]. Another drawback is that most of the heat generated in the multiple quantum well (MQW) has to be dissipated via the sapphire substrate of which the thermal conductivity is only about 35 W/(m·K). Therefore, using the lateral LED chip in a high-power LED will yield an excessively high junction temperature, adding up the risk of failure of the LEDs.

Figure 18.2b shows a flip chip initially imported from the semiconductor packaging technology by Philips Lighting in the early 2000s. However, the commercial products have only appeared until 2012, known as chip scale packages (CSPs). In this structure, the sapphire substrate lies on the top of the chip, making that the light generated in the MQW layer can directly emit through the sapphire without obstructions from the electrodes. Furthermore, both electrodes are soldered on the electric circuit substrate by a reflow or eutectic bonding technique replacing the wire bonding process. The solder layer not only provides an efficient heat dissipation path to transfer the heat from the MQW layer but also allows a large current to pass through. Therefore, CSP has a great potential for high-power LEDs. Nevertheless, the drawback is the CSP industry lags far behind in achieving a mature process for industrialization production. So commercial CSPs are produced with a high cost and low yield, and that is the main reason the CSPs have not yet gained a wide range of applications for the time being.

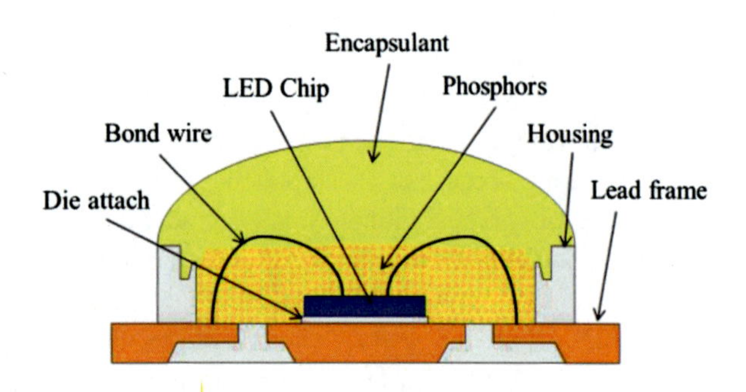

Fig. 18.1 Illustration of a pc-white LED package [7]

Figure 18.2c shows a LED chip with vertical structure invented by Cree. In this structure, the n contact electrode is below the SiC substrate, connecting to the electric circuit substrate via a eutectic bonding technology, whereas the p contact electrode is on the top of the chip, connecting to the electric circuit substrate via wire bond. The vertical LED has the follow advantages: Firstly, the direct bonding between the n contact electrode and electric circuit substrate is very helpful for the heat dissipation inside of the chip. Secondly the current vertically flows through the chip, dramatically reducing the current crowding phenomenon. Therefore, the CSL is not necessarily needed in the vertical LED. Moreover, SiC is a perfect material for substrate by its superior electric and thermal properties. Especially the thermal conductivity of the SiC substrate is as high as 490 W/(m·K). However, the manufacturing cost of the SiC substrate is also quite higher compared to that of the sapphire substrate. For the time being, a new manufacturing process is developed by using silicon (Si) as the substrate in lateral and vertical LEDs. Compared to sapphire, Si has much higher electrical and thermal conductivities for the improvement of current spreading and therefore more suitable for substrate.

Reliability problems of the above three types of LED chips are basically caused by several reasons, such as

Fig. 18.2 Illustration of LED chip structures. (**a**) Lateral LED. (**b**) Flip chip LED. (**c**) Vertical LED

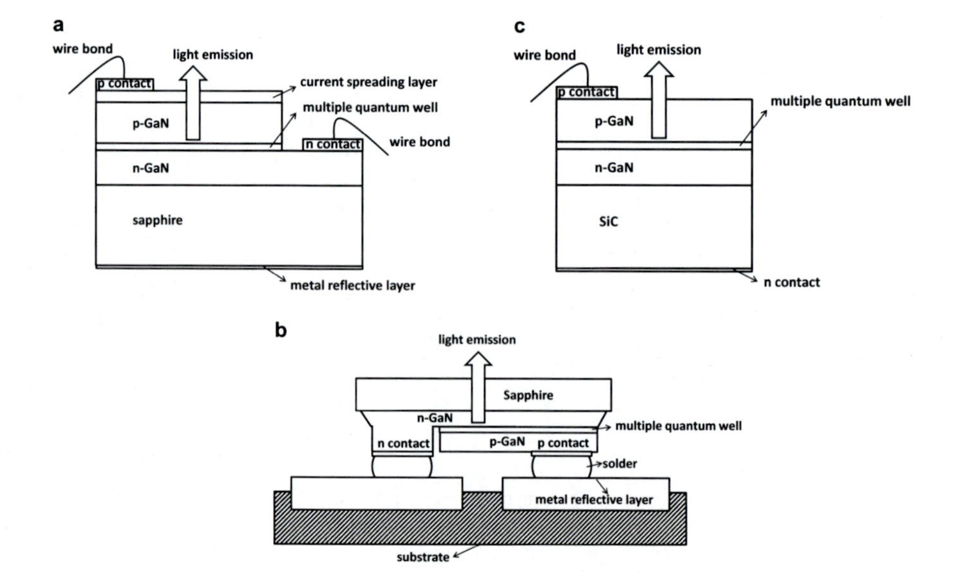

material defects in the substrate and epitaxial layers, dopant degradation, electro-metallization, and manufacturing damage such as electrostatic discharge (ESD) [4–9]. Then driving under the influence of current, temperature, and moisture, failures of luminous flux depreciation, current leakages, and even sudden death might occur as presented in below [10–14].

1. Temperature and current usually affect the LED chip by reducing the internal quantum efficiency (IQE) and lifetime. High environmental temperature and driving current result in a high junction temperature of the chip, producing a sharp temperature gradient within the LED chip. As a result, the thermal stress and strain are generated because of the different thermal expansion inside of the chip to deteriorate its IQE and reliability. In extreme cases, micro-cracks will be initiated at the areas with high thermal strains. Another negative outcome of the high temperature and current is the degradation of ohmic contact between ITO and p-GaN, which will bring a more serious current crowding effect and therefore aggravate the light output degradation of the chip. This is because the degradation of ohmic contact will give rise to an increase of parasitic resistance, leading to a decrease of forward current for a constant input voltage.
2. There are a numerous number of material defects in GaN epitaxy because of the lattice mismatch in between the substrate (i.e., sapphire, SiC, Si, etc.) and epitaxy. During the operating process, these defects react as non-radiative recombination centers and charge carriers tunneling paths to form low-resistance ohmic passage. Increase of the non-radiative recombination centers is one of the root causes of the light output degradation of LEDs. When high-energy electrons enter the MQW layer, they can break the Ga-N bonds to generate nitrogen vacancy (N-V) centers. Increase of the newly formed N-V centers gives more possibilities of the non-radiative recombination, leading to the reduction of the light output power of the LEDs. Even under an extremely low current density over the MQW layer, non-radiative recombination centers will be increased to degrade the light output power of the LED. While an excessive current density will increase the risk of metallic electromigration, directly breaking the chips as catastrophic failures in severe cases. Since it takes a while to reach a balance on the N-V centers, the light output power of the chip is usually degraded in a gradual process.
3. ESD makes an instantaneously released high voltage applied on the chip to cause two types of failures: the hard and soft failures. The hard failure occurs when the instantaneously released voltage is high enough to induce chip short circuit via the dielectric failure or throughout current under a super high current density. Compared to the vertical LEDs, lateral LEDs are more likely to suffer short circuit because the p contact and n contact electrodes are located in the same side of the chip. If the instantaneously released voltage produced by ESD is not high enough, the damage (so-called soft failures) of the LED might be imperceptible. These soft failures will not increase the failure probability of the chip during its service, resulting in an unstable fluctuation to the chip product yield rate.
4. The chip damage generated in the manufacturing process will result in rapid light output degradation or catastrophic failure to the chip. Under such circumstance, micro-cracks are initiated from the internal strain induced by the improper operation in the processes such as substrate dicing, polishing, and die bonding. Together with the effects by thermal stress resulting from temperature inequality and material mismatch, high current density, material defects, etc., these micro-cracks grow quickly until the chip is completely failed.

18.3 Phosphors

Today, the phosphors used in pc-white LEDs have become a huge family including garnet, silicate, nitride, oxynitrides, nitridosilicate, and so on. Garnet-based phosphors refer to the phosphors with Ce^{3+}-doped luminescent materials that have a general structural formula of $[X]_3\{Y\}_2(Z)_3O_{12}$ (X = Ca, Mg, Fe, Mn, Y, Lu; Y = Al, Fe, Cr; Z = Si, Ge) [15], where [], {}, () denote dodecahedral, octahedral, and tetrahedral coordination, respectively [16, 17]; therefore, they have a very stiff cubic structure [18]. Several phosphors with the garnet structure have been developed for pc-white LEDs. The mostly widely used one is yellow- emitting Ce^{3+} activated yttrium aluminum garnet $Y_3Al_2Al_3O_{12}$: Ce^{3+} (YAG: Ce), because of its excitation spectrum matching perfectly with the blue LED chips. The first use of YAG in pc-white LED was published by Nakamura in 1995 [19]. After decades of development, it has been regarded as one of most promising and efficient phosphors for pc-white LEDs [20].

Silicate-based phosphors are another type of phosphors with tunable wide range emission wavelengths [21, 22]. Categorized by their compositions, the silicate-based phosphors can be divided into a binary system, such as $2MO\text{-}SiO_2$ (M = Ca, Sr, Ba) [23], and a ternary system, such as $MO\text{-}NO\text{-}SiO_2$ (M = Ca, Sr, Ba, N = Li, Mg, Al, Zn) [24]. Some ternary system phosphors, such as $MMg_2Si_2O_7$, $CaAl_2Si_2O_8$, and $M_3MgSi_2O_8$ (M = Ca, Sr, Ba), have low absorption of blue light, which makes this kind of phosphor less frequently used in the pc-white LEDs [25, 26]. By mixing with different alkaline earth metal ions, the silicate-based phosphors emit different color lights. Figure 18.3 shows the emission peaks of an M_2SiO_4: Eu (M = Ca, Sr, Ba) phosphor affected by the alkaline earth metal ions. It can be seen that emission peaks of the phosphors Ca_2SiO_4, Sr_2SiO_4, and Ba_2SiO_4 are 515 nm, 575 nm, and 505 nm, respectively. The other emission peaks of among the 515 nm, 575 nm and 505 nm can be obtained by mixing the two neighboring Ca, Sr, Ba ions.

Nitride and oxynitride-based phosphors are recently developed for pc-white LEDs [27]. In terms of their compositions, nitrides can be divided into three types: (1) binary, (2) ternary, and (3) quaternary. However, binary covalent nitrides, such as GaN, BN, and AlN, can't be easily considered as host lattices for phosphors in white LEDs, because they do not have suitable crystal sites for activators [28]. The ternary, quaternary, and multiparty covalent nitride compounds are interesting because of their unique and rigid crystal structures, availability of suitable crystal sites for activators, and their structural versatility, which enable the doping of rare-earth ions to provide useful photoluminescence. SiAlON oxynitride green phosphor is expected to have excellent chemical and thermal stabilities because their basic crystal structure is built on rigid tetrahedral networks. SiAlON includes α-SiAlON and β-SiAlON. α-SiAlON is derived from α-Si_3N_4 structure by equivalent substitution of Al and O for Si and N, and its chemical composition can be written as α-$M_{m/v}Si_{12-(m+n)}Al_{m+n}O_nN_{16-n}$ (M = Ca, Li, Mg, Y, m/v $\leq$ 2.0) [29–31]. α-SiAlON has a hexagonal crystal structure and the $P3_{1c}$ space group. Because of the better luminescence property and particle morphology, researches on the SiAlON mainly focus on the α-SiAlON [32]. β-SiAlON is derived from β-Si_3N_4 structure by equivalent substitution of Al and O for Si and N, and its chemical composition can be written as $Si_{6-z}Al_zO_zN_{8-z}$ (z represents the number of Al-O pairs substituting for Si-N pairs and z = 0 ~ 4.2) [33]. β-SiAlON has a hexagonal crystal structure and the $P6_3$ space group.

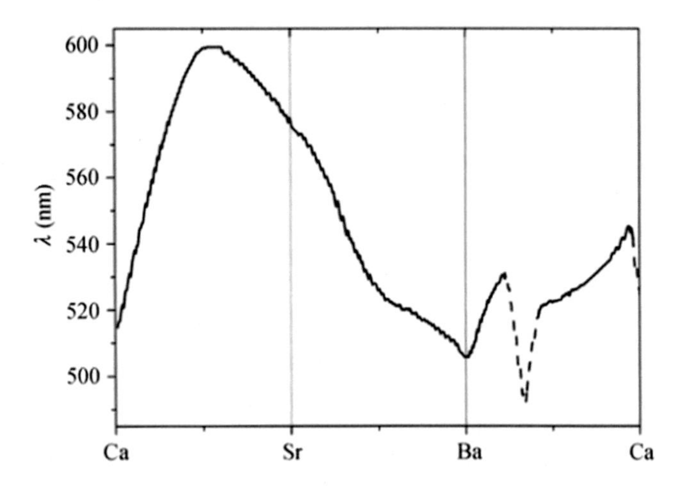

Fig. 18.3 Emission peaks of an M_2SiO_4: Eu (M = Ca, Sr, Ba) phosphor [22]

Nitridosilicate-based phosphors are very promising phosphors with much higher IQEs compared to the traditional YAG phosphors. There are mainly two types of nitridosilicate which are alkaline-earth silicone nitrides, such as $M_2Si_5N_8$ (M = Ca, Sr, Ba), and alkaline-earth aluminum silicon nitrides, such as $MAlSiN_3$ (M = Mg, Ca, Sr). Their activators include Ce^{3+}, Eu^{2+}, Mn^{2+}, and Tm^{3+}, and Eu^{2+} is generally the most active one in many cases [34]. The luminescence properties of the nitridosilicate phosphors depend not only on the type of activator ion but also on the chemical composition of the host lattice. Table 18.1

Table 18.1 Luminescent properties of the nitridosilicate-based phosphor [34]

Type	Host lattice	Activator	Absorption wavelength	FWHM[a]	Peak wavelength	IQE(%)[b]
$M_2Si_5N_8$: B^{n+}	$Ca_2Si_5N_8$	Eu^{2+}	250~580	104	605 ~ 615	55
		Ce^{3+}	250~450	95	470	–
		Mn^{2+}	250~550	70	599	–
	$Sr_2Si_5N_8$	Eu^{2+}	~450	100	~615	80
		Mn^{2+}	250~550	60	606	–
		Tm^{3+}	~460	84	612	–
		Ce^{3+}	230~550	170	553	–
	$Ba_2Si_5N_8$	Eu^{2+}	350~600	100	593 ~ 633	67.8
		Mn^{2+}	250~550	60	567	–
		Ce^{3+}	220~450	–	457	–
M $SiAlN_3$:A^{n+}	$MgSiAlN_3$	Eu^{2+}	200~500	200	620 ~ 710	–
	$CaSiAlN_3$	Eu^{2+}	~450	100	670	80
		Ce^{3+}	200~550	150	570 ~ 603	70
		Mn^{2+}	220~600	100	627	–
	$SrSiAlN_3$	Eu^{2+}	200~600	80	610	–
	(Sr, Ca)$SiAlN_3$	Eu^{2+}	250~600	80	610 ~ 650	80
Oxide (YAG)	$Y_3Al_5O_{12}$	Ce^{3+}	300~525	100	532	26.4

[a]*FWHM* full width at half maximum
[b]*IQE* internal quantum efficiency

enumerates the chemical compositions and luminescent properties of the nitridosilicate-based phosphor, compared with the commonly used YAG phosphors.

Thermal quenching which describes the luminescence reduction caused by the temperature increase is the major reliability problem of phosphors. The Arrhenius equation, as shown in Eqs. (18.1) and (18.2), is always widely used to explain the thermal quenching effects of phosphors.

$$I(T) = \frac{I_0}{1 + C\exp(-E/KT)} \quad (18.1)$$

$$\ln\frac{I_0}{I(T)} = -\frac{E}{KT} - \ln C \quad (18.2)$$

where I_0 and $I(T)$ are the luminous intensity at 0 and $T(K)$, respectively, and K is the Boltzmann constant (8.629 × 10-5 eV), while C is a rate constant for the thermal activated escape. E is thermal activation energy for the thermal quenching process, which indicates the thermal energy required to promote an electron from the lowest excited state to the host lattice conduction band. The lower value of E means a more rapid non-radiative rate at a given temperature [35].

The peak wavelength of the emission spectrum is another significant material characteristic for phosphors, which represents the wavelength where the strongest relative light radiation can be achieved. With an increase of temperature, the peak wavelength of the emission spectrum might shift toward to the long wavelength direction (i.e., red-shifted phenomenon) or short wavelength direction (i.e., blue-shifted phenomenon). The red-shifted phenomenon can be explained by the temperature-dependence luminescence effect with the semi-empirical Varshni equation as shown in Eq. (18.3) [36].

$$E(T) = E_0 - \frac{aT^2}{T + b} \quad (18.3)$$

where E_0 is the energy difference at 0 K, $E(T)$ is the energy difference between the excited states and ground states at temperature T, and a and b are Varshni coefficients.

Usually, Eq. 18.3 can be used to explain the change of peak wavelength for most phosphor powders under heat treatments. At the high temperature, the bond length between the luminescent center and its ligand is increased to result in a decreased crystal field. Furthermore, the distortion of the highly symmetrical luminescent center is expected to be occurred in the host, which reduces its symmetry and makes the so-called Jahn-Teller effect dominant. These two effects result in the splitting of degenerate excited state or ground state, and the emission peak is red-shifted with increasing of temperature [37].

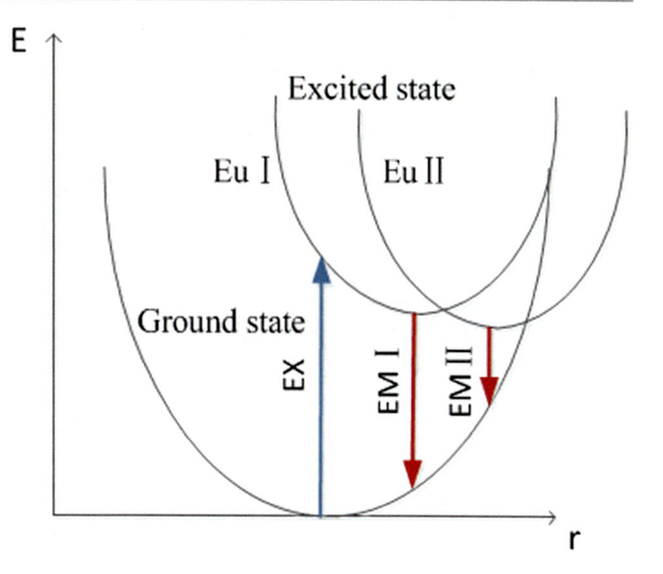

Fig. 18.4 Two emissions of Eu ion in configuration coordinate diagram [18]

Sometimes, the blue-shifted phenomenon might occur because of the tunnel effect of activator ions from the excited states of low-energy emission band to the excited states of high-energy emission band, which can be described in the configuration coordinate diagram [18] in Fig. 18.4. The two emissions of activator Eu ions in $(Sr,Ba)_3SiO_5$ and $CaAlSiN_3$ arise from each ground state to the potential energy surface of excited state. It is observed that the higher-energy emission (EM I) is more likely to happen at higher temperature, while the low-energy emission (EM II) is dominant at low temperature. As the excited state of Eu ion corresponds to the emission spectra, the blue shift behavior is observed with increasing of temperature. Moreover, the blue shift of $(Sr,Ba)_3SiO_5$ is small, while the $CaAlSiN_3$ almost keeps unchanged, probably due to the low Eu ion content in these two phosphor powders [38].

The full width at half maximum (FWHM), which is the distance in between two wavelengths with respect to the half peak amplitude of the emission spectrum, usually is used to characterize the spectral purity of a phosphor material [39]. FWHM always standards for the spectral purity and also is used to measure the dispersion of the energy which contributes to luminescence. The temperature dependence of FWHMs of the emission spectrum of the phosphor follows Boltzmann distribution function shown by Eq. (18.4):

$$\Gamma(T) = \Gamma_0\sqrt{\coth(hw/2kT)}, \ \Gamma_0 = \sqrt{8\ln 2}(hw)\sqrt{S} \quad (18.4)$$

in which $\Gamma(T)$ is the FWHM at temperature T, Γ_0 is the FWHM at 0 K, and $h\omega$ is the energy of the lattice vibration that interacts with the electronic transitions and S is the Huang-Rhys coupling factor that measures the strength of the electron-lattice coupling. If S < 1, the coupling is weak, if

$1 < S < 5$, the coupling is intermediate, and if $S > 5$, the coupling is strong [76]. Furthermore, the Stokes shift (ΔS) expressed by Eq. (18.5) is an important feature for the photoluminescence properties of phosphors [40].

$$\Delta S = (2S - 1)\, hw \tag{18.5}$$

18.4 Encapsulant and Phosphor/Encapsulant Composite

18.4.1 Encapsulant

Within a pc-white LED, the encapsulant plays an important role on not only preventing the LED chip from the natural environmental deterioration but also enhancing the light extraction efficiency due to its intermediate refractive index (RI) in between the air (i.e., about 1) and LED chip (i.e., about 2.7). Generally speaking, the higher RI of an encapsulant is, the higher light extraction efficiency of a pc-white LED will have. To achieve a higher RI of the encapsulant, a common way is to add nanoparticles (e.g., TO_2 nano-particles) with higher RI into it [41–44]. For instance, Liu et al. [45] increased the RI of silicone encapsulant from 1.54 to 1.62 by adding 5 wt% in-house-synthesized TO_2 nanoparticles. Nevertheless, a loss of transparency of the TO_2/silicone composite was observed. The relative transmittance at 589 nm is found to reduce from 100% to 84% but still much higher compared to the other commercial TO_2/silicone composites with 5 wt.% TO_2 nanoparticles. Later on, they refined the TO_2 nanoparticles synthesis process to get a highly dispersed TO_2/silicone composite. Then they can further increase the RI of the silicone encapsulant up to 1.73 by adding 10 wt% TO_2 nanoparticles. Meanwhile, because of a high dispersion of TO_2 nanoparticles, the transparency of the TO_2/silicone composite was greatly improved. At 589 nm, the relative transmittance remained at a high level of at least 93% [46].

Regarding reliability issues of the encapsulant, the most critical problem is the yellowing phenomena caused by either thermal aging or UV/blue light exposure [47–49]. Yellowing reduces the transparency of encapsulant and finally decreases light extraction from the pc-white LED [50]. The yellowing issue has been often found in high-power LED packages with epoxy encapsulant [51]. In high-power LED packages, there are large amounts of heat generated and accumulated inside the package, especially around the LED chip, resulting in yellowing of epoxy encapsulant surrounding the LED chip. The yellowing of the epoxy results in a significant loss of light output over time. Typically, this phenomenon is due to the aromatic group or free radical in the epoxy encapsulant [52]. Hsu et al. [53] found that the epoxy could have dissociated easily to form free radicals at high temperatures. As a result, epoxies form many different types of chromophoric structure, including the yellow quinone structure [54]. The presence of free radicals enhances the formation of these types of chromophoric group. Furthermore, a very small amount of chromophoric group can cause a dramatic yellowing of the polymer. On the other hand, some researchers [55] found that the epoxy materials could become yellow due to the effects of moisture as the LED samples were aged at high-temperature and high relative humidity environment.

Compared to the epoxy-based encapsulant, silicone-based encapsulant exhibits much better thermal stability and therefore becomes a preferred candidate used in pc-white LEDs [56, 57]. Among different silicone-based materials, a phenyl-silicone has an improved refractive index due to the high polarizability of the phenyl groups [58]. But at high temperatures, the transparency of the phenyl-silicone encapsulant can be reduced by yellowing, since the phenyl groups are susceptible to thermal oxidation and cleavage from the silicone backbone [59]. Bae et al. [60] observed a yellowing of the silicone encapsulant at an elevated temperature (180 °C) due to its high catalyst content (200 ppm) and the presence of unreacted groups, probably due to the thermal oxidation and cleavage from the silicone backbone of the un-crosslinked vinyl groups or phenyl groups. The linear silicone bonds are cut during thermal pyrolysis, forming ring structures. During the pyrolysis process, the phenyl radical is cleaved from the silicone chain and causes the discoloration to yellow [61]. Moreover, the authors also indicated that high non-polymerized methacrylate groups can lower the optical transmittance and thermal stability. This behavior is interpreted by the thermal degradation of methacrylate groups in the silicone hybrimers. The non-polymerized methacrylate groups can be the source of yellowing in thermal degradation [62].

Carbonization is a more severe problem than yellowing for the encapsulant working in high-power pc-white LEDs under a high-temperature environment [63]. The most apparent evidence is that the encapsulant in the vicinity of the LED chip or lead frame is carbonized [8]. The carbonization decreases the encapsulant's insulation resistance, significantly inhibiting its ability to provide electrical insulation between adjacent bond wires and leads. This phenomenon can initiate a runaway process in which current through the encapsulant increases as its insulation resistance decreases, leading to joule heating which in turn decreases the insulation resistance and can eventually cause combustion of the encapsulant.

The reasons to cause encapsulant carbonization are multifold. A major root cause is the joule heating due to the high resistances of the Ohmic contacts, n-type and p-type cladding layers, and degradation of heat conduction materials

[64–68]. Another potential mechanism is the local high temperature yielded by phosphor self-heating [69, 70]. The phosphor self-heating is a side effect of phosphor emission in which the excited energy is not released by light but heat due to non-radiative transfer and Stokes shift [28]. Recent studies reveal that as small as 3-mW heat generation on a 20-μm diameter spherical phosphor particle might lead to excessive temperatures in the local area of encapsulant varied from 195 to 316 °C [71–73]. The third possible reason is that the encapsulant absorbs the excessive radiation by the blue light emitted from the LED chip. Under a synergistic effect of short wavelength blue light, high temperature, and oxygen, the silicone material is very easily degraded, causing an aggravation on the degradation of the pc-white LED.

Figure 18.5 shows the top of cross section of pc-white LED packages which were stressed in HAST tests with and without driving current, respectively. It was found that silicone carbonization appeared only when the current was applied. It was speculated that the silicone carbonization was due to either the self-heating of phosphors or over-absorption of blue lights, which may induce extremely high temperature in silicone encapsulant.

Barton et al. [63] conducted a number of tests on pc-white LEDs under various driving currents to get different junction temperatures. They stated that the rapid degradation observed at higher currents was correlated to carbonization of the plastic encapsulation material on the chip surface, which leads to the formation of a conductive path across the LED and subsequently to the destruction of the diode itself. Zhou et al. [75] studied the reliability of a high-power LED package and found that the thermal overstress on the epoxy along the interface owing to the high junction temperature is the main factor to degrade the epoxy gradually. EDAX results illuminate that the percentage of C and O elements in failed samples is higher than that in the normal one. The presence of oxygen observed on the surface of degraded LED die is associated with the darkening. As a result, a darkening area made up of carbon as well its oxide occurs on the LED die surface and depresses the luminous efficiency.

Huang et al. reported a new degradation mechanism of the encapsulant carbonization in pc-white LED samples under a highly accelerated temperature and humidity stress test (HAST) [74]. They found that the over-absorption of blue lights by silicone bulk is considered as the root cause for silicone carbonization. This is mainly due to the scattering effect of water particles inside the silicone materials, which can increase the encapsulant temperature significantly. A finite element simulation also verified that the temperature of the encapsulant was as high as 298.4 °C when the absorption of the blue light is 80%, as indicated in Fig. 18.6.

18.4.2 Phosphor/Encapsulant Composite

When used in pc-white LEDs, phosphors and encapsulant are usually mixed as a whole to undertake the tasks of blue light

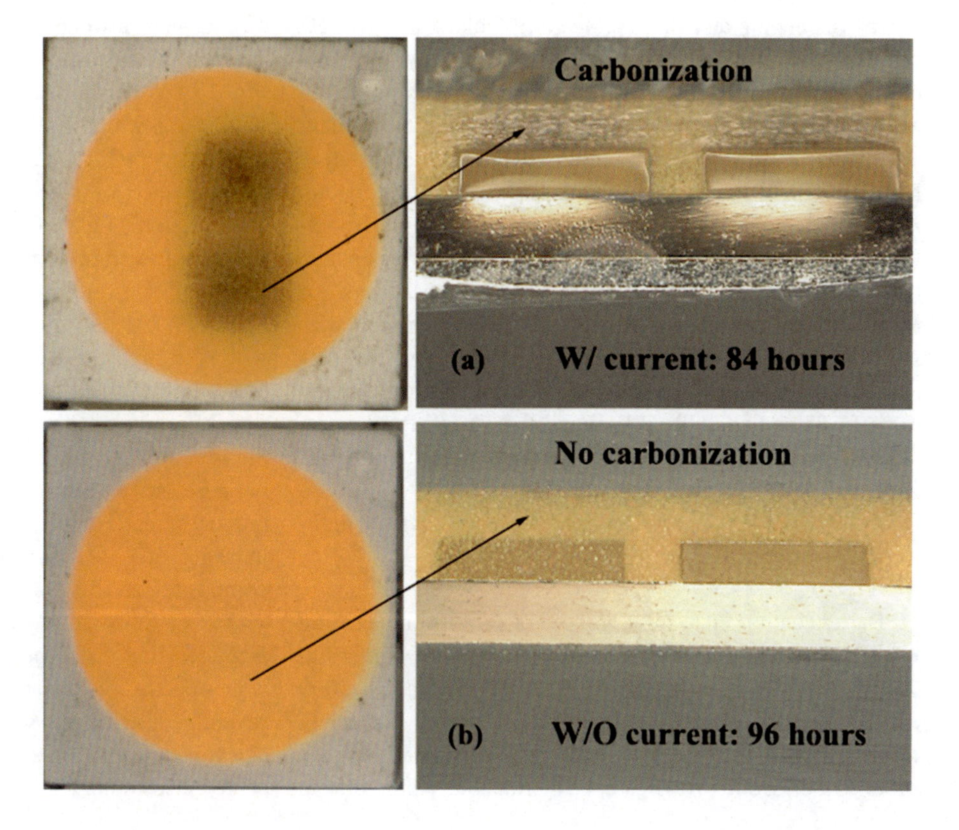

Fig. 18.5 Top and cross section of pc-white LEDs after aging tests with different conditions: (**a**) HAST test with driving current; (**b**) HAST test without driving current [74]

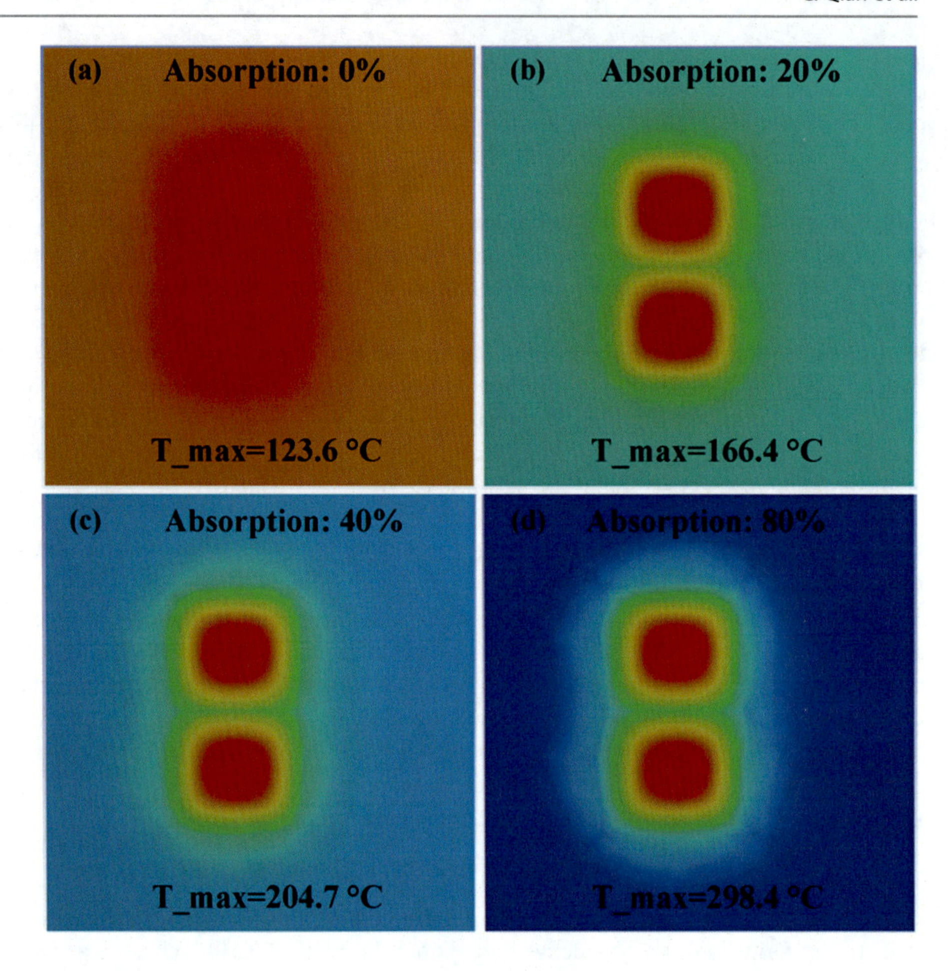

Fig. 18.6 Simulated temperature distributions of the encapsulant in pc-white LEDs [74]

down-conversion and chip protection. This brings a new problem that the failures existing in the phosphors and encapsulant will interact with each other resulting in an acceleration of the deterioration of the phosphor/encapsulant composite. In this subsection, this synergistic interaction between phosphor and encapsulant under high temperature, high temperature and blue light exposure, and high-temperature and high humidity conditions are investigated through the tests on three widely used types of phosphors and corresponding composites which are made by mixing with silicone encapsulant. Table 18.2 lists the test samples consisting of three pristine phosphors (herein called Type G, Type O and Type R, respectively) and three phosphor/silicone composites (herein called G + silicone, O + silicone, and R + silicone, respectively).

A. *Phosphor-silicone interaction effects under high-temperature condition*

The excitation and emission spectra of all samples were measured with a gradually increased temperature from 30 to 150 °C in an increment of 20 °C by using an EX-1000 exciting spectra and thermal quenching analyzer. As shown in Fig. 18.7a–d, the changes of emission spectra of seven test samples are used to investigate the thermal quenching effects of phosphor powders, phosphor/silicone composites, and the interactions between phosphor powders and silicone. In Fig. 18.7d, the spectrum of silicone is the excitation spectra of Xe lamp which go through silicone film and then received by the acceptor; thus it can be used to indirectly reflect the transmittance of silicone. It is obviously shown that the transmission light decreases with the increasing of temperatures from 30 to 150 °C.

As shown in Fig. 18.8, the luminous intensities of Types G, O, and R phosphor powders gradually decrease with the temperature rising. At 150 °C, the emission intensities of three selected three monochromatic phosphor powders were declined to the 98.6%, 77.0%, and 88.2% of the initial ones, respectively, which indicates that high temperature has a great influence on the emission intensity of monochromatic phosphor powders. While the emission intensity of G + silicone, O + silicone, and R + silicone composites at 150 °C only can remain 95.9%, 66.6%, and 78.8% of those at 30 °C, which are much more seriously degraded than those of three phosphor powders. The emission spectra of phosphor/silicon composites were more severely deteriorated with the similar trend; therefore, we can conclude that the transmission of the silicone can significantly affect the emission spectra of the phosphor/silicon

Table 18.2 A summary of phosphor/encapsulant composites

Name	Host crystal structure	Density [g/cm^3]	Particle Size [um]	CIE1931 color coordinates	Pristine phosphor	Phosphor/silicone composite
Type G (green)	$Lu_3Al_5O_{12}$	4.8	12	X = 0.338 Y = 0.576	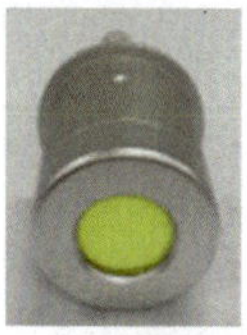	
Type O (orange)	Sr_3SiO_5	4.76	16.5	X = 0.553 Y = 0.440	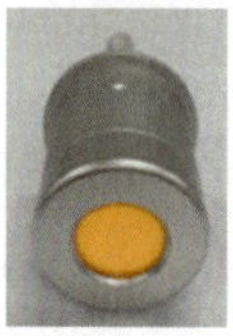	
Type R (red)	$CaAlSiN_3$	3.1	15.5	X = 0.643 Y = 0.356		

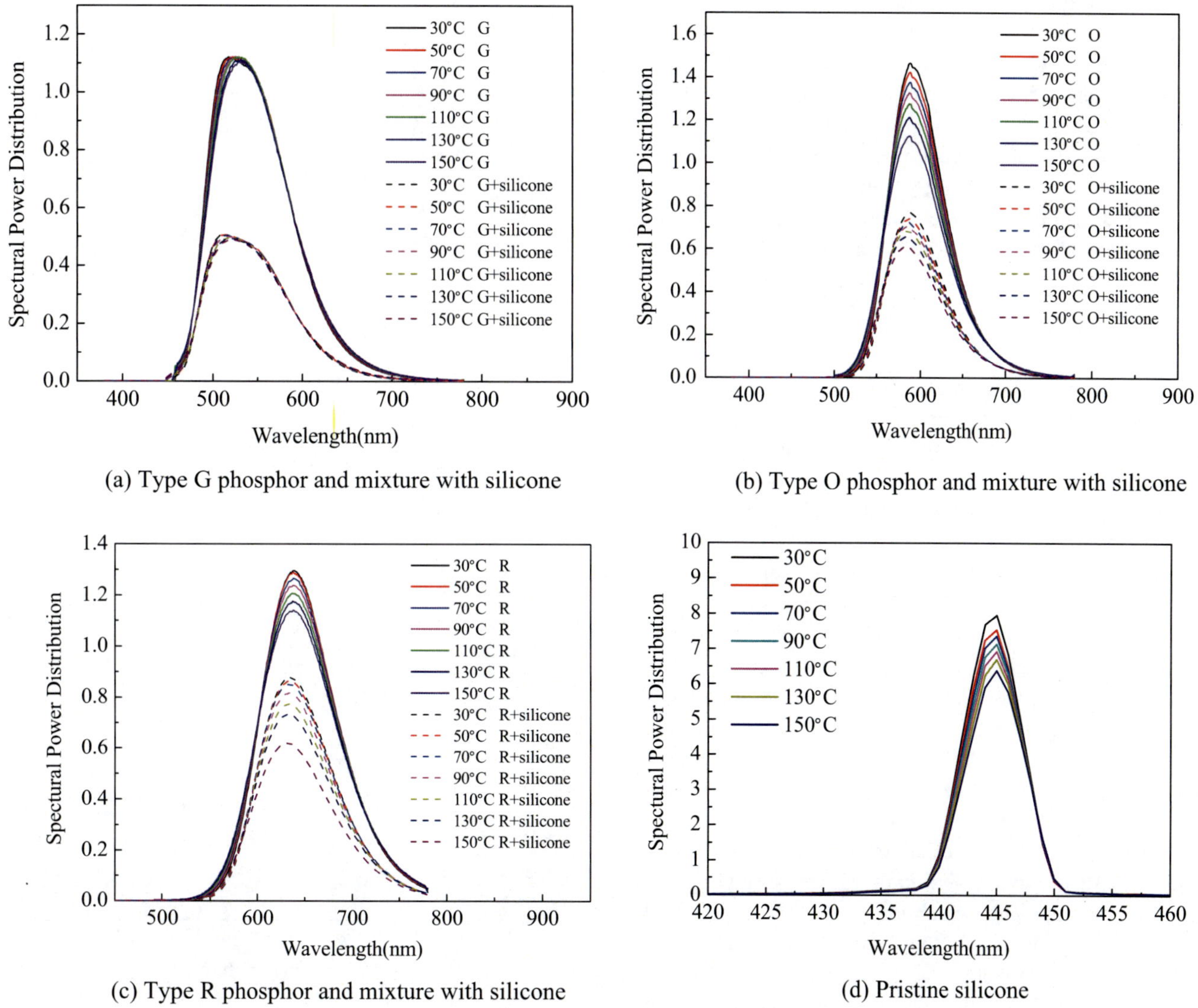

(a) Type G phosphor and mixture with silicone

(b) Type O phosphor and mixture with silicone

(c) Type R phosphor and mixture with silicone

(d) Pristine silicone

Fig. 18.7 Emission spectra of test samples treated under temperatures from 30 to 150 °C. (**a**) Type G phosphor and mixture with silicone. (**b**) Type O phosphor and mixture with silicone. (**c**) Type R phosphor and mixture with silicone. (**d**) Pristine silicone

composites under the high-temperature condition. Figure 18.8 also shows that the luminous intensity of Type O phosphor and its composite declines much more seriously under high temperature, while the Type G phosphor and its composite maintains relatively more stable.

Figure 18.9 shows the temperature dependence on normalized peak wavelengths of the pristine phosphors. In details, the peak position of emission spectra of Type G phosphor is red-shifted with increasing temperatures, and the detail peak wavelength changes from 517 nm at 30 °C to 530 nm at 150 °C, whereas the peak position of the Type O phosphor shows a blue-shifted trend and that of the Type R phosphor almost keeps relative steady at high temperature.

The normalized FWHMs of three emission bands changed with increasing temperatures are given in Fig. 18.10. As shown, the FWHMs of Type O phosphor at 30 and 150 °C are 72.4 and 79.8 nm, broadening 6.4 nm, and that of Type R phosphor is 8.3 nm, while the Type G phosphor is relatively stable. Figure 18.11 shows the temperature dependence of the FWHMs of the emission bands fitted by the Boltzmann distribution function Eq. (18.4). Furthermore, according to Eqs. (18.4) and (18.5), the parameters S, *hw*, and ΔS of the three pristine phosphors are calculated and listed in Table 18.3. It is found that Type G phosphor has the weakest electron-phonon coupling and the lowest Stokes shift, which can indicate that the crystal structure of Type G phosphor is more rigid than that of Type O and R phosphors. Therefore, the change of FWHM in Type G phosphor is smaller.

B. *Phosphor-silicone interaction effects under long-term accelerated aging test*

Next, the reliability of the phosphor/silicone composites was studied under three accelerated aging test conditions (i.e., 85°C, 85°C & blue light exposure and 85°C & 85% RH, respectively).

Firstly, to assess the long-term thermal stabilities of phosphor/silicone composites used in pc-white LEDs, three prepared phosphor/silicone composites were aged under 85 °C for more than 4000 h. The emission spectra of aged phosphor/silicone composites were measured at room temperature. Meanwhile, the transmission of pristine silicone aged at the same conditions was also measured as a reference. As shown in Fig. 18.12, the normalized emission peaks of the selected phosphor/silicone composites aged in 85 °C show a rapid decrease within the initial 500 h, reaching 91.9%, 91.1%, and 93.5% at 500 h, respectively. Then, they fluctuate slightly in the following period. While the peak wavelengths and FWHMs of the prepared samples keep steady regardless of the high-temperature condition, which may represent that the lattice structure and the band gap of the phosphor powders are relatively stable under 85 °C condition. Figure 18.12d shows the transmission change of silicone aged under the 85 °C condition according to exposure time.

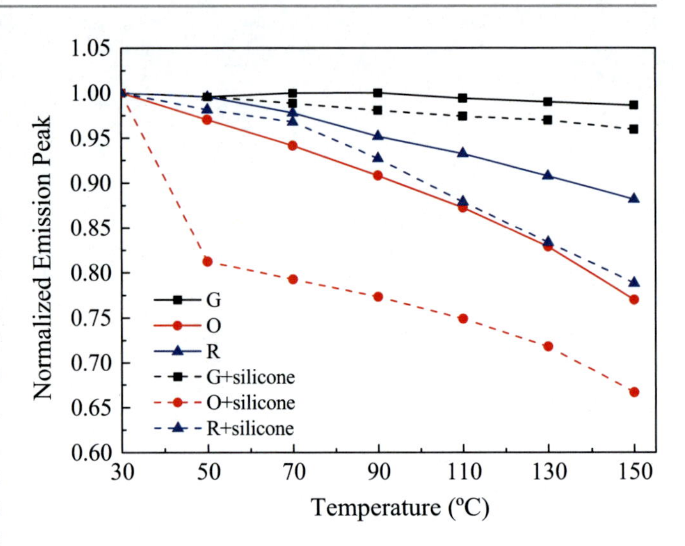

Fig. 18.8 Normalized emission peaks of emission spectra of test samples vs. temperature

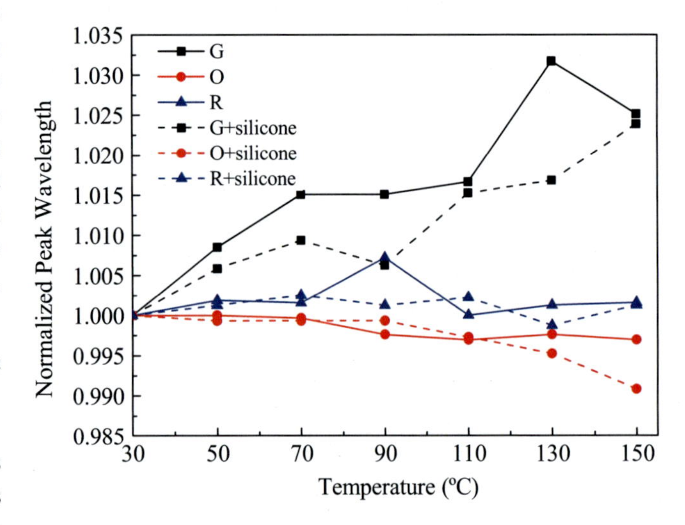

Fig. 18.9 Normalized peak wavelengths of test samples vs. temperatures

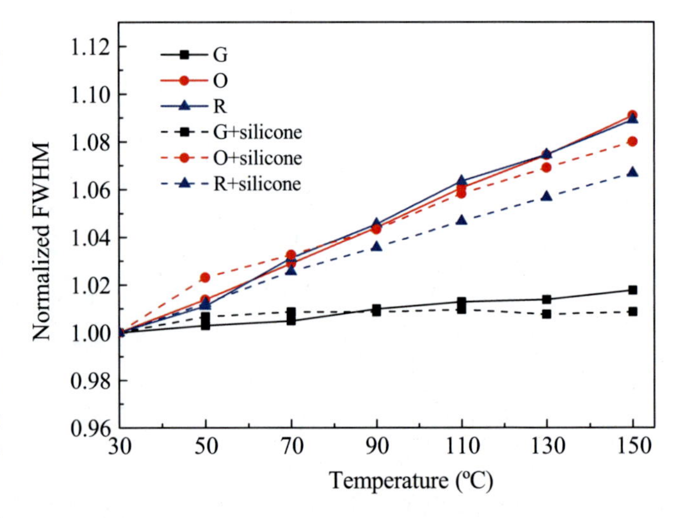

Fig. 18.10 Normalized FWHMs of emission spectra of test samples vs. temperatures

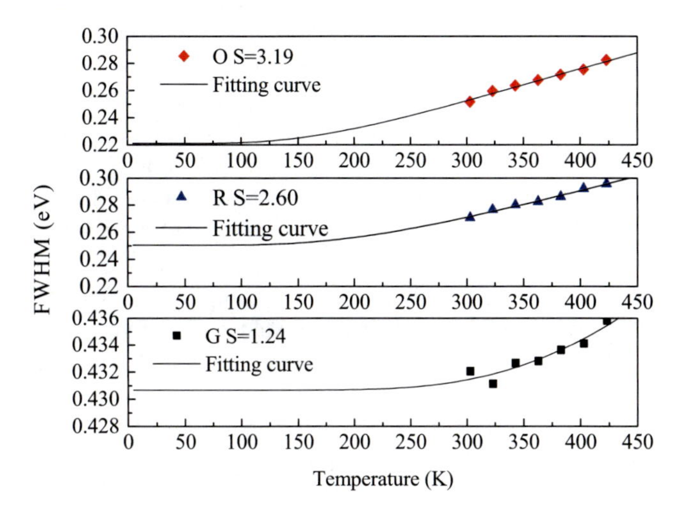

Fig. 18.11 Temperature dependence on FWHMs of the pristine phosphors

The transmission of silicone in the shorter wavelength range declined more severely than that in the longer wavelength range. In the initial aging, the transmission of silicone in the short wavelength has a rapid decrease which may cause the sudden decrease of emission intensities of phosphor/silicone composites. After aging under 85 °C condition, the silicone film shows a slight yellowing, which can be ascribed to the oxidation.

Table 18.3 Electron-lattice coupling parameters of the pristine phosphors

Phosphors	S	*hw*	ΔS
Type G	1.236	0.1645	0.242
Type O	3.194	0.0525	0.283
Type R	2.607	0.0659	0.278

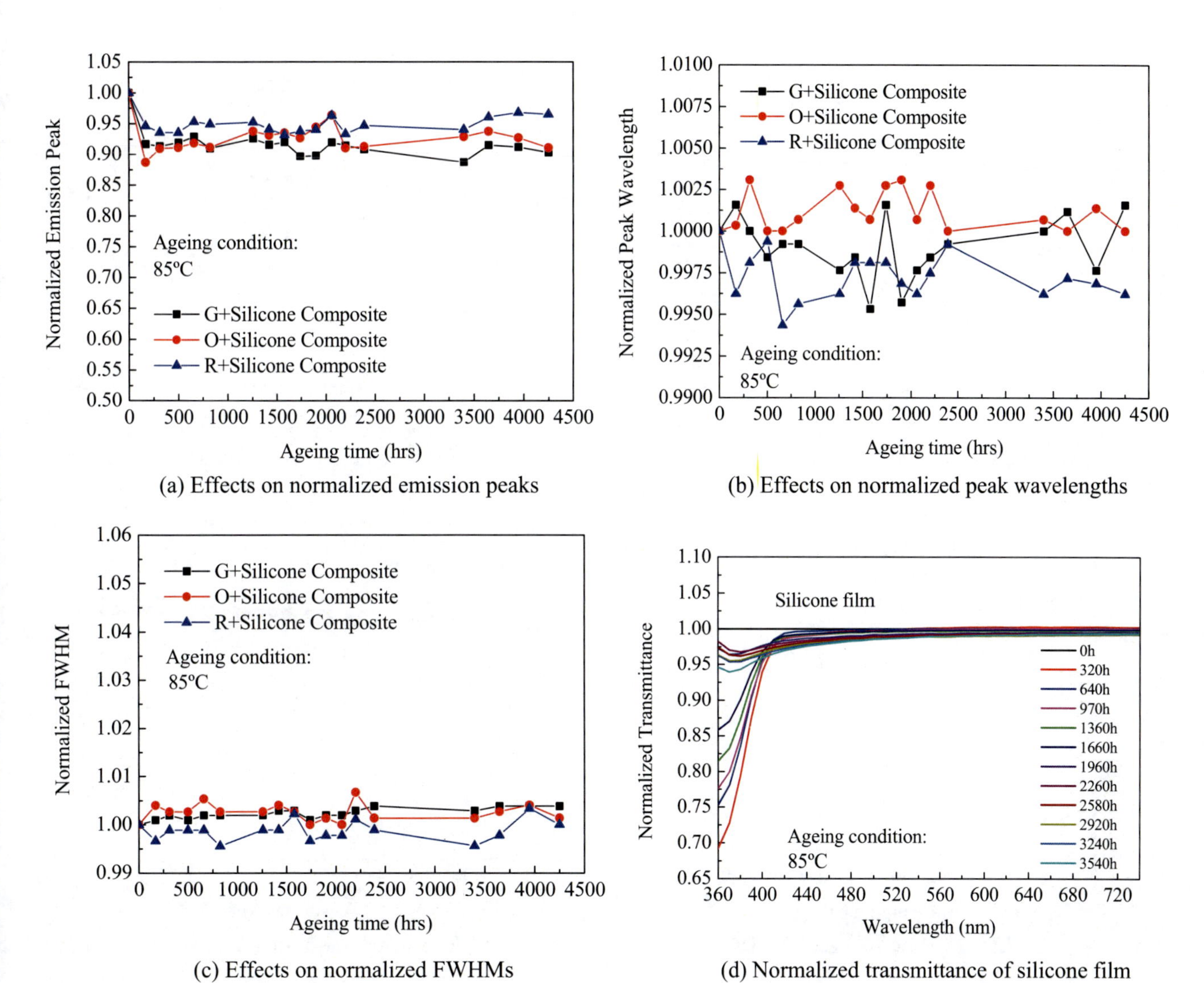

Fig. 18.12 Time dependence on spectral features of the phosphor/silicone composite samples under 85 °C. (**a**) Effects on normalized emission peaks. (**b**) Effects on normalized peak wavelengths. (**c**) Effects on normalized FWHMs. (**d**) Normalized transmittance of silicone film

Then, Fig. 18.13a–d shows time dependences of the emission peaks, peak wavelengths, and FWHMs of the phosphor/silicone composites and silicone aged under the condition of both blue light exposure and high temperature (85 °C). The results show that the feature change trend of the emission spectra under the blue light exposure and 85 °C are similar as that under the 85 °C condition. However, by comparing Fig. 18.12d and Fig. 18.13d, the transmission of the silicone aged under the blue light exposure and 85 °C decreases more severely than that aged under 85 °C condition, which indicates that the blue light irradiation can accelerate the degradation of silicone. Generally, this result may attribute to the photolysis phenomenon, in which the silicone absorbs the blue light and the oxidation reaction can be speeded up to accelerate its degradation.

Finally, Fig. 18.14a–d shows time dependences of the emission peaks, peak wavelengths, and FWHMs of phosphor/silicone composites and silicone film aged under the condition of both high temperature (85 °C) and high humidity (85%RH). Comparing to the aging test results in Figs. 18.12 and 18.13, it can be found that the selected phosphor/silicone composites show the most rapid degradation under the 85 °C and 85% RH condition. The emission peaks of the G + silicone, O + silicone, and R + silicone composites reduce to 89%, 61%, and 82% of their initial values at 3000 h, while all FWHMs increase and their peak wavelengths remain relatively stable. From Fig. 18.14d, the transmission of silicone dropped most significantly under the 85 °C and 85% RH condition. As the moisture can accelerate the yellowing rate of silicone, its oxidation effect will be

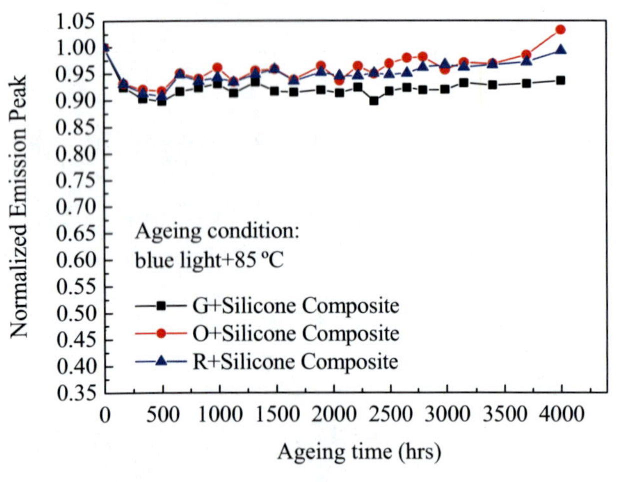

(a) Effects on normalized emission peaks

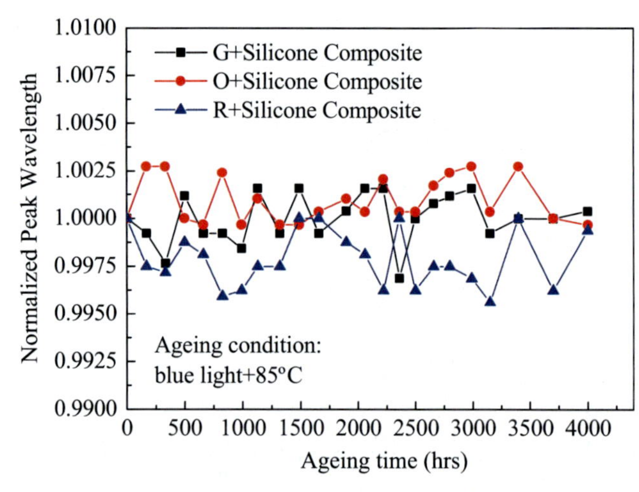

(b) Effects on normalized peak wavelengths

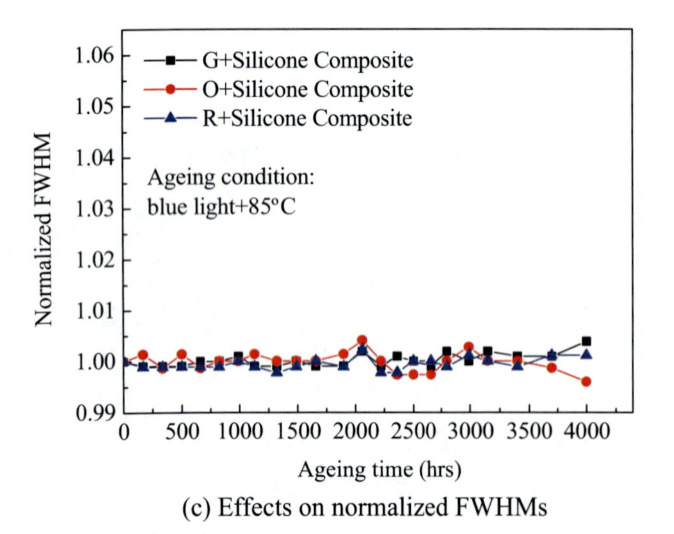

(c) Effects on normalized FWHMs

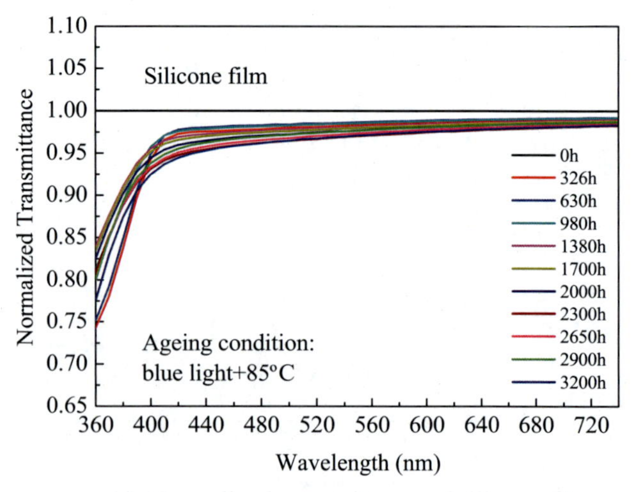

(d) Normalized transmittance of silicone film

Fig. 18.13 Time dependence on spectral features of the phosphor/silicone composite and silicone film samples under 85 °C and blue light exposure. (**a**) Effects on normalized emission peaks. (**b**) Effects on normalized peak wavelengths. (**c**) Effects on normalized FWHMs. (**d**) Normalized transmittance of silicone film

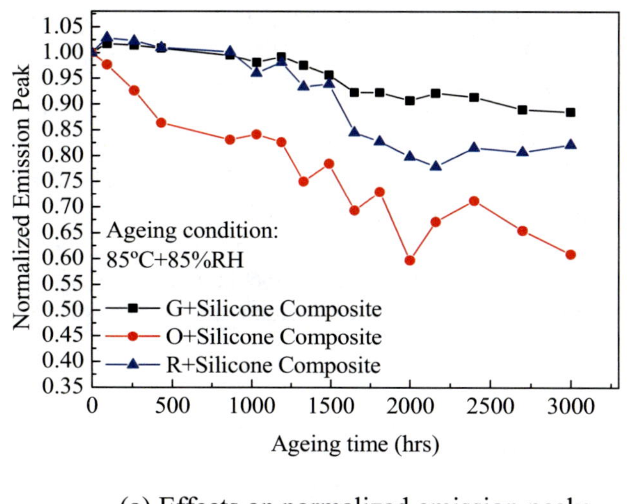

(a) Effects on normalized emission peaks

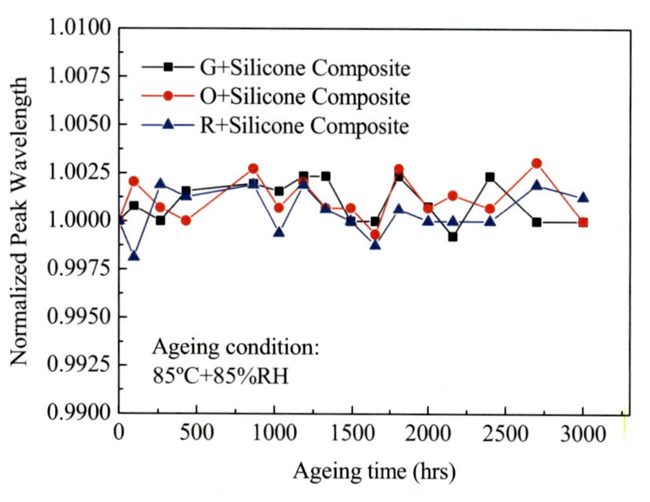

(b) Effects on normalized peak wavelengths

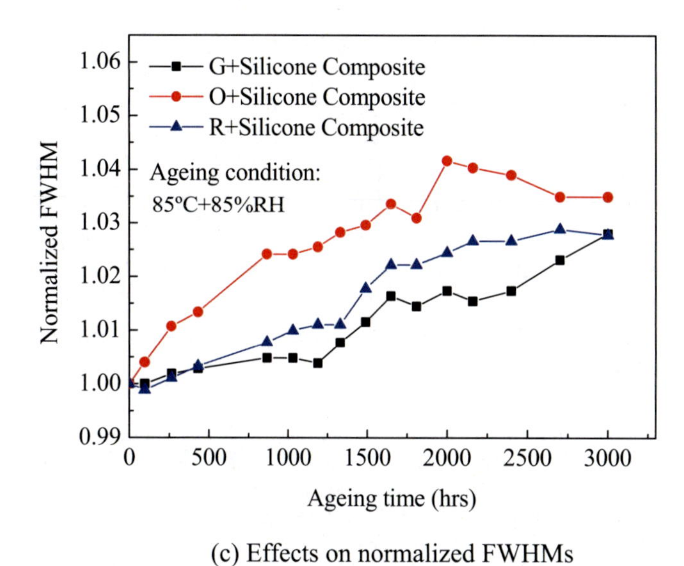

(c) Effects on normalized FWHMs

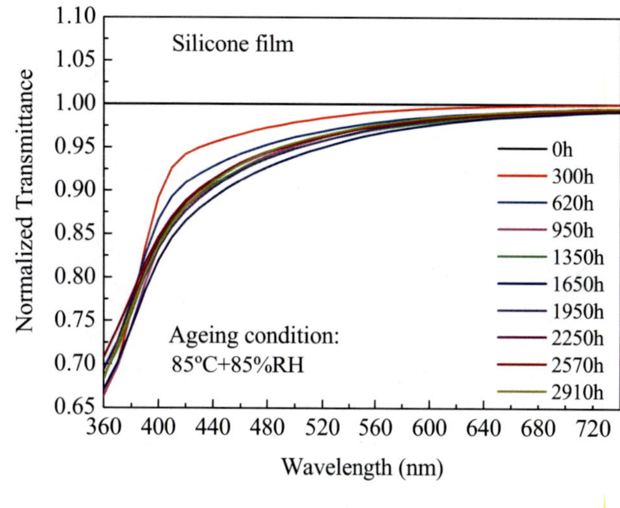

(d) Normalized transmittance of silicone film

Fig. 18.14 Time dependence on spectral features of the phosphor/silicone composite and silicone film samples under 85 °C and 85% RH. (**a**) Effects on normalized emission peaks. (**b**) Effects on normalized peak wavelengths. (**c**) Effects on normalized FWHMs. (**d**) Normalized transmittance of silicone film

increased to severely decrease the transmission. The root cause may be related to degradations of both phosphors and silicone when the moisture emerged in phosphor/silicone composites can dissolve the ions of phosphors [77–80]. In details, the average grain size, lattice strain, and the surface morphology of the phosphor powders are significantly changed, resulting in a variance of the emission spectra of phosphor powders. In addition, the high temperature and high humidity can also accelerate the degradation of transmission of silicone which is confirmed in Fig. 18.14d.

18.5 Lead Frame

Under the actual usage, toxic chemical elements (i.e., S, Cl, etc.) and moistures can be often permeated into the pc-white LED package. These elements will deteriorate the mechanical and optical properties of the lead frame because of the following reactions occurred:

$$4Ag + 2H_2S + O_2 \rightarrow 2Ag_2S + 2H_2O \quad (18.6)$$

$$2Ag + SO_2 + O_2 \rightarrow Ag_2SO_4 \quad (18.7)$$

As a consequent, the lead frames are easily contaminated during use operations [81, 82]. The sulfur contamination will sometimes result in an electrically open failure in the pc-white LED package but more often initiate the yellowing and carbonization of the encapsulants as mentioned in the preceding subsection.

During thermal storage or aging, copper diffusion may also happen in the lead frame of the pc-white LED packages [83]. As driven by temperature, the copper is gradually moving up from the PCB board to the lead frame. Meanwhile, the contamination ions (mainly chlorine) in the air will react with Cu and Ag on the surface of lead frame. Thermodynamically, the diffused Cu can nucleate as particles after reach certain supersaturation degree. The particles pile up when more nuclei generate. Under thermal aging condition, the particles are very active, which can react with the oxygen in atmosphere. Finally, these particles can form as bands so as to decrease the surface energy. The decrease of optical reflectance is consistent with the formation and growth of the Cu particles, as well as the transformation of the surface morphology.

18.6 Interface

Delamination is always reported as one of factors which cause the light output degradation of pc-white LED packages [84]. Thermal stress and moisture absorption are considered as the main causes for this delamination [85]. The mismatching coefficients of thermal expansions (CTEs) and moisture expansions (CMEs) will induce thermal stress on the interface in between neighboring components within the LED package. Under such circumstance, delamination failures may occur in between chip surface and encapsulant, in between chip and die attach, and in between lead frame and encapsulant [82, 86–88].

Luo et al. [89] reported that the moisture-induced stress was much lower than the thermal-induced stress in the same pc-white LED package because it only takes 1.5 min to reach a temperature balance but 46 days to reach a moisture balance. E. H. Wong et al. [90] studied the mechanics and impact of hygroscopic swelling of polymeric materials in electronic packaging and found that the hygroscopic stress induced through moisture conditioning was significant compared to the thermal stress during solder reflow. It is because differential swelling occurs between the polymeric and non-polymeric materials, as well as among the polymeric materials constituting the electronic packages. This differential swelling induces hygroscopic stress in the package that adds to the thermal stress at high reflow temperature, raising the susceptibility of package to popcorn cracking.

Tan, et al. [84] used numerical simulations to show that voids and delamination in the die-attach layer would greatly enhance the thermal resistance and decrease the light extraction efficiency of a pc-white LED, depending on the sizes and locations of the defects distributed in package. They also stated that delamination between the LED chip and phosphor layer decreased the LED's relative light extraction efficiency when the ratio of the delamination length to the chip's length increased. However, the decrease amount of light is only 4% when the ratio of delamination length to chip size reaches to 65%.

18.7 Concluding Remarks

A pc-white LED package is a small but complex optoelectronic device consisting of a number of components, each of which is made of different materials and exhibits its own failure features under certain conditions, for instance, a combination of temperature, humidity, driving current, and perhaps some contamination conditions. This chapter gives a brief introduction on the key components in a pc-white LED, as well as reliability failures which are concerned in those components. Competition and synergism among these abovementioned reliability failures usually lead to the final failures of LEDs, such as luminous flux depreciation, color shift, and catastrophic failures.

Acknowledgment The authors would also like to acknowledge the support of China Association for Science and Technology (CAST) Program of International Collaboration Platform for Science and Technology Organizations in Belt and Road Countries (Grant No. 2020ZZGJB072014), the National Natural Science Foundation of China (Grant No. 51805147, 61673037) and the Six Talent Peaks Project in Jiangsu Province (Grant No. GDZB-017).

References

1. Liu, S., Luo, X.: LED Packaging for Lighting Applications. Wiley (Asia), Singapore (2011)
2. van Driel, W.D., Fan, X.J.: Solid State Lighting Reliability: Components to Systems. Springer, New York (2013)
3. Chang, M.H., et al.: Light emitting diodes reliability review. Microelectron. Reliab. **52**, 762–782 (2012)
4. Fu, J., et al.: Degradation and corresponding failure mechanism for GaN-based LEDs. AIP Adv. **6**, 055219 (2016)
5. Huh, C., et al.: Improvement in light-output efficiency of InGaN/GaN multiple-quantum well light-emitting diodes by current blocking layer. J. Appl. Phys. **92**, 2248–2250 (2002)
6. Palakurthy, S., et al.: Design and comparative study of lateral and vertical LEDs with graphene as current spreading layer. Superlattices Microstruct. **86** (2015)
7. Cao, X.A., et al.: Defect generation in InGaN/GaN light-emitting diodes under forward and reverse electrical stresses. Microelectron. Reliab. **43**(12), 1987–1991 (2003)

8. Meneghini, M., et al.: A review on the physical mechanisms that limit the reliability of GaN-based LEDs. IEEE Trans. Electron Dev. **57**, 108–118 (2010)
9. Vinson, J.E., Liou, J.J.: Electrostatic discharge in semiconductor devices: protection techniques. Proc. IEEE. **88**(12), 1878–1902 (2000)
10. Lin, Y., et al.: Defects dynamics during ageing cycles of InGaN blue light-emitting diodes revealed by evolution of external quantum efficiency – current dependence. Opt. Express. **23** (15), A979 (2015)
11. Aladov, A.V., et al.: Thermal resistance and nonuniform distribution of electroluminescence and temperature in high-power AlGaInN light-emitting diodes. St. Petersburg Polytech. Univ. J Phys Math. **1**, 151–158 (2015)
12. Qian, C., et al.: Electro-Optical Simulation of a GaN Based Blue LED Chip, Proceedings of International Conference on Thermal, Mechanical and Multi-Physics Simulation and Experiments in Microelectronics and Microsystems (2016)
13. Hwu, F.S., et al.: A numerical study of thermal and electrical effects in a vertical LED Chip. J. Electrochem. Soc. **157**(1), H31–H37 (2010)
14. Jayawardena, A., Narendran, N.: Analysis of electrical parameters of InGaN-based LED packages with aging. Microelectron. Reliab. **66**, 22–31 (2016)
15. Pavese, A., et al.: X-ray single-crystal diffraction study of pyrope in the temperature-range 30–973 K. Am. Mineral. **80**, 457–464 (1995)
16. Meagher, E.P.: The crystal structures of pyrope and grossularite at elevated temperatures. Am. Mineral. **60**, 218–228 (1975)
17. Chenavas, J., Joubert, J.C., Marezio, M.: On the crystal symmetry of the garnet structure. Less Common Metals. **62**, 373–380 (1978)
18. Ye, S., et al.: Phosphors in phosphor-converted white light-emitting diodes recent advances in materials, techniques and properties. Mater. Sci. Eng R-Rep. **71**, 1–34 (2010)
19. Nakamura, S., Chichibu, S.F.: Introduction to Nitride Semiconductor Blue Laser and Light Emitting Diodes. Taylor & Francis, London (2000)
20. Chen, W.Q., et al.: Synthesis and photoluminescence properties of YAG:Ce3+ phosphor using a liquid-phase precursor method. J. Lumin. **147**, 304–309 (2014)
21. Kim, E.H., et al.: InGaN/GaN white light-emitting diodes embedded with europium silicate thin film phosphor. IEEE J. Quantum Electron. **46**, 1381–1387 (2010)
22. Chen, Y.B., et al.: Low thermal quenching and high-efficiency Ce3 +, Tb3+−co-doped Ca3Sc2Si3O12 green phosphor for white light-emitting diodes. J. Lumin. **131**, 1589–1593 (2011)
23. Kim, J.S., et al.: Temperature-dependent emission spectra of M2SiO4: Eu2+ (M = ca, Sr, Ba) phosphors for green and greenish white LEDs. Solid State Commun. **133**, 445–448 (2005)
24. Kim, J.S., et al.: Temperature-dependent emission spectrum of Ba3MgSi2O8: Eu2+, Mn2+ phosphor for white-light-emitting diode. Electrochem. Solid State Lett. **8**, H65–H67 (2005)
25. Kim, J.S., et al.: Luminescent and thermal properties of full-color emitting X3MgSi2O8: Eu2+, Mn2+ (X = Ba, Sr, Ca) phosphors for white LED. J. Lumin. **122**, 583–586 (2007)
26. Shimomura, Y., et al.: Photoluminescence and crystal structure of green-emitting Ca3Sc2Si3O12: Ce3+ phosphor for white light emitting diodes. J. Electrochem. Soc. **154**, J35–J38 (2007)
27. Teng, X.M., Zhuang, W.D., Hu, Y.S., Huang, X.W.: Luminescence properties of nitride red phosphor for LED. J. Rare Earth. **26**, 652–655 (2008)
28. Xie, R.J., Hirosaki, N.: Silicon-based oxynitride and nitride phosphors for white LEDs—a review. Sci. Technol. Adv. Mater. **8** (7–8), 588–600 (2007)
29. He, X.H., et al.: Dependence of luminescence properties on composition of rare-earth activated (oxy)nitrides phosphors for white-LEDs applications. J. Mater. Sci. **44**, 4763–4775 (2009)
30. Cao, G.Z., et al.: Effects of the characteristics of silicon-nitride powders on the preparation of alpha′-sialon ceramics. J. Mater. Sci. Lett. **11**, 1685–1686 (1992)
31. Xie, R.J., Hirosaki, N., Mitomo, M.: Oxynitride/nitride phosphors for white light-emitting diodes (LEDs). J. Electroceram. **21**, 370–373 (2008)
32. Liu, L.H., et al.: Facile synthesis of Ca-alpha-SiAlON:Eu2+ phosphor by the microwave sintering method and its photoluminescence properties. Chin. Sci. Bull. **58**, 708–712 (2013)
33. Kim, D.H., Ryu, J.H., Cho, S.Y.: Light emitting properties of SiAlON: Eu2+ green phosphor. Appl. Phys. A. **102**, 79–83 (2011)
34. Chung, S.L., et al.: Phosphors based on nitridosilicates: synthesis methods and luminescent properties. Curr. Opin. Chem. Eng. **3**, 62–67 (2014)
35. Yazdan Mehr, M., van Driel, W.D., Zhang, G.Q.: Accelerated life time testing and optical degradation of remote phosphor plates. Microelectron. Reliab. **54**, 1544–1548 (2014)
36. Qin, J.L., et al.: Temperature-dependent luminescence characteristic of SrSi2O2N2:Eu2+ phosphor and its thermal quenching behavior. J. Mater. Sci. Technol. **30**, 290–294 (2014)
37. Sun, J.F., Shi, Y.M., Zhang, W.L.: Luminescent and thermal properties of a new green emitting Ca6Sr4(Si2O7)3Cl2:Eu2+ phosphor for near-UV light-emitting di-odes. Mater. Res. Bull. **47**, 400–404 (2012)
38. Li, G.H., et al.: Carbothermal synthesis of CaAlSiN3:Eu2+ red-emitting phosphors and the photoluminescent properties. J. Mater. Sci. Mater. Electron. **26**, 10201–10206 (2015)
39. Cooke, D.W., et al.: Oscillator strengths, Huang-Rhys parameters, and vibrational quantum energies of cerium-doped gadolinium oxyorthosilicate. J. Appl. Phys. **87**, 7793–7797 (2000)
40. Fukuda, Y., et al.: Luminescence properties of Eu2+−doped green-emitting Sr-sialon phosphor and its application to white light-emitting diodes. Appl. Phys. Express. **2** (2009)
41. Cheng, Y., et al.: A review on high refractive index nanocomposites for optical applications. Recent Patents Mater. Sci. **4**(1), 15–27 (2011)
42. Monson, T.C., Huber, D.L.: High refractive index TiO2 nanoparticle/silicone composites. Nanosci. Nanotechnol. 46–47
43. Nussbaumer, R.J., et al.: Polymer-TiO2 nanocomposites: a route towards visually transparent broadband UV filters and high refractive index materials. Macromol. Mater. Eng. **288**(1), 44–49 (2003)
44. Tao, P., et al.: TiO2 nanocomposites with high refractive index and transparency. J. Mater. Chem. **21**(46), 18623–18629 (2011)
45. Liu, Y., et al.: High refractive index and transparency nanocomposites as encapsulant for high brightness LED packaging. Electronic Components and Technology Conference. IEEE, pp. 553–556 (2013)
46. Tuan, C.-C., et al.: Ultra-high refractive index LED encapsulant, IEEE. Electronic Components and Technology Conference. IEEE, pp. 447–451 (2014)
47. Song, B.-M., Han, B.: Analytical/experimental hybrid approach based on spectral power distribution for quantitative degradation analysis of phosphor converted LED. IEEE Trans. Device Mater. Reliab. **14**(1), 365–374 (2014)
48. Meneghini, M., et al.: Degradation mechanisms of high-power LEDs for lighting applications: an overview. IEEE Trans. Ind. Appl. **50**(1), 78–85 (2014)
49. Fan, J., Yung, K.C., Pecht, M.: Prognostics of chromaticity state for phosphor-converted white light emitting diodes using an unscented Kalman filter approach. IEEE Trans. Device Mater. Reliab. **14**(1), 564–573 (2014)
50. Barton, D.L., et al.: Life tests and failure mechanisms of GaN/AlGaN/InGaN light-emitting diodes. Optoelectron. High-Power Lasers Appl. 17–27 (1998)
51. Lin, Y.H., et al.: Development of high-performance optical silicone for the packaging of high-power LEDs. IEEE Trans. Compon. Packaging Technol. **33**(4), 761–766 (2010)

52. Lin, C.H., et al.: Development of UV stable LED encapsulants. In: Microsystems, Packaging, Assembly and Circuits Technology Conference, 2009. IMPACT 2009. 4th International, pp. 565–567 (2009)
53. Hsu, C.W., et al.: Effect of thermal aging on the optical, dynamic mechanical, and morphological properties of phenylmethylsilicone-modified epoxy for use as an LED encapsulant. Mater. Chem. Phys. **134**(2), 789–796 (2012)
54. Grassie, N., Guy, M.I., Tennent, N.H.: Degradation of epoxy polymers: 2—mechanism of thermal degradation of bisphenol-a diglycidyl ether. Polym. Degrad. Stabil. **13**(1), 11–20 (1985)
55. Huang, J., et al.: Degradation mechanisms of mid-power white-light LEDs under high-temperature–humidity conditions. IEEE Trans. Device Mater. Reliab. **15**(2), 220–228 (2015)
56. Bahadur, M., et al.: Silicone materials for LED packaging. SPIE Optics+ Photonics. 63370F–63370F-7 (2006)
57. Huang, W., et al.: Studies on UV-stable silicone–epoxy resins. J. Appl. Polym. Sci. **104**(6), 3954–3959 (2007)
58. Atkins, G.R., Krolikowska, R.M., Samoc, A.: Optical properties of an ormosil system comprising methyl-and phenyl-substituted silica. J. Non-Cryst. Solids. **265**(3), 210–220 (2000)
59. Hench, L.L., Ulrich, D.R.: Ultrastructure Processing of Ceramics, Glasses, and Composites. Wiley, New York (1984)
60. Bae, J.-y., et al.: Sol–gel synthesized linear oligosilicone-based hybrid material for a thermally-resistant light emitting diode (LED) encapsulant. RSC Adv. **3**(23), 8871–8877 (2013)
61. Kim, J.-S., Yang, S., Bae, B.-S.: Thermally stable transparent sol−gel based silicone hybrid material with high refractive index for light emitting diode (LED) encapsulation. Chem. Mater. **22**(11), 3549–3555 (2010)
62. Kim, J.-S., Yang, S., Bae, B.-S.: Thermal stability of sol–gel derived methacrylate oligosilicone-based hybrids for LED encapsulants. J. Sol-Gel Sci. Technol. **53**(2), 434–440 (2010)
63. Barton, D.L., et al.: Single-quantum well InGaN green light emitting diode degradation under high electrical stress. Microelectron. Reliab. **39**(8), 1219–1227 (1999)
64. Meneghesso, G., et al.: Reliability of visible GaN LEDs in plastic package. Microelectron. Reliab. **43**(9–11), 1737–1742 (2003)
65. Meneghini, M., et al.: Reversible degradation of ohmic contacts on p-GaN for application in high-brightness LEDs. IEEE Trans. Electron Devices. **54**(12), 3245–3251 (2007)
66. Jung, E., Kim, H.: Rapid optical degradation of GaN-based light-emitting diodes by a current-crowding-induced self-accelerating thermal process. IEEE Trans. Electron Devices. **61**(3), 825–830 (2014)
67. Yan, B., et al.: Influence of die attach layer on thermal performance of high power light emitting diodes. IEEE Trans. Compon. Packaging Technol. **33**(4), 722–727 (2010)
68. You, J., He, Y., Shi, F.: Thermal management of high power LEDs: Impact of die attach materials. In: IMPACT, pp. 239–242 (2007)
69. Luo, X., et al.: Phosphor self-heating in phosphor converted light emitting diode packaging. Int. J. Heat Mass Transf. **58**(1–2), 276–281 (2013)
70. Meneghini, M., et al.: Thermally activated degradation of remote phosphors for application in LED lighting. IEEE Trans. Device Mater Reliab. **13**(1), 316–318 (2013)
71. Ye, H., et al.: Thermal analysis of phosphor in high brightness LED. In: ICEPT-HDP, Guilin, pp. 1535–1539 (2012)
72. B. Yan, et al.: Influence of phosphor configuration on thermal performance of high power white LED array. In: ISAPM, Irvine, California, pp. 274–289 (2013)
73. Juntunen, E., et al.: Effect of phosphor encapsulant on the thermal resistance of a high-power COB LED module. IEEE Trans. Compon. Pack. A. **3**(7), 1148–1154 (2013)
74. Huang, J., et al.: Rapid degradation of mid-power white-light LEDs in saturated moisture conditions. IEEE Trans. Device Mater. Reliab. **15**(4), 478–485 (2015)
75. L. Zhou, et al.: Analysis of delamination and darkening in high power LED packaging. In: 16th IEEE International Symposium on the Physical and Failure Analysis of Integrated Circuits, 2009. IPFA 2009. pp. 656–660 (2009)
76. Fang, Y.C., et al.: Two-step synthesis of SrSi2O2N2: Eu2+ green Oxynitride phosphor: electron-phonon coupling and thermal quenching behavior. J. Electrochem. Soc. **158**, J246–J249 (2011)
77. Choi, M., et al.: Direct correlation between reliability and pH changes of phosphors for white light-emitting diodes. Microelectron. Reliab. **54**, 2849–2852 (2014)
78. Yoon, M.J., et al.: Comparison of YAG: Eu phosphors synthesized by supercritical water in batch and continuous reactors. Korean J. Chem. Eng. **24**, 877–880 (2007)
79. Yang, X.X., et al.: EDTA-mediated morphology and tunable optical properties of Eu3+−doped NaY(MoO4)(2) phosphor. J. Mater. Sci. Mater. Electron. **26**, 6659–6666 (2015)
80. Liu, H.T., et al.: Study on white-light emission and light-emitting mechanism of hydrothermally synthesized YVO4:1 mol.%Dy3+,x mol.%Eu3+ phosphor powders. J. Rare Earths. **34**, 113–117 (2016)
81. Jung, E., Kim, M.S., Kim, H.: Analysis of contributing factors for determining the reliability characteristics of GaN-based white light-emitting diodes with dual degradation kinetics. IEEE Trans. Electron Devices. **60**(1), 186–191 (2013)
82. Mura, G., et al.: Sulfur-contamination of high power white LED. Microelectron. Reliab. **48**(8–9), 1208–1211 (2008)
83. Zhang, L., et al.: Study on Ag-plated Cu lead frame and its effect to LED performance under thermal aging. IEEE Trans. Device Mater. Reliab. **14**(4), 1022–1030 (2014)
84. Tan, L., et al.: Effects of defects on the thermal and optical performance of high-brightness light-emitting diodes. IEEE Trans Electron. Packag. Manuf. **32**(4), 233–240 (2009)
85. Hu, J., Yang, L., Whan Shin, M.: Mechanism and thermal effect of delamination in light-emitting diode packages. Microelectron. J. **38** (2), 157–163 (2007)
86. Luo, X., Wu, B., Liu, S.: Effects of moist environments on LED module reliability. IEEE Trans. Device Mater. Reliab. **10**(2), 182–186 (2010)
87. Kim, H.H., et al.: Thermal transient characteristics of die attach in high power LED PKG. Microelectron. Reliab. **48**(3), 445–454 (2008)
88. Gladkov, A., Bar-Cohen, A.: Parametric dependence of fatigue of electronic adhesives. IEEE Trans. Compon. Packag. Technol. **22**(2), 200–208 (1999)
89. Ries, M.D., et al.: Attachment of solder ball connect (SBC) packages to circuit cards. IBM J. Res. Dev. **37**(5), 597–608 (1993)
90. Wong, E.H., et al.: The mechanics and impact of hygroscopic swelling of polymeric materials in electronic packaging. J. Electron. Packag. **124**(2), 122 (2002)

Part VI

Energy Harvesting

19 Mechanical Energy Harvesting Using Wurtzite Nanowires

Xudong Wang and Zhong Lin Wang

19.1 Introduction

Facing current global energy shortage, seeking for renewable energies is essential for releasing the heavy reliance on mineral-based energy and improving the cleanness of our environment. Solar, mechanical (mostly wind), and bioenergies are common substitutions and have been more and more widely used. Meanwhile, scientific research is intensively focused on these areas for improving the energy efficiency and lowering the cost. Based on the nature of those energy sources, advantages and limitations are obvious for each of them. The type of energy source to be chosen has to be decided on the basis of specific applications. For instance, the supply of solar energy is endless, while it does not apply to dark areas or underwater/underground. Mechanical energy more universally exists on the earth but mostly in a lower energy level, such as friction, body movement, wave, and acoustic. Considering small electronics or advanced sensing systems that do not have high-energy requirement, the mechanical energy would be a perfect power source for them to operate in a "self-sufficient" mode. Small-level mechanical energy harvesting was typically realized using piezoelectric materials, which generates electric potential when subject to mechanical deformation. However, due to the limitation of micro-fabrication, its power density is rather low, and a big mass is usually required for improving the sensitivity to low-level vibrations. Using piezoelectric wurtzite ZnO nanowire (NW) arrays, we have developed a novel technology for converting nanoscale mechanical energy into electric energy. The operation mechanism of the electric generator relies on the unique coupling of piezoelectric and semiconducting dual properties of ZnO as well as the elegant rectifying function of the Schottky barrier formed between the metal electrode and the NW. The nanodevice developed under this concept is named as *nanogenerator* (NG). The details of this development and corresponding mechanisms will be addressed in this chapter.

X. Wang (✉)
Department of Materials Science and Engineering, University of Wisconsin-Madison, Madison, WI, USA
e-mail: xudong@engr.wisc.edu

Z. L. Wang
School of Materials Science and Engineering, Georgia Institute of Technology, Atlanta, GA, USA
e-mail: zhong.wang@mse.gatech.edu

19.2 Mechanical to Electrical Energy Conversion by Wurtzite Nanowires

19.2.1 Signal ZnO Wires

The mechanical energy to electrical energy conversion process was first revealed on a ZnO wire under atomic force microscope (AFM) by directly correlating the manipulation process of the wire and the electric output voltage/current signal [1]. In the experiment, a long ZnO wire that was large enough to be seen under an optical microscope was used. One end of the ZnO wire was affixed on a silicon substrate by silver paste, while the other end was left free. The substrate was intrinsic silicon, so its conductivity was rather poor. The wire was laid on the substrate but kept at a small distance from the substrate to eliminate the friction with the substrate except at the affixed side (Fig. 19.1a). The measurements were performed by AFM using a Si tip coated with Pt film coating, which had a tetrahedral shape with an apex angle of 70° and height of 14 μm and a spring constant of 1.42 N/m. The measurements were done in AFM contact mode under a constant normal force of 5 nN between the tip and sample surface, and the scan area was $70 \times 70\ \mu m^2$.

Both the topography (feedback signal from the scanner) and the corresponding output voltage (V) images across a load were recorded simultaneously when the AFM tip was scanned across a wire/belt. The topography image reflects the

C. P. Wong et al. (eds.), *Nano-Bio- Electronic, Photonic and MEMS Packaging*, https://doi.org/10.1007/978-3-030-49991-4_19

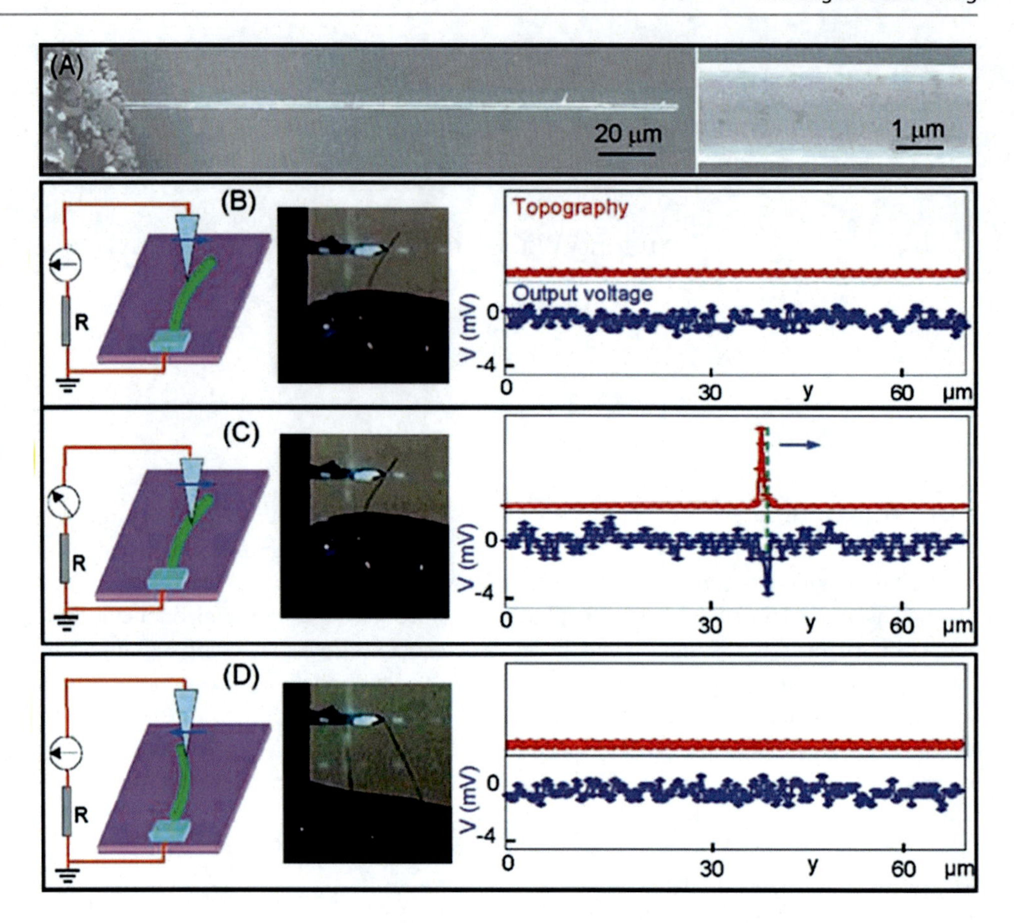

Fig. 19.1 In situ observation of the process for converting mechanical energy into electric energy by a piezoelectric ZnO belt. (**a**) SEM images of a ZnO belt with one end affixed by silver paste onto a silicon substrate and the other end free. (**b**, **c**, and **d**) Three characteristic snapshots and the corresponding topography (*top curve*) and output voltage (*bottom curve*) images when the tip scanned across the middle section of the belt. The schematic illustration of the experimental condition is shown at the left-hand side, with the scanning direction of the tip indicated by an *arrowhead*

change in normal force perpendicular to the substrate, which shows a bump only when the tip scans over the wire. The output voltage between the conductive tip and the ground was continuously monitored as the tip scans over the wire. No external voltage was applied in any stage of the experiment. The output data to be presented were obtained from snapshots of the characteristic events observed during the experiments.

The AFM tip scanned line by line at a speed of 105.57 μm/s perpendicular to the wire and pushed the wire at its middle section. When subjected to a displacement force, one side of the wire is stretched (left-hand side), and the other side (the right-hand side) is compressed. The objective of this experiment is to observe if the piezoelectric discharge occurs when the tip touches either the stretched side or the compressed side or both sides. The topography image directly captured whether the tip passed over the belt or not because it was a representation of the normal height received by the cantilever. When the tip pushed the wire but did not go over and across it, as judged by the flat output signal in the topography image (Fig. 19.1b), no voltage output was produced, indicating the stretched side produced no piezoelectric discharge event. Once the tip goes over the belt and touches the compressed side, as indicated by a peak in the topography image, a sharp voltage output peak is observed (Fig. 19.1c). By analyzing the positions of the peaks observed in the topography image and the output voltage image, we noticed that the discharge occurred after the tip nearly finished crossing the wire. This clearly indicates that the compressed side was responsible for producing the negative piezoelectric discharge voltage. For another wire, when the tip retracted from the right-hand side to the left-hand side (Fig. 19.1d), no output voltage was detected because the tip just touched the stretched side of the belt without crossing it.

As a summary of the above experimental observations, there are three key results. First, piezoelectric discharge is observed only when the AFM tip touches the end of the bent wire. Second, the piezoelectric discharge occurs only when the AFM tip touches the compressed side of the wire, and there is no voltage output if the tip touches the stretched side of the wire. Third, the piezoelectric discharge gives negative output voltage as measured from the load R_L. Finally, in reference to the topography image, the voltage output event occurs when the AFM tip has almost finished crossing the width of the wire/belt, which means that the discharge event is delayed to the last half of the contact between the tip and the wire.

19.2.2 Aligned ZnO Nanowire Arrays

Vertically aligned ZnO NW array is a perfect self-assembled nanostructure that allows series of macroscopic manipulation and operation. Aligned ZnO NWs used in the experiments were grown on a-Al_2O_3 substrate using Au particles as a catalyst, by the vapor–liquid–solid (VLS) process [2]. An epitaxial relation between ZnO and a-Al_2O_3 allows a thin, continuous layer of ZnO to form at the substrate, which serves as a large electrode connecting the NWs with a metal electrode for transport measurement (Fig. 19.2a) [3]. The NW grows along the [0001] direction and has side surfaces of {0110} (Fig. 19.2b). Most of the Au particles at the tips of the NWs either evaporate during the growth or fall off during scanning by the AFM tip. For most of the NWs, the growth front is free of Au particles or has a small hemispherical Au particle that covers only a fraction of the top (inset, Fig. 19.2b). For the purpose of our measurements, we have grown NW arrays that have relatively less density and shorter length (0.2–0.5 mm), so that the AFM tip can exclusively reach one NW without touching another.

Similar measurements were performed by AFM using a Si tip coated with Pt film [4]. The rectangular cantilever had a calibrated normal spring constant of 0.76 N/m (Fig. 19.2c). In the AFM contact mode, a constant normal force of 5 nN was maintained between the tip and the sample surface. The tip scanned over the top of the ZnO NW, and the tip's height was adjusted according to the surface morphology and local contacting force. The thermal vibration of the NWs at room temperature was negligible. For the electric contact at the bottom of the nanowires, silver paste (uncured) was applied to connect the (large) ZnO film on the substrate surface with the measurement circuit. The output voltage across an outside load of resistance $R_L = 500$ megohms was continuously monitored as the tip scanned over the nanowires (note the defined polarity of the voltage signal). No external voltage was applied in any stage of the experiment.

Experimentally, both the topography (feedback signal from the scanner) (Fig. 19.2a) and the corresponding output voltage (V_L) images across the load (Fig. 19.2b) were recorded simultaneously when the AFM tip scanned over the aligned NW arrays. In contact mode, as the tip scanned over the vertically aligned NWs, the NWs were bent consecutively. The bending distance was directly recorded in the topography image, from which the maximum bending deflection distance and the elastic modulus of the NW as well as the density of NWs that have been scanned by the tip were directly derived [5].

In the V_L image, many sharp output peaks (like discharge peaks) were observed. These peaks, typically about 4–50 times the noise level, are rather sharp and narrow, and sometimes one or two pixels represent one voltage peak because of

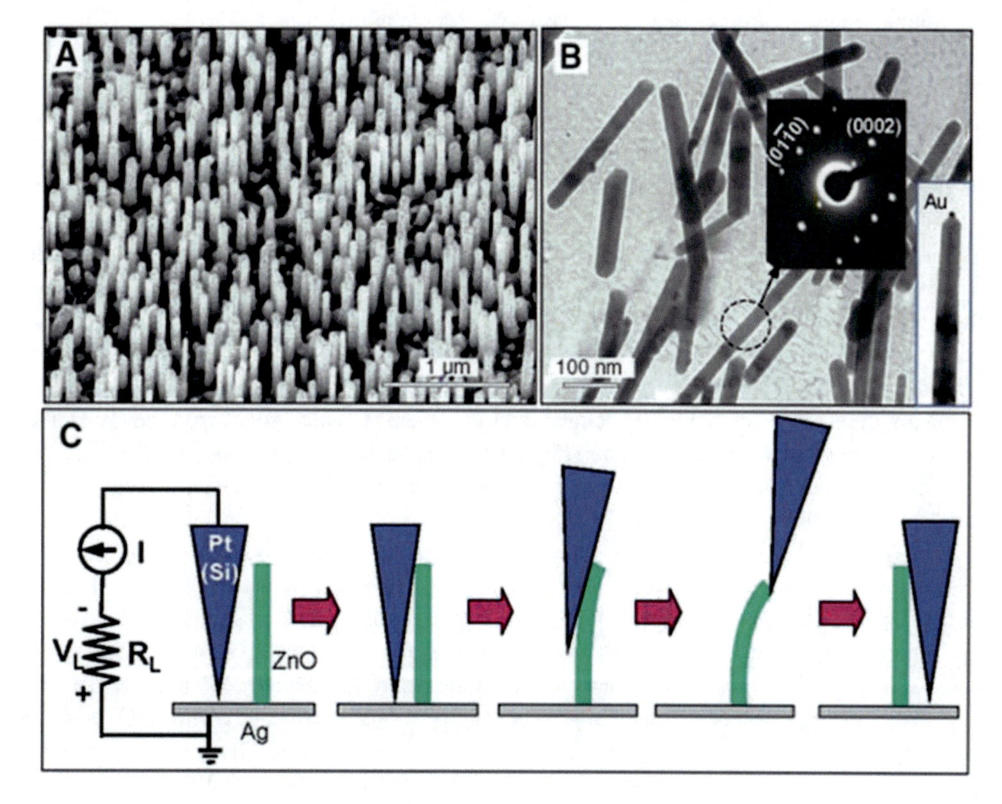

Fig. 19.2 Experimental design for converting nanoscale mechanical energy into electrical energy by a vertical ZnO NW. (**a**) SEM images of aligned ZnO NWs grown on a-Al_2O_3 substrate. (**b**) TEM images of ZnO NWs, showing the typical structure of the NW without an Au particle or with a small Au particle at the *top*. *Inset at center*: an electron diffraction pattern from a NW. *Inset at right*: image of a NW with an Au particle. (**c**) Experimental setup and procedures for generating electricity by deforming a PZ NW with a conductive AFM tip

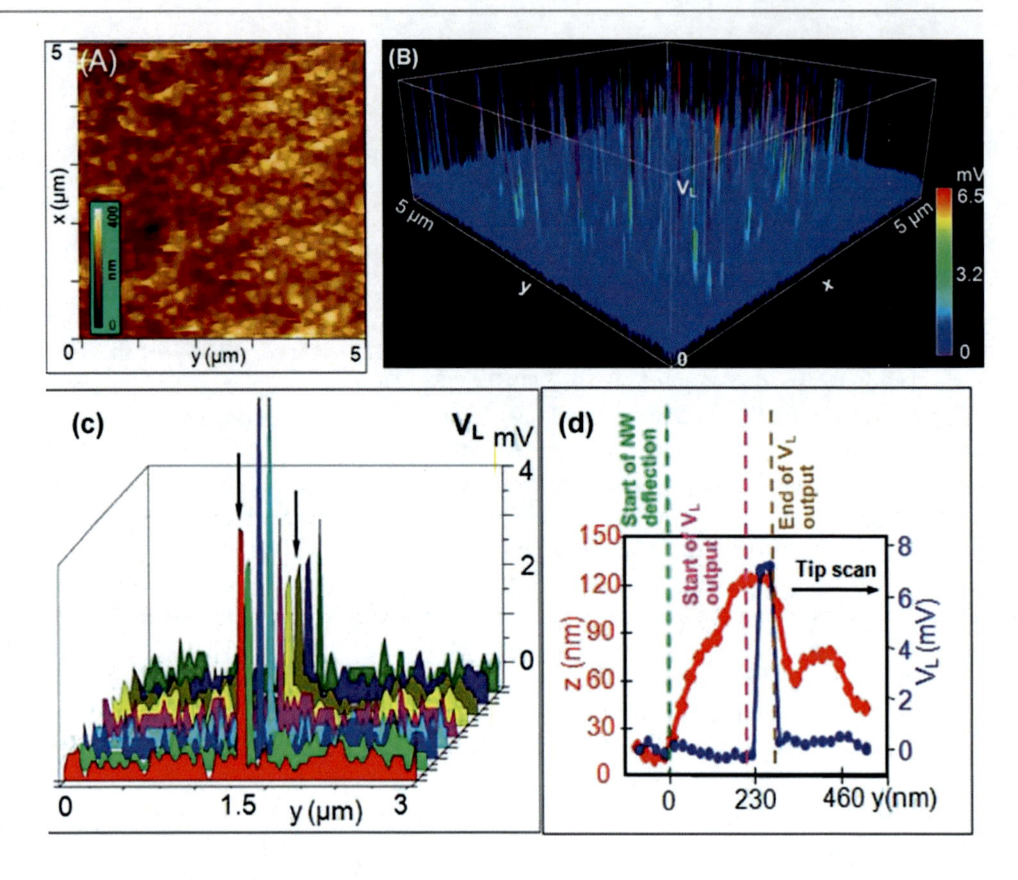

Fig. 19.3 Electromechanically coupled discharging process of aligned piezoelectric ZnO NWs observed in contact mode. Topography image (**a**) and corresponding output voltage image (**b**) of the NW arrays. (**c**) A series of line profiles of the voltage output signal when the AFM tip scanned across a vertical NW at a time interval of 1 min. (**d**) Line profiles from the topography (*diamond marks*) and output voltage (*circle marks*) images across a NW

the limited scanning speed of the AFM tip, so that the color distribution in the plot is not easy to display. By reducing the scan range and increasing the scan frequency, more complete profiles of the discharge peaks were captured. Most of the voltage peaks are ~6–9 mV in height. The density of NWs contacted by the tip is counted from Fig. 19.3a to be ~20 μm^{-2}, and the average density of NWs whose voltage output events had been captured by the tip in Fig. 19.3b is ~8 μm^{-2}; thus ~40% of the NWs were contacted.

The location of the voltage peak is directly registered at the site of the NW. A time series of the voltage output line profiles across one NW acquired at a time interval of 1 min is shown in Fig. 19.3c. Because the dwell time for each data point (or pixel) is 2 ms, which is longer than the average lifetime of the voltage peak of ~0.6 ms (Fig. 19.3d), the peak at which V_L reached the maximum was possibly missed by the "slow" scanning tip, so that V_L shows a chopped top (arrows in Fig. 19.3c).

A sharp peak can be identified continuously at the location of the NW and in the NW output voltage in each scan of the tip. When the tip started to deflect the NW, no voltage output was observed (Fig. 19.3d); V_L was detected when the deflection of the NW approached its maximum. When the NW was released by the AFM tip, V_L dropped to zero, indicating that the output of piezoelectricity was detected toward the end of the AFM scan over the NW.

19.2.3 CdS Nanowires

As another semiconductor material with wurtzite crystal structure, CdS also exhibited the ability for mechanical energy conversion [6]. Vertically CdS NW arrays were grown through a hydrothermal process following the procedures reported in literature with minor modifications [7]. The morphology of the hydrothermally grown CdS NWs is shown by a SEM image (Fig. 19.4a). These NWs were 150 nm in diameter and several micrometers in length. The x-ray diffraction (XRD) pattern of the as-grown CdS NWs is shown in Fig. 19.4b. Also included in Fig. 19.4b are the XRD patterns of the reference cubic zinc blende (ZB) and hexagonal wurtzite (WZ) phases of CdS crystals for comparison purposes. All of the diffraction peaks, except the one located at 26.4°, can be indexed to the WZ phase. The diffraction peak at 2θ of 26.4° is a combined contribution from WZ (0002) and ZB (111) planes. The coexistence of the two phases can be clearly seen from the high-resolution transmission electron microscopy (HRTEM) image shown in the inset of Fig. 19.4a. The NW is composed of alternating ZB and WZ phases, with alternating atomic layer stacking sequence from ABC to AB along the growth direction of the NWs. The growth direction was identified to be along ZB (111) and WZ <0001>.

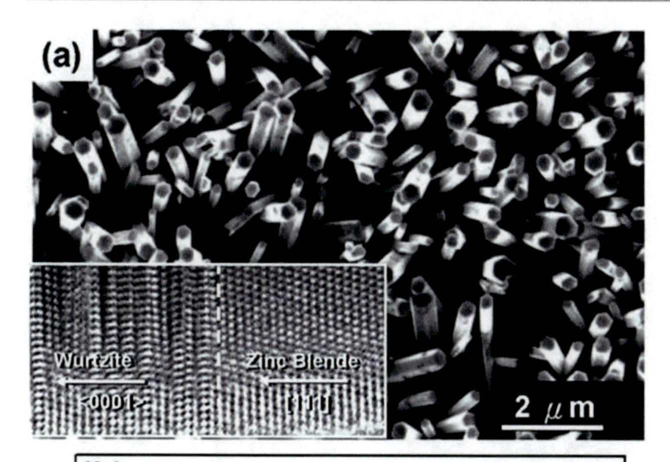

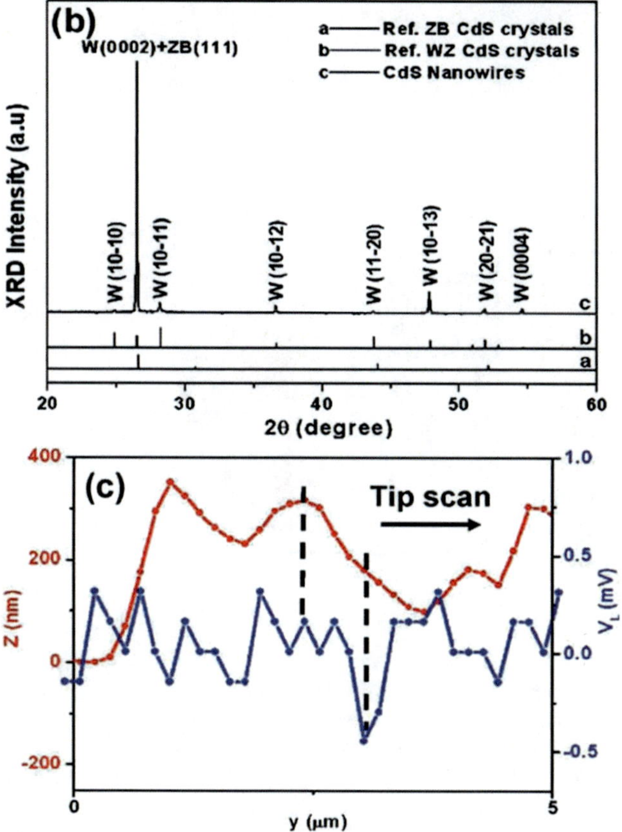

Fig. 19.4 (**a**) Top view SEM image of the CdS NWs prepared from the hydrothermal process, *inset* is an HRTEM image of a single CdS nanowire. (**b**) An XRD spectrum of the CdS NW arrays. (**c**) AFM line profiles of topography (*top*) and output voltage (*bottom*) scanned across CdS NWs

The energy conversion experiments were conducted through the same manner as the measurements on ZnO NW arrays presented in Sect. 19.2.2. Similarly, the process for generating the electric current can be derived from a detailed analysis of the output voltage peak and the topological profile received by the tip when scanned across a NW. During the tip scans, no voltage output signals were observed if the tip touched only the stretched side of the NW and did not lift up to go beyond the central line of the NW to reach the compressed side. A negative voltage output signal was detected when the tip went beyond the NW to reach the compressed side of the NW, as shown in the topography and output voltage images of Fig. 19.4c. This is clearly indicated by a delayed output in voltage signal in reference to the surface profile image of the NW. The typical magnitude of the output voltage was $\simeq$(0.5–1) mV, where the negative sign means that the generated current flowed from the tip to the NW. The low-voltage output is likely to be resulting from the phase transition between ZB and WZ phases along the growth direction of the NW, because WZ phase is piezoactive, while ZB phase is not. However, the presence of the ZB phase along the growth direction is disadvantageous to the piezoelectronic performance of the CdS NWs.

19.3 Energy Conversion Mechanisms

The mechanical to electrical energy conversion phenomenon relies on the piezoelectric and semiconducting coupled property of wurtzite crystals, which controls the charge creating, preserving, and releasing processes [8]. For a vertical, straight ZnO NW (Fig. 19.5a), the deflection of the NW by AFM tip, creates a strain field, with the outer surface being tensile (positive strain ε) and inner surface compressive (negative strain ε) (Fig. 19.5b). A piezoelectric field E_z along the NW is then created inside the NW volume by the piezoelectric effect, with the piezoelectric field closely parallel to the NW at the outer surface and antiparallel to the NW at the inner surface (Fig. 19.5c). As a result, across the width of the NW at the top end, the electric potential distribution from the compressed to the stretched side surface is approximately between V_s^- (negative) to V_s^+ (positive).

In order for releasing the piezoelectric charge, a Schottky barrier formed between the NW and AFM tip is the key controlling factor. In the first step, the AFM conductive tip that induces the deformation is in contact with the tensile surface of positive potential V_s^+ (Fig. 19.5d, e). The Pt metal tip has a potential of nearly zero, $V_m = 0$. So the metal tip ZnO interface is negatively biased for $\Delta V = V_m - V_s^- < 0$. With consideration of the n-type semiconductor characteristic of the as-synthesized ZnO NWs, the Pt metal ZnO semiconductor (M–S) interface in this case is a reverse-biased Schottky diode (Fig. 19.5e) (a Schottky contact is a metal–semiconductor contact at which a potential barrier is formed, so that it behaves like a diode that only allows the current to flow from metal to semiconductor), and little current flows across the interface. This is the process of creating, separating, and accumulating the charges. In the second step, when the AFM tip is in contact with the compressed side of the NW (Fig. 19.5f), the metal tip ZnO interface is

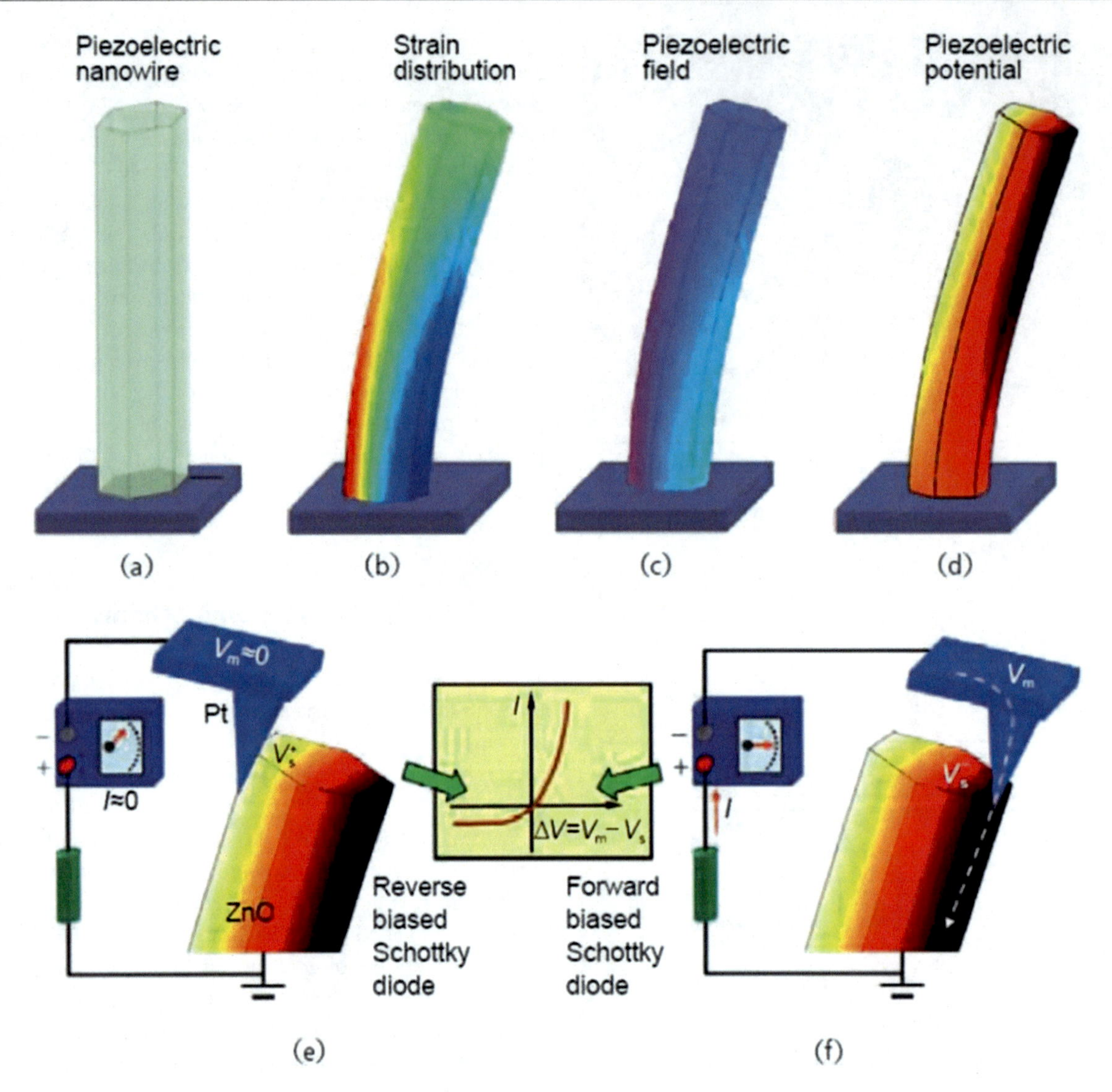

Fig. 19.5 (**a**) Schematic definition of a NW and the coordination system. (**b**) Longitudinal strain ε_z distribution in the NW after being deflected by an AFM tip from the side. (**c**) The corresponding longitudinal piezoelectric-induced electric field E_z distribution in the NW. (**d**) Potential distribution in the NW as a result of the piezoelectric effect. (**e and f**) Contacts between the AFM tip and the semiconductor ZnO NW [*boxed* area in (**d**)] at two reversed local contact potentials (positive and negative), showing reverse- and forward-biased Schottky rectifying behavior, respectively. The *inset* shows a typical current–voltage (*I–V*) relation characteristic of a metal–semiconductor (n-type) Schottky barrier

positively biased for $\Delta V = V_m - V_s^- > 0$. The M–S interface in this case is a positively biased Schottky diode, and it produces a sudden increase in the output electric current. The current is the result of ΔV-driven flow of electrons from the semiconductor ZnO NW to the metal tip. This is the charge releasing process.

Therefore, the theoretical background for such an energy conversion process is based on a voltage drop created across the cross section of the NW when it is laterally deflected, with the tensile side surface in positive voltage and compressive side in negative voltage. To quantify the piezoelectric potential generated at the two side surfaces of the NW is essential to predict the energy conversion efficiency and the output power. We have applied the perturbation theory for calculating the piezoelectric-potential distribution in a nanowire as pushed by a lateral force at the tip [9]. The analytical solution given under the first-order approximation produces a result that is within 6% from the fully numerically calculated result using a finite element method. Under a simplified condition, where we assume that the nanowire has a cylindrical shape with a uniform cross section of diameter $2a$ and length l, the maximum potential at the surface of the NW is given by equation

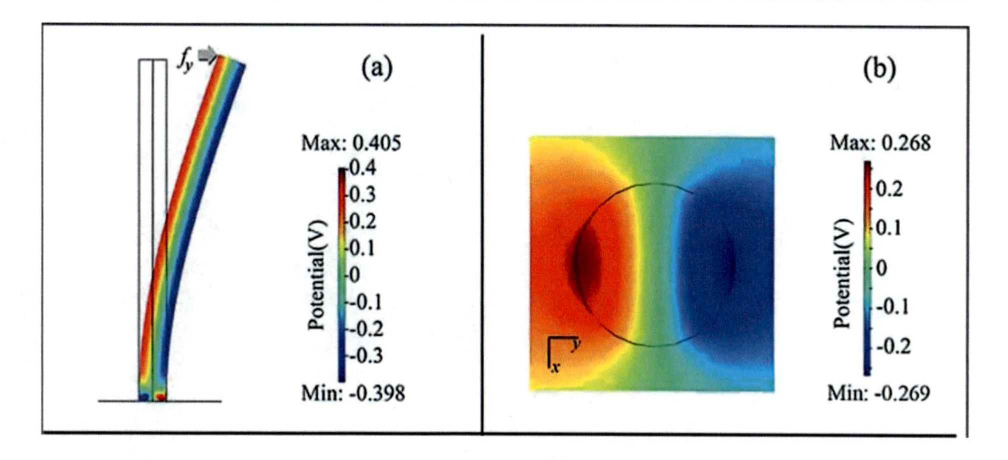

Fig. 19.6 Potential distribution for a ZnO nanowire with $d = 50$ nm and $l = 600$ nm at a lateral bending force of 80 nN. (**a**) and (**b**) are side and cross-sectional (at $z = 300$ nm) output of the piezoelectric potential in the NW given by finite element calculation

$$\varphi_{\max}^{(T,C)} = \pm\frac{3}{4(\kappa_0 + \kappa_{\perp})}[e_{33} - 2(1+\nu)e_{15} - 2\nu e_{31}] \times \frac{a^3}{l^3}\nu_{\max} \quad (19.1)$$

where ϕ is the electric potential; κ_0 and $\kappa_{\perp}$ are the dielectric constants of vacuum and ZnO crystal along its c-plane, respectively; e_{33}, e_{15}, and e_{31} are the linear piezoelectric coefficients; ν is Poisson ratio; and $\nu_{\max}$ is the maximum deflection at the NW's tip. This equation clearly shows that the electrostatic potential is directly related to the aspect ratio of the NW instead of its dimensionality.

For a typical NW that was grown through the vapor–liquid–solid (VLS) process [10], the diameter is $d = 50$ nm and length is $l = 600$ nm. When it is bent 145 nm to the right by an 80 nm lateral force, which is a common scanning situation in AFM, ±0.3 V piezoelectric potential can be induced on the two side surfaces, as shown in Fig. 19.6b, the calculated potential distribution across the NW cross section. Figure 19.6a is the potential distribution along the bent NW generated by finite element calculation using the above equation. Calculation also shows that the piezoelectric potential in the NW almost does not depend on the z-coordination along the NW. Therefore the potential is uniform along z-direction except for regions very close to the ends of the NW. This means that the NW is approximately like a "parallel plate capacitor." The maximum potential at the surface of the NW is directly proportional to the lateral displacement of the NW and inversely proportional to the length-to-diameter aspect ratio of the NW. For a larger size NW with $d = 300$ nm and length $l = 2$ μm, the surface piezoelectric potential can reach ±0.6 V, when it is deflected by a 1000 nN force. From this calculation, it can be concluded that the maximum potential at the surface of the NW is directly proportional to the lateral displacement of the NW and inversely proportional to the cube of length-to-diameter ratio of the NW. The voltage that can be generated in a conventional ZnO NW is in the range of several tenths of volts, which is high enough to drive the M-S Schottky diode at the interface between the AFM tip and the ZnO NW, as the key for nanogenerator operation.

19.4 Direct Current Nanogenerator

AFM-based experiments successfully showed the potential in applying piezoelectric and semiconducting wurtzite NWs for mechanical energy harvesting. Supported by theoretical calculation, the voltage that can be generated by a NW can be sufficiently high for realistic applications. However, in order to convert this concept into an applicable device for power generation, a few challenges need to be addressed. First, sophisticated AFM needs to be eliminated in deflecting the NWs so that the power generation can be achieved by an adaptable, mobile, and cost-effective approach over a larger scale. Second, all of the NWs are required to generate electricity simultaneously and continuously, and all the electricity can be effectively collected and outputted. Finally, the energy to be converted into electricity has to be provided in a form of wave/vibration from the environment, so the NG can operate "independently" and wirelessly. In addressing those challenges, an aligned NW array-based NG model has been developed, which is operated by ultrasonic waves and outputs a direct current (DC) electric signal – known as DC NG [11].

19.4.1 The DC Nanogenerator Model

The DC NG is built on vertically aligned ZnO NW arrays, and the structure is schematically shown in Fig. 19.7a. The ZnO NW array was covered by a zigzag Si electrode coated with Pt. The Pt coating not only enhanced the conductivity of the electrode but also created a Schottky contact at the interface with ZnO. The NWs were grown on either GaN substrates (Fig. 19.7b) or sapphire substrates that were

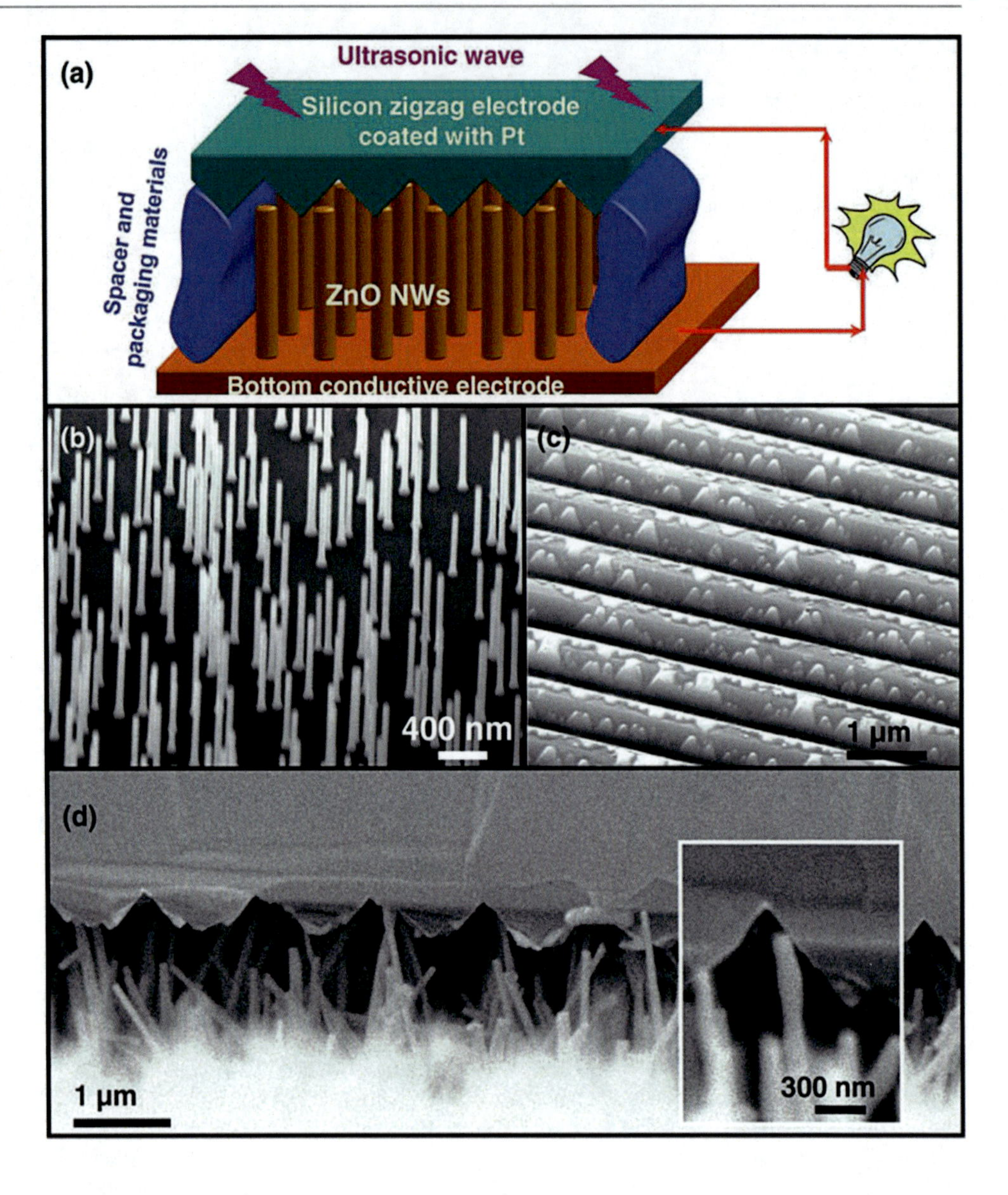

Fig. 19.7 Nanogenerators driven by an ultrasonic wave. (**a**) Schematic diagram showing the design and structure of the nanogenerator. (**b**) Low-density aligned ZnO NWs grown on a GaN substrate (**c**) Zigzag trenched electrode coated with Pt. (**d**) Cross-sectional SEM image of the nanogenerator. *Inset*: A typical NW that is forced by the electrode to bend

covered by a thin layer of ZnO film [3], which served as a common electrode for directly connecting the NWs with an external circuit. The density of the NWs was ~10/μm^2, and the height and diameter were ~1.0 μm and ~40 nm, respectively. The top electrode was composed of parallel zigzag trenches fabricated on a (001)-orientated Si wafer [12] and coated with a thin layer of Pt (200 nm in thickness) (Fig. 19.6c). The electrode was placed above the NW arrays and manipulated by a probe station under an optical microscope to achieve precise positioning; the spacing was controlled by soft-polymer stripes between the electrode and the NWs at the four sides. The assembled device was sealed at the edges to prevent the penetration of liquid. A cross-sectional image of the packaged NW arrays is shown in Fig. 19.7d; it displays a "lip-teeth" relationship between the NWs and the electrode. Some NWs are in direct contact with the top electrode, but some are located between the teeth of the electrode. The inclined NWs in the SEM image were primarily caused by the cross-sectioning of the packaged device.

In this design, the zigzag trenches on the top electrode act as an array of aligned AFM tips. Fig. 19.8a–c shows four possible configurations of contact between a NW and the zigzag electrode. When subject to the excitation of an ultrasonic wave, the zigzag electrode can move down and push the NW, which leads to lateral deflection of NW I. This, in turn, creates a strain field across the width of NW I, with the NW's outer surface being in tensile strain and its inner surface in compressive strain. The inversion of strain across the NW results in an inversion of piezoelectric field E_z along the NW, which produces a piezoelectric-potential inversion from V^- (negative) to V^+ (positive) across the NW (Fig. 19.8b). When the electrode makes contact with the stretched surface of the NW, which has a positive

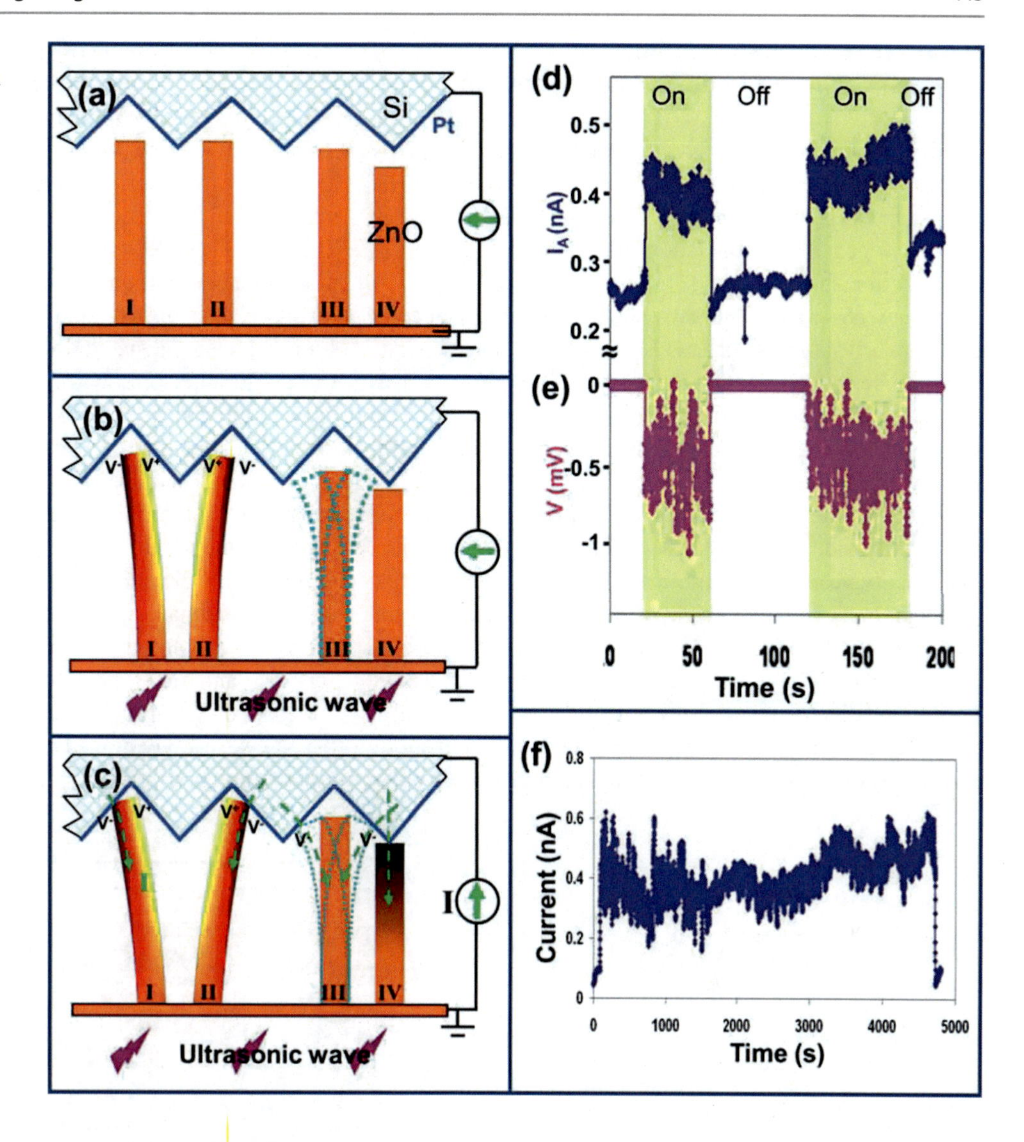

Fig. 19.8 (**a–c**) The mechanism of the nanogenerator driven by an ultrasonic wave. (**d**) and (**e**) Current and voltage measured on the nanogenerator, respectively, when the ultrasonic wave was turned on and off. (**f**) Continuous current output of the nanogenerator for an extended period of time

piezoelectric potential, the Pt metal–ZnO semiconductor interface is a reversely biased Schottky barrier, resulting in little current flowing across the interface. This is the process of creating, separating, preserving, and accumulating charges. With further pushing by the electrode, the bent NW I will reach the other side of the adjacent tooth of the zigzag electrode (Fig. 19.8c). In such a case, the electrode is also in contact with the compressed side of the NW, where the metal–semiconductor interface is a forward-biased Schottky barrier, resulting in a sudden increase in the output electric current flowing from the top electrode into the NW. This is the discharge process. Analogous to the situation described for NW I, the same processes apply to the charge output from NW II. NW III is chosen to elaborate on the vibration/resonance induced by an ultrasonic wave. When the compressive side of NW III is in contact with the electrode, the same discharge process as that for NW I occurs, resulting in the flow of current from the electrode into the NW (Fig. 19.8c). NW IV, which is short in height, is forced (without bending) into compressive strain by the electrode. In such a case, the piezoelectric voltage can also be created at the top of the NW, thus contributing to the electricity output.

The current and voltage outputs of the NG are shown in Fig. 19.8d, e, respectively, with the ultrasonic wave being turned on and off regularly. A jump of ~0.15 nA was observed when the ultrasonic wave was turned on, and the current immediately fell back to the baseline once the ultrasonic wave was turned off. Correspondingly, the voltage signal exhibited a similar on and off trend but with a negative output of ≃0.5 mV. The size of the NG is ~2 mm^2 in effective substrate surface area. The number of NWs that were actively contributing to the observed output current is estimated to be 250–1000 in the current experimental design. The NG worked continuously for an extended period of time of beyond 1 h (Fig. 19.8f).

19.4.2 Nanogenerator in Fluid and Three-Dimensional Integration

The first prototype of the NG has been demonstrated by indirect driving using ultrasonic waves through a metal frame suspended in air. An improved design has been thereafter developed for the operation in biocompatible fluid as driven by ultrasonic waves [13].

The new NG was modified from our original model composed by vertically aligned ZnO NWs and a Pt-coated zigzag top electrode, as described in Sect. 19.4.1. The NG core was completely packaged by a polymer to prevent the infiltration of biofluid into the NG. The polymer also has certain flexibility to remain free of relative movement between ZnO NWs and the top of the electrode. As shown in Fig. 19.9a, the NG was placed inside a container filled with 0.9% NaCl solution, which is a typical biocompatible solution. The container had a diameter of 11 cm and a height of 9.5 cm. The substrate and the top electrode were connected by waterproof extension cord to the outside of the container and marked as the positive and the negative electrodes, respectively. The definition of the two electrodes is according to the flow of current in the external circuit, which is always from the bottom substrate to the top zigzag electrode owing to the rectifying effect of the Schottky barrier between ZnO and Pt in the core of the NG. An ultrasonic stimulation was applied through the bottom of the container with a frequency of 41 kHz. The ultrasonic wave transported through the fluid and triggered the vibration of the electrode and NWs to generate electricity. The size of the NG was about 2 mm^2, on which there are more than one million NWs.

Inside the fluid container, the output of NG exhibited a strong correspondence to the local intensity of the ultrasonic wave. The ultrasonic wave source was placed at the center beneath the container. Once the ultrasonic wave was excited inside the container, it was reflected by the container's wall and the water surface. As a result, the wave intensity was enhanced in a certain region inside the fluid. A NG was placed inside the fluid and held by a clamp that can be freely moved in any direction to trace the enhanced ultrasonic wave, while the output current was continuously monitored. First, the NG was placed at the center region above the water surface, and the ultrasonic wave was kept on. The corresponding current signal is shown in Fig. 19.9b. Then the NG was slowly moved into the fluid along the z-direction (depth direction), and a jump in current for ~1 nA was

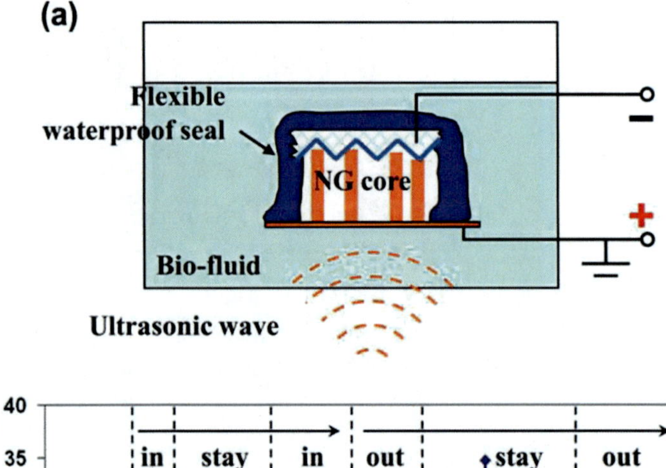

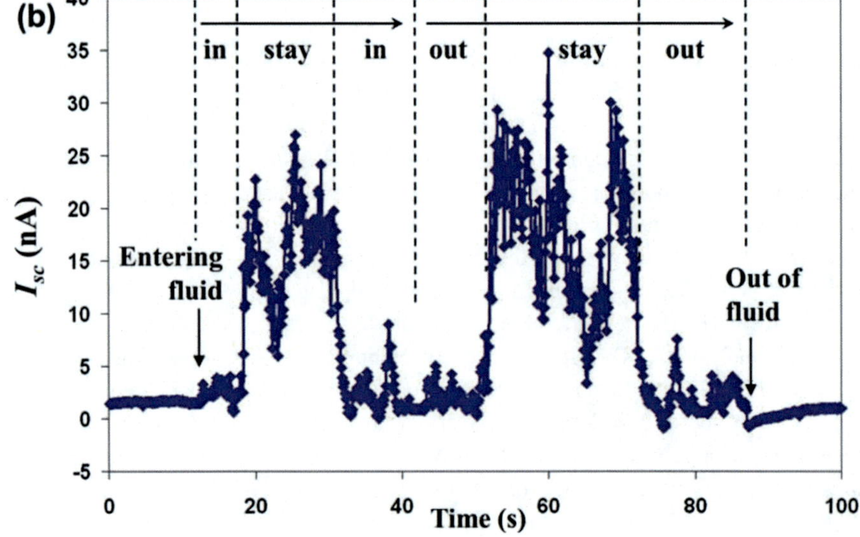

Fig. 19.9 (**a**) Schematic of a NG that operates in biofluid and the definition of its two electrodes. (**b**) Short-circuit current signal measured during the movement of NG along z-direction (from water surface to the bottom and then back to surface)

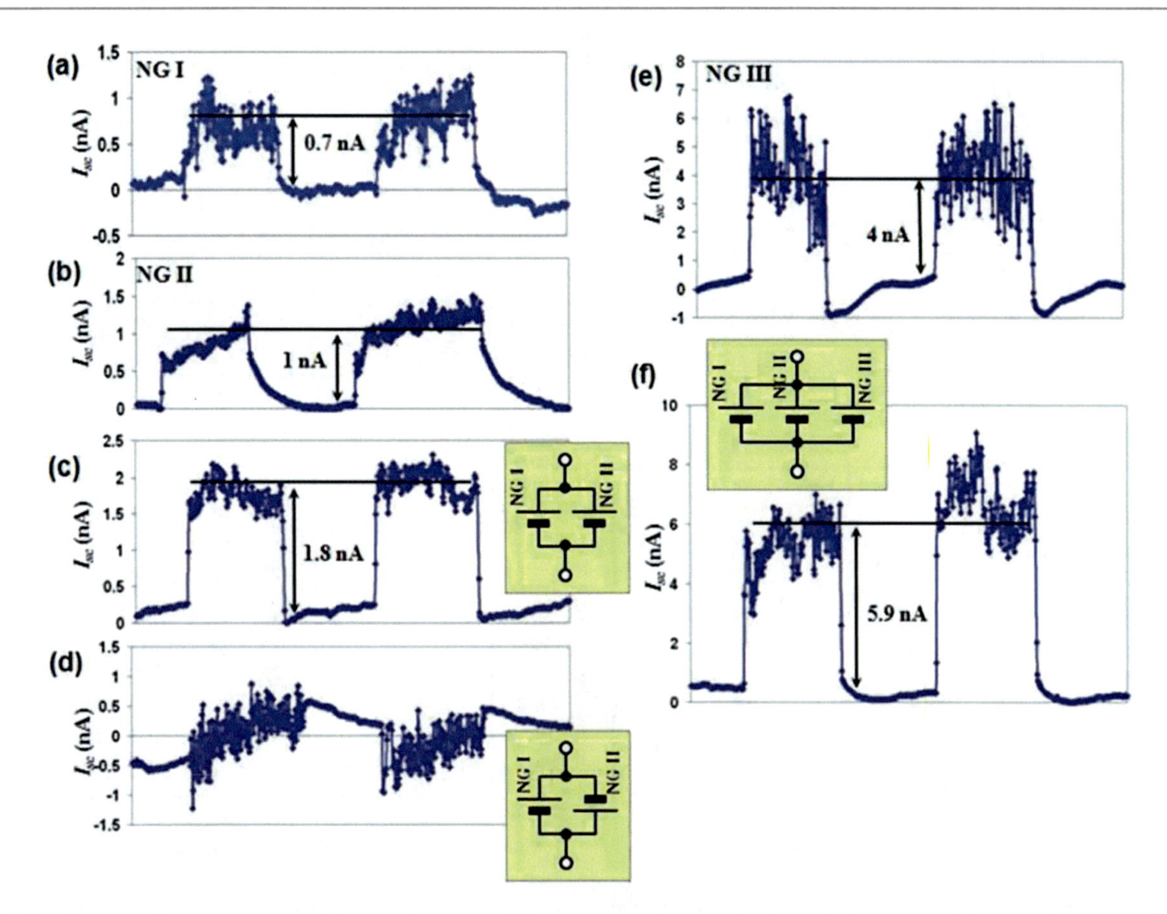

Fig. 19.10 Short-circuit current measured from an integrated NG system. (**a**) and (**b**) Current signal measured from two individuals NG I and II. (**c**) and (**d**) Current signal measured from parallel- and antiparallel-connected NG I and NG II, respectively; the connection configurations are schematically shown in the *insets*. (**e**) Current signal measured from NG III with a better performance. (**f**) Current signal measured from parallel-connected NG I, II, and III; *inset* shows the connection configuration

immediately detected once the NG touched the water surface. When the NG reached ~3.3 cm below the water surface, the current quickly jumped to ~20 nA. The output can be kept at such a high level as long as the NG stayed at this depth. After 15 s of steady high output, the NG was moved further down, and the current dropped back to the 1–2 nA level again. When the bottom of the container was reached, the NG was pulled upward to the water surface. The same 20–25 nA high current output was observed again once the NG reached 3.3 cm depth. The current signal dropped back to its baseline after the NG was pulled out of water.

Raising the output of the NG is essential to advance this technology into an applicable power source. If we take each NG as a battery, the most straightforward way to increase the current/voltage is to put them in parallel/serial. This has been demonstrated experimentally. As shown in parts a and b of Fig. 19.10, two NGs I and II have been tested under the same experimental conditions. NG I exhibited an average I_{sc} of ~0.7 nA, and NG II showed a lower noisy signal of 1 nA. These two NGs were then connected in parallel (inset of Fig. 19.10c) and tested under the same condition again. The resultant output current reached an average of ~1.8 nA, which is the sum of the two individual outputs (Fig. 19.10c). This concept was further proved by antiparallel-connecting NG I and NG II (inset of Fig. 19.10d). Since the magnitudes of the I_{sc} of the two NGs were very close, the total current was mostly cancelled out by the "head-to-tail" connection in parallel, and the received signal was around the baseline (Fig. 19.10d), just as expected.

If more NWs were located in the activate position when packaging the top zigzag electrode, the output current can be improved by several times. As shown in Fig. 19.10e, with a better assembly, NG III exhibited an average I_{sc} of ~4 nA. Once NG III was added to the parallel connection with NG I and NG II (inset of Fig. 19.10f), their total output current was the sum of the three NGs (5.9 nA ≈ 0.7 + 1 + 4 nA), as shown in Fig. 19.10f. This series of experiments demonstrated that the short-circuit current can be effectively improved by parallel-connecting multiple NGs. This sets the platform for developing three-dimensionally stacked NGs.

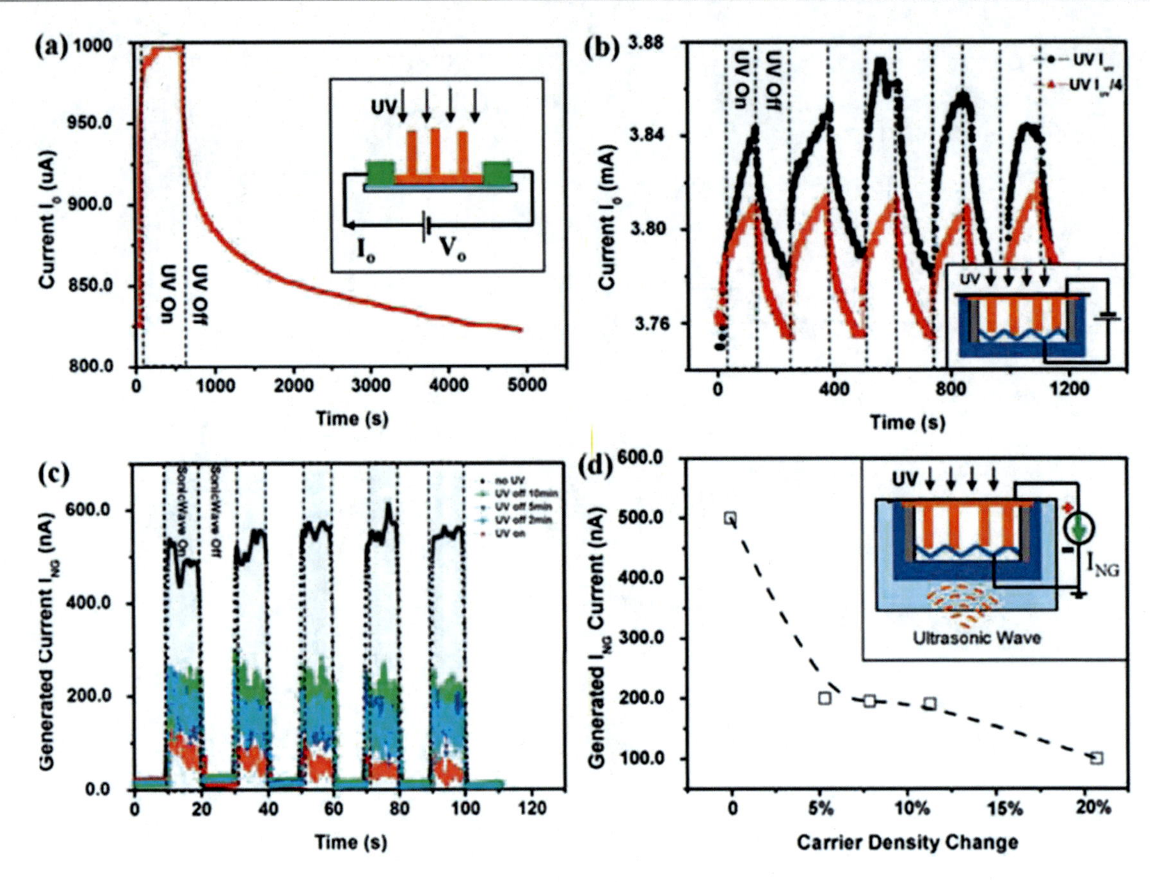

Fig. 19.11 Tuning the transport property of ZnO and an assembled NG by UV excitation. (**a**) Response of the current transported through a thin ZnO film with NWs on top when subjecting to UV illumination. The *inset* is the measurement setup. (**b**) Current transported through an assembled NG when the UV light was tuned on and off. The *black* and *gray* curves represent the response curve at full UV intensity I_{UV} and $I_{UV}/4$, respectively. The *inset* is the measurement setup. (**c**) Current generated by the NG when the ultrasonic wave was turned on and off under different illumination conditions of UV light. (**d**) Plot of the NG-generated current as a function of the increased percentage of carrier density in the NWs. The *inset* is the measurement setup

19.4.3 Performance Analysis: The Carrier Density

The operation mechanism of NG requires that the conductivity and carrier density of the ZnO NW are adequately low in the first step to preserve the piezoelectric-potential distribution in the NW from being "neutralized" by the freely flowing charge carriers, which are electrons for n-type ZnO, but they need to be high enough to transport the current under the driving of the piezoelectric potential in the charge releasing process [14]. To evaluate the performance of NG in response to carrier density, two experiments have been carried out. Figure 19.11a shows the UV-induced carrier density change of the vertically aligned ZnO NWs and the underlying ZnO thin film by connecting two electrodes at the two ends of the film, as shown in the inset. By applying a constant voltage V_0, the transport current was monitored when the UV was turned on or off. The conductance of the film responded rapidly to UV and reached saturation after ~300 s. However, the current decayed very slowly after the UV was turned off, possibly due to the following reason. The UV light can effectively create electron–hole pairs in ZnO and greatly increase the carrier density. After turning off the UV, the electrons and holes would recombine and thus decrease the conductivity. However, some carriers may be trapped in the surface/vacancy states in ZnO, which greatly delayed the electron–hole recombination rate, possibly resulting in a long decay in conductivity.

The long-lasting tail in the recovery of conductivity in Fig. 19.11a may possibly indicate that ZnO can hold the charges for an extended period of time before they are completely and entirely neutralized/screened by external electrons. One may suggest that the piezoelectric charges and potential created in a NW may hold for an equivalent length of time before being neutralized/screened by external electrons, so that it can effectively gate the charge carriers as proposed for nanopiezotronics [15], giving sufficient time for experimental observation of the effect of the piezoelectric potential across a NW. Such length of time may be even longer than tens to hundreds of seconds.

The second experiment was to characterize the UV-induced carrier density change of an assembled NG, as shown in the inset in Fig. 19.11b. By reducing the relative UV intensity from I_{UV} to $\sim I_{UV}/4$, the conductance of the NG changed by ~2%. The conductivity also sharply responded to the on and off of the UV, clearly showing that the UV can effectively increase the carrier density in the NG.

After observing the increased conductivity in NG under UV irradiation, we investigated the performance of the NG for generating electricity. Figure 19.11c is the generated current (I_{NG}) of a NG when subject to ultrasonic wave excitation by changing the UV irradiating condition. When the UV light was off for an extended period of time, the generated current was ~500 nA for a NG with 6 mm^2 in size, corresponding to an output current density of 8.3 μA/cm^2. The key to obtaining a high output current density is to have NWs that are uniform in size and with patterned distribution on substrate that matches the design of the electrode. As soon as the UV light was turned on, the generated current dropped to as low as 30–80 nA. At the same time, we monitored the I_{NG} after the UV was turned off for 2, 5, and 10 min, and the generated current increased to ~180, 190, and 205 nA, respectively. Such a gradual increase in the generated current corresponds to the slow decay in conductivity as shown in Fig. 19.11b. This set of experiments apparently shows that an increase in conductivity, e.g., carrier density, reduces the output current of the NG. By extracting the relative change in carrier density from Fig. 19.11a and the correspondingly measured NG output I_{NG} from Figs. 19.11c, 19.11d illustrates the monotone decrease relationship between the increased percentage in carrier density and the output current I_{NG}. This result indicates that the carrier density in ZnO is one of the key characteristics that dictate the output power of a NG. A too high carrier density would screen the piezoelectric charges, resulting in a lower or even vanishing output current. However, a too low-carrier density would increase the inner resistance and greatly decrease the output current. Therefore, there should be an optimum choice of NW carrier density or conductivity to maximize the NG output.

19.4.4 Performance Analysis: The Schottky Barrier

The Schottky contact between the metal contact and ZnO NW is another key factor to the current generation process [14]. Investigating how it affects the performance of NG will provide effective guidance in designing and fabricating high-output NGs. We characterized the *I–V* characteristic of a set of NGs to illustrate the effects of Schottky barrier for current generation. Figure 19.12a shows the *I–V* characteristics of a NG that did generate output current (top inset of Fig. 19.12a). The experimental setup for the measurement is shown in the inset at the lower right corner in Fig. 19.12a. The NG exhibits a Schottky-like *I–V* characteristic, which means that the NG has different responses for forward and reverse biases. When being illuminated by UV, the NG showed an increased conductance, and its Schottky-like rectifying effect was largely reduced due to the increased conductance. Alternatively, for a "defective" NG that did not produce current in responding to ultrasonic wave, its *I–V* characteristic was clearly ohmic even prior to UV illumination (Fig. 19.12b). By examining over 10 NGs, their *I–V* characteristics reliably indicated whether the NGs were effective for producing current. This study establishes a criterion for identifying the working NG from the defective ones.

To examine the role played by the Schottky barrier in the NG, we used the AFM-based manipulation and measurement system as used in our first study for demonstrating the piezoelectric NG3 (see Fig. 19.13a). When a 100 nm thick Pt-coated Si tip was used for scanning the NWs in contact mode, voltage peaks were observed (Fig. 19.13b), and the output voltages were in the order of ≃11 mV (the negative sign means that the current flowed from the grounded end through the external load). By changing the tip to an Al–In (30 nm/30 nm) alloy-coated Si tip, the ZnO NWs showed no piezoelectric output (Fig. 19.13c). In order to understand the two distinct performances of the two types of scanning tips, we have measured their corresponding *I–V* characteristics with the ZnO NWs. To ensure the stability of the contact, we used a large electrode that was in contact with a group of NWs, as shown in Fig. 19.13d. The Pt–ZnO contact clearly presented a Schottky diode (Fig. 19.13e), whereas the Al–In alloy–ZnO was an ohmic contact. In reference to the piezoelectric output presented in Fig. 19.13b, c, we conclude that a Schottky contact between the metal electrode and ZnO is a must for a working NG.

19.4.5 Output Improvement

The amplitude of the NG's output is largely dependent on the number of NWs that involves in the power conversion. In order to maximize the number of NWs that would contribute to the power generation process while minimizing the parasitic capacitance that could bring down the voltage output, the NW arrays were carefully chosen so that the variance of the average length of the NW was minimal [16].

The NG was then placed in the water bath to measure the closed-circuit current and open-circuit voltage. The ultrasonic wave generator was periodically turned on every other 15 s. Measurements were taken under both connection polarities to rule out any possible artifact caused by the measurement setup. Figure 19.14a shows the closed circuit current when the ultrasonic wave was turned on and off. The

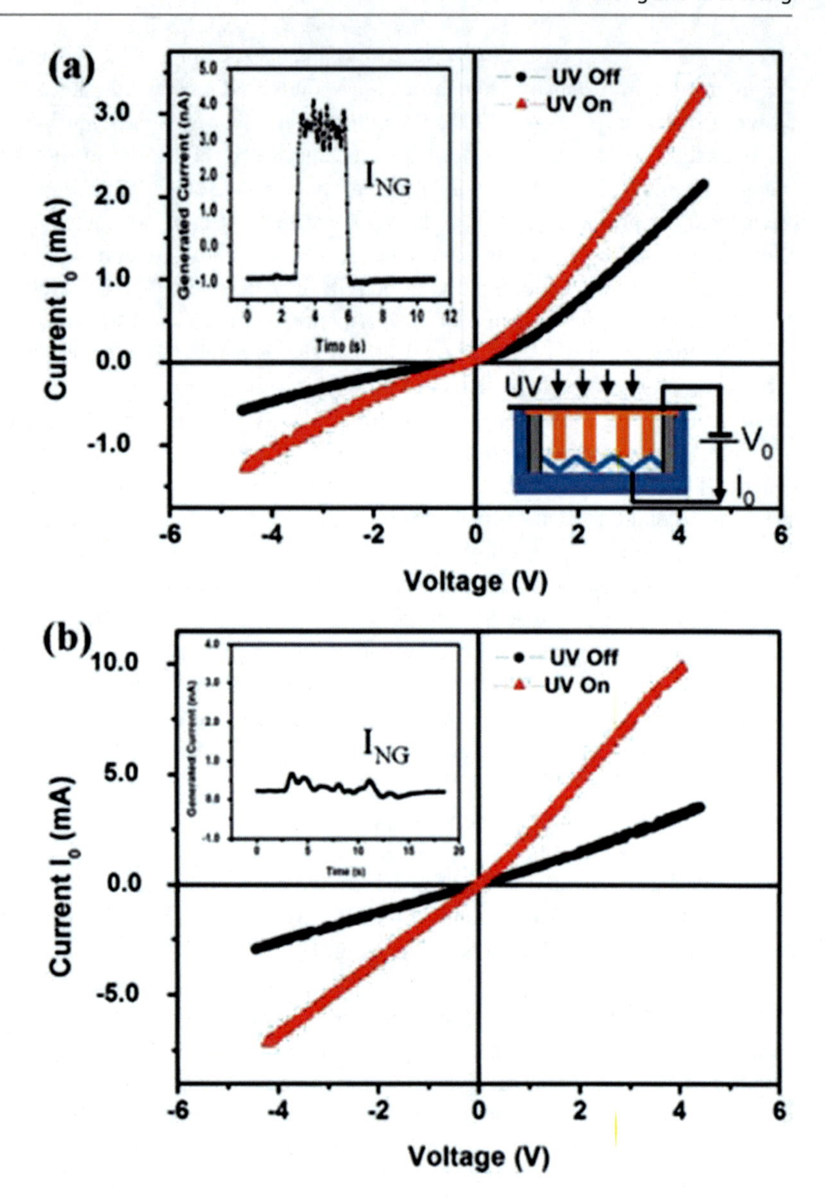

Fig. 19.12 *I–V* characteristic of an assembled NG for identifying its performance for producing current. (**a**) For a NG that is actively generating current, as shown in the *inset*, its *I–V* curve when the ultrasonic wave was off clearly shows a Schottky diode behavior. UV light not only increased the carrier density (or conductivity) but also might reduce the barrier height. (**b**) For a "defective" NG that did not produce current, the *I–V* curve shows a clearly ohmic behavior. The UV light clearly increased the conductivity

data clearly indicate that the current output was originated from the NG as a result of ultrasonic wave excitation, as the output coincided with the working cycle of the ultrasonic wave generator. Furthermore, the output signal switched in sign from positive to negative when the measurement polarity was switched from forward to reversed connection. Finally, there is no destabilization of the output amplitude over time in the graph. A similar pattern in the open-circuit voltage output was also observed, as shown in Fig. 19.14b. Both the current and the voltage outputs exhibit high levels for this type of NG, with a current of ~500 nA and voltage of ~10 mV. Considering the effective area of the NG (6 mm^2), it is equivalent to a current generation density of ~8.3 μA/cm^2, which is ~20 times higher than previously reported [13]. A power generation density of ~83 nW/cm^2 is reported, which shows a great potential to power nanosensors [17].

However, even the highest output signal was still more than 10 times smaller than the output voltage calculated by the bending theory of a single NW. To understand the reason, we need to keep in mind that a realistic NG cannot be simply treated as a single voltage source but has to be treated as multiple sources along with a capacitance and an internal resistance. As Fig. 19.15 shows, each voltage source V_1, V_2, ... V_n corresponds to one of the NWs that generates electric power. For each of the.

NWs, the output consists of multiple pulse outputs:

$$V_i(t) = \sum_{t_{p,i}} V * \left(t - t_{p,i}\right) \tag{19.2}$$

where $V^*(t)$ is a pulse generated by one instance of power release process between the NW and the top electrode. $t_{p,i}$ is

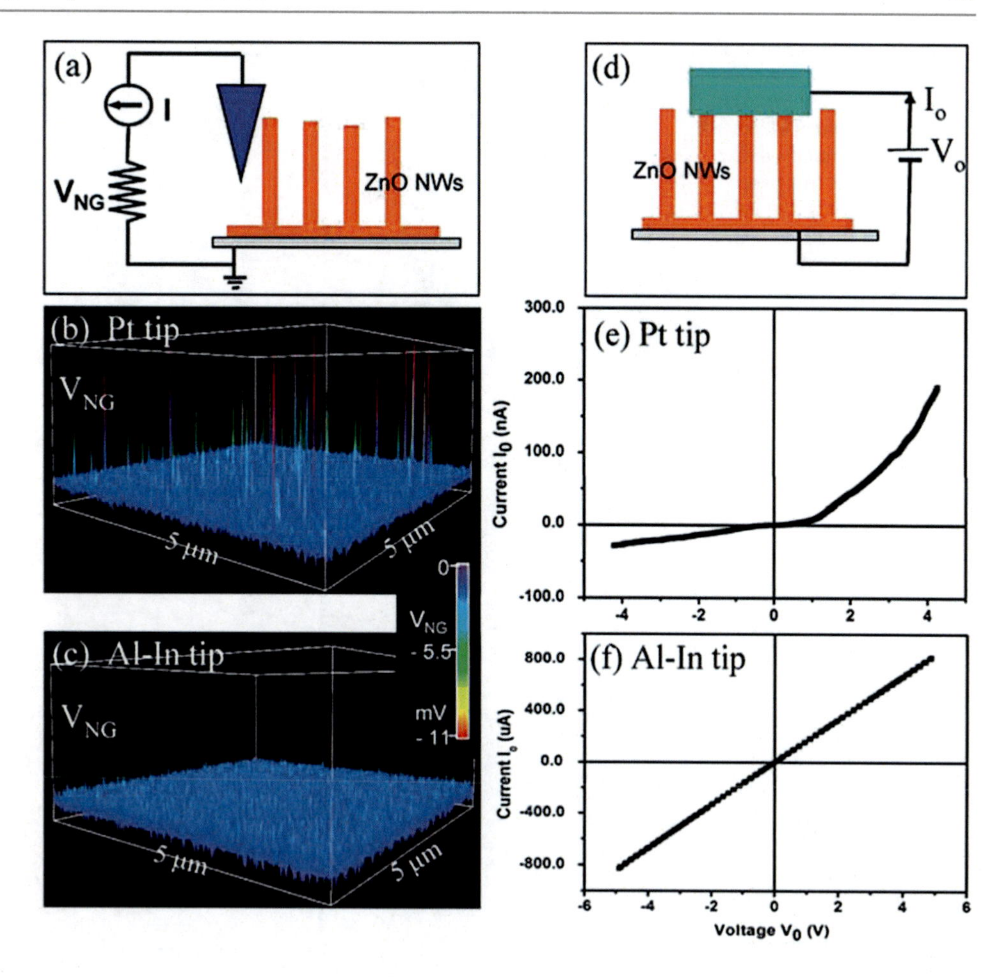

Fig. 19.13 (**a**) AFM-based measurement setup for correlating the relationship between the metal–ZnO contact and the NG output. (**b**) Output potential generated by ZnO NW array when scanned by a Pt-coated Si tip. (**c**) No output potential is generated by ZnO NW array when scanned by an Al–In coated Si tip. (**d**) Experimental setup for characterizing the *I–V* transport property of metal–ZnO NW contact. (**e**) *I–V* curve of a Pt–ZnO NW contact, showing Schottky diode effect. (**f**) *I–V* curve of an alloyed Al/In-ZnO NW contact, showing ohmic behavior

the time when such a pulse p happens for the ith NW. $V^*(t)$ can be approximated as a Gauss package:

$$V * (t) = V_0 \exp\left(-\frac{t^2}{2\tau^2}\right) \tag{19.3}$$

where $V_0 = 0.3$ V is the output of a single pulse, as calculated with the bending theory of single NWs, and τ is the width of each output pulse. When driven by ultrasonic waves at ~40 kHz, τ can be as small as 0.1 µs.

The internal resistance is due to the leakage of electric current through the NWs from the substrate to the top electrode. Suppose a single NW has an average resistance of $R_1 = R_2 = \cdots = R_n = R_0$, experiment shows that $R_0 \sim 10$ MΩ. r is the resistance due to the nanowires that do not generate power. Those NWs that touch the electrode thus give a bypass to the output. Experiment shows that $r \sim 100$ kΩ.

The capacitance between the top electrode and the substrate is

$$C = \varepsilon_{\text{Eff}} \frac{A}{d} \tag{19.4}$$

where A is the area of the electrode and the d is the spacing between the top electrode and substrate. ε_{Eff} is the effective permittivity, which can be approximated to be the permittivity of the vacuum as an estimation. Plugging $\varepsilon_{\text{Eff}} = \varepsilon_r \varepsilon_0 = 7.77 \times 8.85 \times 10^{-12}$ F/m, $A = 2$ mm^2, $d = 1$ µm, we have $C = 0.14$ nF.

Now the timescale of the R–C impedance circuit is

$$\tau_{\text{RC}} \sim rC = 100\ \text{k}\Omega \times 0.14\ \text{nF} = 14\ \mu\text{S} \tag{19.5}$$

In solving the system in Fig. 19.15, we make three assumptions:

(A) The resistance r is a constant that does not change with time. Experimentally, the fluctuation of r is observed very small, meaning that the number of NWs touching the top electrode does not change very much when sonicating.

(B) The change of the capacitance C can be neglected. In the realistic situation, C does change with time; but such a change does not give rise to a significant effect in a NG application. As a simple theory illustrating how the output voltage is decreased, we do not consider the variation of the capacitance C.

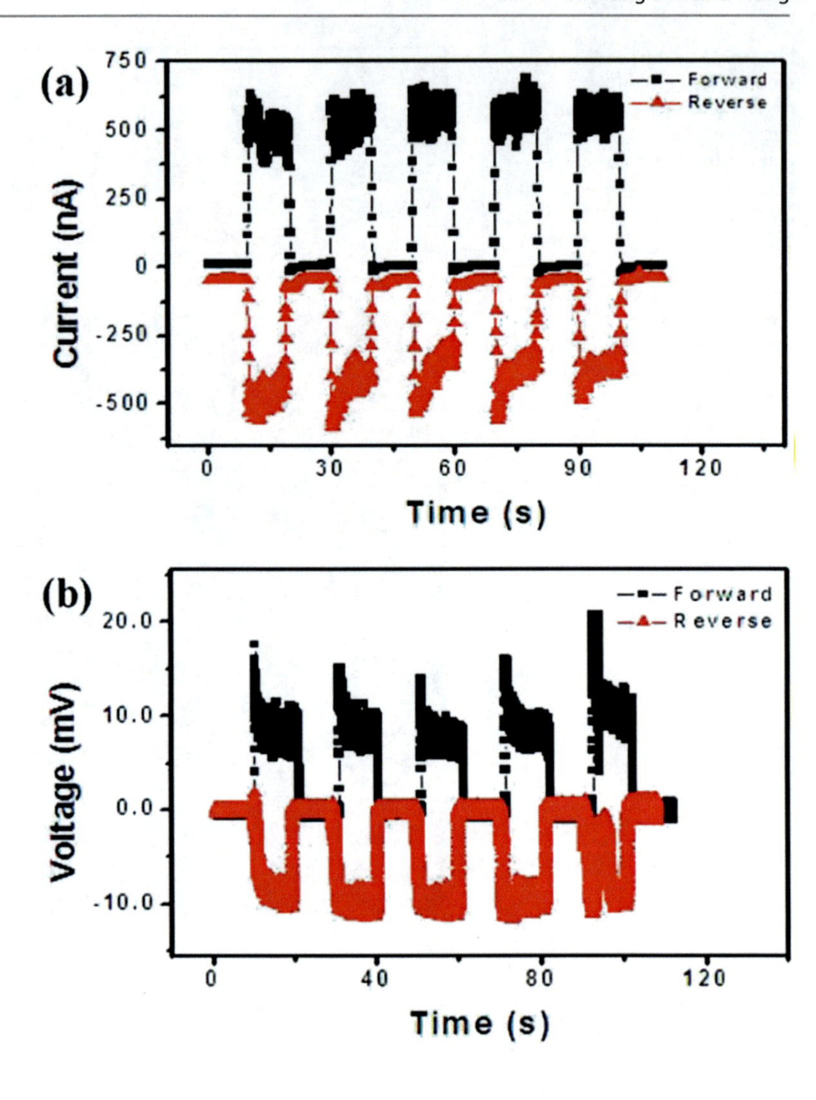

Fig. 19.14 Performance of a high-output NG when periodically excited by ultrasonic wave. (**a**) Closed-circuit current output and (**b**) open-circuit voltage output measured at forward polarity (*black line*) and reversed polarity (*gray line*) connection with the measurement system

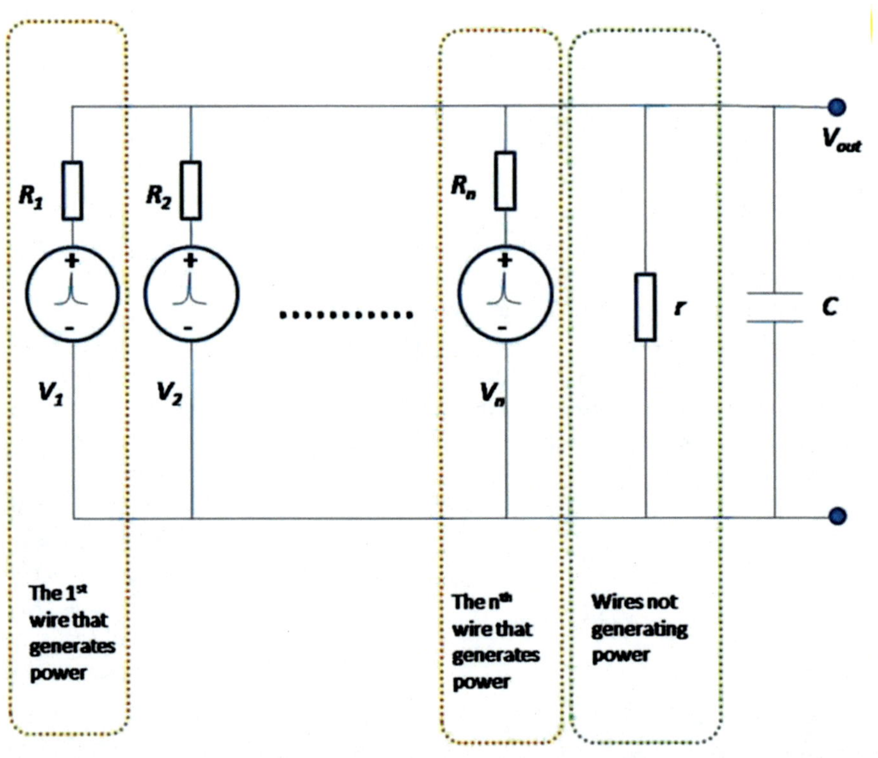

Fig. 19.15 Equivalent circuit of a realistic nanogenerator

(C) The NWs do not interfere with each other. That is, the *i*th wire and *j*th wire work independently. Therefore, we can first calculate the output V_{out} due to each NW and then linearly sum up the contributions of all the NWs.

By Kirchoff's law and the above assumptions, we integrate the output from a single pulse of a single NW; the output voltage is found to be

$$V_{out}^{total}(t) = ri_r^{total}(t) = \sum_i \sum_{t_{p,i}<t} \frac{\sqrt{2\pi} V_0 \tau}{R_0 C} \exp\left(-\frac{t - t_{p,i}}{\tau_{RC}}\right) \quad (19.6)$$

The power generation process of each NW can be assumed as a Poisson-type random process. Instead of solving the random process rigorously, we are only interested in the averaged DC output of the system. Now let us assume that there are n NWs generating power. For each of the NWs, it takes an average time of T between every two instances of power release, namely, the period of each generation process is T. Taking a long time $T_1 \gg T$ to calculate the average, it is found that the average voltage output is

$$\overline{V_{out}^{total}} = \sqrt{2\pi} \frac{\tau}{T} \frac{n}{n'} V_0 \quad (19.7)$$

where n' is the total number of NWs in a NG.

Equation (19.7) shows that we need to improve the ratio n/n' in order to improve the output voltage. Increasing τ/T would also improve the performance, which means that we should increase the time during which the power release happens or shorten the generation period. Both of these two ways to improve the performance require better control over the device morphology.

19.5 Flexible Nanogenerator

19.5.1 ZnO Nanowires on Flexible Substrate

NGs presented previously in this chapter were all based on ZnO NWs that were grown on ceramic and semiconducting substrates, which are hard and brittle and cannot be used in applications that require a foldable or flexible power source. By growing ZnO NW arrays on a flexible plastic substrate, we demonstrate the first successful flexible power source built on conducting polymer films [18]. This approach has two specific advantages: it uses a cost-effective, large-scale, wet-chemistry strategy to grow ZnO NW arrays at temperatures lower than 80 °C, and the growth of aligned ZnO NW arrays can occur on a large assortment of flexible plastic substrates.

The aligned ZnO NWs arrays were grown on a plastic film using a solution-based method [19]. Serving as one of the electrodes for a later electrical connection and also functioning as uniform nucleation sites for NW growth, a thin, ca. 100 nm, layer of Au was deposited on the plastic substrate before the hydrothermal growth of ZnO NW arrays. Through this method, aligned ZnO NWs can be grown uniformly covering an area of 2 inch plastic wafer (Fig. 19.16a) or even larger. In order for the manipulation of AFM, the density of ZnO NWs was controlled to be low by adjusting the reaction temperature, solution concentration, and pH value. Figure 19.16b shows a SEM image of sparsely grown ZnO NW arrays on a plastic substrate. The NWs have a density of ca. 1 μm^{-2}.

A similar AFM scanning process was applied on the NW arrays, and the voltage signal was recorded simultaneously. Figure 19.16c is a 3D plot of a topographic AFM scan of a 40 × 40 μm area of the ZnO NW array. The scanning direction is from the front to the back, which can be seen from the raised linear traces on the flexible plastic substrate. Because of the flexibility of the plastic substrate and in spite of the firm adhesion to the Ag paste, the substrate surface profile under tip contact was somewhat wavy. Relative to the plastic substrate, the NW array was revealed to have a height distribution of 0.5–2.0 μm. The corresponding 3D plot of the voltage output image is shown in Fig. 19.16d. There are a number of sharp peaks ranging from 15 to 25 mV that represent voltage outputs. By counting the pulse numbers shown here with respect to the number of NWs in the topography profile in Fig. 19.16c, the ratio of voltage peaks to the number of available NWs is 90:150 or about 60%, which suggests that the discharge events of the NWs captured by the AFM tip correspond to at least 50% of the NWs. In fact, because of the limitation in the data-collecting speed of the AFM, which has a data step size of ca. 156.2 nm in this measurement, the discharge peaks of some NWs were missed, possibly due to poor contact between the tip and the NWs and/or due to multiple contacts with neighboring NWs. It is suggested that both suitable bonding strength between the ZnO NWs and the polymer substrate and a uniform density distribution of NWs in the array might be very important in terms of improving the piezoelectric discharge efficiency.

19.5.2 Power Fiber

Fabric is another type of widely used soft material, which can be an excellent media for harvesting mechanical energy. By growing ZnO NWs around Kevlar 129 fibers, which have high strength, modulus, toughness, and thermal stability, we demonstrated the power generation ability of a textile fiber system [20]. The preliminary research was based on a

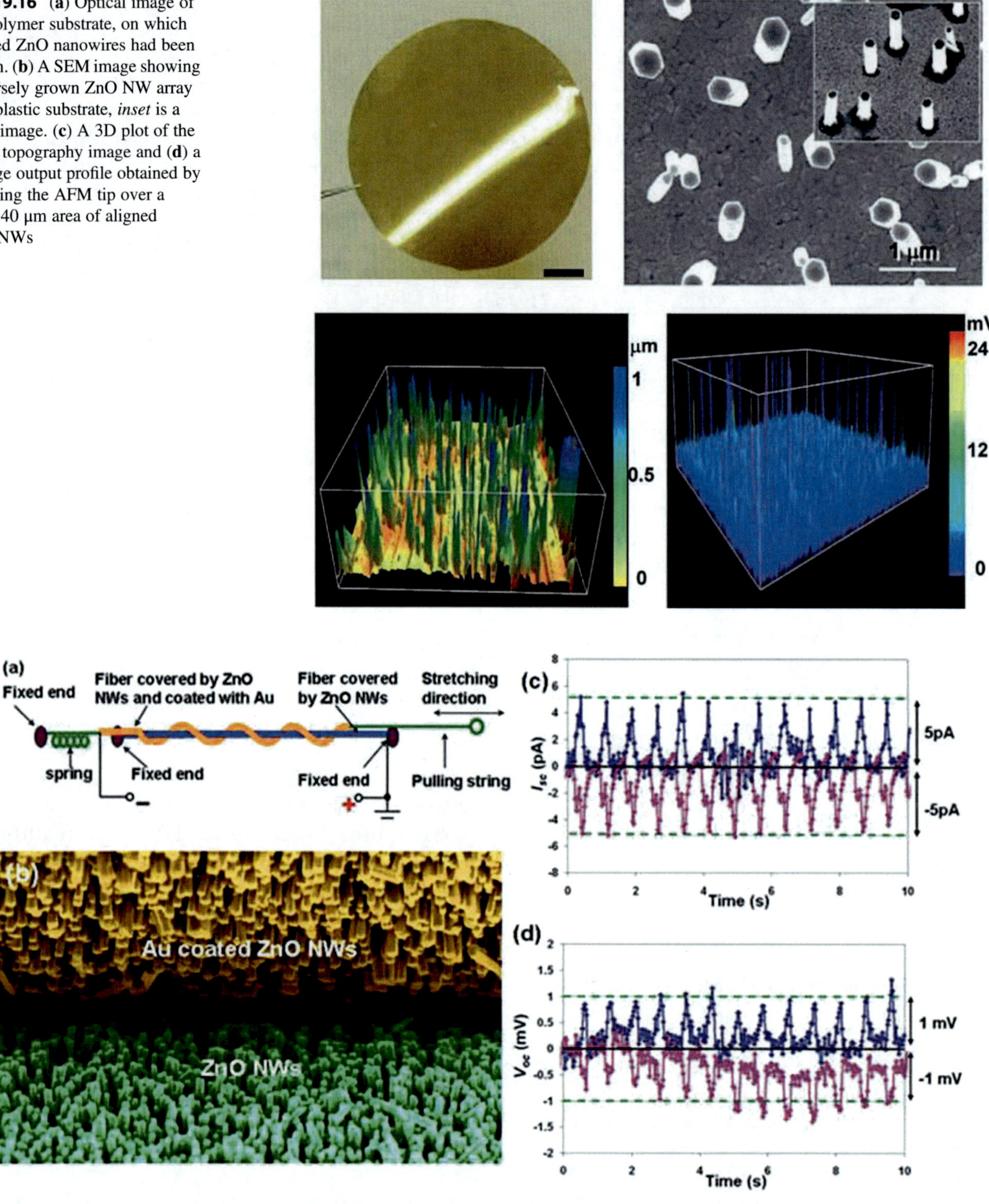

Fig. 19.16 (**a**) Optical image of the polymer substrate, on which aligned ZnO nanowires had been grown. (**b**) A SEM image showing a sparsely grown ZnO NW array on a plastic substrate, *inset* is a tilted image. (**c**) A 3D plot of the AFM topography image and (**d**) a voltage output profile obtained by scanning the AFM tip over a 40 × 40 μm area of aligned ZnO NWs

Fig. 19.17 (**a**) Schematic experimental setup of the fiber-based NG. (**b**) SEM image at the "teeth-to-teeth" interface of two fibers covered by NWs with/without gold. (**c**) and (**d**) I_{sc} and V_{oc} of a double-fiber NG

double-fiber model as shown in Fig. 19.17a. Two fibers, one coated with a 300-nm-thick gold layer and the other as-grown, were entangled to form the core for power generation. The relative brushing between the two fibers was simulated by pulling/releasing the string using an external rotor with a controlled frequency. In this design, the gold-coated ZnO NWs acted as an array of scanning metal tips that deflected the ZnO NWs rooted at the other fiber; a coupled piezoelectric and semiconducting property resulted in a process of charge creation and accumulation and charge release. The gold coating completely covered the ZnO NWs and formed a continuous layer along the entire fiber. Once the two fibers were firmly entangled together, some of the gold-coated NWs penetrated slightly into the spaces between the uncoated NWs rooted at the other fiber, as shown by the interface image in Fig. 19.17b. Thus, when there was a relative sliding/deflection between them, the bending of the uncoated ZnO NWs produced a piezoelectric potential across their width, and the Au-coated NWs acted as the "zigzag" electrode (as for the DC NG) for collecting and transporting the charges.

The short-circuit current (I_{sc}) and open-circuit voltage (V_{oc}) were measured to characterize the performance of the fiber NGs. The pulling and releasing of the gold-coated fiber were accomplished by a motor at a controlled frequency. The resulting electric signals detected at 80 r.p.m. are shown in Fig. 19.17c, d. A "switching polarity" testing method was applied during the entire measurement to rule out system artifacts. When the current meter was forward-connected to the NG, which means that the positive and negative probes were connected to the positive and negative electrodes of the NG as defined in Fig. 19.17a, respectively, ~5 pA-positive current pulses were detected at each pulling-releasing cycle (blue curve in Fig. 19.17c). Negative current pulses with the same amplitude were received (pink curve in Fig. 19.17c) when the current meter was reverse-connected, with its positive and negative probes connected to the negative and positive electrodes of the NG, respectively. The small output current (~5 pA) is attributed mainly to the large loss in the fiber due to an extremely large inner resistance ($R_i \approx 250$ MΩ). Open-circuit voltage was also measured by the switching polarity method. Corresponding positive and negative voltage signals were received when the voltage meter was forward- and reverse-connected to the fiber NG, respectively (Fig. 19.17d). The amplitude of the voltage signal in this case was ~1–3 mV

After demonstrating the electricity generation principle, a few approaches were investigated to increase the output power and prototype integration. To simulate a practical fabric made of yarn, a single yarn made of six fibers was tested; three fibers were covered with NWs and coated with Au, and three were only covered with NWs. All of the gold-coated fibers were movable in the testing (Fig. 19.18b). At a motor speed of 80 r.p.m., an average current of ~0.2 nA was achieved (Fig. 19.18a), which is ~30–50 times larger than the output signal from a single-fiber NG: this is due to the substantially increased surface contact area among the fibers. The width of each pulse was broadened, apparently due to the unsynchronized movement and a relative delay in outputting current among the fibers.

Reducing the inner resistance of the fiber and the NWs was found to be effective for enhancing the output current. By depositing a conductive layer directly onto the fiber before depositing the ZnO seeds, the inner resistance of the NG was reduced from ~1 GΩ to ~1 kΩ; thus the output current I_{sc} of a double-fiber NG was increased from ~4 pA to ~4 nA (Fig. 19.18c). The current I_{sc} is approximately inversely proportional to the inner resistance of the NG (inset in Fig. 19.18c). This study shows an effective approach for increasing the output current.

The textile-fiber-based NG has demonstrated the following innovative advances in comparison with the DC NGs. First, using ZnO NWs grown on fibers, it is possible to fabricate flexible, foldable, wearable, and robust power sources in any shape (such as a "power shirt"). Second, the output electricity can be dramatically enhanced using a bundle of fibers as a yarn, which is the basic unit for fabrics. The optimum output power density from textile fabrics can be estimated on the basis of the data we have reported, and an output density of 20–80 mW per square meter of fabric is expected. Third, the NG operates at low frequency, in the range of conventional mechanical vibration, footsteps, and heartbeats, greatly expanding the application range of NGs. Last, as the ZnO NW arrays were grown using chemical synthesis at 80 °C on a curved substrate, we believe that our method should be applicable to growth on any substrate, so the fields in which the NGs can be applied and integrated may be greatly expanded. This work establishes a methodology for scavenging light-wind energy and body-movement energy using fabrics.

19.6 Prospect

Relying on the unique piezoelectric and semiconducting coupled property, ZnO NWs have been successfully applied for harvesting nanoscale mechanical energy and converting it into electricity. The NGs have been made on solid substrates for harvesting high-frequency acoustic wave or on flexible soft materials that are able to harvest low-frequency perturbations, such as footsteps, wind blows, or even heartbeats. The development of NGs exhibits unique advantages: first, the NW-based NGs can be subjected to extremely large deformation, so they can be used for flexible electronics as a flexible/foldable power source. Second, the large degree of deformation that can be withstood by the NWs is likely to result in a larger volume density of power output. Third, ZnO NW NGs can directly produce current as a result of their enhanced conductivity with the presence of oxygen vacancies. Fourth, the flexibility of the polymer substrate used for growing ZnO NWs/NBs makes it feasible to accommodate the flexibility of human muscles so that we can

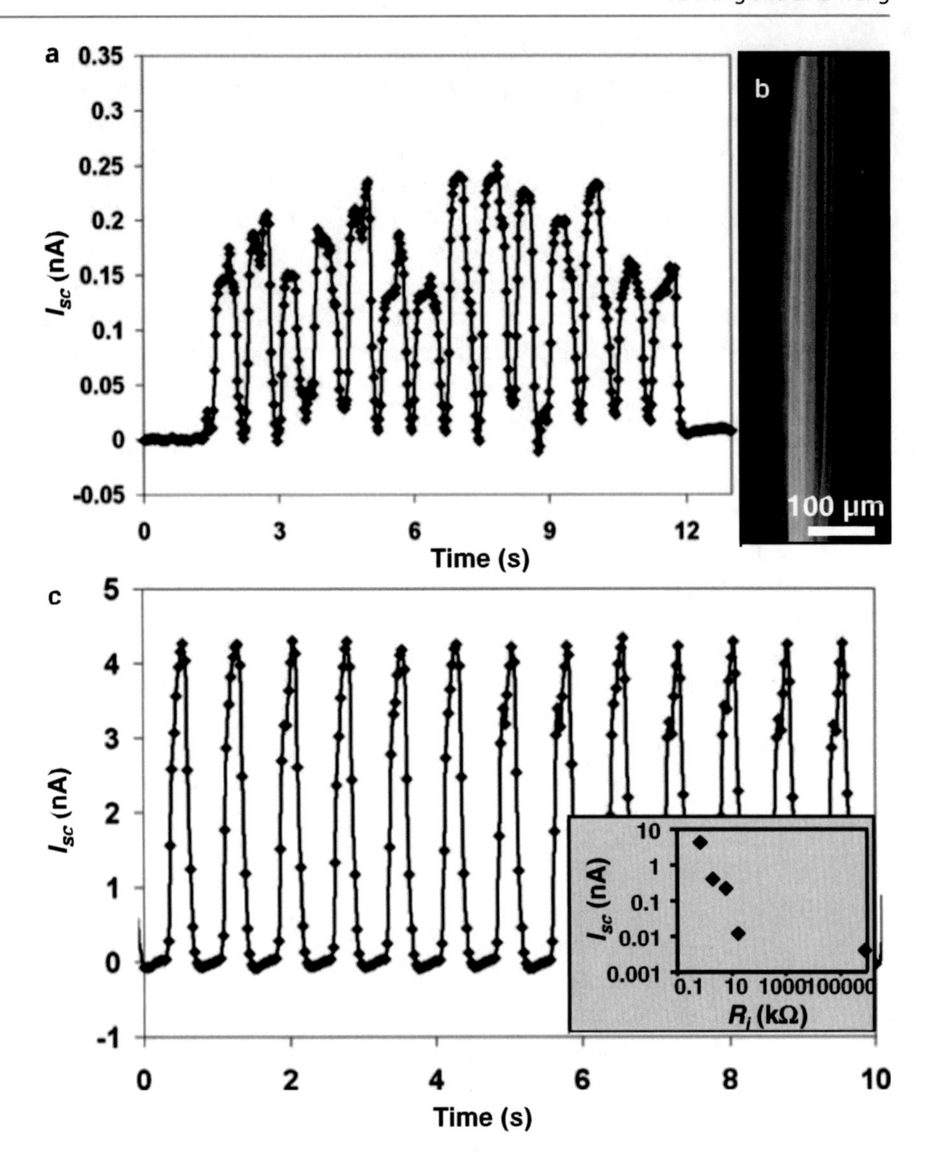

Fig. 19.18 (**a**) I_{sc} of a multi-fiber NG. (**b**) SEM image of the multi-fiber NG composed of six ZnO NW-coated fibers, three of which were gold-coated. (**c**) Enhancement of I_{sc} by reducing the inner resistance of the NG

use the mechanical energy (body movement, muscle stretching) in the human body to generate electricity. Finally, ZnO is a biocompatible and biosafe material [21]; it has great potential as an implantable power source within the human body.

The following development will be focused on improving the output power of a current NG model, while a full article addressing some concerns about NG's performance [22] can be found elsewhere [23]. Combining with the existing knowledge on the relationship between the ZnO NW arrays and the performance of NG, we summarize the key factors that are important to boost up the voltage output as follows:

(A) Increase the number of active NWs for participating in electricity generation. This is crucial in determining the voltage output performance. There are two possible approaches. One is to utilize ZnO NW arrays with uniform size, especially uniform in length. The other approach is to pattern the NW arrays according to the dimension and shape of the top electrode. An increase in output voltage will be accompanied with a simultaneous increase in output current.

(B) Increase the charges generated by individual NW during the deflection process. This may be possible by increasing the magnitude of the external excitation, because the magnitude of the generated voltage is proportional to the deflection of the NW.

(C) Increase the total charge to be the output by a NW to the external load. This requires a great decrease in the contact resistance between the metal electrode and the NW. In a general NG device, the sum of the contact resistance and inner resistance of the wire was estimated to be ∼35 MΩ. This large resistance dissipates a large voltage at the contact. It is thus essential to reduce the contact resistance to receive larger output voltage.

(D) Optimize the electric conductivity of ZnO NW. The relationship between the carrier density and the performance of the NG indicates that a too high conductivity destroys the Schottky contact at the interface, while a too low conductivity consumes too much voltage. It is expected that by tuning the carrier density to an optimal value, both the current and the voltage outputs can be significantly increased. However, the optimum value of the conductivity needs to be modeled with consideration of the charge transport dynamics.

(E) Decrease the capacitance between individual NW and the top electrode, as well as the system capacitance. The size of the electrodes can make large contribution to the system capacitance. The other factor is the capacitance of the measurement circuit.

By addressing the above challenges, we expect to have >80% NWs involved in electricity generation. Under this ideal situation, the power density per unit substrate area is $\sim$1–2 nW/μm^2, i.e., 0.1–0.2 W/cm^2, which is large enough for powering regular small electronic devices.

Acknowledgments This work was supported by DOE BES, NSF, DARPA, and NASA. Thanks to the contributions from our group members: Dr. Jinhui Song, Dr. Yong Qin, Dr. Jin Liu, Dr. Jun Zhou, Dr. Rusen Yang, Yifan Gao, Peng Fei, Dr. Puxian Gao, Dr. Jr-Hau He, Dr. Changshi Lao, Dr. Yi-Feng Lin, and Wenjie Mai.

References

1. Song, J.H., Zhou, J., Wang, Z.L.: Piezoelectric and semiconducting coupled power generating process of a single ZnO Belt/wire. A technology for harvesting electricity from the environment. Nano Lett. **6**, 1656–1662 (2006)
2. Wang, X.D., Summers, C.J., Wang, Z.L.: Large-scale hexagonal-patterned growth of aligned ZnO nanorods for nano-optoelectronics and nanosensor arrays. Nano Lett. **4**, 423–426 (2004)
3. Wang, X.D., Song, J.H., Summers, C.J., Ryou, J.H., Li, P., Dupuis, R.D., Wang, Z.L.: Density- controlled growth of aligned ZnO nanowires sharing a common contact: a simple, low- cost, and mask-free technique for large-scale applications. J. Phys. Chem. B. **110**, 7720–7724 (2006)
4. Wang, Z.L., Song, J.H.: Piezoelectric Nanogenerators based on zinc oxide nanowire arrays. Science. **312**, 242–246 (2006)
5. Song, J.H., Wang, X.D., Riedo, E., Wang, Z.L.: Elastic property of vertically aligned nanowires. Nano Lett. **5**, 1954–1958 (2005)
6. Lin, Y.F., Song, J.H., Ding, Y., Liu, S.Y., Wang, Z.L.: Piezoelectric nanogenerator using CdS nanowires. Appl. Phys. Lett. **92**, 022105 (2008)
7. Cao, B., Jiang, Y., Wang, C., Wang, W., Wang, L., Niu, M., Zhang, W., Li, Y., Lee, S.-T.: Synthesis and lasing properties of highly ordered CdS nanowire arrays. Adv. Funct. Mater. **17**, 1501–1506 (2007)
8. Wang, Z.L., Wang, X.D., Song, J.H., Liu, J., Gao, Y.F.: Piezoelectric nanogenerators for self- powered nanodevices. IEEE Pervasive Computing. **7**, 49–55 (2008)
9. Gao, Y.F., Wang, Z.L.: Electrostatic potential in a bent piezoelectric nanowire. The fundamental theory of nanogenerator and nanopiezotronics. Nano Lett. **7**, 2499–2505 (2007)
10. Wang, X.D., Song, J., Li, P., Ryou, J.H., Dupuis, R.D., Summers, C. J., Wang, Z.L.: Growth of uniformly aligned ZnO nanowire heterojunction arrays on GaN, AlN, and Al0.5Ga0.5 N substrates. J. Am. Chem. Soc. **127**, 7920–7923 (2005)
11. Wang, X.D., Song, J.H., Liu, J., Wang, Z.L.: Direct-current nanogenerator driven by ultrasonic waves. Science. **316**, 102–105 (2007)
12. Frühauf, J., Krönert, S.: Wet etching of silicon gratings with triangular profiles. Microsyst. Technol. **11**, 1287–1291 (2005)
13. Wang, X.D., Liu, J., Song, J.H., Wang, Z.L.: Integrated nanogenerators in biofluid. Nano Lett. **7**, 2475–2479 (2007)
14. Liu, J., Fei, P., Song, J.H., Wang, X.D., Lao, C.S., Tummala, R., Wang, Z.L.: Carrier density and Schottky barrier on the performance of DC nanogenerator. Nano Lett. **8**, 328–332 (2008)
15. Wang, Z.L.: Nanopiezotronics. Adv. Mater. **19**, 889–892 (2007)
16. Liu, J., Fei, P., Zhou, J., Tummala, R., Wang, Z.L.: Toward high output-power nanogenerator. Appl. Phys. Lett. **92**, 173105 (2008)
17. Yu, C., Hao, Q., Saha, S., Shi, L., Kong, X., Wang, Z.L.: Integration of metal oxide nanobelts with microsystems for nerve agent detection. Appl. Phys. Lett. **86**, 063101 (2005)
18. Gao, P.G., Song, J.H., Liu, J., Wang, Z.L.: Nanowire piezoelectric nanogenerators on plastic substrates as flexible power sources for nanodevices. Adv. Mater. **19**, 67–72 (2007)
19. Vayssieres, L.: Growth of arrayed nanorods and nanowires of ZnO from aqueous solutions. Adv. Mater. **15**, 464–466 (2003)
20. Qin, Y., Wang, X.D., Wang, Z.L.: Microfibre-nanowire hybrid structure for energy scavengings. Nature. **451**, 809–813 (2008)
21. Zhou, J., Xu, N.S., Wang, Z.L.: Dissolving behavior and stability of ZnO wires in biofluids: a study on biodegradability and biocompatibility of ZnO nanostructures. Adv. Mater. **18**, 2432–2435 (2006)
22. Alexe, M., Senz, S., Schubert, M.A., Hesse, D., Gosele, U.: Energy harvesting using nanowires? Adv. Mater. **20**, 4021 (2008)
23. Wang, Z.L.: Energy harvesting using piezoelectric nanowires – a correspondence on 'energy harvesting using nanowires?' By Alexe et al. Adv. Mater. **21**, 1311–1315 (2009)

1D Nanowire Electrode Materials for Power Sources of Microelectronics

20

Jaephil Cho

20.1 Introduction

One-dimensional nanostructural materials have been intensively investigated for possible applications related to their magnetic properties and optical properties [1–4]. The superior characteristics of these materials over bulk counterparts originate mainly from their limited crystal dimensions with a very high surface-to-bulk ratio. Presently, the most popular power source for the mobile electronics is lithium-ion batteries for which the market is expected to grow to $\sim$10 billion in 2009. On the other hand, lithium-ion batteries for possible applications in microelectronics (sensor nods (Fig. 20.1), active RF-ID (radio-frequency identification)) need different electrode requirements. For instance, a micro-sensor nod may use the thin-film type Li-ion battery attached to the back panel for power supply, and electrode materials may need to use nanostructured morphology. Especially, these electronics require longer use time and fast charge time, and, in this regard, electrode materials for power sources should have a high surface area, which can provide higher electrode/electrolyte contact areas, which can lead to shorter diffusion paths with the particles and more facile diffusion for electrons and Li-ions. In addition, the reduced strain of intercalation and contributions from charge storage at the surface can also contribute to the Li capacity, compared with bulk counterparts.

In this study, 1D anode (SnO_2, $Sn_{78}Ge_{22}$@carbon) and cathode ($Li_{0.88}[Li_{0.18}Co_{0.33}Mn_{0.49}]O_2$) wires prepared by different synthetic methods were reviewed. These nanowires demonstrated superior capacity and rate capability (capacity retention at different current rates).

J. Cho (✉)
School of Energy Engineering, Ulsan National Institute of Science & Technology, Ulsan, South Korea
e-mail: jpcho@unist.ac.kr

20.2 SnO_2 Nanowires

Due to the fact that SnO2 is an n-type semiconductor with a wide band gap (E_g = 3.6 eV), it has been intensively investigated in the areas of gas sensors, solar cells, and catalysts [5–8]. Furthermore, it has been investigated for an anode material for lithium batteries due to its high capacity (>600 mAh/g) compared with graphite (372 mAh/g) [5–8]. However, similar to all other lithium-reactive materials, it showed a volume change that was greater than 200% during the lithium alloy/dealloy process. This causes cracking and crumbling, resulting in electrical disconnection from the copper current collector. Eventually, rapid capacity fading occurs. In order to minimize such a drastic volume change, morphology controls of SnO_2 by a synthetic method, involving nanotubes [9], hollow nanotubes [10], nanowires [11], mesopores [12], and nanorods [13], have been reported.

To date, several preparative approaches utilizing a supramolecular templating mechanism have been reported for the preparation of mesoporous tin oxide [12, 14]. Upon removal of the surfactant, however, the mesoporous structure was destroyed or was somewhat disordered. Moreover, direct synthesis of these types of mesoporous materials using surfactants is often difficult; compared with silica, the surfactant/oxide composite precursors are often more susceptible to a lack of condensation, redox reactions, or phase transitions accompanied by thermal breakdown of the structural integrity [15]. Accordingly, an alternative method via a hard template method using a silica template with *p6mm* symmetry for preparing SnO_2 nanowire is used. It contains two-dimensional parallel cylindrical pores arranged with hexagonal symmetry.

The silica templates were prepared in an aqueous solution using Pluronic 123 ($EO_{20}PO_{70}EO_{20}$, MW = 5800, Aldrich) according to previously reported methods [15] that were modified. For preparing a silica template, 30 g of P123 and 30 g of *n*-butanol (Aldrich, 99.9%) were dissolved in 2000 g of distilled water and 60 g of 30 wt% HCl solution under

C. P. Wong et al. (eds.), *Nano-Bio- Electronic, Photonic and MEMS Packaging*, https://doi.org/10.1007/978-3-030-49991-4_20

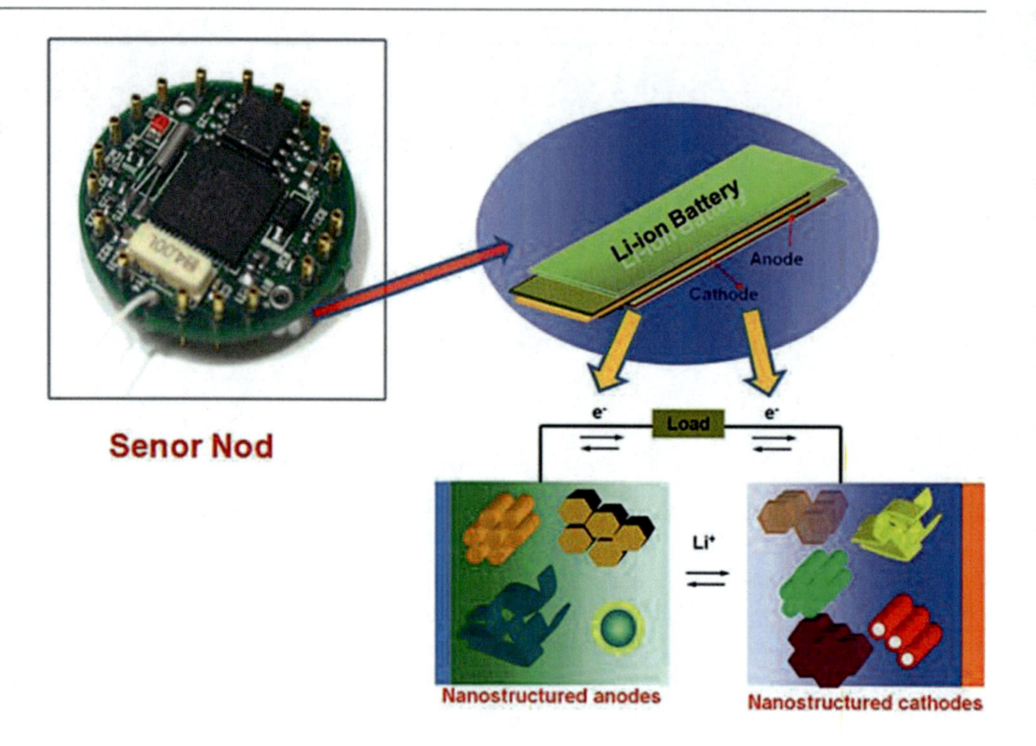

Fig. 20.1 Micro-sensor nod in which thin-film type Li-ion battery is implemented. The battery can be attached to the back panel of the nod and can use nanostructured cathode and anodes, such as nanowires and nanorods

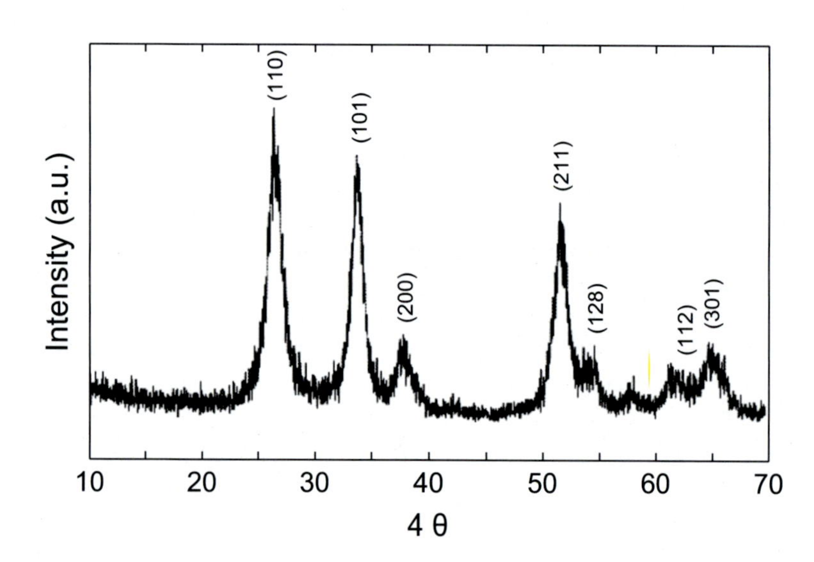

Fig. 20.2 X-ray diffraction patterns of nanowire SnO_2

stirring at 40 °C; 30 g of tetraethoxysilane (TEOS) was added into the mixed solution at 40 °C under stirring, and the remainder of the synthesis was identical to that of SBA-15.

Calcined silica templates at 550 °C for 3 h were used to prepare mesostructural material and nanowire SnO_2; 18 g of $SnCl_45H_2O$ (99%, Aldrich) was dissolved in 20 g of distilled water. This solution was mixed with 20 g of the silica templates and stirred at room temperature until all of the solution was adsorbed. After drying in an oven at 120 °C, this process (impregnation and drying) was repeated one additional time to obtain the SnO_2 precursor/SiO_2. The composite was further annealed at 400 °C for 3 h, and the silica template was removed from the SnO_2/SiO_2 composite by treating with dilute 1 M NaOH solution for 30 min and washing with distilled water and ethanol several times. Finally, the powers were vacuum-dried at 120 °C overnight.

Figure 20.2 shows high-angle XRD pattern of the SnO_2 nanowires and confirms the formation of nanocrystalline SnO_2, and the average crystal domain size was estimated as 6.5 ± 0.2 nm. ICP-MS results of the SnO_2 nanowire showed that residual Si from SiO_2 was 90 ppm, but residual Na from NaOH was not detected in both samples.

The nanowire morphology of SnO_2 prepared from the template is shown in Fig. 20.3a, b. An individual nanowire

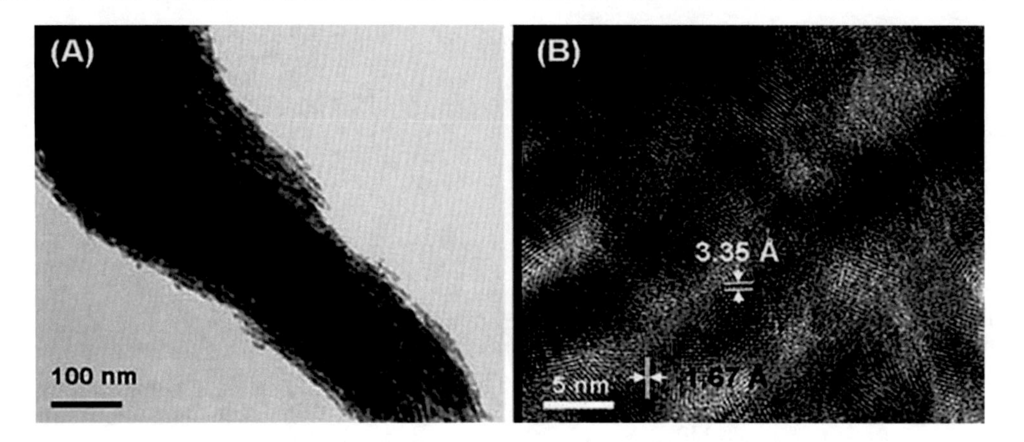

Fig. 20.3 TEM images of SnO_2 nanowires (**a**) along the (110) direction of the *p6mm* structure, (**b**) HREM image of (**a**). (Copyright 2008, Royal Society of Chemistry: reproduced with permission from [16])

has the diameter of about 6 nm and > 3 μm in length, as evidenced by the TEM (Fig. 20.3a). In general, metal oxides prepared from the template led to aggregated nanowires, and our nanowire morphology is quite similar to that of nanowire CeO_2 [15]. A high-resolution TEM image (Fig. 20.3b) shows that the nanowires consist of nanocrystalline domains that are not structurally coherent and that lattice fringes exist with *d* spacing values of 3.35 and 1.67 Å, corresponding to the (110) and (220) planes of SnO_2.

Figure 20.4 shows the cycle-life performance of the nanowire SnO_2 after 1, 10, 20, 30, 40, and 50 cycles between 1.2 and 0 V at a rate of 0.2 C (= 80 mA/g). The first discharge and the charge capacity of the nanowires were 1400 and 795 mAh/g, respectively (the irreversible capacity ratio is 43%), and the capacity retention rate after 50 cycles is 85% at a rate of 0.2 C between 0 and 1.2 V. The high level of irreversible capacity during the first cycle is associated with the formation of an irreversible inactive Li_2O phase and possible side reactions with the electrolytes during the decomposition of SnO_2 ($SnO_2 + 4Li^+ + 4e^- \rightarrow Li_2O + Sn$) [11].

Compared with self-catalysis-grown SnO_2 nanowires, which has about 100 nm diameter that showed capacity decrease from 910 to 220 mAh/g at current density of 100 mA/g after 50 cycles [11], the capacity retention of the nanowires obtained from the template shows significantly improved value (680 mAh/g). This improvement may be related to that the spaces between the nanowires act as the buffer layers during the lithium reactions. As the SnO_2 nanowires are aligned with different crystal orientations, an anisotropic volume change during the lithium alloy and dealloy processes can be expected. Although nanoscale materials can result in better capacity retention compared to micron-sized particles due to a more homogenous volume change and reduced absolute volume change between Sn and Li_xSn [17], nanoparticles would nonetheless aggregate into larger particles, resulting in pulverization and electrical disconnection with the current collector. Accordingly, a 50-cycled electrode shows some fragmentation of the nanowires into aggregated nanoparticles ~5 nm in size, inducing a capacity decay (Fig. 20.5). A TEM image shows the formation of tetragonal Sn with different lattice fringes of (200) and (211), corresponding to 2.95 and 1.99 Å, respectively (Fig. 20.5).

When the cell is cycled at high rates from 1, 3, 5, and 10 C (with a discharge rate fixed at 0.2 C) (Fig. 20.4c), the charge capacity decreases slowly out to 3 C, after which it decreases rapidly, showing 697, 500, and 250 mAh/g at 3, 5, and 10 C rates, respectively; the capacity retention at a 10 C rate, compared with 1 C, is 31%. In addition, capacity recovery after eight cycles at 10 C rate is 68% of the capacity at 1 C rate. This result indicates that nanowires may undergo a higher degree of nanowire pulverization than those cycled at lower C rate.

20.3 $Sn_{78}Ge_{22}$@Carbon Nanowires

One of the most intensively studied anode materials for lithium batteries is Sn due to its relatively high theoretical capacity (991 mAh/g) and high electronic conductivities (~10^4 s/cm) compared to other lithium-reactive metals [18]. In contrast to the metal hosts, lithium alloys Li_xM possess a highly ionic character and, thus, are usually very brittle. Mechanical stresses related to volume changes in excess of 200% induce pulverization and aggregation as well as loss of the electrical inter-phase contact [19]. Although such a drastic volume change cannot be removed completely, the degree of the volume change can be reduced. In this regard, two methods have been proposed; the first involves forming an alloy with a ductile active metal so that it acts as a buffer for volume expansion [20]. The second method is similar to the first: bulk- or nanosized metals are coated with carbon or are dispersed in the carbon matrix [21, 22]. Here, carbon and inactive conducting metal media act as electrical connector with the Cu current collector when the particles pulverized. These methods utilize zero-dimensional bulk particles or nanoparticles.

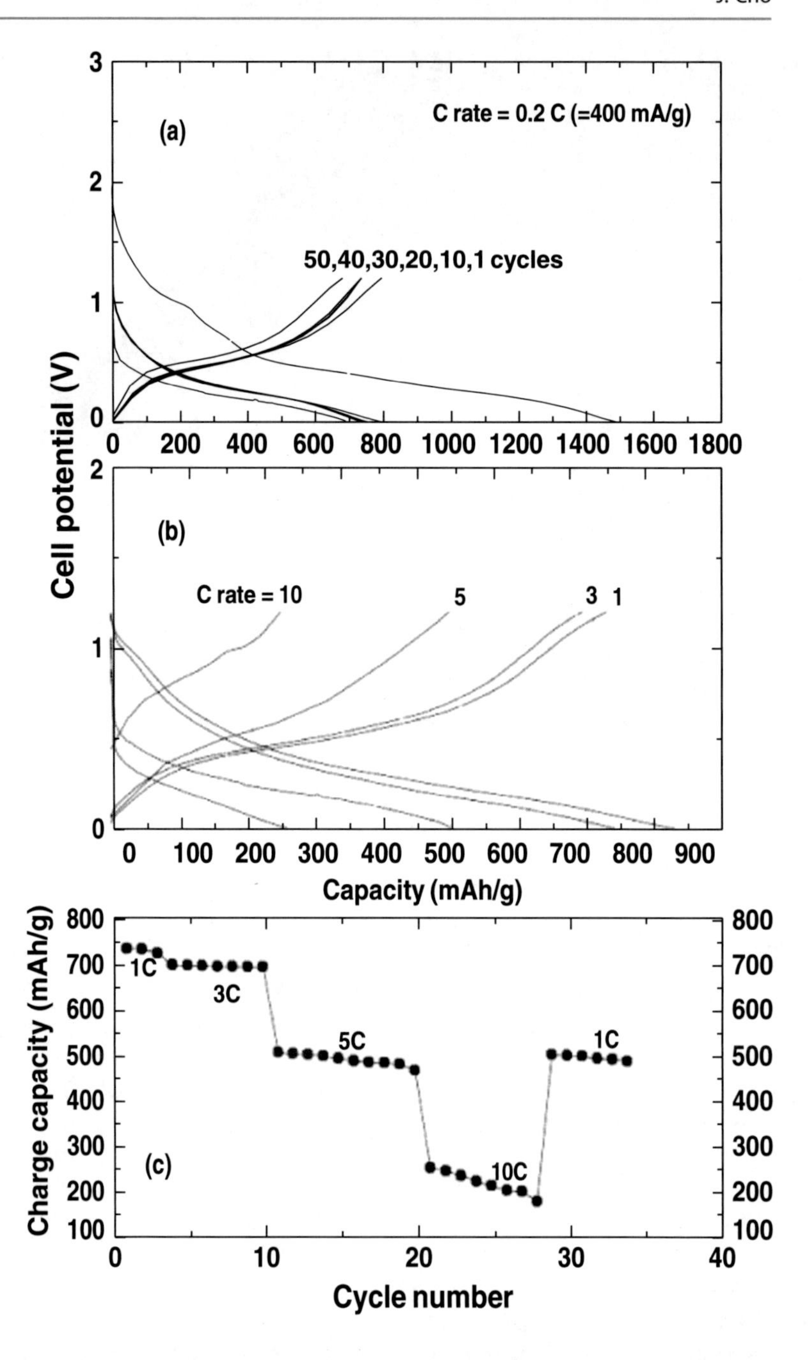

Fig. 20.4 (**a**) Voltage profiles of the nanowire SnO2 anode in coin-type half-cells during the first, fifth, 10th, 20th, 30th, and 50th cycles between 1.2 and 0 V. The charge C rate was maintained at 0.2 C (= 40 mA/g). (**b**) Voltage profiles of the nanowire SnO_2 anode in coin-type half-cells during each first cycle at different discharges of 1, 3, 5, and 10 C. The charge C rate was maintained at 0.2 C (= 40 mA/g)

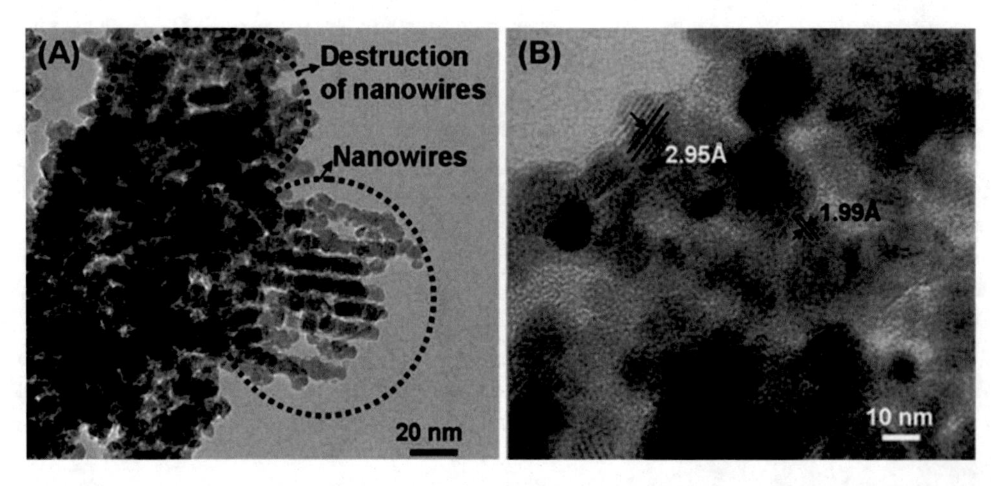

Fig. 20.5 TEM image of (**a**) the nanowire SnO_2 electrode after 50 cycles and (**b**) a high-resolution image of (**a**). (Copyright 2008, Royal Society of Chemistry: reproduced with permission from [16])

Syntheses and electrochemical studies of 1D lithium-reactive metal nanowires are rarely reported, and Si nanowires prepared by a laser ablation method have shown a reversible capacity <500 mAh/g [23]. General synthetic methods for preparing metallic nanowires use vapor–liquid–solid (VLS) [24, 25], solution–liquid–solid (SLS), or vapor–solid (VS) reactions [26, 27].

In this method, for synthesizing the $Sn_{78}Ge_{22}$@carbon nanowires, all the experimental procedures were carried out in purified Ar atmosphere except for the washing process. Sodium naphthalide solution was prepared from 5.4 g (0.23 mol) of sodium (Aldrich, 99.9%) and 19.38 g (0.15 mol) of naphthalene (99%, Aldrich), stirred in 100 ml 1.2-dimethoxyethane for 24 h; 3.978 ml of anhydrous $SnCl_4$ (Aldrich, 99.999%) and 0.684 ml of $GeCl_2$ (Aldrich, 99.99%) in a dried 50 ml of 1.2-dimethoxyethane (Aldrich, 99%) were thoroughly mixed in a flask under a dry argon atmosphere in a glove box for 3 h in order to get the atomic ratio of 85:15 (Sn:Ge) (atom%). This mixed solution was stirred for 24 h, and then 60 ml of butyllithium (n-C_4H_9Li) (Aldrich, 99%) was added. This solution was stirred at room temperature for 4 h and naphthalene being removed by using a rotating evaporator at 110 °C. The product was washed with distilled water seven times and finally vacuum-dried at 120 °C for 48 h. The obtained product, which was a viscous liquid with a black color, was annealed at 600 °C for 5 h under a vacuum atmosphere. The obtained nanoparticles were analyzed with an inductively coupled plasma-mass spectrometry (ICP-MS) to ensure the stoichiometry.

Our simple synthetic approach uses clusters of the butyl-capped $Sn_{78}Ge_{22}$ nanoparticles, and vacuum annealing of these clusters tuned into $Sn_{78}Ge_{22}$@carbon core-shell nanowires (Fig. 20.6a). Figure 20.6b exhibits a TEM image of butyl-capped $Sn_{78}Ge_{22}$ nanoparticles before annealing. With these nanoparticles, particles of 1 μm in size are connected to each other. The electron diffraction (inset of b) pattern confirms the formation of the amorphous phase. Typically, Ge [28] or $Sn_{0.9}Si_{0.1}$ [29] nanoparticles obtained from similar synthetic methods are of a similar particle size, while the particles are unconnected. Furthermore, the particles in the image consist of clusters of nanosized particles (~5–10 nm) capped with butyl groups. Hence, a similar occurrence can be expected with the sample before annealing. After thermal annealing at 600 °C in a vacuum, the morphology before annealing changed completely, showing branched nanowires with diameters in a range of 50–100 nm and lengths that were several micrometers.

As $Sn_{78}Ge_{22}$ particles before annealing consist of many nanoparticles capped with butyl groups, these particles can be grown in different directions when the butyl groups are burnt out, as shown in Fig. 20.6c. Inductively coupled plasma mass spectroscopy (ICP-MS) of the nanowires after annealing confirmed the formation of $Sn_{77}Ge_{23}$, and the carbon content determined by a CHS analyzer in the nanowires was 5 wt%.

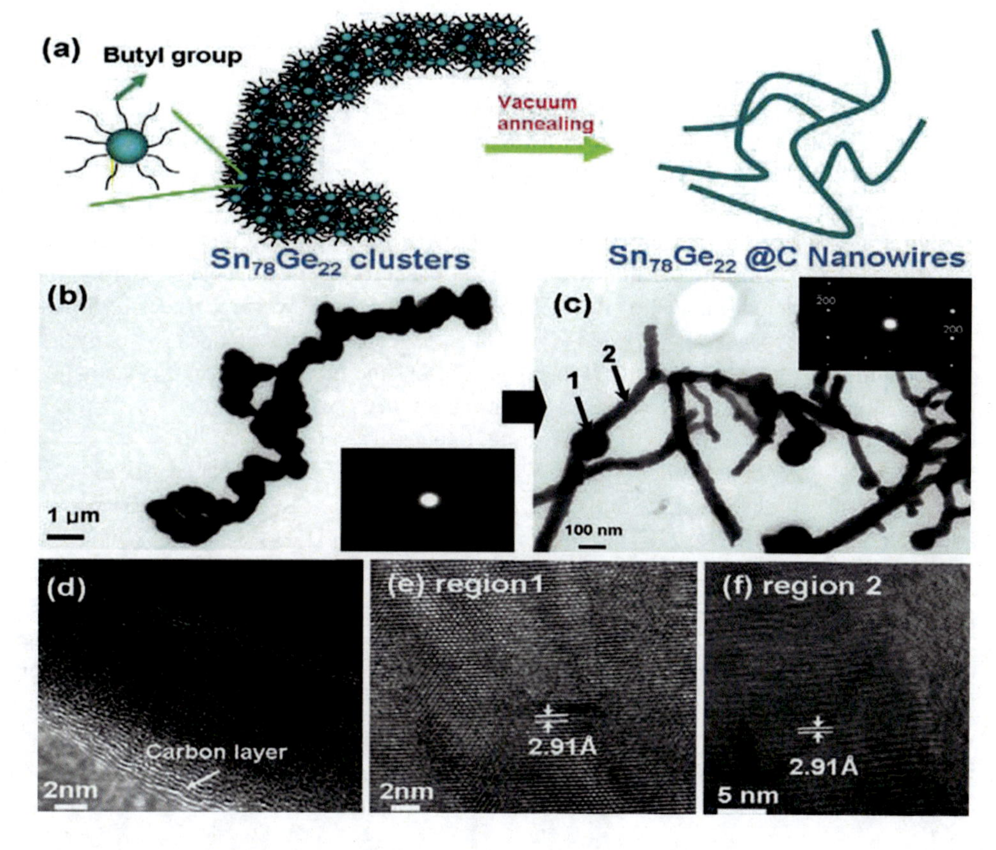

Fig. 20.6 (**a**) Schematic diagram for $Sn_{78}Ge_{22}$@carbon nanowires growth from clusters of butyl-capped Sn78Ge22 nanoparticles and TEM images of the Sn78Ge22@carbon nanowires (**b**) before thermal annealing and (**c**) after thermal annealing. Insets are electron diffraction patterns of (**b**) and (**c**). (**d**) TEM images of Region 1 in (**c**). (**e**) and (**f**) FE-SEM images of the nanowires. (Copyright 2008, American Chemical Society: reproduced with permission from [30])

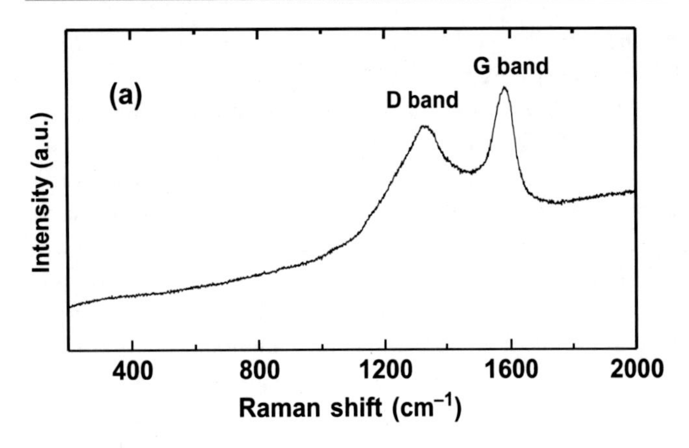

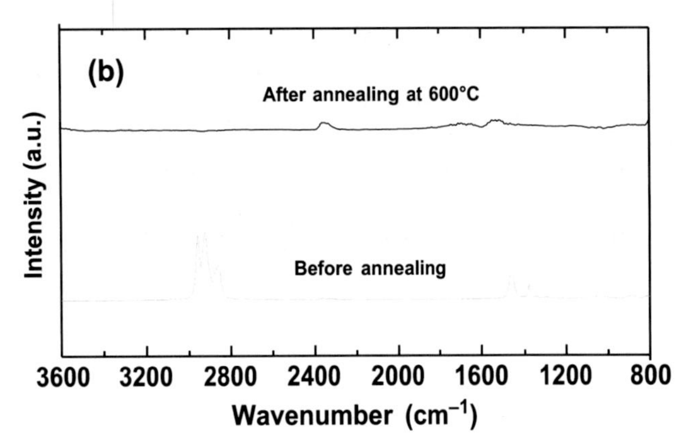

Fig. 20.7 Raman scattering of the $Sn_{78}Ge_{22}$@carbon nanowires after thermal analysis and FT-IR spectra of $Sn_{78}Ge_{22}$@carbon nanowires before and after thermal annealing

Figure 20.6d is an expanded TEM image of Region 2, in which an amorphous carbon layer with a thickness of ~2 nm is shown. This carbon layer was formed from the decomposition of C_4H_9 groups during annealing.

To investigate the ordering degree of amorphous carbon, Raman scattering of the annealed $Ge_{78}Sn_{22}$ nanowire was performed (Fig. 20.7a), and the dimensional ration of the D and G band was estimated, as the ratio of D/G was found to be 2.13. The two peaks at ~1360 and ~1580 cm^{-1} are assigned to the (disordered band) D band and (graphene band) G band, respectively [31]. This is far larger than well-ordered graphite at 0.09 [6]. This result implies the formation of amorphous carbon. Figure 20.7b shows the FT-IR spectra of butyl-capped $Sn_{87}Ge_{22}$ particles before and after annealing at 600 °C. The three peaks at 2955, 2924, and 2853 cm^{-1} fall where expected for C–H stretches of butyl groups, which is in good agreement with the results of an earlier study [32]. The peaks at 1376 and 1456 cm^{-1} are the positions expected for the symmetric and asymmetric bends of the butyl group, respectively [31]. The nanowire annealed at 600 °C for 5 h showed no traces of butyl groups, indicating complete transformation into amorphous carbon. Typically, oxidized SnO_2 or GeO_2 are observed between 800 and 750 cm^{-1}; these are attributed to metal-O stretch [31]. In this case, no peaks in the range were observed, indicating no oxygen contamination.

Figure 20.8 shows the first discharge and charge capacities of the nanowires, exhibiting 1247 and 1107 mAh/g, respectively, corresponding to Coulombic efficiency of 88%. This value is an improved value compared to previous capped Sn [22] or Ge [23] nanoparticles that were measured below 85%. Moreover, such a Coulombic efficiency was superior to that of Si nanowires that show the Coulombic efficiency below 50% [23]. Figure 20.8b shows that charge capacities at 1, 3, 6, and 8 C were 1107, 1075, 1060, and 1034 mAh/g with a capacity retention at 8 C of 93%.

Thus far, very few studies have reported a high-rate stability of nanoalloy particles. The excellent rate capability of the core-shell nanowires is attributed to the increased electrolyte contact area and the reduced lithium diffusion length that facilitates further Li reactions with the nanowire. For the 0D particles, repetitive particle aggregation and pulverization leads to the formation of a new surface, thus forming a new solid electrolyte interface (SEI) layer. Accordingly, part of the active metal becomes isolated by surface films formed on the fresh active surface by reactions with Li-metal compounds and electrolytes, resulting in rapid capacity fading [33]. However, due to the nature of the nanowire/carbon with the branched morphology and carbon cell, direct contact among the nanotubes can be minimized, and less aggregation is expected. After finishing the rate capability test of the nanowire/carbon nanowire, a morphology change of the nanowire/carbon core shell occurred. As shown in Fig. 20.9a, the original branched nanowires were separated from the nanoparticles and nanowire. Furthermore, the separated nanowire has stacking faults, and microtwins are visible with their fault plane (101) parallel to the axis of the nanowire (Fig. 20.9b). These defects are not observed in the pristine nanowires and may be associated with continuous volume change during the lithium dealloy/alloy process. However, respective Sn and Ge atoms segregation was not observed during cycling, as evidenced by the line mapping of Sn and Ge in the nanowires (Fig. 20.9c).

20.4 Layered $Li_{0.88}[Li_{0.18}Co_{0.33}Mn_{0.49}]O_2$ Nanowires

Recently, Mn-rich lithium metal oxides such as $Li[Li_{1/3-2x/3}Mn_{2/3-x/3}M_x]O2$ cathodes are currently receiving significant interest as cathode materials for Li-ion batteries. These materials can provide capacities greater than 200 mAh/g if initially charged to 4.5 V or higher [34–40]. In spite of the capacity advantage, a rapid capacity fade occurs at higher C rates, for instance, over 50% capacity fade was observed when the current was increased from 20 to 200 mA/g in Li

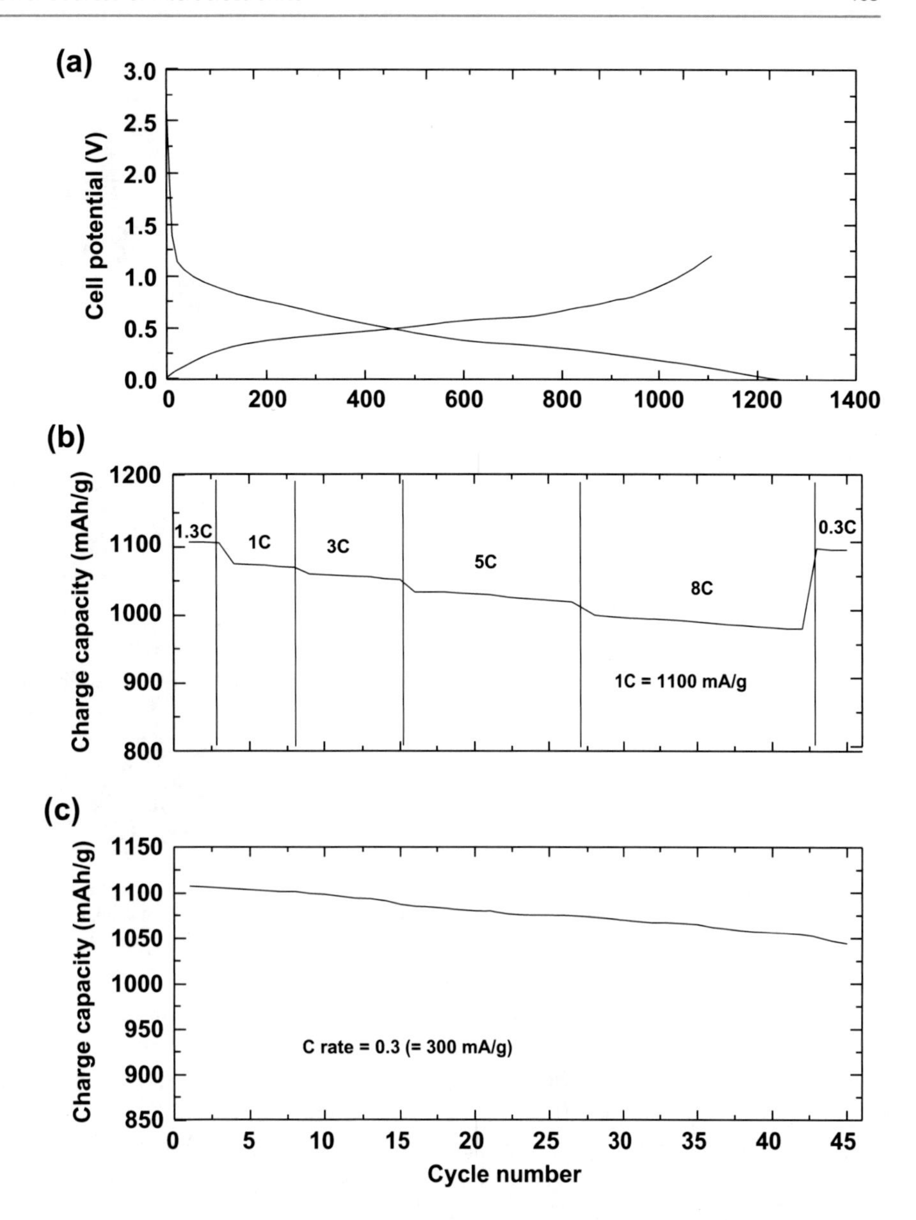

Fig. 20.8 (**a**) Rate capabilities of $Sn_{78}Ge_{22}$@carbon nanowires at different C rates from 0.3, 1, 3, 5, and 8 C (1 C = 800 mA/g) in a coin-type half-cell. Discharge rate was fixed at 0.3 C. (**b**) Plot of rate capabilities as a function of charge capacity during extended cycling. (**c**) Plot of cycle number as a function of the charge capacity of $Sn_{78}Ge_{22}$@carbon nanowires

$[Ni_{0.2}Li_{0.2}Mn_{0.6}]O_2$ [41]. However, Cho et al. recently reported that $Li[Ni_{0.41}Li_{0.08}Mn_{0.51}]O_2$ nanoplates exhibited greatly improved rate capabilities at a 3 C rate compared with bulk counterpart prepared via a coprecipitation method [42].

For preparing the nanowire cathode, a solution containing 7.2 g of $KMnO_4$ and 200 ml of distilled water was slowly stirred for 30 min at 40 °C and added to 2.1 g of fumaric acid which resulted in a rapid, exothermic reaction, forming a brown gel. This gel was further annealed at 400 °C and 700 °C for 6 and 12 h, respectively, and the resulting dark black powder was thoroughly washed with water six times, followed by vacuum drying at 200 °C overnight. Inductively coupled plasma mass spectrometry (ICP-MS) analysis of the sample confirmed $K_{0.3}MnO_2$. As-prepared K-birnessite was mixed with $Co(NO_3)_6 6H_2O{:}K_{0.3}MnO_2$ with a weight ratio of 8:1 in 100 ml of distilled water, kept in an autoclave for 5 days at 200 °C, and finally washed with water six times in order to remove residues that did not participate in the reaction and dissolved K ions. ICP-MS analysis of the sample confirmed formation of $Co_{0.4}Mn_{0.6}O_2$. Finally, $LiNO_3H_2O$ and $Co_{0.40}Mn_{0.60}O_2$ were combined with weight ratios of 4:1, were mixed in 100 ml of distilled water, and transferred in an autoclave, maintaining it at 200 °C for 2 days. As-prepared powder was rinsed with water and dried under vacuum at 120 °C. Cathodes for battery test cells were made from the active material (~20 mg), super P carbon black (MMM, Belgium), and a polyvinylidene fluoride (PVdF) binder (Kureha Company, Japan) at a weight ratio of 90:5:5. A cathode slurry was prepared by thoroughly mixing an *N*-methyl-2-pyrrolidone (NMP) solution with the PVdF, the carbon black, and the powdery cathode-active material. The electrodes were prepared by coating the cathode-slurry onto

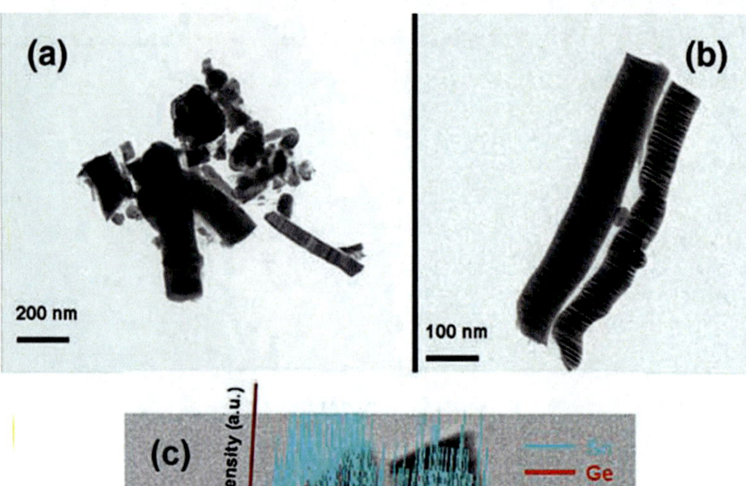

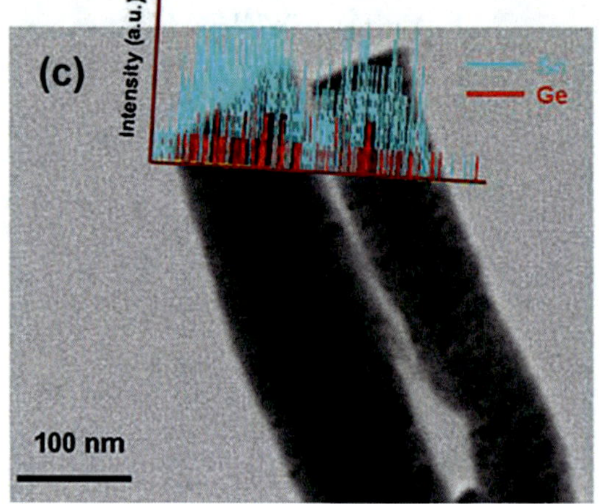

Fig. 20.9 (**a** and **b**) TEM images of $Sn_{78}Ge_{22}$@carbon nanowires after the rate capability test in Fig. 20.8b. (**c**) The EDS (energy-dispersive spectroscopy) line scan across the particle. (Copyright 2008, American Chemical Society: reproduced with permission from [30])

an Al foil, followed by drying at 130 °C for 20 min. Coin-type battery-test cells (size 2016) containing a cathode, a Li-metal anode, and a microporous polyethylene separator were prepared in a helium-filled glove box. The electrolyte used was 1 M $LiPF_6$ with ethylene carbonate/diethylene carbonate/ethyl-methyl carbonate (EC/DEC/EMC) (30:30:40 vol.%) (Cheil Industries, Korea). After addition of the electrolyte, the test cells were aged at room temperature for 24 h before commencing the electrochemical tests.

Figure 20.10a shows an XRD pattern of the as-prepared $Co_{0.4}Mn_{0.6}O_2$ and $Li_{0.88}[Li_{0.18}Co_{0.33}Mn_{0.49}]O_2$. The XRD pattern of the $Co_{0.40}Mn_{0.60}O_2$ is dominated by two major peaks at about 12° (110) and about 25° (200) scattering angles, which is similar to that of previously reported $Ni_{0.45}Mn_{0.55}O_2$ obtained from *K*-birnessite [42]. After a reaction with Li at 200 °C, the $Li_{0.88}[Li_{0.18}Co_{0.33}Mn_{0.49}]O_2$ sample clearly shows the formation of a hexagonal layered structure with a R-3 m space group. The lattice constants of *a* and *c* were estimated as 2.834 Å and 14.208 Å, respectively. Very weak superlattice reflections at approximately 20° and 24° are known to correspond to the ordering of the Li and Co and Mn ions in transition metal sites of the layered lattice [36, 37]. Moreover, preferred orientation along the (003) plane was observed. The ICP-MS results showed the formation of $Li_{1.3}Co_{0.4}Mn_{0.6}O_{4.9}$; when normalized to the $LiMnO_2$, the compound had a formula $Li_{0.88}[Li_{0.18}Co_{0.33}Mn_{0.49}]O_2$.

Figure 20.11 shows SEM and TEM images of the as-prepared $Co_{0.4}Mn_{0.6}O_2$ nanowires (a) and $Li_{0.88}[Li_{0.18}Co_{0.33}Mn_{0.49}]O_2$ nanowires (b, d, and e), and $Co_{0.4}Mn_{0.6}O_2$ nanowires have a diameter of 40 nm and are >1 μm in length. After reaction with the Li nitrate, the dimension of the nanowires increases. The $Li_{0.88}[Li_{0.18}Co_{0.33}Mn_{0.49}]O_2$ nanowires have a diameter of 100 nm and a length of >3 μm. An inset of Fig. 20.10b is an enlarged image of the edge parts of the nanowires, confirming that the particles consisted of aggregated individual nanowires. Energy-dispersive X-ray spectrometry (EDS) of the $Li_{0.88}[Li_{0.18}Co_{0.33}Mn_{0.49}]O_2$ nanowires shows both the presence of Co and Mn. The ICP-MS results confirmed stoichiometry of $Li_{0.88}[Li_{0.18}Co_{0.33}Mn_{0.49}]O_2$. The TEM image of Fig. 20.11d shows that the diameter of the nanowires is about 50 nm. An HREM image of the nanowires (image e) clearly shows the formation of the layered structure with the lattice fringe of the (003) plane corresponding to 0.46 nm. The Brunauer–Emmett–Teller (BET) surface area of the $Li_{0.88}[Li_{0.18}Co_{0.33}Mn_{0.49}]O_2$ nanowires was 50 m^2/g.

Figure 20.12 shows the charge and discharge curves between 4.8 and 2 V at rates of 0.2, 1, 3, 5, 10, and 15 C (1 C = 240 mA/g), corresponding to capacities of 245, 242, 238, 230, 225, and 220 mAh/g, respectively (charge rate was fixed at 1 C). In general, the loading level of the active cathode material critically affects the performance of the

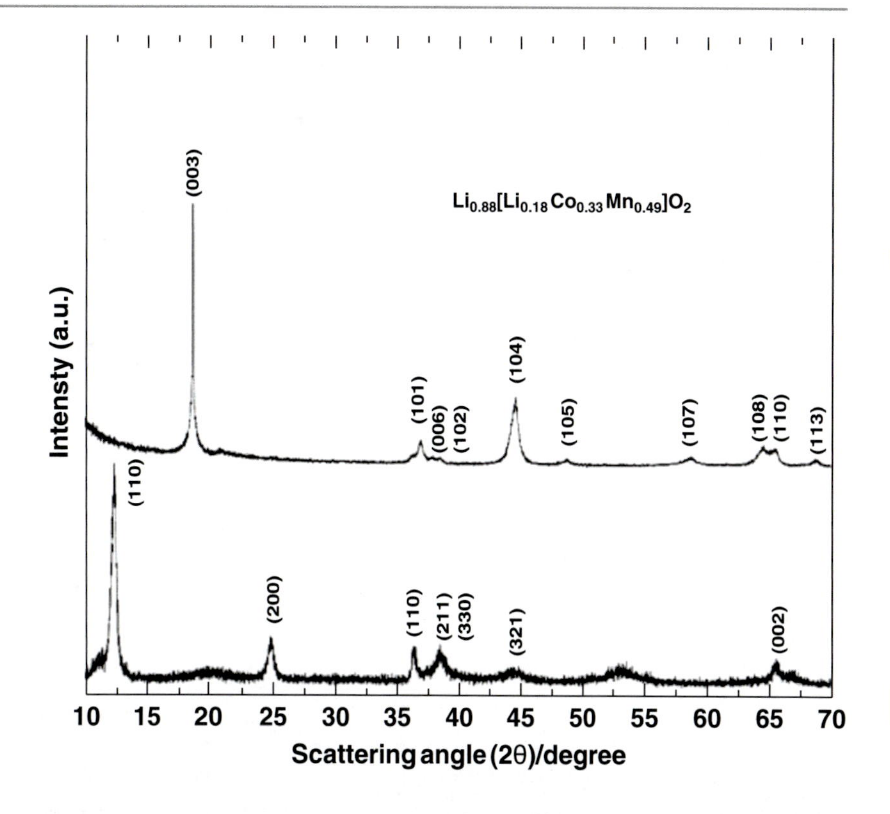

Fig. 20.10 Powder XRD patterns of $Co_{0.4}Mn_{0.6}O_2$ and $Li_{0.88}[Li_{0.18}Co_{0.33}Mn_{0.49}]O_2$ nanowires

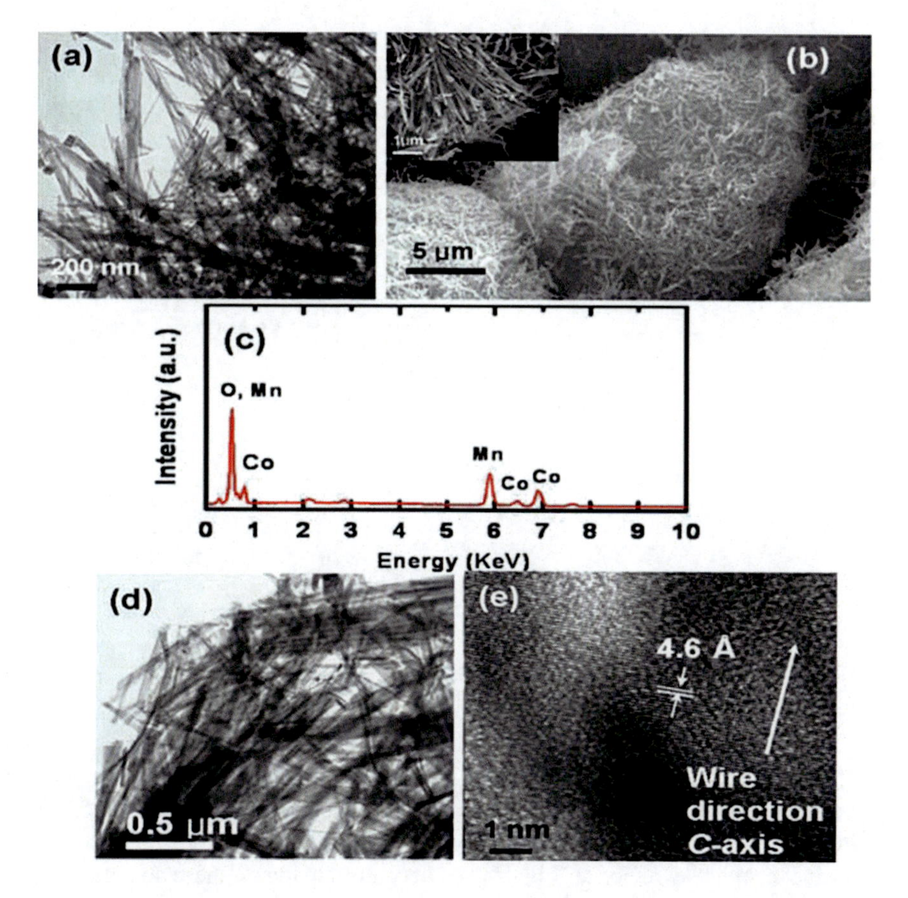

Fig. 20.11 TEM images of (**a**) $Co_{0.4}Mn_{0.6}O_2$ nanowires and (**b** and **d**) SEM and TEM images of the $Li_{0.88}[Li_{0.18}Co_{0.33}Mn_{0.49}]O_2$ nanowires, respectively. An inset of 2b is an expanded image of 2b. (**c**) Corresponding EDS spectrum of image b and (**e**) high-resolution TEM image of a single nanowire of (**d**). (Copyright 2008, American Chemical Society: reproduced with permission from [43])

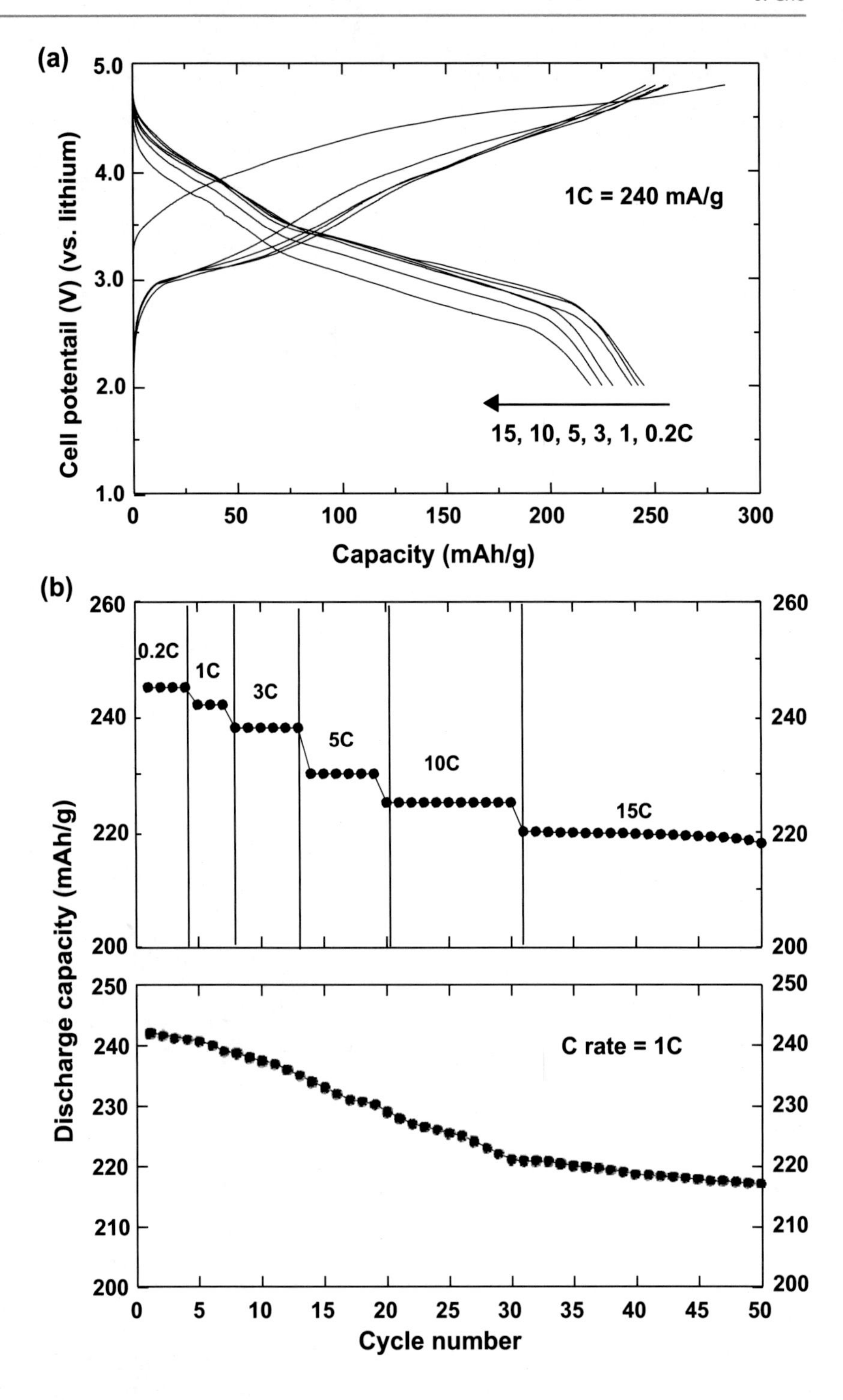

Fig. 20.12 (**a**) Voltage curves of $Li_{0.88}[Li_{0.18}Co_{0.33}Mn_{0.49}]O_2$ nanowires in a coin-type half-cell at different C rates of 0.2, 1, 3, 5, 10, and 15 C (1 C = 240 mA/g) between 2 and 4.8 V. The charge rate was fixed at 1 C, and the discharge capacities vs. cycle number were tested at different C rates. Discharge capacity vs. cycle number of the Li $[Li_{0.08}Co_{0.33}Mn_{0.5}]O_2$ nanowires in a coin-type half-cell at a 1 C rate between 2 and 4.8 V

cathode. In this case, a cathode material loading level of 20 mg/cm^2 was used.

After the first charge process, the plateau vanished in the subsequent cycles. Although a similar irreversible voltage plateau during the first charge process has been reported with Li_2MnO_3-based oxides, Li_2MnO_3-based oxides showed a distinct flat plateau region at ~4.5 V [44]. In contrast to this, $Li_{0.88}[Li_{0.18}Co_{0.33}Mn_{0.49}]O_2$ has an inclined plateau at ~4.6 V. The presence of the inclined plateau at 4.5 V (as shown in Fig. 20.12a) instead of the flat plateau in $Li_{0.88}[Li_{0.18}Co_{0.33}Mn_{0.49}]O_2$ nanowires is also observed in $Li[Co_x Li_{1/3-x/3}Mn_{2-x2/3}]O_2$ and $Li[Cr_x Li_{1/3-x/3}Mn_{2-x2/3}]O_2$, in which the flat plateau turn into a inclined plateau with increasing Co or Cr contents [38, 39]. This is because more transition metals (Co, Cr) are available to be involved in the redox reaction, and oxygen loss will not be required for samples with higher transition metal content to remove all the Li from the Li layer [39].

During the first cycle, the charge and discharge capacities are 283 and 245 mAh/g, respectively, showing 87% columbic efficiency. After 50 cycles at a rate of 1 C, capacity retention was 92%. In particular, the capacity retention at a 15 C rate was 90%, compared to that at a 0.2 C rate. The capacity retention of the nanowires even after repetitive cycling at a 15 C rate was 98%. The rate capability of the cathode with the irregularly distributed particles was determined by the larger particles, and therefore it is hampered by the larger particles with longer Li-ion diffusion distance. For instance, $Li[Ni_{0.20}Li_{0.2}Mn_{0.6}]O_2$ nanoparticles with the primary particles were narrowly distributed in the range of 80–200 nm and with large aggregates of 1–20 μm [41] showed the first discharge capacity of ~280 mAh/g at 20 mA/g, but its discharge capacity significantly decreased to 210 mAh/g at 400 mA/g. However, our nanowires have a uniform diameter of about 50 nm and shows first discharge capacity of 245 mAh/g at 48 mA/g, and its capacity is 220 mAh/g at 3000 mA/g. The capacity retention of the $Li[Ni_{0.41}Li_{0.08}Mn_{0.51}]O_2$ nanoparticle with 20–200 nm sizes (BET surface area was 24 m^2/g) at a current rate of 1200 mA/g rate is 89%, while that of the nanowire is 94% at the same current rate. Even though 6% difference is small in lower current rates, higher current rates can make a huge difference. Accordingly, the capacity retention of the nanowires is expected to be much higher than that of the nanoparticles at higher current rate than 1200 mA/g. This indicates that the nanowires with higher surface areas likely lead to higher electrode/electrolyte contact areas. This coupled with shorter diffusion paths into the lattices, and reduced strain of intercalation may lead to improved cycling performance compared to nanoparticles [45].

20.5 Conclusions

We have reviewed 1D anode and cathode materials to obtain high surface area of the nanostructured electrode, which makes the current density more effective than for the conventional bulk electrode. The high polarization was believed to result from slow ionic and electronic diffusions in the active material or interfaces, but manipulation of the morphology provided versatile strategies toward improving the electrochemical properties. Better rate capabilities were obtained as there was a decrease in diffusion distance between Li^+ and electrons within the nanostructured electrode.

Acknowledgments This research was supported by the converging Research Center Program through the National Research Foundation (NRF) of Korea funded by the Ministry of Education, Science and Technology.

References

1. Peng, X., Manna, L., Yang, W., Wickham, J., Scher, E., Kadavanich, A.: Nature. **404**, 59 (2000)
2. Jun, Y., Lee, S.-M., Kang, N.-J., Cheon, J.: J. Am. Chem. Soc. **123**, 5150 (2001)
3. Manna, L., Scher, E.C., Alivisatos, A.P.: J. Am. Chem. Soc. **122**, 12700 (2000)
4. Jun, Y.-W., Choi, J.-S., Cheon, J.: Angew. Chem. Int. Ed. **45**, 3414 (2006)
5. Liu, Z., Zhang, D., Han, S., Li, C., Tang, T., Jin, W., Liu, X., Lei, B., Zhou, C.: Adv. Mater. **15**, 1754 (2003)
6. Liu, Y., Dong, J., Liu, M.: Adv. Mater. **16**, 353 (2004)
7. Dai, Z.R., Pan, Z.W., Wang, Z.L.: J. Am. Chem. Soc. **124**, 8673 (2002)
8. Hu, J.Q., Ma, X.L., Shang, N.G., Xie, Z.Y., Wong, N.B., Lee, C.S., Lee, S.T.: J. Phys. Chem. B. **106**, 3823 (2002)
9. Wang, Y., Lee, J.Y.: J. Phys. Chem. B. **108**, 17832 (2004).; Y. Wang, H. C. Zeng, J. Y. Lee, Adv. Mater. **2006**, 18, 645; Y. Wang, J. Y. Lee, H. C. Zeng, Chem. Mater. **2005**, 17, 3899
10. Wang, Y., Su, F., Lee, J.Y., Zhao, X.S.: Chem. Mater. **18**, 1347 (2006).; X. W. Lou, Y. Wang, C. Yuan, J. Y. Lee, L. A. Archer, *Adv. Mater.* 2006, 18, 2325; S. Han, B. Jang, T. Kim, S. M. Oh, T. Hyeon, *Adv. Funct. Mater.* **2005**, 15, 1845
11. Park, M.-S., Wang, G.-X., Kang, Y.-M., Wexler, D., Dou, S.-X., Liu, H.-K.: Angew. Chem. Int. Ed. **46**, 750 (2007)
12. Chen, F., Liu, M.: Chem. Commun. 1829 (**1999**)
13. Zhang, D.-F., Sun, L.-D., Yin, J.-L., Yan, C.-H.: Adv. Mater. **12**, 1022 (2003)
14. Ulagappan, N., Rao, C.N.R.: Chem. Commun. 1685 (**1996**).; K. G. Severin, T. M. Abdel-Fattah, T. J. Pinnavaia, Chem. Commun. 1998, 1471
15. Laha, S.C., Ryoo, R.: Chem. Commun. 2138 (**2003**).; K. P. Gierszal, T.-W. Kim, R. Ryoo, M. Jaroniec, J. Phys. Chem. B. **2005**, 109, 8774
16. Kim, H., Cho, J.: J. Mater. Chem. **18**, 771 (2008)
17. Coutney, I.A., Dahn, J.R.: J. Electrochem. Soc. **144**, 2045 (1997).; I. A. Courtney, J. S. Tse, O. Mao, J. Hafner, J. R. Dahn, Phys. Rev. B **1998**, 58, 15583
18. Nazri, G.-A., Pistoia, G.: Lithium Batteries Science and Technology. Kluwer, Boston (2004)
19. Besenhard, J.O., Yang, J., Winter, M.: J. Power Sources. **68**, 87 (1997)
20. Shin, H.C., Liu, M.: Adv. Funct. Mater. **15**, 582 (2005).; Y. Xia, T. Sakai, M. Wada, H. Yoshinaga, J. Electrochem. Soc. **2001**, 148, 471; S. D. Beattie, J. R. Dahn, J. Electrochem. Soc. **2003**, 150, A894; O. Mao, J. R. Dahn, J. Electrochem. Soc. **1999**, 146, 414
21. Kim, I.-S., Blomgren, G.E., Kumta, P.N.: Electrochem. Solid- State Lett. **7**, A44 (2004).; G. X. Wang, J. Yao, H. K. Liu, Electrochem. Solid- State Lett. **2004**, 7, A250; L. K. Lee, Y. S. Jung, S. M. Oh, J. Am. Chem. Soc. **2003**, 125, 5652; M. Noh, Y. Kim, M. G. Kim, H. Lee, Y. Kwon, Y. Lee, J. Cho, Chem. Mater. 2005, 17, 3320
22. Y. Kwon, M. G. Kim, Y. Kim, Y. Lee, J. Cho, J. Electrochem. Solid-State Lett. 2006, 9, A34. ; M. Noh, Y. Kwon, H. Lee, J. Cho, Y. Kim, M. G. Kim, Chem. Mater. **2005**, 17, 1926; H. Kim, J. Cho, J. Electrochem. Soc. **2007**, 154, A462; H. Kim, J. Cho, Electrochim. Acta **2007**, 52, 4197
23. Gao, B., Sinha, S., Fleming, L., Zhou, O.: Adv. Mater. **13**, 816 (2001)
24. Himpsel, F.J., Jung, T., Kirakosian, A., Lin, J.L., Petrovykh, D.Y., Rausher, H., Viernow, J.: MRS Bull. **24**, 20 (1999)
25. Wagner, R.S., Ellis, W.C.: Appl. Phys. Lett., vol. 4, p. 89 (1964).; G. Cao, Nanostructure and Nanomaterials, Imperial College Press, London, 2004

26. Trentler, T.J., Kickman, K.M., Goel, S.C., Viano, S.M., Gibbons, P. C., Buhro, W.E.: Science. **270**, 1791 (1995)
27. Yang, P., Lieber, C.M.: Science. **273**, 1836 (1996).; P. Yang, C. M. Lieber, J. Mater. Res. 1997, 12, 2981
28. Lee, H., Kim, H., Doo, S.-G., Cho, J.: J. Electrochem. Soc. **154**, A343 (2007)
29. Kwon, Y., Kim, H., Doo, S.-G., Cho, J.: Chem. Mater. **19**, 982 (2007)
30. Lee, H., Cho, J.: Nano Lett. **7**, 2638 (2007)
31. Yang, C.-S., Li, Q., Kauzlarich, S.M., Phillips, B.: Chem. Mater. **12**, 983 (2000)
32. Ferrai, A.C., Roberston, R.: J. Phys. Chem. B. **61**, 14095 (2000)
33. Aurbach, D., Nimberger, A., Markosky, B., Levi, E., Sominski, E., Gedanken, A.: Chem. Mater. **14**, 4155 (2002)
34. Lu, Z., Dahn, J.R.: J. Electrochem. Soc. **149**, A815 (2002)
35. Armstrong, A.R., Holzapfel, M., Novak, P., Johnson, C.S., Kang, S.-H., Thackeray, M.M., Bruce, P.G.: J. Am. Chem. Soc. **128**, 8694 (2006)
36. Park, S.-H., Kang, S.-H., Johnson, C.S., Amine, K., Thackeray, M. M.: Electrochem. Commun. **9**, 262 (2007)
37. Thackeray, M.M., Johnson, C.S., Vaughey, J.T., Li, N., Hackney, S. A.: J. Mater. Chem. **15**, 2257 (2005)
38. Johnson, C.S., Li, N., Lefief, C., Thackeray, M.M.: Electrochem. Commun. **9**, 787 (2007)
39. Hong, Y.-S., Park, Y.J., Wu, X., Ryu, K.S., Chang, S.H.: Electrochem. Solid State Lett. **6**, A166 (2003)
40. Ammundsen, B., Paulsen, J.: J. Adv. Mater. **13**, 943 (2001)
41. Hong, Y.-S., Park, Y.J., Ryu, K.S., Chang, S.H., Kim, M.G.: J. Mater. Chem. **14**, 1424 (2004)
42. Cho, J., Kim, Y., Kim, M.G.: J. Phys. Chem. C. **111**, 1186 (2007)
43. Lee, Y., Kim, M.G., Cho, J.: Nano Lett. **8**, 957 (2008)
44. Park, Y.J., Hong, Y.-S., Wu, X., Kim, M.G., Ryu, K.S., Chang, S. H.: J. Electrochem. Soc. **151**, A720 (2004)
45. Kim, C., Noh, M., Choi, M., Cho, J., Park, B.: Chem. Mater. **17**, 3297 (2005)

Part VII

Energy Storage (LIB Battery)

Graphene-Based Materials with Tailored Nanostructures for Lithium-Ion Batteries

21

Cuiping Han, Hongfei Li, Jizhang Chen, Baohua Li, and C. P. (Ching-Ping) Wong

21.1 Introduction

Nowadays, the scarcity of fossil fuels and growing environmental concern are driving scientists and engineers to exploit sustainable, clean, and highly efficient technologies to supply and store energy. Due to the relatively high energy density (100–200 Wh kg^{-1}), long cycle life, non-memory effect, and low self-discharging, lithium-ion batteries (LIBs) have become one of the most popular rechargeable batteries [1]. After their commercialization in the 1990s, LIBs soon found resounding application in a myriad of consumer electronic devices such as cell phones, laptops, and cameras, etc. Moreover, they have also become one of the most preferred power sources for electric and hybrid electric vehicles (EVs and HEVs) and are promising for future smart grid storage [1, 2].

Figure 21.1a shows the working mechanism of a LIB that consists of anode (negative electrode, e.g., graphite), cathode (positive electrode, e.g., $LiFePO_4$), electrolyte, and separator [3]. The electrode materials are coated on current collectors, whereas copper foil is used as the current collector for the anode, and aluminum foil is used as the current collector for the cathode. A porous thin film of polypropylene or polyethylene is served as the separator. During the discharge process, Li-ions that are stored in the anode will transport through the separator to the cathode via the Li-ion conducting electrolyte, thus generating current in the external circuit. Upon charging, the external power source will force the Li-ion to be deintercalated from the cathode and move back to the anode again through the electrolyte.

Depending on the way of packaging, there are mainly three different cell designs of LIB—cylindrical, prismatic, and pouch type. Cylindrical cell is named according to its shape and size. Typical 18,650 cell is one of the most popular cell packages, which has a diameter of 18 mm and height of 65 mm. They hold volumetric energy densities of 600-–650 Wh kg^{-1}, which are ~20% higher than those of prismatic and pouch counterparts [4]. As shown in Fig. 21.1b, cylindrical cells contain spiral wounded electrodes sealed within aluminum or stainless steel can. There are also safety protection components in most cylindrical cells, including positive thermal coefficient (PTC) switch, pressure activated disconnect, and gas release vent [5], which can greatly enhance the safety properties when exposed to excess current, heat, and/or pressure. Compared with cylindrical cells, the size of prismatic and pouch cells is easily customized for the final product. The prismatic cells are formed into elliptic spiral and are packaged by aluminum or stainless steel cans (Fig. 21.1c). The metallic casings of cylindrical and prismatic cells can provide solid structures for the active material, handling shock, vibration, and nail penetration issues [5]. In contrast, the pouch cells are prepared by stacking the electrode plates and are packaged by aluminum/polymer composite film (Fig. 21.1d). Different from the rigid and heavy metal cans used for cylindrical and prismatic cells, the aluminum/polymer composite film is widely used as the package envelop for pouch cell due to its excellent properties such as lightweight, good plasticity, easiness of shaping, and high barrier levels to gas, light, and moisture, thus enabling the pouch cell with lightweight, high energy density, and high dimensional freedom [6]. The aluminum/polymer composite film shows a typical polymer-aluminum-polymer laminated structure, in which the innermost cast polypropylene layer is the heat-sealing layer, the intermediate aluminum layer is a moisture and air barrier, and the outmost polyamide layer protects the cell from damages [7]. The comparison of various cell format parameters is shown in Table 21.1.

Though commercial LIBs can be manufactured into different format, efficient electrode materials are the hard core for the performance improvement of LIBs. Currently, commercialized LIBs are not the best in class in terms of safety, power density, rate capability, low temperature

C. Han (✉) · H. Li · J. Chen · B. Li · C. P. Wong
College of Materials and Engineering, Shenzhen University, Shenzhen Key Laboratory of Special Functional Materials, Shenzhen, China
e-mail: hancuiping06@126.com

C. P. Wong et al. (eds.), *Nano-Bio- Electronic, Photonic and MEMS Packaging*, https://doi.org/10.1007/978-3-030-49991-4_21

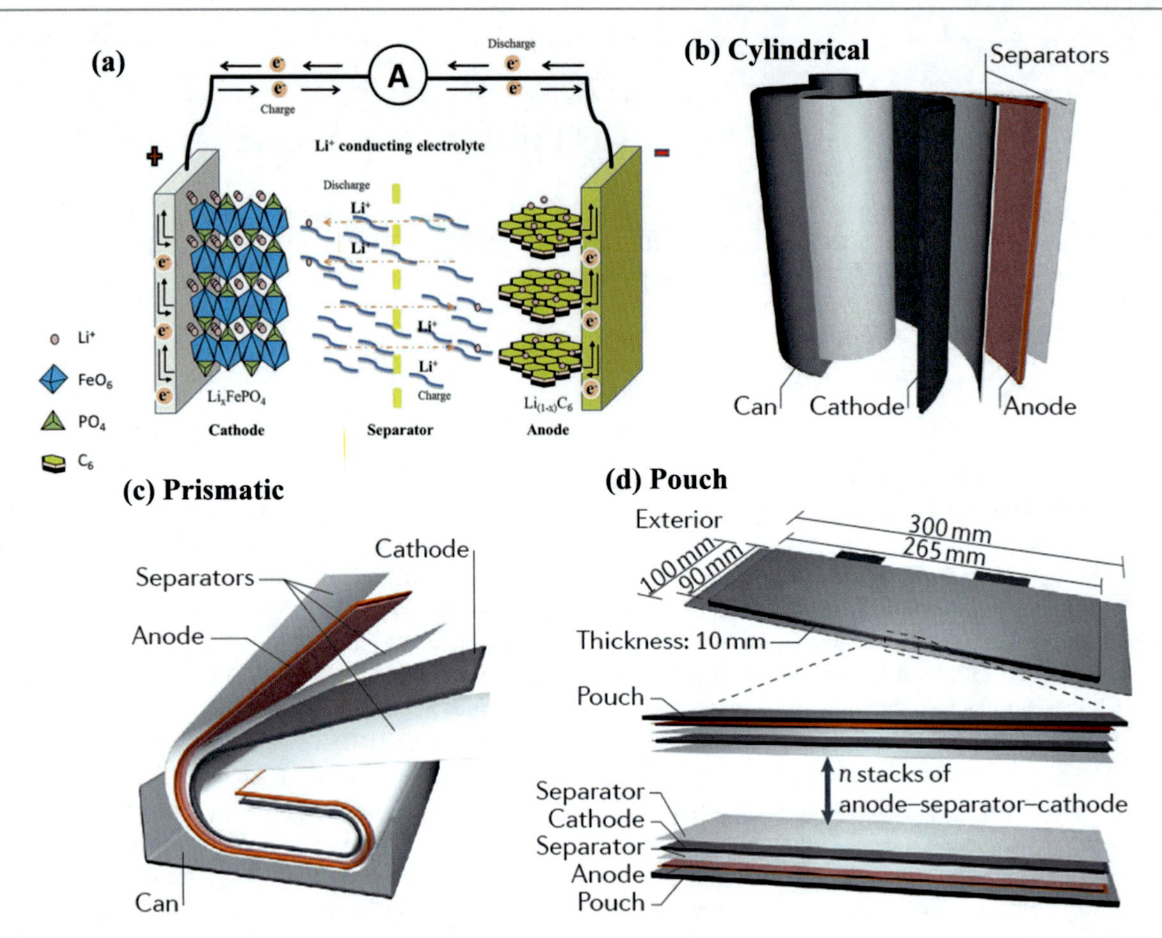

Fig. 21.1 (**a**) Schematic illustration of a typical lithium-ion battery. Reproduced from Ref. [3] with permission from the Royal Society of Chemistry. Three representative commercial cell structures: (**b**) cylindrical-type cell, (**c**) prismatic-type cell, (**d**) pouch-type cell. (Reproduced from Ref. [4] with permission from Springer Nature)

Table 21.1 Comparison of various cell format parameters

	Small cylindrical	Large cylindrical	Prismatic	Pouch
Shape	Contained in metal casing	Contained in metal casing	Contained in semihard plastic or metal casing	Contained in aluminum soft bag
Connections	Welded nickel or copper strips or plates	Threaded stud for nut or threaded hole for bolt	Thread hole for bolt	Tabs that are clamped, welded, or soldered
Retention against expansion	Inherent from cylindrical shape	Inherent from cylindrical shape	Requires retaining plates at ends of battery	Requires retaining plates at ends of battery
Appropriateness for small projects	Poor: high design effort, requires welding, labor intensive	Good: some design effort	Excellent: little design effort	Very poor: design effort too high
Appropriateness for production runs	Good: welded connections are reliable	Good	Excellent	Excellent
Field replacement	Not possible	Possible	Possible	Not possible
Delamination	Not possible	Not possible	Possible	Highly possible
Compressive force holding	Excellent	Excellent	Poor	Extremely poor
Thermal management	Not favorable	Not favorable	Favorable	Favorable
Heat dissipation	Poor	Poor	Fair	Good
Local stress	No	No	No	Yes
Safety	Good, integrated with PTC	Good, integrated with PTC	Good, integrated with PTC	Poor, no safety feature included
Ease of assembly	Poor	Poor	Excellent	Poor
Heat shrink wrapping	Yes	Yes	Depend on casing material	No

Reproduced from Ref. [5] with permission from Elsevier

performance, cost, etc. The current EV battery cells have a specific energy of 200–250 Wh kg^{-1}, but an aggressive goal to reach 500 Wh kg^{-1} with a >10-year life has been proposed by the Battery500 Consortium [1]. To better satisfy the requirements for high power applications, such as future electrical vehicles and stationary grid storage, LIBs with large capacity, high power density, enhanced rate capability, long-term cycle stability, and low cost are urgently needed. Now the dominant anode material for commercial LIBs is graphite due to its low working potential and affordable price. In the past decades, metal oxides and alloys with higher specific capacities and enhanced safety properties have been widely investigated. For example, Si is the second-most abundant element in the earth's crust and is environmentally benign. The alloying reaction between Si and Li can give a final product of $Li_{4.4}Si$, delivering a theoretical capacity as high as 4200mAhg^{-1}, which is significantly higher than that of graphite (372 mAh g^{-1}). As for the cathode, the current LIB is based on lithium metal oxides such as $LiCoO_2$, $LiFePO_4$, $LiMn_2O_4$, $LiNi_xCo_yMn_zO_2$ (NMC), etc. A wide variety of nanostructured cathode materials including high-Ni NCM, $Li_3V_2(PO_4)_3$, V_2O_5, etc. have been explored with increased storage capacity, improved redox potential, and extended cycle life. Further increase in energy and power density relies on advanced electrode materials. With this regard, it is necessary to develop novel high-performance electrode materials with high electric conductivity for fast electron transport, well-designed morphologies for reduced Li-ion diffusion pathway, and large accessible surface area for better electrode/electrolyte contact.

Graphene has found wide application in lithium-ion battery technology due to its unique virtues like large theoretical specific surface area (2630 m^2g^{-1}), superior electric conductivity (~2000 S cm^{-1}), broad electrochemical window, high flexibility, high thermal conductivity (4840–5300 W m^{-1} K^{-1}), and excellent mechanical stability [8–10]. It can not only act as an anode material independently but also serve as a scaffold to support a wide range of electroactive materials such as transitional metal anodes, metal oxides, and cathodes by different configurations (encapsulated, mixed, wrapped, anchored, sandwiched, layered) [10]. With the incorporation of graphene, the obtained composites demonstrate greatly enhanced electrochemical performance due to largely improved electrical conductivity and Li-ion diffusion, well-accommodated volume expansion/contraction, and suppressed particle agglomeration during cycling. Besides, its superior thermal conductivity may facilitate the dissipation of the heat generated during cycling. Therefore, graphene-containing nanomaterials are promising for high-performance LIB applications. In this chapter, recent advances in the rational design and preparation of graphene-containing composites containing a variety of electrochemical active materials (e.g., transition metal anodes, metal oxide anodes, cathodes, etc.) for LIBs will be summarized. Particular emphasis will be put on the synthetic routes, structural configurations, electrochemical performances, and the lithium storage properties of the hybrids.

21.2 Graphene-Based Material as Anodes

21.2.1 Graphene Nanostructures as Anodes

Now the dominate anode material for commercial LIBs is graphite due to its low working potential and affordable price. It takes an insertion/extraction working principle during which lithium-ions insert into/extract from the planes of graphite sheets [11]. The specific lithium storage capacity of graphite is 372 mAh g^{-1} with the formation of LiC_6. Due to the intriguing characteristics like large surface area, superior electric conductivity, excellent mechanical flexibility, as well as rich surface chemistry, graphene was studied as an anode for LIBs [12]. It is proposed that lithium-ions can be adsorbed on both sides of a single layered graphene sheet by the formation of Li_2C_6 with a theoretical capacity of 744 mAh g^{-1}, which is much higher than graphite anode [13, 14].

Intensive research has been devoted to achieve high Li storage performances of graphene, which is affected by a series of factors including electrical conductivity, specific surface area, porosity, defects, interlayer spacing, surface chemistry, etc. Controlled design and synthesis of graphene with diminished layer number, decreased sheet size, increased edge sites, and defects all can be highly beneficial for the enhancement of cyclic performance and rate capability of graphene anode [15]. Yoo et al. [16] tested the lithium storage properties of graphene prepared by hydrazine hydrate reduction of exfoliated graphite oxide. It delivers a specific capacity of 540 mAhg^{-1} at a current density of 0.05Ag^{-1}, which is much larger than that of 372mAhg^{-1} of graphite. The incorporation of carbon nanotubes (CNT) and fullerene (C_{60}) as interlayer spacers further increases the specific capacity to 730 and 784 mAhg^{-1}, respectively. Lian et al. [17] reported graphene sheets with a curled morphology that achieved an initial charge capacity as high as 1264 mAhg^{-1} at a current density of 0.1Ag^{-1} due to reduced layers (~4 layers) and relatively large specific surface area (492.5 m^2g^{-1}). Wang et al. [13] prepared graphene nanosheets with flower petal-like morphology through hydrazine hydrate reduction of graphene oxide under refluxing at 100 °C. It delivers an initial discharge capacity of 945 mAhg^{-1} and a reversible charge capacity of 650 mAhg^{-1} under 1C rate (1C = 744 mAhg^{-1}).

Doping of graphene sheets with heteroatoms such as N, P, S, and B is able to introduce defects and/or disorders into graphene framework, which can effectively contribute to provide more Li storage active sites, as well as enhance the electrical conductivity [18–22]. For example, Ma et al. [18] successfully prepared P and N co-doped graphene nanomesh by introducing $(NH_4)_3PO_4$ as both P and N source into the MgO-templated chemical vapor deposition process. The as-prepared graphene has a P and N doping concentrations of ~0.6 at% and 2.6 at% and shows a high reversible capacity of 2250 mAh g^{-1} at 0.05 A g^{-1} and 750 $mAhg^{-1}$ at 1 Ag^{-1}, which is much larger than that of the undoped graphene. Similarly, a capacity of 1043 mAh g^{-1} for N-doped graphene and 1549 mAh g^{-1} for B-doped graphene at a current density of 0.05 Ag^{-1} was achieved in Wu's report [19]. More importantly, the rate capability is greatly improved after doping with N and B atoms (~199 mAh g^{-1} for the N-doped graphene and ~235 mAh g^{-1} for the B-doped graphene under 25 Ag^{-1}). First principle calculations based on density functional theory studied the effect of nitrogen-doping configurations (i.e., pyridinic, pyrrolic, graphitic doping) on the Li storage properties of graphene and found that pyridinic-N-doped graphene is more suitable for Li storage due to a stronger bonding with lithium [23]. This calculation result coincides well with Hu's work on N-doped graphene, which is prepared by thermal reduction of graphene oxide with ammonia hydroxide [21]. The high-resolution N1s XPS spectrum of as-prepared N-doped graphene suggests that pyridinic-N doping is the dominant doping configuration. Electrochemical measurements showed that the pyridinic-N-doped graphene showed a reversible capacity of 453 mA h g^{-1} after 550 cycles at a current density of 2 A g^{-1}, which is 90% higher than that of undoped graphene.

Chemical reduction of graphene oxide is one of the most commonly used strategies to prepare graphene in large scale. However, the as-prepared graphene usually suffers from restacking due to van der Waals attraction, as well as poor electrical conductivity due to the presence of oxygen functional groups. With this regard, highly conductive carbon nanotubes or carbon nanofibers are introduced to prevent the restacking of graphene sheets and improve its electrical conductivity [16, 24–26]. Vinayan et al. [26] prepared a multiwall carbon nanotube (MWNT)/graphene composite by mixing the negatively charged MWNT with the positively charged graphene (Fig. 21.2). The negatively charged MWNTs not only prevent the restacking of graphene sheets but also contribute to a higher electric conductivity. The hybrid structure gives a reversible discharge capacity of 768 $mAhg^{-1}$ after 100 cycles at 0.09 Ag^{-1}. Furthermore, anchoring of metal oxides into the graphene matrix can achieve a synergy between the high capacity of metal oxides and the large surface area of graphene [27, 28]. As an

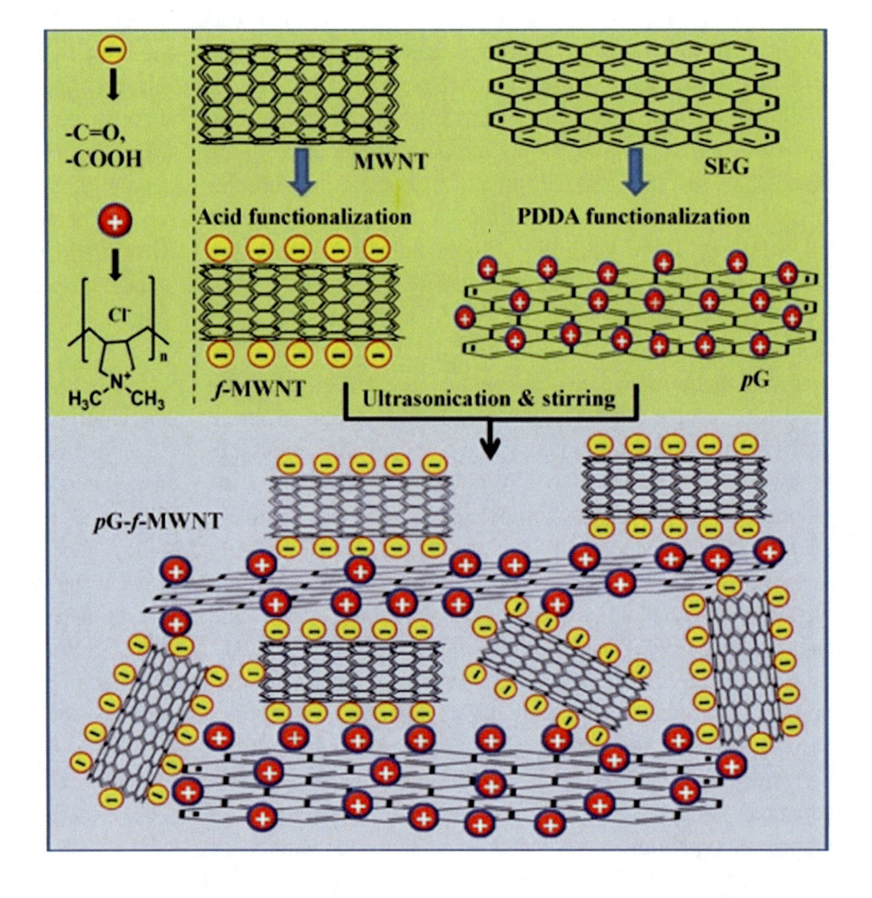

Fig. 21.2 Scheme outlining the synthesis procedure adopted for synthesizing the graphene/MWNT nanocomposite. (Reproduced from Ref. [26] with permission from the Royal Society of Chemistry)

example, the combination of iron oxides, carbon nanotubes, and graphene sheets into one composite (G-CNT-Fe) enables a large number of Li-ions to be inserted and stored in the hierarchical architecture, rendering a specific capacity of 1024 mAh g^{-1} after 45 cycles at 0.1 Ag^{-1} [28].

Though great improvements have been achieved in graphene anode for LIBs, several challenges still remain [26, 29], such as the high irreversible capacity loss during the initial several cycles due to the chemical reaction between lithium-ions and the surface oxygen-containing functional groups of graphene, as well as the formation of a solid electrolyte interface (SEI) layer accumulated from the decomposition products of electrolytes at lower potentials (below 1.0 V vs Li/Li^{+}). Secondly, it is difficult for graphene-based full battery to give a stable and flat output voltage because of the inconspicuous voltage plateaus observed during discharge/charge process. Moreover, the limited rate capability and great capacity decay also hinds its application as an anode for LIBs.

21.2.2 Alloying Graphene Hybrids

The electrochemical alloying reaction of lithium with elements (Si, Sb, Ge) or metals (Sn, In, Cd) in organic carbonate electrolytes offers a high theoretical lithium storage capacity (Eq. 21.1) [30, 31]. For instance, Si is the second-most abundant element in the earth's crust and is environmentally benign. The alloying reaction between Si and Li can give a final product of $Li_{4.4}Si$, delivering a theoretical capacity as high as 4200 mAh g^{-1}, which is significantly higher than that of graphite (372mAhg^{-1}) (Eq. 21.2) [21]. The theoretical capacity for other alloying anodes, such as Ge, Sn, and Sb, are also very attractive (1623 mAh g^{-1}, 994 mAhg^{-1}, and 660 mAh g^{-1}, respectively) (Eqs. 21.3, 21.4, and 21.5) [32]. Unfortunately, their practical applications in reversible LIBs are severely hindered by the rapid capacity decay due to material pulverization, disintegration of the composite electrodes, and unstable formation of solid electrolyte interface (SEI) layer, with all of these originating from the large volume expansion upon lithium insertion (>300% for Si, 260% for Sn, 370% for Ge, 200% for Sb.) and volume contraction upon lithium extraction [31–34].

$$\text{General: } M + xLi^{+} + xe^{-} \rightarrow Li_{x}M \qquad (21.1)$$

$$\text{For Si: } Si + 4.4Li^{+} + 4.4e^{-} \rightarrow Li_{4.4}Si \qquad (21.2)$$

$$\text{For Sn: } Sn + 4.4Li^{+} + 4.4e^{-} \rightarrow Li_{4.4}Sn \qquad (21.3)$$

$$\text{For Ge: } Ge + 4.4Li^{+} + 4.4e^{-} \rightarrow Li_{4.4}Ge \qquad (21.4)$$

$$\text{For Sb: } Sb + 3Li^{+} + 3e^{-} \rightarrow Li_{3}Sb \qquad (21.5)$$

Encapsulations of these metals with graphene have been proven to be promising in circumventing these problems due to its exceptional mechanical, chemical, and electronical properties, and high flexibility. In the hybrids, graphene sheets form an electric conductive framework to facilitate electron transfer and provide large surface area and possibly a nanoporous structure with large void spaces. Meanwhile, the flexible confinement of graphene and void spaces could effectively accommodate the large volume expansion/contraction during Li insertion/extraction and prevent the detachment of electrode materials from the current collector and assist the formation of a stable solid electrolyte interface (SEI) film on surface of the electrode [35]. Though pulverization may happen to larger electrode particles, the fractured small pieces can remain well trapped within the conducting network, thus preserving good electron transport among fractured particles. As a result, the graphene/metal composite electrode retains a promoted Li storage capacity, good cycling stability, and high rate capability (Table 21.2) [36, 37]. For instance, Ko et al. [38] reported a graphene/Si composite with amorphous Si particles (5–10 nm) uniformly anchored on graphene backbone. The hybrid material showed a remarkable initial coulombic efficiency of 92.5% and a high reversible specific capacity of 2858 mAhg^{-1} at 56 mAg^{-1}.

Preventing the agglomeration of nanoparticles and graphene during the preparation of graphene/metal hybrids is challenging and critical for the performance improvement in LIBs. Commercially available Si nanoparticles have a high tendency to aggregate together and are hard to be redispersed in aqueous solutions due to its surface hydrophobicity, thus leading to inhomogeneous mixing of Si and graphene and limited performance improvement. Various approaches have been devoted to achieve uniform anchoring of Si nanoparticles on surface of graphene sheets. Wong et al. [39] found that commercial Si nanoparticles could be well dispersed in organic solvents like N-methylpyrrolidone (NMP). Therefore, a homogeneous Si/graphene composite could be easily obtained by using NMP as solvents for both Si nanoparticles and graphene. Yang et al. [40] successfully grafted Si nanoparticles on surface of p-phenylenediamine functionalized graphene. Zhou et al. [41] developed an electrostatic self-assembly approach by first endowing Si nanoparticles with positively charged poly(diallydimethylammonium chloride) (PDDA) followed by mixing with negatively charged graphene oxides. Li et al. [42] constructed a conformal graphene cage around microsized Si particles using a chemical vapor deposition (CVD) method. The large void spaces between graphene cage and Si particles can effectively buffer the volume expansion during lithiation. Meanwhile, the graphene cage helps retain electrical conductivity of the electrode and contributes to the formation of a stable SEI layer around it. This graphene encapsulated Si

Table 21.2 Electrochemical performances of the graphene/metal hybrids for LIBs reported in the literature

Graphene/metal hybrids	Metal size (nm)	Metal content (wt%)	Voltage range (V vs Li/Li$^+$)	Current density (Ag^{-1})	Specific capacity (mAhg^{-1})	ICE (%)	Capacity retention (%) /cycles/ current density (Ag^{-1})	Refs.
Amorphous Si nanoparticle/G	5–10	82	0.01–1.5	28	1148	92.5	No fading/1000/ 14	[38]
In situ grown Si nanowire/G	20–50 in diameter	/	0.005–1.5	0.42	1386 (30th)	57.3	62.3/2–30/0.42	[46]
Porous Si nanowire/ graphene	10,000 in length	45	0.005–0.8	0.105	2470	64	87/20/0.105	[47]
Si nanowire/RGO	60 in diameter	90.9	0.01–1.5	10.8	825	78	91.8/100/1.2	[48]
3D porous Si/G	300	82.9	0.01–1.2	2.1	697.8	/	/	[49]
3D porous Si/G	10–100	/	0.01–2.0	2	1000	/	81/first few-200/2	[50]
Pyrolyzed PANI-grafted Si/G	<50	66	0.01–2.0	2	~900 (300th)	73.2	76/300/2	[57]
Si nanoparticle/G	50–100	72	0.002–1.5	12	700	60	76/120/1	[58]
Si/RGO aerogel	~100	46	0.001–2.5	5	705	67.2	83.4/2–50/0.5	[59]
Si/honeycomb rGO composite film	20–50	/	0.01–1.5	1	491	47.4	~66/2–50/0.05	[60]
Multilayered Si nanoparticle/ RGO	~100	76	0.01–2.0	24	700	73	80/100/2.4	[61]
Carbon-coated Si/G	<200	63.9	0.02–1.2	0.3	902 (100th)	57.3	68/2–100/0.3	[62]
Carbon-coated G/Si	10–15	~70	0.01–1.0	0.05	3.2 mAh cm^{-2} (100th)	64	/	[63]
Electrospun CNF/Si/G	90–120	42	0–1.5	1	567	71.2	91/50/0.1	[64]
Electrospun CNFs/ SiNPs@rGO	<50	/	0.005–2.5	0.89	1048 (200th)	62	~69/2–200/0.89	[65]
3D Sn/G	5–30	46.8	0.003–3.0	10	270	~69	96.3/2–1000/2	[66]
Sn/N-doped graphene	/	70	0.005–2.0	2	307	65	70/10–100/0.1	[67]
Carbon-coated Sn/G	60–100	77.9	0.01–3.0	3.75	286	/	52.9/100/0.75	[44]
Sandwich-structured G/Sn/ G	~50	60.1	0.005–2.0	1.6	265	66.5	~74/10–60/0.05	[45]
Sn nanopillar array/G	150–200 in diameter	70	0.002–3.0	5	408	/	~72/30/0.05	[51]
Carbon-coated Sn nanocable/RGO	200	61	0.005–2.5	0.1	~630 (50th)	66.9	~46/50/0.1	[52]
Carbon-coated Ge/RGO	70 ± 30	79.3	0.01–1.5	20	720	43	~100/21–600/1	[68]
Ge/GN sponge	2–5	76.4	0.01–2.2	6.4	556	93.3	61.5/50/0.1	[69]
Sandwich-structured C/Ge/ G	20–30	89.9	0.01–1.5	Charge at 20C Discharge at 0.4C	746.3	/	86.4/2–160/charge at 1C, discharge at 0.4C (1C = 1600mAhg^{-1})	[70]
Ge/few layer graphene	150–300	50	0–1.5	0.2	846.3 (50th)	67	86/50/0.2	[71]
Graphene-coated Ge nanowire	46	98.3	0.001–1.5	4.8	1059 (200th)	69	90/24–200/4.8	[53]
Ge nanowire/G	/	75	0.01–3.0	0.139	600 (20th)	/	/	[54]

ICE initial coulombic efficiency

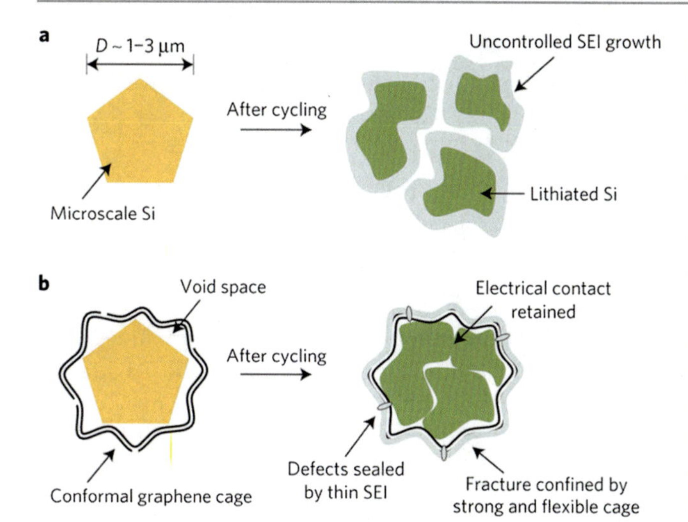

Fig. 21.3 Design and structure of graphene cage encapsulation of Si macroparticle. (Reproduced from Ref. [42] with permission from Springer Nature)

microparticle delivered a high reversible capacity of~3300 mAh g^{-1} at C/20 and good capacity retention of over 85% for 300 cycles (Fig. 21.3).

Metallic tin has a low melting point (232 °C) and prefers to grow into large particles during high-temperature calcination processes, thus making it challenging to be engineered into various nanostructures when hybridized with graphene sheets [43, 44]. The construction of an additional surface carbon coating layer around Sn could effectively suppress the growth and aggregation at high temperature. In the work done by Luo et al. [45], a glucose-derived carbon layer was coated on surface of graphene/SnO_2 precursor. After calcinations, metallic Sn nanoplates reduced from SnO_2 were uniformly distributed on graphene sheets. The glucose-derived carbon layer not only helps direct the growth of metallic Sn but also contributes to buffer the volume expansion of Sn during cycling. The composite gives a capacity of over 590 mAh g^{-1} after 60 cycles at 0.05 Ag^{-1}. Wang et al. [44] reported that graphene-supported SnO_2 nanoparticles can be reduced to metallic Sn (60–100 nm) followed by coating with a surface carbon layer (around 50 nm thick) simultaneously by using a CVD method. The hybrid electrode has an initial discharge capacity as high as 1069 $mAhg^{-1}$ at a current density of 0.075 Ag^{-1}.

Meanwhile, nanostructuring of the metallic elements in the graphene/metal hybrid also helps to enhance both cycle stability and rate capability. Various structures, such as one-dimensional (21.1D) Si nanowire [46–48], three-dimensional (3D) porous Si sphere [49, 50], 1D Sn nanopillar [51], 1D Sn nanocable [52], and 1D Ge nanowire [53, 54], have been reported. The nanostructured metallic elements could well buffer the volume expansion, while graphene ensures a conductive network for lithium-ions and electron transport. Ren et al. [48] reported the direct growth of [111] facet-oriented single crystalline Si nanowires (NWs) on reduced graphene oxide (RGO) nanosheets via an Au-seeded CVD method. The as-grown Si NWs allow efficient volume change on the vertical direction of the Si NWs, and the RGO substrate facilitates electron transfer between the Si NWs. The synergy of these two components contributes to a high initial charge capacity of 2428 mA h g^{-1} at 300 mAg^{-1} and a capacity retention of 91.8% over 100 cycles at 1.2 Ag^{-1}. Chockla et al. [54] reported the growth of Ge nanowires and Si nanowires using solution-liquid-solid (SLS) growth method and supercritical fluid-liquid-solid (SFLS) growth method, respectively. The obtained nanowires were further mixed with RGO in toluene followed by vacuum filtration. They found that the incorporated RGO contributes to improve the Li storage capacity and stabilize the cycling performance.

Though graphene/metal composites exhibit many unique synergistic effects, a theoretical study of Li intercalation mechanism in these hybrid materials could provide more understanding of the properties and performance of graphene/metal composite and assist the optimization of best configuration [55, 56]. Hwang et al. [55], examined the lithiation behavior of silicon-graphene composites using first principle calculation based on density functional theory (DFT). They predicted that the interfacial Li-ions exhibit five times higher mobility along the Si/graphene interface than the bulk Si, and Li atoms are mostly incorporated into the Si matrix rather than at the Si/graphene interface, leading to a voltage profile that mostly resembles those of pure Si. Moreover, it remains challenging to increase the low initial coulombic efficiency and reduce the irreversible capacity loss caused by the formation of solid electrolyte interface (SEI) layer and the consumption of lithium by the reaction with surface functional groups of graphene.

21.2.3 Graphene-Metal Oxide Hybrids

Transition metal oxide anodes, typically offering a capacity larger than that of graphite, have gained much attention due to their low cost and easy preparation [72]. Generally, the electrochemical reaction with lithium follows three different mechanisms: Li insertion/extraction reaction, conversion (redox) reaction, alloying/de-alloying reaction [73, 74].

Typical insertion-type metal oxides are compounds with a two-dimensional (2D) layered structure or a 3D network structure that can reversibly intercalate/deintercalate Li-ion into/out of the host lattice (Eqs. 21.6, 21.7, and 21.8). Insertion-type electrodes include titanium dioxides (anatase, rutile, bronze), Nb_2O_5, vanadium oxides (VO, V_2O_3, VO_2, V_2O_5), layered MoO, $Li_4Ti_5O_{12}$, $MgTi_2O_5$, $LiTiNbO_5$, $TiNb_2O_7$, etc. [75]. The volume change accompanying this

Li intercalation process is comparatively small (less than 4% for anatase TiO_2 and nearly 0% for $Li_4Ti_5O_{12}$), which ensures the structural integrity of the host lattice and renders high coulombic efficiency and long cycling life [76, 77]. However, the specific capacities of this type of electrodes are usually limited by available vacancies presented in the host lattice (e.g., theoretical specific capacities for TiO_2 and $Li_4Ti_5O_{12}$ are 336 and 175mAhg^{-1}, respectively). Besides, their electrochemical performance is greatly hindered by their poor intrinsic electrical conductivity and low lithium-ion diffusion coefficient [78]. Thus, the incorporation of conductive substances like graphene is necessary and important for the performance improvement.

$$\text{Insertion: } Li_mM_xO_y + nLi^+ + ne^- \leftrightarrow Li_{m+n}M_xO_y \quad (\text{M is Ti, Nb, Mo}) \tag{21.6}$$

$$\text{For } TiO_2\text{: } TiO_2 + nLi^+ + ne^- \leftrightarrow Li_nTiO_2 \ (n \le 1) \tag{21.7}$$

$$\text{For } Li_4Ti_5O_{12}\text{: } Li_4Ti_5O_{12} + nLi^+ + ne^- \leftrightarrow Li_{4+n}Ti_5O_{12} \ (n \le 5) \tag{21.8}$$

The conversion reaction is characterized by the formation and decomposition of Li_2O along with the reduction and oxidation of metal nanoparticles (Eqs. 21.9, 21.10, 21.11, and 21.12). Conversion-type anodes generally deliver a high theoretical specific capacity because the reduction reaction of transition metal oxides to their metallic state involves multiple electrons. As a consequence, a wide range of metal oxides including binary oxides (Co_xO_y, Mn_xO_y, Fe_xO_y, Cu_xO, NiO, RuO_2, MoO_2, MoO_3, Cr_2O_3, etc.) and ternary oxides ($CoMn_2O_4$, $CoFe_2O_4$, $NiFe_2O_4$, $CuFe_2O_4$, $MnCo_2O_4$, $FeCo_2O_4$, $NiCo_2O_4$, $CuCo_2O_4$, etc.) have been explored [79, 80]. Unfortunately, drastic volume changes are observed during cycling, which lead to material pulverization and eventually disintegration of the electrode upon extended cycles. Besides, conversion-type metal oxides also suffer from low electrical conductivity, large voltage hysteresis, insufficient coulombic efficiency, and short cycle life [79].

$$\text{Conversion: } M_xO_y + 2yLi^+ + 2ye^- \leftrightarrow yLi_2O + xM \ (\text{M is Mn, Fe, Co, Ni, Cu, Cr, Mo, Ru}) \tag{21.9}$$

$$\text{For Co} - \text{based oxides } (CoO, Co_3O_4)\text{: } Co_xO_y + 2yLi^+ + 2ye^- \leftrightarrow xCo + yLi_2O \tag{21.10}$$

$$\text{For Mn}-\text{based oxides } (MnO, Mn_3O_4, Mn_2O_3, MnO_2)\text{: } Mn_xO_y + 2yLi^+ + 2ye^- \leftrightarrow xMn + yLi_2O \tag{21.11}$$

$$\text{For Fe} - \text{based oxides } (FeO, Fe_3O_4, Fe_2O_3)\text{: } Fe_xO_y + 2yLi^+ + 2ye^- \leftrightarrow xFe + yLi_2O \tag{21.12}$$

As discussed above, the alloying reaction of lithium with elements or semimetals like Si, Sb, Sn, Ge, In, and Cd offers a high Li storage capacity. Since metal oxides are more cheap and easy fabrication, their corresponding oxides (SnO, SnO_2, $ZnFe_2O_4$, $ZnCo_2O_4$, Sb_2O_3, GeO_2, $MGeO_3$ (M = Cu, Fe, and Co), etc.) are also explored as Li storage materials [43, 81]. During the lithium intercalation process, metal oxides are firstly reduced to their metallic state followed by the formation of Li alloys (Eqs. 21.13, 21.14, 21.15, and 21.16). The generated Li_2O accompanying the reduction of metal oxides serves as a buffering domain for the large volume variation during the subsequent alloying process, thereby contributing to maintain the Li storage capacity. In comparison, graphene-containing composites afford not only a buffering matrix for the large volume variation but also a conductive scaffold for electrons and lithium-ion transport. Thereby, the graphene/metal oxide hybrids exhibit improved specific capacity, rate capability, and cycle stability regardless of the Li storage mechanisms.

$$\text{Conversion: } M_xO_y + 2yLi^+ + 2ye^- \leftrightarrow yLi_2O + xM \ (\text{M is Si, Sb, Sn, Zn, In, Bi, Cd, Ge}) \tag{21.13}$$

$$\text{Alloying: } M + zLi^+ + ze^- \leftrightarrow Li_zM \tag{21.14}$$

$$\text{For } SnO_2\text{: } SnO_2 + 4Li^+ + 4e^- \leftrightarrow Sn + 2Li_2O \tag{21.15}$$

$$Sn + 4.4Li^+ + 4.4e^- \leftrightarrow Li_{4.4}Sn \tag{21.16}$$

Generally, the fabrication of graphene/metal oxides is mainly classified as ex situ and in situ hybridization. In the ex situ hybridization, graphene oxide (GO) and metal oxide are firstly prepared by different approaches and then mixed together for the loading of metal oxide, after which the GO is reduced into graphene either using chemical or thermal reduction. In this nanostructure, graphene serves as a conductive network to promote efficient electron transfer. Through simple mechanical mixing, the $Li_4Ti_5O_{12}$/graphene hybrid exhibits improved conductivity and reduced polarization.

As a consequence, a specific capacity of 122 mAh g^{-1} at 30 C and a capacity loss of less than 6% over 300 cycles at 20 C were achieved with 5 wt.% graphene [82]. But the performance improvement is restricted by low degree of hybridization. The solution mixing strategy, such as electrostatic assembly, can realize the well decoration of nanocrystals on graphene [100, 104, 122]. In Zhao's report [122], the prefabricated hollow SnO_2-Fe_2O_3 nanoparticles (h-SnO_2-Fe_2O_3) are first positively charged by grafting with aminopropyltrimethoxysilane (APS), and GO is negatively charged as determined by zeta potential test. Driven by the electrostatic interaction, the h-SnO_2-Fe_2O_3 nanoparticles are tightly encapsulated by GO when mixed together. The obtained hybrid architecture generated a capacity of 550 mAh g^{-1} at 1000 mA g^{-1}. Other GO-wrapped metal oxides, such as Cu_2O, α-Fe_2O_3, SnO_2, Co_3O_4, and $Ni(OH)_2$, with varied particle size, morphology, and configuration were successfully obtained through electrostatic assembly [100, 112].

In comparison with the ex situ way, the in situ hybridization can achieve more uniform hybridization. One widely used method is hydrothermal/solvothermal preparation. GO with rich surface functional groups can be well dispersed in aqueous or organic solvents. Meanwhile, metal salt precursor dissolved in the same solvent undergoes hydrolysis or decomposition during hydrothermal/solvothermal process, resulting in the direct growth of metal oxides on the reactive sites of graphene sheets. Dong et al. [84] demonstrated a $Li_4Ti_5O_{12}$/graphene composite architecture consisting of graphene sheets dispersed between and inside $Li_4Ti_5O_{12}$ particles. The inside graphene can facilitate electron transfer within $Li_4Ti_5O_{12}$ particles, while the interparticle graphene, together with a thin carbon coating layer, promotes efficient electron transport among $Li_4Ti_5O_{12}$ particles. The presence of graphene also plays an important role in reducing the particle size and preventing aggregation of as-grown metal oxides. As an example, in the Co_3O_4/graphene composite reported by Cheng [89], Co_3O_4 nanoparticles with a size of 10–30 nm are homogeneously anchored on the thin graphene layers. By contrast, only microsized Co_3O_4 aggregates can be formed without the presence of graphene sheets. Atomic layer deposition is used to achieve in situ grafting of both amorphous and crystalline SnO_2 on graphene nanosheets matrix. The SnO_2/graphene composite exhibits a sandwich-like structure and demonstrates a capacity of 793 $mAhg^{-1}$ after 150 cycles at $0.4Ag^{-1}$ [126].

As summarized in Table 21.3, the graphene/metal oxide composites occur in varied configurations, including encapsulated, mixed, wrapped, anchored, sandwich-like, and layered structure (Fig. 21.4) [10]. Though the performance is related with the nanostructures of metal oxides,

Table 21.3 Electrochemical performances of the graphene/metal oxide hybrids for LIBs reported in the literature

G/M_xO_y hybrids	M_xO_y size (nm)	M_xO_y content (wt%)	Voltage range (V vs Li/Li$^+$)	Current density (Ag^{-1})	Specific capacity ($mAhg^{-1}$)	ICE (%)	Capacity retention(%)/ cycles/current density (Ag^{-1})	Refs.
$Li_4Ti_5O_{12}$/G	100–400	95	0.8–2.5	30C	122	/	94.8%/300/20C (1C = $175mAhg^{-1}$)	[82]
$Li_4Ti_5O_{12}$/G	30	84.5	1.0–2.5	60C	82.5	/	96.9%/100/1C 97.5%/100/10C 97.1%/100/60C (1C = $175mAhg^{-1}$)	[83]
$Li_4Ti_5O_{12}$/G	100–300	98.8	1.0–2.5	30C	123.5	/	94.6%/100/10C (1C = $175mAhg^{-1}$)	[84]
Mesoporous TiO_2/G aerogels	15–20	64.8	1.0–3.0	5	99	/	/	[85]
TiO_2 nanotube/G	20 in diameter, 200 in length	84.5	1.0–3.0	10C	118	/	79.5%/150/5C (1C = $337mAhg^{-1}$)	[86]
N-doped TiO_2 nanotubes/N-doped G	7–10 in diameter	89.7	1.0–3.0	5	90	/	/	[87]
TiO_2(B) nanosheets/G	10 in thickness, 100x120 in size	77.8	1.0–3.0	1C 40C	275 200	~82	80%/1000/40C (1C = $335mAhg^{-1}$)	[88]
Co_3O_4/G	10–30	75.4	0.01–3.0	0.05	~935 (30th)	68.6	Slight increase /30 / 0.05	[89]
Co_3O_4 hollow sphere/G	~270	76.2	0.01–3.0	5	259	69.3	81.6%/100/1	[90]
Co_3O_4 nanorods/G	30 in diameter, 1000–2000 in length	80	0.01–3.0	1	1090	70.4	Slight increase /40/0.1	[91]
Co_3O_4 nanosheets/G	/	/	0.05–3.0	5C	130	76.6	63.9%/50/0.2C (1C = $891mAhg^{-1}$)	[92]

(continued)

Table 21.3 (continued)

G/M_xO_y hybrids	M_xO_y size (nm)	M_xO_y content (wt%)	Voltage range (V vs Li/Li^+)	Current density (Ag^{-1})	Specific capacity ($mAhg^{-1}$)	ICE (%)	Capacity retention(%)/ cycles/current density (Ag^{-1})	Refs.
MnO nanosheets/G	10 in thickness, 50–300 in diameter	82.6	0.01–3.0	3	625.8	/	96%/400/2	[93]
N-doped MnO/G	120–200	91.1	0.01–3.0	5	202	69.5	Slight increase /90/0.1	[94]
Mn_3O_4/G platelet	200–450	65.7	0.01–3.0	0.075	720 (100th)	/	53.3%/100/0.072	[95]
Mn_3O_4/RGO	10–20	90	0.1–3.0	1.6	390	/	/	[96]
MnO_2 nanorods/PEDOT/G	5 in diameter, 50 in length	/	0.2–3.0	0.4	698		51.7%/15/0.05	[97]
MnO_2 nanoflakes/G	100–200	~50	1.8–4.0	0.2	230 (150th)	/	/	[98]
MnO_2 nanotubes/G film	70–80 in diameter	~50	0.01–3.0	1.6	208	/	/	[99]
α-Fe_2O_3/G	30–50	72	0.01–3.0	0.8	620	66	/	[100]
α-Fe_2O_3 nanorods/G	15–30 in diameter, 120–300 in length	58.3	0.005–3.0	1	210.7	64.7	34.7%/30/0.1	[101]
α-Fe_2O_3 nanodisks/RGO	200 in diameter	75.6	0.005–3.0	10	337	91.6	85.6%/50/0.2	[102]
α-Fe_2O_3 nanospindles/RGO	60 in diameter, 120 in length	75.6	0.005–3.0	0.1 0.5 1 5	969(100th) 589(100th) 368(100th) 336(100th)	65	62.3%/100/0.1 65.6%/100/0.5 37.5%/100/1 48.1%/100/5	[103]
Hollow Fe_3O_4/G	~220	90.2	0.005–3.0	0.1	~900 (50th)	70	Nearly unvarying/50 / 0.1Ag^{-1}	[104]
Fe_3O_4 nanorods/G	11 in diameter	75	0.01–3.0	4.864	569	65.8	/	[105]
Fe_3O_4/G nanoscroll	/	50	0.01–3.0	5	300	73	58.7%/50/0.1	[106]
Fe_3O_4/G foam	30	80	0.01–3.0	60C	190	65.1	Slight increase /500/1C	[107]
CuO nanorods/G	50 in diameter, 300 in length	/	0.001–3.0	5C	262	64.7	74%/50/0.1C (1C = 674$mAhg^{-1}$)	[108]
CuO nanosheets/G	300	50	0.01–3.0	0.067	736.8 (50th)	91.6	/	[109]
Hollow CuO/Cu_2O-G	100–200 nanosphere	/	0.005–3.0	5	183	65	45%/60/0.2	[110]
Cu_2O/graphene	150–400	~75	0.01–3.0	0.1	599.8	94	94%/50/0.1	[111]
Cu_2O@rGO	1000	~83	0.01–3.0	0.1 0.2 0.5 1	1097 403 323 226	65.8	~31%/200/1	[112]
Ni/NiO-G	Several nanometers	<96	0.001–3.0	3	700	73	74.7%/300/1.5	[113]
NiO NPs/G	200	68.5	0.01–3.0	4	152	65.1	76.6%/100/0.1	[114]
SnO_2/G nanoribbon	<10	~80	0.01–2.5	2	580	74.3	~73%/50/0.1	[115]
SnO_2@graphene@graphene	80–160 in diameter	90	0.01–2.0	4	212.8	61.6	60%/120/0.08	[116]
Mesoporous SnO_2/G	3–4	78.8	0.01–3.0	0.782	621.5	69.4	53.1%/50/0.078	[117]
SnO_2 nanorods/G	8–10 in diameter, 90–120 in length	69.5	0.01–3.0	2	583.3	/	96.2/100/0.2	[118]
SnO_2 nanorods/G	10–20 in diameter, 100–200 in length	44	0.01–3.0	0.1	710 (50th)	/	61.9%/50/0.1	[119]
Flowerlike SnO_2/G	40	61	0.01–2.5	0.2	658.4 (50th)	46.4	/	[120]

(continued)

Table 21.3 (continued)

G/M_xO_y hybrids	M_xO_y size (nm)	M_xO_y content (wt%)	Voltage range (V vs Li/Li^+)	Current density (Ag^{-1})	Specific capacity ($mAhg^{-1}$)	ICE (%)	Capacity retention(%)/ cycles/current density (Ag^{-1})	Refs.
Hollow $SnO_2@Co_3O_4@$graphene foam	300	~67	0.01–3.0	0.2	815.2 (100th)	/	~34%/100/0.2	[121]
Hollow SnO_2-$Fe_2O_3@RGO$	~320	~67	0.01–3.0	0.2	1160.9(5th)	61.3	52%/100/0.2	[122]
$Fe_2O_3@SnO_2/G$	/	64.5	0.01–3.0	2	535	60.9	~49%/200/0.1	[123]
Al_2O_3 coated ZnO/G	/	53	0.01–3.0	1	415	53.1	60.8%/2–100/0.1	[124]
Porous carbon/ZnO/G	~3.5	47	0.01–3.0	0.978	420	62.8	/	[125]

ICE initial coulombic efficiency

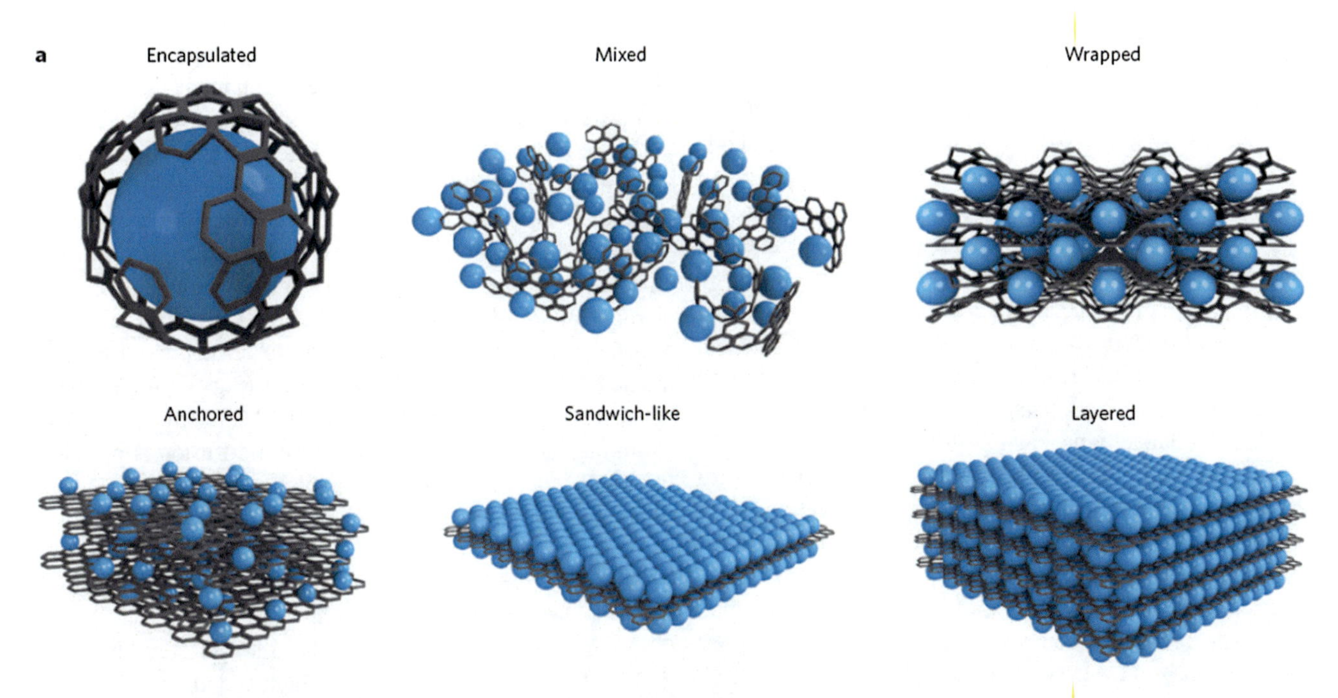

Fig. 21.4 Schematic of the different structures of graphene composite electrode materials. All models (except where specifically indicated) refer to composites in which graphene and the active material are synthesized through one-pot processes. Encapsulated: Single active material particles are encapsulated by graphene, which acts as either an active (for example, LIB anodes) or an inactive (for example, LIB cathodes) component. Mixed: Graphene and active materials are synthesized separately and mixed mechanically during the electrode preparation. In this structure, graphene may serve as an inactive conductive matrix (for example, LIB cathodes) or an active material (for example, LIB anodes). Wrapped: The active material particles are wrapped by multiple graphene sheets. This structure well-represents pseudocapacitor electrodes, in which graphene is the active material, as well as metal-ion battery cathodes, where graphene is an inactive component. Anchored: This is the most common structure for graphene composites, in which electroactive nanoparticles are anchored to the graphene surface. This structure is very relevant for metal-ion battery anodes and pseudocapacitors, where graphene serves as an active material, as well as for metal-ion battery cathodes and in LSBs, where graphene acts as an inactive component. Sandwich-like model: Graphene is used as a template to generate active material/graphene sandwich structures. This graphene-composite model, although not widespread, is used for LIB cathodes. Layered model: Active material nanoparticles are alternated with graphene sheets to form a composite layered structure, which has been proposed for use in metal-ion battery anodes and cathodes. (Reproduced from Ref. [10] with permission from Springer Nature)

configurations of the composite, and mass loading of metal oxides, there are no available data and direct evidences identifying which structural model enables better performance than others [127]. Generally speaking, the nanostructured metal oxides can better accommodate the volume variation, shorten the lithium-ion diffusion pathway, and provide enlarged reaction area, which contributes to augmented electrochemical reaction kinetics. Therefore, the preparation of metal oxides with varied nanostructures, such as mesoporous nanosphere [85, 90, 117], nanotube [86, 87], nanorod [91, 101, 105, 108, 118, 119], nanosheet [92, 109], nanodisk [102], nanospindle [103], and nanoflower [120], have been extensively studied due to better accommodated volume variation and shortened lithium-ion diffusion pathway. Moreover, the incorporated graphene plays multi-functional roles in the composite: during material preparation and electrode fabrication, graphene serves as a matrix to support active materials and prevent particle agglomeration [127]; upon cycling, the flexible and conductive graphene functions as an elastic matrix to alleviate the large volume variation and provides a conductive network for efficient electron transport [127]. As a result, the graphene/metal oxide composite demonstrates an overall improved electrochemical performance in terms of capacity, rate capability, and cycle stability.

21.3 Graphene-Based Material as Cathodes

Presently, the development of high-energy and high-power LIBs is restricted by the relatively low capacity of the cathode materials [128]. Therefore, a wide variety of nanostructured cathode materials including $LiCoO_2$, $LiFePO_4$, $LiMn_2O_4$, $Li_3V_2(PO_4)_3$, V_2O_5, etc. have been explored with increased storage capacity, improved redox potential, and extended cycle life [129]. The fundamental Li storage mechanisms of some typical cathodes are illustrated in Eqs. 21.17, 21.18, 21.19, 21.20, and 21.21. In general, the Li storage reaction kinetics of these cathode materials are limited by their low electrical conductivity and slow Li-ion diffusion (Table 21.4). Similar to anode materials, graphene is proposed as an exceptionally suitable conductive agent in cathodes due to its key properties, including excellent electrical conductivity, augmented electrons and lithium-ion transportation, and reduced particle agglomeration during cycling [128]. These cathode composites can also form wrapped [130], anchored [131], encapsulated [132], layered [133], mixed [134], and sandwich-like architectures [135], as demonstrated in anode materials. Even without being actively involved in the electrochemical reaction, the overall performance of cathodes shows considerable improvement with the addition of graphene.

$$LiCoO_2 \leftrightarrow Li^+ + e^- + CoO_2 \quad (21.17)$$

$$LiFePO_4 \leftrightarrow Li^+ + e^- + FePO_4 \quad (21.18)$$

$$LiMn_2O_4 \leftrightarrow Li^+ + e^- + Mn_2O_4 \quad (21.19)$$

$$Li_3V_2(PO_4)_3 \leftrightarrow V_2(PO_4)_3 + 3Li^+ + 3e^- \quad (21.20)$$

$$xLi^+ + xe^- + V_2O_5 \leftrightarrow Li_xV_2O_5\ (0 \leq x \leq 3) \quad (21.21)$$

Olivine-type $LiFePO_4$ has gained great interest as Li storage cathode because of its relatively large capacity (170 $mAhg^{-1}$) among cathodes, low raw material cost, low toxicity, and safety. Their key limitation is the extremely low electrical conductivity (10^{-9} S cm^{-1}) [138–140] and sluggish Li-ion diffusion (10^{-14}–10^{-16} cm^2 s^{-1} as reported by Prosini et al. [141]), which results in poor high rate performance and greatly hampered application in high-power LIBs. To enhance the performance, conductive graphene nanosheets have been used as a scaffold to grow and/or anchor the insulating $LiFePO_4$ materials. Guo et al. [135] reported the fabrication of sandwich-like $LiFePO_4$/graphene hybrid nanosheets. In this structure, the sandwiched thin $LiFePO_4$ nanosheets shorten the distance for Li-ion diffusion and the graphene carbon, which is derived from the in situ carbonization of dodecylamine, and enable fast electron transport, both of which contribute to an improved performance in fast charge/discharge process (Fig. 21.5). A capacity of 115 mA h g^{-1} at a high rate of 10 C is achieved. Apart from the electrical conductivity, Li-ion diffusion is of equal importance and is reported to be hindered when $LiFePO_4$ is fully and tightly wrapped by graphene [150, 151]. Therefore, a proper configuration should provide a balance between increased electron conduction and fast Li-ion diffusion and help the delivery of a high specific capacity [152].

Vanadium oxides, such as VO_2 and V_2O_5, are attractive LIB cathode materials due to their high abundance, low cost, good safety, and most importantly high theoretical capacity (320 $mAhg^{-1}$ for VO_2(B) and 443 $mAhg^{-1}$ for V_2O_5) compared to other commonly studied cathode materials such as $LiCoO_2$ (140 $mAhg^{-1}$), $LiMn_2O_4$ (148 $mAhg^{-1}$), and $LiFePO_4$ (170 $mAhg^{-1}$). However, it has been challenging to achieve decent electrochemical performance in aspects of specific capacity, rate capability, and cycling life due to the low electrical conductivity, low Li-ion diffusion, and irreversible phase transition upon deep discharge [153, 154]. To

Table 21.4 Properties of some of the reported cathode materials in LIBs

Cathode	Crystal structure	Theoretical capacity ($mAhg^{-1}$)	Electrical conductivity (S cm^{-1})	Refs.	Li-ion diffusion coefficient (cm^2 s^{-1})	Refs.
$LiCoO_2$	Layered	140	~10^{-3}	[136]	10^{-9}	[137]
$LiFePO_4$	Olivine	170	10^{-9}	[138–140]	10^{-14}–10^{-16}	[141]
$LiMn_2O_4$	Spinel	148	10^{-4}–10^{-5}	[142, 143]	10^{-9}–10^{-10}	[144]
$Li_3V_2(PO_4)_3$	Monoclinic	197	10^{-7}	[145]	10^{-9} to 10^{-10}	[146]
V_2O_5	Layered	443	10^{-2}–10^{-3}	[147]	10^{-12}–10^{-13}	[148, 149]

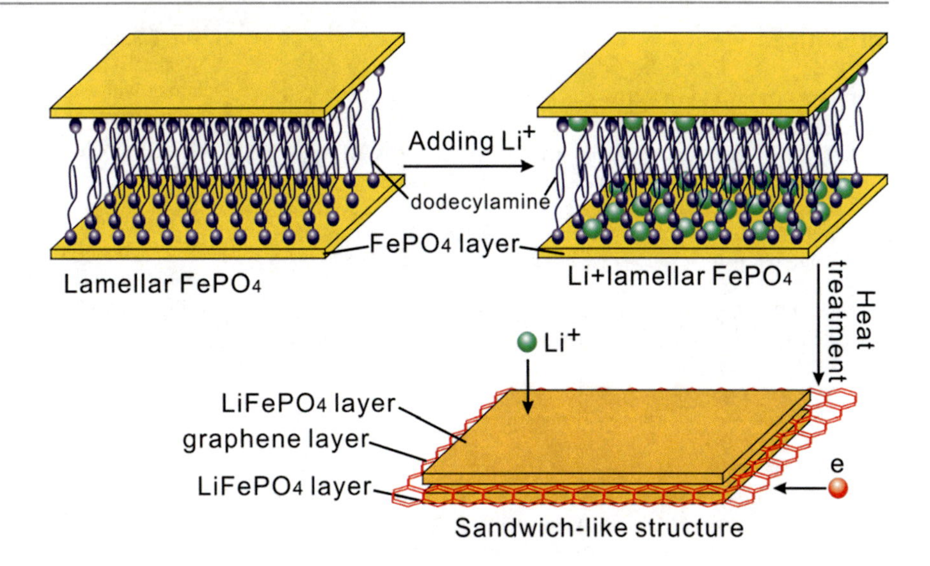

Fig. 21.5 The proposed formation mechanism of $LiFePO_4$/graphene hybrid nanosheets with a sandwich-like structure. (Reproduced from Ref. [135] with permission from the Royal Society of Chemistry)

address the above issues, nanostrucutred V_xO_y/graphene composites have been developed with varied performance improvement. By using hydrothermal method, single-crystalline VO_2 nanoribbons were successfully grown on graphene nanosheets. The as-prepared VO_2 nanoribbons and graphene form a 3D interpenetrating architecture and deliver a remarkable capacity of 190 $mAhg^{-1}$ after 1000 cycles at 190C, corresponding to a retention rate of >90% [155]. A free-standing paper electrode consisting of ultrathin (~10 nm) V_2O_5 nanowire and graphene sheets was prepared by vacuum filtration and exhibited a remarkable rate and cycling performance [156]. The paper electrode with 15 wt% V_2O_5 achieved a reversible capacity of 414.4 $mAhg^{-1}$ at 0.1 Ag^{-1} and 255.5 $mAhg^{-1}$ at 1 Ag^{-1} and a capacity retention of 79.8% over 200 cycles at 0.1 Ag^{-1} based on the mass of V_2O_5 only. When cycled at higher current densities of 1 Ag^{-1} for 3000 times and 10 Ag^{-1} for 100,000 times, the remaining capacities were 191.5 $mAhg^{-1}$ and 94.4 $mAhg^{-1}$, respectively, while the pristine V_2O_5 nanowire lost its capacity after only 150 cycles at 0.1 Ag^{-1}.

Other graphene/cathode composites, like $LiMn_2O_4$/graphene composite, were also investigated. With 5 wt% graphene and 10 wt% acetylene black (AB) as conductive additives, the specific capacities and cycling performance of $LiMn_2O_4$ were obviously enhanced [157]. Bak et al. [158] reported the fabrication of nano-sized $LiMn_2O_4$/RGO hybrid via a microwave-assisted hydrothermal method at 200 °C for 30 min. By using RGO as a template, the $LiMn_2O_4$ particles were uniformly anchored on this conductive substrate. The hybrid structure delivered an outstanding discharge capacity of 117 mAh g^{-1} and 101 mAh g^{-1} at charge/discharge rates of 50 C rate and 100 C rate, respectively.

21.4 Conclusion and Outlooks

In summary, graphene and its composites have achieved great success for use in both the anode and cathode of LIBs. The current state-of-the-art and most recent advances in the preparation of graphene-containing electrodes and their Li storage properties are summarized. The rich surface functional groups make it possible to integrate with varied electrode materials, ranging from element electrodes, transition metals, oxide anodes, to oxides cathodes. Basically, the electrochemical performance of the composite electrodes shows considerable improvement due to the following aspects: firstly, the introduced graphene can serve as a scaffold to separate and support the active material and prevent them from agglomeration during material preparation; secondly, the flexible graphene functions as an elastic matrix to accommodate the volume variation, thus promoting the formation of a stable SEI layer and helping to maintain electrode integrity during cycling; finally, the interconnected graphene provides a conductive network for efficient electron transfer and Li-ion diffusion.

However, there are several challenges facing graphene-containing electrodes: (a) Novel structures and material morphologies need to be rationally designed to enrich the family of graphene-based composites, and elucidate the influence of structure and material composition on the lithium storage property. (b) Efforts should be devoted to the deep understanding of the corelation between the architecture and the electrochemical performance for graphene-based composite electrodes, which will provide guideline for the future material design and synthesis. (c) Green, cost-effective, and large-scale production approaches for high-quality graphene are urgently needed.

References

1. Chu, S., Cui, Y., Liu, N.: The path towards sustainable energy. Nat. Mater. **16**(1), 16–22 (2017)
2. Grey, C.P., Tarascon, J.M.: Sustainability and in situ monitoring in battery development. Nat. Mater. **16**(1), 45–56 (2017)
3. Lee, H., Yanilmaz, M., Toprakci, O., Fu, K., Zhang, X.: A review of recent developments in membrane separators for rechargeable lithium-ion batteries. Energy Environ. Sci. **7**(12), 3857–3886 (2014)
4. Choi, J.W., Aurbach, D.: Promise and reality of post-lithium-ion batteries with high energy densities. Nat. Rev. Mater. **1**, 16013 (2016)
5. Saw, L.H., Ye, Y., Tay, A.A.O.: Integration issues of lithium-ion battery into electric vehicles battery pack. J. Clean. Prod. **113**, 1032–1045 (2016)
6. Kim, T.-H., Park, J.-S., Chang, S.K., Choi, S., Ryu, J.H., Song, H.-K.: The current move of Lithium ion batteries towards the next phase. Adv. Energy Mater. **2**(7), 860–872 (2012)
7. Xia, X.-F., Gu, Y.-Y., Xu, S.-A.: Titanium conversion coatings on the aluminum foil AA 8021 used for lithium–ion battery package. Appl. Surf. Sci. **419**, 447–453 (2017)
8. Stoller, M.D., Park, S., Zhu, Y., An, J., Ruoff, R.S.: Graphene-based ultracapacitors. Nano Lett. **8**(10), 3498–3502 (2008)
9. Chen, K., Song, S., Liu, F., Xue, D.: Structural design of graphene for use in electrochemical energy storage devices. Chem. Soc. Rev. **44**(17), 6230–6257 (2015)
10. Raccichini, R., Varzi, A., Passerini, S., Scrosati, B.: The role of graphene for electrochemical energy storage. Nat. Mater. **14**(3), 271–279 (2015)
11. de las Casas, C., Li, W.: A review of application of carbon nanotubes for lithium ion battery anode material. J. Power Sources. **208**, 74–85 (2012)
12. Zhu, Y., Murali, S., Cai, W., Li, X., Suk, J.W., Potts, J.R., Ruoff, R.S.: Graphene and graphene oxide: synthesis, properties, and applications. Adv. Mater. **22**(35), 3906–3924 (2010)
13. Wang, G., Shen, X., Yao, J., Park, J.: Graphene nanosheets for enhanced lithium storage in lithium ion batteries. Carbon. **47**(8), 2049–2053 (2009)
14. Dahn, J.R., Zheng, T., Liu, Y., Xue, J.S.: Mechanisms for lithium insertion in carbonaceous materials. Science. **270**(5236), 590–593 (1995)
15. Li, X., Hu, Y., Liu, J., Lushington, A., Li, R., Sun, X.: Structurally tailored graphene nanosheets as lithium ion battery anodes: an insight to yield exceptionally high lithium storage performance. Nanoscale. **5**(24), 12607–12615 (2013)
16. Yoo, E., Kim, J., Hosono, E., Zhou, H.-s., Kudo, T., Honma, I.: Large reversible Li storage of graphene nanosheet families for use in rechargeable lithium ion batteries. Nano Lett. **8**(8), 2277–2282 (2008)
17. Lian, P., Zhu, X., Liang, S., Li, Z., Yang, W., Wang, H.: Large reversible capacity of high quality graphene sheets as an anode material for lithium-ion batteries. Electrochim. Acta. **55**(12), 3909–3914 (2010)
18. Ma, X., Ning, G., Qi, C., Xu, C., Gao, J.: Phosphorus and nitrogen dual-doped few-layered porous graphene: a high-performance anode material for Lithium-ion batteries. Acs Appl. Mater. Interfaces. **6**(16), 14415–14422 (2014)
19. Wu, Z.-S., Ren, W., Xu, L., Li, F., Cheng, H.-M.: Doped graphene sheets as anode materials with superhigh rate and large capacity for lithium ion batteries. ACS Nano. **5**(7), 5463–5471 (2011)
20. Reddy, A.L.M., Srivastava, A., Gowda, S.R., Gullapalli, H., Dubey, M., Ajayan, P.M.: Synthesis of nitrogen-doped graphene films for lithium battery application. ACS Nano. **4**(11), 6337–6342 (2010)
21. Hu, T., Sun, X., Sun, H., Xin, G., Shao, D., Liu, C., Lian, J.: Rapid synthesis of nitrogen-doped graphene for a lithium ion battery anode with excellent rate performance and super-long cyclic stability. Phys. Chem. Chem. Phys. **16**(3), 1060–1066 (2014)
22. Yan, Y., Yin, Y.-X., Xin, S., Guo, Y.-G., Wan, L.-J.: Ionothermal synthesis of sulfur-doped porous carbons hybridized with graphene as superior anode materials for lithium-ion batteries. Chem. Commun. **48**(86), 10663–10665 (2012)
23. Ma, C., Shao, X., Cao, D.: Nitrogen-doped graphene nanosheets as anode materials for lithium ion batteries: a first-principles study. J. Mater. Chem. **22**(18), 8911–8915 (2012)
24. Cohn, A.P., Oakes, L., Carter, R., Chatterjee, S., Westover, A.S., Share, K., Pint, C.L.: Assessing the improved performance of freestanding, flexible graphene and carbon nanotube hybrid foams for lithium ion battery anodes. Nanoscale. **6**(9), 4669–4675 (2014)
25. Zhong, C., Wang, J.-Z., Wexler, D., Liu, H.-K.: Microwave autoclave synthesized multi-layer graphene/single-walled carbon nanotube composites for free-standing lithium-ion battery anodes. Carbon. **66**(0), 637–645 (2014)
26. Vinayan, B.P., Nagar, R., Raman, V., Rajalakshmi, N., Dhathathreyan, K.S., Ramaprabhu, S.: Synthesis of graphene-multiwalled carbon nanotubes hybrid nanostructure by strengthened electrostatic interaction and its lithium ion battery application. J. Mater. Chem. **22**(19), 9949–9956 (2012)
27. Park, K.H., Lee, D., Kim, J., Song, J., Lee, Y.M., Kim, H.-T., Park, J.-K.: Defect-free, size-tunable graphene for high-performance lithium ion battery. Nano Lett. **14**(8), 4306–4313 (2014)
28. Lee, S.H., Sridhar, V., Jung, J.H., Karthikeyan, K., Lee, Y.S., Mukherjee, R., Koratkar, N., Oh, I.K.: Graphene--nanotube--iron hierarchical nanostructure as lithium ion battery anode. ACS Nano. **7**(5), 4242–4251 (2013)
29. Xu, C., Xu, B., Gu, Y., Xiong, Z., Sun, J., Zhao, X.S.: Graphene-based electrodes for electrochemical energy storage. Energy Environ. Sci. **6**(5), 1388–1414 (2013)
30. Palacin, M.R.: Recent advances in rechargeable battery materials: a chemist's perspective. Chem. Soc. Rev. **38**(9), 2565–2575 (2009)
31. Wu, H., Cui, Y.: Designing nanostructured Si anodes for high energy lithium ion batteries. Nano Today. **7**(5), 414–429 (2012)
32. Chan, C.K., Zhang, X.F., Cui, Y.: High capacity Li ion battery anodes using Ge nanowires. Nano Lett. **8**(1), 307–309 (2007)
33. Beaulieu, L.Y., Eberman, K.W., Turner, R.L., Krause, L.J., Dahn, J.R.: Colossal reversible volume changes in Lithium alloys. Electrochem. Solid-State Lett. **4**(9), A137–A140 (2001)
34. Kasavajjula, U., Wang, C., Appleby, A.J.: Nano- and bulk-silicon-based insertion anodes for lithium-ion secondary cells. J. Power Sources. **163**(2), 1003–1039 (2007)
35. He, Y.-S., Gao, P., Chen, J., Yang, X., Liao, X.-Z., Yang, J., Ma, Z.-F.: A novel bath lily-like graphene sheet-wrapped nano-Si composite as a high performance anode material for Li-ion batteries. RSC Adv. **1**(6), 958–960 (2011)
36. Zhou, X., Bao, J., Dai, Z., Guo, Y.-G.: Tin nanoparticles impregnated in nitrogen-doped graphene for Lithium-ion battery anodes. J. Phys. Chem. C. **117**(48), 25367–25373 (2013)
37. Liu, X.H., Zhong, L., Huang, S., Mao, S.X., Zhu, T., Huang, J.Y.: Size-dependent fracture of silicon nanoparticles during lithiation. ACS Nano. **6**(2), 1522–1531 (2012)
38. Ko, M., Chae, S., Jeong, S., Oh, P., Cho, J.: Elastic a-silicon nanoparticle backboned graphene hybrid as a self-compacting anode for high-rate lithium ion batteries. ACS Nano. **8**(8), 8591–8599 (2014)
39. Wong, D.P., Tseng, H.-P., Chen, Y.-T., Hwang, B.-J., Chen, L.-C., Chen, K.-H.: A stable silicon/graphene composite using solvent exchange method as anode material for lithium ion batteries. Carbon. **63**, 397–403 (2013)

40. Yang, S., Li, G., Zhu, Q., Pan, Q.: Covalent binding of Si nanoparticles to graphene sheets and its influence on lithium storage properties of Si negative electrode. J. Mater. Chem. **22**(8), 3420–3425 (2012)
41. Zhou, X., Yin, Y.-X., Wan, L.-J., Guo, Y.-G.: Self-assembled nanocomposite of silicon nanoparticles encapsulated in graphene through electrostatic attraction for lithium-ion batteries. Adv. Energy Mater. **2**(9), 1086–1090 (2012)
42. Li, Y., Yan, K., Lee, H.-W., Lu, Z., Liu, N., Cui, Y.: Growth of conformal graphene cages on micrometre-sized silicon particles as stable battery anodes. Nat. Energy. **1**, 15029 (2016)
43. Chen, J.S., Archer, L.A., Wen Lou, X.: SnO_2 hollow structures and TiO_2 nanosheets for lithium-ion batteries. J. Mater. Chem. **21**(27), 9912–9924 (2011)
44. Wang, D., Li, X., Yang, J., Wang, J., Geng, D., Li, R., Cai, M., Sham, T.-K., Sun, X.: Hierarchical nanostructured core-shell Sn@C nanoparticles embedded in graphene nanosheets: spectroscopic view and their application in lithium ion batteries. Phys. Chem. Chem. Phys. **15**(10), 3535–3542 (2013)
45. Luo, B., Wang, B., Li, X., Jia, Y., Liang, M., Zhi, L.: Graphene-confined Sn nanosheets with enhanced lithium storage capability. Adv. Mater. **24**(26), 3538–3543 (2012)
46. Lu, Z., Zhu, J., Sim, D., Shi, W., Tay, Y.Y., Ma, J., Hng, H.H., Yan, Q.: In situ growth of Si nanowires on graphene sheets for Li-ion storage. Electrochim. Acta. **74**, 176–181 (2012)
47. Wang, X.-L., Han, W.-Q.: Graphene enhances Li storage capacity of porous single-crystalline silicon nanowires. Acs Appl. Mater. Inter. **2**(12), 3709–3713 (2010)
48. Ren, J.-G., Wang, C., Wu, Q.-H., Liu, X., Yang, Y., He, L., Zhang, W.: A silicon nanowire-reduced graphene oxide composite as a high-performance lithium ion battery anode material. Nanoscale. **6** (6), 3353–3360 (2014)
49. Wu, P., Wang, H., Tang, Y., Zhou, Y., Lu, T.: Three-dimensional interconnected network of graphene-wrapped porous silicon spheres: in situ Magnesiothermic-reduction synthesis and enhanced Lithium-storage capabilities. Acs Appl. Mater. Inter. **6** (5), 3546–3552 (2014)
50. Ge, M., Rong, J., Fang, X., Zhang, A., Lu, Y., Zhou, C.: Scalable preparation of porous silicon nanoparticles and their application for lithium-ion battery anodes. Nano Res. **6**(3), 174–181 (2013)
51. Ji, L., Tan, Z., Kuykendall, T., An, E.J., Fu, Y., Battaglia, V., Zhang, Y.: Multilayer nanoassembly of Sn-nanopillar arrays sandwiched between graphene layers for high-capacity lithium storage. Energy Environ. Sci. **4**(9), 3611–3616 (2011)
52. Luo, B., Wang, B., Liang, M.H., Ning, J., Li, X.L., Zhi, L.J.: Reduced graphene oxide-mediated growth of uniform tin-core/carbon-sheath coaxial nanocables with enhanced lithium ion storage properties. Adv. Mater. **24**(11), 1405–1409 (2012)
53. Kim, H., Son, Y., Park, C., Cho, J., Choi, H.C.: Catalyst-free direct growth of a single to a few layers of graphene on a germanium nanowire for the anode material of a lithium battery. Angew. Chem. Int. Ed. **52**(23), 5997–6001 (2013)
54. Chockla, A.M., Panthani, M.G., Holmberg, V.C., Hessel, C.M., Reid, D.K., Bogart, T.D., Harris, J.T., Mullins, C.B., Korgel, B.A.: Electrochemical lithiation of graphene-supported silicon and germanium for rechargeable batteries. J. Phys. Chem. C. **116**(22), 11917–11923 (2012)
55. Chou, C.-Y., Hwang, G.S.: Role of interface in the lithiation of silicon-graphene composites: a first principles study. J. Phys. Chem. C. **117**(19), 9598–9604 (2013)
56. Odbadrakh, K., McNutt, N.W., Nicholson, D.M., Rios, O., Keffer, D.J.: Lithium diffusion at Si-C interfaces in silicon-graphene composites. Appl. Phys. Lett. **105**(5), (2014)
57. Li, Z.-F., Zhang, H., Liu, Q., Liu, Y., Stanciu, L., Xie, J.: Novel pyrolyzed polyaniline-grafted silicon nanoparticles encapsulated in graphene sheets as Li-ion battery anodes. Acs Appl. Mater. Inter. **6** (8), 5996–6002 (2014)
58. Wen, Y., Zhu, Y., Langrock, A., Manivannan, A., Ehrman, S.H., Wang, C.: Graphene-bonded and -encapsulated Si nanoparticles for lithium ion battery anodes. Small. **9**(16), 2810–2816 (2013)
59. Park, S.-H., Kim, H.-K., Ahn, D.-J., Lee, S.-I., Roh, K.C., Kim, K.-B.: Self-assembly of Si entrapped graphene architecture for high-performance Li-ion batteries. Electrochem. Commun. **34**, 117–120 (2013)
60. Tang, H., Tu, J.-p., Liu, X.-y., Zhang, Y.-j., Huang, S., Li, W.-z., Wang, X.-l., Gu, C.-d.: Self-assembly of Si/honeycomb reduced graphene oxide composite film as a binder-free and flexible anode for Li-ion batteries. J. Mater. Chem. A. **2**(16), 5834–5840 (2014)
61. Chang, J., Huang, X., Zhou, G., Cui, S., Hallac, P.B., Jiang, J., Hurley, P.T., Chen, J.: Multilayered Si nanoparticle/reduced graphene oxide hybrid as a high-performance lithium-ion battery anode. Adv. Mater. **26**(5), 758–764 (2014)
62. Zhou, M., Cai, T., Pu, F., Chen, H., Wang, Z., Zhang, H., Guan, S.: Graphene/carbon-coated Si nanoparticle hybrids as high-performance anode materials for Li-ion batteries. Acs Appl. Mater. Inter. **5**(8), 3449–3455 (2013)
63. Yi, R., Zai, J., Dai, F., Gordin, M.L., Wang, D.: Dual conductive network-enabled graphene/Si-C composite anode with high areal capacity for lithium-ion batteries. Nano Energy. **6**, 211–218 (2014)
64. Xu, Z.-L., Zhang, B., Kim, J.-K.: Electrospun carbon nanofiber anodes containing monodispersed Si nanoparticles and graphene oxide with exceptional high rate capacities. Nano Energy. **6**, 27–35 (2014)
65. Shin, J., Park, K., Ryu, W.-H., Jung, J.-W., Kim, I.-D.: Graphene wrapping as a protective clamping layer anchored to carbon nanofibers encapsulating Si nanoparticles for a Li-ion battery anode. Nanoscale. **6**(21), 12718–12726 (2014)
66. Qin, J., He, C., Zhao, N., Wang, Z., Shi, C., Liu, E.-Z., Li, J.: Graphene networks anchored with Sn@graphene as Lithium ion battery anode. ACS Nano. **8**(2), 1728–1738 (2014)
67. Zhou, X.S., Bao, J.C., Dai, Z.H., Guo, Y.G.: Tin nanoparticles impregnated in nitrogen-doped graphene for Lithium-ion battery anodes. J. Phys. Chem. C. **117**(48), 25367–25373 (2013)
68. Yuan, F.-W., Tuan, H.-Y.: Scalable solution-grown high-germanium-nanoparticle-loading graphene nanocomposites as high-performance lithium-ion battery electrodes: an example of a graphene-based platform toward practical full-cell applications. Chem. Mater. **26**(6), 2172–2179 (2014)
69. Qin, J., Wang, X., Cao, M., Hu, C.: Germanium quantum dots embedded in N-doping graphene matrix with sponge-like architecture for enhanced performance in lithium-ion batteries. Chem-Eur. J. **20**(31), 9675–9682 (2014)
70. Li, D., Seng, K.H., Shi, D., Chen, Z., Liu, H.K., Guo, Z.: A unique sandwich-structured C/Ge/graphene nanocomposite as an anode material for high power lithium ion batteries. J. Mater. Chem. A. **1**(45), 14115–14121 (2013)
71. Ouyang, L.Z., Guo, L.N., Cai, W.H., Ye, J.S., Hu, R.Z., Liu, J.W., Yang, L.C., Zhu, M.: Facile synthesis of Ge@FLG composites by plasma assisted ball milling for lithium ion battery anodes. J. Mater. Chem. A. **2**(29), 11280–11285 (2014)
72. Poizot, P., Laruelle, S., Grugeon, S., Dupont, L., Tarascon, J.M.: Nano-sized transition-metal oxides as negative-electrode materials for lithium-ion batteries. Nature. **407**(6803), 496–499 (2000)
73. Reddy, M., Subba Rao, G., Chowdari, B.: Metal oxides and oxysalts as anode materials for Li ion batteries. Chem. Rev. **113** (7), 5364–5457 (2013)
74. Wu, Z.-S., Zhou, G., Yin, L.-C., Ren, W., Li, F., Cheng, H.-M.: Graphene/metal oxide composite electrode materials for energy storage. Nano Energy. **1**(1), 107–131 (2012)

75. Guan-Nan Zhu, Y.-G.W.: Yong-Yao Xia: Ti-based compounds as anode materials for Li-ion batteries. Energy Environ. Sci. **5**, 6652–6667 (2012)
76. Han, C., He, Y.-B., Wang, S., Wang, C., Du, H., Qin, X., Lin, Z., Li, B., Kang, F.: Large polarization of $Li_4Ti_5O_{12}$ lithiated to 0 V at large charge/discharge rates. Acs Appl. Mater. Inter. **8**(29), 18788–18796 (2016)
77. Du, G., Guo, Z., Zhang, P., Li, Y., Chen, M., Wexler, D., Liu, H.: SnO_2 nanocrystals on self-organized TiO2 nanotube array as three-dimensional electrode for lithium ion microbatteries. J. Mater. Chem. **20**(27), 5689–5694 (2010)
78. Deng, D., Kim, M.G., Lee, J.Y., Cho, J.: Green energy storage materials: nanostructured TiO_2 and Sn-based anodes for lithium-ion batteries. Energy Environ. Sci. **2**(8), 818–837 (2009)
79. Cabana, J., Monconduit, L., Larcher, D., Palacín, M.R.: Beyond intercalation-based Li-ion batteries: the state of the art and challenges of electrode materials reacting through conversion reactions. Adv. Mater. **22**(35), E170–E192 (2010)
80. Deng, Y., Wan, L., Xie, Y., Qin, X., Chen, G.: Recent advances in Mn-based oxides as anode materials for lithium ion batteries. RSC Adv. **4**(45), 23914–23935 (2014)
81. Wu, S., Han, C., Iocozzia, J., Lu, M., Ge, R., Xu, R., Lin, Z.: Germanium-based nanomaterials for rechargeable batteries. Angew. Chem. Int. Ed. **55**(28), 7898–7922 (2016)
82. Shi, Y., Wen, L., Li, F., Cheng, H.-M.: Nanosized $Li_4Ti_5O_{12}$/graphene hybrid materials with low polarization for high rate lithium ion batteries. J. Power Sources. **196**(20), 8610–8617 (2011)
83. Ding, Y., Li, G.R., Xiao, C.W., Gao, X.P.: Insight into effects of graphene in $Li_4Ti_5O_{12}$/carbon composite with high rate capability as anode materials for lithium ion batteries. Electrochim. Acta. **102**, 282–289 (2013)
84. Dong, H.-Y., He, Y.-B., Li, B., Zhang, C., Liu, M., Su, F., Lv, W., Kang, F., Yang, Q.-H.: Lithium titanate hybridized with trace amount of graphene used as an anode for a high rate lithium ion battery. Electrochim. Acta. **142**(0), 247–253 (2014)
85. Qiu, B., Xing, M., Zhang, J.: Mesoporous TiO_2 nanocrystals grown in situ on graphene aerogels for high photocatalysis and lithium-ion batteries. J. Am. Chem. Soc. **136**(16), 5852–5855 (2014)
86. Tang, Y., Liu, Z., Lu, X., Wang, B., Huang, F.: TiO_2 nanotubes grown on graphene sheets as advanced anode materials for high rate lithium ion batteries. RSC Adv. **4**(68), 36372–36376 (2014)
87. Li, Y., Wang, Z., Lv, X.-J.: N-doped TiO_2 nanotubes/N-doped graphene nanosheets composites as high performance anode materials in lithium-ion battery. J. Mater. Chem. A. **2**(37), 15473–15479 (2014)
88. Etacheri, V., Yourey, J.E., Bartlett, B.M.: Chemically bonded TiO2-bronze nanosheet/reduced graphene oxide hybrid for high-power lithium ion batteries. ACS Nano. **8**(2), 1491–1499 (2014)
89. Wu, Z.-S., Ren, W., Wen, L., Gao, L., Zhao, J., Chen, Z., Zhou, G., Li, F., Cheng, H.-M.: Graphene anchored with Co_3O_4 nanoparticles as anode of lithium ion batteries with enhanced reversible capacity and cyclic performance. ACS Nano. **4**(6), 3187–3194 (2010)
90. Sun, H., Sun, X., Hu, T., Yu, M., Lu, F., Lian, J.: Graphene-wrapped mesoporous cobalt oxide hollow spheres anode for high-rate and long-life lithium ion batteries. J. Phys. Chem. C. **118**(5), 2263–2272 (2014)
91. Tao, L., Zai, J., Wang, K., Zhang, H., Xu, M., Shen, J., Su, Y., Qian, X.: Co_3O_4 nanorods/graphene nanosheets nanocomposites for lithium ion batteries with improved reversible capacity and cycle stability. J. Power Sources. **202**, 230–235 (2012)
92. Sun, H., Liu, Y., Yu, Y., Ahmad, M., Nan, D., Zhu, J.: Mesoporous Co_3O_4 nanosheets-3D graphene networks hybrid materials for high-performance lithium ion batteries. Electrochim. Acta. **118**, 1–9 (2014)
93. Sun, Y., Hu, X., Luo, W., Xia, F., Huang, Y.: Reconstruction of conformal nanoscale MnO on graphene as a high-capacity and long-life anode material for lithium ion batteries. Adv. Funct. Mater. **23**(19), 2436–2444 (2013)
94. Zhang, K., Han, P., Gu, L., Zhang, L., Liu, Z., Kong, Q., Zhang, C., Dong, S., Zhang, Z., Yao, J., et al.: Synthesis of nitrogen-doped MnO/graphene nanosheets hybrid material for lithium ion batteries. Acs Appl. Mater. Inter. **4**(2), 658–664 (2012)
95. Lavoie, N., Malenfant, P.R.L., Courtel, F.M., Abu-Lebdeh, Y., Davidson, I.J.: High gravimetric capacity and long cycle life in Mn_3O_4/graphene platelet/LiCMC composite lithium-ion battery anodes. J. Power Sources. **213**, 249–254 (2012)
96. Wang, H., Cui, L.-F., Yang, Y., Casalongue, H.S., Robinson, J.T., Liang, Y., Cui, Y., Dai, H.: Mn_3O_4-graphene hybrid as a high-capacity anode material for lithium ion batteries. J. Am. Chem. Soc. **132**(40), 13978–13980 (2010)
97. Guo, C.X., Wang, M., Chen, T., Lou, X.W., Li, C.M.: A hierarchically nanostructured composite of MnO_2/conjugated polymer/graphene for high-performance lithium ion batteries. Adv. Energy Mater. **1**(5), 736–741 (2011)
98. Li, J., Zhao, Y., Wang, N., Ding, Y., Guan, L.: Enhanced performance of a MnO_2-graphene sheet cathode for lithium ion batteries using sodium alginate as a binder. J. Mater. Chem. **22**(26), 13002–13004 (2012)
99. Yu, A., Park, H.W., Davies, A., Higgins, D.C., Chen, Z., Xiao, X.: Free-standing layer-by-layer hybrid thin film of graphene-MnO_2 nanotube as anode for lithium ion batteries. J. Phys. Chem. Lett. **2** (15), 1855–1860 (2011)
100. Wei, D., Liang, J., Zhu, Y., Yuan, Z., Li, N., Qian, Y.: Formation of graphene-wrapped nanocrystals at room temperature through the colloidal coagulation effect. Part. Part. Syst. Charact. **30**(2), 143–147 (2013)
101. Zhao, B., Liu, R., Cai, X., Jiao, Z., Wu, M., Ling, X., Lu, B., Jiang, Y.: Nanorod-like Fe_2O_3/graphene composite as a high-performance anode material for lithium ion batteries. J. Appl. Electrochem. **44**(1), 53–60 (2014)
102. Qu, J., Yin, Y.-X., Wang, Y.-Q., Yan, Y., Guo, Y.-G., Song, W.-G.: Layer structured alpha-Fe_2O_3 Nanodisk/reduced graphene oxide composites as high-performance anode materials for lithium-ion batteries. Acs Appl. Mater. Inter. **5**(9), 3932–3936 (2013)
103. Bai, S., Chen, S., Shen, X., Zhu, G., Wang, G.: Nanocomposites of hematite (alpha-Fe_2O_3) nanospindles with crumpled reduced graphene oxide nanosheets as high-performance anode material for lithium-ion batteries. RSC Adv. **2**(29), 10977–10984 (2012)
104. Chen, D., Ji, G., Ma, Y., Lee, J.Y., Lu, J.: Graphene-encapsulated hollow Fe_3O_4 nanoparticle aggregates as a high-performance anode material for lithium ion batteries. Acs Appl. Mater. Inter. **3** (8), 3078–3083 (2011)
105. Hu, A., Chen, X., Tang, Y., Tang, Q., Yang, L., Zhang, S.: Self-assembly of Fe_3O_4 nanorods on graphene for lithium ion batteries with high rate capacity and cycle stability. Electrochem. Commun. **28**, 139–142 (2013)
106. Zhao, J., Yang, B., Zheng, Z., Yang, J., Yang, Z., Zhang, P., Ren, W., Yan, X.: Facile preparation of one-dimensional wrapping structure: graphene nanoscroll-wrapped of Fe_3O_4 nanoparticles and its application for lithium-ion battery. Acs Appl. Mater. Inter. **6**(12), 9890–9896 (2014)
107. Luo, J., Liu, J., Zeng, Z., Ng, C.F., Ma, L., Zhang, H., Lin, J., Shen, Z., Fan, H.J.: Three-dimensional graphene foam supported Fe_3O_4 lithium battery anodes with long cycle life and high rate capability. Nano Lett. **13**(12), 6136–6143 (2013)
108. Wang, Q., Zhao, J., Shan, W., Xia, X., Xing, L., Xue, X.: CuO nanorods/graphene nanocomposites for high-performance lithium-ion battery anodes. J. Alloys Compd. **590**, 424–427 (2014)

109. Liu, Y., Wang, W., Gu, L., Wang, Y., Ying, Y., Mao, Y., Sun, L., Peng, X.: Flexible CuO nanosheets/reduced-graphene oxide composite paper: binder-free anode for high-performance lithium-ion batteries. Acs Appl. Mater. Inter. **5**(19), 9850–9855 (2013)
110. Zhou, X., Shi, J., Liu, Y., Su, Q., Zhang, J., Du, G.: Microwave-assisted synthesis of hollow CuO-Cu_2O nanosphere/graphene composite as anode for lithium-ion battery. J. Alloys Compd. **615**, 390–394 (2014)
111. Xu, Y.T., Guo, Y., Song, L.X., Zhang, K., Yuen, M.M.F., Xu, J.B., Fu, X.Z., Sun, R., Wong, C.P.: Co-reduction self-assembly of reduced graphene oxide nanosheets coated Cu_2O sub-microspheres core-shell composites as lithium ion battery anode materials. Electrochim. Acta. **176**, 434–441 (2015)
112. Xu, Y.T., Guo, Y., Li, C., Zhou, X.Y., Tucker, M.C., Fu, X.Z., Sun, R., Wong, C.P.: Graphene oxide nano-sheets wrapped Cu_2O microspheres as improved performance anode materials for lithium ion batteries. Nano Energy. **11**, 38–47 (2015)
113. Choi, S.H., Ko, Y.N., Lee, J.-K., Kang, Y.C.: Rapid continuous synthesis of spherical reduced graphene ball-nickel oxide composite for lithium ion batteries. Sci. Rep. **4** (2014)
114. Zhuo, L., Wu, Y., Zhou, W., Wang, L., Yu, Y., Zhang, X., Zhao, F.: Trace amounts of water-induced distinct growth behaviors of NiO nanostructures on graphene in CO_2-expanded ethanol and their applications in lithium-ion batteries. Acs Appl. Mater. Inter. **5**(15), 7065–7071 (2013)
115. Lin, J., Peng, Z., Xiang, C., Ruan, G., Yan, Z., Natelson, D., Tour, J.M.: Graphene nanoribbon and nanostructured SnO_2 composite anodes for lithium ion batteries. ACS Nano. **7**(7), 6001–6006 (2013)
116. Zhu, J., Zhang, G., Yu, X., Li, Q., Lu, B., Xu, Z.: Graphene double protection strategy to improve the SnO_2 electrode performance anodes for lithium-ion batteries. Nano Energy. **3**, 80–87 (2014)
117. Yang, S., Yue, W., Zhu, J., Ren, Y., Yang, X.: Graphene-based mesoporous SnO_2 with enhanced electrochemical performance for lithium-ion batteries. Adv. Funct. Mater. **23**(28), 3570–3576 (2013)
118. Han, Q., Zai, J., Xiao, Y., Li, B., Xu, M., Qian, X.: Direct growth of SnO_2 nanorods on graphene as high capacity anode materials for lithium ion batteries. RSC Adv. **3**(43), 20573–20578 (2013)
119. Xu, C., Sun, J., Gao, L.: Direct growth of monodisperse SnO_2 nanorods on graphene as high capacity anode materials for lithium ion batteries. J. Mater. Chem. **22**(3), 975–979 (2012)
120. Guo, Q., Qin, X.: Flower-like SnO_2 nanoparticles grown on graphene as anode materials for lithium-ion batteries. J. Solid State Electrochem. **18**(4), 1031–1039 (2014)
121. Zhao, B., Huang, S.Y., Wang, T., Zhang, K., Yuen, M.M.F., Xu, J. B., Fu, X.Z., Sun, R., Wong, C.P.: Hollow SnO_2@Co_3O_4 core-shell spheres encapsulated in three-dimensional graphene foams for high performance supercapacitors and lithium-ion batteries. J. Power Sources. **298**, 83–91 (2015)
122. Zhao, B., Xu, Y.T., Huang, S.Y., Zhang, K., Yuen, M.M.F., Xu, J. B., Fu, X.Z., Sun, R., Wong, C.P.: 3D RGO frameworks wrapped hollow spherical SnO_2@Fe_2O_3 mesoporous nano-shells: fabrication, characterization and lithium storage properties. Electrochim. Acta. **202**, 186–196 (2016)
123. Liu, S., Wang, R., Liu, M., Luo, J., Jin, X., Sun, J., Gao, L.: Fe_2O_3@SnO_2 nanoparticle decorated graphene flexible films as high-performance anode materials for lithium-ion batteries. J. Mater. Chem. A. **2**(13), 4598–4604 (2014)
124. Yu, M., Wang, A., Wang, Y., Li, C., Shi, G.: An alumina stabilized ZnO-graphene anode for lithium ion batteries via atomic layer deposition. Nanoscale. **6**(19), 11419–11424 (2014)
125. Hsieh, C.-T., Lin, C.-Y., Chen, Y.-F., Lin, J.-S.: Synthesis of ZnO@graphene composites as anode materials for lithium ion batteries. Electrochim. Acta. **111**, 359–365 (2013)
126. Li, X., Meng, X., Liu, J., Geng, D., Zhang, Y., Banis, M.N., Li, Y., Yang, J., Li, R., Sun, X., et al.: Tin oxide with controlled morphology and crystallinity by atomic layer deposition onto graphene nanosheets for enhanced lithium storage. Adv. Funct. Mater. **22** (8), 1647–1654 (2012)
127. Yang, Y., Han, C., Jiang, B., Iocozzia, J., He, C., Shi, D., Jiang, T., Lin, Z.: Graphene-based materials with tailored nanostructures for energy conversion and storage. Mater. Sci. Eng. R Rep. **102**, 1–72 (2016)
128. Wu, S., Xu, R., Lu, M., Ge, R., Iocozzia, J., Han, C., Jiang, B., Lin, Z.: Graphene-containing nanomaterials for lithium-ion batteries. Adv. Energy Mater. **5**(21), 1500400–n/a (2015)
129. Kucinskis, G., Bajars, G., Kleperis, J.: Graphene in lithium ion battery cathode materials: a review. J. Power Sources. **240**, 66–79 (2013)
130. Ma, R., Lu, Z., Wang, C., Wang, H.-E., Yang, S., Xi, L., Chung, J. C.Y.: Large-scale fabrication of graphene-wrapped FeF_3 nanocrystals as cathode materials for lithium ion batteries. Nanoscale. **5**(14), 6338–6343 (2013)
131. Han, S., Wang, J., Li, S., Wu, D., Feng, X.: Graphene aerogel supported $Fe_5(PO_4)_4(OH)_3{\cdot}2H_2O$ microspheres as high performance cathode for lithium ion batteries. J. Mater. Chem. A. **2** (17), 6174–6179 (2014)
132. Fei, H., Peng, Z., Yang, Y., Li, L., Raji, A.-R.O., Samuel, E.L.G., Tour, J.M.: $LiFePO_4$ nanoparticles encapsulated in graphene nanoshells for high-performance lithium-ion battery cathodes. Chem. Commun. **50**(54), 7117–7119 (2014)
133. Hu, J., Lei, G., Lu, Z., Liu, K., Sang, S., Liu, H.: Alternating assembly of Ni-Al layered double hydroxide and graphene for high-rate alkaline battery cathode. Chem. Commun. **51**(49), 9983–9986 (2015)
134. Ma, R., Dong, Y., Xi, L., Yang, S., Lu, Z., Chung, C.: Fabrication of LiF/Fe/graphene nanocomposites as cathode material for lithium-ion batteries. Acs Appl. Mater. Inter. **5**(3), 892–897 (2013)
135. Guo, X., Fan, Q., Yu, L., Liang, J., Ji, W., Peng, L., Guo, X., Ding, W., Chen, Y.: Sandwich-like $LiFePO_4$/graphene hybrid nanosheets: in situ catalytic graphitization and their high-rate performance for lithium ion batteries. J. Mater. Chem. A. **1**(38), 11534–11538 (2013)
136. Molenda, J., Stokłosa, A., Bak, T.: Modification in the electronic structure of cobalt bronze Li_xCoO_2 and the resulting electrochemical properties. Solid State Ionics. **36**(1), 53–58 (1989)
137. Barker, J., Pynenburg, R., Koksbang, R., Saidi, M.Y.: An electrochemical investigation into the lithium insertion properties of Li_xCoO_2. Electrochim. Acta. **41**(15), 2481–2488 (1996)
138. Chung, S.-Y., Bloking, J.T., Chiang, Y.-M.: Electronically conductive phospho-olivines as lithium storage electrodes. Nat. Mater. **1**(2), 123–128 (2002)
139. Chung, S.-Y., Chiang, Y.-M.: Microscale measurements of the electrical conductivity of doped $LiFePO_4$. Electrochem. Solid-State Lett. **6**(12), A278–A281 (2003)
140. Xu, Y.-N., Chung, S.-Y., Bloking, J.T., Chiang, Y.-M., Ching, W.Y.: Electronic structure and electrical conductivity of undoped $LiFePO_4$. Electrochem. Solid-State Lett. **7**(6), A131–A134 (2004)
141. Prosini, P.P., Lisi, M., Zane, D., Pasquali, M.: Determination of the chemical diffusion coefficient of lithium in $LiFePO_4$. Solid State Ionics. **148**(1–2), 45–51 (2002)
142. Shimakawa, Y., Numata, T., Tabuchi, J.: Verwey-type transition and magnetic properties of the $LiMn_2O_4$Spinels. J. Solid State Chem. **131**(1), 138–143 (1997)
143. Kawai, H., Nagata, M., Kageyama, H., Tukamoto, H., West, A.R.: 5 V lithium cathodes based on spinel solid solutions $Li_2Co_{1+X}Mn_{3-X}O_8$: $-1 \leq X \leq 1$. Electrochim. Acta. **45**(1–2), 315–327 (1999)

144. Wakihara, M., Guohua, L., Ikuta, H., Uchida, T.: Chemical diffusion coefficients of lithium in $LiM_yMn_{2-y}O_4$ (M = Co and Cr). Solid State Ionics. **86–88, Part 2**(0), 907–909 (1996)
145. Yin, S.C., Strobel, P.S., Grondey, H., Nazar, L.F.: $Li_{2.5}V_2(PO_4)_3$: a room-temperature analogue to the fast-ion conducting high-temperature γ-phase of $Li_{2.5}V_2(PO_4)_3$. Chem. Mater. **16**(8), 1456–1465 (2004)
146. Rui, X.H., Ding, N., Liu, J., Li, C., Chen, C.H.: Analysis of the chemical diffusion coefficient of lithium ions in $Li_{2.5}V_2(PO_4)_3$ cathode material. Electrochim. Acta. **55**(7), 2384–2390 (2010)
147. Livage, J.: Vanadium pentoxide gels. Chem. Mater. **3**(4), 578–593 (1991)
148. Potiron, E., Le Gal La Salle, A., Verbaere, A., Piffard, Y., Guyomard, D.: Electrochemically synthesized vanadium oxides as lithium insertion hosts. Electrochim. Acta. **45**(1–2), 197–214 (1999)
149. Lantelme, F., Mantoux, A., Groult, H., Lincot, D.: Electrochemical study of phase transition processes in Lithium insertion in V_2O_5 electrodes. J. Electrochem. Soc. **150**(9), A1202–A1208 (2003)
150. Wei, W., Lv, W., Wu, M.-B., Su, F.-Y., He, Y.-B., Li, B., Kang, F., Yang, Q.-H.: The effect of graphene wrapping on the performance of $LiFePO_4$ for a lithium ion battery. Carbon. **57**, 530–533 (2013)
151. Su, F.-Y., You, C., He, Y.-B., Lv, W., Cui, W., Jin, F., Li, B., Yang, Q.-H., Kang, F.: Flexible and planar graphene conductive additives for lithium-ion batteries. J. Mater. Chem. **20**(43), 9644–9650 (2010)
152. Hu, L.-H., Wu, F.-Y., Lin, C.-T., Khlobystov, A.N., Li, L.-J.: Graphene-modified $LiFePO_4$ cathode for lithium ion battery beyond theoretical capacity. Nat. Commun. **4** (2013)
153. Nethravathi, C., Rajamathi, C.R., Rajamathi, M., Gautam, U.K., Wang, X., Golberg, D., Bando, Y.: N-doped graphene-VO_2(B) nanosheet-built 3D flower hybrid for lithium ion battery. Acs Appl. Mater. Inter. **5**(7), 2708–2714 (2013)
154. Liu, H., Yang, W.: Ultralong single crystalline V_2O_5 nanowire/graphene composite fabricated by a facile green approach and its lithium storage behavior. Energy Environ. Sci. **4**(10), 4000–4008 (2011)
155. Yang, S., Gong, Y., Liu, Z., Zhan, L., Hashim, D.P., Ma, L., Vajtai, R., Ajayan, P.M.: Bottom-up approach toward single-crystalline VO_2-graphene ribbons as cathodes for ultrafast lithium storage. Nano Lett. **13**(4), 1596–1601 (2013)
156. Lee, J.W., Lim, S.Y., Jeong, H.M., Hwang, T.H., Kang, J.K., Choi, J.W.: Extremely stable cycling of ultra-thin V_2O_5 nanowire-graphene electrodes for lithium rechargeable battery cathodes. Energy Environ. Sci. **5**(12), 9889–9894 (2012)
157. Jiang, R., Cui, C., Ma, H.: Using graphene nanosheets as a conductive additive to enhance the rate performance of spinel LiMn2O4 cathode material. Phys. Chem. Chem. Phys. **15**(17), 6406–6415 (2013)
158. Bak, S.-M., Nam, K.-W., Lee, C.-W., Kim, K.-H., Jung, H.-C., Yang, X.-Q., Kim, K.-B.: Spinel $LiMn_2O_4$/reduced graphene oxide hybrid for high rate lithium ion batteries. J. Mater. Chem. **21**(43), 17309–17315 (2011)

Part VIII

Energy Storage (Supercapacitor)

Fe-Based Anode Materials for Asymmetric Supercapacitors

22

Jizhang Chen, Cuiping Han, and C. P. (Ching-Ping) Wong

22.1 Introduction

Supercapacitors, also known as electrochemical capacitors, are nowadays considered as one of the most important energy storage devices, principally owing to their high power and excellent cyclability [1]. At present, supercapacitors have a variety of applications, e.g., powering tools, backup power systems, and start/stop systems [2, 3]. According to different charge storage mechanisms, supercapacitors can be classified into two types, namely, electrochemical double-layer capacitors (EDLCs) and pseudocapacitors. In EDLCs, energy storage is realized through electrosorption of electrolyte ions onto the surface of the electrode material, typically activated carbon (AC; see Fig. 22.1a–c). Since the formation of double layer (non-faradaic process) is very rapid and theoretically does not involve any mechanical strain or structural stress, ELDCs can be charged and discharged in a very short period of time while simultaneously delivering extremely high specific powder (up to 10 kW kg^{-1}) and extraordinary life span (>500,000 cycles). A notable advantage of EDLCs is that they are maintenance-free. Different from EDLCs, the capacitance of pseudocapacitors arises from surface and near-surface redox reactions of pseudocapacitive electrode materials (see, e.g., Fig. 22.1d–f, taking Fe_2O_3). Theoretically, the faradaic process enables pseudocapacitive materials to possess one order of magnitude higher theoretical-specific capacitance than that of EDLC materials. Besides pseudocapacitance, the electrosorption of electrolyte ions at the electrode/electrolyte interface also contributes to the total capacitance of pseudocapacitors, whereas it only accounts for a small portion.

Currently, nearly all supercapacitor manufacturers are using organic electrolytes, typically tetraethylammonium tetrafluoroborate ($TEABF_4$) dissolved in propylene carbonate (PC) or acetonitrile (AN). However, there exist several drawbacks for organic electrolytes, such as relatively high cost, inferior ionic conductivity, as well as serious safety issues associated with flammability, volatility, and toxicity of organic solvents. By contrast, aqueous electrolytes have a series of advantages over organic electrolytes, such as low cost, high ionic conductivity, environmentally benignness, and high electrothermal safety. In particular, the high ionic conductivity of aqueous electrolytes (e.g., 0.6 and 0.8 S cm^{-1} for 6 M KOH and 1 M H_2SO_4, respectively) is in favor of high specific capacitance and high specific power [3–5]. In respect of electrothermal safety, it is an important parameter for supercapacitors, since supercapacitors are frequently operated at high rates that might lead to thermal runaway. Moreover, it requires no special atmosphere during device manufacturing if aqueous electrolytes are used. Unfortunately, aqueous electrolytes suffer from low electrochemical stability window (ESW) due to water splitting that theoretically occurs at 1.23 V. As the specific energy of EDLCs is proportional to the square of the maximum operating voltage and the operating voltage of aqueous EDLCs (~1.0 V) is much lower than that of organic EDLCs (2.5–2.8 V), the specific energy of aqueous EDLCs is much lower than that of organic EDLCs. In terms of electrochemical evaluations on Fe-based pseudocapacitive materials, aqueous electrolytes are used in nearly all reports. The strategies to widen the ESW of aqueous electrolytes will be discussed in the aftermentioned sections.

As pseudocapacitive materials are usually poor in electronic and ionic transport properties, the interior bulk of them oftentimes remains inactive during redox reactions (see Fig. 22.1d), which is the reason why their experimentally measured capacitance values are oftentimes much lower than theoretical ones. In addition, the capacitance retention of pseudocapacitive materials is much lower than that of EDLC materials, due to large mechanical strain and structural distortion during redox reactions. In order to enhance the utilization efficiency and capacitance retention of pseudocapacitive materials, three effective strategies have been

J. Chen (✉) · C. Han · C. P. Wong
College of Materials Science and Engineering, Nanjing Forestry University, Nanjing, China

C. P. Wong et al. (eds.), *Nano-Bio- Electronic, Photonic and MEMS Packaging*, https://doi.org/10.1007/978-3-030-49991-4_22

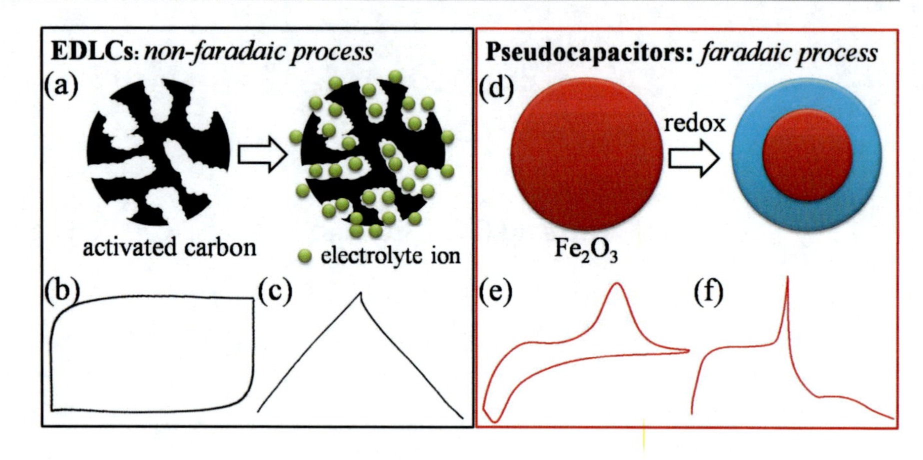

Fig. 22.1 Illustrations of the charge storage in (**a**) EDLCs and (**d**) pseudocapacitors and typical (**b**, **e**) CV curves and (**c**, **f**) GCD curves of (**b**, **c**) EDLC materials and (**e**, **f**) pseudocapacitive materials

proposed. The first strategy is to construct nanostructures, which offer three major advantages over the micrometer-sized counterparts: (1) increasing the electrode/electrolyte contact area per unit mass, therefore creating more reaction sites; (2) shortening transport paths for both electrolyte ions and electrons, therefore realizing higher pseudocapacitance and better rate capability; and (3) accommodating the mechanical strain and structural distortion upon redox reactions in a more efficient manner, therefore providing better cyclability. The second strategy extends the first one by introducing a porous configuration, which renders large interfacial contact with the electrolyte and thus allows fast ionic transfers at the interface. The third strategy is to build heterostructures by integrating pseudocapacitive materials with conductive skeletons (e.g., grapheme, carbon nanofibers (CNFs), carbon nanotubes (CNTs), porous carbon), which could in principle supply efficient pathways for electron transports and buffer strain/stress, thereby reducing electrochemical polarization, enhancing pseudocapacitive kinetics, and improving cycling stability.

Thanks to the large specific capacitance, suitable redox potential window, natural abundance, low cost, non-toxicity, and environmentally benignness, Fe-based pseudocapacitive materials (including Fe_2O_3, Fe_3O_4, FeOOH, FeS_2, etc.) are very promising as anode materials for asymmetric supercapacitors (ASCs) [6–9]. This chapter reviews recent research progress and development of Fe-based pseudocapacitive materials from charge storage mechanisms to synthesis methods and from three-electrode evaluations to full-cell devices.

22.2 Charge Storage Mechanisms

Some researchers insisted that RuO_2 and MnO_2 are pseudocapacitive materials when using aqueous electrolytes, since they show similar cyclic voltammetry (CV) and galvanostatic charging/discharging (GCD) profiles to that of EDLC materials, while Fe-based electrode materials are not pseudocapacitive materials, but are battery-type materials, due to the existences of redox peaks in their CV curves and plateaus in their GCD curves (see typical CV and GCD curves of Fe_2O_3 in the alkaline aqueous electrolyte in Fig. 22.1e, f, reported by Wong et al. [6]). In fact, Fe-based electrode materials can also show EDLC-like electrochemical profiles in other electrolytes and in other reports [10–12]. Moreover, both batteries and pseudocapacitors store charges through redox reactions, as a result, it would be misleading if separating batteries and pseudocapacitors only by curve profiles [13–15]. Actually, the main difference between batteries and pseudocapacitors is that the former is kinetically controlled by ion diffusion within the bulk of electrode material, while the latter is surface-controlled. A classic power equation $i = a\nu^b$ has been utilized as a powerful tool to distinguish batteries and pseudocapacitors (or EDLCs), where i is the response peak current and ν is the scan rate of CV measurements [13–15]. In the equation, a and b are adjustable values, representing charge storage ability and rate capability, respectively. Specifically, b value of 0.5 indicates the charge storage is controlled by semi-infinite linear diffusion of ions within the crystalline framework of the active material (battery-type), while b value of 1.0 indicates the charge storage is a surface process (supercapacitor-type). Owing to the inevitable resistances and impedances in the test system, b value would be lower than the theoretical value, especially at high rates (>20 mV s^{-1}). That is to say, the electronic conductivity of the active material, the ionic conductivity of the electrolyte, and the electrode/electrolyte interface properties should be seriously taken into account when concerning b values. For electrode materials, if the electronic conductivity is low, the particle size is large, or the mass loading is high, b value would be pushed to a lower value, meaning that a b value lower than 1.0 is reasonable for pseudocapacitive materials. Wong et al. report b value of 0.811 from 5 to 50 mV s^{-1} for pyrite FeS_2 nanobelts, even higher than RuO_2 and MnO_2 in most of the reports [10]. In fact, the electrochemical process of RuO_2, MnO_2, and Fe-based materials are all affected by ion diffusion, and nano-engineering is a prerequisite to achieve high performances. Benefiting from the short transport paths for both ions and electrons, nanostructured electrode materials are less limited by ion diffusions and electron transports, as a result, the rate capability (i.e., *b* value) can be improved, approaching that of EDLC materials. Consequently, Fe-based materials are pseudocapacitive, as RuO_2 and MnO_2.

22.2.1 Charge Storage Mechanisms of Fe_3O_4

O'Neill et al. evaluated the performance of FeO_x in different electrolytes, i.e., 0.5 M Na_2SO_3, Li_2SO_4, Na_2SO_4, and KCl aqueous electrolytes, and found that the capacitance in the Na_2SO_3 electrolyte (312 F g^{-1} at 2 mV s^{-1}) is much higher than that in other electrolytes [16]. The improved capacitance can be ascribed to the redox reactions of SO_3^{2-} ions adsorbed onto the surface of FeO_x. That is, the pseudocapacitive behavior is owing to Fe^{3+}/Fe^{2+} reversible reaction and the following reactions [17]:

$$2SO_3^{2-} + 3H_2O + 4e^- \leftrightarrow S_2O_3^{2-} + 6OH^- \quad (22.1)$$

$$SO_3^{2-} + 3H_2O + 8e^- \leftrightarrow 2S^{2-} + 6OH^- \quad (22.2)$$

Wu et al. and Chen et al. ascribed the pseudocapacitance of Fe_3O_4 in the Na_2SO_3 aqueous electrolyte to the surface redox reaction of sulfur in the forms of sulfate (SO_4^{2-}) and sulfite (SO_3^{2-}) anions, together with the redox reaction between Fe^{2+} and Fe^{3+} accompanied by the intercalation of SO_3^{2-} to balance the extra charge within the iron oxide layers (this proposed mechanism lacks experimental supports) [18, 19]:

$$FeO + SO_3^{2-} \leftrightarrow FeSO_4 + 2e^- \quad (22.3)$$

$$2Fe^{II}O + SO_3^{2-} \leftrightarrow \left(Fe^{III}O\right)^+ SO_3^{2-} \left(Fe^{III}O\right)^+ + 2e^- \quad (22.4)$$

Feng et al. investigated FeOOH in 1 M LiOH aqueous electrolyte after electrochemical cycles and found the original FeOOH was transformed into Fe_3O_4 on the basis of XRD and XPS characterizations [20]. That is, FeOOH can be electrochemically converted to Fe_3O_4 (denoted as e-Fe_3O_4). The structure and composition changes of e-Fe_3O_4 at different GCD states were studied. The e-Fe_3O_4 electrodes were taken out from the electrolyte immediately after being charged or discharged to a certain potential (see Fig. 22.2a), then rinsed with de-ionized water, and dried at 60 °C under vacuum. The FTIR spectra of e-Fe_3O_4 at different states are presented in Fig. 22.2b. The broad peak at 585 cm^{-1} corresponds to the Fe-O stretching vibration mode of Fe_3O_4, while two other peaks at 880 and 796 cm^{-1} can be assigned to the Fe-O-H bending vibration, probably originating from the residual FeOOH in e-Fe_3O_4. As there are no obvious absorption peaks in the region of 3200–3600 cm^{-1} that belong to the stretching vibration of free -OH or hydrogen-bonded -OH, the adsorption of OH^-

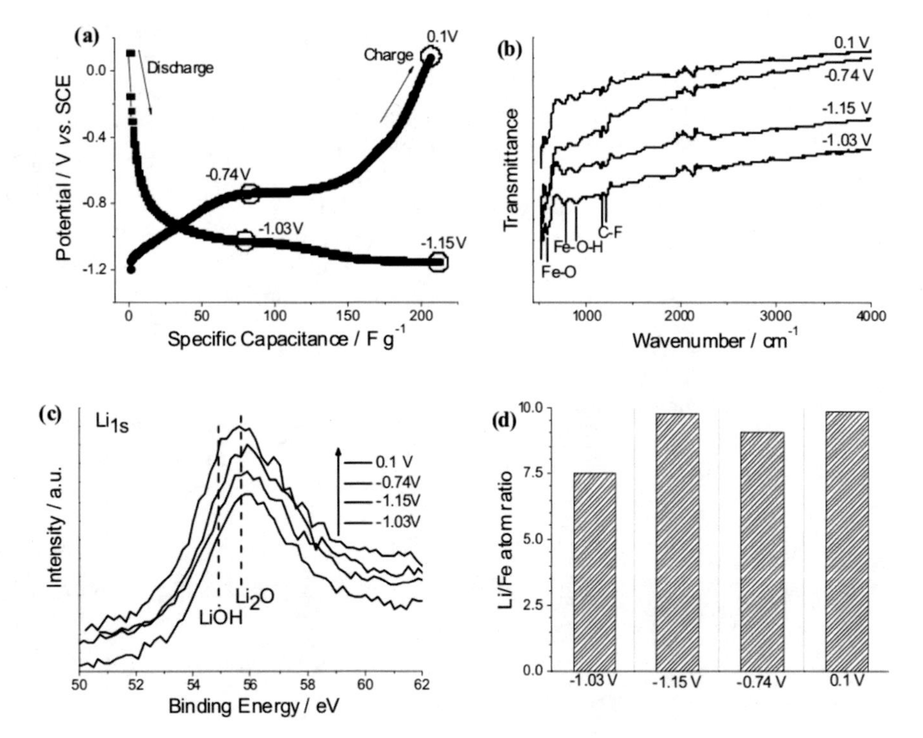

Fig. 22.2 (**a**) Typical GCD curves of e-Fe_3O_4 in 1 M LiOH aqueous electrolyte at 0.5 A g^{-1}. (**b**) FTIR spectra, (**c**) Li 1 s XPS spectra, and (**d**) the relative atomic ratios of Li/Fe in e-Fe_3O_4 at different GCD states [20]

on the surface of e-Fe_3O_4 is not the major contributor for the charge storage. In Fig. 22.2c, the existing state and content of Li^+ were investigated by XPS. The binding energy of Li 1 s is close to that of Li_2O instead of LiOH, excluding the existence of LiOH electrolyte in e-Fe_3O_4. Consequently, the Li 1 s spectra indicate that the adsorbed or intercalated Li^+ cations have a strong interaction with e-Fe_3O_4 (especially O^{2-} anions). Figure 22.2d shows the elemental composition of e-Fe_3O_4 at different states, obtained from XPS spectra. The results reveal that a large number of Li^+ cations reside in the interior of e-Fe_3O_4, which can be attributed to the nanoparticle structure of e-Fe_3O_4 and the inherent ability of Fe_3O_4 to adsorb cations. Another point was found that the crystalline structure of Fe_3O_4 keeps stable during cycles (evidenced by XRD and XPS measurements), which is different from the transition mechanism of Fe_3O_4 when used as the anode material for Li-ion batteries, where metallic Fe is formed upon lithiation. Thus, it was concluded that the pseudocapacitive behavior of Fe_3O_4 comes from fast adsorption/intercalation of Li^+.

22.2.2 Charge Storage Mechanisms of Fe_2O_3

It is reasonable to speculate that the pseudocapacitive behavior of Fe_2O_3 is similar to that proposed for Fe_3O_4, whereas it lacks sufficient experimental supports.

Yu et al. used ex situ XRD and Raman characterizations to explore the charge storage mechanism of α-Fe_2O_3 in 2 M Li_2SO_4 aqueous electrolyte [21]. It was found that there existed characteristic peaks of Li+-intercalated compounds in the XRD and Raman spectra of the electrode material after cycles. Thus, it was concluded that a Li+-intercalated compound is formed on the surface of the electrode material, which is associated with fast electrochemical reactions between α-Fe_2O_3 and Li_2SO_4.

Mohapatra et al. attributed the pseudocapacitive mechanism of Fe_2O_3 in 0.1 M Na_2SO_4 aqueous electrolyte to the following reversible redox reaction (an assumption) [22]:

$$Fe_2O_3 + OH^- \leftrightarrow Fe_2O_3OH + e^- \quad (22.5)$$

Wang et al. believed Fe_2O_3 in the KOH electrolyte stores charges via the following redox reaction (an assumption) [23]:

$$Fe_2O_3 + 3H_2O + 2e^- \leftrightarrow 2Fe(OH)_2 + 2OH^- \quad (22.6)$$

Wong et al. proposed that the charge storage mechanism of Fe_2O_3 in the KOH electrolyte is as follows (an assumption) [6]:

$$Fe_2O_3 + 2K^+ + 2e^- \leftrightarrow K_2Fe_2O_3 \quad (22.7)$$

22.2.3 Charge Storage Mechanisms of FeOOH

Long et al. used X-ray absorption spectroscopy (XAS) and extended X-ray absorption fine-structure (EXAFS) analysis to investigate the charge storage mechanism of FeO_x in 2.5 M Li_2SO_4 aqueous electrolyte and found the pseudocapacitance of the FeOOH coating originated from the reversible Fe^{3+}/Fe^{2+} redox couple, whereas the identity of the charge-compensating cation involved in the charge storage process is as yet undetermined [24].

Xue et al. proposed that the pseudocapacitive mechanism of γ-FeOOH is due to (note that this is short of experimental supports) [25]:

$$FeOOH + H_2O + e^- \leftrightarrow Fe(OH)_2 + OH^- \quad (22.8)$$

Hsu et al. performed systematic in situ XAS studies to elucidate the charge storage mechanism and the evolutions of Fe oxidation state in γ-FeOOH in Li_2SO_4 aqueous electrolyte during GCD cycles [26]. Figure 22.3a presents in situ X-ray absorption near-edge structure (XANES) spectra of the γ-FeOOH electrode, as well as that of Fe_3O_4 and Fe_2O_3. As the potential varied, a clear energy shift of the adsorption peak can be recognized. In Fig. 22.3b, the absorption threshold energy (E_0) obtained from the first inflection point on the absorption edge is linearly correlated with the oxidation state of Fe in the γ-FeOOH. Thus, the average oxidation state of Fe in the γ-FeOOH can be determined according to E_0 derived from in situ XANES spectra, as shown in Fig. 22.3c, indicating that the pseudocapacitance of the FeOOH comes from the reversible Fe^{3+}/Fe^{2+} redox couple. Figure 22.3d exhibits the Fourier-transformed (FT) magnitude of Fe K-edge EXAFS spectra of the γ-FeOOH electrode measured at several potentials in a sequence. The first FT maximum at ~1.5 Å corresponds to the Fe-O bond within a $[FeO_6]$ octahedral unit, while the second FT maximum at ~2.7 Å is attributed to the Fe-Fe interatomic distance between neighboring $[FeO_6]$ units. As the potential varied from −0.1 to −0.8 V, the Fe-O interatomic distance increased progressively, resulting from the reduction from Fe^{3+} to Fe^{2+}. It can also be observed that the Fe-Fe interatomic distance

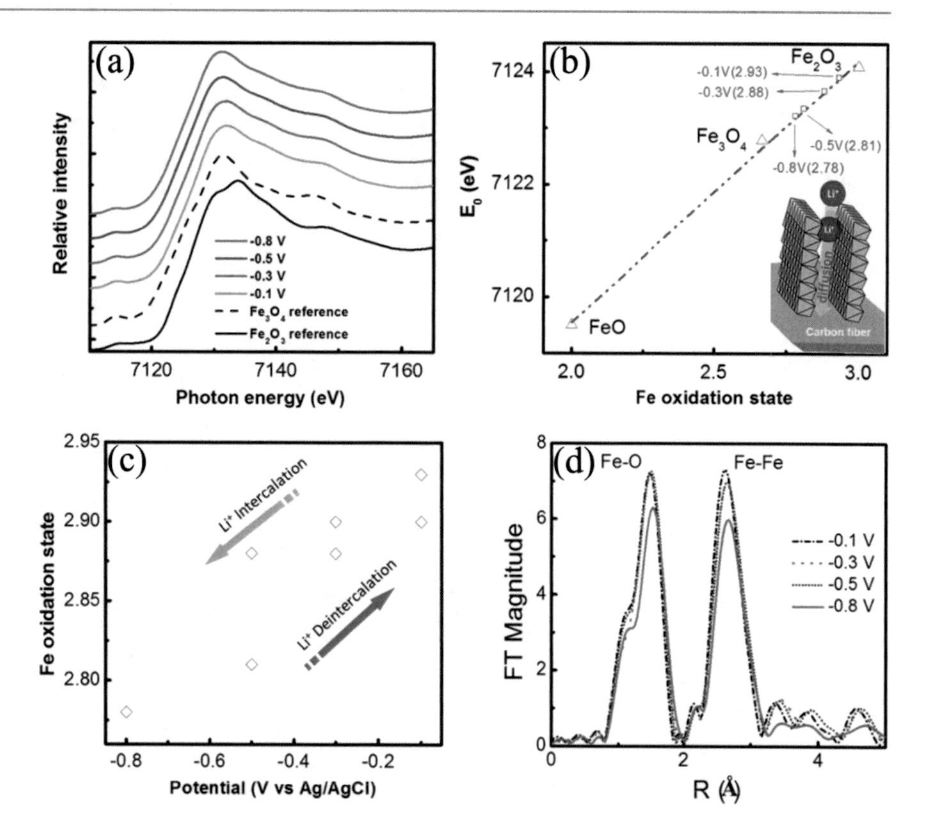

Fig. 22.3 (**a**) In situ Fe K-edge XANES spectra of the γ-FeOOH electrode in Li_2SO_4 aqueous electrolyte with respect to the potential from −0.8 to −0.1 V. (**b**) Relation between the edge position and Fe average oxidation state for γ-FeOOH at different potentials. (**c**) Evolutions of the Fe oxidation state with respect to the potential. (**d**) In situ Fe K-edge EXAFS spectra of the γ-FeOOH electrode in Li_2SO_4 aqueous electrolyte with respect to the potential from −0.8 to −0.1 V [26]

was increased, implying that the $[FeO_6]$ octahedral units were expanded, which was associated with Li^+ intercalation. In addition, it is seen that the intensity of the FT Fe-O maximum declined with varying the potential from −0.1 to −0.8 V, indicative of a distortion of γ-FeOOH during the intercalation process. Therefore, Hsu et al. concluded that the redox transition of Fe^{3+}/Fe^{2+} is charge compensated by the reversible intercalation and de-intercalation of Li^+ into and from 2D layered channels between the $[FeO_6]$ octahedral units, that is, the pseudocapacitive mechanism of γ-FeOOH in Li_2SO_4 aqueous electrolyte is based on [26]:

$$\text{Fe(III)OOH} + \text{Li}^+ + \text{e}^- \leftrightarrow \text{LiFe(II)OOH} \quad (22.9)$$

22.3 Synthesis Methods of Fe-Based Pseudocapacitive Materials

22.3.1 Hydrothermal Method

Hydrothermal method is well-known for synthesizing nanostructured or highly crystalline materials from hot-temperature (80–260 °C) aqueous solutions under high vapor pressure in a Teflon-lined stainless steel autoclave. It is possible to design and synthesize materials with specified morphology, size, structure, and crystallinity through altering certain experimental conditions, including reaction precursors, temperature, time, solvent, and surfactant. The disadvantages of hydrothermal method, however, are the inevitable usage of expensive autoclaves and the impossibility of in situ observing reaction process inside the autoclave during hydrothermal synthesis. With hydrothermal methods, Fe-based materials with various morphologies and structures have been reported.

22.3.1.1 Fe_2O_3

Tong et al. used a precursor solution containing $FeCl_3$, $NaNO_3$, and HCl (pH = ~1.5) to synthesize β-FeOOH nanorods after hydrothermal reaction at 95 °C for 6 h [11]. Then, β-FeOOH was converted to oxygen-deficient α-Fe_2O_3 nanorods after thermal annealing at 400 °C under N_2 atmosphere. The oxygen vacancies in Fe_2O_3 can serve as shallow donors, therefore effectively improving the electronic property and enhancing the electrochemical performance. The product exhibited a capacitance of 89 F g^{-1} at 0.5 A cm^{-2} in 3 M KCl aqueous electrolyte.

Yu et al. synthesized α-Fe_2O_3 nanoneedles grown on carbon cloth (CC) from a hydrothermal reaction [21]. The precursor solution was comprised of $Fe(NO_3)_3$ and Na_2SO_4 dissolved in deionized water, and CC was used as the substrate. The product obtained after 120 °C for 6 h delivered a capacitance of 1.78 F cm^{-2} at 2.0 mA cm^{-2} in 2 M Li_2SO_4 aqueous electrolyte.

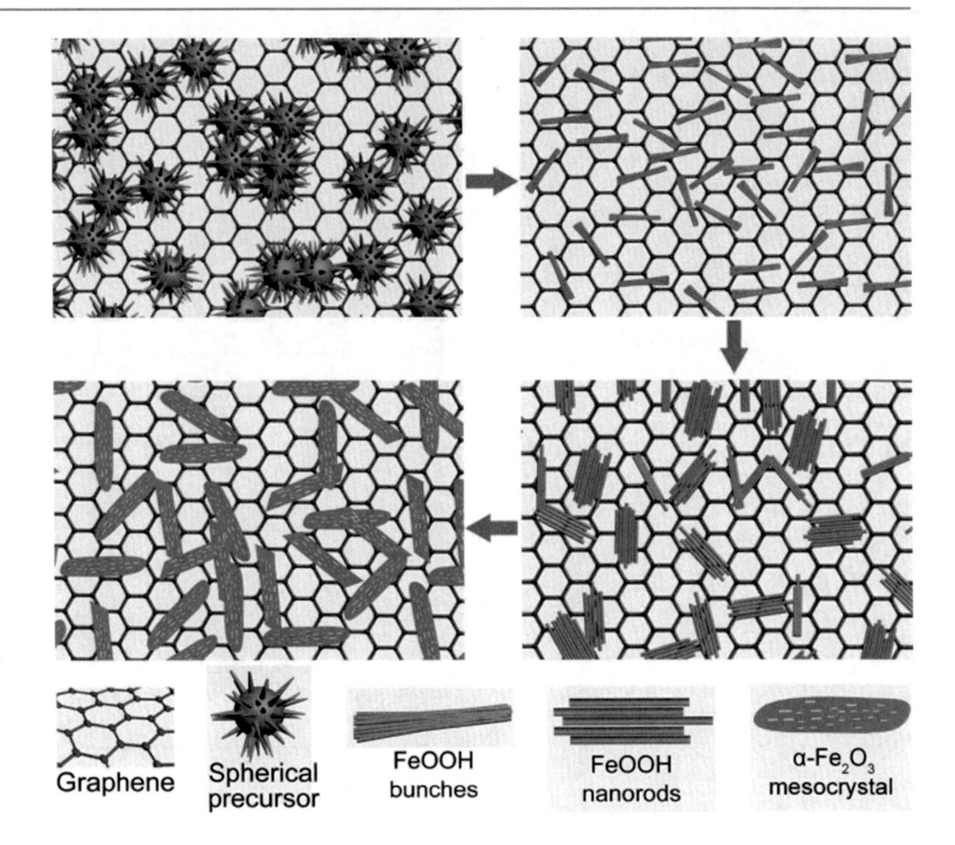

Fig. 22.4 Schematic illustration of the formation mechanism of α-Fe_2O_3 mesocrystals on graphene sheets and the corresponding shape-evolution process [28]

Guo et al. reported a composite of three-dimensional (3D) ultra-porous graphene-like carbon supporting well-dispersed and ultrafine α-Fe_2O_3 quantum dots (QDs), synthesized by a hydrothermal reaction at 100 °C for 0.5 h, from an aqueous precursor solution consisting of 3D carbon, $FeCl_3$, and ethanol [27]. The product showed a capacitance of 945 F g^{-1} at 1 A g^{-1} in 2 M KOH aqueous electrolyte.

Gao et al. fabricated graphene/α-Fe_2O_3 mesocrystals nanocomposite by a hydrothermal method [28]. The aqueous precursor solution was composed of $FeSO_4$ and graphene oxide (GO) and maintained at 160 °C for 24 h. The growth of α-Fe_2O_3 mesocrystals onto graphene nanosheets was ascribed to the synergic effects of dissolution-recrystallization, oriented attachment, and Ostwald ripening mechanisms. A possible growth mechanism was proposed based on the time-dependent experiments, as shown in Fig. 22.4. Owing to sufficient oxygen-containing functional groups on the basal planes and edges of GO nanosheets, GO aqueous solution was negatively charged, as a result, GO was favorably bound to Fe^{2+} cations via electrostatic interactions. During hydrothermal reaction, spherical particles with numerous protuberances were formed through a homogeneous nucleation process. As the reaction went on, the reaction solution turned to be acidic due to the retention of H^+ or the loss of OH^-, which might make the spherical precursor dissolve and recrystallize to form renascent FeOOH bunches. Owing to the continuous growth of FeOOH bunches, the FeOOH nanorod agglomerates were subsequently formed onto graphene nanosheets. Finally, the FeOOH nanorods were epitaxially fused together and transformed into α-Fe_2O_3 mesocrystals by the mechanisms of oriented attachment and Ostwald ripening. The product gave a capacitance of 306.9 F g^{-1} at 3 A g^{-1} in 1 M Na_2SO_4 aqueous electrolyte.

22.3.1.2 Fe_3O_4

Yan et al. used an aqueous solution containing GO, $FeCl_3$, $FeCl_2$, and $NH_3{\cdot}H_2O$ as the precursor solution to synthesize reduced graphene oxide (rGO)/Fe_3O_4 nanoparticles nanocomposite after hydrothermal reaction at 150 °C for 2 h [29]. The product delivered a capacitance of 890 F g^{-1} at 1 A g^{-1} in 1 M KOH aqueous electrolyte.

Yang et al. fabricated Fe_3O_4 nanoparticles (~5 nm in average) anchored on rGO by a hydrothermal reaction at 180 °C for 12 h [30]. The precursor solution was composed of GO, $FeCl_3$, glucose, and NaOH, with a pH value of ~7. The obtained product exhibited a capacitance of 241 F g^{-1} at 1 A g^{-1} in 1 M KOH aqueous electrolyte.

22.3.1.3 Others

Xue et al. obtained FeOOH nanorods through a low-temperature hydrothermal method (100 °C), only using $FeCl_3$ aqueous precursor solution without any addition of surfactants or templates [31]. The product fabricated from 0.2 M $FeCl_3$ showed the highest specific capacitance, i.e., 714.8 F g^{-1} at 1 A g^{-1} in 1 M KOH aqueous electrolyte.

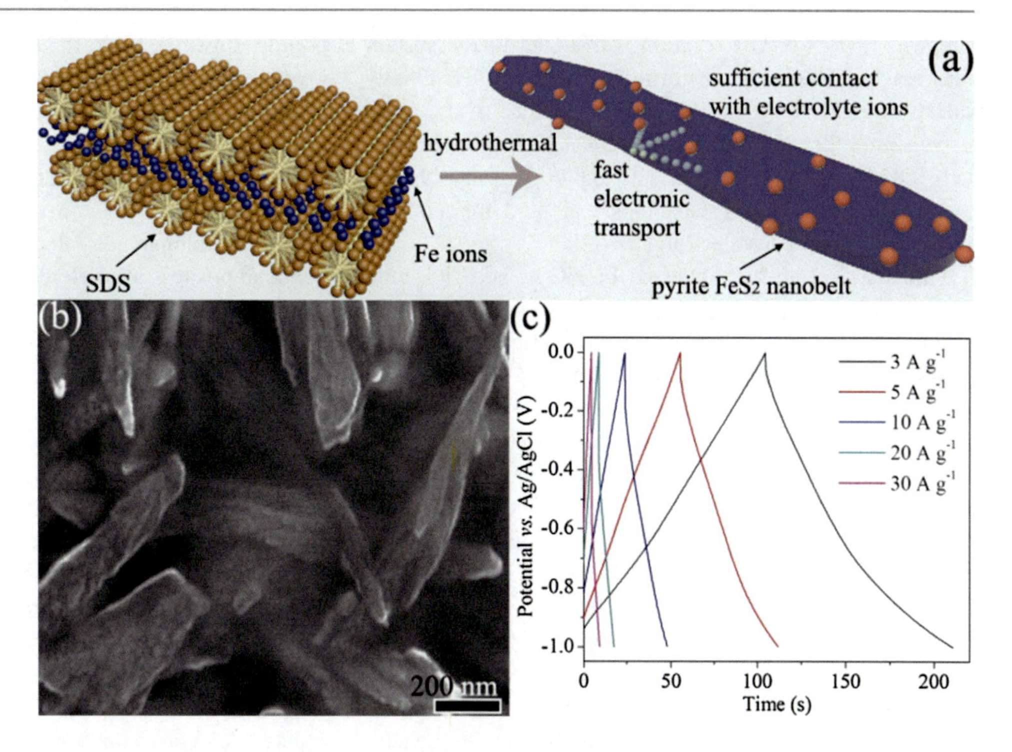

Fig. 22.5 (**a**) Schematic illustration for the formation, (**b**) SEM image, and (**c**) GCD curves at various current densities of pyrite FeS_2 nanobelts [10]

Hu et al. synthesized FeS_2 nanospheres from a hydrothermal process at 200 °C for 24 h, using an aqueous precursor solution containing $Fe(NO_3)_3$, ethylenediamine (EN), and carbon disulfide (CS_2) [32]. The product gave a capacitance of 484 F g^{-1} at 5 mV s^{-1} in 1 M LiCl aqueous electrolyte.

Wong et al. fabricated pyrite FeS_2 nanobelts via a facile hydrothermal method by introducing sodium dodecyl sulfate (SDS) anodic surfactant as the soft template, as schematically illustrated in Fig. 22.5a [10]. The nanobelt morphology was formed with the assistance of SDS micelles and can guarantee sufficient contact with electrolyte ions and shorten paths for charge transports. The product delivered a capacitance of 317.8 F g^{-1} at 3 A g^{-1} in 1 M Na_2SO_4 aqueous electrolyte.

22.3.2 Solvothermal Method

Solvothermal method is very similar to hydrothermal method. The only difference between hydrothermal method and solvothermal method is that the precursor solution of the latter is usually not aqueous. Solvothermal method integrates the benefits of hydrothermal method and sol-gel method, thus allowing for precise control of the products.

Chen et al. used CNFs as the conductive matrix, $FeCl_3$ and sodium acetate (NaAc) as reactants, and ethylene glycol (EG) as the solvent to conduct solvothermal method at 160 °C for 18 h [18]. CNFs were derived from polyacrylonitrile (PAN) nanofibers via carbonization in N_2 at 1000 °C. The obtained CNFs/highly dispersed Fe_3O_4 nanosheets nanocomposite delivered a capacitance of 135 F g^{-1} at 0.42 A g^{-1} in 1 M Na_2SO_3 aqueous electrolyte.

Golberg et al. fabricated carbon nanosheets (CNSs)/ultrathin nanoporous Fe_3O_4 nanocomposite by the solvothermal reaction of $FeCl_3$ and NH_4HCO_3 in EG solvent at 185 °C [33]. With the assistance of NH_4HCO_3 in the EG system, the chelated iron precursor with a lamellar structure can be formed, therefore resulting in intriguing morphology of the product, which exhibited a capacitance of 163.4 F g^{-1} at 1 A g^{-1} in 1 M Na_2SO_3 aqueous electrolyte.

Sarkar et al. synthesized 1D porous α-Fe_2O_3 nanoribbons via a solvothermal method ($FeCl_3$, urea, and oleylamine in EG and ethanol mixed solvent, reacting at 180 °C for 12 h) and subsequent heat treatment at 500 °C [34]. Detailed time-dependent TEM and HRTEM characterizations were used to find that the 1D nanoribbon structure was formed by the aggregation of porous α-Fe_2O_3 nanoparticles. The product gave a capacitance of 145 F g^{-1} at 1 A g^{-1} in 1 M KOH aqueous electrolyte.

Fu et al. used $FeCl_3$ and CNFs activated by concentrated HNO_3 solution as the precursors, and EG as the solvent, to synthesize CNFs/Fe_3O_4 spherulites nanocomposite after solvothermal reaction at 200 °C for 16 h [35]. The obtained product delivered a capacitance of 225 F g^{-1} at 1 A g^{-1} in 3 M KOH aqueous electrolyte.

22.3.3 Chemical Solution Deposition

Cao et al. synthesized rGO/Fe_3O_4 nanocrystals through chemical solution deposition by using GO as the precursor [36]. Typically, an aqueous solution of $Fe(NO_3)_3$ was added to the GO suspension, resulting in the formation of GO/Fe^{3+}

composite. Then NaOH solution was added to the suspension, making GO/Fe^{3+} transform into GO/FeOOH. Finally, $NaBH_4$ aqueous solution was added dropwise to the above suspension, followed by heat treatment at 120 °C for 12 h. As such, GO/FeOOH was reduced to rGO/Fe_3O_4 by $NaBH_4$. The product delivered a capacitance of 48.8 mF cm^{-2} in saturated KCl aqueous electrolyte.

Mohapatra et al. homogenized $Fe(NO_3)_3$ and EG under vigorous stirring and then heated the mixed solution at the desired temperature for a certain interval of time, producing a brown precipitate, i.e., nanostructured α-Fe_2O_3 [22]. EG not only acted as the solvent but also played a dominate role in controlling the surface morphology and facet orientation of α-Fe_2O_3 during the crystal growth. The influences of different parameters on the morphology, structure, and electrochemical property of α-Fe_2O_3 were investigated. The α-Fe_2O_3 synthesized at the Fe/EG ratio of 1: 2 gave the highest capacitance of 450 F g^{-1} in 0.1 M Na_2SO_4 aqueous electrolyte. Mohapatra et al. also reported flowery-shaped α-FeOOH@Fe_2O_3 core-shell nanoparticles through a ligand mediation-precipitation approach in a semi-aqueous-organic medium [37]. Typically, $Fe(NO_3)_3$ was stirred with the addition of ethylene glycol monomethyl ether (MEG) to form a homogeneous solution, followed by heat treatment for a certain period and subsequent calcination at 400 °C for 4 h in air. The obtained product exhibited a capacitance of 200 F g^{-1} at 5 mA g^{-1} in 0.1 M Na_2SO_4 aqueous electrolyte.

Xia et al. first dissolved $FeCl_3 \cdot 6H_2O$ into ethanol and then added rGO to the solution, followed by 10 min ultrasonication; finally NH_4HCO_3 was added to the suspension with continuous stirring for 8 h [38]. $FeCl_3$ can be converted to FeOOH on the basis of the following chemical reaction:

$$FeCl_3 + 3NH_4HCO_3 \rightarrow FeOOH + 3CO_2 + H_2O + 3NH_4Cl \quad (22.10)$$

Thanks to the uniform mixing of soluble reactants, the above reaction proceeded homogeneously and therefore generated uniform amorphous FeOOH QDs (~2 nm). Besides, the oxy-containing functional groups in rGO can tightly anchor FeOOH QDs through covalent chemical bondings. The obtained rGO/FeOOH QDs showed a capacitance of 365 F g^{-1} at 10 mV s^{-1} in 1 M Li_2SO_4 aqueous electrolyte.

Kolekar et al. deposited $ZnFe_2O_4$ nanoflake thin films onto stainless steel (SS) substrates using a chemical solution deposition method [39]. The bath solution consisted of $ZnCl_2$, $FeCl_2$, mono-ethanolamine (MEA), and ammonia with a pH value of ~10 and was maintained at 55 °C. Then ultrasonically cleaned SS substrates were immersed into the solution and rotated at 60 rpm speed with a gear motor. The pseudocapacitive performances of the obtained product were investigated in different alkaline aqueous electrolytes, i.e., 1 M LiOH, 1 M NaOH, and 1 M KOH, and their binary combinations of LiOH-NaOH, NaOH-KOH, and LiOH-KOH. It was found that $ZnFe_2O_4$ gave the highest capacitance of 433 F g^{-1} at 2 mA cm^{-2} in the KOH electrolyte.

Purushothaman et al. fabricated mesh-like C/Fe_2O_3 nanocomposites at different annealing temperatures and analyzed the relationship between the annealing temperature and electrochemical property [40]. In a typical synthesis, dextran was added into $Fe(NO)_3$ solution under constant stirring, and ammonia solution was added dropwise to maintain the pH at ~10. After vigorous stirring for 4 h, dark brown precipitate appeared in the solution and was subsequently annealed at 400, 500, and 600 °C to obtain mesh-like C/Fe_2O_3 nanocomposites, which are denoted as FCC4, FCC5, and FCC6, respectively, depending on the annealing temperature. The formation mechanism of the mesh-like structure is shown in Fig. 22.6a. The formation of $Fe(OH)_2$ chains was initiated by the chain-like structure of dextran. Dextran was decomposed into carbon during the annealing process, leading to the formation of C/Fe_2O_3 nanocomposite. The burnt off carbon varied with the annealing temperature, resulting in morphological changes of the nanocomposite. Figure 22.6b shows specific capacitances of FCC4, FCC5, and FCC6 in 2 M KOH aqueous electrolyte as a function of the current density. FCC5 revealed the best performance among all the samples, i.e., 315 F g^{-1} at 2 mV s^{-1}.

22.3.4 Thermal Decomposition of Fe-Based Precursor

Fe-based pseudocapacitive materials can also be produced through thermal decomposition of Fe-based precursors at a high temperature [6, 41, 42]. Zhi et al. reported porous carbon/Fe_3O_4 nanoparticles composite via calcination of Fe-based metal organic framework (Fe-MOF) [42]. The product was composed of porous carbon and Fe_3O_4 nanoparticles, resulting from well-controlled incomplete annealing (500 °C under N_2 atmosphere) of the MOF. A capacitance of 139 F g^{-1} was achieved for this product at 0.5 A g^{-1} in 1 M KOH aqueous electrolyte.

Wong et al. designed and fabricated hierarchical graphene/Fe_2O_3 nanocomposite comprised of interconnected Fe_2O_3 with porous structure anchored on the graphene scaffold [6]. Figure 22.7a describes the schematic (a template-assisted nanocasting process) for fabricating the abovementioned nanocomposite. In step I, cetyl trimethyl ammonium bromide (CTAB) was used as the pore former, and tetraethyl orthosilicate (TEOS) was used as the Si source, to anchor porous SiO_2 onto the graphene scaffold through a solution method, followed by calcination at 800 °C under Ar

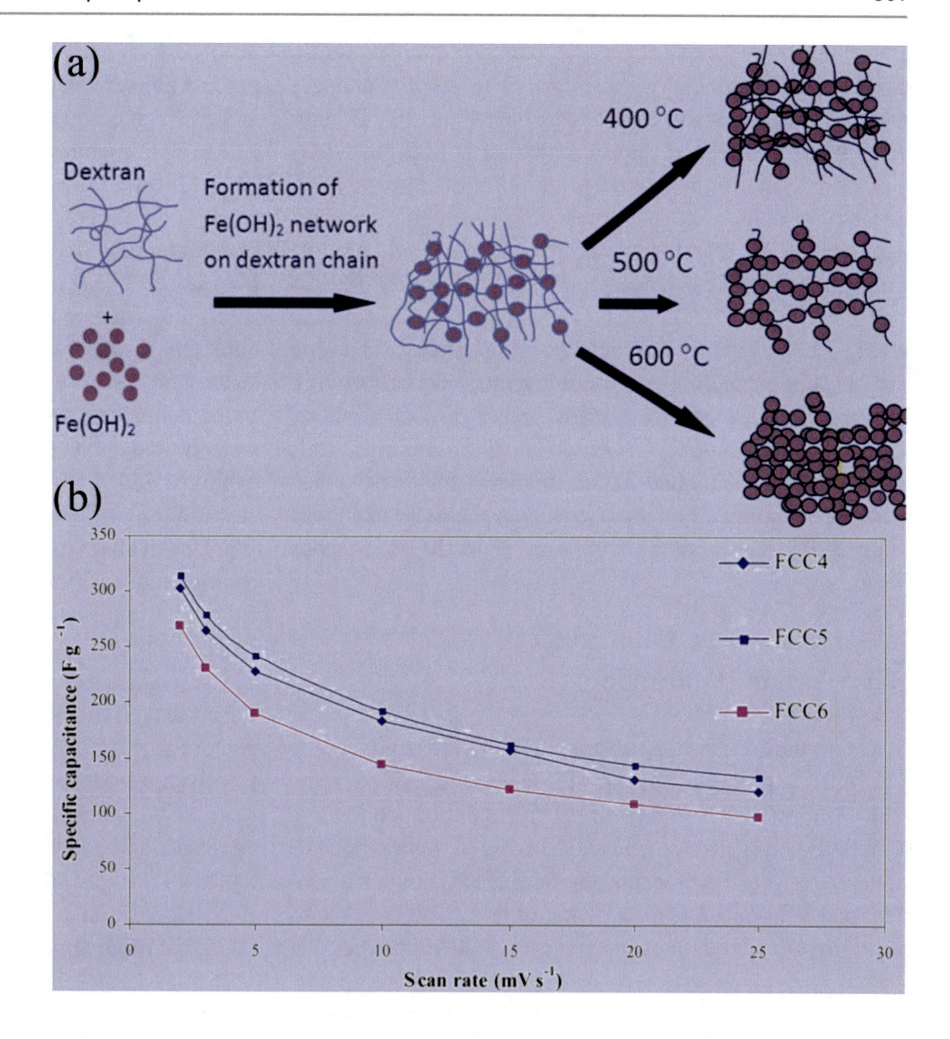

Fig. 22.6 (**a**) Formation mechanism of C/Fe_2O_3 nanocomposites with different annealing temperatures. (**b**) Specific capacitances of different C/Fe_2O_3 nanocomposites at various current densities [40]

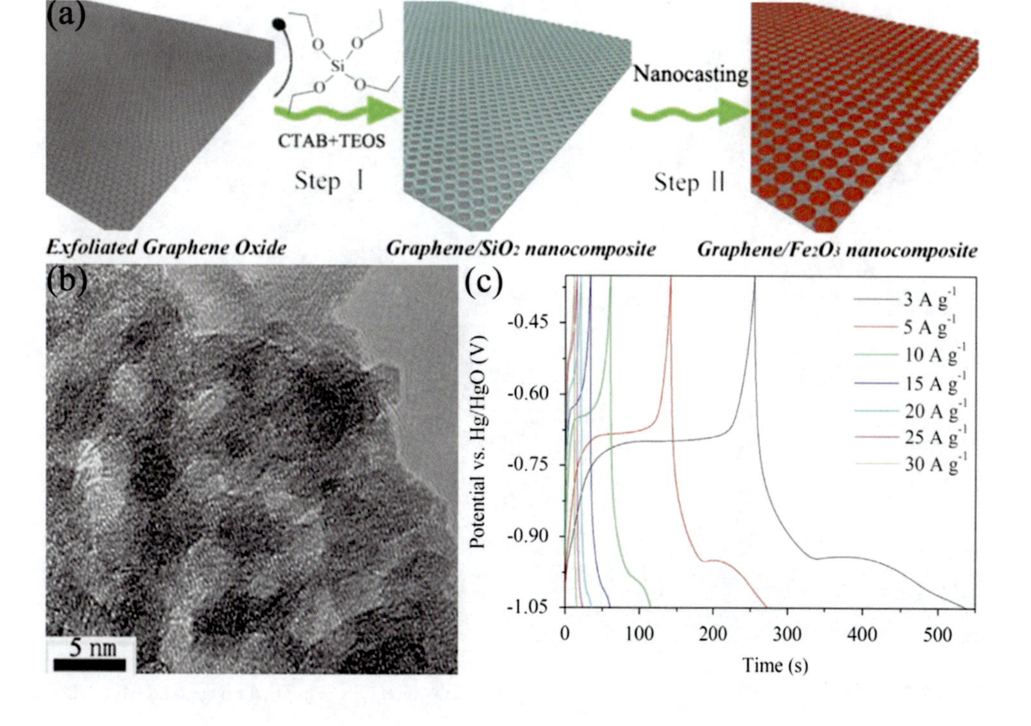

Fig. 22.7 (**a**) Schematic for the fabrication process, (**b**) HRTEM image, and (**c**) GCD curves as a function of the current density of graphene/porous Fe_2O_3 nanocomposite [6]

atmosphere. In step II, the obtained graphene/SiO_2 nanocomposite was added to an ethanol solution containing $Fe(NO_3)_3$, and the suspension was stirred at 50 °C until ethanol was evaporated. After calcination at 350 °C, Fe_2O_3 was formed due to thermal decomposition of Fe-containing precursor and filled into the pores of SiO_2. Finally, the SiO_2 was etched by NaOH, resulting in the final product (graphene/porous Fe_2O_3 nanocomposite), whose HRTEM is shown in Fig. 22.7b. The GCD curves of the nanocomposite in Fig. 22.7c reveal distinct charging and discharging plateaus, corresponding to reversible redox reaction between Fe^{2+} and Fe^{3+}. Combining the advantages of nanostructure, porous configuration, and conductive matrix, this nanocomposite exhibited much higher capacitance than most of reported Fe-based pseudocapacitive materials, being 1095 F g^{-1} at 3 A g^{-1} in 3 M KOH aqueous electrolyte.

22.3.5 Electrodeposition

Hsu et al. used CC as the substrate to fabricate lepidocrocite γ-FeOOH nanosheets via one-step electrodeposition route [26]. The CC was immersed in the solution containing 0.01 M $Fe(NH_4)_2(SO_4)_2$ and 0.04 M NaAc, and γ-FeOOH nanosheets were electrodeposited onto the CC under a constant potential of 0.7 V for a period of time. The nanosheets were smooth and homogeneous with a low thickness of 30–50 nm and an average length of ~1.4 μm. The obtained product delivered a capacitance of 310.3 F g^{-1} at 0.13 A g^{-1} in 1 M Li_2SO_4 aqueous electrolyte.

Lu et al. synthesized graphene/γ-Fe_2O_3 nanocomposite via an electrodeposition method [43]. Typically, graphene powers were mixed with poly(vinylidene fluoride) (PVDF) binder at a ratio of 10: 1, and then suspended in 1-methyl-2-pyrrolidone (NMP), followed by drop-casting on a graphite substrate. Then γ-Fe_2O_3 was potentiostatically deposited onto the graphite substrate using 25 mM K_2FeO_4 as the electrolyte, resulting in graphene/γ-Fe_2O_3 nanocomposite. In the composite, the γ-Fe_2O_3 nanocrystals of ~5 nm in size were hosted by the porous graphene film. The composite exhibited 224 F g^{-1} at 25 mV s^{-1} in 1 M Na_2SO_3 aqueous electrolyte.

Wang et al. prepared core/shell ZnO/Ag/FeOOH electrode by electrodeposition of FeOOH onto Ag-decorated ZnO nanorod arrays [44]. Figure 22.8a illustrates the fabrication process of ZnO/Ag/FeOOH. First, ZnO nanorods were grown on the Ni foam through a wet chemical process. Then, the Ni foam was soaked in a solution containing glucose and $AgNO_3$ at room temperature for Ag decoration. Finally, electrodeposition was carried out using the treated Ni foam as the working electrode, a carbon rod as the counter electrode, and a solution containing 0.1 M $Fe(NH_4)_2(SO_4)_2$, 0.1 M Na_2SO_4, and 0.2 M NaAc as the electrolyte. After electrodeposition at a constant current of 0.125 mA cm^{-2} for 10 min at room temperature, ZnO/Ag/FeOOH was obtained.

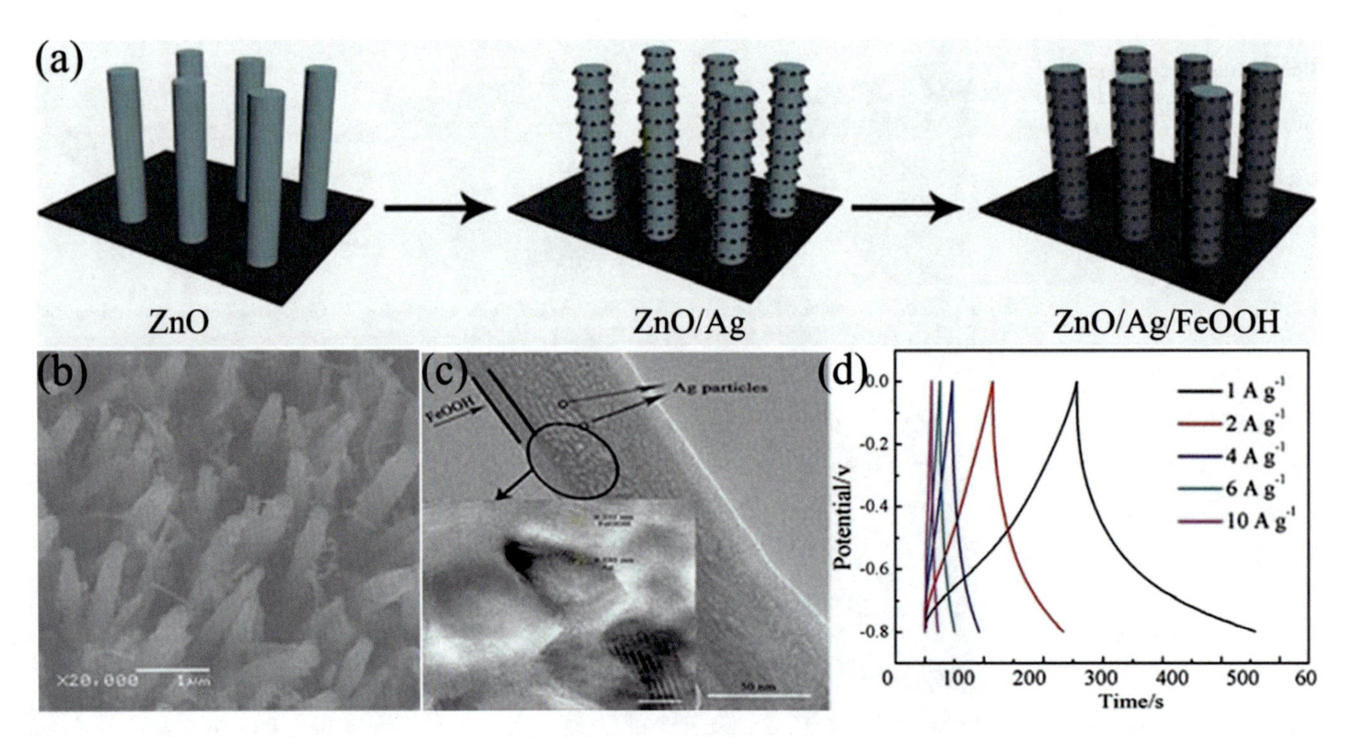

Fig. 22.8 (**a**) Schematic illustration of the fabrication, (**b**) SEM image, (**c**) HRTEM image, and (**d**) GCD curves of the ZnO/Ag/FeOOH electrode [44]

Figure 22.8b exhibits SEM image of ZnO/Ag/FeOOH, demonstrating the product possesses nanorod morphology. The HRTEM image of ZnO/Ag/FeOOH is presented in Fig. 22.8c. In the ZnO/Ag/FeOOH, ZnO nanorods have a diameter ~60 nm, Ag particles are homogeneously distributed on ZnO nanorods, and FeOOH shell with a thickness of ~10 nm is well-distributed on the surface of ZnO/Ag nanorods. ZnO nanorod arrays function as the skeleton for loading active material, thus providing a large specific area for easy access of electrolyte ions, while the Ag decoration improves the electronic conductivity. As a result, the product gave a capacitance of 376.6 F g^{-1} at 1A g^{-1} in 0.5 M Li_2SO_4 aqueous electrolyte, as shown in Fig. 22.8d.

22.3.6 Anodization

Misra et al. conducted anodization in an EG solution with 0.5% NH_4F and 3% H_2O, for synthesizing iron oxide nanotubes [45]. The anodization system was comprised of a DC power source, an iron coupon anode, and a platinum cathode. After applying a constant voltage of 50 V for 30 min at 15 °C, iron oxide nanotubes were produced on the iron substrate. Then the anodized product was further annealed in air at a temperature ranging from 200 to 700 °C for 2 h. The influence of annealing temperature on the pseudocapacitive performance of iron oxide was studied, and it was found the product annealed at 300 °C revealed the best performance, i.e., 364 mF cm^{-2} at 1 mA cm^{-2} in 1 M Li_2SO_4 aqueous electrolyte.

Wijayantha et al. anodized steel in a 10 M NaOH solution containing NH_4F with the initial voltage and current being 3.7 V and 1.5 A, respectively, resulting in highly porous sponge-like Fe_2O_3. The porous structure of Fe_2O_3 resulted from the oxidation of Fe(0) to Fe(III) and the subsequent dissolution of Fe(III) species into the F^--containing electrolyte:

$$2Fe + 3H_2O \rightarrow Fe_2O_3 + 3H_2 \quad (22.11)$$

$$Fe_2O_3 + 12F^- + 6H^+ \rightarrow 2[FeF_6]^{3-} + 3H_2O \quad (22.12)$$

The anodized electrode showed a capacitance of 18 mF cm^{-2} in 1 M NaOH aqueous electrolyte.

22.3.7 Other Synthesis Methods

Long et al. used electroless deposition for the growth of FeO_x onto fiber-supported carbon nanofoam paper [24]. The carbon nanofoam was first infiltrated with KOH and then soaked in 25 mM K_2FeO_4 and 9 M KOH solution. K_2FeO_4 is a strong oxidant, whereas its solution is stable only in alkaline solutions with a narrow pH range. Consequently, 9 M KOH was used to stabilize K_2FeO_4. The redox reaction between K_2FeO_4 and carbon generated nanostructured FeO_x both on the interior and exterior of the carbon matrix. The obtained carbon nanofoam/FeO_x composite delivered a capacitance of 343 F g^{-1} at 5 mV s^{-1} in 2.5 M Li_2SO_4 aqueous electrolyte.

Sivakumar et al. used an electrospinning technique to fabricate α-Fe_2O_3 porous fibers [46]. In the synthesis, iron (III) acetylacetonate (Fe(acac)$_3$), polyvinylpyrrolidone (PVP), and acetic acid were utilized as the Fe precursor, spinning aid, and reaction controller, respectively. After electrospinning under an applying potential of 10 kV, the electrospun fiber was further annealed at 500 °C in air, producing α-Fe_2O_3 porous fibers with a particle size of ~21 nm. The product gave a capacitance of 256 F g^{-1} at 1 mV s^{-1} in 1 M LiOH aqueous electrolyte.

Chen et al. reported core-shell structured Fe_2O_3@C nanoparticles through oxidizing Fe@C nanoparticles that were prepared by a direct current carbon arc discharge method [47]. During the arc discharge, a cylindrical graphite was used as the cathode, and a mixture of graphite powder and Fe powder (weight ratio of 1:1) was used as the anode. The arc between the cathode and anode was initiated at a current of 100 A and voltage of 50 V. The obtained material was further calcined at 750 °C for 2 h in air to produce Fe_2O_3@C, followed by activation with KOH at 750 °C for 2 h under N_2 atmosphere. The activated product exhibited a capacitance of 612 F g^{-1} at 0.5 A g^{-1} in 5 M NaOH aqueous electrolyte.

Yao et al. prepared C/Fe_xC_y composite via a high-temperature solid-state method (calcining the filter paper containing ferric citrate at 990 °C under Ar atmosphere) [48]. The Fe_xC_y showed polyhedron shape with an average size of ~50 nm and were distributed on the wire-like carbon randomly. According to the XRD characterization, the Fe_xC_y was a hybrid of $Fe_{1.88}C_{0.12}$ and $Fe_{15.1}C$. The C/Fe_xC_y composite showed a capacitance of 274.7 F g^{-1} at 2.5 A g^{-1} in 6 M KOH aqueous electrolyte.

Fu et al. fabricated hollow and porous Fe_2O_3 micro-rods through a ball-milling method combined with chemical treatment [49]. The schematic diagram of the fabrication process is shown in Fig. 22.9a. The kg-scale mill chunks were obtained directly from the rolling steel plant. Firstly, the chunks were ball-milled for 4 h to produce powders with a mean size of ~2 mm. This step was aimed at increasing surface area of the raw material, so as to ensure fast and complete reaction in the next step. Secondly, the ball-milled mill scale powders were dispersed into an aqueous solution containing oxalic acid. During the chemical reaction between the mill scale and oxalic acid, the mill scale dissolved progressively, and a new and finer-scale solid phase was formed

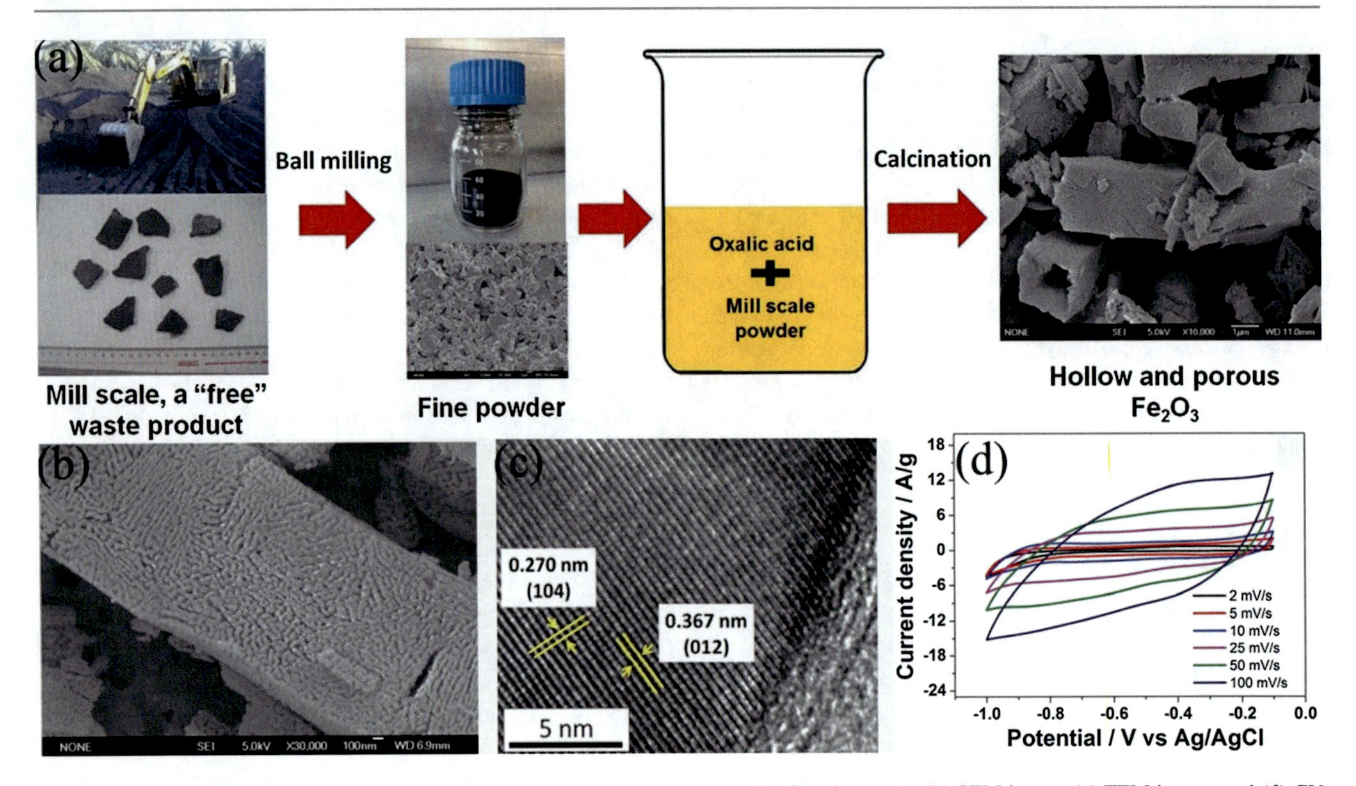

Fig. 22.9 (**a**) Schematic diagram of converting mill scale to hollow and porous Fe_2O_3 micro-rods. (**b**) SEM image, (**c**) TEM image, and (**d**) CV curves of the product [49]

as the sediment progressively. Finally, the obtained sediment was calcined in a furnace, generating hollow and porous Fe_2O_3 micro-rods, as shown in Fig. 22.9b, c. The CV curves of the product are shown in Fig. 22.9d, revealing a capacitance of 346 F g^{-1} at 2 mV s^{-1} in 0.5 M Na_2SO_3 aqueous electrolyte.

22.4 Doping of Fe-Based Pseudocapacitive Materials

Doping (i.e., introducing external elements into the crystalline framework of host inorganic materials to form a solid solution) has been a widely used mythology to tune different properties. For energy storage materials, doping is usually utilized to enhance charge transport properties.

Lu et al. reported Ti^{4+}-doped Fe_2O_3 prepared by hydrothermal reaction at 120 °C and the following sintering at 300 °C under N_2 atmosphere [50]. Ti^{4+} is an electron donor that can reduce Fe^{3+} to Fe^{2+}, significantly enhancing the donor density of Fe_2O_3. Compared to the pristine Fe_2O_3, the Raman spectrum of Ti-Fe_2O_3 was negatively shifted and became more broadened, implying more oxygen vacancies after Ti-doping. In XPS spectra, the existence of Fe^{2+} was observed after Ti-doping, corresponding to oxygen vacancies. The results demonstrated greatly improved electronic conductivity after Ti-doping, which contributed to a better performance, being 1.15 F cm^{-2} (311 F g^{-1}) at 1 mA cm^{-2} in 5 M LiCl aqueous electrolyte.

Wu et al. prepared α-Fe_2O_3 nanorods that were doped by multiple heteroatoms (C, N, and S) and self-assembled into super-long hollow tubes. The α-Fe_2O_3 nanorods were synthesized using the chicken eggshell membrane as both the template and dopant via hydrothermal and calcination techniques. The product delivered a capacitance of 330 g^{-1} at 0.5 A g^{-1} in 6 M KOH aqueous electrolyte.

Yu et al. synthesized metal-like β-FeO(OH,F) nanorods through a hydrothermal process at 120 °C and post-annealing at 400 °C [12]. For the synthesis of typical sample, Fe $(NO_3)_3$, NH_4F, and urea with a molar ratio of 2:2:5 were used as the raw materials in the hydrothermal reaction. Sample-1 and Sample-2 were synthesized when the molar ratios were 2:4:5 and 2:1:5, respectively. The control sample was synthesized by a same process to that of typical sample except that NH_4F was replaced by NH_4Cl. XRD patterns of different samples are shown in Fig. 22.10a. The XRD pattern of typical sample was similar to the pattern from the standard JCPDS card of β-FeO(OH,Cl), indicating the formation of β-FeO(OH,F). The smaller diameter of F^- than that of Cl^- made the lattice parameters of β-FeO(OH,F) slightly smaller than that of β-FeO(OH,Cl). With twice amounts of F^-, Sample-1 had a lower crystallization, owing to disorders or defects within the lattices originating from excess F^- dopants. With half amounts of F^-, Sample-2 had a low

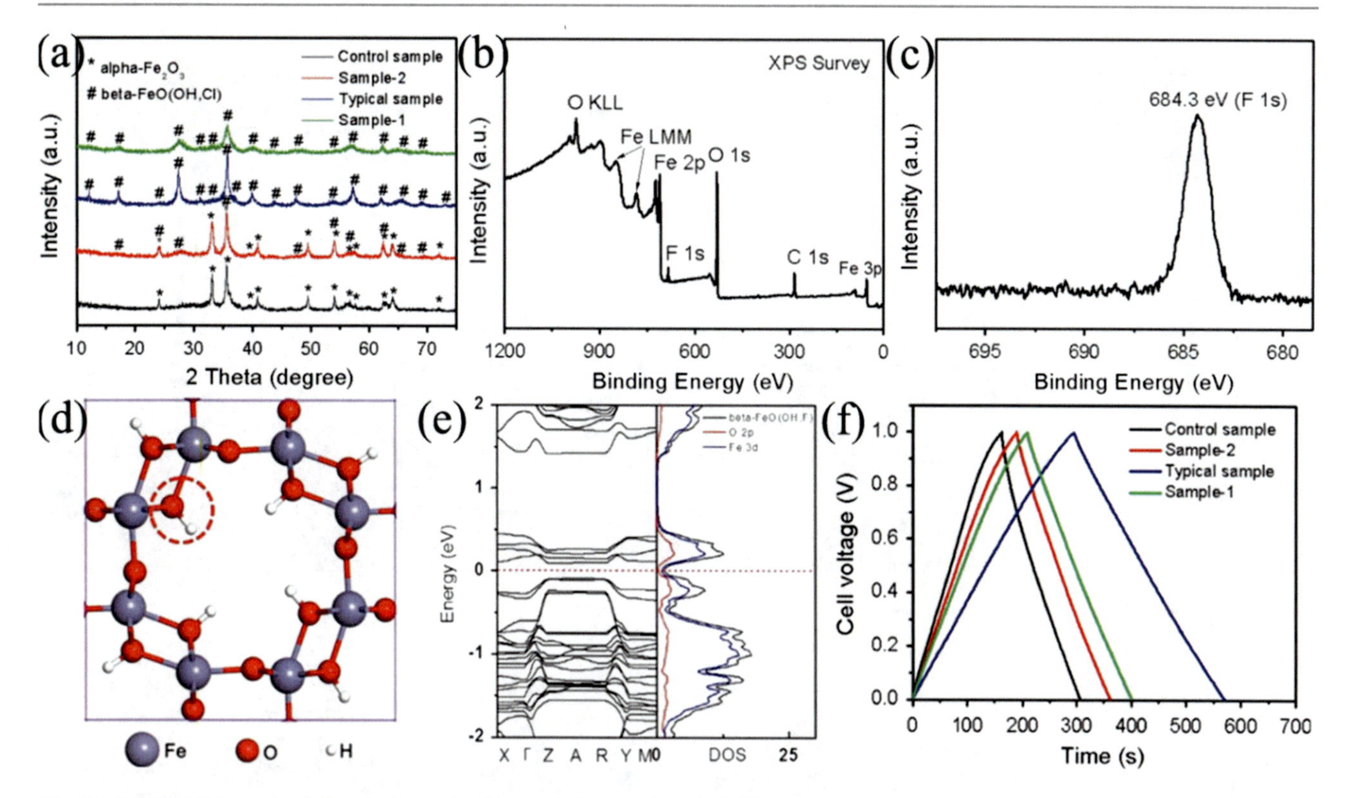

Fig. 22.10 (**a**) XRD patterns of typical sample, Sample-1, Sample-2, and control sample. (**b**) XPS survey spectrum and (**c**) high-resolution XPS F 1 s spectrum of typical sample. (**d**) Computational model for the β-FeOOH (the marked -OH illustrates the F-doping site). (**e**) Calculated DOS plots for β-FeO(OH,F). (f) GCD curves of different samples [12]

peak intensity of FeO(OH,F) and some characteristic peaks belonging to α-Fe_2O_3, which arose from the Fe^{3+} cations not coordinated with F^-. As for control sample, only α-Fe_2O_3 existed. The results suggested that the strong complexation between Fe^{3+} and F^- made the formation of β-FeO(OH,F) more favorable than other phases. The XPS survey spectrum in Fig. 22.10b and high-resolution XPS F 1 s spectrum in Fig. 22.10c of typical sample further manifest successful doping of F^-. Yu et al. also performed density functional theory (DFT) calculations based on the model in Fig. 22.10d and found the band gap of F-doped β-FeOOH was merely ~0.2 eV (see calculated density of state (DOS) plots in Fig. 22.10e). Such a metal-like behavior renders β-FeO(OH, F) promising for charge storage. The β-FeO(OH,F)-based supercapacitor gave a high capacitance of 1.12 F cm^{-2} at 1.0 mA cm^{-2} in 6 M KOH aqueous electrolyte, as shown in Fig. 22.10f.

22.5 Heterostucture

Combining advantages of different materials, heterostructure is an effective strategy to improve charge storage. Considering carbon-based matrixes are common and have been discussed by the above sections, this section is mainly focused on materials other than carbon for constituting heterostructures with Fe-based pseudocapacitive materials.

Arul et al. synthesized CeO_2/Fe_2O_3 composite nanospindles through co-precipitation of $Ce(NO_3)_3$ and Fe $(NO_3)_2$ in an ammonia hydroxide solution with a pH of ~11 [51]. It was speculated that integrating Fe_2O_3 with CeO_2 can improve the redox stability. The composite revealed a capacitance of 142.6 F g^{-1} at 5 mV s^{-1}, together with excellent capacitance retention of ~95% after 1000 cycles.

Tour et al. used Ta_2O_5 nanotubes grown on the tantalum foil as the framework to support carbon-onion-coated Fe_2O_3 nanoparticles, so as to fabricate 3D heterostructured thin-film electrode [52]. First, Ta_2O_5 nanotubes were grown by an anodization method using tantalum foils as the substrate, followed by annealing at 750 °C under Ar atmosphere. Then, Fe_2O_3 was deposited into Ta_2O_5 nanotubes by a solution-based method. Afterward, carbon onions were deposited through an in situ CVD process, during which Fe_2O_3 was reduced to metallic Fe. Finally, the sample was annealed at 300 °C in air to convert metallic Fe back to Fe_2O_3. In the heterostructured composite, carbon onions acted as microelectrodes to separate Ta_2O_5 and Fe_2O_3, thus forming a nanoscale 3D sandwich structure. As such, space charge layers can be formed at the phase boundaries, providing additional energy storage by charge separation.

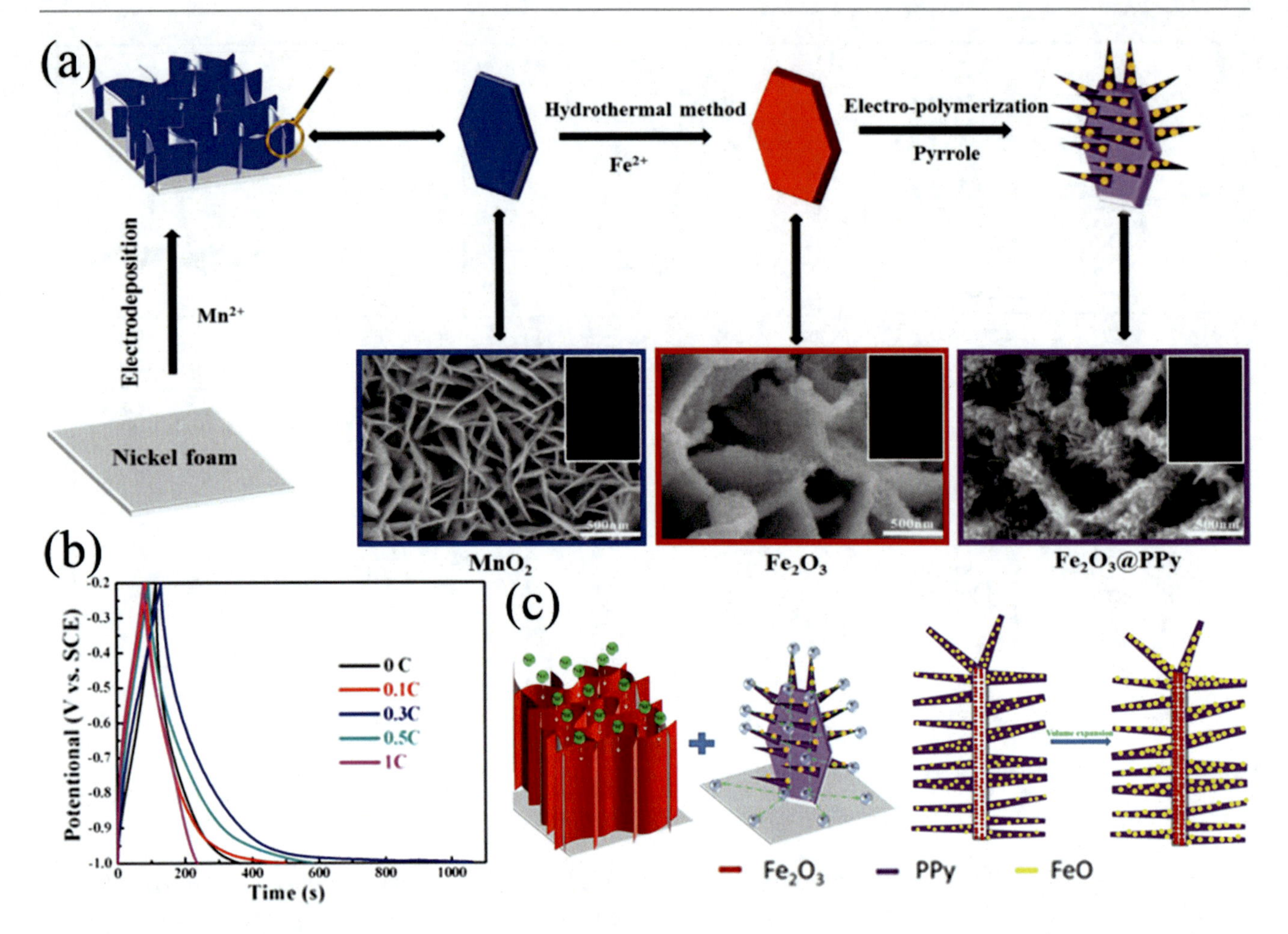

Fig. 22.11 (**a**) Fabrication process of core-branch Fe_2O_3@PPy. (**b**) GCD curves of neat Fe_2O_3 and Fe_2O_3@PPy with different amounts of PPy at 1 A g^{-1}. (**c**) Schematic illustration of charge transport paths within Fe_2O_3@PPy [53]

Therefore, the obtained thin-film electrode delivered good supercapacitor performance (~18.2 mF cm^{-2}).

Arbiol et al. reported core-branch honeycomb-like Fe_2O_3 nanoflakes@polypyrrole (PPy) nanoleaves arrays that were grown on the Ni foam substrate, which combined 1D leaves-like PPy, 2D mesoporous Fe_2O_3 nanoflakes, and 3D macroporous substrate [53]. The fabrication process is depicted in Fig. 22.11a, employing MnO_2 nanoflakes as the sacrificed template to synthesize honeycomb-like Fe_2O_3 nanoflakes. Typically, $FeSO_4$ was dissolved in a mixed solution of EG and water, and a Ni foam containing MnO_2 nanoflakes was immersed in this solution for hydrothermal reaction at 120 °C, producing $Fe(OH)_3$, which was transformed into α-Fe_2O_3 after calcination at 400 °C in air. Subsequently, PPy was deposited onto α-Fe_2O_3 via electropolymerization in a solution containing pyrrole, $LiClO_4$, and sodium dodecyl sulfate (SDS). The as-prepared Fe_2O_3@PPy exhibited a maximum capacitance of 1168 F g^{-1} at 1 A g^{-1} in 0.5 M Na_2SO_4 aqueous electrolyte, as shown in Fig. 22.11b. Although this capacitance was rather high, water splitting definitely largely contributed to this capacitance, since the discharging time was considerably larger than the corresponding charging time and there existed large tags at the low potential of discharging curves. The charge transport paths within Fe_2O_3@PPy are schematically illustrated in Fig. 22.11c. The honeycomb-like structure of Fe_2O_3 backbones provides large surface area accessible to electrolyte ions, while the wrapped PPy branches functioned as "superhighways" for efficient transports of electrons, therefore improving charge storage performances.

Boulmedais et al. fabricated a ternary hybrid material consisting of nanostructured Fe_3O_4 and poly (3,4-ethylenedioxy thiophene)/poly(styrene sulfonate) (PEDOT/PSS) grafted onto few-layer graphene (FLG) using hydrothermal reaction followed by step-by-step spin-coating [54]. Graphene and PEDOT/PSS guaranteed high electronic conductivity. PEDOT/PSS also acted as a binder to ensure cohesion to the hybrid material. The hybrid material gave a capacitance of 153 F g^{-1} at 0.1 A g^{-1} in 0.5 M Na_2SO_3 aqueous electrolyte.

Konstantinov et al. decorated α-Fe_2O_3 nanorods uniformly onto PEDOT/PSS-functionalized rGO scaffold,

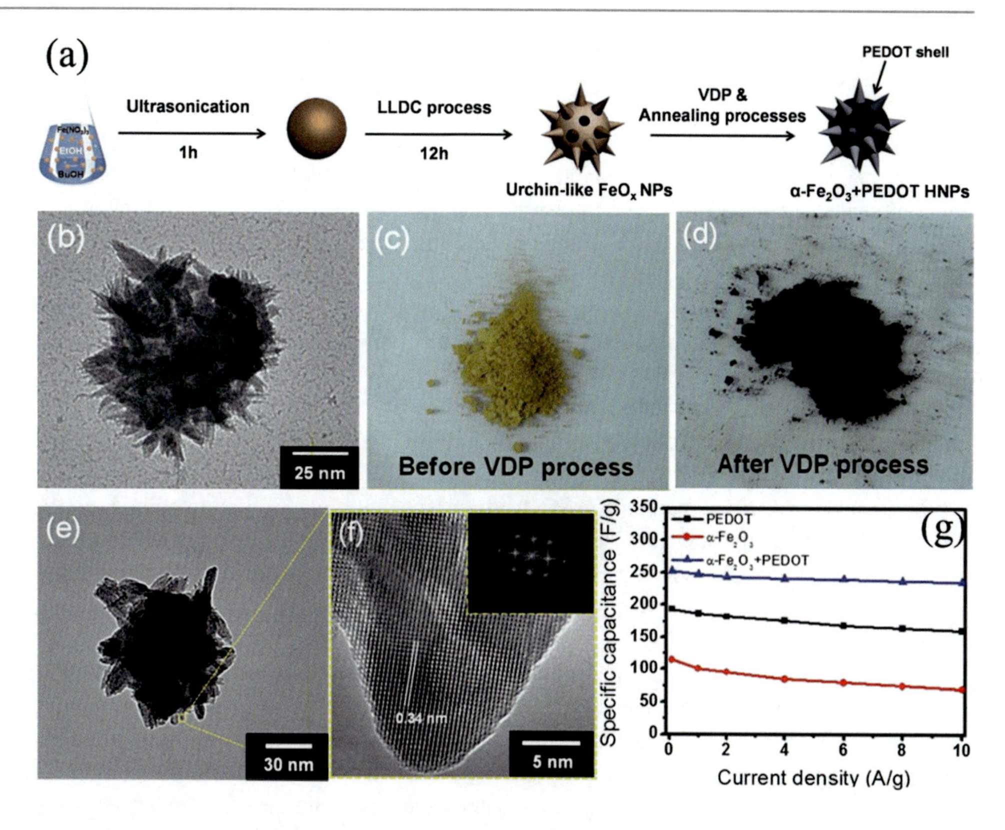

Fig. 22.12 (**a**) Schematic illustration of the synthesis of α-Fe_2O_3 + PEDOT HNPs. (**b**) TEM image of FeO_x NPs obtained after the LPL. Optical micrographs of (**c**) FeO_x NPs before VDP process and (**d**) α-Fe_2O_3 + PEDOT HNPs after VDP process. (**e**) TEM and (**f**) HRTEM images of α-Fe_2O_3 + PEDOT HNPs. (**g**) Specific capacitances of different sample at various current densities [56]

realizing 3D interconnected layer-by-layer (LBL) architecture [55]. Such an interpenetrating network ensured highly conductive pathways for facile transports of electrons and electrolyte ions. Benefiting from the synergistic effects of ternary components, this hybrid material showed a high capacitance of 875 F g^{-1} at 5 mV s^{-1} in 1 M KOH aqueous electrolyte.

Jang et al. synthesized hierarchical α-Fe_2O_3/PEDOT core-shell hybrid nanoparticles (HNPs) through sonochemical and liquid-liquid diffusion-assisted crystallization (LLDC) methods followed by vapor deposition polymerization (VDP) [56]. The overall fabrication procedure is illustrated in Fig. 22.12a. In the first step, highly uniform FeO_x nanospheres were obtained by template-free sonochemical and LLDC methods. FeO_x particles were observed to be urchin-like with a diameter of ~60 nm (see Fig. 22.12b). In the second step, VDP was carried out in a chemical vapor deposition (CVD) chamber at 100 °C, for PEDOT coating. The successful VDP was indicated by the color change from yellow to black after VDP process (see Fig. 22.12c, d). In the third step, the sample was annealed at 300 °C, converting FeO_x (comprised of mixed phases with amorphous structures) to α-Fe_2O_3. No changes in the structure or morphology were observed for the product after VDP and annealing processes, as shown in Fig. 22.12e. Figure 22.12f demonstrates a uniform PEDOT layer with the thickness being ~2.5 nm was formed on the surface of crystalline α-Fe_2O_3. PEDOT is a highly stable and conductive polymer, which can not only effectively improve the electronic conductivity but also act as a buffering layer to prevent α-Fe_2O_3 from destruction/degradation during cycling. Consequently, the α-Fe_2O_3/PEDOT possessed better performances than the control samples (see Fig. 22.12g, e.g., 252.8 F g^{-1} at 0.1 A g^{-1} in 1 M H_2SO_4 aqueous electrolyte.

22.6 Aqueous Asymmetric Supercapacitors

Aqueous ASCs are composed of two different supercapacitor electrode materials with complementary potential windows using an aqueous electrolyte. Owing to the high overpotential for O_2/H_2 evolutions of pseudocapacitive materials, the devices can be operated up to 1.5 or 1.6 V, higher than the ESW of water (1.23 V). Up to now, different aqueous ASC systems have been reported using Fe-based anode materials coupled with different cathode materials, e.g., NiO nanoflakes [23], $Ni(OH)_2$ [57], 3D graphene/NiOOH/Ni_3S_2 [58], $Ni_3(PO_4)_2$ [59], manganous hexacyanoferrate (MnHCF) [60], $Co(OH)_2$ nanosheets [61], CoNi-layered double hydroxides (LDHs) [6], MnO_2 [62], $NiCo_2S_4$ [63], and $CoMoO_4$ [64].

Fan et al. used graphene foam (GF)/CNT hybrid film as the substrate to grow porous Fe_2O_3 nanorods and $Ni(OH)_2$ nanosheets, employed as the anode material and cathode material, respectively [57]. The assembled ASC delivered high specific energy/power (100.7 Wh kg^{-1} at

287 W kg^{-1}, 70.9 Wh kg^{-1} at 1.4 kW kg^{-1}) and high capacitance retention of 89.1% after 1000 cycles in 6 M KOH aqueous electrolyte.

Lin et al. used $FeCl_3$, $FeCl_2$, GO, and ammonia solution as the hydrothermal precursors to synthesize rGO/Fe_3O_4 nanocomposite as the anode material for assembling aqueous ASC with 3D graphene/NiOOH/Ni_3S_2 cathode [58]. The ASC showed a specific energy of 82.5 Wh kg^{-1} at a power density of 930 W kg^{-1} in 1 M KOH aqueous electrolyte.

Crosnier et al. synthesized nanosized $FeWO_4$ via a polyol-mediated route and used this material to assemble aqueous ASC with MnO_2 [62]. The ASC gave a capacitance of 9 F g^{-1} (based on the total active materials) in 5 M $LiNO_3$ aqueous electrolyte.

Wong et al. assembled aqueous ASC using graphene/porous Fe_2O_3 nanocomposite and CNT/CoNi-layered double hydroxide (LDH) nanocomposite as the anode material and cathode material, respectively [6]. The device delivered a high specific capacitance of 252.4 F g^{-1} when the current density was 0.5 A g^{-1} with the open circuit voltage being 1.5 V in 3 M KOH aqueous electrolyte. A high specific energy (up to 98.0 W h kg^{-1}) was obtained under a specific power of 465.9 W kg^{-1}. When the specific power was increased to very high at 22,862 W kg^{-1}, a specific energy of 9.0 W h kg^{-1} can still be maintained. Wong et al. also used nitrogen-doped hierarchically porous carbon foam as the mechanical support for graphene/porous Fe_2O_3 nanocomposite and $NiCo_2S_4$, based on which, an aqueous ASC was assembled [63]. The ASC device exhibited a high specific energy of 93.9 Wh kg^{-1} (or 3.55 mWh cm^{-3}) and a high specific power of 21.1 kW kg^{-1} (or 799 mW cm^{-3}). In another report, Wong et al. fabricated amorphous nanostructured fish-scale-like FeOOH and flower-like CoNi-DH nanosheets onto Ni foams through facile and scalable one-step electrodeposition [65]. The fabrications of FeOOH and CoNi-DH, as well the assembly of aqueous ASC, are illustrated in Fig. 22.13a. The SEM image of FOOH is shown in Fig. 22.13b, revealing intriguing fish-scale-like morphology. The aqueous ASC consisting of FeOOH anode, CoNi-DH cathode, and 3 M KOH aqueous electrolyte exhibited nonlinear GCD curves, with a high capacitance of 849 mF cm^{-2} at 5 mA cm^{-2} and a low equivalent series resistance (R_{ESR}) of ~2.22 Ω cm^{-2}, as shown in Fig. 22.13c. Figure 22.13d displays gravimetric Ragone plots of the ASC compared with other reported advanced ASCs. The ASC assembled by Wong et al. gave high specific energy ranging from 86.4 to 22.8 W h kg^{-1} as the specific power varied from 1832 to 11,628 W kg^{-1}, superior to many other reports [65]. The ASC also possessed a high energy density of 0.7 mW h cm^{-3} at high power density of 15.3 mW cm^{-3} (including the volumes of two Ni foam substrates).

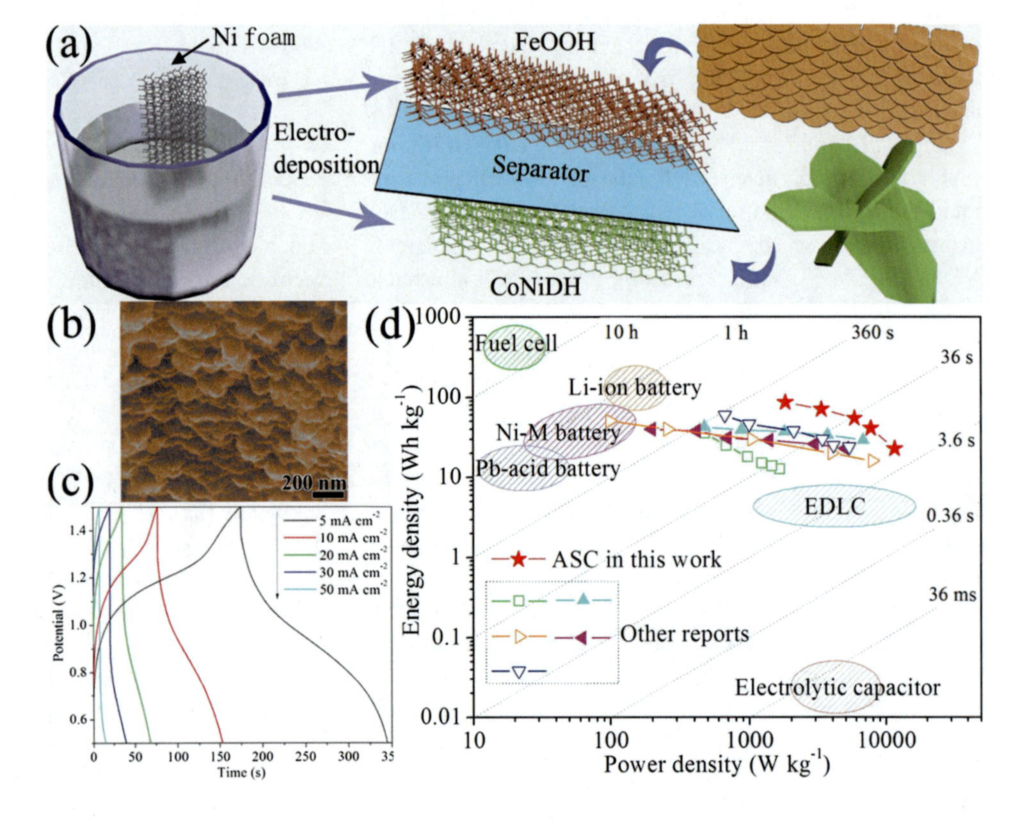

Fig. 22.13 (**a**) Schematic illustration for the fabrications of nanostructured FeOOH and CoNi-DH. (**b**) SEM image of FeOOH. (**c**) GCD curves and (**d**) Ragone plots of specific energy vs. specific power of the aqueous FeOOH//CoNi-DH ASC [65]

22.7 Solid-State Asymmetric Supercapacitors

With the growing demands for flexible, bendable, and wearable personal electronics, it is a technological trend to develop flexible and lightweight power sources. In order to conform to such trend, it is a prerequisite to replace aqueous electrolytes with solid or quasi-solid polymer electrolytes. Meantime, strict sealing and housing requirements can be avoided if using polymer electrolytes. Owing to their high ionic conductivity (~10^{-3} to 10^{-4} S cm^{-1}, much higher than ~10^{-7} to 10^{-8} S cm^{-1} of solid polymer electrolytes) [66], gel polymer electrolytes (GPEs) are currently the most investigated electrolytes for solid-state power sources. Typically, a GPE consists of a polymer matrix as the host and a liquid electrolyte (e.g., aqueous electrolyte, organic electrolyte, ionic liquids) as the guest. As for aqueous GPEs (also called hydrogel electrolytes), poly (vinyl alcohol) (PVA, a linear polymer) is the most used polymer matrix, owing to its facile preparation, low cost, high hydrophilicity, non-toxicity, great electrochemical stability, and good mechanical properties [66, 67]. In practical applications, PVA is blended with different aqueous electrolytes, such as H_2SO_4, H_3PO_3, KOH, and LiCl solutions. The ionic conductivity of PVA-based GPEs is at the magnitude of 10^{-3} S cm^{-1}. However, the ESW of aqueous PVA-based GPEs is limited by water splitting, since PVA matrixes are swollen with aqueous electrolytes.

Mai et al. developed a flexible solid-state ASC with an extended operating voltage window of 1.6 V, using α-MnO_2 nanowires (NWs) and amorphous Fe_2O_3 nanotubes (NTs) grown on CCs as two electrodes and PVA/LiCl gel as the electrolyte [68]. The synthesis procedures of MnO_2 NWs and Fe_2O_3 NTs are schematically shown in Fig. 22.14a. A hydrothermal method was used to deposit MnO_2 NWs onto the CC. And a sacrificial template-assisted hydrolysis approach was used to grow Fe_2O_3 NTs onto the CC. SEM images of MnO_2 and Fe_2O_3 are displayed in Fig. 22.14b, c, respectively, indicating the CCs were covered by MnO_2 NWs with 100–150 nm in diameter and 2–4 μm in length and Fe_2O_3 NTs with 100–200 nm in outer diameter, respectively. The assembled ASC exhibited high energy density of 0.55 mWh cm^{-3}, high power density of 139.1 mW cm^{-3}, as well as good rate capability, as shown in Fig. 22.14d. Moreover, a commercial blue light-emitting diode (LED) can be easily lightened up by the tandem device comprised of two ASCs connected in series.

Tong et al. assembled Fe_2O_3//MnO_2 solid-state ASC on the basis of CC substrates and PVA/LiCl gel electrolyte [11]. The device delivered an energy density of 0.41 mWh cm^{-3} at a current density of 0.5 mA cm^{-2} and still remained 0.35 mWh cm^{-3} at 6 mA cm^{-2}.

Lokhande et al. demonstrated flexible and bendable all-solid-state thin-film ASC, using nanostructured α-MnO_2 and α-Fe_2O_3 thin films grown on the flexible SS substrate as two electrodes and PVA/$LiClO_4$ gel as electrolyte [69]. The ASC possessed a high potential window and great cycling stability of 2500 cycles, together with high specific energy of 41 Wh kg^{-1} at 2.1 kW kg^{-1}.

Wei et al. used wrinkled CNT/MnO_2 hybrid film (prepared by CVD method) and wrinkled CNT/Fe_2O_3 hybrid film (prepared by solution precipitation) as the cathode and

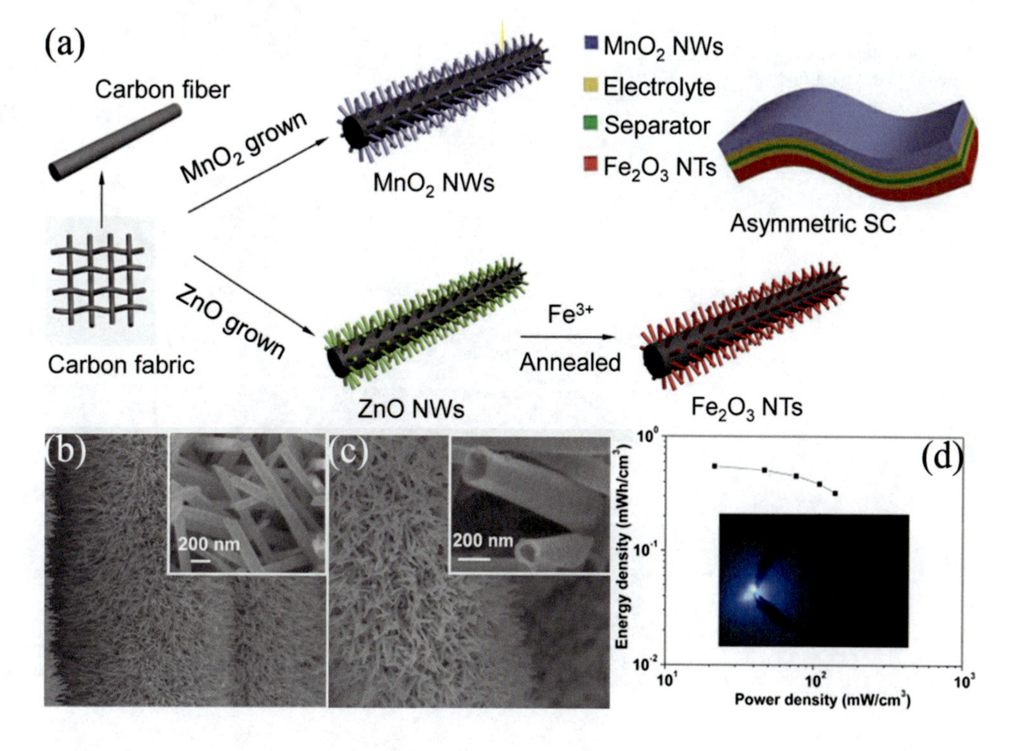

Fig. 22.14 (**a**) Schematic illustration of the synthesis procedures of MnO_2 NWs and Fe_2O_3 NTs grown on CCs and the assembly of solid-state ASC devices. SEM images of (**b**) MnO_2 NWs and (**c**) Fe_2O_3 NTs. (**d**) Ragone plots of the ASC device [68]

anode, respectively, to constitute stretchable all-solid-state ASC using PVA/Na_2SO_4 gel electrolyte [70]. An optimized ASC can be reversibly cycled at a high voltage while simultaneously showing high specific energy of 45.8 W h kg^{-1} at 410 W kg^{-1} and great cycling stability of 98.9% capacitance retention after 10,000 cycles.

Hu et al. presented a flexible fiber-shaped ASC using PEDOT@MnO_2 and C@Fe_3O_4 composites as two electrodes and PVA/LiCl gel as the electrolyte [71]. The PEDOT@MnO_2 and C@Fe_3O_4 electrodes were prepared by electrodeposition onto SS fibers, followed by annealing at 500 °C under Ar atmosphere. The fabricated ASC exhibited high areal-specific capacitance of 60 mF cm^{-2} and large specific energy of 0.0335 mW h cm^{-2}.

Wang et al. used Au-coated CC as the flexible substrate to electrochemically grow PPy nanowires, which were converted to CNFs via annealing at 800 °C under N_2 atmosphere, and the CNFs were used as the matrixes to grow needlelike Fe_3O_4 and flake-like NiO via hydrothermal reaction and chemical bath deposition, respectively [72]. The assembled flexible quasi-solid ASC demonstrated high energy density of 5.2 mWh cm^{-3} or 94.5 Wh Kg^{-1}, high power density of 0.64 W cm^{-3} or 11.8 kW Kg^{-1}, together with excellent cycling stability up to 2600 cycles in the PVA/KOH electrolyte.

Lokhande et al. fabricated all-solid-state ASCs using Fe_2O_3 and CuO thin-film electrodes and carboxymethyl cellulose (CMC)/Na_2SO_4 gel electrolyte [73]. The thin-film electrodes were prepared by using SS as the substrate and using synthesis methods of successive ionic layer adsorption and reaction (SILAR) and chemical bath deposition (CBD). A specific capacitance of 79 F g^{-1} can be obtained at 2 mA cm^{-2}, and specific energy of 23 W h kg^{-1} and specific power of 19 kW kg^{-1} can be achieved for this ASC.

Zhang et al. reported an ASC device consisting of Fe_3O_4 anode, Co_2AlO_4@MnO_2 cathode, and PVA/KOH electrolyte [74]. The growth procedures of active materials onto Ni foams are illustrated in Fig. 22.15a. Fe_3O_4 nanoflakes were prepared by a hydrothermal reaction at 120 °C and subsequent annealing at 350 °C, while Co_2AlO_4@MnO_2 nanosheets were fabricated by two-step hydrothermal processes. The SEM images in Fig. 22.15b show that cross-linked Fe_3O_4 nanosheets (0.5–1 mm in length, 30–50 nm in thickness) grew vertically as a dense film on the Ni foam. And the TEM image indicates Fe_3O_4 nanoflakes were comprised of numerous nanoparticles and nanopores. The as-designed ASC device possessed an extended operating voltage window of 1.6 V with a specific capacitance of 99.1 F g^{-1} at 2 A g^{-1}. The Ragone plots in Fig. 22.15c reveal that the ASC delivered a specific energy of 35.2 Wh kg^{-1} at a specific power of 800 W kg^{-1}. When the specific power was increased to 8033 W kg^{-1}, the ASC still maintained a specific energy of 24.1 Wh kg^{-1}. In addition, the ASC was able to retain 92.4% capacitance retention after 5000 cycles.

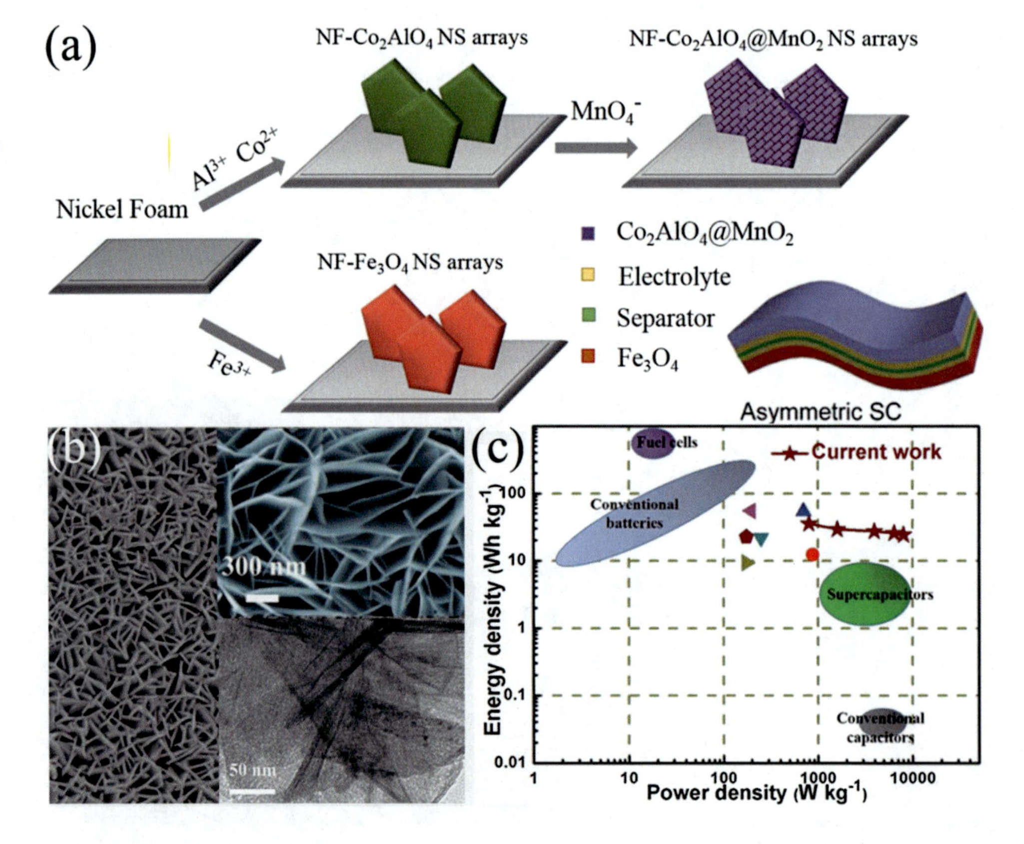

Fig. 22.15 (**a**) Schematic illustration of the synthesis procedures of Co_2AlO_4@MnO_2 and Fe_3O_4 on Ni foams. (**b**) SEM and TEM images of Fe_3O_4 grown on the Ni foam. (**c**) Ragone plots of the ASC device [74]

22.8 Conclusion and Outlook

This chapter summarizes the latest progress with respect to Fe-based pseudocapacitive materials, covering charge storage mechanisms, synthesis methods, performance optimizations (nanostructures, porous structures, heterostructures, and dopings), as well as both aqueous and solid-state ASCs. Despite significant improvements made by some researchers, Fe-based pseudocapacitive materials still lack competitiveness in electrochemical performances when compared to well-developed Li-ion battery systems. Thanks to outstanding advantages of high specific power and low cost, Fe-based pseudocapacitive materials have great potential to be high-performance electrode materials for next-generation energy storage devices. An outlook on performance improvements of Fe-based ASCs is made as follows.

22.8.1 Fundamental Studies on Charge Storage Mechanisms

Although many reports have proposed some charge storage mechanisms for Fe-based pseudocapacitive materials, most of them are only an assumption (without experimental supports) and have yet to be verified. For example, the charge storage mechanisms of Fe_2O_3 and Fe_3O_4 in the KOH electrolyte were ascribed to what is hard to happen within the potential window in aqueous electrolytes [57, 72, 75]:

$$Fe_2O_3 + 3H_2O + 6e^- \rightarrow 2Fe + 6OH^- \quad (22.13)$$

$$Fe_3O_4 + 4H_2O + 8e^- \rightarrow 3Fe + 8OH^- \quad (22.14)$$

It is also certainly impossible that redox reactions arising from SO_3^{2-} in the electrolyte contribute largely to the total capacitance of Fe-based pseudocapacitive materials [16–19], since such mechanism is short of experimental evidences and cannot explain much higher capacitances when using alkaline electrolytes in many other reports. Some other reports conducted characterizations on the electrodes after cycling and proposed some mechanisms based on the characterization results [20, 24, 26], whereas the corresponding capacitances (<325 F g^{-1}) are rather low, making these mechanisms not convincing enough. And the extraordinary content of Li in the active material was far from what the practical capacitance corresponded to [20]. In addition, the electrochemical profiles of Fe-based pseudocapacitive materials in alkaline electrolytes have a big difference with that in neutral electrolytes, which has not been well dealt with so far. Therefore, the underlying charge storage mechanisms of Fe-based pseudocapacitive materials require further fundamental studies, so that better performance can be realized on the basis of new understandings.

22.8.2 Expanding Operating Voltage of Aqueous Electrolytes

Although the specific energy of Fe-based ASCs is much higher than that of EDLCs, it is lower than that of Li-ion batteries (~150 Wh kg^{-1}). In general, the specific energy is greatly influenced by two factors, namely, specific capacitance and operating voltage. The attempts to improve the specific capacitance of Fe-based pseudocapacitive materials have been described in the above sections, whereas there have been few reports tackling the strategies to increase the operating voltage of Fe-based ASCs. On the other side, the operating voltage of aqueous electrolytes and aqueous GPEs is limited by water splitting, which results in a low ESW (theoretically 1.23 V). It should also be noted that the practical operating voltage of some electrochemical energy storage devices can exceed the theoretical ESW owing to O_2/H_2 evolution over-potential, which depends on the chemical nature of electrodes and electrolytes. For example, Pb-acid batteries can be reversibly operated above 2 V in the concentrated H_2SO_4 electrolyte. This section clarifies some typical strategies to increase the operating voltage of aqueous EDLCs, which might bring up hints for increasing the operating voltage of Fe-based ASCs.

Whitacre et al. investigated the role of electrolyte acidity on hydrogen uptake in AC and found the di-hydrogen evolution over-potential is much higher in neutral electrolytes (pH = 4–8) than in strong acidic and alkaline ones [76]. When operated at a relatively high voltage, the nascent hydrogen is generated from the aqueous solution as Eqs. (22.15) or (22.16), while at the same time, the surface pH shits to a higher value. The generated hydrogen adsorbs onto the carbon as Eq. (22.17) and/or recombines to evolve di-hydrogen as Eqs. (22.18), (22.19), or (22.20). As a consequence of pH increment, the hydrogen adsorption and diffusion of adsorbed hydrogen into the matrix of AC are more favored than H_2 evolution, which significantly promotes the storage of mono-hydrogen while preventing the recombination to di-hydrogen:

$$H_3O^+ + e^- \rightarrow H + H_2O \text{ (acid electrolytes)} \quad (22.15)$$

$$H_2O + e^- \rightarrow H + OH^- \text{ (neutral and basic electrolytes)} \quad (22.16)$$

$$C + H \rightarrow CH_{ad} \quad (22.17)$$

$$CH_{ad} + H_2O + e^- \rightarrow H_2 + OH^- + C \quad (22.18)$$

$$2H \rightarrow H_2 \quad (22.19)$$

$$CH_{ad} + CH_{ad} \rightarrow H_2 + 2C \quad (22.20)$$

Belanger et al. used 1 M Li_2SO_4 aqueous electrolyte (pH = 6.5) for AC-based EDLCs and realized a high operating voltage of 1.6 V with high coulombic efficiency (99%) and low steady self-discharge profile [77]. During the water reduction at high voltage, the generation and accumulation of OH^- in the pores of the AC anode make the surface pH value higher than 10, contributing to high over-potential for di-hydrogen evolution.

Frackowiak et al. investigated EDLC performances in three types of alkali metal (Li, Na, K) sulfate aqueous electrolytes and found the capacitance is much higher in the Li_2SO_4 electrolyte than that in Na_2SO_4 and K_2SO_4 electrolytes [78]. As alkali metal ions are strongly solvated in aqueous solutions with a diameter increase of the hydrated complex in the order of K^+ (3.34 Å) < Na^+ (3.59 Å) < Li^+ (3.81 Å), the mobilities of them increase in the order of $Li^+ < Na^+ < K^+$. When the pores within AC are well-saturated with the electrolyte, only a small distance is required for electrolytes ions to be adsorbed on or desorbed from AC, that is, double-layer-related charge storage does not require very high ion mobility. Instead, the double-layer capacitance would be decreased if the mobility of the electrolyte ion is too high. In Na_2SO_4 and K_2SO_4 electrolytes, Na^+ and K^+ would migrate into the electrolyte bulk rapidly upon charging, whereas their return to the electrode/electrolyte interface would not be so fast. Frackowiak et al. also investigated EDLC performances when using different concentrations (0.1–2.5 M) of Li_2SO_4 aqueous electrolytes and found 1 M concentration is the most appropriate [78]. If the concentration is 0.1 or 0.5 M, the ionic conductivity of the electrolyte becomes lower, resulting in unsatisfying capacitances. If the concentration is 1.5 or 2.5 M, the solvated complexes turn to be steric hindrance, aggravating fast charge transfer. Given that the molar concentration of H_2O molecules in water is 55.56 M, all H_2O molecules in the electrolyte are more or less "bonded" to Li^+ or SO_4^{2-} if assuming that each Li^+ is surrounded by 27 molecules of H_2O and each SO_4^{2-} by 12–16 molecules of H_2O. As the solvation energy for Li^+ or SO_4^{2-} is strong, in the order of 160–220 kJ mol^{-1}, the energy required to break the bonds in the solvation shell is competitive to the energy of water decomposition at voltages from 1.6 to 2.2 V. Therefore, EDLCs using 1 M Li_2SO_4 aqueous electrolyte can be operated up to 2.2 V without any significant capacitance fade during 15,000 cycles [78]. In another report, the voltage can be up to 1.8 – 2.0 V through capacitance unequalization approach [79].

22.8.3 Developing Redox Electrolytes

As the capacitance of EDLCs is limited by the double layer, introducing redox active species into aqueous electrolytes has proven to be an effective strategy to enhance the total capacitance. Blanco et al. added quinone/hydroquinone to the H_2SO_4 aqueous electrolyte of AC-based EDLCs and found that the redox-type reactions at the anode and the pseudocapacitive hydrogen electrosorption at the cathode bring in extra capacitances [80]. As a result, the specific capacitance of the anode was as large as 5017 F g^{-1}, while that of the cathode was 477 F g^{-1} (initially 290 F g^{-1}).

Wu et al. introduced p-phenylenediamine into the KOH aqueous electrolyte, and as a consequence, the AC-based EDLC exhibited much higher electrode-specific capacitance (605 F g^{-1}) than that (144 F g^{-1}) using the neat KOH electrolyte [81]. In addition, the cyclability of the EDLC was superior after the introduction of p-phenylenediamine, with high capacitance retention of 94.5% after 4000 cycles.

Ishikawa et al. proposed a pretreatment strategy for AC-based EDLCs through impregnating pores of the cathode into bromine-containing water before cell assembly [82]. The obtained EDLCs achieved high operating voltage of 1.8 V in an aqueous electrolyte system, as well as considerably improved capacitance compared to that in a conventional electrolyte. Consequently, the specific energy of the aqueous device was 1.2 times higher than that of conventional 2.7 V EDLCs with organic electrolytes.

22.8.4 Exploring High ESW Electrolytes

The electrolytes used in the above sections, namely, aqueous electrolytes and aqueous GPEs, suffer from narrow ESW due to water splitting, which inevitably lower the specific energy and specific power of Fe-based ASCs. Thus, finding a suitable high ESW electrolyte might be an effective strategy. During the past years, iongels that contain ionic liquids in polymer matrixes have attracted great attention as the electrolyte for high-voltage electrochemical energy storage devices, thanks to some intriguing characteristics of ionic liquids, e.g., high ESW, non-volatility, non-flammability, and high thermal stability [83]. Different polymer matrixes, such as PTFE, PVdF/-HFP, polysaccharides, and polymeric ionic liquids, have been investigated to accommodate ionic liquids to develop all-solid-state EDLCs. Using iongels, EDLCs can be operated at a voltage of ~3.5 V, much higher than that using organic electrolytes (~2.7 V) and using aqueous electrolytes (~1.0 V). As iongels are non-volatile, sealing is not necessary, which is beneficial to the versatility of the devices. Although iongels have gained great success in EDLCs, they have not been reported for Fe-based ASCs. Therefore, studies on the compatibility of iongels with Fe-based anode materials are highly desirable.

References

1. Huang, P., Lethien, C., Pinaud, S., Brousse, K., Laloo, R., Turq, V., Respaud, M., Demortiere, A., Daffos, B., Taberna, P.L., Chaudret, B., Gogotsi, Y., Simon, P.: On-chip and freestanding elastic carbon films for micro-supercapacitors. Science. **351**, 691–695 (2016)
2. Miller, J.R.: Engineering electrochemical capacitor applications. J. Power Sources. **326**, 726–735 (2016)
3. Wang, Y., Song, Y., Xia, Y.: Electrochemical capacitors: mechanism, materials, systems, characterization and applications. Chem. Soc. Rev. **45**, 5925–5950 (2016)
4. Kötz, R., Carlen, M.: Principles and applications of electrochemical capacitors. Electrochim. Acta. **45**, 2483–2498 (2000)
5. Zhong, C., Deng, Y., Hu, W., Qiao, J., Zhang, L., Zhang, J.: A review of electrolyte materials and compositions for electrochemical supercapacitors. Chem. Soc. Rev. **44**, 7484–7539 (2015)
6. Chen, J., Xu, J., Zhou, S., Zhao, N., Wong, C.-P.: Template-grown graphene/porous Fe_2O_3 nanocomposite: a high-performance anode material for pseudocapacitors. Nano Energy. **15**, 719–728 (2015)
7. Nithya, V.D., Arul, N.S.: Review on α-Fe_2O_3 based negative electrode for high performance supercapacitors. J. Power Sources. **327**, 297–318 (2016)
8. Nithya, V.D., Sabari, A.N.: Progress and development of Fe_3O_4 electrodes for supercapacitors. J. Mater. Chem. A. **4**, 10767–10778 (2016)
9. Zeng, Y., Yu, M., Meng, Y., Fang, P., Lu, X., Tong, Y.: Iron-based supercapacitor electrodes: advances and challenges. Adv. Energy Mater. **6**, 1601053 (2016)
10. Chen, J., Zhou, X., Mei, C., Xu, J., Zhou, S., Wong, C.-P.: Pyrite FeS_2 nanobelts as high-performance anode material for aqueous pseudocapacitor. Electrochim. Acta. **222**, 172–176 (2016)
11. Lu, X., Zeng, Y., Yu, M., Zhai, T., Liang, C., Xie, S., Balogun, M. S., Tong, Y.: Oxygen-deficient hematite nanorods as high-performance and novel negative electrodes for flexible asymmetric supercapacitors. Adv. Mater. **26**, 3148–3155 (2014)
12. Chen, L.-F., Yu, Z.-Y., Wang, J.-J., Li, Q.-X., Tan, Z.-Q., Zhu, Y.-W., Yu, S.-H.: Metal-like fluorine-doped β-FeOOH nanorods grown on carbon cloth for scalable high-performance supercapacitors. Nano Energy. **11**, 119–128 (2015)
13. Brezesinski, T., Wang, J., Tolbert, S.H., Dunn, B.: Ordered mesoporous alpha-MoO_3 with iso-oriented nanocrystalline walls for thin-film pseudocapacitors. Nat. Mater. **9**, 146–151 (2010)
14. Augustyn, V., Come, J., Lowe, M.A., Kim, J.W., Taberna, P.L., Tolbert, S.H., Abruna, H.D., Simon, P., Dunn, B.: High-rate electrochemical energy storage through Li^+ intercalation pseudocapacitance. Nat. Mater. **12**, 518–522 (2013)
15. Kim, H.-S., Cook, J.B., Lin, H., Ko Jesse, S., Tolbert Sarah, H., Ozolins, V., Dunn, B.: Oxygen vacancies enhance pseudocapacitive charge storage properties of MoO_{3-x}. Nat. Mater. **16**, 454–460 (2017)
16. O'neill, L., Johnston, C., Grant, P.S.: Enhancing the supercapacitor behaviour of novel Fe_3O_4/FeOOH nanowire hybrid electrodes in aqueous electrolytes. J. Power Sources. **274**, 907–915 (2015)
17. Wang, S.-Y., Ho, K.-C., Kuo, S.-L., Wu, N.-L.: Investigation on capacitance mechanisms of Fe_3O_4 electrochemical capacitors. J. Electrochem. Soc. **153**, A75–A80 (2006)
18. Mu, J., Chen, B., Guo, Z., Zhang, M., Zhang, Z., Zhang, P., Shao, C., Liu, Y.: Highly dispersed Fe_3O_4 nanosheets on one-dimensional carbon nanofibers: synthesis, formation mechanism, and electrochemical performance as supercapacitor electrode materials. Nanoscale. **3**, 5034 (2011)
19. Wu, N.-L., Wang, S.-Y., Han, C.-Y., Wu, D.-S., Shiue, L.-R.: Electrochemical capacitor of magnetite in aqueous electrolytes. J. Power Sources. **113**, 173–178 (2003)
20. Qu, Q., Yang, S., Feng, X.: 2D sandwich-like sheets of iron oxide grown on graphene as high energy anode material for supercapacitors. Adv. Mater. **23**, 5574–5580 (2011)
21. Chen, L.-F., Yu, Z.-Y., Ma, X., Li, Z.-Y., Yu, S.-H.: In situ hydrothermal growth of ferric oxides on carbon cloth for low-cost and scalable high-energy-density supercapacitors. Nano Energy. **9**, 345–354 (2014)
22. Barik, R., Mohapatra, M.: Solvent mediated surface engineering of α-Fe_2O_3 nanomaterials: facet sensitive energy storage materials. CrystEngComm. **17**, 9203–9215 (2015)
23. Tang, Q., Wang, W., Wang, G.: The perfect matching between the low-cost Fe_2O_3 nanowire anode and the NiO nanoflake cathode significantly enhances the energy density of asymmetric supercapacitors. J. Mater. Chem. A. **3**, 6662–6670 (2015)
24. Sassin, M.B., Mansour, A.N., Pettigrew, K.A., Rolison, D.R., Long, J.W.: Electroless deposition of conformal nanoscale iron oxide on carbon nanoarchitectures for electrochemical charge storage. ACS Nano. **4**, 4505–4514 (2010)
25. Chen, X., Chen, K., Wang, H., Xue, D.: A colloidal pseudocapacitor: direct use of $Fe(NO_3)_3$ in electrode can lead to a high performance alkaline supercapacitor system. J. Colloid Interf. Sci. **444**, 49–57 (2015)
26. Chen, Y.-C., Lin, Y.-G., Hsu, Y.-K., Yen, S.-C., Chen, K.-H., Chen, L.-C.: Novel iron oxyhydroxide lepidocrocite nanosheet as ultrahigh power density anode material for asymmetric supercapacitors. Small. **10**, 3803–3810 (2014)
27. Li, Y., Zhang, H., Wang, S., Lin, Y., Chen, Y., Shi, Z., Li, N., Wang, W., Guo, Z.: Facile low-temperature synthesis of hematite quantum dots anchored on a three-dimensional ultra-porous graphene-like framework as advanced anode materials for asymmetric supercapacitors. J. Mater. Chem. A. **4**, 11247–11255 (2016)
28. Yang, S., Song, X., Zhang, P., Sun, J., Gao, L.: Self-assembled alpha-Fe_2O_3 mesocrystals/graphene nanohybrid for enhanced electrochemical capacitors. Small. **10**, 2270–2279 (2014)
29. Shi, W., Zhu, J., Sim, D.H., Tay, Y.Y., Lu, Z., Zhang, X., Sharma, Y., Srinivasan, M., Zhang, H., Hng, H.H., Yan, Q.: Achieving high specific charge capacitances in Fe_3O_4/reduced graphene oxide nanocomposites. J. Mater. Chem. **21**, 3422–3427 (2011)
30. Li, L., Gao, P., Gai, S., He, F., Chen, Y., Zhang, M., Yang, P.: Ultra small and highly dispersed Fe_3O_4 nanoparticles anchored on reduced graphene for supercapacitor application. Electrochim. Acta. **190**, 566–573 (2016)
31. Chen, K., Chen, X., Xue, D.: Hydrothermal route to crystallization of FeOOH nanorods via $FeCl_3 \cdot 6H_2O$: effect of Fe^{3+} concentration on pseudocapacitance of iron-based materials. CrystEngComm. **17**, 1906–1910 (2015)
32. Javed, M.S., Jiang, Z., Zhang, C., Chen, L., Hu, C., Gu, X.: A high-performance flexible solid-state supercapacitor based on Li-ion intercalation into tunnel-structure iron sulfide. Electrochim. Acta. **219**, 742–750 (2016)
33. Liu, D., Wang, X., Wang, X., Tian, W., Liu, J., Zhi, C., He, D., Bando, Y., Golberg, D.: Ultrathin nanoporous Fe_3O_4-carbon nanosheets with enhanced supercapacitor performance. J. Mater. Chem. A. **1**, 1952–1955 (2013)
34. Sarkar, D., Mandal, M., Mandal, K.: Design and synthesis of high performance multifunctional ultrathin hematite nanoribbons. ACS Appl. Mater. Interf. **5**, 11995–12004 (2013)
35. Fu, C., Mahadevegowda, A., Grant, P.S.: Fe_3O_4/carbon nanofibres with necklace architecture for enhanced electrochemical energy storage. J. Mater. Chem. A. **3**, 14245–14253 (2015)
36. Li, B., Cao, H., Shao, J., Qu, M., Warner, J.H.: Superparamagnetic Fe_3O_4 nanocrystals@graphene composites for energy storage devices. J. Mater. Chem. **21**, 5069–5075 (2011)
37. Barik, R., Jena, B.K., Dash, A., Mohapatra, M.: In situ synthesis of flowery-shaped α-FeOOH/Fe_2O_3 nanoparticles and their phase dependent supercapacitive behaviour. RSC Adv. **4**, 18827 (2014)

38. Liu, J., Zheng, M., Shi, X., Zeng, H., Xia, H.: Amorphous FeOOH quantum dots assembled mesoporous film anchored on graphene nanosheets with superior electrochemical performance for supercapacitors. Adv. Funct. Mater. **26**, 919–930 (2016)
39. Vadiyar, M.M., Bhise, S.C., Patil, S.K., Kolekar, S.S., Chang, J.-Y., Ghule, A.V.: Comparative study of individual and mixed aqueous electrolytes with $ZnFe_2O_4$ nano-flakes thin film as an electrode for supercapacitor application. ChemistrySelect. **1**, 959–966 (2016)
40. Sethuraman, B., Purushothaman, K.K., Muralidharan, G.: Synthesis of mesh-like Fe_2O_3/C nanocomposite via greener route for high performance supercapacitors. RSC Adv. **4**, 4631–4637 (2014)
41. Lin, Y., Wang, X., Qian, G., Watkins, J.J.: Additive-driven self-assembly of well-ordered mesoporous carbon/iron oxide nanoparticle composites for supercapacitors. Chem. Mater. **26**, 2128–2137 (2014)
42. Meng, W., Chen, W., Zhao, L., Huang, Y., Zhu, M., Huang, Y., Fu, Y., Geng, F., Yu, J., Chen, X., Zhi, C.: Porous Fe_3O_4/carbon composite electrode material prepared from metal-organic framework template and effect of temperature on its capacitance. Nano Energy. **8**, 133–140 (2014)
43. Chen, H.-C., Wang, C.-C., Lu, S.-Y.: γ-Fe_2O_3/graphene nanocomposites as a stable high performance anode material for neutral aqueous supercapacitors. J. Mater. Chem. A. **2**, 16955–16962 (2014)
44. Yang, S., Li, Y., Xu, T., Li, Y., Fu, H., Cheng, K., Ye, K., Yang, L., Cao, D., Wang, G.: FeOOH electrodeposited on Ag decorated ZnO nanorods for electrochemical energy storage. RSC Adv. **6**, 39166–39171 (2016)
45. Sarma, B., Jurovitzki, A.L., Smith, Y.R., Ray, R.S., Misra, M.: Influence of annealing temperature on the morphology and the supercapacitance behavior of iron oxide nanotube (Fe-NT). J. Power Sources. **272**, 766–775 (2014)
46. Binitha, G., Soumya, M.S., Madhavan, A.A., Praveen, P., Balakrishnan, A., Subramanian, K.R.V., Reddy, M.V., Nair, S.V., Nair, A.S., Sivakumar, N.: Electrospun α-Fe_2O_3 nanostructures for supercapacitor applications. J. Mater. Chem. A. **1**, 11698–11704 (2013)
47. Ye, Y., Zhang, H., Chen, Y., Deng, P., Huang, Z., Liu, L., Qian, Y., Li, Y., Li, Q.: Core–shell structure carbon coated ferric oxide (Fe_2O_3@C) nanoparticles for supercapacitors with superior electrochemical performance. J. Alloy. Compd. **639**, 422–427 (2015)
48. Yan, M., Yao, Y., Wen, J., Fu, W., Long, L., Wang, M., Liao, X., Yin, G., Huang, Z., Chen, X.: A facile method to synthesize Fe_xC_y/C composite as negative electrode with high capacitance for supercapacitor. J. Alloy. Compd. **641**, 170–175 (2015)
49. Fu, C., Mahadevegowda, A., Grant, P.S.: Production of hollow and porous Fe_2O_3 from industrial mill scale and its potential for large-scale electrochemical energy storage applications. J. Mater. Chem. A. **4**, 2597–2604 (2016)
50. Zeng, Y., Han, Y., Zhao, Y., Zeng, Y., Yu, M., Liu, Y., Tang, H., Tong, Y., Lu, X.: Advanced Ti-doped Fe_2O_3@PEDOT Core/Shell anode for high-energy asymmetric supercapacitors. Adv. Energy Mater. **5**, 1402176 (2015)
51. Arul, N.S., Mangalaraj, D., Ramachandran, R., Grace, A.N., Han, J. I.: Fabrication of CeO_2/Fe_2O_3 composite nanospindles for enhanced visible light driven photocatalysts and supercapacitor electrodes. J. Mater. Chem. A. **3**, 15248–15258 (2015)
52. Yang, Y., Peng, Z., Wang, G., Ruan, G., Fan, X., Li, L., Fei, H., Hauge, R.H., Tour, J.M.: Three-dimensional thin film for lithium-ion batteries and supercapacitors. ACS Nano. **8**, 7279–7287 (2014)
53. Tang, P.-Y., Han, L.-J., Genç, A., He, Y.-M., Zhang, X., Zhang, L., Galán-Mascarós, J.R., Morante, J.R., Arbiol, J.: Synergistic effects in 3D honeycomb-like hematite nanoflakes/branched polypyrrole nanoleaves heterostructures as high-performance negative electrodes for asymmetric supercapacitors. Nano Energy. **22**, 189–201 (2016)
54. Pardieu, E., Pronkin, S., Dolci, M., Dintzer, T., Pichon, B.P., Begin, D., Pham-Huu, C., Schaaf, P., Begin-Colin, S., Boulmedais, F.: Hybrid layer-by-layer composites based on a conducting polyelectrolyte and Fe_3O_4 nanostructures grafted onto graphene for supercapacitor application. J. Mater. Chem. A. **3**, 22877–22885 (2015)
55. Islam, M.M., Cardillo, D., Akhter, T., Aboutalebi, S.H., Liu, H.K., Konstantinov, K., Dou, S.X.: Liquid-crystal-mediated self-assembly of porous α-Fe_2O_3 nanorods on PEDOT:PSS-functionalized graphene as a flexible ternary architecture for capacitive energy storage. Part. Part. Sys. Char. **33**, 27–37 (2016)
56. Park, J.W., Na, W., Jang, J.: Hierarchical core/shell Janus-type α-Fe_2O_3/PEDOT nanoparticles for high performance flexible energy storage devices. J. Mater. Chem. A. **4**, 8263–8271 (2016)
57. Liu, J., Chen, M., Zhang, L., Jiang, J., Yan, J., Huang, Y., Lin, J., Fan, H.J., Shen, Z.X.: A flexible alkaline rechargeable Ni/Fe battery based on graphene foam/carbon nanotubes hybrid film. Nano Lett. **14**, 7180–7187 (2014)
58. Lin, T.W., Dai, C.S., Hung, K.C.: High energy density asymmetric supercapacitor based on NiOOH/Ni_3S_2/3D graphene and Fe_3O_4/graphene composite electrodes. Sci. Rep. **4**, 7274 (2014)
59. Li, J.-J., Liu, M.-C., Kong, L.-B., Wang, D., Hu, Y.-M., Han, W., Kang, L.: Advanced asymmetric supercapacitors based on $Ni_3(PO_4)_2$@GO and Fe_2O_3@GO electrodes with high specific capacitance and high energy density. RSC Adv. **5**, 41721–41728 (2015)
60. Lu, K., Li, D., Gao, X., Dai, H., Wang, N., Ma, H.: An advanced aqueous sodium-ion supercapacitor with a manganous hexacyanoferrate cathode and a Fe_3O_4/rGO anode. J. Mater. Chem. A. **3**, 16013–16019 (2015)
61. Li, N., Zhi, C., Zhang, H.: High-performance transparent and flexible asymmetric supercapacitor based on graphene-wrapped amorphous FeOOH nanowire and $co(OH)_2$ nanosheet transparent films produced at air-water interface. Electrochim. Acta. **220**, 618–627 (2016)
62. Goubard-Bretesché, N., Crosnier, O., Buvat, G., Favier, F., Brousse, T.: Electrochemical study of aqueous asymmetric $FeWO_4$/MnO_2 supercapacitor. J. Power Sources. **326**, 695–701 (2016)
63. Chen, J., Xu, J., Zhou, S., Zhao, N., Wong, C.-P.: Nitrogen-doped hierarchically porous carbon foam: a free-standing electrode and mechanical support for high-performance supercapacitors. Nano Energy. **25**, 193–202 (2016)
64. Guan, C., Liu, J., Wang, Y., Mao, L., Fan, Z., Shen, Z., Zhang, H., Wang, J.: Iron oxide-decorated carbon for supercapacitor anodes with ultrahigh energy density and outstanding cycling stability. ACS Nano. **9**, 5198–5207 (2015)
65. Chen, J., Xu, J., Zhou, S., Zhao, N., Wong, C.-P.: Amorphous nanostructured FeOOH and co–Ni double hydroxides for high-performance aqueous asymmetric supercapacitors. Nano Energy. **21**, 145–153 (2016)
66. Choudhury, N.A., Sampath, S., Shukla, A.K.: Hydrogel-polymer electrolytes for electrochemical capacitors: an overview. Energy Environ. Sci. **2**, 55–67 (2009)
67. Senthilkumar, S.T., Wang, Y., Huang, H.: Advances and prospects of fiber supercapacitors. J. Mater. Chem. A. **3**, 20863–20879 (2015)
68. Yang, P., Ding, Y., Lin, Z., Chen, Z., Li, Y., Qiang, P., Ebrahimi, M., Mai, W., Wong, C.P., Wang, Z.L.: Low-cost high-performance solid-state asymmetric supercapacitors based on MnO_2 nanowires and Fe_2O_3 nanotubes. Nano Lett. **14**, 731–736 (2014)
69. Chodankar, N.R., Dubal, D.P., Gund, G.S., Lokhande, C.D.: Bendable all-solid-state asymmetric supercapacitors based on MnO_2 and Fe_2O_3 thin films. Energ. Technol. **3**, 625–631 (2015)
70. Gu, T., Wei, B.: High-performance all-solid-state asymmetric stretchable supercapacitors based on wrinkled MnO_2/CNT and Fe_2O_3/CNT macrofilms. J. Mater. Chem. A. **4**, 12289–12295 (2016)

71. Sun, J., Huang, Y., Fu, C., Huang, Y., Zhu, M., Tao, X., Zhi, C., Hu, H.: A high performance fiber-shaped PEDOT@MnO_2//C@Fe_3O_4 asymmetric supercapacitor for wearable electronics. J. Mater. Chem. A. **4**, 14877–14883 (2016)
72. Guan, C., Zhao, W., Hu, Y., Ke, Q., Li, X., Zhang, H., Wang, J.: High-performance flexible solid-state Ni/Fe battery consisting of metal oxides coated carbon cloth/carbon nanofiber electrodes. Adv. Energy Mater. **6**, 1601034 (2016)
73. Shinde, A.V., Chodankar, N.R., Lokhande, V.C., Lokhande, A.C., Ji, T., Kim, J.H., Lokhande, C.D.: Highly energetic flexible all-solid-state asymmetric supercapacitor with Fe_2O_3 and CuO thin films. RSC Adv. **6**, 58839–58843 (2016)
74. Li, F., Chen, H., Liu, X.Y., Zhu, S.J., Jia, J.Q., Xu, C.H., Dong, F., Wen, Z.Q., Zhang, Y.X.: Low-cost high-performance asymmetric supercapacitors based on Co_2AlO_4@MnO_2 nanosheets and Fe_3O_4 nanoflakes. J. Mater. Chem. A. **4**, 2096–2104 (2016)
75. Wang, H., Liang, Y., Gong, M., Li, Y., Chang, W., Mefford, T., Zhou, J., Wang, J., Regier, T., Wei, F., Dai, H.: An ultrafast nickel-iron battery from strongly coupled inorganic nanoparticle/nanocarbon hybrid materials. Nat. Commun. **3**, 917 (2012)
76. Chun, S.-E., Whitacre, J.F.: Investigating the role of electrolyte acidity on hydrogen uptake in mesoporous activated carbons. J. Power Sources. **242**, 137–140 (2013)
77. Abbas, Q., Ratajczak, P., Babuchowska, P., Comte, A.L., Belanger, D., Brousse, T., Beguin, F.: Strategies to improve the performance of carbon/carbon capacitors in salt aqueous electrolytes. J. Electrochem. Soc. **162**, A5148–A5157 (2015)
78. Fic, K., Lota, G., Meller, M., Frackowiak, E.: Novel insight into neutral medium as electrolyte for high-voltage supercapacitors. Energy Environ. Sci. **5**, 5842–5850 (2012)
79. Chae, J.H., Chen, G.Z.: 1.9V aqueous carbon–carbon supercapacitors with unequal electrode capacitances. Electrochim. Acta. **86**, 248–254 (2012)
80. RoldáN, S., Granda, M., MenéNdez, R., Santamarí, A.R., Blanco, C.: Mechanisms of energy storage in carbon-based supercapacitors modified with a quinoid redox-active electrolyte. J. Phys. Chem. C. **115**, 17606–17611 (2011)
81. Wu, J., Yu, H., Fan, L., Luo, G., Lin, J., Huang, M.: A simple and high-effective electrolyte mediated with p-phenylenediamine for supercapacitor. J. Mater. Chem. **22**, 19025 (2012)
82. Yamazaki, S., Ito, T., Murakumo, Y., Naitou, M., Shimooka, T., Yamagata, M., Ishikawa, M.: Hybrid capacitors utilizing halogen-based redox reactions at interface between carbon positive electrode and aqueous electrolytes. J. Power Sources. **326**, 580–586 (2016)
83. Tiruye, G.A., Muñoz-Torrero, D., Palma, J., Anderson, M., Marcilla, R.: Performance of solid state supercapacitors based on polymer electrolytes containing different ionic liquids. J. Power Sources. **326**, 560–568 (2016)

Part IX

Metrology

Nanoscale Deformation and Strain Analysis by AFM–DIC Technique

23

Y. F. Sun and John H. L. Pang

23.1 Introduction

Digital image correlation (DIC) deformation measurements employ optical image correlation and deformation analysis and provide a noncontact, full-field displacement measurement technique. The digital image correlation (DIC) technique can be employed with different digital imaging instruments for macro/micro/nanodeformation measurements. The deformation measurement at the submicron scale is a challenging need for micro-to-nanoscale deformation analysis in characterizing packaging and interconnection materials.

The DIC method was first developed by Peters et al. [1–3]. The DIC technique has great capabilities in adapting to digital images obtained from different imaging tools. For example, combined with optical microscope, digital image correlation can be used for deformation measurements at micrometer scale. With the aid of scanning electron microscope (SEM)/scanning probe microscope (SPM), the submicron- and nanoscale resolution of deformation measurements by DIC can be applied if distortions in SEM [4–6] or scan drifts in SPM [7, 8] could be calibrated and corrected.

With the shrinking size of the packages and increasing number of components integrated into the packages, more challenges are attributed to experimental deformation/strain analysis for the design and reliability study of such packages. There is a potential to employ digital image correlation technique to handle AFM (a typical scanning probe microscope) images in expectation of the nanoscale deformation measurement with the aid of high spatial resolution of AFM. There are few successful nanoscale applications [9, 10] of digital image correlation with the aid of atomic force microscope (AFM). The failures in the use of AFM–DIC technique result from the AFM scanner drift effect. For example, a drift of 1 nm when scanning a length of 100 nm causes a false strain of 1%, which can flood the actual deformation to be measured. In order to apply AFM–DIC technique successfully for deformation measurements, a method to compensate for the AFM scanner drifts is needed [11, 12].

The spurious distortion of AFM images results not only from the nonlinear motions of the AFM piezoelectric scanner when responding to the applied voltage but also from its hysteresis and creep effects, even from the complex cross coupling between x, y, and z motions. The nonlinearity corrections of AFMs have been investigated in some studies [4–8] and proved to be effective for the compensation of scanner's nonlinearity. The nonlinearity [13] of AFMs can be reduced within 10% by software solutions and to less than 1% by hardware solutions with external sensors [14–16] to positioning the scanner. A 1% deviation may be unnoticed or can be ignored by a common user who examines the topograph of micro/nanostructures under AFM. However, the deviation must be taken into account for the application of digital image correlation to the accurate deformation measurement. In fact, the scanner drifts can easily be observed by digital image correlation of AFM images acquired at the same load condition.

The digital image correlation (DIC) is applied to obtain the displacements of the center point of a subimage on the reference image taken before deformation, when the subimage matches another subimage on the deformed image taken after deformation. The match of a subimage is to be found by maximizing the cross-correlation coefficient between intensity patterns of two subimages [11]:

Y. F. Sun (✉) · J. H. L. Pang
School of Mechanical & Aerospace Engineering, Nanyang Technological University, Singapore, Singapore
e-mail: sunyaofeng@pmail.ntu.edu.sg; mhlpang@ntu.edu.sg

C. P. Wong et al. (eds.), *Nano-Bio- Electronic, Photonic and MEMS Packaging*, https://doi.org/10.1007/978-3-030-49991-4_23

$$C(\mathbf{p}) = \frac{\sum_{y=y_0-N/2}^{y_0+N/2} \sum_{x=x_0-M/2}^{x_0+M/2} g(x,y)h(x+u(x,y,\mathbf{p}), y+v(x,y,\mathbf{p}))}{\sqrt{\sum_{y=y_0-N/2}^{y_0+N/2} \sum_{x=x_0-M/2}^{x_0+M/2} g^2(x,y) \sum_{y=y_0-N/2}^{y_0+N/2} \sum_{x=x_0-M/2}^{x_0+M/2} h^2(x+u(x,y,\mathbf{p}), y+v(x,y,\mathbf{p}))}} \tag{23.1}$$

where the coordinate ($x0$, $y0$) represents the center point of a subimage with its size $M \times N$; $g(x, y)$ is the light intensity at the point (x, y) in the reference image; and $h(x + u(x, y, \mathbf{p}), y + v(x, y, \mathbf{p}))$ is the light intensity in the deformed image at the match position of the point (x, y). The intensity at a sub-pixel position is obtained by using bicubic spline interpolation scheme: $I(x,y) = \sum_{n=0}^{3} \sum_{m=0}^{3} \alpha_{mn} x^m y^n$. The coefficients amn are determined from the intensity values of pixel locations. $u(x, y, \mathbf{p})$ and $v(x, y, \mathbf{p})$ are displacement functions which define how each of the subimage points located in the reference image at (x, y) is mapped to the deformed image. First-order displacement functions are often used:

$$\begin{aligned} u(x,y,\mathbf{p}) &= p_0 + p_2(x - x_0) + p_4(y - y_0) \\ v(x,y,\mathbf{p}) &= p_1 + p_3(x - x_0) + p_5(y - y_0) \end{aligned} \tag{23.2}$$

The parameter vector $\mathbf{p}(p0, p1, p2, p3, p4, p5)$ can be explicitly denoted $\mathbf{p}\left(u(x_0, y_0), v(x_0, y_0), \frac{\partial u}{\partial x}\Big|_{x_0,y_0}, \frac{\partial v}{\partial x}\Big|_{x_0,y_0}, \frac{\partial u}{\partial y}\Big|_{x_0,y_0}, \frac{\partial v}{\partial y}\Big|_{x_0,y_0} \right)$ (where $u(x_0, y_0)$ and $v(x_0, y_0)$ are the displacements of the subimage center in x and y directions, respectively). To find the resultant vector $\mathbf{p}$ that maximizes the correlation coefficient (Eq. (23.1)), the Newton–Raphson method is used to iteratively find the solution with a given approximation $\mathbf{p}_{\text{old}}$ to the solution. The next improved approximation $\mathbf{p}_{\text{new}}$ can be obtained using the following Newton–Raphson iteration formula:

$$\nabla\nabla C(\mathbf{p}_{\text{old}})(\mathbf{p}_{\text{new}} - \mathbf{p}_{\text{old}}) = -\nabla C(\mathbf{p}_{\text{old}}) \tag{23.3}$$

Newton–Raphson method is known for its fast convergence capability. Usually, two or three iterations are enough to solve the parameter vector $\mathbf{p}$, which represents the displacements and displacement gradients at the subimage center. It is practical that the displacement gradients above are used for the direct calculation of strain components in uniform deformation measurements instead of additional operations of displacement field smoothing and derivative calculation.

In what follows, image correlation results use u to represent the displacement in x/horizontal direction of images and v in y/vertical direction of images.

Figure 23.1 shows the displacement field obtained from the correlation of two AFM images of size 512×512 $pixel^2$ or 2×2 μm^2 from two continuous scans without any load. Although the drift effect or the poor reproducibility has also been mentioned in literature, the method to compensate the AFM nonlinearity and thus derive the accurate displacement field has not been given out.

23.2 Atomic Force Microscope (AFM)

The AFM is a typical member in the scanning probe microscope family, the powerful modern research tools to investigate the morphology, and the local properties of the sample surface from the atomic to the micron level. The experimental system (see Fig. 23.2) studied in this work is a Digital

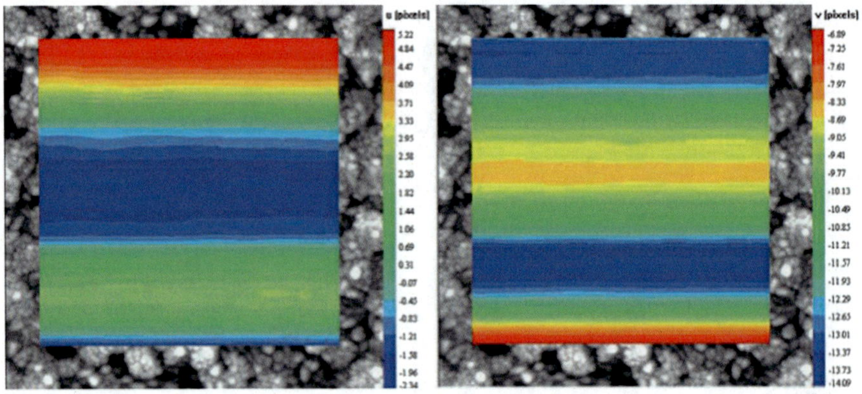

Fig. 23.1 Artificial displacement fields between two AFM images scanned at the same condition: (**a**) u field, (**b**) v field

(a) (b)

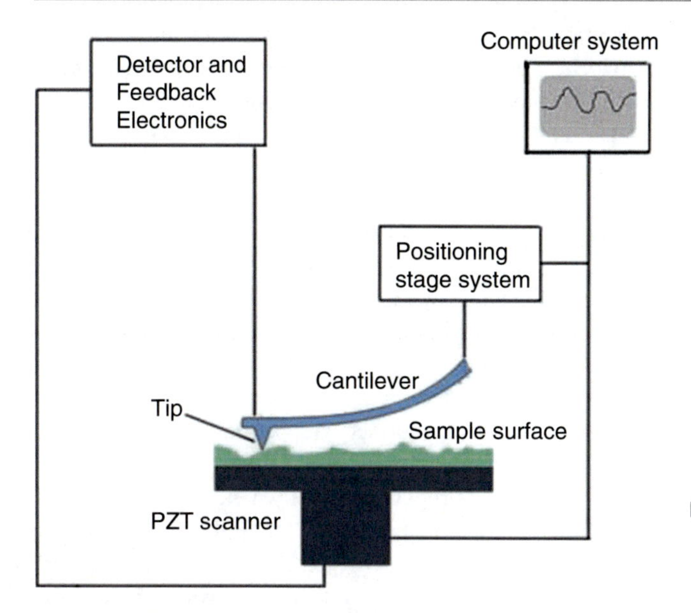

Fig. 23.2 Schematic of the experimental AFM system

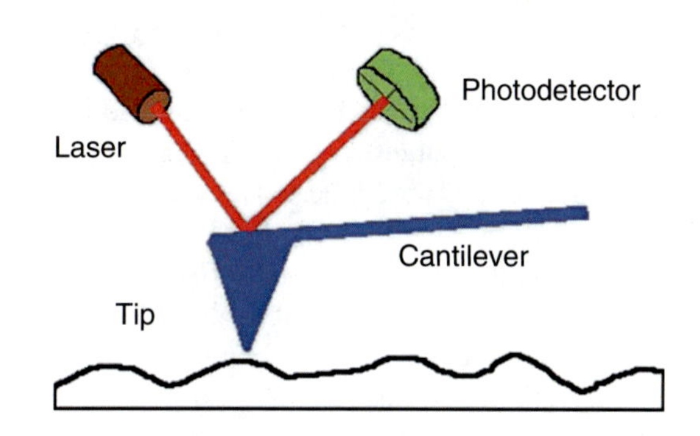

Fig. 23.3 Position-sensitive photo-detector system in AFM

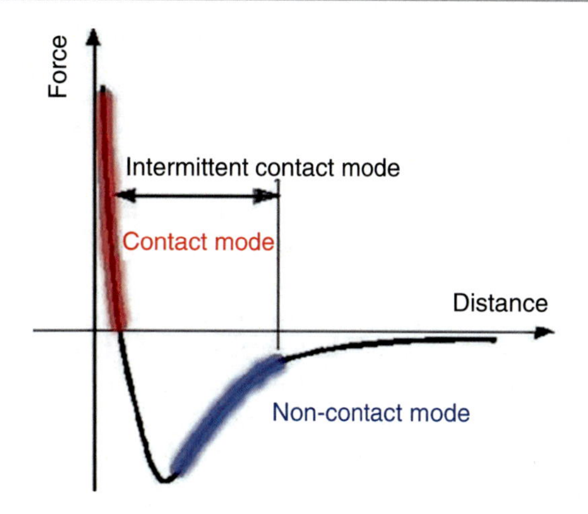

Fig. 23.4 Curve of van der Waals force vs. distance and AFM modes

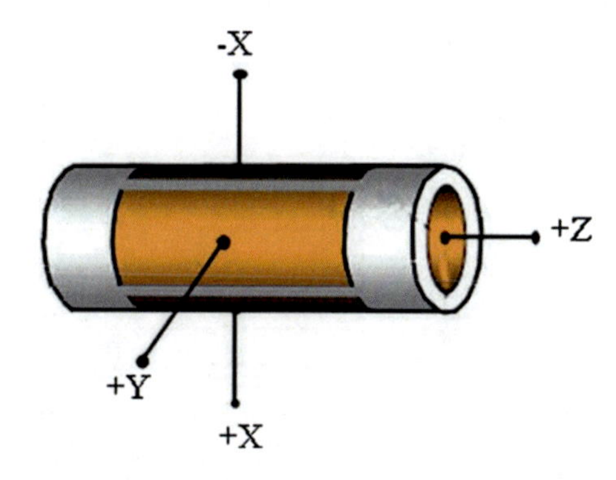

Fig. 23.5 Tube-shaped piezoscanner in AFM. Voltage applied to the x- and y-axis produces the bend of the tube. Voltage applied to the z-axis produces the extension of the tube

Instrument-3000 AFM. A sharp tip, typically less than 5 μm and often less than 10 nm in diameter at the apex, is located at the free end of a cantilever that is 100–500 μm long.

The interactive forces between the tip and the sample surface cause the cantilever to deflect. A position-sensitive photodetector (see Fig. 23.3) measures the cantilever deflections as the sample is scanned under the tip. The measured cantilever deflections allow a computer to generate a map of surface topography.

The interactive force primarily contributed to the deflection of an AFM cantilever is the van der Waals force. The relationship between the van der Waals force and the tip-to-sample separation [13] is shown in Fig. 23.4. Depending on the distance from the sample surface that the probe tip is held, an AFM can be operated at contact, noncontact, and intermittent contact modes. The selection of AFM modes is dependent on specific applications. Detailed information is available on the operation manual from AFM vendors.

Scanning a sample surface in an AFM is performed by the piezoelectric scanner (Fig. 23.5). The scanner acts as a fine positioning stage to move the sample under the probe tip and thus makes a high spatial resolution of AFM possible. Similar to moving an electronic beam on the screen in the cathode ray tube of a television, the AFM control unit drives the scanner in a type of raster pattern as shown in Fig. 23.6. The tip goes along the first line in forward (fast-scan direction), and then back, then moves across the next line (slow-scan direction perpendicular to the fast scan direction), and so forth. During the scanning, data can be collected separately as the probe moves from left to right (the "trace") and from right to left (the "retrace"). The data are sampled digitally at equally spaced intervals scheduled in the fast-scan direction and the slow-scan direction. The distance between the sample points is called the step size. The step size is calculated as the full-scan size divided by the number of sample points per line. As a rule, the number of scan lines equals the number of sample points per line. Thus, the data set from a square grid of measurements, called the AFM frame, consists of a square matrix of data elements, whose size is typically 256 × 256 or 512 × 512. The data elements are the cantilever deflection that reflects the surface topography of the scanned area.

The AFM frame can be visualized on a computer as a three-dimensional topography or two-dimensional color/grayscale image (Fig. 23.7a, b), where the frame data elements fill image pixels.

23.3 AFM Scanner Imperfections

While an ideally square grid of measurements is scheduled for the scanner, virtually there are a lot of difficulties in achieving perfect tip-over-sample motions in the plane of the scanned sample surface, as well as in the out-of-plane direction. Besides external influences, such as electronic noises, vibrations, acoustic noises, and instability of tip-sample contact, the imperfections of a piezoelectric scanner itself cause the distortions in the AFM images. Here scanner imperfections, including intrinsic nonlinearity, hysteresis, creep, and cross coupling, are discussed in the following sections.

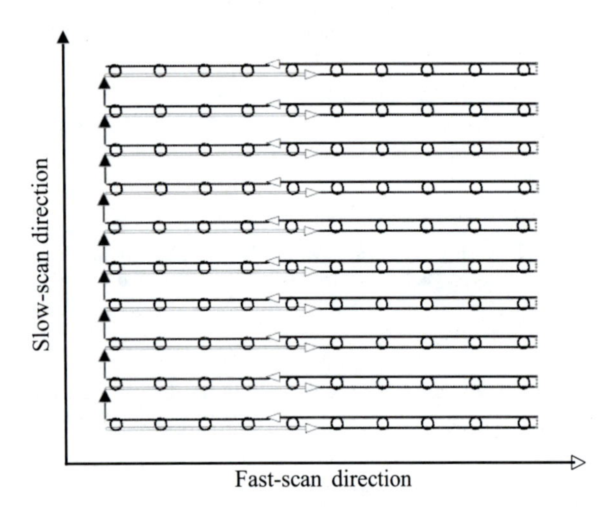

Fig. 23.6 Schematic description of scanner motion and point sampling

Figure 23.8 shows the intrinsic nonlinearity and hysteretic behavior of a piezoelectric scanner. The intrinsic nonlinearity describes a nonlinear extension responding to the applied voltage, and the hysteretic behavior of a piezoscanner displays differences in forward and reverse motions. To make things worse, the nonlinearity and hysteresis are also dependent on the range and even the rate of change of applied voltage. The intrinsic nonlinearity causes the distortion of the measurement grid shown in Fig. 23.8.

The sample points are not equally spaced when the scanner does not move linearly with applied voltage. The hysteretic behavior causes the forward and reverse scan directions to behave differently. A surface with periodic structures, such as gratings, is generally used to calibrate an AFM. AFM images of such a structure are shown in Fig. 23.9. Both scans are in the down direction. Notice the differences in the spacing, size, and shape of the pits between the bottom and the top of each image. The effect of the hysteresis loop on each scan direction in *xy* plane is demonstrated.

As for this nonlinear relationship, the correction solution in the AFM is the use of a calibration routine by applying a nonlinear voltage (see Fig. 23.10) in real time to produce a linear scan along *x*- and *y*-axis in both tracing and retracing scan directions. The feature distortion in Fig. 23.9 can be reduced or eliminated by the application of such a solution, as shown in Fig. 23.11a.

A quantitative investigation of the effect of hysteresis correction can be done by correlating two AFM images scanned in the tracing and retracing directions. Figure 23.11b shows the displacement field in the tracing/retracing direction from the correlation of such a pair of images. The result indicates good linearity of scanner motions along fast-scan direction. However, in fact the procedure does not work well for the linearity correction in slow-scan direction, which can be proved by the results (Fig. 23.1) from correlating two "trace" images of the same area. The almost horizontal contour bands shown in Fig. 23.1b show large artificial deformation in slow-scan/vertical direction,

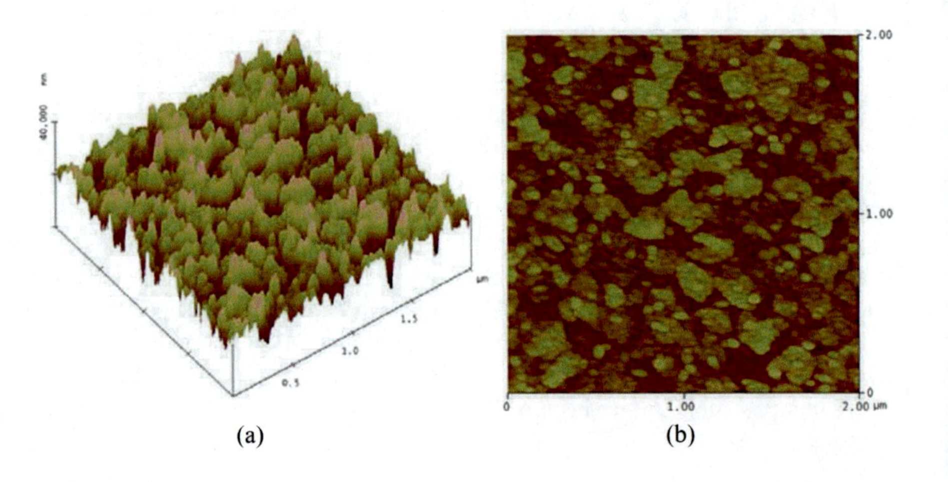

Fig. 23.7 AFM frame visualized as (**a**) a 3D topography or (**b**) 2D color/grayscale image

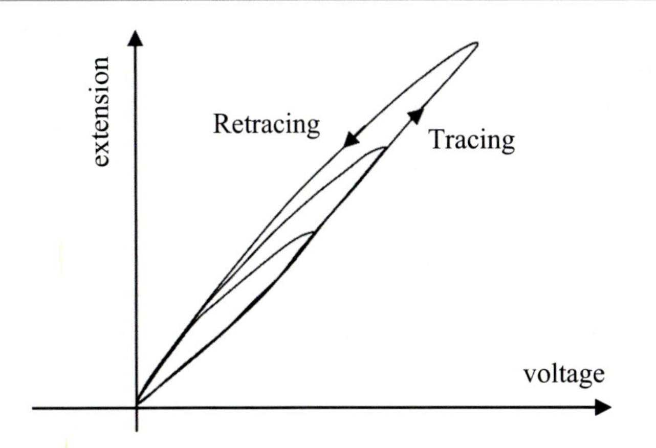

Fig. 23.8 Intrinsic nonlinearity and hysteresis curves of a piezoscanner for various peak voltages or scan sizes. The hysteresis loop is related to the scan size

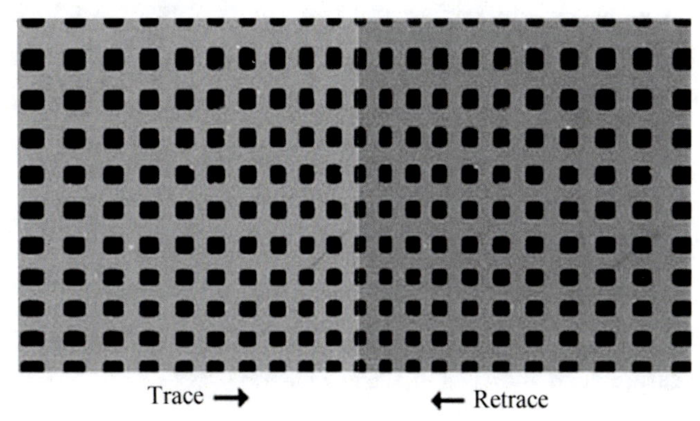

Fig. 23.9 100 × 100 μm scans in the forward (trace) and reverse (retrace) directions of a two-dimensional 10 μm pitch calibration grating without linearity correction [17]

Fig. 23.10 Nonlinear voltage waveform applied to the piezo electrodes to produce linear scanner movement

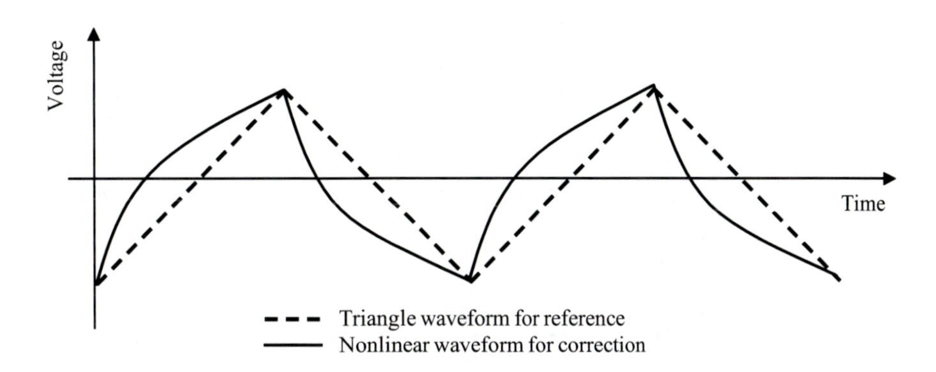

Fig. 23.11 Effect of hysteresis correction: (**a**) 100 × 100 μm scans of the same two-dimensional 10 μm pitch calibration grating with linearity correction, (**b**) correlation between the images in the tracing and retracing directions

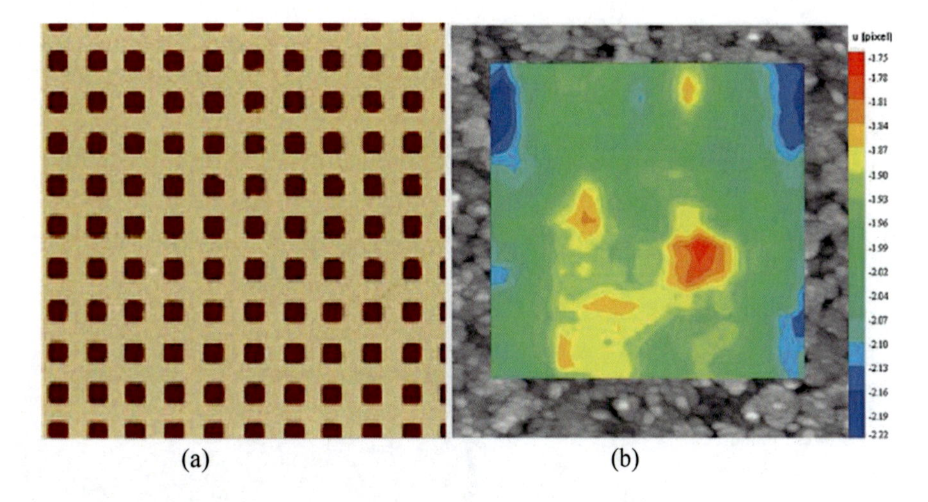

but the contour bands of the same type in Fig. 23.1a denote tiny artificial deformation within a fast-scan/horizontal line. The effectiveness disparity of nonlinearity correction between fast-scan and slow-scan directions ascribed to the disparity of scanner motions in both directions, which is shown in Fig. 23.6.

Another drawback of a piezoscanner is the so-called creep or drift, a delay in the response to sudden change of the applied voltage. The schematic diagram of applied voltage change versus the scanner motions in xy plane is shown in Fig. 23.12. Strong influence [18] of the creep occurs on the initial stage of the scanning process (see Fig. 23.13), when making a change in x and y offsets (see Fig. 23.14) or when using the frame up and frame down commands to bring along abrupt voltage change applied to the scanner.

Last but not the least, cross coupling exerts influence during AFM scanning. The cross coupling refers to the interaction between *x*-, *y*-, and *z*-axis scanner movements. The basis of geometric cross coupling is that the scanner is constructed by combining operated piezo electrodes for *x*, *y*, and *z* into a single tube, and the movements of the scanner in *xy* plane are achieved by bending the tube at the free end. The influences of cross coupling increase with scan size and surface roughness. Higher surface smoothness can reduce the effect of the cross coupling on in-plane deformation measurements by AFM image correlation.

23.4 Reconstruction of AFM Images

Considering that Figs. 23.1 and 23.13 show the accuracy of the scan along fast-scan/horizontal direction much better than that in slow-scan/vertical direction, it occurred to the author that the AFM image would have better quality if the image data in two directions were both from the accurate scan along fast-scan direction, and such an image can be produced by combining two AFM images of the same area scanned at the angle of 0° and 90°, respectively. Exactly speaking, new image can be generated from the reconstruction of the image scanned at the angle of 0° whose distortions are determined by digital image correlation with the image scanned at the angle of 90°. The detailed procedure for AFM image reconstruction is introduced as follows:

1. Acquire one AFM image at the scan angle of zero degree at some load condition.
2. Acquire the other AFM image at the scan angle of 90° at the same load condition and rotate the image 90° clockwise.
3. With the image from step 1 as the reference image and the image from step 2 as the deformed image, correlate the two AFM images by digital image correlation technique and obtain the displacement field.
4. Determine the spurious displacements in the reference image based on the displacement field obtained from step 3.
5. Reconstruct the reference AFM image according to the spurious displacements calculated from step 4.

In the procedure, steps 4 and 5 are the core of the method and are illustrated in the following paragraphs.

Considering that the *u*- and *v*-displacement contour bands are both in almost horizontal and parallel pattern in Figs. 23.1 and 23.13, it is reasonable to assign common *u*- and *v*-displacements to all points on a same horizontal/scan line. With this understanding, it can be imagined that the normal grid (Fig. 23.15a) scanned horizontally will become distorted as schematically shown in Fig. 23.15b, likewise, the normal grid scanned vertically will become distorted as schematically shown in Fig. 23.15c. When Fig. 23.15a, b overlap, the offsets of horizontal scan lines will be exposed as shown in Fig. 23.15d, where u^r_j and v^r_j denote the horizontal and vertical offsets of horizontal scan line (row) *j* scanned in the reference image from its ideal position. Likewise, the overlapping of Fig. 23.15a, c will show up the offsets of

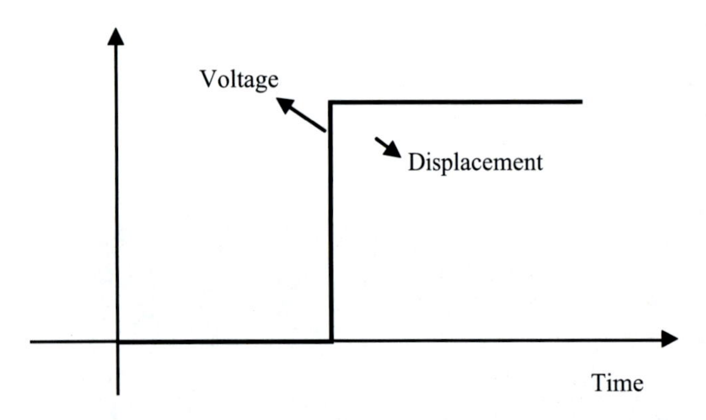

Fig. 23.12 Schematic time diagram of applied voltage change (*blue line*) vs. the scanner motion along x-axis (*red line*)

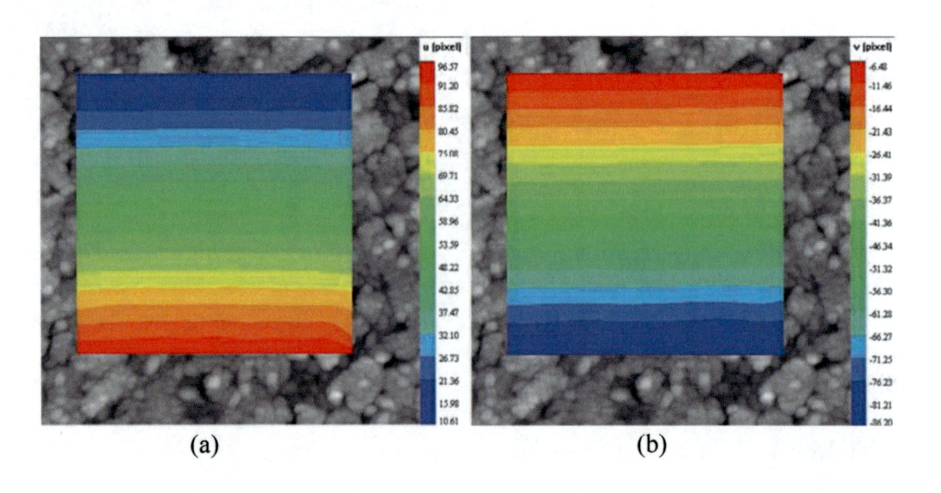

Fig. 23.13 Artificial displacement fields in AFM images due to creep: (**a**) *u* field in the fast-scan direction, (**b**) *v* field in the slow-scan direction

vertical scan lines, as shown in Fig. 23.15e, where u_i^d and v_i^d denote the horizontal and vertical offsets of vertical scan line (column) i scanned in the deformed image from its ideal position. Now if a point is located both on the horizontal line j and the vertical line i in the normal grid, the point will have the position $\left(i+u_j^r, j+v_j^r\right)$ in the reference image, and have the position $\left(i+u_i^d, j+v_i^d\right)$ in the deformed image. If Ux,y and Vx,y denote the horizontal and vertical displacements between a point at the location (x, y) in the reference image and its match point in the deformed image, the displacements of the point at $\left(i+u_j^r, j+v_j^r\right)$ in the reference image can be denoted as $U_{i+u_j^r,j+v_j^r}$ r inhorizontal direction and in $V_{i+u_j^r,j+v_j^r}$ vertical direction and expressed as:

Fig. 23.14 Effect of creep from 300 nm offsets in x- and y-axis

$$\begin{aligned} U_{i+u_j^r,j+v_j^r} &= \left(i+u_i^d\right)-\left(i+u_j^r\right)=u_i^d-u_j^r \\ V_{i+u_j^r,j+v_j^r} &= \left(j+v_i^d\right)-\left(j+v_j^r\right)=v_i^d-v_j^r \end{aligned} \tag{23.4}$$

If the number of horizontal lines and of vertical lines is n ($j = 1, 2,.. .n$) and m ($i = 1, 2,.. .m$), respectively, the number of the unknowns (terms on the right-hand side of Eq. (23.4)) is $2n + 2\,m$. Compared with the number of the knowns (terms on the left-hand side of Eq. (23.4)) mn, we know that there is redundancy for the solution of Eq. (23.4). The redundancy can be used to reduce the deviation effect of measured displacements by image correlation. Consider random errors that can usually be minimized with the use of the

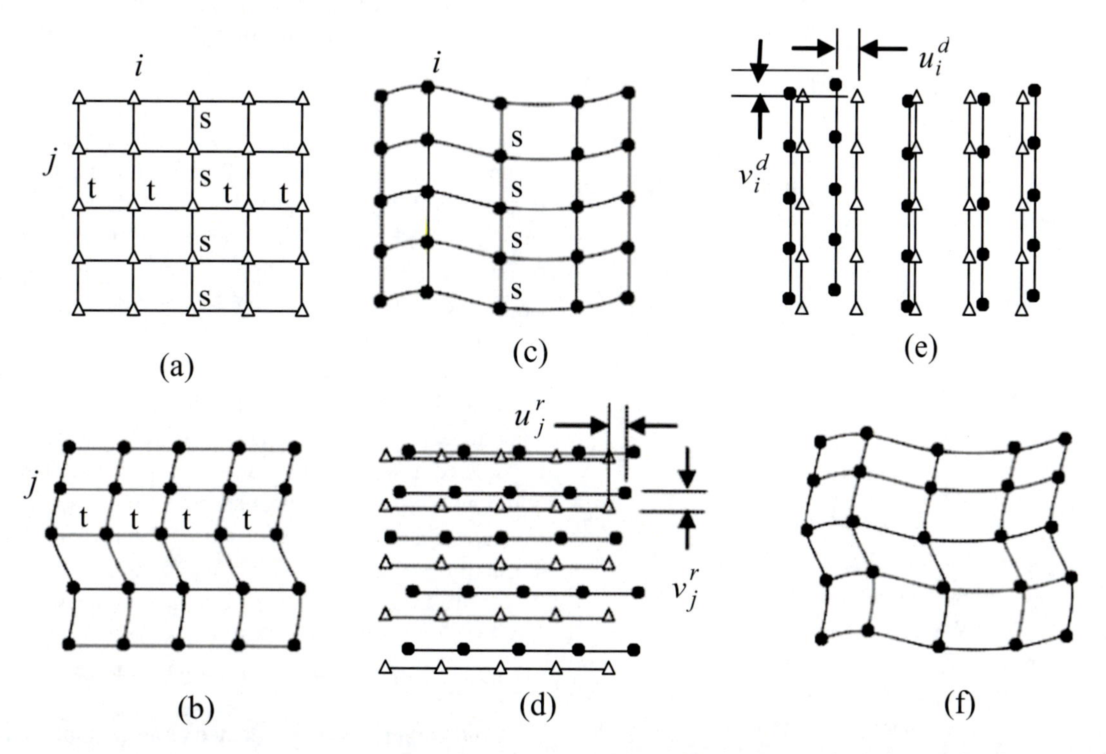

Fig. 23.15 Schematic diagram for AFM image reconstruction: (**a**) normal grid to be scanned by the AFM, the grid with row spacing s and column spacing t, j as row index, i as column index, (**b**) distorted grid scanned at the scan angle of 0° and used as reference image in DIC, (**c**) distorted grid scanned at the scan angle of 90°, used as deformed image in DIC, (**d**) overlapping (**a**) and (**b**) shows displacements between corresponding rows, (**e**) overlapping (**a**) and (**c**) shows displacements between corresponding columns, (**f**) correlation between the two distorted grids (**b**) and (**c**)

average approach. Sums of the displacements of all m points on scan line j lead to:

$$\sum_{i=1}^{m} U_{i+u^r_j,j+v^r_j} = \sum_{i=1}^{m} u^d_i - mu^r_j$$
$$\sum_{i=1}^{m} V_{i+u^r_j,j+v^r_j} = \sum_{i=1}^{m} v^d_i - mv^r_j \quad (23.5)$$

Equation (23.5) can be rewritten as:

$$u^r_j = -\frac{1}{m}\sum_{i=1}^{m} U_{i+u^r_j,j+v^r_j} + \frac{1}{m}\sum_{i=1}^{m} u^d_i$$
$$v^r_j = -\frac{1}{m}\sum_{i=1}^{m} V_{i+u^r_j,j+v^r_j} + \frac{1}{m}\sum_{i=1}^{m} v^d_i \quad (23.6)$$

The subtraction of displacements between two horizontal lines $j + k$ (k is an integer) and j will cancel out the unknowns u^d_i and v^d_i and is shown as follows:

$$u^r{}_{j+k} - u^r_j = -\frac{1}{m}\sum_{i=1}^{m}\left(U_{i+u^r{}_{j+k},j+k+v^r{}_{j+k}} - U_{i+u^r_j,j+v^r_j}\right)$$
$$v^r{}_{j+k} - v^r_j = -\frac{1}{m}\sum_{i=1}^{m}\left(V_{i+u^r{}_{j+k},j+k+v^r{}_{j+k}} - V_{i+u^r_j,j+v^r_j}\right) \quad (23.7)$$

To reconstruct the reference image, a reference/anchor line is needed. Provided that the central horizontal line jc is selected as the reference line, then $u^r_{jc} = 0$ and $v^r_{jc} = 0$. The displacements of horizontal line $jc + 1$ in the reference image can be rewritten based on Eq. (23.7):

$$u^r_{jc+1} = u^r_{jc+1} - u^r_{jc} = -\frac{1}{m}\sum_{i=1}^{m}\left(U_{i+u^r_{jc+1},jc+1+v^r_{jc+1}} - U_{i,jc}\right)$$
$$v^r_{jc+1} = v^r_{jc+1} - v^r_{jc} = -\frac{1}{m}\sum_{i=1}^{m}\left(V_{i+u^r_{jc+1},j+1+v^r_{jc+1}} - V_{i,jc}\right) \quad (23.8)$$

u^r_{jc+1} and v^r_{jc+1} in Eq. (23.8) can be solved by iteration technique by taking the initial guesses of them as zeros. The solutions are found when the iteration converges. The convergence is reached when the u^r_{jc+1} and v^r_{jc+1} ifferences between the current and the last estimates of them, respectively, are both lower than their convergence limits, 0.01 pixel. Likewise, we can obtain u^r_{jc-1} and v^r_{jc-1}, u^r_{jc+2} and v^r_{jc+2}, and so on. When the displacements of all horizontal lines scanned in the reference image are determined, the reference image can be reconstructed to remove its distortion.

The reconstruction of the reference image is done based on the following expression:

$$\bar{f}(i,j) = f\left(i + u^r_j, j + v^r_j\right)$$
$$(i = 1, 2, \ldots, m;\ j = 1, 2, \ldots, n.) \quad (23.9)$$

where $\bar{f}$ denotes the reconstructed image profile and f denotes the reference image profile.

In the implementation of the above algorithms for reconstruction of AFM images, the spurious displacements at a grid of control points located on the image are obtained from DIC. The values of the rest points are calculated using interpolation schemes, such as bilinear interpolation, which is to obtain the values of a point by linear interpolation between its four nearest neighbors. As shown in Fig. 23.16, this correction function has been added in the original DIC software.

23.5 Verification and Buckling Test

Figure 23.17 shows the displacement field between two AFM images scanned at the angles of 0° and 90°, respectively, but under the same conditions. Either the displacement u field or v field displays the same pattern along a scan line, which agrees well with the supposed (see Fig. 23.15d).

The second verification case is zero-deformation measurement. Two images reconstructed from two pairs of AFM images scanned at the angle of 0° and 90° are correlated using digital image correlation technique. The displacement field between one pair of images scanned at the angle of 0° and 90° is shown in Fig. 23.17. The displacement field between both images scanned at the angle of 0° is shown in Fig. 23.18a, b and the displacement field between two reconstructed images in Fig. 23.18c, d. The sample ranges of displacements u and v are 65.7 and 68.8 nm before image correction and 2.1 and 2.7 nm after image correction. In this instance, the maximum image distortion was decreased from 4.3% to 0.16% (correlated image area is 1.6 × 1.6 μm). Therefore, the correction method to AFM images is effective to cancel out the distortion due to the AFM scanner drifts. In addition, a comparison of measured strains before and after image correction has been listed in Table 23.1. After image correction, the average remaining strains, ε_x, ε_y, and ε_{xy}, are 76.6, −124, and −2.83 $\mu\varepsilon$ and their standard deviations 0.00148, 0.00180, and 0.00124, respectively, which shows the promise of accurate deformation measurements by AFM image correlation.

A buckling test was conducted as an example of deformation measurements by AFM image correlation. The buckling specimen (see Fig. 23.19) is a thin PCB (printed circuit board) with cross copper lines located in the centerlines of the plate. The size of the specimen is 56 × 48 × 0.42

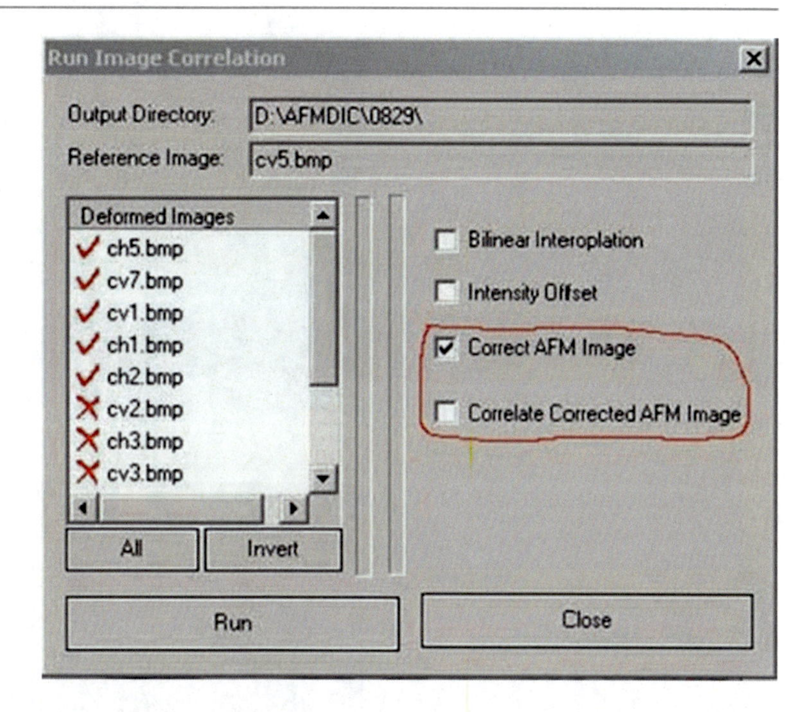

Fig. 23.16 Software implementation for AFM image correction

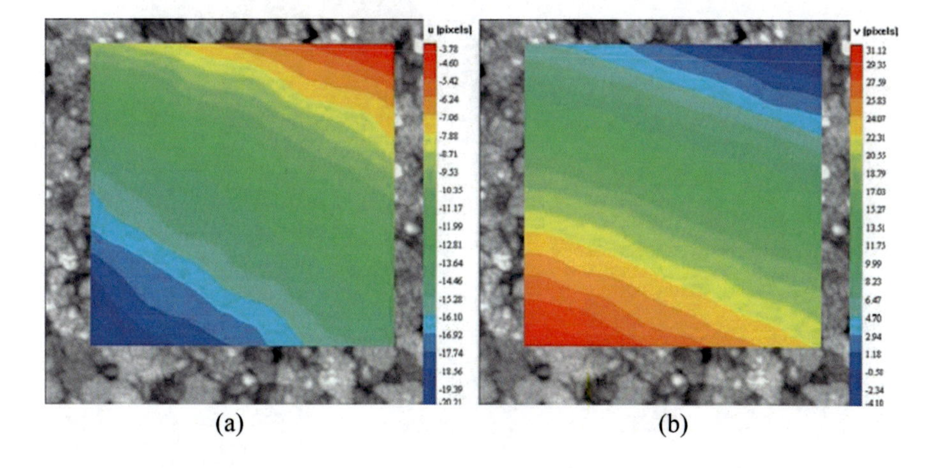

Fig. 23.17 Displacement (3.906 nm/*pixel*) fields between two AFM images scanned at 0° and 90°: (**a**) *u* field, (**b**) *v* field

mm. The center area of the plate to be scanned was sputtered with nanoparticles to obtain AFM images with high contrast pattern. Nanoindentation marks of 1 μm size were shaped for the assistance in the positioning of scan areas between two loading states. A simple fixture was designed, and the schematic diagram of loading fixture was shown in Fig. 23.20. The fixture was required to be aligned with the AFM scan direction and be tightly fixed on the AFM stage. The specimen was first scanned at the lengthwise buckling state because of the ease to unload the specimen buckling during the test. An area of $10 \times 10\ \mu m^2$ in the center of the plate surface was selected to be scanned considering that the smaller area will increase the difficulty in the search for the same area to be scanned at two separate scans. It is noted that the operation of probe offset was not used in this test because false variational displacements along fast scan axis have been found between AFM images scanned with probe offset and without probe offset. This nonlinearity of AFM scan has been calibrated and shown earlier.

The area of interest was scanned by AFM images at the scan angle of 0° and 90° when the specimen was pushed 1.5 mm at one end to buckling deformation and was then unloaded. Acquired images were reconstructed by following the procedure mentioned earlier, and digital image correlation was conducted on the reconstructed images. The displacements *u* and *v* in the direction of specimen length and width have been shown in Fig. 23.21a, b.

The tensile deformation shown in Fig. 23.21a is almost uniform. Finite element analysis (FEA) of the buckling test using ABAQUS shows that the strains in the center area of the buckling plate are uniform. Measured strains by digital image correlation and calculated strains by FEA are listed in Table 23.2. The results show the average strain in *x* direction measured by DIC agrees well with that calculated by FEA.

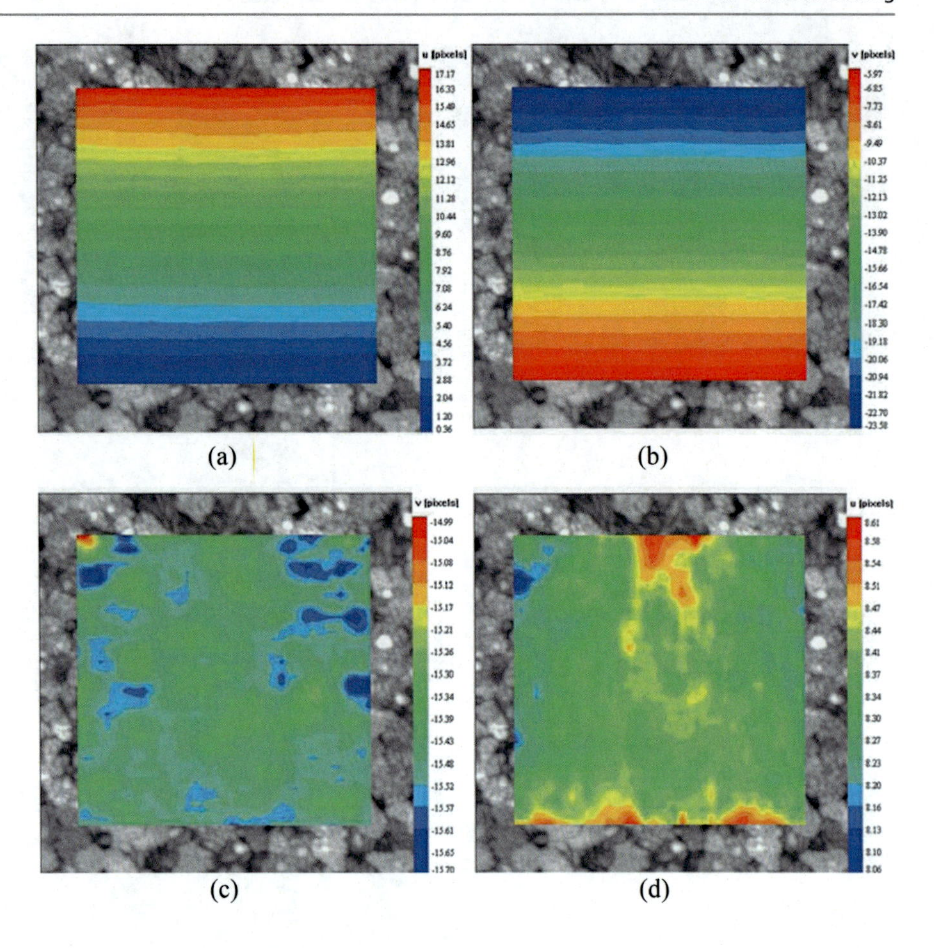

Fig. 23.18 u- and v-displacement (3.906 nm/*pixel*) fields in the zero-deformation test: (**a–b**) u, v fields before image correction, (**c–d**) u, v fields after image correction.

Table 23.1 Remaining strains in the zero-deformation test

	Before correction			After correction		
Strains	ε_x	ε_y	ε_{xy}	ε_x	ε_y	ε_{xy}
Mean(ε)	/	/	/	7.66e–5	–1.24e–4	–2.83e–6
Min(ε)	−3.78e−3	3.01e–2	−3.18e–2	–4.92e–3	–7.69e–3	–4.25e–3
Max(ε)	5.33e–3	5.48e–2	−1.16e–2	5.50e–3	6.78e–3	3.97e–3
σ (ε)	/	/	/	1.48e–3	1.80e–3	1.24e–3

Fig. 23.19 Thin plate (printed circuit board) for the buckling test

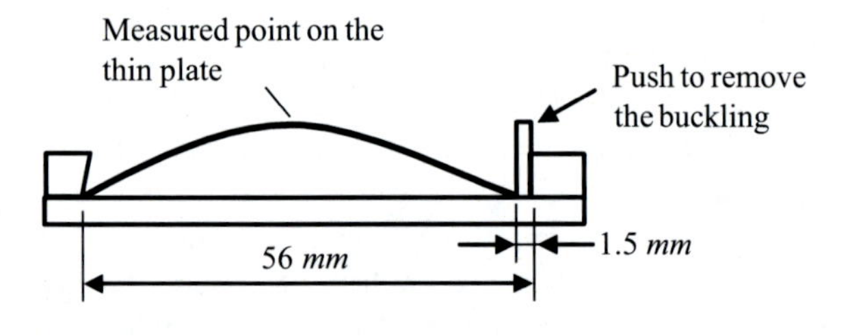

Fig. 23.20 Schematic diagram of the fixture for the buckling test

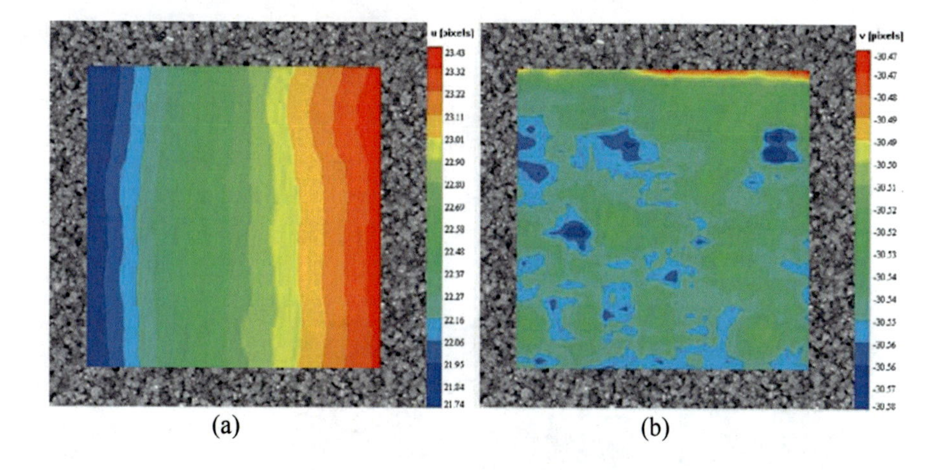

Fig. 23.21 *u* (**a**) and *v* (**b**) displacement (19.53 nm/*pixel*) fields in the buckling test

Table 23.2 Strains measured by DIC and from FEAσ

Strains		ε_x	ε_y	ε_{xy}
Measured by DIC	Mean(ε)	3.96e–3	1.36e–5	2.35e–5
FEA solution	σ (ε)	6.96e–4	3.91e–4	3.58e–4
		3.88e–3	6.59e–8	2.20e–9

The standard deviations of measured strains in Table 23.2 are less than those in Table 23.1. The difference can be attributed to the different scan sizes in two tests. Figure 23.21b shows a displacement deviation of 0.11 *pixels* or 2.2 nm which is close to that in the zero-deformation test. The minimum strain that can be identified in this instance is estimated at hundreds of micro-strain.

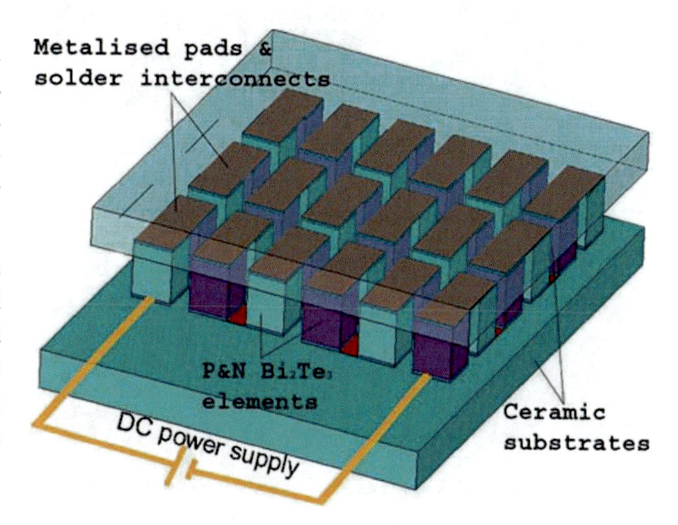

Fig. 23.22 Structure of solid-state thermoelectric module

23.6 AFM–DIC Experiments

It is a challenge to conduct thermomechanical tests under AFM. Thermal loading can be exerted on a specimen readily by conducting a thermomechanical test. The issues of thermal drift of AFM scanner must be resolved, and this led to the design of the thermomechanical experiments on a micro thermoelectric cooler under AFM.

A micro thermoelectric cooler (TEC) was used in this work. A micro TEC is a tiny solid-state heat pump, which is a critical component in many applications of integrated electronic packaging assemblies as it provides precise cooling and temperature stabilization. A typical TEC module comprises of two ceramic substrates that serve as a foundation and electrical insulation for p-type and n-type bismuth telluride (Bi_2Te_3) pellets that are connected electrically in series and thermally in parallel between the substrates (Fig. 23.22). Metallized pads are attached to the ceramic substrates to maintain the electrical connections inside the module. Solder is typically used at the connection joints to enhance the electrical connections and hold the module together. The prototype of the micro TECs with outer dimension of *L*3.8 × *W*3.0 × *H*1.0 mm is shown in Fig. 23.23. The dimension details of the micro TEC are listed in Table 23.3.

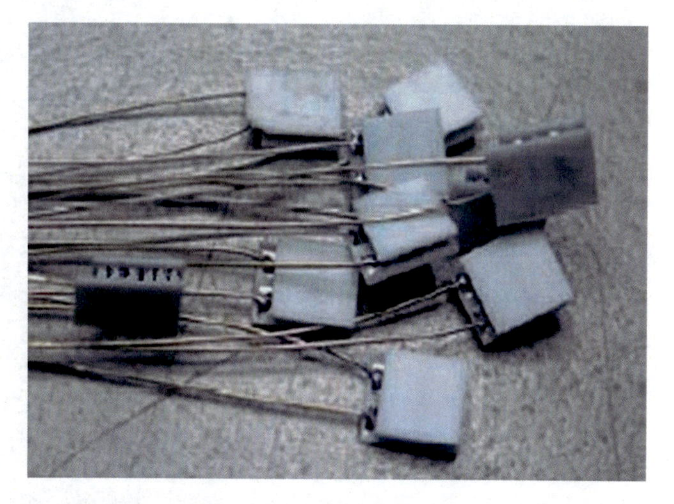

Fig. 23.23 Prototype of micro thermoelectric cooler

Table 23.3 Dimensions of the micro TEC module

Components	Dimensions
Top ceramic substrate	3.0 × 3.0 × 0.28 mm
Bottom ceramic substrate	3.8 × 3.0 × 0.28 mm
TE elements	0.3 × 0.3 × 0.4 mm
TE element pitch	0.45 × 0.45 mm
Solder interconnections	0.3 × 0.3 × 0.02 mm

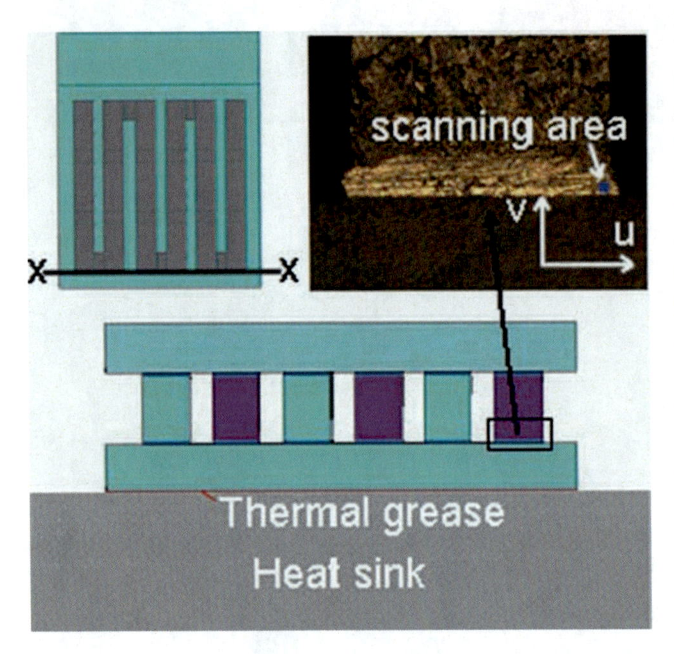

Fig. 23.24 Sample preparation and experimental test assembly

After the specimen check, the hot junction side of the good TEC module was attached to a copper heat sink, and thermal grease was filled between them (see Fig. 23.24). Two electric wires were soldered to each terminal of the micro TEC and then connected to a direct-current power supply. Before the deformation measurement by digital image correlation with the aid of AFM, a calibration check of the assembly performance was done to obtain a configuration of input current, the surface temperature, or temperature difference for use in AFM–DIC experiments. For the calibration, a fine thermal couple was attached to the top surface of the TEC module. An input current of 0.8 A and a temperature difference of 50°C under an ambient temperature of 25°C were determined for the following AFM–DIC experiments. In other words, in our experiments, the device working at its cooling stage was subjected to a temperature gradient of 25 to –25°C from the bottom substrate to the top substrate, while working at its heating stage was subjected to a reverse temperature gradient of 25–75°C from the bottom substrate to the top substrate. The multimode AFM (Fig. 23.25) of Dimension 3000 produced by digital instrument was used in this work. The assembly was then fixed on the stage of AFM with the region of interest facing the AFM scanner tip. The electric wires of the TEC module were fastened on the stage by the tape, squeezed tightly by the AFM shelter, and connected to a DC power supply.

The scanning area (shown in Fig. 23.24) was located at the right bottom corner of the solder interconnection, which is connected to the bottom substrate. During the experiments,

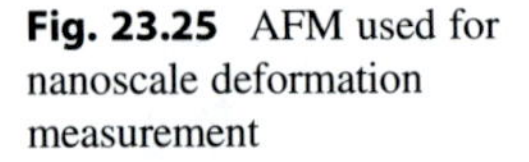

Fig. 23.25 AFM used for nanoscale deformation measurement

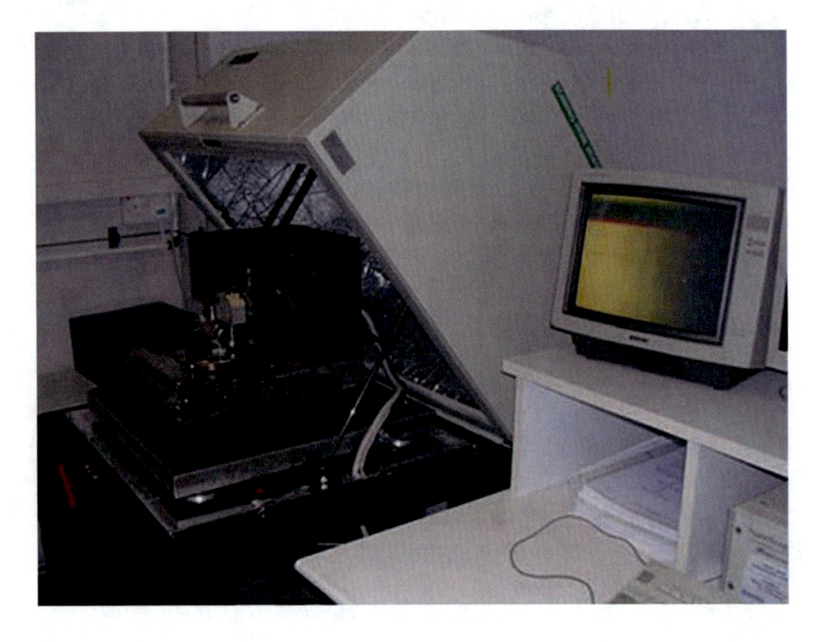

the bottom substrate maintained at the temperature of 25°C, and so the scanning area remained at a temperature of nearly 25°C, which allowed neglecting the thermal effect on the AFM scanner. The tapping mode of AFM was used to scan the specimen. All acquired AFM images were 5 × 5 μm in size with 512 pixel resolution along each scan line. In the present test method, the topography of the area of interest was recorded via AFM at both the horizontal scan and vertical scan directions. The two AFM images were used to reconstruct the topography of the scanned area in order to reduce the drift of AFM scanning. The detail of AFM image correction [11] for deformation measurements was presented in the previous section.

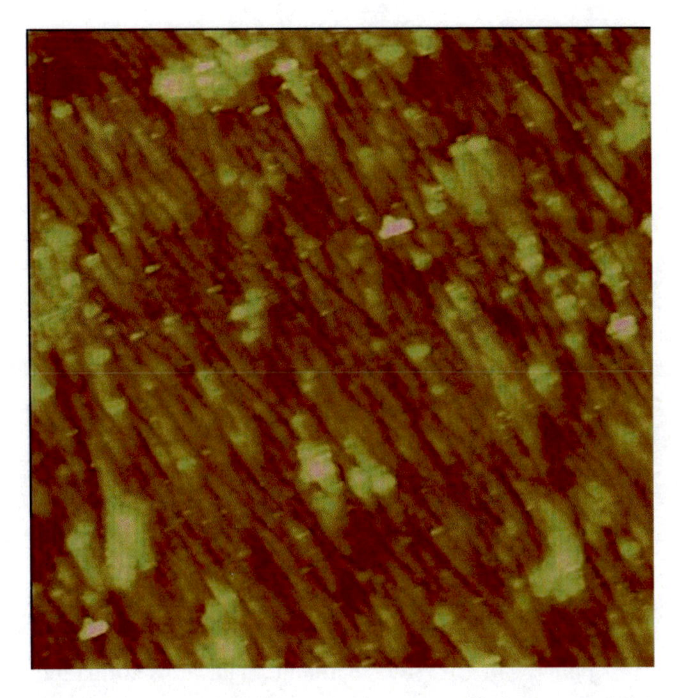

Fig. 23.26 AFM image of 5 × 5 μm size captured for image correlation analysis

Once the scan parameters of AFM scan were determined, the region of interest in the solder interconnection of the sample was positioned and focused under the AFM. Without the current input, the 5 × 5 μm region of interest was initially scanned in horizontal and vertical directions with the tapping mode, and thus a pair of AFM images before deformation was captured and later correlated to obtain a corrected AFM image as the reference image for DIC analysis. Next, a constant current of 0.8 A was applied to the micro TEC, which was thus operating at its cooling stage. When the steady-state condition was reached after approximately 2 min, the region of interest was scanned again with the same AFM settings, and a pair of AFM images after deformation was captured. The pair of AFM images was later correlated to obtain a corrected AFM image, which carried the deformation information due to the TEC's operating at its cooling stage. After that, the input current polarity was changed, and so the micro TEC was operating at its heating stage. In other words, the hot side and the cool side of the TEC changed. Under this condition, the region of interest was scanned twice in the horizontal and vertical directions to obtain another pair of AFM images, which carried the deformation information due to the TEC's operating at its heating stage. A typical AFM image of the scanned region captured in this test is shown in Fig. 23.26.

The AFM images captured before and after deformation were correlated to obtain the deformation using the AFM–DIC software developed. The resultant displacement fields due to the TEC's operating at its cooling stage and at its heating stage are shown in Figs. 23.27 and 23.28, respectively, where u and v represent the displacement fields in x/

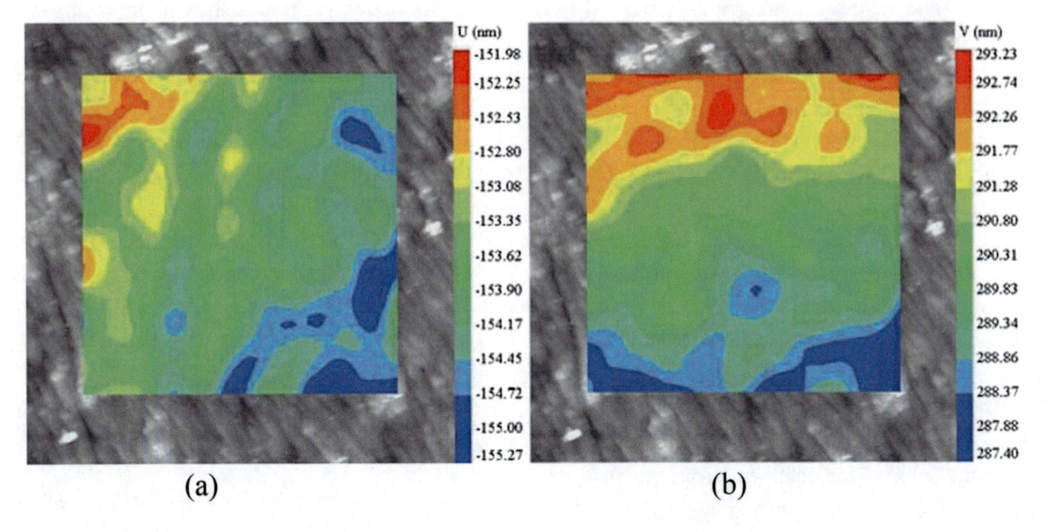

Fig. 23.27 Displacement fields of the observed region in the solder interconnection when the TEC is working at its cooling stage: (**a**) u field; (**b**) v field

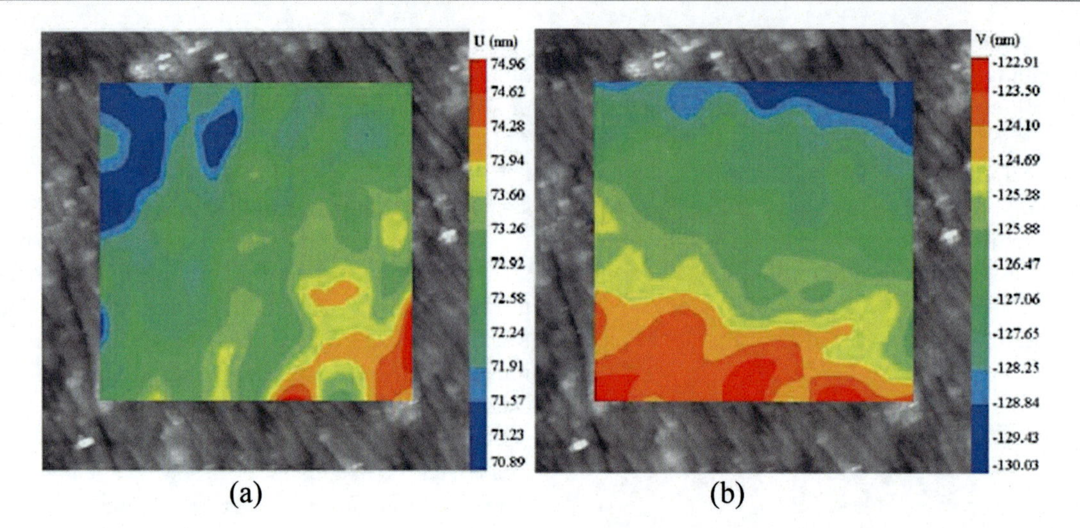

Fig. 23.28 Displacement fields of the observed region in the solder interconnection when the TEC is working at its heating stage: (**a**) u field; (**b**) v field

horizontal and y/vertical direction of images, respectively (see Fig. 23.24). Figure 23.27 shows the range of u displacements, which is 3.29 nm, is less than that of v-displacements, which is 5.83 nm, and the u-displacement contour bands lean to the right, while the v-displacement contour bands are almost horizontal.

Thus, it can be estimated that the average strain in vertical direction $\overline{\varepsilon}_y$ is several times larger than the average horizontal strain component $\overline{\varepsilon}_x$ and the shear strain component $\overline{\varepsilon}_{xy}$. Similarly, Fig. 23.28 shows that the range of u-displacements, which is 4.07 nm, is less than that of v-displacements, which is 7.12 nm, and the u-displacement contour bands incline from the vertical more than the v-displacement contour bands incline from the horizontal.

The heterogeneity of the displacement fields shown in Figs. 23.27 and 23.28 did not completely result from the actual deformation. The displacement errors from AFM–DIC analysis can contribute much to the heterogeneous displacement fields when the actual deformation is small. The gradients of strains on the small scanned area of 5 × 5 μm are expected to be small. The average strains derived from linearly fitting the displacement fields can be very accurate. The calculation of average strains shows that $\overline{\varepsilon}_x$, $\overline{\varepsilon}_y$, and $\overline{\varepsilon}_{xy}$ are 0.042%, 0.126%, and 0.005%, respectively, when the TEC was operating at its cooling stage. The calculated average strains $\overline{\varepsilon}_x$, $\overline{\varepsilon}_y$, and $\overline{\varepsilon}_{xy}$ are 0.035, –0.155, and –0.07%, respectively, when the TEC was operating at its heating stage. Clearly, compared with $\overline{\varepsilon}_x$ and $\overline{\varepsilon}_{xy}$, the vertical strain component $\overline{\varepsilon}_y$ is dominant in the observed region. When the operation of the TEC module was switched from the cooling mode to the heating mode, $\overline{\varepsilon}_y$ changed from positive to negative, which means that the observed region of solder interconnection was subjected to cyclic stresses between tension and compression.

23.7 Conclusions

A method was developed to reconstruct AFM images for nanodeformation measurements by digital image correlation. The developed method is to generate a corrected digital image from two AFM images scanned at the angle of 0° and 90°, respectively. The proposed method was shown valid for the correction of distorted AFM images by the calibration test, which shows the displacement deviation in a reconstructed image is less than 0.16%, compared with 4.3% in the original AFM image. The AFM–DIC method was applied for in situ thermomechanical deformation measurements on solder interconnection of a micro thermoelectric cooler. The AFM–DIC measurements can verify finite element analysis, which was used to study the behavior of the whole device. It was shown that cyclic strains occurred at the top solder interconnections when the TEC worked at its switching mode between cooling stage and heating stage.

The improvement of AFM–DIC technique is desirable considering that the continuous miniaturization of microelectronic devices and interconnections more and more demand experimental strain/stress analysis of micro- and nanoscale components for material characterization and structure reliability analysis. The advance of AFM, calibration algorithm, and DIC algorithm will all contribute toward the robustness of AFM–DIC technique.

References

1. Peters, W.H., Ranson, W.F.: Digital imaging techniques in experimental stress analysis. Opt. Eng. **21**, 427–432 (1982)
2. Sutton, M.A., Wolters, W.J., Peters, W.H., Ranson, W.F., McNeil, S.R.: Determination of displacements using an improved digital image correlating method. Image. Vis. Comput. **1**, 133–139 (1983)

3. Chu, T.C., Ranson, W.F., Sutton, M.A., Peters, W.H.: Applications of digital image correlation techniques to experimental mechanics. Exp. Mech. **25**, 234–244 (1985)
4. Murray, S., Gillham, C.J., Windle, A.H.: Characterization and correction of distortions encountered in scanning electron micrographs. J. Phys. E: Sci. Instrum. **6**, 381–384 (1973)
5. Suganuma, I.: A novel method for automatic measurement and correction of astigmatism in the SEM. J. Phys. E: Sci. Instrum. **20**, 67–73 (1987)
6. Garnaes, J., Nielsen, L., Dirscherl, K., Jorgensen, J.F., Kasmussen, J.B., Lindelof, P.E., Sorensen, C.B.: Two-dimensional nanometer-scale calibration based on one-dimensional gratings. Appl. Phys. A Mater. Sci. Process. **66**, S831–S835 (1998)
7. Dirscherl, K., Jogensen, J.F., Sorensen, M.P.: Modeling the hysteresis of a scanning probe microscope. J. Vac. Sci.Technol. B. **18**, 621–624 (2000)
8. Stoll, E.P.: Correction of geometrical distortions in scanning tunneling and atomic force microscopes caused by piezo hysteresis and nonlinear feedback. Rev. Sci. Instrum. **65**, 2864–2869 (1994)
9. Chasiotis, I., Knauss, W.G.: A new microtensile tester for the study of MEMS materials with the aid of atomic force microscopy. Exp. Mech. **42**, 51–57 (2002)
10. Chang, S., Wang, C.S., Xiong, C.Y., Fang, J.: Nanoscale in-plane displacement evaluation by AFM scanning and digital image correlation processing. Nanotechnology. **16**, 344–349 (2005)
11. Sun, Y.F., Pang, J.H.L.: AFM image reconstruction for deformation measurements by digital image correlation. Nanotechnology. **17**, 933–939 (2006)
12. Sun, Y.F., Pang, J.H.L., Fan, W.: Nanoscale deformation measurement of microscale interconnection assemblies by a digital image correlation technique. Nanotechnology. **18**, 1–8 (2007)
13. Howland, R., Benatar, L.; A practical guide to scanning probe microscopy, http://web.mit.edu/cortiz/www/AFMGallery/PracticalGuide.pdf (2000)
14. Barrett, R.C., Quate, C.F.: Optical scan-correction system applied to atomic force microscopy. Rev. Sci. Instrum. **62**, 1393–1399 (1991)
15. Griffith, J.E., et al.: A scanning tunneling microscope with a capacitance-based position monitor. J. Vac. Sci. Technol. B. **8**, 2023–2027 (1990)
16. Yamada, H., Fujii, T., Nakayama, K.: Linewidth measurement by a new scanning tunneling microscope. Jpn. J. Appl. Phys. **28**, 2402–2404 (1989)
17. D. I. V. M. Group, AFM/LFM instruction manual (Version 4.22ce), Digital Instruments (1999)
18. Helena, J., Bruck, H.A.: A new method for characterizing nonlinearity in scanning probe microscopes using digital image correlation. Nanotechnology. **16**, 1849–1855 (2005)

Part X

Computational

Molecular Dynamics Applications in Packaging 24

Yao Li, Jeffery A. Hinkley, and Karl I. Jacob

24.1 Molecular Dynamics Procedure

Molecular dynamics (MD) is a simulation methodology based on the well-familiar Newton's law, relating force with the rate of change of momentum, namely:

$$\vec{F} = m\vec{a} \tag{24.1}$$

If all the atoms and/or molecules in the simulation box are treated as classical particles obeying Newton's equations, then the positions and velocities of each atom can be calculated knowing the positions and velocities at a time immediately before the current time, once the forces acting on each atom are computed. Forces are derived from a conservative potential function, also known as force fields, arising from the interactions between the chemical units present in the simulation box.

Assuming that we have N particles in the system, we have $6N$ independent variables: positions $r1$, $r2$, ..., rN and velocities $v1$,$v1$, ... ,vN. Derived from the classical dynamics, positions, and velocities of particles are related as

$$\frac{\mathrm{d}\vec{r}_i}{\mathrm{d}t} = \vec{v}i \tag{24.2}$$

and

$$\frac{\mathrm{d}\vec{v}_i}{\mathrm{d}t} = \frac{\vec{f}_i}{m_i} = \frac{\mathrm{d}^2\vec{r}_i}{\mathrm{d}t^2} \tag{24.3}$$

where fi is the force vector acting on particle i. In order to implement Eqs. (24.2) and (24.3) in a computer simulation, numerical integration is employed instead of analytical integration. If the position of particle i at time t is known as $ri(t)$, using the Taylor's expansion, the position of the particle at time $t + \Delta t$ is given by [1–3]

$$\vec{r}_i(t+\Delta t) = \vec{r}_i(t) + \Delta t \cdot \dot{\vec{r}}_i(t) + \frac{(\Delta t)^2}{2}\ddot{\vec{r}}_i(t) + \frac{(\Delta t)^3}{6}\overrightarrow{\dddot{r}_i}(t) + O\left((\Delta t)^4\right) \tag{24.4}$$

Substituting Eqs. (24.2) and (24.3) into Eq. (24.4), we can get

$$\vec{r}_i(t+\Delta t) = \vec{r}_i(t) + \Delta t \cdot \vec{v}_i(t) + \frac{(\Delta t)^2}{2} \cdot \frac{\vec{f}_i(t)}{m_i} + \frac{(\Delta t)^3}{6}\overrightarrow{\dddot{r}_i}(t) + O\left((\Delta t)^4\right) \tag{24.5}$$

Thus, the position of atoms at time $t + \Delta t$ can be calculated by knowing the position at t and the forces at t, provided the time interval Δt is short. A similar relation can be derived for calculating the velocity at a given time, knowing the velocity at a short interval prior to that time.

Another assumption made by MD is the reversibility of time, i.e., if we trace particles' trajectories in the reverse time direction, all particles will follow their original trajectories. Under this assumption, we can obtain the particle's position at time $t - \Delta t$ as

Y. Li · K. I. Jacob (✉)
Department of Polymer, Textile and Fiber Engineering, Georgia Institute of Technology, Atlanta, GA, USA
e-mail: yli6@gatech.com; jacob@gatech.edu

J. A. Hinkley
Langley Research Center, NASA, Hampton, VA, USA
e-mail: jeffrey.a.hinkley@nasa.gov

C. P. Wong et al. (eds.), *Nano-Bio- Electronic, Photonic and MEMS Packaging*, https://doi.org/10.1007/978-3-030-49991-4_24

$$\vec{r}_i(t-\Delta t)=\vec{r}_i(t)-\Delta t\cdot\vec{v}_i(t)+\frac{(\Delta t)^2}{2}\cdot\frac{\vec{f}_i(t)}{m_i} -\frac{(\Delta t)^3}{6}\overrightarrow{\dddot{r}}_i(t)+O\left((\Delta t)^4\right) \tag{24.6}$$

Adding the Eqs. (24.5) and (24.6) and omitting the higher-order derivatives will give us

$$\vec{r}_i(t+\Delta t)+\vec{r}_i(t-\Delta t)=2\vec{r}_i(t)+(\Delta t)^2\cdot\frac{\vec{f}_i(t)}{m_i} \tag{24.7}$$

Thus, as long as we know the positions at least at time t and preferably at $t-\Delta t$ as well as the net force acting on the particles at time t, we can compute the positions at time $t+\Delta t$. Equation (24.7) is the widely used Verlet algorithm in MD simulation, which is similar to the finite difference numerical procedure.

Since higher-order terms are neglected, a requirement for Eq. (24.7) to be use- ful is to have very small time interval Δt, comparable to the characteristic time scale of actual atoms' movement. A typical time step of 1–10 fs is adopted in most MD simulations.

The remaining unknown in Eq. (24.7) is *fi*(t), which is the gradient of the potential energy U field, given by [4]

$$\vec{f}_i(t)=-\nabla_i U(t) \tag{24.8}$$

The interaction potential energy U in an MD simulation for polymers usually consists of a number of bonded and nonbonded terms, as given by

$$U=\sum^{N_{\text{bond}}}U_{\text{bond}}+\sum^{N_{\text{angle}}}U_{\text{angle}}+\sum^{N_{\text{torsion}}}U_{\text{torsion}}+\sum^{N_{\text{inversion}}}U_{\text{inversion}} +\sum_{i=1}^{N-1}\sum_{j>i}^{N}U_{\text{vdw}}+\sum_{i=1}^{N-1}\sum_{j>i}^{N}U_{\text{electrostatic}} \tag{24.9}$$

Since these energy terms may involve two or more particles, to compute *fi*(t), we need to go through all the energy terms, calculate corresponding net force acting on particle i by other particles. The potential energy function or the coefficients used in this expression are generally called force field in MD. Besides the functional form of the potentials, force field defines a set of parameters for the interactions involved in each type of particle that contribute to the potential energy. Both the functional form and parameters can be obtained from empirical, semi-empirical, or ab initio methods. In Eq. (24.9), the first four terms stand for interactions between bonded particles, i.e., bond stretching *U*bond, bond angle bending *U*angle, dihedral angle torsion *U*torsion, and inversion interaction *U*inversion; while the last two terms are interactions between unbonded particles, i.e., van der Waals energy *U*vdw and electrostatic energy *U*electrostatic; *N*bond, *N*angle, *N*torsion, and *N*inversion are the total numbers of these specific interactions in the simulated system; i and j in the van der Waals and electrostatic terms indicate the pair-wise particles involved in the interaction [5]. A number of force fields have been developed for different types of systems. Popular force fields include AMBER [6], widely used for proteins and DNA; CHARMM [7], widely used for both small molecules and macromolecules; EAM [8], broadly used for fcc metals; COMPASS [9], developed specially for polymers; CFF [10], adapted to a broad variety of organic compounds, to name a few. Thus, a conservative force field is assumed, arising from the interactions of a given atom with other atoms.

Subtracting Eq. (24.6) from Eq. (24.5) will give the following equation:

$$\vec{r}_i(t+\Delta t)-\vec{r}_i(t-\Delta t)=2\Delta t\cdot\vec{v}_i(t) +O\left((\Delta t)^3\right) \tag{24.10}$$

Neglecting the last term and dividing the Eq. (24.10) by $2\Delta t$ leads to

$$\vec{v}_i(t)=\frac{\vec{r}_i(t+\Delta t)-\vec{r}_i(t-\Delta t)}{2\Delta t} \tag{24.11}$$

The total kinetic energy then can be calculated as

$$\vec{E}_{\text{kinetic}}(t)=\sum_{i=1}^{N}\frac{1}{2}m_i\vec{v}_i^{\,2}(t) \tag{24.12}$$

Many thermodynamics properties (temperature, pressure, etc.) of the simulated system can also be derived from the time-dependent position and velocity data of atoms based on statistical mechanics. For an isolated system, the total energy and linear momentums are conserved.

MD simulations are carried out within the framework of an appropriate ensemble, such as microcanonical ensemble (NVE), canonical ensemble (NVT), and isothermal–isobaric ensemble (NPT), depending upon which thermodynamic variables are fixed. For example, in the isothermal–isobaric ensemble, the number of moles of the substance (N), pressure (P), and temperature (T) are kept constant. There are many subtle procedures involved in keeping the average value of these variables fixed during the simulation, which is achieved through various "thermostats." Since the number of interactions between atoms increase with the number of atoms present in the system, for a system containing

N particles, the number of times forces are calculated at a given time step vary at least with N^2 [2]. This will limit the number of particles one could simulate and the simulation time one can afford, even with the availability of larger and faster massively parallel clusters. Although there are many exceptions, the highest number of particles or atoms generally studied in MD is in the order of millions of atoms, and the simulation time frame has been within nanosecond duration. Even with a very large number of atoms in a system, many of those atoms will reside on the surface of the simulation box, giving rise to the influence by surface energy in the simulation of such nanosystems. Thus, if the interest is to study the bulk properties, the surface effects must be removed, which is achieved by using the periodic boundary condition. For periodic boundary conditions, the same simulation box or unit cell is repeated in every direction from the primary simulation box to fill the entire domain. When periodic boundary conditions are applied, an atom that moves from the primary simulation cell to the next cell will reappear like an atom entering from the opposite side of the primary simulation cell as all cells are images of the same cell.

MD is a convenient tool to study the static and dynamic attributes of materials. Researchers have been applying MD to most materials science areas, including metals, small molecules, polymers, biomaterials, etc.

24.2 Structures and Properties of Epoxy

The earliest materials used in IC packaging were ceramic flat packs. During the development of the IC industry, the need for encapsulation shifted packaging materials from ceramic flat packs to plastic flat packs. Encapsulation plastics primarily used in the last step of IC packaging are epoxy resins. In addition, epoxies also serve as strong adhesives applicable to most interfaces. Flexible epoxy resins can be excellent electrical insulators, good conductors of heat, offer reliable protection to moisture and strong adhesion to metals. Detailed studies about the thermal and mechanical properties of epoxy resins are crucial to choosing suitable epoxies for packaging applications.

With MD simulations, when done with proper forcefields, one can predict how material properties will change by changing functional groups, backbone structures, crosslink densities, and other structural parameters. There are a variety of epoxies compounds in the market, and the resulting variability of epoxy structures by using various epoxy compounds is highly useful for practical applications, but a challenging problem for simulations. Epoxy resins are crosslinked polymer networks without a universal long-range repeat unit. Construction of a realistic epoxy initial structure is the first step in the simulation process. If the initial structure is not a realistic one with energies close to the minimum energy for that simulation cell, then MD results could be affected, or simulation time will be significantly higher for realistic results. In order to avoid the lengthy simulation time requirement, an initial configuration should be generated as close to the final lowest energy state as possible. One of the major problems of generating initial structures for epoxy resins is the complexity of crosslinking reactions between epoxy groups and curing agents. The reaction process is even more complex considering the presence of various types of epoxy groups and curing agents and possible side reactions.

Recently, new algorithms were developed to build realistic initial configurations of epoxy polymer networks at different length scales accessible to MD. The first category of these methods adopt "step growth" algorithm to create networks from basic repeat units. The assumption made here is that those repeating units are identical and therefore can be replicated and extended in three dimensions. Starting from one central unit, surrounding units are grown from the reactive points of the central unit such that each head atom should be linked to a tail atom. This method is straightforward and easy to implement. However, this procedure only works for certain networks, and the network produced is too regular and flawless to be realistic. Spatial hindrance encountered in the growth process may be another problem if the repeating unit is short or spatially compact [11].

In the second approach, segments between fixed crosslinking junctions are generated first according to the junction-to-junction distance of segments between them. This technique is appropriate for networks with uncertain number of repeating units. If atoms at crosslinking points are the same and branches growing from the crosslinking points are chemically identical, efficiency of this approach can be very good. Linkages at each crosslinking point should be determined, while the structures of each branch are different. The computational cost is expected to be very high when the network is complex and large and, as in the previous case, presence of defects and unreacted functional groups, are not easy to introduce using this approach [12].

Another way to generate initial epoxy structures is to mimic the actual polymerization process. Generally, as the first step monomers are created in a simulation box, and an equilibrium simulation is performed to blend and relax monomers. Then, using a crosslinking criteria (e.g., the distance between nearby atoms is within a chosen limit) adopted to generate bonds between possible reactive groups, crosslinking is implemented. This type of method can build either all atom models or coarse grain models. They are able to simulate networks with multifunctional groups and capture the complexity of these networks. Usually, an imperfect network will be generated [13]. All these methodologies to generate initial polymer network configurations rely on using a valid force field for polymers, although reasonable configurations can also be created using other methods such as Monte Carlo methods using less rigorous force fields. COMPASS and Dreiding [14] are the two major popular force fields commonly used for polymers, among others.

Once the initial structure is generated, then a series of MD simulations are carried oriate conditions of temperature and pressure (or stresses). The data generated from these simulations, namely, position and velocity of each atom in the simulation cell at various time steps, can be used to calculate a number of material properties. Many reports have introduced different methodology to simulate polymer networks, for understanding the structure and confirmation of polymers as well as to extract physical and mechanical properties of a polymer. Generation of initial structures under various conditions is generally enough to study the conformational aspects of polymers, for example, Yarovsky et al. [15] developed a way to construct low-molecular-weight water-soluble epoxy resins. Recently, Xu et al. [16] conducted atomic simulations of epoxy resins using an alternate MM (molecular mechanics)/MD method. A four-step procedure (monomer construction, monomer equilibrium, crosslinking, and the final network equilibrium) was employed in these papers to understand the conformational aspects of polymer chains. Reasonable agreements between computational results and experimental results have shown the validity of these methods. Although many researchers use custom codes for MD simulations, industrial research and even undergraduate instructions are usually done with commercial simulations codes [17], such as the molecular simulation package Cerius2 (now Material Studio) [18] from Accelrys to study conformational aspects and to predict material properties. However, the size of the simulation cell is generally limited when such commercial software is used mainly with PCs, and control of the detailed crosslinking process is also difficult to implement. In the last few years, researchers tried to combine and modify previous methodologies to build polymer networks, taking efficiency and robustness as a criteria. Komarov et al. [19] and Varshney et al. [20] both successfully deployed different mechanisms to construct highly crosslinked, relatively large-scale polymer networks with fairly consistent properties with experiments. Figure 24.1 is a typical example of the initial structure of a simulation box filled with crosslinked epoxy resins.

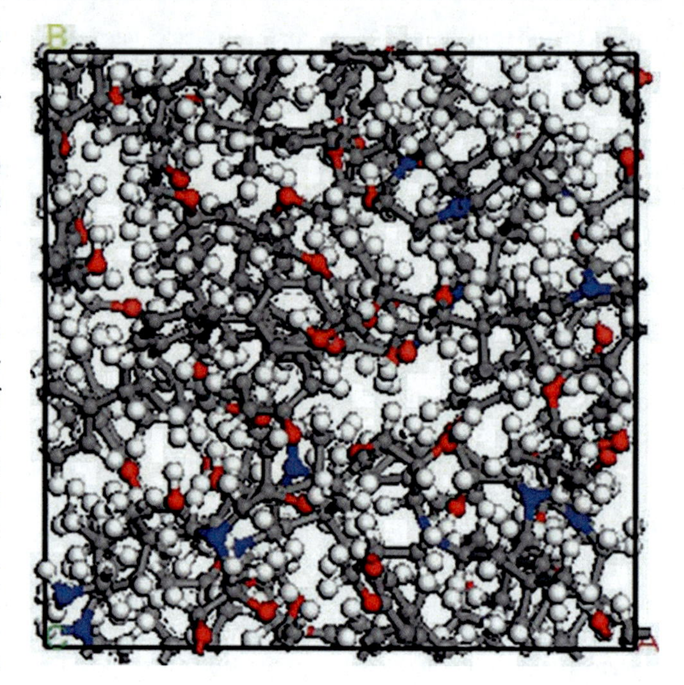

Fig. 24.1 Epoxy resins model system: stick–ball model, balls with different grades of black represent oxygen, carbon, hydrogen, and nitrogen. (Reprinted from Ref. [16], copyright 2006, with permission from Elsevier)

There are a number of material properties that are relevant in the IC packaging area. Thermal stability of epoxy resins is the one of the most critical properties. Glass transition temperature (*Tg*) is one physical parameter that we can obtain easily from MD simulation, for epoxy with various compositions. To obtain the glass transition, a simulated annealing process can be performed with molecular to find the volume of the simulation box for various temperatures. The temperature at which a sudden change in the slope of the volume and temperature relationship occurs indicates the *Tg*. Figure 24.2 is an example showing the change of volume as a function of temperature, where the intersection point of the two fitting lines gives the *Tg*. Another attribute that is important in packaging systems is the volume thermal expansion coefficient (VTEC), which can be obtained from the volume–temperature relationship given in Fig. 24.2. VTEC is different above and below the *Tg* for the same sample. Linear thermal expansion coefficient is related to VTEC by a 1/3 factor for isotropic materials.

Important mechanical properties, such as bulk modulus, Young's modulus, and Poisson's ratio, also can be computed with molecular simulations. This can be done in two different ways, using molecular mechanics (MM) or with MD simulations. In the first approach, boundary atoms in the opposite sides are moved by a very small amount and the force that must be applied on those planes to keep the entire system at the minimum energy configuration can be calculated. Then by uniformly increasing the axial strain in the material (by moving the boundary layers progressively and calculating the corresponding force), the force–displacement relationship can be obtained. There are various modifications to this approach, such as instead of moving only the boundary atoms, all atoms are moved, in the direction the nearest boundary was moved, proportional to their position from the center of the simulation box such that a uniform displacement is induced in the material from the displacement of the boundary atoms. Such a modification could allow one to

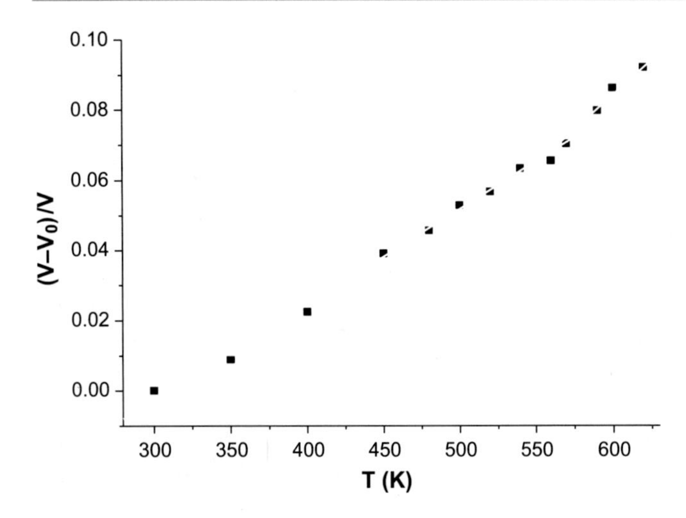

Fig. 24.2 Volume change vs. temperature of diaminodiphenyl methane/4,4'-diaminodiphenylsulfone (TGDDM/DDS) system

apply larger displacements on the boundary layer, consequently larger strains, and thus fewer simulations. In the MD approach, stresses are applied on the boundary and MD simulations can be carried out with constant pressure constraint until internal equilibrium is reached, thus the corresponding strain can be obtained. Then the force is increased and the simulation process is continued. There are a number of coupling procedures, to connect applied stress with the simulation procedure, such as loose coupling technique [21] and Parrinello–Rahman technique [22]. As an example, a stress–strain curve drawn from a simulated uniaxial tensile test of a model sample is shown in Fig. 24.3. From this curve, the Young's modulus or the tangent or secant modulus at any strain can be calculated easily. Mechanical strength can also be calculated; but to obtain strength, these simulations must be repeated many times since strength is more a statistical property sensitive to polymer confirmation in the simulation cell.

As implied earlier, crosslink densities that relate directly to the polymer morphology will have a significant influence on the thermal and mechanical properties. Usually, higher crosslink densities lead to higher *Tg* and better mechanical strength. M. Stevens et al. [23–26] manipulated connectivity of epoxy networks to investigate its influence on the strength of interfaces. It was also found that mechanical properties were dependent on the degree of curing. The factors affecting properties of epoxies as adhesives will be discussed again later.

24.3 Moisture Diffusion

Considering the environment ICs usually encounter, its necessary to investigate the durability aspects of the system, which relates to the resistance of insulating materials to

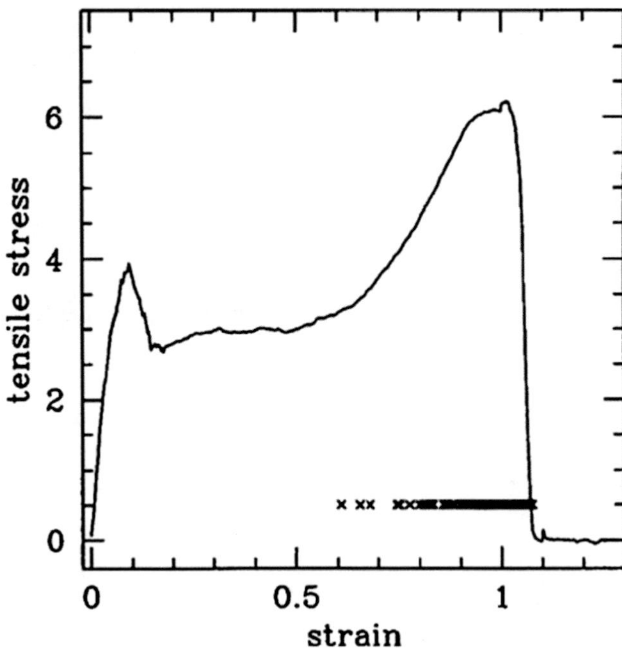

Fig. 24.3 Tensile stress–strain curve for the fully bonded network. The points mark the strains at which bond breaking occurs. Stress is in LJ units. (Reprinted with permission from Ref. [23]. Copyright 2001 American Chemical Society)

water intake and diffusion. Presence of water could create swelling, affect thermal expansion, and could cause hydrothermal degradation of the polymeric material. Experimentally, it is relatively easy to measure the diffusion coefficient of small molecules through bulk material once the material is synthesized, although for interfaces it may pose some difficulties. However, compared to experiments, MD is an effective approach to get an insight into the diffusion behavior at an atomistic and fundamental level, if such materials are unavailable for diffusion experiments or we want to design and synthesize an epoxy composition with certain moisture transport properties. In this case, the power of MD is more on narrowing down the candidate materials with specific properties that needs to be synthesized for further experimental verification. The basic MD approach for moisture diffusion in polymers is fairly straightforward [27–29]:

According to Fick's law,

$$\vec{J} = -D\nabla c \qquad (24.13)$$

where J is the flux, D is the diffusion coefficient, and c is the concentration of the diffuser.

Conservation of mass gives

$$\frac{\partial c}{\partial t} + \nabla \cdot \vec{J} = 0 \qquad (24.14)$$

Combining Eqs. (24.13) and (24.14), we can get

$$\frac{\partial c}{\partial t} = D\nabla^2 c \quad (24.15)$$

Solution diffusion equation in three dimensions is given by

$$c\left(\vec{r}, t\right) = (2\pi Dt)^{3/2} \exp\left[-\frac{r^2}{2Dt}\right] \quad (24.16)$$

The approach first used by Einstein will lead to the following equation:

$$\left\langle r^2(t) \right\rangle = 6Dt \quad (24.17)$$

If we define $\Delta \vec{r}_i(t) = \vec{r}_i(t) - \vec{r}_i(0)$, then

$$\begin{aligned} \left\langle \Delta r^2(t) \right\rangle &= \frac{1}{N} \sum_{i=1}^{N} \left\langle \Delta r_i^2(t) \right\rangle \\ &= \frac{1}{N} \sum_{i=1}^{N} \left\langle \left[\vec{r}_i(t) - \vec{r}_i(0)\right]^2 \right\rangle \end{aligned} \quad (24.18)$$

Substituting Eq. (24.18) into Eq. (24.17) gives

$$D = \lim_{t\to\infty} \frac{1}{6N} \sum_{i=1}^{N} \left\langle \left[\vec{r}_i(t) - \vec{r}_i(0)\right]^2 \right\rangle \quad (24.19)$$

This equation can be applied to get the diffusion coefficient D, using the positions of diffusing molecules in a medium. For the problem of interest, in Eqs. (24.18) and (24.19), N is the number of penetrating water molecules in the epoxy; $ri(t)$ is the position vector of water molecule i at time t; () means an ensemble average.

Diffusion of water molecules in polymers was studied by many researchers [30–37]. Diffusion coefficient is mainly related to temperature, relative humidity, and the structure of materials. In MD, one can compute mean square displacement (MSD) and plot it as a function of time. From this plot, diffusion coefficient can be derived from the slope of a linear fitting to data points according to Eq. (24.17) or (24.19). Figure 24.4 shows an example for this method. Although the principle is fairly straightforward, the implementation and analysis of the result can be involved when using this method. Since there are various regions in the mean- squared displacement–time plot, one has to be careful in picking the appropriate region for calculating the diffusion coefficient. For example, diffusion coefficient calculated from the ballistic region may not be of any significance. In addition, many diffusion processes are non-Fickian; thus care must be taken in the interpretation of data to obtain meaningful results.

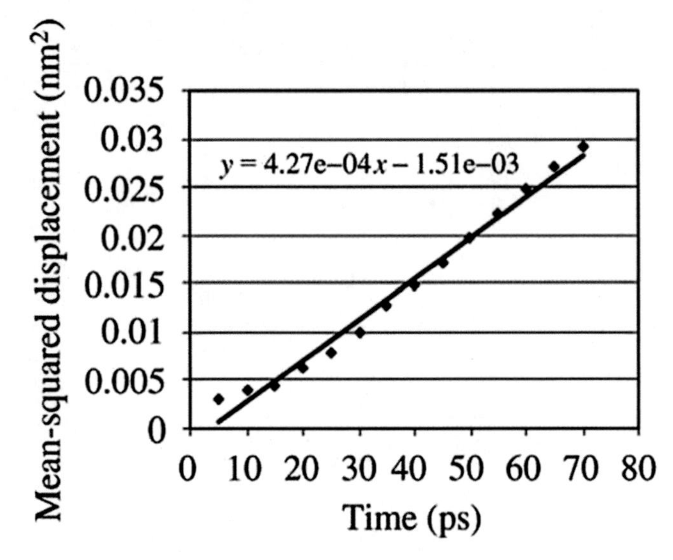

Fig. 24.4 Mean-squared displacements of water molecules against time in the epoxy molding compound (EMC) bulk and the fitted lines. (Reprinted from Ref. [30] with permission from Koninklijke Brill NV)

For polymers, the total amount and the distribution of free volume cavities will directly influence how fast small molecules can diffuse. Diffusion coefficients of same penetrant molecules in polydimethylsiloxane (PDMS) is three times larger than in amorphous polyethylene (PE) due to larger free volume and broader free volume distribution in PDMS [35]. Since most polymers stay in a glassy state at room temperature, the diffusion of small molecules always take a much longer time than in a simple liquid. Interactions with water molecules, such as hydrogen bonding, can significantly alter the diffusion process. This procedure is very useful in obtaining various parameters associated with the diffusion of water molecules, as shown by E. Dermitzaki et al. [38], who by systematically changing the structures of epoxy resins found out that the polarity is the dominant factor for water uptake.

Size effect of materials also plays a role in water transport. Penetrating capabilities of small molecules through bulky materials are believed to differ from that in thin films or membranes. A comparison between interface and bulk models of simulated rubbery and glassy polymers addressed by D. Hofmann et al. [32] revealed an experimentally important pervaporation feature of membranes, which means a combination of membrane permeation and evaporation. Comparison between epoxy molding compound (EMC) bulky samples and EMC/Cu interfaces [30] shows that water molecules are more prone to penetrate in thin interfaces. The diffusion coefficient of moisture through interfaces increased by one magnitude compared to that in bulk. Such behavior demands researchers to pay more attention to study and prevent moisture diffusion particularly at thin interfaces, and this is becoming an important issue as electronic devices are miniaturized.

Some small penetrants (such as water) in a polymer have aggregative attributes and intend to form large clusters. In one of the earliest papers in this field, Yoshinori Tamai et al. [35] studied the mechanism of small molecules diffusing through an amorphous polymer membrane. Diffusion of different penetrants (water, methane, and ethanol) in different polymers (PE and PDMS) was conducted. Diffusion coefficient of nonaggregated penetrant species coincided with experimental value; however, aggregation of water and ethanol in PDMS reduced diffusion coefficient by more than 1 magnitude. It suggested that water clusters forming in polymers played an important role in diffusion. Results by Fukuda et al. [31] also validated this observation. This study was also performed with MD simulation with a simulation time of about 10 ns. The diffusion coefficient of a water cluster containing 10 molecules in polyethylene was detected to be 7.1×10^{-7} cm^2/s. Relevant to the topic on aggregation, the relative humidity affects the diffusion behavior through the for- mation of water aggregates. At a low moisture concentration, the number of water clusters in polymer was believed to be few. Higher relative humidity leads to a low moisture concentration gradient and more water aggregates. These two results will contribute to a low diffusion coefficient [30].

The result from water uptake can be divided into two categories: change in physical properties and loss of mechanical strength. It was discussed earlier that water update by epoxy will decrease glass transition temperatures [38]. Larger amount of moisture in epoxies means a larger reduction of glass transition temperature, which can change the state of some materials from glassy to rubbery. It was observed experimentally that the joint strength of epoxy adhesives was substantially weakened when exposed to moisture above a critical relative humidity for a long time [39]. However, the influence of moisture exposure on the mechanical properties of epoxies has not been adequately captured through MD simulation.

Since moisture diffusion will directly determine the reliability of ICs in long periods of time, it is worthwhile to clearly understand the mechanism of moisture penetration at a deeper level. Besides the diffusion coefficient, further studies on the mechanical behavior of materials, coupling of water molecular diffusion and polymer swelling, combination of diffusion, and chemical degradation of polymers are possible areas for the application of MD techniques.

24.4 Interconnection Alloys

Several materials are used as electrical interconnects between electronics devices in electronics packaging. A major part of these interconnect materials are metals with low melting point, because melting metals at high temperature will cause serious damage to the IC devices. β-Tin (β-Sn) and lead (Pb) are the two most commonly used materials among the metals with low melting temperature. While used as interconnects in the packaging area, both Sn and Pb have poor electrical conductivity. Another problem with Pb is its harmful impact to the environment. More and more lead-free alloys of tin and other metals with high electrical conductivities are used as interconnects. Copper and silver are the best two electrical conductors; while the drawback with them is the high melting temperature.

One solution to the problem addressed in the previous paragraph is to mix tin and silver/copper at a proper ratio to obtain a reasonable melting temperature and electrical conductivity. However, a commonly used Sn/Ag solder with a ratio of 96.5/3.5 has a melting point of 221 °C [40], higher than temperature threshold of some electronic components. Considering nanosize effects on material properties, more and more researchers [40–45] tried to reduce the melting temperature by decreasing the size of metals. This is an area of considerable interest in nanoscale simulations. H. Dong et al. [40] employed MD simulation in conjunction with modified embedded atom method (MEAM) to investigate the intermix ability of nano-Ag/Sn alloy. MEAM is a modified approach based on embedded atom method (EAM) describing the energies between atoms, particularly for bcc metallic atoms. In these simulations, two 4-nm droplets of Ag and Sn were generated, equilibrated, and placed together at 500 K, a temperature that is higher than the melting temperature of Sn and lower than that of Ag. The initial configurations of Ag and Sn are shown in Fig. 24.5, where Ag maintained its ordered lattice structure and Sn had a liquid-like structure.

A 3 ns simulation time for interdiffusion suggested that no substantial intermix took place except collapse of both particles (Fig. 24.6). If the temperature was elevated to 1000 K, a total interdiffusion was observed. Using analytical tools and the calculated activation energy for interdiffusion, time need for intermix at 500 K was estimated to be 118 ns. This time scale is too large for MD simulation but easily feasible for practical solder application, but with increasing temperature this problem also falls within the capabilities of MD. Such MD simulations can be used to understand temperature dependence of the degree of mixing, speed of interdiffusion, and even activation energy.

Another approach to replace the current Sn/Pb solder is by using conductive polymer/metal composites, which are often called electrically conductive adhesive (ECA). Two examples of ECA systems are shown in Fig. 24.7. One concern is the electrical conductivity of these composites, which is not only related to the conductive polymer matrix but also with the dispersion state of metal fills. Interfaces between metal fills will significantly degrade the electrical conductivity. To interconnect isolated metal fills, one has to heat the polymer/metal composite assembly to fuse metal fills. The maximum temperature one can heat such a part is limited by the degradation temperature of polymers. Again, decreasing the size of metal particles is a viable means to suppress the melting temperature of metals. MEAM/EAM-coupled MD simulations were conducted by H. Dong et al. [43] to investigate the time evolution of Cu nanoparticles between two substrates in the same temperature range as discussed before. Results showed that a more ordered structure than the original one was formed under the applied strain, which might improve conductivity of ECA, although experimental verification is needed to validate this result.

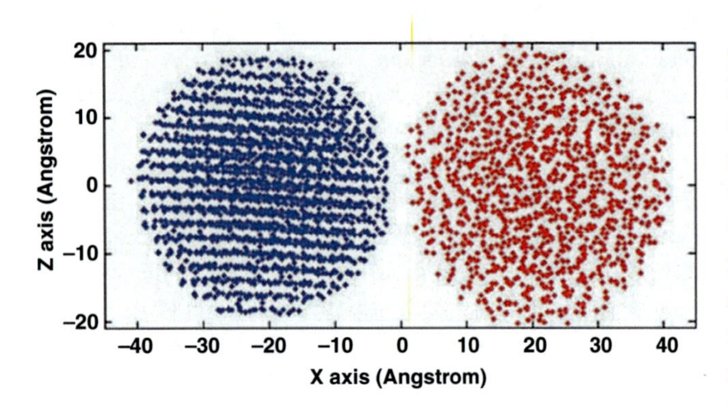

Fig. 24.5 Initial configurations of Ag and Sn. (Reprinted from Ref. [40] with permission from Institute of Physics)

Besides thermal properties of interconnect alloys, another problem to solve is the failure mechanism of Sn-based alloys. Many experimental investigations were reported recently, but a fundamental understanding of the universal failure mechanism for these systems is still lacking. MD simulations provide a convenient and powerful guiding tool to investigate the failure mechanism at the atomic level. W. Wang et al. [46] investigated uniaxial tension of nanocracked β-Sn single crystal using the MD method. The sample was prepared by first generating ideal β-Sn single crystal and then manually removing some atoms to create defects. Morse potential was used to compute the pairwise interatomic forces. Uniaxial tension applied along different directions was shown to induce change in dislocation mechanisms. In Fig. 24.8,

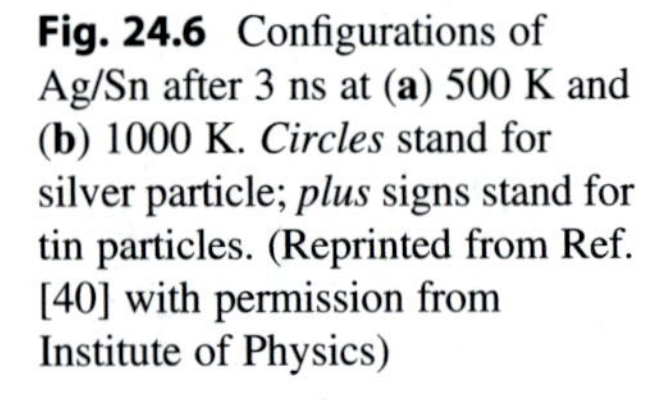

Fig. 24.6 Configurations of Ag/Sn after 3 ns at (**a**) 500 K and (**b**) 1000 K. *Circles* stand for silver particle; *plus* signs stand for tin particles. (Reprinted from Ref. [40] with permission from Institute of Physics)

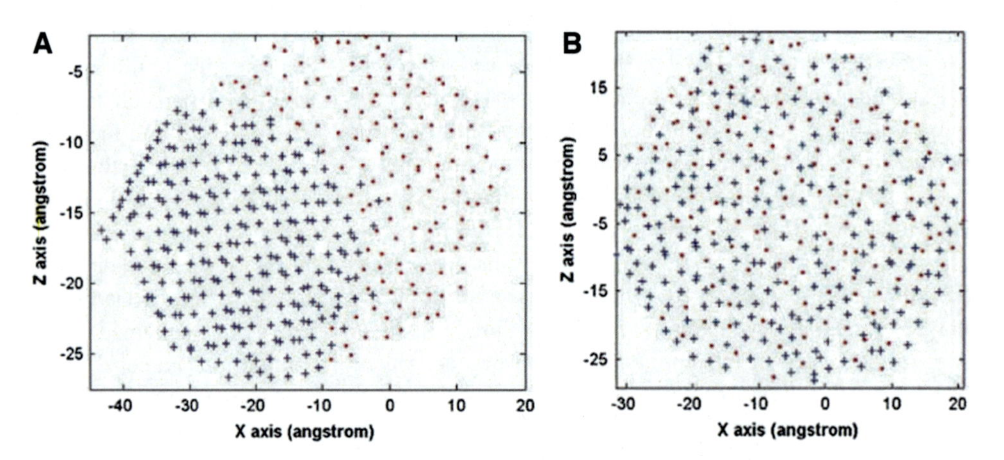

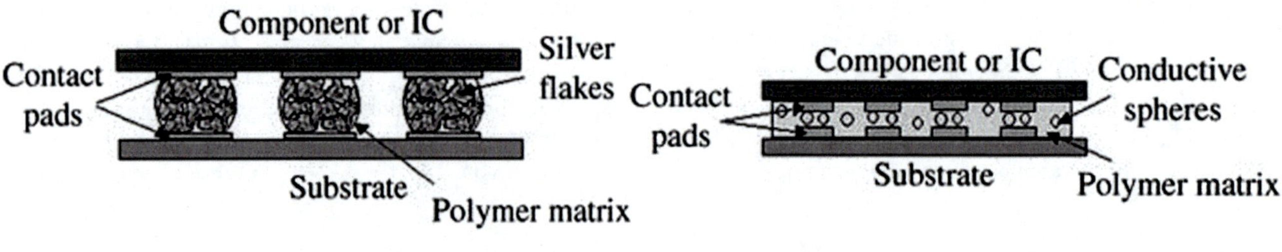

Fig. 24.7 Schematic diagram of electrically conductive adhesives (ECAs). (Reprinted from Ref. [43] with permission from Koninklijke Brill NV)

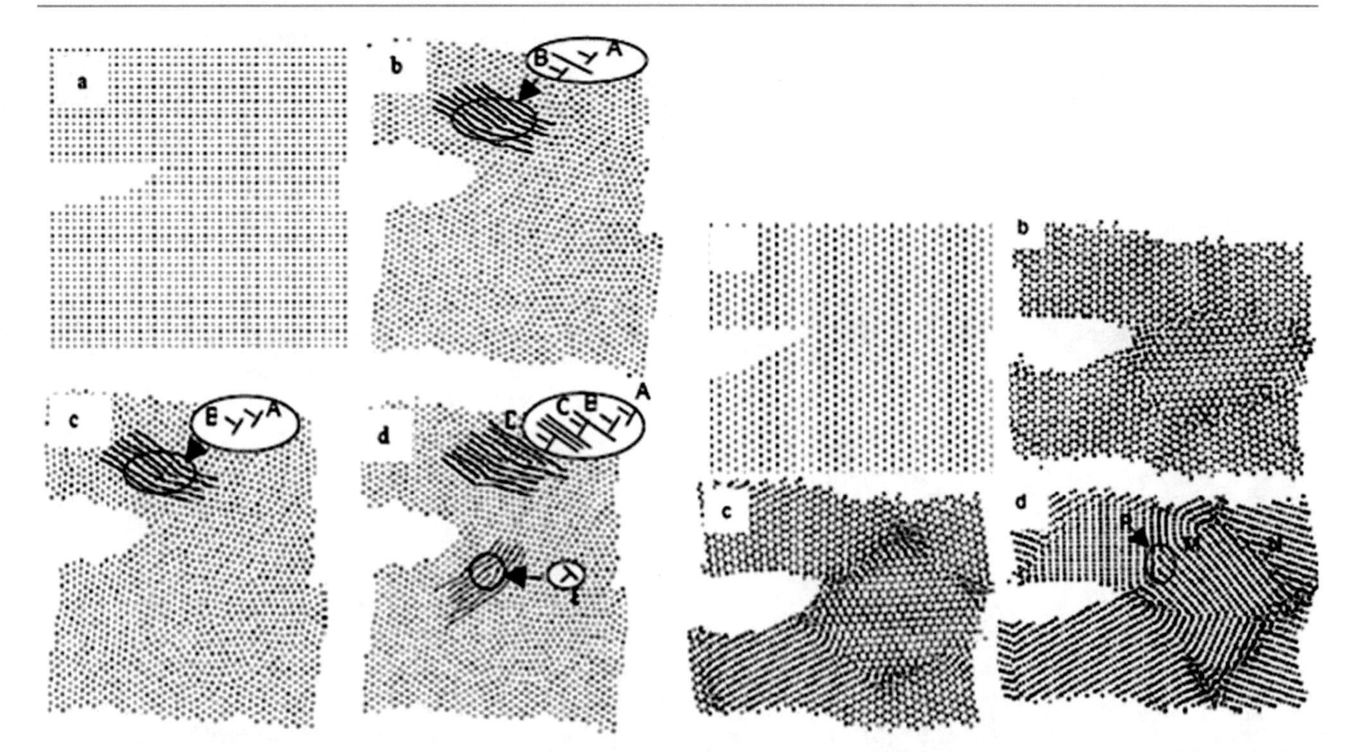

Fig. 24.8 *Left*: Simulation plots of the first case: applied tensile force along (010) direction (**a**) initial, (**b**) after relaxation, (**c**) after 4000 fs, (**d**) after 5000 fs. *Right*: Simulation plots of the second case: applied tensile force along (001) direction (**a**) initial, (**b**) after relaxation, (**c**) after 5000 fs, (**d**) after another 5000 fs without loading. (Reprinted with permission from Ref. [46], Copyright 2005 IEEE)

simulation plots are given for two different tension cases. In the first case (force along the (010) direction), only a one-slip system was observed, which means a few dislocations. In the second case, two-slip system was observed almost simultaneously. But the disorder induced by atomic positions from the two slips would initiate sub-boundaries to minimize energy. These results were similar to the experiments conducted in TEM.

24.5 Modification of Interface

The adhesives at the interface between multilayers are expected to provide suf- ficiently strong adhesion to undergo thermal cycling, moisture transport, and chemical and radiation degradation. However, interactions between plastics and metals are observed in general to be weak, which is a significant challenge for electronic packaging. Efforts were made to improve joint adhesion by introducing strong connections between polymers and metals. Among all the physical and chemical interactions, metallic bonds, ionic bonds, as well as covalent bonds are three candidates for adhesive purpose. When making modifications for the adhesive layers, it is convenient to examine the effectiveness of newly introduced interfacial connections using calculations with MD [47–50]. Using MD simulations, with appropriate modifications, one can also observe bond formation and bond breakage at the interfaces (under appropriate conditions the system is subjected to) and compare changes in chemical bonds and to calculate the strength of these bonded joints. Although the procedure appears straightforward, there are a number of difficulties in implementing MD to study interface bonding, but still the results are expected to provide qualitative and sometimes quantitative patterns. Normally, the interactions between plastics and metals were believed to be the van der Waals and the weaker type of electrostatic forces. These pair forces are dependent on distances between neighboring atoms for a specific system. However, some researchers [51] argued that the bonding between epoxy adhesive and metal was strongly related to the content of metal oxides, although the mechanism was unclear. Development of the force field involved in simulating the interface between dissimilar materials is a major challenge. Unless appropriate force fields are used, the results will not be representative for the system we are investigating.

Fan et al. [52–55] (cf. Figs. 24.9 and 24.10) tracked the interfacial energies of epoxy/copper system in a thermal cycling test via MD method. Interfacial energies were calculated from the difference between total energy of combined copper/epoxy system and the sum of energies of copper and epoxy considered separately. The bonding energy was found to decrease with a decreasing percentage of cuprous oxide, and additional complications were found from voids occurring at the epoxy/copper interface. So, the key factor

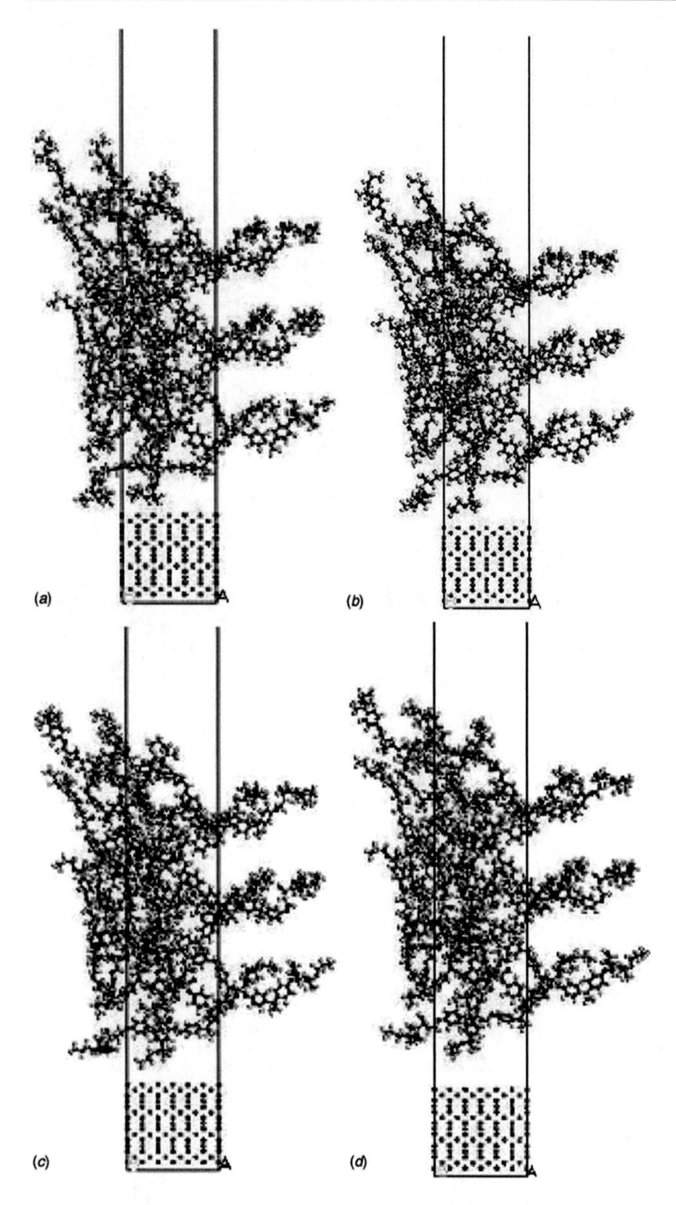

Fig. 24.9 Snapshots of the system after (**a**) 0 thermal cycle, (**b**) 600 thermal cycles, (**c**) 1200 thermal cycles, and (**d**) 1800 thermal cycles from the MD simulations. (Reprinted from Ref. [53] with permission from ASME)

governing the interfacial adhesion strength is still under considerable debate.

In general, addition of a third layer between the two layers in an interface (like between epoxy and copper) can create stronger interactions between the neighboring layers. If suitable force fields are available, this problem provides a fertile area for MD simulations, in order to understand the enhancement of strength with the nature of the newly added third layer. An alternative way to enhance interactions is to make an epoxy/copper composite instead of pure epoxy to provide metallic interactions.

Wong et al. [54] used a self-assembly monolayer (SAM) to enhance joint adhesion in the epoxy/copper system. Four different types of thiols/disulfides, copper surface modifiers, were chosen as the SAM candidates. Their structures are shown in Fig. 24.11.

Change of potential energies before and after a covalent bond forming between SAM and epoxy was used as a criterion to determine how many interfacial bonds were formed. SAM density on the copper substrate was estimated by the same method. Interfacial bonding energies between SAM A and SAM C modified copper and epoxy showed a significant enhancement of adhesion, which was also consistent with shear tests (Fig. 24.12). Problems like this are suitable for MD simulations, to understand how certain functional groups could enhance the adhesion and the resulting structure of the adhesive layer.

24.6 Thermal Conductivity

Fast technological development in the semiconductor industry and tougher new requirements governed by the market place have reduced the size of chips to nanometer scale, that allows to pack more dense chips in a given area. Since many of the devices containing these chips are continuously used, these chips and associated connective circuits will produce a significant amount of heat. Performance of IC chips will be weakened if the heat produced during the process is not dissipated quickly. As described below, heat conduction can also be studied using the MD approach. Theoretically, the parameter that describes the speed of thermal dissipation is called thermal conductivity λ, which is defined by the following equation:

$$\lambda = -\frac{J}{\Delta T} \tag{24.20}$$

where J is the heat flux, which is defined as the amount of energy transferred in a given time through a given surface area, and ΔT is the temperature gradient. But this Fourier law is valid only for macroscopic materials. When the size of devices is reduced to nanoscale, thermal conductivity is believed to depend not only on the properties of materials but also the size of the materials. Thus, considering the complexities involved on the dependence of properties on the size, this is an active area of research. It is relatively easy to measure the thermal conductivity of bulk materials, but finding size-dependent thermal conductivity coefficients for nanosized materials pose some experimental difficulties. So, instead of using numerous experiments, it may be preferable to use simulation methods to study thermal transport in IC chips. MD deals with the smallest scale simulation of heat

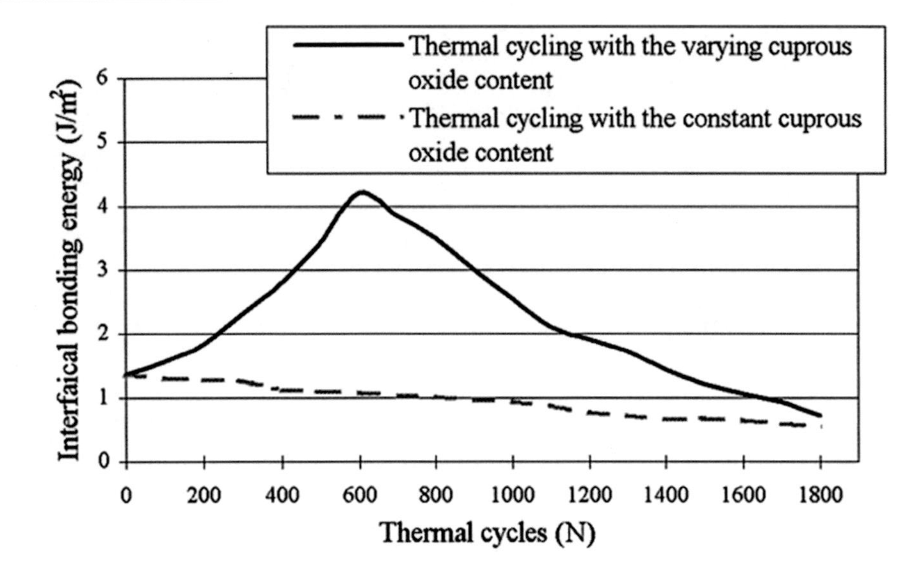

Fig. 24.10 Interfacial bonding energy against thermal cycles. (Reprinted from Ref. [53] with permission from ASME)

Fig. 24.11 Chemical structure for four types of SAMs. (Reprinted with permission from Ref. [54], Copyright 2008 IEEE)

HClH$_2$N S—S NH$_2$HCl

SAM A

SAM C

HO—(CH$_2$)$_2$—SH

SAM O

CH$_3$—(CH$_2$)$_{15}$—SH

SAM E

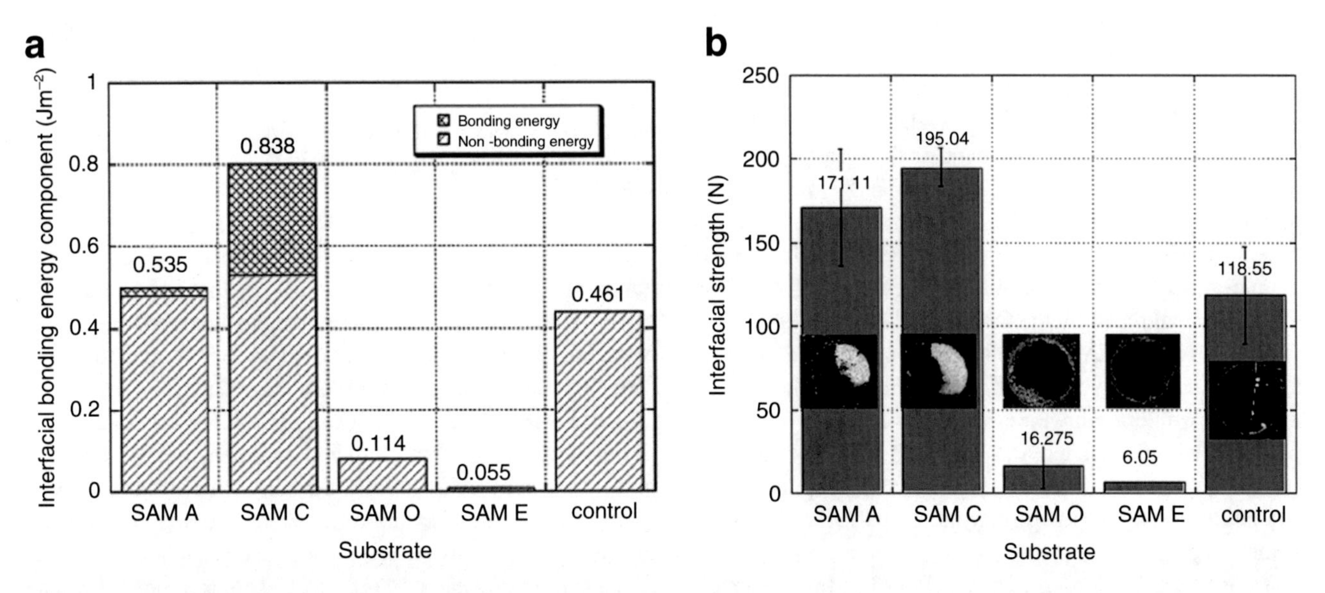

Fig. 24.12 (**a**) Energy component of the interfacial bonding energy predicted by MD, (**b**) interfacial bonding strength data evaluated by the button shear test. (Reprinted with permission from Ref. [54], Copyright 2008 IEEE)

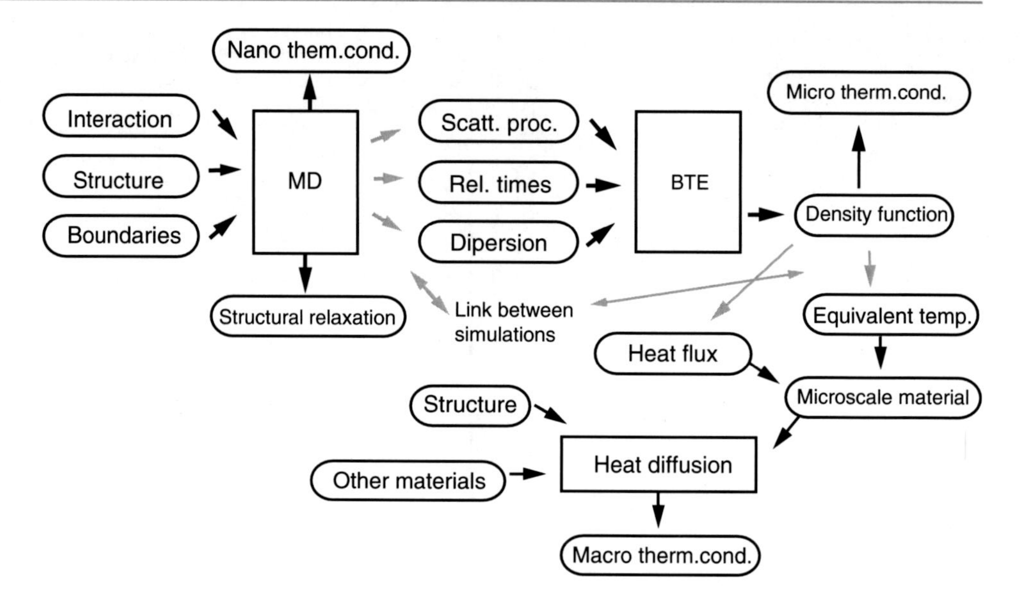

Fig. 24.13 Input and output quantities required in different models. (Reprinted from Ref. [56] with permission from Springer Science+Business Media)

dissipation compared to other continuum models as well as Boltzmann transport equation (BTE) models. Dependence of various parameters in MD and their relationship between heat-transfer phenomenon is summarized in the sketch given in Fig. 24.13.

In MD simulations, the thermal conductivity can be computed either using nonequilibrium MD (NEMD) or equilibrium MD (EMD) approach. In NEMD, the thermal conductivity is computed by adding a thermal gradient or a heat flux to the heat carrier; while in EMD one calculates the thermal conductivity from current thermal fluctuation in the system using the so-called Green–Kubo method [57].

Schelling et al. [57] measured thermal conductivities of silicon films using the methods mentioned above. In NEMD, a temperature gradient was added in a single well-defined direction. The potentials (force fields) in silicon were described with Stillinger and Weber model. Thermal conductivities were found to depend on the system size (finite system effect) using either NEMD or EMD method. But reliable results could be derived by extrapolation to infinite system size. It is well known that the thermal conductivity of a thin film is lower than that of a bulk (see Fig. 24.14), which was further validated by the work of Schelling et al.

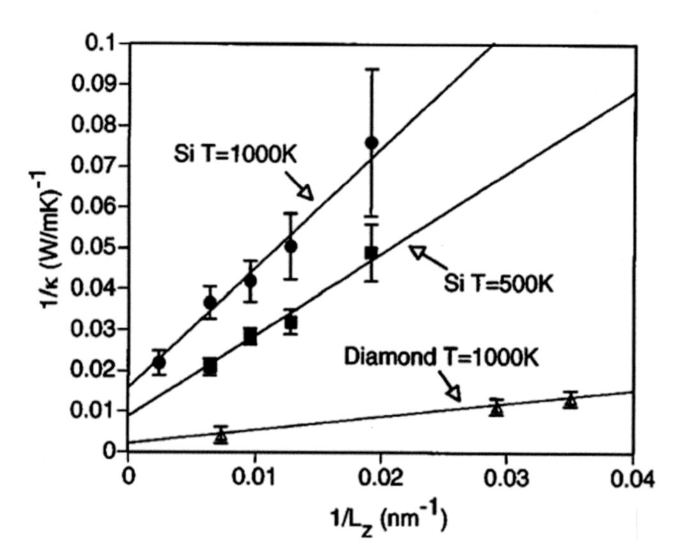

Fig. 24.14 System size dependence of 1/k on 1/Lz. (Reprinted with permission from Ref. [57], Copyright 2002 by the American Physical Society)

The other NEMD method to compute thermal conductivities is by imposing heat flux on the system reversely. MüllerPlathe [58] used this kind of NEMD to measure thermal conductivities. The advantage is relatively easy to implement. Also, the system was compatible with simulation boxes with or without periodic boundary conditions.

Yang et al. [59] proposed a MD/FEM (finite element method) coupled methodology to investigate heat conductivities through a copper/aluminum interface. The advantage of coupled method is making implication of boundary conditions easier, and this method can extract nanoscale aspects of certain regions in the bulk properties of the whole system, since the domain covered by MD is embedded in the finite element domain that applies to larger elements and larger domains. This approach is particularly effective when a nanoscale second phase (or different material) is embedded in the larger domain. Although this is a powerful technique in linking domains with different length scales, linking the MD and finite element domains is one of the challenges in this process. Various approaches have been proposed. As an example, Fig. 24.15 shows a method where the intermediate region between MD and finite element is covered by both FEM meshes and MD particles, to keep the continuity from the smaller MD regions to the finite element region.

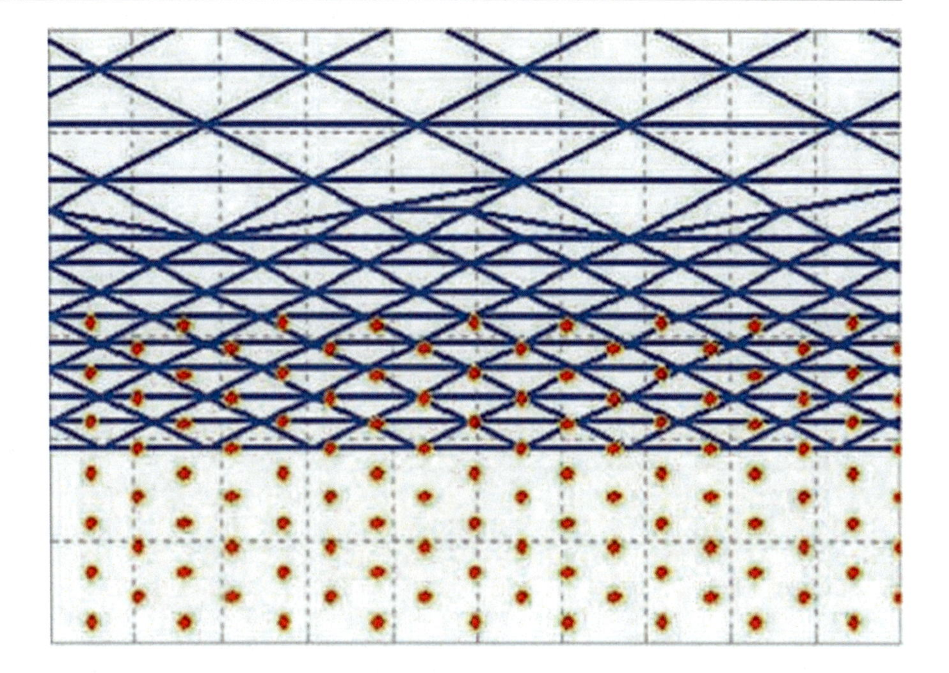

Fig. 24.15 The configuration of the coupled region. (Reprinted from Ref. [59] with permission from Springer Science+Business Media)

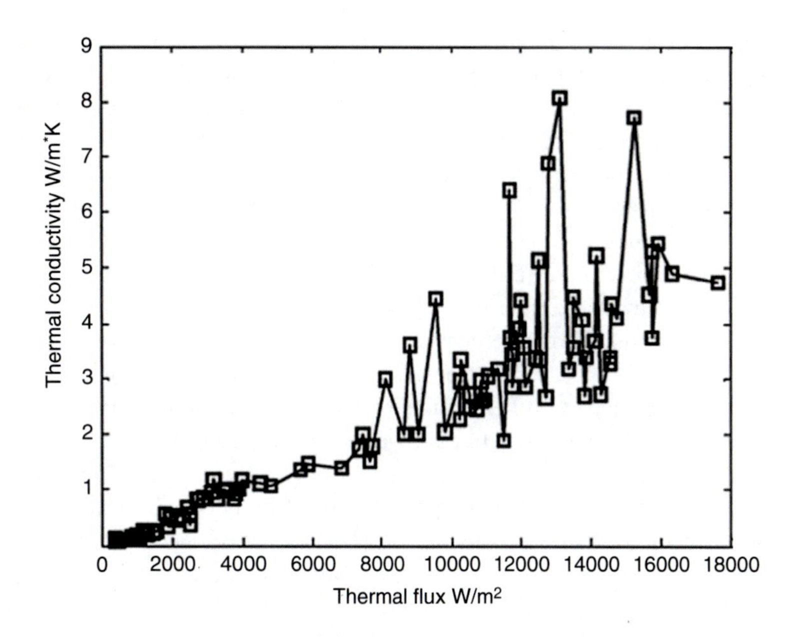

Fig. 24.16 The relation of thermal conductivity and thermal flux. (Reprinted from Ref. [59] with permission from Springer Science+Business Media)

Using such an approach, the thermal conductivity was calculated for a Cu/Al interface and the results are summarized in Fig. 24.16. An approximately linear relationship between thermal conductivities and thermal flux was found during the primary stage. When the thermal flux achieved certain critical point, thermal conductivities would fluctuate intensely, showing the instability and dependence of thermal conductivity on thermal flux.

24.7 Applications of Carbon Nanotubes

Carbon nanotubes (CNTs) have been extensively studied [60–71] as a potential alternative/additive of refills in composites used in the electronic industry due to its excellent mechanical strength, heat dissipation and electrical transport, and low thermal expansion. Since theoretically calculated

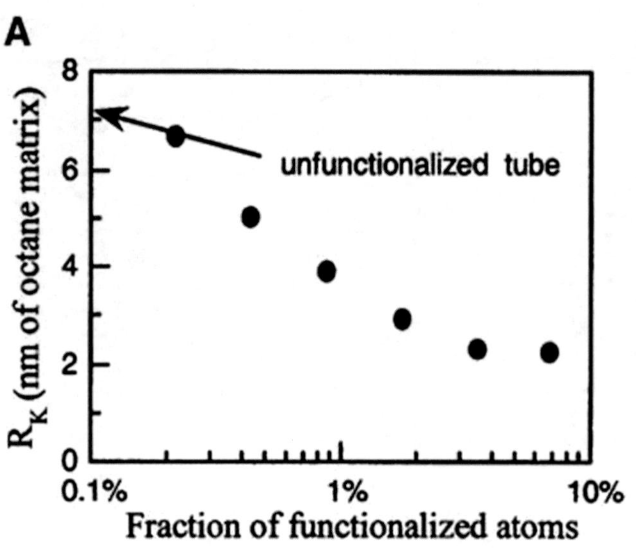

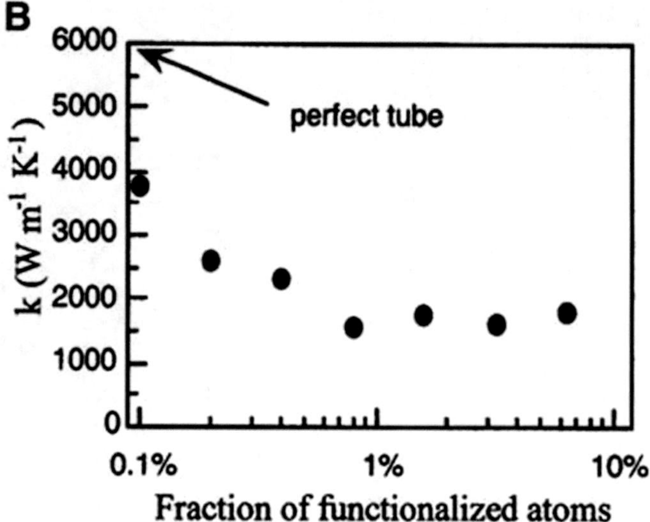

Fig. 24.17 (**a**) Interfacial resistance in units of equivalent matrix thickness vs. fraction of tube carbon atoms with covalently attached octane molecules, (**b**) carbon nanotube conductivity vs. fraction of functionalized tube carbon atoms, obtained from MD simulations. (Reprinted with permission from Ref. [71], Copyright 2004, American Institute of Physics)

thermal conductivity of multiwall carbon nanotubes (MWNT) at room temperature can be as high as 6600 W/mK [72], researchers are paying more and more attention to promising applications of CNTs as components of thermal interface materials (TIMs). However, the actual thermal conductivity depends on a number of parameters and varies from tens to thousands W/mK. For isolated CNTs, those factors governing thermal conductivity include length of CNTs [66, 67], temperature [60], defects [61], and other factors. For a CNT-based composite, a possible challenge to deal with is the high thermal resistance at the interfaces between fillers and matrix. An experimentally measured thermal conductivity coefficient is often smaller than the value predicted by mixing rules. MD is an effective way to help to understand the enhanced heat dissipation behaviors of TIMs. Algorithms to compute thermal conductivity are described in the previous subtopic. Some results of CNT-based composites obtained from MD simulation are presented here and compared with experimental measurement.

It has been shown that defects in isolated CNTs will scatter phonons (the quantized modes of crystal vibration associated with heat transfer) and decrease intrinsic tube conductivity [60]. As a major type of TIMs, CNT/polymer composites have higher thermal conductivity than polymer bulks, lower elastic modulus, and thermal expansion coefficient than metals. However, the interface between the tubes and soft matrix plays an important role in the overall performance of TIM. Increasing the number of covalent bonding between CNTs and polymer will reduce interface resistance and provide durable adhesion but also induce defects into CNTs, since this is generally done through incorporating functionalized groups to the nanotubes that are capable of forming covalent bonds. MD studies of CNTs with varied degree of chemical functionalization from different research groups validated each other – namely, functionalization of CNTs reduces thermal conductivity and interfacial resistance. For example, Shenogin et al. [71] grafted octane molecules to single- wall carbon nanotubes (SWNT) to investigate dependency of thermal conductivity and the MD simulation results are shown in Fig. 24.17. It can be seen that CNT with 10% functionalized atoms has a thermal conductivity of about half of that of perfect tubes, while interface resistance is also reduced by a factor of about 2. Similar results are also reported by other researchers [62, 69]. Although the percentages of conductivity and resistance degraded by functionalized atoms vary case by case, the trends drawn from these researches are identical. Hence, the problem is how to find an optimal percentage of functionalized atoms to balance conductivity and resistance.

Precise experimental manipulation of arraying or growing CNTs makes it possible to align vertically to the surface. Figure 24.18 gives an example of a pure CNT TIM with controllable thickness. When a pure CNT TIM was placed between two aluminum surfaces, the CNT TIM had less thermal contact resistance than in the case of direct contact or with graphite coating. Observed from experiments, the thermal conductivity was also enhanced by more than three times [70] with this construction. Under this situation, CNTs are linked to planar surfaces with points contacts. The size effects of the contact points, and many other aspects associated with CNT growth and behavior can be studied by MD.

Saha et al. [73] built two planar silicon substrates with constriction to mimic the point contact situation using MD,

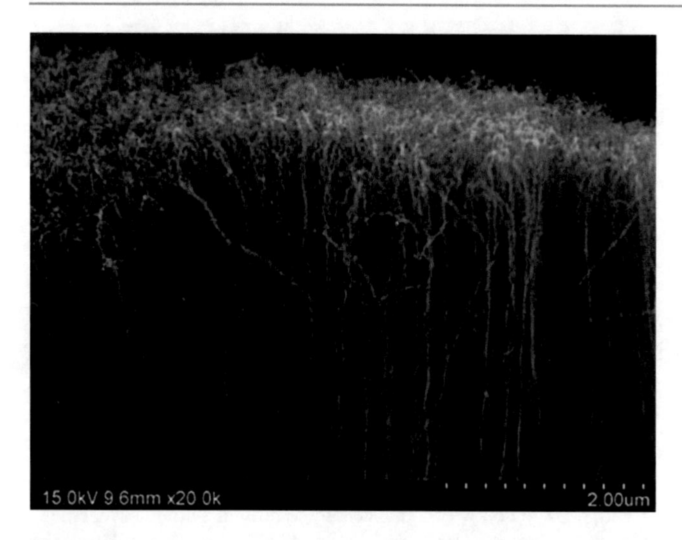

Fig. 24.18 Scanning electron microscopy image of low-density aligned CNT. (Reprinted from Ref. [70], Copyright 2007, with permission from Elsevier)

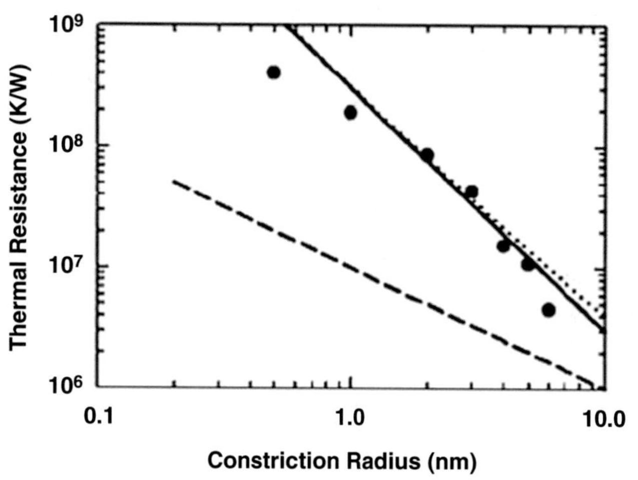

Fig. 24.19 The MD thermal resistance (*filled circles*) measured as a function of constriction radius. Also plotted are theoretically calculated curves (*solid line*: *R*b, *dashed line*: *R*d, *dotted line*: *R*). (Reprinted with permission from Ref. [73], Copyright 2007, American Institute of Physics)

where the focus was how the radius of constriction influences thermal resistance with other conditions fixed. Theoretically, the total thermal resistance R is a summation of two parts: the diffusive resistance R_d and the ballistic resistance R_b, which can be calculated based on the following two equations:

$$R_\mathrm{d} = 1/(2ka) \tag{24.21}$$

$$R_\mathrm{b} = 4l/\left(3\pi ka^2\right) \tag{24.22}$$

where k is thermal conductivity, a is radius of constriction, and l is phonon mean free path (phonon group velocity and phonon mean free path are assumed independent of frequency). The general trend is that a larger radius will lead to a lower resistance (Fig. 24.19). The pattern for the variation of thermal resistance from the theoretical approach (Eqs. 24.21 and 24.22) is also shown in the figure. MD with appropriate idealizations can be used to study properties such as thermal resistance.

24.8 Summary

In this chapter, we reviewed applications of molecular dynamics techniques in the electronic packaging area. In general, MD has been used to help to understand events at the atomistic length scales, such as atomistic mechanisms of diffusion, adhesion, heat conduction and failure, predict material properties, and design new structures. MD can be applied to almost all types of materials in all states, although simulating certain phenomenon in solid state is more difficult than in the liquid state. It can generally compute both statistical and dynamic properties and explain physical and mechanical behaviors. As long as initial configurations are correctly generated and proper force fields are selected, MD will give fairly good results compared to experiments. However, the interpretation of MD output to obtain relevant parameters has to be done properly; otherwise, the results may not be useful. In general, at least useful qualitative patterns can be obtained and in many problems quantitative information can be extracted by the judicious application of MD. The advantage of MD is that it is less expensive and much more convenient than experiments, if the objective is materials design. When the scale of interest comes to nanoscale, MD can explore interactions at atomistic level, which is not easily achieved through experiments. MD can also be employed to measure parameters associated with a phenomenon, which cannot easily be characterized experimentally. However, MD does have some shortcomings. First of all, it is a molecular modeling method that cannot deal with mesoscopic and macroscopic simulations. Second, generally chemical reactions cannot be modeled by classical MD, although some aspects of it can be done through specific modifications. Third, results of MD rely on the accuracy of the force fields and initial configurations. Currently, researchers try to combine MD with other simulation methods (such as finite element methods, quantum mechanics simulation) to overcome some of the drawbacks of the classical MD, although that could remove some of the difficulties of classical MD, but it can create its own difficulties. Also, significant effort has been devoted to increase the efficiency and effectiveness of MD algorithms.

The main application of MD methods in "nanopackaging," where there are nanoscale components, is to give another tool to understand conformational aspects, properties, and in aiding the design of new materials such as insulators, conductors, adhesives, and solders. Currently, MD simulations are also routinely applied to study nanoparticle/polymer composites [74], lead-free solder [75], and nanometals [76, 77], that are particularly relevant to the packaging area. In summary, applications of MD techniques in the packaging area are many and growing. We have provided only an outline of the applications of MD in this area due to space limitations, and the literature contains vast amount of information about the procedure as well as applications of MD in the packaging area.

Acknowledgment We gratefully acknowledge the financial support from NASA Langley Research Center.

References

1. Alder, B.J., Wainwright, T.E.: Studies in molecular dynamics. 1. General method. J. Chem. Phys. **31**(2), 459–466 (1959)
2. Frenkel, D., Smit, B.: Understanding Molecular Simulation: From Algorithms to Applications, 2nd edn, p. xxii, 638 p. Academic, San Diego (2002)
3. Haile, J.M.: Molecular Dynamics Simulation: Elementary Methods, p, xvii, 489 p. Wiley, New York (1992)
4. Rapaport, D.C.: The Art of Molecular Dynamics Simulation, 2nd edn, p, xiii, 549 p. Cambridge University Press, Cambridge/New York (2004)
5. Sadus, R.J.: Molecular Simulation of Fluids: Theory, Algorithms, and Object-Orientation, 1st edn, p, xxvii, 523 p. Elsevier, Amsterdam/New York (1999)
6. Ponder, J.W., Case, D.A.: Force fields for protein simulations. Protein Simul. **66**, 27–86 (2003)
7. MacKerell, A.D., Bashford, D., Bellott, M., Dunbrack, R.L., Evanseck, J.D., Field, M.J., Fischer, S., Gao, J., Guo, H., Ha, S., Joseph-McCarthy, D., Kuchnir, L., Kuczera, K., Lau, F.T.K., Mattos, C., Michnick, S., Ngo, T., Nguyen, D.T., Prodhom, B., Reiher, W.E., Roux, B., Schlenkrich, M., Smith, J.C., Stote, R., Straub, J., Watanabe, M., Wiorkiewicz-Kuczera, J., Yin, D., Karplus, M.: All-atom empirical potential for molecular modeling and dynamics studies of proteins. J. Phys. Chem. B. **102**(18), 3586–3616 (1998)
8. Daw, M.S., Foiles, S.M., Baskes, M.I.: The embedded-atom method – a review of theory and applications. Mater. Sci. Rep. **9** (7–8), 251–310 (1993)
9. Sun, H.: COMPASS: an ab initio force-field optimized for condensed-phase applications – overview with details on alkane and benzene compounds. J. Phys. Chem. B. **102**(38), 7338–7364 (1998)
10. Niketic, S.R., Rasmussen, K.: The Consistent Force Field: A Documentation, p. ix, 212 p. Springer, Berlin/New York (1977)
11. Hamerton, I., Heald, C.R., Howlin, B.J.: Molecular modelling of the physical and mechani- cal properties of two polycyanurate network polymers. J. Mater. Chem. **6**(3), 311–314 (1996)
12. Leung, Y.K., Eichinger, B.E.: Computer-simulation of end-linked elastomers. 1. Trifunctional networks cured in the bulk. J. Chem. Phys. **80**(8), 3877–3884 (1984)
13. Cheng, K.C., Chiu, W.Y.: Monte-Carlo simulation of polymer network formation with com- plex chemical-reaction mechanism – kinetic approach on curing of epoxides with amines. Macromolecules. **27**(12), 3406–3414 (1994)
14. Mayo, S.L., Olafson, B.D., Goddard, W.A.: Dreiding – a generic force-field for molecular simulations. J. Phys. Chem. **94**(26), 8897–8909 (1990)
15. Yarovsky, I., Evans, E.: Atomistic simulation of the sol formation during synthesis of organic/inorganic hybrid materials. Mol. Simul. **28**(10–11), 993–1004 (2002)
16. Wu, C.F., Xu, W.J.: Atomistic molecular modelling of crosslinked epoxy resin. Polymer. **47**(16), 6004–6009 (2006)
17. Fan, H.B., Yuen, M.M.F.: Material properties of the cross-linked epoxy resin compound predicted by molecular dynamics simulation. Polymer. **48**(7), 2174–2178 (2007)
18. http://accelrys.com/products/materials-studio/
19. Komarov, P.V., Chiu, Y.T., Chen, S.M., Khalatur, P.G., Reineker, P.: Highly cross-linked epoxy resins: an atomistic molecular dynamics simulation combined with a mapping/reverse mapping procedure. Macromolecules. **40**(22), 8104–8113 (2007)
20. Varshney, V., Patnaik, S.S., Roy, A.K., Farmer, B.L.: A molecular dynamics study of epoxy- based networks: cross-linking procedure and prediction of molecular and material properties. Macromolecules. **41**(18), 6837–6842 (2008)
21. Brown, D., Clarke, J.H.R.: A loose-Coupling, Constant-pressure, molecular-dynamics algorithm for use in the modeling of polymer materials. Comput. Phys. Commun. **62**(2–3), 360–369 (1991)
22. Parrinello, M., Rahman, A.: Polymorphic transitions in single-crystals – a new molecular- dynamics method. J. Appl. Phys. **52** (12), 7182–7190 (1981)
23. Stevens, M.J.: Interfacial fracture between highly cross-linked polymer networks and a solid surface: effect of interfacial bond density. Macromolecules. **34**(8), 2710–2718 (2001)
24. Stevens, M.J.: Manipulating connectivity to control fracture in network polymer adhesives. Macromolecules. **34**(5), 1411–1415 (2001)
25. Tsige, M., Lorenz, C.D., Stevens, M.J.: Role of network connectivity on the mechanical properties of highly cross-linked polymers. Macromolecules. **37**(22), 8466–8472 (2004)
26. Tsige, M., Stevens, M.J.: Effect of cross-linker functionality on the adhesion of highly cross- linked polymer networks: a molecular dynamics study of epoxies. Macromolecules. **37**(2), 630–637 (2004)
27. Drozdov, A.D., Christiansen, J.D., Gupta, R.K., Shah, A.P.: Model for anomalous moisture diffusion through a polymer-clay nanocomposite. J. Polym. Sci. B Polym. Phys. **41**(5), 476–492 (2003)
28. Marsh, L.L., Lasky, R., Seraphim, D.P., Springer, G.S.: Moisture solubility and diffusion in epoxy and epoxy-glass composites. IBM J. Res. Dev. **28**(6), 655–661 (1984)
29. Yu, Y.T., Pochiraju, K.: Three-dimensional simulation of moisture diffusion in polymer composite materials. Polym.-Plast. Technol. Eng. **42**(5), 737–756 (2003)
30. Fan, H.B., Chan, E.K.L., Wong, C.K.Y., Yuen, M.M.F.: Investigation of moisture diffusion in electronic packages by molecular dynamics simulation. J. Adhes. Sci. Technol. **20**(16), 1937–1947 (2006)
31. Fukuda, M., Kuwajima, S.: Molecular-dynamics simulation of moisture diffusion in polyethylene beyond 10 ns duration. J. Chem. Phys. **107**(6), 2149–2159 (1997)
32. Hofmann, D., Fritz, L., Ulbrich, J., Schepers, C., Bohning, M.: Detailed-atomistic molec- ular modeling of small molecule diffusion and solution processes in polymeric membrane materials. Macromol. Theory Simul. **9**(6), 293–327 (2000)
33. Lin, Y.C., Chen, X.: Investigation of moisture diffusion in epoxy system: experiments and molecular dynamics simulations. Chem. Phys. Lett. **412**(4–6), 322–326 (2005)
34. Soles, C.L., Yee, A.F.: A discussion of the molecular mechanisms of moisture transport in epoxy resins. J. Polym. Sci. Part B Polym. Phys. **38**(5), 792–802 (2000)

35. Tamai, Y., Tanaka, H., Nakanishi, K.: Molecular simulation of permeation of small penetrants through membranes. 1. Diffusion-coefficients. Macromolecules. **27**(16), 4498–4508 (1994)
36. Vanlandingham, M.R., Eduljee, R.F., Gillespie, J.W.: Moisture diffusion in epoxy systems. J. Appl. Polym. Sci. **71**(5), 787–798 (1999)
37. Zannideffarges, M.P., Shanahan, M.E.R.: Diffusion of water into an epoxy adhesive – com- parison between bulk behavior and adhesive joints. Int. J. Adhes. Adhes. **15**(3), 137–142 (1995)
38. Dermitzaki, E., Wunderle, B., Bauer, J., Walter, H., Michel, B.: Structure property correlation of epoxy resins under the influence of moisture and comparison of diffusion coefficient with MD-simulations. In: 9th International Conference on Thermal, Mechanical and Multiphysics Simulation and Experiments in Micro-Electronics and Micro-Systems. EuroSimE (2008)
39. Comyn, J., Groves, C.L., Saville, R.W.: Durability in high humidity of glass-to-lead alloy joints bonded with an epoxide adhesive. Int. J. Adhes. Adhes. **14**(1), 15–20 (1994)
40. Dong, H., Fan, L.H., Moon, K.S., Wong, C.P., Baskes, M.I.: MEAM molecular dynamics study of lead free solder for electronic packaging applications. Model. Simul. Mater. Sci. Eng. **13**(8), 1279–1290 (2005)
41. Dong, H., Moon, K.S., Wong, C.P.: Molecular dynamics study on the coalescence of Cu nanoparticles and their deposition on the Cu substrate. J. Electron. Mater. **33**(11), 1326–1330 (2004)
42. Dong, H., Moon, K.S., Wong, C.P.: Molecular dynamics study of nanosilver particles for low-temperature lead-free interconnect applications. J. Electron. Mater. **34**(1), 40–45 (2005)
43. Dong, H., Zhang, Z.Q., Wong, C.P.: Molecular dynamics study of a nano-particle joint for potential lead-free anisotropic conductive adhesives applications. J. Adhes. Sci. Technol. **19**(2), 87–94 (2005)
44. Dong, H., Fan, L., Moon, K., Wong, C.P.: Molecular dynamics simulation of lead free sol- der for low temperature reflow applications. In: 55th Electronic Components and Technology Conference, pp. 983–987 (2005)
45. Dong, H., Moon, K., Wong, C.P.: Molecular dynamics study on coalescence of silver (Ag) nanoparticles and their deposition on gold (Au) substrates. In: 9th International Symposium on Advanced Packaging Materials: Processes, Properties and Interfaces, pp. 152–157 (2004)
46. Wang, W., Ding, Y., Wang, C.: Molecular dynamics (MD) simulation of uniaxial tension of β-Sn single crystals with nanocracks. In: 6th International Conference on Electronic Packaging Technology (2005)
47. Iwamoto, N., Pedigo, J.: Property trend analysis and simulations of adhesive formulation effects in the microelectronics packaging industry using molecular modeling. In: 48th IEEE Electronic Components and Technology Conference, USA, pp. 1241–1246 (1998)
48. Iwamoto, N.: Applying polymer process studies using molecular modeling. In: 4th International Conference on Adhesive Joining and Coating Technology in Electronics Manufacturing, pp. 182–187 (2000)
49. Iwamoto, N.: Advancing materials using interfacial process and reliability simulations on the molecular level. In: International Symposium on Advanced Packaging Materials: Processes, Properties and Interfaces, pp. 14–17 (2000)
50. Mustoe, G.G.W., Nakagawa, M., Lin, X., Iwamoto, N.: Simulation of particle compaction for conductive adhesives using discrete element modeling. In: 49th Electronic Components and Technology Conference, pp. 353–359 (1999)
51. Su, Y.Y., Shemenski, R.M.: The role of oxide structure on copper wire to the rubber adhesion. Appl. Surf. Sci. **161**(3–4), 355–364 (2000)
52. Chan, E.K.L., Fan, H., Yuen, M.M.F.: Effect of interfacial adhesion of copper/epoxy under different moisture level. In: 7th International Conference on Thermal, Mechanical and Multiphysics Simulation and Experiments in Micro-Electronics and Micro-Systems. EuroSimE (2006)
53. Fan, H.B., Chan, E.K.L., Wong, C.K.Y., Yuen, M.M.F.: Molecular dynamics simulation of thermal cycling test in electronic packaging. J. Electron. Packag. **129**(1), 35–40 (2007)
54. Wong, C.K.Y., Fan, H.B., Yuen, M.M.F.: Interfacial adhesion study for SAM induced cova- lent bonded copper-EMC interface by molecular dynamics simulation. IEEE Trans. Compon. Packag. Technol. **31**(2), 297–308 (2008)
55. Wong, C.K., Fan, H., Yuen, M.M.F.: Investigation of adhesion properties of Cu-EMC interface by molecular dynamic simulation. In: 6th International Conference on Thermo, Mechanical and Multiphysics Simulation and Experiments in Micro-Electronics and Micro-Systems. EuroSimE (2005)
56. Heino, P.: Simulations of nanoscale thermal conduction. Microsyst. Technol. Micro- Nanosyst. Inf. Storage Process. Syst. **15**(1), 75–81 (2009)
57. Schelling, P.K., Phillpot, S.R., Keblinski, P.: Comparison of atomic-level simulation methods for computing thermal conductivity. Phys. Rev. B. **65**(14), 144306 (2002)
58. MullerPlathe, F.: A simple nonequilibrium molecular dynamics method for calculating the thermal conductivity. J. Chem. Phys. **106**(14), 6082–6085 (1997)
59. Yang, P., Liao, N.B.: Research on characteristics of interfacial heat transport between two kinds of materials using a mixed MD-FE model. Appl. Phys. A Mater. Sci. Process. **92**(2), 329–335 (2008)
60. Bi, K.D., Chen, Y.F., Yang, J.K., Wang, Y.J., Chen, M.H.: Molecular dynamics simulation of thermal conductivity of single-wall carbon nanotubes. Phys. Lett. A. **350**(1–2), 150–153 (2006)
61. Che, J.W., Cagin, T., Goddard, W.A.: Thermal conductivity of carbon nanotubes. Nanotechnology. **11**(2), 65–69 (2000)
62. Fan, H.B., Zhang, K., Yuen, M.M.F.: Thermal performance of carbon nanotube-based com- posites investigated by molecular dynamics simulation. In: 57th Electronic Components and Technology Conference, Reno, NV, pp. 269–272 (2007)
63. Grujicic, M., Cao, G., Gersten, B.: Atomic-scale computations of the lattice contribution to thermal conductivity of single-walled carbon nanotubes. Mater. Sci. Eng. B Solid State Mater. Adv. Technol. **107** (2), 204–216 (2004)
64. Lopez, M.J., Rubio, A., Alonso, J.A.: Deformations and thermal stability of carbon nanotube ropes. IEEE Trans. Nanotechnol. **3**(2), 230–236 (2004)
65. Ma, P.C., Tang, B.Z., Kim, J.K.: Effect of CNT decoration with silver nanoparticles on electrical conductivity of CNT-polymer composites. Carbon. **46**(11), 1497–1505 (2008)
66. Maruyama, S.: A molecular dynamics simulation of heat conduction of a finite length single- walled carbon nanotube. Microscale Thermophys. Eng. **7**(1), 41–50 (2003)
67. Mingo, N., Broido, D.A.: Length dependence of carbon nanotube thermal conductivity and the "problem of long waves". Nano Lett. **5** (7), 1221–1225 (2005)
68. Ngo, Q., Cruden, B.A., Cassell, A.M., Sims, G., Meyyappan, M., Li, J., Yang, C.Y.: Thermal interface properties of Cu-filled vertically aligned carbon nanofiber arrays. Nano Lett. **4**(12), 2403–2407 (2004)
69. Padgett, C.W., Brenner, D.W.: Influence of chemisorption on the thermal conductivity of single-wall carbon nanotubes. Nano Lett. **4** (6), 1051–1053 (2004)
70. Shaikh, S., Lafdi, K., Silverman, E.: The effect of a CNT interface on the thermal resistance of contacting surfaces. Carbon. **45**(4), 695–703 (2007)
71. Shenogin, S., Bodapati, A., Xue, L., Ozisik, R., Keblinski, P.: Effect of chemical function- alization on thermal transport of carbon nanotube composites. Appl. Phys. Lett. **85**(12), 2229–2231 (2004)

72. Berber, S., Kwon, Y.K., Tomanek, D.: Unusually high thermal conductivity of carbon nanotubes. Phys. Rev. Lett. **84**(20), 4613–4616 (2000)
73. Saha, S.K., Shi, L.: Molecular dynamics simulation of thermal transport at a nanometer scale constriction in silicon. J. Appl. Phys. **101**(7), 074304-1–074304-7 (2007)
74. Starr, F.W., Schroder, T.B., Glotzer, S.C.: Molecular dynamics simulation of a polymer melt with a nanoscopic particle. Macromolecules. **35**(11), 4481–4492 (2002)
75. Sellers, M.S., Schultz, A.J., Kofke, D.A., Basaran, C.: Molecular dynamics modeling of grain boundary diffusion in Sn-Ag-Cu solder. In: AIChe Annual Meeting, Salt Lake City, Utah (2007)
76. Heino, P., Ristolainen, E.: Molecular dynamics study of thermally induced shear strain in nanoscale copper. IEEE Trans. Adv. Packag. **22**(3), 510–514 (1999)
77. Liu, H.H., Jiang, E.Y., Bai, H.L., Wu, P., Li, Z.Q., Sun, C.Q.: The kinetics and modes of gold nanowire breaking. J. Comput. Theor. Nanosci. **5**(7), 1450–1453 (2008)

25 Nano-scale and Atomistic-Scale Modeling of Advanced Materials

Ruo Li Dai, Wei-Hsin Liao, Chun-Te Lin, Kuo-Ning Chiang, and Shi-Wei Ricky Lee

25.1 Modeling of Carbon Nanotube Composites for Vibration Damping

Damping materials play an important role in vibration damping and control systems. Vibration can be reduced via damping materials by transferring the kinetic energy into heat. One of the widely used damping materials is viscoelastic material (VEM). In general, the VEM damping effects are significant for being used in structures due to the energy dissipation when the structures are vibrating. However, the stiffness of traditional viscoelastic materials is low; thus large shear strain could be caused while having significant energy dissipation. The softness property may constrain the VEMs from being used in the structures with high strength-to-weight ratio, where the strength and/or transmissibility issues are crucial. On the other hand, epoxy resin (which is also a VEM, but has a highly crosslinked 3D network) is much stiffer than the VEMs and could be used to enhance structural stiffness; but it dissipates little energy from the structure. Considering the above situations, it is desirable to develop alternative means for significant vibration damping, while the stiffness and strength of the material/structure are high. It could be possible to make it when composites made up of carbon nanotubes and epoxy resins are used.

People have expected that carbon nanotubes would bring us a revolution in scientific research since the multiwalled carbon nanotubes were discovered [1]. Carbon nanotubes (CNTs) are cylinder allotropes of carbon with diameter in the scale of nanometers. This results in a nano-structure where the length-to-diameter ratio might exceed 10,000. Carbon nanotubes are one of the strongest and stiffest materials known, in terms of tensile strength and elastic modulus, respectively.

A multiwalled carbon nanotube was tested to have a tensile strength of 63 GPa. For comparison, high-carbon steel has a tensile strength of approximately 1.2 GPa. CNTs have very high elastic moduli, on the order of 1 TPa [2]. Since carbon nanotubes have a low density of 1.3–1.4 g/cm^3, its specific strength of up to 48,462 kN·m/kg is the best of known materials, compared with high-carbon steel's 154 kN·m/kg.

It was recognized that CNTs might be the ultimate reinforcing materials for developing composites with excellent properties, such as exceptionally high stiff- ness, strength, and resilience, as well as superior electrical and thermal properties [3]. Researches have been conducted in the field of nano-structures enhanced polymer resin composites. Most of those researches focused on the strength, toughness, thermal conduction, and electrical conduction enhancement [4–7].

Recently, researchers have started to pay attention to the damping enhancement of the composites due to the CNT addition. The unique damping properties of CNT-based structures, such as densely packed CNT thin films [8–11] and polymeric composites with dispersed CNT fillers [12–16], have been of great interest. Multiwalled carbon nanotube (MWNT) thin films were fabricated using catalytic chemical vapor deposition (CVD) of xylene and ferrocene mixture precursor [10, 11]. The nanotube films were employed as inter-layers within the composite to reinforce the interfaces between composite plies and enhance laminate stiffness as well as structural damping. Zhou et al. [17] conducted experimental and analytical investigations on the energy dissipation of polymeric composites containing carbon nanotube fillers. Their specimens were fabricated by directly mixing single-walled carbon nanotubes into two types of polymers. The loss factors of the composites were characterized while the loading was applied in the normal direction. It was also reported that poor load transfer between nanotubes and surrounding polymer chains may detriment to high stiffness and strength but greatly enhance the structural damping

R. L. Dai (✉) · W.-H. Liao · C.-T. Lin · K.-N. Chiang · S.-W. R. Lee
Department of Mechanical and Automation Engineering, The Chinese University of Hong Kong, Shatin, NT, Hong Kong

C. P. Wong et al. (eds.), *Nano-Bio- Electronic, Photonic and MEMS Packaging*, https://doi.org/10.1007/978-3-030-49991-4_25

[14]. Dai and Liao [18] experimentally studied the damping effect caused by CNT addition. General procedures for fabricating CNT/epoxy composites were given as follows. A set of material testing equipment was constructed. The fabricated composite specimens were evaluated under different loading conditions by varying excitation frequencies and amplitudes. Storage modulus and loss factors were measured for both pure epoxy specimen and CNT composite specimens. In particular, CNT addition was randomly dispersed in the composite fabrication, and shear loading was applied during the damping measurement. Experimental results showed that the CNT additives greatly enhanced the damping capacity of pure epoxy resins. This fabricated CNT/epoxy composite has the same level of storage modulus as epoxy resin and a similar level of loss factor as compared to VEMs, while strain is over a certain level. It was concluded by Dai and Liao that for sufficiently high strain level, CNT composite could provide more damping than traditional damping materials. The observed phenomena matched the anticipation and the results were promising for potential applications when both damping and stiffness are desired.

Zhou et al. [17] first developed a "stick-slip" damping model for composites containing well-dispersed single-wall carbon nanotubes (SWNTs) and performed damping measurements on CNT-based composites. SWNTs were assumed to be in parallel bundles referred to as "nanotube ropes" or "nanoropes." In each bundle, the SWNTs were held together by relatively weak van der Waals forces. Based on this assumption, a micromechanical model for the composites with aligned nanoropes was developed. Analytical results showed strain-dependent damping enhancement due to "stick-slip" motion at the interfaces of the SWNTs and the resin. Zhou et al. [17] and Liu et al. [15, 16] presented an analysis on the structural damping characteristics of polymeric composites containing dilute, randomly oriented nanoropes. The SWNT rope was modeled as a closely packed lattice consisting of seven nanotubes in hexagonal array. The resin was described as a viscoelastic material using two models: Maxwell model and three-element standard solid model. The composite was modeled as a three-phase system consisting of a resin, a resin sheath acting as a shear transfer zone, and SWNT ropes. The analytical results indicated that the loss factor of the composite is sensitive to stress magnitude. It was illustrated that the "stick-slip" friction is the main contribution for the total loss factor of CNT-based composites even with a small amount of nanotubes/ropes. Odegard et al. [19] also gave an analytical model to simulate nanotube-reinforced polymer composites. Liu et al. [20] used the boundary element method (BEM) for analyzing carbon nanotube-based composites. It should be noted that the damping performance of CNT composites was not considered in by the last two research groups.

In this chapter, a finite element method (FEM)-based model for CNT composite is developed. A composite unit cell model is built. The FEM-based composite shows the detailed sliding surface status developed between CNT and base epoxy matrix, while none of previous models does. Under the shear loading, simulation is carried out to investigate interfacial sliding under different strain level and CNT orientation. Loss factors can be obtained from the ratio of energy dissipation out of energy input per cycle. The simulation results are compared with experimental data. Based on the FEM model, further parametric studies are performed. Several parameters such as CNT dimension and CNT alignment orientation are discussed with respect to composite damping capacity (loss factor).

25.2 Nanotube Composite Modeling

25.2.1 Modeling Platform and Assumptions

The finite element method (FEM) is used for modeling the CNT composites. The chosen platform is commercial FEM software ANSYS 10.0/Multiphysics. Three- dimensional solid element SOLID185 is used. SOLID185 is defined by eight nodes having 3° of freedom at each node: translations in *x*-, *y*-, and *z*-directions [21]. This element can be used in relatively large deformation cases, and there is no redundant degree of freedom that brings extra computational load.

It was assumed that the energy dissipation in CNT composite mostly comes from interfacial sliding friction between CNT and base matrix [17]. In addition, the base matrix (i.e., epoxy resin) itself also attributes to the composite damping, which is assumed to be constant with respect to strain and the loss factor of the base matrix equals to that of the pure epoxy resin. Several assumptions are made in order to model CNT composites:

(1) Damping consists of two parts: contribution from epoxy resin matrix and contribution from CNT/matrix interfacial sliding.
(2) CNT and base matrix are perfectly bonded at initial state before loading is applied. As the interfacial strain increases with applied shear loading, sliding occurs when the critical strain is reached.
(3) Interfacial friction transfers mechanical energy to thermal energy thus dissipates energy out of system.
(4) The CNTs in composite are assumed to be well aligned.
(5) CNT segments are well separated, and each single CNT segment is isolated within a small epoxy resin cubic, which is called "composite unit cell." The dimension of this unit cell is calculated from CNT volume ratio. This assumption would be acceptable when the damping

mainly comes from CNT/epoxy sliding but not CNT/CNT friction.

(6) CNTs in matrix are randomly oriented and uniformly distributed so that we can average the energy loss over all CNT segments.

Based on assumptions (4) and (5), the CNTs in composite are modeled as straight segments isolated in composite unit cells. It has been reported that while ultrasonic process is used for dispersing the CNTs into the epoxy resin, the sonication energy breaks the CNTs into shorter segments. The ultrasonic process will decrease CNT aspect ratio, while the dispersibility is increased [22]. For CNT composite fabricated in this research, it can be observed from a SEM image given in Fig. 25.1 that the CNT dispersion quality is good in general. The aspect ratio of CNTs is also found to be decreased. Under such condition, the model with straight CNT segments isolated in composite unit cells seems acceptable.

25.2.2 Finite Element Model

The composite unit cell (cubic shaped) containing a single CNT segment is modeled in 3D as shown in Fig. 25.2. Extra blocks are mounted on the top and bottom of this unit cell in the model in order to facilitate applied loading. The dimension of this composite unit cell is obtained from CNT weight ratio. Higher CNT weight ratio gives smaller CNT composite unit cell. Detailed dimension and properties are given in Table 25.1. The CNT segment is modeled as a hollow tube with rounded ends and a built-in thin sheath tube in order to simulate the contact/bonding surface between CNT and matrix (see Fig. 25.3). This sheath layer plays a very important role in damping simulation. Some of the overstressed elements in this layer will be deleted in order to simulate the interfacial sliding mechanism.

For finite element modeling, hexahedron elements are preferred over tetrahedron elements because fewer well-formed hexahedron elements are required than tetrahedron elements for the same accuracy. In another word, the size of the model and the computational effort could be reduced by using hexahedron elements. However, the mesh generation may be more complicated or even impossible for some of the topologies [21]. In this model, an all-hexahedron mesh is generated for the CNT segment, the sheath layer, and the blocks, which are fixed on the top and bottom sides of the composite unit cell. Tetrahedron and prism-shaped elements are used for meshing the epoxy part of the unit cell. This treatment greatly reduces the calculation time as compared with an all-tetrahedron meshed model. Figure 25.4 shows the comparison between different meshing strategies. The model shown in the top two figures is generated by an all-tetrahedron meshing, which contains 81,531 elements, while the model shown in the bottom two is generated by all-hexahedron meshing, which has 56,968 elements. In addition, the mesh quality is improved by using all-hexa mesh as the element shape is more regular and the mesh density in the

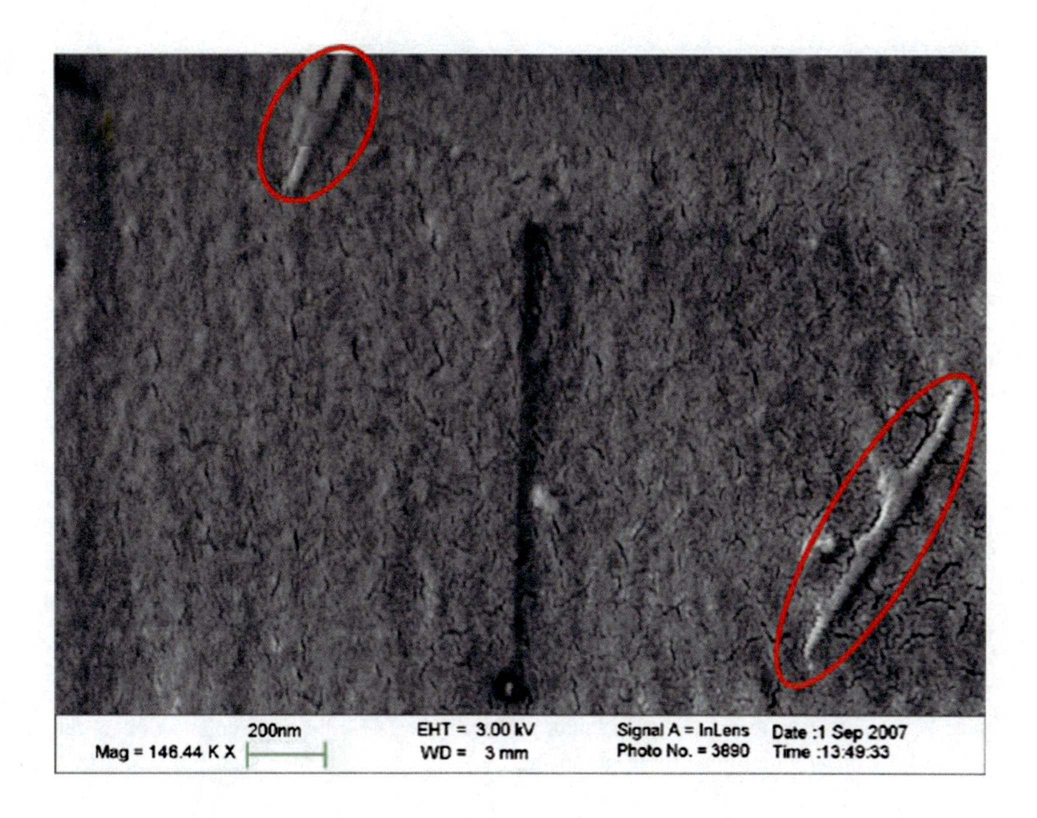

Fig. 25.1 CNT segments (*circled*) in matrix

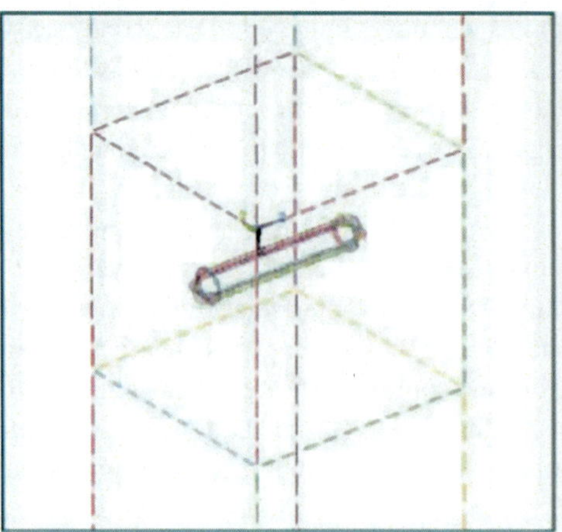
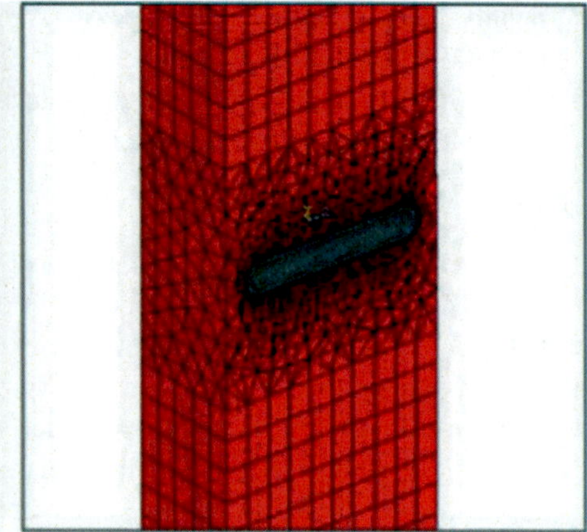

Fig. 25.2 Finite element model of composite unit cell

Table 25.1 Model parameters

CNT segment		Epoxy resin unit cell		Sheath layer	
Length	100 nm	Cubic edge length	300 nm	Thickness	2 nm
Outer diameter	20 nm	Young's modulus	3 GPa	Young's modulus	3 GPa
Thickness	4 nm	Poisson's ratio	0.37	Poisson's ratio	0.37
Young's modulus	1000 GPa	Loading	Shear	Critical sliding strain	0.0016
Poisson's ratio	0.3				

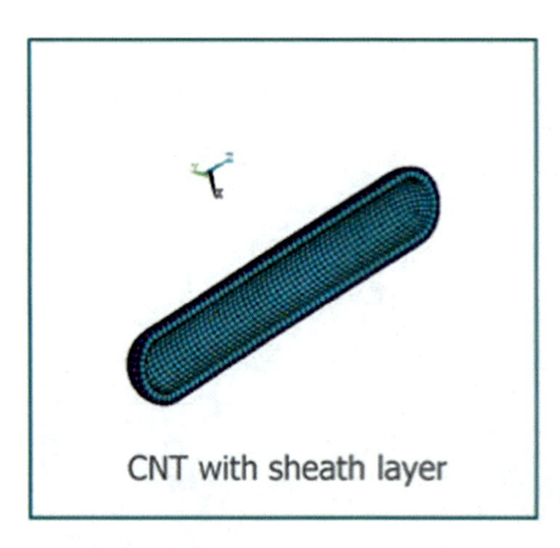

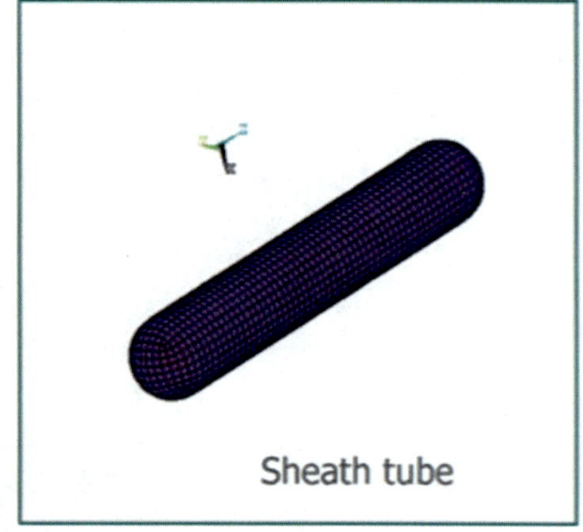

Fig. 25.3 CNT model with sheath layer

epoxy part is even finer with fewer elements. As a result, it takes 4.5 h using the all-tetra meshed model for obtaining a solution while the model with all-hexa mesh runs a little bit longer than 3 h (the results of the two models are close to each other). The solver convergence of hexahedral versus tetrahedral meshes has been studied. Cifuentes et al. [23] concluded that the results obtained with quadratic tetrahedral elements, compared to bilinear hexahedral elements, were equivalent in terms of both accuracy and CPU time. Bussler et al. [24] reported that better accuracy can be achieved by using the same order hexahedral elements over tetrahedrons. Therefore, the model containing less tetrahedron elements would have better solver convergence and accuracy when material nonlinearity or large deformation is significant.

The Young's modulus and Poisson's ratio of the sheath layer elements are set to be the same as the regular epoxy resin but a failure strain (0.0016) is set in order to simulate the interfacial sliding mechanism. The value of this critical sliding strain can be determined from experiments and numerical analysis, which will be discussed in the next section.

The flowchart for calculating composite loss factor is given in Fig. 25.5. The model is first constructed with boundary conditions and initial loading applied. The model is then solved by using ANSYS^{QR} implicit sparse solver and DYNA^{QR} explicit solver for highly nonlinear case such as large displacement. Observing the results, we can pick up the sheath elements over the critical sliding strain. These elements are deleted from system in order to simulate the

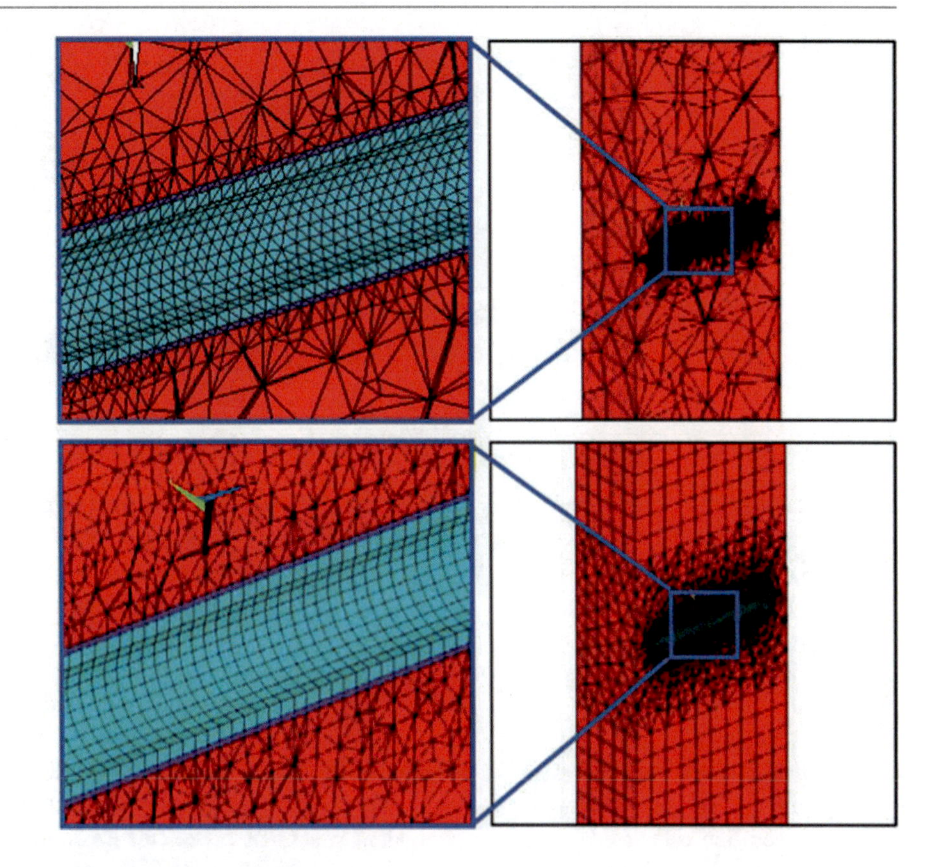

Fig. 25.4 Comparison between different meshing strategies (*Top*: all-tetra mesh, *Bottom*: all-hexa mesh except the epoxy part of composite unit cell)

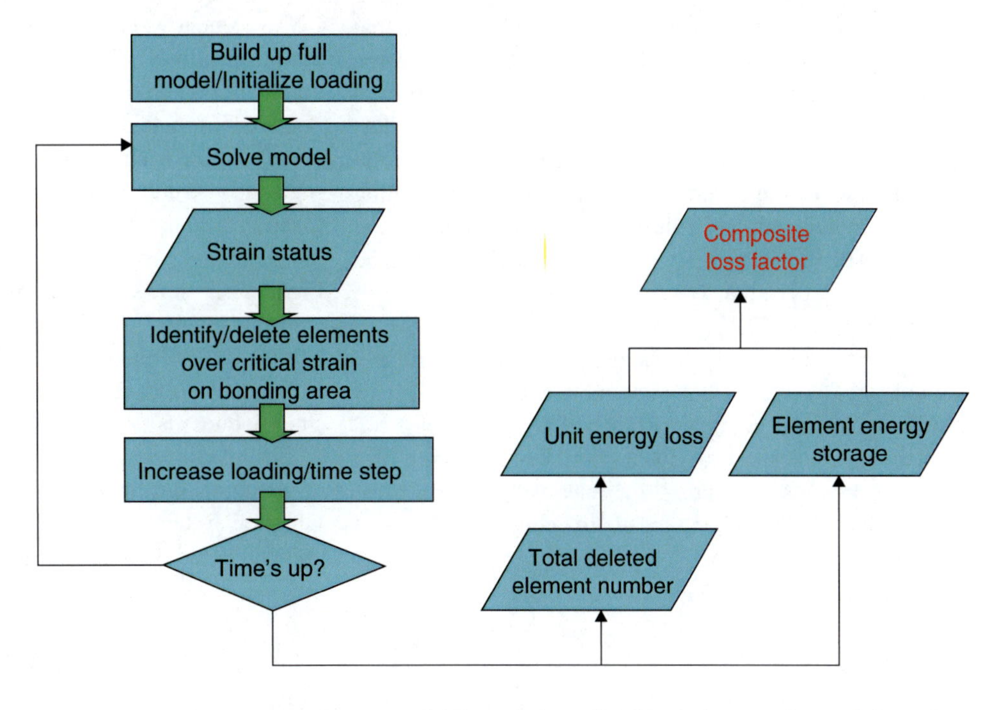

Fig. 25.5 Flow chart for calculating composite loss factor

debonding between nanotube segment and base epoxy matrix. Then, as clock steps up, another set of loading is applied, the model is solved again and more elements are deleted in the new round of calculation. After one period of cyclic loading applied, the total number of deleted elements is counted. The number of deleted elements is used for calculating energy loss. The total element energy storage (element strain energy) during this time interval is also calculated from the model. By using these two energies, we can calculate composite loss factor. More specifically, from simulated

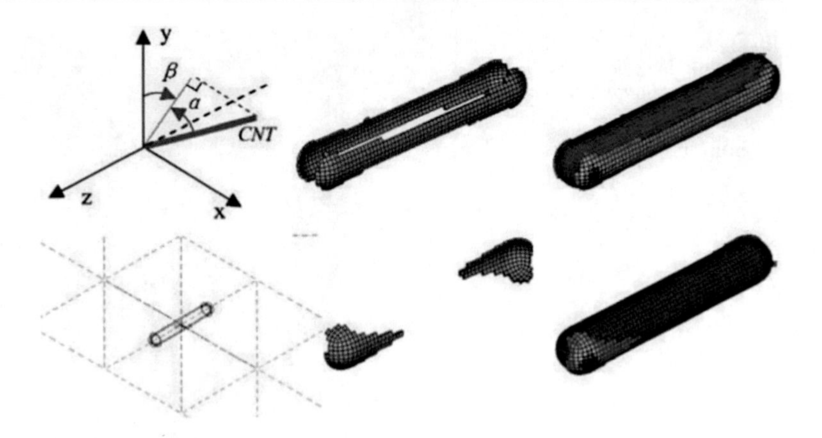

Fig. 25.6 Deleted sheath elements during CNT/epoxy sliding

strain/stress distribution, the stored energy (energy input) of a composite unit cell per loading cycle, *U*cell can be calculated as follows:

$$U_{\text{cell}} = \frac{1}{2}\sum_{n} \sigma_{\text{cell}}\varepsilon_{\text{cell}} V \tag{25.1}$$

where *V*, *n*, σ cell, and εcell represent the volume of element in composite unit cell, the total number of elements in composite unit cell, the calculated von Mises equivalent stress, and strain, respectively [21].

As CNT is much stiffer than base matrix, it is assumed that the strain near CNT/matrix interface is larger than the average strain in composite unit cell. Once the critical sliding strain is determined, the sheath elements over this critical strain during each loading cycle are detected and removed from the model in order to simulate the debonding process and sliding effect. As mentioned, the critical strain in this simulation is chosen as 0.0016; further discussions on critical strain will be given in the next section. The number of deleted sheath tube elements is counted after each loading cycle in order to calculate the cyclic energy dissipation. The zenith angle and azimuth angle of the CNT segment are denoted by two parameters α and β as shown in Fig. 25.6 to describe CNT orientation with respect to global coordinate system. After the solver finishes computing, the energy loss during one cycle is calculated by adding the strain energy of all deleted elements

$$U_{\text{loss}} = \frac{1}{2}\sum_{m} \sigma_{\text{cr}}\varepsilon_{\text{cr}} v \tag{25.2}$$

where *v*, *m*, σ cr, and εcr are volume of sheath element, number of deleted sheath elements, critical sliding stress, and critical sliding strain, respectively.

25.2.3 Simulation Results

In this simulation, the interfacial sliding between CNT segments and base epoxy resin is modeled. The model shows when and where the sliding occurs on CNT/matrix interface and how the sliding develops. These conditions were not considered by previous researches.

Figure 25.6 shows interfacial sliding examples between CNT segment and base matrix in two different CNT orientations. The upper two show the sliding status while CNT segment is perpendicular to loading surface ($\alpha = 90°$, $\beta = 90°$). The lower two illustrate the sliding status, while CNT segment's zenith angle and azimuth angle are 30 and 90°, respectively. The sheath layer elements shown in the middle of Fig. 25.6 are the ones over critical strain, which are removed from the model in order to simulate the sliding effect during each loading cycle. More studies on CNT segment's zenith and azimuth alignment angles will be given in the next section. Other than CNT orientation, interfacial sliding area is also related to the applied strain level. Figure 25.7 shows CNT segment sliding status under different strain levels, i.e., 0.0001, 0.0002, 0.0003, 0.0005, 0.0006, 0.0007, 0.0008, and 0.0012, respectively, from (a) to (h). The sliding area is developed along with strain level – higher strain induces larger sliding area.

The loss factor is the ratio of energy dissipated as the result of damping during one cycle to the total energy of the vibration system under cyclic loading. The loss factor of the unit cell could be obtained as follows:

$$\eta = \frac{U_{\text{loss}}}{U_{\text{cell}}} = \frac{\frac{1}{2}\sum_{m}\sigma_{\text{cr}}\varepsilon_{\text{cr}} v}{\frac{1}{2}\sum_{n}\sigma_{\text{cell}}\varepsilon_{\text{cell}} V} = \frac{\sum_{m} E_{\text{expoxy}}\varepsilon_{\text{cr}}^{2} v}{\sum_{n} E_{\text{cell}}\varepsilon_{\text{cell}}^{2} V} \tag{25.3}$$

Shear loading is applied on the composite unit cell. CNT orientation parameters α and β range from 0 to 90° while the step size is 7.5°, which leads to $13 \times 13 = 169$ different cases built in the first octant for each loading level. The angular

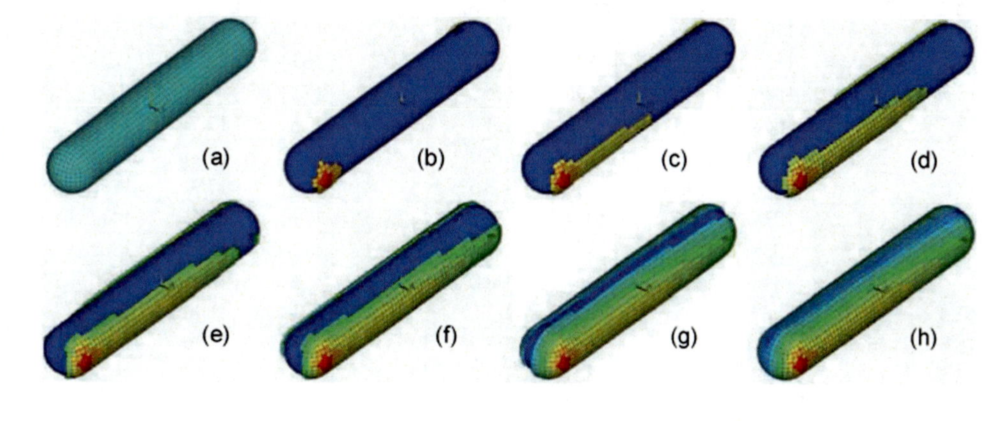

Fig. 25.7 CNT/epoxy interfacial sliding status under different strain level ($\alpha = 7.5°$, $\beta = 90°$, εcell = [0.0001, 0.0012]). (**a**) 0.0001, (**b**) 0.0002, (**c**) 0.0003, (**d**) 0.0005, (**e**) 0.0006, (**f**) 0.0007, (**g**) 0.0008, and (**h**) 0.0012

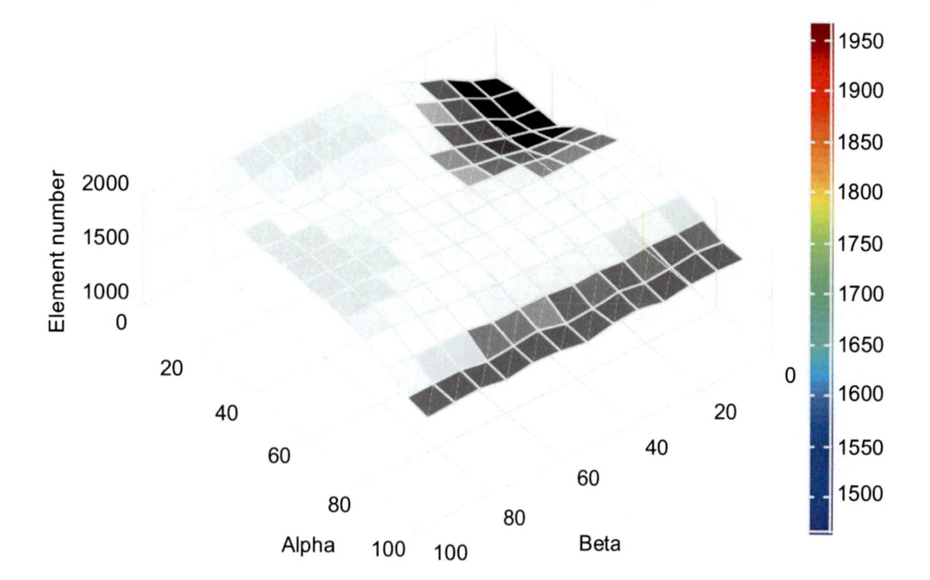

Fig. 25.8 Number of deleted sheath elements versus CNT orientation while εcell = 0.0007

range covers one-eighth of the 3D coordinate system so that it is sufficient for our study due to geometric symmetry. The number of deleted sheath elements is recorded in terms of α and β under different strain level. In this model, the total number of elements in the sheath layer $n$max = 1973, which corresponds to the maximum number of elements involved in interfacial sliding. In other words, at most 1973 elements could be deleted during sliding to result in the largest possible energy dissipation. In general, higher strain level leads to larger sliding area that induces more energy dissipation; different CNT alignment causes different sliding area size and location under the same loading. A 3D graph is given in Fig. 25.8 to show the number of deleted sheath elements versus CNT orientation when εcell = 0.0007. The related data are given in Table 25.2.

For each loading level, based on simulation results for 169 cases (represent 13 × 13 different CNT orientations in the first octant), the average number of sliding sheath elements is calculated. The average energy loss of each composite unit cell can be further obtained. The average number of sliding elements, number of maximum sliding elements, number of minimum sliding elements, and standard deviation of sliding number (w.r.t. CNT segment orientation) associated with strain range [0.0002, 0.001] are given in Table 25.3. The loss factor of the CNT composite thus can be calculated. The calculated loss factor of the CNT composite is compared to the experimental data as shown in Fig. 25.9. We can see from this figure that the calculated and measured loss factors of the CNT composite are in the similar trend and level with respect to strain.

In this simulation, the CNT/matrix interfacial critical sliding strain is set to 0.0016. The critical strain is determined from experimental data. However, from available experimental data and other references [2, 12, 16], the critical strain can only be estimated. A range of critical strain is varied to find out which one gives the loss factor curve most fit to experimental data. Simulation results for different critical strain values are shown in Fig. 25.10, where critical strain value is set to 0.0012, 0.0014, 0.0016, and 0.0018. We can see from this figure that the simulation with εc = 0.0016 fits experimental data the best, and thus this value is used for further simulation in this research.

Table 25.2 Number of deleted sheath elements versus CNT orientation while εcell = 0.0007

$\beta\alpha$	0	7.5	15	22.5	30	37.5	45	52.5	60	67.5	75	82.5	90
0	1453	1407	1382	1365	1452	1632	1726	1753	1747	1705	1645	1639	1572
7.5	1445	1423	1406	1394	1505	1600	1694	1746	1721	1700	1642	1604	1604
15	1578	1540	1529	1508	1552	1612	1713	1754	1745	1717	1641	1610	1607
22.5	1707	1668	1616	1579	1588	1650	1718	1736	1718	1707	1643	1623	1573
30	1812	1797	1706	1673	1659	1701	1731	1760	1736	1734	1632	1581	1589
37.5	1938	1898	1818	1750	1716	1729	1747	1750	1750	1721	1648	1612	1574
45	1945	1935	1874	1811	1745	1754	1784	1780	1777	1748	1644	1598	1555
52.5	1953	1938	1899	1827	1777	1775	1810	1810	1805	1760	1671	1573	1566
60	1929	1896	1845	1851	1840	1840	1849	1849	1835	1773	1659	1602	1583
67.5	1887	1783	1796	1852	1882	1924	1911	1896	1870	1813	1661	1582	1566
75	1639	1761	1783	1856	1903	1941	1955	1936	1901	1830	1711	1631	1610
82.5	1622	1637	1776	1887	1955	1961	1957	1957	1926	1844	1689	1617	1609
90	1617	1627	1690	1965	1963	1966	1963	1957	1929	1850	1672	1601	1566

Table 25.3 Number of deleted sheath elements under different strain level

Strain	0.0002	0.0003	0.0004	0.0005	0.0006	0.0007	0.0008	0.0009	0.001
Average	92	441	957	1333	1586	1725	1810	1911	1961
Max	352	746	1174	1755	1945	1966	1973	1973	1973
Min	0	157	707	1033	1220	1365	1463	1577	1694
Std. dev.	95	167	136	159	155	143	123	84	55
Loss factor	0.095	0.141	0.159	0.149	0.133	0.117	0.104	0.096	0.089

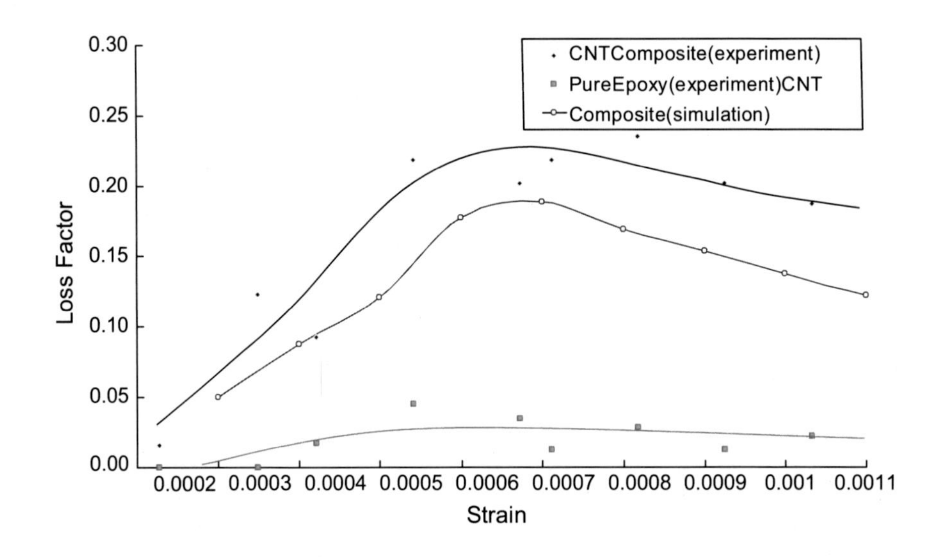

Fig. 25.9 Comparison of loss factor between simulation and experiment

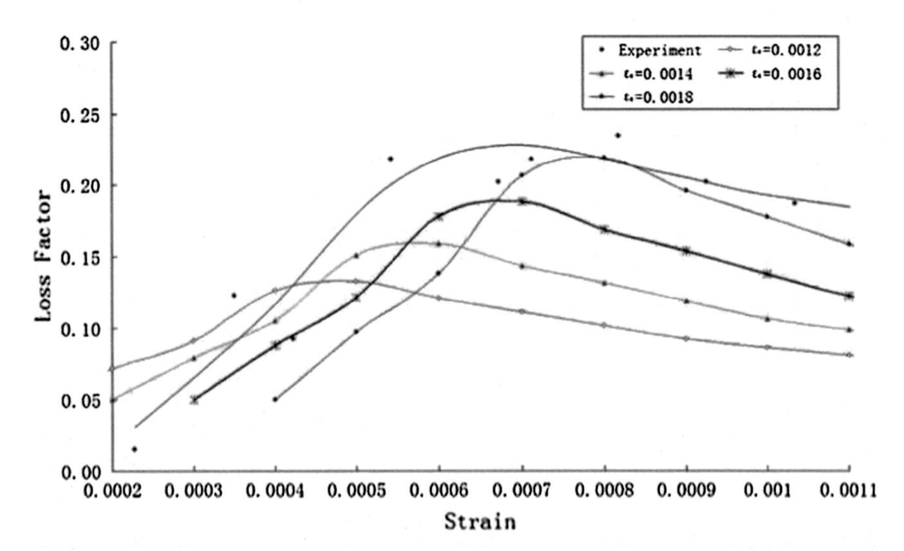

Fig. 25.10 Simulated composite loss factors with different critical strains

Figure 25.9 shows that simulated loss factor is smaller than the one measured from experiment for a given strain. It should be noted that only the contributions from epoxy resin matrix and CNT/matrix interfacial sliding were considered when modeling damping, as stated in the first assumption. Some other mechanisms could also dissipate energy. CNT/CNT stick-slip motion could be one of these mechanisms while energy dissipation from bubbles/cavities inside epoxy matrix might be another one. These mechanisms could be further considered for more accurate models.

25.2.4 Parametric Study on CNT Composites

The validated model could be used for predicting CNT composite dynamic properties and guiding composite fabrication. Several factors could affect the energy dissipation ability of the CNT composite, such as the dimensions of CNTs and the alignment orientations of CNTs inside matrix. These key factors will be studied in order to provide some references for material selection and composite fabrication.

There are several commercially available carbon nanotubes. A brief summary on geometrical properties of some major MWNTs is given in Table 25.4. As shown in the table, typical diameter of commercially available MWNTs ranges from 5 nm to 50 nm while the length ranges from 200 nm to 20 μm. For the studies in this research, CNT diameter range [10, 50 nm] and length range [100, 600 nm] are set as they cover most commonly used MWNTs. Another reason for the chosen CNT length range is that an ultrasonic process was performed during CNT dispersion. High- energy sonication could break CNTs down to small segments, thus the selection of the length range [100, 600 nm] would be reasonable.

Table 25.4 Geometrical properties of commercial MWNTs

Supplier	Type	Length	Diameter
NanoLab	MWNT 1	1–5 μm	15 ± 5 nm
	MWNT 2	5–20 μm	15 ± 5 nm
	MWNT 3	1–5 μm	30–15 nm
	MWNT 4	5–20 μm	30–15 nm
	MWNT 1	200 nm	6.5 nm
Rosseter	MWNT 2	300 nm	12 nm
	MWNT 3	500 nm	20 nm
SunNano	MWNT	400 nm–1 μm	10–30 nm
n-Tec	MWNT	2 μm	2–50 nm

CNT models with different CNT length while keeping the same CNT diameter are built. Loss factor versus strain with respect to different CNT length are shown in Fig. 25.11. Corresponding data are given in Table 25.5. These simulation results are obtained based on the same epoxy matrix and CNT weight ratio while assuming randomly oriented dispersion of CNTs in composite. It is observed that, given the same CNT diameter, CNT addition with shorter segment length induces slightly higher composite damping while strain is over certain level (around 0.00065 for the case shown in Fig. 25.11) and longer CNTs bring out higher damping before this strain level. However, composite loss factor curves with different CNT length are close to each other in general. In other words, composite loss factor is not sensitive to CNT length. As it is hard to keep CNTs with high aspect ratio than instead of low one, this observation would keep the cost of composite fabrication relatively low.

Another study is performed with respect to CNT diameter in range [10, 50 nm]. Other geometrical parameters including CNT segment length, diameter, and the dimension of composite unit cell are kept the same. Loss factor versus strain with different CNT diameter are shown in Fig. 25.12 and corresponding data are given in Table 25.6. It can be seen that the effect of CNT diameter on the composite loss factor is significant. More specifically, larger CNT diameter gives more energy dissipation for a single composite unit cell as increasing CNT diameter enlarges the surface area of the

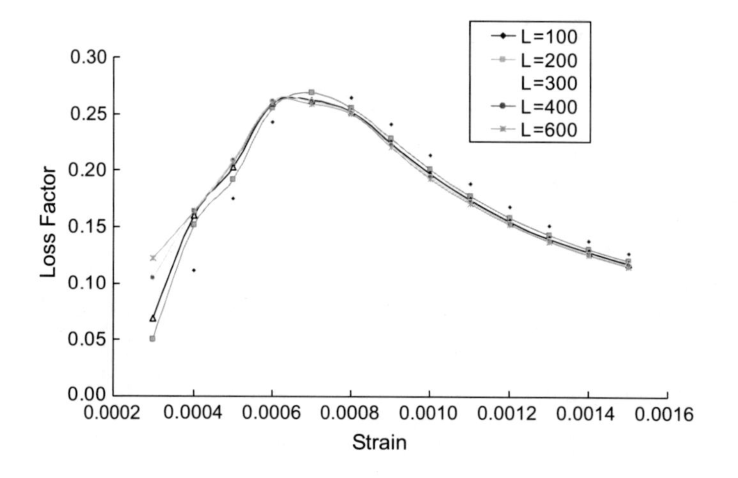

Fig. 25.11 Loss factor versus strain with respect to different CNT length (CNT diameter $D = 20$ nm; L represents CNT segment length in nanometer)

Table 25.5 Calculated loss factors with different CNT length (CNT diameter = 20 nm)

Global strain	Loss factor (L100)	Del. elem. (L100)	Loss factor (L200)	Del. elem. (L200)	Loss factor (L300)	Del. elem. (L300)	Loss factor (L400)	Del. elem. (L400)	Loss factor (L600)	Del. elem. (L600)
0.0001	0.050	0	0.050	0	0.069	2	0.105	8	0.122	16
0.0002	0.112	9	0.152	31	0.160	51	0.165	71	0.163	105
0.0003	0.175	44	0.192	102	0.203	167	0.209	233	0.207	345
0.0004	0.243	131	0.256	284	0.260	436	0.262	589	0.260	874
0.0005	0.270	241	0.269	480	0.262	693	0.261	918	0.259	1360
0.0006	0.265	338	0.256	640	0.252	939	0.251	1243	0.250	1855
0.0007	0.241	397	0.229	731	0.224	1064	0.223	1401	0.221	2078
0.0008	0.213	428	0.201	780	0.197	1130	0.195	1484	0.193	2196
0.0009	0.188	446	0.178	814	0.174	1179	0.172	1546	0.171	2286
0.0010	0.168	460	0.159	835	0.155	1208	0.154	1584	0.152	2341
0.0011	0.152	469	0.143	851	0.140	1231	0.139	1612	0.137	2383
0.0012	0.138	476	0.131	865	0.128	1251	0.127	1641	0.126	2425
0.0013	0.127	483	0.120	877	0.118	1270	0.117	1666	0.116	2458
0.0014	0.118	488	0.112	886	0.110	1284	0.109	1683	0.108	2478
0.0015	0.110	491	0.105	892	0.103	1291	0.102	1690	0.101	2488

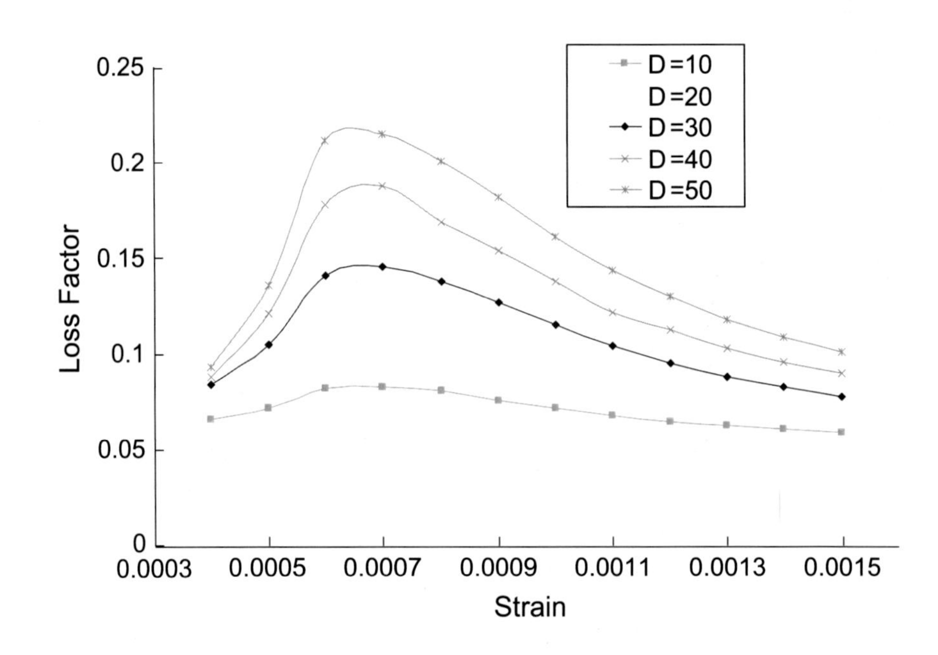

Fig. 25.12 Loss factor versus strain with respect to different CNT diameters (D represents CNT diameter and CNT segment length $L = 200$ nm)

Table 25.6 Calculated loss factors with different CNT diameter (CNT length = 200 nm)

Global strain	Loss factor (D10)	Del. elem. (D10)	Loss factor (D20)	Del. elem. (D20)	Loss factor (D30)	Del. elem. (D30)	Loss factor (D40)	Del. elem. (D40)	Loss factor (D50)	Del. elem. (D50)
0.0004	0.066	18	0.077	31	0.084	39	0.088	43	0.093	49
0.0005	0.072	56	0.090	102	0.105	143	0.121	189	0.136	233
0.0006	0.082	146	0.111	284	0.141	437	0.178	643	0.212	845
0.0007	0.083	235	0.115	480	0.146	726	0.188	1100	0.215	1358
0.0008	0.081	321	0.111	640	0.138	951	0.169	1333	0.201	1755
0.0009	0.076	365	0.102	731	0.127	1123	0.154	1555	0.182	2041
0.0010	0.072	399	0.093	780	0.115	1226	0.138	1688	0.161	2200
0.0011	0.068	412	0.085	814	0.104	1275	0.122	1733	0.144	2303
0.0012	0.065	420	0.080	835	0.095	1305	0.113	1830	0.130	2372
0.0013	0.063	422	0.075	851	0.088	1328	0.103	1866	0.118	2421
0.0014	0.061	423	0.071	865	0.083	1345	0.096	1889	0.109	2459
0.0015	0.059	424	0.069	877	0.078	1357	0.090	1906	0.101	2488

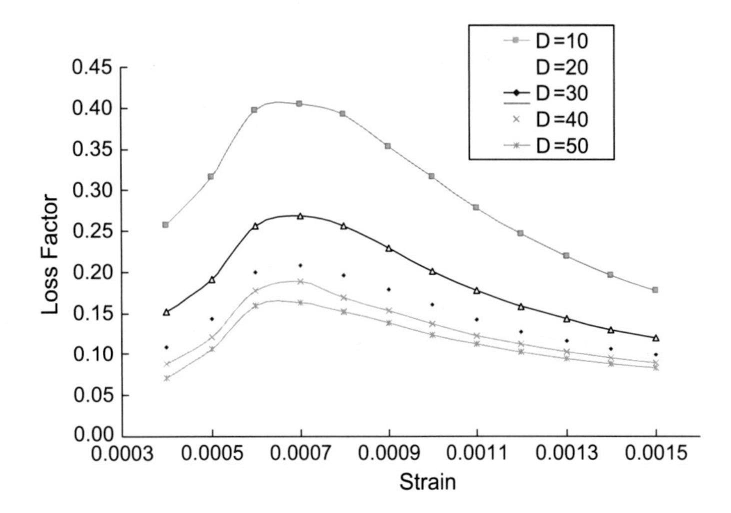

Fig. 25.13 Loss factor versus strain with respect to different CNT diameters (*D* represents CNT diameter and CNT segment length $L = 200$ nm, CNT weight ratio is fixed for each case)

Table 25.7 Calculated loss factors with different CNT diameter, fixed CNT weight ratio (CNT length = 200 nm)

Global strain	Loss factor (D10)	Del. elem. (D10)	Loss factor (D20)	Del. elem. (D20)	Loss factor (D30)	Del. elem. (D30)	Loss factor (D40)	Del. elem. (D40)	Loss factor (D50)	Del. elem. (D50)
0.0004	0.258	18	0.152	31	0.109	39	0.088	43	0.072	49
0.0005	0.316	56	0.192	102	0.143	143	0.121	189	0.107	233
0.0006	0.397	146	0.256	284	0.200	437	0.178	643	0.160	845
0.0007	0.404	235	0.269	480	0.208	726	0.188	1100	0.163	1358
0.0008	0.392	321	0.256	640	0.196	951	0.169	1333	0.152	1755
0.0009	0.353	365	0.229	731	0.179	1123	0.154	1555	0.139	2041
0.0010	0.317	399	0.201	780	0.160	1226	0.138	1688	0.124	2200
0.0011	0.279	412	0.178	814	0.143	1275	0.122	1733	0.112	2303
0.0012	0.247	420	0.159	835	0.128	1305	0.113	1830	0.102	2372
0.0013	0.219	422	0.143	851	0.116	1328	0.103	1866	0.095	2421
0.0014	0.196	423	0.131	865	0.107	1345	0.096	1889	0.088	2459
0.0015	0.178	424	0.120	877	0.099	1357	0.090	1906	0.083	2488

CNT segment. Larger composite loss factor is thus induced when the number of CNT segments in composite is the same.

Additional analysis on the loss factor with different CNT diameters is performed, while CNT weight ratio is fixed at 2%. The results are shown in Fig. 25.13 and Table 25.7. It can be observed from the figure that thinner CNTs induce higher composite damping. The maximum composite loss factor with 10 nm CNT additives is 0.4, while the one with 50 nm CNT additives is 0.16. The reason is as follows. Although a smaller diameter gives smaller surface area for a single CNT segment, a larger total sliding surface is induced as the number of CNT segments in composite increases when CNT diameter decreases given the same CNT weight ratio. Therefore, if the weight ratio of CNT additives is the same, CNTs with smaller diameter can be chosen to obtain higher composite damping. If cost is not a big concern, we can even use SWNTs, which are with a much smaller diameter (only a few nanometers).

As mentioned, CNT/epoxy interfacial sliding status is related to CNT alignment orientation in matrix. A model is used here to study the relationship between CNT orientation and composite damping ability. Figure 25.14 shows simulation results regarding CNT/matrix interfacial sliding status in different CNT segment orientations, while CNT segments are aligned on *x*–*z* plane, swept from negative *z* toward positive *x* ($\alpha = [0°, 90°]$, $\beta = 90°$). As we can see from this figure, interfacial sliding occurs along the CNT longitudinal direction while $\alpha = 0°$. The sliding surface location moves to the both ends of the segment while CNT rotating around *y* axis with $\alpha = 52.5°$.

As observed in the simulation results, the effect of CNT alignment on the composite damping varies under different strain level. The sliding element numbers versus CNT orientation are illustrated in Fig. 25.15 for strain range [0, 0.0012]. As increasing the applied load, interesting phenomena are observed. The CNT/matrix interfacial sliding could be categorized into five different stages with respect to the average strain in composite unit cell.

1. For very low strain level (Fig. 25.15a), no sliding occurs between CNTs and matrix. At this stage, composite loss factor is small and corresponding to that of pure epoxy.

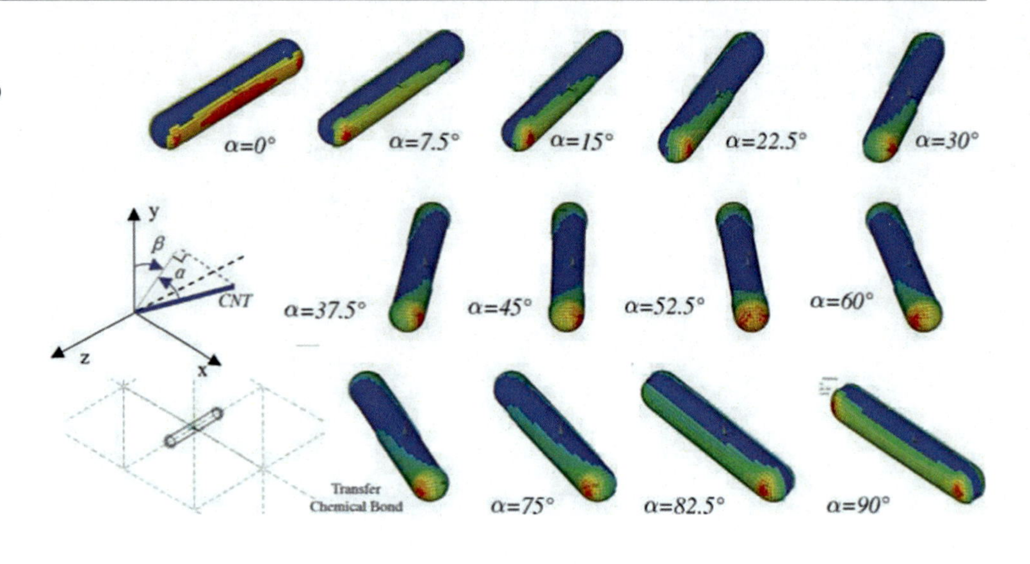

Fig. 25.14 CNT/epoxy interfacial sliding status by varying α ($\alpha = [0°, 90°]$, $\beta = 90°$)

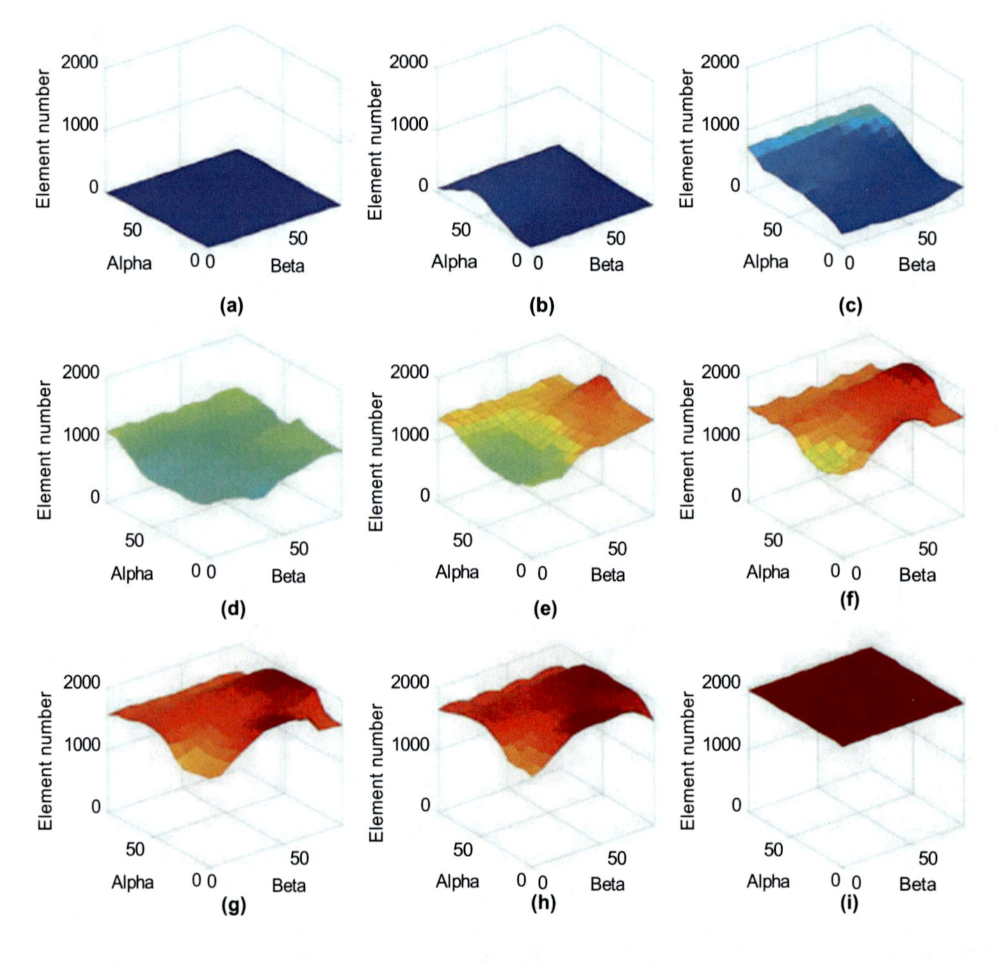

Fig. 25.15 Sliding element number versus CNT orientation (εcell = (**a**) 0, (**b**) 0.0002, (**c**) 0.0003, (**d**) 0.0004, (**e**) 0.0005, (**f**) 0.0006, (**g**) 0.0007, (**h**) 0.0008, and (**i**) 0.0012

2. For low strain level (Fig. 25.15b), sheath layer strain could become larger than the critical strain and interfacial sliding occurs only on CNT segments in some orientations. In this case, the orientation with $\alpha \approx 50°$ and $\beta = 0°$ (all CNT segments are in x–y plane) gives higher loss factor.
3. For medium strain level (Fig. 25.15c, d, and e), sliding occurs on all-oriented CNT segments. We can see from these figures that, under strain range εcell = [0.0003, 0.0005], CNTs in composite are partially debonded (the numbers of sliding sheath elements are less than 1973). Meanwhile, CNTs with orientation $\alpha = 90°$ dissipate relatively more energy than other orientations.
4. For high strain level (Fig. 25.15f, g, and h), sliding occurs on all oriented CNT segments. For some CNT

orientations, the CNT segment is fully debonded. It is observed from these figures that, for some CNT orientations ($\beta = 90°$, $\alpha \approx 50°$ and $\beta \approx 50°$, $\alpha = 90°$), more sheath layer elements are deleted to result in higher damping effect. It should be noted that, although not all the CNTs are fully debonded, composite loss factor could reach its maximum at this stage as observed in Fig. 25.9 (strain about 0.0007). This is because the relationship between interfacial sliding area and strain is not linear. The energy dissipation (which is related to sliding area) at this stage is already close to the maximum while the total energy input (which is related to strain) can still be further increased for even higher strain level (stage 5).

5. For very high strain level (Fig. 25.15i), all CNT segments are fully debonded, nearly the full sheath layer is deleted (over 1940 elements are deleted). The interfacial sliding is complete. In this case, regardless of the CNT alignment, the energy loss of composite unit cell reaches its upper bound and the composite gives maximum energy loss *U*loss. However, as total energy input *U*cell keeps going up as the strain level is further increased beyond the energy loss upper bound, the loss factor η calculated from Eq. (25.3) would drop.

Based on the aforementioned discussion, we can conclude that if the strain is above a certain level (0.0012 in this case), as all CNTs inside composite are thoroughly debonded, CNT alignment is not a concern as the composite already provides the maximum energy dissipation.

25.3 Modeling of Nano-scale Single Crystal Silicon with Atomistic–Continuum Mechanics

Single crystal silicon (SCS) is one of the most common structure materials used in microelectromechanical system (MEMS). Rapid development in the field of MEMS, particularly in the design of mechanical devices, requires an accurate value of the mechanical properties that include Young's modulus, Poisson's ratio, and tensile strength. Tests of MEMS materials have been carried out for evaluating their mechanical properties, such as tensile tests [25–29], bending tests [30–32]. However, the material properties have been estimated only on a micrometer to millimeter-scale structure because of difficulties in fabrication of a nano-scale test specimens and problems associated with measuring ultra-small physical phenomena in an experiment.

Although, the material testing techniques can provide much knowledge, they may not be adequate for the accurate determination of the material properties. Even if these testing methods were possible, the equipment cost would be significant in performing nano-scale tensile testing. An alternate method would be to investigate the nature of deformation and the material properties of nano-scale devices by simulation techniques. Recently, in the nano- or atomistic regime, the molecular dynamics (MD) simulation technique has been successfully applied to a variety of phenomena including solid fracture, surface friction, crack propagation, material testing [33], and machining [34]. The essence of the MD simulation method is the numerical solution of Newton's equation on motion for an ensemble of atoms. To render atomistic simulation studies practical, a classical or semi-classical potential function from which interatomic force can be derived is necessary. This is accomplished by using an appropriate empirical potential energy function that satisfies material properties criteria that include the lattice constant, compressibility, elastic constant, cohesive energy. The interatomic forces, involve attractive and repulsive, are defined by an appropriate empirical potential energy function, such as Morse potential for metal, Stillinger–Weber potential for single crystal silicon. The MD simulations can be likened to the dynamic response of numerous nonlinear spring-mass systems under an applied load, velocity, or displacement conditions. From this point of view, the mass represents the atoms (ions) linked one-with-other by spring elements to represent the interatomic bonds (interatomic force). However, MD simulations are limited to very small systems over very short times. Furthermore, in traditional continuum mechanics, the FEM is widely used to model and simulate the mechanical behaviors of solid/discrete body [49], it is a mature technology after decades of development. Compared with MD simulation, the continuum mechanics is considered very efficient and can provide results quickly within an accurate range. Moreover, many researchers successfully employed the traditional continuum theory that directly incorporates the inter-atomic potential or MD algorithmic into the constitutive model of the solid [35–39]. According to their results, the continuum mechanics theory is applicable on the nanometer scale.

Therefore, this investigation will adopt a hybrid method which combines FEM and the interatomic potential function to explore the mechanical properties of nano- scale single crystal silicon (SCS) under uniaxial tensile loading condition. In the present approach, the minimum potential energy criterion will be applied to individual atoms for obtaining the force–displacement relation of atoms, once the force–displacement curve is determined the analysis could follow the conventional FEM steps to complete the solution procedures. This novel approach is called the atomistic–continuum mechanics (ACM, [48]). Mechanical properties such as the Young's modulus, coefficient of thermal expansion (CTE), and Poisson's ratio are evaluated from the ACM simulation and compare with the experimental data or other simulation results in the literature.

25.4 Numerical Simulation Strategy of Atomic and Nano-structures

This section briefly reviews several numerical simulation strategies of the nano- scale materials and nano-structures. Similarly, the benefits and disadvantages of the simulation methods would be addressed. The ab initio method, quantum mechanics (QM), Monte Carlo (MC), molecular mechanics (MM), molecule dynamics (MD), and the ACM methods are included.

25.4.1 Ab Initio Method

The ab initio method is computational methodology that can be used without any experimental parameters. In general, the Schrödinger equation is used to calculate the quantum description of the atomic systems. Except for some special cases, one normally assumes the Bohr–Oppenheimer approximation in which the motions of nuclei are suppressed.

The time-dependent Schrödinger equation would be expressed as

$$\left(-\frac{1}{2}\nabla^2 - \sum_i\sum_a \frac{Z_a}{|r_i - R_a|} + \sum_{i,j}\frac{1}{|r_i - r_j|}\right) \times |\Psi\rangle = E|\Psi\rangle \tag{25.4}$$

where *ri* and *rj* are the positions of the electrons and *Ra* and *Za* are the positions and the atomic number of the nuclei, *E* is the energy, and Ψ is the wave function. The energy may be computed by the Schrödinger equation which in the time-dependent Bohr–Oppenheimer approximation is

$$H\Psi(r_1, r_2, \cdots, r_N) = E\Psi(r_1, r_2, \cdots, r_N) \tag{25.5}$$

The wave function Ψ depends on the position and spins of the *N* electrons. The Hamiltonian operator, *H*, consists of a sum of three terms: the kinetic energy, the interaction with the external potential (*V*ext), and the electron–electron interaction (*V*ee). Therefore, to compare Eqs. (25.4) and (25.5), the Hamiltonian operator could be expressed as

$$H = -\frac{1}{2}\nabla^2 - \sum_i\sum_a \frac{Z_a}{|r_i - R_a|} + \sum_{i,j}\frac{1}{|r_i - r_j|} \tag{25.6}$$

To resolve the issue, the Hartree–Fock theory is by far the most frequently quoted ab initio methods. This theory allows for the Pauli exclusion principle while lets each electron have its own orbital.

The ab initio method also exhibits important limitations. The major drawback of this method is its intense computational cost, which limits the modeled systems to within a certain length and time.

25.4.2 Monte Carlo Simulation

Related to molecular dynamics are Monte Carlo methods which randomly move to a new trajectory. If this conformation has a lower energy it is accepted, if not, an entirely new conformation is generated. This process is continued until a set of low-energy conformers has been generated in a certain number of times.

25.4.3 Molecular Mechanics

Molecular mechanics is a classical mechanical model that represents a molecule as a group of atoms held together by elastic bonds. Molecular mechanics looks at the bonds as springs, which can be stretched, compressed, and bent at the bond angles. Interactions between non-bonded atoms also are considered. The sum of all these forces is called the force field of the molecule. To make a molecular mechanics calculation, a force field is chosen, and suitable molecular structure values are constructed. Molecular mechanics is a valuable tool for predicting geometries of molecules.

25.4.4 Molecule Dynamics

The molecular dynamics is developed to study of the properties of atoms system with empirical interatomic potential functions. Thus, the interaction between two atoms is represented by a potential energy. Based on the electronic database or by alternatively using experimental measurements of specific properties, an approximate empirical potential function can be constructed. According to classic Newtonian mechanics, the dynamic evolution of all atoms can be determined by numerical integration. In principle, once the positions and velocities of atoms in the finite region within the simulation core are known, all thermodynamic properties can be readily extracted. A successful simulation depends on three major factors:

1. The computational implementation of the MD method.
2. The construction of accurate interatomic potential energy.
3. The analysis of massive data resulting from computer simulation.

On the other hand, how can MD use Newton's law to move atoms when everyone knows that the system at the atomistic level obeys quantum law rather than classic laws? A simple validation of the classic approximation is based on the de Broglie thermal wavelength [40], which is defined as

$$\Lambda\sqrt{\frac{h^2}{2\pi M k_B T}} \tag{25.7}$$

where h represents Plank's constant, M is the atomic mass, kB denotes the Boltzmann's constant, and T is the temperature. The classic approximation is justified if $\Lambda < a$, where a is the mean nearest neighbor separation of atoms. As such, the quantum effect must be considered to demonstrate the behaviors of nano-scale. For example, the de Broglie thermal wavelength of the silicon atom is the 0.19 Å whose value is smaller than the nearest neighbor separation of 2.53 Å. Therefore, classical mechanics is appropriate to study the motion of the Si atom system in this statement.

The following presents the critical concepts in the numerical implementation of molecular dynamics methods. The MD is used to describe the solution of classic equations of motion for a set of atoms. The microscopic coordinates *rj* and momentum *Pi* of the molecular computed over time may then be used to obtain thermodynamic and transport properties through statistical mechanics. It is assumed that the motion of atoms is governed by Newton's equations of motion of classic mechanics, as given by

$$\frac{dr_i}{dt} = \frac{P_i}{m_i} \tag{25.8}$$

$$\frac{dP_i}{dt} = F_i \tag{25.9}$$

where *mi* and *Fi* are the mass and force experienced by atom *i* as given by

$$F_i = -\frac{\partial V}{\partial r_i} = -\nabla V \tag{25.10}$$

where V is the potential function. Therefore, all the forces between the particles have been calculated using Eq. (25.10). The Newton's equation of motion can be integrated to calculate a new velocity and consequently a next position for each atom.

For a system of N atoms, Eqs. (25.8) and (25.9) represent 6 N first-order differential equations that require solving. These equations are always solved using finite different methods such as Verlet algorithms and predictor–corrector algorithms which are the two most common methods. The basic Verlet method is based on position $r(t)$, acceleration $a(t)$, and the position $r(t–\delta t)$ from the previous step. The equations for advancing the position, velocity, and acceleration are

$$r(t+\delta t) = 2r(t) - r(t-\delta t) + \delta t^2 a(t) \tag{25.11}$$

$$a(t) = -(1/m)\nabla(r(t)) \tag{25.12}$$

$$v(t) = \frac{r(t+\delta t) - r(t-\delta t)}{2\delta t} \tag{25.13}$$

where acceleration is substituted using Newton's equation of motion to give force divided by mass. The application of the Verlet algorithm (central difference) to MD simulation is completed in two steps. First, the force is calculated using position $r(t)$ and second, positions $r(t)$ and $r(t–\delta t)$, along with the force determined in step one, are used to calculate the position at $r(t + \delta t)$. These steps are repeated to integrate the motion over a particular time length. To ensure accuracy, the Leap-Forg algorithm (modified central difference) is used for a slight modification of the Verlet algorithm.

$$\begin{aligned} r(t+\delta t) &= r(t) + v\left(t+\frac{1}{2}\delta t\right)\delta t \\ v\left(t+\frac{1}{2}\delta t\right) &= v\left(t-\frac{1}{2}\delta t\right) + a\delta t \end{aligned} \tag{25.14}$$

Accordingly, a basic MD simulation is usually a period of about a few femtoseconds (fs) since the intermolecular or interatomic vibration.

Molecular dynamics is a very powerful technique but has limitations [40].

1. Classical description of atomic motion: According to the de Broglie thermal wavelength of Eq. (25.7), the classic approximation is justified if $\Lambda < a$, where a is the mean nearest neighbor separation of atoms. As such, the quantum effect must be considered to demonstrate the behaviors of nano-scale.
2. Realism of forces: In the molecular dynamics, atoms interact with each other. These interactions originate forces with act upon atoms, and atoms move under the action of these instantaneous forces. As the atoms move, their positions change and force changes as well. According to the abovementioned, forces are usually obtained as the gradient of a potential energy function,

depending on the positions of the particles. The realism of the simulation therefore depends on the ability of the potential chosen to reproduce the behavior of the material under the conditions at which the simulation is run.

3. Time and size limitation:
   - Time scale
     The maximum time step of integration in MD simulation is defined by the fastest motion in the system. Vibration frequencies in a molecular system are up to 100 THz which corresponds to a period of about 10 fs. Optical phonon frequencies are 10 THz, period of about 100 fs. Therefore, a typical time step in MD simulation is on the order of a femtosecond.
     Using modern computers it is possible to calculate 10^6–10^8 time steps. Therefore, it can only simulate processes that occur within 1–100 ns.
   - Size scale
     The size of the computational model is limited by the number of atoms that can be included in the simulation, typically 10^4–10^6 atoms. This corresponds to the size of the computational cell on the order of tens of nanometer.

25.4.5 Atomistic–Continuum Mechanics

ACM is a novel atomistic level numerical methodology based on the classic finite element theory. Figure 25.16 shows the schematic of the atomistic–continuum mechanics. In this figure, a simple model for a diamond crystal is that of an orderly array of spheres which represents the atoms linked with one another by elastic bonds (nonlinear spring elements) to represent the interatomic bonds (interatomic force). To move one atom aside from its equilibrium position entails stretching/compressing or attractive/repulsive behavior. If one atom is pulled to increase the distance between the atoms then the spring will exert an attractive force on the atom and try to attract it back to its equilibrium position. If the atoms are pushed together then the spring will exert a repulsive force on the atom and restore the atom to its equilibrium position.

25.5 Concept of Atomistic–Continuum Mechanics

Potential functions are developed to simplify the complexity of quantum mechanic- based computation such as ab initio calculations. These functions return a value of energy based on the conformation of the molecules or atoms. Some potentials such as Morse potential [35] and embedded-atom method (EAM [19, 50]) have been proposed to describe the potential energy of metal atoms, and they have been successfully applied to estimate the material properties of metals at room temperature in many articles [41–43]. For covalent bonds the most commonly used potential functions are the reactive empirical bond order (REBO) potential [44] for carbon and hydrocarbon, Cornell potential [45] for bio-structure, Stillinger–Weber potential and Tersoff potential [46] for silicon, germanium family. The Stillinger–Weber potential is selected as a benchmark for silicon model in this research.

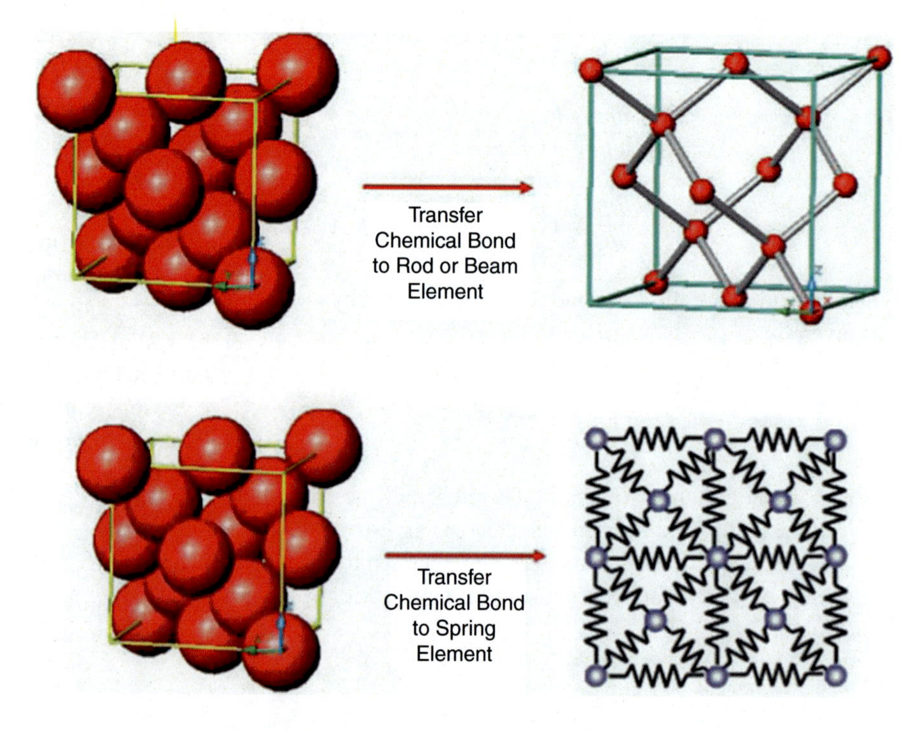

Fig. 25.16 Schematic of the atomistic–continuum mechanics (ACM)

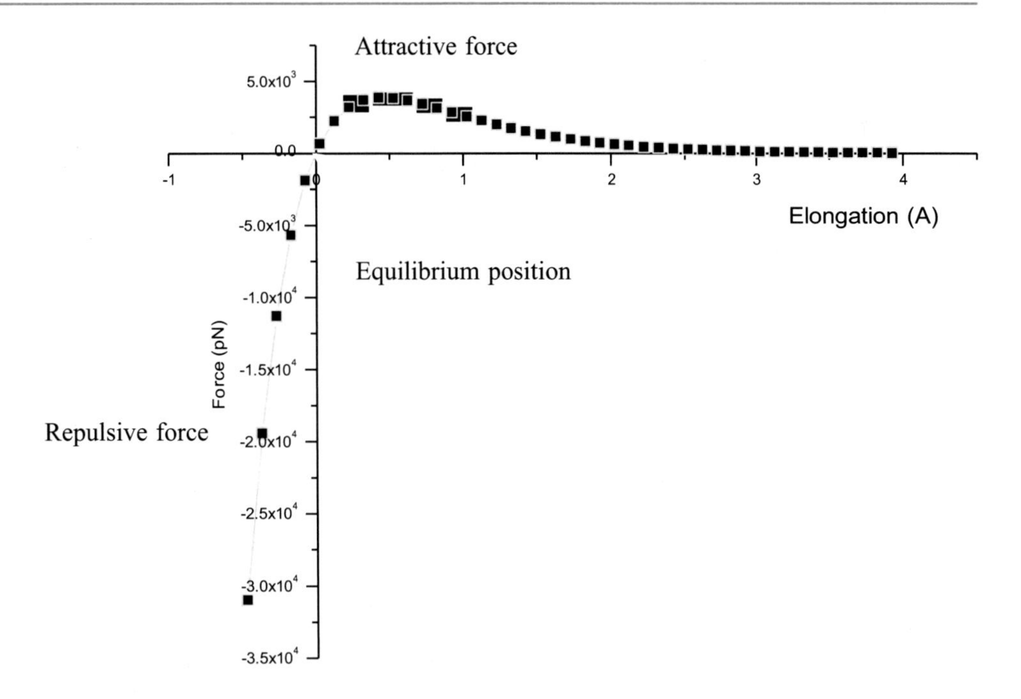

Fig. 25.17 The force–elongation curve (interatomic potential function for silicon)

In general the force–elongation (or so-called force–displacement) curve between two atoms in a crystal is considered to have the form shown in Fig. 25.17.

The minimization of the total potential energy is considered due to the ACM. Therefore, this section also introduces the principle of minimum potential energy.

The total potential energy of an elastic body $\prod$ is defined as

$$\prod = U - W \tag{25.15}$$

where U is the strain energy and W is the energy of the external loads, respectively. The strain energy of a linear elastic body is defined as

$$U = \frac{1}{2}\int_V \{\varepsilon\}^T [E] \{\varepsilon\} dV \tag{25.16}$$

The energy of the external loads can be stated as

$$W = \int_V \{u\}^T \{F\} dV + \{u\}^T \{P\} \tag{25.17}$$

where the $\{\varepsilon\}$ represents the strain filed, $[E]$ represents the material matrix, $\{u\}$ represents the displacement field, $\{F\}$ is the body force, $[D]$ represents the nodal of degree-of-freedom (D.O.F) of the structure, and $\{P\}$ is the load applied to D.O.F by external agencies.

Equations (16) and (25.17) which could be rewritten as the finite element form for a 3D analysis were given by

$$U = \frac{1}{2}\int_V [D]^T [k] [D] dV \tag{25.18}$$

$$W = \int_V [D]^T [N]^T \{F\} dV + [D]^T \{P\} \tag{25.19}$$

where $[k]$ is the element stiffness matrix ($[k] = \int_V [B^T][E][B]\, dV$), the $[B]$ is the strain–displacement matrix, and $[N]$ represents the shape function matrix. The minimization of total potential energy with respect to the nodal displacement requires that

$$\frac{\partial \prod}{\partial [D]} = 0 \tag{25.20}$$

According to Eq. (25.20), the overall structure can now be obtained as

$$[k][D] = \{R\} \tag{25.21}$$

where $[k] = \sum_{i=1}^{n} [k]_i$ and $[R] = \{P\} + \sum_{i=1}^{n} [N]_i \{F\}$. The required solution for the nodal displacement and reaction force can be obtained after solving Eq. (25.21).

Therefore, the atomistic–continuum mechanics is a methodology that the forces are usually obtained as the gradient of a potential energy function, depending on the positions of the particles. The realism of the simulation therefore depends on the ability of the potential chosen to reproduce the behavior of the material under the conditions at which the simulation is run.

Table 25.8 The characteristic comparison of many-particles system simulations

	Characteristic	Quantum mechanics included	Degree of freedom of particles	The max. Size of many particle systems
Ab initio (QM)	To solve the Schrödinger equation for many particle systems	Yes	High	30–100 atoms
MC	New particle configurations are created randomly step by step	No	Case by case	Case by case
MM	With potential functions, to solve static molecular configuration near equilibrium position	No	Medium	10^5–10^6 atoms
MD	Works through the solution of Newton's equations of motion	No	Medium	10^4–10^6 atoms
ACM	Transfer interatomic relationship to continuum mechanics, and then solve problem by the finite element method	No	Low	10^5–10^7 atoms

The feasible atomistic–continuum mechanics should be established, making it capable of including the chemical bond force in the analysis of the single crystal silicon finite element model. In this investigation, the Stillinger–Weber and Tersoff potentials were used to describe the chemical bond force between the silicon atoms.

Compared with the ab initio method and the molecular dynamics, the ACM based on the classical finite element theory has several benefits. First, the ACM model could predict accurate physical and chemical behaviors of atom and molecule systems with the million or billion atoms. Second, the ACM numerical method based on the finite element theory could reduce the computational time of the simulation. However, the ACM method has two disadvantages. First, the potential function is based on known properties. If one is interested in the properties of a new type of molecules, an appropriate force field probably may not be available for that type of molecules. Second, because ACM models set at molecules as a group of springs, it cannot be used to predict electronic properties of molecules. This novel simulation method to investigate nano-scale materials is not limited to any materials; it can also be applied to other nano-structured materials once the interatomic potential and the atomic structure of the material are known. Table 25.8 lists the characteristic comparison of ab initio method, Monte Carlo simulation, molecular mechanics, molecular dynamics, and atomistic–continuum mechanics.

25.6 Interatomic Potential Energy for Silicon

The potential energy function considered for the interaction between the silicon atoms in this investigation is a general form of Stillinger–Weber (S–W) potential. The Stillinger–Weber potential is a combination of two-body and three-body potentials, *f2* and *f3*. The two-body potential describes the formation of a chemical bond between two atoms. The three-body potential represents the angle between two bonds. The two-body potential has a Lennard-Jones type form:

$$f_2(r) - \{ {}^{A(Br^{-p} - r^{-q}) \exp\left[(r-a)^{-1}\right], r < a}_{0, r \geq a} \quad (25.22)$$

Note that *f2(r)* is a function of the rescaled bond length *r* only, and it vanishes without discontinuities at $r = a$.

The three-body potential *f3* is given by *w*

$$f_3(r_i, r_j, r_k) = h(r_{ij}, r_{ik}, \theta_{jik}) + h(r_{ji}, r_{jk}, \theta_{ijk}) + h(r_{ki}, r_{kj,}\theta_{ikj}) \quad (25.23)$$

where θ *jik* is the angle between *rj* and *rk* bond at vertex atom *i*. Provided that both *rj* and *rk* are less than the previously introduced cutoff *a*, it has the following form:

$$h(r_{ij}, r_{ik}, \theta_{jik}) = \lambda \exp\left[\gamma(r_{ij} - a)^{-1} + \gamma(r_{ik} - a)^{-1}\right] \cdot \leq \left(\cos\,\theta_{jik} + 1/3\right)^2 \quad (25.24)$$

Original the parameters of the S–W potential function were fitting to the simulation results of the total energy and the lattice constant to the experiment data. The energy and the distance units are deduced from the observed atomic energy and lattice constant at 0 K in the silicon diamond structure: $\sigma = 0.21$ nm. The parameters *A*, *B*, *p*, *q*, *a*, λ, and γ are constant used in the Stillinger–Weber potential function for single crystal silicon (see Table 25.9 for detail) [36].

25.7 Methodology of Atomistic–Continuum Mechanics

Single crystal silicon forms a covalently bonded structure, the diamond cubic structure. Silicon, with its four covalent bonds, coordinates itself tetrahedrally, and these tetrahedrons make up the diamond-cubic structure. This structure can also be represented as two interpenetrating face-centered cubic (FCC) lattices, one displaced (1/4, 1/4/, 1/4) with respect to

Table 25.9 Parameters used in the Stillinger–Weber potential function for silicon [36]

Parameter		Parameter	
A	7.049556277	a	1.8
B	0.602224558	Λ	21
p	4	Γ	1.2
q	0		

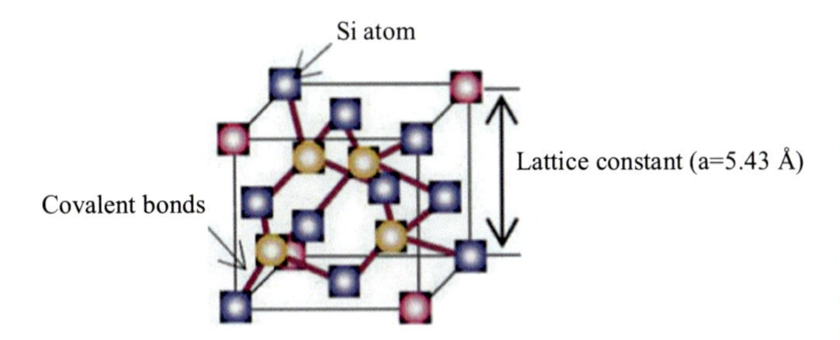

Fig. 25.18 Single crystal silicon structure

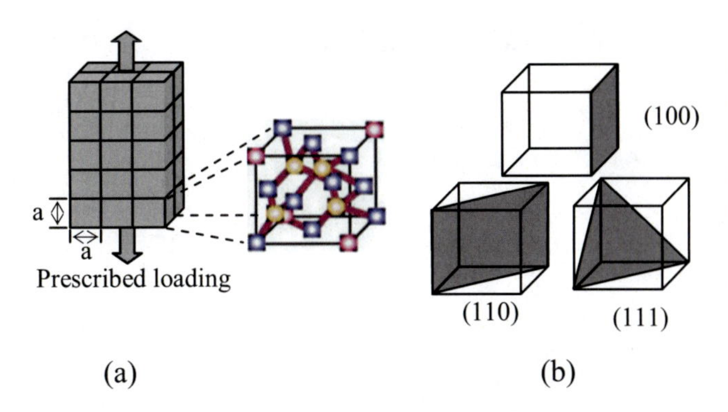

Fig. 25.19 (**a**) Schematic of the model used in the ACM simulations of uniaxial tensile testing of SCS (a = 5.43 Å: lattice constant). (**b**) The different crystallographic planes include (100), (110), and (111)

the other, as shown in Fig. 25.18. Figure 25.19a is a schematic of the model used in the ACM simulations of uniaxial tensile testing of SCS. The top and bottom surface are free from any prescribed loads (displacement), as in conventional tensile testing. Furthermore, this study also investigated the different crystallographic plane, whose prescribed loading direction axes were aligned in plane of (100), (110), and (111), respectively. Figure 25.19b displayed the different crystallographic plane of SCS.

According to the fundamentals of ACM simulation, the covalent bond (inter- atomic force) between the two atoms was replaced by the nonlinear spring element. Moreover, in order to describe completely the two-body and three-body

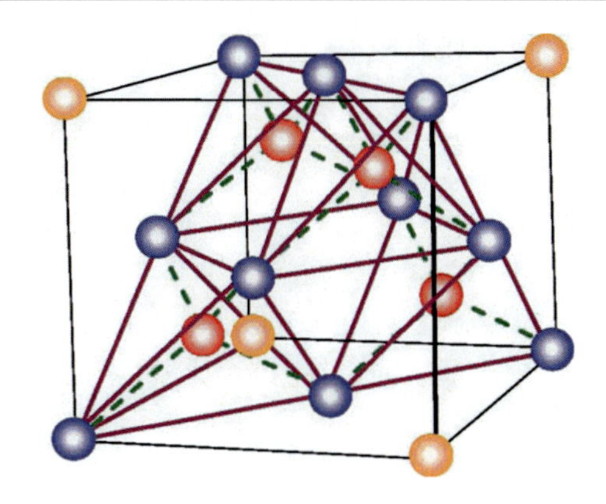

Fig. 25.20 The modification of single crystal silicon model. (*Solid line*: three-body term of S–W potential function; *Dotted line*: two-body term of S–W potential function)

interaction in ACM simulation, this investigation was set to two nonlinear spring elements of two kinds of the different behaviors. One was illustrated the stretch motion, another was defined the angle variation behavior between the two atoms and two covalent bonds, i.e., two-body and three-body term of the Stillinger–Weber potential function, respectively. Therefore, the original single crystal silicon structure in Fig. 25.18 was reconstructed the new model in ACM simulation, as shown in Fig. 25.20. Figure 25.21 showed that the force–elongation curves of the nonlinear spring elements of the two-body and three-body potential function. The relationship between the force and elongation was derived from the gradient of the potential function with respect to the position of atom, i.e.,

$$F(r) = -\nabla f(r) \tag{25.25}$$

25.8 Atomistic–Continuum Mechanics Models

In order to study the mechanical properties of the nano-scale single crystal silicon, the finite element software was adopted to simulate its behavior. In this investigation, the commercial software ANSYS^{QR} was applied to analyze the mechanical performance of the nano-scale single crystal silicon.

The half of finite element model of nano-scale single crystal silicon of the (110) crystallographic plane were established in Fig. 25.22, since the single crystal silicon is a half symmetry. The boundary conditions of the top surface were the prescribed displacement along y-direction, the x- and z-direction fixed and the y-direction fixed on the bottom surface. Table 25.10 gives the dimension of ACM models, loading, etc.

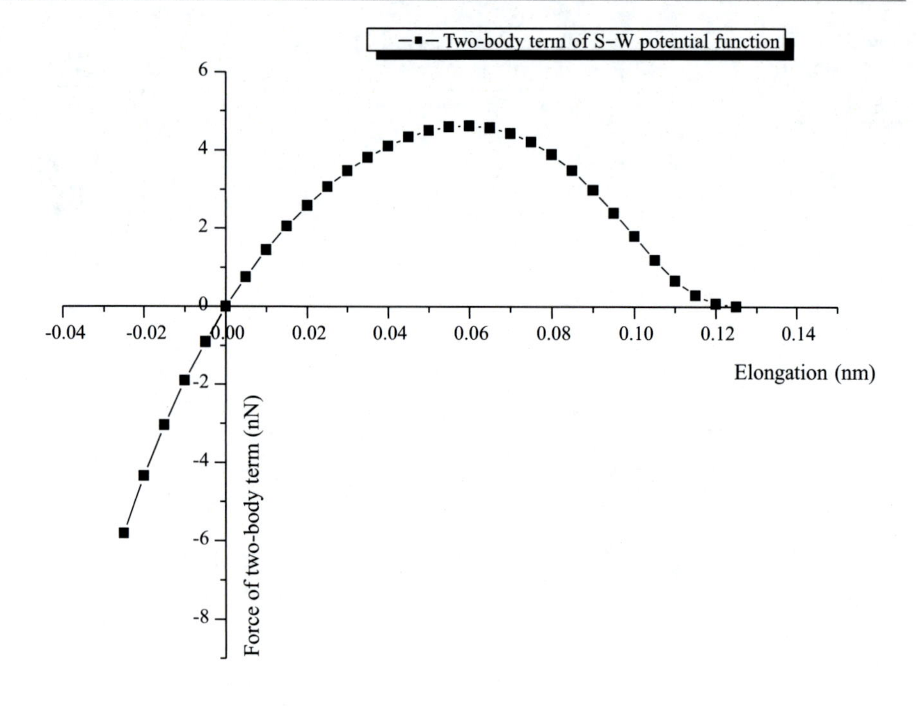

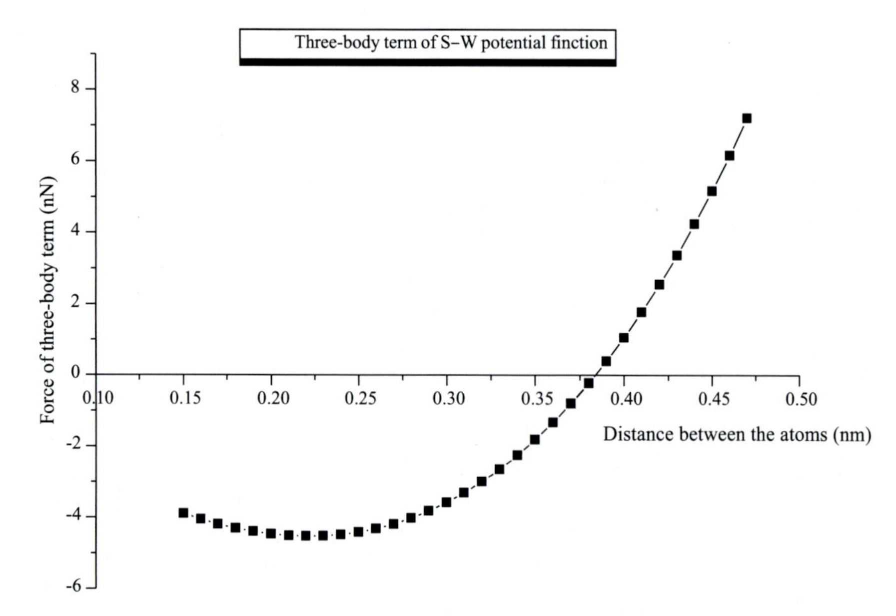

Fig. 25.21 Force–elongation curves of the nonlinear spring elements of the two-body and three-body term of S–W potential function

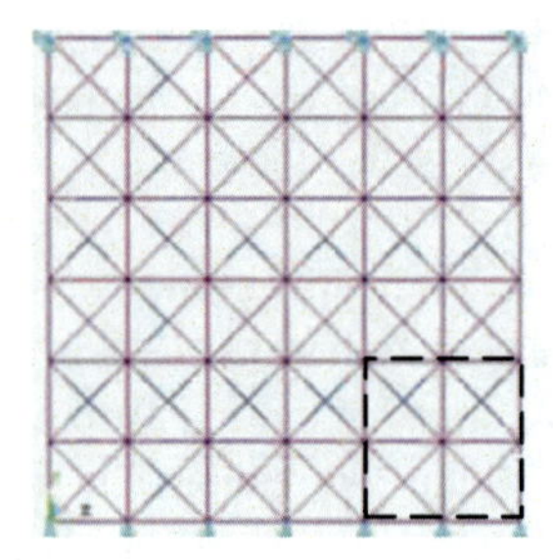

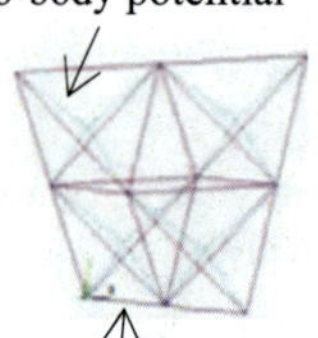

Fig. 25.22 The half of finite element model of nano-scale single crystal silicon of the (110) crystallographic plane

Table 25.10 ACM simulations conditions used for the tensile loading

Materials	Single crystal silicon
Potential used	Stillinger–Weber potential
Simulation model	Three dimension
Simulation size	3 × 5 × 3a (*a* is the lattice constant)
Prescribed displacement	0.000543–0.00543 nm
Strain	0.1–1%
Crystallographic plane	(100), (110), and (111)

25.9 Results and Discussion

In the following, the mechanical behaviors of the nano-scale single crystal silicon on different crystallographic plane under various prescribed displacement will be presented.

25.9.1 Young's Modulus Estimation

Table 25.11 listed the simulated Young's modulus compared with those values in the literature measured for bulk silicon [25, 26, 30, 47]. It could be observed that the estimated Young's modulus of SCS along the (100), (110), and (111) plane are in agree with the values reported in the literature based on macro- and microscale experimental work. The errors of the Young's modulus on the different crystal plane were 5, 4, and 5%, respectively. There are two main reasons to explain why the Young's modulus of simulation results agreed with the experiment data. First, the single crystal silicon was a defect free, no impurity structure as well as its atoms were also arranged regular and perfect. Therefore, the silicon material was identical in behavior no matter in simulation method or experiment testing. Second, the parameters of Stillinger–Weber potential function (listed in Table 25.11) were accurate, since the parameters fitted and deduced from the experiment data.

Table 25.11 Young's modulus estimation of SCS (GPa)

Direction	(100)	(110)	(111)
Brantley [47]	130.2	169.9	187.5
Wilson and Beck [30, 31]	130	161	–
Yi et al. [26]	–	169.2	–
Sato [25]	125	140	180
Present study	121.8	153	174.6

25.9.2 Effect of Prescribed Displacement (Strain) on the Tensile Response

ACM simulations of uniaxial tension were conducted at various prescribed displacement for SCS to investigate the effect of strain on the mechanical properties.

Figure 25.23 showed the effect of various prescribed displacement (strain) on the Young's modulus for nano-scale SCS. It could be observed that the Young's modulus was nearly constant and relatively insensitive to the strain. According to the literature [33] also revealed that the elastic modulus for silicon was not affected by the loading rate. The foregoing results were laid on the strain lower than 1%. However, while the strain exceeded 1%, the SCS would probably have reached the fracture strain and then the SCS structure would be directly fractured. Since the SCS was a brittle material, there was no necking effect before failure, unlike the metal materials.

25.10 Summary and Conclusions

25.10.1 Modeling of Carbon Nanotube Composites for Vibration Damping

In this research, an FEM composite unit cell model was developed, and the loss factor of the CNT composite was calculated and compared with the experimental results. The simulation curve fits well in general with experimental data.

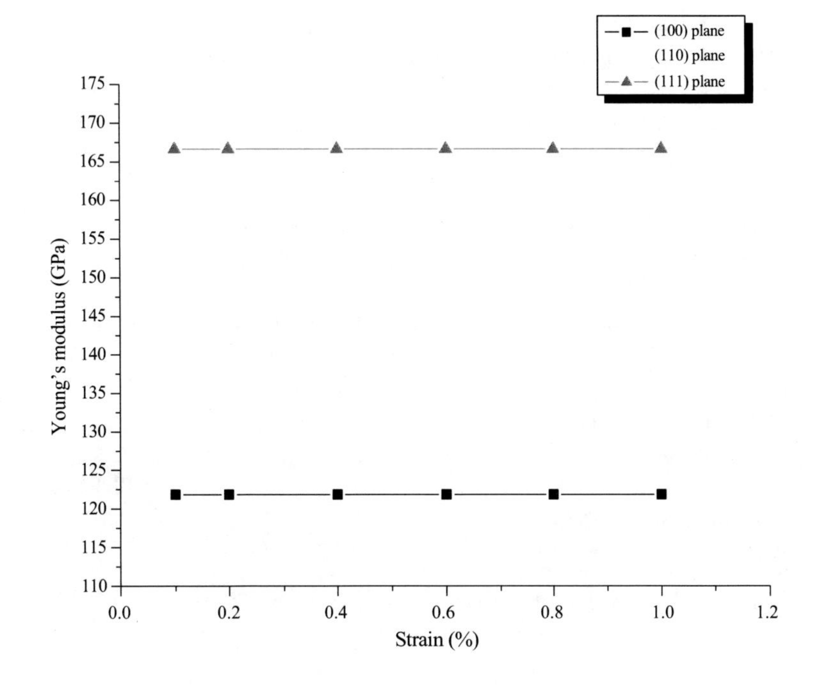

Fig. 25.23 Effect of various prescribed displacement (strain) on the Young's modulus for SCS

It could be observed from simulation results that a composite loss factor is related to the strain level, CNT dimension, CNT weight ratio, and alignment of CNT segments. As the location, size, and shape of CNT/epoxy interfacial sliding area could be simulated by using this model, before composite fabrication, CNT alignment for higher damping under different loading level could be identified. Furthermore, the upper bound of the CNT composite loss factor could be estimated by using the developed model. In parametric studies, several factors are discussed while composite energy dissipation ability is taken into consideration. Parameters such as CNT length, diameter, and alignment orientation are investigated. It is found that the composite loss factor is insensitive to CNT segment length but highly sensitive to CNT diameter. The interfacial sliding developed between CNT and matrix could be categorized into five different stages under different strain level, while the CNT alignment orientation for larger composite loss factor was identified.

25.10.2 Atomistic–Continuum Mechanics Models

This research presents a novel mechanics for coupling continuum mechanics and interatomic potential function (Stillinger–Weber potential) to predict the mechanical behavior of nano-scale single crystal silicon under uniaxial tensile loading. According to this methodology, once the interatomic potential and the atomic structure of the material are known, a systematic approach, which does not involve any parameter fitting, has been proposed. The measured Young's modulus of SCS were 121.8, 153, and 174.6 GPa along the (100), (110), and (111) crystallographic plane, respectively, which are in reasonable agreement with the results reports in the literature based on macro- and microscale experiment. The effect of strain through the variation of prescribed displacement on the measured mechanical properties was investigated. It could be observed that the Young's modulus were nearly constant and relatively insensitive to the strain (0.1–1%).

This novel simulation mechanics to investigate the nano-scale materials was not limited to single crystal silicon; it can also be applied to other nano-structured materials once the interatomic potential and the atomic structure of the material are known. In the future, the author would explore the size effects of the nano-scale single crystal silicon using the atomistic–continuum mechanics with Stillinger–Weber potential function.

Acknowledgments The work presented in this chapter was sponsored by the Research Grants Council of the Hong Kong Special Administrative Region, China (Project No. CUHK4144/07E), the Research Boost Program of National Tsing Hua University (Project No. 98N2914E1), and the National Nano Science and Technology Project of National Science Council, Taiwan (Project No. NSC98-2120-M-194- 002). The authors wish to acknowledge these supports.

References

1. Iijima, S.: Helical microtubules of graphitic carbon. Nature. **354**, 56–58 (1991)
2. Yu, M.F., Lourie, O., Dyer, M.J.: Strength and breaking mechanism of multiwalled carbon nanotubes under tensile load. Science. **287**, 637–640 (2000)
3. Thostenson, E.T., Ren, Z.F., Chou, T.W.: Advances in the science and technology of carbon nanotubes and their composites: a review. Compos. Sci. Technol. **61**, 1899–1912 (2001)
4. Ajayan, P.M., Stephan, O., Colliex, C., Trauth, D.: Aligned carbon nanotube arrays formed by cutting a polymer resin-nanotube composite. Science. **256**, 1212–1214 (1994)
5. Qian, D., Dickey, E., Andrews, C.R., Rantell, T.: Load transfer and deformation mechanisms in carbon nanotube-polystyrene composites. Appl. Phys. Lett. **76**, 2868–2870 (2000)
6. Ruan, S.L., Gao, P., Yang, X.G., Yu, T.X.: Toughening high performance ultrahigh molecular weight polyethylene using multiwalled carbon nanotubes. Polymer. **44**, 5643–5654 (2003)
7. Lau, K.T., Shi, S.Q., Zhou, L.M., Cheng, H.M.: Micro-hardness and flexural properties of randomly-oriented carbon nanotube composites. J. Compos. Mater. **37**, 365–376 (2003)
8. Koratkar, N., Wei, B., Ajayan, P.M.: Carbon nanotube films for damping applications. Adv. Mater. **14**, 997–1000 (2002)
9. Koratkar, N., Wei, B., Ajayan, P.M.: Multifunctional structural reinforcement featuring carbon nanotube films. Compos. Sci. Technol. **63**, 1525–1531 (2003)
10. Lass, E.A., Koratkar, N.A., Ajayan, P.M., Wei, B.Q., Keblinski, P.: Damping and stiffness enhancement in composite systems with carbon nanotubes films. Proc. Mater. Res. Soc. Symp. **740**, 371–376 (2002)
11. Lass, E., Ajayan, P., Koratkar, N.: Engineered connectivity in carbon nanotube films for damping applications. Proc. SPIE. **5052**, 141–150 (2003)
12. Wong, M., Paramsothy, M., Xu, X.J., Ren, Y., Li, S., Liao, K.: Physical interactions at carbon nanotube-polymer interface. Polymer. **44**, 7757–7764 (2003)
13. Zhou, X., Shin, E., Wang, K.W., Bakis, C.E.: Interfacial damping characteristics of carbon nanotube based composites. Compos. Sci. Technol. **64**, 2425–2437 (2004)
14. Suhr, J., Koratkar, N., Keblinski, P., Ajayan, P.: Viscoelasticity in carbon nanotube composites. Nat. Mater. **4**, 134–137 (2005)
15. Liu, A., Huang, J.H., Wang, K.W., Bakis, C.E.: Effects of interfacial friction on the damping characteristics of composites containing randomly oriented carbon nanotube ropes. J. Intell. Mater. Syst. Struct. **17**, 217–229 (2006)
16. Liu, A., Wang, K.W., Bakis, C.E., Huang, J.H.: Analysis of damping characteristics of a vis- coelastic polymer filled with randomly oriented single-wall nanotube ropes. Proc. SPIE Int. Symp. Smart Struct. Mater. **6169**, 275–291 (2006)
17. Zhou, X., Wang, K.W., Bakis, C.E.: Effects of interfacial friction on composites containing nanotube ropes. In: Proceedings of 44th AIAA/ASME/ASCE/AHS/ASC structures, structural dynamics, and materials conference 2004; AIAA–2004–1784
18. Dai, R.L., Liao, W.H.: Experimental studies on carbon nanotube composites for vibration damping. In: Proceedings of 17th international conference on adaptive structures and technologies 2006
19. Odegard, G.M., Gates, T.S., Wise, K.E., Park, C., Siochi, E.J.: Constitutive modeling of nanotube-reinforced polymer composites. ICASE Report 2002;NASA/CR-2002-211760:No. 2002–2027

20. Liu, Y.J., Chen, X.L.: Continuum models of carbon nanotube-based composites using the boundary element method. Electron. J. Bound. Elem. **1**, 316–335 (2003)
21. Inc ANSYS Theory Reference 2005;10.0
22. Andrews, R., Wiesenberger, M.C.: Carbon nanotube polymer composites. Curr. Opin. Solid State Mater. Sci. **8**, 31–37 (2004)
23. Cifuentes, A.O., Kalbag, A.: A performance study of tetrahedral and hexahedral elements in 3-D finite element structural analysis. Finite Elem. Anal. Des. **12**, 313–318 (1992)
24. Bussler, M., Ramesh, A.: The eight-node hexahedral element in FEA of part designs. Foundry Manag. Technol. **121**(11), 26–28 (1993)
25. Sato, K., Yoshioka, T., Ando, T., Ahilida, M., Kawabata, T.: Tensile testing of silicon film having different crystallographic orientation carrier out on a silicon chip. Sens. Actuator A. **70**, 148–152 (1998)
26. Yi, T., Lu, L., Kim, C.J.: Microscale material testing of single crystalline silicon: process effects on surface morphology and tensile strength. Sens. Actuator A. **83**, 172–178 (2000)
27. Sharpe, W.N., Yuan, B., Edwards, R.L.: A new technique for measuring the mechanical properties of thin films. J. Microelectromech. Syst. **6**, 193 (1997)
28. Li, X., Kasai, T., Nakao, S., Ando, T., Shikida, M., Sato, K., Tanaka, H.: Anisotropy in fracture of single crystal silicon film characterized under uniaxial tensile condition. Sens. Actuator A. **117**, 143–150 (2005)
29. Tsuchiya, T., Tabata, O., Sakata, J., Taga, Y.: Specimen size effect ontensile strength of surface- micromachined polycrystalline silicon thin film. J. Microelectromech. Syst. **7**, 106 (1998)
30. Wilson, C.J., Beck, P.A.: Fracture testing of bulk silicon microcantilever beams subjected to a side load. J. Microelectromech. Syst. **5**, 142–150 (1996)
31. Wilson, C.J., Oremggi, A., Narbutovskih, M.: Fracture testing of silicon microcantilever beams. J. Appl. Phys. **79**, 2386 (1996)
32. Namazu, T., Isono, Y., Tanaka, T.: Evaluation of size effect on mechanical properties of single crystal silicon by nanoscale bending test using AFM. J. Microelectromech. Syst. **9**, 450–459 (2000)
33. Komanduri, R., Chandrasekaran, N., Raff, L.M.: Molecular dynamics simulation of uniaxial tension at nanoscale of semiconductor materials for micro-electro-mechanical systems (MEMS) application. Mater. Sci. Eng. A. **34**, 58–67 (2003)
34. Komanduri, R., Raff, L.M.: A review on the molecular dynamics simulation of machining at the atomic scale. Proc. Inst. Mech. Eng. Part B. **215**, 1639–1672 (2001)
35. Girifalca, L.A., Weizer, V.G.: Application of morse potential function to cubic metals. Phys. Rev. **114**, 687–690 (1959)
36. Stillinger, F.H., Weber, T.A.: Computer simulation of local order in condensed phases of silicon. Phys. Rev. B. **31**, 5262–5271 (1985)
37. Zhang, P., Huang, Y., Geubelle, P.H., Klvin, P.A., Hwang, K.C.: The elastic modulus of single-wall carbon nanotubes: a continuum analysis incorporating interatomic potentials. Int. J. Solid Struct. **39**, 3893–3906 (2002)
38. Xiao, S.P., Belytschko, T.: A bridging domain method for coupling continua with molecular dynamics. Comput. Methods Appl. Mech. Eng. **193**, 1645–1669 (2004)
39. Wagner, G.J., Liu, W.K.: Coupling of atomistic and continuum simulations using a bridging scale decomposition. J. Comput. Phys. **190**, 249–274 (2003)
40. Ercolessi, F.A.P.: Dynamics. website: http://www.sissa.it/furio/
41. Chiang, K.N., Chou, C.Y., Wu, C.J., Yuan, C.A.: Prediction of the bulk elastic constant of metals using atomic-level single-lattice analytical method. Appl. Phys. Lett. **88**, 171904 (2006)
42. Jeng, Y.R., Tan, C.M.: Computer simulation of tension experiments of a thin film using atomic model. Phys. Rev. B. **65**, 174107 (2002)
43. Foiles, S.M., Daw, M.S.: Calculation of the thermal expansion of metals using the embedded- atom method. Phys. Rev. B. **38**, 12643 (1988)
44. Brenner, D.W., Shenderova, O.A., Harrison, J.A., Stuart, S.J., Ni, B., Sinnott, S.B.: A second- generation reactive empirical bond order potential energy expression for hydrocarbons. J. Phys. Condens. Matter. **14**, 783–802 (2002)
45. Cornell, W.D., Cieplak, P., Bayly, C.I., Gould, I.R., Merz, K.M., Ferguson, D.M., Spellmeyer, D.C., Fox, T., Caldwell, J.W., Kollman, P.A.: A second generation force filed for the simulation of proteins, nucleic, acids, and organic molecules. J. Am. Chem. Soc. **117**, 5179–5197 (1995)
46. Tersoff, J.: New empirical model for the structure properties of silicon. Phys. Rev. Lett. **56**, 632–635 (1986)
47. Brantly, W.A.: Calculated elastic constants for stress problems associated with semiconductor devices. J. Appl. Phys. **44**, 534–535 (1973)
48. Chiang, K.N., Yuan, C.A., Han, C.N., Chou, C.Y., Cui, Y.: Mechanical characteristic of ssDNA/dsDNA molecule under external loading. Appl. Phys. Lett. **88**, 023902–023903 (2006)
49. Chiang, K.N., Chang, C.H., Peng, C.T.: Local-strain effects in Si/SiGe/Si islands on oxide. Appl. Phys. Lett. **87**, 191901–191903 (2005)
50. Daw, M.S., Baskes, M.I.: Embedded-atom method: derivation and application to impurities, surfaces, and other defects in metals. Phys. Rev. B. **29**, 6443 (1984)

Index

C. P. Wong et al. (eds.), *Nano-Bio- Electronic, Photonic and MEMS Packaging*, https://doi.org/10.1007/978-3-030-49991-4

Printed in the United States
by Baker & Taylor Publisher Services